T0134711

# Lecture Notes in Artificial Intelligence 12977

Subseries of Lecture Notes in Computer Science

More information about this subseries at http://www.springer.com/series/1244

Nuria Oliver · Fernando Pérez-Cruz ·
Stefan Kramer · Jesse Read ·
Jose A. Lozano (Eds.)

# Machine Learning and Knowledge Discovery in Databases

## Research Track

European Conference, ECML PKDD 2021
Bilbao, Spain, September 13–17, 2021
Proceedings, Part III

*Editors*
Nuria Oliver
ELLIS - The European Laboratory
for Learning and Intelligent Systems
Alicante, Spain

Avda Universidad, San Vicente del Raspeig
Alicante, Spain

Vodafone Institute for Society
and Communications
Berlin, Germany

Data-Pop Alliance
New York, USA

Stefan Kramer
Johannes Gutenberg University of Mainz
Mainz, Germany

Jose A. Lozano
Basque Center for Applied Mathematics
Bilbao, Spain

Fernando Pérez-Cruz
ETHZ and EPFL
Zürich, Switzerland

Jesse Read
École Polytechnique
Palaiseau, France

ISSN 0302-9743 ISSN 1611-3349 (electronic)
Lecture Notes in Artificial Intelligence
ISBN 978-3-030-86522-1 ISBN 978-3-030-86523-8 (eBook)
https://doi.org/10.1007/978-3-030-86523-8

LNCS Sublibrary: SL7 – Artificial Intelligence



This Springer imprint is published by the registered company Springer Nature Switzerland AG
The registered company address is: Gewerbestrasse 11, 6330 Cham, Switzerland

# Preface

This edition of the European Conference on Machine Learning and Principles and Practice of Knowledge Discovery in Databases (ECML PKDD 2021) has still been affected by the COVID-19 pandemic. Unfortunately it had to be held online and we could only meet each other virtually. However, the experience gained in the previous edition joined to the knowledge collected from other virtual conferences allowed us to provide an attractive and engaging agenda.

ECML PKDD is an annual conference that provides an international forum for the latest research in all areas related to machine learning and knowledge discovery in databases, including innovative applications. It is the leading European machine learning and data mining conference and builds upon a very successful series of ECML PKDD conferences. Scheduled to take place in Bilbao, Spain, ECML PKDD 2021 was held fully virtually, during September 13–17, 2021. The conference attracted over 1000 participants from all over the world. More generally, the conference received substantial attention from industry through sponsorship, participation, and also the industry track.

The main conference program consisted of presentations of 210 accepted conference papers, 40 papers accepted in the journal track and 4 keynote talks: Jie Tang (Tsinghua University), Susan Athey (Stanford University), Joaquin Quiñonero Candela (Facebook), and Marta Kwiatkowska (University of Oxford). In addition, there were 22 workshops, 8 tutorials, 2 combined workshop-tutorials, the PhD forum, and the discovery challenge. Papers presented during the three main conference days were organized in three different tracks:

- Research Track: research or methodology papers from all areas in machine learning, knowledge discovery, and data mining.
- Applied Data Science Track: papers on novel applications of machine learning, data mining, and knowledge discovery to solve real-world use cases, thereby bridging the gap between practice and current theory.
- Journal Track: papers that were published in special issues of the Springer journals Machine Learning and Data Mining and Knowledge Discovery.

We received a similar number of submissions to last year with 685 and 220 submissions for the Research and Applied Data Science Tracks respectively. We accepted 146 (21%) and 64 (29%) of these. In addition, there were 40 papers from the Journal Track. All in all, the high-quality submissions allowed us to put together an exceptionally rich and exciting program.

The Awards Committee selected research papers that were considered to be of exceptional quality and worthy of special recognition:

- Best (Student) Machine Learning Paper Award: Reparameterized Sampling for Generative Adversarial Networks, by Yifei Wang, Yisen Wang, Jiansheng Yang and Zhouchen Lin.

- First Runner-up (Student) Machine Learning Paper Award: “Continual Learning with Dual Regularizations”, by Xuejun Han and Yuhong Guo.
- Best Applied Data Science Paper Award: “Open Data Science to fight COVID-19: Winning the 500k XPRIZE Pandemic Response Challenge”, by Miguel Angel Lozano, Oscar Garibo, Eloy Piñol, Miguel Rebollo, Kristina Polotskaya, Miguel Angel Garcia-March, J. Alberto Conejero, Francisco Escolano and Nuria Oliver.
- Best Student Data Mining Paper Award: “Conditional Neural Relational Inference for Interacting Systems”, by Joao Candido Ramos, Lionel Blondé, Stéphane Armand and Alexandros Kalousis.
- Test of Time Award for highest-impact paper from ECML PKDD 2011: “Influence and Passivity in Social Media”, by Daniel M. Romero, Wojciech Galuba, Sitaram Asur and Bernardo A. Huberman.

We would like to wholeheartedly thank all participants, authors, Program Committee members, area chairs, session chairs, volunteers, co-organizers, and organizers of workshops and tutorials for their contributions that helped make ECML PKDD 2021 a great success. We would also like to thank the ECML PKDD Steering Committee and all sponsors.

September 2021

Jose A. Lozano
Nuria Oliver
Fernando Pérez-Cruz
Stefan Kramer
Jesse Read
Yuxiao Dong
Nicolas Kourtellis
Barbara Hammer

# Organization

## General Chair

| | |
|---|---|
| Jose A. Lozano | Basque Center for Applied Mathematics, Spain |

## Research Track Program Chairs

| | |
|---|---|
| Nuria Oliver | Vodafone Institute for Society and Communications, Germany, and Data-Pop Alliance, USA |
| Fernando Pérez-Cruz | Swiss Data Science Center, Switzerland |
| Stefan Kramer | Johannes Gutenberg Universität Mainz, Germany |
| Jesse Read | École Polytechnique, France |

## Applied Data Science Track Program Chairs

| | |
|---|---|
| Yuxiao Dong | Facebook AI, Seattle, USA |
| Nicolas Kourtellis | Telefonica Research, Barcelona, Spain |
| Barbara Hammer | Bielefeld University, Germany |

## Journal Track Chairs

| | |
|---|---|
| Sergio Escalera | Universitat de Barcelona, Spain |
| Heike Trautmann | University of Münster, Germany |
| Annalisa Appice | Università degli Studi di Bari, Italy |
| Jose A. Gámez | Universidad de Castilla-La Mancha, Spain |

## Discovery Challenge Chairs

| | |
|---|---|
| Paula Brito | Universidade do Porto, Portugal |
| Dino Ienco | Université Montpellier, France |

## Workshop and Tutorial Chairs

| | |
|---|---|
| Alipio Jorge | Universidade do Porto, Portugal |
| Yun Sing Koh | University of Auckland, New Zealand |

## Industrial Track Chairs

| | |
|---|---|
| Miguel Veganzones | Sherpa.ia, Portugal |
| Sabri Skhiri | EURA NOVA, Belgium |

## Award Chairs

| | |
|---|---|
| Myra Spiliopoulou | Otto-von-Guericke-University Magdeburg, Germany |
| João Gama | University of Porto, Portugal |

## PhD Forum Chairs

| | |
|---|---|
| Jeronimo Hernandez | University of Barcelona, Spain |
| Zahra Ahmadi | Johannes Gutenberg Universität Mainz, Germany |

## Production, Publicity, and Public Relations Chairs

| | |
|---|---|
| Sophie Burkhardt | Johannes Gutenberg Universität Mainz, Germany |
| Julia Sidorova | Universidad Complutense de Madrid, Spain |

## Local Chairs

| | |
|---|---|
| Iñaki Inza | University of the Basque Country, Spain |
| Alexander Mendiburu | University of the Basque Country, Spain |
| Santiago Mazuelas | Basque Center for Applied Mathematics, Spain |
| Aritz Pèrez | Basque Center for Applied Mathematics, Spain |
| Borja Calvo | University of the Basque Country, Spain |

## Proceedings Chair

| | |
|---|---|
| Tania Cerquitelli | Politecnico di Torino, Italy |

## Sponsorship Chair

| | |
|---|---|
| Santiago Mazuelas | Basque Center for Applied Mathematics, Spain |

## Web Chairs

| | |
|---|---|
| Olatz Hernandez Aretxabaleta | Basque Center for Applied Mathematics, Spain |
| Estíbaliz Gutièrrez | Basque Center for Applied Mathematics, Spain |

## ECML PKDD Steering Committee

| | |
|---|---|
| Andrea Passerini | University of Trento, Italy |
| Francesco Bonchi | ISI Foundation, Italy |
| Albert Bifet | Télécom ParisTech, France |
| Sašo Džeroski | Jožef Stefan Institute, Slovenia |
| Katharina Morik | TU Dortmund, Germany |
| Arno Siebes | Utrecht University, The Netherlands |
| Siegfried Nijssen | Université Catholique de Louvain, Belgium |

## Program Committee

### Guest Editorial Board, Journal Track

| | |
|---|---|
| Peer Kröger | Ludwig Maximilian University of Munich |
| Meelis Kull | University of Tartu |
| Michel Lang | TU Dortmund University |
| Helge Langseth | Norwegian University of Science and Technology |
| Oswald Lanz | FBK |
| Mark Last | Ben-Gurion University of the Negev |
| Kangwook Lee | University of Wisconsin-Madison |
| Jurica Levatic | IRB Barcelona |
| Thomar Liebig | TU Dortmund |
| Hsuan-Tien Lin | National Taiwan University |
| Marius Lindauer | Leibniz University Hannover |
| Marco Lippi | University of Modena and Reggio Emilia |
| Corrado Loglisci | Università degli Studi di Bari |
| Manuel Lopez-Ibanez | University of Malaga |
| Nuno Lourenço | University of Coimbra |
| Claudio Lucchese | Ca' Foscari University of Venice |
| Brian Mac Namee | University College Dublin |
| Gjorgji Madjarov | Ss. Cyril and Methodius University |
| Davide Maiorca | University of Cagliari |
| Giuseppe Manco | ICAR-CNR |
| Elena Marchiori | Radboud University |
| Elio Masciari | Università di Napoli Federico II |
| Andres R. Masegosa | Norwegian University of Science and Technology |
| Ernestina Menasalvas | Universidad Politécnica de Madrid |
| Rosa Meo | University of Torino |
| Paolo Mignone | University of Bari "Aldo Moro" |
| Anna Monreale | University of Pisa |
| Giovanni Montana | University of Warwick |
| Grègoire Montavon | TU Berlin |
| Katharina Morik | TU Dortmund |
| Animesh Mukherjee | Indian Institute of Technology, Kharagpur |
| Amedeo Napoli | LORIA Nancy |
| Frank Naumann | University of Adelaide |
| Thomas Dyhre | Aalborg University |
| Bruno Ordozgoiti | Aalto University |
| Rita P. Ribeiro | University of Porto |
| Pance Panov | Jozef Stefan Institute |
| Apostolos Papadopoulos | Aristotle University of Thessaloniki |
| Panagiotis Papapetrou | Stockholm University |
| Andrea Passerini | University of Trento |
| Mykola Pechenizkiy | Eindhoven University of Technology |
| Charlotte Pelletier | Université Bretagne Sud |
| Ruggero G. Pensa | University of Torino |
| Nico Piatkowski | TU Dortmund |
| Dario Piga | IDSIA Dalle Molle Institute for Artificial Intelligence Research - USI/SUPSI |

| | |
|---|---|
| Gianvito Pio | Università degli Studi di Bari "Aldo Moro" |
| Marc Plantevit | LIRIS - Université Claude Bernard Lyon 1 |
| Marius Popescu | University of Bucharest |
| Raphael Prager | University of Münster |
| Mike Preuss | Universiteit Leiden |
| Jose M. Puerta | Universidad de Castilla-La Mancha |
| Kai Puolamäki | University of Helsinki |
| Chedy Raïssi | Inria |
| Jan Ramon | Inria |
| Matteo Riondato | Amherst College |
| Thomas A. Runkler | Siemens Corporate Technology |
| Antonio Salmerón | University of Almería |
| Joerg Sander | University of Alberta |
| Roberto Santana | University of the Basque Country |
| Michael Schaub | RWTH Aachen |
| Lars Schmidt-Thieme | University of Hildesheim |
| Santiago Segui | Universitat de Barcelona |
| Thomas Seidl | Ludwig-Maximilians-Universitaet Muenchen |
| Moritz Seiler | University of Münster |
| Shinichi Shirakawa | Yokohama National University |
| Jim Smith | University of the West of England |
| Carlos Soares | University of Porto |
| Gerasimos Spanakis | Maastricht University |
| Giancarlo Sperlì | University of Naples Federico II |
| Myra Spiliopoulou | Otto-von-Guericke-University Magdeburg |
| Giovanni Stilo | Università degli Studi dell'Aquila |
| Catalin Stoean | University of Craiova |
| Mahito Sugiyama | National Institute of Informatics |
| Nikolaj Tatti | University of Helsinki |
| Alexandre Termier | Université de Rennes 1 |
| Kevin Tierney | Bielefeld University |
| Luis Torgo | University of Porto |
| Roberto Trasarti | CNR Pisa |
| Sébastien Treguer | Inria |
| Leonardo Trujillo | Instituto Tecnológico de Tijuana |
| Ivor Tsang | University of Technology Sydney |
| Grigorios Tsoumakas | Aristotle University of Thessaloniki |
| Steffen Udluft | Siemens |
| Arnaud Vandaele | Université de Mons |
| Matthijs van Leeuwen | Leiden University |
| Celine Vens | KU Leuven Kulak |
| Herna Viktor | University of Ottawa |
| Marco Virgolin | Centrum Wiskunde & Informatica |
| Jordi Vitrià | Universitat de Barcelona |
| Christel Vrain | LIFO – University of Orléans |
| Jilles Vreeken | Helmholtz Center for Information Security |

| | |
|---|---|
| Willem Waegeman | Ghent University |
| David Walker | University of Plymouth |
| Hao Wang | Leiden University |
| Elizabeth F. Wanner | CEFET |
| Tu Wei-Wei | 4paradigm |
| Pascal Welke | University of Bonn |
| Marcel Wever | Paderborn University |
| Man Leung Wong | Lingnan University |
| Stefan Wrobel | Fraunhofer IAIS, University of Bonn |
| Zheng Ying | Inria |
| Guoxian Yu | Shandong University |
| Xiang Zhang | Harvard University |
| Ye Zhu | Deakin University |
| Arthur Zimek | University of Southern Denmark |
| Albrecht Zimmermann | Université Caen Normandie |
| Marinka Zitnik | Harvard University |

**Area Chairs, Research Track**

| | |
|---|---|
| Fabrizio Angiulli | University of Calabria |
| Ricardo Baeza-Yates | Universitat Pompeu Fabra |
| Roberto Bayardo | Google |
| Bettina Berendt | Katholieke Universiteit Leuven |
| Philipp Berens | University of Tübingen |
| Michael Berthold | University of Konstanz |
| Hendrik Blockeel | Katholieke Universiteit Leuven |
| Juergen Branke | University of Warwick |
| Ulf Brefeld | Leuphana University Lüneburg |
| Toon Calders | Universiteit Antwerpen |
| Michelangelo Ceci | Università degli Studi di Bari "Aldo Moro" |
| Duen Horng Chau | Georgia Institute of Technology |
| Nicolas Courty | Université Bretagne Sud, IRISA Research Institute Computer and Systems Aléatoires |
| Bruno Cremilleux | Université de Caen Normandie |
| Philippe Cudre-Mauroux | University of Fribourg |
| James Cussens | University of Bristol |
| Jesse Davis | Katholieke Universiteit Leuven |
| Bob Durrant | University of Waikato |
| Tapio Elomaa | Tampere University |
| Johannes Fürnkranz | Johannes Kepler University Linz |
| Eibe Frank | University of Waikato |
| Elisa Fromont | Université de Rennes 1 |
| Stephan Günnemann | Technical University of Munich |
| Patrick Gallinari | LIP6 - University of Paris |
| Joao Gama | University of Porto |
| Przemyslaw Grabowicz | University of Massachusetts, Amherst |

| | |
|---|---|
| Eyke Hüllermeier | Paderborn University |
| Allan Hanbury | Vienna University of Technology |
| Daniel Hernández-Lobato | Universidad Autónoma de Madrid |
| José Hernández-Orallo | Universitat Politècnica de València |
| Andreas Hotho | University of Wuerzburg |
| Inaki Inza | University of the Basque Country |
| Marius Kloft | TU Kaiserslautern |
| Arno Knobbe | Universiteit Leiden |
| Lars Kotthoff | University of Wyoming |
| Danica Kragic | KTH Royal Institute of Technology |
| Sébastien Lefèvre | Université Bretagne Sud |
| Bruno Lepri | FBK-Irst |
| Patrick Loiseau | Inria and Ecole Polytechnique |
| Jorg Lucke | University of Oldenburg |
| Fragkiskos Malliaros | Paris-Saclay University, CentraleSupelec, and Inria |
| Giuseppe Manco | ICAR-CNR |
| Dunja Mladenic | Jozef Stefan Institute |
| Katharina Morik | TU Dortmund |
| Sriraam Natarajan | Indiana University Bloomington |
| Siegfried Nijssen | Université catholique de Louvain |
| Andrea Passerini | University of Trento |
| Mykola Pechenizkiy | Eindhoven University of Technology |
| Jaakko Peltonen | Aalto University and University of Tampere |
| Marian-Andrei Rizoiu | University of Technology Sydney |
| Céline Robardet | INSA Lyon |
| Maja Rudolph | Bosch |
| Lars Schmidt-Thieme | University of Hildesheim |
| Thomas Seidl | Ludwig-Maximilians-Universität München |
| Arno Siebes | Utrecht University |
| Myra Spiliopoulou | Otto-von-Guericke-University Magdeburg |
| Yizhou Sun | University of California, Los Angeles |
| Einoshin Suzuki | Kyushu University |
| Jie Tang | Tsinghua University |
| Ke Tang | Southern University of Science and Technology |
| Marc Tommasi | University of Lille |
| Isabel Valera | Saarland University |
| Celine Vens | KU Leuven Kulak |
| Christel Vrain | LIFO - University of Orléans |
| Jilles Vreeken | Helmholtz Center for Information Security |
| Willem Waegeman | Ghent University |
| Stefan Wrobel | Fraunhofer IAIS, University of Bonn |
| Min-Ling Zhang | Southeast University |

### Area Chairs, Applied Data Science Track

| | |
|---|---|
| Francesco Calabrese | Vodafone |
| Michelangelo Ceci | Università degli Studi di Bari "Aldo Moro" |
| Gianmarco De Francisci Morales | ISI Foundation |
| Tom Diethe | Amazon |
| Johannes Fünkranz | Johannes Kepler University Linz |
| Han Fang | Facebook |
| Faisal Farooq | Qatar Computing Research Institute |
| Rayid Ghani | Carnegie Mellon Univiersity |
| Francesco Gullo | UniCredit |
| Xiangnan He | University of Science and Technology of China |
| Georgiana Ifrim | University College Dublin |
| Thorsten Jungeblut | Bielefeld University of Applied Sciences |
| John A. Lee | Université catholique de Louvain |
| Ilias Leontiadis | Samsung AI |
| Viktor Losing | Honda Research Institute Europe |
| Yin Lou | Ant Group |
| Gabor Melli | Sony PlayStation |
| Luis Moreira-Matias | University of Porto |
| Nicolò Navarin | University of Padova |
| Benjamin Paaßen | German Research Center for Artificial Intelligence |
| Kitsuchart Pasupa | King Mongkut's Institute of Technology Ladkrabang |
| Mykola Pechenizkiy | Eindhoven University of Technology |
| Julien Perez | Naver Labs Europe |
| Fabio Pinelli | IMT Lucca |
| Zhaochun Ren | Shandong University |
| Sascha Saralajew | Porsche AG |
| Fabrizio Silvestri | Facebook |
| Sinong Wang | Facebook AI |
| Xing Xie | Microsoft Research Asia |
| Jian Xu | Citadel |
| Jing Zhang | Renmin University of China |

### Program Committee Members, Research Track

| | |
|---|---|
| Hanno Ackermann | Leibniz University Hannover |
| Linara Adilova | Fraunhofer IAIS |
| Zahra Ahmadi | Johannes Gutenberg University |
| Cuneyt Gurcan Akcora | University of Manitoba |
| Omer Deniz Akyildiz | University of Warwick |
| Carlos M. Alaíz Gudín | Universidad Autónoma de Madrid |
| Mohamed Alami | Ecole Polytechnique |
| Chehbourne Abdullah Alchihabi | Carleton University |
| Pegah Alizadeh | University of Caen Normandy |

| | |
|---|---|
| Reem Alotaibi | King Abdulaziz University |
| Massih-Reza Amini | Université Grenoble Alpes |
| Shin Ando | Tokyo University of Science |
| Thiago Andrade | INESC TEC |
| Kimon Antonakopoulos | Inria |
| Alessandro Antonucci | IDSIA |
| Muhammad Umer Anwaar | Technical University of Munich |
| Eva Armengol | IIIA-SIC |
| Dennis Assenmacher | University of Münster |
| Matthias Aßenmacher | Ludwig-Maximilians-Universität München |
| Martin Atzmueller | Osnabrueck University |
| Behrouz Babaki | Polytechnique Montreal |
| Rohit Babbar | Aalto University |
| Elena Baralis | Politecnico di Torino |
| Mitra Baratchi | University of Twente |
| Christian Bauckhage | University of Bonn, Fraunhofer IAIS |
| Martin Becker | University of Würzburg |
| Jessa Bekker | Katholieke Universiteit Leuven |
| Colin Bellinger | National Research Council of Canada |
| Khalid Benabdeslem | LIRIS Laboratory, Claude Bernard University Lyon I |
| Diana Benavides-Prado | Auckland University of Technology |
| Anes Bendimerad | LIRIS |
| Christoph Bergmeir | University of Granada |
| Alexander Binder | UiO |
| Aleksandar Bojchevski | Technical University of Munich |
| Ahcène Boubekki | UiT Arctic University of Norway |
| Paula Branco | EECS University of Ottawa |
| Tanya Braun | University of Lübeck |
| Katharina Breininger | Friedrich-Alexander-Universität Erlangen Nürnberg |
| Wieland Brendel | University of Tübingen |
| John Burden | University of Cambridge |
| Sophie Burkhardt | TU Kaiserslautern |
| Sebastian Buschjäger | TU Dortmund |
| Borja Calvo | University of the Basque Country |
| Stephane Canu | LITIS, INSA de Rouen |
| Cornelia Caragea | University of Illinois at Chicago |
| Paula Carroll | University College Dublin |
| Giuseppe Casalicchio | Ludwig Maximilian University of Munich |
| Bogdan Cautis | Paris-Saclay University |
| Rémy Cazabet | Université de Lyon |
| Josu Ceberio | University of the Basque Country |
| Peggy Cellier | IRISA/INSA Rennes |
| Mattia Cerrato | Università degli Studi di Torino |
| Ricardo Cerri | Federal University of Sao Carlos |
| Alessandra Cervone | Amazon |
| Ayman Chaouki | Institut Mines-Télécom |

| | |
|---|---|
| Paco Charte | Universidad de Jaén |
| Rita Chattopadhyay | Intel Corporation |
| Vaggos Chatziafratis | Stanford University |
| Tianyi Chen | Zhejiang University City College |
| Yuzhou Chen | Southern Methodist University |
| Yiu-Ming Cheung | Hong Kong Baptist University |
| Anshuman Chhabra | University of California, Davis |
| Ting-Wu Chin | Carnegie Mellon University |
| Oana Cocarascu | King's College London |
| Lidia Contreras-Ochando | Universitat Politècnica de València |
| Roberto Corizzo | American University |
| Anna Helena Reali Costa | Universidade de São Paulo |
| Fabrizio Costa | University of Exeter |
| Gustavo De Assis Costa | Instituto Federal de Educação, Ciância e Tecnologia de Goiás |
| Bertrand Cuissart | GREYC |
| Thi-Bich-Hanh Dao | University of Orleans |
| Mayukh Das | Microsoft Research Lab |
| Padraig Davidson | Universität Würzburg |
| Paul Davidsson | Malmö University |
| Gwendoline De Bie | ENS |
| Tijl De Bie | Ghent University |
| Andre de Carvalho | Universidade de São Paulo |
| Orphée De Clercq | Ghent University |
| Alper Demir | İzmir University of Economics |
| Nicola Di Mauro | Università degli Studi di Bari "Aldo Moro" |
| Yao-Xiang Ding | Nanjing University |
| Carola Doerr | Sorbonne University |
| Boxiang Dong | Montclair State University |
| Ruihai Dong | University College Dublin |
| Xin Du | Eindhoven University of Technology |
| Stefan Duffner | LIRIS |
| Wouter Duivesteijn | Eindhoven University of Technology |
| Audrey Durand | McGill University |
| Inês Dutra | University of Porto |
| Saso Dzeroski | Jozef Stefan Institute |
| Hamid Eghbalzadeh | Johannes Kepler University |
| Dominik Endres | University of Marburg |
| Roberto Esposito | Università degli Studi di Torino |
| Samuel G. Fadel | Universidade Estadual de Campinas |
| Xiuyi Fan | Imperial College London |
| Hadi Fanaee-T. | Halmstad University |
| Elaine Faria | Federal University of Uberlandia |
| Fabio Fassetti | University of Calabria |
| Kilian Fatras | Inria |
| Ad Feelders | Utrecht University |

| | |
|---|---|
| Songhe Feng | Beijing Jiaotong University |
| Àngela Fernández-Pascual | Universidad Autónoma de Madrid |
| Daniel Fernández-Sánchez | Universidad Autónoma de Madrid |
| Sofia Fernandes | University of Aveiro |
| Cesar Ferri | Universitat Politécnica de Valéncia |
| Rémi Flamary | École Polytechnique |
| Michael Flynn | University of East Anglia |
| Germain Forestier | Université de Haute Alsace |
| Kary Främling | Umeå University |
| Benoît Frénay | Université de Namur |
| Vincent Francois | University of Amsterdam |
| Emilia Gómez | Joint Research Centre - European Commission |
| Luis Galárraga | Inria |
| Esther Galbrun | University of Eastern Finland |
| Claudio Gallicchio | University of Pisa |
| Jochen Garcke | University of Bonn |
| Clément Gautrais | KU Leuven |
| Yulia Gel | University of Texas at Dallas and University of Waterloo |
| Pierre Geurts | University of Liège |
| Amirata Ghorbani | Stanford University |
| Heitor Murilo Gomes | University of Waikato |
| Chen Gong | Shanghai Jiao Tong University |
| Bedartha Goswami | University of Tübingen |
| Henry Gouk | University of Edinburgh |
| James Goulding | University of Nottingham |
| Antoine Gourru | Université Lumière Lyon 2 |
| Massimo Guarascio | ICAR-CNR |
| Riccardo Guidotti | University of Pisa |
| Ekta Gujral | University of California, Riverside |
| Francesco Gullo | UniCredit |
| Tias Guns | Vrije Universiteit Brussel |
| Thomas Guyet | Institut Agro, IRISA |
| Tom Hanika | University of Kassel |
| Valentin Hartmann | Ecole Polytechnique Fédérale de Lausanne |
| Marwan Hassani | Eindhoven University of Technology |
| Jukka Heikkonen | University of Turku |
| Fredrik Heintz | Linköping University |
| Sibylle Hess | TU Eindhoven |
| Jaakko Hollmén | Aalto University |
| Tamas Horvath | University of Bonn, Fraunhofer IAIS |
| Binbin Hu | Ant Group |
| Hong Huang | UGoe |
| Georgiana Ifrim | University College Dublin |
| Angelo Impedovo | Università degli studi di Bari "Aldo Moro" |

| | |
|---|---|
| Nathalie Japkowicz | American University |
| Szymon Jaroszewicz | Institute of Computer Science, Polish Academy of Sciences |
| Saumya Jetley | Inria |
| Binbin Jia | Southeast University |
| Xiuyi Jia | School of Computer Science and Technology, Nanjing University of Science and Technology |
| Yuheng Jia | City University of Hong Kong |
| Siyang Jiang | National Taiwan University |
| Priyadarshini Kumari | IIT Bombay |
| Ata Kaban | University of Birmingham |
| Tomasz Kajdanowicz | Wroclaw University of Technology |
| Vana Kalogeraki | Athens University of Economics and Business |
| Toshihiro Kamishima | National Institute of Advanced Industrial Science and Technology |
| Michael Kamp | Monash University |
| Bo Kang | Ghent University |
| Dimitrios Karapiperis | Hellenic Open University |
| Panagiotis Karras | Aarhus University |
| George Karypis | University of Minnesota |
| Mark Keane | University College Dublin |
| Kristian Kersting | TU Darmstadt |
| Masahiro Kimura | Ryukoku University |
| Jiri Klema | Czech Technical University |
| Dragi Kocev | Jozef Stefan Institute |
| Masahiro Kohjima | NTT |
| Lukasz Korycki | Virginia Commonwealth University |
| Peer Kröger | Ludwig Maximilian University of Münich |
| Anna Krause | University of Würzburg |
| Bartosz Krawczyk | Virginia Commonwealth University |
| Georg Krempl | Utrecht University |
| Meelis Kull | University of Tartu |
| Vladimir Kuzmanovski | Aalto University |
| Ariel Kwiatkowski | Ecole Polytechnique |
| Emanuele La Malfa | University of Oxford |
| Beatriz López | University of Girona |
| Preethi Lahoti | Aalto University |
| Ichraf Lahouli | Euranova |
| Niklas Lavesson | Jönköping University |
| Aonghus Lawlor | University College Dublin |
| Jeongmin Lee | University of Pittsburgh |
| Daniel Lemire | LICEF Research Center and Université du Québec |
| Florian Lemmerich | University of Passau |
| Elisabeth Lex | Graz University of Technology |
| Jiani Li | Vanderbilt University |
| Rui Li | Inspur Group |
| Wentong Liao | Lebniz University Hannover |

| | |
|---|---|
| Jiayin Lin | University of Wollongong |
| Rudolf Lioutikov | UT Austin |
| Marco Lippi | University of Modena and Reggio Emilia |
| Suzanne Little | Dublin City University |
| Shengcai Liu | University of Science and Technology of China |
| Shenghua Liu | Institute of Computing Technology, Chinese Academy of Sciences |
| Philipp Liznerski | Technische Universität Kaiserslautern |
| Corrado Loglisci | Università degli Studi di Bari “Aldo Moro” |
| Ting Long | Shanghai Jiaotong University |
| Tsai-Ching Lu | HRL Laboratories |
| Yunpu Ma | Siemens AG |
| Zichen Ma | The Chinese University of Hong Kong |
| Sara Madeira | Universidade de Lisboa |
| Simona Maggio | Dataiku |
| Sara Magliacane | IBM |
| Sebastian Mair | Leuphana University Lüneburg |
| Lorenzo Malandri | University of Milan Bicocca |
| Donato Malerba | Università degli Studi di Bari “Aldo Moro” |
| Pekka Malo | Aalto University |
| Robin Manhaeve | KU Leuven |
| Silviu Maniu | Université Paris-Sud |
| Giuseppe Marra | KU Leuven |
| Fernando Martínez-Plumed | Joint Research Centre - European Commission |
| Alexander Marx | Max Plank Instiute for Informatics and Saarland University |
| Florent Masseglia | Inria |
| Tetsu Matsukawa | Kyushu University |
| Wolfgang Mayer | University of South Australia |
| Santiago Mazuelas | Basque center for Applied Mathematics |
| Stefano Melacci | University of Siena |
| Ernestina Menasalvas | Universidad Politécnica de Madrid |
| Rosa Meo | Università degli Studi di Torino |
| Alberto Maria Metelli | Politecnico di Milano |
| Saskia Metzler | Max Planck Institute for Informatics |
| Alessio Micheli | University of Pisa |
| Paolo Mignone | Università degli studi di Bari “Aldo Moro” |
| Matej Mihelčić | University of Zagreb |
| Decebal Constantin Mocanu | University of Twente |
| Nuno Moniz | INESC TEC and University of Porto |
| Carlos Monserrat | Universitat Politécnica de Valéncia |
| Corrado Monti | ISI Foundation |
| Jacob Montiel | University of Waikato |
| Ahmadreza Mosallanezhad | Arizona State University |
| Tanmoy Mukherjee | University of Tennessee |
| Martin Mundt | Goethe University |

| | |
|---|---|
| Mohamed Nadif | Université de Paris |
| Omer Nagar | Bar Ilan University |
| Felipe Kenji Nakano | Katholieke Universiteit Leuven |
| Mirco Nanni | KDD-Lab ISTI-CNR Pisa |
| Apurva Narayan | University of Waterloo |
| Nicolò Navarin | University of Padova |
| Benjamin Negrevergne | Paris Dauphine University |
| Hurley Neil | University College Dublin |
| Stefan Neumann | University of Vienna |
| Ngoc-Tri Ngo | The University of Danang - University of Science and Technology |
| Dai Nguyen | Monash University |
| Eirini Ntoutsi | Free University Berlin |
| Andrea Nuernberger | Otto-von-Guericke-Universität Magdeburg |
| Pablo Olmos | University Carlos III |
| James O'Neill | University of Liverpool |
| Barry O'Sullivan | University College Cork |
| Rita P. Ribeiro | University of Porto |
| Aritz Pèrez | Basque Center for Applied Mathematics |
| Joao Palotti | Qatar Computing Research Institute |
| Guansong Pang | University of Adelaide |
| Pance Panov | Jozef Stefan Institute |
| Evangelos Papalexakis | University of California, Riverside |
| Haekyu Park | Georgia Institute of Technology |
| Sudipta Paul | Umeå University |
| Yulong Pei | Eindhoven University of Technology |
| Charlotte Pelletier | Université Bretagne Sud |
| Ruggero G. Pensa | University of Torino |
| Bryan Perozzi | Google |
| Nathanael Perraudin | ETH Zurich |
| Lukas Pfahler | TU Dortmund |
| Bastian Pfeifer | Medical University of Graz |
| Nico Piatkowski | TU Dortmund |
| Robert Pienta | Georgia Institute of Technology |
| Fábio Pinto | Faculdade de Economia do Porto |
| Gianvito Pio | University of Bari "Aldo Moro" |
| Giuseppe Pirrò | Sapienza University of Rome |
| Claudia Plant | University of Vienna |
| Marc Plantevit | LIRIS - Universitè Claude Bernard Lyon 1 |
| Amit Portnoy | Ben Gurion University |
| Melanie Pradier | Harvard University |
| Paul Prasse | University of Potsdam |
| Philippe Preux | Inria, LIFL, Universitè de Lille |
| Ricardo Prudencio | Federal University of Pernambuco |
| Zhou Qifei | Peking University |
| Erik Quaeghebeur | TU Eindhoven |

| | |
|---|---|
| Kristina Yordanova | University of Rostock |
| Tetsuya Yoshida | Nara Women's University |
| Guoxian Yu | Shandong University, China |
| Sha Yuan | Tsinghua University |
| Valentina Zantedeschi | INSA Lyon |
| Albin Zehe | University of Würzburg |
| Bob Zhang | University of Macau |
| Teng Zhang | Huazhong University of Science and Technology |
| Liang Zhao | University of São Paulo |
| Bingxin Zhou | University of Sydney |
| Kenny Zhu | Shanghai Jiao Tong University |
| Yanqiao Zhu | Institute of Automation, Chinese Academy of Sciences |
| Arthur Zimek | University of Southern Denmark |
| Albrecht Zimmermann | Université Caen Normandie |
| Indre Zliobaite | University of Helsinki |
| Markus Zopf | NEC Labs Europe |

**Program Committee Members, Applied Data Science Track**

| | |
|---|---|
| Mahdi Abolghasemi | Monash University |
| Evrim Acar | Simula Research Lab |
| Deepak Ajwani | University College Dublin |
| Pegah Alizadeh | University of Caen Normandy |
| Jean-Marc Andreoli | Naver Labs Europe |
| Giorgio Angelotti | ISAE Supaero |
| Stefanos Antaris | KTH Royal Institute of Technology |
| Xiang Ao | Institute of Computing Technology, Chinese Academy of Sciences |
| Yusuf Arslan | University of Luxembourg |
| Cristian Axenie | Huawei European Research Center |
| Hanane Azzag | Université Sorbonne Paris Nord |
| Pedro Baiz | Imperial College London |
| Idir Benouaret | CNRS, Université Grenoble Alpes |
| Laurent Besacier | Laboratoire d'Informatique de Grenoble |
| Antonio Bevilacqua | Insight Centre for Data Analytics |
| Adrien Bibal | University of Namur |
| Wu Bin | Zhengzhou University |
| Patrick Blöbaum | Amazon |
| Pavel Blinov | Sber Artificial Intelligence Laboratory |
| Ludovico Boratto | University of Cagliari |
| Stefano Bortoli | Huawei Technologies Duesseldorf |
| Zekun Cai | University of Tokyo |
| Nicolas Carrara | University of Toronto |
| John Cartlidge | University of Bristol |
| Oded Cats | Delft University of Technology |
| Tania Cerquitelli | Politecnico di Torino |

| | |
|---|---|
| Prithwish Chakraborty | IBM |
| Rita Chattopadhyay | Intel Corp. |
| Keru Chen | GrabTaxi Pte Ltd. |
| Liang Chen | Sun Yat-sen University |
| Zhiyong Cheng | Shandong Artificial Intelligence Institute |
| Silvia Chiusano | Politecnico di Torino |
| Minqi Chong | Citadel |
| Jeremie Clos | University of Nottingham |
| J. Albert Conejero Casares | Universitat Politécnica de Vaécia |
| Evan Crothers | University of Ottawa |
| Henggang Cui | Uber ATG |
| Tiago Cunha | University of Porto |
| Padraig Cunningham | University College Dublin |
| Eustache Diemert | CRITEO Research |
| Nat Dilokthanakul | Vidyasirimedhi Institute of Science and Technology |
| Daizong Ding | Fudan University |
| Kaize Ding | ASU |
| Michele Donini | Amazon |
| Lukas Ewecker | Porsche AG |
| Zipei Fan | University of Tokyo |
| Bojing Feng | National Laboratory of Pattern Recognition, Institute of Automation, Chinese Academy of Science |
| Flavio Figueiredo | Universidade Federal de Minas Gerais |
| Blaz Fortuna | Qlector d.o.o. |
| Zuohui Fu | Rutgers University |
| Fabio Fumarola | University of Bari "Aldo Moro" |
| Chen Gao | Tsinghua University |
| Luis Garcia | University of Brasília |
| Cinmayii Garillos-Manliguez | University of the Philippines Mindanao |
| Kiran Garimella | Aalto University |
| Etienne Goffinet | Laboratoire LIPN-UMR CNRS |
| Michael Granitzer | University of Passau |
| Xinyu Guan | Xi'an Jiaotong University |
| Thomas Guyet | Institut Agro, IRISA |
| Massinissa Hamidi | Laboratoire LIPN-UMR CNRS |
| Junheng Hao | University of California, Los Angeles |
| Martina Hasenjaeger | Honda Research Institute Europe GmbH |
| Lars Holdijk | University of Amsterdam |
| Chao Huang | University of Notre Dame |
| Guanjie Huang | Penn State University |
| Hong Huang | UGoe |
| Yiran Huang | TECO |
| Madiha Ijaz | IBM |
| Roberto Interdonato | CIRAD - UMR TETIS |
| Omid Isfahani Alamdari | University of Pisa |

| | |
|---|---|
| Anna Nguyen | Karlsruhe Institute of Technology |
| Hao Niu | KDDI Research, Inc. |
| Inna Novalija | Jožef Stefan Institute |
| Tsuyosh Okita | Kyushu Institute of Technology |
| Aoma Osmani | LIPN-UMR CNRS 7030, Université Paris 13 |
| Latifa Oukhellou | IFSTTAR |
| Andrei Paleyes | University of Cambridge |
| Chanyoung Park | KAIST |
| Juan Manuel Parrilla Gutierrez | University of Glasgow |
| Luca Pasa | Università degli Studi Di Padova |
| Pedro Pereira Rodrigues | University of Porto |
| Miquel Perelló-Nieto | University of Bristol |
| Beatrice Perez | Dartmouth College |
| Alan Perotti | ISI Foundation |
| Mirko Polato | University of Padua |
| Giovanni Ponti | ENEA |
| Nicolas Posocco | Eura Nova |
| Cedric Pradalier | GeorgiaTech Lorraine |
| Giulia Preti | ISI Foundation |
| A. A. A. Qahtan | Utrecht University |
| Chuan Qin | University of Science and Technology of China |
| Dimitrios Rafailidis | University of Thessaly |
| Cyril Ray | Arts et Metiers Institute of Technology, Ecole Navale, IRENav |
| Wolfgang Reif | University of Augsburg |
| Kit Rodolfa | Carnegie Mellon University |
| Christophe Rodrigues | Pôle Universitaire Léonard de Vinci |
| Natali Ruchansky | Netflix |
| Hajer Salem | AUDENSIEL |
| Parinya Sanguansat | Panyapiwat Institute of Management |
| Atul Saroop | Amazon |
| Alexander Schiendorfer | Technische Hochschule Ingolstadt |
| Peter Schlicht | Volkswagen |
| Jens Schreiber | University of Kassel |
| Alexander Schulz | Bielefeld University |
| Andrea Schwung | FH SWF |
| Edoardo Serra | Boise State University |
| Lorenzo Severini | UniCredit |
| Ammar Shaker | Paderborn University |
| Jiaming Shen | University of Illinois at Urbana-Champaign |
| Rongye Shi | Columbia University |
| Wang Siyu | Southwestern University of Finance and Economics |
| Hao Song | University of Bristol |
| Francesca Spezzano | Boise State University |
| Simon Stieber | University of Augsburg |

Laurens Stoop Utrecht University
Hongyang Su Harbin Institute of Technology
David Sun Apple
Weiwei Sun Shandong University
Maryam Tabar Pennsylvania State University
Anika Tabassum Virginia Tech
Garth Tarr University of Sydney
Dinh Van Tran University of Padova
Sreekanth Vempati Myntra
Herna Viktor University of Ottawa
Daheng Wang University of Notre Dame
Hongwei Wang Stanford University
Wenjie Wang National University of Singapore
Yue Wang Microsoft Research
Zhaonan Wang University of Tokyo and National Institute of Advanced Industrial Science and Technology
Michael Wilbur Vanderbilt University
Roberto Wolfler Calvo LIPN, Université Paris 13
Di Wu Chongqing Institute of Green and Intelligent Technology
Gang Xiong Chinese Academy of Sciences
Xiaoyu Xu Chongqing Institute of Green and Intelligent Technology
Yexiang Xue Purdue University
Sangeeta Yadav Indian Institute of Science
Hao Yan Washington University in St. Louis
Chuang Yang University of Tokyo
Yang Yang Northwestern University
You Yizhe Institute of Information Engineering, Chinese Academy of Sciences
Alexander Ypma ASML
Jun Yuan The Boeing Company
Mingxuan Yue University of Southern California
Danqing Zhang Amazon
Jiangwei Zhang Tencent
Xiaohan Zhang Sony Interactive Entertainment
Xinyang Zhang University of Illinois at Urbana-Champaign
Yongxin Zhang Sun Yat-sen University
Mia Zhao Airbnb
Tong Zhao University of Notre Dame
Bin Zhou National University of Defense Technology
Bo Zhou Baidu
Louis Zigrand Université Sorbonne Paris Nord

## Sponsors

Google ASML

amazon

# Contents – Part III

## Generative Models



## Search and optimization

**Supervised Learning**

**Text Mining and Natural Language Processing**

**Image Processing, Computer Vision and Visual Analytics**

Generative Models

# Deep Conditional Transformation Models

Philipp F. M. Baumann[1(✉)], Torsten Hothorn[2], and David Rügamer[3]

[1] KOF Swiss Economic Institute, ETH Zurich, Zurich, Switzerland
baumann@kof.ethz.ch

[2] Epidemiology, Biostatistics and Prevention Institute, University of Zurich, Zurich, Switzerland
torsten.hothorn@uzh.ch

[3] Department of Statistics, LMU Munich, Munich, Germany
david.ruegamer@stat.uni-muenchen.de

**Abstract.** Learning the cumulative distribution function (CDF) of an outcome variable conditional on a set of features remains challenging, especially in high-dimensional settings. Conditional transformation models provide a semi-parametric approach that allows to model a large class of conditional CDFs without an explicit parametric distribution assumption and with only a few parameters. Existing estimation approaches within this class are, however, either limited in their complexity and applicability to unstructured data sources such as images or text, lack interpretability, or are restricted to certain types of outcomes. We close this gap by introducing the class of deep conditional transformation models which unifies existing approaches and allows to learn both interpretable (non-)linear model terms and more complex neural network predictors in one holistic framework. To this end we propose a novel network architecture, provide details on different model definitions and derive suitable constraints as well as network regularization terms. We demonstrate the efficacy of our approach through numerical experiments and applications.

**Keywords:** Transformation models · Distributional regression · Normalizing flows · Deep learning · Semi-structured regression

## 1 Introduction

Recent discussions on the quantification of uncertainty have emphasized that a distinction between aleatoric and epistemic uncertainty is useful in classical machine learning [18,38]. Moreover, this distinction was also advocated in the deep learning literature [4,20]. While epistemic uncertainty describes the uncertainty of the model and can be accounted for in a Bayesian neural network, aleatoric uncertainty [12] can be captured by modeling an outcome probability distribution that has a stochastic dependence on features (i.e., conditional on features). Apart from non-parametric estimation procedures, four fundamental approaches in statistics exist that allow to model the stochastic dependence between features and the outcome distribution [14]. First, parametric models


N. Oliver et al. (Eds.): ECML PKDD 2021, LNAI 12977, pp. 3–18, 2021.
https://doi.org/10.1007/978-3-030-86523-8_1

where additive functions of the features describe the location, scale and shape (LSS) of the distribution [34] or where these features are used in heteroscedastic Bayesian additive regression tree ensembles [31]. Second, quantile regression models [1,23,28] that directly model the conditional quantiles with a linear or non-linear dependence on feature values. Third, distribution regression and transformation models [3,9,27,35,43] that have response-varying effects on the probit, logit or complementary log-log scale. Finally, hazard regression [25] which estimates a non-proportional hazard function conditional on feature values. Parallel to this, various approaches in machine learning and deep learning have been evolved to model the outcome distribution through input features. A prominent example is normalizing flows (see, e.g. [22,30,33,41]), used to learn a complex distribution of an outcome based on feature values. Normalizing flows start with a simple base distribution $F_Z$ and transform $F_Z$ to a more complex target distribution using a bijective transformation of the random variable coming from the base distribution. The most recent advances utilize monotonic polynomials [19,32] or splines [5,6,29] to learn this transformation. As pointed out recently by several authors [21,39], normalizing flows are conceptually very similar to transformation models. However, normalizing flows, in contrast to transformation models, usually combine a series of transformations with limited interpretability of the influence of single features on the distribution of the target variable. In this work, we instead focus on conditional transformation approaches that potentially yield as much flexibility as normalizing flows, but instead of defining a generative approach, strive to build interpretable regression models without too restrictive parametric distribution assumptions.

### 1.1 Transformation Models

The origin of transformation models (TM) can be traced back to [2] studying a parametric approach to transform the variable of interest $Y$ prior to the model estimation in order to meet a certain distribution assumption of the model. Many prominent statistical models, such as the Cox proportional hazards model or the proportional odds model for ordered outcomes, can be understood as transformation models. Estimating transformation models using a neural network has been proposed by [39]. However, [39] only focus on a smaller subclass of transformation models, we call (linear) shift transformation models and on models that are not interpretable in nature. Recently, fully parameterized transformation models have been proposed [15,16] which employ likelihood-based learning to estimate the cumulative distribution function $F_Y$ of $Y$ via estimation of the corresponding transformation of $Y$. The main assumption of TM is that $Y$ follows a known, log-concave error distribution $F_Z$ after some monotonic transformation $h$. CTMs specify this transformation function conditional on a set of features $\boldsymbol{x}$:

$$\mathbb{P}(Y \leq y|\boldsymbol{x}) = F_{Y|\boldsymbol{x}}(y) = F_Z(h(y|\boldsymbol{x})). \tag{1}$$

The transformation function $h$ can be decomposed as $h(y|\boldsymbol{x}) := h_1 + h_2$, where $h_1$ and $h_2$ can have different data dependencies as explained in the following. When

$h_1$ depends on $y$ as well as $\boldsymbol{x}$, we call the CTM an *interacting CTM*. When $h_1$ depends on $y$ only, we call the model a *shift CTM*, with shift term $h_2$. When $h_2$ is omitted in an interacting CTM, we call the CTM a *distributional CTM*. In general, the bijective function $h(y|\boldsymbol{x})$ is unknown a priori and needs to be learned from the data. [16] study the likelihood of this transformation function and propose an estimator for the most likely transformation. [16] specify the transformation function through a flexible basis function approach, which, in the unconditional case $h(y)$ (without feature dependency), is given by $h(y) = \boldsymbol{a}(y)^\top \boldsymbol{\vartheta}$ where $\boldsymbol{a}(y)$ is a matrix of evaluated basis functions and $\boldsymbol{\vartheta}$ a vector of basis coefficients which can be estimated by maximum likelihood. For continuous $Y$ Bernstein polynomials [8] with higher order $M$ provide a more flexible but still computationally attractive choice for $\boldsymbol{a}$. That is,

$$\boldsymbol{a}(y)^\top \boldsymbol{\vartheta} = \frac{1}{(M+1)} \sum_{m=0}^{M} \vartheta_m f_{Be(m+1, M-m+1)}(\tilde{y}) \tag{2}$$

where $f_{Be(m,M)}$ is the probability density function of a Beta distribution with parameters m, M and a normalized outcome $\tilde{y} := \frac{y-l}{u-l} \in [0,1]$ with $u > l$ and $u, l \in \mathbb{R}$. In order to guarantee monotonicity of the estimate of $F_{Y|\boldsymbol{x}}$, strict monotonicity of $\boldsymbol{a}(y)^\top \boldsymbol{\vartheta}$ is required. This can be achieved by restricting $\vartheta_m > \vartheta_{m-1}$ for $m = 1, \dots, M+1$. When choosing $M$, the interplay with $F_Z$ should be considered. For example, when $F_Z = \Phi$, the standard Gaussian distribution function, and $M = 1$, then $\hat{F}_Y$ will also belong to the family of Gaussian distributions functions. Further, when we choose $M = n - 1$ with $n$ being the number of independent observations, then $\hat{F}_Y$ is the non-parametric maximum likelihood estimator which converges to $F_Y$ by the Glivenko-Cantelli lemma [13]. As a result, for small $M$ the choice of $F_Z$ will be decisive, while TMs will approximate the empirical cumulative distribution function well when $M$ is large independent of the choice of $F_Z$. Different choices for $F_Z$ have been considered in the literature (see, e.g., [16]), such as the standard Gaussian distribution function ($\Phi$), the standard logistic distribution function ($F_L$) or the minimum extreme value distribution function ($F_{MEV}$).

In CTMs with structured additive predictors (STAP), features considered in $h_1$ and $h_2$ enter through various functional forms and are combined as an additive sum. The STAP is given by

$$\eta_{struc} = s_1(\boldsymbol{x}) + \ldots + s_k(\boldsymbol{x}) \tag{3}$$

with $s_1, \dots, s_k$ being partial effects of one or more features in $\boldsymbol{x}$. Common choices include linear effects $\boldsymbol{x}^\top \boldsymbol{w}$ with regression coefficient $\boldsymbol{w}$ and non-linear effects based on spline basis representation, spatial effects, varying coefficients, linear and non-linear interaction effects or individual-specific random effects [7]. Structured additive models have been proposed in many forms, for example in additive (mixed) models where $\mathbb{E}(Y|\boldsymbol{x}) = \eta_{struc}$.

### 1.2 Related Work and Our Contribution

The most recent advances in transformation models [14,17,21] learn the transformation functions $h_1$ an $h_2$ separately, using, e.g., a model-based boosting algorithm with pre-specified base learners [14]. Very recent neural network-based approaches allow for the joint estimation of both transformation functions, but do either not yield interpretable models [39] or are restricted to STAP with ordinal outcomes [24].

Our framework combines the existing frameworks and thereby extends approaches for continuous outcomes to transformation models able to 1) learn more flexible and complex specifications of $h_1$ and $h_2$ simultaneously 2) learn the CDF without the necessity of specifying the (type of) feature contribution a priori, 3) retain the interpretability of the structured additive predictor in $h_1$ and $h_2$ 4) estimate structured effects in high-dimensional settings due to the specification of the model class within a neural network 5) incorporate unstructured data source such as texts or images.

## 2 Model and Network Definition

We now formally introduce the deep conditional transformation model (DCTM), explain its network architecture and provide details about different model definitions, penalization and model tuning.

### 2.1 Model Definition

Following [14], we do not make any explicit parameterized distribution assumption about $Y$, but instead assume

$$\mathbb{P}(Y \leq y|\boldsymbol{x}) = F_Z(h(y|\boldsymbol{x})) \tag{4}$$

with error distribution $F_Z : \mathbb{R} \mapsto [0,1]$, an a priori known CDF that represents the data generating process of the transformed outcome $h(Y|\boldsymbol{x})$ conditional on some features $\boldsymbol{x} \in \chi$. For tabular data, we assume $\boldsymbol{x} \in \mathbb{R}^p$. For unstructured data sources such as images, $\boldsymbol{x}$ may also include multidimensional inputs. Let $f_Z$ further be the corresponding probability density function of $F_Z$. We model this transformation function conditional on some predictors $\boldsymbol{x}$ by $h(y|\boldsymbol{x}) = h_1 + h_2 = \boldsymbol{a}(y)^\top \boldsymbol{\vartheta}(\boldsymbol{x}) + \beta(\boldsymbol{x})$, where $\boldsymbol{a}(y)$ is a (pre-defined) basis function $\boldsymbol{a} : \Xi \mapsto \mathbb{R}^{M+1}$ with $\Xi$ the sample space and $\boldsymbol{\vartheta} : \chi_\vartheta \mapsto \mathbb{R}^{M+1}$ a conditional parameter function defined on $\chi_\vartheta \subseteq \chi$. $\boldsymbol{\vartheta}$ is parameterized through structured predictors such as splines, unstructured predictors such as a deep neural network, or the combination of both and $\beta(\boldsymbol{x})$ is a feature dependent distribution shift. More specifically, we model $\boldsymbol{\vartheta}(\boldsymbol{x})$ by the following additive predictor:

$$\boldsymbol{\vartheta}(\boldsymbol{x}) = \sum_{j=1}^{J} \boldsymbol{\Gamma}_j \boldsymbol{b}_j(\boldsymbol{x}), \tag{5}$$

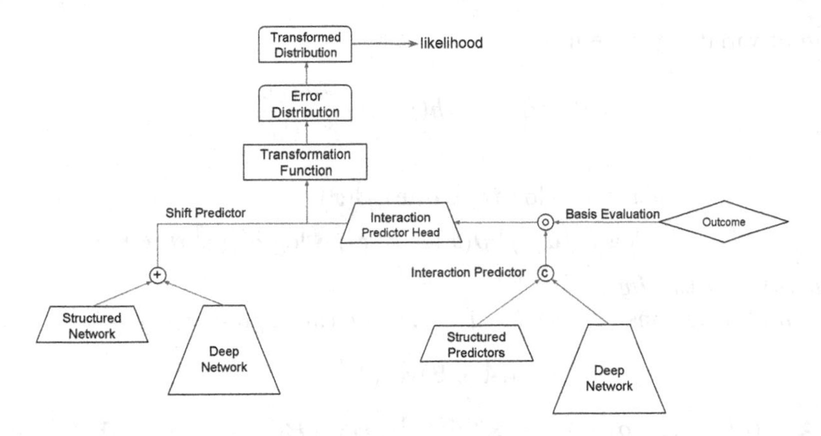


**Fig. 1.** Architecture of a deep conditional transformation model. Both the shift and interaction predictor can potentially be defined by a structured network including linear terms, (penalized) splines or other structured additive regression terms and deep neural network defined by an arbitrary network structure. While the shift predictor ($\mathcal{C}\boldsymbol{\Psi}$) is a sum of both subnetwork predictions, the interaction predictor ($\mathcal{A} \odot \mathcal{B}$) is only multiplied with a final 1-hidden unit fully-connected layer (network head, vec($\boldsymbol{\Gamma}$)) after the structured predictors and latent features of the deep neural network are combined with the basis evaluated outcome. The shift and interaction network part together define the transformation function, which transforms the error distribution and yields the final likelihood used as loss function.

with $\boldsymbol{\Gamma}_j \in \mathbb{R}^{(M+1)\times O_j}, O_j \geq 1$, being joint coefficient matrices for the basis functions in $\boldsymbol{a}$ and the chosen predictor terms $\boldsymbol{b}_j : \chi_{b_j} \mapsto \mathbb{R}^{O_j}, \chi_{b_j} \subseteq \chi$. We allow for various predictor terms including an intercept (or bias term), linear effects $\boldsymbol{b}_j(\boldsymbol{x}) = x_{k_j}$ for some $k_j \in \{1, \ldots, p\}$, structured non-linear terms $\boldsymbol{b}_j(\boldsymbol{x}) = G(x_{k_j})$ with some basis function $G : \mathbb{R} \mapsto \mathbb{R}^q, q \geq 1$ such as a B-spline basis, bivariate non-linear terms $\boldsymbol{b}_j(\boldsymbol{x}) = G'(x_{k_j}, x_{k_{j'}})$ using a tensor-product basis $G' : \mathbb{R} \times \mathbb{R} \mapsto \mathbb{R}^{q'}, q' \geq 1$ or neural network predictors $\boldsymbol{b}_j(\boldsymbol{x}) = d(\boldsymbol{x}_{k_j})$, which define an arbitrary (deep) neural network that takes (potentially multidimensional) features $\boldsymbol{x}_{k_j} \in \chi$. The network will be used to learn latent features representing the unstructured data source. These features are then combined as a linear combination when multiplied with $\boldsymbol{\Gamma}_j$. The same types of predictors can also be defined for the shift term $\beta(\boldsymbol{x}) = \sum_{j=1}^{J} \boldsymbol{c}_j(\boldsymbol{x})^\top \boldsymbol{\psi}_j$, which we also defined as an additive predictor of features, basis functions or deep neural networks times their (final) weighting $\boldsymbol{\psi}_j$.

The final model output for the transformation of $y$ is then given by

$$\boldsymbol{a}(y)^\top \boldsymbol{\vartheta}(\boldsymbol{x}) = \boldsymbol{a}(y)^\top \boldsymbol{\Gamma} \boldsymbol{B}, \tag{6}$$

with $\boldsymbol{\Gamma} = (\boldsymbol{\Gamma}_1, \ldots, \boldsymbol{\Gamma}_J) \in \mathbb{R}^{(M+1)\times P}, P = \sum_{j=1}^{J} O_j$ the stacked coefficient matrix combining all $\boldsymbol{\Gamma}_j$s and $\boldsymbol{B} \in \mathbb{R}^P$ a stacked vector of the predictor terms $b_j(\boldsymbol{x})$s. Based on model assumption (4) we can define the loss function based on the change of variable theorem

$$f_Y(y|\boldsymbol{x}) = f_Z(h(y|\boldsymbol{x})) \cdot \left|\frac{\partial h(y|\boldsymbol{x})}{\partial y}\right|$$

as

$$\begin{aligned} \ell(h(y|\boldsymbol{x})) &= -\log f_Y(y|\boldsymbol{\vartheta}(\boldsymbol{x}), \boldsymbol{\beta}(\boldsymbol{x})) \\ &= -\log f_Z(\boldsymbol{a}(y)^\top \boldsymbol{\vartheta}(\boldsymbol{x}) + \beta(\boldsymbol{x})) - \log[\boldsymbol{a}'(y)^\top \boldsymbol{\vartheta}(\boldsymbol{x})] \end{aligned} \tag{7}$$

with $\boldsymbol{a}'(y) = \partial \boldsymbol{a}(y)/\partial y$.

For $n$ observations $(y_i, \boldsymbol{x}_i), i = 1, \ldots, n$, we can represent (6) as

$$(\boldsymbol{\mathcal{A}} \odot \boldsymbol{\mathcal{B}})\text{vec}(\boldsymbol{\Gamma}^\top) \tag{8}$$

with $\boldsymbol{\mathcal{A}} = (\boldsymbol{a}(y_1), \ldots, \boldsymbol{a}(y_n))^\top \in \mathbb{R}^{n \times (M+1)}$, $\boldsymbol{\mathcal{B}} = (\boldsymbol{B}_1, \ldots, \boldsymbol{B}_n)^\top \in \mathbb{R}^{n \times P}$, vectorization operator vec$(\cdot)$ and the row-wise tensor product (also known as transpose Kathri-Rao product) operator $\odot$. Similar, the distribution shift can be written in matrix form as $\boldsymbol{C\Psi}$ with $\boldsymbol{C} \in \mathbb{R}^{n \times Q}$ consisting of the stacked $\boldsymbol{c}_j(\boldsymbol{x})$s and $\boldsymbol{\Psi} = (\boldsymbol{\psi}_1^\top, \ldots, \boldsymbol{\psi}_J^\top)^\top \in \mathbb{R}^Q$ the stacked vector of all shift term coefficients. A schematic representation of an exemplary DCTM is given in Fig. 2.

### 2.2 Network Definition

Our network consists of two main parts: a feature transforming network (FTN) part, converting $\boldsymbol{X} = (\boldsymbol{x}_1^\top, \ldots, \boldsymbol{x}_n^\top)^\top \in \mathbb{R}^{n \times p}$ to $\boldsymbol{\mathcal{B}}$ and an outcome transforming network (OTN) part, transforming $\boldsymbol{y} = (y_1, \ldots, y_n)^\top \in \mathbb{R}^n$ to $h(\boldsymbol{y}|\boldsymbol{X}) \in \mathbb{R}^n$. In the OTN part the matrix $\boldsymbol{\Gamma}$ is learned, while the FTN part only contains additional parameters to be learned by the network if some feature(s) are defined using a deep neural network. In other words, if only structured linear effects or basis function transformations are used in the FTN part, $\boldsymbol{\Gamma}$ contains all trainable parameters. Figure 1 visualizes an exemplary architecture.

After the features are processed in the FTN part, the final transformed outcome is modeled using a conventional fully-connected layer with input $\boldsymbol{\mathcal{A}} \odot \boldsymbol{\mathcal{B}}$, one hidden unit with linear activation function and weights corresponding to vec$(\boldsymbol{\Gamma})$. The deep conditional transformation model as visualized in Fig. 1 can also be defined with one common network which is split into one part that is added to the shift predictor and one part that is used in the interaction predictor.

### 2.3 Penalization

$L_1$-, $L_2$-penalties can be incorporated in both the FTN and OTN part by adding corresponding penalty terms to the loss function. We further use smoothing penalties for structured non-linear terms by regularizing the respective entries in $\boldsymbol{\Psi}$ and $\boldsymbol{\Gamma}$ to avoid overfitting and easier interpretation. Having two smoothing directions, the penalty for $\boldsymbol{\Gamma}$ is constructed using a Kronecker sum of individual marginal penalties for anisotropic smoothing

$$\boldsymbol{D}_{\Gamma} = \lambda_a \boldsymbol{D}_a \oplus \lambda_b \boldsymbol{D}_b,$$

where the involved tuning parameters $\lambda_a, \lambda_b$ and penalty matrices $\boldsymbol{D}_a, \boldsymbol{D}_b$ correspond to the direction of $y$ and the features $\boldsymbol{x}$, respectively. Note, however, that for $\boldsymbol{\Gamma}$, the direction of $y$ usually does not require additional smoothing as it is already regularized through the monotonicity constraint [16]. The corresponding penalty therefore reduces to

$$\boldsymbol{D}_{\Gamma} = \boldsymbol{I}_P \otimes (\lambda_b \boldsymbol{D}_b) \tag{9}$$

with the diagonal matrix $\boldsymbol{I}_P$ of size $P$. These penalties are added to the negative log-likelihood defined by (7), e.g.,

$$\ell_{pen} = \ell(h(y|\boldsymbol{x})) + \text{vec}(\boldsymbol{\Gamma})^\top \boldsymbol{D}_{\Gamma} \text{vec}(\boldsymbol{\Gamma})$$

for a model with penalized structured effects only in $\boldsymbol{\mathcal{B}}$. As done in [37] we use the Demmler-Reinsch orthogonalization to relate each tuning parameter for smoothing penalties to its respective degrees-of-freedom, which allows a more intuitive setting of parameters and, in particular, allows to define equal amount of penalization for different smooth terms. Leaving the least flexible smoothing term unpenalized and adjusting all others to have the same amount of flexibility works well in practice.

### 2.4 Bijectivitiy and Monotonocity Constraints

To ensure bijectivity of the transformation of each $y_i$, we use Bernstein polynomials for $\boldsymbol{\mathcal{A}}$ and constraint the coefficients in $\boldsymbol{\Gamma}$ to be monotonically increasing in each column. The monotonicity of the coefficients in $\boldsymbol{\Gamma}$ can be implemented in several ways, e.g., using the approach by [11] or [39] on a column-basis. Note that this constraint directly yields monotonically increasing transformation functions if $P = 1$, i.e., if no or only one feature is used for $h_1$. If $P > 1$, we can ensure monotonicity of $h_1$ by using predictor terms in $\boldsymbol{\mathcal{B}}$ that are non-negative. A corresponding proof can be found in the Supplementary Material (Lemma 1)[1]. Intuitively the restriction can be seen as an implicit positivity assumption on the learned standard deviation of the error distribution $F_Z$ as described in the next section using the example of a normal distribution. Although non-negativity of predictor terms is not very restrictive, e.g., allowing for positive linear features, basis functions with positive domain such as B-splines or deep networks with positivity in the learned latent features (e.g., based on a ReLU activation function), the restriction can be lifted completely by simply adding a positive constant to $\boldsymbol{\mathcal{B}}$.

### 2.5 Interpretability and Identifiability Constraints

Several choices for $M$ and $F_Z$ will allow for particular interpretation of the coefficients learned in $\boldsymbol{\Psi}$ and $\boldsymbol{\Gamma}$. When choosing $F_Z = \Phi$ and $M = 1$, the DCTM effectively learns an additive regression model with Gaussian error distribution,

[1] https://github.com/PFMB/DCTMs.

i.e., $Y|\boldsymbol{x} \sim N(\tilde{\beta}(\boldsymbol{x}), \sigma_s^2)$. The unstandardized structured additive effects in $\tilde{\beta}(\boldsymbol{x})$ can then be divided by $\sigma_s$ yielding $\beta(\boldsymbol{x})$. Therefore $\beta(\boldsymbol{x})$ can be interpreted as shifting effects of normalized features on the transformed response $\mathbb{E}(h_1(y)|\boldsymbol{x})$. For $M > 1$, features in $\beta(\boldsymbol{x})$ will also affect higher moments of $Y|\boldsymbol{x}$ through a non-linear $h_1$, leading to a far more flexible modeling of $F_{Y|\boldsymbol{x}}$. Smooth monotonously increasing estimates for $\beta(\boldsymbol{x})$ then allow to infer that a rising $\boldsymbol{x}$ leads to rising moments of $Y|\boldsymbol{x}$ independent of the choice for $F_Z$. Choosing $F_Z = F_{MEV}$ or $F_Z = F_L$ allows $\beta(\boldsymbol{x})$ to be interpreted as additive changes on the log-hazard ratio or on the log-odds ratio, respectively. The weights in $\boldsymbol{\Gamma}$ determine the effect of $\boldsymbol{x}$ on $F_{Y|\boldsymbol{x}}$ as well as whether $F_{Y|\boldsymbol{x}}$ varies with the values of $y$ yielding a response-varying distribution [3] or not. In general, structured effects in $\boldsymbol{\Gamma}$ are coefficients of the tensor product $\boldsymbol{\mathcal{A}} \odot \boldsymbol{\mathcal{B}}$ and can, e.g., be interpreted by 2-dimensional contour or surface plots (see, e.g., Fig. 1 in the Supplementary Material).

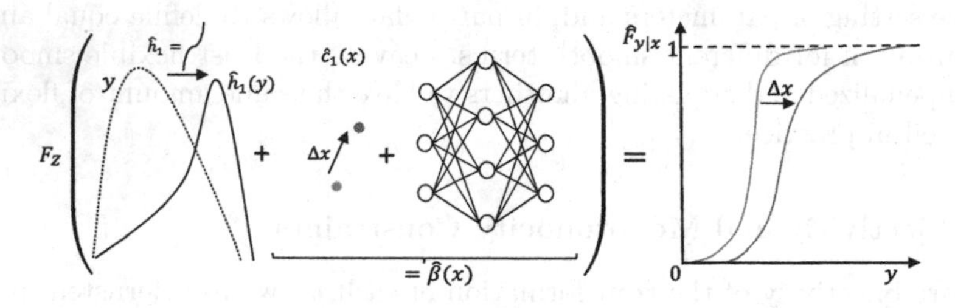


**Fig. 2.** Schematic representation of an exemplary DCTM with a learned transformation $\hat{h}_1$ for the outcome $y$. The shift term $\hat{\beta}(x)$ is composed of an estimated smooth term $\hat{c}_1(x) = c_1(x)\hat{\psi}_1$ for $x$ and a neural network predictor. An increase in $x$ is indicated by $\Delta x$ with corresponding effect on $\hat{F}_{Y|x}$ through $\hat{h}_2 = \hat{\beta}(x)$ on the right hand side of the equation.

In order to ensure identifiability and thus interpretability of structured effects in $h_1$ and $h_2$, several model definitions require the additional specifications of constraints. If certain features in $\boldsymbol{\mathcal{B}}$ or $\boldsymbol{\mathcal{C}}$ are modeled by both a flexible neural network predictor $d(\boldsymbol{x})$ and structured effects $s(\boldsymbol{x})$, the subnetwork $d(\boldsymbol{x})$ can easily assimilate effects $s(\boldsymbol{x})$ is supposed to model. In this case, identifiability can be ensured by an orthogonalization cell [37], projecting the learned effects of $d(\boldsymbol{x})$ in the orthogonal complement of the space spanned by features modeled in $s(\boldsymbol{x})$. Further, when more than one smooth effect or deep neural network is incorporated in either $\boldsymbol{\mathcal{B}}$ or $\boldsymbol{\mathcal{C}}$, these terms can only be learned up to an additive constants. To solve this identifiability issue we re-parameterize the terms and learn these effects with a sum-to-zero constraint. As a result, corresponding effects can only be interpreted on a relative scale. Note that this is a limitation of additive models per se, not our framework.

## 3 Numerical Experiments

We now demonstrate the efficacy of our proposed framework for the case of a shift CTM, a distributional CTM and an interacting CTM based on a general data generating process (DGP).

*Data Generating Process.* The data for the numerical experiments were generated according to $g(y) = \eta(\boldsymbol{x}) + \epsilon(\boldsymbol{x})$ where $g : \mathbb{R}^n \mapsto \mathbb{R}^n$ is bijective and differentiable, $\eta(\boldsymbol{x})$ is specified as in (3) and $\epsilon \sim F_Z$ with $F_Z$ being the error distribution. We choose $\epsilon(\boldsymbol{x}) \sim N(0, \sigma^2(\boldsymbol{x}))$ where $\sigma^2(\boldsymbol{x}) \in \mathbb{R}^+$ is specified as in (3) so that we can rewrite the model as

$$F_Z\left(\frac{g(y) - \eta(\boldsymbol{x})}{\sigma(\boldsymbol{x})}\right) = F_Z\left(h_1 + h_2\right). \tag{10}$$

From (1) and our model definition, (10) can be derived by defining $h_1$ as $g(y)\sigma^{-1}(\boldsymbol{x})$ and $h_2$ as $-\eta(\boldsymbol{x})\sigma^{-1}(\boldsymbol{x})$. We finally generate $y$ according to $g^{-1}(\eta(\boldsymbol{x}) + \epsilon(\boldsymbol{x}))$ with $\epsilon(\boldsymbol{x}) \sim N(0, \sigma^2(\boldsymbol{x}))$. We consider different specification for $g$, $\eta$, $\sigma$ and the order of the Bernstein polynomial $M$ for different samples sizes $n$ (see Supplementary Material).

*Evaluation.* To assess the estimation performance, we compute the relative integrated mean squared error (RIMSE) between $\hat{h}_1$, evaluated on a fine grid of $(y_i,\ \sigma(\boldsymbol{x}_i))$ value pairs, with the true functional form of $h_1$ as defined by the data generating process. For the estimation performance of $h_2$, we evaluate the corresponding additive predictor by calculating the mean squared error (MSE) between estimated and true linear coefficients for linear feature effects and the RIMSE between estimated and true smooth non-linear functions for non-linear functional effects. We compare the estimation against transformation boosting machines (TBM) [14] that also allow to specify structured additive predictors. Note, however, that TBMs only implement either the shift (TBM-Shift) or distributional CTM (TBM-Distribution), but do not allow for the specification of an interacting CTM with structured predictors, a novelty of our approach. In particular, only the TBM-Shift comes with an appropriate model specification such that it can be used for comparison in some of the DGP defined here.

*Results.* We first discuss the 4 out of 10 specifications of the true DGP where $h_1$ is not learned through features and thus allows for a direct comparison of TBM-Shift and DCTMs. For $h_1$, we find that, independent of the size of the data set and the order of the Bernstein polynomial, DCTMs provide a viable alternative to TBM-Shift, given the overlap between the (RI)MSE distributions and the fact that the structured effects in DCTMs are not tuned extensively in these comparisons. For $h_2$, DCTMs outperform TBM-Shift in all 16 configurations for $M/n$ among the 4 DGPs depicted in Fig. 3 when taking the mean or the median across the 20 replications. The simulation results for the 6 remaining DGPs can be found in the Supplementary Material. For $h_1$ and $h_2$, the

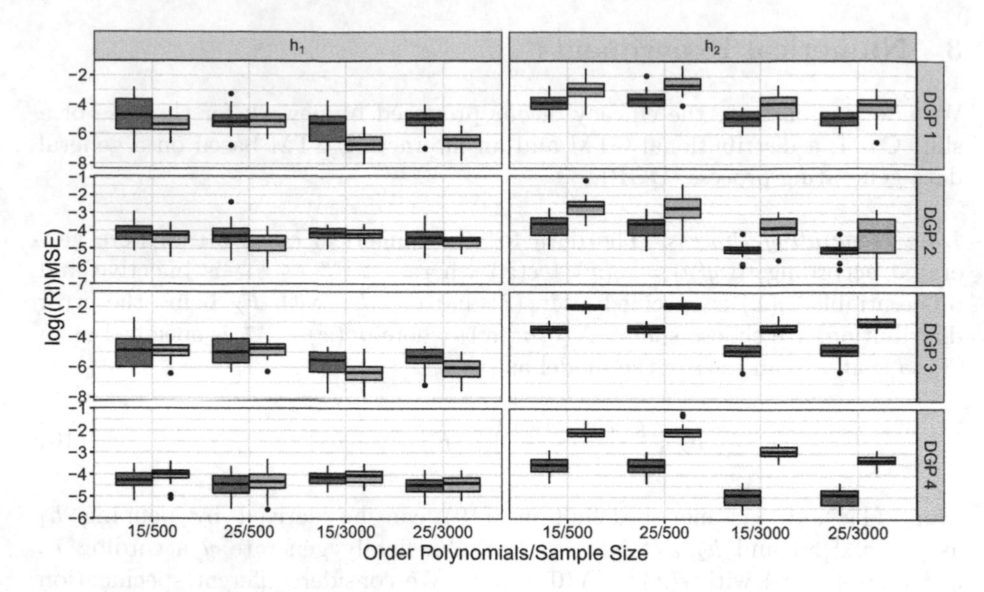


**Fig. 3.** Comparison of the logarithmic (RI)MSEs between TBM-Shift (yellow) and DCTM (blue) for different data generating processes (DGP in rows) as well as different orders of the Bernstein polynomial and the sample size ($M/n$ on the x-axis) for 20 runs. The specification of the DGPs can be found in the Supplementary Material and for this figure are based on $\sigma_1$ with alternating $g \in \{g_1, g_2\}$ and $\eta \in \{\eta_1, \eta_2\}$. (Color figure online)

results for the majority of specifications reveal that DCTMs benefit from lower order Bernstein polynomials independent of the sample size. When only unstructured model components were specified, DCTM's estimation of $h_1$ benefits from Bernstein polynomial with higher order. This holds regardless of $g$. Figure 1 in the Supplementary Material exemplary depicts the estimation performance of DCTMs for one DGP setting.

# 4 Application

We now demonstrate the application of DCTMs by applying the approach to a movie reviews and a face data set.

## 4.1 Movie Reviews

The Kaggle movies data set consists of $n = 4442$ observations. Our goal is to predict the movies' revenue based on their production budget, popularity, release date, runtime and genre(s). Figure 1 in the Supplementary Material depicts the revenue for different genres. We deliberately do not log-transform the response, but let the transformation network convert a standard normal distribution (our error distribution) to fit to the given data.

*Model Description.* First, we define a DCTM solely based on a structured additive predictor (i.e. no deep neural net predictor) as a baseline model which we refer to as the "Structured Model". The structured additive predictor includes the binary effect for each element of a set of 20 available genres ($x_0$) as well as smooth effects (encoded as a univariate thin-plate regression splines [42]) for the popularity score ($x_1$), for the difference of release date and a chosen date in the future in days ($x_2$), for the production budget in US dollars ($x_3$) and the run time in days ($x_4$):

$$\sum_{r=1}^{20}\beta_r I(r \in x_{0,i}) + s_1(x_{1,i}) + s_2(x_{2,i}) + s_3(x_{3,i}) + s_4(x_{4,i}). \quad (11)$$

This linear predictor (11) is used to define the structured component in the shift term $\beta(\boldsymbol{x})$. For the interaction term, the STAP consists of all the genre effects and the resulting design matrix $\boldsymbol{\mathcal{B}}$ is then combined with the basis of a Bernstein polynomial $\boldsymbol{\mathcal{A}}$ of order $M = 25$. We compare this model with three deep conditional transformation models that use additional textual information of each movie by defining a "Deep Shift Model", a "Deep Interaction Model" and a "Deep Combination Model". The three models all include a deep neural network as input either in the shift term, in the interaction term or as input for both model parts, respectively. As deep neural network we use an embedding layer of dimension 300 for 10000 unique words and combine the learned outputs by flatting the representations and adding a fully-connected layer with 1 unit for the shift term and/or 1 units for the interaction term on top. As base distribution we use a logistic distribution, i.e., $F_Z(h) = F_L(h) = (1 + \exp(-h))^{-1}$.

*Comparisons.* We use 20% of the training data as validation for early stopping and define the degrees-of-freedom for all non-linear structured effects using the strategy described in Sect. 2.3. We compare our approach again with the shift and distributional TBM (TBM-Shift and TBM-Distribution, respectively) as state-of-the-art baseline. We run both models with the predictor specification given in (11). For TBM, we employ a 10-fold bootstrap to find the optimal stopping iteration by choosing the minimum out-of-sample risks averaged over all folds. Finally we evaluate the performance on the test data for both algorithms.

*Results.* The non-linear estimations of all models show a similar trend for the four structured predictors. Figure 4 depicts an example for the estimated partial effects in the $h_2$ term of each model. The resulting effects in Fig. 4 can be interpreted as functional log-odds ratios due to the choice $F_Z = F_L$. For example, the log-odds for higher revenue linearly increase before the effect stagnates for three of the four model at a level greater than 150 million USD. Table 1 shows (Movie Reviews column) the mean predicted log-scores [10], i.e., the average log-likelihood of the estimated distribution of each model when trained on 80% of the data (with 20% of the training data used as validation data) and evaluated on the remaining 20% test data. Results suggest that deep extensions with movie descriptions as additional predictor added to the baseline model can improve over the TBM, but do not achieve as good prediction results as the purely structured DCTM model in this case. Given the small amount of data, this result is

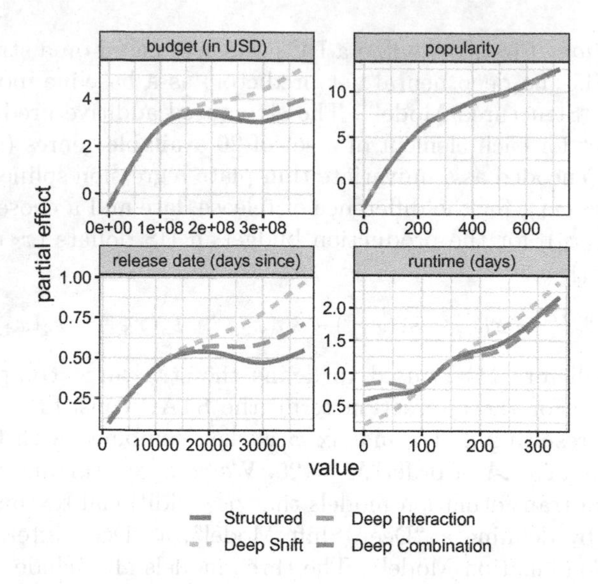


**Fig. 4.** Estimated non-linear partial effect of the 4 available numerical features for $h_2$ (in each sub-plot) based on the four different DCTM models (colors). (Color figure online)

not surprising and showcases a scenario, where the potential of the structured model part outweighs the information of a non-tabular data source. The flexibility of our approach in this case allows to seamlessly traverse different model complexities and offers a trade-off between complexity and interpretability.

**Table 1.** Average result (standard deviation in brackets) over different training/test-splits on the movie reviews (left) and UTKFace data set. Values correspond to negative predicted log-scores (PLS; smaller is better) for each model with best score in bold.

| | Model | Movie reviews | UTKFace |
|---|---|---|---|
| DCTM | Structured | **19.26** (0.18) | 3.98 (0.02) |
| | Deep Shift | 19.32 (0.20) | 3.81 (0.52) |
| | Deep Interaction | 19.69 (0.22) | 3.79 (0.21) |
| | Deep Combination | 19.67 (0.19) | **3.37** (0.09) |
| TBM | Shift | 23.31 (0.83) | 4.25 (0.02) |
| | Distribution | 22.38 (0.31) | 4.28 (0.03) |

## 4.2 UTKFace

The UTKFace dataset is a publicly available image dataset with $n = 23708$ images and additional tabular features (age, gender, ethnicity and collection date). We use this data set to investigate DCTMs in a multimodal data setting.

*Model Description.* Our goal is to learn the age of people depicted in the images using both, the cropped images and the four tabular features. As in the previous section we fit the four different DCTM models, all with the same structured additive predictor (here effect for race, gender and a smooth effect for the collection date) and add a deep neural network predictor to the $h_1$ (Deep Interaction), $h_2$ (Deep Shift), to both (Deep Combination) or only fit the structured model without any information of the faces (Structured). The architecture for the faces consists of three CNN blocks (see Supplementary Material for details) followed by flattening operation, a fully-connected layer with 128 units with ReLU activation, batch normalization and a dropout rate of 0.5. Depending on the model, the final layer either consists of 1 hidden unit (Deep Shift, Deep Interaction) or 2 hidden units (Deep Combination).

*Comparisons.* The baseline model is a two-stage approach that first extracts latent features from the images using a pre-trained VGG-16 [40] and then uses these features together with the original tabular features in a TBM-Shift/-Distribution model to fit a classical structured additive transformation model. We again compare the 4 DCTM models and 2 baseline models using the PLS on 30% test data and report model uncertainties by repeating the data splitting and model fitting 4 times. For the DCTMs we use early stopping based on 20% of the train set used for validation. For TBM models we search for the best stopping iteration using a 3-fold cross-validation. The results in Table 1 (UTKFace column) suggest that our end-to-end approach works better than the baseline approach and that the DCTM benefits from a combined learning of $h_1$ and $h_2$ through the images.

### 4.3 Benchmark Study

We finally investigate the performance of our approach by comparing its density estimation on four UCI benchmark data sets (Airfoil, Boston, Diabetes, Forest Fire) against parametric alternatives. We use a deep distributional regression approach (DR) [37], a Gaussian process (GP) and a GP calibrated with an isotonic regression (IR) [26]. We adapt the same architecture as in DR to specifically examine the effect of the proposed transformation. To further investigate the impact of the polynomials' order M (i.e., flexibility of the transformation vs. risk of overfitting), we run the DCTM model with $M \in \{1, 16, 32, 64\}$ (DCTM-$M$). We also include a normalizing flow baseline with a varying number of radial flows $\mathcal{M}$ (NF-$\mathcal{M}$; [36]). This serves as a reference for a model with more than one transformation and thus potentially more expressiveness at the expense of the feature-outcome relationship being not interpretable. Details for hyperparameter specification can be found in the Supplementary Material.

Results (Table 2) indicate that our approach performs similar to alternative methods. For two data sets, the greater flexibility of the transformation yields superior performance compared to methods without transformation (DR, GP, IR), suggesting that the transition from a pure parametric approach to a more flexible transformation model can be beneficial. For the other two data sets,

**Table 2.** Comparison of neg. PLS (with standard deviation in brackets) of different methods (rows; best-performing model in bold, second best underlined) on four different UCI repository datasets (columns) based on 20 different initializations of the algorithms.

| | Airfoil | Boston | Diabetes | Forest F. |
|---|---|---|---|---|
| DR | 3.11 (0.02) | 3.07 (0.11) | **5.33** (0.00) | 1.75 (0.01) |
| GP | 3.17 (6.82) | 2.79 (2.05) | 5.35 (5.76) | 1.75 (7.09) |
| IR | 3.29 (1.86) | 3.36 (5.19) | 5.71 (2.97) | **1.00** (1.94) |
| DCTM-1 | 3.07 (0.01) | 2.97 (0.03) | 5.44 (0.02) | 1.83 (0.02) |
| DCTM-16 | 3.07 (0.02) | 2.76 (0.02) | <u>5.34</u> (0.01) | 1.30 (0.12) |
| DCTM-32 | 3.08 (0.02) | <u>2.71</u> (0.03) | 5.39 (0.02) | <u>1.08</u> (0.15) |
| DCTM-64 | 3.08 (0.03) | **2.66** (0.05) | 5.37 (0.01) | 1.41 (1.03) |
| NF-1 | 3.04 (0.22) | 2.98 (0.20) | 5.59 (0.10) | 1.77 (0.02) |
| NF-3 | **2.88** (0.17) | 2.76 (0.14) | 5.54 (0.11) | 1.76 (0.02) |
| NF-5 | <u>2.90</u> (0.18) | 2.81 (0.20) | 5.47 (0.09) | 1.77 (0.12) |

DCTM's performance is one standard deviation apart from the best performing model. For Airfoil the even greater flexibility of a chain of transformations (NF-$\mathcal{M}$ in comparison to DCTM-$M$) improves upon the result of DCTMs.

## 5 Conclusion and Outlook

We introduced the class of deep conditional transformation models which unifies existing fitting approaches for transformation models with both interpretable (non-)linear model terms and more complex predictors in one holistic neural network. A novel network architecture together with suitable constraints and network regularization terms is introduced to implement our model class. Numerical experiments and applications demonstrate the efficacy and competitiveness of our approach.

**Acknowledgement.** This work has been partly funded by SNF grant 200021-184603 from the Swiss National Science Foundation (Torsten Hothorn) and the German Federal Ministry of Education and Research (BMBF) under Grant No. 01IS18036A (David Rügamer).

## References

1. Athey, S., Tibshirani, J., Wager, S., et al.: Generalized random forests. Ann. Stat. **47**(2), 1148–1178 (2019)
2. Box, G.E., Cox, D.R.: An analysis of transformations. J. Roy. Stat. Soc. Ser. B (Methodol.) **26**(2), 211–243 (1964)
3. Chernozhukov, V., Fernández-Val, I., Melly, B.: Inference on counterfactual distributions. Econometrica **81**(6), 2205–2268 (2013)

4. Depeweg, S., Hernandez-Lobato, J.M., Doshi-Velez, F., Udluft, S.: Decomposition of uncertainty in Bayesian deep learning for efficient and risk-sensitive learning. In: International Conference on Machine Learning, pp. 1184–1193. PMLR (2018)
5. Durkan, C., Bekasov, A., Murray, I., Papamakarios, G.: Cubic-spline flows. arXiv preprint arXiv:1906.02145 (2019)
6. Durkan, C., Bekasov, A., Murray, I., Papamakarios, G.: Neural spline flows. In: Wallach, H., Larochelle, H., Beygelzimer, A., d' Alché-Buc, F., Fox, E., Garnett, R. (eds.) Advances in Neural Information Processing Systems, vol. 32. Curran Associates, Inc. (2019)
7. Fahrmeir, L., Kneib, T., Lang, S., Marx, B.: Regression: Models, Methods and Applications. Springer, Heidelberg (2013). https://doi.org/10.1007/978-3-642-34333-9
8. Farouki, R.T.: The Bernstein polynomial basis: a centennial retrospective. Comput. Aided Geom. Des. **29**(6), 379–419 (2012)
9. Foresi, S., Peracchi, F.: The conditional distribution of excess returns: an empirical analysis. J. Am. Stat. Assoc. **90**(430), 451–466 (1995)
10. Gelfand, A.E., Dey, D.K.: Bayesian model choice: asymptotics and exact calculations. J. Roy. Stat. Soc. Ser. B (Methodol.) **56**(3), 501–514 (1994)
11. Gupta, M., et al.: Monotonic calibrated interpolated look-up tables. J. Mach. Learn. Res. **17**(109), 1–47 (2016)
12. Hora, S.C.: Aleatory and epistemic uncertainty in probability elicitation with an example from hazardous waste management. Reliab. Eng. Syst. Saf. **54**(2–3), 217–223 (1996)
13. Hothorn, T.: Most likely transformations: the mlt package. J. Stat. Softw. Articles **92**(1), 1–68 (2020)
14. Hothorn, T.: Transformation boosting machines. Stat. Comput. **30**(1), 141–152 (2019). https://doi.org/10.1007/s11222-019-09870-4
15. Hothorn, T., Kneib, T., Bühlmann, P.: Conditional transformation models. J. R. Stat. Soc. Ser. B Stat. Methodol. **76**, 3–27 (2014)
16. Hothorn, T., Möst, L., Bühlmann, P.: Most likely transformations. Scand. J. Stat. **45**(1), 110–134 (2018)
17. Hothorn, T., Zeileis, A.: Predictive distribution modeling using transformation forests. J. Comput. Graph. Stat. 1–16 (2021). https://doi.org/10.1080/10618600.2021.1872581
18. Hüllermeier, E., Waegeman, W.: Aleatoric and epistemic uncertainty in machine learning: a tutorial introduction. arXiv preprint arXiv:1910.09457 (2019)
19. Jaini, P., Selby, K.A., Yu, Y.: Sum-of-squares polynomial flow. CoRR (2019)
20. Kendall, A., Gal, Y.: What uncertainties do we need in Bayesian deep learning for computer vision? In: Advances in Neural Information Processing Systems, pp. 5574–5584 (2017)
21. Klein, N., Hothorn, T., Kneib, T.: Multivariate conditional transformation models. arXiv preprint arXiv:1906.03151 (2019)
22. Kobyzev, I., Prince, S., Brubaker, M.: Normalizing flows: an introduction and review of current methods. IEEE Trans. Pattern Anal. Mach. Intell. 1 (2020). https://doi.org/10.1109/tpami.2020.2992934
23. Koenker, R.: Quantile Regression. Economic Society Monographs, Cambridge University Press, Cambridge (2005)
24. Kook, L., Herzog, L., Hothorn, T., Dürr, O., Sick, B.: Ordinal neural network transformation models: deep and interpretable regression models for ordinal outcomes. arXiv preprint arXiv:2010.08376 (2020)

25. Kooperberg, C., Stone, C.J., Truong, Y.K.: Hazard regression. J. Am. Stat. Assoc. **90**(429), 78–94 (1995)
26. Kuleshov, V., Fenner, N., Ermon, S.: Accurate uncertainties for deep learning using calibrated regression. In: Proceedings of the 35th International Conference on Machine Learning, vol. 80, pp. 2796–2804 (2018)
27. Leorato, S., Peracchi, F.: Comparing distribution and quantile regression. EIEF Working Papers Series 1511, Einaudi Institute for Economics and Finance (EIEF) (2015)
28. Meinshausen, N.: Quantile regression forests. J. Mach. Learn. Res. **7**, 983–999 (2006)
29. Müller, T., McWilliams, B., Rousselle, F., Gross, M., Novák, J.: Neural importance sampling (2019)
30. Papamakarios, G., Nalisnick, E., Rezende, D.J., Mohamed, S., Lakshminarayanan, B.: Normalizing flows for probabilistic modeling and inference (2019)
31. Pratola, M., Chipman, H., George, E.I., McCulloch, R.: Heteroscedastic BART via multiplicative regression trees. J. Comput. Graph. Stat. **29**, 405–417 (2019)
32. Ramasinghe, S., Fernando, K., Khan, S., Barnes, N.: Robust normalizing flows using Bernstein-type polynomials (2021)
33. Rezende, D., Mohamed, S.: Variational inference with normalizing flows. In: Proceedings of Machine Learning Research, vol. 37, pp. 1530–1538 (2015)
34. Rigby, R.A., Stasinopoulos, D.M.: Generalized additive models for location, scale and shape. J. Roy. Stat. Soc. Ser. C (Appl. Stat.) **54**(3), 507–554 (2005)
35. Rothe, C., Wied, D.: Misspecification testing in a class of conditional distributional models. J. Am. Stat. Assoc. **108**(501), 314–324 (2013)
36. Rothfuss, J., et al.: Noise regularization for conditional density estimation (2020)
37. Rügamer, D., Kolb, C., Klein, N.: Semi-Structured Deep Distributional Regression: Combining Structured Additive Models and Deep Learning. arXiv preprint arXiv:2002.05777 (2020)
38. Senge, R., et al.: Reliable classification: learning classifiers that distinguish aleatoric and epistemic uncertainty. Inf. Sci. **255**, 16–29 (2014)
39. Sick, B., Hathorn, T., Dürr, O.: Deep transformation models: Tackling complex regression problems with neural network based transformation models. In: 2020 25th International Conference on Pattern Recognition (ICPR), pp. 2476–2481 (2021). https://doi.org/10.1109/ICPR48806.2021.9413177
40. Simonyan, K., Zisserman, A.: Very deep convolutional networks for large-scale image recognition. In: Bengio, Y., LeCun, Y. (eds.) 3rd International Conference on Learning Representations, ICLR 2015, San Diego, CA, USA, 7–9 May 2015, Conference Track Proceedings (2015)
41. Tabak, E.G., Turner, C.V.: A family of nonparametric density estimation algorithms. Commun. Pure Appl. Math. **66**(2), 145–164 (2013)
42. Wood, S.N.: Thin plate regression splines. J. R. Stat. Soc. Ser. B (Stat. Methodol.) **65**(1), 95–114 (2003)
43. Wu, C.O., Tian, X.: Nonparametric estimation of conditional distributions and rank-tracking probabilities with time-varying transformation models in longitudinal studies. J. Am. Stat. Assoc. **108**(503), 971–982 (2013)

# Disentanglement and Local Directions of Variance

Alexander Rakowski[1(✉)] and Christoph Lippert[1,2]

[1] Hasso Plattner Institute for Digital Engineering, University of Potsdam, Potsdam, Germany
{alexander.rakowski,christoph.lippert}@hpi.de
[2] Hasso Plattner Institute for Digital Health at Mount Sinai, New York, USA

**Abstract.** Previous line of research on learning disentangled representations in an unsupervised setting focused on enforcing an uncorrelated posterior. These approaches have been shown both empirically and theoretically to be insufficient for guaranteeing disentangled representations. Recent works postulate that an implicit PCA-like behavior might explain why these models still tend to disentangle, exploiting the structure of variance in the datasets. Here we aim to further verify those hypotheses by conducting multiple analyses on existing benchmark datasets and models, focusing on the relation between the structure of variance induced by the ground-truth factors and properties of the learned representations. We quantify the effects of global and local directions of variance in the data on disentanglement performance using proposed measures and seem to find empirical evidence of a negative effect of local variance directions on disentanglement. We also invalidate the robustness of models with a global ordering of latent dimensions against the local vs. global discrepancies in the data.

**Keywords:** Disentanglement · Variational autoencoders · PCA

## 1 Introduction

Given the growing sizes of modern datasets and the costs of manual labeling, it is desirable to provide techniques for learning useful low-dimensional representations of data in an unsupervised setting. Disentanglement has been postulated as an important property of the learned representations [2,12,28]. Current state-of-the-art approaches utilize Variational Autoencoders (VAEs) [18], a powerful framework both for variational inference and generative modeling.

The primary line of work introduced models that were claimed to disentangle either by controlling the bottleneck capacity [5,13] or explicitly imposing a factorising prior [6,17,19]. However, the validity of these claims has been challenged with the *Impossibility Result* [22,23], showing the unidentifiability of disentangled representations in a purely unsupervised setting. An exhaustive empirical study from the same work further revealed that these models indeed


N. Oliver et al. (Eds.): ECML PKDD 2021, LNAI 12977, pp. 19–34, 2021.
https://doi.org/10.1007/978-3-030-86523-8_2

fail to provide a consistent performance regardless of any tested hyperparameter setting. Instead, the scores exhibit a relatively high variance - disentanglement seems to happen "at random" - or at least cannot be attributed to the proposed techniques.

Meanwhile another line of research emerged, pointing to similarities between VAEs and Principal Component Analysis (PCA) [27] as a possible explanation to why disentanglement can happen in VAE-learned representations [33]. It was further postulated that the **global** structure of variance in the data is an inductive bias that is being exploited, while local discrepancies in the directions of variance in the data should have an adverse effect [41]. An example of different amounts of variance caused by the same underlying factor is shown in Fig. 1.

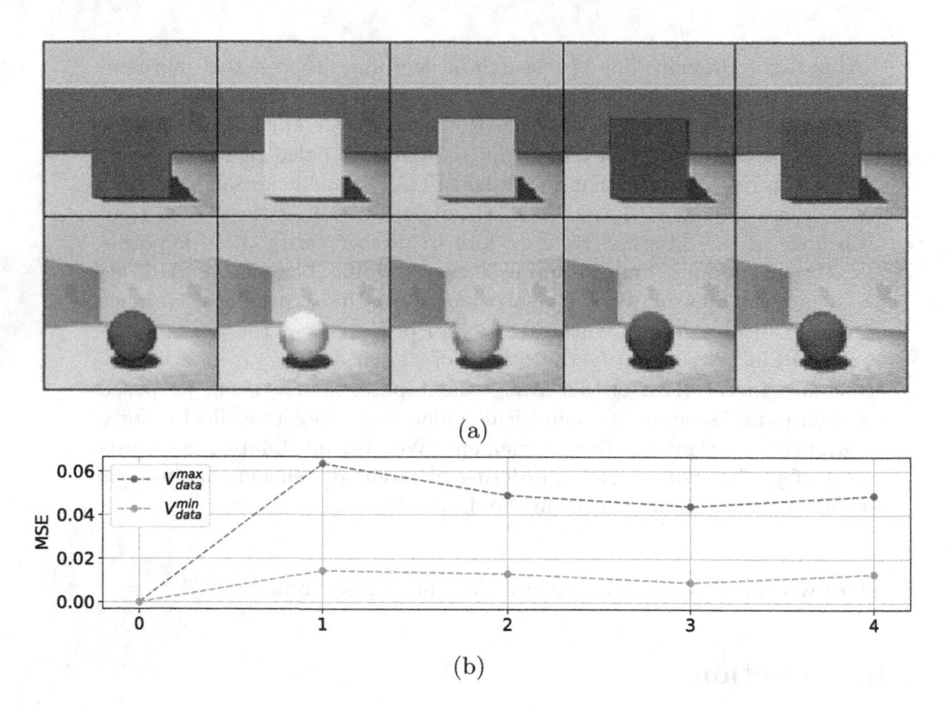


**Fig. 1.** Example of different local impacts of a ground-truth factor in the Shapes3D dataset. **a)**: Top and bottom rows show images generated by interpolating over a single underlying factor - the color of the foreground object - for points belonging to hyperplanes in the latent space where this factor induces the most and the least variance respectively. **b)**: Mean Squared Error between consecutive pairs of these images - blue values correspond to the region with the most induced variance, the orange ones to the one with least variance. (Color figure online)

In this work we aim to gain more insight and possibly further validate these claims by performing empirical analyses of properties of both the datasets and trained models. Our contributions are as follows: a) We formulate questions regarding the relations between the structure of variance in data, learned encodings and disentanglement. b) We define measures to quantify these properties and use them to perform statistical analyses to answer these questions c) We design synthetic datasets with hand-controlled specific structures of variance and employ models with more explicit connections to PCA. d) We identify strong connections between the defined measures and model properties and performance - in particular, we seem to find evidence for the negative effect of local variance directions on disentanglement. e) Contrary to the hypothesis from [41] we do not observe benefits of employing models with a global PCA-like behavior in the presence of such local discrepancies.

## 2 Related Work

Arguably one of the most important works on unsupervised disentanglement is the study of [22], which showed the limitations of current approaches, necessitating a shift towards either weak supervision [3,14,16,24,36] or investigating (potentially implicit) inductive biases [11,31,41].

Connections between autoencoders and PCA have been investigated, among others, in [1,29,33,41] and between PCA modifications (e.g., pPCA, RPCA) in [7,20,25]. [37] note the bias of VAEs towards PCA solutions and illustrate it with examples of toy data where this bias is beneficial and ones where it is not. Furthermore, the framework of Deep Restricted Kernel Machines [38] provides a Kernel PCA [34] interpretation of autoencoders. Such models with imposed orthogonality constraints on the latent space encodings are shown to be a competitive alternative for VAEs in terms of disentanglement [40].

We build largely upon the findings of the recent work of [41], who postulate a global structure of variance in the datasets as an existing inductive bias. While providing an insightful perspective on the problem, they aim to validate this claim somewhat indirectly, generating artificial data from trained neural networks. The use of such complex non-linear models can obfuscate and change the actual problem being analyzed. Instead, we work directly on the original data, analyzing the local and global variance structures with quantities computed either analytically or estimated using Monte Carlo procedures. [8] is another work connected to ours, as they compute similarities between learned encodings of multiple trained models. These quantities are then used to arguably successfully select in an unsupervised manner models that disentangle, based on the assumption that "disentangled representations are all alike". We are not aware, to the best of our knowledge, of other works that conduct an empirical analysis of properties of both data and models, in the context of disentanglement.

## 3 Disentanglement, PCA and VAEs

### 3.1 Preliminaries

**Variational Autoencoders.** Introduced in [18], VAEs are a framework for performing variational inference using latent variable models. The objective is to find parameters $\boldsymbol{\theta}$ that maximize the Evidence Lower Bound (ELBO), which we will denote as $\mathcal{L}(\boldsymbol{\theta}, \mathbf{x})$, which lower-bounds the otherwise intractable marginal log-likelihood of the $N$ observed data points $\mathbf{x}^{(i)}$:

$$\sum_i^N \log p_{\boldsymbol{\theta}}(\mathbf{x}^{(i)}) \geq \sum_i^N \mathbb{E}_{q_{\boldsymbol{\theta}}(\mathbf{z}|\mathbf{x}^{(i)})}[\log p_{\boldsymbol{\theta}}(\mathbf{x}^{(i)}|\mathbf{z})] - D_{KL}(q_{\boldsymbol{\theta}}(\mathbf{z}|\mathbf{x}^{(i)})||p(\mathbf{z})) \quad (1)$$

In practice the prior distribution over the latent variables $p(\mathbf{z})$ is taken to be the PDF of a standard Normal distribution $\mathcal{N}(\mathbf{0}, \mathbf{I})$, while $q_{\boldsymbol{\theta}}$ and $p_{\boldsymbol{\theta}}$ are modeled with a neural network model each (encoder and decoder), optimized with a gradient ascent algorithm of choice using the objective function:

$$-\mathcal{L}(\boldsymbol{\theta}, \mathbf{x}) = \mathcal{L}_{Rec}(\mathbf{x}', \mathbf{x}) + \beta \mathcal{L}_{KLD}(\boldsymbol{\mu}, \boldsymbol{\sigma}^2) \quad (2)$$

where $\boldsymbol{\mu}, \boldsymbol{\sigma} = Enc_\theta(\mathbf{x})$ is the local parameterization of $q_{\boldsymbol{\theta}}(\mathbf{z}|\mathbf{x})$, $\mathbf{x}' = Dec_\theta(\mathbf{z})$ is the reconstructed input and $\mathbf{z} \sim \mathcal{N}(\boldsymbol{\mu}, \boldsymbol{\sigma}^2\mathbf{I})$ is its sampled latent representation. $\mathcal{L}_{Rec}$ is usually based on Gaussian or Bernoulli priors, yielding the mean squared error (MSE) or binary cross-entropy (BCE) losses. The second term, $\mathcal{L}_{KLD}(\cdot, \cdot)$, is the Kullback-Leibler divergence between the (approximate) posterior and prior distributions. For the standard Normal prior it can be computed in closed form:

$$\mathcal{L}_{KLD}(\boldsymbol{\mu}, \boldsymbol{\sigma}^2) = \sum_{j=1}^{d} [\boldsymbol{\mu}_j^2 + \boldsymbol{\sigma}_j^2 - \log \boldsymbol{\sigma}_j^2 - 1] \quad (3)$$

This can also be understood as a regularization term, forcing the distribution $q_{\boldsymbol{\theta}}(\mathbf{z}|\mathbf{x})$ of latent codes $\mathbf{z}$ around each data point $\mathbf{x}$ to match the (uninformative) prior. The $\beta$ term in Eq. 2 acts as a hyperparameter controlling the tradeoff between reconstruction quality and regularization, introduced with the $\beta$-VAE model [13]. It is worth highlighting, that even though both $Enc_\theta$ and $Dec_\theta$ are deterministic functions, $\mathcal{L}_{Rec}(\mathbf{x}', \mathbf{x})$ contains hidden stochasticity induced by sampling $\mathbf{z} \sim Enc_\theta(\mathbf{x})$ which in turn generates $\mathbf{x}' = Dec_\theta(\mathbf{z})$.

**Disentanglement.** This property is usually defined with respect to the *ground-truth factors of variation* - a set of $d$ random variables $\mathbf{G} = (G_1, \ldots, G_d)$ which generate the (usually higher-dimensional) data $\mathbf{X}$ via an unknown function $f : \mathbb{R}^d \mapsto \mathbb{R}^k$. It is commonly assumed that the factors are mutually independent, i.e., coming from a factorized distribution, e.g., $\mathcal{N}(\mathbf{0}, \mathbf{I})$. The task is then to learn a representation $r : \mathbb{R}^k \mapsto \mathbb{R}^l$, where each dimension is dependent on at most one factor $G_i$. There exist several metrics for measuring disentanglement of representations [6,9,13,17,19,32], however the majority of them was found to be strongly correlated [22].

Thus for brevity we report only the Mutual Information Gap (MIG) [6] throughout this work, as it has the advantage of being deterministic and having relatively few hyperparameters. Informally, it first computes the mutual information between each factor and latent dimension of the representation, and stores these values in a $d \times l$ matrix. For each $G_i$, its corresponding MIG is then taken to be the difference (gap) between the two highest normalized entries of the $i$-th row. Intuitively, if a factor is encoded with a single latent, this value will be close to 1, and close to 0 otherwise (for an entangled representation).

**Impossibility Result.** In a fully unsupervised setting we never observe the true values of $\mathbf{g}$ and only have access to the corresponding points in the data space $\mathbf{x} = f(\mathbf{g})$ generated from them. Variational autoencoders are arguably the state-of-the-art method in this setting, where the representation $r$ is defined by the encoder network, with its corresponding posterior distribution $q_{\boldsymbol{\theta}}(\mathbf{z}|\mathbf{x})$. However, the main difficulty lies assessing whether a representation is disentangled, without having access to $\mathbf{g}$. There exists a potentially infinite number of solutions yielding the same marginal $p(\mathbf{x})$ and $p(\mathbf{z})$, which VAEs are optimized to match, but different conditional $p_{\boldsymbol{\theta}}(\mathbf{x}|\mathbf{z})$ and $q_{\boldsymbol{\theta}}(\mathbf{z}|\mathbf{x})$ [22]. This is due to the rotational invariance of the prior distribution - applying a rotation to $\mathbf{z}$ in the encoder does not change $p(\mathbf{z})$. By undoing the rotation in the decoder, the marginal $p(\mathbf{x})$ remains unchanged, keeping the value of the ELBO objective (Eq. 1) intact.

### 3.2 Disentanglement in a PCA Setting

Unsupervised disentanglement is thus fundamentally impossible in a general setting. It is believed however, that certain inductive biases might help to overcome this difficulty. In particular, a PCA-like behavior of VAEs has been postulated as one [33,41]. Recall that the PCA objective can be formulated as minimizing the squared reconstruction error between the original and reconstructed data:

$$\min_{\mathbf{W}} \|\mathbf{X} - \mathbf{W}\mathbf{W}^{\top}\mathbf{X}\|_2^2, \quad s.t.\ \forall i : \|\mathbf{W}_i\| = 1 \tag{4}$$

where $\mathbf{W}$ is usually taken to be of a smaller rank than $\mathbf{X}$. We can draw the analogy between disentanglement and PCA by treating $\mathbf{W}^{\top}$ and $\mathbf{W}$ as (linear) encoder and decoder networks respectively, and $\mathbf{z} = \mathbf{W}^{\top}\mathbf{x}$ as the low-dimensional representations of the data. A PCA model could thus be able to disentangle, given that the underlying factors $G$ that generated the data coincide with the principal components [12]. An important benefit of this approach is the potential to escape the problem of non-identifiability of the solution highlighted in Sect. 3.1, due to the imposed ordering of principal components. Adopting this view allows us to reason about the conditions regarding the data for identifiability of a potentially disentangled representation. For the solution of PCA to be unique, the data covariance matrix $\mathbf{X}^{\top}\mathbf{X}$ must have non-degenerate eigenvalues. This translates to the assumption, that each $G_i$ should have a different impact on the data. While it is unlikely that these values will be exactly the same in real-life scenarios, one can still recover differing solutions due to noise in the data or stochasticity of optimization algorithms. If the alignment with PCA directions were in fact an inductive bias beneficial for disentangling, we could ask the following:

**Question 1.** *Do models disentangle better on datasets where the ground-truth factors of variation contribute to the observed data with different magnitudes?*

### 3.3 PCA Behavior in Variational Autoencoders

The relation between autoencoders and PCA has been studied from several different angles, under the assumption of linear networks. Just the $L_2$

reconstruction objective is already enough for a linear autoencoder to learn the subspace spanned by the PCA solution [1,4], while the actual principal components can be identified by performing Singular Value Decomposition of the model's weights [29]. The variational mechanisms of VAEs provide further connections to PCA. Specifically, the variance components $\boldsymbol{\sigma}^2$ promote local orthogonality of the decoder [33] and even alignment with local the principal components [41] around a point $\mathbf{x}$. While insightful, these findings hold under several simplifying assumptions, such as linearity of the decoder and varying $\boldsymbol{\sigma}_i^2$, which must not necessarily be true in practice. The optimal solution also depends on the value $\beta$, via its effect on $\mathcal{L}_{KLD}$. However indirect, these analogies to PCA are argued to be crucial for disentanglement [26,33,40,41], leading us to ask:

**Question 2.** *Given the assumption from Question 1, will models with a more explicit PCA behavior achieve better disentanglement scores (e.g., MIG)?*

Another caveat of the above is that it is a **local** (wrt. the data space) effect, as in the standard VAE framework values of $(\boldsymbol{\sigma}^{(i)})^2$ depend on $\mathbf{x}^{(i)}$. In case of non-linearly generated data these locally uncovered directions must not match the global ones. It is postulated in [41] that a consistent structure of variance in the dataset is an inductive bias allowing VAEs to disentangle.

**Question 3.** *Are local vs. global discrepancies present in the benchmark datasets? Do they have an influence on disentanglement scores obtained by the models?*

If the assumption about the negative effect of non-global PCA behavior in VAEs were indeed true, a natural solution would be to employ models which order the latent dimensions globally [25,35].

**Question 4.** *Are models explicitly imposing a global ordering of latents robust to the inconsistency of variance in the data?*

We describe two such models employed in our study in Sect. 5.2.

# 4 Measuring Induced Variance and Consistency

## 4.1 Ground-Truth Factor Induced Variance

Since the impact of the $j$-th ground-truth factor on the generated data cannot be computed analytically in the benchmark datasets, we define it instead as the average per-pixel variance induced by interventions on that factor:

$$V^i_{data}(\mathbf{g}) = \frac{1}{k}\sum_{j}^{k} \operatorname*{Var}_{g_i \sim p(g_i)} (p(\mathbf{x}_j|\mathbf{g}_{(\mathbf{g}_i=g_i)})), \tag{5}$$

which is computed locally on realizations of $\mathbf{g}$. By marginalizing over $\mathbf{z}$ we can obtain a global measure of influence: $\overline{V}^i_{data} = \int V^i_{data}(\mathbf{g})p(\mathbf{g})d\mathbf{g}$. We can estimate the values of $\overline{V}^i_{data}$ via a Monte Carlo procedure: we sample $N$ points in the ground-truth space, and for each of them generate additional points by iterating over all possible values of $\mathbf{g}_i$ while keeping all the other factors $\mathbf{g}_{j \neq i}$ fixed. Using these values we define the "spread" of per-factor contributions as:

$$S_{data} = \operatorname*{Var}_{i \in [d]}(\overline{V}^i_{data}) \tag{6}$$

over the (finite) set of all per-factor values of $\overline{V}^i_{data}$ for a given dataset.

## 4.2 Local Directions of Variance

Building up on the definition of the per-factor induced variance we can define the consistency of a factor's impact on the data:

$$C^i_{data} = \operatorname*{Var}_{\mathbf{g} \sim p(\mathbf{g})}(V^i_{data}(\mathbf{g})) \tag{7}$$

In contrast to $\overline{V}_{data}$, which measures the average impact over the data, $C_{data}$ quantifies the variability of it. For example, data generated by a (noiseless) linear transformation would be considered perfectly consistent ($\forall i : C^i_{data} = 0$). This must not be true however for more complex data - changes in color of a particular object will induce a smaller total change in regions of the data space where the said object has a smaller size or is occluded - see Fig. 1. To estimate these values empirically we use an analogous procedure as in Sect. 4.1.

## 4.3 Consistency of Encodings

To quantify which latent dimensions of a model encode a certain ground-truth factor $i$ we measure the amount of variance induced in each dimension $j$ by changes in that factor around a $\mathbf{g}$:

$$V^{i,j}_{enc}(\mathbf{g}) = \operatorname*{Var}_{g_i \sim p(g_i)}(q(\mathbf{z}_j|\mathbf{x})p(\mathbf{x}|\mathbf{g}_{(\mathbf{g}_i = g_i)})) \tag{8}$$

We estimate these values empirically in a similar manner as $V_{data}$ and $C_{data}$ (Sects. 4.1, 4.2), with an additional step of passing the obtained points in the data (image) space through a trained encoder network. Due to the non-linear nature of the networks modeling $q(\mathbf{z}|\mathbf{x})$ there is no guarantee that a given factor will be mapped to the same latent dimensions **globally** over the dataset. Variance of $V_{enc}$ over the dataset can then be used as a proxy measure of the consistency of encodings:

$$C^i_{enc} = \frac{1}{l}\sum_{j=1}^{l} C^{i,j}_{enc} = \frac{1}{l}\sum_{j=1}^{l} \operatorname*{Var}_{\mathbf{g} \sim p(\mathbf{g})}(V^{i,j}_{enc}(\mathbf{g})) \tag{9}$$

High values of $C^i_{enc}$ might indicate that the $i$-th factor is encoded differently over different parts of the data space - perhaps due to differing local vs. global structures of variance. Note that this is irrespective of whether the representation is disentangled or not - instead it can be thought of as consistency in preserving the same rotation of the latent space wrt. the ground-truth factors.

# 5 Experimental Setup

## 5.1 Datasets

We constructed two families of synthetic datasets to investigate the effects of the spread of per-factor induced variances $S_{data}$ and consistency of per-factor induced variances $C_{data}$ on disentanglement and properties of VAE models.

**Synthetic Data with Varying $\overline{V}_{data}$.** In the former case we generate higher-dimensional data $\mathbf{x} \in \mathbb{R}^k$ from low-dimensional ground-truth factors $\mathbf{z} \in \mathbb{R}^d$ via random linear mappings $\mathbf{W} \in \mathbb{R}^{d \times k}$ with unit-norm columns:

$$\mathbf{x} = \mathbf{z}\mathbf{W} = \sum_i \mathbf{z}_i \mathbf{W}_i; \ \|\mathbf{W}_i\| = 1; \ \mathbf{z} \sim \mathcal{N}(\mathbf{0}, \boldsymbol{\sigma}^2 \mathbf{I}) \tag{10}$$

This is a non-trivial task in terms of disentangling, since (linear) ICA results require at most one factor to be Gaussian for identifiability [10,15]. Values of $\overline{V}^i_{data}$ can be computed in closed form, since the data points are defined as linear transformations of Gaussian random variables. We create several different datasets by varying the diversity of entries of $\boldsymbol{\sigma}^2$, ranging from all factors having the same variance ($\sigma_1^2 = \sigma_2^2 = \ldots = \sigma_d^2$) to all of them being different ($\sigma_1^2 < \sigma_2^2 < \ldots < \sigma_d^2$). Note that for this dataset we have $\forall i : C^i_{data} = 0$ due to the linearity of the data-generating process.

**Synthetic Data with Non-global Variance Structure.** The second variant is constructed to have differing local principal directions. We define three ground-truth factors: two zero-centered Normal random variables $z_1, z_2$ with $\sigma_1^2 < \sigma_2^2$, and a third variable $z_3 \sim \mathcal{U}\{1, k\}$. We then construct the data vectors $\mathbf{x} \in \mathbb{R}^k$ the following way:

$$\mathbf{x}_i = \begin{cases} z_1, & \text{if } i \leq z_3 \\ z_2, & \text{otherwise} \end{cases} \tag{11}$$

Thus $z_3$ defines the number of entries in $\mathbf{x}$ equal to $z_1$, while the rest is filled with $z_2$. Because of that it also changes the local principal components of $\mathbf{x}$ (and their corresponding eigenvalues). There are regions of high $z_3$ where it is more profitable for the encoder to transmit $z_1$ with more precision than $z_2$ - even though the latter induces more variance in $\mathbf{x}$ globally ($\overline{V}^2_{data} > \overline{V}^1_{data}$). Thus, contrary to the previous dataset, we have $C^1_{data}, C^2_{data} > 0$.

**Benchmark Data.** We also used commonly established "benchmark" datasets, first employed together in the study of [22]: *dSprites*, *noisy-*, *color-* and *scream-dSprites*, *NORB*, *Cars3D* and Shapes3D [13,17,21,22,30]. They all consist of $64 \times 64$ sized images, created in most cases artificially, with known corresponding ground-truth factors used to generate each image.

### 5.2 Models

**Baseline.** As a baseline for the experiments we took the $\beta$-VAE model. For the benchmark datasets we used pretrained models from the study of [22], which are publicly available for download[1] (apart from models trained on the Shapes3D dataset).

**Global Variance VAE.** Introduced in [25], this model uses a global variance vector i.e., $\boldsymbol{\sigma}^2$ is not a function of $\mathbf{x}$. More formally, the posterior $q_{\boldsymbol{\theta}}$ is defined as:

$$q_{\boldsymbol{\theta}}(\mathbf{z}|\mathbf{x}) = \mathcal{N}(\boldsymbol{\mu}, \boldsymbol{\sigma}^2\mathbf{I}) = \mathcal{N}(Enc_\theta(\mathbf{x}), \boldsymbol{\sigma}^2\mathbf{I}) \tag{12}$$

where $\boldsymbol{\sigma}^2 \perp\!\!\!\perp \mathbf{X}$. Note that the variance components are still learned from the data, just kept as constant parameters after training. For linear networks, the global solution of this model coincides with the solution of pPCA [39].

**Hierarchical Non-linear PCA (h-NLPCA).** Introduced in [35], this model imposes an explicit ordering of the latent dimensions in terms of their contribution to reconstruction. This is done by using a modified reconstruction loss:

$$\mathcal{L}_{Rec\_H}(\theta, \mathbf{x}) = \frac{1}{l}\sum_{i}^{l} \mathcal{L}^{i}_{Rec\_H}(\theta, \mathbf{x}) = \frac{1}{l}\sum_{i}^{l} \mathcal{L}_{Rec}(Dec_\theta(Enc_\theta(\mathbf{x}) \odot \boldsymbol{\delta}_i)), \tag{13}$$

where $\boldsymbol{\delta}_i$ denotes a "masking" vector with $l - i$ leading 1's and $i$ trailing 0's. $\mathcal{L}^{i}_{H\_Rec}$ thus measures the reconstruction loss when allowing information to be encoded only in the first $i$ latent dimensions, forcing each $\mathbf{z}_i$ to be more beneficial for reconstruction than $\mathbf{z}_{i+1}$.

## 6 Results

### 6.1 The Effect of Different Per-Factor Contributions

**Synthetic Data.** We created several variants of the dataset defined in Sect. 5.1 with increasing values of $S_{data}$. There is barely any change in disentanglement scores wrt. $S_{data}$ for $\beta$-VAE and Global Variance models. On the other hand, the h-NLPCA models exhibit a much stronger relation, outperforming the baseline the more the higher the dataset's $S_{data}$ (see Fig. 2). This might indicate that the explicit ordering of dimensions in h-NLPCA is beneficial for settings with high $S_{data}$. Even more notable, however, is the fact that even these models underperform against a simple PCA baseline.

[1] github.com/google-research/disentanglement_lib.

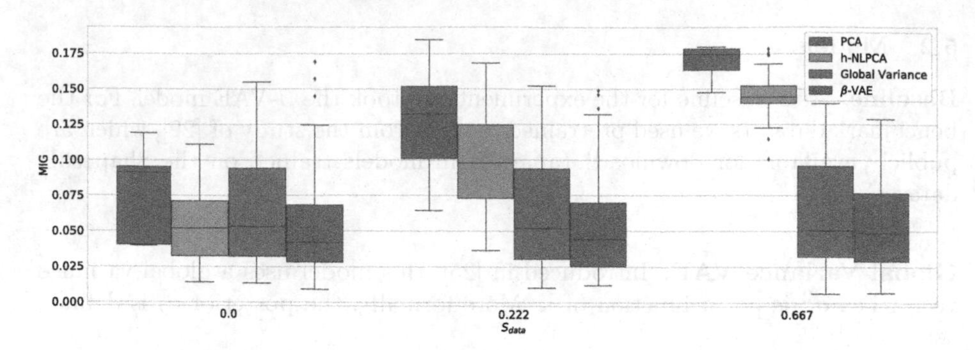


**Fig. 2.** MIG scores obtained on synthetic data with controlled $S_{data}$. Models imposing an explicit ordering of latent dimensions wrt. impact on reconstruction (PCA and h-NLPCA) perform best, especially when $S_{data} > 0$.

**Benchmark Data.** This relation is also visible on the benchmark datasets. Figure 3 shows the Pearson correlation of estimated values of $S_{data}$ with obtained MIG scores. Interestingly, while $\beta$-VAE models exhibit an arguably strong correlation (>0.70) for smaller values of $\beta$, increasing the regularization strength seems to consistently weaken this relation. This seems contradictory with the regularization loss term being postulated as the mechanism inducing PCA-like behavior in VAEs. The other models exhibit a somewhat reverse relation: while still present, the correlations are weaker, but they grow with increased $\beta$-s.

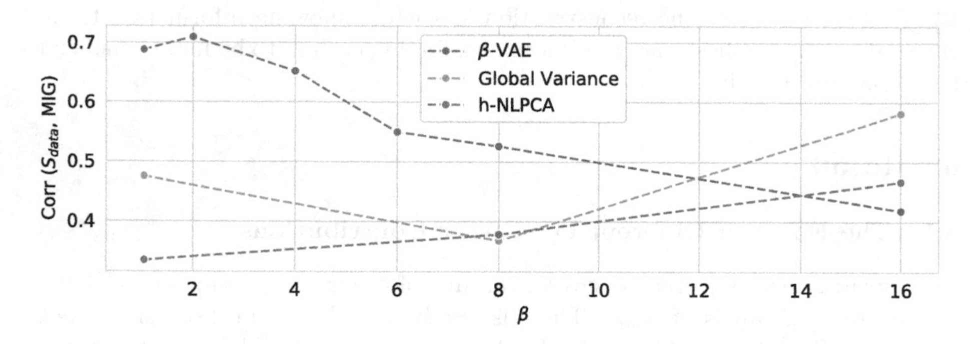


**Fig. 3.** Pearson correlation coefficient between $S_{data}$ for each benchmark dataset with MIG scores obtained on it (y-axis), for different values of $\beta$ (x-axis). There is an arguably strong correlation, which decreases with higher $\beta$-s for $\beta$-VAE while increasing for the other models.

### 6.2 The Effect of Non-global Variance Structure in the Data

**Does Inconsistency of Variance Structure in Data Correlate with Inconsistency of Encodings?** We observed prevailing correlations between the consistency of per-factor induced variance in the dataset $C^i_{data}$ and the corresponding consistency of learned encodings of that factor $C_{enc}^{i,\cdot}$, with the strongest values obtained by the Global Variance models. Table 1 shows values of the per-dataset Pearson correlation coefficient averaged over all $\beta$ values and random seeds. It seems that factors whose $V_{data}$ vary across the data tend to also be encoded with different latent dimensions, depending on the location in the data space. Figure 4 shows the mean correlation over all datasets wrt. values of $\beta$. While almost constant for Global Variance models, these correlations steadily decrease with higher $\beta$-s. One can see that for smaller values ($\leq$8) this effect is much stronger - up to a coefficient of over 0.7. This could mean that stronger regularization could be beneficial for alleviating non-global mappings of factor-latent being caused by non-global variance structure in the data.

**Table 1.** Per-dataset correlation between consistency of per-factor induced variance in data $C_{data}^{i}$ and variance in dimensions of learned representations $C_{enc}^{i,\cdot}$ for models trained on the benchmark datasets. While differing in strength, it is clearly present in almost all cases, the only exception being Cars3D for the h-NLPCA models.

| Model | dSprites | Color-dS | Noisy-dS. | Scream-dS. | Norb | Cars3D | Shapes3D | Mean |
|---|---|---|---|---|---|---|---|---|
| $\beta$-VAE | 0.302 | 0.598 | 0.432 | 0.501 | 0.447 | 0.462 | 0.314 | 0.437 |
| Global Var. | 0.710 | 0.862 | 0.409 | 0.647 | 0.287 | 0.501 | 0.379 | 0.542 |
| h-NLPCA | 0.307 | 0.558 | 0.247 | 0.407 | 0.339 | −0.124 | 0.635 | 0.338 |

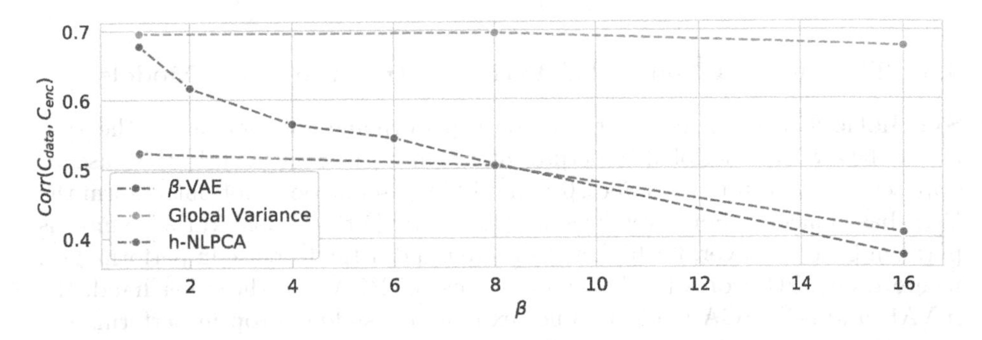


**Fig. 4.** Correlations between $C_{data}^{i}$ and $C_{enc}^{i,\cdot}$ (y-axis) averaged over all datasets, for different values of $\beta$ (x-axis). For models with a non-global variance structure a clear negative relation is visible.

**Do These Inconsistencies Correlate with Disentanglement Scores?** A perhaps more interesting question is whether the above effects translate to disentanglement of learned representations. In Sect. 6.1 we analyzed the effects of global, per-dataset, variance structures. Here we are interested in a more fine-grained, per-factor effect caused by the discrepancies between local and global directions of variance.

First we investigate whether disentanglement is correlated with inconsistency of encodings. This could reveal an indirect effect of $C_{data}$ on disentanglement, propagated through its correlation with $C_{enc}$ seen in the previous section. Table 2 shows per-dataset correlations between a model's $C_{enc}$ and its obtained MIG score - we observe negative correlations in most settings, albeit weaker than these from the previous section. While present for all model families, they are strongest for $\beta$-VAE.

We also look for a direct relation with $C_{data}$. To account for the correlation we observed in Sect. 6.1 (see Fig. 3), we first fit an ordinary least squares model on each dataset's values of $S_{data}$ and MIG scores across different $\beta$ values, and compute the correlations between $C_{data}$ on the residuals. While weaker than with $S_{data}$, there are still negative correlations present - models tend to perform worse on datasets with higher average $C_{data}$ (see Fig. 5). Interestingly, increasing $\beta$ seems to amplify this effect.

**Table 2.** Correlation between the MIG scores and $C_{enc}$ for models trained on the benchmark datasets. Negative correlations seem to be most prevalent for the $\beta$-VAE models.

| Model | dSprites | Color-dS. | Noisy-dS. | Scream-dS. | Norb | Cars3D | Shapes3D | Mean |
|---|---|---|---|---|---|---|---|---|
| $\beta$-VAE | −0.612 | −0.439 | −0.320 | −0.754 | 0.058 | −0.054 | −0.657 | −0.397 |
| Global Var | −0.110 | −0.251 | −0.194 | −0.532 | 0.069 | −0.033 | −0.207 | −0.180 |
| h-NLPCA | −0.103 | −0.068 | −0.081 | −0.035 | −0.105 | −0.410 | −0.337 | −0.163 |

## 6.3 The Effect of Non-global Variance Structure in the Models

**Synthetic Data.** In Fig. 6 we compare performance of models on the synthetic data with non-global structures of variance (see Sect. 5.1). In this case, as opposed to the results from Sect. 6.1, the PCA baseline does not outperform the VAE-based models. For lower $\beta$ values ($\{10^{-1}, 10^{0}\}$) the Global Variance models perform best. However, for higher $\beta$-s there is a drastic decrease in performance, where their MIG scores fall below even those of PCA. On the other hand, the $\beta$-VAE and h-NLPCA models do not exhibit this sudden drop in performance. This might suggest that when the directions of variance change over the dataset, the ability to adapt the variance structure locally improves robustness against tighter bottlenecks.

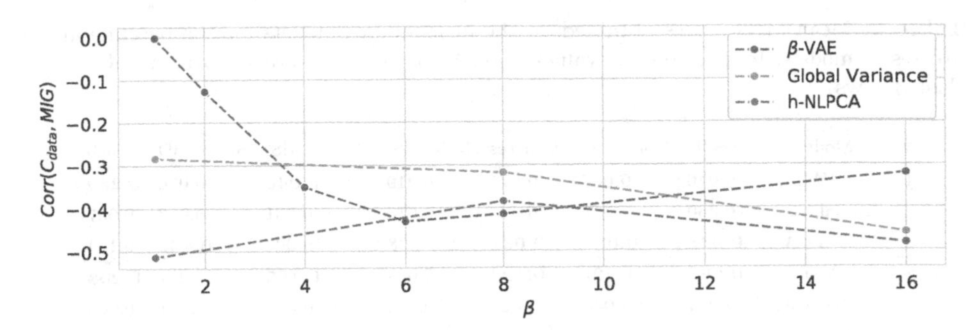


**Fig. 5.** Correlation between $C_{data}$ of benchmark datasets and obtained MIG scores (y-axis) for different values of $\beta$ (x-axis). At least a weak negative correlation is present for all tested models, often growing with higher $\beta$-s.

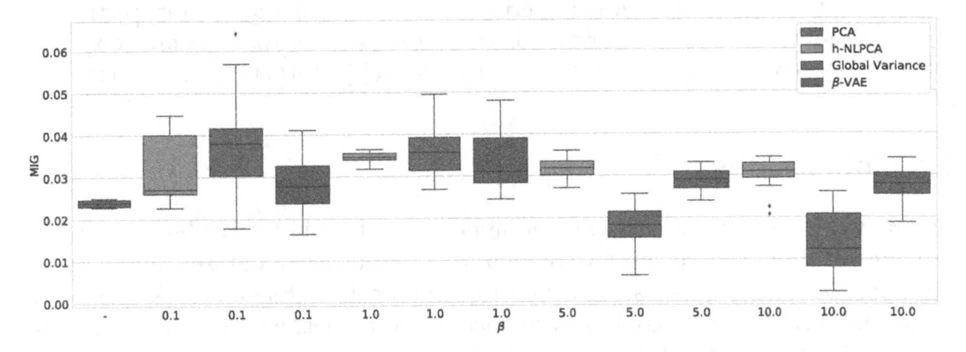


**Fig. 6.** MIG scores (y-axis) obtained on the dataset with non-global variance structure for different values of $\beta$ (x-axis). Notable is the sudden drop in performance for the Global Variance VAE for higher $\beta$-s.

**Benchmark Data.** We also perform a similar analysis on the benchmark data - the obtained MIG scores for each dataset and $\beta$ value are reported in Table 3. There seems to be no clear advantage of employing the proposed models over $\beta$-VAE. Especially the Global Variance VAE seems to drastically underperform, regardless of the hyperparameter setting. h-NLPCA models achieve results more comparable to the baseline. They seem to have an advantage with weak regularization - perhaps this is due to the PCA-like behavior being induced not only by the KL divergence, but also by the modified reconstruction loss.

**Table 3.** Mean MIG scores obtained on the benchmark datasets for the baseline and proposed models, for different $\beta$ values (first 9 rows) and averaged over all of them (last 3 rows).

| $\beta$ | Model | Cars3D | Color-dS. | dSprites | Noisy-dS. | Scream-dS. | Shapes3D | Norb |
|---|---|---|---|---|---|---|---|---|
| 1 | $\beta$-VAE | 0.046 | 0.067 | 0.075 | 0.019 | **0.040** | 0.076 | **0.250** |
| | Global Var. | 0.030 | 0.027 | 0.025 | 0.015 | 0.031 | 0.053 | 0.057 |
| | h-NLPCA | **0.068** | **0.080** | **0.087** | **0.108** | 0.022 | **0.432** | 0.154 |
| 8 | $\beta$-VAE | **0.099** | 0.127 | **0.124** | 0.068 | **0.165** | 0.291 | **0.208** |
| | Global Var. | 0.031 | 0.048 | 0.029 | 0.029 | 0.049 | 0.091 | 0.066 |
| | h-NLPCA | 0.092 | **0.148** | 0.103 | **0.087** | 0.018 | **0.492** | 0.130 |
| 16 | $\beta$-VAE | **0.121** | **0.148** | **0.243** | 0.080 | **0.105** | **0.465** | **0.188** |
| | Global Var. | 0.049 | 0.030 | 0.027 | 0.034 | 0.042 | 0.112 | 0.158 |
| | h-NLPCA | 0.101 | 0.106 | 0.094 | **0.087** | 0.001 | 0.397 | 0.187 |
| – | $\beta$-VAE | **0.089** | **0.114** | **0.147** | 0.056 | **0.103** | 0.277 | **0.215** |
| | Global Var. | 0.037 | 0.035 | 0.027 | 0.026 | 0.041 | 0.085 | 0.094 |
| | h-NLPCA | 0.087 | 0.111 | 0.094 | **0.094** | 0.014 | **0.440** | 0.157 |

## 7 Conclusions

In this work we approach an existing hypothesis of an inductive bias for disentanglement, from the perspective of analyzing properties of datasets and models directly. Reviewing the existing connections between VAEs and PCA we stated questions regarding expected performance of the models wrt. structure of variance in the datasets, and defined quantifiable measures used to answer them.

Models seem to disentangle better on datasets where the per ground-truth factor induced variances vary stronger (Question 1). However there doesn't seem to be a clear benefit of exploiting this relation with models with stronger connections to PCA (Question 2). We also find that local vs. global discrepancies of variance structure are indeed present in the datasets, and are negatively correlated with both consistency of encodings and disentanglement (Question 3). Surprisingly, contrary to the assumption from [41], models with a global ordering of latents seem to be less robust against these discrepancies (Question 4).

We note that albeit seemingly strong in some cases, these correlations should not be taken as exhaustive or causal explanations of the mechanisms governing variational autoencoders and disentanglement. Instead, they are meant to empirically (in-)validate previously stated assumptions and point to new intuitions. Specifically, we see potential in further analyzing the connection to PCA and the role of local directions of variance.

## References

1. Baldi, P., Hornik, K.: Neural networks and principal component analysis: learning from examples without local minima. Neural Netw. **2**(1), 53–58 (1989)
2. Bengio, Y., Courville, A., Vincent, P.: Representation learning: a review and new perspectives. IEEE Trans. Pattern Anal. Mach. Intell. **35**(8), 1798–1828 (2013)

3. Bouchacourt, D., Tomioka, R., Nowozin, S.: Multi-level variational autoencoder: learning disentangled representations from grouped observations. In: Proceedings of the AAAI Conference on Artificial Intelligence, vol. 32 (2018)
4. Bourlard, H., Kamp, Y.: Auto-association by multilayer perceptrons and singular value decomposition. Biol. Cybern. **59**(4), 291–294 (1988)
5. Burgess, C.P., et al.: Understanding disentangling in beta-VAE. arXiv preprint arXiv:1804.03599 (2018)
6. Chen, T.Q., Li, X., Grosse, R.B., Duvenaud, D.K.: Isolating sources of disentanglement in variational autoencoders. In: Advances in Neural Information Processing Systems, pp. 2610–2620 (2018)
7. Dai, B., Wang, Y., Aston, J., Hua, G., Wipf, D.: Connections with robust PCA and the role of emergent sparsity in variational autoencoder models. J. Mach. Learn. Res. **19**(1), 1573–1614 (2018)
8. Duan, S., et al.: Unsupervised model selection for variational disentangled representation learning. In: International Conference on Learning Representations (2019)
9. Eastwood, C., Williams, C.K.: A framework for the quantitative evaluation of disentangled representations (2018)
10. Eriksson, J., Koivunen, V.: Identifiability and separability of linear ICA models revisited. In: Proceedings of ICA, vol. 2003, pp. 23–27 (2003)
11. Gondal, M.W., et al.: On the transfer of inductive bias from simulation to the real world: a new disentanglement dataset. Adv. Neural Inf. Process. Syst. **32**, 15740–15751 (2019)
12. Goodfellow, I., Bengio, Y., Courville, A.: Deep Learning. MIT Press, Cambridge (2016)
13. Higgins, I., et al.: Early visual concept learning with unsupervised deep learning. arXiv preprint arXiv:1606.05579 (2016)
14. Hosoya, H.: Group-based learning of disentangled representations with generalizability for novel contents. In: IJCAI, pp. 2506–2513 (2019)
15. Hyvärinen, A.: Survey on independent component analysis (1999)
16. Khemakhem, I., Kingma, D., Monti, R., Hyvarinen, A.: Variational autoencoders and nonlinear ICA: a unifying framework. In: International Conference on Artificial Intelligence and Statistics, pp. 2207–2217. PMLR (2020)
17. Kim, H., Mnih, A.: Disentangling by factorising. In: International Conference on Machine Learning, pp. 2649–2658. PMLR (2018)
18. Kingma, D.P., Welling, M.: Auto-encoding variational Bayes. arXiv preprint arXiv:1312.6114 (2013)
19. Kumar, A., Sattigeri, P., Balakrishnan, A.: Variational inference of disentangled latent concepts from unlabeled observations. In: International Conference on Learning Representations (2018)
20. Kunin, D., Bloom, J., Goeva, A., Seed, C.: Loss landscapes of regularized linear autoencoders. In: Chaudhuri, K., Salakhutdinov, R. (eds.) Proceedings of the 36th International Conference on Machine Learning. Proceedings of Machine Learning Research, vol. 97, pp. 3560–3569. PMLR, 09–15 June 2019
21. LeCun, Y., Huang, F.J., Bottou, L.: Learning methods for generic object recognition with invariance to pose and lighting. In: Proceedings of the 2004 IEEE Computer Society Conference on Computer Vision and Pattern Recognition 2004. CVPR 2004. vol. 2, pp. II–104. IEEE (2004)
22. Locatello, F., et al.: Challenging common assumptions in the unsupervised learning of disentangled representations. In: International Conference on Machine Learning, pp. 4114–4124. PMLR (2019)

23. Locatello, F., et al.: A commentary on the unsupervised learning of disentangled representations. In: Proceedings of the AAAI Conference on Artificial Intelligence, vol. 34, pp. 13681–13684 (2020)
24. Locatello, F., et al.: Weakly-supervised disentanglement without compromises. In: International Conference on Machine Learning, pp. 6348–6359. PMLR (2020)
25. Lucas, J., Tucker, G., Grosse, R.B., Norouzi, M.: Don't blame the ELBO! A linear VAE perspective on posterior collapse. In: Wallach, H., Larochelle, H., Beygelzimer, A., d'Alché-Buc, F., Fox, E., Garnett, R. (eds.) Advances in Neural Information Processing Systems, vol. 32. Curran Associates, Inc. (2019)
26. Pandey, A., Fanuel, M., Schreurs, J., Suykens, J.A.: Disentangled representation learning and generation with manifold optimization. arXiv preprint arXiv:2006.07046 (2020)
27. Pearson, K.: LIII. On lines and planes of closest fit to systems of points in space. London Edinburgh Dublin Philos. Mag. J. Sci. **2**(11), 559–572 (1901)
28. Peters, J., Janzing, D., Schölkopf, B.: Elements of Causal Inference: Foundations and Learning Algorithms. The MIT Press, Cambridge (2017)
29. Plaut, E.: From principal subspaces to principal components with linear autoencoders. arXiv preprint arXiv:1804.10253 (2018)
30. Reed, S.E., Zhang, Y., Zhang, Y., Lee, H.: Deep visual analogy-making. In: Advances in Neural Information Processing Systems, pp. 1252–1260 (2015)
31. Ren, X., Yang, T., Wang, Y., Zeng, W.: Rethinking content and style: exploring bias for unsupervised disentanglement. arXiv preprint arXiv:2102.10544 (2021)
32. Ridgeway, K., Mozer, M.C.: Learning deep disentangled embeddings with the f-statistic loss. In: Advances in Neural Information Processing Systems, pp. 185–194 (2018)
33. Rolinek, M., Zietlow, D., Martius, G.: Variational autoencoders pursue PCA directions (by accident). In: Proceedings of the IEEE/CVF Conference on Computer Vision and Pattern Recognition, pp. 12406–12415 (2019)
34. Schölkopf, B., Smola, A., Müller, K.-R.: Kernel principal component analysis. In: Gerstner, W., Germond, A., Hasler, M., Nicoud, J.-D. (eds.) ICANN 1997. LNCS, vol. 1327, pp. 583–588. Springer, Heidelberg (1997). https://doi.org/10.1007/BFb0020217
35. Scholz, M., Vigário, R.: Nonlinear PCA: a new hierarchical approach. In: ESANN, pp. 439–444 (2002)
36. Shu, R., Chen, Y., Kumar, A., Ermon, S., Poole, B.: Weakly supervised disentanglement with guarantees. In: International Conference on Learning Representations (2019)
37. Stühmer, J., Turner, R., Nowozin, S.: Independent subspace analysis for unsupervised learning of disentangled representations. In: International Conference on Artificial Intelligence and Statistics, pp. 1200–1210. PMLR (2020)
38. Suykens, J.A.: Deep restricted kernel machines using conjugate feature duality. Neural Comput. **29**(8), 2123–2163 (2017)
39. Tipping, M.E., Bishop, C.M.: Probabilistic principal component analysis. J. R. Stat. Soc. Ser. B (Stat. Methodol.) **61**(3), 611–622 (1999)
40. Tonin, F., Patrinos, P., Suykens, J.A.: Unsupervised learning of disentangled representations in deep restricted kernel machines with orthogonality constraints. arXiv preprint arXiv:2011.12659 (2020)
41. Zietlow, D., Rolinek, M., Martius, G.: Demystifying inductive biases for *beta*-VAE based architectures. arXiv preprint arXiv:2102.06822 (2021)

# Neural Topic Models for Hierarchical Topic Detection and Visualization

Dang Pham(✉) and Tuan M. V. Le(✉)

Department of Computer Science, New Mexico State University, Las Cruces, USA
{dangpnh,tuanle}@nmsu.edu

**Abstract.** Given a corpus of documents, hierarchical topic detection aims to learn a topic hierarchy where the topics are more general at high levels of the hierarchy and they become more specific toward the low levels. In this paper, we consider the joint problem of hierarchical topic detection and document visualization. We propose a joint neural topic model that can not only detect topic hierarchies but also generate a visualization of documents and their topic structure. By being able to view the topic hierarchy and see how documents are visually distributed across the hierarchy, we can quickly identify documents and topics of interest with desirable granularity. We conduct both quantitative and qualitative experiments on real-world large datasets. The results show that our method produces a better hierarchical visualization of topics and documents while achieving competitive performance in hierarchical topic detection, as compared to state-of-the-art baselines.

## 1 Introduction

Given a corpus of documents, hierarchical topic detection aims to learn a topic hierarchy where the topics are more general at high levels of the hierarchy and they become more specific toward the low levels. Flat topic models such as LDA [5] are not designed to detect topic hierarchies. Therefore, several hierarchical topic models including the nested Chinese restaurant process (nCRP) [3,4], the nested hierarchical Dirichlet process (nHDP) [24] have been proposed to overcome this limitation. These models can learn the latent hierarchical structure of topics and they have a wide variety of applications such as language modeling [10], entity disambiguation [14], and sentiment analysis [1,17]. More recently, there has been an increasing interest in neural approaches for topic modeling. Several flat neural topic models have been proposed for document modeling [7,28], and supervised topic modeling [29]. For detecting topic hierarchies, we have neural methods such as TSNTM [11]. While traditional hierarchical topic models often use inference algorithms like collapsed Gibbs sampling or stochastic variational inference, TSNTM is trained using the autoencoding variational Bayes (AEVB) [18], which scales to large datasets.

Besides topic modeling, visualization is also an important tool for the analysis of text corpora. Topic modeling with visualization can provide users with


N. Oliver et al. (Eds.): ECML PKDD 2021, LNAI 12977, pp. 35–51, 2021.
https://doi.org/10.1007/978-3-030-86523-8_3

an effective overview of the text corpus which could help users discover useful insights without going through each document. Therefore, in this work, we investigate neural approaches for the joint problem of hierarchical topic detection and visualization. We propose a joint neural topic model that can not only detect topic hierarchies but also generate a visualization of documents and their topic structure. By being able to view the topic hierarchy as well as how documents are distributed across the hierarchy, users can quickly identify documents and topics of interest with desirable granularity. There are several types of visualization for visualizing topic hierarchies and documents including scatter plots [8], Sankey diagram [15], Sunburst diagram [26], and tag cloud [30]. In this work, we are interested in scatter plot visualization where documents, topics, and the topic hierarchy are embedded in a 2-d or 3-d visualization space. The joint problem of hierarchical topic detection and visualization can be formally stated as follows.

**Problem.** Let $\mathcal{D} = \{\mathbf{w}_n\}_{n=1}^{\mathcal{N}}$ denote a finite set of $\mathcal{N}$ documents and let $\mathcal{V}$ be a finite vocabulary from these documents. A document $n$ is represented as a vector of word counts $\mathbf{w}_n \in \mathcal{R}^{|\mathcal{V}|}$. Given visualization dimension $d$: 1) For hierarchical topic modeling, we want to find a hierarchy structure of latent topics where each node in the hierarchy is a topic $z$ and $\beta_z$ is its word distribution. The hierarchy can have an infinite number of branches and nodes (topics). The most general topic is at the root node and more specific topics are at the leaf nodes. We also find topic distributions of documents that are collectively denoted as $\boldsymbol{\Theta} = \{\theta_n\}_{n=1}^{\mathcal{N}}$; 2) For visualization, we want to find $d$-dimensional visualization coordinates for $\mathcal{N}$ documents $\boldsymbol{X} = \{x_n\}_{n=1}^{\mathcal{N}}$, and all $Z$ topics $\boldsymbol{\Phi} = \{\phi_z\}_{z=1}^{Z}$ such that the distances between documents, topics in the visualization space reflect the topic-document distributions $\boldsymbol{\Theta}$ as well as properties of the topic hierarchy.

There are three aspects considered in the stated problem. In the first aspect, we want to infer the latent topics in the text corpus. In the second aspect, we also want to organize these topics into a hierarchy. Finally, we want to visualize documents and their topics in the same visualization space for visual analysis. Most of the joint approaches so far only focus on one or two aspects. LDA [5] can learn topics but not their structure. nCRP [4], TSNTM [11] or other hierarchical topic models can both learn topics and organize them into a hierarchy. However, these topic models do not generate a visualization of documents and their topics. Therefore, recent topic models such as PLSV [12] and its variants [21,22] are proposed to jointly infer topics and visualization using a single objective function. However, since they are flat topic models, they cannot learn or visualize the topic hierarchy.

In this paper, we aim to propose a neural hierarchical topic model, namely *HTV*, that jointly addresses all three aspects of the problem. In our approach, documents and topics are embedded in the same 2-d or 3-d visualization space. We introduce the path and level distributions over an infinite tree, and parameterize them by document and topic coordinates. To possibly create an unbounded topic tree, we use a doubly-recurrent neural network (DRNN) [2] to generate topic embeddings. Our contributions are as follows:

- We propose *HTV*[1], a novel visual hierarchical neural topic model for hierarchical topic detection and visualization.
- We develop an AEVB inference for our model that involves using a doubly-recurrent neural network (DRNN) over an infinite tree and parameterizing the path and level distributions by document and topic coordinates. We also introduce the use of graph layout objective function of the Kamada-Kawai (KK) algorithm for visualizing the topic tree in our model.
- We conduct extensive experiments on several real-world datasets. The experimental results show that our method produces a better hierarchical visualization of topics and documents while achieving competitive performance in hierarchical topic detection, as compared to state-of-the-art baselines.

# 2 Visual and Hierarchical Neural Topic Model

## 2.1 Generative Model

In this section, we present the generative process of our proposed model. As shown in Fig. 1, the topic hierarchy can be considered as a tree where each node is a topic. The tree could have an infinite number of branches and levels. The topic at the root is the most general and topics at the leaf nodes are more specific. To sample a topic for each word $w_{nm}$ in a document $n$, a path $c_{nm}$ from the root to a leaf node and a level $l_{nm}$ are drawn. Let $\beta_{c_{nm}[l_{nm}]}$ be the topic distribution of the topic in the path $c_{nm}$ and at level $l_{nm}$. The word $w_{nm}$ is then drawn from the multinomial distribution Mult $\left(\beta_{c_{nm}[l_{nm}]}\right)$. The full generative process of *HTV* is as follows:

1. For each document $n = 1, \cdots, \mathcal{N}$:
   (a) Draw a document coordinate: $x_n \sim \text{Normal}(\mathbf{0}, \gamma \boldsymbol{I})$
   (b) Obtain a path distribution: $\pi_n = f_\pi(x_n, \boldsymbol{\Phi})$
   (c) Obtain a level distribution: $\delta_n = f_\delta(x_n, \boldsymbol{\Phi})$
   (d) For each word $w_{nm}$ in document $n$:
      (i) Draw a path: $c_{nm} \sim \text{Mult}(\pi_n)$
      (ii) Draw a level: $l_{nm} \sim \text{Mult}(\delta_n)$
      (iii) Draw a word: $w_{nm} \sim \text{Mult}\left(\beta_{c_{nm}[l_{nm}]}\right)$

Here $\boldsymbol{\Phi} = \{\phi_z\}_{z=1}^{Z}$ are coordinates of all topics in the tree, $x_n$ is the coordinate of a document $n$. As in [4,11], for each document $n$, besides topic distribution $\theta_n$, we associate it with a path distribution $\pi_n$ over all the paths from the root to the leaf nodes, and a level distribution $\delta_n$ over all tree levels. To possibly model the topic tree with an infinite number of branches and levels, nCRP [4] assumes that the level distribution is drawn from a stick-breaking construction:

$$\eta_l \sim \text{Beta}(1, \alpha), \delta_l = \eta_l \prod_{i=1}^{l-1} (1 - \eta_i), \tag{1}$$

[1] The source code is available at https://github.com/dangpnh2/htv.

and the path distribution is drawn from a nested stick-breaking construction as follows:

$$v_z \sim \text{Beta}(1, \varphi), \pi_z = \pi_{par(z)} v_z \prod_{z', z' \in ls(z)} (1 - v_{z'}) \tag{2}$$

here $l$ is one of the levels, $z$ is one of the topics in the topic tree, $par(z)$ is the parent topic of $z$, and $ls(z)$ represents the set of $z$'s left siblings. $\eta_l$ and $v_z$ are stick proportions of level $l$ and topic $z$ respectively. In our model, since we also want to visualize the topic tree, we need to formulate a way to encode these stick breaking constructions into the visualization space to make sure that the tree can grow unbounded. We introduce two functions $f_\pi(x_n, \boldsymbol{\Phi})$ and $f_\delta(x_n, \boldsymbol{\Phi})$ that are parameterized by document and topic visualization coordinates for computing the path distribution and the level distribution respectively.

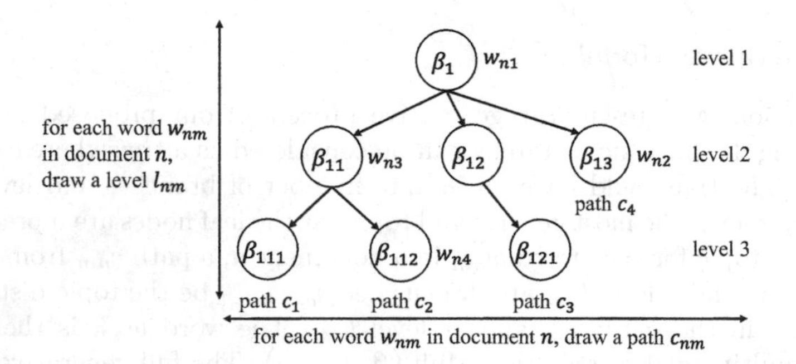


**Fig. 1.** Steps to sample a topic for each word $w_{nm}$ in a document $n$. For each word $w_{nm}$, a path $c_{nm}$ (from the root to a leaf node) and a level $l_{nm}$ are sampled. The topic assigned to $w_{nm}$ is $\beta_{c_{nm}[l_{nm}]}$. In this example, for the word $w_{n3}$, assume that the path $c_1$ and the level 2 are drawn. Topic $\beta_{11}$ is then assigned to the word $w_{n3}$

## 2.2 Parameterizing Path Distribution and Level Distribution

In this section, we explain how path distribution and level distribution are parameterized by document and topic visualization coordinates. From Eq. 2, generally for all topics that are children of a parent node $p$, this will hold:

$$\sum_{z, z \in child(p)} \pi_z = \pi_p \Longleftrightarrow \sum_{z, z \in child(p)} \frac{\pi_z}{\pi_p} = 1 \tag{3}$$

here $child(p)$ represents the set of all children of the parent node $p$. Let $\tau_z = \frac{\pi_z}{\pi_p} = \frac{\pi_z}{\pi_{par(z)}}$. To encode the nested stick breaking construction of the path distribution into the visualization, we parameterize $\tau_z$ of each document $n$ as a function of the distance between $x_n$ and $\phi_z$ as follows:

$$\tau_{zn} = \frac{\rho(\|x_n - \phi_z\|)}{\sum_{z', z' \in child(par(z))} \rho(\|x_n - \phi_{z'}\|)} \tag{4}$$

here *child(par(z))* represents the set of all children of parent of $z$, the denominator is for normalization so that (3) still holds, and $\rho$ is a radial basis function (RBF) which can have different forms such as Gaussian: $\exp(-\frac{1}{2}r^2)$, or Inverse quadratic: $\frac{1}{1+r^2}$ where $r = \|x - \phi_z\|$ is the distance from $x_n$ to topic coordinate $\phi_z$[2]. Equation 4 with Gaussian $\rho$ is also used in PLSV to encode the topic distribution in the visualization space [12]. As shown in [25], Inverse quadratic consistently produces good performance and in some cases it gives better results. Therefore, we choose to use Inverse quadratic in our experiments. Equation 4 becomes:

$$\tau_{zn} = \frac{\frac{1}{1+\|x_n-\phi_z\|^2}}{\sum_{z',z'\in child(par(z))}\frac{1}{1+\|x_n-\phi_{z'}\|^2}} \tag{5}$$

As we can see from the above formula, when the document $n$ is close to the topic $z$ in the visualization space, the numerator will be high and thus $\tau_{zn}$ and $\pi_{zn} = \tau_{zn}\pi_{par(z)n}$ will be high. Therefore, in step (1)(d)(i) of the generative process, the words in document $n$ tend to be assigned to the paths that going through topic $z$.

Note that $\pi_n$ is the path distribution of a document $n$. It is easy to see that the number of paths is equal to the number of leaf nodes in the topic tree. Therefore, $\pi_{in}$ of the leaf node $i$ is the path proportion of the path that goes to the leaf node $i$ and it is computed as follows:

$$\pi_{in} = \tau_{in}\pi_{par(i)n} = \prod_{z,z\in path(i)} \tau_{zn} \tag{6}$$

here note that $\pi_{root} = 1$ and $path(i)$ represents all the nodes that lie on the path from the root to the leaf node $i$. From (6), $\pi_n$ is then a function of $x_n, \boldsymbol{\Phi}$, i.e., $\pi_n = f_\pi(x_n, \boldsymbol{\Phi})$, which is used in step (1)(b) of the generative process.

Similarly, we also parameterize the level distribution of a document $n$ as a function of $x_n$ and topic coordinates $\boldsymbol{\Phi}$:

$$\delta_{ln} = \frac{\frac{1}{1+\min\{\|x_n-\phi_z\|^2, \forall z \text{ in level } l\}}}{\sum_{l'=1}^{L}\frac{1}{1+\min\{\|x_n-\phi_{z'}\|^2, \forall z' \text{ in level } l'\}}} \tag{7}$$

where $\min\{\|x_n - \phi_z\|^2, \forall z \text{ in level } l\}$ is the minimum distance between a document $n$ and all topics in the $l$-th level. From (7), $\delta_n$ is a function of $x_n, \boldsymbol{\Phi}$, i.e., $\delta_n = f_\delta(x_n, \boldsymbol{\Phi})$, which is used in step (1)(c) of the generative process. Based on $\pi_n$ and $\delta_n$, the topic distribution $\theta_n$ can be derived as: $\theta_{zn} = (1 - \sum_{l=1,l\neq l_z}^{L}\delta_{ln})(\sum_{c:c_l=z}\pi_{cn})$, where $l_z$ is the level of topic $z$.

### 2.3 Parameterizing Word Distribution

Let $t_z \in \mathbb{R}^H$ be the embedding of topic $z$ and $U \in \mathbb{R}^{V\times H}$ be the embeddings of words. The word distribution of topic $z$ is computed as: $\beta_z = \text{softmax}(\frac{U\cdot t_z^T}{\kappa^{\frac{1}{l_z}}})$,

[2] $r$ is Euclidean distance in our experiments.

where $\kappa^{\frac{1}{l_z}}$ is the temperature value that controls the sparsity of $\beta_z$. When the level $l_z$ is deeper, the probability distribution over words $\beta_z$ is sparser [11]. To possibly create an unbounded topic tree, as in [11] we use a doubly-recurrent neural network (DRNN) [2] to generate topic embeddings. A DRNN consists of two RNNs that respectively model the ancestral (parent-to-children) and fraternal (sibling-to-sibling) flows of information in the topic tree. More specifically, the hidden state $h_z$ of the topic $z$ is given by:

$$h_z = \tanh(W_h(\tanh(W_p h_{par(z)} + b_p) + \tanh(W_s h_{z-1} + b_s)) + b_h) \quad (8)$$

where $\tanh(W_p h_{par(z)} + b_p)$ and $\tanh(W_s h_{z-1} + b_s)$ can be considered as the ancestral and fraternal hidden states. The output topic embedding $t_z$ is computed based on $h_z$ as: $t_z = W h_z + b$. To increase the diversity of topics in the tree while allowing parent-children correlations, as in [11] we apply the following tree-specific diversity regularizer to the final objective function (Sect. 2.6):

$$L_{td} = \sum_{z \notin Leaf} \sum_{i,j \in Child(z); i \neq j} \left( \frac{\bar{t}_{zi}^{\top} \cdot \bar{t}_{zj}}{||\bar{t}_{zi}|| \, ||\bar{t}_{zj}||} - 1 \right)^2 \quad (9)$$

where $\bar{t}_{zi} = t_i - t_z$, $Leaf$ and $Child(z)$ denote the set of the topics with no children and the children of the $z$ topic, respectively.

### 2.4 Visualizing the Topic Tree

Our model also aims to visualize the topic tree. While the model can learn the topic visualization coordinates, it does not guarantee that the edges connecting topics do not cross each other. Therefore, to ensure that we have a visually appealing layout of the topic tree (e.g., the number of crossing edges is minimized), we employ the graph layout objective function of the Kamada-Kawai (KK) algorithm [13] and use it to regularize the topic coordinates in our final objective function (Sect. 2.6). The layout objective function of the KK algorithm is specified as: $L_{kk} = \sum_{i \neq j} \frac{1}{2}(\frac{d_{i,j}}{s_{i,j}} - 1)^2$, where, in our case, $d_{i,j} = ||\phi_i - \phi_j||$ is the Euclidean distance between topics $i$,$j$ in the visualization space, $s_{i,j}$ is the graph-theoretic distance of the shortest-path between topics $i$, $j$ in the tree. The weight of the edge connecting topics $i$, $j$ is computed based on the cosine distance between topic embeddings $t_i$, $t_j$, i.e., $\text{weight}(i,j) = 1 + \text{cosine\_dist}(t_i, t_j)$. Intuitively, two connected topics that are not similar will result in a longer edge.

### 2.5 Dynamically Growing the Topic Tree

We explain how the model dynamically updates the tree using heuristics. For each topic $z$, we estimate the proportion of the words in the corpus belonging to topic $z$: $p_z = \frac{\sum_{n=1}^{N} M_n \hat{\theta}_{zn}}{\sum_{n=1}^{N} M_n}$, where $M_n$ is the number of words in document $n$. We compare $p_z$ with the level-dependent pruning and adding thresholds to determine whether $z$ should be removed or a new child topic should be added to

$z$ for refining it. We use the adding threshold defined as $\max(a, \frac{1}{min\_child*2^{l-1}})$ and the pruning threshold defined as $\max(b, \frac{1}{max\_child*2^{l-1}})$. Here, $min\_child$ and $max\_child$ can be interpreted as the expected minimum and maximum numbers of children. The max function is to ensure that the thresholds are not too small when the number of levels is increasing[3]. For a topic $z$ that has $p_z$ greater than the adding threshold: 1) If $z$ is a non-leaf node, one child is added to $z$; (2) If $z$ is a leaf node, two children are added. This is the case where the model grows the tree by increasing the number of levels. Finally, if the sum of the proportions of all descendants of topic $z$, i.e., $\sum_{j \in \text{Des}(z)} p_j$ is smaller than the pruning threshold then $z$ and its descendants are removed.

### 2.6 Autoencoding Variational Inference

In this section, we present the inference of our model based on AEVB. The marginal likelihood of a document is given by:

$$\begin{aligned} p(\mathbf{w}_n|\boldsymbol{\Phi}, \boldsymbol{\beta}, \gamma) &= \int_x \Big\{\prod_m \sum_{c,l} p(w_{nm}|\boldsymbol{\beta}_{c[l]}) p(c|x, \boldsymbol{\Phi}) p(l|x, \boldsymbol{\Phi})\Big\} p(x|\gamma) dx \\ &= \int_x \Big\{\prod_m \sum_z p(w_{nm}|\boldsymbol{\beta}_z)\theta_{zn}\Big\} p(x|\gamma) dx \end{aligned} \tag{10}$$

where $p(c|x, \boldsymbol{\Phi})$, $p(l|x, \boldsymbol{\Phi})$ are the path distribution $\pi_n$ and the level distribution $\delta_n$ respectively. They are computed as in Eq. 6 and Eq. 7. $\theta_n$ is the topic distribution and $\theta_{zn} = (1 - \sum_{l=1, l \neq l_z}^{L} \delta_{ln})(\sum_{c: c_l = z} \pi_{cn})$. Based on the AEVB framework, we have the following lower bound to the marginal log likelihood (ELBO) of a document:

$$\mathcal{L}(\eta|\gamma, \boldsymbol{\Phi}, \boldsymbol{\beta}) = -\mathbb{D}_{\text{KL}}\left[q(x|\mathbf{w}_n, \eta) \| p(x|\gamma)\right] + \mathbb{E}_{q(x|\mathbf{w}_n, \eta)}\left[\log\left(\theta_n \boldsymbol{\beta}\right) \mathbf{w}_n^T\right] \tag{11}$$

where $q(x|\mathbf{w}_n, \eta) = \text{Normal}(\boldsymbol{\mu}_n, \boldsymbol{\Sigma}_n)$ is the variational distribution and $\boldsymbol{\mu}_n$, diagonal $\boldsymbol{\Sigma}_n \in \mathbb{R}^d$ are outputs of the encoding feed forward neural network with variational parameters $\eta$. The whole inference network architecture including the DRNN of *HTV* is shown in Fig. 8. To estimate the expectation w.r.t $q(x|\mathbf{w}_n, \eta)$ in Eq. 11, we sample an $\hat{x}$ from the posterior $q(x|\mathbf{w}_n, \eta)$ by using reparameterization trick, i.e., $\hat{x} = \boldsymbol{\mu}_n + \boldsymbol{\Sigma}_n^{1/2}\hat{\boldsymbol{\epsilon}}$ where $\hat{\boldsymbol{\epsilon}} \sim \text{Normal}(\mathbf{0}, \boldsymbol{I})$ [18]. For the whole corpus, the lower bound is then approximated as:

$$\begin{aligned} \mathcal{L}(\Omega) = \sum_{n=1}^{N} \Bigg[ &-\frac{1}{2}\left(\text{tr}\left((\gamma \boldsymbol{I})^{-1}\boldsymbol{\Sigma}_n\right) + (-\boldsymbol{\mu}_n)^T (\gamma \boldsymbol{I})^{-1} (-\boldsymbol{\mu}_n) - d + \log\frac{|\gamma \boldsymbol{I}|}{|\boldsymbol{\Sigma}_n|}\right) \\ &+ \log\left(\hat{\theta}_n \boldsymbol{\beta}\right) \mathbf{w}_n^T \Bigg] \end{aligned} \tag{12}$$

Adding the tree-specific diversity regularizer (Eq. 9) and the KK layout regularizer, we have the final objective function:

$$\mathcal{L} = \mathcal{L}(\Omega) + \lambda_{td} * L_{td} + \lambda_{kk} * L_{kk} \tag{13}$$

[3] In the experiments, we set $a = b = 0.01$.

## 3 Experiments

**Datasets.** In our experiments, we use four real-world datasets: 1) BBC[4] consists of 2225 documents from BBC News [9]. It has 5 classes: business, entertainment, politics, sport, and tech; 2) REUTERS[5] contains 7674 newswire articles from 8 categories [6]; 3) 20 NEWSGROUPS[6] contains 18251 newsgroups posts from 20 categories; 4) WEB OF SCIENCE[7] contains the abstracts and keywords of 46,985 published papers from 7 research domains: CS, Psychology, Medical, ECE, Civil, MAE, and Biochemistry [19]. All datasets are preprocessed by stemming and removing stopwords. The vocabulary sizes are 2000, 3000, 3000, and 5000 for BBC, REUTERS, 20 NEWSGROUPS, and WEB OF SCIENCE respectively. **Comparative Baselines.** We compare our proposed model with the following baselines: 1) *LDA-VAE*[8]: LDA with variational auto-encoder (VAE) inference [27]; 2) *PLSV-VAE*[9]: PLSV using VAE inference with Inverse quadratic RBF [25]; 3) *nCRP*[10]: A hierarchical topic model based on the nested Chinese restaurant process with collapsed Gibbs sampling [3]; 4) *TSNTM*[11]: A hierarchical neural topic model using VAE inference [11]; 5) *HTV* (our model): A novel joint model for both hierarchical topic modeling and visualization with VAE inference.

*LDA-VAE*, *nCRP*, and *TSNTM* are methods for topic modeling but they do not produce visualization. Therefore, for these methods, we use t-SNE [23][12] to embed the documents' topic proportions for visualization. In contrast, *PLSV-VAE* and *HTV* are joint methods for both topic modeling and visualization. Although *PLSV-VAE* is a flat topic model that does not detect topic hierarchies, for completeness we will compare our method with it. In our experiments, VAE-based methods are trained by AdaGrad with 2000 epochs, learning rate 0.01, batch size 512, and dropout with probability $p = 0.2$. For *TSNTM* and *HTV*, we use 256-dimensional word and topic embeddings, and $\kappa = 0.1$ for computing temperature value in $\beta_z$. The adding and pruning thresholds of *TSNTM* are 0.01 and 0.005 respectively. In *HTV*, we experimentally set (*min_child*, *max_child*) as (5, 10) for BBC, (5, 15) for REUTERS, and (6, 15) for both 20 NEWSGROUPS and WEB OF SCIENCE. These values work well for these datasets. We set the regularization parameters as $\lambda_{td} = 0.1$ and $\lambda_{kk} = 1000$, which consistently produces good performance across all datasets. Smaller $\lambda_{kk}$ would result in more crossing edges. We initialize the tree with 3 levels where each node has 3 children. The maximum level is set to 4.

[4] https://mlg.ucd.ie/datasets/bbc.html.
[5] https://ana.cachopo.org/datasets-for-single-label-text-categorization.
[6] https://scikit-learn.org/0.19/datasets/twenty_newsgroups.html.
[7] https://data.mendeley.com/datasets/9rw3vkcfy4/6.
[8] Its implementation is at https://github.com/akashgit/autoencoding_vi_for_topic_models.
[9] We use the implementation at https://github.com/dangpnh2/plsv_vae.
[10] We use the implementation at https://github.com/blei-lab/hlda.
[11] We use the implementation at https://github.com/misonuma/tsntm.
[12] https://github.com/DmitryUlyanov/Multicore-TSNE.

Different from hierarchical methods, *PLSV-VAE* and *LDA-VAE* need the number of topics to be specified before training. For a fair comparison, we set the number of topics in *PLSV-VAE* and *LDA-VAE* to be equal to the number of topics generated by *HTV* for each run. Regarding *nCRP*, we set its hyper-parameters as follows: $\gamma = 0.01$, the Dirichlet parameter $\eta = 5$, and the GEM parameters are set as $\pi = 10$ and $m = 0.5$. All the experiment results are averaged across 10 runs on a system with 64GB memory, an Intel(R) Xeon(R) CPU E5-2623v3, 16 cores at 3.00GHz. The GPU in use on this system is NVIDIA Quadro P2000 GPU with 1024 CUDA cores and 5 GB GDDR5.

### 3.1 Tree-Structure and Visualization Quantitative Evaluation

We evaluate the quality of the tree structure using document specialization in the visualization space and two other metrics: node specialization and hierarchical affinity that are also used in [11,16].

**Document Specialization in the Visualization Space.** In this task, we measure the quality of hierarchical visualization of documents and topics. A good hierarchical visualization should put general documents close to general topics and the farther the documents are from the root, the more specific they are. We quantify this aspect by finding the top $5\%, 10\%, \ldots$ of all documents that are the closest to the root topic in the visualization space. For each such set of documents, we compute the average cosine similarity between each document and the vector of the entire corpus. As in [11,16], the vector of the entire corpus is computed based on the frequencies of the words and is considered as the most general topic. We would expect that the average cosine similarity will be high for documents near the root and it will be decreasing when farther away. Since *PLSV-VAE*, *LDA-VAE*, *nCRP*, and *TSNTM* do not visualize topics, we use the average of all documents coordinates as the root. Figure 2 shows the average cosine similarity (i.e., doc specialization as in the figure) by the methods for different top $k\%$ of documents. The high steepness of the curve by our model *HTV* indicates that the documents are organized better into hierarchies in the visualization where the most general documents are near the root and they become increasingly specific when farther away.

**Classification in Visualization Space.** We show that while producing better hierarchical visualization, our method still generates a high quality scatterplot visualization in terms of $k$-NN accuracy in the visualization space. $k$-NN accuracy is widely used to evaluate the quality of the visualization [23,25]. In this evaluation approach, a $k$-NN classifier is used to classify documents using their visualization coordinates. A good visualization should group documents with the same label together and hence yield a high classification accuracy in the visualization space. Figure 3 shows $k$-NN accuracy of all models across datasets. This figure shows that visualization by *HTV* is as good as other methods. This will be further confirmed when we look at the example visualizations in Sect. 3.3.

**Node Specialization.** A good tree structure should have the general topics near the root and topics become more specific toward the low levels. To quantify

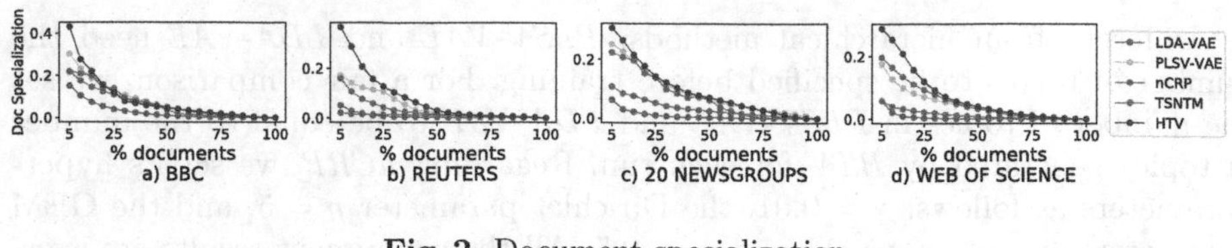


**Fig. 2.** Document specialization

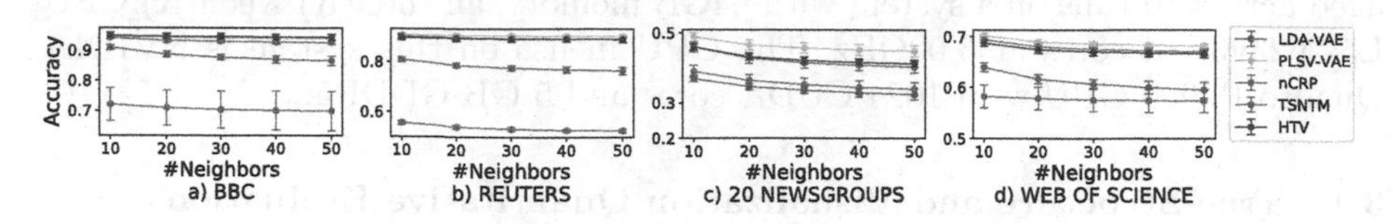


**Fig. 3.** $k$-NN accuracy in the visualization space

this aspect, we rely on node specialization that measures the specialization score as the cosine distance between the word distribution of each topic and the vector of the entire corpus [11]. Since the entire corpus vector is regarded as the most general topic, more specific topics should have higher cosine distances. Table 1 shows the average cosine distance of all topics at each level. We only compare our model with hierarchical methods in this task. Except for *TSNTM* in BBC, the specialization scores for each model increase as the level increases.

**Hierarchical Affinity.** Another characteristic of a good tree structure is that a parent topic should be more similar to its children than the topics descended from the other parents. As in [11], we compute the average cosine similarity between a node to its children and non-children nodes. Table 2 shows the average cosine similarity over the topics of all models. The higher score over child nodes indicates that a parent is more similar to its child nodes. We only show the results of hierarchical methods in this task. All three models infer child topics similar to their parents.

### 3.2 Topic Coherence and Running Time Comparison

We evaluate the quality of topic models produced by all methods in terms of topic coherence. The objective is to show that while generating better hierarchical visualization quality, *HTV* also achieves competitive performance on topic coherence. For topic coherence, we use the Normalized Pointwise Mutual Information (NPMI) [20] estimated based on a large external corpus. We use Wikipedia 7-gram dataset created from the Wikipedia dump data as of June 2008 version[13]. Table 3 shows the average Normalized Pointwise Mutual Information (NPMI [20]) over all topics for all models. The NPMI scores of *HTV* over all datasets are comparable to all baselines. Comparing to hierarchical methods *nCRP* and *TSNTM*, *HTV* can find slightly better topics. For running time, since

[13] https://nlp.cs.nyu.edu/wikipedia-data/.

**Table 1.** Topic specialization scores. Except *TSNTM* method in BBC from level 3 to level 4, the scores increase as the level increases for all models

| Dataset | Model | Level 1 | Level 2 | Level 3 | Level 4 |
|---|---|---|---|---|---|
| BBC | *nCRP* | 0.188 | < 0.529 | < 0.792 | < 0.845 |
| | *TSNTM* | 0.321 | < 0.528 | < 0.557 | > 0.516 |
| | *HTV* | 0.339 | < 0.579 | < 0.722 | < 0.831 |
| Reuters | *nCRP* | 0.097 | < 0.612 | < 0.815 | < 0.882 |
| | *TSNTM* | 0.315 | < 0.535 | < 0.563 | < 0.566 |
| | *HTV* | 0.450 | < 0.561 | < 0.739 | < 0.877 |
| 20 Newsgroups | *nCRP* | 0.097 | < 0.612 | < 0.847 | < 0.894 |
| | *TSNTM* | 0.247 | < 0.456 | < 0.538 | < 0.561 |
| | *HTV* | 0.447 | < 0.452 | < 0.672 | < 0.802 |
| Web of Science | *nCRP* | 0.148 | < 0.606 | < 0.814 | < 0.870 |
| | *TSNTM* | 0.306 | < 0.439 | < 0.511 | < 0.518 |
| | *HTV* | 0.411 | < 0.431 | < 0.671 | < 0.754 |

**Table 2.** Hierarchical affinity. Except *TSNTM* method in BBC from level 3 to level 4, the scores increase as the level increases for all models

| Dataset | Model | Child | Non-Child |
|---|---|---|---|
| BBC | *nCRP* | 0.146 | 0.063 |
| | *TSNTM* | 0.201 | 0.171 |
| | *HTV* | 0.127 | 0.060 |
| Reuters | *nCRP* | 0.139 | 0.095 |
| | *TSNTM* | 0.254 | 0.188 |
| | *HTV* | 0.151 | 0.070 |
| 20 Newsgroups | *nCRP* | 0.138 | 0.095 |
| | *TSNTM* | 0.238 | 0.194 |
| | *HTV* | 0.146 | 0.081 |
| Web of Science | *nCRP* | 0.140 | 0.089 |
| | *TSNTM* | 0.275 | 0.205 |
| | *HTV* | 0.143 | 0.081 |

**Table 3.** Average NPMI of all topics over 10 runs

| model | BBC | Reuters | 20 Newsgroups | Web of Science |
|---|---|---|---|---|
| *LDA* | 0.091 | 0.051 | 0.95 | 0.094 |
| *PLSV-VAE* | 0.095 | 0.054 | 0.095 | 0.099 |
| *nCRP* | 0.043 | 0.039 | 0.031 | 0.053 |
| *TSNTM* | 0.090 | 0.053 | 0.092 | 0.094 |
| *HTV*(Our model) | 0.091 | 0.052 | 0.094 | 0.099 |

**Table 4.** Running time (in seconds) of three models: *nCRP*, *TSNTM*, and *HTV*

| Dataset | *nCRP* | *TSNTM* | *HTV* |
|---|---|---|---|
| BBC | 84120 | 6008 | 7223 |
| REUTERS | 31300 | 2132 | 2247 |
| 20 NEWSGROUPS | 7079 | 1011 | 882 |
| WEB OF SCIENCE | 5535 | 294 | 295 |

*HTV* uses VAE inference, it scales well to large datasets. As shown in Table 4, it runs much faster than *nCRP* and has comparable running time to *TSNTM*.

### 3.3 Visualization Qualitative Evaluation

Figures 4, 5, and 6 show visualization examples by *HTV*, *PLSV-VAE*, and *TSNTM* on REUTERS, 20 NEWSGROUPS, and WEB OF SCIENCE respectively. Each colored point represents a document, and the larger points with black border are topics (only in *PLSV-VAE* and *HTV*). In *HTV*, the red point with black border is the root topic and the green points with black border are the level 2 topics, finally, the blue and yellow points with black border represent level 3 and level 4 topics. It is clear that *HTV* can find good document clusters as compared to *PLSV-VAE* and *TSNTM* with t-SNE. Moreover, *HTV* learns a topic tree for each dataset and visualizes it in the visualization space using KK layout objective function. This helps to minimize the crossing edges as seen in all the visualization examples. In contrast, *TSNTM* does not visualize topics, and for *PLSV-VAE*, it does not infer the topic hierarchy. Therefore, it is difficult to tell the relationship between topics in the visualization. In Fig. 7, we show the visualization of documents along with the generated topics by *HTV* on WEB

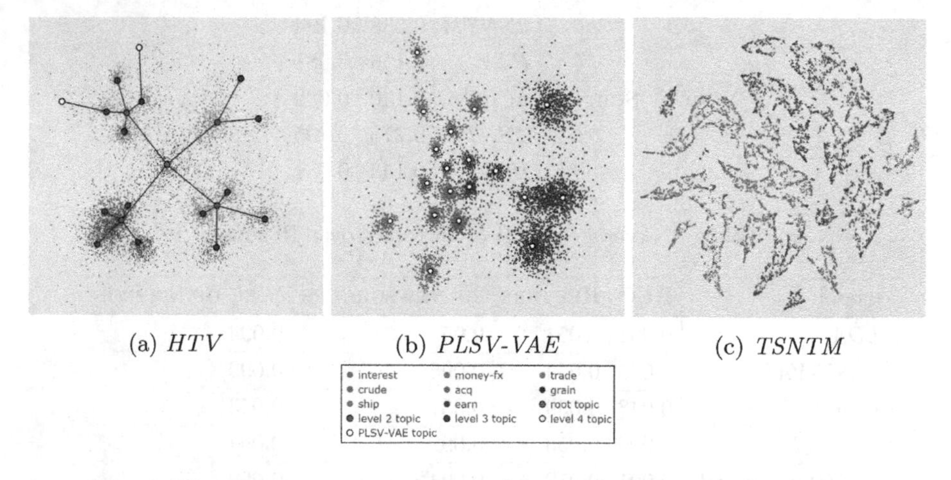


**Fig. 4.** Visualization of REUTERS by a) *HTV* b) *PLSV-VAE* c) *TSNTM*

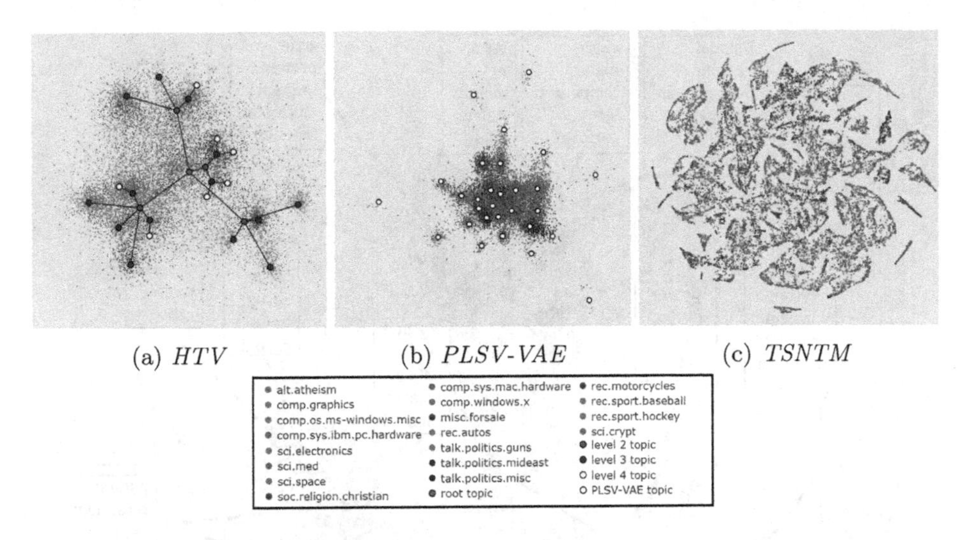


**Fig. 5.** Visualization of 20 NEWSGROUPS by a) *HTV* b) *PLSV-VAE* c) *TSNTM*

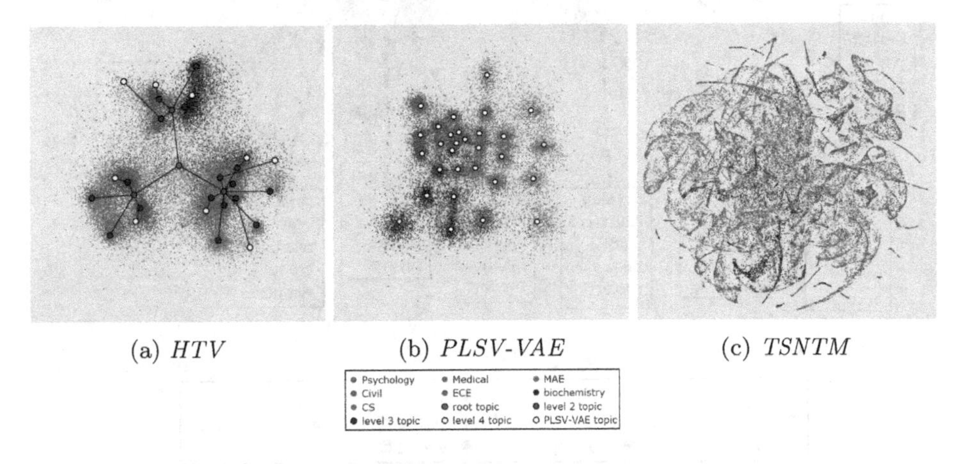


**Fig. 6.** Visualization of WEB OF SCIENCE by a) *HTV* b) *PLSV-VAE* c) *TSNTM*

OF SCIENCE. The inferred topic tree has three branches. The root topic has top 5 words: "method, studi, data, measur, img" which are very general words in sciences domain. As we can see, topics at the lower levels are more specific. For example, topics on levels 3, 4 in the top branch are very specific. They are topics in Civil, MAE, Biochemistry domains such as "water, are, model, system, soil", "model, engin, experiment, properti, flow, fluid", and "gene, protein, sequenc, molecular, function". The layout of topics and their structure show that our model can extract the topic hierarchy and visualize it along with the documents.

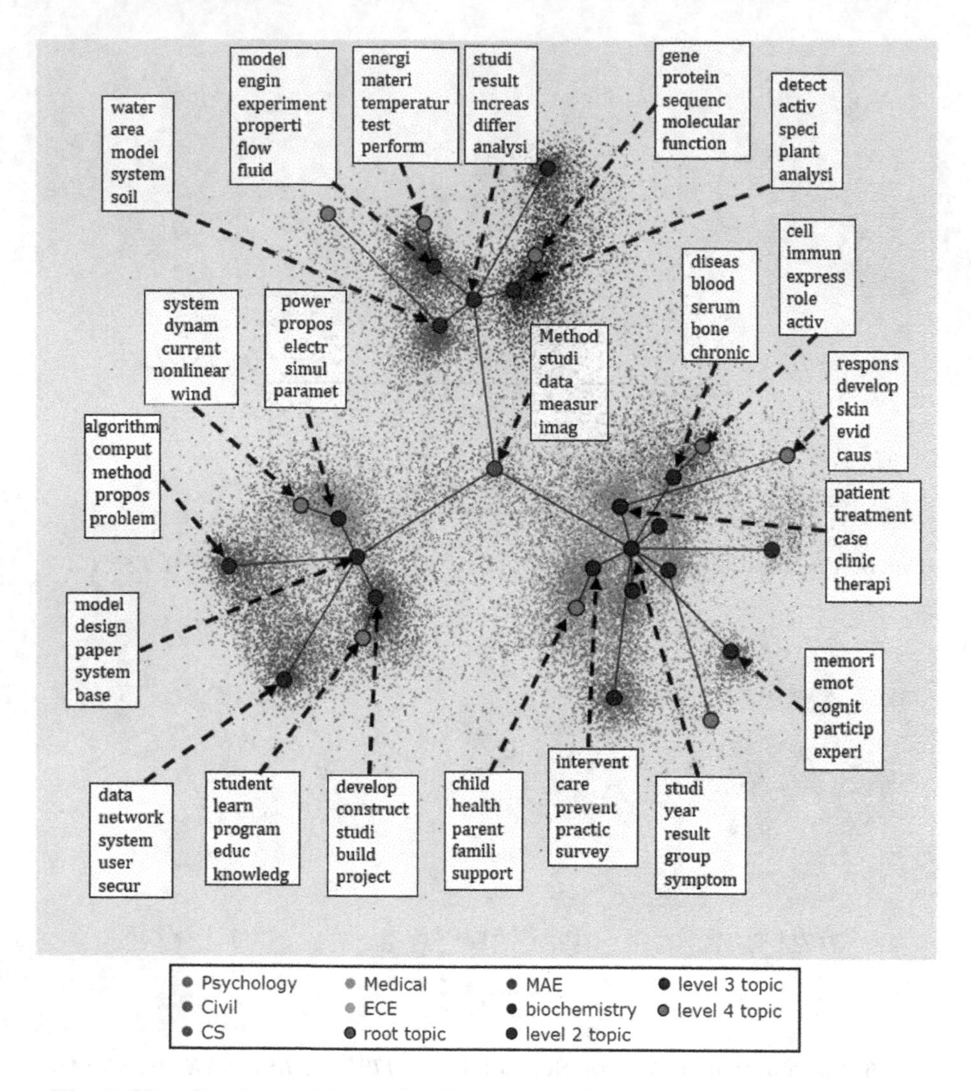


**Fig. 7.** Visualization and hierarchical topics found by *HTV* on Web of Science

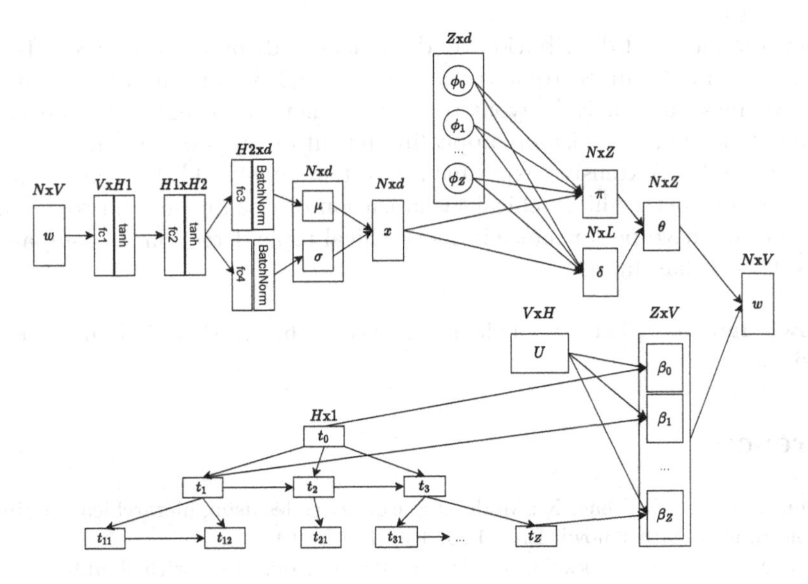


**Fig. 8.** The inference network architecture of *HTV*

## 4 Related Work

Hierarchical structure is an effective way to organize topics as it helps users to understand and explore the structure of topics. Flat topic models such as LDA [5] are not designed to detect topic hierarchies. Therefore, several hierarchical topic models including the nested Chinese restaurant process (nCRP) [3,4], the nested hierarchical Dirichlet process (nHDP) [24] have been proposed to overcome this limitation. Recently, there has been an increasing interest in neural approaches for topic modeling [7,28,29]. For detecting topic hierarchies, we have neural methods such as TSNTM [11], which is a neural extension of nCRP. TSNTM parameterizes the topic distribution over an infinite tree by a doubly-recurrent neural network (DRNN). TSNTM is trained using AEVB, making it scale well to larger datasets than the nCRP-based model.

All of the above methods work well for topic modeling but they are not designed for visualization tasks. Therefore, several works including the pioneering model PLSV [12] and its variants [21,22] have been proposed to jointly perform topic modeling and visualization. PLSV is a flat topic model where a generative model is used to generate both topics and visualization. Recently, *PLSV-VAE* [25] proposes using AEVB for scalable inference in PLSV. These joint models are not for hierarchical topic detection. To the best of our knowledge, our model is the first joint model for detecting topic hierarchies and visualization.

## 5 Conclusion

In this paper, we propose *HTV*, a visual hierarchical neural topic model for jointly detecting topic hierarchies and visualization. We parameterize the path distribution and level distribution by document and topic coordinates. To possibly create an unbounded topic tree, we use a DRNN to generate topic embeddings. We make use of KK layout objective function to regularize the model, ensuring that we have a visually appealing layout of the topic tree in the visualization space. Our extensive experiments on four real-world datasets show that *HTV* generates better hierarchical visualization of documents and topics while gaining competitive performance in hierarchical topic detection, as compared to state-of-the-art baselines.

**Acknowledgments.** This research is sponsored by NSF #1757207 and NSF #1914635.

## References


1. Almars, A., Li, X., Zhao, X.: Modelling user attitudes using hierarchical sentiment-topic model. Data Knowl. Eng. **119**, 139–149 (2019)
2. Alvarez-Melis, D., Jaakkola, T.: Tree-structured decoding with doubly-recurrent neural networks. In: ICLR (2017)
3. Blei, D.M., Griffiths, T.L., Jordan, M.I.: The nested Chinese restaurant process and Bayesian nonparametric inference of topic hierarchies. J. ACM (JACM) **57**(2), 1–30 (2010)
4. Blei, D.M., Griffiths, T.L., Jordan, M.I., Tenenbaum, J.B.: Hierarchical topic models and the nested Chinese restaurant process. In: Advances in Neural Information Processing Systems, vol. 16, no. 16, pp. 17–24 (2004)
5. Blei, D.M., Ng, A.Y., Jordan, M.I.: Latent Dirichlet allocation. J. Mach. Learn. Res. **3**, 993–1022 (2003)
6. Cardoso-Cachopo, A.: Improving methods for single-label text categorization. Ph.D. thesis, Instituto Superior Tecnico, Universidade Tecnica de Lisboa (2007)
7. Chen, Y., Zaki, M.J.: KATE: K-competitive autoencoder for text. In: Proceedings of the 23rd ACM SIGKDD International Conference on Knowledge Discovery and Data Mining, pp. 85–94 (2017)
8. Choo, J., Lee, C., Reddy, C.K., Park, H.: UTOPIAN: user-driven topic modeling based on interactive nonnegative matrix factorization. IEEE Trans. Visual Comput. Graph. **19**(12), 1992–2001 (2013)
9. Greene, D., Cunningham, P.: Practical solutions to the problem of diagonal dominance in kernel document clustering. In: Proceedings of the 23rd International Conference on Machine learning (ICML 2006), pp. 377–384. ACM Press (2006)
10. Guo, D., Chen, B., Lu, R., Zhou, M.: Recurrent hierarchical topic-guided RNN for language generation. In: Proceedings of the 37th International Conference on Machine Learning. Proceedings of Machine Learning Research, vol. 119, pp. 3810–3821 (2020)
11. Isonuma, M., Mori, J., Bollegala, D., Sakata, I.: Tree-structured neural topic model. In: Proceedings of the 58th Annual Meeting of the Association for Computational Linguistics, pp. 800–806 (2020)
12. Iwata, T., Yamada, T., Ueda, N.: Probabilistic latent semantic visualization: topic model for visualizing documents. In: KDD, pp. 363–371 (2008)
13. Kamada, T., Kawai, S.: An algorithm for drawing general undirected graphs. Inf. Process. Lett. **31**, 7–15 (1989)

14. Kataria, S.S., Kumar, K.S., Rastogi, R.R., Sen, P., Sengamedu, S.H.: Entity disambiguation with hierarchical topic models. In: KDD, pp. 1037–1045 (2011)
15. Kim, H., Drake, B., Endert, A., Park, H.: ArchiText: interactive hierarchical topic modeling. IEEE Trans. Visual. Comput. Graph. **27**, 3644–3655 (2020)
16. Kim, J.H., Kim, D., Kim, S., Oh, A.H.: Modeling topic hierarchies with the recursive Chinese restaurant process. In: Proceedings of the 21st ACM International Conference on Information and Knowledge Management (2012)
17. Kim, S., Zhang, J., Chen, Z., Oh, A., Liu, S.: A hierarchical aspect-sentiment model for online reviews. In: AAAI, vol. 27 (2013)
18. Kingma, D.P., Welling, M.: Auto-encoding variational Bayes. In: 2nd International Conference on Learning Representations, ICLR 2014, Banff, AB, Canada, 14–16 April 2014, Conference Track Proceedings (2014)
19. Kowsari, K., et al.: HDLTex: hierarchical deep learning for text classification. In: International Conference on Machine Learning and Applications. IEEE (2017)
20. Lau, J.H., Newman, D., Baldwin, T.: Machine reading tea leaves: automatically evaluating topic coherence and topic model quality. In: Proceedings of the 14th Conference of the European Chapter of the Association for Computational Linguistics, pp. 530–539 (2014)
21. Le, T., Lauw, H.: Manifold learning for jointly modeling topic and visualization. In: Proceedings of the AAAI Conference on Artificial Intelligence, vol. 28 (2014)
22. Le, T.M., Lauw, H.W.: Semantic visualization for spherical representation. In: Proceedings of the 20th ACM SIGKDD International Conference on Knowledge Discovery and Data Mining, pp. 1007–1016 (2014)
23. Van der Maaten, L., Hinton, G.: Visualizing data using t-SNE. J. Mach. Learn. Res. **9**(11), 2579–2605 (2008)
24. Paisley, J., Wang, C., Blei, D.M., Jordan, M.I.: Nested hierarchical Dirichlet processes. IEEE Trans. Pattern Anal. Mach. Intell. **37**(2), 256–270 (2014)
25. Pham, D., Le, T.: Auto-encoding variational Bayes for inferring topics and visualization. In: Proceedings of the 28th International Conference on Computational Linguistics (2020)
26. Smith, A., Hawes, T., Myers, M.: Hiearchie: visualization for hierarchical topic models. In: Proceedings of the Workshop on Interactive Language Learning, Visualization, and Interfaces, pp. 71–78 (2014)
27. Srivastava, A., Sutton, C.A.: Autoencoding variational inference for topic models. In: ICLR (2017)
28. Wang, R., et al.: Neural topic modeling with bidirectional adversarial training. In: Proceedings of the 58th Annual Meeting of the Association for Computational Linguistics, pp. 340–350 (2020)
29. Wang, X., Yang, Y.: Neural topic model with attention for supervised learning. In: International Conference on Artificial Intelligence and Statistics, pp. 1147–1156. PMLR (2020)
30. Yang, Y., Yao, Q., Qu, H.: VISTopic: a visual analytics system for making sense of large document collections using hierarchical topic modeling. Visual Inform. **1**(1), 40–47 (2017)

# Semi-structured Document Annotation Using Entity and Relation Types

Arpita Kundu(✉), Subhasish Ghosh, and Indrajit Bhattacharya

TCS Research, Kolkata, India
{arpita.kundu1,g.subhasish,b.indrajit}@tcs.com

**Abstract.** Semi-structured documents such as web-pages and reports contain text units with complex structure connecting these. We motivate and address the problem of annotating such semi-structured documents using a knowledge graph schema with entity and relation types. This poses significant challenges not addressed by the existing literature. The latent document structure needs to be recovered, and paths in the latent structure need to be jointly annotated with entities and relationships. We present a two stage solution. First, the most likely document structure is recovered by structure search using a probabilistic graphical model. Next, nodes and edges in the recovered document structure are jointly annotated using a probabilistic logic program, considering logical constraints as well as uncertainty. We additionally discover new entity and relation types beyond those in the specified schema. We perform experiments on real webpage and complex table data to show that our model outperforms existing table and webpage annotation models for entity and relation annotation.

**Keywords:** Semi-structured documents · Document annotation · Knowledge discovery · Graphical model · Structure search · Probabilistic logic program

## 1 Introduction

Semi-structured or richly structured documents, such as web-pages, reports and presentations, are among the most abundant information sources created by humans for human consumption. These consist of text units connected by spatial structure, and the spatial structures bear evidence of the semantic relations between the text units. There has been a lot of recent interest in extracting information from such semi-structured documents by annotating these with entities and relations from a knowledge graph [9,10,18].

However, document structures in the real world are more complex and interesting than those addressed in the literature. One view of such documents is that of complex tables, with cells that span multiple rows and columns. But the table annotation literature so far has only considered simple tables [3,8,17,20]. The second view is that of html structures as in web-pages [9,10]. Again, this area of research has ignored much of the complexity of such structures.


N. Oliver et al. (Eds.): ECML PKDD 2021, LNAI 12977, pp. 52–68, 2021.
https://doi.org/10.1007/978-3-030-86523-8_4

First, a typical document is not about a single entity and its properties, but about multiple related entities. Secondly, the observed visual structure may not be the true semantic structure. Text units are not always related to the closest neighbors, and the observed structure often leaves out header text units. Thirdly, groups of documents often share a document structure template (or DST), such as distinct structures for movie pages, actor pages, studio pages, etc. in the movies domain. However, the DSTs are not directly observed, and the structure of individual documents may contain minor variations within the theme.

We consider the problem of annotating such semi-structured documents using a knowledge graph schema that specifies entity types and binary relation types between these entity types. The structure of real schemas can be complex. In general, these are directed multi-graphs. In contrast, existing literature [9,10] only considers schemas that are collections of disconnected star-shapes, each about an entity and its properties. Our observation is that, for complex document structures and complex schemas, the nature of the annotation is also complex. Not all nodes in the document structures correspond to entity types. As a result, individual edges in the document structures cannot be annotated with relation types in the schema. Instead, relation types correspond to document structure paths. This leads to structural constraints on the annotations, which calls for joint annotation of the nodes and paths in the document structure with entity and relation types. Finally, we highlight the task of entity type and relation type discovery. Available knowledge graph schemas for a domain are precise, but far from complete. A typical document corpus mentions many entity and relation types not specified in the schema. Such new entity and relation types need to be discovered from the corpus and incorporated into the schema.

To address these challenges, we propose a two-stage approach. The first stage uses a probabilistic graphical model with structural parameters to capture the complexity of DSTs and variations within a DST. We use a novel structure search algorithm using this model to identify the latent structure and DST for each document. The second stage annotates recovered document structures with entity and relation types. Since this stage requires logical rules to capture background knowledge about constraints while accommodating uncertainty of annotation, we use probabilistic logic programming [13,14], which naturally supports both. We also propose an iterative algorithm for discovering new entity and relation types by analyzing patterns in the unannotated parts of the document structure.

Using experiments on three real datasets of different types, we demonstrate the flexibility of our approach and improvements in annotation accuracy over state-of-the-art webpage and table annotation models.

## 2 Related Work

Some of the work on extracting information from semi-structured documents [9,10,18] focus specifically on web-pages [9,10]. These assume documents to be about a single entity so that the document structure is a star. No latent structure different from the observed structure is considered. Though these use a notion of structural templates, documents within a template are assumed to have exactly the same structure. The knowledge graph schema is also assumed to be a star with the topic entity at the center, so that only properties of the topic entity are annotated as relations. Independent annotation of individual document edges is sufficient in this setting.

Annotation of web table using a knowledge graph has received a lot of attention in recent years [3,8,17,20]. These only address simple table structures, whereas the document structures for us are DAGs. These either use graphical models [8,17] to reason jointly using the table structure, or reduce tables to documents and use deep models for sequence data for annotation [3,20]. These consider the table structure to be directly observed. These also do not consider shared templates for tables in a corpus.

While we have used ProbLog for probabilistic logic programming, other existing frameworks could have been used [1,15,16]. Markov logic network (MLN) [15] combine (undirected) graphical models with logic, which is also our end-to-end goal. However, the existing Alchemy implementation of MLNs cannot perform the sophisticated structure search required by our task.

## 3 Problem Statement

We begin by defining semi-structured documents and the problem of annotating such documents with a knowledge graph schema.

**Document, Structure and Template:** We have a set of semi-structured documents $D$: webpages, reports, slide decks, etc. Each document $d_i$ has a set of text units. In Fig. 1, the first column shows two example documents and their text units. Each text unit has a sequence of tokens $w_{ij}$.

The documents have visible structure connecting the text units based on their relative locations within the document. The second column in Fig. 1 shows the corresponding observed structures for the two documents in the first column. Each text unit is a node in this structure. Node 1 is immediately above nodes 2 and 3, while node 2 is on the immediate left of node 3, which explains the edges between them. We consider the observed document structure to be a directed acyclic graph (DAG), where $\pi_{ijj'}$ is a boolean variable that indicates whether node $j$ is a parent of node $j'$ in document $i$. The parents may be obtained directly from html constructs of webpages, or indirectly from the spatial layout of the text units. To accommodate this, a node can additionally have two dimensional location coordinates $l_{ij} = (x_{ij}, y_{ij})$. The examples illustrate the alternative view of such documents as tables, with integer location coordinates. But observe that the tables are complex - text units can span multiple rows or columns.

The observed structure may not reveal the true semantic relationships. In both example documents, the 'Details' and 'Movie Details' have 'Stars' as parents. Semantically, both should have 'Movie' nodes as their parents. The third column shows the latent structures for the two documents. The latent structure is also a DAG. Let $\pi^*_{ijj'}$ denote the latent parents. Each latent node has its corresponding text $w^*_{ij}$ and location $l^*_{ij}$.

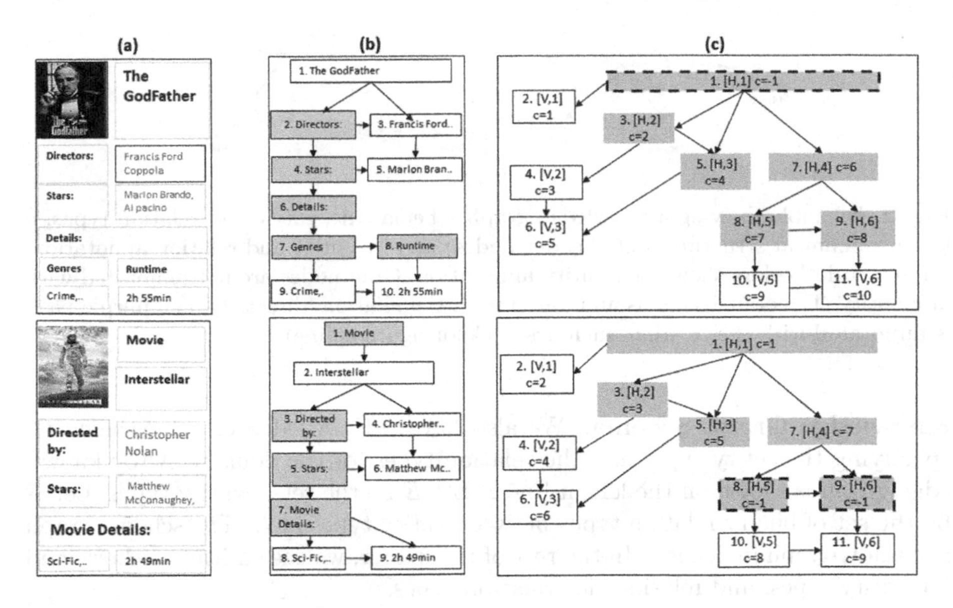


**Fig. 1.** Example showing (a) two semi-structured documents, (b) their observed structures, and (c) latent structures. Header nodes are shaded gray and node types are mentioned as $[t^p_{ij}, t^s_{ij}]$ inside latent nodes. Corresponding observed nodes ($c_{ij}$) are also mentioned inside latent nodes. Latent nodes missing in the observed structure ($c = -1$) are marked with dashed borders. Both documents follow the 'Movie' template.

Additionally, the nodes are of different types. At the highest level, these are headers ($H$) or values ($V$). The variable $t^p_{ij} \in \{H, V\}$ denotes the primary type of the node. The nodes also have secondary types $t^s_{ij} \in \mathbb{N}$ for both headers and values. These represent different types of values nodes - long textual descriptions, short texts, numbers, etc., and accordingly different types of header nodes associated with these different types of value nodes.

Not all nodes in the latent structure have corresponding nodes in the observed structure. Specifically, header nodes are often meant to be understood from context. Let $c_{ij}$ denotes the corresponding observed node for the $j^{th}$ latent node. Latent nodes that do not have corresponding observed nodes have $c_{ij} = -1$.

Finally, latent structures are often shared between groups of documents in a corpus. Both our example documents are about 'Movies'. As a result, even though they have different observed structures, their latent structures are the same. Other documents about 'Actors' or 'Production Houses' may again have similar structure, different from those of 'Movie' documents. We use $T_i \in \mathbb{N}$ to denote the document structure template (DST in short) of the $i^{th}$ document. Note that two documents with the same DST do not necessarily have identical latent structures. For instance, not all 'Movie' documents may mention run-times. However, they have the same type for the root node of their latent structures, and very similar latent structures overall.

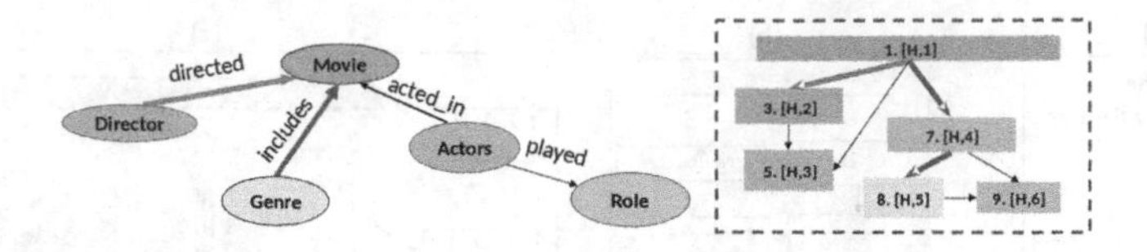


**Fig. 2.** Example showing a knowledge graph schema with entity and relation types, a latent document structure (only header nodes) and its entity and relation annotation. Latent nodes' colors show their entity annotation. Gray nodes are not annotated with any entity. The orange edge is annotated with the relation 'directed'. The purple path is annotated with the relation 'includes'. (Color figure online)

**Knowledge Graph Schema:** We also have a Knowledge Graph Schema $G$, specifying the entity types and the relation types for the domain. A toy knowledge graph is shown on the left in Fig. 2. Let $E$ be the set of entity types. Let $R$ be the set of binary relation types between entity types in $E$. The schema graph is a multi-graph in general. In the rest of this paper, we use entities as shorthand for entity types, and relations for relation types.

**Annotation:** An annotation is a mapping between the latent document structure and the entities and relations. Only header latent nodes ($t^p_{ij} = H$) can correspond to entities. For a latent node, $e_{ij} \in E \cup \{-1\}$ denotes its *entity annotation*. The right side of Fig. 2 shows the annotation of the common latent structure of our two example documents in Fig. 1(c), with the knowledge graph schema on the left. Some header nodes may not correspond to entities in the given schema $G$. Such nodes have $e_{ij} = -1$.

Next is *relation annotation*. In the simplest case, an edge in the latent structure corresponds to a relation. However, relations cannot always be mapped to individual edges. For example, node 1 corresponding to 'Movie' is connected to node 8 corresponding to 'Genre' by the *path* $(1, 7, 8)$. Therefore, the entire path corresponds to relation 'includes' between 'Genre' and 'Movie'. This is allowed only because the intermediate node 7 does not correspond to any entity. A latent path $p_{il}$ in the $i^{th}$ document is a sequence of latent nodes of header type ($t^p_{ij} = H$). Let $r_{il} \in R \cup \{-1\}$ denote the *relation annotation* of path $p_{il}$. Path annotations have to be consistent with entity annotations. If $p_{il} = r$ and the relation $r$ is between entities $e_1$ and $e_2$, then the first node $j$ in the path must have $e_{ij} = e_1$ and the last node $j'$ must have $e_{ij'} = e_2$. As a consequence, in a path annotated with any relation ($p_{il} \neq -1$), an internal node $j''$ cannot be annotated with an entity (i.e. must have $e_{ij''} = -1$). Additionally, path annotations in a document have to be consistent with each other. The same edge cannot appear in two different paths with different relation annotations.

There may be many consistent entity and relation annotations for a document latent structure. Note that annotating every node and edge with $-1$ is a trivially consistent annotation but one that should ideally have a very low probability. On the other hand, the annotation shown in the example in Fig. 2 looks very likely and should have a high probability. Thus, annotation involves logical constraints as well as uncertainty.

**Structured Document Annotation Problem:** We now define the annotation problem for a semi-structured corpus. We are given a knowledge graph schema $G$ with entities $E$ and relationships $R$. We are also given a training corpus $D$ of semi-structured documents. For these, the observed document structures are given. In addition, we assume the latent document structure $\pi^*$, and the annotations of the latent nodes and paths with entities and relations in the KG are also available for training. Based on the entity annotations, the primary types of some nodes are known to be header, but secondary node types $t^s$, or the document templates $T$ are not given. The number of secondary node types and the number of templates may also not be known.

Now, given only the observed document structure of a test document $d$, the task is two fold. (A) **Structure recovery:** Identify the latent document structure $\pi^*_d$, along with the primary and secondary node types $t^p_d$ and $t^s_d$ and the document template $T_d$. (B) **Annotation:** Annotate the recovered latent structure with entities and relations. Note that the relation annotations for different paths needs to be performed jointly along with the entity annotations, because of the constraints on them.

Additionally, we consider the problem of **entity and relation discovery**. Here, we assume that only a subset $G^o$ of the KG schema containing a subset of the entities $E^o$ and relations $R^o$ are observed during training. As a result, the training annotations now only contain the observed entities and relations. No annotations are available for the remaining entities $E - E^o$ and relations $R - R^o$. In our example, the complete schema may contain entity 'Duration' and relation 'has_runtime' between 'Movie' and 'Duration', but this is not in the training schema $G^o$. Now, an additional task is to recover such latent entities and relations from the training corpus, and annotate the test documents not only with the observed entities and relations, but the latent ones as well.

## 4 Proposed Approach

Our proposed approach has three stages. As shown in Fig. 1, the first stage takes the observed structure of a document and uses a probabilistic graphical model to recover the latent structure, node types and the DST by structure search. As shown in Fig. 2, the second stage takes the latent structure and node types, and uses a probabilistic logic program to annotate the nodes and the paths satisfying logical constraints. These two stages are supervised. The final stage analyzes patterns in un-annotated parts of the latent structures to discover new entities and relations.

### 4.1 Document Structure Recovery Using Generative PGM

We define a joint probability distribution over the observed and latent structure variables, and then present a structure recovery algorithm using the model.

**Generative Model:** The model is iid over documents. We factorize the joint distribution over the variables for each document into 5 major components and provide these with generative semantics as follows:

$$T \sim Cat(\alpha)$$
$$t, \pi^* \sim P(t, \pi^* \mid T; \beta, \mu) \qquad w^*, l^* \sim P(w^*, l^* \mid T, t, \pi^*; h(t), \lambda, \tau, \gamma)$$
$$c, \pi \sim P(c, \pi \mid T, t, \pi^*; \phi) \qquad w, l \sim P(w, l \mid c_i, w^*, l^*)$$

The first step samples the DST $T$. The second and third steps sample the node types $t$ and the latent structure $\pi^*$, content $w^*$ and location $l^*$ conditioned on the DST $T$. The fourth and fifth steps sample the observed structure $\pi$ and correspondence $c$ conditioned on the latent structure and the DST, and finally the content and location of the observed nodes conditioned on those for the latent nodes and the correspondence. Next, we define each of these steps in further detail and the parameters involved in each.

When the number of unique DSTs is known ahead of time, the DST $T$ follows a categorical distribution $Cat(\alpha)$ shared by all documents. When the number of DSTs is a variable, following Bayesian non-parametric methods, we instead use a Chinese Restaurant Process (CRP) prior $T \sim CRP(\alpha')$, so that the number of DSTs grows slowly with increasing number of documents [12].

Sampling of the latent structure starts by creating a latent root node, setting its parent $\pi_0^* = -1$ and sampling its node type $t_0$ from a DST-specific categorical distribution $Cat(\beta_T)$. The rest of the latent structure is created recursively. For each remaining node $i$ with type $t$, a child is created (or not) for each unique node type $t'$ by first sampling from a Bernoulli $Ber(\mu_{t,t'}\beta_{T,t'})$ to decide if a child of type $t'$ is to be created, and then, if true, creating a new node $j$ with $\pi_j^* = i$ and $t_j = t'$. The pattern for parent type $t$ and child type $t'$ is captured by $\mu_{t,t'}$, and that for a DST $T$ and a node type $t'$ by $\beta_{T,t'}$. When the number of types is not known in advance, we use the Indian Buffet Process (IBP) $t' \sim IBP(\mu'\beta')$ as the appropriate Bayesian non-parametric distribution [6], so that the number of node types also grows slowly with the number of documents.

For sampling of the content $w^*$ of latent nodes, we make use of the multinomial mixture model with one mixture component for each node type. To make use of precomputed word embeddings, we combine generative mixture models and embeddings [5,11]. However, instead of using word embeddings, we modify the LFF-DMM model [11] to make use of sequence embeddings, since many text units, such as for 'movie plots', have long sequential content. We first create sequence embeddings of all text units, fine-tuned using BERT [4]. Then the (unnormalized) probability of a text unit content $w$ with embedding $h(w)$ for a node type $t$ with embedding $h(t)$ is defined as the dot product $h(w)h(t)^T$ of the two. The specific content $w_j^*$ of the latent node $j$ is sampled from a categorical distribution over possible text unit contents in the data for the node type $t$, obtained by applying soft-max on the unnormalized probabilities.

When the data has node locations $l_j = (x_j, y_j)$, we model the location of the $j^{th}$ node as a relative shift $(\Delta_j^x, \Delta_j^y)$ from the location of its parent node. Viewing the documents as complex tables, the relative shifts are integers. We model the shifts of a node of type $t'$ with parent type $t$ using Poisson distributions:

$\Delta_j^x \sim Poi(\tau_{tt'}^x)$, $\Delta_j^y \sim Poi(\tau_{tt'}^y)$. Spans of nodes are similarly modeled using Poisson distributions $Poi(\gamma_{t'}^x)$, $Poi(\gamma_{t'}^y)$.

The final step is the sampling of the observed structure conditioned on the latent structure, the DST and node types. The latent nodes are traversed in the topological order of their generation. For the $j^{th}$ latent node of type $t$, whether it is observed is decided by sampling from a Bernoulli $Ber(\phi_t)$. If it is not observed, then it is skipped and $c_j = -1$. Else, an observed node $j'$ is created with $c_j = j'$. The parent is recorded by setting $\pi_{j'c(\pi_j^*)} = 1$ if $\pi_j^*$ is observed, or setting $\pi_{j'c(a_j^*)} = 1$, where $a_j^*$ is the nearest ancestor of latent node $j$ which is observed. An observed node $j'$ gets its content $w_j$ and location $l_j$ from its corresponding latent node.

**Inference:** Now that we have defined the generative model, or equivalently the joint distribution over the observed and the latent variables in a document, we now address the task of inferring the latent structure based on the observed structure, using estimates of the parameters, $\beta$, $\mu$, $\tau$, $\gamma$ and $\phi$ and the embeddings $h(t)$ for the different node types. We will address the task of parameter estimation after describing our inference algorithm. Specifically, the inference task is to identify, as in the third column of Fig. 1, the most likely DST $T$, latent nodes $N^*$ and parents $\pi^*$, node types $t = (t^p, t^s)$, their content $w^*$ and correspondences $c$ between latent and observed nodes, given, as in the second column, the observed nodes $N$, their parents $\pi$, location $l$ and content $w$. For the inference setting, we assume the number of DSTs and node types to be known. The main challenge in recovering the latent structure is that the number of latent nodes is unknown. Additional latent nodes, which do not have corresponding observed nodes, may need to be introduced into the latent structure.

**Algorithm 1:** Document Structure Inference

**Input**: Observed document $N, \pi, w, l$ ; parameters $\beta$, $\mu$, $\tau$, $\phi$, $\{h(t)\}$
1 $N^* = N$, $c_j = j \; \forall j$, $(\pi^*, w^*, l^*) = (\pi, w, l)$, Initialize $t, T$
2 **repeat**
  // *Update node types, parents and DST*
3   **for** $j \in N^*$ **do**
4     $t_j = \arg\max_t P(t_j = t \mid T, t, \pi^*, w_j^*, l_j^*)$
5     $\pi_j^* = \arg\max_{j'} P(\pi_j^* = j' \mid T, t, l^*)$
6   $T = \arg\max_k P(T = k \mid t, \pi^*)$
7   **for** *each missing header sub-type* $t'$ *in document:* **do**
8     $L = P(N, \pi, w, l, N^*, T, \pi^*, t, w^*, c)$ // *Record likelihood before insertion*
9     $N^* = N^* \cup \{n'\}$, $t_{n'} = t'$, $\pi_{n'}^* = 0$, $c_{n'} = -1$
10     For $j \in N^*$: $\pi_j^* = \arg\max_{j'} P(\pi_j^* = j' \mid T, t, l^*)$
11     $L' = P(N, \pi, w, l, N^*, T, \pi^*, t, w^*, c)$ // *Check likelihood after insertion*
12     If $L' < L$, revert to earlier $N^*, \pi^*$
13     Else $w^*, l^* \sim P(w^*, l^* \mid T, t, \pi^*; h(t'), \lambda, \tau, \gamma)$
14 **until** *convergence*

The high-level inference algorithm is shown in Algorithm 1. The inference algorithm first initializes all the latent variables - the latent structure is set to the initial structure, node types are set based on their 'local' content ($w$) and location ($l$) features, and the DST is assigned randomly. Then it iteratively updates these variables until convergence. It is essential for the inference algorithm to be iterative since the assignment to the latent variables depend on each other even when the parameters are known. Our algorithm follows the spirit of belief propagation [19], which is known to converge in two passes for polytrees. However, our setting is complicated by the search over new nodes.

The first part of the update, reassigns the node type $t$ and parent $\pi^*$ for each of the existing current nodes and the document template, based on the *current assignment of all the other variables*. Updates are made using the posterior distributions for these variables given the other assignments. All the posterior distributions are categorical. The node type depends on the content and location, as well as the node types of the parent and all child nodes, and the DST. The parent depends on the DST, the types of all nodes as well as their locations. The DST depends on the types and parents of all nodes. The variables are set to the mode of the posterior distribution. In each iteration, this takes $O(N(N+n_t)+n_T)$ time, where $N$ is the current number of nodes in the document, $n_t$ is the number of node types, and $n_T$ is the number of DSTs.

The second part in each update iteration is the search over new nodes. The algorithm identifies a missing node type in the document, iteratively finds the best position to introduce it into the document by reassigning parents of all nodes, and finally decides to add it if this leads to an improvement in the overall likelihood. In general, this takes $O(n_t N^2)$ time. The structural constraints in our specific setting simplify the search to some extent. Specifically, (a) only nodes with primary type header can be missing and (b) header nodes can only be children of header nodes. When only a few document nodes are missing, the complexity reduces to $O(N^2)$. Some domains may have some additional constraints, such as (c) a header node has at most $K$ children of a specific header node type, and (d) there are at most $K'$ nodes corresponding to a header node type in a document. In our experimental domains, both $K$ and $K'$ were 1. We also found that the algorithm converges in a few iterations in such settings.

**Learning:** The parameter learning task is to estimate parameters ($\alpha$, $\mu$, $\beta$, $\tau$, $\gamma$, $\phi$) and the embeddings $h(t)$ for each node type from the training documents. We assume that each training document has the entity and relation annotations over the true structure of the document. For the purpose of training the graphical model for structure recovery, this means that the true nodes $N^*$ and the true parents $\pi^*$ for these are provided. Additionally, since only header nodes can correspond to entities, the primary nodes types for a subset of nodes are implicitly known to be header. However, the primary types for the remaining nodes are not known. Further, the secondary node type for any of the nodes, as well as the DST for the documents are not known. Since the number of distinct node types and distinct DSTs are also typically difficult to provide as inputs, these are also assumed to be unknown.

We use a hard version of the expectation maximization (EM) algorithm for parameter learning. In each iteration, the E-step performs inference for the latent variables, which are the node types $t$ and the DST $T$. These are done following the first part of Algorithm 1. The only difference is that new types and new DSTs need to be considered, beyond those already encountered in the training data. These are done using the CRP for the DST and the IBP for the node types, both of which reserve a constant probability for unseen types in their prior. The M-step takes maximum a posteriori (MAP) estimates of all the parameters using current assignments to the latent variables. For the embedding $h(t)$ of each node type $t$, instead of the true MAP, for simplicity we take the average of the embeddings of the text units currently assigned to the specific type $t$. The iterations continue until there are no changes to the assignments.

### 4.2 Document Structure Annotation Using PLP

At this stage, for each document, we have the output of structure recovery, which includes $N^*$, $\pi_j^*$, $t_j$ and $T$. We also have the knowledge graph schema with the entities $E$ and binary relations $R$ defined over these. The task is to find the entity annotation $e_j \in E \cup \{-1\}$ for each header node $n_j$ and the relation annotation $r_l \in R \cup \{-1\}$ for each header path $p_l$, as in Fig. 2. The goal is to find the most likely consistent annotation.

The probability of a consistent annotation is determined by the parameters $\theta_{et}$, which is the probability of entity label $e$ for a node of type $t$, and $\psi_{re_1e_2}$, which is the probability of a relation label $r$ between entity pair $(e_1, e_2)$. The joint probability of a document annotation is the product of the individual entity and relation annotations. All inconsistent annotations have zero probability.

To address this problem, we define a valid joint probability distribution over entity and relation annotations that follow logical rules encoding the constraints. A natural framework for this is probabilistic (first order) logic programming (PLP) [13,14], which combines first order logic programming such as Prolog, with probabilistic semantics. Specifically, we use ProbLog [2,13], but other frameworks [1,15,16] can be substituted without much difficulty.

A probabilistic logic program is composed of a set of ground facts (or predicates), a set of uncertain facts, and a set of first order logic rules encoding background knowledge over the predicates. These define a distribution over possible worlds of assignments to the remaining (query) predicates. The probabilities can be learnt from evidence facts. We first define our reduction of the entity and relation annotation task to a probabilistic logic program, and then describe the inference and learning algorithms.

**Probabilistic Logic Program for Annotation:** For document structure annotation, we set the input document header structure as the ground facts: $\text{edge}(n_j, n_{j'}) = 1$ if $\pi_j^* = j'$ or $\pi_{j'}^* = j$, and $\text{type}(t, n_j) = 1$ if $t_j = t$. The uncertain facts are $eLabel(e, t)$ for node with type $t$ having entity label $e$, and $rLabel(r, e_1, e_2)$ for entity pair $e_1$, $e_2$ getting relation label $r$. These are specified using the parameters that determine the probability of an annotation, using the notion of annotated disjunctions, as follows: $\theta_{et}$ :: $eLabel(e,t)$; .... and $\psi_{re_1e_2}$ :: $rLabel(r,e_1,e_2)$; ..... The entity and relation annotations form the query predicates eAnnot$(e,n_j)$ and rAnnot$(r,p_l)$.

The rules capture background knowledge about constraints to define the consequences of assignments:

$$\text{eAnnot}(e,n) := \text{type}(t,n), \text{eLabel}(e,t).$$
$$\text{rAnnot}(r,p) := \text{path}(p), \text{eAnnot}(e_1, head(p)), \text{eAnnot}(e_2, last(p)), \text{rLabel}(r,e_1,e_2).$$
$$\text{path}(p) := len(p,2), \text{edge}(\text{head}(p), \text{last}(p)).$$
$$\text{path}(p) := \text{path}(tail(p)), \text{edge}(\text{head}(p), \text{head}(\text{tail}(p))), \text{eAnnot}(\text{head}(\text{tail}(p)), -1).$$

The first rule defines the probability of an entity annotation of a node $n$ with entity $e$, in terms of the type $t$ of the node, which is a ground fact, and the entity label $e$ for a node of type $t$, which is an uncertain fact. Similarly, the second rule defines the probability of a relation annotation of a path $p$ with relation $r$, in terms of $p$ being a logical path, the probability of entity label $e_1$ for the head node of $p$, the probability of entity label $e_2$ for the last node of $p$, and the probability of relation label $r$ between entities $e_1$ and $e_2$. The final two rules define logical paths. We represent paths as lists, and for notational convenience assume functions head$(l)$, tail$(l)$ and last$(l)$ for lists. The simplest path has length 2 and consists of an edge between the head and the last node. Alternatively, $p$ is a logical path if its tail is a logical path, there is an edge between the head of $p$ and the head of the tail, and the head node of the tail has entity annotation $-1$.

**Inference:** The inference task uses known estimates of the probabilities $\theta$ (for eLabel $(e,t)$) and $\psi$ (for rLabel$(r,e_1,e_2)$) and the ground predicates edge$(n,n')$ and type$(t,n)$ to infer the query predicates eAnnot$(e,n)$ and rAnnot$(r,p)$. To do this in ProbLog, we execute the query rAnnot$(_,_)$ to get the probability of different relation annotations for each path. We annotate each path with the relation which has the highest probability. We also annotate the first and last node of each path with the first and last entity of the selected relation type.

**Parameter Learning:** The learning task is to estimate the values of parameters $\theta_{et}$ and $\psi_{re_1e_2}$ from training data containing entity annotations for header nodes and relation annotations for header paths. This is done by providing to ProbLog as evidence ground facts eAnnot$(e,n)$ and rAnnot$(r,p)$ prepared from training data.

### 4.3 Entity and Relation Discovery

We finally describe our approach for entity and relation discovery, given the latent header structure and the entity and relation annotations using the existing (incomplete) knowledge graph. Our discovery algorithm is iterative. It identifies as candidates for a new entity, a sets of nodes $C_t^e$ with the same secondary header type $t$ but not yet assigned to any entity. The algorithm takes the largest such set and assigns all those nodes to a new entity id. Next, it looks for new relations between existing entities including the newly discovered one. It identifies as candidates for a new relation, a sets of header-type paths $C^r_{e,e'}$ with the same entities $e$ and $e'$ assigned to the head and last entities but with the intermediate nodes not assigned to any entity. It takes the largest such set and assigns all those paths to a new relation. This continues while the sizes of best candidate entity set and the best candidate relation set are larger than $\epsilon_e$ and $\epsilon_r$ respectively, which are the entity and relation discovery thresholds.

# 5 Experiments

We now report experiments for evaluating our proposed model, which we name **PGM-PLP**. We first describe our datasets and baselines for comparison.

*Datasets:* We use three real datasets of different types.

**Web Complex:** This dataset is created from the website of a large multinational company. This is our primary dataset as this contains all discussed complexities of semi-structured documents. This has 640 webpages following 6 DSTs. We convert the html of the pages to a DAG structure containing the locations of the text units. The DAG structure of a webpage contains 6–18 nodes. We have also annotated each webpage with the gold-standard structure. Nodes are annotated using 40 node types (20 for headers, 20 for values). The schema has 11 entity types and 13 relation types. Nodes and paths in the gold-standard structure are annotated with entities and relations. A webpage has 3–6 nodes with entity annotations and 2–6 paths with relation annotations. We use 15% of the pages as training, 15% for validation and 70% as test.

**SWDE:** This is a benchmark webpage dataset for entity annotation [10]. It contains gold annotations of 4–5 predicates of a single entity for 4 verticals, with 10 websites in each vertical and 200–2000 pages for each site. The schema has a star shape, with a central entity and 4–5 secondary entities. We use the same train-test setting as [7]. We converted the DOM tree structure of each webpage to DAG structure with only the text units as nodes. This dataset does not differentiate between observed and true structures, and also does not identify DSTs.

**Tab Complex:** This is a proprietary dataset containing real tables from the pharmaceuticals industry. The tables are complex with 2-level hierarchy in value cells under some columns. We use this dataset to demonstrate that our solution is not restricted to webpages, but can handle document structure with only spatial locations of text units and no associated html. This has 76 tables of 9 types. The number of tables for each type ranges from 4 to 14. The total number of cells per table ranges from 13 to 8292, corresponding to 36 different cell types. The gold-standard structure is the same as the observed structure. The associated knowledge graph schema has 18 entity types and 26 relation types. Each table contains gold-standard entity and relation annotations. A table has 2–5 entity

**Table 1.** Entity (E) and relation (R) annotation performance on Web Complex, Tab Complex and SWDE for PGM-PLP, CERES and Limaye. Limaye cannot handle html structure in SWDE.

| | Web complex | | Tab complex | | SWDE | | | | | | | |
|---|---|---|---|---|---|---|---|---|---|---|---|---|
| | | | | | Movie | | NBA Player | | University | | Book | |
| | E | R | E | R | E | R | E | R | E | R | E | R |
| Limaye | 0.81 | 0.61 | 0.93 | 0.86 | – | – | – | – | – | – | – | – |
| CERES | 0.65 | 0.28 | 0.64 | 0.31 | 1.00 | 1.00 | 1.00 | 1.00 | 1.00 | 1.00 | 0.95 | 0.92 |
| PGM-PLP | **0.97** | **0.96** | **0.95** | **0.92** | 1.00 | 1.00 | 1.00 | 1.00 | 0.99 | 0.99 | **1.00** | **1.00** |

annotations and 1–6 relation annotations. We use 15% of the data as training, 15% for validation, and remaining for test.

*Models:* We next describe the models used for comparison.

**PGM-PLP:** This is our proposed model that recovers latent structure as well as the DST, followed by entity and relation annotation. The model hyper-parameters - the CRP parameters $\alpha'$, the IBP parameters $\mu'$, $\beta'$ and the entity and relation discovery thresholds $\epsilon_e$ and $\epsilon_r$ - are tuned using the validation sets.

**Ceres:** CERES [10] is the state-of-the-art on entity information extraction from web-pages. Though it can be trained using distant supervision, for fair comparison we provide it with the same training data as the other models. It does not recover latent structure and also does not consider DSTs for webpages. As no implementation is publicly available, we have used our own implementation. Since CERES can only handle star schemas, we provide it with the best possible star approximation of the schemas. It uses html and text features for annotation. However, since html features are not available for Web Complex and Tab Complex, its entity annotation classifier only uses text-based features.

**Limaye:** This is a common baseline for annotation of simple tables [8] that uses joint inference using a probabilistic graphical model. It performs comparably with more recent table annotation models [17]. Since semi-structured documents can alternatively be represented as (complex) tables, we use this as the representative of table annotation approaches. Note that this approach does not recognize DSTs or latent structures. Since Limaye, like other table annotation approaches, can only handle simple tables, we approximate our datasets as simple tables as faithfully as possible to create the input for Limaye, as follows. We define a complex table as a tree where each leaf node is a simple table, and each internal node has the structure of a simple table but each cell points to another internal or leaf node. Sibling tables (internal or leaf) do not necessarily have the same dimensions. We simplify such tables in a bottom-up fashion. An internal node is simplified only when all its children are simple tables. We order the cells of an internal (table) node and merge its children sequentially and pairwise. The merge operation for two tables of different dimensions takes all combination of the rows of the two tables, and columns from both.

*Annotation Experiments:* We first present results for entity and relation annotation in Table 1. We use micro-averaged F1, which is the weighted average of the F1 scores for the different classes (entities or relations). Limaye is defined only for tables and cannot annotate html structures as in SWDE.

We observe that for Web Complex, which has the most complex document structure and knowledge graph schema, PGM-PLP significantly outperforms both Limaye and CERES. This reflects the usefulness of structure and DST recovery, and joint annotation. Limaye performs much better than CERES since it is able to handle arbitrary schemas, while CERES takes in a star-approximation. Unlike Web Complex, Tab Complex does not have any latent structure different from the observed structure. As a result, performance improves for both Limaye and CERES, with CERES still lagging Limaye. However, PGM-PLP is still significantly better than CERES and also outperforms Limaye. The smaller gap between PGP-PLP and Limaye is also because of lower structural complexity in Tab Complex compared to Web Complex. For SWDE, both PGM-PLP and CERES are able to annotate almost perfectly when provided with the same training setting as in [7].

**Table 2.** Performance for identification of parent (F1), node type (NMI) and DST (NMI) for PGM and its three ablations.

| | Parent | Node type | DST |
|---|---|---|---|
| PGM(-T) | 0.82 | 0.73 | 0.64 |
| PGM(-L) | 0.97 | 0.85 | 0.88 |
| PGM(-S) | 0.78 | 0.82 | 0.79 |
| PGM | **0.98** | **0.95** | **0.99** |

**Table 3.** Entity (E) and relation (R) discovery performance (NMI) for PGM-PLP with varying percentage of hidden entities during training.

| | Web complex | | Tab Complex | |
|---|---|---|---|---|
| % Hidden | E | R | E | R |
| 30 | 0.96 | 0.95 | 0.92 | 0.91 |
| 60 | 0.95 | 0.95 | 0.91 | 0.91 |
| 100 | 0.95 | 0.94 | 0.88 | 0.90 |

*Document Structure Recovery:* Next, we report performance for document structure recovery. Specifically, we evaluate the goodness of parent identification of nodes, node type identification (at the secondary level), and DST identification. Note that the first is a binary classification task, so that we evaluate using F1. The second and the third are clustering tasks, with no supervision provided during training. Accordingly, we use normalized mutual information (NMI) over pair-wise clustering decisions for evaluation. The other models do not address this task. Therefore, we compare against ablations of our own model. This task does not involve the PLP stage. So we use ablations of PGM. Ablation **-T** does not use the textual content, **-L** does not use the locations of the text units, and **-S** does not use structural patterns between parent and child types as captured by the parameter $\beta$. We report performance only for Web Complex, which has gold standards for all of these aspects of the document structure.

From Table 2, we can see importance of each of these aspects for structure recovery. PGM is able to recover the structure almost perfectly. Not surprisingly, text content has the biggest individual impact, most of all on the DST. Structural pattern has the next biggest impact. Location has the smallest individual impact, but still results in a significant difference in performance.

*Entity and Relation Discovery:* We next report performance for entity and relation discovery. In this experiment, we hide some entities and all their relations from the training schema and the training annotations. We evaluate how accurately occurrences of these entities and relations are recognized in the test documents. This is again a clustering task, since these entity and relation labels are not seen during training. Accordingly, we evaluate performance using NMI. Note that the other approaches cannot address this task. So we evaluate only PGM-PLP. Since SWDE has a simple star schema to begin with, we evaluate only for Web Complex and Tab Complex. Table 3 records performance for varying percentages of hidden entities during training. Performance is extremely robust for Web Complex. Even when no initial schema is provided, entities and relations are detected very accurately based on the recovered document structure properties. This is based on the accurate recovery of the secondary header types and clear path patterns between these in the document templates.

*Ablation Study for Annotation:* Finally, after presenting the usefulness of our overall model, we report performance of different ablations of PGP-PLP for the end task of entity and relation annotation. **PGM(-S)-PLP** performs only partial structure recovery. It identifies node types and DSTs but keeps the original observed structure. Its second stage is the same as in PGM-PLP. **PGM-LP** has the same first stage as PGM-PLP. However, instead of a probabilistic logic program in the second stage, it uses a deterministic logic program, specifically ProLog, with no uncertain facts. In Table 4, we record performance of these two ablated versions along with that of the full model for all three datasets. PGM(-S)-PLP performs the worst for Web Complex because of its structural complexity, where structure recovery has the biggest impact on both entity and relation annotation. In contrast, it almost matches the performance of the full model for SWDE, where the structural complexity is the least. On the other hand, PGM-LP takes a very significant hit across all three datasets. This shows

**Table 4.** Ablation study for entity (E) and relation (R) detection on 3 datasets. PGM(-S)-PLP keeps original document structure. PGM-LP uses Prolog instead of ProbLog.

| | Web complex | | Tab complex | | SWDE | | | | | | | |
|---|---|---|---|---|---|---|---|---|---|---|---|---|
| | | | | | Movie | | NBA player | | University | | Book | |
| | E | R | E | R | E | R | E | R | E | R | E | R |
| PGM(-S)-PLP | 0.41 | 0.11 | 0.88 | 0.80 | 1.00 | 1.00 | 1.00 | 1.00 | 0.99 | 0.99 | 0.99 | 0.98 |
| PGM-LP | 0.21 | 0.10 | 0.35 | 0.20 | 0.00 | 0.00 | 0.31 | 0.20 | 0.00 | 0.00 | 0.24 | 0.10 |
| PGM-PLP | 0.97 | 0.96 | 0.95 | 0.92 | 1.00 | 1.00 | 1.00 | 1.00 | 0.99 | 0.99 | 1.00 | 1.00 |

the importance of accounting for uncertainty when annotating the recovered structures, and that of learning the parameters of this model.

## 6 Conclusions

We have formalized the general problem of annotation of complex semi-structured documents with a knowledge graph schema containing entity and relationship types. We have addressed the problem using a two-stage solution that performs structure recovery using a probabilistic graphical model, followed by joint annotation of nodes and edges of the structure using a probabilistic logic program. We also address the discovery of new entity and relation types not in the schema. We show the usefulness of our model by experiments on real data for both web-pages and complex tables, and we show that we outperform both state-of-the-art table and web-page annotation approaches.

## References


1. Bach, S.H., Broecheler, M., Huang, B., Getoor, L.: Hinge-loss Markov random fields and probabilistic soft logic. J. Mach. Learn. Res. (JMLR) **18**, 1–67 (2017)
2. De Raedt, L., Kersting, K.: Probabilistic inductive logic programming. In: De Raedt, L., Frasconi, P., Kersting, K., Muggleton, S. (eds.) Probabilistic Inductive Logic Programming. LNCS (LNAI), vol. 4911, pp. 1–27. Springer, Heidelberg (2008). https://doi.org/10.1007/978-3-540-78652-8_1
3. Deng, L., Zhang, S., Balog, K.: Table2Vec: neural word and entity embeddings for table population and retrieval. In: SIGIR (2019)
4. Devlin, J., Chang, M., Lee, K., Toutanova, K.: BERT: pre-training of deep bidirectional transformers for language understanding. In: NAACL HLT (2019)
5. Dieng, A.B., Ruiz, F.J.R., Blei, D.M.: Topic modeling in embedding spaces. Trans. Assoc. Comput. Linguist. **8**, 439–453 (2020)
6. Griffiths, T.L., Ghahramani, Z.: The Indian buffet process: an introduction and review. J. Mach. Learn. Res. **12**(32), 1185–1224 (2011)
7. Gulhane, P., et al.: Web-scale information extraction with vertex. In: ICDE (2011)
8. Limaye, G., Sarawagi, S., Chakrabarti, S.: Annotating and searching web tables using entities, types and relationships. Proc. VLDB Endow. **3**(1–2), 1338–1347 (2010)
9. Lockard, C., Shiralkar, P., Dong, X.L.: OpenCeres: when open information extraction meets the semi-structured web. In: NAACL-HLT (2019)
10. Lockard, C., Dong, X.L., Einolghozati, A., Shiralkar, P.: Ceres: distantly supervised relation extraction from the semi-structured web. Proc. VLDB Endow. **11**(10), 1084–1096 (2018)
11. Nguyen, D.Q., Billingsley, R., Du, L., Johnson, M.: Improving topic models with latent feature word representations. Trans. ACL **3**, 299–313 (2015)
12. Orbanz, P., Teh, Y.: Modern Bayesian nonparametrics. In: NIPS Tutorial (2011)
13. Raedt, L.D., Kimmig, A.: Probabilistic programming. In: Tutorial at IJCAI (2015)
14. Raedt, L.D., Poole, D., Kersting, K., Natarajan, S.: Statistical relational artificial intelligence: logic, probability and computation. In: Tutorial at Neurips (2017)
15. Richardson, M., Domingos, P.: Markov logic networks. Mach. Learn. **62**(1–2), 107–136 (2006)

16. Sato, T., Kameya, Y.: Parameter learning of logic programs for symbolic-statistical modeling. J. Artif. Intell. Res. **15**, 391–454 (2001)
17. Takeoka, K., Oyamada, M., Nakadai, S., Okadome, T.: Meimei: an efficient probabilistic approach for semantically annotating tables. In: AAAI (2019)
18. Wu, S., et al.: Fonduer: Knowledge base construction from richly formatted data. In: SIGMOD (2018)
19. Yedidia, J.S., Freeman, W.T., Weiss, Y.: Understanding belief propagation and its generalizations, pp. 239–269. Morgan Kaufmann Publishers Inc. (2003)
20. Zhang, S., Balog, K.: Ad hoc table retrieval using semantic similarity. In: WWW (2018)

# Learning Disentangled Representations with the Wasserstein Autoencoder

Benoit Gaujac[1(✉)], Ilya Feige[2], and David Barber[1]

[1] University College London, London, UK
{benoit.gaujac.16,david.barber}@ucl.ac.uk
[2] Faculty, London, UK
ilya@faculty.ai

**Abstract.** Disentangled representation learning has undoubtedly benefited from objective function surgery. However, a delicate balancing act of tuning is still required in order to trade off reconstruction fidelity versus disentanglement. Building on previous successes of penalizing the total correlation in the latent variables, we propose TCWAE (Total Correlation Wasserstein Autoencoder). Working in the WAE paradigm naturally enables the separation of the total-correlation term, thus providing disentanglement control over the learned representation, while offering more flexibility in the choice of reconstruction cost. We propose two variants using different KL estimators and analyse in turn the impact of having different ground cost functions and latent regularization terms. Extensive quantitative comparisons on data sets with known generative factors shows that our methods present competitive results relative to state-of-the-art techniques. We further study the trade off between disentanglement and reconstruction on more-difficult data sets with unknown generative factors, where the flexibility of the WAE paradigm leads to improved reconstructions.

**Keywords:** Wasserstein Autoencoder · Variational Autoencoder · Generative modelling · Representation learning · Disentanglement learning

## 1 Introduction

Learning representations of data is at the heart of deep learning; the ability to interpret those representations empowers practitioners to improve the performance and robustness of their models [4,32]. In the case where the data is underpinned by independent latent generative factors, a good representation should encode information about the data in a semantically meaningful manner with statistically independent latent variables encoding for each factor. [4] define a disentangled representation as having the property that a change in one dimension corresponds to a change in one factor of variation, while being relatively invariant to changes in other factors. While many attempts to formalize


N. Oliver et al. (Eds.): ECML PKDD 2021, LNAI 12977, pp. 69–84, 2021.
https://doi.org/10.1007/978-3-030-86523-8_5

this concept have been proposed [8,9,15], finding a principled and reproducible approach to assess disentanglement is still an open problem [24].

Recent successful unsupervised learning methods have shown how simply modifying the ELBO objective, either re-weighting the latent regularization terms or directly regularizing the statistical dependencies in the latent, can be effective in learning disentangled representation. [16] and [6] control the information bottleneck capacity of Variational Autoencoders (VAEs, [20,29]) by heavily penalizing the latent regularization term. [7] perform ELBO surgery to isolate the terms at the origin of disentanglement in $\beta$-VAE, improving the reconstruction-disentanglement trade off. [10] further improve the reconstruction capacity of $\beta$-TCVAE by introducing structural dependencies both between groups of variables and between variablcs within each group. Alternatively, directly regularizing the aggregated posterior to the prior with density-free divergences [41] or moments matching [21], or simply penalizing a high Total Correlation (TC) [39]) in the latent [19] has shown good disentanglement performance.

In fact, information theory has been a fertile ground to tackle representation learning. [1] re-interpret VAEs from an Information Bottleneck view [34], re-phrasing it as a trade off between sufficiency and minimality of the representation, regularizing a pseudo TC between the aggregated posterior and the true conditional posterior. Similarly, [12] use the principle of total Correlation Explanation (CorEX) [37] and maximize the mutual information between the observation and a subset of anchor latent points. Maximizing the mutual information (MI) between the observation and the latent has been broadly used [3,17,27,36], showing encouraging results in representation learning. However, [36] argued that MI maximization alone cannot explain the disentanglement performance of these methods.

Building on developments in Optimal Transport (OT) [35,38] introduced the Wasserstein Autoencoder (WAE), an alternative to VAEs for learning generative models. WAE maps the data into a (low-dimensional) latent space while regularizing the averaged encoding distribution. This is in contrast with VAEs where the posterior is regularized at each data point, and allows the encoding distribution to capture significant information about the data while still matching the prior when averaged over the whole data set. Interestingly, by directly regularizing the aggregated posterior, WAE hints at more explicit control on the way the information is encoded, and thus better disentanglement. The reconstruction term of the WAE allows for any cost function on the observation space and non-deterministic decoders. This removes the need of random decoders for which the log density is tractable and derivable (with regard to the decoder parameters) as it is the case in VAEs. The combination of non-degenerated Gaussian decoders with the Kullback-Leibler (KL) divergence acting as the posterior regularizer is viewed as the cause of the samples blurriness in VAEs [5].

Few works have sought to use WAE for disentanglement learning. Initial experiments from [31] showed encouraging results but did not fully leverage the flexibility found in WAEs. The authors simply applied the original WAE objective with cross-entropy reconstruction cost without studying the impact of using different reconstruction terms and latent regularisation functions. Mirroring the KL-based TC, [40] introduced Wasserstein Total Correlation (WTC). Resorting to the triangle inequality, they decomposed the latent regularization term of the WAE into the sum of two WTC terms, obtaining an upper bound of the WAE objective. However, WTC lacks an information-theoretic interpretation such as the one offered by the TC and the authors resorted to the dual formulation of the 1-Wasserstein distance to approximate the WTC, adversarially training two critics.

In this work[1], following the success of regularizing the TC in disentanglement, we propose to use the KL divergence as the latent regularization function in the WAE. We introduce the Total Correlation WAE (TCWAE) with an explicit dependency on the TC of the aggregated posterior. We study separately the impact of using a different ground cost function in the reconstruction term and the impact of a different composition of the latent regularization term. Performing extensive comparisons with successful methods on a number of data sets, we found that TCWAEs achieve competitive disentanglement performance while improving modelling performance by allowing flexibility in the choice of reconstruction cost.

## 2 Importance of Total Correlation in Disentanglement

### 2.1 Total Correlation

The TC of a random vector $Z \in \mathcal{Z}$ under $P$ is defined by

$$\mathbf{TC}(Z) \triangleq \sum_{d=1}^{d_Z} H_{P_d}(Z_d) - H_P(Z) \tag{1}$$

where $p_d(z_d)$ is the marginal density over only the $d$th component $z_d$, and $H_P(Z) \triangleq -\mathbb{E}_P \log p(Z)$ is the Shannon differential entropy, which encodes the information contained in $Z$ under $P$. Since

$$\sum_{d=1}^{d_Z} H_{P_d}(Z_d) \leq H_P(Z) \tag{2}$$

with equality when the marginals $Z_d$ are mutually independent, the TC can be interpreted as the loss of information when assuming mutual independence of the $Z_d$; namely, it measures the mutual dependence of the marginals. Thus, in the context of disentanglement learning, we seek a low TC of the aggregated posterior, $p(z) = \int_{\mathcal{X}} p(z|x)\, p(x)\, dx$, which forces the model to encode the data into statistically independent latent codes. High MI between the data and the latent is then obtained when the posterior, $p(z|x)$, manages to capture relevant information from the data.

[1] Implementation details and additional results can be found in the Appendix of [13].

### 2.2 Total Correlation in ELBO

We consider latent generative models $p_\theta(x) = \int_{\mathcal{Z}} p_\theta(x|z)\,p(z)\,dz$ with prior $p(z)$ and decoder network, $p_\theta(x|z)$, parametrized by $\theta$. VAEs approximate the intractable posterior $p_\theta(z|x)$ by introducing an encoding distribution (the encoder), $q_\phi(z|x)$, and learning simultaneously $\theta$ and $\phi$ when optimizing the variational lower bound, or ELBO, defined in Eq. (3) for a single observation $x$:

$$\mathcal{L}_{\mathrm{ELBO}}(x,\theta,\phi) \triangleq \underset{Q_\phi(Z|X=x)}{\mathbb{E}}\big[\log p_\theta(x|Z)\big] - \mathbf{KL}\Big(Q_\phi(Z|X=x)\;\|\;P(Z)\Big) \tag{3}$$

Following [18], we treat the data index, $n$, as a uniform random variable over $\{1,\dots,N\}$: $p(n) = \frac{1}{N}$. We can then define respectively the posterior, *joint* posterior and *aggregated* posterior:

$$q_\phi(z|n) = q_\phi(z|x_n), \quad q_\phi(z,n) = q_\phi(z|n)\,p(n), \quad q_\phi(z) = \sum_{n=1}^{N} q_\phi(z|n)\,p(n) \tag{4}$$

Using this notation, we decompose the KL term in Eq. (3) into the sum of ⓘ, an *index-code mutual information* term, and ⓘⓘ, a *marginal KL* term, with:

$$\text{ⓘ} = \mathbf{KL}\Big(Q_\phi(Z,N)\;\|\;Q_\phi(Z)P(N)\Big) \quad \text{and} \quad \text{ⓘⓘ} = \mathbf{KL}\Big(Q_\phi(Z)\;\|\;P(Z)\Big) \tag{5}$$

The index-code mutual information (index-code MI) represents the MI between the data and the latent under the joint distribution $q(z,n)$, and the marginal KL enforces the aggregated posterior to match the prior. The marginal KL term plays an important role in disentanglement. Indeed, it pushes the encoder network to match the prior when *averaged*, as opposed to matching the prior for each data point. Combined with a factorized prior $p(z) = \prod_d p_d(z_d)$, as it is often the case, the aggregated posterior is forced to factorize and align with the axis of the prior. More explicitly, the marginal KL term in Eq. (5) can be decomposed as sum of a *Total Correlation* term and a *dimension-wise KL* term:

$$\text{ⓘⓘ} = \mathbf{TC}\Big(Q_\phi(Z)\Big) + \sum_{d=1}^{d_{\mathcal{Z}}} \mathbf{KL}\Big(Q_{\phi,d}(Z_d)\;\|\;P_d(Z_d)\Big) \tag{6}$$

Thus maximizing the ELBO implicitly minimizes the TC of the aggregated posterior, enforcing the aggregated posterior to disentangle as [16] and [6] observed when strongly penalizing the KL term in Eq. (3). [7] leverage the KL decomposition in Eq. (6) by refining the heavy latent penalization to the TC only. However, the index-code MI term of Eq. (5) seems to have little to no role in disentanglement, potentially harming the reconstruction performance as we will see in Sect. 4.1.

## 3 Is WAE Naturally Good at Disentangling?

### 3.1 WAE

The Kantorovich formulation of the OT between the true-but-unknown data distribution $P_D$ and the model distribution $P_\theta$, for a given cost function $c$, is defined by:

$$\mathrm{OT}_c(P_D, P_\theta) = \inf_{\Gamma \in \mathcal{P}(P_D, P_\theta)} \mathop{\mathbb{E}}_{\Gamma(X,\tilde{X})} \left[c(X, \tilde{X})\right] \tag{7}$$

where $\mathcal{P}(P_D, P_\theta)$ is the space of all couplings of $P_D$ and $P_\theta$; namely, the space of joint distributions $\Gamma$ on $\mathcal{X} \times \mathcal{X}$ whose densities $\gamma$ have marginals $p_D$ and $p_\theta$. [35] derive the WAE objective by constraining this space but then relaxing the hard constraint on the marginal using a Lagrange multiplier (see Appendix A of [13]):

$$W_{\mathcal{D},c}(\theta, \phi) \triangleq \mathop{\mathbb{E}}_{P_D(X)} \mathop{\mathbb{E}}_{Q_\phi(Z|X)} \mathop{\mathbb{E}}_{P_\theta(\tilde{X}|Z)} \left[c(X, \tilde{X})\right] + \lambda \mathcal{D}\Big(Q_\phi(Z) \parallel P(Z)\Big) \tag{8}$$

where $\mathcal{D}$ is any divergence function and $\lambda$ a relaxation parameter. The decoder, $p_\theta(\tilde{x}|z)$, and the encoder, $q_\phi(z|x)$, are optimized simultaneously by dropping the closed-form minimization over the encoder network, with standard gradient descent methods.

Similarly to the ELBO, objective (8) consists of a reconstruction term and a latent regularization term, preventing the latent codes from drifting away from the prior. However, WAE naturally and explicitly penalizes the aggregate posterior as opposed to VAE, where the dependency to the TC is only implicit and competes with other terms, especially, the index-code MI. This motivates, following Sect. 2.2, the use of WAE in disentanglement learning.

Another important difference lies in the functional form of the reconstruction cost in the reconstruction term. Indeed, WAE allows for more flexibility in the reconstruction term with any cost function allowed, and in particular, it allows for cost functions better suited to the data at hand and for the use of deterministic decoder networks [11,35]. This can potentially result in an improved reconstruction-disentanglement trade off as we empirically find in Sects. 4.2 and 4.1.

### 3.2 TCWAE

For notational simplicity, we drop the explicit dependency of the distributions on their parameters $\theta$ and $\phi$ throughout this section.

We chose the divergence function $\mathcal{D}$ in Eq. (8) to be the KL divergence and assume a factorized prior (*e.g.* $p(z) = \mathcal{N}(0_{d_\mathcal{Z}}, \mathcal{I}_{d_\mathcal{Z}})$), thus obtaining the same decomposition as in Eq. (6). Similarly to [7], we use different parameters, $\beta$ and $\gamma$, for each term in the decomposition in Eq. (6) of the latent regularization term from Eq. (8), obtaining our TCWAE objective:

$$\begin{aligned} W_{TC} \triangleq & \mathop{\mathbb{E}}_{P(X_n)} \mathop{\mathbb{E}}_{Q(Z|X_n)} \Big[ \mathop{\mathbb{E}}_{P(\tilde{X}_n|Z)} [c(X_n, \tilde{X}_n)] \Big] \\ & + \beta \mathbf{KL}\Big(Q(Z) \parallel \prod_{d=1}^{d_\mathcal{Z}} Q_d(Z_d)\Big) + \gamma \sum_{d=1}^{d_\mathcal{Z}} \mathbf{KL}\Big(Q_d(Z_d) \parallel P_d(Z_d)\Big) \end{aligned} \tag{9}$$

TCWAE is identical to the WAE objective with KL divergence when $\lambda = \beta = \gamma$; and provides an upper bound with $\min(\beta, \gamma) = \lambda$.

Equation (9) can be directly related to the $\beta$-TCVAE objective of [7]:

$$\mathcal{L}_{\beta TC} \triangleq \mathop{\mathbb{E}}_{P(X_n)} \mathop{\mathbb{E}}_{Q(Z|X_n)} \Big[ -\log p(X_n|Z) \Big] + \alpha \mathbf{KL}\Big( Q(Z,N) \parallel Q(Z)P(N) \Big) + \beta \mathbf{KL}\Big( Q(Z) \parallel \prod_{d=1}^{d_Z} Q_d(Z_d) \Big) + \gamma \sum_{d=1}^{d_Z} \mathbf{KL}\Big( Q_d(Z_d) \parallel P_d(Z_d) \Big) \tag{10}$$

As mentioned above, the differences are the absence of index-code MI and a different reconstruction cost function. Setting $\alpha = 0$ in Eq. (10) makes the two latent regularizations match but breaks the inequality provided by the ELBO (Eq. (3)). Matching the two reconstruction terms would be possible if we could find a ground cost function $c$ such that $\mathbb{E}_{P(\tilde{X}_n|Z)} c(x_n, \tilde{X}_n) = -\log p(x_n|Z)$ for all $x_n$.

### 3.3 Estimators

With the goal of being grounded by information theory and earlier works on disentanglement, using the KL as the latent divergence function, as opposed to other sample-based divergences [28,35], presents its own challenges. Indeed, the KL terms are intractable; in particular, we will need estimators to approximate the entropy terms. We propose to use two estimators, one based on importance weight-sampling [7], the other on adversarial estimation using the density-ratio trick [19].

*TCWAE-MWS.* [7] propose to estimate the intractable terms $\mathbb{E}_Q \log q$ and $\mathbb{E}_{Q_d} \log q_d$ within the KL in Eq. (9) with Minibatch-Weighted Sampling (MWS). Considering a batch of observation $\{x_1, \ldots x_{N_{\text{batch}}}\}$, they sample the latent codes $z_i \sim Q(Z|x_i)$ and compute:

$$\mathop{\mathbb{E}}_{Q(Z)} \log q(Z) \approx \frac{1}{N_{\text{batch}}} \sum_{i=1}^{N_{\text{batch}}} \log \frac{1}{N \times N_{\text{batch}}} \sum_{j=1}^{N_{\text{batch}}} q(z_i|x_j) \tag{11}$$

This estimator, while being easily computed from samples, is a biased estimator of $\mathbb{E}_q \log q(Z)$. [7] also proposed an unbiased version, the Minibatch-Stratified Sampling (MSS). However, they found that it did not result in improved performance, and thus, they and we choose to use the simpler MWS estimator. We call the resulting algorithm the TCWAE-MWS. Other sampled-based estimators of the entropy or the KL divergence have been proposed [10,30]. However, we choose the solution of [7] for i) its simplicity and ii) the similarities between the TCWAE and $\beta$-TCVAE objectives.

*TCWAE-GAN.* A different approach, similar in spirit to the WAE-GAN originally proposed by [35], is based on adversarial-training. While [35] use the adversarial training to approximate the JS divergence, [19] use the density-ratio trick and adversarial training to estimate the intractable terms in Eq. (9). The density-ratio trick [26,33] estimates the KL divergence as:

$$\mathbf{KL}\Big(Q(Z) \parallel \prod_{d=1}^{d_Z} Q_d(Z_d)\Big) \approx \mathop{\mathbb{E}}_{Q(Z)} \log \frac{D(Z)}{1 - D(Z)} \tag{12}$$

where $D$ plays the same role as the discriminator in GANs and outputs an estimate of the probability that $z$ is sampled from $Q(Z)$ and not from $\prod_{d=1}^{d_Z} Q_d(Z_d)$. Given that we can easily sample from $q(z)$, we can use Monte-Carlo sampling to estimate the expectation in Eq. (12). The discriminator $D$ is adversarially trained alongside the decoder and encoder networks. We call this adversarial version the TCWAE-GAN.

## 4 Experiments

We perform a series of quantitative and qualitative experiments, isolating in turn the effect of the ground cost and latent regularization functions in TCWAE. We found that while the absence of index-code MI in TCWAEs does not impact the disentanglement performance, using the square Euclidean distance as our ground cost function improves the reconstruction while retaining the same disentanglement. Finally, we compare our methods with the benchmarks $\beta$-TCVAE and FactorVAE, both quantitatively on toy data sets and qualitatively more challenging data sets with unknown generative factors.

### 4.1 Quantitative Analysis: Disentanglement on Toy Data Sets

In this section we train the different methods on the dSprites [25], 3D shapes [19] and smallNORB [22] data sets. We use the Mutual Information Gap (MIG, [7]), the factorVAE metric [19] and the Separated Attribute Predictability score (SAP, [21]) to asses the disentanglement performances (see [24] for the implementation). We assess the reconstruction performances with the Mean Square Error (MSE) of the reconstructions. See Appendix B of [13] for more details on the experimental setups.

*$\gamma$ Tuning.* Mirroring [7,19], we first tune $\gamma$, responsible for the dimensionwise-KL regularization, subsequently focusing on the role of the TC term in the disentanglement performance. We trained the TCWAEs with six different values for each parameter, resulting in thirty-six different models. The heat map for the different data sets can be seen in Appendix C of [13]. We observe that while $\beta$ controls the trade off between reconstruction and disentanglement, $\gamma$ affects the range achievable when varying $\beta$. We choose $\gamma$ with the best overall mean scores when aggregated over all the different $\beta$ (see Appendix C of [13] for more details). The scores violin plots for the different $\gamma$ as well as the chosen values for each method and data set are given in Appendix C of [13]. These values are fixed in all the following experiments.

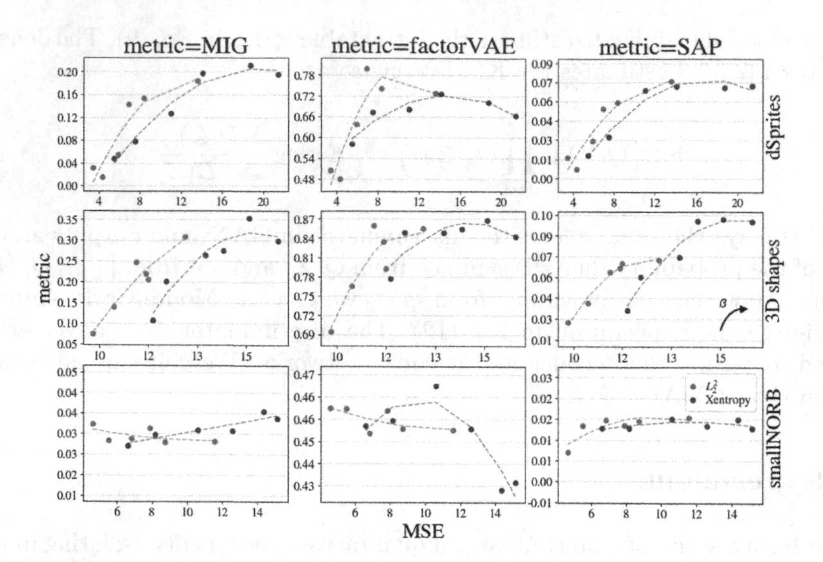


**Fig. 1.** Disentanglement versus reconstruction for TCWAE-MWS with square euclidean distance (red) and cross-entropy (blue) cost. Points represent different models corresponding to different $\beta$ (quadratic regression represented by the dashed lines). Low MSE and high scores (top-left corner) is better. Higher TC penalisation (bigger $\beta$) results in higher MSE and higher scores. (Color figure online)

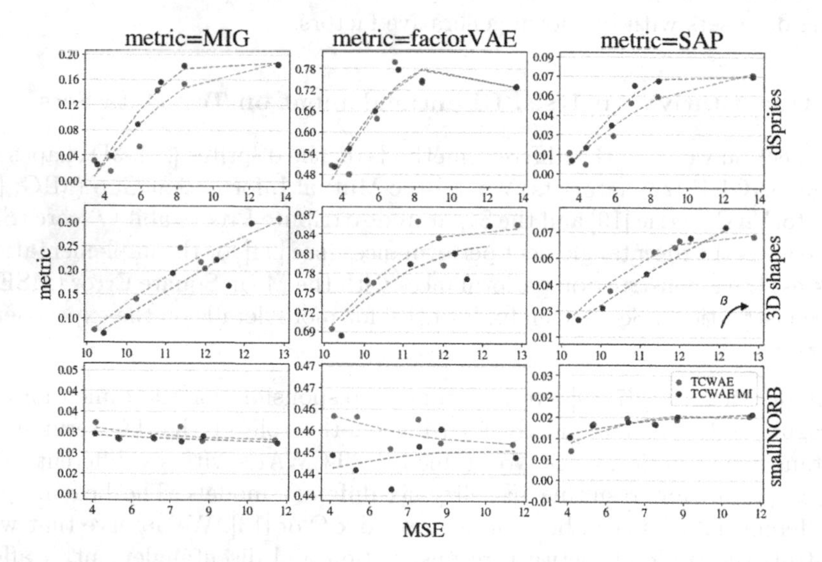


**Fig. 2.** Disentanglement versus reconstruction for TCWAE-MWS with (blue) and without code-index MI (red). The Points represent different models corresponding to different $\beta$ (quadratic regression represented by the dashed lines). Low MSE and high scores (top-left corner) is better. Higher TC penalisation (bigger $\beta$) results in higher MSE and higher scores. (Color figure online)

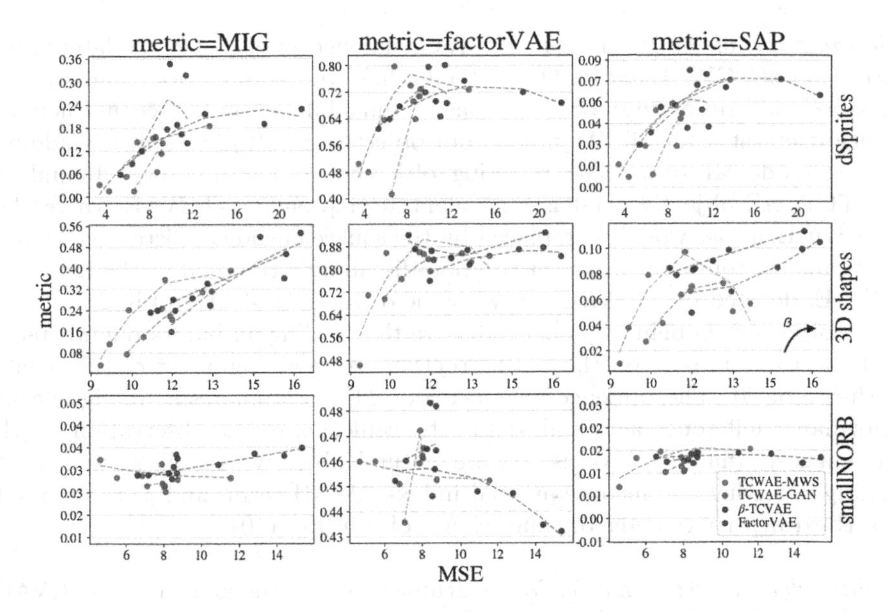


**Fig. 3.** Disentanglement versus reconstruction for TCWAE-MWS (red), TCWAE-GAN (green), $\beta$-TCVAE (blue) and FactorVAE (purple). The Points represent different models corresponding to different $\beta$ (quadratic regression represented by the dashed lines). Low MSE and high scores (top-left corner) is better. Higher TC penalisation (bigger $\beta$) results in higher MSE and higher scores. (Color figure online)

*Ablation Study of the Ground Cost Function.* We study the impact of using the square euclidean distance as the ground cost function as opposed to the cross-entropy cost as in the ELBO (Eq. (3)). When using the cross-entropy, the TCWAE objective is equivalent that in Eq. (10) with $\alpha = 0$; in this case, TCWAE-MWS simply boils down to $\beta$-TCVAE with no index-code MI term.

We train two sets of TCWAE-MWS, one with the square euclidean distance and one with the cross-entropy, for different $\beta$ and compare the reconstruction-disentanglement trade off by plotting the different disentanglement scores versus the MSE in Fig. 1.

We see that the square euclidean distance provides better reconstructions on all data sets as measured by the MSE (points on the left side of the plots have lower MSE). Models trained with the square euclidean distance show similar disentanglement performance for smallNORB and better ones for dSprites. In the case of 3D shapes, some disentanglement performance has been traded away for better reconstruction. We would argue that this simply comes from the range of parameters used for both $\beta$ and $\gamma$. These results show that the flexibility in choice of the ground cost function of the reconstruction term of TCWAE offers better reconstruction-disentanglement trade off by improving the reconstruction while exhibiting competitive disentanglement.

*Ablation Study of the Index-Code MI.* As mentioned in Sect. 3.2, the latent regularisation of TCWAE and $\beta$-TCVAE only differ by the presence or absence of index-code MI. Here, we show that it has minimal impact on the reconstruction-disentanglement trade off. We modify our objective Eq. (9) by explicitly adding an index-code MI term. The resulting objective is now an upper bound of the (TC)WAE objective and is equivalent to a pseudo $\beta$-TCVAE where the reconstruction cost would be replaced by the square euclidean distance. We use $\alpha = \gamma$ and we compare the reconstruction-disentangle trade off of the modified TCWAE, denoted by TCWAE MI, with the original TCWAE in Fig. 2.

All other things being equal, we observe that adding an index-code MI term has little to no impact on the reconstruction (vertical alignment of the points for the same $\beta$). The presence of index-code MI does not result in significant performance difference across all data sets, conforms to the observation of [7]. This suggests that TCWAE is a more natural objective for disentanglement learning removing the inconsequential index-code MI term, and thus the need for a third hyperparameter to tune (choice of $\alpha$ in Eq. (10)).

*Disentanglement Performance.* We benchmark our methods against $\beta$-TCVAE [7] and FactorVAE [19] on the three different data sets. As [7] we take $\alpha = 1$ in $\beta$-TCVAE while for $\gamma$, we follow the same method than for TCWAE (see Appendix C of [13]). As previously, we visualise the reconstruction-disentanglement trade off by plotting the different disentanglement scores against the MSE in Fig. 3.

As expected, the TC term controls the reconstruction-disentanglement trade off as more TC penalization leads to better disentanglement scores but higher reconstruction cost. However, similarly to their VAE counterparts, too much penalization on the TC deteriorates the disentanglement as the poor quality of the reconstructions prevents any disentanglement in the generative factors.

With the range of parameters chosen, both the TCWAE-GAN and FactorVAE reconstruction seem to be less sensitive to the TC regularization as shown by the more concentrated MSE scores. TCWAE models also exhibit smaller MSE than VAEs methods for almost all the range of selected parameters. Thus, while it remains difficult to assert the dominance of one method, both because the performance varies from one data set to another and from one metric to another within each data set[2], we argue that TCWAE improves the reconstruction performance (smaller MSE) while retaining competitive disentanglement performance (see for example the scores violin plots for the different data sets provided in Appendix C of [13]). In Table 1, we report, for each method, the best $\beta$ taken to be the one achieving an overall best ranking over the disentanglement scores (see Appendix B of [13] for more details). TCWAEs achieve competitive disentanglement performance.

For each data set, we plot the latent traversals to visually asses the learned representation in Fig. 4. More specifically, we encode one observation and traverse the latent dimensions one at the time (rows) and reconstruct the resulting latent traversals (columns). Only the active latent dimension are shown.

---

[2] See [24] for the importance of inductive biases in disentanglement learning.

**Table 1.** Reconstruction and disentanglement scores (± one standard deviation). Best scores are in bold while second best are underlined.

| Method | MSE | MIG | factorVAE | SAP |
|---|---|---|---|---|
| *(a) dSprites* | | | | |
| TCWAE MWS | $13.6 \pm 1.2$ | $0.18 \pm .09$ | $\underline{0.73 \pm .08}$ | $\underline{0.076 \pm .004}$ |
| TCWAE GAN | $\mathbf{8.6 \pm 1.1}$ | $0.16 \pm .06$ | $0.69 \pm .10$ | $0.058 \pm .022$ |
| $\beta$-TCVAE | $18.57 \pm 0.6$ | $\underline{0.19 \pm .06}$ | $0.72 \pm .09$ | $\underline{0.076 \pm .006}$ |
| FactorVAE | $\underline{9.9 \pm 0.5}$ | $\mathbf{0.34 \pm .06}$ | $\mathbf{0.80 \pm .05}$ | $\mathbf{0.083 \pm .023}$ |
| *(b) 3D shapes* | | | | |
| TCWAE MWS | $\underline{13.1 \pm 0.9}$ | $0.31 \pm .12$ | $\underline{0.85 \pm .07}$ | $0.073 \pm .014$ |
| TCWAE GAN | $\mathbf{11.8 \pm 0.8}$ | $0.36 \pm .26$ | $\underline{0.85 \pm .12}$ | $0.097 \pm .065$ |
| $\beta$-TCVAE | $16.2 \pm 0.9$ | $\underline{0.45 \pm .16}$ | $\mathbf{0.93 \pm .07}$ | $\mathbf{0.113 \pm .015}$ |
| FactorVAE | $16.8 \pm 3.0$ | $\mathbf{0.53 \pm .17}$ | $0.84 \pm .03$ | $\underline{0.105 \pm .014}$ |
| *(c) smallNORB* | | | | |
| TCWAE MWS | $\mathbf{4.6 \pm 0.0}$ | $0.032 \pm .00$ | $0.46 \pm .02$ | $0.008 \pm .002$ |
| TCWAE GAN | $\underline{7.9 \pm 0.2}$ | $0.027 \pm .00$ | $\underline{0.47 \pm .01}$ | $\underline{0.014 \pm .003}$ |
| $\beta$-TCVAE | $12.80 \pm 0.1$ | $\mathbf{0.034 \pm .00}$ | $0.45 \pm .01$ | $\mathbf{0.017 \pm .002}$ |
| FactorVAE | $8.7 \pm 0.1$ | $\underline{0.033 \pm .00}$ | $\mathbf{0.48 \pm .02}$ | $\mathbf{0.017 \pm .002}$ |

**Table 2.** MSE and FID scores for 3D chairs and CelebA (± one standard deviation). Best scores are in bold while second best are underlined.

| Method | MSE | Rec. FID | Samples FID |
|---|---|---|---|
| *(a) 3D chairs* | | | |
| TCWAE-MWS | $38.2 \pm 0.1$ | $\mathbf{61.3 \pm 0.4}$ | $\mathbf{72.6 \pm 0.5}$ |
| TCWAE-GAN | $\mathbf{36.6 \pm 0.3}$ | $\underline{61.6 \pm 1.4}$ | $\underline{74.6 \pm 1.7}$ |
| $\beta$-TCVAE | $54.0 \pm 0.5$ | $74.9 \pm 1.7$ | $85.1 \pm 0.1$ |
| FactorVAE | $\underline{36.7 \pm 0.7}$ | $62.8 \pm 1.0$ | $98.9 \pm 6.6$ |
| *(b) CelebA* | | | |
| TCWAE-MWS | $\mathbf{150.5 \pm 0.7}$ | $75.2 \pm 0.2$ | $\underline{82.8 \pm 0.0}$ |
| TCWAE-GAN | $186.4 \pm 18.9$ | $\mathbf{72.4 \pm 3.5}$ | $\mathbf{73.3 \pm 2.6}$ |
| $\beta$-TCVAE | $\underline{182.4 \pm 0.3}$ | $84.5 \pm 0.2$ | $91.5 \pm 0.7$ |
| FactorVAE | $237.6 \pm 10.7$ | $\underline{74.0 \pm 1.4}$ | $83.3 \pm 4.5$ |

TCWAEs manage to capture and disentangle most of the generative factors in every data sets. The exception being the discrete generative factor representing the shape category in dSprites and 3D shapes and the instance category in smallNORB. This probably comes from the fact that the models try to learn discrete factors with continuous Gaussian variables. Possible fixes would be to use more structured priors and posteriors with hierarchical latent model, for example

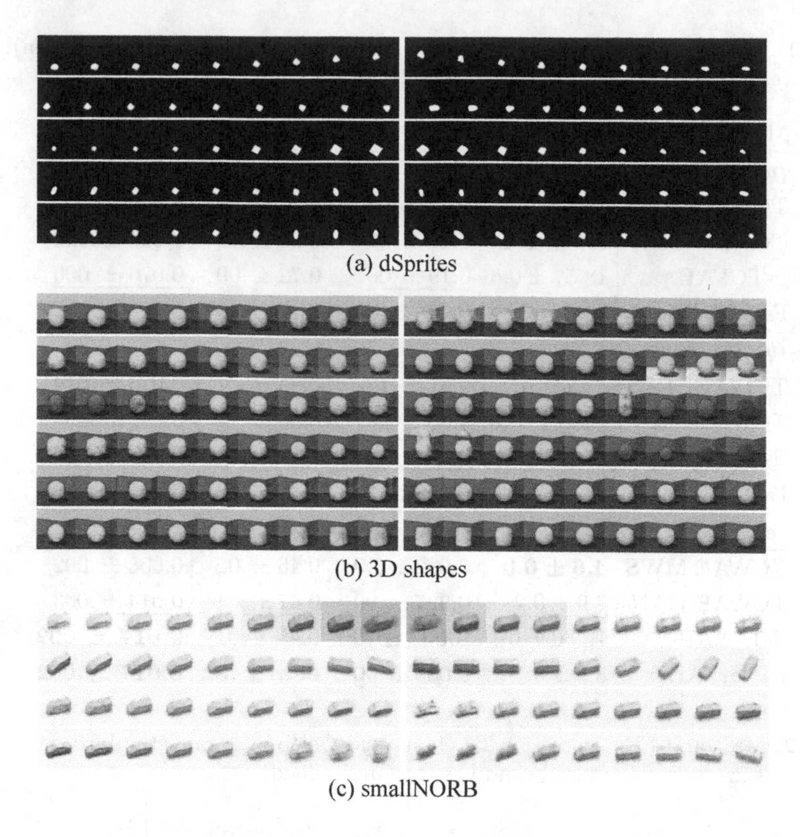


**Fig. 4.** Active latent traversals for each model. The parameters are the same as the ones used in Table 1. Subplots on the left column show traversals for TCWAE-MWS while subplots on the right show traversals TCWAE-GAN. Traversal range is $[-2, 2]$.

[10], but is out-of-scope for this work. Model reconstructions and samples for the different data sets are given in Appendix C of [13]. Visual inspection re-enforces the justification for a different reconstruction cost function as TCWAE present qualitatively crispier reconstructions and samples.

### 4.2 Qualitative Analysis: Disentanglement on Real-World Data Sets

We train our methods on 3D chairs [2] and CelebA [23] whose generative factors are not known and qualitatively find that TCWAEs achieve good disentanglement. Figure 5 shows the latent traversals of four different factors learned by the TCWAEs on 3D chairs, with reconstructions, samples and additional features discovered by the model shown in Appendix D of [13]. Similarly for CelebA, we plot 2 different features found by TCWAEs in Fig. 6 with the reconstructions, model samples additional features given in Appendix D of [13]. Visually, TCWAEs manage to capture different generative factors while retaining good reconstructions and samples. This confirms our intuition that the flexibility offered in the construction of the reconstruction term, mainly the possibility to chose the reconstruction cost function and use deterministic decoders, improves the reconstruction-disentanglement trade off.

In order to further assess the quality of the reconstructions and samples, we compute the MSE of the reconstructions and the FID scores [14] of the reconstructions and samples. Results are reported in Table 2. While the fact that TCWAEs indeed presents better MSE is not a surprise (the reconstruction cost actually minimises the MSE in TCWAEs), TCWAEs also present better reconstruction FID scores. More interesting from a generative modeling point of view, TCWAEs also present better samples than their VAE-counterparts when looking at the samples FID scores. This is visually confirmed when inspecting the model samples given in Appendix D of [13]. These results show the benefit of using a different ground cost function, allowing better reconstruction and generative performance while retaining good disentanglement performance.

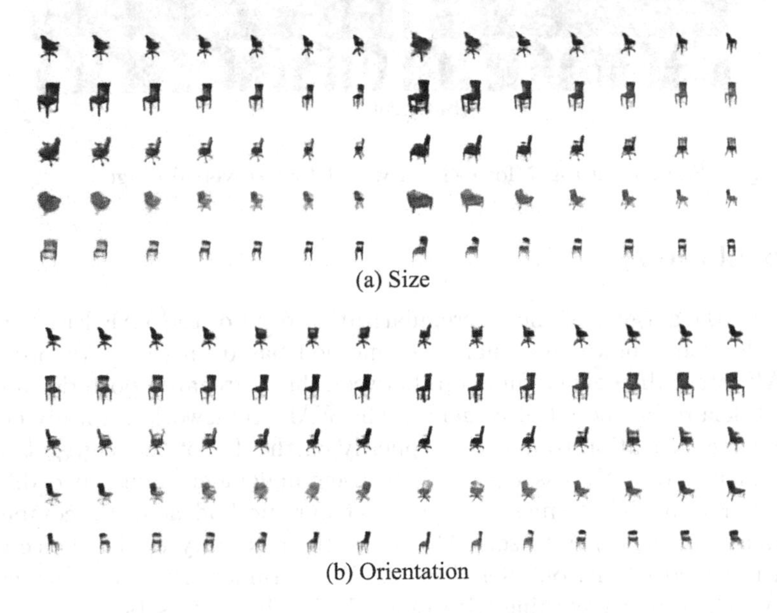

(a) Size

(b) Orientation

**Fig. 5.** Latent traversals for TCWAE-MWS (left column) and TCWAE-GAN (right column) on 3D chairs. Within each subplot, each line corresponds to one input data point. We vary evenly the encoded latent codes in the interval $[-2, 2]$.

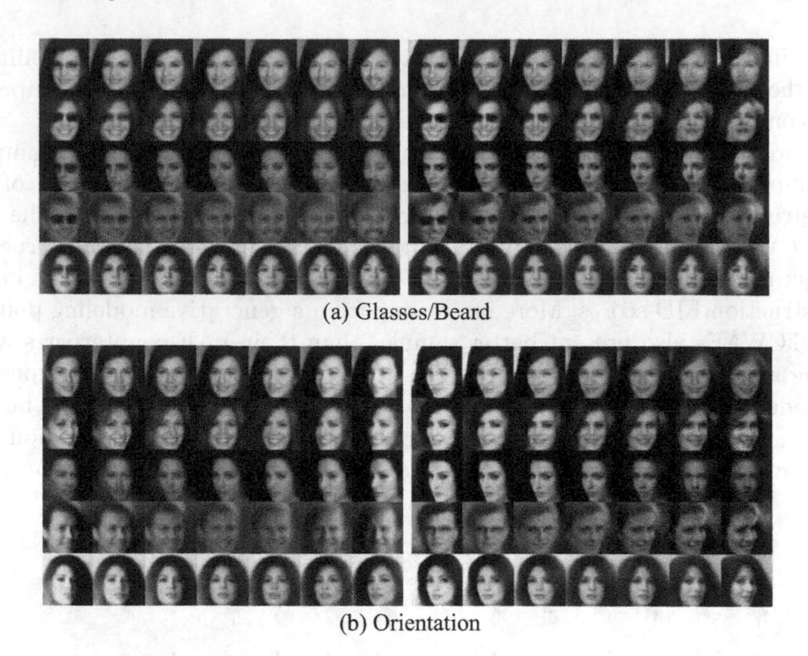


**Fig. 6.** Same than Fig. 5 for CelebA with latent traversal range $[-4, 4]$.

## 5 Conclusion

Inspired by the surgery of the KL regularization term of the VAE ELBO objective, we develop a new disentanglement method based on the WAE objective. The WAE latent divergence function, between the aggregated posterior and the prior, is taken to be the KL divergence. The WAE framework naturally enables the latent regularization to depend explicitly on the TC of the aggregated posterior, which is directly associated with disentanglement. Using two different estimators for the KL terms, we show that our method achieves competitive disentanglement on toy data sets. Moreover, the flexibility in the choice of the reconstruction cost function offered by the WAE framework makes our method more compelling when working with more challenging data sets.

## References

1. Achille, A., Soatto, S.: Information dropout: learning optimal representations through noisy computation. IEEE Trans. Pattern Anal. Mach. Intell. **40**, 2897–2905 (2018)
2. Aubry, M., Maturana, D., Efros, A., Russell, B., Sivic, J.: Seeing 3D chairs: exemplar part-based 2D–3D alignment using a large dataset of CAD models. In: CVPR (2014)
3. Bachman, P., Hjelm, R.D., Buchwalter, W.: Learning representations by maximizing mutual information across views. In: Advances in Neural Information Processing Systems (2019)

4. Bengio, Y., Courville, A., Vincent, P.: Representation learning: a review and new perspectives. IEEE Trans. Pattern Anal. Mach. Intell. **35**, 1798–1828 (2013)
5. Bousquet, O., Gelly, S., Tolstikhin, I., Simon-Gabriel, C.-J., Schoelkopf, B.: From optimal transport to generative modeling: the VEGAN cookbook. arXiv:1705.07642 (2017)
6. Burgess, C.P., et al.: Understanding disentangling in $\beta$-VAE. arXiv:804.03599 (2018)
7. Chen, R.T.K., Li, X., Grosse, R., Duvenaud, D.: Isolating sources of disentanglement in VAEs. In: Advances in Neural Information Processing Systems (2018)
8. Do, K., Tran, T.: Theory and evaluation metrics for learning disentangled representations. arXiv:1908.09961 (2019)
9. Eastwood, C., Williams, C.K.I.: A framework for the quantitative evaluation of disentangled representations. In: International Conference on Learning Representations (2018)
10. Esmaeili, B., et al.: Structured disentangled representations. In: AISTATS (2018)
11. Frogner, C., Zhang, C., Mobahi, H., Araya, M., Poggio, T.A.: Learning with a Wasserstein loss. In: Advances in Neural Information Processing Systems (2015)
12. Gao, S., Brekelmans, R., Ver Steeg, G., Galstyan, A.: Auto-encoding total correlation explanation. In: International Conference on Artificial Intelligence and Statistics (2019)
13. Gaujac, B., Feige, I., Barber, D.: Learning disentangled representations with the Wasserstein autoencoder. arXiv:2010.03459 (2020)
14. Heusel, M., Ramsauer, H., Unterthiner, T., Nessler, B., Hochreiter, S.: GANs trained by a two time-scale update rule converge to a local Nash equilibrium. In: Advances in Neural Information Processing Systems (2017)
15. Higgins, I., et al.: Towards a definition of disentangled representations. arXiv:1812.02230 (2018)
16. Higgins, I., et al.: beta-VAE: learning basic visual concepts with a constrained variational framework. In: International Conference on Learning Representations (2017)
17. Hjelm, R.D., et al.: Learning deep representations by mutual information estimation and maximization. In: International Conference on Learning Representations (2019)
18. Hoffman, M.D., Johnson, M.J.: ELBO surgery: yet another way to carve up the variational evidence lower bound. In: NIPS Workshop on Advances in Approximate Bayesian Inference (2016)
19. Kim, H., Mnih, A.: Disentangling by factorising. In: International Conference on Machine Learning (2018)
20. Kingma, D.P., Welling, M.: Auto-encoding variational Bayes. In: International Conference on Learning Representations (2014)
21. Kumar, A., Sattigeri, P., Balakrishnan, A.: Variational inference of disentangled latent concepts from unlabeled observations. In: International Conference on Learning Representations (2018)
22. LeCun, Y., Huang, F.J., Bottou, L.: Learning methods for generic object recognition with invariance to pose and lighting. In: IEEE Computer Society Conference on Computer Vision and Pattern Recognition (2004)
23. Liu, Z., Luo, P., Wang, X., Tang, X.: Deep learning face attributes in the wild. In: International Conference on Computer Vision (2015)
24. Locatello, F., et al.: Challenging common assumptions in the unsupervised learning of disentangled representations. In: International Conference on Machine Learning (2019)

25. Matthey, L., Higgins, I., Hassabis, D., Lerchner, A.: dSprites: Disentanglement testing Sprites dataset (2017). https://github.com/deepmind/dsprites-dataset/
26. Nguyen, X., Wainwright, M.J., Michael, I.J.: Estimating divergence functionals and the likelihood ratio by penalized convex risk minimization. In: Advances in Neural Information Processing Systems 20 (2008)
27. van den Oord, A., Li, Y., Vinyals, O.: Representation learning with contrastive predictive coding. arXiv:1807.03748 (2018)
28. Patrini, G., et al.: Sinkhorn autoencoders. arXiv:1810.01118 (2018)
29. Rezende, D.J., Mohamed, S., Wierstra, D.: Stochastic backpropagation and approximate inference in deep generative models. In: International Conference on Machine Learning (2014)
30. Rubenstein, P., Bousquet, O., Djolonga, J., Riquelme, C., Tolstikhin, I.: Practical and consistent estimation of f-divergences. In: Advances in Neural Information Processing Systems (2019)
31. Rubenstein, P.K., Schoelkopf, B., Tolstikhin, I.: Learning disentangled representations with Wasserstein auto-encoders. In: ICLR Workshop (2018)
32. van Steenkiste, S., Locatello, F., Schmidhuber, J., Bachem, O.: Are disentangled representations helpful for abstract visual reasoning? In: Advances in Neural Information Processing Systems (2019)
33. Sugiyama, M., Suzuki, T., Kanamori, T.: Density ratio matching under the Bregman divergence: a unified framework of density ratio estimation. Ann. Inst. Stat. Math. **64**, 1009–1044 (2011). https://doi.org/10.1007/s10463-011-0343-8
34. Tishby, N., Pereira, F.C., Bialek, W.: The information bottleneck method. In: Annual Allerton Conference on Communication, Control and Computing (1999)
35. Tolstikhin, I., Bousquet, O., Gelly, S., Schoelkopf, B.: Wasserstein auto-encoders. In: International Conference on Learning Representations (2018)
36. Tschannen, M., Djolonga, J., Rubenstein, P.K., Gelly, S., Lucic, M.: On mutual information maximization for representation learning. In: International Conference on Learning Representations (2020)
37. Ver Steeg, G., Galstyan, A.: Discovering structure in high-dimensional data through correlation explanation. In: Advances in Neural Information Processing Systems (2014)
38. Villani, C.: Optimal Transport: Old and New. GL, vol. 338. Springer, Heidelberg (2008). https://doi.org/10.1007/978-3-540-71050-9
39. Watanabe, S.: Information theoretical analysis of multivariate correlation. IBM J. Res. Dev. **4**, 66–82 (1960)
40. Xiao, Y., Wang, W.Y.: Disentangled representation learning with Wasserstein total correlation. arXiv:1912.12818 (2019)
41. Zhao, S., Song, J., Ermon, S.: InfoVAE: balancing learning and inference in variational autoencoders. In: AAAI Conference on Artificial Intelligence (2019)

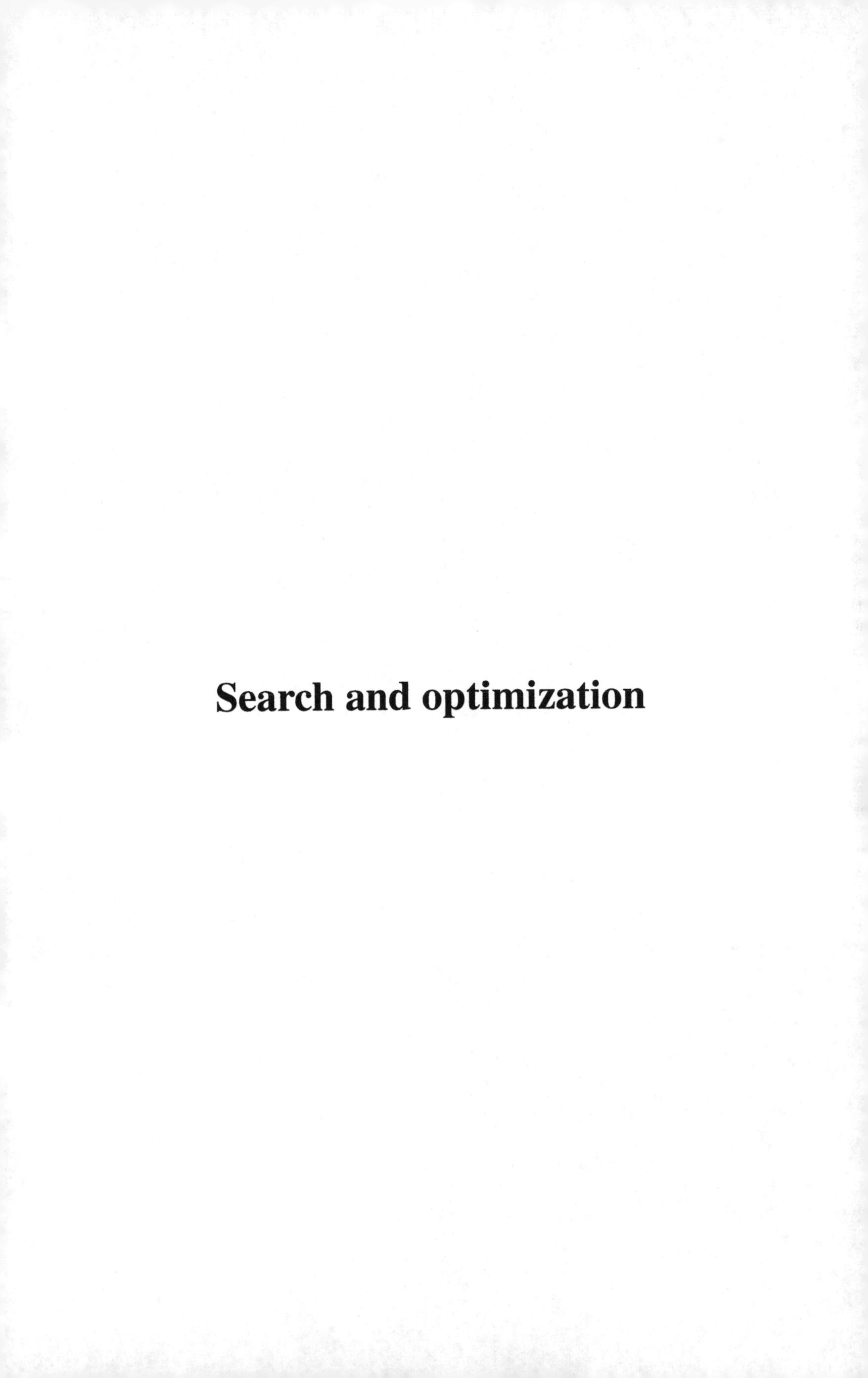

# Search and optimization

# Which Minimizer Does My Neural Network Converge To?

Manuel Nonnenmacher[1,2](✉), David Reeb[1], and Ingo Steinwart[2]

[1] Bosch Center for Artificial Intelligence (BCAI), 71272 Renningen, Germany
manuel.nonnenmacher@de.bosch.com

[2] Institute for Stochastics and Applications, University of Stuttgart, 70569 Stuttgart, Germany

**Abstract.** The loss surface of an overparameterized neural network (NN) possesses many global minima of zero training error. We explain how common variants of the standard NN training procedure change the minimizer obtained. First, we make explicit how the size of the initialization of a strongly overparameterized NN affects the minimizer and can deteriorate its final test performance. We propose a strategy to limit this effect. Then, we demonstrate that for adaptive optimization such as AdaGrad, the obtained minimizer generally differs from the gradient descent (GD) minimizer. This adaptive minimizer is changed further by stochastic mini-batch training, even though in the non-adaptive case, GD and stochastic GD result in essentially the same minimizer. Lastly, we explain that these effects remain relevant for less overparameterized NNs. While overparameterization has its benefits, our work highlights that it induces sources of error absent from underparameterized models.

**Keywords:** Overparameterization · Optimization · Neural networks

## 1 Introduction

Overparameterization is a key ingredient in the success of neural networks (NNs), thus modern NNs have become ever more strongly overparameterized. As much as this has helped increase NN performance, overparameterization has also caused several puzzles in our theoretical understanding of NNs, especially with regards to their good optimization behavior [36] and favorable generalization properties [37]. In this work we shed light on the optimization behavior, identifying several caveats.

More precisely, we investigate how the obtained minimizer can change depending on the NN training procedure – we consider common techniques like adjusting the initialization size, the use of adaptive optimization, and stochastic gradient descent (SGD).

These training choices can have significant impact on the final test performance, see Fig. 1. While some of these peculiar effects had been observed experimentally [34,38],


N. Oliver et al. (Eds.): ECML PKDD 2021, LNAI 12977, pp. 87–102, 2021.
https://doi.org/10.1007/978-3-030-86523-8_6

We explain and quantify them in a general setting. Note further that this effect is absent from the more commonly studied underparameterized models, whose minimizer is generically unique ([31] and App. A (see [26])).

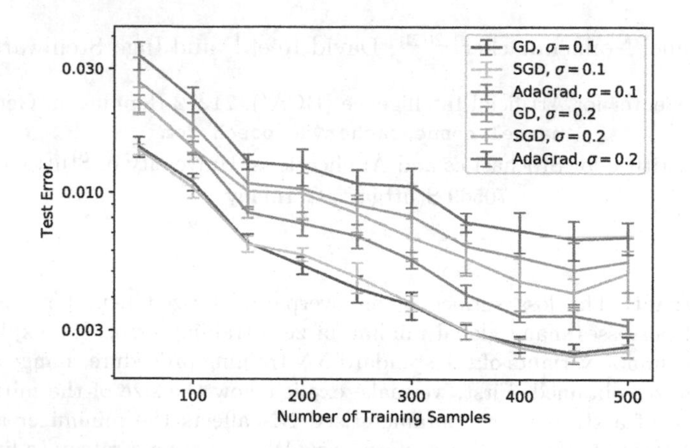


**Fig. 1.** The test performance of an overparameterized NN depends considerably on the optimization method (GD, SGD, AdaGrad) and on the initialization size $\sigma$, even though all nets have been trained to the same low empirical error of $10^{-5}$. Shown are results on MNIST 0 vs. 1 under squared loss in 5 repetitions of each setting, varying the degree of overparameterization by changing the training set size. Our theoretical results explain and quantify these differences between training choices.

Our analysis makes use of the improved understanding of the training behavior of strongly overparameterized NNs which has been achieved via the Neural Tangent Kernel (NTK) [15,20,22]. Through this connection one can show that overparameterized NNs trained with gradient descent (GD) converge to a minimizer which is an interpolator of low complexity w.r.t. the NTK [7]. We also extend our analysis to less overparameterized NNs by using linearizations at later training times instead of the NTK limit (Sect. 6).

Our contributions are as follows:

- We explain quantitatively how the size of initialization impacts the trained overparameterized NN and its test performance. While the influence can generally be severe, we suggest a simple algorithm to detect and mitigate this effect (Sect. 3).
- We prove that the choice of adaptive optimization method changes the minimizer obtained and not only the training trajectory (Sect. 4). This can significantly affect the test performance. As a technical ingredient of independent interest we prove that strongly overparameterized NNs admit a linearization under adaptive training similar to GD and SGD training [1,14,22].
- We show that the batch size of mini-batch SGD affects the minimizer of adaptively trained NNs, in contrast to the non-adaptive setting, where the SGD minimizer is virtually the same as in full-batch (GD) training (Sect. 5).

- Our theoretical findings are confirmed by extensive experiments on different datasets, where we investigate the effect of the changed minimizer on the test performance (Sect. 7).

## 2 Background

Throughout, the $N$ training points $\mathcal{D} = \{(x_i, y_i)\}_{i=1}^N \subset B_1^d(0) \times \mathbb{R}$ have inputs from the $d$-dimensional unit ball $B_1^d(0) := \{x \in \mathbb{R}^d : \|x\|_2 \leq 1\}$, and one-dimensional outputs for simplicity. $\mathcal{D}$ is called *non-degenerate* if $x_i \nparallel x_j$ for $i \neq j$. We often view the training inputs $X = (x_1, \ldots, x_N)^T \in \mathbb{R}^{N \times d}$ and corresponding labels $Y = (y_1, \ldots, y_N)^T \in \mathbb{R}^N$ in matrix form. For any function $g$ on $\mathbb{R}^d$, we define $g(X)$ by row-wise application of $g$.

The output of a fully-connected NN with $L$ layers and parameters $\theta$ is denoted $f_\theta^{\text{NN}}(x) = h^L(x)$ with

$$\begin{aligned} h^l(x) &= \frac{\sigma}{\sqrt{m_l}} W^l a(h^{l-1}) + \sigma b^l \quad \text{for } l = 2, \ldots, L, \\ h^1(x) &= \frac{\sigma}{\sqrt{m_1}} W^1 x + \sigma b^1, \end{aligned} \tag{1}$$

where $a : \mathbb{R} \to \mathbb{R}$ is the activation function, applied component-wise. We assume $a$ either to have bounded second derivative [14] or to be the ReLU function $a(h) = \max\{h, 0\}$ [1]. Layer $l$ has $m_l$ neurons ($m_0 = d$, $m_L = 1$); we assume $m_l \equiv m$ constant for all hidden layers $l = 1, \ldots, L-1$ for simplicity.[1] $W^l \in \mathbb{R}^{m_l \times m_{l-1}}$ and $b^l \in \mathbb{R}^{m_l}$ are the NN weights and biases, for a total of $P = \sum_{l=1}^L (m_{l-1} + 1)m_l$ real parameters in $\theta = [W^{1:L}, b^{1:L}]$. We keep the parameter scaling $\sigma$ as an explicit scalar parameter, that can be varied [20]. The parametrization (1) together with a *standard normal initialization* $W_{i,j}^l, b_i^l \sim \mathcal{N}(0, 1)$ (sometimes with *zero biases* $b_i^l = 0$) is the *NTK-parametrization* [14]. This is equivalent to the standard parametrization (i.e., no prefactors in (1)) and initialization $W_{i,j}^l \sim \mathcal{N}(0, \sigma^2/m_l)$, $b_i^l \sim \mathcal{N}(0, \sigma^2)$ [18] in a NN forward pass, while in gradient training the two parametrizations differ by a width-dependent scaling factor of the learning rate [20,22].

We mostly consider the squared error $|\hat{y} - y|^2/2$ and train by minimizing its empirical loss

$$L_{\mathcal{D}}(f_\theta) = \frac{1}{2N} \sum_{(x,y) \in \mathcal{D}} |f_\theta(x) - y|^2 = \frac{1}{2N} \|f_\theta(X) - Y\|_2^2.$$

We train by *discrete* update steps, i.e. starting from initialization $\theta = \theta_0$ the parameters are updated via the discrete iteration $\theta_{t+1} = \theta_t - \eta U_t[\theta_t]$ for $t = 0, 1, \ldots$, where $U_t$ is some function of the (past and present) parameters and $\eta$ the learning rate. Gradient descent (GD) training uses the present loss

[1] For unequal hidden layers, the infinite-width limit (below) is $\min_{1 \leq l \leq L-1}\{m_l\} \to \infty$ [22].

gradient $U_t[\theta_t] = \nabla_\theta L_\mathcal{D}(f_\theta)|_{\theta=\theta_t}$; for adaptive and stochastic gradient methods, see Sects. 4 and 5.

A central object in our study is the feature map $\phi(x) := \nabla_\theta f_\theta^{\mathrm{NN}}(x)\big|_{\theta=\theta_0} \in \mathbb{R}^{1\times M}$ associated with a NN $f_\theta^{\mathrm{NN}}$ at its initialization $\theta_0$. With this we will consider the NN's linearization around $\theta_0$ as

$$f_\theta^{\mathrm{lin}}(x) \;:=\; f_{\theta_0}^{\mathrm{NN}}(x) + \phi(x)(\theta - \theta_0). \tag{2}$$

On the other hand, $\phi$ gives rise to the so-called neural tangent kernel (NTK) $K : \mathbb{R}^d \times \mathbb{R}^d \to \mathbb{R}$ by $K(x, x') := \phi(x)\phi(x')^T$ [20,22]. We use the associated kernel norm to define the *minimum complexity interpolator* of the data $\mathcal{D}$:

$$f^{\mathrm{int}} := \operatorname*{arg\,min}_{f\in\mathcal{H}_K} \|f\|_{\mathcal{H}_K} \qquad \text{subject to } \; Y = f(X), \tag{3}$$

where $\mathcal{H}_K$ is the reproducing kernel Hilbert space (RKHS) associated with $K$ [7]. Its explicit solution is $f^{\mathrm{int}}(x) = \phi(x)\phi(X)^T \left(\phi(X)\phi(X)^T\right)^{-1} Y$ (App. B (see [26])). Here, $\phi(X)\phi(X)^T = K(X, X)$ is invertible for "generic" $X$ and $\theta_0$ in the overparameterized regime $P \geq N$. Technically, we always assume that the infinite-width kernel $\Theta(x, x') := \lim_{m\to\infty} K(x, x')$ has positive minimal eigenvalue $\lambda_0 := \lambda_{\min}(\Theta(X, X)) > 0$ on the data (this $\lim_{m\to\infty}$ exists in probability [20]). $\lambda_0 > 0$ holds for non-degenerate $\mathcal{D}$ and standard normal initialization with zero biases [14]; for standard normal initialization (with normal biases) it suffices that $x_i \neq x_j$ for $i \neq j$.

With these prerequisites, we use as a technical tool the fact that a strongly overparameterized NN stays close to its linearization during GD training (for extensions, see Theorem 7 More precisely, the following holds:

**Lemma 1** [1,22]. *Denote by $\theta_t$ and $\tilde{\theta}_t$ the parameter sequences obtained by gradient descent on the NN $f_\theta^{\mathrm{NN}}$ (1) and on its linearization $f_{\tilde{\theta}}^{\mathrm{lin}}$ (2), respectively, starting from the same initialization $\theta_0 = \tilde{\theta}_0$ and with sufficiently small step size $\eta$. There exists some $C = \mathrm{poly}(1/\delta, N, 1/\lambda_0, 1/\sigma)$ such that for all $m \geq C$ and for all $x \in B_1^d(0)$ it holds with probability at least $1 - \delta$ over the random initialization $\theta_0$ that: $\sup_t |f_{\theta_t}^{\mathrm{NN}}(x) - f_{\tilde{\theta}_t}^{\mathrm{lin}}(x)|^2 \leq O(1/m)$.*

## 3 Impact of Initialization

In this section we quantify theoretically how the initialization $\theta_0$ influences the final NN trained by gradient descent (GD), and in particular its test error or risk. As a preliminary result, we give an analytical expression for the GD-trained NN, which becomes exact in the infinite-width limit:

**Theorem 1.** *Let $f^{\mathrm{NN}}$ be the fully converged solution of an L-layer ReLU-NN (1), trained by gradient descent under squared loss on non-degenerate data $\mathcal{D} = (X, Y)$. There exists $C = \mathrm{poly}(1/\delta, N, 1/\lambda_0, 1/\sigma)$ such that whenever there are*

*$m \geq C$ neurons in each hidden layer, it holds for any $x \in B_1^d(0)$ that, with probability at least $1-\delta$ over standard normal initialization $\theta_0$,*

$$f^{\text{NN}}(x) = \phi(x)\phi(X)^T\left(\phi(X)\phi(X)^T\right)^{-1}Y + \frac{1}{L}\phi(x)(\mathbb{1} - P_{\mathcal{F}})\theta_0 + O\Big(\frac{1}{\sqrt{m}}\Big), \tag{4}$$

*where $P_{\mathcal{F}} := \phi(X)^T\left(\phi(X)\phi(X)^T\right)^{-1}\phi(X)$ is the projector onto the data feature subspace.*

Results similar to Theorem 1 have been found in other works before [20,21,38]. Our result has a somewhat different form compared to them and makes the dependence on the initialization $\theta_0$ more explicit. For this, we simplified the $Y$-independent term by using the property $f_\theta^{\text{NN}}(x) = \frac{1}{L}\langle\theta, \nabla_\theta f_\theta^{\text{NN}}(x)\rangle$ of ReLU-NNs (Lemma 4 (see [26])). Further, our Theorem 1 is proven for discrete, rather than continuous, update steps. The proof in App. F.1 (see [26]) first solves the dynamics of the linearized model (2) iteratively and recovers the first two terms in the infinite training time limit. Finally, Lemma 1 gives $O(1/\sqrt{m})$-closeness to $f^{\text{NN}}(x)$ in the strongly overparameterized regime.

The expression (4) for the converged NN has two main parts. The first term is just the minimum complexity interpolator $f^{\text{int}}(x) = \phi(x)\phi(X)^T(\phi(X)\phi(X)^T)^{-1}Y$ from Eq. (3), making the solution interpolate the training set $\mathcal{D}$ perfectly. Note, $f^{\text{int}}(x)$ is virtually independent of the random initialization $\theta_0$ in the strongly overparameterized regime since $\phi(x)\phi(x')^T$ converges in probability as $m \to \infty$ (Sect. 2); intuitively, random $\theta_0$'s yield similarly expressive features $\phi(x) \in \mathbb{R}^P$ as $P \to \infty$.

The second term in (4), however, depends on $\theta_0$ explicitly. More precisely, it is proportional to the part $(\mathbb{1}-P_{\mathcal{F}})\theta_0$ of the initialization that is orthogonal to the feature subspace $\mathcal{F} = \text{span}\{\phi(x_1), \ldots, \phi(x_N)\}$ onto which $P_{\mathcal{F}}$ projects. This term is present as GD alters $\theta_t \in \mathbb{R}^P$ only along the $N$-dimensional $\mathcal{F}$. It vanishes on the training inputs $x = X$ due to $\phi(X)(\mathbb{1} - P_{\mathcal{F}}) = 0$.

Our main concern is now the extent to which the *test* error is affected by $\theta_0$ and thus in particular by the second term $\phi(x)(\mathbb{1}-P_{\mathcal{F}})\theta_0/L$. Due to its $Y$-independence, this term will generally harm test performance. While this holds for large initialization scaling $\sigma$, small $\sigma$ suppresses this effect:

**Theorem 2.** *Under the prerequisites of Theorem 1 and fixing a test set $\mathcal{T} = (X_{\mathcal{T}}, Y_{\mathcal{T}})$ of size $N_{\mathcal{T}}$, there exists $C = \text{poly}(N, 1/\delta, 1/\lambda_0, 1/\sigma, N_{\mathcal{T}})$ such that for $m \geq C$ the test error $L_{\mathcal{T}}(f^{\text{NN}})$ of the trained NN satisfies the following bounds, with probability at least $1-\delta$ over the standard normal initialization with zero biases,*

$$\begin{aligned} \sqrt{L_{\mathcal{T}}(f^{\text{NN}})} &\geq \sigma^L J(X_{\mathcal{T}}) - \sqrt{L_{\mathcal{T}}(f^{\text{int}})} - O\Big(\frac{1}{\sqrt{m}}\Big), \\ \sqrt{L_{\mathcal{T}}(f^{\text{NN}})} &\leq \sqrt{L_{\mathcal{T}}(f^{\text{int}})} + O\Big(\sigma^L + \frac{1}{\sqrt{m}}\Big), \end{aligned} \tag{5}$$

*where $J(X_{\mathcal{T}})$ is independent of the initialization scaling $\sigma$ and $J(X_{\mathcal{T}}) > 0$ holds almost surely. With standard normally initialized biases, the same bounds hold with both $\sigma^L$ replaced by $\sigma$.*

The lower bound in (5) shows that big initialization scalings $\sigma$ leave a significant mark $\sim \sigma^L$ on the the test error, while the final *training* error of $f^{\mathrm{NN}}$ is always 0 due to strong overparameterization. This underlines the importance of good initialization schemes, as they do not merely provide a favorable starting point for training, but impact which minimizer the weights converge to. To understand the scaling, note that the features scale with $\sigma$ like $\phi_\sigma(x) = \sigma^L \phi_1(x)$ (the behavior is more complex for standard normal biases, see App. F.2 (see [26]). The first term in (4) is thus invariant under $\sigma$, while the second scales as $\sim \sigma^L$.

The main virtue of the upper bound in (5) is that the harmful influence of initialization can be reduced by adjusting $\sigma$ and $m$ simultaneously. To show this, App. F.3 (see [26]) takes care to bound the second term in (4) on the test set $\|\phi(X_{\mathcal{T}})(1-P_{\mathcal{F}})\theta_0\|/\sqrt{N_{\mathcal{T}}} \leq O(\sigma^L)$ *independently* of $\phi$'s dimension $P$, which would grow with $m$. Note further that the kernel interpolator $f^{\mathrm{int}}$ in (5) with loss $L_{\mathcal{T}}(f^{\mathrm{int}})$ was recently found to be quite good empirically [5] and theoretically [4,24].

Based on these insights into the decomposition (4) and the scaling of $L_{\mathcal{T}}(f^{\mathrm{NN}})$ with $\sigma$, we suggest the following algorithm to mitigate the potentially severe influence of the initialization on the test error in large NNs as much as possible: *(a)* randomly sample an initialization $\theta_0$ and train $f^{\mathrm{NN}'}$ using a standard scaling $\sigma' \simeq O(1)$ (e.g. [18]); *(b)* train $f^{\mathrm{NN}''}$ with the same $\theta_0$ and a somewhat smaller $\sigma'' < \sigma'$ (e.g. by several ten percent); *(c)* compare the losses on a validation set $\mathcal{V}$: if $L_{\mathcal{V}}(f^{\mathrm{NN}''}) \approx L_{\mathcal{V}}(f^{\mathrm{NN}'})$ then finish with step (e), else if $L_{\mathcal{V}}$ decreases by a significant margin then continue; *(d)* repeat from step (b) with successively smaller $\sigma''' < \sigma''$ until training becomes impractically slow; *(e)* finally, return the trained $f^{\mathrm{NN}}$ with smallest validation loss.

It is generally not advisable to start training (or the above procedure) with a too small $\sigma < O(1)$ due to the vanishing gradient problem which leads to slow training, even though the upper bound in (5) may suggest very small $\sigma$ values to be beneficial from the viewpoint of the test error. A "antisymmetrical initialization" method was introduced in [38] to reduce the impact of the initialization-dependence on the test performance by doubling the network size; this however also increases the computational cost.

Note further that the above theorems do not hold exactly anymore for $\sigma$ too small due to the $m \geq \mathrm{poly}(1/\sigma)$ requirement; our experiments (Fig. 1 and Sect. 7) however confirm the predicted $\sigma$-scaling even for less strongly overparameterized NNs.

## 4 Impact of Adaptive Optimization

We now explain how the choice of adaptive optimization method affects the minimizer to which overparameterized models converge. The discrete weight update step for adaptive gradient training methods is

$$\theta_{t+1} = \theta_t - \eta D_t \nabla_\theta L_{\mathcal{D}}(\theta)\big|_{\theta=\theta_t}, \tag{6}$$

where the "adaptive matrices" $D_t \in \mathbb{R}^{P\times P}$ are prescribed by the method. This generalizes GD, which is obtained by $D_t \equiv \mathbb{1}$, and includes AdaGrad [16] via $D_t = \left(\mathrm{diag}(\sum_{u=0}^t g_u g_u^T)\right)^{-1/2}$ with the loss gradients $g_t = \nabla_\theta L(\theta)\big|_{\theta_t} \in \mathbb{R}^P$, as well as RMSprop [19] and other adaptive methods. We say that a sequence of adaptive matrices concentrates around $D \in \mathbb{R}^{P\times P}$ if there exists $Z \in \mathbb{R}$ such that $\|D_t - D\|_{\mathrm{op}}/D_{\max} \le Z/\sqrt{m}$ holds for all $t \in \mathbb{N}$, where $D_{\max} := \sup_t \|D_t\|_{\mathrm{op}}$; this is the simplest assumption under which we can generalize Theorem 1, but in general we only need that the linearization during training holds approximately (App. F.4 (see [26])).

The following result gives a closed-form approximation to strongly overparameterized NNs during training by Eq. (6). Note that this overparameterized adaptive case was left unsolved in [31]. The closed-form expression allows us to illustrate via explicit examples that the obtained minimizer can be markedly different from the GD minimizer, see Example 1 below.

**Theorem 3.** *Given a NN $f_\theta^{\mathrm{NN}}$ (1) and a non-degenerate training set $\mathcal{D} = (X, Y)$ for adaptive gradient training (6) under squared loss with adaptive matrices $D_t \in \mathbb{R}^{P\times P}$ concentrated around some $D$, there exists $C = \mathrm{poly}(N, 1/\delta, 1/\lambda_0, 1/\sigma)$ such that for any width $m \ge C$ of the NN and any $x \in B_1^d(0)$ it holds with probability at least $1-\delta$ over the random initialization that*

$$\begin{aligned} f_{\theta_t}^{\mathrm{NN}}(x) &= \phi(x)A_t\left[\mathbb{1} - \prod_{u=t-1}^{0}\left(\mathbb{1} - \frac{\eta}{N}\phi(X)D_u\phi(X)^T\right)\right]\left(Y - f_{\theta_0}^{\mathrm{NN}}(X)\right), \\ &\quad + f_{\theta_0}^{\mathrm{NN}}(x) + \phi(x)B_t + O(\frac{1}{\sqrt{m}}), \end{aligned} \tag{7}$$

*where* $A_t = D_{t-1}\phi(X)^T\left(\phi(X)D_{t-1}\phi(X)^T\right)^{-1}$ *and*

$$B_t = \sum_{v=2}^{t}(A_{v-1} - A_v)\cdot\left[\mathbb{1} - \prod_{w=v-2}^{0}\left(\mathbb{1} - \frac{\eta}{N}\phi(X)D_w\phi(X)^T\right)\right]\cdot\left(Y - f_{\theta_0}^{\mathrm{NN}}(X)\right).$$

To interpret this result, notice that on the training inputs $X$ we have $\phi(X)A_t = \mathbb{1}$ and therefore $\phi(X)B_t = 0$, so that the dynamics $f_{\theta_t}^{\mathrm{NN}}(X)$ on the training data simplifies significantly: When the method converges, i.e. $\prod_{u=t-1}^{0}(\ldots) \to 0$ as $t \to \infty$, we have $f_{\theta_t}^{\mathrm{NN}}(X) \to Y + O(1/\sqrt{m})$, meaning that the training labels are (almost) perfectly interpolated at convergence. Even at convergence, however, the interpolating part $f_{\theta_0}^{\mathrm{NN}}(x) + \phi(x)A_t(Y - f_{\theta_0}^{\mathrm{NN}}(X))$ depends on $D_t$ (via $A_t$) for test points $x$. In addition to that, the term $\phi(x)B_t$ in (7) is "path-dependent" due to the sum $\sum_{v=2}^{t}(\ldots)$ and takes account of changes in the $A_t$ (and thus, $D_t$) matrices during optimization, signifying changes in the geometry of the loss surface. The proof of Theorem 3 in App. F.4 (see [26]) is based on a generalization of the linearization Lemma 1 for strongly overparameterized NNs to adaptive training methods (Theorem 7 in App. G (see [26])).

We make the dependence of the trained NN (7) on the choice of adaptive method explicit by the following example of AdaGrad training.

*Example 1.* Perform adaptive optimization on a ReLU-NN with adaptive matrices $D_t$ according to the AdaGrad prescription with

$$g_t = a_t g \quad \text{where} \quad g \in \mathbb{R}^P, \; a_t \in \mathbb{R}, \tag{8}$$

i.e. we assume that the gradients point all in the same direction $g$ with some decay behavior set by the $a_t$. We choose this simple setting for illustration purposes, but it occurs e.g. for heavily overparameterized NNs with $N = 1$ training point, where $g = \phi(x_1)^T$ (a similar setting was used in [36]). The adaptive matrices are then $D_t = D\left(\sum_{i=0}^t a_i^2\right)^{-1/2}$ with $D = \left(\text{diag}(gg^T)\right)^{-1/2}$. As $D_t$ evolves only with scalar factors, $A_t = D\phi(X)^T\left(\phi(X)D\phi(X)^T\right)^{-1}$ is constant and thus $B_t = 0$. We can then explicitly evaluate Theorem 3 and use similar arguments as in Theorem 2 to collect the $f_{\theta_0}^{\text{NN}}$-terms into $O(\sigma^L)$, to write down the minimizer at convergence:

$$f^{\text{NN}}(x) = \phi(x)D\phi(X)^T\left(\phi(X)D\phi(X)^T\right)^{-1} Y + O\left(\sigma^L\right) + O\left(1/\sqrt{m}\right). \tag{9}$$

This example shows explicitly that the minimizer obtained by adaptive gradient methods in overparameterized NNs can be different from the GD minimizer, which results by setting $D = \mathbb{1}$. This difference is not seen on the training inputs $X$, where Eq. (9) always evaluates to $Y$, but only at test points $x$. In fact, any function of the form $f^{\text{NN}}(x) = \phi(x)w + O(\sigma^L + 1/\sqrt{m})$ that interpolates the data, $\phi(X)w = Y$, can be obtained as the minimizer by a judicious choice of $D$. In contrast to the toy example in Wilson et al. [34], our Example 1 does not require a finely chosen training set and is just a special case of the more general Theorem 3.

Another way to interpret Example 1 is that this adaptive method converges to the interpolating solution of minimal complexity w.r.t. a kernel $K_D(x, x') = \phi(x)D\phi(x')^T$ different from the kernel $K(x, x') = \phi(x)\phi(x')^T$ associated with GD (see Sect. 2 and Theorem 1). Thus, unlike in Sect. 3 where the disturbing term can in principle be diminished by initializing with small variance, adaptive training directly changes the way in which we interpolate the data and is harder to control.

Note that the situation is different for underparameterized models, where in fact the same (unique) minimizer is obtained irrespectively of the adaptive method chosen (see App. A [26] and [31]).

## 5 Impact of Stochastic Optimization

Here we investigate the effect SGD has on the minimizer to which the NN converges. The general update step of adaptive mini-batch SGD with adaptive matrices $D_t \in \mathbb{R}^{P \times P}$ is given by

$$\theta_{t+1} = \theta_t - \eta D_t \nabla_\theta L_{\mathcal{D}_{B_t}}(\theta)\big|_{\theta=\theta_t}, \tag{10}$$

where $\mathcal{D}_{B_t}$ contains the data points of the $t$-th batch $B_t$. Further, we denote by $X_{B_t}$ the corresponding data matrix obtained by zeroing all rows outside $B_t$. Ordinary SGD corresponds to $D_t = \mathbb{1}$.

The main idea behind the results of this section is to utilize the fact that ordinary SGD can be written in the form of an adaptive update step by using the adaptive matrix $D_{B_t} := \frac{N}{|B_t|} (\phi(X_{B_t}))^T \left(\phi(X)\phi(X)^T\right)^{-1} \phi(X)$ in Eq. (6). Combining this re-writing of mini-batch SGD with the adaptive update rule, we can write Eq. (10) as $\theta_{t+1} = \theta_t - \eta D_t D_{B_t} \nabla_\theta L_{\mathcal{D}}(\theta)|_{\theta=\theta_t}$. Now applying a similar approach as for Theorem 3 leads to the following result:

**Theorem 4 (informal).** *Let $f^{\mathrm{NN}}_{adGD}$ and $f^{\mathrm{NN}}_{adSGD}$ be fully trained strongly over-parameterized NNs trained on the empirical squared loss with adaptive GD and adaptive SGD, respectively. Then, for sufficiently small learning rate $\eta$ it holds with high probability that:*

*(a) If $D_t = \mathrm{const}$, then $\left|f^{\mathrm{NN}}_{adGD}(x) - f^{\mathrm{NN}}_{adSGD}(x)\right| \leq O(1/\sqrt{m})$.*
*(b) If $D_t$ changes during training, the minimizers $f^{\mathrm{NN}}_{adGD}$ and $f^{\mathrm{NN}}_{adSGD}$ differ by a path- and batch-size-dependent contribution on top of the $O(1/\sqrt{m})$ linearization error.*

Part (a) shows that GD and SGD lead to basically the same minimizer if the adaptive matrices $D_t$ do not change during training. This is the case in particular for vanilla (S)GD, where $D_t = \mathbb{1}$. Part (b) on the other hand shows that for adaptive methods with varying adaptive matrices, the two NN minimizers obtained by GD and mini-batch SGD differ by a path-dependent contribution, where the path itself can be influenced by the batch size. We expect this effect to be smaller for more overparameterized models since then the adaptive matrices are expected to be more concentrated. For the formal version of Theorem 4 see App. C (see [26]).

One of the prerequisites of Theorem 4 is a small learning rate, but it is straightforward to generalize the results to any strongly overparameterized NN (in the NTK-regime) with a learning rate schedule such that the model converges to a minimizer of zero training loss (App. F.5 (see [26])).

## 6 Beyond Strong Overparameterization

The previous three sections explain the impact of common training techniques on the obtained minimizer for strongly overparameterized NNs. In practice, NNs are usually less overparameterized or trained with a large initial learning rate, both of which can cause $\phi(X)$, and thus also $K(X, X)$, to change appreciably during training. This happens especially during the initial stages of training, where weight changes are most significant. The question thus arises how the theoretical results of Sects. 3, 4 and 5 transfer to less overparameterized NNs. (Note that experimentally, the effects do still appear in less overparameterized NNs, see Fig. 1.)

The basis of our theoretical approach is the validity of Lemma 1 and its cousins like Theorem 7 , which build on the fact that the weights of a strongly overparameterized NN do not change significantly during training, i.e. $\|\theta_t - \theta_0\|_2$ remains small for all $t > 0$. For less overparameterized NNs this does not hold in general. One can circumvent this by selecting a later training iteration $T > 0$ such that $\|\theta_t - \theta_T\|_2$ is small enough for all $t > T$. One can always find such $T$, assuming that $\theta_t$ converges as $t \to \infty$. Next, to proceed with a similar analysis as before, we linearize the NN around iteration $T$ instead of Eq. (2):

$$f_\theta^{\mathrm{lin},T}(x) \;:=\; f_{\theta_T}^{\mathrm{NN}}(x) + \phi_T(x)(\theta - \theta_T), \tag{11}$$

where $\phi_T(x) := \nabla_\theta f_\theta^{\mathrm{NN}}(x)\big|_{\theta=\theta_T}$ are the NN features at training time $T$, assumed such that $\|\theta_t - \theta_T\|_2$ is sufficiently small for all $t > T$. Assuming further that $\lambda_0^T := \lambda_{\min}(\phi_T(X)\phi_T(X)^T) > 0$, gives straightforward adaptations of Thms. 1, 3, and 4 with features $\phi_T(x)$ and valid at times $t \geq T$. The main difference is that the results are no longer probabilistic but rather conditional statements.

To demonstrate how this observation can be applied, assume now that we are given two versions $\theta_T$ and $\theta'_T$ of a NN, trained with two different training procedures up to iteration $T$. In case that $\theta_T$ and $\theta'_T$ are (significantly) different, then the adapted version of either Theorem 1 or 3 shows that both models generally converge to different minimizers; this effect would persist even if the two NNs were trained by the same procedure for $t > T$. On the contrary, if $\theta_T$ and $\theta'_T$ were the same (or similar) at iteration $T$, then Theorem 4 suggests that the minimizers will nevertheless differ when $\theta_T$ is updated with adaptive (S)GD and $\theta'_T$ with vanilla (S)GD. While our results only consider the impact on the minimizer obtained after an initial training period, there may exist further effects not considered in our analysis during the initial training steps $t < T$, where the features change.

## 7 Experiments

Here we demonstrate our theoretical findings experimentally. We perform experiments on three different data sets, MNIST (here), Fashion-MNIST (App. E (see [26])), and CIFAR10 (Fig. 5), using their 0 and 1 labels. The first experiment (Fig. 2) investigates the effect which the initialization size can have on the test performance of NNs, confirming the results of Sect. 3 qualitatively and quantitatively. Additionally, the behavior of the test (validation) error with $\sigma$ demonstrates the effectiveness of the error mitigation algorithm described in Sect. 3. The second experiment (Fig. 3) demonstrates the significant difference in test performance between NNs trained with vanilla GD and the adaptive optimization methods AdaGrad and Adam (Sect. 4). The third experiment (Fig. 4) illustrates that for non-adaptive SGD there is only weak dependency on the batch-size and ordering of the datapoints, whereas for adaptive optimization with mini-batch SGD the dependence is noticeable (Sect. 5).

These first three experiments are run with one and two hidden-layer NNs of different widths $m$. In line with our framework and with other works on

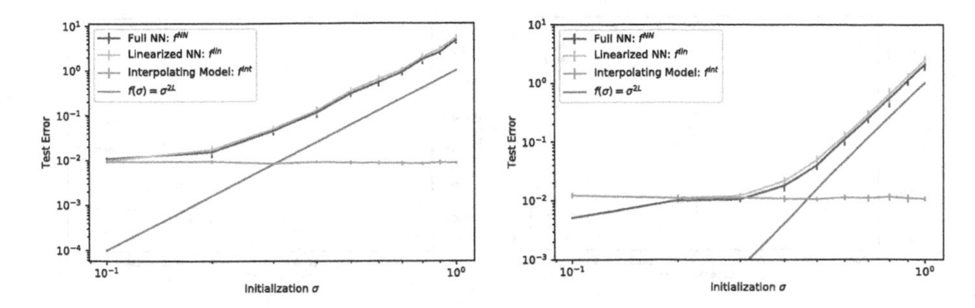


**Fig. 2. Impact of initialization.** The impact of initialization on the test performance of trained overparameterized NNs is illustrated for $L = 2$ (width $m = 4000$, left) and $L = 3$ layers ($m = 2000$, right). At small initialization size $\sigma$, the trained $f^{\mathrm{NN}}$ is close to the interpolating model $f^{\mathrm{int}}$ (Sect. 3), which is virtually independent of the initialization $\theta_0$ and $\sigma$. At larger $\sigma$, the test error grows strongly $\sim \sigma^{2L}$, with depth-dependent exponent (Theorem 2). All NNs were trained to the same low empirical error and are close to their linearizations $f^{\mathrm{lin}}$, verifying that indeed we are in the overparameterized limit. Our results underline that initialization does not merely provide a favorable starting point for optimization, but can strongly impact NN test performance.

overparameterized NNs we minimize the weights of the NN w.r.t. the empirical squared loss on a reduced number of training samples ($N = 100$), to make sure that overparameterization $m \gg N$ is satisfied to a high degree. We train all the NNs to a very low training error ($< 10^{-5}$) and then compare the mean test error from 10 independently initialized NNs with error bars representing the standard deviation[2].

In addition to the effects of the three settings described in the previous paragraph, Fig. 1 illustrates that similar effects appear in less overparameterized settings as well. Furthermore, while some of the theoretical conditions are not satisfied for modern architectures such as ResNets or networks trained with the cross-entropy loss, Fig. 5 shows that comparable effects appear in these settings as well.

## 8 Related Work

The fact that NN initialization can influence the average test performance was pointed out in [38] and [36], who investigated this experimentally, but did not quantify it. The method suggested by [38] to reduce this effect doubles the number of NN parameters, significantly increasing the computational cost. Another method to reduce this effect was suggested by [11] with the "doubling trick", which was shown to potentially harm the training dynamics [38]. [35] also investigates the impact of different initialization scalings but require deep and wide

[2] More details on the settings needed to reproduce the experiments can be found in App. D (see [26]).

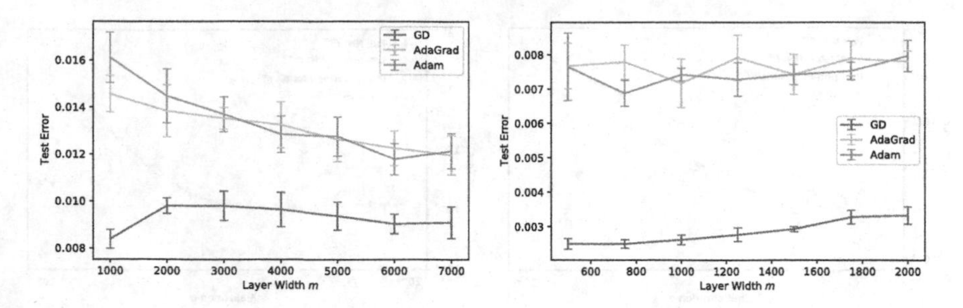


**Fig. 3. Impact of adaptive optimization.** It is shown that the test performance depends strongly on the choice of adaptive optimization method, for a variety of overparameterized NNs with $L = 2$ (left) and $L = 3$ layers (right) over different widths $m$. Even though all models were trained to the same low empirical error, there is a significant performance gap between vanilla GD and the adaptive methods, underlining the results of Sect. 4. Thus, while adaptive methods for overparameterized NNs may improve the training dynamics, they also affect the minimizers obtained.

NNs, while we only requires wide NNs. Furthermore, [34] and [36] observed a significant test performance gap between GD and adaptive optimization methods for overparameterized NNs. While [34] does provide an analytical toy example for this in a linear model, our analysis of NNs is general and holds for generic training data. [31] attempts to explain such gaps for overparameterized linear models, but lacks the explicit expressions Eqs. (7), (9) we find. [28] looks into the implicit bias of AdaGrad only for the cross-entropy loss and [3] investigates the generalization behaviour of natural gradient descent under noise. While we on the other hand investigate the impact of general adaptive training methods for the squared loss on the minimizer obtained.

The convergence of the training error under SGD in comparison to GD has been the subject of many works, recently also in the context of strongly overparameterized NNs [1,10,25,30]. We, on the other hand, investigate whether the minimizer found by SGD is similar to the GD minimizer on a test set. Another concept closely linked to this is the implicit bias of (S)GD on the trained NN, certain aspects of which have been investigated in [9,17,27,29,32]. Our work elucidates the implicit bias caused by the random NN initialization and by the optimizer.

We use interpolating kernel methods [7] and the idea of an overparameterized learning regime which does not suffer from overfitting [6,8,25]. Bounds on their test performance have been derived by [24] and [4]. [5] on the other hand demonstrates their good performance experimentally, but does not investigate how the scale of NN initialization or nonzero NN biases influence the test behavior (our Sect. 3). While [33] also explores different minimum norm solutions they only consider linear models and different loss functions and do not focus on the impact of different training techniques as we do.

Our analysis uses the Neural Tangent Kernel (NTK) limit [20], where the NN width scales polynomially with the training set size. In this regime, it is

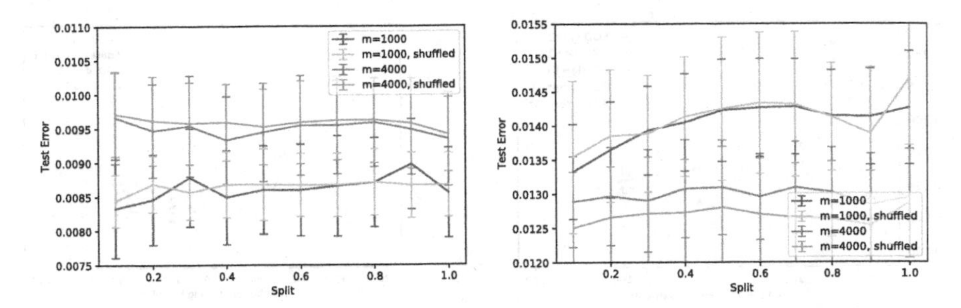


**Fig. 4. Impact of stochastic training.** Without adaptive optimization (left), SGD-trained NNs have basically the same test error as GD-trained ones (split = 1.0), for a large range of mini-batch sizes and both with and without shuffling the training set (Theorem 4(a)). In contrast to this, when using adaptive optimization methods (AdaGrad, right) the test-performance becomes dependent on the batch-size (Theorem 4(b)). The mini-batch size is given as the split ratio of the training set size, with vanilla GD corresponding to 1.0. The NNs shown have $L = 2$ layers and widths $m$ as indicated.

found that NNs converge to arbitrarily small training error under (S)GD. The first works to investigate this were [12], [23] and [15]. Later, [1], [2], and [14] extended these results to deep NNs, CNNs and RNNs. The recent contribution by [22], building upon [21] and [13], explicitly solves the training dynamics.

## 9 Discussion

We have explained theoretically how common choices in the training procedure of overparameterized NNs leave their footprint in the obtained minimizer and in the resulting test performance. These theoretical results provide an explanation for previous experimental observations [34,36], and are further confirmed in our dedicated experiments.

To identify and reduce the harmful influence of the initialization on the NN test performance, we suggest a new algorithm motivated by the bounds in Theorem 2. The potentially harmful influence of adaptive optimization on the test performance, however, cannot be reduced so easily if one wants to keep the beneficial effects on training time. Indeed, current adaptive methods experimentally seem to have worse test error than SGD ([34] and Fig. 3). Therefore, adaptive methods for overparameterized models should be designed not only to improve convergence to any minimum, but this minimum should also be analyzed form the perspective of statistical learning theory.

While our theory applies to NNs trained with squared loss, we believe the same effects to appear in NNs trained with cross-entropy loss (see Fig. 5). And while we show that in the less overparameterized regime, the minimizer is still dependent on training choices, it remains an open question to disentangle how exactly the learned data-dependent features contribute to the effects explained in our work.

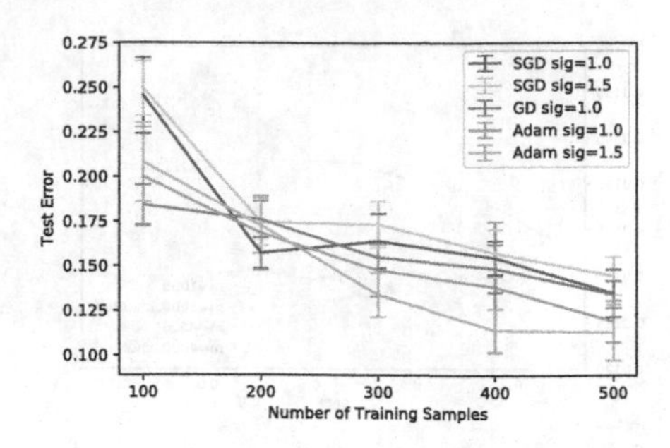


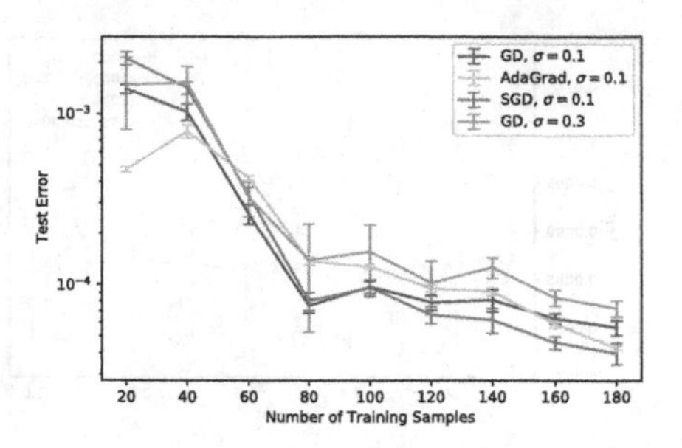


**Fig. 5. ResNet networks and cross-entropy loss.** The left figure shows the test performance of ResNet-20 on the Cifar10 data set using "airplane" vs. "automobile" labels (encoded as 0 and 1) trained with squared loss, for different optimization methods (GD, SGD, AdaGrad) and initialization sizes $\sigma$, over the training set size. The figure on the right shows the same setting as in Fig. 1 but instead of the squared loss the network is trained using the cross-entropy loss. The shown results are averaged over 5 repetitions for each setting and all models are trained to the same low empirical error of $10^{-5}$.

## References


1. Allen-Zhu, Z., Li, Y., Song, Z.: A convergence theory for deep learning via over-parameterization. In: International Conference on Machine Learning, pp. 242–252 (2019)
2. Allen-Zhu, Z., Li, Y., Song, Z.: On the convergence rate of training recurrent neural networks. In: Advances in Neural Information Processing Systems, pp. 6673–6685 (2019)
3. Amari, S., Ba, J., Grosse, R.B., Li, X., Nitanda, A., Suzuki, T., Wu, D., Xu, J.: When does preconditioning help or hurt generalization? In: International Conference on Learning Representations (2021). https://openreview.net/forum?id=S724o4_WB3
4. Arora, S., Du, S., Hu, W., Li, Z., Wang, R.: Fine-grained analysis of optimization and generalization for overparameterized two-layer neural networks. In: International Conference on Machine Learning, pp. 322–332 (2019)
5. Arora, S., Du, S.S., Hu, W., Li, Z., Salakhutdinov, R.R., Wang, R.: On exact computation with an infinitely wide neural net. In: Advances in Neural Information Processing Systems, pp. 8139–8148 (2019)
6. Belkin, M., Hsu, D.J., Mitra, P.: Overfitting or perfect fitting? risk bounds for classification and regression rules that interpolate. In: Advances in Neural Information Processing Systems, pp. 2300–2311 (2018)
7. Belkin, M., Ma, S., Mandal, S.: To understand deep learning we need to understand kernel learning. In: International Conference on Machine Learning, pp. 541–549 (2018)
8. Belkin, M., Rakhlin, A., Tsybakov, A.B.: Does data interpolation contradict statistical optimality? In: The 22nd International Conference on Artificial Intelligence and Statistics, pp. 1611–1619 (2019)
9. Bietti, A., Mairal, J.: On the inductive bias of neural tangent kernels. In: Advances in Neural Information Processing Systems, pp. 12873–12884 (2019)

10. Borovykh, A.: The effects of optimization on generalization in infinitely wide neural networks. In: ICML Workshop (2019)
11. Chizat, L., Oyallon, E., Bach, F.: On lazy training in differentiable programming. In: Advances in Neural Information Processing Systems, pp. 2937–2947 (2019)
12. Daniely, A.: SGD learns the conjugate kernel class of the network. In: Guyon, I., et al. (eds.) Advances in Neural Information Processing Systems, vol. 30, pp. 2422–2430. Curran Associates, Inc (2017)
13. De Matthews, A., Hron, J., Rowland, M., Turner, R., Ghahramani, Z.: Gaussian process behaviour in wide deep neural networks. In: 6th International Conference on Learning Representations, ICLR 2018-Conference Track Proceedings (2018)
14. Du, S., Lee, J., Li, H., Wang, L., Zhai, X.: Gradient descent finds global minima of deep neural networks. In: International Conference on Machine Learning, pp. 1675–1685 (2019)
15. Du, S.S., Zhai, X., Poczos, B., Singh, A.: Gradient descent provably optimizes over-parameterized neural networks. In: International Conference on Learning Representations (2019). https://openreview.net/forum?id=S1eK3i09YQ
16. Duchi, J., Hazan, E., Singer, Y.: Adaptive subgradient methods for online learning and stochastic optimization. J. Mach. Learn. Res. **12**(Jul), 2121–2159 (2011)
17. Glorot, X., Bengio, Y.: Understanding the difficulty of training deep feedforward neural networks. In: Proceedings of the Thirteenth International Conference on Artificial Intelligence and Statistics, pp. 249–256 (2010)
18. He, K., Zhang, X., Ren, S., Sun, J.: Delving deep into rectifiers: Surpassing human-level performance on imagenet classification. In: Proceedings of the IEEE International Conference on Computer Vision, pp. 1026–1034 (2015)
19. Hinton, G.: Lecture 6 of the online course "neural networks for machine learning". Lecture 6 of the online course "Neural Networks for Machine Learning" (2012). https://www.cs.toronto.edu/~tijmen/csc321/slides/lecture_slides_lec6.pdf
20. Jacot, A., Gabriel, F., Hongler, C.: Neural tangent kernel: convergence and generalization in neural networks. In: Advances in Neural Information Processing Systems, pp. 8571–8580 (2018)
21. Lee, J., Sohl-Dickstein, J., Pennington, J., Novak, R., Schoenholz, S., Bahri, Y.: Deep neural networks as gaussian processes. In: International Conference on Learning Representations (2018). https://openreview.net/forum?id=B1EA-M-0Z
22. Lee, J., et al.: Wide neural networks of any depth evolve as linear models under gradient descent. In: Advances in Meural Information Processing Systems, pp. 8570–8581 (2019)
23. Li, Y., Liang, Y.: Learning overparameterized neural networks via stochastic gradient descent on structured data. In: Advances in Neural Information Processing Systems, pp. 8157–8166 (2018)
24. Liang, T., Rakhlin, A.: Just interpolate: Kernel "ridgeless" regression can generalize. To appear in The Annals of Statistics (preprint arXiv:1808.00387) (2019)
25. Ma, S., Bassily, R., Belkin, M.: The power of interpolation: understanding the effectiveness of SGD in modern over-parametrized learning. In: International Conference on Machine Learning, pp. 3325–3334 (2018)
26. Nonnenmacher, M., Reeb, D., Steinwart, I.: Which minimizer does my neural network converge to? arXiv preprint arXiv:2011.02408 (2020)
27. Oymak, S., Soltanolkotabi, M.: Overparameterized nonlinear learning: gradient descent takes the shortest path? In: International Conference on Machine Learning, pp. 4951–4960 (2019)
28. Qian, Q., Qian, X.: The implicit bias of adagrad on separable data. arXiv preprint arXiv:1906.03559 (2019)

29. Rahaman, N., et al.: On the spectral bias of neural networks. In: International Conference on Machine Learning, pp. 5301–5310 (2019)
30. Sankararaman, K.A., De, S., Xu, Z., Huang, W.R., Goldstein, T.: The impact of neural network overparameterization on gradient confusion and stochastic gradient descent. arXiv preprint arXiv:1904.06963 (2019)
31. Shah, V., Kyrillidis, A., Sanghavi, S.: Minimum norm solutions do not always generalize well for over-parameterized problems. arXiv preprint arXiv:1811.07055 (2018)
32. Soudry, D., Hoffer, E., Nacson, M.S., Gunasekar, S., Srebro, N.: The implicit bias of gradient descent on separable data. J. Mach. Learn. Res. **19**(1), 2822–2878 (2018)
33. Vaswani, S., Babanezhad, R., Gallego, J., Mishkin, A., Lacoste-Julien, S., Roux, N.L.: To each optimizer a norm, to each norm its generalization. arXiv preprint arXiv:2006.06821 (2020)
34. Wilson, A.C., Roelofs, R., Stern, M., Srebro, N., Recht, B.: The marginal value of adaptive gradient methods in machine learning. In: Advances in Neural Information Processing Systems, pp. 4148–4158 (2017)
35. Xiao, L., Pennington, J., Schoenholz, S.: Disentangling trainability and generalization in deep neural networks. In: International Conference on Machine Learning, pp. 10462–10472. PMLR (2020)
36. Zhang, C., Bengio, S., Hardt, M., Mozer, M.C., Singer, Y.: Identity crisis: memorization and generalization under extreme overparameterization. In: International Conference on Learning Representations (2020). https://openreview.net/forum?id=B1l6y0VFPr
37. Zhang, C., Bengio, S., Hardt, M., Recht, B., Vinyals, O.: Understanding deep learning requires rethinking generalization. In: International Conference on Learning Representations (2017). https://openreview.net/forum?id=Sy8gdB9xx
38. Zhang, Y., Xu, Z.Q.J., Luo, T., Ma, Z.: A type of generalization error induced by initialization in deep neural networks. In: Mathematical and Scientific Machine Learning, pp. 144–164. PMLR (2020)

# Information Interaction Profile of Choice Adoption

Gaël Poux-Médard(✉), Julien Velcin, and Sabine Loudcher

Université de Lyon, Lyon 2, ERIC UR 3083,
5 avenue Pierre Mendès France, F69676 Bron Cedex, France
{gael.poux-medard,julien.velcin,sabine.loudcher}@univ-lyon2.fr

**Abstract.** Interactions between pieces of information (*entities*) play a substantial role in the way an individual acts on them: adoption of a product, the spread of news, strategy choice, etc. However, the underlying interaction mechanisms are often unknown and have been little explored in the literature. We introduce an efficient method to infer both the entities interaction network and its evolution according to the temporal distance separating interacting entities; together, they form the *interaction profile*. The interaction profile allows characterizing the mechanisms of the interaction processes. We approach this problem *via* a convex model based on recent advances in multi-kernel inference. We consider an ordered sequence of exposures to entities (URL, ads, situations) and the actions the user exerts on them (share, click, decision). We study how users exhibit different behaviors according to *combinations* of exposures they have been exposed to. We show that the effect of a combination of exposures on a user is more than the sum of each exposure's independent effect–there is an interaction. We reduce this modeling to a non-parametric convex optimization problem that can be solved in parallel. Our method recovers state-of-the-art results on interaction processes on three real-world datasets and outperforms baselines in the inference of the underlying data generation mechanisms. Finally, we show that interaction profiles can be visualized intuitively, easing the interpretation of the model.

## 1 Introduction

When told in the year 2000 that the XX$^{th}$ century was the century of physics and asked whether he agrees that the next one would be the century of biology, Stephen Hawkins answered that he believed the XXI$^{th}$ century would be the century of complexity. Be it a reasoned forecast or a tackle to promote scientific multidisciplinarity, there has been an undeniable growing interest for complex systems in research over the past decades. A complex system can be defined as a system composed of many components that interact with each other. Their study often involves network theory, a branch of mathematics that aims at modeling those interactions –that can be physical, biological, social, etc. A significant point of interest is understanding how information spreads along the edges of


N. Oliver et al. (Eds.): ECML PKDD 2021, LNAI 12977, pp. 103–118, 2021.
https://doi.org/10.1007/978-3-030-86523-8_7

a network–with a particular interest in social networks. If the social network skeleton (edges, nodes) plays a significant role in such processes, recent studies pointed out that the interaction between spreading entities might also play a non-trivial role in it [2,23]. The histograms presented Fig. 1 illustrate this finding: the probability for a piece of information to be adopted (or spread) varies according to the exposure to another one at a previous time. We refer to this figure as the *interaction profile*. The study of this quantity is a novel perspective: the interaction between pieces of information has been little explored in the literature, and no previous work aims at unveiling trends in the information interaction mechanisms.

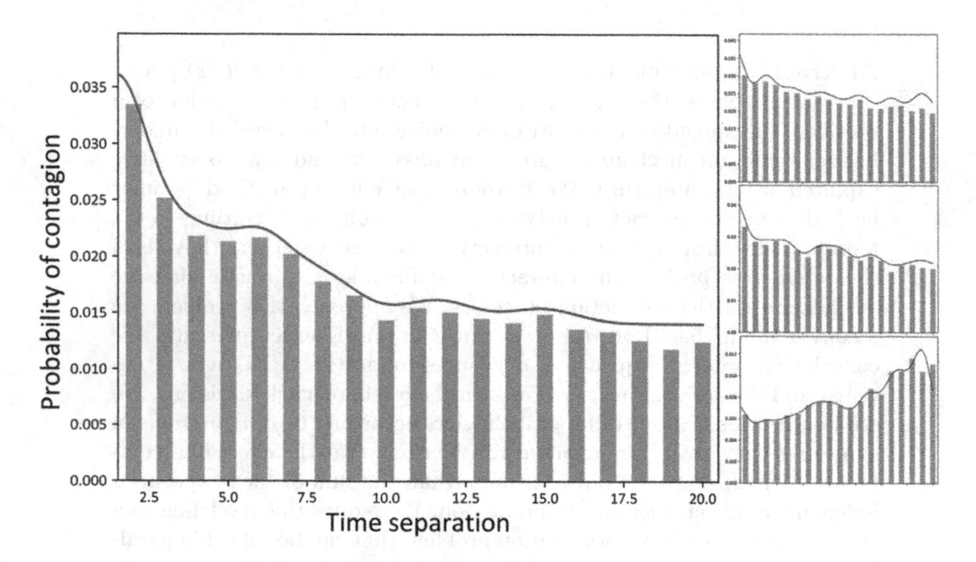


**Fig. 1. Interaction profiles between pairs of entities**—Examples of interaction profiles on Twitter; here is shown the effect of URL shortening services migre.me (left), bit.ly (right-top), tinyurl (right-middle) and t.co (right-bottom) on the probability of tweeting a t.co URL and its evolution in time. This interaction profile shows, for instance, that there is an increased probability of retweet for a t.co URL when it appears shortly after a migre.me one (interaction). This increase fades when the time separation grows (no more interaction). In blue, the interaction profile inferred by our model.

The study of interactions between pieces of information (or entities) has several applications in real-world systems. We can mention the fields of recommender systems (the probability of adoption is influenced by what a user saw shortly prior to it), news propagation and control (when to expose users to an entity in order to maximize its spreading probability [20]), advertising (same reasons as before [3]), choice behavior (what influences one's choice and how [4]).

In the present work, we propose to go one step further and to unveil the mechanisms at stake within those interacting processes by inferring information interaction profiles. Let us imagine, for instance, that an internet user is exposed to an ad at time $t_1$ and to another ad for a similar product at time $t_2 > t_1$. Due to the semantic similarity between the two exposures, we suppose that the exposure to the first one influences the user's sensitivity (likeliness of a click) to the second one a time $t_2 - t_1$ later. Modeling this process involves quantifying the influence an ad exerts on the other and how it varies with the time separation between the exposures. The reunion of those quantities form what we call the *interaction profile* –illustrated Fig. 1), that is, the influence an exposure exerts on the adoption (click, buy, choice, etc.) of another one over time.

Following this idea, we introduce an efficient method to infer both the entities (ad, tweets, products, etc.) interaction network and its evolution according to the temporal distance separating the interacting entities (the influence of an entity A on entity B will not be the same depending on whether A appeared 10 min or 10 h before B). Together they form the interaction profile.

First, we develop a method for inferring this interaction profile in a continuous-time setup using multi-kernel inference methods [5]. Then we show that the inference of the parameters boils down to a convex optimization problem for kernel families obeying specific properties. Moreover, the problem can be subdivided into as many subproblems as entities, which can be solved in parallel. The convexity of the problem guarantees convergence to the likelihood's global optimum for each subproblem and, therefore, to the problem's optimal likelihood. We apply the model to investigate the role of interaction profiles on synthetic data and in various corpora from different fields of research: advertisement (the exposure to an ad influences the adoption of other ads [3]), social dilemmas (the previous actions of one influences another's actions [4]) and information spread on Twitter (the last tweets read influence what a user retweets [12])[1]. Finally, we provide analysis leads and show that our method recovers state-of-the-art results on interaction processes on each of the three datasets considered.

### 1.1 Contributions

The main contributions of this paper are the following:

- We introduce the interaction profile, which is the combination of both the interaction network between entities and its evolution with the interaction time distance, according to the inferred kernels. The interaction profile is a powerful tool to understand how interactions take place in a given corpus (see Fig. 6) and has not been developed in the literature. Its introduction in research is the main contribution of the present work.
- We design a convex non-parametric algorithm that can be solved in parallel, baptized InterRate. InterRate automatically infers the kernels that account the best for information interactions in time within a given kernel family. Its output is the aforementioned interaction profile.

[1] Supplementary materials, codes and datasets can be found at https://github.com/GaelPouxMedard/InterRate.

- We show that InterRate yields better results than non-interacting or non-temporal baseline models on several real-world datasets. Furthermore, our model can recover several conclusions about the datasets from state-of-the-art works.

## 2 Related Work

Previous efforts in investigating the role of interactions in information diffusion have shown their importance in the underlying spreading processes. Several works study the interaction of information with users' attention [11], closely linked to information overload concepts [18], but not the interaction between the pieces of information themselves. On the other hand, whereas most of the modeling of spreading processes are based on either no competition [15,19] or perfect competition [17] assumption, it has been shown that relaxing this hypothesis leads to a better description of competitive spread [2] –with the example of Firefox and Chrome web browsers, whose respective popularities are correlated. According to this finding, a significant effort has been done in elaborating complex processes to *simulate* interaction [17,22] on real-world networks.

However, fewer works have been developed to tackle interaction in information spread from a machine learning point of view. The correlated cascade model [23] aims to infer an interacting spreading process's latent diffusion network. In this work, the interaction is modeled by a hyper-parameter $\beta$ tuning the intensity of interactions according to an exponentially decaying kernel. In their conclusion, the authors formulate the open problem of learning several kernels and the interaction intensity parameter $\beta$, which we address in the present work.

To our knowledge, the attempt the closest to our task to model the interaction intensity parameter $\beta$ is Clash of the contagions [12]; this aims to predict retweets on Twitter based on tweets seen by a user. This model estimates the probability of retweet for a piece of information, given the last tweets a user has been exposed to, according to their relative position in the Twitter feed. The method suffers various flaws (scalability, non-convexity). It also defines interactions based on an arguable hypothesis made on the prior probability of a retweet (in the absence of interactions) that makes its conclusions about interactions sloppy. It is worth noting that in [12], the authors outline the problem of the inference of the interaction profile but do so without searching for global trends such as the one shown in Fig. 1. Recent works address the various flaws observed in [12] and suggest a more general approach to the estimation of the interaction intensity parameters [16]. The latter model develops a scalable algorithm that correctly accounts for interacting processes but neglects the interactions' temporal aspect. To take back the Twitter case study, it implies that in the case of a retweet at time $t$, a tweet appearing at $t_1 \ll t$ in the news feed has the same influence on the retweet as a tweet that appeared at $t_2 \approx t$. A way to relax this assumption is to integrate a temporal setting in the interaction network inference problem.

In recent years, temporal networks inference has been a subject of interest. Significant advances have been made using survival theory modeling applied to partial observations of independent cascades of contagions [6, 7]. In this context, an infected node tries to contaminate every other node at a rate that is tuned by $\beta$. While this work is not directly linked to ours, it has been a strong influence on the interaction profile inference problem we develop here; the problems are different, but the methodology they introduce deserves to be mentioned for its help in the building of our model (development and convexity of the problem, analogy between interaction profile and hazard rate). Moreover, advances in network inference based on the same works propose a multi-kernel network inference method that we adapted to the problem we tackle here [5]. Inspired by these works, we develop a flexible approach that allows for the inference of the best interaction profile from several candidate kernels.

# 3 InterRate

## 3.1 Problem Definition

We illustrate the process to model in Fig. 2. It runs as follows: a user is first exposed to a piece of information at time $t_0$. The user then chooses whether to act on it at time $t_0 + t_s$ (an act can be a retweet, a buy, a booking, etc.); $t_s$ can be interpreted as the "reaction time" of the user to the exposure, assumed constant. The user is then exposed to the next piece of information a time $\delta t$ later, at $t_1 = t_0 + \delta t$ and decides whether to act on it a time $t_s$ later, at $t_1 + t_s$, and so on. Here, $\delta t$ is the time separating two consecutive exposures, and $t_s$ is the reaction time, separating the exposure from the possible contagion. In the remaining of the paper, we refer to the user's action on an exposure (tweet appearing in the feed, exposure to an ad, etc.) as a contagion (retweet or the tweet, click on an ad, etc.).

This choice of modeling comes with several hypotheses. **First**, the pieces of information a user is exposed to appear independently from each other. It is the main difference between our work and survival analysis literature: the pseudo-survival of an entity is conditioned by the random arrival of pieces of information. Therefore, it cannot be modeled as a point process. This assumption holds in our experiments on real-world datasets, where users have no influence on what information they are exposed to. **Second** hypothesis, the user is contaminated solely on the basis of the previous exposures in the feed [12, 23]. **Third**, the reaction time separating the exposure to a piece of information from its possible contagion, $t_s$, is constant (i.e. the time between a read and a retweet in the case of Twitter). Importantly, this hypothesis is a deliberate simplification of the model for clarity purposes; relaxing this hypothesis is straightforward by extending the kernel family, which preserves convexity and time complexity. Note that this simplification does not always hold, as shown in recent works concluding that response time can have complex time-dependent mechanisms [21].

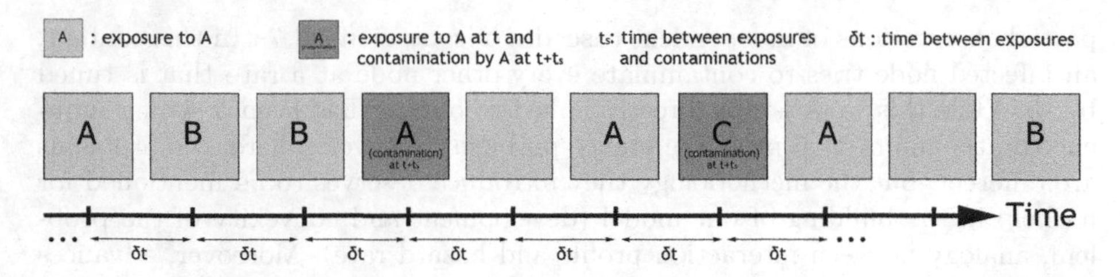


**Fig. 2. Illustration of the interacting process**—Light orange squares represent the exposures, dark orange squares represent the exposures that are followed by contagions and empty squares represent the exposures to the information we do not consider in the datasets (they only play a role in the distance between exposures when one considers the order of appearance as a time feature). A contagion occurs at a time $t_s$ after the corresponding exposure. Each new exposure arrives at a time $\delta t$ after the previous one. Contagion takes place with a probability conditioned by all previous exposures. In the example, the contagion by A at time $t + t_s$ depends on the effect of the exposure to A at times $t$ and $t - 3\delta t$, and to B at times $t - \delta t$ and $t - 2\delta t$. (Color figure online)

### 3.2 Likelihood

We define the likelihood of the model of the process described in Fig. 2. Let $t_i^{(x)}$ be the exposure to $x$ at time $t_i$, and $t_i^{(x)} + t_s$ the time of its possible contagion. Consider now the instantaneous probability of contagion (*hazard function*) $H(t_i^{(x)} + t_s|t_j^{(y)}, \beta_{xy})$, that is the probability that a user exposed to the piece of information $x$ at time $t_i$ is contaminated by $x$ at $t_i + t_s$ given an exposure to $y$ at time $t_j \leq t_i$. The matrix of parameters $\beta_{ij}$ is what the model aims to infer. $\beta_{ij}$ is used to characterize the interaction profile between entities. We define the set of exposures preceding the exposure to $x$ at time $t_i$ (or history of $t_i^{(x)}$) as $\mathcal{H}_i^{(x)} \equiv \{t_j^{(y)} \leq t_i^{(x)}\}_{j,y}$. Let $\mathcal{D}$ be the whole dataset such as $\mathcal{D} \equiv \{(\mathcal{H}_i^{(x)}, t_i^{(x)}, c_{t_i}^{(x)})\}_{i,x}$. Here, c is a binary variable that account for the contagion ($c_{t_i}^{(x)} = 1$) or non-contagion ($c_{t_i}^{(x)} = 0$) of $x$ at time $t_i + t_s$. The likelihood for one exposure in the sequence given $t_j^{(y)}$ is:

$$\begin{aligned} L(\beta_{xy}|\mathcal{D}, t_s) &= P(\mathcal{D}|\beta_{xy}, t_s) \\ &= \underbrace{H(t_i^{(x)} + t_s|t_j^{(y)}, \beta_{xy})^{c_{t_i}^{(x)}}}_{\text{contagion at } t_i^{(x)}+t_s \text{ due to } t_j^{(y)}} \cdot \underbrace{(1 - H(t_i^{(x)} + t_s|t_j^{(y)}, \beta_{xy}))^{(1-c_{t_i}^{(x)})}}_{\text{Survival at } t_i^{(x)}+t_s \text{ due to } t_j^{(y)}} \end{aligned}$$

The likelihood of a sequence (as defined Fig. 2) is then the product of the previous expression over all the exposures that happened before the contagion event $t_i^{(x)} + t_s$ e.g. for all $t_j^{(y)} \in \mathcal{H}_i^{(x)}$. Finally, the likelihood of the whole dataset $\mathcal{D}$ is the product of $L(\beta_x|\mathcal{D}, t_s)$ over all the observed exposures $t_i^{(x)}$. Taking the logarithm of the resulting likelihood, we get the final log-likelihood to maximize:

$$\ell(\beta|\mathcal{D}, t_s) = \sum_{\mathcal{D}} \sum_{t_j^{(y)} \in \mathcal{H}_i^{(x)}} c_{t_i}^{(x)} \log\left(H(t_i^{(x)} + t_s | t_j^{(y)}, \beta_{xy})\right) + (1 - c_{t_j}^{(y)}) \log\left(1 - H(t_i^{(x)} + t_s | t_j^{(y)}, \beta_{xy})\right) \tag{1}$$

### 3.3 Proof of Convexity

The convexity of a problem guarantees to retrieve its optimal solution and allows using dedicated fast optimization algorithms.

**Proposition 1.** *The inference problem* $\min_\beta -\ell(\beta|\mathcal{D}, t_s)$ $\forall \beta \geq 0$*, is convex in all of the entries of* $\beta$ *for any hazard function that obeys the following conditions:*

$$\begin{cases} H'^2 \geq H''H \\ H'^2 \geq -H''(1-H) \\ H \in \,]0;1[ \end{cases} \tag{2}$$

*where* $'$ *and* $''$ *denote the first and second derivative with respect to* $\beta$*, and* $H$ *is the shorthand notation for* $H(t_i^{(x)} + t_s | t_j^{(y)}, \beta_{xy})$ $\forall i, j, x, y$.

*Proof.* The negative log-likelihood as defined in Eq. 1 is a summation of $-\log H$ and $-\log(1-H)$; therefore $H \in \,]0;1[$. The second derivative of these expressions according to any entry $\beta_{mn}$ (noted $''$) reads:

$$\begin{cases} (-\log H)'' = \left(\frac{-H'}{H}\right)' = \frac{H'^2 - H''H}{H^2} \\ (-\log(1-H))'' = \left(\frac{H'}{1-H}\right)' = \frac{H'^2 + H''(1-H)}{(1-H)^2} \end{cases} \tag{3}$$

The convexity according to a single variable holds when the second derivative positive, which leads to Eq. 2. The convexity of the problem then follows from composition rules of convexity. □

A number of functions obey the conditions of Eq. 2, such as the exponential ($e^{-\beta t}$), Rayleigh ($e^{-\frac{\beta}{2}t^2}$), power-law ($e^{-\beta \log t}$) functions, and any log-linear combination of those [5]. These functions are standard in survival theory literature [8].

The final convex problem can then be written $\min_{\beta \geq 0} -\ell(\beta|\mathcal{D}, t_s)$. An interesting feature of the proposed method is that the problem can be subdivided into N convex subproblems that can be solved independently (one for each piece of information). To solve the subproblem of the piece of information $x$, that is to find the vector $\beta_x$, one needs to consider only the subset of $\mathcal{D}$ where $x$ appears. Explicitly, each subproblem consists in maximizing Eq. 1 over the set of observations $\mathcal{D}^{(x)} \equiv \{(\mathcal{H}_i^{(x)}, t_i^{(x)}, c_{t_i}^{(x)})\}_i$.

# 4 Experimental Setup

## 4.1 Kernel Choice

**Gaussian RBF Kernel Family (IR-RBF).** Based on [5], we consider a log-linear combination of Gaussian radial basis function (RBF) kernels as hazard function. We also consider the time-independent kernel needed to infer the base probability of contagion discussed in the section "Background noise in the data" below. The resulting hazard function is then:

$$\log H(t_i^{(x)} + t_s | t_j^{(y)}, \beta_{ij}) = -\beta_{ij}^{(bg)} - \sum_{s=0}^{S} \frac{\beta_{ij}^{(s)}}{2}(t_i + t_s - t_j - s)^2$$

The parameters $\beta^{(s)}$ of Rayleigh kernels are the amplitude of a Gaussian distribution centered on time $s$. The parameter S represents the maximum time shift we consider. In our setup, we set S = 20. We think it is reasonable to assume that an exposition does not significantly affect a possible contagion 20 steps later. The parameter $\beta_{ij}^{(bg)}$ corresponds to the time-independent kernel –base probability of contagion by i. The formulation allows the model to infer complex distributions from a reduced set of parameters whose interpretation is straightforward.

**Exponentially Decaying Kernel (IR-EXP).** We also consider an exponentially decaying kernel that can be interpreted as a modified version of a multivariate Hawkes point process (or self-exciting point process). A multivariate Hawkes process is a point process where different classes of objects exert an influence on the point process of the others [9]. The most common (and historical) form of such a process considers a time-decaying exponential function to model a class's influence on others. However, in our setup, we cannot consider a Hawkes process: we infer the variation of contagion probabilities in time conditionally on the earlier random exposure to a piece of information. The studied process is therefore not rigorously self-exciting. We consider the following form for the hazard function and refer to this modeling as IR-EXP instead of a Hawkes process:

$$\log H(t_i^{(x)} + t_s | t_j^{(y)}, \beta_{xy}) = -\beta_{ij}^{(bg)} - \beta_{ij}(t_i + t_s - t_j)$$

Where $\beta_{ij}^{(bg)}$ once again accounts for the background noise in the data discussed further in this section.

## 4.2 Parameters Learning

Datasets are made of sequences of exposures and contagions, as shown in Fig. 2. To assess the robustness of the proposed model, we apply a 5-folds cross-validation method. After shuffling the dataset, we use 80% of the sequences as a training set and the 20% left as a test set. We repeat this slicing five times, taking care that an interval cannot be part of the test set more than once. The optimization is made in parallel for each piece of information via the convex optimization module for Python CVXPY.

We also set the time separating two exposures $\delta t$ as constant. It means that we consider only the order of arrival of exposures instead of their absolute arrival time. The hypothesis that the order of exposures matters more than the absolute exposure times has already been used with success in the literature [12]. Besides, in some situations, the exact exposure time cannot be collected, while the exposures' order is known. For instance, in a Twitter corpus, we only know in what order a user read her feed, unlike the exact time she read each of the posts. However, from its definition, our model works the same with non-integer and non-constant $\delta t$ in datasets where absolute time matters more than the order of appearance.

### 4.3 Background Noise in the Data

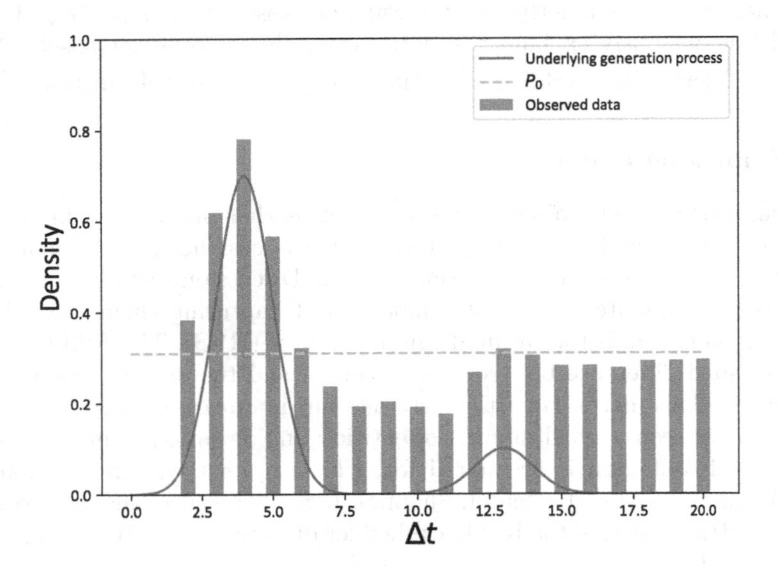


**Fig. 3. Underlying generation process vs observed data**—The red curve represents the underlying probability of contagion by C given an exposure observed $\Delta t$ steps before C. The orange bars represent the observed probability of such events. We see that there is a noise $P_0(C)$ in the observed data. The underlying generation process can then only be observed in the dataset when its effect is larger than some threshold $P_0(C)$. (Color figure online)

Because the dataset is built looking at all exposure-contagion correlations in a sequence, there is inherent noise in the resulting data. To illustrate this, we look at the illustrated example Fig. 2 and consider the exposure to C leading to a contagion happening at time $t$. We assume that in the underlying interaction process, the contagion by C at time $t+t_s$ took place only because C appeared at time $t$. However, when building the dataset, the contagion by C is also attributed to A appearing at times $t-\delta t$, $t-3\delta t$ and $t-6\delta t$, and to B appearing at times $t-4\delta t$ and $t-5\delta t$. It induces a noise in the data. In general, for any contagion in the dataset, several observations (pair exposure-contagion) come from the random presence of entities unrelated to this contagion.

We now illustrate how this problem introduces noise in the data. In Fig. 3, we see that the actual underlying data generation process (probability of a contagion by C given an exposure present $\Delta t$ step earlier) does not exactly fit the collected resulting data: the data gathering process induces a constant noise whose value is noted $P_0(C)$ –that is the average probability of contagion by C. Thus the interaction effect can only be observed when its associated probability of contagion is larger than $P_0(C)$. Consequently, the performance improvement of a model that accounts for interactions may seem small compared with a baseline that only infers $P_0(C)$. That is we observe in the experimental section. However, in this context, a small improvement in performance shows an extended comprehension of the underlying interacting processes at stake (see Fig. 3, where the red line obviously explains the data better than a constant baseline). Our method efficiently infers $P_0(C)$ *via* a time-independent kernel function $\beta_{i,j}^{P_0(i)}$.

### 4.4 Evaluation Criteria

The main difficulty in evaluating these models is that interactions might occur between a small number of entities only. It is the case here, where many pairs of entities have little to no interaction (see the Discussion section). This makes it difficult to evaluate how good a model is at capturing them. To this end, our principal metric is the residual sum of squares (**RSS**). The RSS is the sum of the squared difference between the observed and the expected frequency of an outcome. This metric is particularly relevant in our case, where interactions may occur between a small number of entities: any deviation from the observed frequency of contagion is accounted for, which is what we aim at predicting here. We also consider the Jensen-Shannon (**JS**) divergence; the JS divergence is a symmetric version of the Kullback–Leibler divergence, which makes it usable as a metric [14].

We finally consider the best-case F1-score (**BCF1**) of the models, that is, the F1-score of the best scenario of evaluation. It is not the standard F1 metric (that poorly distinguishes the models since few interactions occur), although its computation is similar. Explicitly, it generalizes F1-score for comparing probabilities instead of comparing classifications; the closer to 1, the closer the inferred and observed probabilities. It is derived from the best-case confusion matrix, whose building process is as follows: we consider the set of every information that appeared before information i at time $t_i$ in the interval, that we denote $\mathcal{H}_i$. We then compute the contagion probability of i at time $t_i+t_s$ to every exposure event $t_j^{(y)} \in \mathcal{H}_i$. Confronting this probability with the observed frequency f of contagions of i at time $t_i+t_s$ given $t_j^{(y)}$ among N observations, we can build the best-case confusion matrix. In the best case scenario, if out of N observations the observed frequency is f and the predicted frequency is p, the number of True

Positives is $N \times \min\{p, f\}$, the number of False Positives is $N \times \min\{p - f, 0\}$, the number of True Negatives is $N \times \min\{1 - p, 1 - f\}$, the number of False Positives is $N \times \min\{f - p, 0\}$.

Finally, when synthetic data is considered, we also compute the mean squared error of the $\beta$ matrix inferred according to the $\beta$ matrix used to generate the observations, that we note MSE $\beta$.

We purposely ignore evaluation in prediction because, as we show later, interactions influence quickly fades over time: probabilities of contagion at large times are mainly governed by the background noise discussed in previous sections. Therefore, it would be irrelevant to evaluate our approach's predictive power on the whole range of times where it does not bring any improvement over a naive baseline (see Fig. 1). A way to alleviate this problem would be to make predictions only when interactions effects are above/below a certain threshold (at short times, for instance). However, such an evaluation process would be debatable. Here, we choose to focus on the descriptive aspect of InterRate.

### 4.5 Baselines

**Naive Baseline.** For a given piece of information i, the contagion probability is defined as the number of times this information is contaminated divided by its number of occurrences.

**Clash of the Contagions.** We use the work presented in [12] as a baseline. In this work, the authors model the probability of a retweet given the presence of a tweet in a user's feed. This model does not look for trends in the way interactions take place (it does not infer an interaction profile), considers discrete time steps (while our model works in a continuous-time framework), and is optimized via a non-convex SGD algorithm (which does not guarantee convergence towards the optimal model). More details on implementation are provided in SI file.

**IMMSBM.** The Interactive Mixed-Membership Stochastic Block Model is a model that takes interactions between pieces of information into account to compute the probability of a (non-)contagion [16]. Note that this baseline does not take the position of the interacting pieces of information into account (time-independent) and assumes that interactions are symmetric (the effect of A on B is the same as B on A).

**ICIR.** The Independent Cascade InterRate (ICIR) is a reduction of our main IR-RBF model to the case where interactions are not considered. We consider the same dataset, enforcing the constraint that off-diagonal terms of $\beta$ are null. The (non-)contagion of a piece of information i is then determined solely by the previous exposures to i itself.

# 5 Results

## 5.1 Synthetic Data

| | | RSS | JS div. | BCF1 | MSE $\beta$ |
|---|---|---|---|---|---|
| Synth-20 | IR-RBF | 18.4151 | 0.002 28 | 0.919 | 0.001 |
| | ICIR | 139.5926 | 0.009 98 | 0.827 | 0.016 |
| | Naive | 145.5132 | 0.010 38 | 0.822 | |
| | CoC | 123.0583 | 0.009 38 | 0.822 | |
| | IMMSBM | 222.0555 | 0.017 29 | 0.727 | |
| Synth-5 | IR-RBF | 0.1169 | 0.000 22 | 0.974 | 0.005 |
| | ICIR | 8.2661 | 0.008 12 | 0.850 | 0.019 |
| | Naive | 10.0264 | 0.009 96 | 0.821 | |
| | CoC | 0.1154 | 0.000 20 | 0.976 | |
| | IMMSBM | 11.6936 | 0.013 62 | 0.769 | |

**Fig. 4. Experimental results**—Darker is better (linear scale). Our model outperforms all of the baselines in almost every dataset for every evaluation metric. The standard deviations of the 5 folds cross-validation are negligible and reported in SI file.

We generate synthetic data according to the process described in Fig. 2 for a given $\beta$ matrix using the RBF kernel family. First, we generate a random matrix $\beta$, whose entries are between 0 and 1. A piece of information is then drawn with uniform probability and can result in a contagion according to $\beta$, the RBF kernel family and its history. We simulate the outcome by drawing a random number and finally increment the clock. The process then keeps on by randomly drawing a new exposure and adding it to the sequence. We set the maximum length of intervals to 50 steps and generate datasets of 20,000 sequences.

We present in Fig. 4 the results of the various models with generated interactions between 20 (Synth-20) and 5 (Synth-5) entities. The interactions are generated using the RBF kernel, hence the fact we are not evaluating the IR-EXP model –its use would be irrelevant. The InterRate model outperforms the proposed baselines for every metric considered. It is worth noting that performances of non-interacting and/or non-temporal baselines are good on the JS divergence and F1-score metrics due to the constant background noise $P_0$. For cases where interactions do not play a significant role, IMMSBM and Naive models perform well by fitting only the background noise. By contrast, the RSS metric distinguishes very well the models that are better at modeling interactions.

Note that while the baseline [12] yields good results when few interactions are simulated (Synth-5), it performs as bad as the naive baseline when this number increases (Synth-20). This is due to the non-convexity of the proposed model, which struggles to reach a global maximum of the likelihood even after 100 runs (see supplementary materials for implementation details).

### 5.2 Real Data

| | | RSS | JS div. | BCF1 |
|---|---|---|---|---|
| Twitter | IR-RBF | 0.0015 | 0.000 06 | 0.983 |
| | IR-EXP | 0.0011 | 0.000 05 | 0.986 |
| | ICIR | 0.0137 | 0.000 63 | 0.961 |
| | Naive | 0.0161 | 0.000 73 | 0.938 |
| | CoC | 0.0017 | 0.000 07 | 0.957 |
| | IMMSBM | 0.0147 | 0.000 68 | 0.954 |
| PD | IR-RBF | 1.1268 | 0.007 58 | 0.979 |
| | IR-EXP | 1.5526 | 0.008 67 | 0.966 |
| | ICIR | 3.5359 | 0.018 23 | 0.938 |
| | Naive | 3.6527 | 0.019 15 | 0.945 |
| | CoC | 1.2409 | 0.008 09 | 0.974 |
| | IMMSBM | 20.3773 | 0.087 01 | 0.767 |
| Ads | IR-RBF | 0.0043 | 0.000 04 | 0.981 |
| | IR-EXP | 0.0030 | 0.000 03 | 0.985 |
| | ICIR | 0.0983 | 0.000 85 | 0.966 |
| | Naive | 0.1453 | 0.001 26 | 0.913 |
| | CoC | 0.0045 | 0.000 05 | 0.974 |
| | IMMSBM | 0.0155 | 0.000 15 | 0.954 |

**Fig. 5. Experimental results**—Darker is better (linear scale). Our model outperforms all of the baselines in almost every dataset for every evaluation metric. The standard deviations of the 5 folds cross-validation are negligible and reported in SI file.

We consider 3 real-world datasets. For each dataset, we select a subset of entities that are likely to interact with each other. For instance, it has been shown that the interaction between the various URL shortening services on Twitter is non-trivial [23]. The datasets are a **Twitter** dataset (104,349 sequences, exposure are tweets and contagions are retweets) [10], a Prisoner's dilemma dataset (**PD**) (2,337 sequences, exposures are situations, contagions are players choices) [1,13] and an **Ads** dataset (87,500 sequences, exposures are ads, contagions are clicks on ads) [3]. A detailed description of the datasets is presented in SI file, section Datasets.

The results on real-world datasets are presented in Fig. 5. We see that the IMMSBM baseline performs poorly on the PD dataset: either considering the time plays a consequent role in the probability of contagion, or interactions are not symmetric. Indeed, the core hypothesis of the IMMSBM is that the effect of exposition A on B is the same as B on A, whichever is the time separation between them. In a prisoner's dilemma game setting, for instance, we expect that a player does not react in the same way to defection followed by cooperation as to cooperation followed by defection, a situation for which the IMMSBM does not account. When there are few entities, the CoC baseline performs as good as IR,

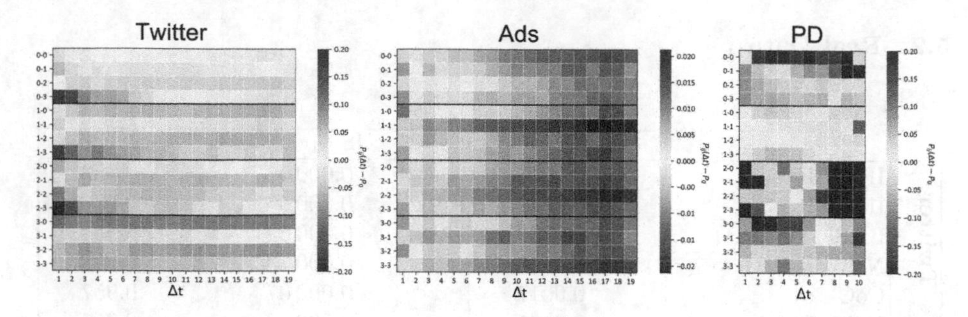


**Fig. 6. Visualization of the interaction profiles**—Intensity of the interactions between every pair of entities according to their time separation (one line is one pair's interaction profile, similar to Fig. 1 seen "from the top"). A positive intensity means that the interaction helps the contagion, while a negative intensity means it blocks it. The key linking numbers on the y-axis to names for each dataset is provided in the Datasets section.

but fails when this number increases; this is due mainly to the non-convexity of the problem that does not guarantee convergence towards the optimal solution. Overall, the InterRate models yield the best results on every dataset.

# 6 Discussion

In Fig. 6, we represent the interaction intensity over time for every pair of information considered in every corpus fitted with the RBF kernel model. The intensity of the interactions is the inferred probability of contagion minus the base contagion probability in any context: $P_{ij}(t) - P_0(i)$. Therefore, we can determine the characteristic range of interactions, investigate recurrent patterns in interactions, whether the interaction effect is positive or negative, etc. Overall, we understand why the EXP kernel performs as good as the RBF on the Twitter and Ads datasets: interactions tend to have an exponentially decaying influence over time. However, this is not the case on the PD dataset: the effect of a given interaction is very dependent on its position in the history (pike on influence at $\Delta t = 3$, shift from positive to negative influence, etc.).

In the Twitter dataset, the most substantial positive interactions occur before $\Delta \mathrm{t} = 3$. This finding agrees with previous works, which stated that the most informative interactions within Twitter URL dataset occur within the 3 time steps before the possible retweet [12]. We also find that the vast majority of interactions are weak, matching with previous study's findings [12,16]. However, it seems that tweets still exert influence even a long time after being seen, but with lesser intensity.

In the Prisoner's Dilemma dataset, players' behaviors are heavily influenced by the previous situations they have been exposed to. For instance, in the situation where both players cooperated in the previous round (pairs 2-x, $3^{rd}$ section in Fig. 6-PD). The probability that the player defects is then significantly increased if both players cooperated or if one betrayed the other exactly two rounds before but decreased if it has been two rounds that players both cooperate.

Finally, we find that the interactions play a lesser role in the clicks on ads. We observe a slightly increased probability of click on every ad after direct exposure to another one. We also observe a globally decreasing probability of click when two exposures distant in time, which agrees with previous work's findings [3]. Finally, the interaction profile is very similar for every pair of ads; we interpret this as a similarity in users' ads perception.

We showed that for each of the considered corpus, considering the interaction profile provides an extended comprehension of choice adoption mechanisms and retrieves several state-of-the-art conclusions. The proposed graphical visualization also provides an intuitive view of how the interaction occurs between entities and the associated trends, hence supporting its relevance as a new tool for researchers in a broad meaning.

## 7 Conclusion

We introduced an efficient convex model to investigate the way interactions happen within various datasets. Interactions modeling has been little explored in data science, despite recent clues pointing to their importance in modeling real-world processes [12,23]. Unlike previous models, our method accounts for both the interaction effects and their influence over time (the interaction profile). We showed that this improvement leads to better results on synthetic and various real-world datasets that can be used in different research fields, such as recommender systems, spreading processes, and human choice behavior. We also discussed the difficulty of observing significant interaction profiles due to the data-gathering process's inherent noise and solved the problem by introducing a time-independent kernel. We finally proposed a way to easily explore the results yielded by our model, allowing one to read the interaction profiles of any couple of entities quickly.

In future works, we will explore the way interactions vary over time and work on identifying recurrent patterns in interaction profiles. It would be the next step for an extended understanding of the role and nature of interacting processes in real-world applications.

## References

1. Bereby-Meyer, Y., Roth, A.E.: The speed of learning in noisy games: partial reinforcement and the sustainability of cooperation. Am. Econ. Rev. **96**(4), 1029–1042 (2006)
2. Beutel, A., Prakash, B.A., Rosenfeld, R., Faloutsos, C.: Interacting viruses in networks: can both survive? In: SIGKDD, pp. 426–434 (2012)
3. Cao, J., Sun, W.: Sequential choice bandits: learning with marketing fatigue. In: AAAI-19 (2019)

4. Cobo-López, S., Godoy-Lorite, A., Duch, J., Sales-Pardo, M., Guimerà, R.: Optimal prediction of decisions and model selection in social dilemmas using block models. EPJ Data Sci. **7**(1), 1–13 (2018). https://doi.org/10.1140/epjds/s13688-018-0175-3
5. Du, N., Song, L., Smola, A., Yuan, M.: Learning networks of heterogeneous influence. In: NIPS (2012)
6. Gomez-Rodriguez, M., Balduzzi, D., Schölkopf, B.: Uncovering the temporal dynamics of diffusion networks. In: ICML, pp. 561–568 (2011)
7. Gomez-Rodriguez, M., Leskovec, J., Schoelkopf, B.: Structure and dynamics of information pathways in online media. In: WSDM (2013)
8. Gomez-Rodriguez, M., Leskovec, J., Schölkopf, B.: Modeling information propagation with survival theory. In: ICML, vol. 28, p. III-666-III-674 (2013)
9. Hawkes, A.: Point spectra of some mutually exciting point processes. JRSS. Series B 33 (1971). https://doi.org/10.1111/j.2517-6161.1971.tb01530.x
10. Hodas, N.O., Lerman, K.: The simple rules of social contagion. Sci. Rep. **4**(4343), 1–7 (2014)
11. Weng, L., Flammini, A., Vespignani, A., Menczer, F., et al.: Competition among memes in a world with limited attention. Nature Sci. Rep. **2**, 335 (2012). https://doi.org/10.1038/srep00335
12. Myers, S., Leskovec, J.: Clash of the contagions: cooperation and competition in information diffusion. In: 2012 IEEE 12th International Conference on Data Mining, pp. 539–548 (2012)
13. Nay, J.J., Vorobeychik, Y.: Predicting human cooperation. PLoS One **11**(5), e0155656 (2016)
14. Nielsen, F.: On a generalization of the Jensen-Shannon divergence and the Jensen-Shannon centroid. Entropy **22**, 221 (2020). https://doi.org/10.3390/e22020221
15. Poux-Médard, G., Pastor-Satorras, R., Castellano, C.: Influential spreaders for recurrent epidemics on networks. Phys. Rev. Res. **2**, 023332 (2020). https://doi.org/10.1103/PhysRevResearch.2.023332
16. Poux-Médard, G., Velcin, J., Loudcher, S.: Interactions in information spread: quantification and interpretation using stochastic block models. RecSys (2021)
17. Prakash, B., Beutel, A., Rosenfeld, R., Faloutsos, C.: Winner takes all: competing viruses or ideas on fair-play networks. In: WWW (2012). https://doi.org/10.1145/2187836.2187975
18. Rodriguez, M., Gummadi, K.P., Schölkopf, B.: Quantifying information overload in social media and its impact on social contagions. In: ICWSM (2014)
19. Senanayake, R., O'Callaghan, S., Ramos, F.: Predicting spatio-temporal propagation of seasonal influenza using variational gaussian process regression. In: AAAI, pp. 3901–3907 (2016)
20. Vosoughi, S., Roy, D., Aral, S.: The spread of true and false news online. Science **359**, 1146–1151 (2018). https://doi.org/10.1126/science.aap9559
21. Yu, L., Cui, P., Song, C., Zhang, T., Yang, S.: A temporally heterogeneous survival framework with application to social behavior dynamics. In: KDD 2017, pp. 1295–1304 (2017). https://doi.org/10.1145/3097983.3098189
22. Zhu, Z., et al.: Cooperation and competition among information on social networks. Nature Sci. Rep. **3103**, 12160 (2020). https://doi.org/10.1038/s41598-020-69098-5
23. Zarezade, A., Khodadadi, A., Farajtabar, M., Rabiee, H.R., Zha, H.: Correlated cascades: compete or cooperate. In: AAAI (2017)

# Joslim: Joint Widths and Weights Optimization for Slimmable Neural Networks

Ting-Wu Chin[1(✉)], Ari S. Morcos[2], and Diana Marculescu[1,3]

[1] Department of ECE, Carnegie Mellon University, Pittsburgh, PA, USA
tingwuc@alumni.cmu.edu

[2] Facebook AI Research, Menlo Park, CA, USA

[3] Department of ECE, The University of Texas at Austin, Austin, TX, USA

**Abstract.** Slimmable neural networks provide a flexible trade-off front between prediction error and computational requirement (such as the number of floating-point operations or FLOPs) with the same storage requirement as a single model. They are useful for reducing maintenance overhead for deploying models to devices with different memory constraints and are useful for optimizing the efficiency of a system with many CNNs. However, existing slimmable network approaches either do not optimize layer-wise widths or optimize the shared-weights and layer-wise widths independently, thereby leaving significant room for improvement by joint width and weight optimization. In this work, we propose a general framework to enable joint optimization for both width configurations and weights of slimmable networks. Our framework subsumes *conventional and NAS-based slimmable methods* as special cases and provides flexibility to improve over existing methods. From a practical standpoint, we propose Joslim, an algorithm that jointly optimizes both the widths and weights for slimmable nets, which outperforms existing methods for optimizing slimmable networks across various networks, datasets, and objectives. Quantitatively, improvements up to 1.7% and 8% in top-1 accuracy on the ImageNet dataset can be attained for MobileNetV2 considering FLOPs and memory footprint, respectively. Our results highlight the potential of optimizing the channel counts for different layers *jointly* with the weights for slimmable networks. Code available at https://github.com/cmu-enyac/Joslim.

**Keywords:** Model compression · Slimmable neural networks · Channel optimization · Efficient deep learning

## 1 Introduction

Slimmable neural networks have been proposed with the promise of enabling multiple neural networks with different trade-offs between prediction error and

**Electronic supplementary material** The online version of this chapter (https://doi.org/10.1007/978-3-030-86523-8_8) contains supplementary material, which is available to authorized users.


N. Oliver et al. (Eds.): ECML PKDD 2021, LNAI 12977, pp. 119–134, 2021.
https://doi.org/10.1007/978-3-030-86523-8_8

the number of floating-point operations (FLOPs), *all at the storage requirement of only a single neural network* [42]. This is in stark contrast to channel pruning methods [4,14,39] that aim for a small standalone model. Slimmable neural networks are useful for applications running on mobile and other resource-constrained devices. As an example, the ability to deploy multiple versions of the same neural network alleviates the maintenance overhead for applications which support a number of different mobile devices with different memory and storage constraints, as only one model needs to be maintained. On the other hand, slimmable networks can bee critical for designing an efficient system that runs multiple CNNs. Specifically, an autonomous robot may execute multiple CNNs for various tasks at the same time. When optimizing the robot's efficiency (overall performance *vs.* computational costs), it is unclear which CNNs should be trimmed by how much to achieve an overall best efficiency. As a result, methods based on trial-and-error are necessary for optimizing such a system. However, if trimming the computational requirement of any CNN requires re-training or fine-tuning, this entire process will be impractically expensive. In this particular case, if we replace each of the CNNs with their respective slimmable versions, optimizing a system of CNNs becomes practically feasible as slimmable networks can be slimmed without the need of re-training or fine-tuning.

A slimmable neural network is trained by simultaneously considering networks with different widths (or filter counts) using a single set of shared weights. The width of a child network is specified by a real number between 0 and 1, which is known as the "width-multiplier" [17]. Such a parameter specifies how many filters per layer to use proportional to the full network. For example, a width-multiplier of $0.35\times$ represents a network with 35% of the channel counts of the full network for all the layers. While specifying child networks using a single width-multiplier for all the layers has shown empirical success [40,42], such a specification neglects that different layers affect the network's output differently [43] and have different FLOPs and memory footprint requirements [13], which may lead to sub-optimal results. As an alternative, neural architecture search (NAS) methods such as BigNAS [41] optimizes the layer-wise widths for slimmable networks, however, a sequential greedy procedure is adopted to optimize the widths and weights. As a result, the optimization of weights is not adapted to the optimization of widths, thereby leaving rooms for improvement by joint width and weight optimization.

In this work, we propose a framework for optimizing slimmable nets by formalizing it as minimizing the area under the trade-off curve between prediction error and some metric of interest, *e.g.*, memory footprint or FLOPs, with alternating minimization. Our framework subsumes both the universally slimmable networks [40] and BigNAS [41] as special cases. The framework is general and provides us with insights to improve upon existing alternatives and justifies our new algorithm Joslim, the first approach that jointly optimizes both shared-weights and widths for slimmable nets. To this end, we demonstrate empirically the superiority of the proposed algorithm over existing methods using various datasets, networks, and objectives. We visualize the algorithmic differences between the proposed method and existing alternatives in Fig. 1.

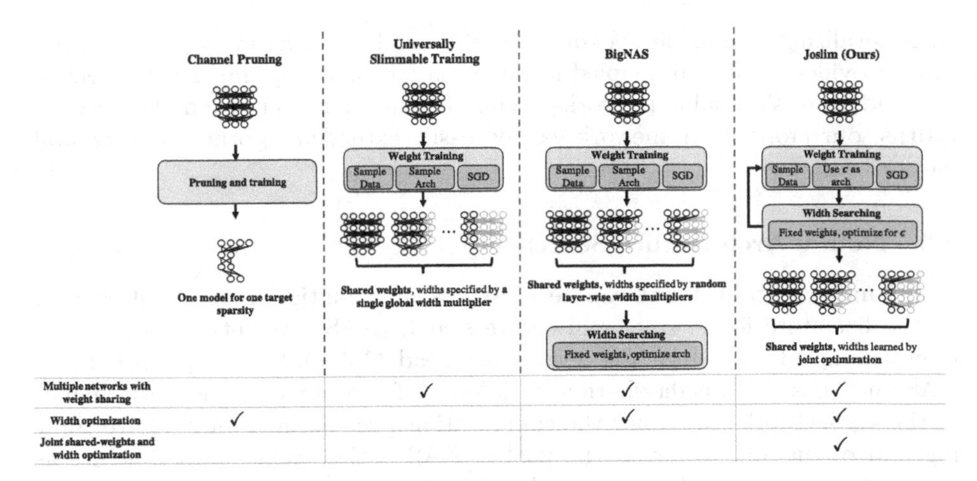


**Fig. 1.** Schematic overview comparing our proposed method with existing alternatives and channel pruning. Channel pruning has a fundamentally different goal compared to ours, *i.e.*, training slimmable nets. Joslim jointly optimizes both the widths and the shared weights.

The contributions of this work are as follows:

- We propose a general framework that enables the joint optimization of the widths and their corresponding shared weights of a *slimmable net*. The framework is general and subsumes existing algorithms as special cases.
- We propose Joslim, an algorithm that jointly optimizes the widths and weights of slimmable nets. We show empirically that Joslim outperforms existing methods on various networks, datasets, and objectives. Quantitatively, improvements up to 1.7% and 8% in top-1 accuracy on ImageNet are attained for MobileNetV2 considering FLOPs and memory footprint, respectively.

## 2 Related Work

### 2.1 Slimmable Neural Networks

Slimmable neural networks [42] enable multiple sub-networks with different compression ratios to be generated from a single network with one set of weights. This allows the network FLOPs to be dynamically configurable at run-time without increasing the storage requirement of the model weights. Based on this concept, better training methodologies have been proposed to enhance the performance of slimmable networks [40]. One can view a slimmable network as a dynamic computation graph where the graph can be constructed dynamically with different accuracy and FLOPs profiles. With this perspective, one can go beyond changing just the width of the network. For example, one can alter the network's sub-graphs [31], network's depth [5,18,19,21], and network's kernel sizes and input resolutions [6,35,36,41]. Complementing prior work primarily focusing on generalizing slimmable networks to additional architectural paradigms, our work provides the first principled formulation for jointly optimizing the weights and widths of slimmable networks. While our analysis focuses on the network widths, our proposed framework can be easily extended to other architectural parameters.

### 2.2 Neural Architecture Search

A slimmable neural network can be viewed as an instantiation of weight-sharing. In the literature for neural architecture search (NAS), weight-sharing is commonly adopted to reduce the search overhead [2,4,15,22,33,39]. Specifically, NAS methods use weight-sharing as a proxy for evaluating the performance of the sub-networks to reduce the computational requirement of iterative training and evaluation. However, the goal of NAS is the resulting architecture as opposed to both shared-weights and architecture. Exceptions are BigNAS [41] and Once-for-all (OFA) networks [6]; however, in neither case the architecture and shared-weights are jointly optimized. Specifically, both BigNAS and OFA employ a two-stage paradigm where the shared-weights are optimized before the architectures are optimized. This makes the trained weights oblivious to the optimized architectures.

While slimmable networks are inherently multi-objective, multi-objective optimization has also been adopted in NAS literature [7,11,12,26,37]. However, a crucial difference of the present work compared to these papers is that we are interested in learning a single set of weights from which multiple FLOP configurations can be used (as in slimmable networks) rather than finding architectures independently for each FLOP configuration that can be trained from scratch freely. Put another way, in our setting, both shared-weights and the searched architecture are optimized jointly, whereas in prior work, only searched architectures were optimized.

When it comes to joint neural architecture search and weight training, ENAS [29] and TuNAS [3] can both be seen as joint optimization. However, in stark contrast to our work, their search is dedicated to a single network of a single computational requirement (e.g., FLOPs) while our method is designed to obtain the weights that work for various architectures across a wide range of computational requirements.

### 2.3 Channel Pruning

Reducing the channel or filter counts for a pre-trained model is also known as channel pruning. In channel pruning, the goal is to find a single small model that maximizes the accuracy while satisfying some resource constraints by optimizing the layer-wise channel counts [4,8,9,16,20,23–25,27,38,39]. While channel pruning also optimizes for non-uniform widths, the goal of channel pruning is crucially different from ours. The key difference is that channel pruning is concerned with a single pruned model while slimmable neural networks require a set of models to be trained using weight sharing. Nonetheless, we compare our work with pruning methods that conduct greedy channel pruning since they naturally produce in a sequence of models that have different FLOPs. In particular, we compare our work with AutoSlim [39] in Appendix E.1 and demonstrate the effectiveness of our proposed Joslim.

# 3 Methodology

In this work, we are interested in jointly optimizing the network widths and network weights. Ultimately, when evaluating the performance of a slimmable neural network, we care about the trade-off curve between multiple objectives, *e.g.*, theoretical speedup and accuracy. This trade-off curve is formed by evaluating the two objectives at multiple width configurations using the same shared-weights. Viewed from this perspective, both the widths and shared-weights should be optimized in such a way that the resulting networks have a better trade-off curve (*i.e.*, larger area under curve). This section formalizes this idea and provides an algorithm to solve it in an approximate fashion.

## 3.1 Problem Formulation

Our goal is to find both the weights and the width configurations that optimize the area under the trade-off curve between two competing objectives, *e.g.*, accuracy and inference speed. Without loss of generality, we use cross entropy loss as the accuracy objective and FLOPs as the inference speed objective throughout the text for clearer context. Note that FLOPs can also be replaced by other metrics of interest such as memory footprint. Since in this case both objectives are better when lower, the objective for the optimizing slimmable nets becomes to *minimize* the area under curve. To quantify the area under curve, one can use a Riemann integral. Let $w(c)$ be a width configuration of $c$ FLOPs, one can quantify the Riemann integral by evaluating the cross entropy loss $L_{\mathcal{S}}$ on the training set $\mathcal{S}$ using the shared weights $\boldsymbol{\theta}$ for the architectures that spread uniformly on the FLOPs-axis between a lower bound $l$ and an upper bound $u$ of FLOPs: $\{\boldsymbol{a}|\boldsymbol{a} = w(c), c \in [l, u]\}$. More formally, the area under curve $\mathbb{A}$ is characterized as

$$\mathbb{A}(\boldsymbol{\theta}, w) \stackrel{\text{def}}{=} \int_{l}^{u} L_{\mathcal{S}}\left(\boldsymbol{\theta}, w(c)\right) dc \tag{1}$$

$$\approx \sum_{i=0}^{N} L_{\mathcal{S}}\left(\boldsymbol{\theta}, w(c_i)\right) \delta, \tag{2}$$

where Eq. 2 approximates the Riemann integral with the Riemann sum using $N$ architectures that are spread uniformly on the FLOPs-axis with a step size $\delta$. With a quantifiable area under curve, our goal for optimizing slimmable neural networks becomes finding both the shared-weights $\boldsymbol{\theta}$ and the architecture function $w$ to minimize their induced area under curve:

$$\arg\min_{\boldsymbol{\theta},w} \mathbb{A}(\boldsymbol{\theta}, w) \approx \arg\min_{\boldsymbol{\theta},w} \sum_{i=0}^{N} L_{\mathcal{S}}\left(\boldsymbol{\theta}, w(c_i)\right)\delta \quad (3)$$

$$= \arg\min_{\boldsymbol{\theta},w} \frac{1}{N} \sum_{i=0}^{N} L_{\mathcal{S}}\left(\boldsymbol{\theta}, w(c_i)\right) \quad (4)$$

$$\approx \arg\min_{\boldsymbol{\theta},w} \mathbb{E}_{c\sim U(l,u)} L_{\mathcal{S}}\left(\boldsymbol{\theta}, w(c)\right), \quad (5)$$

where $U(l, u)$ denotes a uniform distribution over a lower bound $l$ and an upper bound $u$. Note that the solution to Eq. 5 is the shared-weight vector and a set of architectures, which is drastically different from the solution to the formulation used in the NAS literature [22,34], which is an architecture.

## 3.2 Proposed Approach: Joslim

Since both the shared-weights $\boldsymbol{\theta}$ and the architecture function $w$ are optimization variables of two natural groups, we start by using alternating minimization:

$$w^{(t+1)} = \arg\min_{w} \mathbb{E}_{c\sim U(l,u)} L_{\mathcal{S}}\left(\boldsymbol{\theta}^{(t)}, w(c)\right) \quad (6)$$

$$\boldsymbol{\theta}^{(t+1)} = \arg\min_{\boldsymbol{\theta}} \mathbb{E}_{c\sim U(l,u)} L_{\mathcal{S}}\left(\boldsymbol{\theta}, w^{(t+1)}(c)\right). \quad (7)$$

In Eq. 6, we maintain the shared-weights $\boldsymbol{\theta}$ fixed and for each FLOPs between $l$ and $u$, we search for a corresponding architecture that minimizes the cross entropy loss. This step can be seen as a multi-objective neural architecture search given a fixed set of pre-trained weights, and can be approximated using smart algorithms such as multi-objective Bayesian optimization [28] or evolutionary algorithms [10]. However, even with smart algorithms, such a procedure can be impractical for every iteration of the alternating minimization.

In Eq. 7, one can use stochastic gradient descent by sampling from a set of architectures that spread uniformly across FLOPs obtained from solving Eq. 6. However, training such a weight-sharing network is practically 4× the training time of the largest standalone subnetwork [40] (it takes 6.5 GPU-days to train a slimmable ResNet18), which prevents it from being adopted in the alternating minimization framework.

To cope with these challenges, we propose targeted sampling, local approximation, and temporal sharing to approximate both equations.

**Targeted Sampling.** We propose to sample a set of FLOPs to approximate the expectation in Eqs. 6 and 7 with empirical estimates. Moreover, the sampled FLOPs are shared across both steps in the alternating minimization so that one does not have to solve for the architecture function $w$ (needed for the second step), but only solve for a set of architectures that have the corresponding FLOPs. Specifically, we approximate the expectation in both Eqs. 6 and 7 with the sample mean:

$$c_i^{(t)} \sim U(l,u) \ \forall\, i = 1,\ldots,M \tag{8}$$

$$w^{(t+1)} \approx \arg\min_{w} \frac{1}{M}\sum_{i=1}^{M} L_{\mathcal{S}}\left(\boldsymbol{\theta}^{(t)}, w(c_i^{(t)})\right) \tag{9}$$

$$\boldsymbol{\theta}^{(t+1)} \approx \arg\min_{\boldsymbol{\theta}} \frac{1}{M}\sum_{i=1}^{M} L_{\mathcal{S}}\left(\boldsymbol{\theta}, w^{(t+1)}(c_i^{(t)})\right). \tag{10}$$

From Eq. 9 and 10, we can observe that at any timestamp $t$, we only query the architecture function $w^{(t)}$ and $w^{(t+1)}$ at a fixed set of locations $c_i \ \forall\, i = 1,\ldots,M$. As a result, instead of solving for the architecture function $w$, we solve for a fixed set of architectures $\mathcal{W}^{(t+1)}$ at each timestamp as follows:

$$\mathcal{W}^{(t+1)} := \{w^{(t+1)}(c_i),\ldots,w^{(t+1)}(c_M)\} \tag{11}$$

where

$$\begin{aligned} w^{(t+1)}(c_i) = \arg\min_{\boldsymbol{a}} L_{\mathcal{S}}&\left(\boldsymbol{\theta}^{(t)}, \boldsymbol{a}\right) \\ \text{s.t. FLOPs}(\boldsymbol{a}) &= c_i. \end{aligned} \tag{12}$$

With these approximations, for each iteration in the alternating minimization, we solve for $M$ architectures with targeted FLOPs as opposed to solving for the entire approximate trade-off curve.

**Local Approximation.** To reduce the overhead for solving Eq. 10, we propose to approximate it with a few steps of gradient descent. Specifically, instead of training a slimmable neural network with sampled architectures until convergence in each iteration of alternating minimization (Eq. 10), we propose to perform $K$ steps of gradient descent:

$$\begin{aligned} &\boldsymbol{x}^{0} \overset{\text{def}}{=} \boldsymbol{\theta}^{(t)} \\ &\boldsymbol{x}^{(k+1)} \overset{\text{def}}{=} \boldsymbol{x}^{(k)} - \eta\frac{1}{M}\sum_{i=1}^{M} \nabla_{\boldsymbol{x}^{(k)}} L_{\mathcal{S}}\left(\boldsymbol{x}^{(k)}, \mathcal{W}_i^{(t+1)}\right) \\ &\boldsymbol{\theta}^{(t+1)} \approx \boldsymbol{x}^{(K)}, \end{aligned} \tag{13}$$

where $\eta$ is the learning rate. Larger $K$ indicates better approximation with higher training overhead.

**Temporal Sharing.** Since we use local approximation, $\boldsymbol{\theta}^{(t+1)}$ and $\boldsymbol{\theta}^{(t)}$ would not be drastically different. As a result, instead of performing constrained neural architecture search *from scratch* (*i.e.*, solving for Eq. 12) in every iteration of the alternating minimization, we propose to share information across the search procedures in different iterations of the alternation.

To this end, we propose to perform temporal sharing for multi-objective Bayesian optimization with random scalarization (MOBO-RS) [28] to solve

Eq. 12. MOBO-RS itself is a sequential model-based optimization algorithm, where one takes a set of architectures $\mathcal{H}$, builds models (typically Gaussian Processes [30]) to learn a mapping from architectures to cross entropy loss $g_{\text{CE}}$ and FLOPs $g_{\text{FLOPs}}$, scalarizes both models into a single objective with a random weighting $\lambda$ ($\lambda$ controls the preference for cross entropy and FLOPs), and finally optimizes the scalarized model to obtain a new architecture and stores the architecture back to the set $\mathcal{H}$. This entire procedure repeats for $T$ iterations for one MOBO-RS.

To exploit temporal similarity, we propose MOBO-TS2, which stands for multi-objective Bayesian optimization with targeted scalarization and temporal sharing. Specifically, we propose to let $T = 1$ and share $\mathcal{H}$ across alternating minimization. Additionally, we modify the random scalarization with targeted scalarization where we use binary search to search for the $\lambda$ that results in the desired FLOPs. As such, $\mathcal{H}$ grows linearly with the number of alternations. In such an approximation, for each MOBO in the alternating optimization, we reevaluate the cross-entropy loss for each $\boldsymbol{a} \in \mathcal{H}$ to build faithful GPs. We further provide theoretical analysis for approximation via temporal similarity for Bayesian optimization in Appendix D.

**Joslim.** Based on this preamble, we present our algorithm, Joslim, in Algorithm 1. In short, Joslim has three steps: (1) build surrogate functions (*i.e.*, GPs) and acquisition functions (*i.e.*, UCBs) using historical data $\mathcal{H}$ and their function responses, (2) sample $M$ target FLOPs and solve for the corresponding widths (*i.e.*, $\boldsymbol{a}$) via binary search with the scalarized acquisition function and store them back to $\mathcal{H}$, and (3) perform $K$ gradient descent steps using the solved widths. The first two steps solve Eq. 12 with targeted sampling and temporal sharing, and the final step solves Eq. 10 approximately with local approximation. In the end, to obtain the best widths, we use non-dominated sorting based on the training loss and FLOPs for $a \in \mathcal{H}$.

### 3.3 Relation to Existing Approaches

For direct comparison with our work we consider the universally slimmable neural networks [40], which uses a single width multiplier to specify the widths of a slimmable network and NAS-based approaches such as OFA [6] and BigNAS [41], which have decoupled widths and weights optimization. To demonstrate the generality of the proposed framework, we show how these previously published works are special cases of our framework.

**Slim.** Universally slimmable networks [40], or Slim for short, is a special case of our framework where the widths are not optimized but pre-specified by a single global width multiplier. This corresponds to solving Eq. 5 with $w$ given as a function that returns the width that satisfies some FLOPs by controlling a single global width multiplier. Our framework is more general as it introduces the freedom for optimizing the widths of slimmable nets.

**Algorithm 1:** Joslim

**Input** : Model parameters $\boldsymbol{\theta}$, lower bound for width-multipliers $w_0 \in [0, 1]$, number of full iterations $F$, number of gradient descent updates $K$, number of $\lambda$ samples $M$
**Output** : Trained parameter $\boldsymbol{\theta}$, approximate Pareto front $\mathcal{N}$

```
1  H = {}    (Historical minimizers a)
2  for i = 1...F do
3      x, y = sample_data()
4      u_CE, u_FLOPs = L_CE(H; θ, x, y), FLOPs(H)
5      g_CE, g_FLOPs = GP_UCB( H, u_CE, u_FLOPs )
6      widths = []
7      for m = 1...M do
8          a = MOBO_TS2( g_CE, g_FLOPs, H )    (Algorithm 2)
9          widths.append(a)
10     end
11     H = H ∪ widths         (update historical data)
12     widths.append(w_0)
13     for j = 1...K do
14         SlimmableTraining( θ, widths )
15         (line 3-16 of Algorithm 1 in [40])
16     end
17     N=nonDominatedSort(H, u_CE, u_FLOPs)
18 end
```

**Algorithm 2:** MOBO-TS2

**Input** : Acquisition functions $g_{\text{CE}}, g_{\text{FLOPs}}$, historical data $\mathcal{H}$, search precision $\epsilon$
**Output** : channel configurations $\boldsymbol{a}$

```
1  c = Uniform( l, u )    (Sample a target FLOPs)
2  λ_FLOPs, λ_min, λ_max = 0.5, 0, 1
3  while |(FLOPs(a) − c) / FullModelFLOPs| > ε do        // binary search
4      c = arg min_c Scalarize( λ_FLOPs, g_CE, g_FLOPs )
5      if FLOPs(a) > c then
6          λ_min = λ_FLOPs
7          λ_FLOPs = (λ_FLOPs + λ_max)/2
8      else
9          λ_max = λ_FLOPs
10         λ_FLOPs = (λ_FLOPs + λ_min)/2
11     end
12 end
```

**OFA and BigNAS.** OFA and BigNAS use the same approach when it comes to the channel search space[1]. They are also a special case of our framework

[1] Since we only search for channel counts, the progressive shrinking strategy proposed in OFA does not apply. As a result, both OFA and BigNAS have the same approach.

where the optimization of the widths and the shared-weights are carried out greedily. Specifically, BigNAS first trains the shared-weights by random layer-wise width multipliers. After convergence, BigNAS performs evolutionary search to optimize the layer-wise width multipliers considering both error and FLOPs. This greedy algorithm can be seen as performing one iteration of alternating minimization by solving Eq. 7 followed by solving Eq. 6. From this perspective, one can observe that the shared-weights $\boldsymbol{\theta}$ are not jointly optimized with the widths. Our framework is more general and enables joint optimization for both widths and weights.

As we demonstrate in Sect. 4.2, our comprehensive empirical analysis reveals that Joslim is superior to either approach when compared across multiple networks, datasets, and objectives.

# 4 Experiments

## 4.1 Experimental Setup

For all the Joslim experiments in this sub-section, we set $K$ such that Joslim only visits 1000 width configurations throughout the entire training ($|\mathcal{H}| = 1000$). Also, we set $M$ to be 2, which follows the conventional slimmable training method [40] that samples two width configurations in between the largest and the smallest widths. As for binary search, we conduct at most 10 binary searches with $\epsilon$ set to 0.02, which means that the binary search terminates if the FLOPs difference is within a two percent margin relative to the full model FLOPs. On average, the procedure terminates by using 3.4 binary searches for results on ImageNet. The dimension of $\boldsymbol{a}$ is network-dependent and is specified in Appendix A and the training hyperparameters are detailed in Appendix B. To arrive at the final set of architectures for Joslim, we use non-dominated sort based on the training loss and FLOPs for $\boldsymbol{a} \in \mathcal{H}$.

## 4.2 Performance Gains Introduced by Joslim

We consider three datasets: CIFAR-10, CIFAR-100, and ImageNet. To provide informative comparisons, we verify our implementation for the conventional slimmable training with the reported numbers in [40] using MobileNetV2 on ImageNet. Our results follow closely to the reported numbers as shown in Fig. 3a, which makes our comparisons on other datasets convincing.

We compare to the following baselines:

- **Slim**: the conventional slimmable training method (the universally slimmable networks by [40]). We select 40 architectures uniformly distributed across FLOPs and run a non-dominated sort using training loss and FLOPs.
- **BigNAS**: disjoint optimization that first trains the shared-weights, then uses search methods to find architectures that work well given the trained weights (similar to OFA [6]). To compare fairly with Joslim, we use MOBO-RS for the search. After optimization, we run a non-dominated sort for all the visited architectures $\mathcal{H}$ using training loss and FLOPs.

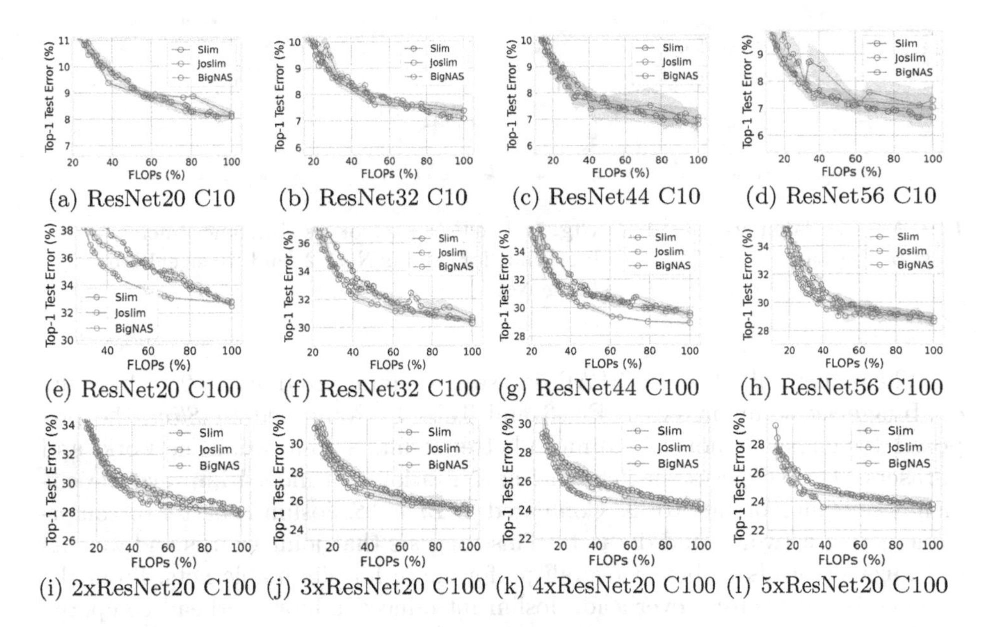


(a) ResNet20 C10 (b) ResNet32 C10 (c) ResNet44 C10 (d) ResNet56 C10
(e) ResNet20 C100 (f) ResNet32 C100 (g) ResNet44 C100 (h) ResNet56 C100
(i) 2xResNet20 C100 (j) 3xResNet20 C100 (k) 4xResNet20 C100 (l) 5xResNet20 C100

**Fig. 2.** Comparisons among Slim, BigNAS, and Joslim. C10 and C100 denote CIFAR-10/100. We perform three trials for each method and plot the mean and standard deviation. $n$xResNet20 represents a $n$ times wider ResNet20.

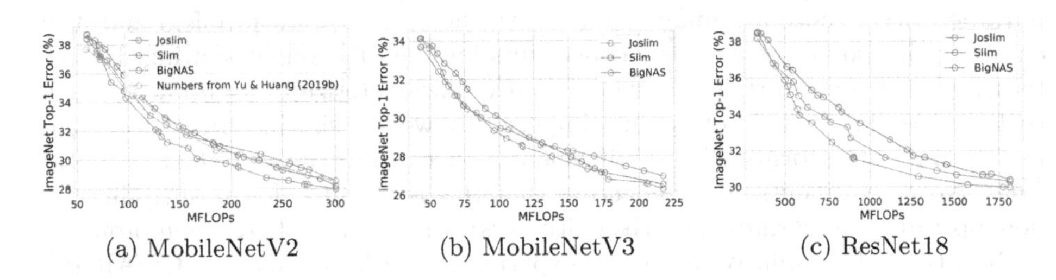


(a) MobileNetV2 (b) MobileNetV3 (c) ResNet18

**Fig. 3.** Comparisons among Slim, BigNAS, and Joslim on ImageNet.

**Table 1.** Comparing the top-1 accuracy among Slim, BigNAS, and Joslim on ImageNet. Bold represents the highest accuracy of a given FLOPs.

| MobileNetV2 | | | | MobileNetV3 | | | | ResNet18 | | | |
|---|---|---|---|---|---|---|---|---|---|---|---|
| MFLOPs | Slim | BigNAS | Joslim | MFLOPs | Slim | BigNAS | Joslim | MFLOPs | Slim | BigNAS | Joslim |
| 59 | 61.4 | 61.3 | **61.5** | 43 | 65.8 | **66.3** | 65.9 | 339 | 61.5 | 61.5 | **61.8** |
| 84 | 63.0 | 63.1 | **64.6** | 74 | 68.1 | 68.1 | **68.8** | 513 | 63.4 | 64.2 | **64.5** |
| 102 | 64.7 | **65.5** | **65.5** | 85 | 69.1 | **70.0** | **70.0** | 650 | 64.7 | 65.6 | **66.5** |
| 136 | 67.1 | 67.5 | **68.2** | 118 | 71.0 | **71.4** | **71.4** | 718 | 65.1 | 66.1 | **67.5** |
| 149 | 67.6 | 68.2 | **69.1** | 135 | 71.5 | 71.5 | **72.1** | 939 | 66.5 | 67.3 | **68.5** |
| 169 | 68.2 | 68.8 | **69.9** | 169 | 72.7 | 72.0 | **72.8** | 1231 | 68.0 | 68.4 | **69.4** |
| 212 | 69.7 | 69.6 | **70.6** | 184 | 73.0 | 72.5 | **73.2** | 1659 | 69.3 | 69.3 | **69.9** |
| 300 | 71.8 | 71.5 | **72.1** | 217 | 73.5 | 73.1 | **73.7** | 1814 | 69.6 | 69.7 | **70.0** |

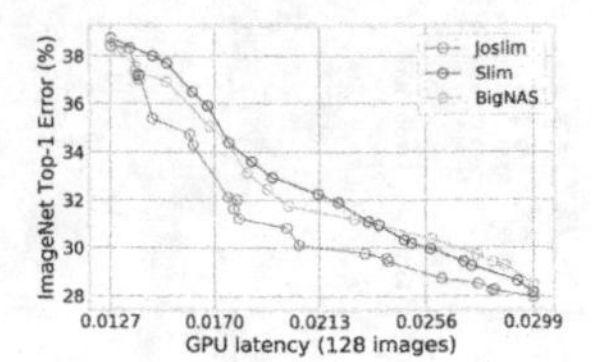


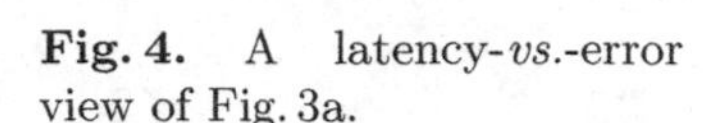

**Fig. 4.** A latency-*vs.*-error view of Fig. 3a.

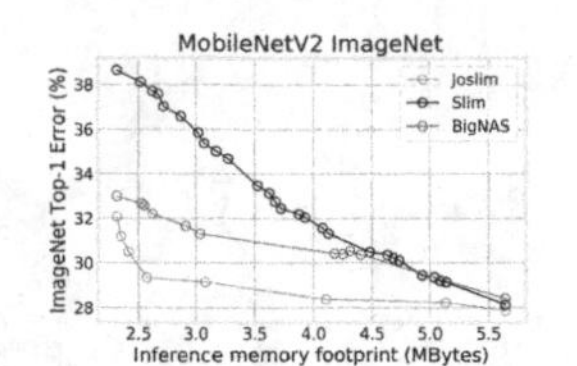


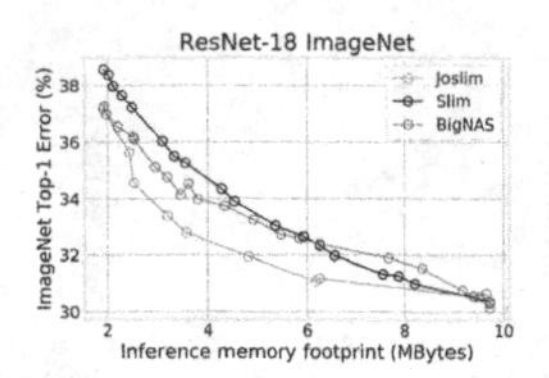


**Fig. 5.** Prediction error *vs.* inference memory footprint for MobileNetV2 and ResNet18 on ImageNet.

The main results for the CIFAR dataset are summarized in Fig. 2 with results on ImageNet summarized in Fig. 3 and Table 1. Compared to *Slim*, the proposed Joslim has demonstrated much better results across various networks and datasets. This suggests that channel optimization can indeed improve the efficiency of slimmable networks. Compared to *BigNAS*, Joslim is better or comparable across networks and datasets. This suggests that joint widths and weights optimization leads to better overall performance for slimmable nets. From the perspective of training overhead, Joslim introduced minor overhead compared to Slim due to the temporal similarity approximation. More specifically, on ImageNet, Joslim incurs approximately 20% extra overhead compared to Slim.

Note that the performance among these three methods are similar for the CIFAR-10 dataset. This is plausible since when a network is more over-parameterized, there are many solutions to the optimization problem and it is easier to find solutions with the constraints imposed by weight sharing. In contrast, when the network is relatively less over-parameterized, compromises have to be made due to the constraints imposed by weight sharing. In such scenarios, Joslim outperforms Slim significantly, as it can be seen in CIFAR-100 and ImageNet experiments. We conjecture that this is because Joslim introduces a new optimization variable (width-multipliers), which allows better compromises to be attained. Similarly, from the experiments with ResNets on CIFAR-100 (Fig. 2e to Fig. 2h), we find that shallower models tend to benefit more from joint channel and weight optimization than their deeper counterparts.

As FLOPs may not necessarily reflect latency improvements since FLOP does not capture memory accesses, we in addition plot latency-*vs.*-error for the data in Fig. 3a in Fig. 4. The latency is measured on a single V100 GPU using a batch size of 128. When visualized in latency, Joslim still performs favorably compared to Slim and BigNAS for MobileNetV2 on ImageNet.

Lastly, we consider another objective that is critical for on-device machine learning, *i.e.*, inference memory footprint [42]. Inference memory footprint decides whether a model is executable or not on memory-constrained devices. We detailed the memory footprint calculation in Appendix E. Since Joslim is general, we can replace the FLOPs calculation with memory footprint calculation to optimize for memory-*vs.*-error. As shown in Fig. 5, Joslim significantly outperform other alternatives. Notably, Joslim outperforms Slim by up to 8% top-1 accuracy for MobileNetV2. Such a drastic improvement comes from the fact that memory footprint depends mostly on the largest layers. As a result, slim-

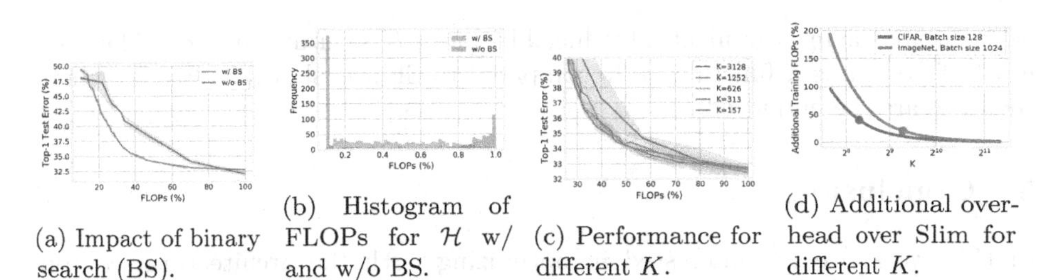


(a) Impact of binary search (BS).

(b) Histogram of FLOPs for $\mathcal{H}$ w/ and w/o BS.

(c) Performance for different $K$.

(d) Additional overhead over Slim for different $K$.

**Fig. 6.** Ablation study for the introduced binary search and the number of gradient descent updates per full iteration using ResNet20 and CIFAR-100. Experiments are conducted three times and we plot the mean and standard deviation.

ming all the layers equally to arrive at networks with smaller memory footprint (as done in Slim) is less than ideal since only one layer contributes to the reduced memory. In addition, when comparing Joslim with BigNAS, we can observe significant improvements as well, *i.e.*, around 2% top-1 accuracy improvements for MobileNetV2, which demonstrates the effectiveness of joint width and weights optimization.

### 4.3 Ablation Studies

In this subsection, we ablate the hyperparameters that are specific to Joslim to understand their impact. We use ResNet20 and CIFAR-100 for the ablation with the results summarized in Fig. 6.

**Binary Search.** Without binary search, one can also consider sampling the scalarization weighting $\lambda$ uniformly from $[0, 1]$, which does not require any binary search and is easy to implement. However, the issue with this sampling strategy is that uniform sampling $\lambda$ does not necessarily imply uniform sampling in the objective space, *e.g.*, FLOPs. As shown in Fig. 6a and Fig. 6b, sampling directly in the $\lambda$ space results in non-uniform FLOPs and worse performance compared to binary search.

**Number of Gradient Descent Steps.** In the approximation, the number of architectures ($|\mathcal{H}|$) is affected by the number of gradient descent updates $K$. In previous experiments for CIFAR, we have $K = 313$, which results in $|\mathcal{H}| = 1000$. Here, we ablate $K$ to $156, 626, 1252, 3128$ such that $|\mathcal{H}| = 2000, 500, 250, 100$, respectively. Given a fixed training epoch and batch size, Joslim produces a better approximation for Eq. 10 but a worse approximation for Eq. 9 with larger $K$. The former is because of the local approximation while the latter is because there are overall fewer iterations put into Bayesian optimization due to temporal sharing. As shown in Fig. 6c, we observe worse results with higher $K$. On the other hand, the improvement introduced by lower $K$ saturates quickly. The overhead of Joslim as a function of $K$ compared to Slim is shown in Fig. 6d where the dots are the employed $K$.

## 5 Conclusion

In this work, we are interested in optimizing both the architectural components and shared-weights of slimmable neural networks. To achieve this goal, we propose a general framework that optimizes slimmable nets by minimizing the area under the trade-off curve between cross entropy and FLOPs (or memory footprint) with alternating minimization. We further show that the proposed framework subsumes existing methods as special cases and provides flexibility for devising better algorithms. To this end, we propose Joslim, an algorithm that jointly optimizes the weights and widths of slimmable nets, which empirically outperforms existing alternatives that either neglect width optimization or conduct widths and weights optimization independently. We extensively verify the effectiveness of Joslim over existing techniques on three datasets (*i.e.*, CIFAR10, CIFAR100, and ImageNet) with two families of network architectures (*i.e.*, ResNets and MobileNets) using two types of objectives (*i.e.*, FLOPs and memory footprint). Our results highlight the importance and superiority in results of jointly optimizing the channel counts for different layers and the weights for slimmable networks.

**Acknowledgement.** This research was supported in part by NSF CCF Grant No. 1815899, NSF CSR Grant No. 1815780, and NSF ACI Grant No. 1445606 at the Pittsburgh Supercomputing Center (PSC).

## References

1. Balandat, M., et al.: BoTorch: a framework for efficient Monte-Carlo Bayesian optimization. In: NeurIPS (2020)
2. Bender, G., Kindermans, P.J., Zoph, B., Vasudevan, V., Le, Q.: Understanding and simplifying one-shot architecture search. In: ICML, pp. 550–559 (2018)
3. Bender, G., et al.: Can weight sharing outperform random architecture search? An investigation with tunas. In: CVPR, pp. 14323–14332 (2020)
4. Berman, M., Pishchulin, L., Xu, N., Medioni, G., et al.: AOWS: adaptive and optimal network width search with latency constraints. In: CVPR (2020)
5. Bolukbasi, T., Wang, J., Dekel, O., Saligrama, V.: Adaptive neural networks for efficient inference. In: ICML (2017)
6. Cai, H., Gan, C., Wang, T., Zhang, Z., Han, S.: Once-for-all: train one network and specialize it for efficient deployment. In: ICLR (2020)
7. Cheng, A.C., et al.: Searching toward pareto-optimal device-aware neural architectures. In: ICCAD, pp. 1–7 (2018)
8. Chin, T.W., Ding, R., Zhang, C., Marculescu, D.: Towards efficient model compression via learned global ranking. In: CVPR (2020)
9. Chin, T.W., Marculescu, D., Morcos, A.S.: Width transfer: on the (in) variance of width optimization. In: CVPR Workshops, pp. 2990–2999 (2021)

10. Deb, K., Pratap, A., Agarwal, S., Meyarivan, T.: A fast and elitist multiobjective genetic algorithm: NSGA-II. IEEE Trans. Evol. Comput. **6**(2), 182–197 (2002)
11. Dong, J.-D., Cheng, A.-C., Juan, D.-C., Wei, W., Sun, M.: DPP-Net: device-aware progressive search for pareto-optimal neural architectures. In: Ferrari, V., Hebert, M., Sminchisescu, C., Weiss, Y. (eds.) ECCV 2018. LNCS, vol. 11215, pp. 540–555. Springer, Cham (2018). https://doi.org/10.1007/978-3-030-01252-6_32
12. Elsken, T., Metzen, J.H., Hutter, F.: Efficient multi-objective neural architecture search via Lamarckian evolution. arXiv preprint arXiv:1804.09081 (2018)
13. Gordon, A., et al.: MorphNet: fast & simple resource-constrained structure learning of deep networks. In: CVPR, pp. 1586–1595 (2018)
14. Guo, S., Wang, Y., Li, Q., Yan, J.: DMCP: differentiable Markov channel pruning for neural networks. In: CVPR, pp. 1539–1547 (2020)
15. Guo, Z., et al.: Single path one-shot neural architecture search with uniform sampling. In: Vedaldi, A., Bischof, H., Brox, T., Frahm, J.-M. (eds.) ECCV 2020. LNCS, vol. 12361, pp. 544–560. Springer, Cham (2020). https://doi.org/10.1007/978-3-030-58517-4_32
16. He, Y., Liu, P., Wang, Z., Hu, Z., Yang, Y.: Filter pruning via geometric median for deep convolutional neural networks acceleration. In: CVPR, pp. 4340–4349 (2019)
17. Howard, A.G., et al.: MobileNets: efficient convolutional neural networks for mobile vision applications. arXiv preprint arXiv:1704.04861 (2017)
18. Huang, G., Chen, D., Li, T., Wu, F., van der Maaten, L., Weinberger, K.Q.: Multi-scale dense networks for resource efficient image classification. In: ICLR (2018)
19. Kaya, Y., Hong, S., Dumitras, T.: Shallow-deep networks: understanding and mitigating network overthinking. In: ICML, Long Beach, CA, June 2019
20. Li, H., Kadav, A., Durdanovic, I., Samet, H., Graf, H.P.: Pruning filters for efficient convnets. arXiv preprint arXiv:1608.08710 (2016)
21. Li, H., Zhang, H., Qi, X., Yang, R., Huang, G.: Improved techniques for training adaptive deep networks. In: ICCV, pp. 1891–1900 (2019)
22. Liu, H., Simonyan, K., Yang, Y.: DARTS: differentiable architecture search. arXiv preprint arXiv:1806.09055 (2018)
23. Liu, Z., et al.: MetaPruning: meta learning for automatic neural network channel pruning. In: ICCV (2019)
24. Liu, Z., Li, J., Shen, Z., Huang, G., Yan, S., Zhang, C.: Learning efficient convolutional networks through network slimming. In: ICCV, pp. 2736–2744 (2017)
25. Louizos, C., Welling, M., Kingma, D.P.: Learning sparse neural networks through $l\_0$ regularization. arXiv preprint arXiv:1712.01312 (2017)
26. Lu, Z., et al.: NSGA-NET: neural architecture search using multi-objective genetic algorithm. In: GECCO, pp. 419–427 (2019)
27. Ma, X., Triki, A.R., Berman, M., Sagonas, C., Cali, J., Blaschko, M.B.: A Bayesian optimization framework for neural network compression. In: ICCV (2019)
28. Paria, B., Kandasamy, K., Póczos, B.: A flexible framework for multi-objective Bayesian optimization using random scalarizations. In: Globerson, A., Silva, R. (eds.) UAI (2019)
29. Pham, H., Guan, M., Zoph, B., Le, Q., Dean, J.: Efficient neural architecture search via parameters sharing. In: ICML, pp. 4095–4104. PMLR (2018)
30. Rasmussen, C.E.: Gaussian processes in machine learning. In: Bousquet, O., von Luxburg, U., Rätsch, G. (eds.) ML 2003. LNCS (LNAI), vol. 3176, pp. 63–71. Springer, Heidelberg (2004). https://doi.org/10.1007/978-3-540-28650-9_4
31. Ruiz, A., Verbeek, J.: Adaptative inference cost with convolutional neural mixture models. In: ICCV, pp. 1872–1881 (2019)

32. Sandler, M., Howard, A., Zhu, M., Zhmoginov, A., Chen, L.C.: MobileNetV2: inverted residuals and linear bottlenecks. In: CVPR, pp. 4510–4520 (2018)
33. Stamoulis, D., et al.: Single-Path NAS: designing hardware-efficient convnets in less than 4 hours. In: ECML-PKDD (2019)
34. Tan, M., et al.: MnasNet: platform-aware neural architecture search for mobile. In: CVPR, pp. 2820–2828 (2019)
35. Wang, D., Gong, C., Li, M., Liu, Q., Chandra, V.: AlphaNet: improved training of supernet with alpha-divergence. In: ICML (2021)
36. Yang, T., Zhu, S., Chen, C., Yan, S., Zhang, M., Willis, A.: MutualNet: adaptive ConvNet via mutual learning from network width and resolution. In: Vedaldi, A., Bischof, H., Brox, T., Frahm, J.-M. (eds.) ECCV 2020. LNCS, vol. 12346, pp. 299–315. Springer, Cham (2020). https://doi.org/10.1007/978-3-030-58452-8_18
37. Yang, Z., et al.: CARS: continuous evolution for efficient neural architecture search. In: CVPR, June 2020
38. Ye, J., Lu, X., Lin, Z., Wang, J.Z.: Rethinking the smaller-norm-less-informative assumption in channel pruning of convolution layers. In: ICLR (2018)
39. Yu, J., Huang, T.: AutoSlim: towards one-shot architecture search for channel numbers. arXiv preprint arXiv:1903.11728, August 2019
40. Yu, J., Huang, T.S.: Universally slimmable networks and improved training techniques. In: ICCV, pp. 1803–1811 (2019)
41. Yu, J., et al.: BigNAS: scaling up neural architecture search with big single-stage models. In: Vedaldi, A., Bischof, H., Brox, T., Frahm, J.-M. (eds.) ECCV 2020. LNCS, vol. 12352, pp. 702–717. Springer, Cham (2020). https://doi.org/10.1007/978-3-030-58571-6_41
42. Yu, J., Yang, L., Xu, N., Yang, J., Huang, T.: Slimmable neural networks. In: ICLRa (2019)
43. Zhang, C., Bengio, S., Singer, Y.: Are all layers created equal? arXiv preprint arXiv:1902.01996 (2019)

# A Variance Controlled Stochastic Method with Biased Estimation for Faster Non-convex Optimization

Jia Bi(✉) and Steve R. Gunn(✉)

Electronics and Computer Science, University of Southampton, Southampton, UK
J.Bi@soton.ac.uk, srg@ecs.soton.ac.uk

**Abstract.** This paper proposed a new technique Variance Controlled Stochastic Gradient (VCSG) to improve the performance of the stochastic variance reduced gradient (SVRG) algorithm. To avoid over-reducing the variance of gradient by SVRG, a hyper-parameter $\lambda$ is introduced in VCSG that is able to control the reduced variance of SVRG. Theory shows that the optimization method can converge by using an unbiased gradient estimator, but in practice, biased gradient estimation can allow more efficient convergence to the vicinity since an unbiased approach is computationally more expensive. $\lambda$ also has the effect of balancing the trade-off between unbiased and biased estimations. Secondly, to minimize the number of full gradient calculations in SVRG, a variance-bounded batch is introduced to reduce the number of gradient calculations required in each iteration. For smooth non-convex functions, the proposed algorithm converges to an approximate first-order stationary point (i.e. $\mathbb{E}\|\nabla f(x)\|^2 \leq \epsilon$) within $\mathcal{O}(min\{1/\epsilon^{3/2}, n^{1/4}/\epsilon\})$ number of stochastic gradient evaluations, which improves the leading gradient complexity of stochastic gradient-based method SCSG ($\mathcal{O}(min\{1/\epsilon^{5/3}, n^{2/3}/\epsilon\})$ [19]. It is shown theoretically and experimentally that VCSG can be deployed to improve convergence.

**Keywords:** Non-convex optimization · Deep learning · Computational complexity

## 1 Introduction

We study smooth non-convex optimization problems which is shown in Eq. 1,

$$\min_{x \in \mathbb{R}^d} f(x), \;\; f(x) := \frac{1}{n} \sum_{i=1}^{n} f_i(x), \tag{1}$$

where each component $f_i(x)(i \in [n])$ is possibly non-convex and Lipschitz ($\mathcal{L} - smooth$) [28,31]. We use $\mathcal{F}_n$ to denote all $f_i(x)$ functions of the form in Eq. 1, and optimize such functions using Incremental First-order (IFO) and Stochastic First-Order (SFO) Oracles, which are defined in Definition 1 and 2 respectively.

**Electronic supplementary material** The online version of this chapter (https://doi.org/10.1007/978-3-030-86523-8_9) contains supplementary material, which is available to authorized users.


N. Oliver et al. (Eds.): ECML PKDD 2021, LNAI 12977, pp. 135–150, 2021.
https://doi.org/10.1007/978-3-030-86523-8_9

**Definition 1.** *[1] For a function $F(x) = \frac{1}{n}\sum_i f_i(x)$, an IFO takes an index $i \in [n]$ and a point $x \in \mathbb{R}^d$, and returns the pair $(f_i(x), \nabla f_i(x))$.*

**Definition 2.** *[26] For a function $F(x) = \mathbb{E}_y f(x, y)$ where $y \sim P$, a SFO returns the stochastic gradient $G(x_k, y_k) = \nabla_x f(x_k, y_k)$ where $y_k$ is a sample drawn i.i.d. from $P$ in the $k_{th}$ call.*

Non-convex optimization is required for many statistical learning tasks ranging from generalized linear models to deep neural networks [19,20]. Many earlier works have focused on the asymptotic performance of algorithms [6,14,32] and non-asymptotic complexity bounds have emerged [19]. To our knowledge, the first non-asymptotic convergence for stochastic gradient descent (SGD) was proposed by [15] with $\mathcal{O}(1/\epsilon^2)$. Full batch gradient descent (GD) is known to ensure convergence with $\mathcal{O}(n/\epsilon)$. Compared with SGD, GD's rate has better dependence on $\epsilon$ but worse dependence on $n$ due to the requirement of computing a full gradient. Variance reduced (VR) methods based on SGD, e.g. Stochastic Variance Reduced Gradient (SVRG) [17], SAGA [12] have been shown to achieve better dependence on $n$ than GD on non-convex problems with $\mathcal{O}(n + (n^{2/3}/\epsilon))$ [27,28]. However, compared with SGD, the rate of VR based methods still have worse dependence on $\epsilon$ unless $\epsilon \ll n^{-2/3}$. Recently, [19] proposed a method called SCSG combining the benefits of SGD and SVRG, which is the first algorithm that achieves a better rate than SGD and is no worse than SVRG with[1] $\mathcal{O}(1/\epsilon^{5/3} \wedge n^{2/3}/\epsilon)$. SNVRG proposed by [34] uses nested variance reduction to reduce the result of SCSG to[2] $\tilde{\mathcal{O}}((1/\epsilon^{3/2}) \wedge (n^{1/2}/\epsilon))$ that outperforms both SGD, GD and SVRG. Further SPIDER [13] proposes their both lower and upper bound as $\mathcal{O}(1/\epsilon^{3/2} \wedge n^{1/2}/\epsilon)$. Recently, [9] provide the lower bound of $\epsilon$-based convergence rate as $\mathcal{O}(1/\epsilon^{3/2})$ which is same with the $\epsilon$-related upper bound of SPIDER. As a result, the $\epsilon$-related convergence rate $\mathcal{O}(1/\epsilon^{3/2})$ is likely to be the best currently. To the best of our knowledge, SPIDER is a leading result of gradient complexity for smooth non-convex optimization by using averaged L-Lipschitz gradients. Their work motivates the research question about whether an algorithm based on SGD or VR-based methods can further reduce the rate of SPIDER when it depends on $\epsilon$ in the regime of modest target accuracy or depends on $n$ in the regime of high target accuracy, respectively.

However, for SGD and VR-based stochastic algorithms, there still exists three challenges. Firstly, stochastic based optimization algorithm do not require a full gradient computation. As a result, SCSG, SNVRG, SPIDER reduce full batch-size from $\mathcal{O}(n)$ to its subset $\mathcal{O}(B)$ where $1 \leq B < n$, which can significantly reduce the computational cost. However, it is challenging to appropriately scale the subset of samples in each stage of optimization to accelerate the convergence and achieve the same accuracy with full samples. Secondly, the variance of SGD is reduced by VR methods since the gradient of SGD is often too noisy to converge. However, VR schemes reduce the ability to escape local minima in later iterations due to a diminishing variance [8,28]. The challenge of SGD and VR methods is, therefore, to control the variance of gradients. ISVRG [8] proposed an approach that uses a hyper-parameter to control the reduced

[1] $(a \wedge b)$ means $\min(a, b)$.
[2] $\tilde{\mathcal{O}}(\cdot)$ hides the logarithmic factors.

variance of optimization, allowing it to have benefits between SGD and SVRG, which further being applied on the Sparse-SVRG method to address sparse data samples problems [7]. Such methods inspired us to address our challenges in this paper. Lastly, there exists a trade-off between biased/unbiased estimation in VR-based algorithms. SVRG is an unbiased estimation that can guarantee to converge but is not efficient to be used in real-world applications. Biased estimation can give a lower upper bound of the mean squared error (MSE) loss function [25], and many works have proposed asymptotically biased optimization with biased gradient estimators to converge to the vicinity of minima, which is an economical alternative to an unbiased version [10,11,16,33]. These methods provide a good insight into the biased gradient search. However, they hold under restrictive conditions, which are very hard to verify for complex stochastic gradient algorithms. Thus, the last challenge is how to balance the unbiased and biased estimator in different stages of the non-convex optimization process.

To address these three challenges in order to accelerate the convergence of non-convex optimization, we propose our method Variance Controlled Stochastic Gradient(VCSG) to control the reduced variance of gradients, scale the subset of full batch samples and choose the biased or unbiased estimator in each iteration. It is still an NP-hard for non-convex optimization that provably guarantee to estimate *approximate* global optimum by using bounding technique [2,4]. Alternatively, our algorithm can fast and *heuristic* turning a relatively good local optima into a global one. Table 1 compares the five methods' theoretical convergence rates, which shows that VCSG has the faster convergence rate than other methods. Here, we did not compare our result to SNVRG and SPIDER since both of their results are under averaged Lipschitz assumption, which is not same with our problem domain. We then show empirically that VCSG has faster rates of convergence than SGD, SVRG and SCSG.

**Table 1.** Comparison of results on SFO Definition 2 and IFO calls Definition 1 of gradient methods for smooth non-convex problems. The best upper bound of SFO in VCSG is still the lower bound that is proven by [9]. The upper bound of IFO in VCSG is better than other methods that use both full or subset of batch samples.

| Algorithms | SFO/IFO calls on Non-convex | Batch size $B$ | Learning rate $\eta$ |
|---|---|---|---|
| GD [23] | $\mathcal{O}(n/\epsilon)$ | $n$ | $\mathcal{O}(L^{-1})$ |
| SGD [15] | $\mathcal{O}(1/\epsilon^2)$ | $n$ | $\mathcal{O}(L^{-1})$ |
| SVRG [3,28] | $\mathcal{O}(n + (n^{2/3}/\epsilon))$ | $n$ | $\mathcal{O}(L^{-1}n^{-2/3})$ |
| SCSG [19] | $\mathcal{O}(1/\epsilon^{5/3} \wedge n^{2/3}/\epsilon)$ | $B(B < n)$ | $\mathcal{O}(L^{-1}(n^{-2/3} \wedge \epsilon^{4/3}))$ |
| VCSG | $\mathcal{O}(\mathbf{1}/\epsilon^{\mathbf{3/2}} \wedge \mathbf{n}^{\mathbf{1/4}}/\epsilon)$ | $B(B < n)$ | $\mathcal{O}(L^{-1} \wedge L^{-1}B^{-1/2})$ |

We summarize and list our main contributions:

- We provide an new method VCSG, a well-balanced VR method for SGD to achieve a competitive convergence rate. We provide a theoretical analysis of VCSG on non-convex problems, which might be the first analysis about controlled variance reduction that can achieve comparable or faster convergence than gradient-based optimization.

- VCSG provides an appropriate sample size in each iteration by the controlled variance reduction, which can significantly save computational cost.
- VCSG balances the trade-off in biased and unbiased estimation, which provides a fast convergence rate.
- We also evaluate VCSG on three different datasets with three deep learning models. It is shown that our method in practice can achieve better performance than other leading results.

## 2 Preliminaries

We use $\|\cdot\|$ to denote the Euclidean norm for brevity throughout the paper. For our analysis, the background that are required to introduce definitions for $L$-smooth and $\epsilon$-accuracy which now are defined in Definition 3 and Definition 4 respectively.

**Definition 3.** *Assume the individual functions $f_i$ in Eq. 1 are $\mathcal{L}$-smooth if there is a constant L such that*

$$\|\nabla f_i(x) - \nabla f_i(y)\| \leq L\|x - y\|, \forall x, y \in \mathbb{R}^d$$

*for some $L < \infty$ and for all $i \in \{1, ..., n\}$.*

We analyze convergence rates for Eq. 1 and apply $\|\nabla f\|^2 \leq \epsilon$ convergence criterion by [22], which the concept of *$\epsilon$-accurate* is defined in Definition 4. Moreover, the minimum IFO/SFO in Definition 1 and 2 to reach an $\epsilon-$accurate solution is denoted by $C_{comp}(\epsilon)$, and its complexity bound is denoted by $\mathbb{E}C_{comp}(\epsilon)$.

**Definition 4.** *A point $x$ is called $\epsilon$-accurate if $\|\nabla f(x)\|^2 \leq \epsilon$. An iterative stochastic algorithm can achieve $\epsilon$-accuracy within $t$ iterations if $\mathbb{E}[\|\nabla f(x^t)\|^2] \leq \epsilon$, where the expectation is over the algorithm.*

We follow part of the work in SCSG. Based on their algorithm settings, we recall that a random variable $N$ has a geometric distribution $N \sim Geom(\gamma)$ if $N$ is supported on the non-negative integrates, which their elementary calculation has been shown as $\mathbb{E}_{N\sim Geom}(\gamma) = \frac{\gamma}{1-\gamma}$. For brevity, we also write $\nabla f_{\mathcal{I}}(x) = \frac{1}{|\mathcal{I}|}\sum_{i\in\mathcal{I}} \nabla f_i(x)$. Note that calculating $\nabla f_{\mathcal{I}}(x)$ incurs $|\mathcal{I}|$ units of computational cost. The minimum IFO complexity to reach an *$\epsilon$-accurate* solution is denoted by $\mathcal{C}_{comp}(\epsilon)$.

To formulate our complexity bound, we define:

$$f^* = \inf_x f(x) \quad \text{and} \quad \triangle_f = f(\tilde{x}_0) - f^* > 0, \tag{2}$$

Further, an upper bound on the variance of the stochastic gradients can be defined as:

$$\mathcal{S}^* = \sup_x \frac{1}{n}\sum_{i=1}^{n} \|\nabla f_i(x) - \nabla f(x)\|^2. \tag{3}$$

## 3 Variance Controlled SVRG with a Combined Unbiased/Biased Estimation

To resolve the first challenge of SG-based optimization, we provide an adjustable schedule of batch size $B < n$, which scales the sample size for optimization. For the second challenge of controlling reduced variance, one method [8] balanced the gradient of SVRG in terms of the stochastic element and its variance to allow the algorithm to choose appropriate behaviors of gradient from stochastic, through reduced variance, to batch gradient descent by introducing a hyper-parameter $\lambda$. Based on this method, we focus on analysing the variance controller $\lambda$ in our case. Towards the last challenge associated with the trade-off between biased and unbiased estimators, we analyze the nature of biased and unbiased estimators in different stages of the non-convex optimization and propose a method that combines the benefits of both biased and unbiased estimator to achieve a fast convergence rate. Firstly, we show a generic form of the batched SVRG in Algorithm 1, which is proposed by [19]. Compared with the SVRG algorithm, the batched SVRG algorithm has a mini-batch procedure in the inner loop and outputs a random sample that instead of an average of the iterates. As seen in the pseudo-code, the batched SVRG method consists of multiple epochs, the batch-size $B_j$ is randomly chosen from the whole samples $n$ in $j$-th epoch and work with mini-batch $b_j$ to generate the total number of updates for inner $k$-th epoch by a geometric distribution with mean equal to the batch size. Finally it outputs a random sample from $\{\tilde{x}_j\}_{j=1}^T$. This is a standard way also proposed by [21], which can save additional overhead by calculating the minimum value of output as $arg\min_{j \le T \|\nabla f(\tilde{x}_j)\|}$.

**Algorithm 1:** Batching SVRG

**input** : Number of epochs $T$, initial iterate $\tilde{x}_0$, step-size $(\eta_j)_{j=1}^T$, batch size $(B_j)_{j=1}^T$, mini-batch sizes $(b_j)_{j=1}^T$.

1 **for** $j = 1$ **to** $T$ **do**
2 Uniformly sample a batch $\mathcal{I}_j \subset \{1, ..., n\}$ with $|\mathcal{I}_j| = B_j$;
3 $g_j \leftarrow \nabla f_{\mathcal{I}_j}(\tilde{x}_{j-1})$;
4 $x_0^{(j)} \leftarrow \tilde{x}_{j-1}$;
5 Generate $\mathcal{N}_j \sim \text{Geom}(B_j/(B_j + b_j))$;
6 **for** $k = 1$ **to** $\mathcal{N}_j$ **do**
7 Randomly select $\tilde{\mathcal{I}}_{k-1} \subset \{1, ..., n\}$ with $|\tilde{\mathcal{I}}_{k-1}| = b_j$;
8 $v_{k-1}^{(j)} \leftarrow \nabla f_{\tilde{\mathcal{I}}_{k-1}}(x_{k-1}^{(j)}) - \nabla f_{\tilde{\mathcal{I}}_{k-1}}(x_0^{(j)}) + g_j$;
9 $x_k^{(j)} \leftarrow x_{k-1}^{(j)} - \eta_j v_{k-1}^{(j)}$;
10 **end**
11 $\tilde{x}_j \leftarrow x_{\mathcal{N}_j}^{(j)}$;
12 **end**

**output**: Sample $\tilde{x}_T^*$ from $\{\tilde{x}_j\}_{j=1}^T$ with $P(\tilde{x}_T^* = \tilde{x}_j) \propto \eta_j B_j / b_j$.

For the two cases of unbiased and biased estimations for the batched SVRG, in the following two sub-sections, we provide two great upper bounds on their convergence for their gradients with two corresponding lower bounds of batch size for each case when their dependency is sample size $n$. Unlike the specific parameter settings in SCSG, we use more general schedules (including learning rate $\eta_j$ and mini-batch size $b_j$), aiming to estimate the best schedules in each stage of optimization for both unbiased and biased estimators, which avoids ad hoc choosing parameters. Proof details are presented in the appendix.

### 3.1 Weighted Unbiased Estimator Analysis

In the first case, we introduce a hyper-parameter $\lambda$ that is applied in a weighted unbiased version of the batched SVRG and is shown in Algorithm 2. Since our method based on SVRG, the $\lambda$ should be within the range $0 < \lambda < 1$ in unbiased and biased cases.

**Algorithm 2:** Batching SVRG with weighted unbiased estimator

1 Replace line number 8 in Alg. 1 with the following line:
$v_{k-1}^{(j)} \leftarrow (1-\lambda)\nabla f_{\tilde{\mathcal{I}}_{k-1}}(x_{k-1}^{(j)}) - \lambda\left(\nabla f_{\tilde{\mathcal{I}}_{k-1}}(x_0^{(j)}) - g_j\right)$;

We now analyse the upper bound of expectation of gradients in a single epoch. Under our settings, we can achieve the upper bound for one-epoch analysis which is shown in Theorem 2.

**Theorem 1.** *Let* $\eta_j L = \gamma(\frac{b_j}{B_j})^\alpha$ $(0 \le \alpha \le 1)$ *and* $B_j \ge b_j \ge B_j^\beta$ $(0 \le \beta \le 1)$ *for all* $j$. *Suppose* $0 < \gamma \le \frac{1}{3}$, *then under Definition 3, the output* $\tilde{x}_j$ *of Algorithm 2 we have*

$$\mathbb{E}\|\nabla f(\tilde{x}_j)\|^2 \le \frac{2L}{\gamma\theta}\cdot(\frac{b_j}{B_j})^{1-\alpha}\mathbb{E}\left(f(\tilde{x}_{j-1}) - f(\tilde{x}_j)\right) + \frac{2\lambda^4 I(B_j < n)\mathcal{S}^*}{\theta B_j^{1-2\alpha}},$$

*where* $I(B_j < n) \ge \frac{n - B_j}{(n-1)B_j}$, $\mathcal{S}$ *is defined in Eq. 3,* $\lambda = \frac{1}{2}$ *and* $\theta = 2(1-\lambda) - (2\gamma B_j^{\alpha\beta-\alpha} + 2B_j^{\beta-1})(1-\lambda)^2 - 1.29(1-\lambda)^2$.

Over all epochs $T$, the output $\tilde{x}_T^*$ that is randomly selected from $(\tilde{x}_j)_{j=1}^T$ should be non-convex and $L$-smooth. Thus, Theorem1 can be telescoped for over all epochs in the following theorem.

**Theorem 2.** *Under all assumptions of Theorem 1,*

$$\mathbb{E}\|\nabla f(\tilde{x}_j)\|^2 \le \frac{(\frac{2L}{\gamma})\triangle_f}{\theta\sum_{j=1}^T b_j^{\alpha-1}B_j^{1-\alpha}} + \frac{2\lambda^4 I(B_j < n)\mathcal{S}^*}{\theta\sum_{j=1}^T B_j^{1-2\alpha}},$$

*where* $\nabla f$ *is defined in Eq. 2.*

### 3.2 Biased Estimator Analysis

In this sub-section we theoretically analyze the performance of the biased estimator, which is shown in Algorithm 3.

**Algorithm 3:** Batching SVRG with biased estimator

1 Replace the line number 8 in Alg 1 with the following line:
$v_{k-1}^{(j)} \leftarrow (1-\lambda)\left(\nabla f_{\tilde{\mathcal{I}}_{k-1}}(x_{k-1}^{(j)}) - \nabla f_{\tilde{\mathcal{I}}_{k-1}}(x_0^{(j)})\right) + \lambda g_j$;

Applying the same schedule of $\eta_j$ and $b_j$ that are used in the unbiased case, we can achieve the results on both one-epoch and all-epoch for this case, which are shown in Theorem 3 and Theorem 4 respectively.

**Theorem 3.** *let* $\eta_j L = \gamma(\frac{b_j}{B_j})^\alpha$ $(0 \le \alpha \le 1)$. *Suppose* $0 < \gamma \le \frac{1}{3}$ *and* $B_j \ge b_j \ge B_j^\beta$ $(0 \le \beta \le 1)$ *for all* $j$, *then under Definition 3, the output* $\tilde{x}_j$ *of Algorithm 3 we have,*

$$\mathbb{E}\|\nabla f(\tilde{x}_j)\|^2 \le \frac{2L}{\gamma\Theta}\cdot(\frac{b_j}{B_j})^{1-\alpha}\mathbb{E}\left(f(\tilde{x}_{j-1}) - f(\tilde{x}_j)\right) + \frac{2(1-\lambda)^2 I(B_j < n)\mathcal{S}^*}{\Theta B_j^{1-2\alpha}},$$

*where* $I(B_j < n) \ge \frac{n-B_j}{(n-1)B_j}$, $\mathcal{S}$ *is defined in Eq. 3,* $0 < \lambda < 1$ *and* $\Theta = 2(1-\lambda) - (2\gamma B_j^{\alpha\beta-\alpha} + 2B_j^{\beta-1} - 4LB_j^{2\alpha-2})(1-\lambda)^2 - 1.16(1-\lambda)^2$.

**Theorem 4.** *Under all assumptions of Theorem 3,*

$$\mathbb{E}\|\nabla f(\tilde{x}_j)\|^2 \le \frac{(\frac{2L}{\gamma})\triangle_f}{\Theta\sum_{j=1}^T b_j^{\alpha-1}B_j^{1-\alpha}} + \frac{2(1-\lambda)^2 I(B_j < n)\mathcal{S}^*}{\Theta\sum_{j=1}^T B_j^{1-2\alpha}},$$

*where* $I(B_j < n) \ge \frac{n-B_j}{(n-1)B_j}$, $0 < \lambda < 1$ *and* $\Theta = 2(1-\lambda) - (2\gamma B_j^{\alpha\beta-\alpha} + 2B_j^{\beta-1} - 4LB_j^{2\alpha-2})(1-\lambda)^2 - 1.16(1-\lambda)^2$.

### 3.3 Convergence Analysis for Smooth Non-convex Optimization

Starting to consider from a constant batch/mini-batch size $1 \le B_j \equiv B \le n$ for some $1 < B \le n$, $b_j = B_j^\beta \equiv B^\beta$ $(0 \le \beta \le 1)$, we can achieve the computational complexity of output from Theorem 2 and 4 that is given as

$$\mathbb{E}\|\nabla f(\tilde{x}_T^*)\|^2 = \mathcal{O}\left(\frac{L\triangle_f}{TB^{1+\alpha\beta-\alpha-\beta}} + \frac{\mathcal{S}^*}{B^{1-2\alpha}}\right), \tag{4}$$

which covers two extreme cases of complexity bounds since the batch-size $B_j$ has two different dependencies.

**Dependence on** $\epsilon$. If $b_j = B^\beta$ when $\beta = 1$ and $1 < B_j \equiv B < n$, the second term of Eq. 4 can be made $\mathcal{O}(\epsilon)$ by setting $B_j^{1-2\alpha} = B = \mathcal{O}\left(\frac{\mathcal{S}^*}{\epsilon}\right)$, where incurs

$\alpha = 0$. And $T(\epsilon) = \left(\frac{L\triangle_f}{\epsilon}\right)$ resulting in the complexity bound is given as $\mathbb{E}C_{comp}(\epsilon) = \mathcal{O}\left(\frac{L\triangle_f B}{\epsilon}\right) = \mathcal{O}\left(\frac{L\triangle_f \mathcal{S}^*}{\epsilon^2}\right)$, which obtains the same with the rate of SGD as shown in Table 1.

**Dependence on n.** If $b_j = 1$ when $\beta = 0$ and $B_j = n$, Eq. 4 can be further alternative as $\mathbb{E}\|\nabla f(\tilde{x}_T^*)\|^2 = \mathcal{O}\left(\frac{L\triangle_f}{Tn^{1-\alpha}} + \frac{\mathcal{S}^*}{n^{1-2\alpha}}\right)$. When $\alpha \leq \frac{1}{2}$, $T(\epsilon)$ can be made as $\mathcal{O}\left(1 + \frac{L\triangle_f}{\epsilon n^{1/2}}\right)$, which yields the complexity bound become as

$$\mathbb{E}C_{comp}(\epsilon) = \mathcal{O}\left(n + \frac{n^{\frac{1}{2}} L\triangle_f}{\epsilon}\right). \tag{5}$$

This upper bound of rate can guarantee to be better than SCSG, as shown in Table 1.

However, both of the above settings are two sub-optimal cases since their extreme setting either the parameter mini-batch size $b_j$ is too large or batch size $B_j$ is too large. We now discuss the batch-size schedules depending on the above two dependencies.

### 3.4 Scaling Batch Samples

**For the case of batch size $B_j$ depending on** $\epsilon$, $B_j = \mathcal{O}\left(\frac{\mathcal{S}^*}{\epsilon}\right)$, $b_j \neq 1$, and learning rate $\eta_j = \frac{\gamma}{L}(\frac{1}{B_j})^{\alpha(1-\beta)}$ where $0 \leq \alpha \leq \frac{1}{2}$. To determine the optimal value of $b_j$ in this case, we compared to the extreme case when $b_j = 1$ and $B_j = n$ that the optimal schedule of learning rate $\eta_j = \frac{\gamma}{L}(\frac{1}{B_j})^{\frac{2}{3}}$ is provided by [3,19,27,28]. Correspondingly in our general form of learning rate, they specified $\alpha = \frac{2}{3}$ and $\beta = 0$. Thus, the learning rate $\eta_j$ has a range which is shown as $\frac{\gamma}{L} \geq \frac{\gamma}{L}(\frac{1}{B_j})^{\frac{2}{3}(1-\beta)} \geq \frac{\gamma}{L}(\frac{1}{B_j})^{\frac{1}{2}}$. As a result, we can estimate the range of $\beta$ as $0 \leq \beta \leq 1/4$. Consequently, $\beta = 1/4$ and $\alpha = 0$ are the optimal values in this case.

After determined the three schedules including $B_j$, $\eta_j$ and $b_j$, we can estimate the optimal value of $\lambda^*$. For the first case that $B_j = \mathcal{O}\left(\frac{\mathcal{S}^*}{\epsilon}\right)$, $b_j = B_j^{\frac{1}{4}}$, $\eta_j = \frac{1}{3L}$, Eq. 4 is specified as

$$\mathbb{E}\|\nabla f(\tilde{x}_T^*)\|^2 = \mathcal{O}\left(\frac{L\triangle_f}{T}\left(\frac{\epsilon}{\mathcal{S}^*}\right)^{\frac{3}{4}} + \epsilon\right). \tag{6}$$

Since in this case batch size depends on $\epsilon$, we more focus on the second term in Eq. 6. As a result, we optimize the second term of $\mathbb{E}\|\nabla f(\tilde{x}_T^*)\|^2$ from both Theorem 2 and 4 in order to achieve lowest upper bound. After comparison the upper bounds in both Theorem 2 and 4, we choose the optimal value of $\lambda^* = \frac{1}{2}$ with the unbiased estimation case, which can provide the lowest upper bound of gradient resulting faster convergence.

**For the case of batch size $B_j$ depending on** $n$, we now analyse the lower bound of batch size $B_j$ in both unbiased and biased estimations. When applying unbiased estimator, for a single epoch, $j$, we define the weighted unbiased variance as $e_j = \lambda\left(\nabla f_{\mathcal{I}_j}(\tilde{x}_{j-1}) - \nabla f(\tilde{x}_{j-1})\right)$. Thus, the gradients in Algorithm 2 can be updated within the $j$-th epoch as $\mathbb{E}_{\tilde{\mathcal{I}}_k} v_k^{(j)} = (1-\lambda)\nabla f(x_k^{(j)}) + e_j$, which reveals the key difference between the batched SVRG and the variance controlled batched SVRG on both unbiased/ biased estimators. Most of the novelty in our analysis lies in dealing with the extra term $e_j$. Since we achieve a lower bound of batch-size by bounding the term $e_j$, we provide the bound of the term $e_j$ as $\mathbb{E}_{\mathcal{I}_j}\|e_j\|^2 \leq \lambda^2 \frac{n-B_j}{nB_j}\mathcal{K}^2 \leq \lambda^2 \frac{n-B_j}{nB_j}\frac{n}{\sqrt{n-1}}\mathcal{S}^* \leq \sigma\rho^{2j}$, where the first inequation follows [5,18] the variance of the norms of gradients $\mathcal{K}^2 \geq \frac{1}{n-1}\sum_{i=1}^{n}[\|\nabla f_i(\tilde{x}_{j-1})\|^2 - \|\nabla f(\tilde{x}_{j-1})\|^2]$, the second inequation follows the Samuelson inequality [24] that $\mathcal{K}^2 \leq \frac{n}{\sqrt{n-1}}\mathcal{S}^*$ where $\mathcal{S}^*$ is shown in Eq. 3, and in the last inequation, there is an upper bound of variance where $\sigma \geq 0$ is a constant for some $\rho < 1$. Thus $B_j$ in unbiased case can be bounded as,

$$B_j \geq \frac{n\mathcal{S}^*}{\mathcal{S}^* + \lambda^2 n^{\frac{1}{2}}\sigma\rho^{2j}}. \tag{7}$$

For batch size in biased case, we use the same approach adopted in the unbiased version. For a single epoch, $j$, we define the biased variance as $e_j = \lambda\nabla f_{\mathcal{I}_j}(\tilde{x}_{j-1}) - (1-\lambda)\nabla f(\tilde{x}_{j-1})$. And we achieve the lower bound of batch-size, which is shown in the following.

$$B_j \geq \begin{cases} \dfrac{n\mathcal{S}^*}{\mathcal{S}^* + (1-\lambda)^2 n^{\frac{1}{2}}\sigma\rho^{2j}}, & \text{if } 0 < \lambda < \dfrac{\sqrt{2}}{2}. \\ \dfrac{n\mathcal{S}^*}{\mathcal{S}^* + (3\lambda^2 - 2\lambda)^2 n^{\frac{1}{2}}\sigma\rho^{2j}}, & \text{if } \dfrac{\sqrt{2}}{2} < \lambda < 1. \end{cases} \tag{8}$$

To estimate the optimal value of $\lambda^*$ in this case that batch size depending on $n$, we specified lower bound of batch size $B_j$ which has two versions of biased and unbiased estimations, $b_j = 1$ and $\eta_j = \frac{1}{3L}(\frac{1}{B_j})^{\frac{1}{2}}$ when optimal value $\alpha = \frac{1}{2}$. Thus Eq. 4 can be specified as

$$\mathbb{E}\|\nabla f(\tilde{x}_T^*)\|^2 = \mathcal{O}\left(\frac{L\triangle_f}{TB_j^{\frac{1}{2}}} + \mathcal{S}^*\right). \tag{9}$$

Due to this case that batch size depending on $n$, we more focus on the first term in Eq. 9. Thus we optimise the first term in the upper bound of $\mathbb{E}\|\nabla f(\tilde{x}_T^*)\|^2$ in both Theorem 2 and 4. After comparison of upper bounds both in unbiased and biased cases, we determine $\lambda^* = 5/8$ with biased estimation that obtain the lowest upper bound.

Consequently, we can achieve the greater complexity bound of Eq. 4 for both biased/unbiased estimations via replacing full sample size $n$ by the batched sample size $B_j$ in Eq. 5, which is shown in Eq. 10.

$$\mathbb{E}C_{comp}(\epsilon) = \mathcal{O}\left(B + \frac{B^{\frac{1}{2}}L\triangle_f}{\epsilon}\right) \tag{10}$$

.

### 3.5 Best of Two Worlds

We have seen in the previous section that the variance controlled SVRG combines the benefits of both SVRG and SGD. We now show these benefits can be made more pronounced by $\lambda^*$ with best combinations between $B_j$ and $b_j$ in different stages of optimization. We introduce our algorithm *VCSG* shown in Algorithm 4.

---

**Algorithm 4:** (Mini-Batch)VCSG

---

**input** : Same input parameters with Alg 1, initial batch size $B_0 = n$ and $b_j = n^{1/4}$.
1 **for** $j = 1$ **to** $T$ **do**
2 Uniformly sample a batch $\mathcal{I}_j \subset \{1, ..., n\}$ with $|\mathcal{I}_j| = B_j$;
3 $B_j \leftarrow \left\{ \frac{12\mathcal{S}_j^*}{\epsilon} \wedge \frac{n\mathcal{S}_j^*}{\mathcal{S}_j^* + 0.14 \cdot n^{\frac{1}{2}} \sigma \rho^{2j}} \right\}$ where $\sigma \geq 0, \rho < 1$;
4 $g_j \leftarrow \nabla f_{\mathcal{I}_j}(\tilde{x}_{j-1})$;
5 $\tilde{x}_0^{(j)} \leftarrow \tilde{x}_{j-1}$;
6 Generate $\mathcal{N}_j \sim \text{Geom}(B_j/(B_j + b_j))$;
7 **for** $k = 1$**to** $\mathcal{N}_j$ **do**
8 **if** $B_j = \mathcal{S}_j^*/\epsilon$ **then**
9 $b_j = B_j^{\frac{1}{4}}; \eta_j = \frac{1}{3L}$;
10 Randomly select $\mathcal{I}_{k-1} \subset \{1, ..., n\}$ with $|\tilde{\mathcal{I}}_{k-1}| = b_j$;
11 $v_{k-1}^{(j)} = \frac{1}{2} \cdot \left( \nabla f_{\tilde{\mathcal{I}}_{k-1}}(x_{k-1}^{(j)}) - \nabla f_{\tilde{\mathcal{I}}_{k-1}}(x_0^{(j)}) + g_j \right)$;
12 **else if** $B_j = B_0$ *or* $B_j = \frac{n\mathcal{S}_j^*}{\mathcal{S}_j^* + 0.14 \cdot n^{\frac{1}{2}} \sigma \rho^{2j}}$ **then**
13 $b_j = 1; \eta_j = \frac{1}{3L}(\frac{1}{B_j})^{\frac{1}{2}}$;
14 Randomly select $\mathcal{I}_{k-1} \subset \{1, ..., n\}$ with $|\tilde{\mathcal{I}}_{k-1}| = b_j$;
15 $v_{k-1}^{(j)} = \frac{3}{8} \cdot (\nabla f_{\tilde{\mathcal{I}}_{k-1}}(x_{k-1}^{(j)}) - \nabla f_{\tilde{\mathcal{I}}_{k-1}}(x_0^{(j)})) + \frac{5}{8} \cdot g_j$;
16 $x_k^{(j)} \leftarrow x_{k-1}^{(j)} - \eta_j v_{k-1}^{(j)}$;
17 $\mathcal{S}_k^{(j)} \leftarrow \| \nabla f_{\tilde{\mathcal{I}}_{k-1}}(\tilde{x}_0^{(j)}) - g_j \|^2$;
18 **end**
19 $\tilde{x}_j \leftarrow x_{\mathcal{N}_j}^{(j)}, \mathcal{S}_j^* \leftarrow \mathcal{S}_{\mathcal{N}_j}^{(j)}$;
20 **end**
**output**: Sample $\tilde{x}_T^*$ from $(\tilde{x}_j)_{j=1}^T$ with $P(\tilde{x}_T^* = \tilde{x}_j) \propto \eta_j B_j / b_j$

---

Following Algorithm 4, we can achieve a general result for VCSG in the following theorem.

**Theorem 5.** *Suppose* $\gamma \leq \frac{1}{3}$. *Let* $B_j = \min\left\{ \frac{\mathcal{S}^*}{\epsilon}, \frac{n\mathcal{S}^*}{\mathcal{S}^* + 0.14 \cdot n^{\frac{1}{2}} \sigma \rho^{2j}} \right\}$, *under Definition 3 and Theorem 2 and 4, the output* $\tilde{x}_T^*$ *in Algorithm 4 satisfies one of two bounds.*

1. *If* $B_j = \frac{\mathcal{S}^*}{\epsilon}$, $b_j = B_j^{\frac{1}{4}}$, $\eta_j = \frac{\gamma}{L}$, $\lambda^* = \frac{1}{2}$, $\theta \approx 0.51$ *with an unbiased estimator,*

$$\mathbb{E}\|\nabla f(\tilde{x}_T^*)\|^2 \leq \frac{\frac{4L}{\gamma}\triangle_f}{\sum_{j=1}^T B_j^{\frac{3}{4}}} + \frac{0.24(I(B_j < n)\mathcal{S}^*}{B_j},$$

2. *If* $B_j = \frac{n\mathcal{S}^*}{\mathcal{S}^* + 0.14 \cdot n^{\frac{1}{2}}\sigma\rho^{2j}}$, $b_j = 1$, $\eta_j = \frac{\gamma}{L}(\frac{1}{B_j})^{\frac{1}{2}}$, $\lambda^* = \frac{5}{8}$, $\Theta \approx 0.59$ *with a biased estimator,*

$$\mathbb{E}\|\nabla f(\tilde{x}_T^*)\|^2 < \frac{\frac{3.4L}{\gamma}\triangle_f}{\sum_{j=1}^T B_j^{\frac{1}{2}}} + 0.48\mathcal{S}^*.$$

Now we discuss how parameters, including $\lambda$, step-size, batch-size, and mini-batch size, work together to control the variance of gradients from stochastic to batch and balance the trade-off between bias/unbiased estimation in batched optimization. Firstly, in very early iterations $B_j$ might choose its first term due to the low variance. In this condition, the small $\lambda$ with relatively large learning rate may help gradients being more stochastic to search more region of problem space, and also can help points escape from bad local minima. During increasing variance, the first term of $B_j$ would be increased as well, resulting $B_j$ will choose its second term. In the second case, both relatively large $\lambda$, small learning rate and the biased estimator work together that can reduce variance to fast converge into a small region of space. In case of the variance that is reduced too small in the second case, $B_j$ will turn to be its first term. We regard this whole process as **Coarse-to-Fine** dynamic searching methods.

To calculate the computational complexity of VCSG, we bring the schedule of batch size $B_j$ into Eq. 10, which is shown in Corollary 1.

**Corollary 1.** *Under parameters setting in Theorem 5,* $B_j \equiv B = \{\frac{\mathcal{S}^*}{\epsilon} \wedge \frac{n\mathcal{S}^*}{\mathcal{S}^*+0.14\cdot n^{(1/2)}\sigma\rho^{2j}}\}$ *then it holds that*

$$\mathbb{E}_{comp}(\epsilon) = \mathcal{O}\left(B + \frac{L\triangle_f}{\epsilon} \cdot B^{\frac{1}{2}}\right).$$

$B = \{\frac{1}{\epsilon} \wedge n^{\frac{1}{2}}\}$ *since assume that* $L\triangle_f, \mathcal{S}^*, \sigma\rho^{2j} = \mathcal{O}(1)$. *Thus, the above bound can be simplified to*

$$\mathbb{E}_{comp}(\epsilon) = \mathcal{O}\left((\frac{1}{\epsilon} \wedge n^{\frac{1}{2}}) + \frac{1}{\epsilon} \cdot (\frac{1}{\epsilon} \wedge n^{\frac{1}{2}})^{\frac{1}{2}}\right) = \mathcal{O}\left(\frac{1}{\epsilon^{\frac{3}{2}}} \wedge \frac{n^{\frac{1}{4}}}{\epsilon}\right).$$

## 4 Application

To experimentally verify our theoretical results and insights, we evaluate VCSG compared with SVRG, SGD, and SCSG on three common DL topologies, including LeNet

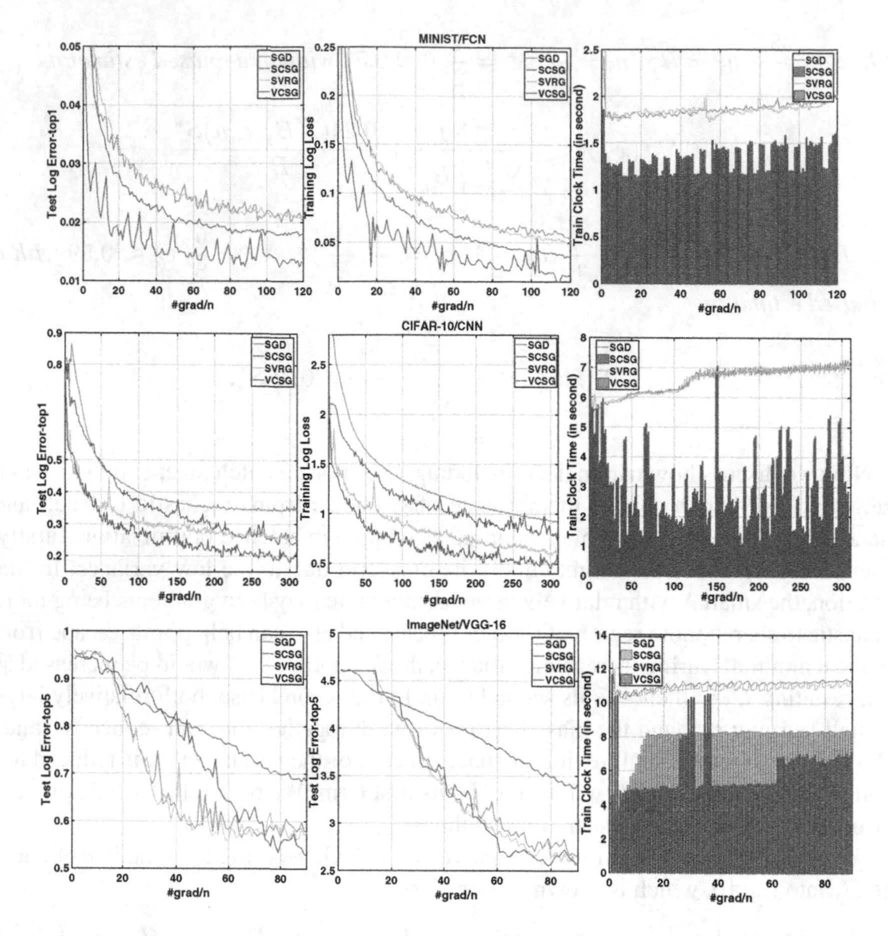


**Fig. 1.** Comparison of rates of convergence in four approaches, including SGD, SVRG, SCSG, and VCSG via test error, training loss and time consumption. Comparatively, we can see that VCSG can converge fastest during all iterations on MINIST and CIFAR-10 data sets. Even though VCSG on the ImageNet data set is slightly slower converging than the other three methods in the beginning, it can significantly decrease after several epochs when the batch-size becomes stable.

(LeNet-300-100 which has two fully connected layers as hidden layers with 300 and 100 neurons respectively, and LeNet-5 which has two convolutional layers and two fully connected layers) and VGG-16 [30] using three datasets including MNIST, CIFAR-10 and tiny ImageNet. Tiny ImageNet contains 200 classes for training each with 500 images and the test set contains 10,000 images. Each image is re-sized to 64 × 64 pixels [29]. We initialize $B_j = B_0 = n$, correspondingly $b_j = b_0 = n^{\frac{1}{4}}$, $\eta_0 = 1/(3Ln^{\frac{1}{2}})$, and $\lambda = \frac{5}{8}$ via using biased estimator in the first epoch. Meanwhile, we choose a scaled SGD as our baseline by multiplying $0.5$ with stochastic gradients, and applied decayed learning rate $\eta_j = \eta_0/(j)$ on SGD. For SVRG, we set-up $\lambda = 0.5$ with fixed learning rate $\eta_j = 1/(3Ln^{\frac{1}{2}})$ in Algorithm 1. The reason we choose SCSG is that our algorithm is inspired from SCSG which is a leading batched SVRG.

Figure 1 compares the performance of four methods, including SGD, SVRG, SCSG, and VCSG, via test log error, training log loss, and training time usage. It has two baselines in all sub-figures, including the performance of SVRG and SGD. The performance of SCSG test error and training loss is smaller than SGD on MNIST and CIFAR-10 data sets, consistent with the experimental results shown in [19]. However, in the ImageNet data set, which is a relatively larger scale application than the previous two data sets, the performance of SCSG becomes worse than SVRG and SGD, which showed weak robustness in our experiments. By contrast, VCSG shown as the green colour in all three datasets, has the lowest test error and training loss among all methods. In the ImageNet data set, both the test error of VCSG is initially higher than SVRG and SGD, but VCSG can reduce the test error and loss dramatically after around 75 epochs. One possible explanation is that the algorithm changes the batch size to the first term resulting in an escape from a local minima by increasing the variance to find a better solution. The right-hand column of Fig. 1 presents the time usage, and it can be seen that SVRG and SGD are similar, having higher training time than the other two methods in all three data sets. In Fig. 2, we use a more visualized format to show the time usage in Fig. 1. We can see in three sub-figures VCSG can achieve the lowest test error over a shorter time. To achieve the 0.025 top-1 test error in the MNIST data set, VCSG only takes 16 s around 2× faster than SCSG, 3× faster than SVRG, and 4× faster than SGD. In CIFAR-10 to achieve 0.3 top-1 test error, VCSG is around 6× faster than SVRG, 4× faster than SCSG and 13× faster than SGD. In the ImageNet data set, to achieve 0.55 top-5 test error, VCSG can be faster than other methods by up to 5×.

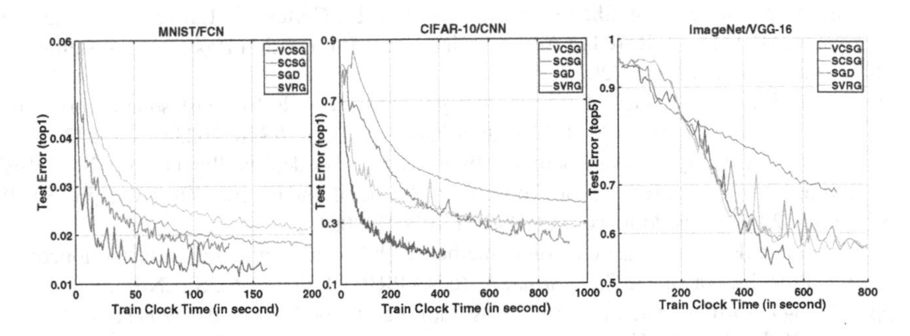


**Fig. 2.** Visualization of test error of four approaches, including SGD, SVRG and SCSG and VCSG against time consumption.

## 5 Discussion

In this paper, we proposed a VR-based optimization $VCSG$ for non-convex problems. We theoretically determined that a hyper-parameter $\lambda$ in each iteration can control the reduced variance of SVRG and balance the trade-off between a biased and an unbiased estimator. Meanwhile, an adjustable batch bounded by controlled reduced variance can work with $\lambda$, step size, and mini-batch to choose an appropriate estimator to converge faster to a stationary point on non-convex problems. Moreover, to verify our theoretical results, our experiments use three datasets on three DL models to present the performance of VCSG via test error/loss and elapsed time and compare these with other leading results. Both theoretical and experimental results show that VCSG can efficiently accelerate convergence. We believe that our algorithm is worthy of further study for non-convex optimization, particularly in deep neural networks training in large-scale applications.

## References

1. Agarwal, A., Bottou, L.: A lower bound for the optimization of finite sums. In: Proceedings of the 32nd International Conference on Machine Learning, ICML 2015, Lille, France, 6–11 July 2015, pp. 78–86. JMLR Workshop and Conference Proceedings, France (2015). http://leon.bottou.org/papers/agarwal-bottou-2015
2. Agarwal, N., Allen-Zhu, Z., Bullins, B., Hazan, E., Ma, T.: Finding approximate local minima faster than gradient descent. In: STOC 2017 - Proceedings of the 49th Annual ACM SIGACT Symposium on Theory of Computing, pp. 1195–1199. Association for Computing Machinery, June 2017. https://doi.org/10.1145/3055399.3055464
3. Allen-Zhu, Z., Hazan, E.: Variance reduction for faster non-convex optimization. In: Balcan, M., Weinberger, K. (eds.) 33rd International Conference on Machine Learning, ICML 2016, pp. 1093–1101. International Machine Learning Society (IMLS), January 2016
4. Allen-Zhu, Z.: Natasha 2: faster non-convex optimization than SGD. In: Bengio, S., Wallach, H., Larochelle, H., Grauman, K., Cesa-Bianchi, N., Garnett, R. (eds.) Advances in Neural Information Processing Systems, vol. 31, pp. 2675–2686. Curran Associates, Inc. (2018)
5. Babanezhad Harikandeh, R., Ahmed, M.O., Virani, A., Schmidt, M., Konečný, J., Sallinen, S.: Stopwasting my gradients: practical SVRG. In: Cortes, C., Lawrence, N., Lee, D., Sugiyama, M., Garnett, R. (eds.) Advances in Neural Information Processing Systems, vol. 28. Curran Associates, Inc. (2015)
6. Bertsekas, D.: A new class of incremental gradient methods for least squares problems. SIAM J. Optim. **7**(4), 913–926 (1997). https://doi.org/10.1137/S1052623495287022
7. Bi, J., Gunn, S.R.: Sparse deep neural networks for embedded intelligence. In: 2018 IEEE 30th International Conference on Tools with Artificial Intelligence (ICTAI), pp. 30–38, November 2018. https://doi.org/10.1109/ICTAI.2018.00016
8. Bi, J., Gunn, S.R.: A stochastic gradient method with biased estimation for faster nonconvex optimization. In: Nayak, A.C., Sharma, A. (eds.) PRICAI 2019. LNCS (LNAI), vol. 11671, pp. 337–349. Springer, Cham (2019). https://doi.org/10.1007/978-3-030-29911-8_26
9. Carmon, Y., Duchi, J.C., Hinder, O., Sidford, A.: Lower bounds for finding stationary points I. Math. Program. **184**(1), 71–120 (2020). https://doi.org/10.1007/s10107-019-01406-y
10. Chen, H., Gao, A.: Robustness analysis for stochastic approximation algorithms. Stoch. Stoch. Rep. **26**(1), 3–20 (1989). https://doi.org/10.1080/17442508908833545
11. Chen, J., Ma, T., Xiao, C.: FastGCN: fast learning with graph convolutional networks via importance sampling. In: International Conference on Learning Representations (2018)
12. Defazio, A., Bach, F., Lacoste-Julien, S.: SAGA: a fast incremental gradient method with support for non-strongly convex composite objectives. In: Proceedings of the 27th International Conference on Neural Information Processing Systems - Volume 1, NIPS 2014, pp. 1646–1654. MIT Press, Cambridge (2014)
13. Fang, C., Li, C.J., Lin, Z., Zhang, T.: SPIDER: near-optimal non-convex optimization via stochastic path-integrated differential estimator. In: Bengio, S., Wallach, H., Larochelle, H., Grauman, K., Cesa-Bianchi, N., Garnett, R. (eds.) Advances in Neural Information Processing Systems 31, pp. 689–699. Curran Associates, Inc. (2018)

14. Gaivoronski, A.A.: Convergence properties of backpropagation for neural nets via theory of stochastic gradient methods. Part 1. Optim. Methods Softw. **4**(2), 117–134 (1994). https://doi.org/10.1080/10556789408805582
15. Ghadimi, S., Lan, G.: Accelerated gradient methods for nonconvex nonlinear and stochastic programming. Math. Program. **156**(1–2), 59–99 (2016)
16. HanFu Chen, L.G., Gao, A.: Convergence and robustness of the Robbins-Monro algorithm truncated at randomly varying bounds. Stoch. Process. Appl. **27**, 217–231 (1987). https://doi.org/10.1016/0304-4149(87)90039-1
17. Johnson, R., Zhang, T.: Accelerating stochastic gradient descent using predictive variance reduction. In: Burges, C.J.C., Bottou, L., Welling, M., Ghahramani, Z., Weinberger, K.Q. (eds.) Advances in Neural Information Processing Systems 26, pp. 315–323. Curran Associates, Inc. (2013)
18. Lohr, S.L.: Sampling: design and analysis. Technometrics **42**(2), 223–224 (2000). https://doi.org/10.2307/1271491
19. Lei, L., Ju, C., Chen, J., Jordan, M.I.: Non-convex finite-sum optimization via SCSG methods. In: Guyon, I., et al (eds.) Advances in Neural Information Processing Systems 30, pp. 2348–2358. Curran Associates, Inc. (2017)
20. McCullagh, P., Nelder, J.A.: Generalized Linear Models. Chapman & Hall/CRC, London (1989)
21. Nemirovski, A., Juditsky, A.B., Lan, G., Shapiro, A.: Robust stochastic approximation approach to stochastic programming. SIAM J. Optim. **19**(4), 1574–1609 (2009)
22. Nesterov, Y.: Introductory Lectures on Convex Optimization: A Basic Course. Springer, New York (2003). https://doi.org/10.1007/978-1-4419-8853-9
23. Nesterov, Y.: Introductory Lectures on Convex Optimization: A Basic Course, 1st edn. Springer, New York (2014)
24. Niezgoda, M.: Laguerre-Samuelson type inequalities. Linear Algebra Appl. **422**(2), 574–581 (2007). https://doi.org/10.1016/j.laa.2006.11.016
25. Percy Liang, F.R.B., Bouchard, G., Jordan, M.I.: Asymptotically optimal regularization in smooth parametric models. In: Advances in Neural Information Processing Systems 22: 23rd Annual Conference on Neural Information Processing Systems 2009. Proceedings of a meeting held 7–10 December 2009, Vancouver, British Columbia, Canada, pp. 1132–1140 (2009)
26. Qu, C., Li, Y., Xu, H.: Non-convex conditional gradient sliding. In: Dy, J., Krause, A. (eds.) Proceedings of the 35th International Conference on Machine Learning. Proceedings of Machine Learning Research, vol. 80, pp. 4208–4217. PMLR, Stockholmsmässan, Stockholm Sweden, 10–15 July 2018. http://proceedings.mlr.press/v80/qu18a.html
27. Reddi, S.J., Sra, S., Póczos, B., Smola, A.: Fast incremental method for smooth nonconvex optimization. In: 2016 IEEE 55th Conference on Decision and Control (CDC), pp. 1971–1977, December 2016. https://doi.org/10.1109/CDC.2016.7798553
28. Reddi, S.J., Hefny, A., Sra, S., Poczos, B., Smola, A.: Stochastic variance reduction for nonconvex optimization. In: Balcan, M.F., Weinberger, K.Q. (eds.) Proceedings of The 33rd International Conference on Machine Learning. Proceedings of Machine Learning Research, vol. 48, pp. 314–323. PMLR, New York, 20–22 June 2016. http://proceedings.mlr.press/v48/reddi16.html
29. Russakovsky, O., et al.: ImageNet large scale visual recognition challenge. Int. J. Comput. Vis. (IJCV) **115**(3), 211–252 (2015). https://doi.org/10.1007/s11263-015-0816-y
30. Simonyan, K., Zisserman, A.: Very deep convolutional networks for large-scale image recognition. In: Bengio, Y., LeCun, Y. (eds.) 3rd International Conference on Learning Representations, ICLR 2015, San Diego, CA, USA, 7–9 May 2015, Conference Track Proceedings (2015)

31. Strongin, R.G., Sergeyev, Y.D.: Global Optimization with Non-Convex Constraints - Sequential and Parallel Algorithms (Nonconvex Optimization and Its Applications Volume 45). Springer, Heidelberg (2000). https://doi.org/10.1007/978-1-4615-4677-1
32. Tseng, P.: An incremental gradient(-projection) method with momentum term and adaptive stepsize rule. SIAM J. Optim. **8**(2), 506–531 (1998)
33. Yifan, H., Siqi, Z., Xin, C., Niao, H.: Biased stochastic first-order methods for conditional stochastic optimization and applications in meta learning. In: Larochelle, H., Ranzato, M., Hadsell, R., Balcan, M.F., Lin, H. (eds.) Advances in Neural Information Processing Systems, vol. 33, pp. 2759–2770. Curran Associates, Inc. (2020)
34. Zhou, D., Xu, P., Gu, Q.: Stochastic nested variance reduced gradient descent for nonconvex optimization. In: Bengio, S., Wallach, H., Larochelle, H., Grauman, K., Cesa-Bianchi, N., Garnett, R. (eds.) Advances in Neural Information Processing Systems 31, pp. 3921–3932. Curran Associates, Inc. (2018)

# Very Fast Streaming Submodular Function Maximization

Sebastian Buschjäger(✉), Philipp-Jan Honysz, Lukas Pfahler, and Katharina Morik

Artificial Intelligence Group, TU Dortmund, Germany
{sebastian.buschjaeger,philipp.honysz, lukas.pfahler,katharina.morik}@tu-dortmund.de

**Abstract.** Data summarization has become a valuable tool in understanding even terabytes of data. Due to their compelling theoretical properties, submodular functions have been the focus of summarization algorithms. Submodular function maximization is a well-studied problem with a variety of algorithms available. These algorithms usually offer worst-case guarantees to the expense of higher computation and memory requirements. However, many practical applications do not fall under this mathematical worst-case but are usually much more well-behaved. We propose a new submodular function maximization algorithm called ThreeSieves that ignores the worst-case and thus uses fewer resources. Our algorithm selects the most informative items from a data-stream on the fly and maintains a provable performance in most cases on a fixed memory budget. In an extensive evaluation, we compare our method against 6 state-of-the-art algorithms on 8 different datasets including data with and without concept drift. We show that our algorithm outperforms the current state-of-the-art in the majority of cases and, at the same time, uses fewer resources.

**Keywords:** Submodular function maximization · Streaming data · Data summarization

## 1 Introduction

In recent years, submodular optimization has found its way into the toolbox of machine learning and data mining. Submodular functions reward adding a new element to a smaller set more than adding the same element to a larger set. This makes them ideal for solving data summarization tasks [22], active learning [28], user recommendation [1], and many other related tasks. In these tasks, the amount of data is often huge and generated in real-time. Consequently, a line of research studies streaming algorithms for maximizing a submodular function.

In this paper, we consider the problem of maximizing a submodular function over a data stream and focus on the task of data summarization. More formally, we consider the problem of selecting $K$ representative elements from a ground set


N. Oliver et al. (Eds.): ECML PKDD 2021, LNAI 12977, pp. 151–166, 2021.
https://doi.org/10.1007/978-3-030-86523-8_10

$V$ into a summary set $S \subseteq V$. To do so, we maximize a non-negative, monotone submodular set function $f: 2^V \to \mathbb{R}_+$ which assigns a utility score to each subset:

$$S^* = \arg\max_{S \subseteq V, |S|=K} f(S) \tag{1}$$

For the empty set, we assume zero utility $f(\emptyset) = 0$. We denote the maximum of $f$ with $OPT = f(S^*)$. A set function can be associated with a marginal gain which represents the increase of $f(S)$ when adding an element $e \in V$ to $S$:

$$\Delta_f(e|S) = f(S \cup \{e\}) - f(S)$$

We call $f$ submodular iff for all $A \subseteq B \subseteq V$ and $e \in V \setminus B$ it holds that

$$\Delta_f(e|A) \geq \Delta_f(e|B)$$

The function $f$ is called monotone, iff for all $e \in V$ and for all $S \subseteq V$ it holds that $\Delta_f(e|S) \geq 0$.

The maximization of a submodular set function is NP-hard [11] and therefore, a natural approach is to find an approximate solution. Table 1 gives an overview of streaming algorithms which have been proposed for solving Eq. 1. To this date, the best performing online algorithms offer an $\mathcal{O}(\frac{1}{2} - \varepsilon)$ approximation ratio where $\varepsilon$ also influences the resource consumption. Even moderate choices for $\varepsilon$ quickly result in unmanageable resource consumption. Feldman et al. showed [12] that this approximation ratio is the best possible for streaming algorithms and that any algorithm with a better worst-case approximation guarantee essentially stores all the elements of the stream (up to a polynomial factor in $K$).

We ask whether we can design an algorithm that – despite the negative result from Feldman et al. – offers a *better* approximation ratio using *fewer* resources. Existing algorithms are designed for the mathematical *worst-case* and thereby have a worst-case approximation guarantee. We note, that this worse-cast is often a pathological case in their mathematical analysis whereas practical applications are usually much more well-behaved. Thus, we propose to *ignore* these pathological cases and derive an algorithm with a *better* approximation guarantee in *most* cases. Our proposed ThreeSieves algorithm estimates the probability of finding a more informative data item on the fly and only adds those items to the summary which are likely to not be 'out-valued' in the future. The resulting algorithm offers a *non-deterministic* approximation ratio of $(1 - \varepsilon)(1 - 1/\exp(1)) > \frac{1}{2} - \varepsilon$ in high probability $(1 - \alpha)^K$, where $\alpha$ is the desired user certainty. It performs $\mathcal{O}(1)$ function queries per item and requires $\mathcal{O}(K)$ memory. Note, that this does not contradict the upper bound of $\frac{1}{2} - \varepsilon$ since our algorithm offers a better approximation quality *in high probability*, but not deterministically for *all* cases. Our contributions are the following:

- The novel ThreeSieves algorithm has an approximation guarantee of $(1 - \varepsilon)(1 - 1/\exp(1))$ in high probability. The fixed memory budget is independent of $\varepsilon$ storing at most $K$ elements and the number of function queries is just one per element.

- For the first time we apply submodular function maximization algorithms to data containing concept drift. We show that our ThreeSieves algorithm offers competitive performance in this setting despite its weaker theoretical guarantee.
- We compare our algorithm against 6 state of the art algorithms on 8 datasets with and without concept drift. To the best of our knowledge, this is the first extensive evaluation of state-of-the-art submodular function maximization algorithms in a streaming setting. We show that ThreeSieves outperforms the current state-of-the-art in many cases while being up to 1000 times faster using a fraction of its memory.

The paper is organized as follows. Section 2 surveys related work, whereas in Sect. 3 we present our main contribution, the ThreeSieves algorithm. Section 4 experimentally evaluates ThreeSieves. Section 5 concludes the paper.

**Table 1.** Algorithms for non-negative, monotone submodular maximization with cardinality constraint $K$. ThreeSieves offers the smallest memory consumption and the smallest number of queries per element in a streaming-setting.

| Algorithm | Approximation ratio | Memory | Queries per Element | Stream | Ref. |
|---|---|---|---|---|---|
| Greedy | $1 - 1/\exp(1)$ | $\mathcal{O}(K)$ | $\mathcal{O}(1)$ | ✗ | [23] |
| StreamGreedy | $1/2 - \varepsilon$ | $\mathcal{O}(K)$ | $\mathcal{O}(K)$ | ✗ | [13] |
| PreemptionStreaming | $1/4$ | $\mathcal{O}(K)$ | $\mathcal{O}(K)$ | ✓ | [4] |
| IndependentSetImprovement | $1/4$ | $\mathcal{O}(K)$ | $\mathcal{O}(1)$ | ✓ | [8] |
| Sieve-Streaming | $1/2 - \varepsilon$ | $\mathcal{O}(K \log K/\varepsilon)$ | $\mathcal{O}(\log K/\varepsilon)$ | ✓ | [2] |
| Sieve-Streaming++ | $1/2 - \varepsilon$ | $\mathcal{O}(K/\varepsilon)$ | $\mathcal{O}(\log K/\varepsilon)$ | ✓ | [16] |
| Salsa | $1/2 - \varepsilon$ | $\mathcal{O}(K \log K/\varepsilon)$ | $\mathcal{O}(\log K/\varepsilon)$ | (✓) | [24] |
| QuickStream | $1/(4c) - \varepsilon$ | $\mathcal{O}(cK \log K \log(1/\varepsilon))$ | $\mathcal{O}(\lceil 1/c \rceil + c)$ | ✓ | [18] |
| ThreeSieves | $(1-\varepsilon)(1 - 1/\exp(1))$ with prob. $(1-\alpha)^K$ | $\mathcal{O}(K)$ | $\mathcal{O}(1)$ | ✓ | This paper |

## 2 Related Work

For a general introduction to submodular function maximization, we refer interested readers to [17] and for a more thorough introduction into the topic of streaming submodular function maximization to [9]. Most relevant to this publication are non-negative, monotone submodular streaming algorithms with cardinality constraints. There exist several algorithms for this problem setting which we survey here. The theoretical properties of each algorithm are summarized in Table 1. A detailed formal description including the pseudo-code of each algorithm is given in the appendix.

While not a streaming algorithm, the Greedy algorithm [23] forms the basis of many algorithms. It iterates $K$ times over the entire dataset and greedily selects that element with the largest marginal gain $\Delta_f(e|S)$ in each iteration.

It offers a $(1-(1/\exp(1))) \approx 63\%$ approximation and stores $K$ elements. StreamGreedy [13] is its adaption to streaming data. It replaces an element in the current summary if it improves the current solution by at-least $\nu$. It offers an $\frac{1}{2} - \varepsilon$ approximation with $\mathcal{O}(K)$ memory, where $\varepsilon$ depends on the submodular function and some user-specified parameters. The optimal approximation factor is only achieved if multiple passes over the data are allowed. Otherwise, the performance of StreamGreedy degrades arbitrarily with $K$ (see Appendix of [2] for an example). We therefore consider StreamGreedy not to be a proper streaming algorithm.

Similar to StreamGreedy, PremptionStreaming [4] compares each marginal gain against a threshold $\nu(S)$. This time, the threshold dynamically changes depending on the current summary $S$ which improves the overall performance. It uses constant memory and offers an approximation guarantee of 1/4. It was later shown that this algorithm is outperformed by SieveStreaming++ (see below) and was thus not further considered our experiments. Chakrabarti and Kale propose in [8] a streaming algorithm also with approximation guarantee of 1/4. Their algorithm stores the marginal gain of each element upon its arrival and uses this 'weight' to measure the importance of each item. We call this algorithm IndependentSetImprovement.

Norouzi-Fard et al. propose in [24] a meta-algorithm for submodular function maximization called Salsa which uses different algorithms for maximization as sub-procedures. The authors argue, that there are different types of data-streams and for each stream type, a different thresholding-rule is appropriate. The authors use this intuition to design a $r$-pass algorithm that iterates $r$ times over the entire dataset and adapts the thresholds between each run. They show that their approach is a $(r/(r+1))^r - \varepsilon$ approximation algorithm. For a streaming setting, i.e. $r = 1$, this algorithm recovers the $1/2 - \varepsilon$ approximation bound. However note, that some of the thresholding-rules require additional information about the data-stream such as its length or density. Since this might be unknown in a real-world use-case this algorithm might not be applicable in all scenarios.

The first proper streaming algorithm with $1/2 - \varepsilon$ approximation guarantee was proposed by Badanidiyuru et al. in [2] and is called SieveStreaming. SieveStreaming tries to estimate the potential gain of a data item before observing it. Assuming one knows the maximum function value $OPT$ beforehand and $|S| < K$, an element $e$ is added to the summary $S$ if the following holds:

$$\Delta_f(e|S) \geq \frac{OPT/2 - f(S)}{K - |S|} \tag{2}$$

Since $OPT$ is unknown beforehand one has to estimate it before running the algorithm. Assuming one knows the maximum function value of a singleton set $m = max_{e \in V} f(\{e\})$ beforehand, then the optimal function value for a set with $K$ items can be estimated by submodularity as $m \leq OPT \leq K \cdot m$. The authors propose to manage different summaries in parallel, each using one threshold from the set $O = \{(1+\varepsilon)^i \mid i \in \mathbb{Z}, m \leq (1+\varepsilon)^i \leq K \cdot m\}$, so that for at least one $v \in O$ it holds: $(1-\varepsilon)OPT \leq v \leq OPT$. In a sense, this approach sieves out elements with marginal gains below the given threshold - hence the authors name their approach SieveStreaming. Note, that this algorithm requires the knowledge of $m = max_{e \in V} f(\{e\})$ before running the algorithm. The authors also present an algorithm to estimate $m$ on the fly which does not alter the theoretical performance of SieveStreaming. Recently, Kazemi et al. proposed in [16] an extension of the SieveStreaming called SieveStreaming++. The authors point out, that the currently best performing sieve $S_v = \arg\max_v\{f(S_v)\}$ offers a better lower bound for the function value and they propose to use $[\max_v\{f(S_v)\}, K \cdot m]$ as the interval for sampling thresholds. This results in a more dynamic algorithm, in which sieves are removed once they are outperformed by other sieves and new sieves are introduced to make use of the better estimation of $OPT$. SieveStreaming++ does not improve the approximation guarantee of SieveStreaming, but only requires $\mathcal{O}(K/\varepsilon)$ memory instead of $\mathcal{O}(K \log K/\varepsilon)$. Last, Kuhnle proposed the QuickStream algorithm in [18] which works under the assumption that a single function evaluation is very expensive. QuickStream buffers up to $c$ elements and only evaluates $f$ every $c$ elements. If the function value is increased by the $c$ elements, thy all are added to the solution. Additionally, older examples are removed if there are more than $K$ items in the solution. QuickStream performs well if the evaluation of $f$ is very costly and if it is ideally independent from the size of the current solution $S$. This is unfortunately not the case in our experiments (see below). Moreover, QuickStream has a guarantee of $1/(4c) - \varepsilon$ that is outperformed by SieveStreaming(++) with similar resource consumption. Thus, we did not consider QuickStream in our experiments.

## 3 The Three Sieves Algorithm

We recognize, that SieveStreaming and its extensions offer a worst-case guarantee on their performance and indeed they can be consider optimal providing an approximation guarantee of $\frac{1}{2} - \varepsilon$ under polynomial memory constraints [12]. However, we also note that this worst case often includes pathological cases, whereas practical applications are usually much more well-behaved. One common practical assumption is, that the data is generated by the same source and thus follows the same distribution (e.g. in a given time frame). In this paper, we want to study these better behaving cases more carefully and present an algorithm which improves the approximation guarantee, while reducing memory and runtime costs in these cases. More formally, we will now assume that the items in the given sample (batch processing) or in the data stream (stream processing) are independent and identically distributed (iid). Note, that we do *not* assume any specific distribution. For batch processing this means, that all items in the data should come from the same (but unknown) distribution and that items should not influence each other. From a data-streams perspective this assumptions means, that the data source will produce items from the same distribution which does not change over time. Hence, we specifically *ignore* concept drift and assume that an appropriate concept drift detection mechanism is in place, so that summaries are e.g. re-selected periodically. We will study streams with drift in more detail in our experimental evaluation. We now use this assumption to derive an algorithm with $(1-\varepsilon)(1-1/\exp(1))$ approximation guarantee in high probability:

SieveStreaming and its extension, both, manage $\mathcal{O}(\log K/\varepsilon)$ sieves in parallel, which quickly becomes unmanageable even for moderate choices of $K$ and $\varepsilon$. We note the following behavior of both algorithms: Many sieves in SieveStreaming have *too small a novelty-threshold* and quickly fill-up with uninteresting events. SieveStreaming++ exploits this insight by removing small thresholds early and thus by focusing on the most promising sieves in the stream. On the other hand, both algorithms manage sieves with *too large a novelty-threshold*, so that they never include any item. Thus, there are only a few thresholds that produce valuable summaries. We exploit this insight with the following approach: Instead of using many sieves with different thresholds we use only a single summary and carefully calibrate the threshold. To do so, we start with a large threshold that rejects most items, and then we gradually reduce this threshold until it accepts some - hopefully the most informative - items. The set $O = \{(1+\varepsilon)^i \mid i \in \mathbb{Z}, m \leq (1+\varepsilon)^i \leq K \cdot m\}$ offers a somewhat crude but sufficient approximation of $OPT$ (c.f. [2]). We start with the largest threshold in $O$ and decide for each item if we want to add it to the summary or not. If we do not add any of $T$ items (which will be discussed later) to $S$ we may lower the threshold to the next smallest value in $O$ and repeat the process until $S$ is full.

The key question now becomes: How to choose $T$ appropriately? If $T$ is too small, we will quickly lower the threshold and fill up the summary before any interesting item arrive that would have exceeded the original threshold. If $T$ is too large, we may reject interesting items from the stream. Certainly, we cannot determine with absolute certainty when to lower a threshold without knowing the rest of the data stream or knowing the ground set entirely, but we can do so with high probability. More formally, we aim at estimating the probability $p(e|f, S, v)$ of finding an item $e$ which exceeds the novelty threshold $v$ for a given summary $S$ and function $f$. Once $p$ drops below a user-defined certainty margin $\tau$

$$p(e|f, S, v) \leq \tau$$

we can safely lower the threshold. This probability must be estimated on the fly. Most of the time, we reject $e$ so that $S$ and $f(S)$ are unchanged and we keep estimating $p(e|f, S, v)$ based on the negative outcome. If, however, $e$ exceeds the current novelty threshold we add it to $S$ and $f(S)$ changes. In this case, we do not have any estimates for the new summary and must start the estimation of $p(e|f, S, v)$ from scratch. Thus, with a growing number of rejected items $p(e|f, S, v)$ tends to become close to 0 and the key question is how many observations do we need to determine – with sufficient evidence – that $p(e|f, S, v)$ will be 0.

The computation of confidence intervals for estimated probabilities is a well-known problem in statistics. For example, the confidence interval of binominal distributions can be approximated with normal distributions, Wilson score intervals, or Jeffreys interval. Unfortunately, these methods usually fail for probabilities near 0 [3]. However, there exists a more direct way of computing a confidence interval for heavily one-sided binominal distribution with probabilities near zero and iid data [15]. The probability of not adding one item in $T$ trials is:

$$\alpha = (1 - p(e|f, S, v))^T \Leftrightarrow \ln(\alpha) = T \ln(1 - p(e|f, S, v))$$

A first order Taylor Approximation of $\ln(1 - p(e|f, S, v))$ reveals that $\ln(1 - p(e|f, S, v)) \approx -p(e|f, S, v)$ and thus $\ln(\alpha) \approx T(-p(e|f, S, v))$ leading to:

$$\frac{-\ln(\alpha)}{T} \approx p(e|f, S, v) \leq \tau \tag{3}$$

Therefore, the confidence interval of $p(e|f, S, v)$ after observing $T$ events is $\left[0, \frac{-\ln(\alpha)}{T}\right]$. The 95% confidence interval of $p(e|f, S, v)$ is $\left[0, -\frac{\ln(0.05)}{T}\right]$ which is approximately $[0, 3/T]$ leading to the term "Rule of Three" for this estimate [15]. For example, if we did not add any of $T = 1000$ items to the summary, then the probability of adding an item to the summary in the future is below 0.003 given a $1 - \alpha = 0.95$ confidence interval. We can use the Rule of Three to quantify the certainty that there is a very low probability for finding a novel item in the data stream after observing $T$ items. Note that we can either set $\alpha$, $\tau$ and use Eq. 3 to compute the appropriate $T$ value. Alternatively, we may directly specify $T$ as a user parameter instead of $\alpha$ and $\tau$, thereby effectively removing one hyperparameter. We call our algorithm ThreeSieves and it is depicted in Algorithm 1. Its theoretical properties are presented in Theorem 1.

```
Input: Stream e_1, e_2, ..., submodular function f and parameters K, T ∈ N_{>0}
Output: A set S with at-most K elements maximizing f(S)
O ← {(1+ε)^i | i ∈ Z, m ≤ (1+ε)^i ≤ K·m}
v ← max(O); O ← O \ {max(O)}; S ← ∅; t ← 0
for next item e do
    if Δ_f(e|S) ≥ (v/2 − f(S))/(K − |S|) and |S| < K then
        S ← S ∪ {e}; t ← 0
    else
        t ← t + 1
        if t ≥ T then
            v ← max(O); O ← O \ {max(O)}; t ← 0
        end
    end
end
return S
```

**Algorithm 1:** ThreeSieves algorithm.

**Theorem 1.** *ThreeSieves has the following properties:*

- *Let $K \in \mathbb{N}_{>0}$ be the maximum desired cardinality and let $1.0 - \alpha$ be the desired confidence interval. Given a fixed groundset $V$ or an infinite data-stream in*

*which each item is independent and identically distributed (iid) it outputs a set $S$ such that $|S| \leq K$ and with probability $(1-\alpha)^K$ it holds for a non-negative, monotone submodular function $f$: $f(S) \geq (1-\varepsilon)(1-1/\exp(1))OPT$*
- *It does 1 pass over the data (streaming-capable) and stores at most $\mathcal{O}(K)$ elements*

*Proof.* The detailed proof can be found in the appendix. The proof idea is as follows: The Greedy Algorithm selects that element with the largest marginal gain in each round. Suppose we know the marginal gains $v_i = \Delta(S|e_i)$ for each element $e_i \in S$ selected by Greedy in each round. Then we can simulate the Greedy algorithm by stacking $K$ copies of the dataset consecutively and by comparing the gain $\Delta_f(e|S)$ of each element $e \in V$ against the respective marginal gain $v_i$. Let $O$ be a set of estimated thresholds with $v_1^*, \ldots, v_K^* \in O$. Let $v_1^*$ denote the first threshold used by ThreeSieves before any item has been added to $S$. By the statistical test of ThreeSieves the $1-\alpha$ confidence interval for $P(v_1 \neq v_1^*)$ is given by $\frac{-\ln(\alpha)}{T}$. Differently coined, it holds with probability $1-\alpha$ that

$$P(v_1 \neq v_1^*) \leq \frac{-\ln(\alpha)}{T} \Leftrightarrow P(v_1 = v_1^*) > 1 - \frac{-\ln(\alpha)}{T}$$

Now we apply the confidence interval $K$ times for each $P(v_j \neq v_j^*)$ individually. Then it holds with probability $(1-\alpha)^K$

$$P\left(v_1 = v_1^*, \ldots, v_K = v_K^*\right) > \left(1 - \frac{-\ln(\alpha)}{T}\right)^K$$

By the construction of $O$ it holds that $(1-\varepsilon)v_i^* \leq v_i \leq v_i^*$ (c.f. [2]). Let $e_K$ be the element that is selected by ThreeSieves after $K-1$ items have already been selected. Let $S_0 = \emptyset$ and recall that by definition $f(\emptyset) = 0$, then it holds with probability $(1-\alpha)^K$:

$$\begin{aligned} f(S_K) &= f(\emptyset) + \sum_{i=1}^{K} \Delta(e_i|S_{K-1}) \geq \sum_{i=1}^{K} (1-\varepsilon)\, v_i^* \\ &= (1-\varepsilon)\, f_G(S_K) \geq (1-\varepsilon)\,(1-1/\exp(1))\, OPT \end{aligned}$$

where $f_G$ denotes the solution of the Greedy algorithm. □

Similar to SieveStreaming, ThreeSieves tries different thresholds until it finds one that fits best for the current summary $S$, the data $V$, and the function $f$. In contrast, however, ThreeSieves is optimized towards minimal memory consumption by maintaining one threshold and one summary at a time. If more memory is available, one may improve the performance of ThreeSieves by running multiple instances of ThreeSieves in parallel on different sets of thresholds. So far, we assumed that we know the maximum singleton value $m = max_{e \in V} f(\{e\})$ beforehand. If this value is unknown before running the algorithm we can estimate it on-the-fly without changing the theoretical guarantees of ThreeSieves.

As soon as a new item arrives with a new $m_{new} > m_{old}$ we remove the current summary and use the new upper bound $K \cdot m_{new}$ as the starting threshold. It is easy to see that this does not affect the theoretical performance of ThreeSieves: Assume that a new item arrives with a new maximum single value $m_{new}$. Then, all items in the current summary have a smaller singleton value $m_{old} < m_{new}$. The current summary has been selected based on the assumption that $m_{old}$ was the largest possible value, which was invalidated as soon as $m_{new}$ arrived. Thus, the probability estimate that the first item in the summary would be 'out-valued' later in the stream was wrong since we just observed that it is being out-valued. To re-validate the original assumption we delete the current summary entirely and re-start the summary selection.

# 4 Experimental Evaluation

In this section, we experimentally evaluate ThreeSieves and compare it against SieveStreaming(++), IndepdendentSetImprovement, Slasa, and Greedy. As an additional baseline we also consider a random selection of items via Reservoir Sampling [27]. We denote this algorithm as Random. We focus on two different application scenarios: In the first experiment, we have given a batch of data and are tasked to compute a comprehensive summary. In this setting, algorithms are allowed to perform multiple passes over the data to the expense of a longer runtime, e.g. as Greedy does. In the second experiment we shift our focus towards the summary selection on streams with concept drift. Here, each item is only seen once and the algorithms do not have any other information about the data-stream. All experiments have been run on an Intel Core i7-6700 CPU machine with 8 cores and 32 GB main memory running Ubuntu 16.04. The code for our experiments is available at https://github.com/sbuschjaeger/SubmodularStreamingMaximization

## 4.1 Batch Experiments

In the batch experiments each algorithm is tasked to select a summary with exactly $K$ elements. Since most algorithms can reject items they may select a summary with less than $K$ elements. To ensure a summary of size $K$, we re-iterate over the entire data-set as often as required until $K$ elements have been selected, but at most $K$ times. We compare the relative maximization performance of all algorithms to the solution of Greedy. For example, a relative performance of 100% means that the algorithm achieved the same performance as Greedy did, whereas a relative performance of 50% means that the algorithm only achieved half the function value of Greedy. We also measure the total runtime and memory consumption of each algorithm. The runtime measurements include all re-runs, so that many re-runs over the data-set result in larger runtimes. To make our comparison implementation independent, we report the *algorithmic* memory consumption in terms of the total number of items stored by each algorithm. For example, the Forestcover dataset contains observations with $d = 10$ features which can be each be represented using a 4 byte `float` variable. Thus, an algorithm that stores 4096 of these observations uses $4096 \cdot 4 \cdot d/1024 = 160$ KB of memory in this specific instance. We evaluate four key questions: First, is ThreeSieves competitive against the other algorithms or will the probabilistic guarantee hurt its practical performance? Second, if ThreeSieves is competitive, how does it related to a Random selection of summaries? Third, how large is the resource consumption of ThreeSieves in comparison? Fourth, how does ThreeSieves behave for different $T$ and different $\varepsilon$?

In total, we evaluate 3895 hyperparameter configurations on the datasets shown in the top group in Table 2. We extract summaries of varying sizes $K \in \{5, 10, \dots, 100\}$ maximizing the log-determinant $f(S) = \frac{1}{2}\log\det(\mathcal{I} + a\Sigma_S)$. Here, $\Sigma_S = [k(e_i, e_j)]_{ij}$ is a kernel matrix containing all similarity pairs of all points in $S$, $a \in \mathbb{R}_+$ is a scaling parameter and $\mathcal{I}$ is the identity matrix. In [26], this function is shown to be submodular. Its function value does not depend on $V$, but only on the summary $S$, which makes it an ideal candidate for summarizing data in a streaming setting. In [5], it is proven that $m = max_{e \in V} f(\{e\}) = 1 + aK$ and that $OPT \leq K \log(1 + a)$ for kernels with $k(\cdot, \cdot) \leq 1$. This property can be enforced for every positive definite kernel with normalization [14]. In our experiments we set $a = 1$ and use the RBF kernel $k(e_i, e_j) = \exp\left(-\frac{1}{2l^2} \cdot ||e_i - e_j||_2^2\right)$ with $l = \frac{1}{2\sqrt{d}}$ where $d$ is the dimensionality of the data. We vary $\varepsilon \in \{0.001, 0.005, 0.01, 0.05, 0.1\}$ and $T \in \{50, 250, 500, 1000, 2500, 5000\}$.

**Table 2.** Data sets used for the experiments.

| Name | Size | Dim | Reference |
|---|---|---|---|
| ForestCover | 286,048 | 10 | [10] |
| Creditfraud | 284,807 | 29 | [21] |
| FACT Highlevel | 200,000 | 16 | [6] |
| FACT Lowlevel | 200,000 | 256 | [6] |
| KDDCup99 | 60,632 | 41 | [7] |
| Stream51 | 150,736 | 2048 | [25] |
| abc | 1,186,018 | 300 | [20] |
| Examiner | 3,089,781 | 300 | [19] |

We present two different sets of plots, one for varying $K$ and one for varying $\varepsilon$. Additional plots with more parameter variations are given in the appendix. Figure 1 depicts the relative performance, the runtime and the memory consumption over different $K$ for a fixed $\varepsilon = 0.001$. For presentational purposes, we selected $T = 500, 1000, 2500, 5000$ for ThreeSieves. In all experiments, we find that ThreeSieves with $T = 5000$ and Salsa generally perform best with a very close performance to Greedy for $K \geq 20$. For smaller summaries with $K < 20$ all algorithms seem to underperform, with Salsa and SieveStreaming performing best. Using $T \leq 1000$ for ThreeSieves seems to decrease the performance on some datasets, which can be explained by the probabilistic nature of the algorithm. We also observe a relative performance above 100 where ThreeSieves performed *better* than Greedy on Creditfraud and Fact Highlevel. Note, that only ThreeSieves showed this behavior, whereas the other algorithms never exceeded Greedy. Expectantly, Random selection shows the weakest performance. SieveStreaming and SieveStreaming++ show identical behavior. Looking at the runtime, please, note the logarithmic scale. Here, we see that ThreeSieves and Random are by far the fastest methods. Using $T = 1000$ offers some performance benefit, but is hardly justified by the decrease in maximization performance, whereas $T = 5000$ is only marginally slower but offers a much better maximization performance. SieveStreaming and SieveStreaming++ have very similar runtime, but are magnitudes slower than Random and ThreeSieves. Last, Salsa is the slowest method. Regarding the memory consumption, please, note again the logarithmic scale. Here, all versions of ThreeSieves use the fewest resources as our algorithm only stores a single summary in all configurations. These curves are identical with Random and IndependentSetImprovement so that only four instead of 7 curves are to be seen. SieveStreaming and its siblings use roughly two magnitudes more memory since they keep track of multiple sieves in parallel.

Now we look at the behavior of the algorithms for different approximation ratios. Again we refer readers to the additional plots with more parameter variations in the appendix. Figure 2 depicts the relative performance, the runtime, and the memory consumption over different $\varepsilon$ for a fixed $K = 50$. We again selected $T = 500, 1000, 2500, 5000$ for ThreeSieves. Note, that for Random, Greedy and IndependentSetImprovement there is now a constant curve since their performance does not vary with $\varepsilon$. Looking at the relative performance we see a slightly different picture than before: For small $\varepsilon \leq 0.05$ and larger $T$ we see that ThreeSieves and Salsa again perform best in all cases. For larger $\varepsilon > 0.05$ the performance of the non-probabilistic algorithms remain relatively stable, but ThreeSieves performance starts to deteriorate. Again we note, that SieveStreaming and SieveStreaming++ show identical behavior. Looking at the runtime and memory consumption we see a similar picture as before: ThreeSieves is by far the fastest method using the fewest resources followed by SieveStreaming(++) and Salsa.

**We Conclude:** In summary, ThreeSieves works best for small $\varepsilon$ and large $T$. The probabilistic nature of the algorithm does not decrease its maximization performance but actually helps it in some cases. In contrast to the other algorithms, the resource consumption and overall runtime of ThreeSieves does not suffer from decreasing $\varepsilon$ or increasing $T$. Additional experiments and comparisons can be found in the appendix. They all depict the same general trend discussed here. In particular, when comparing SieveStreaming(++) and Salsa with ThreeSieves under similar maximization performance (e.g. using $\varepsilon = 0.1$ for SieveStreaming(++) and Salsa, and using $\varepsilon = 0.001$ for ThreeSieves) we still find that ThreeSieves uses magnitudes less resources. Its overall performance is better or comparable to the other algorithms while being much more memory efficient and overall faster.

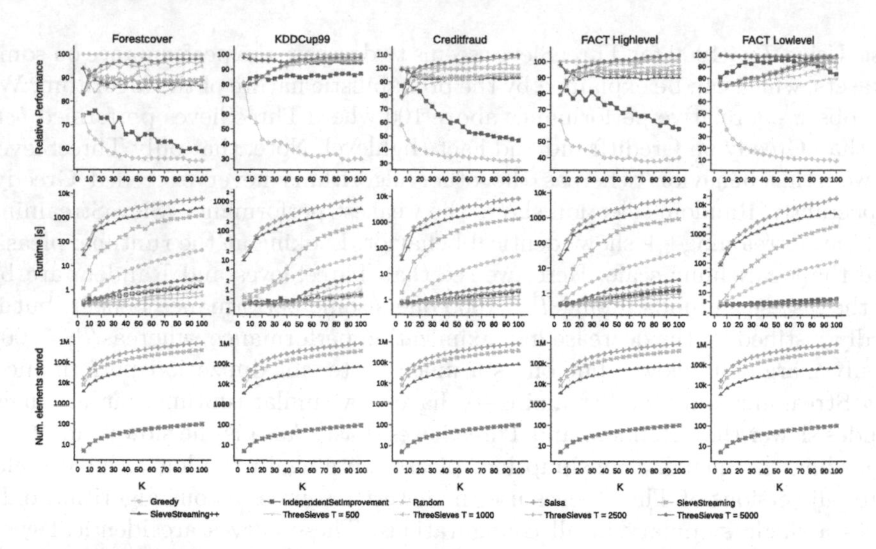


**Fig. 1.** Comparison between SieveStreaming, SieveStreaming++, Salsa, and ThreeSieves for different $K$ values and fixed $\varepsilon = 0.001$. The first row shows the relative performance to Greedy (larger is better), the second row shows the total runtime in seconds (logarithmic scale, smaller is better) and the third row shows the maximum memory consumption (logarithmic scale, smaller is better). Each column represents one dataset.

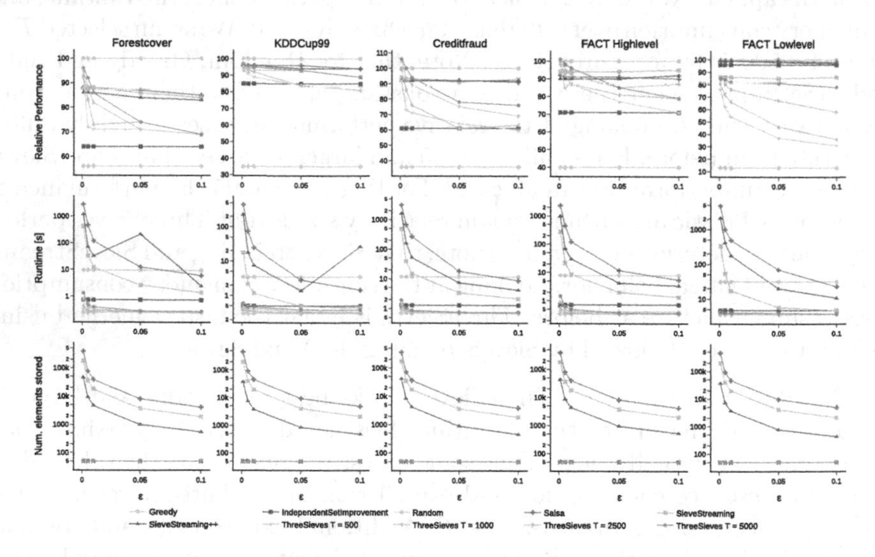


**Fig. 2.** Comparison between IndependentSetImprovement, SieveStreaming, SieveStreaming++, Salsa, Random, and ThreeSieves for different $\varepsilon$ values and fixed $K = 50$. The first row shows the relative performance to Greedy (larger is better), the second row shows the total runtime in seconds (logarithmic scale, smaller is better) and the third row shows the maximum memory consumption (logarithmic scale, smaller is better). Each column represents one dataset.

### 4.2 Streaming Experiments

Now we want to compare the algorithms in a true streaming setting including concept drift. Here, we present each item only once and must decide immediately if it should be added to the summary or not. Since Salsa requires additional information about the stream we excluded it for these experiments. We use two real-world data-sets and one artificial data-set depicted in the bottom group in Table 2: The stream51 dataset [25] contains image frames from a sequence of videos, where each video shows an object of one of 51 classes. Subsequent frames in the video stream are highly dependent. Over the duration of the stream, new classes are introduced. The dataset is constructed such that online streaming classification methods suffer from 'Catastrophic forgetting' [25]. We utilize a pretrained InceptionV3 convolutional neural network that computes 2048-dimensional embeddings of the images. The abc dataset contain news headlines from the Australian news source 'ABC' gathered over 17 years (2003–2019) and the examiner dataset contains news headlines from the Australian news source 'The Examiner' gathered over 6 years (2010–2015). Due to this long time-span we assume that both datasets contain a natural concept drift occurring due to different topics in the news. We use pretrained Glove embeddings to extract 300-dimensional, real-valued feature vectors for all headlines and present them in order of appearance starting with the oldest headline. We ask the following questions: First, will the iid assumption of ThreeSieves hurt its practical performance on data-streams with concept drift? Second, how will the other algorithms with a worst-case guarantee perform in this situation?

In total, we evaluate 3780 hyperparameters on these three datasets. Again, we extract summaries of varying sizes $K \in \{5, 10, \ldots, 100\}$ maximizing the log-determinant with $a = 1$. We use the RBF kernel with $l = \frac{1}{\sqrt{d}}$ where $d$ is the dimensionality of the data. We vary $\varepsilon \in \{0.01, 0.1\}$ and $T \in \{500, 1000, 2500, 5000\}$. Again, we report the relative performance of the algorithms compared to Greedy (executed in a batch-fashion). Figure 3 shows the relative performance of the streaming algorithms for different $K$ with fixed $\varepsilon = 0.1$ (first row) and fixed $\varepsilon = 0.01$ (second row) on the three datasets. On the stream51 dataset we see very chaotic behavior for $\varepsilon = 0.1$. Here, SieveStreaming++ generally seems to be best with a performance around 90–95%. ThreeSieves's performance suffers for smaller $T \leq 1000$ not exceeding 85%. For other configurations with larger $T$ ThreeSieves has a comparable performance to SieveStreaming and IndependentSetImprovement all achieving 85–92%. For $\varepsilon = 0.01$ the behavior of the algorithms stabilizes. Here we find that ThreeSieves with $T = 5000$ shows a similar and sometimes even better performance compared to SieveStreaming(++) beyond 95%. An interesting case occurs for $T = 5000$ and $K = 100$ in which the function value suddenly drops. Here, ThreeSieves rejected more items than were available in the stream and thus returned a summary with less than $K = 100$ items. Somewhere in the middle, we find IndependentSetImprovement reaching 90% performance in the best case. Expectantly, Random selection is the worst in all cases. In general, we find a similar behavior on the other two datasets. For $\varepsilon = 0.1$, SieveStreaming(++) seem to be the best option followed by ThreeSieves with larger $T$ or IndependentSetImprovement. For larger $K$, IndependentSetImprovement is not as good as ThreeSieves and its performance approaches Random quite rapidly. For $\varepsilon = 0.01$ the same general behavior can be observed. SieveStreaming(++) again holds well under concept drift followed by ThreeSieves with larger $T$ followed by IndependentSetImprovement and Random. We conjecture that ThreeSieves's performance could be further improved for larger $T$ and that there seems to be a dependence between $T$ and the type of concept drift occurring in the data.

**We Conclude:** ThreeSieves holds surprisingly well under concept drift, especially for larger $T$. In many cases its maximization performance is comparable with SieveStreaming(++) while being more resource efficient. For smaller $T$ the performance of ThreeSieves clearly suffers, but remains well over the performance of Random or IndependentSetImprovement. For larger $T$ and smaller $\varepsilon$, ThreeSieves becomes more competitive to SieveStreaming(++) while using fewer resources. We conclude that ThreeSieves is also applicable for streaming data with concept drift even though its theoretical guarantee does not explicitly hold in this context.

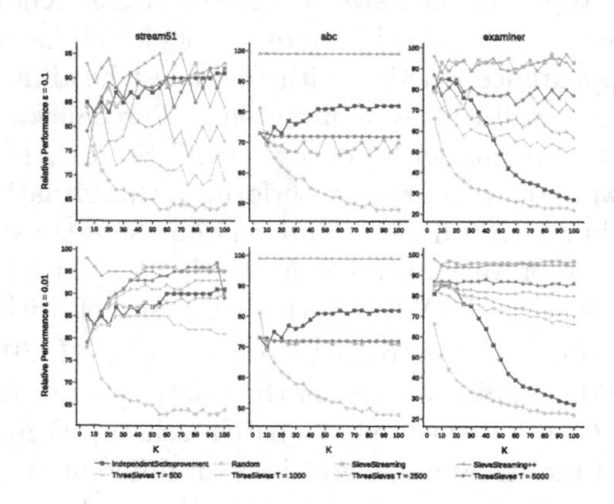


**Fig. 3.** Comparison between IndependentSetImprovement, SieveStreaming, SieveStreaming++, Random and ThreeSieves for different $K$ values and fixed $\varepsilon = 0.1$ (first row) and $\varepsilon = 0.01$ (second row). Each column represents one dataset.

## 5 Conclusion

Data summarization is an emerging topic for understanding and analyzing large amounts of data. In this paper, we studied the problem of on-the-fly data summarization where one must decide immediately whether to add an item to the summary, or not. While some algorithms for this problem already exist, we recognize that these are optimized towards the worst-case and thereby require more resources. We argue that practical applications are usually much more well-behaved than the commonly analyzed worst-cases. We proposed the ThreeSieves algorithm for non-negative monotone submodular streaming function maximization and showed, that – under moderate assumptions – it has an approximation ratio of $(1-\varepsilon)(1-1/\exp(1))$ in high probability. It runs on a fixed memory budget and performs a constant number of function evaluations per item. We compared ThreeSieves against 6 state of the art algorithms on 8 different datasets with and without concept drift. For data without concept drift, ThreeSieves outperforms the other algorithms in terms of maximization performance while using two magnitudes less memory and being up to 1000 times faster. On datasets with concept drift, ThreeSieves outperforms the other algorithms in some cases and offers similar performance in the other cases while being much more resource efficient. This allows for applications, where based on the summary, some action has to be performed. Hence, the novel ThreeSieves algorithm opens up opportunities beyond the human inspection of data summaries, which we want to explore in the future.

**Acknowledgements.** Part of the work on this paper has been supported by Deutsche Forschungsgemeinschaft (DFG) within the Collaborative Research Center SFB 876 "Providing Information by Resource-Constrained Analysis", DFG project number 124020371, SFB project A1, http://sfb876.tu-dortmund.de. Part of the work on this research has been funded by the Federal Ministry of Education and Research of Germany as part of the competence center for machine learning ML2R (01—18038A), https://www.ml2r.de/.

## References

1. Ashkan, A., Kveton, B., Berkovsky, S., Wen, Z.: Optimal greedy diversity for recommendation. IJCAI. **15**, 1742–1748 (2015)
2. Badanidiyuru, A., Mirzasoleiman, B., Karbasi, A., Krause, A.: Streaming submodular maximization: massive data summarization on the fly. In: ACM SIGKDD (2014)
3. Brown, L.D., Cai, T.T., DasGupta, A.: Interval estimation for a binomial proportion. Statist. Sci. **16**(2), 101–117 (2001)
4. Buchbinder, N., Feldman, M., Schwartz, R.: Online submodular maximization with preemption. ACM Trans. Algorithms **15**(3), 1–31 (2019)
5. Buschjäger, S., Morik, K., Schmidt, M.: Summary extraction on data streams in embedded systems. In: ECML Conference Workshop IoT Large Scale Learning from Data Streams (2017)
6. Buschjäger, S., Pfahler, L., Buss, J., Morik, K., Rhode, W.: On-site gamma-hadron separation with deep learning on FPGAs. In: Dong, Y., Mladenić, D., Saunders, C. (eds.) ECML PKDD 2020. LNCS (LNAI), vol. 12460, pp. 478–493. Springer, Cham (2021). https://doi.org/10.1007/978-3-030-67667-4_29
7. Campos, G.O., et al.: On the evaluation of unsupervised outlier detection: measures, datasets, and an empirical study. Data Min. Knowl. Discovery **30**(4), 891–927 (2016). https://doi.org/10.1007/s10618-015-0444-8
8. Chakrabarti, A., Kale, S.: Submodular maximization meets streaming: matchings, matroids, and more. In: Integer Programming and Combinatorial Optimization, pp. 210–221 (2014)

9. Chekuri, C., Gupta, S., Quanrud, K.: Streaming algorithms for submodular function maximization. In: Halldórsson, M.M., Iwama, K., Kobayashi, N., Speckmann, B. (eds.) ICALP 2015. LNCS, vol. 9134, pp. 318–330. Springer, Heidelberg (2015). https://doi.org/10.1007/978-3-662-47672-7_26
10. Dal Pozzolo, A., Caelen, O., Johnson, R.A., Bontempi, G.: Calibrating probability with undersampling for unbalanced classification. In: 2015 IEEE SSCI (2015). https://www.kaggle.com/mlg-ulb/creditcardfraud
11. Feige, U.: A threshold of ln n for approximating set cover. J. ACM **45**(4), 634–652 (1998)
12. Feldman, M., Norouzi-Fard, A., Svensson, O., Zenklusen, R.: The one-way communication complexity of submodular maximization with applications to streaming and robustness. In: 52nd Annual ACM SIGACT STOC, pp. 1363–1374 (2020)
13. Gomes, R., Krause, A.: Budgeted nonparametric learning from data streams. In: ICML, vol. 1, p. 3 (2010)
14. Graf, A.B., Borer, S.: Normalization in support vector machines. In: DAGM Symposium of Pattern Recognition (2001)
15. Jovanovic, B.D., Levy, P.S.: A look at the rule of three. Am. Statist. **51**(2), 137–139 (1997)
16. Kazemi, E., Mitrovic, M., Zadimoghaddam, M., Lattanzi, S., Karbasi, A.: Submodular streaming in all its glory: tight approximation, minimum memory and low adaptive complexity. In: ICML, pp. 3311–3320 (2019)
17. Krause, A., Golovin, D.: Submodular function maximization. Tractability **3**, 71–104 (2014)
18. Kuhnle, A.: Quick streaming algorithms for maximization of monotone submodular functions in linear time. In: International Conference on Artificial Intelligence and Statistics, pp. 1360–1368 (2021)
19. Kulkarni, R.: The examiner - spam clickbait catalog - 6 years of crowd sourced journalism (2017). https://www.kaggle.com/therohk/examine-the-examiner
20. Kulkarni, R.: A million news headlines - news headlines published over a period of 17 years (2017), https://www.kaggle.com/therohk/million-headlines
21. Liu, F.T., Ting, K.M., Zhou, Z.H.: Isolation forest. In: 2008 Eighth IEEE International Conference on Data Mining, pp. 413–422. IEEE (2008). http://odds.cs.stonybrook.edu/forestcovercovertype-dataset/
22. Mirzasoleiman, B., Badanidiyuru, A., Karbasi, A.: Fast constrained submodular maximization: personalized data summarization. In: ICML, pp. 1358–1367 (2016)
23. Nemhauser, G., et al.: An analysis of approximations for maximizing submodular set functions-i. Mathematical Programming (1978). https://doi.org/10.1007/BF01588971
24. Norouzi-Fard, A., Tarnawski, J., Mitrovic, S., Zandieh, A., Mousavifar, A., Svensson, O.: Beyond 1/2-approximation for submodular maximization on massive data streams. In: ICML, pp. 3829–3838 (2018)
25. Roady, R., Hayes, T.L., Vaidya, H., Kanan, C.: Stream-51: streaming classification and novelty detection from videos. In: Proceedings of the IEEE/CVF Conference on Computer Vision and Pattern Recognition (CVPR) Workshops (2020)
26. Seeger, M.: Greedy forward selection in the informative vector machine. Technical report, University of California at Berkeley, Tech. rep. (2004)
27. Vitter, J.S.: Random sampling with a reservoir. ACM Trans. Math. Softw. (TOMS) **11**(1), 37–57 (1985)
28. Wei, K., Iyer, R., Bilmes, J.: Submodularity in data subset selection and active learning. In: International Conference on Machine Learning, pp. 1954–1963 (2015)

# Dep-$L_0$: Improving $L_0$-Based Network Sparsification via Dependency Modeling

Yang Li and Shihao Ji([✉])

Georgia State University, Atlanta, GA, USA
yli93@student.gsu.edu, sji@gsu.edu

**Abstract.** Training deep neural networks with an $L_0$ regularization is one of the prominent approaches for network pruning or sparsification. The method prunes the network during training by encouraging weights to become exactly zero. However, recent work of Gale et al. [11] reveals that although this method yields high compression rates on smaller datasets, it performs inconsistently on large-scale learning tasks, such as ResNet50 on ImageNet. We analyze this phenomenon through the lens of variational inference and find that it is likely due to the independent modeling of binary gates, the mean-field approximation [2], which is known in Bayesian statistics for its poor performance due to the crude approximation. To mitigate this deficiency, we propose a dependency modeling of binary gates, which can be modeled effectively as a multi-layer perceptron (MLP). We term our algorithm Dep-$L_0$ as it prunes networks via a dependency-enabled $L_0$ regularization. Extensive experiments on CIFAR10, CIFAR100 and ImageNet with VGG16, ResNet50, ResNet56 show that our Dep-$L_0$ outperforms the original $L_0$-HC algorithm of Louizos et al. [32] by a significant margin, especially on ImageNet. Compared with the state-of-the-arts network sparsification algorithms, our dependency modeling makes the $L_0$-based sparsification once again very competitive on large-scale learning tasks. Our source code is available at https://github.com/leo-yangli/dep-l0.

**Keywords:** Network sparsification · $L_0$-norm regularization · Dependency modeling

## 1 Introduction

Convolutional Neural Networks (CNNs) have achieved great success in a broad range of tasks. However, the huge model size and high computational price make the deployment of the state-of-the-art CNNs to resource-limited embedded systems (e.g., smart phones, drones and surveillance cameras) impractical. To alleviate this problem, substantial efforts have been made to compress and speed up the networks [6]. Among these efforts, network pruning has been proved to be an effective way to compress the model and speed up inference without losing noticeable accuracy [9,13,24,26,32,36,38,40].

The existing network pruning algorithms can be roughly categorized into two categories according to the pruning granularity: unstructured pruning [9,13,24,36] and structured pruning [8,25,27,31,38,40,43]. As shown in Fig. 1, unstructured pruning


N. Oliver et al. (Eds.): ECML PKDD 2021, LNAI 12977, pp. 167–183, 2021.
https://doi.org/10.1007/978-3-030-86523-8_11

includes weight-level, vector-level and kernel-level pruning, while structured pruning normally refers to filter-level pruning. Although unstructured pruning methods usually lead to higher prune rates than structured ones, they require specialized hardware or software to fully utilize the benefits induced by the high prune rates due to the irregular network structures yielded by unstructured pruning. On the other hand, structured pruning can maintain the regularity of network structures, while pruning the networks effectively, and hence can fully utilize the parallel computing resources of general-purpose CPUs or GPUs. Because of this, in recent years structured pruning has attracted a lot of attention and achieved impressive performances [8,27,31,43]. In this work, we focus on structured pruning, more specifically, filter-level pruning.

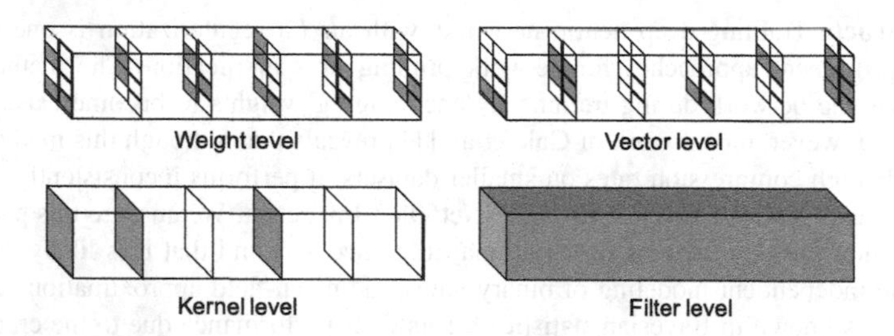


**Fig. 1.** Visualization of the weights of a convolutional filter and different pruning granularities. The red regions highlight the weights that can be pruned by different pruning methods. This paper focuses on filter-level pruning. (Color figure online)

In terms of pruning methods, a simple yet effective strategy is heuristic-based, e.g., pruning the weights based on their magnitudes [13,19,24,25]. Another popular approach is to penalize the model size via sparsity inducing regularization, such as $L_1$ or $L_0$ regularization [30,32,38]. Among them, $L_0$-HC [32] is one of the state-of-the-art pruning algorithms that incorporates $L_0$ regularization for network pruning and has demonstrated impressive performances on many image classification benchmarks (e.g., MNIST, CIFAR10 and CIFAR100). This method attaches a binary gate to each weight of a neural network, and penalizes the complexity of the network, measured by the $L_0$ norm of the weight matrix. However, recent work of Gale et al. [11] reveals that although $L_0$-HC works well on smaller datasets, it fails to prune very deep networks on large-scale datasets, such as ResNet50 on ImageNet. The original $L_0$-HC algorithm was proposed and evaluated on filter-level pruning, while Gale et al. [11] focus on the weight-level pruning. Therefore, it is unclear if the observation of [11] is due to pruning granularity or the deficiency of the $L_0$ regularization based method. To understand this, we evaluate the original $L_0$-HC to sparsify ResNet50 at filter level on ImageNet, and find that it indeed cannot prune ResNet50 without a significant damage of model quality, confirming the observation made by [11]. This indicates that the failure of $L_0$-HC is likely due to the deficiency of the $L_0$-norm based approach. We further analyze $L_0$-HC in the lens of variational inference [2], and find that **the failure is likely due to an over-simplified assumption that models the variational posterior of binary gates to be element-wise independent**. To verify this hypothesis, we propose to incorporate the dependency into the binary gates, and model the gate dependency across

CNN layers with a multi-layer perceptron (MLP). Extensive experiments show that our dependency-enabled $L_0$ sparsification, termed Dep-$L_0$, once again is able to prune very deep networks on large-scale datasets, while achieving competitive or sometimes even better performances than the state-of-the-art pruning methods.

Our main contributions can be summarized as follows:

- From a variational inference perspective, we show that the effectiveness of $L_0$-HC [32] might be hindered by the implicit assumption that all binary gates attached to a neural network are independent to each other. To mitigate this issue, we propose Dep-$L_0$ that incorporates the dependency into the binary gates to improve the original $L_0$-based sparsification method.
- A series of experiments on multiple datasets and multiple modern CNN architectures demonstrate that Dep-$L_0$ improves $L_0$-HC consistently, and is very competitive or sometimes even outperforms the state-of-the-art pruning algorithms.
- Moreover, Dep-$L_0$ converges faster than $L_0$-HC in terms of network structure search, and reduces the time to solution by 20%–40% compared to $L_0$-HC in our experiments.

## 2 Related Work

Model compression [6] aims to reduce the size of a model and speed up its inference at the same time. Recently, there has been a flurry of interest in model compression, ranging from network pruning [13,24,26,28,32], quantization and binarization [4,12], tensor decomposition [7,22], and knowledge distillation [18]. Since our algorithm belongs to the category of network pruning, we mainly focus on reviewing related work in pruning.

***Network Pruning.*** A large subset of pruning methods is heuristic-based, which assigns an importance score to each weight and prune the weights whose importance scores are below a threshold. The importance scores are usually devised according to types of networks, e.g., the magnitude of weights [13,24] (Feed-forward NNs), the $L_1$ or $L_2$ norm of filters [25] (CNNs), and the average percentage of zero activations [19] (CNNs). However, Ye et al. [39] point out that the assumption that weights/filters of smaller norms are less important may not hold in general, challenging the heuristic-based approaches. These methods usually follow a three-step training procedure: training - pruning - retraining in order to achieve the best performance.

Another subset of pruning methods focuses on training networks with sparsity inducing regularizations. For example, $L_2$ and $L_1$ regularizations [30,38] or $L_0$ regularization [32] can be incorporated into the objective functions to train sparse networks. Similarly, Molchanov et al. [34] propose variational dropout, a sparse Bayesian learning algorithm under an improper logscale uniform prior, to induce sparsity. In this framework, network pruning can be performed from scratch and gradually fulfilled during training without separated training stages.

Recently, Gale et al. [11] evaluate three popular pruning methods, including variational dropout [34], $L_0$-HC [32] and magnitude-based pruning [42], on two large-scale benchmarks. They reveal that although $L_0$-HC [32] is more sophisticated and yields state-of-the-art results on smaller datasets, it performs inconsistently on large-scale learning task of ImageNet. This observation motivates our development of Dep-$L_0$. To mitigate the deficiency of $L_0$-HC, we propose dependency modeling, which makes the $L_0$-based pruning once again competitive on large-scale learning tasks.

***Dependency Modelling.*** Even though there are many network pruning algorithms today, most of them (if not all) *implicitly* assume all the neurons of a network are independent to each other when selecting neurons for pruning. There are quite few works exploring the dependency inside neural networks for pruning. The closest one is LookAhead [36], which reinterprets the magnitude-based pruning as an optimization of the Frobenius distortion of a single layer, and improves the magnitude-based pruning by optimizing the Frobenius distortion of multiple layers, considering previous layer and next layer. Although the interaction of different layers is considered, the authors do not model the dependency of them explicitly. To the best of our knowledge, our Dep-$L_0$ is the first to model the dependency of neurons explicitly for network pruning.

# 3 Method

Our algorithm is motivated by $L_0$-HC [32], which prunes neural networks by optimizing an $L_0$ regularized loss function and relaxing the non-differentiable Bernoulli distribution with the Hard Concrete (HC) distribution. Since $L_0$-HC can be viewed as a special case of variational inference under the spike-and-slab prior [32], in this section we first formulate the sparse structure learning from this perspective, then discuss the deficiency of $L_0$-HC and propose dependency modeling, and finally present Dep-$L_0$.

## 3.1 Sparse Structure Learning

Consider a dataset $\mathcal{D} = \{\boldsymbol{x}_i, y_i\}_{i=1}^N$ that consists of $N$ pairs of instances, where $\boldsymbol{x}_i$ is the $i$th observed data and $y_i$ is the associated class label. We aim to learn a model $p(\mathcal{D}|\boldsymbol{\theta})$, parameterized by $\boldsymbol{\theta}$, which fits $\mathcal{D}$ well with the goal of achieving good generalization to unseen test data. In order to sparsify the model, we introduce a set of binary gates $\boldsymbol{z} = \{z_1, \cdots, z_{|\theta|}\}$, one gate for each parameter, to indicate whether the corresponding parameter being kept ($z = 1$) or not ($z = 0$).

This formulation is closely related to the spike-and-slab distribution [33], which is widely used as a prior in Bayesian inference to impose sparsity. Specifically, the spike-and-slab distribution defines a mixture of a delta spike at zero and a standard Gaussian distribution:

$$
\begin{aligned}
p(z) &= \mathrm{Bern}(z|\pi) \\
p(\theta|z=0) = \delta(\theta), &\quad p(\theta|z=1) = \mathcal{N}(\theta|0,1),
\end{aligned} \tag{1}
$$

where $\mathrm{Bern}(\cdot|\pi)$ is the Bernoulli distribution with parameter $\pi$, $\delta(\cdot)$ is the Dirac delta function, i.e., a point probability mass centered at the origin, and $\mathcal{N}(\theta|0,1)$ is the Gaussian distribution with zero mean and unit variance. Since both $\boldsymbol{\theta}$ and $\boldsymbol{z}$ are vectors, we assume the prior $p(\boldsymbol{\theta}, \boldsymbol{z})$ factorizes over the dimensionality of $\boldsymbol{z}$.

In Bayesian statistics, we would like to estimate the posterior of $(\boldsymbol{\theta}, \boldsymbol{z})$, which can be calculated by Bayes' rule:

$$p(\boldsymbol{\theta}, \boldsymbol{z}|\mathcal{D}) = \frac{p(\mathcal{D}|\boldsymbol{\theta}, \boldsymbol{z})p(\boldsymbol{\theta}, \boldsymbol{z})}{p(\mathcal{D})}. \tag{2}$$

Practically, the true posterior distribution $p(\boldsymbol{\theta}, \boldsymbol{z}|\mathcal{D})$ is intractable due to the non-conjugacy of the model likelihood and the prior. Therefore, here we approximate the posterior distribution via variational inference [2]. Specially, we can approximate the true posterior with a parametric variational posterior $q(\boldsymbol{\theta}, \boldsymbol{z})$, the quality of which can be measured by the Kullback-Leibler (KL) divergence:

$$KL[q(\boldsymbol{\theta}, \boldsymbol{z})||p(\boldsymbol{\theta}, \boldsymbol{z}|\mathcal{D})], \tag{3}$$

which is again intractable, but can be optimized by maximizing the variational lower bound of $\log p(\mathcal{D})$, defined as

$$L = \mathbb{E}_{q(\boldsymbol{\theta},\boldsymbol{z})}[\log p(\mathcal{D}|\boldsymbol{\theta}, \boldsymbol{z})] - KL[q(\boldsymbol{\theta}, \boldsymbol{z})||p(\boldsymbol{\theta}, \boldsymbol{z})], \tag{4}$$

where the second term can be further expanded as:

$$\begin{aligned} KL[q(\boldsymbol{\theta}, \boldsymbol{z})||p(\boldsymbol{\theta}, \boldsymbol{z})] &= \mathbb{E}_{q(\boldsymbol{\theta},\boldsymbol{z})}[\log q(\boldsymbol{\theta}, \boldsymbol{z}) - \log p(\boldsymbol{\theta}, \boldsymbol{z})] \\ &= \mathbb{E}_{q(\boldsymbol{\theta},\boldsymbol{z})}[\log q(\boldsymbol{\theta}|\boldsymbol{z}) - \log p(\boldsymbol{\theta}|\boldsymbol{z}) + \log q(\boldsymbol{z}) - \log p(\boldsymbol{z})] \\ &= KL[q(\boldsymbol{\theta}|\boldsymbol{z})||p(\boldsymbol{\theta}|\boldsymbol{z})] + KL[q(\boldsymbol{z})||p(\boldsymbol{z})]. \end{aligned} \tag{5}$$

In $L_0$-HC [32], the variational posterior $q(\boldsymbol{z})$ is factorized over the dimensionality of $\boldsymbol{z}$, i.e., $q(\boldsymbol{z}) = \prod_{j=1}^{|\boldsymbol{\theta}|} q(z_j) = \prod_{j=1}^{|\boldsymbol{\theta}|} \text{Bern}(z_j|\pi_j)$. By the law of total probability, we can further expand Eq. 5 as

$$\begin{aligned} &KL[q(\boldsymbol{\theta}, \boldsymbol{z})||p(\boldsymbol{\theta}, \boldsymbol{z})] \\ &= \sum_{j=1}^{|\boldsymbol{\theta}|} \Big( q(z_j = 0) KL[q(\theta_j|z_j = 0)||p(\theta_j|z_j = 0)] \\ &\quad + q(z_j = 1) KL[q(\theta_j|z_j = 1)||p(\theta_j|z_j = 1)] \Big) + \sum_{j=1}^{|\boldsymbol{\theta}|} KL[q(z_j)||p(z_j)] \\ &= \sum_{j=1}^{|\boldsymbol{\theta}|} q(z_j = 1) KL[q(\theta_j|z_j = 1)||p(\theta_j|z_j = 1)] + \sum_{j=1}^{|\boldsymbol{\theta}|} KL[(q(z_j)||p(z_j)]. \end{aligned} \tag{6}$$

The last step holds because $KL[q(\theta_j|z_j = 0)||p(\theta_j|z_j = 0)] = KL[q(\theta_j|z_j = 0)||\delta(\theta_j)] = 0$.

Furthermore, letting $\boldsymbol{\theta} = \tilde{\boldsymbol{\theta}} \odot \boldsymbol{z}$ and assuming $\lambda = KL[q(\theta_j|z_j = 1)||p(\theta_j|z_j = 1)]$, the lower bound $L$ (4) can be simplified as

$$L = \mathbb{E}_{q(\boldsymbol{z})}[\log p(\mathcal{D}|\tilde{\boldsymbol{\theta}} \odot \boldsymbol{z})] - \sum_{j=1}^{|\boldsymbol{\theta}|} KL\left(q\left(z_j\right)||p\left(z_j\right)\right) - \lambda \sum_{j=1}^{|\boldsymbol{\theta}|} q\left(z_j = 1\right)$$

$$\leq \mathbb{E}_{q(\boldsymbol{z})}[\log p(\mathcal{D}|\tilde{\boldsymbol{\theta}} \odot \boldsymbol{z})] - \lambda \sum_{j=1}^{|\boldsymbol{\theta}|} \pi_j, \tag{7}$$

where the inequality holds due to the non-negativity of KL-divergence.

Given that our model is a neural network $h(\boldsymbol{x}; \tilde{\boldsymbol{\theta}}, \boldsymbol{z})$, parameterized by $\tilde{\boldsymbol{\theta}}$ and $\boldsymbol{z}$, Eq. 7 turns out to be an $L_0$-regularized loss function [32]:

$$\mathcal{R}(\tilde{\boldsymbol{\theta}}, \boldsymbol{\pi}) = \mathbb{E}_{q(\boldsymbol{z})}\left[\frac{1}{N}\sum_{i=1}^{N} \mathcal{L}\left(h(\boldsymbol{x}_i; \tilde{\boldsymbol{\theta}} \odot \boldsymbol{z}), y_i\right)\right] + \lambda \sum_{j=1}^{|\boldsymbol{\theta}|} \pi_j, \tag{8}$$

where $\mathcal{L}(\cdot)$ is the cross entropy loss for classification.

In the derivations above, the variational posterior $q(\boldsymbol{z})$ is assumed to factorize over the dimensionality of $\boldsymbol{z}$, i.e., $q(\boldsymbol{z}) = \prod_{j=1}^{|\boldsymbol{\theta}|} q(z_j)$. This means all the binary gates $\boldsymbol{z}$ are assumed to be *independent* to each other – the mean-field approximation [2]. In variational inference, it is common to assume the prior $p(\boldsymbol{z})$ to be element-wise independent; the true posterior $p(\boldsymbol{z}|\mathcal{D})$, however, is unlikely to be element-wise independent. Therefore, approximating the true posterior by an element-wise independent $q(\boldsymbol{z})$ is a very restrict constraint that limits the search space of admissible $q(\boldsymbol{z})$ and is known in Bayesian statistics for its poor performance [1,2]. We thus hypothesize that this mean-field approximation may be the cause of the failure reported by Gale et al. [11], and the independent assumption hinders the effectiveness of the $L_0$-based pruning method. Therefore, we can potentially improve $L_0$-HC by relaxing this over-simplified assumption and modeling the dependency among binary gates $\boldsymbol{z}$ explicitly.

Specifically, instead of using a fully factorized variational posterior $q(\boldsymbol{z})$, we can model $q(\boldsymbol{z})$ as a conditional distribution by using the chain rule of probability

$$q(\boldsymbol{z}) = q(z_1)q(z_2|z_1)q(z_3|z_1, z_2)\cdots q(z_{|\theta|}|z_1, \cdots, z_{|\theta|-1}),$$

where given an order of binary gates $\boldsymbol{z} = \{z_1, z_2, \cdots, z_{|\theta|}\}$, $z_i$ is dependent on all previous gates $z_{<i}$. With this, Eq. 8 can be rewritten as

$$\mathcal{R}(\tilde{\boldsymbol{\theta}}, \boldsymbol{\pi}) = \lambda \sum_{j=1}^{|\boldsymbol{\theta}|} \pi_j + \mathbb{E}_{q(z_1)\cdots q(z_{|\theta|}|z_1,\cdots,z_{|\theta|-1})}\left[\frac{1}{N}\sum_{i=1}^{N} \mathcal{L}\left(h(\boldsymbol{x}_i; \tilde{\boldsymbol{\theta}} \odot \boldsymbol{z}), y_i\right)\right], \tag{9}$$

which is a dependency-enabled $L_0$ regularized loss function for network pruning. Detailed design of the dependency modeling is to be discussed in later sections.

### 3.2 Group Sparsity

So far we have modeled a sparse network by attaching a set of binary gates $\boldsymbol{z}$ to the network at the weight level. As we discussed in the introduction, we prefer to prune the network at the filter level to fully utilize general purpose CPUs or GPUs. To this end, we consider group sparsity that shares a gate within a group of weights. Let $G = \{g_1, g_2, \cdots, g_{|G|}\}$ be a set of groups, where each element corresponds to a group of weights, and $|G|$ is the number of groups. With the group sparsity, the expected $L_0$-norm of model parameters (the first term of Eq. 9) can be calculated as

$$\mathbb{E}_{q(\boldsymbol{z})}\|\boldsymbol{\theta}\|_0 = \sum_{j=1}^{|\boldsymbol{\theta}|} q(z_j = 1|z_{<j}) = \sum_{k=1}^{|G|} |g_k|\pi_k, \tag{10}$$

where $|g_k|$ denotes the number of weights in group $k$.

In all our experiments, we perform filter-level pruning by attaching a binary gate to all the weights of a filter (i.e., a group). Since modern CNN architectures often contain batch normalization layers [21], in our implementation we make a slight modification that instead of attaching the gates to filters directly, we attach the gates to the feature maps after batch normalization. This is because batch normalization accumulates a moving average of feature statistics for normalization during the training process. Simply attaching a binary gate to the weights of a filter cannot remove the impact of a filter completely when $z = 0$ due to the memorized statistics from batch normalization. By attaching the gates to the feature maps after batch normalization, the impact of the corresponding filter can be completely removed when $z = 0$.

### 3.3 Gate Partition

Modern CNN architectures, such as VGGNet [37], ResNet [14] and WideResNet [41], often come with a large number of weights and filters. For example, VGG16 [37] contains 138M parameters and 4,224 filters. Since we attach a binary gate to each filter, the number of gates would be large and modeling the dependencies among them would lead to a huge computational overhead and optimization issues. To make our dependency modeling more practical, we propose gate partition to simplify the dependency modeling among gates. Specifically, the gates are divided into blocks, and the gates within each block are considered independent to each other, whereas the gates cross blocks are considered dependent. Figure 2 illustrates the difference between an element-wise sequential dependency modeling and a partition-wise dependency modeling. Let's consider $z_1$, $z_2$, $z_3$ and $z_4$ in these two cases. In the element-wise sequential dependency modeling, as shown in Fig. 2(a), $z_2$ is dependent on $z_1$, $z_3$ is dependent on $z_1$ and $z_2$, and so on. As number of gates could be very large, the element-wise sequential modeling would lead to a very long sequence, whose calculation would incur huge computational overhead. Instead, we can partition the gates, as in Fig. 2(b), where $z_1$, $z_2$ and $z_3$ are in block $\boldsymbol{b}_1$ so they are considered independent to each other, while $z_4$ in block $\boldsymbol{b}_2$ is dependent on all $z_1$, $z_2$ and $z_3$.

We formally describe the gate partition as following. Given a set of gates $G = \{g_1, g_2, \cdots, g_{|G|}\}$, let $B = \{b_1, b_2, \cdots, b_{|B|}\}$ be a partition of $G$, where $b_i$ denotes block $i$, and $|B|$ is the total number of blocks. Then we can approximate the variational posterior of $\boldsymbol{z}$ by modeling the distribution over blocks as

$$q(\boldsymbol{z}) \approx q(b_1)q(b_2|b_1)q(b_3|b_1, b_2)\cdots q(b_{|B|}|b_1, \cdots, b_{|B|-1}).$$

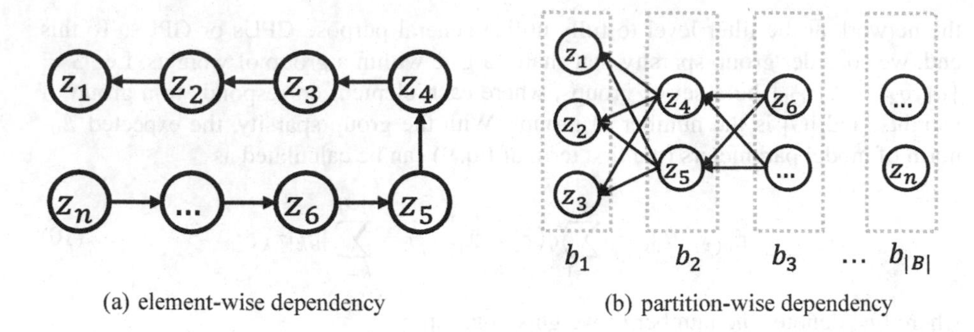


(a) element-wise dependency (b) partition-wise dependency

**Fig. 2.** Illustration of (a) element-wise sequential dependency modeling, and (b) partition-wise dependency modeling.

To reduce the complexity, we can further simplify it as

$$q(\boldsymbol{z}) \approx q(b_1)q(b_2|b_1)q(b_3|b_2)\cdots q(b_{|B|}|b_{|B|-1}), \tag{11}$$

where block $i$ only depends on previous block $i-1$, ignoring all the other previous blocks, i.e., $q(b_i|b_{i-1})$, – the first-order Markov assumption.

In our experiments, we define a layer-wise partition, i.e., a block containing all the filters in one layer. For example, in VGG16 after performing a layer-wise gate partition, we only need to model the dependency within 16 blocks instead of 4,224 gates, and therefore the computational overhead can be reduced significantly.

### 3.4 Neural Dependency Modeling

Until now we have discussed the dependency modeling in a mathematical form. To incorporate the dependency modeling into the original deep network, we adopt neural networks to model the dependencies among gates. Specifically, we choose to use an MLP network as the *gate generator*. With the proposed *layer-wise* gate partition (i.e., attaching an MLP layer to a convolutional layer and a gate to a filter; see Fig. 2(b)), the MLP architecture can model the dependency of gates, as expressed in Eq. 11, effectively.

Formally, we represent the gate generator as an MLP, with $gen_l$ denoting the operation of the $l$th layer. The binary gate $z_{lk}$ (i.e. the $k$th gate in block $l$) can be generated by

$$\begin{aligned} \log \alpha_0 &= \mathbf{1} \\ \log \alpha_l &= gen_l(\log \alpha_{l-1}) \quad \text{with } gen_l(\cdot) = c \cdot \tanh(W_l \cdot), \\ z_{lk} &\sim \mathrm{HC}(\log \alpha_{lk}, \beta), \end{aligned} \tag{12}$$

where $W_l$ is the weight matrix of MLP at the $l$th layer, $c$ is a hyperparameter that bounds the value of $\log \alpha$ in the range of $(-c, c)$, and $\mathrm{HC}(\log \alpha, \beta)$ is the Hard Concrete distribution with the location parameter $\log \alpha$ and the temperature parameter $\beta$ [32][1],

[1] Following $L_0$-HC [32], $\beta$ is fixed to 2/3 in our experiments.

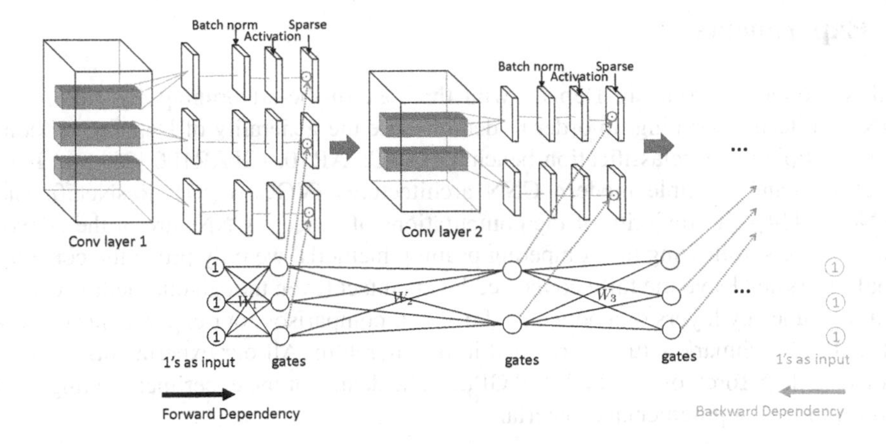


**Fig. 3.** The computational graph of Dep-$L_0$. The original CNN network is shown on the top, and the gate generator network (MLP) is shown at the bottom. Instead of attaching gates directly to filters, we attach gates to the feature maps after batch normalization (as shown by the red dotted lines). The gate can be generated by propagating the generator networks in either forward or backward direction. Both the original network and the gate generator are trained together as the whole pipeline is fully differentiable. $\{W_1, W_2, W_3, \cdots\}$ are the parameters of the MLP gate generator.

which makes the sample $z_{lk}$ differentiable w.r.t. $\log \alpha_{lk}$. We set $c = 10$ as default, which works well in all our experiments.

Figure 3 illustrates the overall architecture of Dep-$L_0$. The original network is shown on the top, and the gate generator (MLP) is shown at the bottom. Here we have a layer-wise partition of the gates, so the gate generator has the same number of layers as the original network. As we discussed in group sparsity, each gate is attached to a filter's output feature map after batch normalization. The input of the gate generator is initialized with a vector of 1's (i.e., $\log \alpha_0 = \mathbf{1}$, such that all the input neurons are activated at the beginning). The values of gates $\boldsymbol{z}$ are generated as we forward propagate the generator. The generated $\boldsymbol{z}$s are then attached to original networks, and the gate dependencies can be learned from the data directly. The whole pipeline (the original network and the gate generator) is fully differentiable as the Hard Concrete distribution (instead of Bernoulli) is used to sample $\boldsymbol{z}$, so that we can use backpropagation to optimize the whole pipeline.

Furthermore, as shown in Fig. 3, the dependencies can be modeled in a backward direction as well, i.e., we generate the gates from the last layer $L$ of MLP first, and then generate the gates from layer $L-1$, and so on. In our experiments, we will evaluate the performance impacts of both forward and backward modeling.

In addition to MLPs, other network architectures such as LSTMs and CNNs can be used to model the gate generator as well. However, neither LSTMs nor CNNs achieves a competitive performance to MLPs in our experiments. Detailed ablation study is provided in the supplementary material[2].

[2] https://arxiv.org/abs/2107.00070.

# 4 Experiments

In this section we compare Dep-$L_0$ with the state-of-the-art pruning algorithms for CNN architecture pruning. In order to demonstrate the generality of Dep-$L_0$, we consider multiple image classification benchmarks (CIFAR10, CIFAR100 [23] and ImageNet [5]) and multiple modern CNN architectures (VGG16 [37], ResNet50, and ResNet56 [14]). As majority of the computations of modern CNNs are in the convolutional layers, following the competing pruning methods, we only prune the convolutional filters and leave the fully connected layers intact (even though our method can be used to prune any layers of a network). For a fair comparison, our experiments closely follow the benchmark settings provided in the literature. All our experiments are performed with PyTorch on Nvidia V100 GPUs. The details of the experiment settings can be found in the supplementary material.

**$L_0$-HC Implementations.** From our experiments, we found that the original $L_0$-HC implementation[3] has a couple issues. First, the binary gates are not properly attached after batch normalization, which results in pruned neurons still having impact after being removed. Second, it only uses one optimizer – Adam for the original network parameters and the hard concrete parameters. We noted that using two optimizers: SGD with momentum for the original network and Adam for the hard concrete parameters works better. Therefore, we fixed these issues of $L_0$-HC for all the experiments and observed improved performance. For a fair comparison, we follow the same experiment settings as in Dep-$L_0$, and tune $L_0$-HC for the best performance.

## 4.1 CIFAR10 Results

We compare Dep-$L_0$ with ten state-of-the-art filter pruning algorithms, including our main baseline $L_0$-HC, in this experiment. Since the baseline accuracies in all the reference papers are different, we compare the performances of all competing methods by their accuracy gains $\Delta_{Acc}$ and their pruning rates in terms of FLOPs and network parameters. For Dep-$L_0$, we evaluate our algorithm with *forward* and *backward* dependency modeling. Table 1 provides the results on CIFAR10. As can be seen, for VGG16, our algorithm (with *backward* dependency modeling) achieves the highest FLOPs reduction of 65.9% on CIFAR10 with only 0.1% of accuracy loss. For ResNet56, our *forward* dependency modeling achieves the highest accuracy gain of 0.2% with a very competitive FLOPs reduction of 45.5%.

Since $L_0$-HC is our main baseline, we highlight the comparison between Dep-$L_0$ and $L_0$-HC in the table. As we can see, Dep-$L_0$ outperforms $L_0$-HC consistently in all the experiments. For VGG16, $L_0$-HC prunes only 39.8% of FLOPs but suffers from a 0.4% of accuracy drop, while our algorithm prunes more (65.9%) and almost keeps the same accuracy (–0.1%). For ResNet56, our algorithm prunes more (45.5% v.s. 44.1%) while achieves a higher accuracy (0.2% vs. –0.5%) than that of $L_0$-HC.

[3] https://github.com/AMLab-Amsterdam/L0_regularization.

**Table 1.** Comparison of pruning methods on CIFAR10. "$\Delta_{Acc}$": '+' denotes accuracy gain; '-' denotes accuracy loss; the worst result is in red. "FLOPs (P.R. %)": pruning ratio in FLOPs. "Params. (P.R. %)": prune ratio in parameters. "-": results not reported in original paper.

| Model | Method | Acc. (%) | $\Delta_{Acc}$ | FLOPs (P.R. %) | Params. (P.R. %) |
|---|---|---|---|---|---|
| VGG16 | Slimming [31] | 93.7→93.8 | +0.1 | 195M (51.0) | 2.30M (88.5) |
| | DCP [43] | 94.0→94.6 | **+0.6** | 109.8M (65.0) | **0.94M (93.6)** |
| | AOFP [8] | 93.4→93.8 | +0.4 | 215M (31.3) | – |
| | HRank [27] | 94.0→93.4 | −0.6 | 145M (53.5) | 2.51M (82.9) |
| | $L_0$-HC (Our implementation) | 93.5→93.1 | −0.4 | 135.6M (39.8) | 2.8M (80.9) |
| | Dep-$L_0$ (forward) | 93.5→93.5 | 0 | 111.9M (64.4) | 2.1M (85.7) |
| | Dep-$L_0$ (backward) | 93.5→93.4 | −0.1 | **107.0M (65.9)** | 1.8M (87.8) |
| ResNet56 | SFP [15] | 93.6→93.4 | −0.2 | 59.4M (53.1) | – |
| | AMC [17] | 92.8→91.9 | −0.9 | 62.5M (50.0) | – |
| | DCP [43] | 93.8→93.8 | 0 | 67.1M (47.1) | **0.25M (70.3)** |
| | FPGM [16] | 93.6→93.5 | −0.1 | 59.4M (52.6) | – |
| | TAS [10] | 94.5→93.7 | −0.8 | **59.5M (52.7)** | – |
| | HRank [27] | 93.3→93.5 | **+0.2** | 88.7M (29.3) | 0.71M (16.8) |
| | $L_0$-HC (Our implementation) | 93.3→92.8 | −0.5 | 71.0M (44.1) | 0.46M (45.9) |
| | Dep-$L_0$ (forward) | 93.3→93.5 | **+0.2** | 69.1M (45.5) | 0.48M (43.5) |
| | Dep-$L_0$ (backward) | 93.3→93.0 | −0.3 | 66.7M (47.4) | 0.49M (42.4) |

**Table 2.** Comparison of pruning methods on CIFAR100. "$\Delta_{Acc}$": '+' denotes accuracy gain; '-' denotes accuracy loss; the worst result is in red. "FLOPs (P.R. %)": pruning ratio in FLOPs. "Params. (P.R. %)": prune ratio in parameters. "-": results not reported in original paper.

| Model | Method | Acc. (P.R. %) | $\Delta_{Acc}$ | FLOPs (P.R. %) | Params. (%) |
|---|---|---|---|---|---|
| VGG16 | Slimming [31] | 73.3→73.5 | +0.2 | 250M (37.1) | 5.0M (75.1) |
| | $L_0$-HC (Our implementation) | 72.2→70.0 | -1.2 | 138M (56.2) | 4.1M (72.5) |
| | Dep-$L_0$ (forward) | 72.2→71.6 | -0.6 | **98M (68.8)** | **2.1M (85.7)** |
| | Dep-$L_0$ (backward) | 72.2→72.5 | **+0.3** | 105M (66.6) | 2.2M (85.0) |
| ResNet56 | SFP [15] | 71.4→68.8 | -2.6 | **59.4M (52.6)** | – |
| | FPGM [16] | 71.4→69.7 | −1.7 | **59.4M (52.6)** | – |
| | TAS [10] | 73.2→72.3 | −0.9 | 61.2M (51.3) | – |
| | $L_0$-HC (Our implementation) | 71.8→70.4 | −1.4 | 82.2M (35.2) | 0.73M (15.2) |
| | Dep-$L_0$ (forward) | 71.8→71.7 | **−0.1** | 87.6M (30.9) | 0.56M (34.9) |
| | Dep-$L_0$ (backward) | 71.8→71.2 | −0.6 | 93.4M (26.3) | **0.52M (39.5)** |

## 4.2 CIFAR100 Results

Experimental results on CIFAR100 are reported in Table 2, where Dep-$L_0$ is compared with four state-of-the-arts pruning algorithms: Slimming [31], SFP [15], FPGM [16] and TAS [10]. Similar to the results on CIFAR10, on this benchmark Dep-$L_0$ achieves the best accuracy gains and very competitive or sometimes even higher prune rates com-

**Table 3.** Comparison of pruning methods on ImageNet. "$\Delta_{Acc}$": '+' denotes accuracy gain; '-' denotes accuracy loss; the worst result is in red. "FLOPs (P.R. %)": pruning ratio in FLOPs. "Params. (P.R. %)": prune ratio in parameters. "–": results not reported in original paper.

| Model | Method | Acc. (%) | $\Delta_{Acc}$ | FLOPs (P.R.%) | Params. (P.R.%) |
|---|---|---|---|---|---|
| ResNet50 | SSS-32 [20] | 76.12 → 74.18 | -1.94 | 2.82B (31.1) | 18.60M (27.3) |
| | DCP [43] | 76.01 → 74.95 | -1.06 | **1.82B (55.6)** | **12.40M (51.5)** |
| | GAL-0.5 [29] | 76.15 → 71.95 | -4.2 | 2.33B (43.1) | 21.20M (17.2) |
| | Taylor-72 [35] | 76.18 → 74.50 | -1.68 | 2.25B (45.0) | 14.20M (44.5) |
| | Taylor-81 [35] | 76.18 → 75.48 | -0.70 | 2.66B (34.9) | 17.90M (30.1) |
| | FPGM-30 [16] | 76.15 → 75.59 | -0.56 | 2.36B (42.2) | – |
| | FPGM-40 [16] | 76.15 → 74.83 | -1.32 | 1.90B (53.5) | – |
| | LeGR [3] | 76.10 → 75.70 | -0.40 | 2.37B (42.0) | – |
| | TAS [10] | 77.46 → 76.20 | -1.26 | 2.31B (43.5) | – |
| | HRank [27] | 76.15 → 74.98 | -1.17 | 2.30B (43.8) | 16.15M (36.9) |
| | $L_0$-HC (Our implementation) | 76.15 → 76.15 | 0 | 4.09B (0.00) | 25.58M (0.00) |
| | Dep-$L_0$ (forward) | 76.15 → 74.77 | -1.38 | 2.58B (36.9) | 16.04M (37.2) |
| | Dep-$L_0$ (backward) | 76.15 → 74.70 | -1.45 | 2.53B (38.1) | 14.34M (43.9) |

pared to the state-of-the-arts. More importantly, Dep-$L_0$ outperforms $L_0$-HC in terms of classification accuracies and pruning rates consistently, demonstrating the effectiveness of dependency modeling.

### 4.3 ImageNet Results

The main goal of the paper is to make $L_0$-HC once again competitive on the large-scale benchmark of ImageNet. In this section, we conduct a comprehensive experiment on ImageNet, where the original $L_0$-HC fails to prune without a significant damage of model quality [11]. Table 3 reports the results on ImageNet, where eight state-of-the-art filter pruning methods are included, such as SSS-32 [20], DCP [43], Taylor [35], FPGM [16], HRank [27] and others. As can be seen, Dep-$L_0$ (forward) prunes 36.9% of FLOPs and 37.2% of parameters with a 1.38% of accuracy loss, which is comparable with other state-of-the-art algorithms as shown in the table.

Again, since $L_0$-HC is our main baseline, we highlight the comparison between Dep-$L_0$ and $L_0$-HC in the table. We tune the performance of $L_0$-HC extensively by searching for the best hyperparameters in a large space. However, even with extensive efforts, $L_0$-HC still fails to prune the network without a significant damage of model quality, confirming the observation made by [11]. On the other hand, our Dep-$L_0$ successfully prunes ResNet50 with a very competitive pruning rate and high accuracy compared to the state-of-the-arts, indicating that our dependency modeling indeed makes the original $L_0$-HC very competitive on the large-scale benchmark of ImageNet – the main goal of the paper.

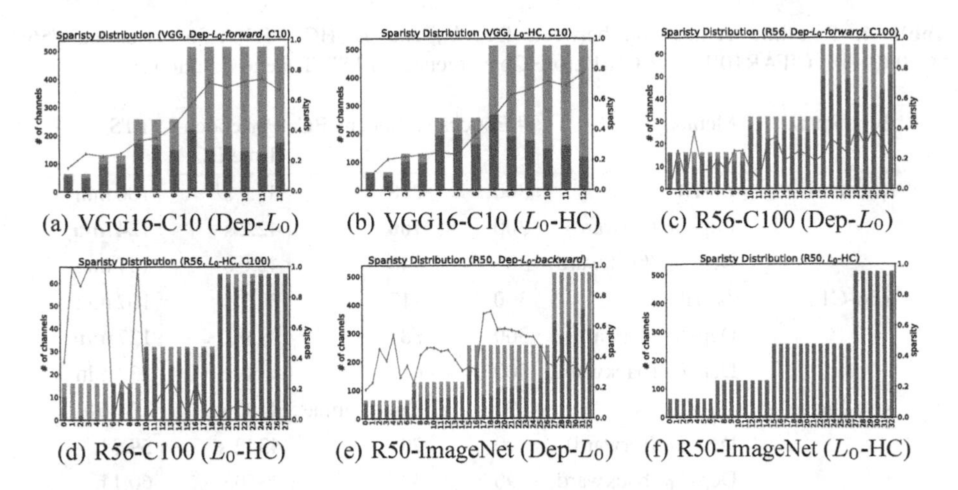


(a) VGG16-C10 (Dep-$L_0$) (b) VGG16-C10 ($L_0$-HC) (c) R56-C100 (Dep-$L_0$)

(d) R56-C100 ($L_0$-HC) (e) R50-ImageNet (Dep-$L_0$) (f) R50-ImageNet ($L_0$-HC)

**Fig. 4.** The layer-wise prune ratios (red curves) of learned sparse structures. The height of a bar denotes the number of filters of a convolutional layer and gray (green) bars correspond to the original (pruned) architecture, respectively. "R50/R56": ResNet 50/56; "C10/C100": CIFAR10/100. (Color figure online)

### 4.4 Study of Learned Sparse Structures

To understand of the behavior of Dep-$L_0$, we further investigate the sparse structures learned by Dep-$L_0$ and $L_0$-HC, with the results reported in Fig. 4. For VGG16 on CIFAR10, Figs. 4(a–b) demonstrate that both algorithms learn a similar sparsity pattern: the deeper a layer is, the higher prune ratio is, indicating that the shallow layers of VGG16 are more important for its predictive performance. However, for deeper networks such as ResNet56 and ResNet50, the two algorithms perform very differently. For ResNet56 on CIFAR100, Figs. 4(c–d) show that Dep-$L_0$ sparsifies each layer by a roughly similar prune rate: it prunes around 20% of the filters in first 12 layers, and around 30% of the filters in the rest of layers. However, on the same benchmark $L_0$-HC tends to prune *all or nothing*: it completely prunes 5 out of 28 layers, but does not prune any filters in other six layers; for the rest of layers, the sparsity produced by $L_0$-HC is either extremely high or low. As of ResNet50 on ImageNet, Figs. 4(e–f) show that the difference between Dep-$L_0$ and $L_0$-HC is more significant: Dep-$L_0$ successfully prunes the model with a roughly similar prune rate across all convolutional layers, while $L_0$-HC fails to prune any filters.

### 4.5 Run-Time Comparison

The main architectural difference between Dep-$L_0$ and $L_0$-HC is the gate generator. Even though the gate generator (MLP) is relatively small compared to the original deep network to be pruned, its existence increases the computational complexity of Dep-$L_0$. Thus, it is worth comparing the run-times of Dep-$L_0$ and $L_0$-HC as well as their convergence rates in terms of sparse structure search. Once a sparse structure is learned by a pruning algorithm, we can extract the sparse network from the original network and

**Table 4.** Run-time comparison between Dep-$L_0$ and $L_0$-HC. "R50/R56": ResNet50/56; "C10/C100": CIFAR10/100. "BC": Before Convergence; "TTS": Time to Solution.

| Benchmark | Method | # Epochs | # Epochs BC | Per-epoch Time BC | TTS |
|---|---|---|---|---|---|
| R56-C10 | $L_0$-HC | 300 | 218 | 37.9 s | 160 min |
| | Dep-$L_0$ (forward) | 300 | **106** | 42.2 s | **124 min** |
| | Dep-$L_0$ (backward) | 300 | 140 | 42.6 s | 141 min |
| R56-C100 | $L_0$-HC | 300 | 117 | 38.7 s | 167 min |
| | Dep-$L_0$ (forward) | 300 | **58** | 43.9 s | **127 min** |
| | Dep-$L_0$ (backward) | 300 | 61 | 43.6 s | 133 min |
| R50-ImageNet | $L_0$-HC | 90 | Fail to prune | 4185 s | 104.6 h |
| | Dep-$L_0$ (forward) | 90 | **30** | 4342 s | **59.5 h** |
| | Dep-$L_0$ (backward) | 90 | 32 | 4350 s | 60.1 h |

continue the training on the smaller structure such that we can reduce the total time to solution (TTS). To this end, we compare Dep-$L_0$ with $L_0$-HC in terms of (1) structure search convergence rate, i.e., how many training epochs are needed for a pruning algorithm to converge to a sparse structure? (2) Per-epoch training time before convergence, and (3) the total time to solution (TTS). The results are reported in Table 4. As can be seen, Dep-$L_0$ (both *forward* and *backward*) converges to a sparse structure in roughly half of the epochs that $L_0$-HC needs (column 4). Even though the per-epoch training time of Dep-$L_0$ is 12% (4%) larger than that of $L_0$-HC on CIFAR10/100 (ImageNet) due to the extra computation of the gate generator (column 5), the total time to solution reduces by 22.5% (43.1%) on the CIFAR10/100 (ImageNet) benchmarks thanks to the faster convergence rates and sparser models induced by Dep-$L_0$ as compared to $L_0$-HC (column 6).

# 5 Conclusion and Future Work

We propose Dep-$L_0$, an improved $L_0$ regularized network sparsification algorithm via dependency modeling. The algorithm is inspired by a recent observation of Gale et al. [11] that $L_0$-HC performs inconsistently in large-scale learning tasks. Through the lens of variational inference, we found that this is likely due to the mean-field assumption in variational inference that ignores the dependency among all the neurons for network pruning. We further propose a dependency modeling of binary gates to alleviate the deficiency of the original $L_0$-HC. A series of experiments are performed to evaluate the generality of our Dep-$L_0$. The results show that our Dep-$L_0$ outperforms the original $L_0$-HC in all the experiments consistently, and the dependency modeling makes the $L_0$-based sparsification once again very competitive and sometimes even outperforms the state-of-the-art pruning algorithms. Further analysis shows that Dep-$L_0$ also learns a better structure in fewer epochs, and reduces the total time to solution by 20%-40%.

As for future work, we plan to explore whether dependency modeling can be used to improve other pruning methods. To the best of our knowledge, there are very few prior works considering dependency for network pruning (e.g., [36]). Our results show that this may be a promising direction to further improve many existing pruning algorithms. Moreover, the way we implement dependency modeling is still very preliminary, which can be improved further in the future.

**Acknowledgment.** We would like to thank the anonymous reviewers for their comments and suggestions, which helped improve the quality of this paper. We would also gratefully acknowledge the support of VMware Inc. for its university research fund to this research.

## References

1. Bishop, C.M.: Pattern Recognition and Machine Learning (Information Science and Statistics). Springer, Heidelberg (2007)
2. Blei, D.M., Kucukelbir, A., McAuliffe, J.D.: Variational inference: a review for statisticians. J. Am. Stat. Assoc. **112**, 859–877 (2017)
3. Chin, T.W., Ding, R., Zhang, C., Marculescu, D.: Legr: filter pruning via learned global ranking. arXiv pp. arXiv-1904 (2019)
4. Courbariaux, M., Hubara, I., Soudry, D., El-Yaniv, R., Bengio, Y.: Binarized neural networks: training deep neural networks with weights and activations constrained to +1 or −1. arXiv preprint arXiv:1602.02830 (2016)
5. Deng, J., Dong, W., Socher, R., Li, L.J., Li, K., Fei-Fei, L.: Imagenet: a large-scale hierarchical image database. In: CVPR, pp. 248–255 (2009)
6. Deng, L., Li, G., Han, S., Shi, L., Xie, Y.: Model compression and hardware acceleration for neural networks: a comprehensive survey. Proc. IEEE **108**, 485–532 (2020)
7. Denton, E.L., Zaremba, W., Bruna, J., LeCun, Y., Fergus, R.: Exploiting linear structure within convolutional networks for efficient evaluation. In: Advances in Neural Information Processing Systems pp. 1269–1277 (2014)
8. Ding, X., Ding, G., Guo, Y., Han, J., Yan, C.: Approximated oracle filter pruning for destructive cnn width optimization. arXiv preprint arXiv:1905.04748 (2019)
9. Ding, X., Ding, G., Zhou, X., Guo, Y., Han, J., Liu, J.: Global sparse momentum sgd for pruning very deep neural networks. In: Advances in Neural Information Processing Systems (2019)
10. Dong, X., Yang, Y.: Network pruning via transformable architecture search. In: Advances in Neural Information Processing Systems, pp. 760–771 (2019)
11. Gale, T., Elsen, E., Hooker, S.: The state of sparsity in deep neural networks. arXiv preprint arXiv:1902.09574 (2019)
12. Gupta, S., Agrawal, A., Gopalakrishnan, K., Narayanan, P.: Deep learning with limited numerical precision. In: International Conference on Machine Learning, pp. 1737–1746 (2015)
13. Han, S., Pool, J., Tran, J., Dally, W.: Learning both weights and connections for efficient neural network. In: Advances in Neural Information Processing Systems, pp. 1135–1143 (2015)
14. He, K., Zhang, X., Ren, S., Sun, J.: Deep residual learning for image recognition. In: CVPR, pp. 770–778 (2016)
15. He, Y., Kang, G., Dong, X., Fu, Y., Yang, Y.: Soft filter pruning for accelerating deep convolutional neural networks. arXiv preprint arXiv:1808.06866 (2018)
16. He, Y., Liu, P., Wang, Z., Hu, Z., Yang, Y.: Filter pruning via geometric median for deep convolutional neural networks acceleration. In: Proceedings of the IEEE Conference on Computer Vision and Pattern Recognition, pp. 4340–4349 (2019)

17. He, Y., Lin, J., Liu, Z., Wang, H., Li, L.J., Han, S.: Amc: automl for model compression and acceleration on mobile devices. In: Proceedings of the European Conference on Computer Vision (ECCV), pp. 784–800 (2018)
18. Hinton, G., Vinyals, O., Dean, J.: Distilling the knowledge in a neural network. arXiv preprint arXiv:1503.02531 (2015)
19. Hu, H., Peng, R., Tai, Y.W., Tang, C.K.: Network trimming: a data-driven neuron pruning approach towards efficient deep architectures. arXiv preprint arXiv:1607.03250 (2016)
20. Huang, Z., Wang, N.: Data-driven sparse structure selection for deep neural networks. In: Proceedings of the European Conference on Computer Vision (ECCV), pp. 304–320 (2018)
21. Ioffe, S., Szegedy, C.: Batch normalization: accelerating deep network training by reducing internal covariate shift. In: International Conference on Machine Learning (ICML) (2015)
22. Jaderberg, M., Vedaldi, A., Zisserman, A.: Speeding up convolutional neural networks with low rank expansions. arXiv preprint arXiv:1405.3866 (2014)
23. Krizhevsky, A., Hinton, G., et al.: Learning multiple layers of features from tiny images (2009)
24. LeCun, Y., Denker, J.S., Solla, S.A.: Optimal brain damage. In: Advances in Neural Information Processing Systems, pp. 598–605 (1990)
25. Li, H., Kadav, A., Durdanovic, I., Samet, H., Graf, H.P.: Pruning filters for efficient convnets. arXiv preprint arXiv:1608.08710 (2016)
26. Li, Y., Ji, S.: $l\_0$-arm: Network sparsification via stochastic binary optimization. In: Joint European Conference on Machine Learning and Knowledge Discovery in Databases, pp. 432–448. Springer, Heidelberg (2019)
27. Lin, M., Ji, R., Wang, Y., Zhang, Y., Zhang, B., Tian, Y., Shao, L.: Hrank: filter pruning using high-rank feature map. In: Proceedings of the IEEE/CVF Conference on Computer Vision and Pattern Recognition, pp. 1529–1538 (2020)
28. Lin, S., Ji, R., Li, Y., Deng, C., Li, X.: Toward compact convnets via structure-sparsity regularized filter pruning. IEEE Trans. Neural Netw. Learn. Syst. **31**(2), 574–588 (2019)
29. Lin, S., et al.: Towards optimal structured cnn pruning via generative adversarial learning. In: Proceedings of the IEEE Conference on Computer Vision and Pattern Recognition, pp. 2790–2799 (2019)
30. Liu, B., Wang, M., Foroosh, H., Tappen, M., Pensky, M.: Sparse convolutional neural networks. In: CVPR, pp. 806–814 (2015)
31. Liu, Z., Li, J., Shen, Z., Huang, G., Yan, S., Zhang, C.: Learning efficient convolutional networks through network slimming. In: Proceedings of the IEEE International Conference on Computer Vision, pp. 2736–2744 (2017)
32. Louizos, C., Welling, M., Kingma, D.P.: Learning sparse neural networks through $l\_0$ regularization. In: International Conference on Learning Representations (ICLR) (2018)
33. Mitchell, T.J., Beauchamp, J.J.: Bayesian variable selection in linear regression. J. Am. Stat. Assoc. **83**(404), 1023–1032 (1988)
34. Molchanov, D., Ashukha, A., Vetrov, D.: Variational dropout sparsifies deep neural networks. arXiv preprint arXiv:1701.05369 (2017)
35. Molchanov, P., Mallya, A., Tyree, S., Frosio, I., Kautz, J.: Importance estimation for neural network pruning. In: Proceedings of the IEEE Conference on Computer Vision and Pattern Recognition, pp. 11264–11272 (2019)
36. Park, S., Lee, J., Mo, S., Shin, J.: Lookahead: a far-sighted alternative of magnitude-based pruning. arXiv preprint arXiv:2002.04809 (2020)
37. Simonyan, K., Zisserman, A.: Very deep convolutional networks for large-scale image recognition. arXiv preprint arXiv:1409.1556 (2014)
38. Wen, W., Wu, C., Wang, Y., Chen, Y., Li, H.: Learning structured sparsity in deep neural networks. In: Advances in Neural Information Processing Systems (2016)

39. Ye, J., Lu, X., Lin, Z., Wang, J.Z.: Rethinking the smaller-norm-less-informative assumption in channel pruning of convolution layers. arXiv preprint arXiv:1802.00124 (2018)
40. You, Z., Yan, K., Ye, J., Ma, M., Wang, P.: Gate decorator: global filter pruning method for accelerating deep convolutional neural networks. In: Advances in Neural Information Processing Systems, pp. 2133–2144 (2019)
41. Zagoruyko, S., Komodakis, N.: Wide residual networks (2016)
42. Zhu, M., Gupta, S.: To prune, or not to prune: exploring the efficacy of pruning for model compression. arXiv preprint arXiv:1710.01878 (2017)
43. Zhuang, Z., et al.: Discrimination-aware channel pruning for deep neural networks. In: Advances in Neural Information Processing Systems, pp. 875–886 (2018)

# Variance Reduced Stochastic Proximal Algorithm for AUC Maximization

Soham Dan(✉) and Dushyant Sahoo

University of Pennsylvania, Philadelphia, USA
{sohamdan,sadu}@seas.upenn.edu

**Abstract.** Stochastic Gradient Descent has been widely studied with classification accuracy as a performance measure. However, these stochastic algorithms are not applicable when non-decomposable pair-wise performance measures are used, such as Area under the ROC curve (AUC), a standard performance metric used when the classes are imbalanced. Several algorithms have been proposed for optimizing AUC as a performance metric, one of the recent being a Stochastic Proximal Gradient Algorithm (SPAM). However, the downside of stochastic gradient descent is that it suffers from high variance leading to very slow convergence. Several variance reduced methods have been proposed with faster convergence guarantees than vanilla stochastic gradient descent to combat this issue. Again, these variance reduced methods are not applicable when non-decomposable performance measures are used. In this paper, we develop a Variance Reduced Stochastic Proximal algorithm for AUC Maximization (VRSPAM) that combines the two areas of analyzing non-decomposable performance metrics with and optimization efforts to guarantee faster convergence. We perform an in-depth theoretical and empirical analysis to demonstrate that our algorithm converges faster than existing state-of-the-art algorithms for the AUC maximization problem.

**Keywords:** Optimization · AUC · Variance reduction

## 1 Introduction

With the wide application of machine learning, there has been significant focus in recent times on applications that involve class imbalance—the case where one of the classes (the majority class) occurs much more frequently than the other class (the minority class) [6]. A concrete example is a medical diagnosis for a rare disease where far fewer instances from the disease class are observed than the healthy class. Traditional classification accuracy is not an appropriate performance metric in this setting, as predicting the majority class will give a high classification accuracy, even if the model always gives the wrong prediction on the minority class. To overcome this drawback, the Area under the ROC

S. Dan and D. Sahoo—Equal Contribution


N. Oliver et al. (Eds.): ECML PKDD 2021, LNAI 12977, pp. 184–199, 2021.
https://doi.org/10.1007/978-3-030-86523-8_12

curve (AUC) [7] is used as a standard metric for quantifying the performance of a binary classifier in this setting. AUC measures the ability of a family of classifiers to correctly rank an example from the positive class with respect to a randomly selected example from the negative class.

Several algorithms have been proposed for AUC maximization in the batch setting, where all the training data is assumed to be available at the beginning [12,25]. However, this assumption is unrealistic in several cases, especially for streaming data analysis, where examples are observed one at a time. For the usual classification accuracy metric, there exists *online algorithms* for such a streaming setting where the per iteration complexity is low [18,21]. However, despite several studies on online algorithms for classification accuracy, the case of maximizing AUC as a performance measure has been looked at only recently [14,26]. The main challenge for optimizing the AUC metric in the online setting is the pairwise nature of the AUC metric which, compared to classification accuracy, does not decompose over individual instances. In the AUC maximization framework, in each step the algorithm needs to pair the current datapoint with all previously observed datapoints leading to $\mathcal{O}(td)$ space and time complexity at step $t$, where the dimension of the instance space is $d$. The problem was not alleviated by the technique of buffering [14,26] since, good generalization performance depends on maintaining a large buffer.

From an optimization perspective, the AUC metric is non-convex and thus hard to optimize. Instead, it is attractive to optimize the convex surrogate, which is consistent, such as the pairwise squared surrogate [1,10,16]. Recently, [24] reformulated the pairwise squared loss surrogate of AUC as a saddle point problem and gave an algorithm that has a convergence rate of $\mathcal{O}(\frac{1}{\sqrt{t}})$. However, they only consider smooth regularization (penalty) terms such as Frobenius norm. Further, their convergence rate is sub-optimal to what stochastic gradient descent (SGD) achieves with classification accuracy as a performance measure $\mathcal{O}(\frac{1}{t})$. [17] improves on this with a stochastic proximal algorithm for AUC maximization, which under assumptions of strong convexity can achieve a convergence rate of $\mathcal{O}(\frac{\log t}{t})$ and has per iteration complexity of $\mathcal{O}(d)$ i.e., one datapoint and applies to general, non-smooth regularization terms.

Although [17] improves convergence for surrogate-AUC maximization, it still suffers from a high variance of the gradient in each iteration. Due to the large variance in random sampling, the stochastic gradient algorithm wastes time bouncing around, leading to worse performance and a slower sub-linear convergence rate of $\mathcal{O}(\frac{1}{t})$ (even if we ignore the $\log(t)$ term). Thus, we have the following trade-off: low per iteration complexity for the stochastic algorithm but slow convergence contrasted with high per iteration complexity and fast convergence for full gradient descent. Thus, it will take longer to get a good approximation of the solution to the AUC optimization problem if we employ the algorithm proposed by [17]. It is precisely this problem that we tackle in this paper: can we design an algorithm for AUC optimization that also enjoys fast convergence (potentially by controlling the variance of the iterates from the stochastic gradient algorithm).

In the relatively well-studied context of classification accuracy, techniques to reduce the variance of SGD have been proposed—SAG [19], SDCA [20], SVRG [13]. While SAG and SDCA require the storage of all the gradients and dual variables respectively, for complex models SVRG enjoys the same fast convergence rates as SDCA and SAG but has a much simpler analysis and does not require storage of gradients. This allows SVRG to be applicable in complex problems where the storage of all gradients would be infeasible.

Several works have explored ways to apply SVRG on classification problems involving a regularizer: the overall objective consists of the sum of a regularizer term and the average of several smooth component function terms in SVRG. Two simple strategies commonly used are the Proximal Full Gradient and the Proximal Stochastic Gradient method. While the Proximal Stochastic Gradient is much faster since it computes only the gradient of a single component function per iteration, it convergences much slower than the Proximal Full Gradient method, alluding to the same trade-off we mentioned earlier. The proximal gradient methods can be viewed as a particular case of splitting methods [2,3]. However, both the proximal methods do not fully exploit the problem structure. Proximal SVRG [22] is an extension of the SVRG [13] technique and can be used whenever the objective function is composed of two terms- the first term is an average of smooth functions (decomposable across the individual instances), and the second term admits a simple proximal mapping. Prox-SVRG needs far fewer iterations to achieve the same approximation ratio than the proximal full and stochastic gradient descent methods. However, all the existing techniques discussed that guarantee faster convergence by controlling the variance, including Prox-SVRG and the proximal full and stochastic gradient descent methods, all have a very restrictive assumption - they require the metric and the loss function to be decomposable over instances (for example, classification accuracy and the corresponding decomposable pointwise surrogate loss functions) and are not directly applicable to non-decomposable pairwise loss functions as in surrogate-AUC optimization (refer to Sect. 2); this is the gap that we close in this paper.

In this paper, we present Variance Reduced Stochastic Proximal algorithm for AUC Maximization (VRSPAM). VRSPAM builds upon previous work for surrogate-AUC maximization by using the SVRG algorithm. We provide theoretical analysis for the VRSPAM algorithm showing that it achieves a linear convergence rate with a fixed step size (much faster than SPAM [17], which has a sub-linear convergence rate and a decreasing step size). Also, the theoretical analysis provided in this paper simplifies the convergence analysis of SPAM. We perform numerical experiments to show that the VRSPAM algorithm converges significantly faster than SPAM.

## 2 AUC Formulation

The AUC score associated with a linear scoring function $g(x) = \mathbf{w}^T x$, is defined as the probability that the score of a randomly chosen positive example is higher than a randomly chosen negative example [5,11] and is denoted by AUC($\mathbf{w}$). If $z = (x, y)$ and $z' = (x', y')$ are drawn independently from an unknown distribution $\mathcal{Z} = \mathcal{X} \times \mathcal{Y}$, then

$$\begin{aligned} \text{AUC}(\mathbf{w}) &= Pr(\mathbf{w}^T x \geq \mathbf{w}^T x' | y = 1, y' = -1) \\ &= \mathbb{E}[\mathbb{I}_{\mathbf{w}^T(x-x') \geq 0} | y = 1, y' = -1] \end{aligned}$$

Since AUC($\mathbf{w}$) in the above form is not convex because of the 0–1 loss, it is a common practice to replace this by a convex surrogate loss. In this paper, we focus on the least square loss which is known to be consistent (consistency of a surrogate loss function w.r.t the AUC metric means that, maximizing the surrogate function also maximizes the AUC).
Let $f(\mathbf{w}) = p(1-p)\mathbb{E}[(1 - \mathbf{w}^T(x - x'))^2 | y = 1, y' = -1]$ and $\Omega$ be the convex regularizer where $p = Pr(y = +1)$ and $1 - p = Pr(y = -1)$ are the class priors. We consider the following objective for surrogate-AUC maximization :

$$\min_{\mathbf{w} \in \mathbb{R}^d} f(\mathbf{w}) + \Omega(\mathbf{w}) \tag{1}$$

The form for $f(\mathbf{w})$ follows from the definition of AUC : expected pairwise loss between a positive instance and a negative instance. Throughout this paper we assume

1. $\Omega$ is $\beta$ strongly convex i.e. for any $\mathbf{w}, \mathbf{w}' \in \mathbb{R}^d$,

$$\Omega(\mathbf{w}) \geq \Omega(\mathbf{w}') + \partial\Omega(\mathbf{w}')^T(\mathbf{w} - \mathbf{w}') + \frac{\beta}{2}\|\mathbf{w} - \mathbf{w}'\|^2$$

2. $\exists M$ such that $\|x\| \leq M \ \forall x \in \mathcal{X}$.

In this paper we have used Frobenius norm $\Omega(\mathbf{w}) = \beta\|\mathbf{w}\|^2$ and Elastic Net $\Omega(\mathbf{w}) = \beta\|\mathbf{w}\|^2 + \nu\|\mathbf{w}\|_1$ as the convex regularizers where $\beta, \nu \neq 0$ are the regularization parameters.

It is important to note that standard stochastic gradient based algorithms cannot be applied to Eq. 1 directly, because of the pairwise nature of $f(\cdot)$. Instead we will use a reformulation that will allows us to apply stochastic gradient descent to find the optimum value of $w$. We write Eq. 1 in a pointwise manner rather than the above pairwise form, as originally proposed in [17], as follows:

$$\min_{\mathbf{w},a,b} \max_{\zeta \in \mathbb{R}} \mathbb{E}[F(\mathbf{w}, a, b, \zeta; z)] + \Omega(\mathbf{w}) \tag{2}$$

where the expectation is with respect to $z = (x, y)$ and

$$\begin{aligned} F(\mathbf{w}, a, b, \zeta; z) = (1-p)(\mathbf{w}^T x - a)^2 \mathbb{I}_{[y=1]} + p(\mathbf{w}^T x - b)^2 \mathbb{I}_{[y=-1]} + \\ 2(1+\zeta)\mathbf{w}^T x(p\mathbb{I}_{[y=-1]} - (1-p)\mathbb{I}_{[y=1]}) - p(1-p)\zeta^2 \end{aligned}$$

Thus, $f(\mathbf{w}) = \min_{a,b} \max_{\zeta \in R} \mathbb{E}[F(\mathbf{w}, a, b, \zeta; z)]$. The optimal choices for $a, b, \zeta$ satisfy :

$$a(\mathbf{w}) = \mathbf{w}^T \mathbb{E}[x|y = 1]$$
$$b(\mathbf{w}) = \mathbf{w}^T \mathbb{E}[x|y = -1]$$
$$\zeta(\mathbf{w}) = \mathbf{w}^T (\mathbb{E}[x'|y' = -1] - \mathbb{E}[x|y = 1])$$

It is important to note here that we differentiate the objective function only with respect to $\mathbf{w}$ and do not compute the gradient with respect to the other parameters $(a, b, \zeta)$ which themselves depend on $\mathbf{w}$. Since, $a, b, \zeta$ are expressible in a closed-form, stochastic gradient algorithms can now be applied to Eq. 2. This is the SPAM algorithm [17].

## 3 Method

In the previous section, we discussed a stochastic gradient based algorithm for AUC maximization, that uses an alternative formulation of the objective, to make it decomposable. However, the SPAM algorithm suffers from very slow convergence in most real world problems which are high dimensional and consists of a large number of instances. The major issue that slows down convergence for SGD is the decay of the step size to 0 as the iteration increase. This is a necessary evil for mitigating the effect of variance introduced by random sampling in SGD. Thus, in this paper we directly attack the variance problem for SGD in the AUC maximization framework. We apply the Prox-SVRG method on the reformulation of AUC to derive the proximal SVRG algorithm for AUC maximization described in Algorithm 1. We store a $\tilde{\mathbf{w}}$ after every $m$ Prox-SGD iterations that is progressively closer to the optimal $\mathbf{w}$ (essentially an estimate of the optimal value of (1). Full gradient $\tilde{\boldsymbol{\mu}}$ is computed whenever $\tilde{\mathbf{w}}$ gets updated i.e. after every $m$ iterations of Prox-SGD:

$$\tilde{\boldsymbol{\mu}} = \frac{1}{n} \sum_{i=1}^{n} G(\tilde{\mathbf{w}}, z_i)$$

where $G(\mathbf{w}; z) = \partial_{\mathbf{w}} F(\mathbf{w}, a(\mathbf{w}), b(\mathbf{w}), \zeta(\mathbf{w}); z)$, $n$ is the number of samples and $\tilde{\boldsymbol{\mu}}$ is used to update next $m$ gradients.
Next $m$ iterations are initialized by $\mathbf{w}_0 = \tilde{\mathbf{w}}$. For each iteration, we randomly pick $i_t \in \{1, ..., n\}$ and compute

$$\hat{\mathbf{w}}_t = \mathbf{w}_{t-1} - \eta \mathbf{v}_t$$

where $\mathbf{v}_t = G(\mathbf{w}_{t-1}, z_{i_t}) - G(\tilde{\mathbf{w}}, z_{i_t}) + \tilde{\boldsymbol{\mu}}$ and then the proximal step is taken

$$\mathbf{w}_t = \mathrm{prox}_{\eta, \Omega}(\hat{\mathbf{w}}_t)$$

Notice that if we take expectation of $G(\tilde{\mathbf{w}}, z_{i_t})$ with respect to $i_t$ we get $\mathbb{E}[G(\tilde{\mathbf{w}}, z_{i_t})] = \tilde{\boldsymbol{\mu}}$. Now if we take expectation of $\mathbf{v}_t$ with respect to $i_t$ conditioned on $\mathbf{w}_{t-1}$, we can get the following:

$$\begin{aligned}\mathbb{E}[\mathbf{v}_t|\mathbf{w}_{t-1}] &= \mathbb{E}[G(\mathbf{w}_{t-1}, z_{i_{t-1}})] - \mathbb{E}[G(\tilde{\mathbf{w}}, z_{i_{t-1}})] + \tilde{\boldsymbol{\mu}} \\ &= \frac{1}{n}\sum_{i=1}^{n} G(\tilde{\mathbf{w}}_{t-1}, z_i)\end{aligned}$$

Hence the modified direction $\mathbf{v}_t$ is the stochastic gradient of $G$ at $\mathbf{w}_{t-1}$. However, the variance $\mathbb{E}\|\mathbf{v}_t - \partial f(\mathbf{w}_{t-1})\|^2$ can be much smaller than $\mathbb{E}\|G(\mathbf{w}_{t-1}, z_{i_t}) - \partial f(\mathbf{w}_{t-1})\|^2$, shown in Sect. 4.1. We will also show that the variance goes to 0 as the algorithm converges. Thus, this is a multi-stage scheme to explicitly reduce the variance of the modified proximal gradient.

**Algorithm 1.** Proximal SVRG for AUC maximization

INPUT Constant step size $\eta$ and update frequency $m$
INITIALIZE $\tilde{\mathbf{w}}_0$
  **for** $s = 1, 2, ...$ **do**
    $\tilde{\mathbf{w}} = \tilde{\mathbf{w}}_{s-1}$
    $\tilde{\boldsymbol{\mu}} = \frac{1}{n}\sum_{i=1}^{n} G(\tilde{\mathbf{w}}, z_i)$
    $\mathbf{w}_0 = \tilde{\mathbf{w}}$
    **for** $t = 1, 2, ..., m$ **do**
      Randomly pick $i_t \in \{1, .., n\}$ and update weight
      $\hat{\mathbf{w}}_t = \mathbf{w}_{t-1} - \eta(G(\mathbf{w}_{t-1}, z_{i_t}) - G(\tilde{\mathbf{w}}, z_{i_t}) + \tilde{\boldsymbol{\mu}})$
      $\mathbf{w}_t = \text{prox}_{\eta\Omega}(\hat{\mathbf{w}}_t)$
    $\tilde{\mathbf{w}}_s = \mathbf{w}_m$

## 4 Convergence Analysis

In this section, we formally analyze the convergence rate of VRSPAM. We first present some lemmas which will be used for proving the Theorem 1 which is the main theorem proving the geometric convergence of Algorithm 1. Lemma 1 states that
$\partial_{\mathbf{w}} F(\mathbf{w}_t, a(\mathbf{w}_t), b(\mathbf{w}_t), \alpha(\mathbf{w}_t); z_t)$ is an unbiased estimator of the true gradient. As we are not calculating the true gradient in VRSPAM, we need the following lemma to prove the convergence result.

**Lemma 1** [17]**.** Let $\mathbf{w}_t$ be given by VRSPAM in Algorithm 1. Then, we have

$$\partial f(\mathbf{w}_t) = \mathbb{E}_{z_t}[\partial_{\mathbf{w}} F(\mathbf{w}_t, a(\mathbf{w}_t), b(\mathbf{w}_t), \alpha(\mathbf{w}_t); z_t)]$$

This lemma is directly applicable in VRSPAM since the proof of the lemma hinges on the objective function reformulation and not on the algorithm specifics.

The next lemma provides an upper bound on the norm of the difference of gradients at different time steps.

**Lemma 2** [17]. Let $\mathbf{w}_t$ be described as above. Then, we have

$$\|G(\mathbf{w}_{t'}, z_t) - G(\mathbf{w}_{t''}, z_t)\| \leq 8M^2\|\mathbf{w}_{t'} - \mathbf{w}_{t''}\|$$

*Proof.*

$$\begin{aligned}\|G(\mathbf{w}_{t'}; z_t) - G(\mathbf{w}_{t''}; z_t)\| &\leq 4M^2p\|\mathbf{w}_{t'} - \mathbf{w}_{t''}\|\mathbb{1}_{[y_t=-1]} \\ &+ 4M^2p\|\mathbf{w}_{t'} - \mathbf{w}_{t''}\|\mathbb{1}_{[y_t=-1]} + 4M^2(1-p)\|\mathbf{w}_{t'} - \mathbf{w}_{t''}\|\mathbb{1}_{[y_t=1]} \\ &+ 4M^2|p - \mathbb{1}_{[y_t=1]}|\|\mathbf{w}_{t'} - \mathbf{w}_{t''}\| \\ &\leq 8M^2\|\mathbf{w}_{t'} - \mathbf{w}_{t''}\|\end{aligned}$$

The proof directly follows by writing out the difference and using the second assumption on the boundedness of $\|x\|$.

We now present and prove a key result that will be necessary in showing convergence in Theorem 1

**Lemma 3.** *Let* $C = \frac{1+128M^4\eta^2}{(1+\eta\beta)^2}$ *and* $D = \frac{128M^4\eta^2}{(1+\eta\beta)^2}$; *if* $\eta \leq \frac{\beta}{128M^4}$ *then* $C^m + DC\frac{C^m-1}{C-1} \leq 1$ *holds true.*

*Proof.* We start with:

$$\begin{aligned}&\eta \leq \frac{\beta}{128M^4} \\ &\Rightarrow 128M^4\eta^2 \leq \eta\beta \\ &\Rightarrow 128M^4\eta^2(2 + 128M^4\eta^2) \leq \eta\beta(2 + 1\eta\beta) \\ &\Rightarrow 128M^4\eta^2 + (128M^4\eta^2)^2 \leq (\eta\beta)^2 + 2\eta\beta - 128M^4\eta^2 \\ &\Rightarrow 128M^4\eta^2 \leq \frac{(1+\eta\beta)^2 - 1 - 128M^4\eta^2}{1 + 128M^4\eta^2} \\ &\Rightarrow 128M^4\eta^2 \leq \frac{1 - \frac{1+128M^4\eta^2}{(1+\eta\beta)^2}}{\frac{1+128M^4\eta^2}{(1+\eta\beta)^2}}\end{aligned}$$

Substituting values of $C$ and $D$ and using the condition that $D \leq 128M^4\eta^2$, we get

$$\begin{aligned}&\Rightarrow D \leq \frac{1-C}{C} \\ &\Rightarrow DC\frac{C^m - 1}{C - 1} \leq 1 - C^m \\ &\Rightarrow C^m + DC\frac{C^m - 1}{C - 1} \leq 1\end{aligned}$$

Now we present and prove the main theorem of this paper that gives the convergence rate of Algorithm 1 and its analysis.

**Theorem 1.** *Consider* VRSPAM *(Algorithm 1) and let* $\mathbf{w}^* = arg\,min_{\mathbf{w}} f(\mathbf{w}) + \Omega(\mathbf{w})$; *if* $\eta < \frac{\beta}{128M^4}$, *then the following inequality holds true*

$$\alpha = C^m + DC\frac{C^m - 1}{C - 1} < 1$$

*and we have the geometric convergence in expectation:*

$$\mathbb{E}[\|\tilde{\mathbf{w}}_s - \mathbf{w}^*\|^2] \le \alpha^s \mathbb{E}[\|\mathbf{w_0} - \mathbf{w}^*\|^2]$$

For proving the above theorem, first we upper bound the variance of the gradient step and show that it approaches zero as $\mathbf{w_s}$ approaches $\mathbf{w}^*$.

### 4.1 Bounding the Variance

In this section, we will derive a bound on the variance of the modified gradient $\mathbf{v}_t = G(\mathbf{w}_{t-1}, z_{i_t}) - G(\tilde{\mathbf{w}}, z_{i_t}) + \tilde{\boldsymbol{\mu}}$. We first present a lemma that will help derive the bound in Lemma 5.

**Lemma 4.** *Consider* VRSPAM *(Algorithm 1), then* $\mathbb{E}[\|\mathbf{v}_t - \partial f(\mathbf{w}^*)]\|^2$ *is upper bounded as:*

$$\mathbb{E}[\|\mathbf{v}_t - \partial f(\mathbf{w}^*)]\|^2] \le 2(8M^2)^2\|\mathbf{w}_{t-1} - \mathbf{w}^*\|^2 + 2(8M^2)^2\|\tilde{\mathbf{w}} - \mathbf{w}^*\|^2$$

*Proof.* Let the variance reduced update be denoted as $\mathbf{v}_t = G(\mathbf{w}_{t-1}, z_{i_t}) - G(\tilde{\mathbf{w}}, z_{i_t}) + \tilde{\boldsymbol{\mu}}$.
As we know $\mathbb{E}[\mathbf{v}_t] = \partial f(\mathbf{w}_{t-1})$, the variance of $\mathbf{v}_t$ can be written as below

$$\begin{aligned}\mathbb{E}[\|\mathbf{v}_t - \partial f(\mathbf{w}^*))\|^2] \le& 2\mathbb{E}[\|G(\mathbf{w}^*, z_{i_t}) - G(\tilde{\mathbf{w}}, z_{i_t}) + \tilde{\boldsymbol{\mu}} - \partial f(\mathbf{w}^*))\|^2] \\ &+ 2\mathbb{E}[\|G(\mathbf{w}_{t-1}, z_{i_t}) - G(\mathbf{w}^*, z_{i_t})\|^2]\end{aligned}$$

Also, $\mathbb{E}[G(\mathbf{w}^*, z_{i_t}) - G(\tilde{\mathbf{w}}, z_{i_t})] = \partial f(\mathbf{w}^*) - \partial f(\tilde{\mathbf{w}})$ from Lemma 1 and using the property that $\mathbb{E}[(X - \mathbb{E}[X])^2] \le \mathbb{E}[X^2]$ we get

$$\begin{aligned}\mathbb{E}[\|\mathbf{v}_t - \partial f(\mathbf{w}^*))\|^2] \le& 2\mathbb{E}[\|G(\mathbf{w}_{t-1}, z_{i_t}) - G(\mathbf{w}^*, z_{i_t})\|^2] \\ &+ 2\mathbb{E}[\|G(\mathbf{w}^*, z_{i_t}) - G(\tilde{\mathbf{w}}, z_{i_t})\|^2]\end{aligned}$$

From Lemma 2, we have $\|G(\mathbf{w}_{t-1}, z_{i_t}) - G(\mathbf{w}^*, z_{i_t})\| \le 8M^2\|\mathbf{w}_{t-1} - \mathbf{w}^*\|$ and $\|G(\mathbf{w}^*, z_{i_t}) - G(\tilde{\mathbf{w}}, z_{i_t})\| \le 8M^2\|\tilde{\mathbf{w}} - \mathbf{w}^*\|$. Using this, we can upper bound the variance of gradient step as:

$$\mathbb{E}[\|\mathbf{v}_t - \partial f(\mathbf{w}^*)]\|^2] \le 2(8M^2)^2\|\mathbf{w}_{t-1} - \mathbf{w}^*\|^2 + 2(8M^2)^2\|\tilde{\mathbf{w}} - \mathbf{w}^*\|^2 \qquad (3)$$

We have the desired result.

We now present the lemma that gives the bound on the variance of modified gradient $\mathbf{v}_t$.

**Lemma 5.** *Consider* VRSPAM *(Algorithm 1), then the variance of the* $\mathbf{v}_t$ *is upper bounded as:*

$$\mathbb{E}[\|\mathbf{v}_t - \partial f(\mathbf{w}_{t-1})]\|^2] \leq 4(8M^2)^2\|\mathbf{w}_{t-1} - \mathbf{w}^*\|^2 + 2(8M^2)^2\|\tilde{\mathbf{w}} - \mathbf{w}^*\|^2$$

*Proof.*

$$\begin{aligned}\mathbb{E}[\|\mathbf{v}_t - \partial f(\mathbf{w}_{t-1})\|^2] &\leq 2\mathbb{E}[\|\mathbf{v}_t - \partial f(\mathbf{w}^*)\|^2] + 2\mathbb{E}[\|\partial f(\mathbf{w}^*) - \partial f(\mathbf{w}_{t-1})]\|^2] \\ &\leq 2(8M^2)^2\|\mathbf{w}_{t-1} - \mathbf{w}^*\|^2 + 2(8M^2)^2\|\tilde{\mathbf{w}} - \mathbf{w}^*\|^2 \\ &+ 2\mathbb{E}[\|G(\mathbf{w}_{t-1}, z_{i_t}) - G(\mathbf{w}^*, z_{i_t})\|^2] \\ &\leq 4(8M^2)^2\|\mathbf{w}_{t-1} - \mathbf{w}^*\|^2 + 2(8M^2)^2\|\tilde{\mathbf{w}} - \mathbf{w}^*\|^2\end{aligned}$$

where the second inequality uses Lemma 4 and last inequality uses Lemma 2.

At convergence, $\tilde{\mathbf{w}} = \mathbf{w}^*$ and $\mathbf{w}_t = \mathbf{w}^*$. Thus, the variance of the updates are bounded and go to zero as the algorithm converges whereas in the case of the SPAM algorithm, the variance of the gradient does not go to zero (which is a characteristic of a stochastic gradient descent based algorithm). We now present the proof of Theorem 1 using the above lemmas.

### 4.2 Proof of Theorem 1

From the first order optimality condition, we can directly write

$$\mathbf{w}^* = \text{prox}_{\eta\Omega}(\mathbf{w}^* - \eta\partial f(\mathbf{w}^*))$$

Using the above we can write

$$\|\mathbf{w}_{t+1} - \mathbf{w}^*\|^2 = \|\,\text{prox}_{\eta\Omega}(\hat{\mathbf{w}}_{t+1}) - \text{prox}_{\eta\Omega}(\mathbf{w}^* - \eta\partial f(\mathbf{w}^*))\|^2$$

Using Proposition 23.11 from [2], we have $\text{prox}_{\eta\Omega}$ is $(1+\eta\beta)$-cocoercieve and for any $\mathbf{u}$ and $\mathbf{w}$ using Cauchy Schwartz we can get the following inequality

$$\|\,\text{prox}_{\eta\Omega}(\mathbf{u}) - \text{prox}_{\eta\Omega}(\mathbf{w})\| \leq \frac{1}{1+\eta\beta}\|\mathbf{u} - \mathbf{w}\|$$

From above we get

$$\begin{aligned}\|\mathbf{w}_{t+1} - \mathbf{w}^*\|^2 &\leq \frac{1}{(1+\eta\beta)^2}\|(\hat{\mathbf{w}}_{t+1}) - (\mathbf{w}^* - \eta\partial f(\mathbf{w}^*))\|^2 \\ &\leq \frac{1}{(1+\eta\beta)^2}\|(\mathbf{w}_t - \mathbf{w}^*) - \eta(G(\mathbf{w}_t, z_{i_{t+1}}) - G(\tilde{\mathbf{w}}, z_{i_{t+1}}) + \tilde{\boldsymbol{\mu}} - \partial f(\mathbf{w}^*))\|^2\end{aligned}$$

Taking expectation on both sides we get

$$\begin{aligned}\mathbb{E}\|\mathbf{w}_{t+1} - \mathbf{w}^*\|^2 \leq \frac{1}{(1+\eta\beta)^2}&(\eta^2\mathbb{E}[\|G(\mathbf{w}_t, z_{i_{t+1}}) - G(\tilde{\mathbf{w}}, z_{i_{t+1}}) + \tilde{\boldsymbol{\mu}} - \partial f(\mathbf{w}^*))\|^2] \\ + \mathbb{E}[\|\mathbf{w}_t - \mathbf{w}^*\|^2] - 2\eta\mathbb{E}[\langle \mathbf{w}_t - \mathbf{w}^*, &G(\mathbf{w}_t, z_{i_{t+1}}) - G(\tilde{\mathbf{w}}, z_{i_{t+1}}) + \tilde{\boldsymbol{\mu}} - \partial f(\mathbf{w}^*)\rangle])\end{aligned} \tag{4}$$

Now, we first bound the last term $T = \mathbb{E}[\langle \mathbf{w}_t - \mathbf{w}^*, G(\mathbf{w}_t, z_{i_{t+1}}) - G(\tilde{\mathbf{w}}, z_{i_{t+1}}) + \tilde{\boldsymbol{\mu}} - \partial f(\mathbf{w}^*)\rangle]$ in Eq. 4. Using Lemma 1 we can write

$$\begin{aligned} T &= \mathbb{E}[\langle \mathbf{w}_t - \mathbf{w}^*, \mathbb{E}_{z_{t+1}}[G(\mathbf{w}_{t-1}, z_{i_{t+1}})] - \mathbb{E}_{z_{t+1}}[G(\tilde{\mathbf{w}}, z_{i_{t+1}})] + \tilde{\boldsymbol{\mu}} - \partial f(\mathbf{w}^*)\rangle] \\ &= \mathbb{E}[\langle \mathbf{w}_t - \mathbf{w}^*, \mathbb{E}_{z_{t+1}}[G(\mathbf{w}_t, z_{i_{t+1}})] - \partial f(\mathbf{w}^*)\rangle] \\ &= \mathbb{E}[\langle \mathbf{w}_t - \mathbf{w}^*, \partial f(\mathbf{w}_t) - \partial f(\mathbf{w}^*)\rangle] \\ &\geq 0 \end{aligned}$$

Now, $\mathbb{E}\|\mathbf{w}_{t+1} - \mathbf{w}^*\|^2$ can be bounded by using above bound and Lemma 4 as below

$$\begin{aligned} \mathbb{E}\|\mathbf{w}_{t+1} - \mathbf{w}^*\|^2 &\leq \frac{1}{(1+\eta\beta)^2}(\mathbb{E}[\|\mathbf{w}_t - \mathbf{w}^*\|^2] \\ &\quad + 2(8M^2)^2\eta^2(\mathbb{E}[\|\mathbf{w}_t - \mathbf{w}^*\|^2] + \mathbb{E}[\|\tilde{\mathbf{w}} - \mathbf{w}^*\|^2])) \\ &\leq \frac{1+128M^4\eta^2}{(1+\eta\beta)^2}\mathbb{E}[\|\mathbf{w}_t - \mathbf{w}^*\|^2] + \frac{128M^4\eta^2}{(1+\eta\beta)^2}\mathbb{E}[\|\tilde{\mathbf{w}} - \mathbf{w}^*\|^2] \end{aligned}$$

Let $C = \frac{1+128M^4\eta^2}{(1+\eta\beta)^2}$ and $D = \frac{128M^4\eta^2}{(1+\eta\beta)^2}$, then after $m$ iterations $\mathbf{w}_t = \tilde{\mathbf{w}}_s$ and $\mathbf{w}_0 = \tilde{\mathbf{w}}_{s-1}$. Substituting this in the above inequality, we get

$$\begin{aligned} \mathbb{E}\|\tilde{\mathbf{w}}_s - \mathbf{w}^*\|^2 &\leq C^m(\mathbb{E}\|\tilde{\mathbf{w}}_{s-1} - \mathbf{w}^*\|^2 + \sum_{i=0}^{m-1}\frac{D}{C^i}\mathbb{E}\|\tilde{\mathbf{w}}_{s-1} - \mathbf{w}^*\|^2) \\ &\leq (C^m + \sum_{i=0}^{m-1}\frac{DC^m}{C^i})\mathbb{E}\|\tilde{\mathbf{w}}_{s-1} - \mathbf{w}^*\|^2 \\ &\leq (C^m + DC^m\frac{1-(1/C^m)}{1-(1/C)})\mathbb{E}\|\tilde{\mathbf{w}}_{s-1} - \mathbf{w}^*\|^2 \\ &\leq (C^m + DC\frac{C^m-1}{C-1})\mathbb{E}\|\tilde{\mathbf{w}}_{s-1} - \mathbf{w}^*\|^2 \\ &\leq \alpha\mathbb{E}\|\tilde{\mathbf{w}}_{s-1} - \mathbf{w}^*\|^2 \end{aligned}$$

where $\alpha = C^m + DC\frac{C^m-1}{C-1}$ is the decay parameter, and $\alpha < 1$ by using Lemma 3. After $s$ steps in outer loop of Algorithm 1, we get $\mathbb{E}\|\tilde{\mathbf{w}}_s - \mathbf{w}^*\|^2 \leq \alpha^s\mathbb{E}\|\mathbf{w}_0 - \mathbf{w}^*\|^2$ where $\alpha < 1$. Hence, we get geometric convergence of $\alpha^s$ which is much stronger than the $\mathcal{O}(\frac{1}{t})$ convergence obtained in [17]. In the next section we derive the time complexity of the algorithm and investigate dependence of $\alpha$ on the problem parameters.

### 4.3 Complexity Analysis

To have $\mathbb{E}\|\tilde{\mathbf{w}}_s - \mathbf{w}^*\|^2 \leq \epsilon$, the number of iterations $s$ must satisfy:

$$s \geq \frac{1}{\log\frac{1}{\alpha}}\log\frac{\mathbb{E}\|\mathbf{w}_0 - \mathbf{w}^*\|^2}{\epsilon}$$

At each stage, the number of gradient evaluations are $n + 2m$ where $n$ is the number of samples and $m$ is the iterations in the inner loop and the complexity is $\mathcal{O}(n+m)(\log(\frac{1}{\epsilon}))$ i.e. Algorithm 1 takes $\mathcal{O}(n+m)(\log(\frac{1}{\epsilon}))$ gradient complexity to achieve accuracy of $\epsilon$. Here, the complexity is dependent on $M$ and $\beta$ as $m$ itself is dependent on $M$ and $\beta$.

Now we find the dependence of $\alpha$ and $m$ on $M$ and $\beta$. Let $\eta = \frac{\theta\beta}{128M^4}$ where $0 < \theta < 1$. Then,

$$\begin{aligned} C &= \frac{1 + 128M^4\eta^2}{(1+\eta\beta)^2} = \frac{1 + \frac{\theta^2\beta^2}{128M^4}}{(1 + \frac{\theta\beta^2}{128M^4})^2} \\ &< \frac{1 + \frac{\theta\beta^2}{128M^4}}{(1 + \frac{\theta\beta^2}{128M^4})^2} \\ &= \frac{1}{(1 + \frac{\theta\beta^2}{128M^4})} \\ &= E \end{aligned}$$

Therefore, $D = \theta(E - E^2)$ and $DC < \theta E^2(1-E)$, and using the above equations we can simplify $\alpha$ as

$$\begin{aligned} \alpha &= C^m + DC\frac{1 - C^m}{1 - C} \\ &< C^m + \theta E^2(1 - E)\frac{1 - C^m}{1 - C} \\ &< C^m + \theta E^2(1 - C^m) \quad \because \frac{1 - E}{1 - C} < 1 \\ &= \theta E^2 + C^m - \theta E^2 C^m \end{aligned}$$

In the above equation, only $C^m - \theta E^2 C^m$ depends on $m$. If we choose $m$ to be sufficiently large then $\alpha = \theta E^2$. An important thing to note here is that $\theta E < C < E$, now if we choose $m \approx 2\frac{\log\theta}{\log E}$ then $\alpha \approx 2\theta E^2$. Thus, the time complexity of the algorithm is

$$\mathcal{O}(n + 2\frac{\log\theta}{\log E})(\log(\frac{1}{\epsilon})) \quad \text{when} \quad m = \Theta(\frac{\log\theta}{\log E}).$$

As the order has inverse dependency on $\log E = \log\frac{128M^4}{128M^4+\theta\beta^2}$, increase in $M$ will result in increase in number of iterations i.e. as the maximum norm of training samples is increased, larger $m$ is required to reach $\epsilon$ accuracy.

**Comparison of Time Complexities:** Now let us compare the time complexities of our algorithm with that of the SPAM algorithm. First, we derive the time complexity of SPAM. We will use Theorem 3 from [17] which states that SPAM achieves the following:

$$\mathbb{E}[\|\mathbf{w}_{T+1} - \mathbf{w}^*\|^2] \leq \frac{t_0}{T}\mathbb{E}[\|\mathbf{w}_{t_0} - \mathbf{w}^*\|^2] + c\frac{\log T}{T}$$

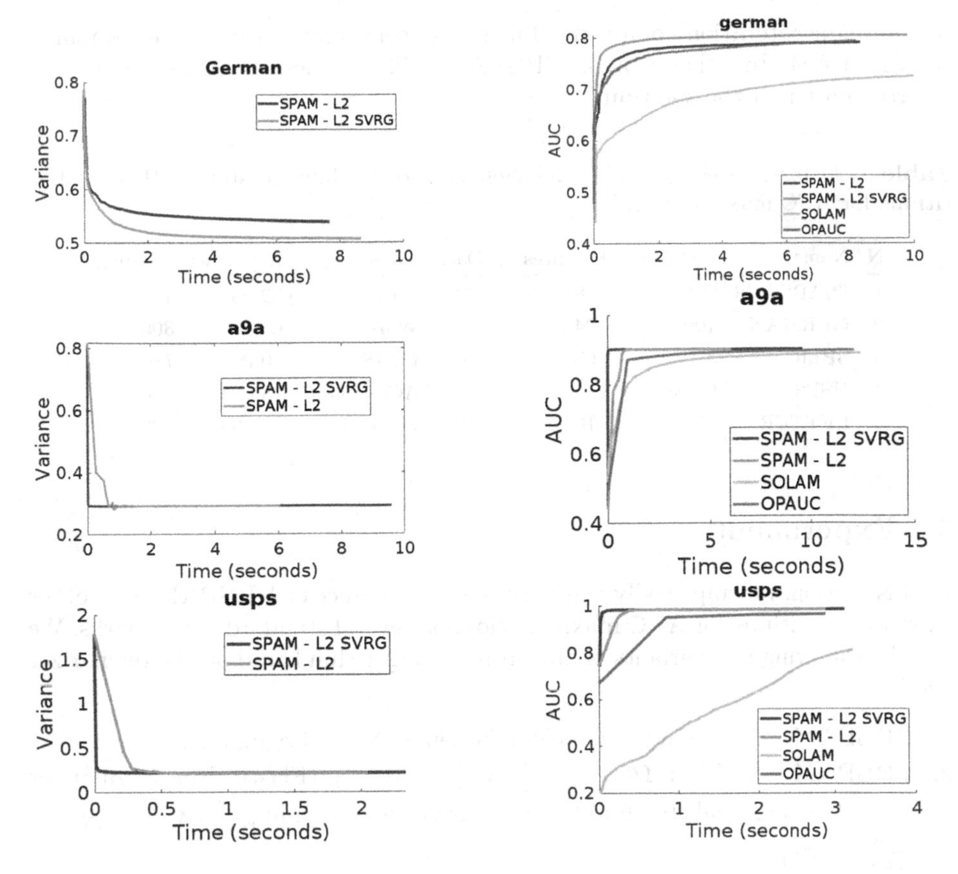


**Fig. 1.** The left column shows that VRSPAM (SPAM-L2-SVRG) has lower variance than SPAM-L2 across different datasets. The right column shows VRSPAM (SPAM-L2-SVRG) converges faster and performs better than existing algorithms on AUC maximization

where $t_0 = \max\big(2, \lceil 1 + \frac{(128M^4+\beta^2)^2}{128M^4\beta^2} \rceil\big)$, $T$ is the number of iterations and $c$ is a constant. Using the averaging scheme developed by [15], the following can be obtained:

$$\mathbb{E}[\|\mathbf{w}_{T+1} - \mathbf{w}^*\|^2] \leq \frac{t_0}{T}\mathbb{E}[\|\mathbf{w}_{t_0} - \mathbf{w}^*\|^2] \tag{5}$$

where

$$\mathbb{E}[\|\mathbf{w}_{t_0} - \mathbf{w}^*\|^2] \leq \frac{2\sigma_*^2}{\tilde{C}_{\beta,M}^2} + \exp\left(\frac{128M^4}{\tilde{C}_{\beta,M}^2}\right) = F,$$

$$\tilde{C}_{\beta,M}^2 = \frac{\beta}{(1 + \frac{\beta^2}{128M^4})^2} \quad \text{and} \quad \mathbb{E}[\|G(\mathbf{w}^*; z) - \partial f(\mathbf{w}^*)\|^2] = \sigma_*^2.$$

Using Eq. 5, the time complexity of the SPAM algorithm can be written as $\mathcal{O}(\frac{t_0 F}{\epsilon})$ i.e. SPAM takes $\mathcal{O}(\frac{t_0 F}{\epsilon})$ iterations to achieve $\epsilon$ accuracy. Thus, SPAM has lower per iteration complexity but slower convergence rate when compared to VRSPAM. In other words, VRSPAM will take less time to get a good approximation of the solution.

**Table 1.** Datasets used for evaluating VRSPAM and the different state-of-the-art algorithms for AUC maximization.

| N | Name | Instances | Features | Data | Name | Instances | Features |
|---|---|---|---|---|---|---|---|
| 1 | DIABETES | 768 | 8 | 6 | A9A | 32,561 | 123 |
| 2 | GERMAN | 1000 | 24 | 7 | W8A | 64,700 | 300 |
| 3 | SPLICE | 3,175 | 60 | 8 | MNIST | 60,000 | 780 |
| 4 | USPS | 9,298 | 256 | 9 | ACOUSTIC | 78,823 | 50 |
| 5 | LETTER | 20,000 | 16 | 10 | IJCNN1 | 141,691 | 22 |

# 5 Experiment

In this section, we empirically compare the performance of VRSPAM with other existing algorithms for AUC maximization, on several standard benchmarks. We use the following two variants of our proposed algorithm based on the regularizer used:

1. VRSPAM $- L^2 : \Omega(\mathbf{w}) = \frac{\beta}{2}\|\mathbf{w}\|^2$ (Frobenius Norm Regularizer)
2. VRSPAM $- NET : \Omega(\mathbf{w}) = \frac{\beta}{2}\|\mathbf{w}\|_2^2 + \beta_1\|\mathbf{w}\|_1$ (Elastic Net Regularizer [27]). The proximal step for elastic net is given as $\arg\min_{\mathbf{w}}\{\frac{1}{2}\|\mathbf{w} - \frac{\hat{\mathbf{w}}_{t+1}}{\eta_t\beta+1}\|^2 + \frac{\eta_t\beta_1}{\eta_t\beta+1}\|\mathbf{w}\|_1\}$

VRSPAM is compared with several baselines: SPAM, SOLAM [24] and one-pass AUC optimization algorithm (OPAUC) [9], which are state-of-the-art methods for AUC maximization. SOLAM was modified to have the Frobenius Norm Regularizer (as in [17]). VRSPAM is compared against OPAUC with the least square loss.

**Table 2.** Comparison of AUC values (mean±std) achieved by the different algorithms on the test data of the different datasets described in Table 1.

| N | VRSPAM-$L^2$ | VRSPAM-NET | SPAM-$L^2$ | SPAM-NET | SOLAM | OPAUC |
|---|---|---|---|---|---|---|
| 1 | .8299±.0323 | **.8305±.0319** | .8272±.0277 | .8085±.0431 | .8128±.0304 | .8309±.0350 |
| 2 | .7902±0386 | .7845±.0398 | .7942±.0388 | .7937±.0386 | .7778±.0373 | **.7978±.0347** |
| 3 | .9640±.0156 | **.9699±.0139** | .9263±.0091 | .9267±.0090 | .9246±.0087 | .9232±.0099 |
| 4 | **.8552±.006** | .8549±.0059 | .8542±.0388 | .8537±.0386 | .8395±.0061 | .8114±.0065 |
| 5 | .9834±.0023 | .9804±.0032 | **.9868±.0032** | .9855±.0029 | .9822±.0036 | .9620±.0040 |
| 6 | **.9003±.0045** | .8981±.0046 | .8998±.0046 | .8980±.0047 | .8966±.0043 | .9002±.0047 |
| 7 | **.9876±.0008** | .9787±.0013 | .9682±.0020 | .9604±.0020 | .9817±.0015 | .9633±.0035 |
| 8 | **.9465±.0014** | .9351±.0014 | .9254±.0025 | .9132±.0026 | .9118±.0029 | .9242±.0021 |
| 9 | .8093±.0033 | .8052±.033 | .8120±.0030 | .8109±.0028 | .8099±.0036 | **.8192±.0032** |
| 10 | **.9750±.001** | .9745±.002 | .9174±.0024 | .9155±.0024 | .9129±.0030 | .9269±.0021 |

All datasets are publicly available from [4] and [8]. Some of the datasets, like MNIST, are multiclass, and we convert them to binary labels by numbering the classes and assigning all the even labels to one class and all the odd labels to another. The results are the mean AUC score and standard deviation of 20 runs on each dataset. All the datasets were randomly divided into training and test splits with 80% and 20% of the data. The parameters $\beta \in 10^{[-5:5]}$ and $\beta_1 \in 10^{[-5:5]}$ for VRSPAM $- L^2$ and VRSPAM $- NET$ are chosen by a 5 fold cross-validation on the training set. All the code is implemented in MATLAB. We measured the algorithm's computational time using an Intel $i7$ CPU with a clock speed of 3538 MHz.

### 5.1 VRSPAM Has Lower Variance

Theoretically, we derived that VRSPAM has lower variance than the baseline SPAM algorithm. Here, we see empirically this holds across the different datasets. In the left column of Fig. 1, we show the variance of the VRSPAM update ($\mathbf{v}_t$) in comparison with the variance of SPAM update $(G(\mathbf{w}_{t-1}, z_{i_{t-1}}))$ . We observe that the variance of VRSPAM is lower than the variance of SPAM and decreases to the minimum value faster, which is in line with Theorem 1.

### 5.2 VRSPAM Has Faster Convergence

Theoretically, we derived that VRSPAM converges faster than the baseline SPAM algorithm. Here, we see empirically this holds across the different datasets. In the right column of Fig. 1, we show the performance of VRSPAM compared to existing methods for AUC maximization. We observe that VRSPAM converges to the maximum value faster than the other methods, and in some cases, this maximum value itself is higher for VRSPAM.

We found that the best results were obtained when the initial weights of VRSPAM were set to be the output generated by SPAM after one iteration, which happens to be standard practice in related problems in optimization [13]. Table 2 summarizes the results of the performance of different algorithms as measured by the AUC metric, across different datasets. AUC values for SPAM-$L^2$, SPAM-NET, SOLAM and OPAUC were taken from [17]. It is seen that in almost all the datasets, one of the two versions of VRSPAM has the best performance and this gain is consistent across multiple runs, as seen by the standard error. This shows that under finite computational time, VRSPAM is able to converge to the global optimum faster than the other algorithms.

## 6 Conclusion

In this paper, we propose a variance reduced stochastic proximal algorithm for AUC maximization (VRSPAM). We theoretically analyze the proposed algorithm and derive a much faster convergence rate of $\mathcal{O}(\alpha^t)$ where $\alpha < 1$ (linear convergence rate), improving upon state-of-the-art methods [17] which have a convergence rate of $\mathcal{O}(\frac{1}{t})$ (sub-linear convergence rate), for strongly convex objective functions with per iteration complexity of one data-point. We gave a theoretical analysis of this and showed empirically VRSPAM converges faster than other methods for AUC maximization.

For future work, it will be interesting to explore if other algorithms used to accelerate SGD can be used in this setting and if they lead to even faster convergence. It is also interesting to apply the proposed methods in practice to non-decomposable performance measures other than AUC. It would be interesting to extend the analysis to a non-convex and non-smooth regularizer using method presented in [23].

## References


1. Agarwal, S.: Surrogate regret bounds for the area under the roc curve via strongly proper losses. In: Conference on Learning Theory, pp. 338–353 (2013)
2. Bauschke, H.H., Combettes, P.L.: Convex Analysis and Monotone Operator Theory in Hilbert Spaces. CBM, Springer, Cham (2017). https://doi.org/10.1007/978-3-319-48311-5
3. Beck, A., Teboulle, M.: A fast iterative shrinkage-threshold algorithm for linear inverse problems. Technion-Israel Institute of Technology, Technical Report (2008)
4. Chang, C.C., Lin, C.J.: Libsvm: a library for support vector machines. ACM Trans. Intell. Syst. Technol. (TIST) **2**(3), 27 (2011)
5. Clémençon, S., Lugosi, G., Vayatis, N., et al.: Ranking and empirical minimization of u-statistics. Ann. Statist. **36**(2), 844–874 (2008)
6. Elkan, C.: The foundations of cost-sensitive learning. In: International Joint Conference on Artificial Intelligence, vol. 17, pp. 973–978. Lawrence Erlbaum Associates Ltd (2001)
7. Fawcett, T.: An introduction to roc analysis. Pattern Recogn. Lett. **27**(8), 861–874 (2006)
8. Frank, A., Asuncion, A.: UCI machine learning repository [http://archive.ics.uci.edu/ml]. University of California, Irvine. School of Information and Computer Science **213**, 2 (2010)
9. Gao, W., Jin, R., Zhu, S., Zhou, Z.H.: One-pass AUC optimization. In: International Conference on Machine Learning, pp. 906–914 (2013)
10. Gao, W., Zhou, Z.H.: On the consistency of AUC pairwise optimization. In: Twenty-Fourth International Joint Conference on Artificial Intelligence (2015)
11. Hanley, J.A., McNeil, B.J.: The meaning and use of the area under a receiver operating characteristic (roc) curve. Radiology **143**(1), 29–36 (1982)
12. Herschtal, A., Raskutti, B.: Optimising area under the roc curve using gradient descent. In: Proceedings of the Twenty-First International Conference on Machine Learning, p. 49. ACM (2004)
13. Johnson, R., Zhang, T.: Accelerating stochastic gradient descent using predictive variance reduction. In: Advances in Neural Information Processing Systems, pp. 315–323 (2013)
14. Kar, P., Sriperumbudur, B.K., Jain, P., Karnick, H.C.: On the generalization ability of online learning algorithms for pairwise loss functions. In: Proceedings of the 30th International Conference on International Conference on Machine Learning, vol. 28, pp. III-441. JMLR. org (2013)

15. Lacoste-Julien, S., Schmidt, M., Bach, F.: A simpler approach to obtaining an o (1/t) convergence rate for the projected stochastic subgradient method. arXiv preprint arXiv:1212.2002 (2012)
16. Narasimhan, H., Agarwal, S.: Support vector algorithms for optimizing the partial area under the roc curve. Neural Comput. **29**(7), 1919–1963 (2017)
17. Natole, M., Ying, Y., Lyu, S.: Stochastic proximal algorithms for AUC maximization. In: International Conference on Machine Learning, pp. 3707–3716 (2018)
18. Orabona, F.: Simultaneous model selection and optimization through parameter-free stochastic learning. In: Advances in Neural Information Processing Systems, pp. 1116–1124 (2014)
19. Roux, N.L., Schmidt, M., Bach, F.R.: A stochastic gradient method with an exponential convergence _rate for finite training sets. In: Advances in Neural Information Processing Systems, pp. 2663–2671 (2012)
20. Shalev-Shwartz, S., Zhang, T.: Stochastic dual coordinate ascent methods for regularized loss minimization. J. Mach. Learn. Res. **14**(Feb), 567–599 (2013)
21. Shalev-Shwartz, S., et al.: Online learning and online convex optimization. Found. Trends® Mach. Learn. **4**(2), 107–194 (2012)
22. Xiao, L., Zhang, T.: A proximal stochastic gradient method with progressive variance reduction. SIAM J. Optim. **24**(4), 2057–2075 (2014)
23. Xu, Y., Qi, Q., Lin, Q., Jin, R., Yang, T.: Stochastic optimization for dc functions and non-smooth non-convex regularizers with non-asymptotic convergence. In: International Conference on Machine Learning, pp. 6942–6951 (2019)
24. Ying, Y., Wen, L., Lyu, S.: Stochastic online AUC maximization. In: Advances in Neural Information Processing Systems, pp. 451–459 (2016)
25. Zhang, X., Saha, A., Vishwanathan, S.: Smoothing multivariate performance measures. J. Mach. Learn. Res. **13**(Dec), 3623–3680 (2012)
26. Zhao, P., Hoi, S.C., Jin, R., Yang, T.: Online AUC maximization. In: Proceedings of the 28th International Conference on International Conference on Machine Learning, pp. 233–240. Omnipress (2011)
27. Zou, H., Hastie, T.: Regularization and variable selection via the elastic net. J. R. Statist. Soc. B Statist. Methodol. **67**(2), 301–320 (2005)

# Robust Regression via Model Based Methods

Armin Moharrer(✉), Khashayar Kamran, Edmund Yeh, and Stratis Ioannidis

Northeastern University, Boston, MA 02115, USA
{amoharrer,kamrank,eyeh,ioannidis}@ece.neu.edu

**Abstract.** The mean squared error loss is widely used in many applications, including auto-encoders, multi-target regression, and matrix factorization, to name a few. Despite computational advantages due to its differentiability, it is not robust to outliers. In contrast, $\ell_p$ norms are known to be robust, but cannot be optimized via, e.g., stochastic gradient descent, as they are non-differentiable. We propose an algorithm inspired by so-called model-based optimization (MBO) [35,36], which replaces a non-convex objective with a convex model function and alternates between optimizing the model function and updating the solution. We apply this to robust regression, proposing SADM, a stochastic variant of the Online Alternating Direction Method of Multipliers (OADM) [48] to solve the inner optimization in MBO. We show that SADM converges with the rate $O(\log T/T)$. Finally, we demonstrate experimentally (a) the robustness of $\ell_p$ norms to outliers and (b) the efficiency of our proposed model-based algorithms in comparison with gradient methods on autoencoders and multi-target regression.

## 1 Introduction

Mean Squared Error (MSE) loss problems are ubiquitous in machine learning and data mining. Such problems have the following form:

$$\min_{\theta} \frac{1}{n} \sum_{i=1}^{n} \|F(\boldsymbol{\theta}; \boldsymbol{x}_i)\|_2^2 + g(\boldsymbol{\theta}), \tag{1}$$

where function $F : \mathbb{R}^d \times \mathbb{R}^m \to \mathbb{R}^N$ captures the contribution of a sample $\boldsymbol{x}_i \in \mathbb{R}^m, i = 1, \ldots, n$, to the objective under the parameter $\boldsymbol{\theta} \in \mathbb{R}^d$ and $g : \mathbb{R}^d \to \mathbb{R}$ is a regularizer. Example applications include training auto-encoders [18,28], matrix factorization [16], and multi-target regression [47].

The MSE loss in (1) is computationally convenient, as the resulting problem is smooth and can thus be optimized efficiently via gradient methods, such as

The authors gratefully acknowledge support from the National Science Foundation (Grants CCF-1750539, IIS-1741197, and CNS-1717213), DARPA (Grant HR0011-17-C-0050), and a research grant from American Tower Corp.


N. Oliver et al. (Eds.): ECML PKDD 2021, LNAI 12977, pp. 200–216, 2021.
https://doi.org/10.1007/978-3-030-86523-8_13

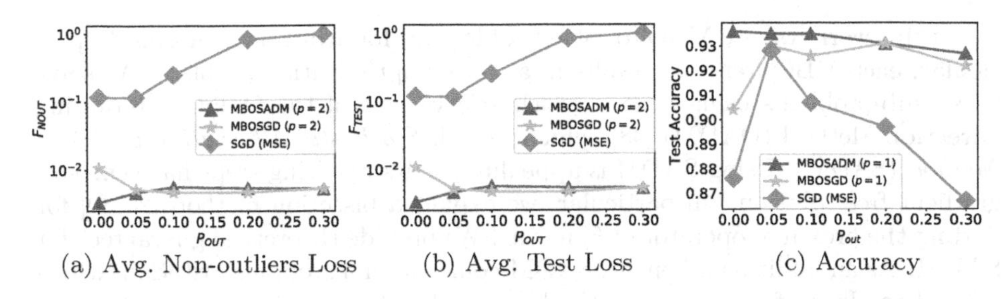


(a) Avg. Non-outliers Loss (b) Avg. Test Loss (c) Accuracy

**Fig. 1.** Robustness of $\ell_p$ norms vs. MSE to outliers introduced to MNIST when training an autoencoder. Figures 1a and 1b show the average loss over the non-outliers and the test set, respectively; values in each figure are normalized w.r.t. the largest value. The test accuracy of a logistic regression on the latent features is shown in Fig. 1c. We see that, under MSE, both for the loss values and classification accuracy are significantly affected by the fraction of outliers $P_{\text{out}}$. Robust embeddings under $p = 1, 2$ norms optimized via our proposed MBO methods exhibit almost constant behavior w.r.t. $P_{\text{out}}$.

stochastic gradient descent (SGD). However, it is well-known that the MSE loss is not robust to *outliers* [8,15,19,20,34], i.e., samples far from the dataset mean. Intuitively, when squaring the error, outliers tend to dominate the objective. To mitigate the effect of outliers, a classic approach is to introduce robustness by replacing the squared error with either the $\ell_2$ norm [8,11,18,28,34,41] or the $\ell_1$ norm [2,5,13,19–21,24,40]. This has been applied to several applications, including feature selection [34,41], PCA [2,8,21,24], K-means clustering [11], training autoencoders [18,28], matrix factorization [5,13,19,20], and regression [40]. Motivated by this approach, we study the following robust variant of Problem (1):

$$\min_{\theta} \frac{1}{n} \sum_{i=1}^{n} \|F(\boldsymbol{\theta}; \boldsymbol{x}_i)\|_p + g(\boldsymbol{\theta}), \tag{2}$$

where $\|\cdot\|_p$ denotes an $\ell_p$ norm ($p \geq 1$). We are particularly interested in cases where $F$ is not affine and, in general, Problem (2) is non-convex. This includes, e.g., feature selection [34], matrix factorization [13,20], auto-encoders [18], and deep multi-target regression [40,47].

A significant challenge behind solving Problem (2) is that its objective is not smooth, precisely because the $\ell_p$ norm is not differentiable at $\mathbf{0} \in \mathbb{R}^N$. For non-convex and non-smooth problems of the form (2), where the objective contains a composite function, *Model-Based Optimization* (MBO) methods [6,7,9,10,23,35] come with good experimental performance as well as theoretical guarantees. In particular, these MBO methods define a convex (but non-smooth) approximation of the main objective, called the *model function*. They then iteratively optimize this model function plus a proximal quadratic term. Under certain conditions, MBO converges to a stationary point of the non-convex problem [23].

In this work, we use MBO to solve Problem (2) for arbitrary $\ell_p$ norms. In particular, each MBO iteration results in a convex optimization problem. We solve these sub-problems using a novel stochastic variant of the Online Alternating Direction Method (OADM) [48], which we call *Stochastic Alternating Direction Method* (SADM). Using SADM is appealing, as its resulting steps have efficient gradient-free solutions; in particular, we exploit a bisection method [25,30] for finding the proximal operator of $\ell_p$ norms. We provide theoretical guarantees for SADM. As an additional benefit, SADM comes with a stopping criterion, which is hard to obtain for gradient methods when the objective is non-smooth [27].

Overall, we make the following contributions:

- We study a general outlier-robust optimization that replaces the MSE with $\ell_p$ norms. We show that such problems can be solved via Model-Based Optimization (MBO) methods.
- We propose SADM, i.e., a stochastic version of OADM, and show that under strong convexity of the regularizer $g$, it converges with a $O(\log T/T)$ rate when solving the sub-problems arising at each MBO iteration.
- We conduct extensive experiments on training auto-encoders and multi-target regression. We show (a) the higher robustness of $\ell_p$ norms in comparison with MSE and (b) the superior performance of MBO, against stochastic gradient methods, both in terms of minimizing the objective and performing downstream classification tasks. In some cases, we see that the MBO variant using SADM obtains objectives that are $29.6\times$ smaller than the ones achieved by the competitors.

The performance of our MBO approach is illustrated in Fig. 1. An autoencoder trained via SGD over the MSE objective is significantly affected by the presence of outliers; in contrast, our MBO methods applied to $\ell_p$ objectives are robust to outliers. These relative benefits are also evident in a downstream classification task over the latent embeddings. The remainder of this paper is organized as follows. We review related work in Sect. 2. We introduce our robust formulation along with its applications in Sect. 3. We describe the instance of MBO applied to our problem in Sect. 4. We introduce SADM and its convergence analysis in Sec. 5 and present our experiments in Sect. 6. We finally conclude in Sect. 7.

## 2 Related Work

**Robustness of $\ell_p$ Norms:** To improve the sensitivity of MSE to outliers, Ding et al. [8] first suggested replacing the MSE with the $\ell_2$ norm in the context of Principal Component Analysis (PCA). This motivated a line of research for developing robust algorithms using the $\ell_2$ norm in different applications, e.g., non-negative matrix factorization [20], feature selection [34,41], training autoencoders [18], and $k$-means clustering [11]. Attaining robustness via the $\ell_1$ norm has also been used in matrix factorization [5,13,19], PCA [2,21,24], and regression [40]. Robustness of the $\ell_1$ norm can be linked to robustness of median to outliers in comparison to average value (see, e.g., Friedman et al. [15]). Our problem includes robust variations considered in, e.g., [13,18,20,34,40], as special cases. However, these earlier algorithms are tailored to specific $\ell_p$ norms and/or do not generalize beyond the studied objective or application (some works, e.g., [34,40], only consider convex problems). In contrast, we unify these variations for different applications as a non-convex and non-smooth problem, and present a general optimization algorithm for arbitrary $\ell_p$ norms.

**Non-smooth/Non-convex Optimization:** Non-smooth and non-convex optimization problems arise in many applications, such as non-negative matrix factorization [16], compressed sensing with non-convex norms [1], and $\ell_p$ norm regularized sparse regression problems [3,33]. A class of non-smooth non-convex optimization problems, known as *weakly convex problems* [45], i.e., problems in which the objective function is the sum of a convex function and a quadratic function, have attracted a lot of attention [6,10,12,22,23,27]. Mai and Johansson [27] provided novel theoretical guarantees on the convergence of stochastic gradient descent with momentum for weakly-convex functions. However, in our experiments in Sect. 6, we show that model-based methods considerably outperform these stochastic gradient methods with momentum.

Our approach falls under the class of *prox-linear* methods [6,9,10,12,22, 23], that solve problems where the objective is a composition of a non-smooth convex function and a smooth function, exactly as in Prob. (2). Such methods iteratively minimize the composition of the non-smooth function with the first-order approximation of the smooth function [6,10,12,23]. Lewis and Wright [23] prove convergence to a stationary point while Drusvyatskiy et al. prove linear convergence [9] and obtain sample complexity guarantees [10]. Ochs et al. [35,36] generalize prox-linear methods by proposing *Model-Based Optimization* (MBO) for both smooth and non-smooth non-convex problems. MBO reduces to a prox-linear method when the objective has a composite form, as in our case. Ochs et al. further considered non-quadratic proximal penalties in sub-problems and complemented MBO with an Armijo-like line search. We leverage both their line search and theoretical guarantees (c.f. Prop. 1); our main technical departure is in solving sub-problems per iteration via SADM, which we discuss next.

**ADMM.** The Alternating Direction Method of Multipliers (ADMM) [4] is a convex optimization algorithm that provides efficient methods for non-smooth problems. Applying ADMM often results in sub-problems that can be solved efficiently via proximal operators [4,39,44]. To speed up ADMM, stochastic variants [26,37,49] have been proposed for minimizing sum-like objectives. These stochastic variants, similar to SGD, update solutions using the gradients of a small batch of terms in the objective, at each iteration. Another group of works proposed online variants of ADMM [17,43,48]. In these variants, the goal is to minimize the summation of loss functions that are revealed by an adversary.

Wang and Banerjee [48] proposed the first online variant of ADMM, termed Online Direction Method of Multipliers (OADM). Here, we propose a stochastic version of OADM, Stochastic Alternating Direction Method (SADM), to solve inner-problems in MBO iterations. SADM is similar to OADM with the difference that functions are sampled uniformly at random and are not given by an adversary. We prove that SADM converges with a $O(\log T/T)$ rate when the regularizer is strongly convex. Other existing stochastic or online ADMM variants either require a smooth objective [26,49] or bounded sub-gradients [37,43], neither of which apply for the inner problems we solve. In contrast, we show that applying SADM results in sub-problems that admit gradient-free efficient solutions via a bisection method for finding proximal operators of $\ell_p$ norms [25,30].

# 3 Robust Regression and Applications

**Notations.** Lowercase boldface letters represent vectors, while capital boldface letters represent matrices. We also use the notation $[n] \triangleq \{1, 2, \ldots, n\}$.

**Robust Regression.** We first extend Prob. (2) to include constraints via:

$$\min_{\boldsymbol{\theta}} \frac{1}{n} \sum_{i \in [n]} \|F(\boldsymbol{\theta}; \boldsymbol{x}_i)\|_p + g(\boldsymbol{\theta}) + \chi_{\mathcal{C}}(\boldsymbol{\theta}), \tag{3}$$

where, again, $F : \mathbb{R}^d \times \mathbb{R}^m \to \mathbb{R}^N$ is smooth, $\|\cdot\|_p$ is the $\ell_p$ norm, $g : \mathbb{R}^d \to \mathbb{R}$ is a convex regularizer such that $\inf g > -\infty$, while $\chi_{\mathcal{C}} : \mathbb{R}^d \to \{0, \infty\}$ is the indicator function of the convex set $\mathcal{C} \subseteq \mathbb{R}^d$. In practice, we are often interested in cases where either the regularizer or the constraint is absent.

**Applications.** For the sake of concreteness, we introduce some applications of Prob. (3). Function $g$ is typically either the lasso (i.e., the $\ell_1$ norm $g(\boldsymbol{\theta}) = \|\boldsymbol{\theta}\|_1$) or ridge regularizer (i.e., the $\ell_2$ norm squared $g(\boldsymbol{\theta}) = \|\boldsymbol{\theta}\|_2^2$). We thus focus on the definition of $F(\cdot;\cdot)$ and constraint set $\mathcal{C}$ in each of these applications.

*Auto-encoders* [18]. Given $n$ data points $\boldsymbol{x}_i \in \mathbb{R}^m$, $i \in [n]$, auto-encoders embed them in a $m'$–dimensional space, $m' \ll m$, as follows. The mapping to $\mathbb{R}^{m'}$ is done by a possibly non-linear function (e.g., a neural network) with $d_{\mathsf{enc}}$ parameters $F_{\mathsf{enc}} : \mathbb{R}^{d_{\mathsf{enc}}} \times \mathbb{R}^m \to \mathbb{R}^{m'}$, called the *encoder*. An inverse mapping, the *decoder* $F_{\mathsf{dec}} : \mathbb{R}^{d_{\mathsf{dec}}} \times \mathbb{R}^{m'} \to \mathbb{R}^m$ with $d_{\mathsf{dec}}$ parameters re-constructs the original points given latent embeddings. Both the encoder and the decoder are trained jointly over a dataset $\{\boldsymbol{x}_i\}_{i=1}^n$ by minimizing the reconstruction error; cast in our robust setting, this amounts to minimizing (3) with

$$F(\boldsymbol{\theta}; \boldsymbol{x}_i) = \boldsymbol{x}_i - F_{\mathsf{dec}}\left(\boldsymbol{\theta}_{\mathsf{dec}}; F_{\mathsf{enc}}(\boldsymbol{\theta}_{\mathsf{enc}}; \boldsymbol{x}_i)\right), \tag{4}$$

where $\boldsymbol{\theta} = [\boldsymbol{\theta}_{\mathsf{dec}}; \boldsymbol{\theta}_{\mathsf{enc}}] \in \mathbb{R}^{d_{\mathsf{enc}} + d_{\mathsf{dec}}}$ comprises the parameters of the encoder and the decoder. Robustness here aims to ameliorate the effect of outliers in the dataset $\{\boldsymbol{x}_i\}_{i=1}^n$. The constraint set can be $\mathbb{R}^d$ (i.e., the problem is unconstrained) or an $\ell_p$-norm ball (i.e., $\{\theta \mid \|\theta\|_p \leq r\}$, for some $r > 0$, $p \geq 1$), when the magnitude of parameters is constrained; this can be used instead of a $\ell_1$ or $\ell_2$ norm regularizer. In stacked denoising autoencoders [46], the encoder and decoder are shallow and satisfy the additional constraint $\theta_{\mathsf{enc}} = \theta_{\mathsf{dec}}$.

*Multi-target Regression* [42]. We are given a set of $n$ data points $\boldsymbol{x}_i \in \mathbb{R}^m$, $i \in [n]$ and the corresponding target labels $\boldsymbol{y}_i \in \mathbb{R}^{m'}$. The goal is to train a (again possibly non-linear) function $f : \mathbb{R}^d \times \mathbb{R}^m \rightarrow \mathbb{R}^{m'}$, with $d$ parameters, to predict target values for a given vector $\boldsymbol{x} \in \mathbb{R}^m$. This maps to Prob. (3) via:

$$F(\boldsymbol{\theta}; \boldsymbol{x}_i, \boldsymbol{y}_i) = \boldsymbol{y}_i - f(\boldsymbol{\theta}; \boldsymbol{x}_i). \tag{5}$$

Robustness in this setting corresponds to ameliorating the effect of outliers in the *label* space, i.e., among labels $\{\boldsymbol{y}_i\}_{i=1}^n$. The constraint set can again be $\mathbb{R}^d$ or defined through an $\ell_p$-norm ball (instead of the corresponding regularizer).

*Matrix Factorization* [38]. Given a matrix $\boldsymbol{X} \in \mathbb{R}^{n\times m}$, the goal is to express it a the product of two matrices $\boldsymbol{G}$,$\boldsymbol{H}$. Cast in our setting, each row $\boldsymbol{x}_i \in \mathbb{R}^m, i \in [n]$, of $\boldsymbol{X}$ is mapped to a lower dimensional sub-space as a vector $\boldsymbol{h}_i \in \mathbb{R}^{m'}$, where the sub-space basis is defined by the rows of the matrix $\boldsymbol{G} \in \mathbb{R}^{m\times m'}$. Function $F$ is then given by $F(\boldsymbol{\theta}; \boldsymbol{x}_i) = \boldsymbol{x}_i - \boldsymbol{G}\boldsymbol{h}_i$, where $\boldsymbol{\theta} = (\boldsymbol{G}, \boldsymbol{H})$ and the rows of the matrix $\boldsymbol{H} \in \mathbb{R}^{n\times m'}$ are the low-dimensional embeddings $\boldsymbol{h}_i$. Robustness here limits sensitivity to outliers in rows; a similar problem can be defined in terms of robustness to outliers in columns. Beyond usual boundedness constraints, additional constraints are introduced in so-called *non-negative matrix factorization* [14,38], where matrices $\boldsymbol{G}$ and $\boldsymbol{H}$ are constrained to be non-negative.

For all three applications, we assume that $F$ is smooth; this requires, e.g., smooth activation functions in deep models. Moreover, in all three examples, Prob. (3) is non-convex and non-smooth, as $\|\cdot\|_p$ is non-differentiable at $\mathbf{0} \in \mathbb{R}^N$.

## 4 Robust Regression via MBO

In this section, we outline how non-smooth, non-convex Prob. (3) can be solved via *model-based optimization* (MBO) [35]. MBO relies on the use of a model function, which is a convex approximation of the main objective. In short, the algorithm proceeds iteratively, approximating function $F(\cdot;\cdot)$ by it's 1st order Taylor expansion at each iteration. This approximation is affine in $\boldsymbol{\theta}$, and results in a convex optimization problem per iteration.

In more detail, cast into our setting, MBO proceeds as follows. Starting with a feasible solution $\boldsymbol{\theta}^0 \in \mathcal{C}$, it performs the following operations in each step $k \in \mathbb{N}$:

$$\tilde{\boldsymbol{\theta}}^k = \arg\min_{\boldsymbol{\theta}} F_{\boldsymbol{\theta}^k}(\boldsymbol{\theta}) + \frac{h}{2}\|\boldsymbol{\theta} - \boldsymbol{\theta}^k\|_2^2, \tag{6a}$$

$$\boldsymbol{\theta}^{k+1} = (1 - \eta^k)\boldsymbol{\theta}^k + \eta^k \tilde{\boldsymbol{\theta}}^k, \tag{6b}$$

where $h > 0$ is a regularization parameter, $\eta^k > 0$ is a step size, and function $F_{\boldsymbol{\theta}^k} : \mathbb{R}^d \rightarrow \mathbb{R}$ is the so-called *model function* at $\boldsymbol{\theta}^k$, defined as:

$$F_{\boldsymbol{\theta}^k}(\boldsymbol{\theta}) \triangleq \frac{1}{n}\sum_{i\in[n]} \|F(\boldsymbol{\theta}^k; \boldsymbol{x}_i) + \boldsymbol{D}_{F_i}(\boldsymbol{\theta}^k)(\boldsymbol{\theta} - \boldsymbol{\theta}^k)\|_p + g(\boldsymbol{\theta}) + \chi_{\mathcal{C}}(\boldsymbol{\theta}), \tag{7}$$

where $\boldsymbol{D}_{F_i}(\boldsymbol{\theta}) \in \mathbb{R}^{N\times d}$ is the Jacobian of $F(\boldsymbol{\theta}; \boldsymbol{x}_i)$ w.r.t. $\boldsymbol{\theta}$. Thus, in each step, MBO replaces $F$ with its 1st-order Taylor approximation and minimizes the

**Algorithm 1.** Model-based Minimization (MBO)

1: **Input: Initial solution $\boldsymbol{\theta}^0 \in \mathbf{dom}F$, iteration number $K$ set $\delta, \gamma \in (0,1)$, and $\tilde{\eta} > 0$**
2: **for** $k \in [K]$ **do**
3: $\quad \tilde{\boldsymbol{\theta}}^k := \arg\min_{\boldsymbol{\theta}} F_{\boldsymbol{\theta}^k}(\boldsymbol{\theta}) + \frac{h}{2}\|\boldsymbol{\theta} - \boldsymbol{\theta}^k\|_2^2$
4: $\quad$ Find $\gamma^k$ via Armijo search rule
5: $\quad \boldsymbol{\theta}^{k+1} := (1-\eta^k)\boldsymbol{\theta}^k + \eta^k \tilde{\boldsymbol{\theta}}^k$
6: **end for**

objective plus a proximal penalty; the resulting $\tilde{\boldsymbol{\theta}}^k$ is interpolated with the current solution $\boldsymbol{\theta}^k$.

The above steps are summarized in Alg. 1. The step size $\eta^k$ is computed via an Armijo-type line search algorithm, which we present in detail in App. A in the extended version [31]. Moreover, the inner-step optimization via (6a) can be inexact; the following proposition shows asymptotic convergence of MBO to a stationary point using an inexact solver (see also App. A in [31]):

**Proposition 1** *(Theorem 4.1 of [35]). Suppose $\boldsymbol{\theta}^*$ is the limit point of the sequence $\boldsymbol{\theta}^k$ generated by Alg. 1. Assume $F_{\boldsymbol{\theta}^k}(\tilde{\boldsymbol{\theta}}^k) + \frac{h}{2}\|\tilde{\boldsymbol{\theta}}^k - \boldsymbol{\theta}^k\|_2^2 - \inf_{\tilde{\boldsymbol{\theta}}} F_{\boldsymbol{\theta}^k}(\tilde{\boldsymbol{\theta}}) + \frac{h}{2}\|\tilde{\boldsymbol{\theta}} - \boldsymbol{\theta}^k\|_2^2 \leq \epsilon^k$, for all iterations $k$, and that $\epsilon^k \to 0$. Then $\boldsymbol{\theta}^*$ is a stationary point of Prob.* (3).

For completeness, we prove Proposition 1 in App. B of the extended version [31], by showing that assumptions of Theorem 4.1 of [35] are indeed satisfied. Problem (6a) is convex but still non-smooth; we discuss how it can be solved efficiently via SADM in the next section.

## 5 Stochastic Alternating Direction Method of Multipliers

After dealing with convexity via MBO, there are still two challenges behind solving the constituent sub-problem (6a). The first is the non-smoothness of $\|\cdot\|_p$; the second is scaling in $n$, which calls for a the use of a stochastic optimization method, akin to SGD (which, however, is not applicable due to the lack of smoothness). We address both through the a novel approach, namely, SADM, which is a stochastic version of the OADM algorithm by Wang and Banerjee [48]. Most importantly, our approach reduces the solution of Prob. (6a) to several gradient-free optimization sub-steps, which can be computed efficiently. In addition, using an SADM/ADMM variant comes with clear stopping criteria, which is challenging for traditional stochastic subgradient methods [27].

### 5.1 SADM

We first describe how our SADM can be applied to solve Prob. (6a). We introduce the following notation to make our exposition more concise:

$$F^{(k)}(\boldsymbol{\theta};\boldsymbol{x}_i) \triangleq \|F(\boldsymbol{\theta}^k;\boldsymbol{x}_i) + \boldsymbol{D}_{F_i}(\boldsymbol{\theta}^k)(\boldsymbol{\theta}-\boldsymbol{\theta}^k)\|_p + \frac{h}{2}\|\boldsymbol{\theta}-\boldsymbol{\theta}^k\|_2^2, \tag{8a}$$

$$F^{(k)}(\boldsymbol{\theta}) \triangleq \frac{1}{n}\sum_{i\in[n]} F^{(k)}(\boldsymbol{\theta},\boldsymbol{x}_i), \tag{8b}$$

$$G(\boldsymbol{\theta}) \triangleq g(\boldsymbol{\theta}) + \chi_{\mathcal{C}}(\boldsymbol{\theta}). \tag{8c}$$

We can then rewrite Prob. (6a) as the following equivalent problem:

$$\text{Minimize} \quad F^{(k)}(\boldsymbol{\theta}_1) + G(\boldsymbol{\theta}_2) \tag{9a}$$

$$\text{subject to:} \quad \boldsymbol{\theta}_1 = \boldsymbol{\theta}_2, \tag{9b}$$

where $\boldsymbol{\theta}_1, \boldsymbol{\theta}_2 \in \mathbb{R}^d$ are auxiliary variables.

Note that the objective in (9a) is equivalent to $F^{(k)}(\boldsymbol{\theta}_1) + G(\boldsymbol{\theta}_2)$. SADM starts with initial solutions, i.e., $\boldsymbol{\theta}_1^0 = \boldsymbol{\theta}_2^0 = \boldsymbol{u}^0 = 0$. At the $t$-th iteration, the algorithm performs the following steps:

$$\boldsymbol{\theta}_1^{t+1} := \arg\min_{\boldsymbol{\theta}_1} F^{(k)}(\boldsymbol{\theta}_1;\boldsymbol{x}_t) + \frac{\rho_t}{2}\|\boldsymbol{\theta}_1 - \boldsymbol{\theta}_2^t + \boldsymbol{u}^t\|_2^2 + \frac{\gamma_t}{2}\|\boldsymbol{\theta}_1 - \boldsymbol{\theta}_1^t\|_2^2, \tag{10a}$$

$$\boldsymbol{\theta}_2^{t+1} := \arg\min_{\boldsymbol{\theta}_2} G(\boldsymbol{\theta}_2) + \frac{\rho_t}{2}\|\boldsymbol{\theta}_1^{t+1} - \boldsymbol{\theta}_2 + \boldsymbol{u}^t\|_2^2, \tag{10b}$$

$$\boldsymbol{u}^{t+1} := \boldsymbol{u}^t + \boldsymbol{\theta}_1^{t+1} - \boldsymbol{\theta}_2^{t+1}, \tag{10c}$$

where variables $\boldsymbol{x}_t$ are sampled uniformly at random from $\{\boldsymbol{x}_i\}_{i=1}^n$, $\boldsymbol{u}^t \in \mathbb{R}^d$ is the dual variable, the $\rho_t, \gamma_t > 0$ are scaling coefficients at the $t$-th iteration. We explain how to set $\rho_t, \gamma_t$ in Thm. 1.

The solution to Problem (10b) amounts to finding the proximal operator of function $G$. In general, given that $g$ is smooth and convex, this is a strongly convex optimization problem and can be solved via standard techniques. Nevertheless, for several of the practical cases we described in Sec. 3 this optimization can be done efficiently with gradient-free methods. For example, in the case where the regularizer $g$ is the either a ridge or lasso penalty, and $\mathcal{C} = \mathbb{R}^d$, it is well-known that proximal operators for $\ell_1$ and $\ell_2$ norms have closed-form solutions [4]. For general $\ell_p$ norms, an efficient (gradient-free) bi-section method due to Liu and Ye [25] (see App. I in [31]) can be used to compute the proximal operator. Moreover, in the absence of the regularizer, the proximal operator for the indicator function $\chi_{\mathcal{C}}$ is equivalent to projection on the convex set $\mathcal{C}$. This again has closed-form solution, e.g., when $\mathcal{C}$ is the simplex [29] or an $\ell_p$-norm ball [25,32]. Problem (10a) is harder to solve; we show however that it can also reduced to the (gradient-free) bisection method due to Liu and Ye [25] in the next section.

### 5.2 Inner ADMM

We solve Problem (10a) using another application of ADMM. In particular, note that (10a) assumes the following general form:

$$\min_{\boldsymbol{x}} \quad \|\boldsymbol{A}\boldsymbol{x}+\boldsymbol{b}\|_p + \lambda\|\boldsymbol{x}-\boldsymbol{c}\|_2^2, \tag{11}$$

where $\boldsymbol{A} = \boldsymbol{D}_{F_t}(\boldsymbol{\theta}^{(k)})$, the constituent parameter vectors are $\boldsymbol{c} = \frac{\rho_t}{\rho_t+\gamma_t+h}(\boldsymbol{\theta}_2^t - \boldsymbol{u}^t) + \frac{\gamma_t}{\rho_t+\gamma_t+h}\boldsymbol{\theta}_1^t + \frac{h}{\rho_t+\gamma_t+h}\boldsymbol{\theta}^{(k)}$, $\boldsymbol{b} = F(\boldsymbol{\theta}^{(k)};\boldsymbol{x}_t) - \boldsymbol{D}_{F_t}(\boldsymbol{\theta}^{(k)})\boldsymbol{\theta}^{(k)}$, and $\lambda = \frac{\rho_t+\gamma_t+h}{2}$.

We solve (11) via ADMM by reformulating it as the following problem:

$$\min \quad \|\boldsymbol{y}\|_p + \lambda\|\boldsymbol{x}-\boldsymbol{c}\|_2^2 \tag{12a}$$

$$\text{s.t} \quad \boldsymbol{A}\boldsymbol{x}+\boldsymbol{b}-\boldsymbol{y} = 0. \tag{12b}$$

The ADMM steps at the $k$-th iteration for (12) are the following:

$$\boldsymbol{y}^{k+1} := \arg\min_{\boldsymbol{y}} \|\boldsymbol{y}\|_p + \rho'/2\|\boldsymbol{y} - \boldsymbol{A}\boldsymbol{x}^k - \boldsymbol{b} + \boldsymbol{z}^k\|_2^2, \tag{13a}$$

$$\boldsymbol{x}^{k+1} := \arg\min_{\boldsymbol{x}} \lambda\|\boldsymbol{x}-\boldsymbol{c}\|_2^2 + \rho'/2\|\boldsymbol{y}^{k+1} - \boldsymbol{A}\boldsymbol{x} - \boldsymbol{b} + \boldsymbol{z}^k\|_2^2, \tag{13b}$$

$$\boldsymbol{z}^{k+1} := \boldsymbol{z}^k + \boldsymbol{y}^{k+1} - \boldsymbol{A}\boldsymbol{x}^{k+1} - \boldsymbol{b}, \tag{13c}$$

where $\boldsymbol{z}^k \in \mathbb{R}^N$ denotes the dual variable at the $k$-th iteration and $\rho' > 0$ is a hyper-parameter of ADMM.

Problem (13a) is again equivalent to computing the proximal operator of the $\ell_p$-norm, which, as mentioned earlier, has closed-form solution for $p = 1, 2$. Moreover, for general $\ell_p$-norms the proximal operator can be computed via the bisection algorithm by Liu and Ye [25]. This bisection method yields a solution with an $\epsilon$ accuracy in $O(\log_2(1/\epsilon))$ rounds [25,30] (see App. I in [31]).

### 5.3 Convergence

To attain the convergence guarantee of MBO given by Proposition 1, we need to solve the inner problem within accuracy $\epsilon^k$ at iteration $k$, where $\epsilon^k \to 0$. As our major technical contribution, we ensure this by proving the convergence of SADM when solving Prob. (6a).

Consider the sequence $\{\boldsymbol{\theta}_1^t, \boldsymbol{\theta}_2^t, \boldsymbol{u}^t\}_{t=1}^T$ generated by our SADM algorithm (10), where $\boldsymbol{x}_t$, $t \in [T]$, are sampled u.a.r. from $\{\boldsymbol{x}_i\}_{i=1}^n$. Let also

$$\bar{\boldsymbol{\theta}}_1^T \triangleq \frac{1}{T}\sum_{t=1}^T \boldsymbol{\theta}_1^t,\ \bar{\boldsymbol{\theta}}_2^T \triangleq \frac{1}{T}\sum_{t=1}^T \boldsymbol{\theta}_2^{t+1}, \tag{14}$$

denote the time averages of the two solutions. Let also $\boldsymbol{\theta}^* = \boldsymbol{\theta}_1^* = \boldsymbol{\theta}_2^*$ be the optimal solution of Prob. (9). Finally, denote by

$$R^T \triangleq F^{(k)}(\bar{\boldsymbol{\theta}}_1^T) + G(\bar{\boldsymbol{\theta}}_2^T) - F^{(k)}(\boldsymbol{\theta}^*) - G(\boldsymbol{\theta}^*) \tag{15}$$

the residual error of the objective from the optimal. Then, the following holds:

**Theorem 1.** *Assume that $\mathcal{C}$ is convex, closed, and bounded, while $g(0) = 0$, $g(\boldsymbol{\theta}) \geq 0$, and $g(\cdot)$ is both Lipschitz continuous and $\beta$-strongly convex over $\mathcal{C}$. Moreover, assume that both the function $F(\boldsymbol{\theta}; \boldsymbol{x}_i)$ and its Jacobian $\boldsymbol{D}_{F_i}(\boldsymbol{\theta})$ are bounded on the set $\mathcal{C}$, for all $i \in [n]$. We set $\gamma_t = ht$ and $\rho_t = \beta t$. Then,*

$$\|\bar{\boldsymbol{\theta}}_1^T - \bar{\boldsymbol{\theta}}_2^T\|_2^2 = O\left(\frac{\log T}{T}\right) \tag{16a}$$

$$\mathbb{E}[R^T] = O\left(\frac{\log T}{T}\right) \tag{16b}$$

$$\mathbb{P}\left(R^T \geq k_1 \frac{\log T}{T} + k_2 \frac{M}{\sqrt{T}}\right) \leq e^{-\frac{M^2}{16}} \quad \text{for all } M > 0, T \geq 3, \tag{16c}$$

*where $k_1, k_2 > 0$ are constants (see (32) in App. C in [31] for exact definitions).*

We prove Theorem 1 in App. C in [31]. The theorem has the following important consequences. First, (16a) implies that the infeasibility gap between $\theta_1$ and $\theta_2$ decreases as $O(\frac{\log T}{T})$ *deterministically.* Second, by (16b) the residual error $R^T$ decreases as $O(\frac{\log T}{T})$ in expectation. Finally, (16c) shows that the tail of the residual error as iterations increase is exponentially bounded. In particular, given a desirable accuracy $\epsilon_k$, (16c) gives the number of iterations necessary be within $\epsilon_k$ of the optimal with any probability $1 - \delta$. Therefore, according to Proposition 1, using SADM will result in convergence of Algorithm 1 with high probability. Finally, we note that, although we write Theorem 1 for updates using only one random sample per iteration, the analysis and guarantees readily extend to the case where a batch selected u.a.r. is used instead. A formal statement and proof can be found in App. E in the extended version [31].

## 6 Experiments

**Algorithms.** We run two variants of MBO; the first one, which we call `MBOSADM`, uses SADM (see Sect. 5) for solving the inner problems (6a). The second one, which we call `MBOSGD`, solves inner problems via a sub-gradient method. We also apply stochastic gradient descent with momentum directly to Prob. (3); we refer to this algorithm as `SGD`. This corresponds to the algorithm by [27], applied to our setting. We also solve the problem instances with an MSE objective using `SGD`, as the MSE is smooth and `SGD` is efficient in this case. Hyperparameters and implementation details are in App. F in [31]. Our code is publicly available.[1]

**Applications and Datasets.** We focus on two applications: training autoencoders and multi-target regression, with a ridge regularizer and $\mathcal{C} = \mathbb{R}^d$. The architectures we use are described in App. F in [31]. For autoencoders, We use `MNIST` and `Fashion-MNIST` to train autoencoders and `SCM1d` [42] for multi target regression. All three datasets, including training and test splits, are also described in App. F in [31].

[1] https://github.com/neu-spiral/ModelBasedOptimization.

**Table 1.** Time and Objective Performance. We report objective and time metrics for under different outlier ratios and different $p$-norms. We observe from the table that `MBOSADM` significantly outperforms other competitors in terms of objective metrics. In terms of running time, `SGD` is generally fastest, due to fast gradient updates. However, we see that the time MBO variants take to get to the same or better objective value (i.e., $T^*$), ware comparable to running time of `SGD`.

| $P_{out}$ | $p$ | MBOSADM | | | | | MBOSGD | | | | | SGD | | | |
|---|---|---|---|---|---|---|---|---|---|---|---|---|---|---|---|
| | | $F_{NOUTL}$ | $F_{OBJ}$ | $F_{TEST}$ | $T$(h) | $T^*$(h) | $F_{NOUTL}$ | $F_{OBJ}$ | $F_{TEST}$ | $T$(h) | $T^*$(h) | $F_{NOUTL}$ | $F_{OBJ}$ | $F_{TEST}$ | $T$(h) |
| MNIST | | | | | | | | | | | | | | | |
| 0.0 | 2.0 | **2.50** | **2.51** | **2.50** | 5.69 | 0.14 | 8.08 | 8.08 | 8.12 | 64.17 | 6.47 | 9.21 | 9.22 | 9.30 | 9.83 |
| 0.0 | 1.5 | **2.63** | **2.63** | **2.63** | 11.67 | 0.79 | 20.19 | 20.20 | 20.39 | 65.67 | 59.98 | 20.35 | 20.36 | 20.57 | 14.71 |
| 0.0 | 1.0 | **3.46** | **3.47** | **3.44** | 17.82 | 3.09 | 102.79 | 102.80 | 104.24 | 81.53 | NA | 102.44 | 102.46 | 103.89 | 11.50 |
| 0.05 | 2.0 | **3.48** | **5.35** | **3.46** | 6.33 | 0.31 | 3.89 | 6.36 | 3.86 | 54.92 | 38.31 | 8.03 | 12.52 | 8.09 | 13.96 |
| 0.05 | 1.5 | **4.10** | **9.74** | **4.08** | 45.32 | 2.08 | 5.86 | 11.70 | 5.82 | 57.03 | 25.12 | 20.34 | 34.69 | 20.57 | 14.60 |
| 0.05 | 1.0 | **5.23** | **20.20** | **5.24** | 44.03 | 5.61 | 27.68 | 73.53 | 27.56 | 32.76 | 9.67 | 102.40 | 236.43 | 103.90 | 11.70 |
| 0.1 | 2.0 | 4.27 | **7.77** | 4.23 | 11.67 | 1.34 | **3.56** | 7.83 | **3.54** | 64.20 | 33.70 | 7.02 | 11.64 | 7.04 | 13.97 |
| 0.1 | 1.5 | **4.18** | **11.84** | **4.17** | 68.74 | 0.29 | 5.50 | 13.77 | 5.45 | 67.04 | 9.88 | 20.34 | 48.79 | 20.57 | 14.04 |
| 0.1 | 1.0 | **5.90** | **36.02** | **5.92** | 37.73 | 8.20 | 30.08 | 109.79 | 30.16 | 39.77 | 6.72 | 102.36 | 368.22 | 103.90 | 11.81 |
| 0.2 | 2.0 | 4.07 | 8.97 | 4.04 | 51.69 | 4.39 | **3.54** | **8.23** | **3.52** | 57.08 | 19.19 | 7.48 | 16.44 | 7.51 | 14.25 |
| 0.2 | 1.5 | **3.90** | **11.58** | **3.89** | 195.69 | 1.56 | 7.00 | 20.63 | 6.95 | 45.46 | 6.44 | 20.36 | 77.78 | 20.59 | 15.06 |
| 0.2 | 1.0 | **3.85** | **28.25** | **3.83** | 36.98 | 5.15 | 40.12 | 224.47 | 40.11 | 19.71 | 2.13 | 102.37 | 639.32 | 103.90 | 8.25 |
| 0.3 | 2.0 | **3.99** | 14.21 | **3.98** | 9.83 | 4.93 | 4.02 | **10.55** | 3.99 | 55.60 | 24.14 | 7.46 | 20.63 | 7.48 | 13.53 |
| 0.3 | 1.5 | 20.55 | **22.56** | 20.78 | 159.92 | 39.24 | **7.22** | 24.30 | **7.16** | 42.65 | 15.09 | 23.90 | 58.52 | 23.89 | 16.25 |
| 0.3 | 1.0 | 102.70 | **99.36** | 104.27 | 51.48 | 0.87 | **56.60** | 438.89 | **56.17** | 20.48 | 3.32 | 102.34 | 910.68 | 103.90 | 8.52 |
| Fashion-MNIST | | | | | | | | | | | | | | | |
| 0.0 | 2.0 | **3.51** | **3.51** | **3.51** | 4.33 | 0.31 | 5.01 | 5.01 | 5.01 | 42.13 | 14.80 | 8.72 | 8.73 | 8.70 | 9.78 |
| 0.0 | 1.5 | **6.13** | **6.14** | **6.14** | 14.35 | 2.18 | 8.87 | 8.88 | 8.89 | 63.39 | 12.80 | 22.62 | 22.63 | 22.56 | 14.70 |
| 0.0 | 1.0 | **10.59** | **10.61** | **10.56** | 29.69 | 2.63 | 41.24 | 41.26 | 41.35 | 50.41 | 3.32 | 224.26 | 224.28 | 224.89 | 9.72 |
| 0.05 | 2.0 | **3.80** | **5.82** | **3.80** | 14.70 | 1.40 | 4.53 | 6.75 | 4.54 | 67.29 | 19.15 | 8.30 | 11.98 | 8.27 | 9.71 |
| 0.05 | 1.5 | **7.38** | 14.57 | **7.40** | 96.73 | 2.56 | 7.91 | **11.97** | 7.93 | 64.25 | 16.16 | 20.88 | 27.22 | 20.83 | 10.94 |
| 0.05 | 1.0 | **16.64** | **30.51** | **16.68** | 43.55 | 16.04 | 65.01 | 109.94 | 65.31 | 29.60 | 9.70 | 158.65 | 227.48 | 158.27 | 9.97 |
| 0.1 | 2.0 | **4.05** | **6.73** | **4.06** | 14.66 | 2.35 | 4.28 | 7.90 | 4.29 | 65.06 | 16.46 | 8.96 | 14.35 | 8.94 | 13.67 |
| 0.1 | 1.5 | 11.08 | 32.69 | 11.10 | 20.07 | NA | **8.46** | **15.23** | **8.49** | 65.78 | 9.92 | 17.98 | 31.41 | 17.95 | 10.63 |
| 0.1 | 1.0 | **9.79** | **27.96** | **9.81** | 35.50 | 2.43 | 58.70 | 126.18 | 58.90 | 45.06 | 2.08 | 235.02 | 452.45 | 234.49 | 13.32 |
| 0.2 | 2.0 | 6.07 | 10.19 | 6.08 | 14.25 | 3.67 | **4.77** | **9.28** | **4.77** | 69.84 | 30.16 | 5.71 | 14.34 | 5.71 | 9.31 |
| 0.2 | 1.5 | 28.51 | 53.69 | 28.49 | 39.42 | NA | **10.94** | **27.12** | **10.97** | 39.11 | 16.09 | 19.36 | 42.97 | 19.36 | 10.87 |
| 0.2 | 1.0 | **10.50** | **27.95** | **10.50** | 94.72 | 6.57 | 140.00 | 390.02 | 140.08 | 17.03 | 3.33 | 204.88 | 644.99 | 205.13 | 14.72 |
| 0.3 | 2.0 | 6.63 | 23.18 | 6.63 | 32.87 | NA | **5.84** | **13.04** | **5.85** | 50.52 | 29.95 | 7.45 | 20.12 | 7.46 | 13.65 |
| 0.3 | 1.5 | **7.08** | **22.51** | **7.10** | 86.27 | 30.02 | 11.09 | 24.73 | 11.12 | 52.41 | 12.08 | 19.52 | 58.26 | 19.56 | 11.05 |
| 0.3 | 1.0 | **14.43** | **50.91** | **14.46** | 95.08 | 19.48 | 404.77 | 893.56 | 404.52 | 9.51 | NA | 410.82 | 522.50 | 411.84 | 10.74 |
| SCMD1d | | | | | | | | | | | | | | | |
| 0.0 | 2.0 | 2.88 | 2.88 | 3.02 | 1.82 | 0.12 | **2.85** | **2.85** | **2.99** | 0.36 | 0.04 | 3.62 | 3.63 | 3.72 | 1.37 |
| 0.0 | 1.5 | 4.23 | 4.24 | **4.39** | 7.22 | 0.43 | **4.22** | **4.23** | 4.44 | 0.36 | 0.04 | 5.47 | 5.47 | 5.60 | 1.58 |
| 0.0 | 1.0 | **9.78** | **9.79** | **10.18** | 7.13 | 0.47 | 9.86 | 9.86 | 10.32 | 0.37 | 0.04 | 12.95 | 12.95 | 13.25 | 1.22 |
| 0.05 | 2.0 | 2.88 | 3.13 | 2.99 | 2.52 | 0.13 | **2.86** | **3.11** | 3.00 | 0.54 | 0.05 | 3.64 | 3.89 | 3.71 | 1.31 |
| 0.05 | 1.5 | 4.23 | 4.61 | **4.37** | 10.23 | 4.61 | **4.22** | **4.59** | 4.46 | 0.50 | 0.05 | 5.50 | 5.87 | 5.61 | 1.23 |
| 0.05 | 1.0 | **9.69** | **10.52** | **10.09** | 0.59 | 0.18 | 9.86 | 10.66 | 10.35 | 0.51 | 0.05 | 13.03 | 13.87 | 13.29 | 1.17 |
| 0.1 | 2.0 | 2.90 | 3.41 | 3.01 | 2.22 | 0.13 | **2.84** | **3.34** | **3.00** | 0.46 | 0.05 | 3.63 | 4.12 | 3.69 | 1.30 |
| 0.1 | 1.5 | 4.23 | 4.99 | 4.42 | 9.42 | 0.68 | **4.18** | **4.90** | **4.40** | 0.50 | 0.05 | 5.52 | 6.11 | 5.62 | 1.15 |
| 0.1 | 1.0 | **9.56** | **10.99** | **10.18** | 9.92 | 0.78 | 9.77 | 11.11 | 10.54 | 0.54 | 0.07 | 13.09 | 13.72 | 13.32 | 1.11 |
| 0.2 | 2.0 | 2.93 | 3.90 | 3.03 | 1.83 | 0.95 | **2.86** | **3.79** | **3.02** | 0.5 | 0.3 | 3.63 | 3.97 | 3.66 | 1.15 |
| 0.2 | 1.5 | 4.23 | 5.60 | **4.37** | 8.17 | 3.60 | **4.21** | **5.48** | 4.47 | 0.36 | 0.2 | 5.50 | 5.83 | 5.56 | 1.17 |
| 0.2 | 1.0 | **9.46** | **11.00** | **10.04** | 6.55 | 1.50 | 9.82 | 11.30 | 10.60 | 0.45 | 0.20 | 13.09 | 13.38 | 13.28 | 1.11 |
| 0.3 | 2.0 | 2.93 | 4.32 | 3.03 | 1.80 | NA | **2.85** | 4.10 | **3.03** | 0.46 | NA | 3.61 | **3.90** | 3.64 | 1.18 |
| 0.3 | 1.5 | 4.25 | 6.05 | **4.44** | 8.19 | NA | **4.21** | 5.73 | 4.49 | 0.50 | NA | 5.43 | **5.69** | 5.43 | 1.18 |
| 0.3 | 1.0 | **9.53** | 11.07 | **10.00** | 6.44 | 2.72 | 9.68 | **11.05** | 10.32 | 0.51 | 0.28 | 12.95 | 13.13 | 12.96 | 1.11 |

**Outliers.** We denote the outliers ratio with $P_{\text{out}}$; each datapoint $\boldsymbol{x}_i$, $i \in [n]$, is independently corrupted with outliers with probability $P_{\text{out}}$. The probability $P_{\text{out}}$ ranges from 0.0 to 0.3 in our experiments. In particular, we corrupt training samples by replacing them with samples randomly drawn from a Gaussian distribution whose mean is $\alpha$ away from the original data and its standard deviation equals that of the original dataset. For MNIST and FashionMNIST, we set $\alpha$ to 1.5 times the original standard deviation, while for SCM1d, we set $\alpha$ to 2.5 times the standard deviation.

**Metrics.** We evaluate the solution obtained by different algorithms by using the following three metrics. The first is $F_{\text{OBJ}}$, the regularized objective of Prob. (3) evaluated over the training set. The other two are: $F_{\text{NOUTL}} \triangleq \frac{\sum_{i \notin \mathcal{S}_{\text{OUTL}}} \|F(\boldsymbol{\theta}; \boldsymbol{x}_i)\|_p}{n - |\mathcal{S}_{\text{OUTL}}|}$, and $F_{\text{TEST}} \triangleq \frac{\sum_{i \in \mathcal{S}_{\text{TEST}}} \|F(\boldsymbol{\theta}; \boldsymbol{x}_i)\|_p}{|\mathcal{S}_{\text{TEST}}|}$, where $\mathcal{S}_{\text{OUTL}}$, $\mathcal{S}_{\text{TEST}}$ are the outlier and test sets, respectively. Metric $F_{\text{NOUTL}}$ measures the robustness of algorithms w.r.t. outliers; ideally, $F_{\text{NOUTL}}$ should remain unchanged as the fraction of outliers increases. Metric $F_{\text{TEST}}$ evaluates the generalization ability of algorithms on unseen (test) data, which also does not contain outliers; ideally, $F_{\text{TEST}}$ be similar $F_{\text{NOUTL}}$. Moreover, we report total running time ($T$) of all algorithms. For the two variants of MBO, we additionally report the time ($T^*$) until the they reach the optimal value attained by SGD (N/A if never reached). Finally, for autoencoders, we also use dataset labels to train a logistic regression classifier over latent embeddings, and also report the prediction accuracy on the test set. Classifier hyperparameters are described in App. G in [31].

### 6.1 Time and Objective Performance Comparison

We evaluate our algorithms w.r.t. both objective and time metrics, which we report for different outlier ratios $P_{\text{out}}$ and $p$-norms in Table 1. By comparing objective metrics, we see that MBOSADM and MBOSGD significantly outperform SGD. SGD achieves a better $F_{\text{OBJ}}$ in only 2 out of 45 cases, i.e., SCM1d dataset for $p = 1.5, 2$ and $P_{\text{out}} = 0.3$; however, even for these two cases, MBOSADM and MBOSGD obtain better $F_{\text{NOUTL}}$ and $F_{\text{TEST}}$ values. In terms of overall running time $T$, SGD is generally faster than MBOSADM and MBOSGD; this is expected, as each iteration of SGD only computes the gradient of a mini-batch of terms in the objective, while the other methods need to solve an inner-problem. Nonetheless, by comparing $T^*$, we see that the MBO variants obtain the same or better objective as SGD in a comparable time. In particular, $T^*$ is less than $T$ for SGD in 33 and 15 cases (out of 45) for MBOSADM and MBOSGD, respectively.

Comparing the performance between MBOSADM and MBOSGD, we first note that MBOSADM has a superior performance w.r.t. all three objective metrics for 25 out of 45 cases. In some cases, MBOSADM obtains considerably smaller objective values; for example, for MNIST and $P_{\text{out}} = 0.0, p = 1$, $F_{\text{NOUT}}$ is 0.03 of the value obtained by MBOSGD (also see Figs. 2c and 2f). However, it seems that in the high-outlier setting $P_{\text{out}} = 0.3$ the performance of MBOSADM deteriorates; this is mostly due to the fact that the high number of outliers adversely affects the convergence of SADM and it takes more iterations to satisfy the desired accuracy.

### 6.2 Robustness Analysis

We further study the robustness of different $p$-norms and MSE to the presence of outliers. For brevity, we only report results for MNIST and for $p = 1, 2$, and MSE. For more results refer to App. H in [31]. To make comparisons between different objectives interpretable, we normalize all values in each figure by the largest value in that figure.

By comparing Figs. 2a and 2d, corresponding to MSE, with other plots in Fig. 2, we see that the loss values considerably increase by adding outliers. For other $p$-norms, we see that SGD generally stays unchanged, w.r.t. outliers. However, the loss for SGD is higher than MBO variants. Loss values for MBOSGD also do not increase significantly by adding outliers. Moreover, we see that, when no outliers are present $P_{\text{out}} = 0.0$, MBOSGD obtains higher loss values. MBOSADM generally achieves the lowest loss values and these values again do not increase with increasing $P_{\text{out}}$; however, for the highest outliers ($P_{\text{out}} = 0.3$), the performance of MBOSADM is considerably worse for $p = 1$. As we emphasize in Sect. 6.1, high number of outliers adversely affects the convergence of SADM, and hence the poor performance of MBOSADM for $P_{\text{out}} = 0.3$.

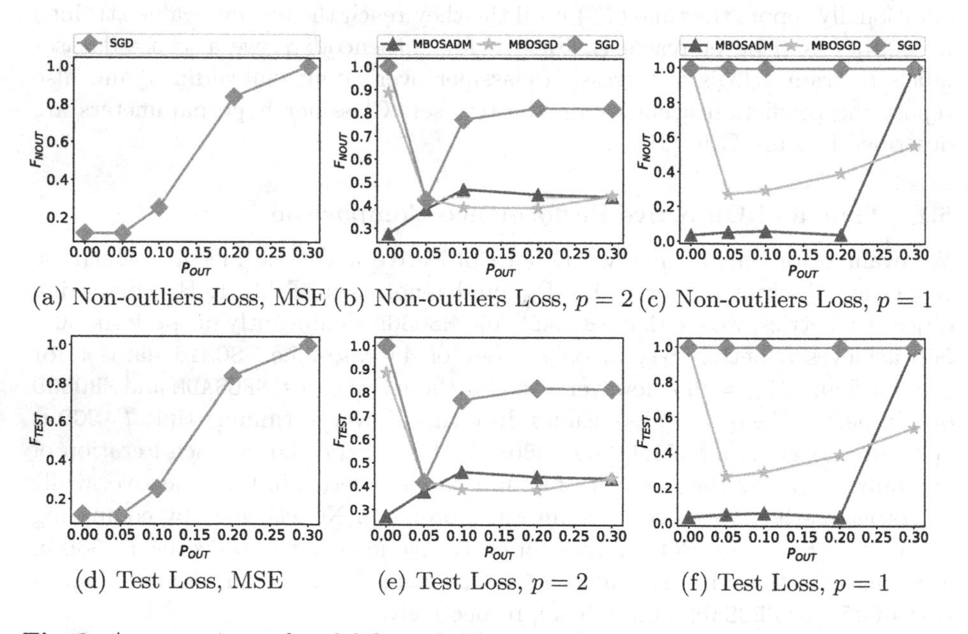


(a) Non-outliers Loss, MSE (b) Non-outliers Loss, $p = 2$ (c) Non-outliers Loss, $p = 1$

(d) Test Loss, MSE (e) Test Loss, $p = 2$ (f) Test Loss, $p = 1$

**Fig. 2.** A comparison of scalability of the non-outlioers loss $F_{\text{NOUT}}$ and the test loss $F_{\text{TEST}}$ for different $p$-norms, w.r.t., outliers fraction $P_{\text{out}}$. We normalize values in each figure by the largest observed value, to make comparisons between different objectives possible. We see that MSE in Figures 2a and 2 are drastically affected by outliers and scale with outliers fraction $P_{\text{out}}$. Other $\ell_p$ norms for different methods in Figures 2b, 2c, 2e, and 2f generally stay unchanged w.r.t. $P_{\text{out}}$. However, MBOSADM in the high outlier regime and $p = 1$ performs poorly.

### 6.3 Classification Performance

Figure 3 shows the quality of the latent embeddings obtained by different trained autoencoders on the downstream classification over `MNIST` and `FashionMNIST`. Additional results are shown in App. G in [31]. We see that MBO variants again outperform `SGD`. For `MNIST`, (reported in Figs. 3a to 3c), we see that `MBOSADM` for $p = 1$ obtains the highest accuracy. Moreover, for `Fashion-MNIST` (reported in Fig. 3d to 3f), we observe that again `MBOSADM` for $p = 1$ outperforms other methods. We also observe that MSE (reported in Figures 3a and 3d) is sensitive to outliers; the corresponding accuracy drastically drops for $P_{\text{out}} \geq 0.1$. An interesting observation is that adding outliers improves the performance of `SGD`; however, we see that `SGD` always results in lower accuracy, except in two cases ($P_{\text{out}} = 0.2$ in Fig. 3b and $P_{\text{out}} = 0.3$ in Fig. 3e).

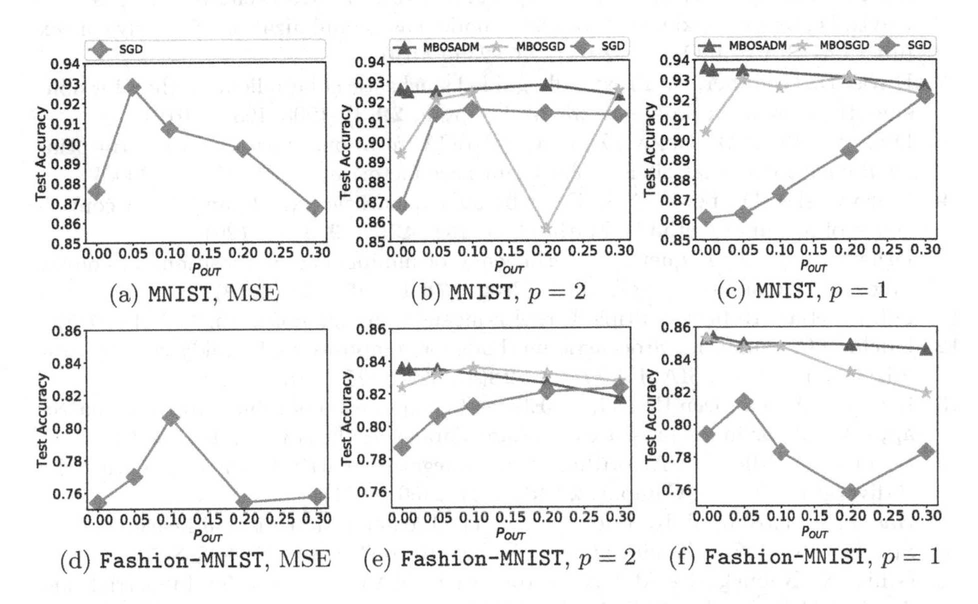


(a) `MNIST`, MSE (b) `MNIST`, $p = 2$ (c) `MNIST`, $p = 1$

(d) `Fashion-MNIST`, MSE (e) `Fashion-MNIST`, $p = 2$ (f) `Fashion-MNIST`, $p = 1$

**Fig. 3.** Classification performance for different methods and datasets. We use the embeddings obtained by auto-encoders trained via different algorithms to train a logistic regression model for classification. We generally observe that `MBOSADM` results in higher accuracy on the test sets. Moreover, we see that MSE is evidently sensitive to outliers, see Figures 3a and 3d for $P_{\text{out}} \geq 0.2$.

## 7 Conclusion

We present a generic class of robust formulations that includes many applications, i.e., auto-encoders, multi-target regression, and matrix factorization. We show that SADM, in combination with MBO, provides efficient solutions for our class of robust problems. Studying other proximal measures described by Ochs et al. [35] is an open area. Moreover, characterizing the sample complexity of our proposed method for obtaining a stationary point, as in MBO variants that use gradient methods [6,10], is an interesting future direction.

## References


1. Attouch, H., Bolte, J., Redont, P., Soubeyran, A.: Proximal alternating minimization and projection methods for nonconvex problems: an approach based on the Kurdyka-Lojasiewicz inequality. Math. Oper. Res. **35**(2), 438–457 (2010)
2. Baccini, A., Besse, P., Falguerolles, A.: A l1-norm PCA and a heuristic approach. Ordinal Symbolic Data Anal. **1**(1), 359–368 (1996)
3. Blumensath, T., Davies, M.E.: Iterative hard thresholding for compressed sensing. Appl. Comput. Harmonic Anal. **27**(3), 265–274 (2009)
4. Boyd, S., Parikh, N., Chu, E., Peleato, B., Eckstein, J.: Distributed optimization and statistical learning via the alternating direction method of multipliers. Found. Trends® Mach. Learn. **3**(1), 1–122 (2011)
5. Croux, C., Filzmoser, P.: Robust factorization of a data matrix. In: COMPSTAT, pp. 245–250. Springer (1998). https://doi.org/10.1007/978-3-662-01131-7_29
6. Davis, D., Drusvyatskiy, D.: Stochastic model-based minimization of weakly convex functions. SIAM J. Optim. **29**(1), 207–239 (2019)
7. Davis, D., Grimmer, B.: Proximally guided stochastic subgradient method for nonsmooth, nonconvex problems. SIAM J. Optim. **29**(3), 1908–1930 (2019)
8. Ding, C., Zhou, D., He, X., Zha, H.: R1-PCA: rotational invariant l 1-norm principal component analysis for robust subspace factorization. In: ICML (2006)
9. Drusvyatskiy, D., Lewis, A.S.: Error bounds, quadratic growth, and linear convergence of proximal methods. Math. Oper. Res. **43**(3), 919–948 (2018)
10. Drusvyatskiy, D., Paquette, C.: Efficiency of minimizing compositions of convex functions and smooth maps. Math. Prog. **178**(1), 503–558 (2019)
11. Du, L., et al.: Robust multiple kernel k-means using l21-norm. In: IJCAI (2015)
12. Duchi, J.C., Ruan, F.: Stochastic methods for composite and weakly convex optimization problems. SIAM J. Optim. **28**(4), 3229–3259 (2018)
13. Eriksson, A., Van Den Hengel, A.: Efficient computation of robust low-rank matrix approximations in the presence of missing data using the l1 norm. In: CVPR (2010)
14. Févotte, C., Idier, J.: Algorithms for nonnegative matrix factorization with the $\beta$-divergence. Neural Comput. **23**(9), 2421–2456 (2011)
15. Hastie, T., Tibshirani, R., Friedman, J.: The Elements of Statistical Learning. SSS, Springer, New York (2009). https://doi.org/10.1007/978-0-387-84858-7
16. Gillis, N.: Nonnegative Matrix Factorization. SIAM - Society for Industrial and Applied Mathematics, Philadelphia (2020)
17. Hosseini, S., Chapman, A., Mesbahi, M.: Online distributed ADMM via dual averaging. In: CDC (2014)
18. Jiang, W., Gao, H., Chung, F.L., Huang, H.: The l2,1-norm stacked robust autoencoders for domain adaptation. In: AAAI (2016)
19. Ke, Q., Kanade, T.: Robust l1 factorization in the presence of outliers and missing data by alternative convex programming. In: CVPR (2005)
20. Kong, D., Ding, C., Huang, H.: Robust nonnegative matrix factorization using l21-norm. In: CIKM (2011)
21. Kwak, N.: Principal component analysis based on l1-norm maximization. IEEE Trans. Pattern Anal. Mach. Intell. **30**(9), 1672–1680 (2008)
22. Le, H., Gillis, N., Patrinos, P.: Inertial block proximal methods for non-convex non-smooth optimization. In: ICML (2020)
23. Lewis, A.S., Wright, S.J.: A proximal method for composite minimization. Math. Prog. **158**(1), 501–546 (2016)

24. Li, X., Pang, Y., Yuan, Y.: l1-norm-based 2DPCA. IEEE Trans. Syst. Man Cybern. Part B Cybern. **40**(4), 1170–1175 (2010)
25. Liu, J., Ye, J.: Efficient l1/lq NormRregularization. arXiv preprint arXiv:1009.4766 (2010)
26. Liu, Y., Shang, F., Cheng, J.: Accelerated variance reduced stochastic ADMM. In: AAAI (2017)
27. Mai, V., Johansson, M.: Convergence of a stochastic gradient method with momentum for non-smooth non-convex optimization. In: ICML (2020)
28. Mehta, J., Gupta, K., Gogna, A., Majumdar, A., Anand, S.: Stacked robust autoencoder for classification. In: NeurIPS (2016)
29. Michelot, C.: A finite algorithm for finding the projection of a point onto the canonical simplex of n. J. Optim. Theor. Appl. **50**(1), 1–6 (1986)
30. Moharrer, A., Gao, J., Wang, S., Bento, J., Ioannidis, S.: Massively distributed graph distances. IEEE Trans. Sig. Inf. Process. Netw. **6**, 667–683 (2020)
31. Moharrer, A., Kamran, K., Yeh, E., Ioannidis, S.: Robust regression via model based methods. arXiv preprint arXiv:2106.10759 (2021)
32. Moreau, J.J.: Décomposition orthogonale d'un espace hilbertien selon deux cônes mutuellement polaires. Comptes rendus hebdomadaires des séances de l'Académie des sciences **255**, 238–240 (1962)
33. Natarajan, B.K.: Sparse approximate solutions to linear systems. SIAM J. Comput. **24**(2), 227–234 (1995)
34. Nie, F., Huang, H., Cai, X., Ding, C.H.: Efficient and robust feature selection via joint l2,1-norms minimization. In: NIPS (2010)
35. Ochs, P., Fadili, J., Brox, T.: Non-smooth non-convex Bregman minimization: unification and new algorithms. J. Optim. Theor. Appl. **181**(1), 244–278 (2019)
36. Ochs, P., Malitsky, Y.: Model function based conditional gradient method with Armijo-like line search. In: Proceedings of the 36th International Conference on Machine Learning (2019)
37. Ouyang, H., He, N., Tran, L., Gray, A.: Stochastic alternating direction method of multipliers. In: International Conference on Machine Learning, pp. 80–88. PMLR (2013)
38. Paatero, P., Tapper, U.: Positive matrix factorization: a non-negative factor model with optimal utilization of error estimates of data values. Environmetrics **5**(2), 111–126 (1994)
39. Peng, Y., Ganesh, A., Wright, J., Xu, W., Ma, Y.: Rasl: robust alignment by sparse and low-rank decomposition for linearly correlated images. IEEE Trans. Pattern Anal. Mach. Intell. **34**(11), 2233–2246 (2012)
40. Pesme, S., Flammarion, N.: Online robust regression via SGD on the l1 loss. In: NeurIPS (2020)
41. Qian, M., Zhai, C.: Robust unsupervised feature selection. In: IJCAI (2013)
42. Spyromitros-Xioufis, E., Tsoumakas, G., Groves, W., Vlahavas, I.: Multi-target regression via input space expansion: treating targets as inputs. Mach. Learn. **104**(1), 55–98 (2016)
43. Suzuki, T.: Dual averaging and proximal gradient descent for online alternating direction multiplier method. In: ICML (2013)
44. Tao, M., Yuan, X.: Recovering low-rank and sparse components of matrices from incomplete and noisy observations. SIAM J. Optim. **21**(1), 57–81 (2011)
45. Vial, J.P.: Strong and weak convexity of sets and functions. Math. Oper. Res. **8**(2), 231–259 (1983)

46. Vincent, P., Larochelle, H., Lajoie, I., Bengio, Y., Manzagol, P.A.: Stacked denoising autoencoders: Learning useful representations in a deep network with a local denoising criterion. J. Mach. Learn. Res. **11**(Dec), 3371–3408 (2010)
47. Waegeman, W., Dembczyński, K., Hüllermeier, E.: Multi-target prediction: a unifying view on problems and methods. Data Min. Knowl. Discovery **33**(2), 293–324 (2019)
48. Wang, H., Banerjee, A.: Online alternating direction method. In: ICML (2012)
49. Zheng, S., Kwok, J.T.: Fast-and-light stochastic ADMM. In: IJCAI, pp. 2407–2613 (2016)

# Black-Box Optimizer with Stochastic Implicit Natural Gradient

Yueming Lyu(✉) and Ivor W. Tsang

Australian Artificial Intelligence Institute, University of Technology Sydney, 15 Broadway, Ultimo, NSW 2007, Australia
Yueming.Lyu@student.uts.edu.au, Ivor.Tsang@uts.edu.au

**Abstract.** Black-box optimization is primarily important for many computationally intensive applications, including reinforcement learning (RL), robot control, etc. This paper presents a novel theoretical framework for black-box optimization, in which our method performs stochastic updates with an implicit natural gradient of an exponential-family distribution. Theoretically, we prove the convergence rate of our framework with full matrix update for convex functions under Gaussian distribution. Our methods are very simple and contain fewer hyper-parameters than CMA-ES [12]. Empirically, our method with full matrix update achieves competitive performance compared with one of the state-of-the-art methods CMA-ES on benchmark test problems. Moreover, our methods can achieve high optimization precision on some challenging test functions (e.g., $l_1$-norm ellipsoid test problem and Levy test problem), while methods with explicit natural gradient, i.e., IGO [21] with full matrix update can not. This shows the efficiency of our methods.

**Keywords:** Black-box optimization · Implicit natural gradient · Stochastic optimization

## 1 Introduction

Given a proper function $f(\boldsymbol{x}) : \mathbb{R}^d \rightarrow \mathbb{R}$ such that $f(\boldsymbol{x}) > -\infty$, we aim at minimizing $f(\boldsymbol{x})$ by using function queries only, which is known as black-box optimization. It has a wide range of applications, such as automatic hyper-parameters tuning in machine learning and computer vision problems [24], adjusting parameters for robot control and reinforcement learning [7,9,17], black-box architecture search in engineering design [27] and drug discovery [20].

Several kinds of approaches have been widely studied for black-box optimization, including Bayesian optimization (BO) methods [8,19,26], evolution strategies (ES) [5,12] and genetic algorithms (GA) [25]. Among them, Bayesian

**Electronic supplementary material** The online version of this chapter (https://doi.org/10.1007/978-3-030-86523-8_14) contains supplementary material, which is available to authorized users.


N. Oliver et al. (Eds.): ECML PKDD 2021, LNAI 12977, pp. 217–232, 2021.
https://doi.org/10.1007/978-3-030-86523-8_14

optimization methods are good at dealing with low-dimensional expensive black-box optimization, while ES methods are better for relatively high-dimensional problems with cheaper evaluations compared with BO methods. ES-type algorithms can well support parallel evaluation, and have drawn more and more attention because of its success in reinforcement learning problems [10,16,23], recently.

CMA-ES [12] is one of state-of-the-art ES methods with many successful applications. It uses second-order information to search candidate solutions by updating the mean and covariance matrix of the likelihood of candidate distributions. Despite its successful performance, the update rule combines several sophisticated components, which is not well understood. Wierstra et al. show that directly applying standard reinforce gradient descent is very sensitive to variance in high precision search for black-box optimization [28]. Thus, they propose Natural evolution strategies (NES) [28] to estimate the natural gradient for black-box optimization. However, they use the Monte Carlo sampling to approximate the Fisher information matrix (FIM), which incurs additional error and computation cost unavoidably. Along this line, [1] show the connection between the rank-$\mu$ update of CMA-ES and NES [28]. [21] further show that several ES methods can be included in an unified framework. Despite these theoretical attempts, the practical performance of these methods is still inferior to CMA-ES. Moreover, these works do not provide any convergence rate analysis, which is the key insight to expedite black-box optimizations.

Another line of research for ES-type algorithms is to reduce the variance of gradient estimators. Choromanski et al. [10] proposed to employ Quasi Monte Carlo (QMC) sampling to achieve more accurate gradient estimates. Recently, they further proposed to construct gradient estimators based on active subspace techniques [9]. Although these works can reduce sample complexity, how does the variance of these estimators influence the convergence rate remains unclear.

To take advantage of second-order information for the acceleration of black-box optimizations, we propose a novel theoretical framework: stochastic Implicit Natural Gradient Optimization (INGO) algorithms, from the perspective of information geometry. Raskutti et al. [22] give a method to compute the Fisher information matrix implicitly using exact gradients, which is impossible for black-box optimization; while our methods and analysis focus on black-box optimization. To the best of our knowledge, we are the first to design stochastic implicit natural gradient algorithms w.r.t natural parameters for black-box optimization. Our methods take a stochastic black-box estimate instead of the exact gradient to update. Theoretically, this update is equivalent to a stochastic natural gradient step w.r.t. natural parameters of an exponential-family distribution. Our contributions are summarized as follows:

- To the best of our knowledge, we are the first to design stochastic implicit natural gradient algorithms w.r.t natural parameters for black-box optimization. We propose efficient algorithms for both continuous and discrete black-box optimization. Our methods construct stochastic black-box update without computing the FIM. Our method can adaptively control the stochastic update

by taking advantage of the second-order information, which is able to accelerate convergence and is primarily important for ill-conditioned problems. Moreover, our methods have fewer hyperparameters and are much simpler than CMA-ES.
- Theoretically, we prove the convergence rate of our continuous optimization methods for convex functions. We also show that reducing variance of the black-box gradient estimators by orthogonal sampling can lead to a small regret bound.
- Empirically, our continuous optimization method achieves a competitive performances compared with the state-of-the-art method CMA-ES on benchmark problems. We find that our method with full matrix update can obtain higher optimization precision compared with IGO [21] on some challenging problems. We further show the effectiveness of our methods on RL control problems. Moreover, our discrete optimization algorithm outperforms a GA method on a benchmark problem.

## 2 Notation and Symbols

Denote $\|\cdot\|_2$ and $\|\cdot\|_F$ as the spectral norm and Frobenius norm for matrices, respectively. Define $\|Y\|_{tr} := \sum_i |\lambda_i|$, where $\lambda_i$ denotes the $i^{th}$ eigenvalue of matrix $Y$. Notation $\|\cdot\|_2$ will also denote $l_2$-norm for vectors. Symbol $\langle \cdot, \cdot \rangle$ denotes inner product under $l_2$-norm for vectors and inner product under Frobenius norm for matrices. Define $\|\boldsymbol{x}\|_C := \sqrt{\langle \boldsymbol{x}, C\boldsymbol{x} \rangle}$. Denote $\mathcal{S}^+$ and $\mathcal{S}^{++}$ as the set of positive semi-definite matrices and the set of positive definite matrices, respectively. Denote $\Sigma^{\frac{1}{2}}$ as the symmetric positive semi-definite matrix such that $\Sigma = \Sigma^{\frac{1}{2}}\Sigma^{\frac{1}{2}}$ for $\Sigma \in \mathcal{S}^+$.

## 3 Implicit Natural Gradient Optimization

### 3.1 Optimization with Exponential-Family Sampling

We aim at minimizing a proper function $f(\boldsymbol{x})$, $\boldsymbol{x} \in \mathcal{X}$ with only function queries, which is known as black-box optimization. Due to the lack of gradient information for black-box optimization, we here present an exponential-family sampling trick to relax any black-box optimization problem. Specifically, the objective is relaxed as the expectation of $f(\boldsymbol{x})$ under a parametric distribution $p(\boldsymbol{x}; \eta)$ with parameter $\eta$, i.e., $J(\boldsymbol{\eta}) := \mathbb{E}_{p(\boldsymbol{x};\eta)}[f(\boldsymbol{x})]$ [28]. The optimal parameter $\boldsymbol{\eta}$ is found by minimizing $J(\boldsymbol{\eta})$ as $\min_{\boldsymbol{\eta}} \left\{\mathbb{E}_{p(\boldsymbol{x};\boldsymbol{\eta})}[f(\boldsymbol{x})]\right\}$ This relaxed problem is minimized when the probability mass is all assigned on the minimum of $f(\boldsymbol{x})$. The distribution $p$ is the sampling distribution for black-box function queries. Note, $p$ can be either continuous or discrete.

In this work, we assume that the distribution $p(\boldsymbol{x}; \boldsymbol{\eta})$ is an exponential-family distribution:

$$p(\boldsymbol{x}; \boldsymbol{\eta}) = h(\boldsymbol{x}) \exp\left\{\langle \phi(\boldsymbol{x}), \boldsymbol{\eta} \rangle - A(\boldsymbol{\eta})\right\}, \tag{1}$$

where $\boldsymbol{\eta}$ and $\phi(\boldsymbol{x})$ are the natural parameter and sufficient statistic, respectively. And $A(\eta)$ is the log partition function defined as $A(\eta) = \log \int \exp\{\langle \phi(\boldsymbol{x}), \boldsymbol{\eta} > h(\boldsymbol{x}) d\boldsymbol{x}$.

It is named as minimal exponential-family distribution when there is a one-to-one mapping between the mean parameter $\boldsymbol{m} := \mathbb{E}_p[\phi(\boldsymbol{x})]$ and natural parameter $\boldsymbol{\eta}$. This one-to-one mapping ensures that we can reparameterize $J(\boldsymbol{\eta})$ as $\tilde{J}(\boldsymbol{m}) = J(\boldsymbol{\eta})$ [3,13]. $\tilde{J}$ is w.r.t parameter $\boldsymbol{m}$, while $J$ is w.r.t parameter $\boldsymbol{\eta}$.

To minimize the objective $\tilde{J}(\boldsymbol{m})$, we desire the updated distribution lying in a trust region of the previous distribution at each step. Formally, we update the mean parameters by solving the following optimization problem.

$$\boldsymbol{m}_{t+1} = \arg\min_{\boldsymbol{m}} \left\langle \boldsymbol{m}, \nabla_{\boldsymbol{m}} \tilde{J}(\boldsymbol{m}_t) \right\rangle + \frac{1}{\beta_t} \mathrm{KL}\left(p_{\boldsymbol{m}} \| p_{\boldsymbol{m}_t}\right), \tag{2}$$

where $\nabla_{\boldsymbol{m}} \tilde{J}(\boldsymbol{m}_t)$ denotes the gradient at $\boldsymbol{m} = \boldsymbol{m}_t$.

The KL-divergence term measures how close the updated distribution and the previous distribution. For an exponential-family distribution, the KL-divergence term in (2) is equal to Bregman divergence between $\boldsymbol{m}$ and $\boldsymbol{m}_t$ [4]:

$$\mathrm{KL}\left(p_{\boldsymbol{m}} \| p_{\boldsymbol{m}_t}\right) = A^*(\boldsymbol{m}) - A^*(\boldsymbol{m}_t) - \langle \boldsymbol{m} - \boldsymbol{m}_t, \nabla_{\boldsymbol{m}} A^*(\boldsymbol{m}_t) \rangle, \tag{3}$$

where $A^*(\boldsymbol{m})$ is the convex conjugate of $A(\boldsymbol{\eta})$. Thus, the problem (2) is a convex optimization problem, and it has a closed-form solution.

### 3.2 Implicit Natural Gradient

**Intractability of Natural Gradient for Black-box Optimization:** Natural gradient [2] can capture information geometry structure during optimization, which enables us to take advantage of the second-order information to accelerate convergence. Direct computation of natural gradient needs the inverse of Fisher information matrix (FIM), which needs to estimate the FIM. The method in [22] provides an alternative way to compute natural gradient without computation of FIM. However, it relies on the exact gradient, which is impossible for black-box optimization.

Hereafter, we propose a novel stochastic implicit natural gradient algorithms for black-box optimization of continuous and discrete variables in Sect. 4 and Section A (in the supplement), respectively. We first show how to compute the implicit natural gradient. In problem Eq. (2), we take the derivative w.r.t $\boldsymbol{m}$, and set it to zero, also note that $\nabla_{\boldsymbol{m}} A^*(\boldsymbol{m}) = \boldsymbol{\eta}$ [22], we can obtain that

$$\boldsymbol{\eta}_{t+1} = \boldsymbol{\eta}_t - \beta_t \nabla_{\boldsymbol{m}} \tilde{J}(\boldsymbol{m}_t). \tag{4}$$

Natural parameters $\boldsymbol{\eta}$ of the distribution lies on a Riemannian manifold with metric tensor specified by the Fisher Information Matrix:

$$\boldsymbol{F}(\boldsymbol{\eta}) := \mathbb{E}_p \left[ \nabla_\eta \log p(\boldsymbol{x}; \boldsymbol{\eta}) \nabla_\eta \log p(\boldsymbol{x}; \boldsymbol{\eta})^\top \right]. \tag{5}$$

For exponential-family with the minimal representation, the natural gradient has a simple form for computation.

**Theorem 1.** *[14, 22] For an exponential-family in the minimal representation, the natural gradient w.r.t $\boldsymbol{\eta}$ is equal to the gradient w.r.t. $\boldsymbol{m}$, i.e.,*

$$\boldsymbol{F}(\boldsymbol{\eta})^{-1}\nabla_{\boldsymbol{\eta}}J(\boldsymbol{\eta}) = \nabla_{\boldsymbol{m}}\tilde{J}(\boldsymbol{m}). \tag{6}$$

**Remark:** Theorem 1 can be easily obtained by the chain rule and the fact $\boldsymbol{F}(\boldsymbol{\eta}) = \frac{\partial^2 A(\eta)}{\partial\eta\partial\eta^\top}$. It enables us to compute the natural gradient implicitly without computing the inverse of the Fisher information matrix. As shown in Theorem 1, the update rule in (4) is equivalent to the natural gradient update w.r.t $\boldsymbol{\eta}$ in (7):

$$\boldsymbol{\eta}_{t+1} = \boldsymbol{\eta}_t - \beta_t\boldsymbol{F}(\boldsymbol{\eta}_t)^{-1}\nabla_{\eta}J(\boldsymbol{\eta}_t). \tag{7}$$

Thus, update rule in (4) selects the steepest descent direction along the Riemannian manifold induced by the Fisher information matrix as natural gradient descent. It can take the second-order information to accelerate convergence.

## 4 Update Rule for Gaussian Sampling

We first present an update method for the case of Gaussian sampling for continuous optimization. For other distributions, we can derive the update rule in a similar manner. We present update methods for discrete optimization in the supplement due to the space limitation.

For a Gaussian distribution $p := \mathcal{N}(\boldsymbol{\mu}, \Sigma)$ with mean $\boldsymbol{\mu}$ and covariance matrix $\Sigma$, the natural parameters $\eta = \{\boldsymbol{\eta}_1, \boldsymbol{\eta}_2\}$ are given as

$$\boldsymbol{\eta}_1 := \Sigma^{-1}\boldsymbol{\mu} \tag{8}$$

$$\boldsymbol{\eta}_2 := -\frac{1}{2}\Sigma^{-1}. \tag{9}$$

The related mean parameters $\boldsymbol{m} = \{\boldsymbol{m}_1, \boldsymbol{m}_2\}$ are given as $\boldsymbol{m}_1 := \mathbb{E}_p[\boldsymbol{x}] = \boldsymbol{\mu}$ and $\boldsymbol{m}_2 := \mathbb{E}_p[\boldsymbol{x}\boldsymbol{x}^\top] = \boldsymbol{\mu}\boldsymbol{\mu}^\top + \Sigma$.

Using the chain rule, the gradient with respect to mean parameters can be expressed in terms of the gradients w.r.t $\boldsymbol{\mu}$ and $\Sigma$ [13,15] as:

$$\nabla_{\boldsymbol{m}_1}\tilde{J}(\boldsymbol{m}) = \nabla_{\boldsymbol{\mu}}\tilde{J}(\boldsymbol{m}) - 2[\nabla_{\Sigma}\tilde{J}(\boldsymbol{m})]\boldsymbol{\mu} \tag{10}$$

$$\nabla_{\boldsymbol{m}_2}\tilde{J}(\boldsymbol{m}) = \nabla_{\Sigma}\tilde{J}(\boldsymbol{m}). \tag{11}$$

It follows that

$$\Sigma_{t+1}^{-1} = \Sigma_t^{-1} + 2\beta_t\nabla_{\Sigma}\tilde{J}(\boldsymbol{m}_t) \tag{12}$$

$$\boldsymbol{\mu}_{t+1} = \boldsymbol{\mu}_t - \beta_t\Sigma_{t+1}\nabla_{\boldsymbol{\mu}}\tilde{J}(\boldsymbol{m}_t). \tag{13}$$

Note that $\tilde{J}(\boldsymbol{m}) = \mathbb{E}_p[f(\boldsymbol{x})]$, the gradients of $\tilde{J}(\boldsymbol{m})$ w.r.t $\mu$ and $\Sigma$ can be obtained by log-likelihood trick as Theorem 2.

**Theorem 2.** *[28] The gradient of the expectation of an integrable function $f(\boldsymbol{x})$ under a Gaussian distribution $p := \mathcal{N}(\boldsymbol{\mu}, \Sigma)$ with respect to the mean $\boldsymbol{\mu}$ and the covariance $\Sigma$ can be expressed as Eq. (14) and Eq. (15), respectively.*

$$\nabla_{\mu}\mathbb{E}_p[f(\boldsymbol{x})] = \mathbb{E}_p\left[\Sigma^{-1}(\boldsymbol{x}-\boldsymbol{\mu})f(\boldsymbol{x})\right] \tag{14}$$

$$\nabla_{\Sigma}\mathbb{E}_p[f(\boldsymbol{x})] = \frac{1}{2}\mathbb{E}_p\left[\left(\Sigma^{-1}(\boldsymbol{x}-\boldsymbol{\mu})(\boldsymbol{x}-\boldsymbol{\mu})^\top\Sigma^{-1}-\Sigma^{-1}\right)f(\boldsymbol{x})\right]. \tag{15}$$

Together Theorem 2 with Eq. (12) and (13), we present the update with only function queries as:

$$\Sigma_{t+1}^{-1} = \Sigma_t^{-1}+\beta_t\mathbb{E}_p\left[\left(\Sigma_t^{-1}(\boldsymbol{x}-\boldsymbol{\mu}_t)(\boldsymbol{x}-\boldsymbol{\mu}_t)^\top\Sigma_t^{-1}-\Sigma_t^{-1}\right)f(\boldsymbol{x})\right] \tag{16}$$

$$\boldsymbol{\mu}_{t+1} = \boldsymbol{\mu}_t - \beta_t\Sigma_{t+1}\mathbb{E}_p\left[\Sigma_t^{-1}(\boldsymbol{x}-\boldsymbol{\mu}_t)f(\boldsymbol{x})\right]. \tag{17}$$

**Remark:** Our method updates the inverse of the covariance matrix instead of the covariance matrix itself.

### 4.1 Stochastic Update

The above gradient update needs the expectation of a black-box function. However, this expectation does not have a closed-form solution. Here, we estimate the gradient w.r.t $\mu$ and $\Sigma$ by Monte Carlo sampling. Equation (16) and (17) enable us to estimate the gradient by the function queries of $f(x)$ instead of $\nabla f(x)$. This property is very crucial for black-box optimization because gradient ($\nabla f(x)$) is not available.

Update rules using Monte Carlo sampling are given as:

$$\Sigma_{t+1}^{-1} = \Sigma_t^{-1} + \frac{\beta_t}{N}\sum_{i=1}^{N}\left[\left(\Sigma_t^{-1}(\boldsymbol{x}_i-\boldsymbol{\mu}_t)(\boldsymbol{x}_i-\boldsymbol{\mu}_t)^\top\Sigma_t^{-1}-\Sigma_t^{-1}\right)f(\boldsymbol{x}_i)\right] \tag{18}$$

$$\boldsymbol{\mu}_{t+1} = \boldsymbol{\mu}_t - \frac{\beta_t}{N}\sum_{i=1}^{N}\left[\Sigma_{t+1}\Sigma_t^{-1}(\boldsymbol{x}_i-\boldsymbol{\mu}_t)f(\boldsymbol{x}_i)\right]. \tag{19}$$

To avoid the numeric ill-scaling problem, we employ a monotonic transformation $h(\cdot)$ as:

$$h(f(\boldsymbol{x}_i)) = \frac{f(\boldsymbol{x}_i)-\widehat{\mu}}{\widehat{\sigma}}. \tag{20}$$

where $\widehat{\mu}$ and $\widehat{\sigma}$ denote mean and stand deviation of function values in a batch of samples. This leads to an unbiased estimator for gradient. The update rule is given as Eq. (21) and Eq. (22). We present our black-box optimization algorithm in Algorithm 1.

$$\Sigma_{t+1}^{-1} = \Sigma_t^{-1} + \beta\sum_{i=1}^{N}\frac{f(\boldsymbol{x}_i)-\widehat{\mu}}{N\widehat{\sigma}}\left(\Sigma_t^{-1}(\boldsymbol{x}_i-\boldsymbol{\mu}_t)(\boldsymbol{x}_i-\boldsymbol{\mu}_t)^\top\Sigma_t^{-1}\right) \tag{21}$$

$$\boldsymbol{\mu}_{t+1} = \boldsymbol{\mu}_t - \beta_t\sum_{i=1}^{N}\frac{f(\boldsymbol{x}_i)-\widehat{\mu}}{N\widehat{\sigma}}\Sigma_{t+1}\Sigma_t^{-1}(\boldsymbol{x}_i-\boldsymbol{\mu}_t). \tag{22}$$

**Algorithm 1.** INGO

**Input:** Number of Samples $N$, step-size $\beta$.
**while** Termination condition not satisfied **do**
Take i.i.d samples $\boldsymbol{z}_i \sim \mathcal{N}(\boldsymbol{0}, \boldsymbol{I})$ for $i \in \{1, \cdots N\}$.
Set $\boldsymbol{x}_i = \boldsymbol{\mu}_t + \Sigma_t^{\frac{1}{2}} \boldsymbol{z}_i$ for $i \in \{1, \cdots N\}$.
Query the batch observations $\{f(\boldsymbol{x}_1), ..., f(\boldsymbol{x}_N)\}$
Compute $\widehat{\sigma} = \text{std}(f(\boldsymbol{x}_1), ..., f(\boldsymbol{x}_N))$.
Compute $\widehat{\mu} = \frac{1}{N} \sum_{i=1}^{N} f(\boldsymbol{x}_i)$.
Set $\Sigma_{t+1}^{-1} = \Sigma_t^{-1} + \beta \sum_{i=1}^{N} \frac{f(\boldsymbol{x}_i) - \widehat{\mu}}{N\widehat{\sigma}} \Sigma_t^{-\frac{1}{2}} \boldsymbol{z}_i \boldsymbol{z}_i^\top \Sigma_t^{-\frac{1}{2}}$.
Set $\boldsymbol{\mu}_{t+1} = \boldsymbol{\mu}_t - \beta \sum_{i=1}^{N} \frac{f(\boldsymbol{x}_i) - \widehat{\mu}}{N\widehat{\sigma}} \Sigma_{t+1} \Sigma_t^{-\frac{1}{2}} \boldsymbol{z}_i$
**end while**

The update of mean $\boldsymbol{\mu}$ in Algorithm 1 is properly scaled by $\Sigma$. Moreover, our method updates the inverse of the covariance matrix instead of the covariance matrix itself, which provides us a stable way to update covariance independent of its scale. Thus, our method can update properly when the algorithm adaptively reduces variance for high precision search. In contrast, directly apply standard reinforce type gradient update is unstable as shown in [28].

### 4.2 Direct Update for $\boldsymbol{\mu}$ and $\boldsymbol{\Sigma}$

We provide an alternative updating equation with simple concept and derivation. The implicit natural gradient algorithms are working on the natural parameter space. Alternatively, we can also directly work on the $\boldsymbol{\mu}$ and $\Sigma$ parameter space. Formally, we derive the update rule by solving the following trust region optimization problem.

$$\boldsymbol{\theta}_{t+1} = \arg\min_{\boldsymbol{\theta}} \langle \boldsymbol{\theta}, \nabla_{\boldsymbol{\theta}} \bar{J}(\boldsymbol{\theta}_t) \rangle + \frac{1}{\beta_t} \text{KL}\left(p_{\boldsymbol{\theta}} \| p_{\boldsymbol{\theta}_t}\right), \tag{23}$$

where $\boldsymbol{\theta} := \{\boldsymbol{\mu}, \Sigma\}$ and $\bar{J}(\boldsymbol{\theta}) := \mathbb{E}_{p(\boldsymbol{x};\boldsymbol{\theta})}[f(\boldsymbol{x})] = J(\boldsymbol{\eta})$.

For Gaussian sampling, the optimization problem in (23) is a convex optimization problem. We can achieve a closed-form update given in Theorem 3:

**Theorem 3.** *For Gaussian distribution with parameter $\boldsymbol{\theta} := \{\boldsymbol{\mu}, \Sigma\}$, problem (23) is convex w.r.t $\boldsymbol{\theta}$. The optimum of problem (23) leads to closed-form update (24) and (25):*

$$\Sigma_{t+1}^{-1} = \Sigma_t^{-1} + 2\beta_t \nabla_{\Sigma} \bar{J}(\boldsymbol{\theta}_t) \tag{24}$$

$$\boldsymbol{\mu}_{t+1} = \boldsymbol{\mu}_t - \beta_t \Sigma_t \nabla_{\boldsymbol{\mu}} \bar{J}(\boldsymbol{\theta}_t). \tag{25}$$

**Remark:** Comparing the update rule in Theorem 3 with Eq. (12) and (13), we can observe that the only difference is in the update of $\boldsymbol{\mu}$. In Eq. (25), the update employs $\Sigma_t$, while the update in Eq. (13) employs $\Sigma_{t+1}$. The update in Eq. (13) takes one step look ahead information of $\Sigma$,

**Algorithm 2.** INGOstep

**Input:** Number of Samples $N$, step-size $\beta$.
**while** Termination condition not satisfied **do**
  Take i.i.d samples $\boldsymbol{z}_i \sim \mathcal{N}(\boldsymbol{0}, \boldsymbol{I})$ for $i \in \{1, \cdots N\}$.
  Set $\boldsymbol{x}_i = \boldsymbol{\mu}_t + \Sigma_t^{\frac{1}{2}} \boldsymbol{z}_i$ for $i \in \{1, \cdots N\}$.
  Query the batch observations $\{f(\boldsymbol{x}_1), ..., f(\boldsymbol{x}_N)\}$
  Compute $\widehat{\sigma} = \text{std}(f(\boldsymbol{x}_1), ..., f(\boldsymbol{x}_N))$.
  Compute $\widehat{\mu} = \frac{1}{N} \sum_{i=1}^{N} f(\boldsymbol{x}_i)$.
  Set $\Sigma_{t+1}^{-1} = \Sigma_t^{-1} + \beta \sum_{i=1}^{N} \frac{f(\boldsymbol{x}_i) - \widehat{\mu}}{N\widehat{\sigma}} \Sigma_t^{-\frac{1}{2}} \boldsymbol{z}_i \boldsymbol{z}_i^\top \Sigma_t^{-\frac{1}{2}}$.
  Set $\boldsymbol{\mu}_{t+1} = \boldsymbol{\mu}_t - \beta \sum_{i=1}^{N} \frac{f(\boldsymbol{x}_i) - \widehat{\mu}}{N\widehat{\sigma}} \Sigma_t^{\frac{1}{2}} \boldsymbol{z}_i$
**end while**

We can obtain the black-box update for $\boldsymbol{\mu}$ and $\Sigma$ by Theorem 3 and Theorem 2. The update rule is given as follows:

$$\Sigma_{t+1}^{-1} = \Sigma_t^{-1} + \beta_t \mathbb{E}_p \left[ \left( \Sigma_t^{-1} (\boldsymbol{x} - \boldsymbol{\mu}_t)(\boldsymbol{x} - \boldsymbol{\mu}_t)^\top \Sigma_t^{-1} - \Sigma_t^{-1} \right) f(\boldsymbol{x}) \right] \tag{26}$$

$$\boldsymbol{\mu}_{t+1} = \boldsymbol{\mu}_t - \beta_t \mathbb{E}_p \left[ (\boldsymbol{x} - \boldsymbol{\mu}_t) f(\boldsymbol{x}) \right]. \tag{27}$$

Using the normalization transformation function $h(f(x)) = (f(x) - \widehat{\mu})/\widehat{\sigma}$, we can obtain Monte Carlo approximation update as

$$\Sigma_{t+1}^{-1} = \Sigma_t^{-1} + \beta \sum_{i=1}^{N} \frac{f(\boldsymbol{x}_i) - \widehat{\mu}}{N\widehat{\sigma}} \left( \Sigma_t^{-1} (\boldsymbol{x}_i - \boldsymbol{\mu}_t)(\boldsymbol{x}_i - \boldsymbol{\mu}_t)^\top \Sigma_t^{-1} \right) \tag{28}$$

$$\boldsymbol{\mu}_{t+1} = \boldsymbol{\mu}_t - \beta_t \sum_{i=1}^{N} \frac{f(\boldsymbol{x}_i) - \widehat{\mu}}{N\widehat{\sigma}} (\boldsymbol{x}_i - \boldsymbol{\mu}_t). \tag{29}$$

We present the algorithm in Algorithm 2 compared with our INGO. The only difference between INGO (Algorithm 1) and INGOstep (Algorithm 2) is the update rule of $\boldsymbol{\mu}$. INGO employs information of $\Sigma_{t+1}$, while INGOstep only uses $\Sigma_t$.

## 5 Convergence Rate

We first show a general framework for continuous optimization in Algorithm 3. Algorithm 3 employs an unbiased estimator ($\widehat{g}_t$) for gradient $\nabla_{\boldsymbol{\mu}} \bar{J}(\boldsymbol{\theta}_t)$. In contrast, it can employ both the unbiased and biased estimators $\widehat{G}_t$ for update. It is worth noting that $\widehat{g}_t$ can be both the first-order estimate (stochastic gradient) and the zeroth-order estimate (function's value-based estimator).

The update step of $\boldsymbol{\mu}$ and $\Sigma$ is achieved by solving the following convex minimization problem.

$$\boldsymbol{m}^{t+1} = \operatorname*{arg\,min}_{\boldsymbol{m} \in \boldsymbol{\mathcal{M}}} \beta_t \langle \boldsymbol{m}, \widehat{v}_t \rangle + \text{KL}\left( p_{\boldsymbol{m}} \| p_{\boldsymbol{m}^t} \right). \tag{30}$$

where $\boldsymbol{m} := \{\boldsymbol{m}_1, \boldsymbol{m}_2\} = \{\boldsymbol{\mu}, \Sigma + \boldsymbol{\mu}\boldsymbol{\mu}^\top\} \in \boldsymbol{\mathcal{M}}$, $\boldsymbol{\mathcal{M}}$ denotes a convex set, and $\widehat{v}_t = \{\widehat{g}_t - 2\widehat{G}_t \boldsymbol{\mu}_t, \widehat{G}_t\}$.

The optimum of problem (30) leads to closed-form update (31) and (32):

$$\Sigma_{t+1}^{-1} = \Sigma_t^{-1} + 2\beta_t \widehat{G}_t \tag{31}$$
$$\boldsymbol{\mu}_{t+1} = \boldsymbol{\mu}_t - \beta_t \Sigma_{t+1} \widehat{g}_t. \tag{32}$$

**General Stochastic Case:** The convergence rate of Algorithm 3 is shown in Theorem 4.

**Theorem 4.** *Given a convex function $f(\boldsymbol{x})$, define $\bar{J}(\boldsymbol{\theta}) := \mathbb{E}_{p(\boldsymbol{x};\boldsymbol{\theta})}[f(\boldsymbol{x})]$ for Gaussian distribution with parameter $\boldsymbol{\theta} := \{\boldsymbol{\mu}, \Sigma^{\frac{1}{2}}\} \in \boldsymbol{\Theta}$ and $\boldsymbol{\Theta} := \{\boldsymbol{\mu}, \Sigma^{\frac{1}{2}} \mid \boldsymbol{\mu} \in \mathcal{R}^d, \Sigma \in \mathcal{S}^+\}$. Suppose $\bar{J}(\boldsymbol{\theta})$ be $\gamma$-strongly convex. Let $\widehat{G}_t$ be positive semi-definite matrix such that $bI \preceq \widehat{G}_t \preceq \frac{\gamma}{2}I$. Suppose $\Sigma_1 \in \mathcal{S}^{++}$ and $\|\Sigma_1\| \leq \rho$, $\mathbb{E}\widehat{g}_t = \nabla_{\boldsymbol{\mu}=\boldsymbol{\mu}_t}\bar{J}$. Assume furthermore $\|\nabla_{\Sigma=\Sigma_t}\bar{J}\|_{tr} \leq B_1$ and $\|\boldsymbol{\mu}^* - \boldsymbol{\mu}_1\|^2_{\Sigma_1^{-1}} \leq R$, $\mathbb{E}\|\widehat{g}_t\|_2^2 \leq \mathcal{B}$ . Set $\beta_t = \beta$, then Algorithm 3 can achieve*

$$\frac{1}{T}\left[\sum_{t=1}^{T}\mathbb{E}f(\boldsymbol{\mu}_t)\right] - f(\boldsymbol{\mu}^*) \leq \frac{2bR + 2b\beta\rho(4B_1+\beta\mathcal{B}) + 4B_1(1+\log T) + (1+\log T)\beta\mathcal{B}}{4\beta bT}$$
$$= \mathcal{O}\left(\frac{\log T}{T}\right). \tag{33}$$

**Remark:** Theorem 4 does not require the function $f(\boldsymbol{x})$ be differentiable. It holds for non-smooth function $f(\boldsymbol{x})$. Theorem 4 holds for convex function $f(\boldsymbol{x})$, as long as $\bar{J}(\boldsymbol{\theta}) := \mathbb{E}_{p(\boldsymbol{x};\boldsymbol{\theta})}[f(\boldsymbol{x})]$ be $\gamma$-strongly convex. Particularly, when $f(\boldsymbol{x})$ is $\gamma$-strongly convex, we know $\bar{J}(\boldsymbol{\theta})$ is $\gamma$-strongly convex [11]. Thus, the assumption here is weaker than strongly convex assumption of $f(\boldsymbol{x})$. Moreover, Theorem 4 does not require the boundedness of the domain. It only requires the boundedness of the distance between the initialization point and an optimal point. Theorem 4 shows that the bound depends on the bound of $\mathbb{E}\|\widehat{g}_t\|_2^2$, which means that reducing variance of the gradient estimators can leads to a small regret bound.

**Black-Box Case:** For black-box optimization, we can only access the function value instead of the gradient. Let $\boldsymbol{z} \sim \mathcal{N}(\boldsymbol{0}, \boldsymbol{I})$, we give an unbiased estimator of $\nabla_{\boldsymbol{\mu}}\bar{J}(\boldsymbol{\theta}_t)$ using function values as

$$\widehat{g}_t = \Sigma_t^{-\frac{1}{2}}\boldsymbol{z}\left(f(\boldsymbol{\mu}_t + \Sigma_t^{\frac{1}{2}}\boldsymbol{z}) - f(\boldsymbol{\mu}_t)\right). \tag{34}$$

The estimator $\widehat{g}_t$ is unbiased, i.e., $\mathbb{E}[\widehat{g}_t] = \nabla_{\boldsymbol{\mu}}\bar{J}(\boldsymbol{\theta}_t)$. The proof of unbiasedness of the estimator $\widehat{g}_t$ is given in Lemma 7 in the supplement. With this estimator, we give the convergence rate of Algorithm 3 for convex black-box optimization as in Theorem 5.

**Theorem 5.** *For a L-Lipschitz continuous convex black box function $f(\boldsymbol{x})$, define $\bar{J}(\boldsymbol{\theta}) := \mathbb{E}_{p(\boldsymbol{x};\boldsymbol{\theta})}[f(\boldsymbol{x})]$ for Gaussian distribution with parameter $\boldsymbol{\theta} :=$*

**Algorithm 3.** General Framework

**Input:** Number of Samples $N$, step-size $\beta$.
**while** Termination condition not satisfied **do**
Construct unbiased estimator $\widehat{g}_t$ of gradient w.r.t $\boldsymbol{\mu}$.
Construct unbiased/biased estimator $\widehat{G}_t \in \mathcal{S}^{++}$ such that $bI \preceq \widehat{G}_t \preceq \frac{\gamma}{2}I$
Set $\Sigma_{t+1}^{-1} = \Sigma_t^{-1} + 2\beta\widehat{G}_t$.
Set $\boldsymbol{\mu}_{t+1} = \boldsymbol{\mu}_t - \beta\Sigma_{t+1}\widehat{g}_t$.
**end while**

$\{\boldsymbol{\mu}, \Sigma^{\frac{1}{2}}\} \in \boldsymbol{\Theta}$ *and* $\boldsymbol{\Theta} := \{\boldsymbol{\mu}, \Sigma^{\frac{1}{2}} \mid \boldsymbol{\mu} \in \mathcal{R}^d, \Sigma \in \mathcal{S}^{+}\}$. *Suppose* $\bar{J}(\boldsymbol{\theta})$ *be* $\gamma$*-strongly convex. Let* $\widehat{G}_t$ *be positive semi-definite matrix such that* $b\boldsymbol{I} \preceq \widehat{G}_t \preceq \frac{\gamma}{2}\boldsymbol{I}$. *Suppose* $\Sigma_1 \in \mathcal{S}^{++}$ *and* $\|\Sigma_1\|_2 \leq \rho$, *Assume furthermore* $\|\nabla_{\Sigma=\Sigma_t}\bar{J}\|_{tr} \leq B_1$ *and* $\|\boldsymbol{\mu}^* - \boldsymbol{\mu}_1\|^2_{\Sigma_1^{-1}} \leq R$, *and set* $\beta_t = \beta$ *and employ estimator* $\widehat{g}_t$ *in Eq. (34), then Algorithm 3 can achieve*

$$\frac{1}{T}\left[\sum_{t=1}^{T}\mathbb{E}f(\boldsymbol{\mu}_t)\right] - f(\boldsymbol{\mu}^*) \leq \frac{bR + b\beta\rho(4B_1 + 2\beta L^2(d+4)^2)}{2\beta bT} + \frac{4B_1(1+\log T) + (1+\log T)\beta L^2(d+4)^2}{4\beta bT} \tag{35}$$

$$= \mathcal{O}\left(\frac{d^2 \log T}{T}\right). \tag{36}$$

**Remark:** Theorem 5 holds for non-differentiable function $f(\boldsymbol{x})$. Thus, Theorem 5 can cover more interesting cases e.g. sparse black box optimization. In contrast, Balasubramanian et al. [6] require function $f(\boldsymbol{x})$ has Lipschitz continuous gradients.

Algorithm 1 employs an unbiased gradient estimator. When further ensure $b\boldsymbol{I} \preceq \widehat{G}_t \preceq \frac{\gamma}{2}\boldsymbol{I}$, Theorem 5 holds for Algorithm 1 Theorem 5 is derived for single sample per iteration. We can reduce the variance of estimators by constructing a set of structured samples that are conjugate of inverse covariance matrix in a batch, i.e., $\boldsymbol{z}_i\Sigma_t^{-1}\boldsymbol{z}_j = 0, i \neq j$. Particularly, when we use $\widehat{\Sigma}_t = \sigma_t\boldsymbol{I}$, sampling $N = d$ orthogonal samples [10] per iteration can lead to a convergence rate $\mathcal{O}\left(\frac{d\log T}{T}\right)$. For $N > d$ samples, we can use the method in [18] with a random rotation to reduce variance. For very large $N$, we can use the construction in Eq. (23) in [18] to transform the complex sampling matrix [29] onto sphere $\mathbb{S}^{d-1}$, then scale samples by i.i.d variables from Chi distribution. This construction has a bounded mutual coherence.

## 6 Optimization for Discrete Variable

**Binary Optimization:** For function $f(\boldsymbol{x})$ over binary variable $\boldsymbol{x} \in \{0,1\}^d$, we employ Bernoulli distribution with parameter $\boldsymbol{p} = [p_1, \cdots, p_d]^\top$ as the underlying distribution, where $p_i$ denote the probability of $x_i = 1$. Let $\boldsymbol{\eta}$ denote the natural parameter, then we know $\boldsymbol{p} = \frac{1}{1+e^{-\eta}}$. The mean parameter is $\boldsymbol{m} = \boldsymbol{p}$.

From Eq. (4), we know that

$$\boldsymbol{\eta}_{t+1} = \boldsymbol{\eta}_t - \beta_t \nabla_{\boldsymbol{p}} \mathbb{E}_{\boldsymbol{p}}[f(\boldsymbol{x})]. \tag{37}$$

Approximate the gradient by Monte Carlo sampling, we obtain that

$$\boldsymbol{\eta}_{t+1} = \boldsymbol{\eta}_t - \beta_t \frac{1}{N} \sum_{n=1}^{N} f(\boldsymbol{x}^n) \boldsymbol{h}^n, \tag{38}$$

where $\boldsymbol{h}_i^n = \frac{1}{p_i} \mathbf{1}(\boldsymbol{x}_i^n = 1) - \frac{1}{1-p_i} \mathbf{1}(\boldsymbol{x}_i^n = 0)$.

In order to achieve stable update, we normalize function value by its mean $\widehat{\mu}$ and standard deviation $\widehat{\sigma}$ in a batch. The normalized update is given as follows:

$$\boldsymbol{\eta}_{t+1} = \boldsymbol{\eta}_t - \beta_t \sum_{n=1}^{N} \frac{f(\boldsymbol{x}^n) - \widehat{\mu}}{N\widehat{\sigma}} \boldsymbol{h}^n. \tag{39}$$

**General Discrete Optimization:** Similarly, for function $f(\boldsymbol{x})$ over discrete variable $\boldsymbol{x} \in \{1, \cdots, K\}^d$, we employ categorical distribution with parameter $\boldsymbol{P} = [\boldsymbol{p}_1, \cdots, \boldsymbol{p}_d]^\top$ as the underlying distribution, where the $ij$-th element of $\boldsymbol{P}$ ($\boldsymbol{P}_{ij}$) denote the probability of $\boldsymbol{x}_i = j$. Let $\boldsymbol{\eta} \in \mathcal{R}^{d \times K}$ denote the natural parameter, then we know $\boldsymbol{P}_{ij} = \frac{e^{\eta_{ij}}}{\sum_{j=1}^{K} e^{\eta_{ij}}}$. The mean parameter is $\boldsymbol{m} = \boldsymbol{P}$.

From Eq. (4), we know that

$$\boldsymbol{\eta}_{t+1} = \boldsymbol{\eta}_t - \beta_t \nabla_{\boldsymbol{P}} \mathbb{E}_{\boldsymbol{P}}[f(\boldsymbol{x})]. \tag{40}$$

Approximate the gradient by Monte Carlo sampling,

$$\boldsymbol{\eta}_{t+1} = \boldsymbol{\eta}_t - \beta_t \frac{1}{N} \sum_{n=1}^{N} f(\boldsymbol{x}^n) \boldsymbol{H}^n, \tag{41}$$

where $\boldsymbol{H}_{ij}^n = \frac{1}{\boldsymbol{P}_{ij}} \mathbf{1}(\boldsymbol{x}_i^n = j)$. We can also normalize the update by the mean $\widehat{\mu}$ and std $\widehat{\sigma}$. More detailed derivation can be found in Appendix H.

# 7 Empirical Study

## 7.1 Evaluation on Synthetic Continuous Test Benchmarks

We evaluate the proposed INGO, INGOstep and Fast-INGO (diagonal case of INGO) by comparing with one of the state-of-the-art method CMA-ES [12] and IGO [21] with full covariance matrix update, and vanilla ES with antithetic gradient estimators [23] on several synthetic benchmark test problems. All the test problems are listed in Table 1 in the supplement.

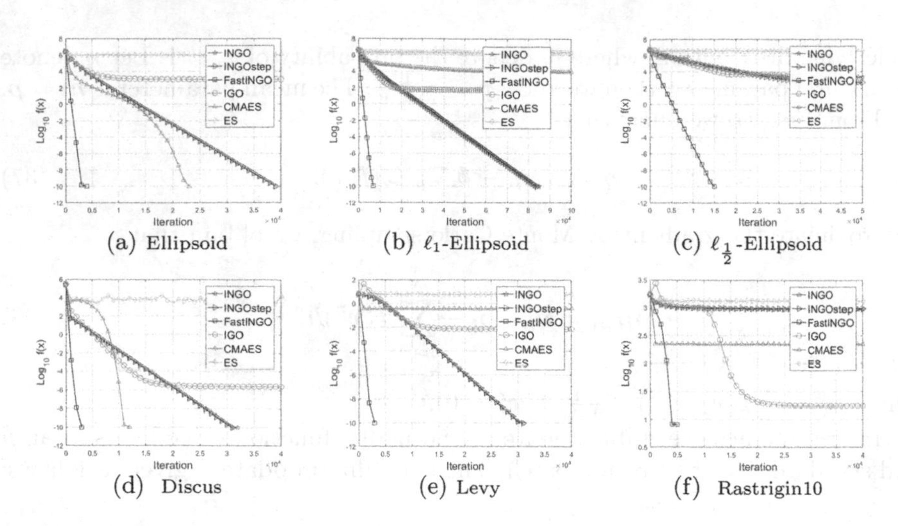


(a) Ellipsoid (b) $\ell_1$-Ellipsoid (c) $\ell_{\frac{1}{2}}$-Ellipsoid

(d) Discus (e) Levy (f) Rastrigin10

**Fig. 1.** Mean value of $f(\boldsymbol{x})$ in $\log_{10}$ scale over 20 independent runs for 100-dimensional problems.

**Parameter Settings:** For INGO, INGOstep and IGO, we use the same normalization transformation $h(f(\boldsymbol{x}_i) = \frac{f(\boldsymbol{x}_i)-\widehat{\mu}}{\widehat{\sigma}}$ and all same hyper-parameters to test the effect of implicit natural gradient. We set step size $\beta = 1/d$ for all of them. For Fast-INGO, we set step size $\beta = 1/\sqrt{d}$, where $d$ is the dimension of the test problems. The number of samples per iteration is set to $N = 2\lfloor 3 + \lfloor 3 \times \ln d \rfloor /2 \rfloor$ for all the methods, where $\lfloor \cdot \rfloor$ denotes the floor function. This setting ensures $N$ to be an even number. We set $\boldsymbol{\sigma}_1 = 0.5 \times \mathbf{1}$ and sample $\boldsymbol{\mu}_1 \sim Uni[\mathbf{0}, \mathbf{1}]$ as the same initialization for all the methods, where $Uni[0, 1]$ denotes the uniform distribution in $[0, 1]$. For ES [23], we use the default step-size hyper-parameters.

The mean value of $f(\boldsymbol{x})$ over 20 independent runs for 100-dimensional problems are show in Fig. 1. From Fig. 1, we can see that INGO, INGOstep and Fast-INGO converge linearly in log scale. Fast-INGO can arrive $10^{-10}$ precision on five cases except the highly non-convex Rastrigin10 problem. Fast-INGO employs the separate structure of the problems, thus it obtains better performance than the other methods with full matrix update. It is worth to note that Fast-INGO is not rotation invariant compared with Full-INGO. The INGO and INGOstep (with full matrix update) can arrive $10^{-10}$ on four cases, while IGO with full matrix update can not achieve high precision. This shows that the update of inverse of covariance matrix is more stable. Moreover, CMA-ES converge linearly in log scale for the convex Ellipsoid problem but slower than Fast-INGO. In addition, CMAES converge slowly on the non-smooth $\ell_1$-Ellipsoid and the non-convex $\ell_{\frac{1}{2}}$-Ellipsoid problem. Furthermore, CMAES fails on the non-convex Levy problem, while INGO, INGOstep and Fast-INGO obtain $10^{-10}$. CMAES converges faster or achieves smaller value than ES. On the non-convex Rastrigin10 problem, all methods fail to obtain $10^{-10}$ precision. Fast-INGO obtains smaller value. The results on synthetic test problems show that methods employing second-order

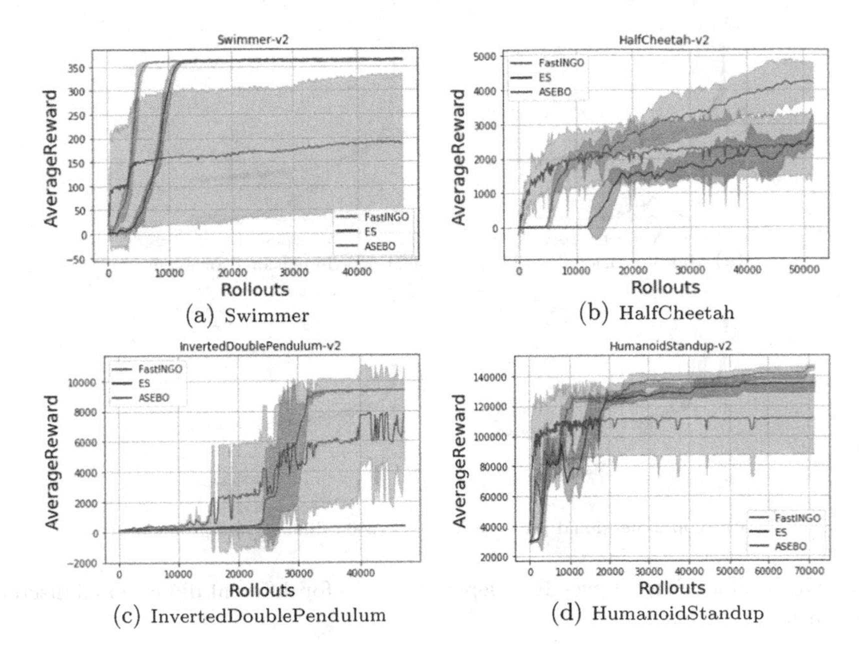

(a) Swimmer
(b) HalfCheetah
(c) InvertedDoublePendulum
(d) HumanoidStandup

**Fig. 2.** Average reward over 5 independent runs on benchmark RL environments

information converge faster than first-order method ES. And employing second-order information is important to obtain high optimization precision, i.e., $10^{-10}$. Moreover, taking stochastic implicit natural gradient update can converge faster than IGO. The test functions are highly ill-conditioned and non-convex; the experimental results show that it is challenging for ES to optimize them well without adaptively update covariance and mean.

### 7.2 Evaluation on RL Test Problems

We further evaluate the proposed Fast-INGO by comparing AESBO [9] and ES with antithetic gradient estimators [23] on MuJoCo control problems: Swimmer, HalfCheetah, HumanoidStandup, InvertedDoublePendulum, in Open-AI Gym environments. CMA-ES is too slow due to the computation of eigendecomposition for high-dimensional problems.

We use one hidden layer feed-forward neural network with tanh activation function as policy architecture. The number of hidden units is set to $h = 16$ for all problems. The goal is to find the parameters of this policy network to achieve large reward. The same policy architecture is used for all the methods on all test problems. The number of samples per iteration is set to $N = 20 + 4\lfloor\lfloor 3 \times \ln d\rfloor/2\rfloor$ for all the methods. For Fast-INGO, we set step-size $\beta = 0.3$. We set $\boldsymbol{\sigma}_1 = 0.1 \times \mathbf{1}$ and $\boldsymbol{\mu}_1 = \mathbf{0}$ as the initialization for both Fast-INGO and ES. For ES [23], we use the default step-size hyper-parameters. Five independent runs are performed. The experimental results are shown in Fig. 2. We can observe

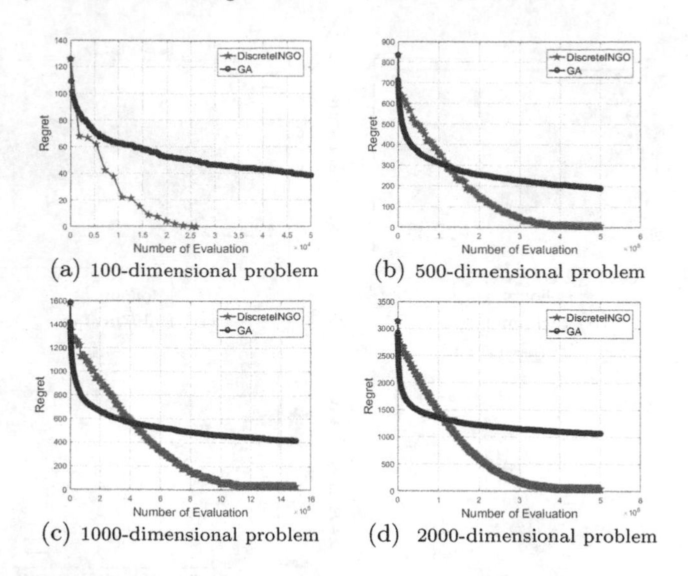


**Fig. 3.** Mean value of regret over 10 independent runs for different dimensional discrete optimization problems

that Fast-INGO increase AverageReward faster than ES on all four cases. This shows that the update using seconder order information in Fast-INGO can help accelerate convergence.

### 7.3 Evaluation on Discrete Test Problems

We evaluate our discrete INGO by comparing with GA method on binary reconstruction benchmark problem, i.e., $f(\boldsymbol{x}) := \|\text{sign}(\boldsymbol{x}-0.5)-\boldsymbol{w}\|_2^2 - \|\text{sign}(\boldsymbol{w})-\boldsymbol{w}\|_2^2$ with $\boldsymbol{x} \in \{0,1\}^d$. We construct $\boldsymbol{w}$ by sampling from standard Gaussian. The dimension $d$ of test problem is set to $\{100, 500, 1000, 2000\}$, respectively. For our discrete INGO, we set the step-size $\beta = 1/d$. The number of samples per iteration is same as INGO, i.e., $N = 20 + 4\lfloor 3 + \lfloor 3 \times \ln d \rfloor /2 \rfloor$.

The experimental results are shown in Fig. 3. We can observe that our discrete INGO achieves much smaller regret compared with GA. Our discrete INGO converges to near zero regret on test problems, while GA decrease very slowly after a short initial greedy phase.

## 8 Conclusions

We proposed a novel stochastic implicit natural gradient frameworks for black-box optimization. Under this framework, we presented algorithms for both continuous and discrete black-box optimization. For Gaussian distribution, we proved the $\mathcal{O}(\log T/T)$ convergence rate of our continuous algorithms with stochastic update for convex function under expectation $\gamma$-strongly convex assumption. We proved $\mathcal{O}\left(d^2 \log T/T\right)$ converge rate for black-box function under same assumptions above. For isometric Gaussian case, we proved the $\mathcal{O}\left(d \log T/T\right)$ converge rate when using $d$ orthogonal samples per iteration, which well supports parallel evaluation. Our method is very simple, and it contains less hyper-parameters than CMA-ES. Empirically, our methods obtain a competitive performance compared with CMA-ES. Moreover, our INGO and INGOstep with full matrix update can achieve high precision on Levy test problem and Ellipsoid problems, while IGO [21] with full matrix update can not. This shows the efficiency of our methods. On RL control problems, our algorithms increase average reward faster than ASEBO [9] and ES, which shows employing second order information can help accelerate convergence. Moreover, our discrete algorithm outperforms than GA on test functions.

**Acknowledgement.** We would like to thank all anonymous reviewers and the area chair for their valuable comments and suggestions. Yueming Lyu was supported by UTS President Scholarship. Ivor Tsang was supported by the Australian Research Council Grant (DP180100106 and DP200101328).

## References

1. Akimoto, Y., Nagata, Y., Ono, I., Kobayashi, S.: Bidirectional relation between CMA evolution strategies and natural evolution strategies. In: Schaefer, R., Cotta, C., Kołodziej, J., Rudolph, G. (eds.) PPSN 2010. LNCS, vol. 6238, pp. 154–163. Springer, Heidelberg (2010). https://doi.org/10.1007/978-3-642-15844-5_16
2. Amari, S.I.: Natural gradient works efficiently in learning. Neural Comput. **10**(2), 251–276 (1998)
3. Amari, S.: Information Geometry and Its Applications. AMS, vol. 194. Springer, Tokyo (2016). https://doi.org/10.1007/978-4-431-55978-8
4. Azoury, K.S., Warmuth, M.K.: Relative loss bounds for on-line density estimation with the exponential family of distributions. Mach. Learn. **43**(3), 211–246 (2001)
5. Back, T., Hoffmeister, F., Schwefel, H.P.: A survey of evolution strategies. In: Proceedings of the Fourth International Conference on Genetic Algorithms, vol. 2. Morgan Kaufmann Publishers, San Mateo (1991)
6. Balasubramanian, K., Ghadimi, S.: Zeroth-order (non)-convex stochastic optimization via conditional gradient and gradient updates. In: Advances in Neural Information Processing Systems, pp. 3455–3464 (2018)
7. Barsce, J.C., Palombarini, J.A., Martínez, E.C.: Towards autonomous reinforcement learning: automatic setting of hyper-parameters using Bayesian optimization. In: Computer Conference (CLEI), 2017 XLIII Latin American, pp. 1–9. IEEE (2017)
8. Bull, A.D.: Convergence rates of efficient global optimization algorithms. J. Mach. Learn. Res. (JMLR) **12**, 2879–2904 (2011)
9. Choromanski, K., Pacchiano, A., Parker-Holder, J., Tang, Y.: From complexity to simplicity: adaptive ES-active subspaces for blackbox optimization. arXiv:1903.04268 (2019)
10. Choromanski, K., Rowland, M., Sindhwani, V., Turner, R.E., Weller, A.: Structured evolution with compact architectures for scalable policy optimization. In: ICML, pp. 969–977 (2018)

11. Domke, J.: Provable smoothness guarantees for black-box variational inference. arXiv preprint arXiv:1901.08431 (2019)
12. Hansen, N.: The CMA evolution strategy: a comparing review. In: Lozano, J.A., Larrañaga, P., Inza, I., Bengoetxea, E. (eds.) Towards a New Evolutionary Computation. STUDFUZZ, vol. 192, pp. 75–102. Springer, Heidelberg (2006). https://doi.org/10.1007/3-540-32494-1_4
13. Khan, M.E., Lin, W.: Conjugate-computation variational inference: converting variational inference in non-conjugate models to inferences in conjugate models. arXiv preprint arXiv:1703.04265 (2017)
14. Khan, M.E., Nielsen, D.: Fast yet simple natural-gradient descent for variational inference in complex models. In: 2018 International Symposium on Information Theory and Its Applications (ISITA), pp. 31–35. IEEE (2018)
15. Khan, M.E., Nielsen, D., Tangkaratt, V., Lin, W., Gal, Y., Srivastava, A.: Fast and scalable Bayesian deep learning by weight-perturbation in adam. In: ICML (2018)
16. Liu, G., et al.: Trust region evolution strategies. In: AAAI (2019)
17. Lizotte, D.J., Wang, T., Bowling, M.H., Schuurmans, D.: Automatic gait optimization with Gaussian process regression. In: IJCAI, vol. 7, pp. 944–949 (2007)
18. Lyu, Y.: Spherical structured feature maps for kernel approximation. In: Proceedings of the 34th International Conference on Machine Learning (ICML), pp. 2256–2264 (2017)
19. Lyu, Y., Yuan, Y., Tsang, I.W.: Efficient batch black-box optimization with deterministic regret bounds. arXiv preprint arXiv:1905.10041 (2019)
20. Negoescu, D.M., Frazier, P.I., Powell, W.B.: The knowledge-gradient algorithm for sequencing experiments in drug discovery. INFORMS J. Comput. **23**(3), 346–363 (2011)
21. Ollivier, Y., Arnold, L., Auger, A., Hansen, N.: Information-geometric optimization algorithms: a unifying picture via invariance principles. J. Mach. Learn. Res. (JMLR) **18**(1), 564–628 (2017)
22. Raskutti, G., Mukherjee, S.: The information geometry of mirror descent. IEEE Trans. Inf. Theory **61**(3), 1451–1457 (2015)
23. Salimans, T., Ho, J., Chen, X., Sidor, S., Sutskever, I.: Evolution strategies as a scalable alternative to reinforcement learning. arXiv preprint arXiv:1703.03864 (2017)
24. Snoek, J., Larochelle, H., Adams, R.P.: Practical Bayesian optimization of machine learning algorithms. In: NeurIPS, pp. 2951–2959 (2012)
25. Srinivas, M., Patnaik, L.M.: Genetic algorithms: a survey. Computer **27**(6), 17–26 (1994)
26. Srinivas, N., Krause, A., Kakade, S.M., Seeger, M.: Gaussian process optimization in the bandit setting: no regret and experimental design. In: ICML (2010)
27. Wang, G.G., Shan, S.: Review of metamodeling techniques in support of engineering design optimization. J. Mech. Des. **129**(4), 370–380 (2007)
28. Wierstra, D., Schaul, T., Glasmachers, T., Sun, Y., Peters, J., Schmidhuber, J.: Natural evolution strategies. J. Mach. Learn. Res. (JMLR) **15**(1), 949–980 (2014)
29. Xu, Z.: Deterministic sampling of sparse trigonometric polynomials. J. Complex. **27**(2), 133–140 (2011)

# More General and Effective Model Compression via an Additive Combination of Compressions

Yerlan Idelbayev[(✉)] and Miguel Á. Carreira-Perpiñán

Department of CSE, University of California, Merced, Merced, USA
{yidelbayev,mcarreira-perpinan}@ucmerced.edu

**Abstract.** Model compression is generally performed by using quantization, low-rank approximation or pruning, for which various algorithms have been researched in recent years. One fundamental question is: what types of compression work better for a given model? Or even better: can we improve by combining compressions in a suitable way? We formulate this generally as a problem of optimizing the loss but where the weights are constrained to equal an additive combination of separately compressed parts; and we give an algorithm to learn the corresponding parts' parameters. Experimentally with deep neural nets, we observe that 1) we can find significantly better models in the error-compression space, indicating that different compression types have complementary benefits, and 2) the best type of combination depends exquisitely on the type of neural net. For example, we can compress ResNets and AlexNet using only 1 bit per weight without error degradation at the cost of adding a few floating point weights. However, VGG nets can be better compressed by combining low-rank with a few floating point weights.

**Keywords:** Additive combinations · Model compression

## 1 Introduction

In machine learning, model compression is the problem of taking a neural net or some other model, which has been trained to perform (near)-optimal prediction in a given task and dataset, and transforming it into a model that is smaller (in size, runtime, energy or other factors) while maintaining as good a prediction performance as possible. This problem has recently become important and actively researched because of the large size of state-of-the-art neural nets, trained on large-scale GPU clusters without constraints on computational resources, but which cannot be directly deployed in IoT devices with much more limited capabilities.

The last few years have seen much work on the topic, mostly focusing on specific forms of compression, such as quantization, low-rank matrix approximation and weight pruning, as well as variations of these. These papers typically


N. Oliver et al. (Eds.): ECML PKDD 2021, LNAI 12977, pp. 233–248, 2021.
https://doi.org/10.1007/978-3-030-86523-8_15

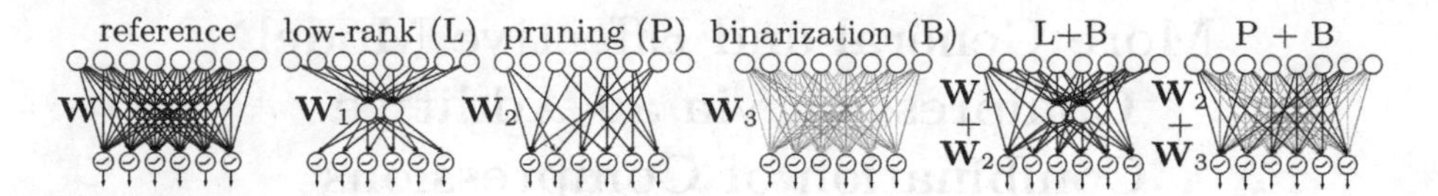


**Fig. 1.** Illustration of compression by additive combination $\mathbf{W} = \mathbf{W}_1 + \mathbf{W}_2 + \mathbf{W}_3$. Black weights are real, red weights are $-1$ and blue weights are $+1$. (Color figure online)

propose a specific compression technique and a specific algorithm to compress a neural net with it. The performance of these techniques individually varies considerably from case to case, depending on the algorithm (some are better than others) but more importantly on the compression technique. This is to be expected, because (just as happens with image or audio compression) some techniques achieve more compression for certain types of signals.

A basic issue is the representation ability of the compression: given an optimal point in model space (the weight parameters for a neural net), which manifold or subset of this space can be compressed exactly, and is that subset likely to be close to the optimal model for a given machine learning task? For example, for low-rank compression the subset contains all matrices of a given rank or less. Is that a good subset to model weight matrices arising from, say, deep neural nets for object classification?

One way to understand this question is to try many techniques in a given task and gain experience about what works in which case. This is made difficult by the multiplicity of existing algorithms, the heuristics often used to optimize the results experimentally (which are compounded by the engineering aspects involved in training deep nets, to start with), and the lack at present of an apples-to-apples evaluation in the field of model compression.

A second way to approach the question which partly sidesteps this problem is to use a common algorithm that can handle any compression technique. While compressing a deep net in a large dataset will still involve careful selection of optimization parameters (such as SGD learning rates), having a common algorithmic framework should put different compression techniques in the same footing. Yet a third approach, which we propose in this paper, is to *combine* several techniques (rather than try each in isolation) while jointly optimizing over the parameters of each (codebook and assignments for quantization, component matrices for low-rank, subset and value of nonzero weights for pruning, etc.).

There are multiple ways to define a combination of compression techniques. One that is simple to achieve is by applying compression techniques sequentially, such as first pruning the weights, then quantizing the remaining nonzero weights and finally encoding them with Huffman codes [13]. This is suboptimal in that the global problem is solved greedily, one compression at a time. The way we propose here is very different: an *additive combination* of compression techniques. For example, we may want to compress a given weight matrix $\mathbf{W}$ as the sum (or linear combination) $\mathbf{W} = \mathbf{W}_1 + \mathbf{W}_2 + \mathbf{W}_3$ of a low-rank matrix

$\mathbf{W}_1$, a sparse matrix $\mathbf{W}_2$ and a quantized matrix $\mathbf{W}_3$. This introduces several important advantages. First, it contains as a particular case each technique in isolation (e.g., quantization by making $\mathbf{W}_1 = \mathbf{0}$ a zero-rank matrix and $\mathbf{W}_2 = \mathbf{0}$ a matrix with no nonzeros). Second, and critically, it allows techniques to help each other because of having complementary strengths. For example, pruning can be seen as adding a few elementwise real-valued corrections to a quantized or low-rank weight matrix. This could result (and does in some cases) in using fewer bits, lower rank and fewer nonzeros and a resulting higher compression ratio (in memory or runtime). Third, the additive combination vastly enlarges the subset of parameter space that can be compressed without loss compared to the individual compressions. This can be seen intuitively by noting that a fixed vector times a scalar generates a 1D space, but the additive combination of two such vectors generates a 2D space rather than two 1D spaces).

One more thing remains to make this possible: a formulation and corresponding algorithm of the compression problem that can handle such additive combinations of arbitrary compression techniques. We rely on the previously proposed "learning-compression (LC)" algorithm [8]. This explicitly defines the model weights as a function (called *decompression mapping*) of low-dimensional compression parameters; for example, the low-rank matrix above would be written as $\mathbf{W}_1 = \mathbf{U}\mathbf{V}^T$. It then iteratively optimizes the loss but constraining the weights to take the desired form (an additive combination in our case). This alternates *learning (L)* steps that train a regularized loss over the original model weights with *compression (C)* steps that compress the current weights, in our case according to the additive compression form.

Next, we review related work (Sect. 2), describe our problem formulation (Sect. 3) and corresponding LC algorithm (Sect. 4), and demonstrate the claimed advantages with deep neural nets (Sects. 5 and 6).

## 2 Related Work

*General Approaches.* In the literature of model and particularly neural net compression, various approaches have been studied, including most prominently weight quantization, weight pruning and low-rank matrix or tensor decompositions. There are other approaches as well, which can be potentially used in combination with. We briefly discuss the individual techniques first. Quantization is a process of representing each weight with an item from a codebook. This can be achieved through fixed codebook schemes, i.e., with predetermined codebook values that are not learned (where only the assignments should be optimized). Examples of this compression are binarization, ternarization, low-precision, fixed-point or other number representations [22,31]. Quantization can also be achieved through adaptive codebook schemes, where the codebook values are learned together with the assignment variables, with algorithms based on soft quantization [1,29] or hard quantization [13]. Pruning is a process of removal of weights (unstructured) or filters and neurons (structured). It can be achieved by salience ranking [13,21] in one go or over multiple refinements, or by using sparsifying norms [9,41]. Low-rank approximation is a process of replacing weights with low-rank [18,20,35,38] or tensors-decomposed versions [28].

*Usage of Combinations.* One of the most used combinations is to apply compressions sequentially, most notably first to prune weights and then to quantize the remaining ones [12,13,13,34,40], which may possibly be further compressed via lossless coding algorithms (e.g., Huffman coding). Additive combination of quantizations [3,39,44], where weights are the sum of quantized values, as well as low-rank + sparse combination [2,43] has been used to compress neural networks. However, these methods rely on optimization algorithms highly specialized to a problem, limiting its application to new combinations (e.g., quantization + low-rank).

# 3 Compression via an Additive Combination as Constrained Optimization

Our basic formulation is that we define the weights as an additive combination of weights, where each term in the sum is individually compressed in some way. Consider for simplicity the case of adding just two compressions for a given matrix[1]. We then write a matrix of weights as $\mathbf{W} = \boldsymbol{\Delta}_1(\boldsymbol{\theta}_1)+\boldsymbol{\Delta}_2(\boldsymbol{\theta}_2)$, where $\boldsymbol{\theta}_i$ is the low-dimensional parameters of the $i$th compression and $\boldsymbol{\Delta}_i$ is the corresponding decompression mapping. Formally, the $\boldsymbol{\Delta}$ maps a compressed representation of the weight matrix $\boldsymbol{\theta}$ to the real-valued, uncompressed weight matrix $\mathbf{W}$. Its intent is to represent the space of matrices that can be compressed via a constraint subject to which we optimize the loss of the model in the desired task (e.g., classification). That is, a constraint $\mathbf{W} = \boldsymbol{\Delta}(\boldsymbol{\theta})$ defines a feasible set of compressed models. For example:

- Low-rank: $\mathbf{W} = \mathbf{U}\mathbf{V}^T$ with $\mathbf{U}$ and $\mathbf{V}$ of rank $r$, so $\boldsymbol{\theta} = (\mathbf{U}, \mathbf{V})$.
- Pruning: $\mathbf{w} = \boldsymbol{\theta}$ s.t. $\|\boldsymbol{\theta}\|_0 \leq \kappa$, so $\boldsymbol{\theta}$ is the indices of its nonzeros and their values.
- Scalar quantization: $w = \sum_{k=1}^{K} z_k c_k$ with assignment variables $\mathbf{z} \in \{0,1\}^K$, $\mathbf{1}^T\mathbf{z} = 1$ and codebook $\mathcal{C} = \{c_1, \dots, c_K\} \subset \mathbb{R}$, so $\boldsymbol{\theta} = (\mathbf{z}, \mathcal{C})$.
- Binarization: $w \in \{-1, +1\}$ or equivalently a scalar quantization with $\mathcal{C} = \{-1, +1\}$.

Note how the mapping $\boldsymbol{\Delta}(\boldsymbol{\theta})$ and the low-dimensional parameters $\boldsymbol{\theta}$ can take many forms (involving scalars, matrices or other objects of continuous or discrete type) and include constraints on $\boldsymbol{\theta}$. Then, our problem formulation takes the form of *model compression as constrained optimization* [8] and given as:

$$\min_{\mathbf{w}} L(\mathbf{w}) \quad \text{s.t.} \quad \mathbf{w} = \boldsymbol{\Delta}_1(\boldsymbol{\theta}_1) + \boldsymbol{\Delta}_2(\boldsymbol{\theta}_2). \tag{1}$$

This expresses in a mathematical way our desire that 1) we want a model with minimum loss on the task at hand ($L(\mathbf{w})$ represents, say, the cross-entropy of

[1] Throughout the paper we use $\mathbf{W}$ or $\mathbf{w}$ or $w$ to notate matrix, vector or scalar weights as appropriate.

the original deep net architecture on a training set); 2) the model parameters $\mathbf{w}$ must take a special form that allows them to be compactly represented in terms of low-dimensional parameters $\boldsymbol{\theta} = (\boldsymbol{\theta}_1, \boldsymbol{\theta}_2)$; and 3) the latter takes the form of an additive combination (over two compressions, in the example). Problem (1) has the advantage that it is amenable to modern techniques of numerical optimization, as we show in Sect. 4.

Although the expression "$\mathbf{w} = \boldsymbol{\Delta}_1(\boldsymbol{\theta}_1) + \boldsymbol{\Delta}_2(\boldsymbol{\theta}_2)$" is an addition, it implicitly is a linear combination because the coefficients can typically be absorbed inside each $\boldsymbol{\Delta}_i$. For example, writing $\alpha\mathbf{U}\mathbf{V}^T$ (for low-rank compression) is the same as, say, $\mathbf{U}'\mathbf{V}^T$ with $\mathbf{U}' = \alpha\mathbf{U}$. In particular, any compression member may be implicitly removed by becoming zero. Some additive combinations are redundant, such as having both $\mathbf{W}_1$ and $\mathbf{W}_2$ be of rank at most $r$ (since rank $(\mathbf{W}_1 + \mathbf{W}_2) \leq 2r$) or having each contain at most $\kappa$ nonzeros (since $\|\mathbf{W}_1 + \mathbf{W}_2\|_0 \leq 2\kappa$).

The additive combination formulation has some interesting consequences. First, *an additive combination of compression forms can be equivalently seen as a new, learned deep net architecture.* For example (see Fig. 1), low-rank plus pruning can be seen as a layer with a linear bottleneck and some skip connections which are learned (i.e., which connections to have and their weight value). It is possible that such architectures may be of independent interest in deep learning beyond compression. Second, while *pruning in isolation means (as is usually understood) the removal of weights from the model, pruning in an additive combination means the addition of a few elementwise real-valued corrections.* This can potentially bring large benefits. As an extreme case, consider binarizing both the multiplicative and additive (bias) weights in a deep net. It is known that the model's loss is far more sensitive to binarizing the biases, and indeed compression approaches generally do not compress the biases (which also account for a small proportion of weights in total). In binarization plus pruning, all weights are quantized but we learn which ones need a real-valued correction for an optimal loss. Indeed, our algorithm is able to learn that the biases need such corrections more than other weights (see corresponding experiment in suppl. mat. [10]).

*Well Known Combinations.* Our motivation is to combine generically existing compressions in the context of model compression. However, some of the combinations are well known and extensively studied. Particularly, low-rank + sparse combination has been used in its own right in the fields of compressed sensing [7], matrix decomposition [45], and image processing [6]. This combination enjoys certain theoretical guarantees [7,11], yet it is unclear whether similar results can be stated over more general additive combinations (e.g., with non-differentiable scheme like quantization) or when applied to non-convex models as deep nets.

*Hardware Implementation.* The goal of model compression is to implement in practice the compressed model based on the $\boldsymbol{\theta}$ parameters, not the original weights $\mathbf{W}$. With an additive combination, the implementation is straightforward and efficient by applying the individual compressions sequentially and cumulatively. For example, say $\mathbf{W} = \mathbf{W}_1 + \mathbf{W}_2$ is a weight matrix in a layer of a deep net and we want to compute the layer's output activations $\sigma(\mathbf{W}\mathbf{x})$ for a given input vector of activations $\mathbf{x}$ (where $\sigma(\cdot)$ is a nonlinearity, such as a ReLU). By the properties of linearity, $\mathbf{W}\mathbf{x} = \mathbf{W}_1\mathbf{x}+\mathbf{W}_2\mathbf{x}$, so we first compute $\mathbf{y} = \mathbf{W}_1\mathbf{x}$ according to an efficient implementation of the first compression, and then we accumulate $\mathbf{y} = \mathbf{y} + \mathbf{W}_2\mathbf{x}$ computed according to an efficient implementation of the second compression. This is particularly beneficial because some compression techniques are less hardware-friendly than others. For example, quantization is very efficient and cache-friendly, since it can store the codebook in registers, access the individual weights with high memory locality, use mostly floating-point additions (and nearly no multiplications), and process rows of $\mathbf{W}_1$ in parallel. However, pruning has a complex, nonlocal pattern of nonzeros whose locations must be stored. Combining quantization plus pruning not only can achieve higher compression ratios than either just quantization or just pruning, as seen in our experiments; it can also reduce the number of bits per weight and (drastically) the number of nonzeros, thus resulting in fewer memory accesses and hence lower runtime and energy consumption.

## 4 Optimization via a Learning-Compression Algorithm

Although optimizing (1) may be done in different ways for specific forms of the loss $L$ or the decompression mapping constraint $\boldsymbol{\Delta}$, it is critical to be able to do this in as generic way as possible, so it applies to any combination of forms of the compressions, loss and model. Following Carreira-Perpinan [8], we apply a penalty method and then alternating optimization. We give the algorithm for the quadratic-penalty method [27], but we implement the augmented Lagrangian one (which works in a similar way but with the introduction of a Lagrange multiplier vector $\boldsymbol{\lambda}$ of the same dimension as $\mathbf{w}$). We then optimize the following while driving a penalty parameter $\mu \to \infty$:

$$Q(\mathbf{w}, \boldsymbol{\theta}; \mu) = L(\mathbf{w}) + \frac{\mu}{2}\|\mathbf{w} - \boldsymbol{\Delta}_1(\boldsymbol{\theta}_1) - \boldsymbol{\Delta}_2(\boldsymbol{\theta}_2)\|^2 \tag{2}$$

by using alternating optimization over $\mathbf{w}$ and $\boldsymbol{\theta}$. The step over $\mathbf{w}$ ("learning (L)" step) has the form of a standard loss minimization but with a quadratic regularizer on $\mathbf{w}$ (since $\boldsymbol{\Delta}_1(\boldsymbol{\theta}_1) + \boldsymbol{\Delta}_2(\boldsymbol{\theta}_2)$ is fixed), and can be done using a standard algorithm to optimize the loss, e.g., SGD with deep nets. The step over $\boldsymbol{\theta}$ ("compression (C)" step) has the following form:

$$\min_{\boldsymbol{\theta}} \|\mathbf{w} - \boldsymbol{\Delta}(\boldsymbol{\theta})\|^2 \Leftrightarrow \min_{\boldsymbol{\theta}_1,\boldsymbol{\theta}_2} \|\mathbf{w} - \boldsymbol{\Delta}_1(\boldsymbol{\theta}_1) - \boldsymbol{\Delta}_2(\boldsymbol{\theta}_2)\|^2. \tag{3}$$

In the original LC algorithm [8], this step (over just a single compression $\boldsymbol{\Delta}(\boldsymbol{\theta})$) typically corresponds to a well-known compression problem in signal processing and can be solved with existing algorithms. This gives the LC algorithm a major advantage: in order to change the compression form, we simply call the corresponding subroutine in this step (regardless of the form of the loss and model). For example, for low-rank compression the solution is given by a truncated SVD, for pruning by thresholding the largest weights, and for quantization

**Algorithm 1.** Pseudocode (quadratic-penalty version)

**input** training data, neural net architecture with weights $\mathbf{w}$
$\mathbf{w} \leftarrow \arg\min_{\mathbf{w}} L(\mathbf{w})$ — reference net
$\boldsymbol{\theta}_1, \boldsymbol{\theta}_2 \leftarrow \arg\min_{\boldsymbol{\theta}_1,\boldsymbol{\theta}_2} \|\mathbf{w} - \boldsymbol{\Delta}_1(\boldsymbol{\theta}_1) - \boldsymbol{\Delta}_2(\boldsymbol{\theta}_2)\|^2$ — init
**for** $\mu = \mu_0 < \mu_1 < \cdots < \infty$
  $\mathbf{w} \leftarrow \arg\min_{\mathbf{w}} L(\mathbf{w}) + \frac{\mu}{2}\|\mathbf{w} - \boldsymbol{\Delta}_1(\boldsymbol{\theta}_1) - \boldsymbol{\Delta}_2(\boldsymbol{\theta}_2)\|^2$ — L step
  **while** alternation does not converge — C step
    $\boldsymbol{\theta}_1 \leftarrow \arg\min_{\boldsymbol{\theta}_1} \|(\mathbf{w} - \boldsymbol{\Delta}_2(\boldsymbol{\theta}_2)) - \boldsymbol{\Delta}_1(\boldsymbol{\theta}_1)\|^2$
    $\boldsymbol{\theta}_2 \leftarrow \arg\min_{\boldsymbol{\theta}_2} \|(\mathbf{w} - \boldsymbol{\Delta}_1(\boldsymbol{\theta}_1)) - \boldsymbol{\Delta}_2(\boldsymbol{\theta}_2)\|^2$
  **if** $\|\mathbf{w} - \boldsymbol{\Delta}_1(\boldsymbol{\theta}_1) - \boldsymbol{\Delta}_2(\boldsymbol{\theta}_2)\|$ is small enough **then** exit the loop
**return** $\mathbf{w}, \boldsymbol{\theta}_1, \boldsymbol{\theta}_2$

by $k$-means. It is critical to preserve that advantage here so that we can handle in a generic way an arbitrary additive combination of compressions. Fortunately, we can achieve this by applying alternating optimization again but now to (3) over $\boldsymbol{\theta}_1$ and $\boldsymbol{\theta}_2$, as follows[2]:

$$\begin{aligned} \boldsymbol{\theta}_1 &= \arg\min_{\boldsymbol{\theta}} \|(\mathbf{w} - \boldsymbol{\Delta}_2(\boldsymbol{\theta}_2)) - \boldsymbol{\Delta}_1(\boldsymbol{\theta})\|^2 \\ \boldsymbol{\theta}_2 &= \arg\min_{\boldsymbol{\theta}} \|(\mathbf{w} - \boldsymbol{\Delta}_1(\boldsymbol{\theta}_1)) - \boldsymbol{\Delta}_2(\boldsymbol{\theta})\|^2 \end{aligned} \quad (4)$$

Each problem in (4) now does have the standard compression form of the original LC algorithm and can again be solved by an existing algorithm to compress optimally according to $\boldsymbol{\Delta}_1$ or $\boldsymbol{\Delta}_2$. At the beginning of each C step, we initialize $\boldsymbol{\theta}$ from the previous C step's result (see Algorithm 1).

It is possible that a better algorithm exists for a specific form of additive combination compression (3). In such case we can employ specialized version during the C step. But our proposed alternating optimization (4) provides a generic, efficient solution as long as we have a good algorithm for each individual compression.

Convergence of the alternating steps (4) to a global optimum of (3) over $(\boldsymbol{\theta}_1, \boldsymbol{\theta}_2)$ can be proven in some cases, e.g., low-rank + sparse [45], but not in general, as one would expect since some of the compression problems involve discrete and continuous variables and can be NP-hard (such as quantization with an adaptive codebook). Convergence can be established quite generally for convex functions [4,36]. For nonconvex functions, convergence results are complex and more restrictive [33]. One simple case where convergence occurs is if the objective in (3) (i.e., each $\boldsymbol{\Delta}_i$) is continuously differentiable and it has a unique minimizer over each $\boldsymbol{\theta}_i$ [5, Proposition 2.7.1]. However, in certain cases the optimization can be solved exactly without any alternation. We give a specific result next.

[2] This form of iterated "fitting" (here, compression) by a "model" (here, $\boldsymbol{\Delta}_1$ or $\boldsymbol{\Delta}_2$) of a "residual" (here, $\mathbf{w} - \boldsymbol{\Delta}_2(\boldsymbol{\theta}_2)$ or $\mathbf{w} - \boldsymbol{\Delta}_1(\boldsymbol{\theta}_1)$) is called backfitting in statistics, and is widely used with additive models [14].

### 4.1 Exactly Solvable C Step

Solution of the C step (Eq. 3) does not need to be an alternating optimization. Below we give an exact algorithm for the additive combination of fixed codebook quantization (e.g., $\{-1,+1\}$, $\{-1,0,+1\}$, etc.) and sparse corrections.

**Theorem 1 (Exactly solvable C step for combination of fixed codebook quantization + sparse corrections).** *Given a fixed codebook $\mathcal{C}$ consider compression of the weights $w_i$ with an additive combinations of quantized values $q_i \in \mathcal{C}$ and sparse corrections $s_i$:*

$$\min_{q,s} \sum_i (w_i - (q_i + s_i))^2 \quad s.t. \quad \|s\|_0 \leq \kappa, \tag{5}$$

*Then the following provides one optimal solution $(q^*, s^*)$: first set $q_i^* = closest(w_i)$ in codebook for each $i$, then solve for $s$: $\min_s \sum_i (w_i - q_i^* - s_i))^2$ s.t. $\|s\|_0 \leq \kappa$.*

*Proof.* Imagine we know the optimal set of nonzeros of the vector $s$, which we denote as $\mathcal{N}$. Then, for the elements not in $\mathcal{N}$, the optimal solution is $s_i^* = 0$ and $q_i^* = closest(w_i)$. For the elements in $\mathcal{N}$, we can find their optimal solution by solving independently for each $i$:

$$\min_{q_i,s_i} (w_i - (q_i + s_i))^2 \quad \text{s.t.} \quad q_i \in \mathcal{C}.$$

The solution is $s_i^* = w_i - q_i$ for arbitrary chosen $q_i \in \mathcal{C}$. Using this, we can rewrite the Eq. 5 as $\sum_{i \notin \mathcal{N}} (w_i - q_i^*)^2$.

This is minimized by taking as set $\mathcal{N}$ the $\kappa$ largest in magnitude elements of $w_i - q_i^*$ (indexed over $i$). Hence, the final solution is: 1) Set the elements of $\mathcal{N}$ to be the $\kappa$ largest in magnitude elements of $w_i - q_i^*$ (there may be multiple such sets, any one is valid). 2) For each $i$ in $\mathcal{N}$: set $s_i^* = w_i - q_i^*$, and $q_i^* =$ any element in $\mathcal{C}$. For each $i$ not in $\mathcal{N}$: set $s_i^* = 0$, $q_i^* = closest(w_i)$ (there may be 2 closest values, any one is valid). This contains multiple solutions. One particular one is as given in the theorem statement, where we set $q_i^* = closest(w_i)$ for every $i$, which is practically more desirable because it leads to a smaller $\ell_1$-norm of $s$.

## 5 Experiments on CIFAR10

We evaluate the effectiveness of additively combining compressions on deep nets of different sizes on the CIFAR10 (VGG16 and ResNets). We systematically study each combination of two or three compressions out of quantization, low-rank and pruning. We demonstrate that the additive combination improves over any single compression contained in the combination (as expected), and is comparable or better than sequentially engineered combinations such as first pruning some weights and then quantizing the rest. We sometimes achieve models that not only compress the reference significantly but also reduce its error. Notably,

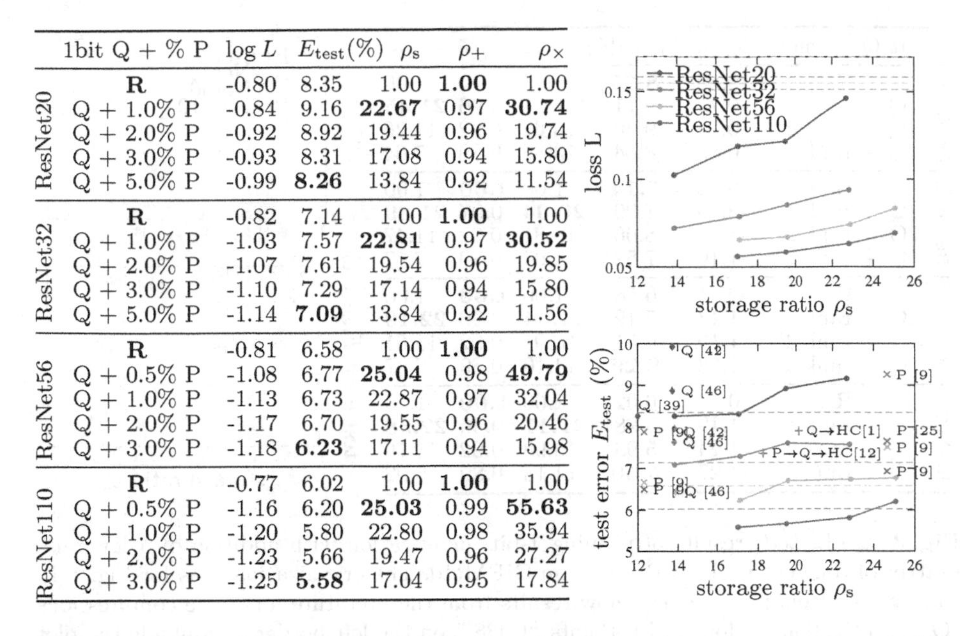

| | 1bit Q + % P | log $L$ | $E_{\text{test}}$(%) | $\rho_s$ | $\rho_+$ | $\rho_\times$ |
|---|---|---|---|---|---|---|
| ResNet20 | **R** | -0.80 | 8.35 | 1.00 | **1.00** | 1.00 |
| | Q + 1.0% P | -0.84 | 9.16 | **22.67** | 0.97 | **30.74** |
| | Q + 2.0% P | -0.92 | 8.92 | 19.44 | 0.96 | 19.74 |
| | Q + 3.0% P | -0.93 | 8.31 | 17.08 | 0.94 | 15.80 |
| | Q + 5.0% P | -0.99 | **8.26** | 13.84 | 0.92 | 11.54 |
| ResNet32 | **R** | -0.82 | 7.14 | 1.00 | **1.00** | 1.00 |
| | Q + 1.0% P | -1.03 | 7.57 | **22.81** | 0.97 | **30.52** |
| | Q + 2.0% P | -1.07 | 7.61 | 19.54 | 0.96 | 19.85 |
| | Q + 3.0% P | -1.10 | 7.29 | 17.14 | 0.94 | 15.80 |
| | Q + 5.0% P | -1.14 | **7.09** | 13.84 | 0.92 | 11.56 |
| ResNet56 | **R** | -0.81 | 6.58 | 1.00 | **1.00** | 1.00 |
| | Q + 0.5% P | -1.08 | 6.77 | **25.04** | 0.98 | **49.79** |
| | Q + 1.0% P | -1.13 | 6.73 | 22.87 | 0.97 | 32.04 |
| | Q + 2.0% P | -1.17 | 6.70 | 19.55 | 0.96 | 20.46 |
| | Q + 3.0% P | -1.18 | **6.23** | 17.11 | 0.94 | 15.98 |
| ResNet110 | **R** | -0.77 | 6.02 | 1.00 | **1.00** | 1.00 |
| | Q + 0.5% P | -1.16 | 6.20 | **25.03** | 0.99 | **55.63** |
| | Q + 1.0% P | -1.20 | 5.80 | 22.80 | 0.98 | 35.94 |
| | Q + 2.0% P | -1.23 | 5.66 | 19.47 | 0.96 | 27.27 |
| | Q + 3.0% P | -1.25 | **5.58** | 17.04 | 0.95 | 17.84 |



**Fig. 2.** Q+P. *Left*: results of running 1-bit quantization with varying amounts of additive pruning (corrections) on ResNets of different depth on CIFAR-10 (with reference nets denoted **R**). We report training loss (logarithms are base 10), test error $E_{\text{test}}$ (%), and ratios of storage $\rho_s$ and floating point additions ($\rho_+$) and multiplications ($\rho_\times$). Boldfaced results are the best for each ResNet depth. *Right*: training loss (*top*) and test error (*bottom*) as a function of the storage ratio. For each net, we give our algorithm's compression over several values of P, thus tracing a line in the error-compression space (reference nets: horizontal dashed lines). We also report results from the literature as isolated markers with a citation: quantization Q, pruning P, Huffman coding HC, and their sequential combination using arrows (e.g., Q → HC). Point "Q [39]" on the left border is outside the plot ($\rho_s < 12$).

this happens with ResNet110 (using quantization plus either pruning or low-rank), even though our reference ResNets were already well trained and achieved a lower test error than in the original paper [15].

We initialize our experiments from reasonably well-trained reference models. We train reference ResNets of depth 20, 32, 56, and 110 following the procedure of the original paper [15] (although we achieve lower errors). The models have 0.26M, 0.46M, 0.85M, and 1.7M parameters and test errors of 8.35%, 7.14%, 6.58% and 6.02%, respectively. We adapt VGG16 [32] to the CIFAR10 dataset (see details in suppl. mat. [10]) and train it using the same data augmentation as for ResNets. The reference VGG16 has 15.2M parameters and achieves a test error of 6.45%.

The optimization protocol of our algorithm is as follows throughout all experiments with minor changes (see suppl. mat. [10]). To optimize the L step we use Nesterov's accelerated gradient method [26] with momentum of 0.9 on minibatches of size 128, with a decayed learning rate schedule of $\eta_0 \cdot a^m$ at the $m$th

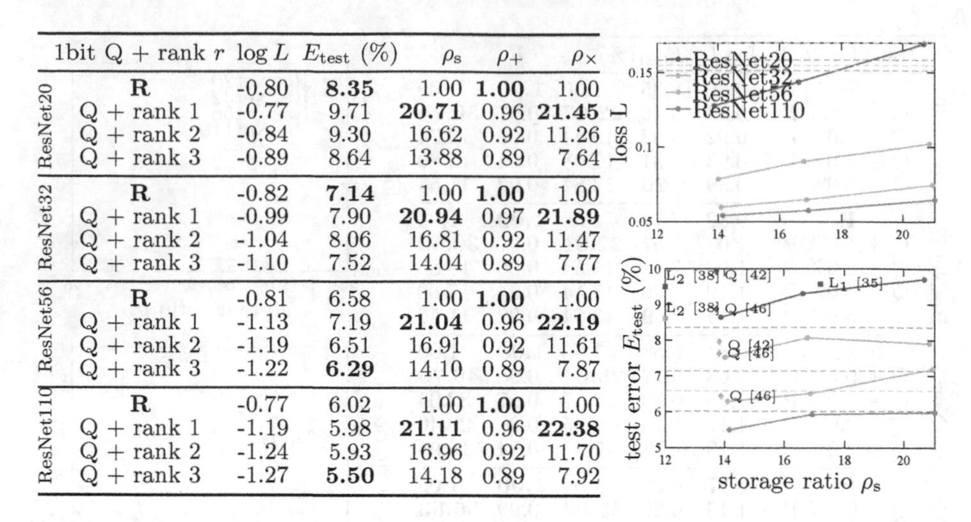

| | 1bit Q + rank $r$ | $\log L$ | $E_{\text{test}}$ (%) | $\rho_s$ | $\rho_+$ | $\rho_\times$ |
|---|---|---|---|---|---|---|
| ResNet20 | **R** | -0.80 | **8.35** | 1.00 | **1.00** | 1.00 |
| | Q + rank 1 | -0.77 | 9.71 | **20.71** | 0.96 | **21.45** |
| | Q + rank 2 | -0.84 | 9.30 | 16.62 | 0.92 | 11.26 |
| | Q + rank 3 | -0.89 | 8.64 | 13.88 | 0.89 | 7.64 |
| ResNet32 | **R** | -0.82 | **7.14** | 1.00 | **1.00** | 1.00 |
| | Q + rank 1 | -0.99 | 7.90 | **20.94** | 0.97 | **21.89** |
| | Q + rank 2 | -1.04 | 8.06 | 16.81 | 0.92 | 11.47 |
| | Q + rank 3 | -1.10 | 7.52 | 14.04 | 0.89 | 7.77 |
| ResNet56 | **R** | -0.81 | 6.58 | 1.00 | **1.00** | 1.00 |
| | Q + rank 1 | -1.13 | 7.19 | **21.04** | 0.96 | **22.19** |
| | Q + rank 2 | -1.19 | 6.51 | 16.91 | 0.92 | 11.61 |
| | Q + rank 3 | -1.22 | **6.29** | 14.10 | 0.89 | 7.87 |
| ResNet110 | **R** | -0.77 | 6.02 | 1.00 | **1.00** | 1.00 |
| | Q + rank 1 | -1.19 | 5.98 | **21.11** | 0.96 | **22.38** |
| | Q + rank 2 | -1.24 | 5.93 | 16.96 | 0.92 | 11.70 |
| | Q + rank 3 | -1.27 | **5.50** | 14.18 | 0.89 | 7.92 |



**Fig. 3.** Q+L. *Left*: results of running 1-bit quantization with addition of a low-rank matrix of different rank on ResNets on CIFAR10. The organization is as for Fig. 2. In the right-bottom plot we also show results from the literature for single compressions (Q: quantization, L: low-rank). Points "L [38]" on the left border are outside the plot ($\rho_s < 12$).

epoch. The initial learning rate $\eta_0$ is one of {0.0007,0.007,0.01}, and the learning rate decay one of {0.94,0.98}. Each L step is run for 20 epochs. Our LC algorithm (we use augmented Lagrangian version) runs for $j$ steps where $j \leq 50$, and has a penalty parameter schedule $\mu_j = \mu_0 \cdot 1.1^j$; we choose $\mu_0$ to be one of $\{5 \cdot 10^{-4}, 10^{-3}\}$. The solution of the C step requires alternating optimization over individual compressions, which we perform 30 times per each step.

We report the training loss and test error as measures of the model classification performance; and the ratios of storage (memory occupied by the parameters) $\rho_s$, number of multiplications $\rho_\times$ and number of additions $\rho_+$ as measures of model compression. Although the number of multiplications and additions is about the same in a deep net's inference pass, we report them separately because different compression techniques (if efficiently implemented) can affect quite differently their costs. We store low-rank matrices and sparse correction values using 16-bit precision floating point values. See our suppl. mat. [10] for precise definitions and details of these metrics.

## 5.1 Q+P: Quantization Plus Pruning

We compress ResNets with a combination of quantization plus pruning. Every layer is quantized separately with a codebook of size 2 (1 bit). For pruning we employ the constrained $\ell_0$ formulation [9], which allows us to specify a single number of nonzero weights $\kappa$ for the entire net ($\kappa$ is the "% P" value in Fig. 2). The C step of Eq. (3) for this combination alternates between running $k$-means (for quantization) and a closed-form solution based on thresholding (for pruning).

**Table 1.** L+P. Compressing VGG16 with low-rank and pruning using our algorithm (top) and by recent works on structured and unstructured pruning. Metrics as in Fig. 2.

| | Model | $E_{\text{test}}$ (%) | $\rho_s$ |
|---|---|---|---|
| | **R** VGG16 | **6.45** | 1.00 |
| ours | rank 2 + 2% P | 6.66 | **60.99** |
| ours | rank 3 + 2% P | 6.65 | 56.58 |
| | pruning [25] | 6.66 | ≈ 24.53 |
| | filter pruning [23] | 6.60 | 5.55 |
| | quantization [30] | 8.00 | 43.48 |

Figure 2 shows our results; published results from the literature are at the bottom-right part, which shows the error-compression space. We are able to achieve considerable compression ratios $\rho_s$ of up to 20× without any degradation in accuracy, and even higher ratios with minor degradation. These results beat single quantization or pruning schemes reported in the literature for these models. The best 2-bit quantization approaches for ResNets we know about [42,46] have $\rho_s \approx 14\times$ and lose up to 1% in test error comparing to the reference; the best unstructured pruning [9,25] achieves $\rho_s \approx 12\times$ and loses 0.8%.

ResNet20 is the smallest and hardest to compress out of all ResNets. With 1-bit quantization plus 3% pruning corrections we achieve an error of 8.26% with $\rho_s = 13.84\times$. To the best of our knowledge, the highest compression ratio of comparable accuracy using only quantization is 6.22× and has an error of 8.25% [39]. On ResNet110 with 1-bit quantization plus 3% corrections, we achieve 5.58% error while still compressing 17×.

Our results are comparable or better than published results where multiple compressions are applied sequentially (Q → HC and P → Q → HC in Fig. 2). For example, quantizing and then applying Huffman coding to ResNet32 [1] achieves $\rho_s = 20.15\times$ with 7.9% error, while we achieve $\rho_s = 22.81\times$ with 7.57% error. We re-emphasize that unlike the "prune then quantize" schemes, our additive combination is different: we quantize all weights and apply a pointwise correction.

### 5.2 Q+L: Quantization Plus Low-Rank

We compress the ResNets with the additive combination of 1-bit quantized weights (as in Sect. 5.1) and rank-$r$ matrices, where the rank is fixed and has the same value for all layers. The convoloutional layers are parameterized by low-rank as in Wen et al. [35]. The solution of the C step (4) for this combination is an alternation between $k$-means and truncated SVD.

Figure 3 shows our results and at bottom-right of Fig. 3 we see that our additive combination (lines traced by different values of the rank $r$ in the error-compression space) consistently improve over individual compression techniques reported in the literature (quantization or low-rank, shown by markers Q or L).

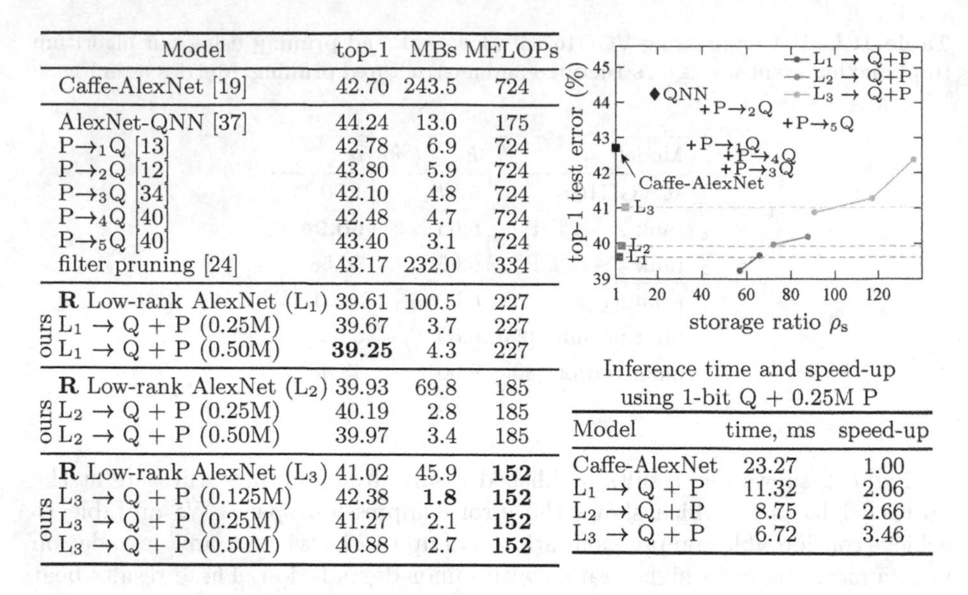

| | Model | top-1 | MBs | MFLOPs |
|---|---|---|---|---|
| | Caffe-AlexNet [19] | 42.70 | 243.5 | 724 |
| | AlexNet-QNN [37] | 44.24 | 13.0 | 175 |
| | $P{\to}_1Q$ [13] | 42.78 | 6.9 | 724 |
| | $P{\to}_2Q$ [12] | 43.80 | 5.9 | 724 |
| | $P{\to}_3Q$ [34] | 42.10 | 4.8 | 724 |
| | $P{\to}_4Q$ [40] | 42.48 | 4.7 | 724 |
| | $P{\to}_5Q$ [40] | 43.40 | 3.1 | 724 |
| | filter pruning [24] | 43.17 | 232.0 | 334 |
| | **R** Low-rank AlexNet ($L_1$) | 39.61 | 100.5 | 227 |
| ours | $L_1 \to Q + P$ (0.25M) | 39.67 | 3.7 | 227 |
| ours | $L_1 \to Q + P$ (0.50M) | **39.25** | 4.3 | 227 |
| | **R** Low-rank AlexNet ($L_2$) | 39.93 | 69.8 | 185 |
| ours | $L_2 \to Q + P$ (0.25M) | 40.19 | 2.8 | 185 |
| ours | $L_2 \to Q + P$ (0.50M) | 39.97 | 3.4 | 185 |
| | **R** Low-rank AlexNet ($L_3$) | 41.02 | 45.9 | **152** |
| ours | $L_3 \to Q + P$ (0.125M) | 42.38 | **1.8** | **152** |
| ours | $L_3 \to Q + P$ (0.25M) | 41.27 | 2.1 | **152** |
| ours | $L_3 \to Q + P$ (0.50M) | 40.88 | 2.7 | **152** |



Inference time and speed-up using 1-bit Q + 0.25M P

| Model | time, ms | speed-up |
|---|---|---|
| Caffe-AlexNet | 23.27 | 1.00 |
| $L_1 \to Q + P$ | 11.32 | 2.06 |
| $L_2 \to Q + P$ | 8.75 | 2.66 |
| $L_3 \to Q + P$ | 6.72 | 3.46 |

**Fig. 4.** Q+P scheme is powerful enough to further compress already downsized models, here, it is used to further compress the low-rank AlexNets [17]. In all our experiments reported here, we use 1-bit quantization with varying amount of pruning. *Left:* We report top-1 validation error, size of the final model in MB when saved to disk, and resulting FLOPs. P—pruning, Q— quantization, L—low-rank. *Top right:* same as the table on the left, but in graphical form. Our compressed models are given as solid connected lines. *Bottom right:* The delay (in milliseconds) and corresponding speed-ups of our compressed models on Jetson Nano Edge GPU.

Notably, the low-rank approximation is not a popular choice for compression of ResNets: Fig. 3 shows only two markers, for the only two papers we know [35, 38]. Assuming storage with 16-bit precision on ResNet20, Wen et al. [35] achieve 17.20× storage compression (with 9.57% error) and Xu et al. [38] respectively 5.39× (with 9.5% error), while our combination of 1-bit quantization plus rank-2 achieves 16.62× (9.3% error).

### 5.3 L+P: Low-Rank Plus Pruning

We compress VGG16 trained on CIFAR10 using the additive combination of low-rank matrices and pruned weights. The reference model has 15.2M parameters, uses 58.17 MB of storage and achieves 6.45% test error. When compressing with L+P scheme of rank 2 and 3% point-wise corrections (Table 1), we achieve a compression ratio of to 60.99× (0.95 MB storage), and the test error of 6.66%.

## 6 Experiments on ImageNet

To demonstrate the power and complementary benefits of additive combinations, we proceed by applying the Q+P combination to *already downsized* models trained on the ILSVRC2012 dataset. We obtain low-rank AlexNets following the work of [17], and compress them further with Q+P scheme. The hyperparameters of the experiments are almost identical to CIFAR10 experiments (Sect. 5) with minor changes, see suppl. mat. [10].

In Fig. 4 (left) we report our results: the achieved top-1 error, the size in megabytes when a compressed model is saved to disk (we use the sparse index compression procedure of Han et al. [13]), and floating point operations required to perform the forward pass through a network. Additionally, we include prominent results from the literature to put our models in perspective. Our Q+P models achieve *significant compression* of AlexNet: we get 136× compression (1.789 MB) without degradation in accuracy and 87× compression with more than 2% improvement in the top-1 accuracy when compared to the Caffe-AlexNet. Recently, Yang et al. [40] (essentially using our Learning-Compression framework) reported 118× and 205× compression on AlexNet with none to small reduction of accuracy. However, as can be found by inspecting the code of Yang et al. [40], these numbers are artificially inflated in that they do not account for the storage of the element indices (for a sparse matrix), for the storage of the codebook, and use a fractional number of bits per element instead of rounding it up to an integer. If these are taken into account, the actual compression ratios become much smaller (52× and 79×), with models of sizes 4.7 MB and 3.1 MB respectively (see left of Fig. 4). Our models outperform those and other results not only in terms of size, but also in terms of inference speed. We provide the runtime evaluation (when processing a single image) of our compressed models on a small edge device (NVIDIA's Jetson Nano) on the right bottom of Fig. 4.

## 7 Conclusion

We have argued for and experimentally demonstrated the benefits of applying multiple compressions as an additive combination. We achieve this via a general, intuitive formulation of the optimization problem via constraints characterizing the additive combination, and an algorithm that can handle *any choice of compression combination* as long as each individual compression can be solved on its own. In this context, pruning takes the meaning of adding a few elementwise corrections where they are needed most. This can not only complement existing compressions such as quantization or low-rank, but also be an interesting way to learn skip connections in deep net architectures. With deep neural nets, we observe that we can find significantly better models in the error-compression space, indicating that *different compression types have complementary benefits*, and that the best type of combination depends exquisitely on the type of neural net. The resulting compressed nets may also make better use of the available hardware. Our codes and models are available at https://github.com/UCMerced-ML/LC-model-compression as part of LC Toolkit [16].

Our work opens up possibly new and interesting mathematical problems regarding the best approximation of a matrix by X, such as when X is the sum of a quantized matrix and a sparse matrix. Also, we do not claim that additive combination is the only or the best way to combine compressions, and future work may explore other ways.

**Acknowledgment.** We thank NVIDIA Corporation for multiple GPU donations.

## References

1. Agustsson, E., et al.: Soft-to-hard vector quantization for end-to-end learning compressible representations. In: Advances in Neural Information Processing Systems (NIPS), vol. 30, pp. 1141–1151 (2017)
2. Alvarez, J.M., Salzmann, M.: Compression-aware training of deep networks. In: Advances in Neural Information Processing Systems (NIPS), vol. 30 (2017)
3. Babenko, A., Lempitsky, V.: Additive quantization for extreme vector compression. In: Proceedings of the 2014 IEEE Computer Society Conf. Computer Vision and Pattern Recognition, CVPR 2014, pp. 931–938 (2014)
4. Beck, A., Tetruashvili, L.: On the convergence of block coordinate descent type methods. SIAM J. Optim. **23**(4), 2037–2060 (2013)
5. Bertsekas, D.P.: Nonlinear Programming, 2nd edn. Athena Scientific, Nashua (1999)
6. Bouwmans, T., Zahzah, E.: Robust principal component analysis via decomposition into low-rank and sparse matrices: an overview, chap. 1. In: Handbook of Robust Low-Rank and Sparse Matrix Decomposition. Applications in Image and Video Processing, pp. 1.1–1.61. CRC Publishers (2016)
7. Candès, E.J., Li, X., Ma, Y., Wright, J.: Robust principal component analysis? J. ACM **58**(3), 11 (2011)
8. Carreira-Perpiñán, M.Á.: Model compression as constrained optimization, with application to neural nets. Part I: general framework. arXiv:1707.01209 (5 July 2017)
9. Carreira-Perpiñán, M.Á., Idelbayev, Y.: "Learning-compression" algorithms for neural net pruning. In: Proceedings of the 2018 IEEE Computer Society Conference Computer Vision and Pattern Recognition, CVPR 2018, pp. 8532–8541 (2018)
10. Carreira-Perpiñán, M.Á., Idelbayev, Y.: Model compression as constrained optimization, with application to neural nets. Part V: combining compressions. arXiv:2107.04380 (2021)
11. Chandrasekaran, V., Sanghavi, S., Parrilo, P.A., Willsky, A.S.: Rank-sparsity incoherence for matrix decomposition. SIAM J. Optim. **21**(2), 572–596 (2010)
12. Choi, Y., El-Khamy, M., Lee, J.: Towards the limit of network quantization. In: Proceedings of the 5th International Conference on Learning Representations, ICLR 2017 (2017)
13. Han, S., Mao, H., Dally, W.J.: Deep compression: compressing deep neural networks with pruning, trained quantization and Huffman coding. In: Proceedings of the 4th International Conference Learning Representations, ICLR 2016 (2016)
14. Hastie, T.J., Tibshirani, R.J.: Generalized Additive Models, vol. 43 in Monographs on Statistics and Applied Probability. Chapman & Hall, London, New York (1990)
15. He, K., Zhang, X., Ren, S., Sun, J.: Deep residual learning for image recognition. In: Proceedings of the 2016 IEEE Computer Society Conference on Computer Vision and Pattern Recognition, CVPR 2016, pp. 770–778 (2016)

16. Idelbayev, Y., Carreira-Perpiñán, M.Á.: A flexible, extensible software framework for model compression based on the LC algorithm. arXiv:2005.07786 (15 May 2020)
17. Idelbayev, Y., Carreira-Perpiñán, M.Á.: Low-rank compression of neural nets: Learning the rank of each layer. In: Proceedings of the 2020 IEEE Computer Society Conference on Computer Vision and Pattern Recognition, CVPR 2020 (2020)
18. Idelbayev, Y., Carreira-Perpiñán, M.Á.: Optimal selection of matrix shape and decomposition scheme for neural network compression. In: Proceedings of the IEEE International Conference on Acoustics, Speech and Signal Processing, ICASSP 2021 (2021)
19. Jia, Y., et al.: Caffe: convolutional architecture for fast feature embedding. arXiv:1408.5093 (20 June 2014)
20. Kim, H., Khan, M.U.K., Kyung, C.M.: Efficient neural network compression. In: Proceedings of the 2019 IEEE Computer Society Conference on Computer Vision and Pattern Recognition, CVPR 2019, pp. 12569–12577 (2019)
21. LeCun, Y., Denker, J.S., Solla, S.A.: Optimal brain damage. In: Touretzky, D.S. (ed.) Advances in Neural Information Processing Systems (NIPS), vol. 2, pp. 598–605. Morgan Kaufmann, San Mateo, CA (1990)
22. Li, F., Zhang, B., Liu, B.: Ternary weight networks. arXiv:1605.04711 (19 November 2016)
23. Li, H., Kadav, A., Durdanovic, I., Graf, H.P.: Pruning filters for efficient ConvNets. In: Proceedings of the 5th International Conference on Learning Representations, ICLR 2017 (2017)
24. Li, J., et al.: OICSR: out-in-channel sparsity regularization for compact deep neural networks. In: Proceedings of the 2019 IEEE Computer Society Conference on Computer Vision and Pattern Recognition, CVPR 2019, pp. 7046–7055 (2019)
25. Liu, Z., Sun, M., Zhou, T., Huang, G., Darrell, T.: Rethinking the value of network pruning. In: Proceedings of the 7th International Conference on Learning Representations, ICLR 2019 (2019)
26. Nesterov, Y.: A method of solving a convex programming problem with convergence rate $\mathcal{O}(1/k^2)$. Sov. Math. Dokl. **27**(2), 372–376 (1983)
27. Nocedal, J., Wright, S.J.: Numerical Optimization. Springer Series in Operations Research and Financial Engineering, 2nd edn. Springer, New York (2006). https://doi.org/10.1007/978-0-387-40065-5
28. Novikov, A., Podoprikhin, D., Osokin, A., Vetrov, D.P.: Tensorizing neural networks. In: Advances in Neural Information Processing Systems (NIPS), vol. 28, pp. 442–450. MIT Press, Cambridge (2015)
29. Nowlan, S.J., Hinton, G.E.: Simplifying neural networks by soft weight-sharing. Neural Comput. **4**(4), 473–493 (1992)
30. Qu, Z., Zhou, Z., Cheng, Y., Thiele, L.: Adaptive loss-aware quantization for multi-bit networks. In: Proceedings of the 2020 IEEE Computer Society Conference on Computer Vision and Pattern Recognition, CVPR 2020 (2020)
31. Rastegari, M., Ordonez, V., Redmon, J., Farhadi, A.: XNOR-Net: ImageNet classification using binary convolutional neural networks. In: Leibe, B., Matas, J., Sebe, N., Welling, M. (eds.) ECCV 2016. LNCS, vol. 9908, pp. 525–542. Springer, Cham (2016). https://doi.org/10.1007/978-3-319-46493-0_32
32. Simonyan, K., Zisserman, A.: Very deep convolutional networks for large-scale image recognition. In: Proceedings of the 3rd International Conference on Learning Representations, ICLR 2015 (2015)
33. Tseng, P.: Convergence of a block coordinate descent method for nondifferentiable minimization. J. Optim. Theor. Appl. **109**(3), 475–494 (2001)

34. Tung, F., Mori, G.: CLIP-Q: deep network compression learning by in-parallel pruning-quantization. In: Proceedings of the 2018 IEEE Computer Society Conference on Computer Vision and Pattern Recognition, CVPR 2018 (2018)
35. Wen, W., Xu, C., Wu, C., Wang, Y., Chen, Y., Li, H.: Coordinating filters for faster deep neural networks. In: Proceedings of the 17th International Conference on Computer Vision, ICCV 2017 (2017)
36. Wright, S.J.: Coordinate descent algorithms. Math. Prog. **151**(1), 3–34 (2016)
37. Wu, J., Leng, C., Wang, Y., Hu, Q., Cheng, J.: Quantized convolutional neural networks for mobile devices. In: Proceedings of the 2016 IEEE Computer Society Conference on Computer Vision and Pattern Recognition, CVPR 2016 (2016)
38. Xu, Y., et al.: TRP: trained rank pruning for efficient deep neural networks. In: Proceedings of the 29th International Joint Conference on Artificial Intelligence, IJCAI 2020 (2020)
39. Xu, Y., Wang, Y., Zhou, A., Lin, W., Xiong, H.: Deep neural network compression with single and multiple level quantization. In: Proceedings of the 32nd AAAI Conference on Artificial Intelligence, AAAI 2018 (2018)
40. Yang, H., Gui, S., Zhu, Y., Liu, J.: Automatic neural network compression by sparsity-quantization joint learning: a constrained optimization-based approach. In: Proceedings of the 2020 IEEE Computer Society Conference on Computer Vision and Pattern Recognition, CVPR 2020, pp. 2175–2185 (2020)
41. Ye, J., Lu, X., Lin, Z., Wang, J.: Rethinking the smaller-norm-less-informative assumption in channel pruning of convolution layers. In: Proceedings of the 6th International Conference on Learning Representations, ICLR 2018 (2018)
42. Yin, P., Zhang, S., Lyu, J., Osher, S., Qi, Y., Xin, J.: BinaryRelax: a relaxation approach for training deep neural networks with quantized weights. SIAM J. Imaging Sci. **11**(4), 2205–2223 (2018). arXiv:1801.06313
43. Yu, X., Liu, T., Wang, X., Tao, D.: On compressing deep models by low rank and sparse decomposition. In: Proceedings of the 2017 IEEE Computer Society Conference on Computer Vision and Pattern Recognition, CVPR 2017, pp. 67–76 (2017)
44. Zhou, A., Yao, A., Guo, Y., Xu, L., Chen, Y.: Incremental network quantization: towards lossless CNNs with low-precision weights. In: Proceedings of the 5th International Conference on Learning Representations, ICLR 2017 (2017)
45. Zhou, T., Tao, D.: GoDec: randomized low-rank & sparse matrix decomposition in noisy case. In: Proceedings of the 28th International Conference on Machine Learning, ICML 2011, pp. 33–40 (2011)
46. Zhu, C., Han, S., Mao, H., Dally, W.J.: Trained ternary quantization. In: Proceedings of the 5th International Conference on Learning Representations, ICLR 2017 (2017)

# Hyper-parameter Optimization for Latent Spaces

Bruno Veloso[1,6(✉)], Luciano Caroprese[5], Matthias König[2], Sónia Teixeira[3,6], Giuseppe Manco[5], Holger H. Hoos[2,4], and João Gama[3,6]

[1] Portucalense University, Porto, Portugal
[2] Leiden University, Leiden, The Netherlands
[3] University of Porto, Porto, Portugal
[4] University of British Columbia, Vancouver, Canada
[5] ICAR-CNR, Arcavacata, Italy
[6] LIAAD-INESC TEC, Porto, Portugal
bruno.m.veloso@inesctec.pt

**Abstract.** We present an online optimization method for time-evolving data streams that can automatically adapt the hyper-parameters of an embedding model. More specifically, we employ the Nelder-Mead algorithm, which uses a set of heuristics to produce and exploit several potentially good configurations, from which the best one is selected and deployed. This step is repeated whenever the distribution of the data is changing. We evaluate our approach on streams of real-world as well as synthetic data, where the latter is generated in such way that its characteristics change over time (concept drift). Overall, we achieve good performance in terms of accuracy compared to state-of-the-art AutoML techniques.

**Keywords:** AutoML · Hyper-parameter optimization · Latent spaces · Nelder-Mead algorithm · SMAC · Recommender systems

## 1 Introduction

In many application scenarios, machine learning systems have to deal with large amounts of continuous data, so-called data streams, whose properties or distribution can change and evolve. This is especially relevant for recommender systems, which usually have to deal with evolving data.

Recommender systems help users to select items among a set of choices, according to their preferences. On e-commerce platforms such as Netflix or Amazon, where the number of items to select is very large, personalized recommendation is almost mandatory.

How these recommendations are made follows one of three types of approaches: i) content-based: items are recommended based on the similarity between their features; ii) collaborative filtering: items are recommended to a user based on known preferences of similar users for those items; iii) hybrid: a


N. Oliver et al. (Eds.): ECML PKDD 2021, LNAI 12977, pp. 249–264, 2021.
https://doi.org/10.1007/978-3-030-86523-8_16

combination of the content-based and collaborative filtering. In this work, we use a collaborative filtering approach. Collaborative filtering is based on a very large sparse users $\times$ items matrix. A typical approach to deal with this huge matrix is *matrix factorization*, which constructs a user embedding matrix and an item embedding matrix, such that users are embedded into the same latent space as items [21]. Through the dot product of the matrices of these embeddings, it is then possible to obtain, by approximation, the ratings (preferences).

The hyper-parameter optimization (HPO) problem seeks to choose optimal settings for the hyper-parameters of a given machine learning system or algorithm, such that the best possible performance is obtained on previously unseen data. The literature suggests approaches such as i) grid search [17], ii) gradient search [24], iii) random search [4] and iv) Bayesian optimization algorithms [13] for hyper-parameter optimization.

In this work, we describe a method for online optimization that can automatically adapt the hyper-parameters of an embedding model to changes in the data. We employ the Nelder-Mead algorithm, which uses a set of operators to produce a potentially good configuration to be deployed. This step is repeated whenever the distribution of the data changes. To evaluate our approach, we developed a synthetic data generator capable of producing realistic data streams with characteristics that change over time (concept drift).

Specifically, we make the following contributions:

- online optimization of the hyper-parameters for matrix factorization on large data sets;
- drift detection, that is, our proposed method reacts to changes in the process producing a given data stream.

The remainder of this paper is structured into five sections. Section 2 provides a systematic literature review on latent space models, automated machine learning in the context of online hyper-parameter optimization, and streaming approaches. Section 3 formulates the problem and describes our proposed solution. Section 4 details our experiments and discusses the results we obtained. Finally, Sect. 5 draws some general conclusions and outlines directions for future work.

## 2 Background and Related Work

In this section, we cover work related to latent variable spaces in recommender systems, online AutoML techniques, and streaming approaches.

***Latent Space Models.*** In this work, we focus on latent variable models, which are historically proven to work effectively in modeling user preferences and providing reliable recommendations (see, *e.g.*, [3] for a survey). Essentially, these approaches embed users and items into latent spaces that translate relatedness to geometrical proximity. Latent embeddings can be used to decompose the large sparse preference matrix [1,35], to devise item similarity [20,27], or more generally, to parameterize probability distributions for item preference [16,29,40] and to sharpen the prediction quality employing meaningful priors.

The recent literature has shifted to more complex models based on deep learning [41], which in principle could show substantial advantages over traditional approaches. For example, neural collaborative filtering (NCF) [15] generalizes matrix factorization to a non-linear setting, where users, items, and preferences are modeled through a simple multilayer perceptron network that exploits latent factor transformations. Despite this progress, collaborative filtering based on simple latent variable modeling still represents a reference framework in the are of recommender systems. Notably, recent studies [30,31] showed that carefully tuned basic matrix factorization models can outperform more complex models based on sophisticated deep learning architectures. This capability, combined with the intrinsic simplicity of the model and the underlying learning process, is the main reason why we focus on them in our work presented here.

***Online AutoML.*** The field of AutoML is generally concerned with automatically constructing machine learning pipelines that efficiently map raw input data to desired outputs [19], such as class labels, and can therefore be seen as an extension of plain *hyper-parameter optimization (HPO)* [9]. The task of automatically constructing a machine learning pipeline is formally modeled by the *combined algorithm selection and hyper-parameter optimization (CASH)* problem, which formalizes the search for the most effective algorithm and its associated hyper-parameter configuration as a joint optimization task [22].

In principle, there are various ways to tackle both the HPO and CASH problems. One of the most prominent approaches is through *sequential model-based algorithm configuration (SMAC)* [18], which is a widely known, freely available, state-of-the-art general-purpose configuration procedure based on sequential model-based (or Bayesian) optimization. The main idea of SMAC is to construct and iteratively update a probabilistic model of target algorithm performance to guide the search for good configurations. In the case of SMAC, this so-called *surrogate model* is implemented using a random forest regressor [5].

AutoML methods have been shown to be able to efficiently assemble and configure full machine learning pipelines (see, *e.g.*, [10]), including automated data pre-processing, feature selection, and hyper-parameter optimization. However, the performance of these systems is usually evaluated in static environments, *i.e.*, on data that does not change over time.

While there exists a vast body of research on processing streams and data in the presence of drift, there has been relatively little work on *online* AutoML, *i.e.*, AutoML methods that can automatically adapt machine learning algorithms to dynamic changes in the data on which they are deployed. Recently, some first attempts have been made to extend AutoML methods to dynamic data streams [8,25]. The main idea behind online AutoML is to not only automatically build a machine learning pipeline, but also to adjust or replace it when its performance degrades due to changes in the data. There are different adaption strategies suggested in the literature, which can be broadly divided into *model replacement* and *model management* strategies [25], where the former globally replace the model with a new one and the latter updates ensemble weights of the initially learned model based on new data.

As our work presented here focuses on a specific algorithm, *i.e.*, matrix factorization, we do not configure a machine learning ensemble, but instead, seek to dynamically optimize the hyper-parameters of the embedding model.

***Streaming Approaches.*** With the development of modern computer architectures and with the substantial increase in the acquisition of sensor data, it becomes evident that offline model training and selection will become largely obsolete in the near future. *React* is one of the first approaches to model selection that explored the emergence of new multi-core architectures to compute several models in parallel [11,12]. This model selection technique implements a tournament-type process to decide which are the most effective hyper-parameter configurations. The major drawback of this technique is the computational cost associated with running multiple model configurations in parallel.

In the literature, we can also find several works proposing incremental model selection for classification tasks. The IL-MS algorithm [23] uses a $k$-fold cross-validation procedure to compute additional support vector machine (SVM) models with different configurations to select the best model. This procedure is run periodically, to minimize the computational cost; however, the evaluation procedure performs a double pass over the data (offline learning), which increases computational cost. A different approach uses meta-learning and weighting strategies to rank a set of heterogeneous ensembles [32,33]. However, this strategy requires the extraction of computationally expensive meta-features, which can pose challenges for stream-based scenarios. More recently, the same authors [34] proposed a way for measuring the score of ensemble members on a recent time window and combining their votes.

The *confStream* algorithm [6,7] was developed to automatically configure a stream-based clustering algorithm using ensembles. Each ensemble has different configurations, which are periodically evaluated and changed, based on the performance on the last window. These performance observations are used to train a regression model, which suggests a set of unknown good configurations for the new ensemble models.

More recently, the problem of hyper-parameter tuning on data streams was formulated as an optimization problem [37,38], using the well-known Nelder-Mead algorithm [26] for optimizing a given loss function. This particular contribution showed to be highly versatile and was applied to three different machine learning tasks: classification [2,36], regression [37], and recommendation [39]. In the particular case of recommendation, the authors adopted an incremental matrix factorization model with a static hyper-parameter configuration, *i.e.*, the hyper-parameters are configured at the beginning of the stream and are subsequently kept unchanged. In the following, we will use embedding models in combination with a concept drift detector, based on the Page-Hinkley test [28], to restart the optimization process.

## 3 Hyper-parameter Optimization for Latent Spaces in Recommender Systems

We start by introducing notation to be used throughout the remainder of this paper. In the following, $u \in U = \{1, \ldots, N\}$ indexes a user and $i \in I = \{1, \ldots, M\}$ indexes an item for which a user can express a preference. Let $R_{u,i} \in \{r_{min}, \ldots, r_{max}\}$ denote the preference (rating) of user $u$ for item $i$. The range $\{r_{min}, \ldots, r_{max}\}$ represents a preference rank: when $R_{u,i} = r_{min}$, user $u$ maximally dislikes item $i$, while $R_{u,i} = r_{max}$ denotes maximal preference for the item. Typical ranges are $\{0, 1\}$ (implicit preference) or $\{1, ..., 5\}$.

The set of all preferences can be represented as a rating matrix $R$, or alternatively, as a set of triplets $R = \{(u_1, i_1, R_1), \ldots, (u_n, i_n, R_n)\}$ where $R_j = R_{u_j, i_j}$ When $N$ and $M$ are large, $R$ only represents a partial and extremely small view of all possible ratings. With an abuse of notation, we shall denote by $(u, i) \in R$ the fact that there exists a triplet $(u, i, R_{u,i})$ in $R$. The underlying learning problem is hence to provide a reliable estimate (completion) $\hat{R}_{u,i}$ for each possible pair $(u, i)$, given the current partial view $R$ not containing the pair.

The standard matrix factorization framework for predicting such values assumes the following. Each user $u$ and item $i$ admit a representation in a $K$-dimensional space. We denote such representations by embedding vectors $\mathbf{p}_u, \mathbf{q}_i \in \mathbb{R}^K$, which represent the row of the embedding matrices $\mathbf{P} \in \mathbb{R}^{N \times K}$ and $\mathbf{Q} \in \mathbb{R}^{M \times K}$. Given user $u$ and item $i$, the corresponding preference can be modeled as a random variable with a fixed distribution whose parameters rely on the embeddings $\mathbf{p}_u$ and $\mathbf{q}_i$. In the following, we assume that $R_{u,i} \sim \mathcal{N}(\mu, \sigma)$, where $\sigma$ is fixed and $\mu = \mathbf{p}_u \cdot \mathbf{q}_i$ [35]. Other modeling choices are possible and do not substantially change the overall framework (see, *e.g.*, [29]). Finally, the learning objective can be specified as finding optimal embedding matrices $\mathbf{P}^*$ and $\mathbf{Q}^*$ that maximize the likelihood of the partial observations $R$, or that minimize the MSE loss:

$$\ell(\mathbf{P}, \mathbf{Q}; R) = \frac{1}{|R|} \cdot \sum_{(u,i) \in R} (R_{u,i} - \mathbf{p}_u \cdot \mathbf{q}_i)^2$$

This optimization problem can be easily solved via stochastic gradient descent, using a given learning rate $\eta$. In general, for a given $R$, the optimal embedding depends on both the embedding size $K$ and the learning rate $\eta$. A proper exploration of the search space induced by these parameters enables the discovery of the most appropriate model for $R$.

An additional assumption that we make in this work is that the partial view $R$ can be continuously updated. That is, either new unknown entries can be disclosed (for example, some users can express preferences for previously unseen items) or former preferences change (for example, due to change of tastes by the user, or due to a more accurate evaluation). Thus, we assume that the history of preferences produces a continuous stream of snapshots $R^{(1)}, R^{(2)}, \ldots, R^{(t)}, \ldots$, where $R^{(t)}$ represents the view of $R$ at time $t$. As already mentioned, snapshots can overlap, *i.e.*, it can happen that both $R^{(n)}_{u,i}$ and $R^{(m)}_{u,i}$ exist for $m \neq n$, and $R^{(n)}_{u,i} \neq R^{(m)}_{u,i}$. We then define $\bigsqcup_{n \leq t} R^{(n)}$ as a merge among

$R^{(1)}, R^{(2)}, \ldots, R^{(t)}$ that preserves recency. That is, by denoting $\bigsqcup_{n \le t} R^{(n)}$ as $V^{(t)}$, we define $V_{u,i}^{(t)} = R_{u,i}^{(t^*)}$, where $t^* = \max\{t \mid (u,i) \in R^{(t)}\}$. If no such $t^*$ exists, then $V_{u,i}^{(t)}$ is undefined. The problem hence becomes: Given the current history $R^{(1)}, R^{(2)}, \ldots, R^{(t)}$, can we predict the missing entries of $\bigsqcup_{n \le t} R^{(n)}$ and provide a reliable estimate $\hat{R}$ of the preferences at time $t$? More specifically, what is the embedding size $K$ and learning rate $\eta$ that produce optimal embeddings $\mathbf{P}, \mathbf{Q}$ minimizing the loss $\ell\left(\mathbf{P}, \mathbf{Q}; \bigsqcup_{n \le t} R^{(n)}\right)$?

### 3.1 The Nelder-Mead Approach

The problem of hyper-parameter tuning we consider can be stated as follows: given a machine learning algorithm *Alg* with a default hyper-parameter configuration *Conf*, represented by $Alg_{Conf}$, and a data stream $S$ consisting of an infinite set of mini-batches *MB*, the goal is to find an optimal configuration $Alg^*_{Conf}$, where $Alg^*_{Conf}$ yields better performance compared to past configurations. In our situation, the model is represented by the embedding matrices $\mathbf{P}$ and $\mathbf{Q}$, *Conf* is represented by the embedding size $K$ and the learning rate $\eta$, and *Alg* is essentially stochastic gradient descent applied to the embedding matrices with learning rate $\eta$.

Here, we adopt a stream-based version of the Nelder-Mead algorithm [37] to find the optimal configuration $Alg^*_{Conf}$. This method consists of two phases: i) exploration, where the algorithm tries different versions of $Alg^*_{Conf}$ to minimize a loss function, using a set of operators based on heuristics; and ii) deployment of the best $Alg^*_{Conf}$ over the following set of *MB* in the stream $S$.

The original Nelder-Mead algorithm requires $n + 1$ configurations *Conf* to optimize a set of $n$ hyper-parameters. Since we try to optimize the learning rate ($\eta$) and embedding size ($K$), we need to maintain three configurations: $B$ (representing the configuration with the best score), $W$ (the configuration with the worst score) and $G$ (with a score in between $B$ and $W$). The stream-based version additionally computes in parallel the $M, R, E, S, C$ auxiliary models for the application of the Nelder-Mead operators. The underlying configurations for these models are obtained by applying four different operations to the configurations $B, G, W$. These are contraction, shrinking, expansion and reflection. Figure 1 illustrates how these models are obtained from $B, G, W$. Essentially, each operation corresponds to modifying the values of $\eta$ and $K$, by either enlarging or narrowing it, and then devising the optimal models corresponding to these modified hyper-parameters.

In the streaming scenario, it is crucial to optimize the model incrementally. Basic stochastic gradient descent (*SGD*) is well-suited to incremental adaptation. This can be achieved by starting from a previous model and performing additional updating steps, possibly while exploiting the new learning rate. However, adapting the embedding dimension $K$ poses several challenges when we try to apply the contraction or expansion operators, because it changes the structure of the model. This means that the *SGD* algorithm cannot exploit a previous

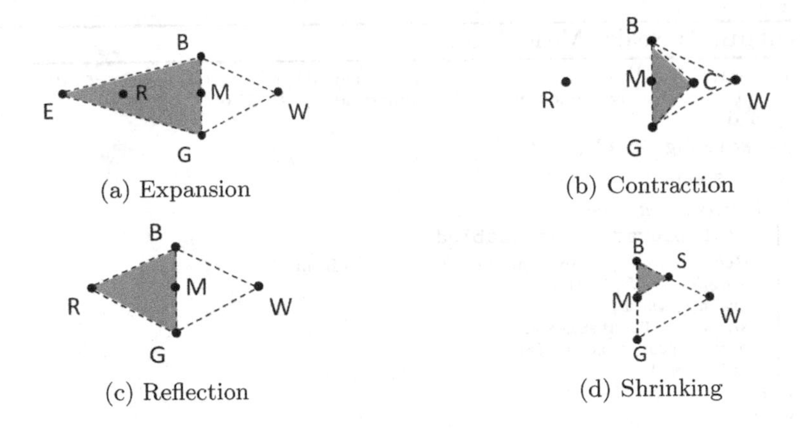


(a) Expansion

(b) Contraction

(c) Reflection

(d) Shrinking

**Fig. 1.** Basic heuristic operations.

model, and consequently has to restart the entire optimization process. To cope with this issue, we adopt two heuristic adaptation strategies:

- If any operation contracts the embedding size, we compute the new embedding matrices from the original ones by dropping some columns. The dropped columns correspond to the latent dimensions exhibiting the lowest variance values. The intuition is that these dimension do not discriminate well among users/items, and hence their contribution is likely to be redundant.
- If, on the other side, the embedding size needs to be expanded, we can keep all previous dimensions and add new ones. To add new dimensions, we create a set of vectors orthogonal to the existing latent vectors. In principle, orthogonal vectors allow for more efficient exploration of the search space by inspecting directions not covered before – for example, compared to randomly generated vectors.

The overall procedure is illustrated in Algorithm 1. It starts with three randomly generated configurations $B, G, W$. Then, the main cycle (lines 2–23) iterates over all time windows (denoted by $t$), while performing two steps:

- In the exploration phase (lines 4–11 and 17–20), the search space is explored to adapt the model with configurations coherent with the changes in the distribution of the data. To detect if a concept drift occurs, we use the Page-Hinkley test (*PHt*) [28] to report a concept drift if the observed prequential loss at some instant is greater than a given user-defined threshold (acceptable magnitude of change). When the mean of new data points exceeds the predefined threshold, an alert is triggered, signaling that a concept drift has been detected.
- In the exploitation phase (lines 13–16), the current best configuration is maintained, and the underlying model is optimized with respect to the current batch. Exploitation only occurs when the differences between the configurations $B, G, W$ are negligible and, consequently, the simplex tends to converge to a single point.

**Algorithm 1:** Nelder-Mead Algorithm

**Input**: A stream $R^{(1)}, R^{(2)}, \ldots, R^{(t)}, \ldots$, where each $R^{(t)}$ is a partial matrix of ratings
1 Initialize the process by randomly creating configurations $B, W, G$
2 **for** $t \geq 1$ **do**
  /* Working on chunk $V^{(t)}$ */
3  $S^{(t)} = \texttt{Sample}\left(\bigsqcup_{n \leq t-1} R^{(n)}\right)$
4  **if** *drift is detected* **then**
   /* Exploration is enabled */
5   Reconfigure $G, W$ by random perturbations from $B$
6   $simplex = \{B, G, W\}$
7   Train($simplex$)
8   $simplex$ = UpdateSimplex($simplex$)
9   $aux$ = UpdateAux($simplex$)
10   Train($aux$)
11  **end**
12  **else**
13   **if** *Convergence is detected on simplex* **then**
    /* Exploitation is enabled */
14    $simplex = \{B\}$
15    $aux = \emptyset$
16   **end**
17   **else**
18    $simplex$ = UpdateSimplex($simplex \cup aux$)
19    $aux$ = UpdateAux($simplex$)
20   **end**
21   Train($simplex \cup aux$)
22  **end**
23 **end**
24 **Function** Train($models$):
25  **for** *each model* $m \in models$ **do**
   /* Update the current model using Gradient Descent */
26   $\mathbf{P}_m^{(t)}, \mathbf{Q}_m^{(t)}, l_m^{(t)} = \texttt{Optimize}(R^{(t)} \sqcup S^{(t)}; \mathbf{P}_m^{(t-1)}, \mathbf{Q}_m^{(t-1)}, \eta_m)$
27  **end**
28 Shoul **Function** UpdateSimplex($models$):
  /* Select best, good and worst model */
29  Re-identify $B, G, W$ from $models$ based on the current associated losses
30  **return** $\{B, G, W\}$
31 **Function** UpdateAux($models$):
  /* Contraction, shrinking, expansion, reflection */
32  Generate auxiliary configurations $M, E, R, S, C$ from $models$
33  **return** $\{M, E, R, S, C\}$

In both situations, the training of the model (line 26) starts from the optimal embeddings for the same configuration, as computed in the preceding time window and adapted according to the previously described contraction and expansion processes. Furthermore, training is performed on the same set of data points as used in the evaluation step (line 3).

## 4 Empirical Evaluation

We conducted an extensive empirical evaluation of the proposed approach, with the goal of answering the following research questions:

**RQ1:** Does the exploitation phase of the Nelder-Mead algorithm converge to high-quality solutions?

**RQ2:** When drift is present in the data, how well does the Nelder-Mead approach adapt to the underlying changes?

**RQ3:** How does the Nelder-Mead algorithm compare to specific adaptations of well-known baseline approaches from the literature for online automatic model learning?

To foster reproducibility, we have publicly released all the data and code required to reproduce our experiments.[1]

### 4.1 Baselines and Evaluation Protocol

Our evaluation was performed on synthetic and real-world datasets. For the evaluation protocol, we considered the temporal data ordering and partitioned each dataset into intervals of the same size. The proposed method was evaluated with the predictive sequential (prequential) evaluation protocol for data streams [14], and we used root-mean-square error (RMSE) as our evaluation metric.

To address the research questions stated above, we considered two baselines. The first of these was aimed at verifying that our algorithm can make correct predictions whenever the entire stream is considered to be resulting from a stationary process. To this end, we considered a statically tuned matrix factorization model. As discussed in [30], a careful setup of the gradient-based algorithm for this basic model can outperform several more sophisticated approaches, including neural collaborative filtering approaches [31]. Hence, it is natural to ask whether the Nelder-Mead approach proposed in Sect. 3 is sufficiently robust to guarantee similar or better performance. To investigate this, we performed an exhaustive hyper-parameter search on the basic matrix factorization model using SMAC over the entire dataset. Hereby, we aimed to find a combination of globally optimal hyper-parameter settings, *i.e.*, settings that perform well over the entire data stream. After completion of this configuration process, these hyper-parameter settings were used to initialize a static model and to start an online training process, where each chunk was processed sequentially using stochastic gradient descent.

As a second baseline, we chose SMAC, a state-of-the-art, general-purpose algorithm configuration procedure that has been widely used for hyper-parameter optimization (see also Sect. 2). The general workflow of SMAC starts with picking a configuration $Alg_{Conf}$ and an instance $\pi$ (a sample of the data). Next, the configurator performs a run of algorithm *Alg* with configuration *Conf* on instance $\pi$ and measures the resulting performance. The information gleaned from such individual target algorithm runs is then used to iteratively build and update a predictive model that provides the basis for identifying promising configurations for subsequent runs, and to thus find configurations that perform well on the given training instances. Once its configuration budget (*e.g.*, the number

[1] https://github.com/BrunoMVeloso/ECMLPKDD2021.

of evaluation runs) is exhausted, SMAC returns its current incumbent $Conf^*$, *i.e.*, the best configuration found so far.

While the Nelder-Mead approach is intrinsically stream-based, the SMAC procedure is static by design. That is, SMAC uses a fixed subset of the data to find an optimal configuration, which is then deployed on the remaining data set. Here, however, we are dealing with a data stream, in which data not only arrives gradually over time, but is also expected to change. Therefore, we adapted the original SMAC procedure to employ a model replacement strategy [25], in which the optimization procedure is re-run from scratch when drift is detected to find better hyper-parameter settings $Conf^*$ and, subsequently, deploy a new instance of *Alg* with $Conf^*$.

The parameters of the model replacement strategy are chosen as follows. First, SMAC finds a configuration $Conf^*$ using the first chunk of the data stream; *i.e.*, the first $n$ mini-batches, where $n$ is the size of a chunk. It should be noted that the value of $n$ has an impact on the effectiveness of the configuration procedure; more precisely, a larger value for $n$ allows for better generalization, but at the same time increases computational cost, as each configuration has to be evaluated on a larger amount of data.

Once drift has been detected, we re-start the configuration procedure using the data chunk in which the drift occurred as the training set. Again, one could increase $n$ or even re-run the configurator using all stored data up to the current batch. However, given an infinite stream of mini-batches, the computational cost incurred by the re-configuration process would grow indefinitely, making this approach infeasible for the scenario studied in this work.

### 4.2 Experiments on Real-World Data

In our first set of experiments, we evaluated our approach using Movielens1M[2], a standard benchmark dataset for collaborative filtering. Movielens is a time-series dataset containing user-item rating pairs along with the corresponding timestamps. The dataset comprises $6K$ users, $4K$ items, and 1 million ratings.

Figure 2 compares the results for this dataset obtained by the automatically tuned matrix factorization model using Nelder-Mead and SMAC, as well as the static matrix factorization model (trained with $K = 35$ and $\eta = 0.0337$).

The graph shows the average prequential loss, computed over a sliding window of fixed size. We can see that the proposed approach consistently outperforms the baselines after an initial burn-in period at the very beginning of the stream. Interestingly, the predictive ability of Nelder-Mead surpasses the static matrix factorization. This answers our first research question: the online version of Nelder-Mead does converge to a high-quality solution.

[2] https://grouplens.org/datasets/movielens/.

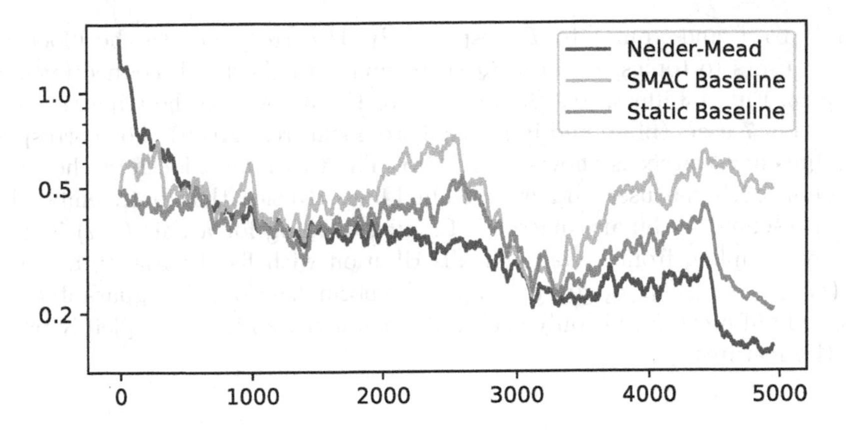


**Fig. 2.** Moving-average error rate for the MovieLens dataset.

### 4.3 Experiments on Synthetic Data

In a second set of experiments, we carried out a careful comparison of the proposed approaches by evaluating their performance in a more controlled way. We tested our framework on datasets produced by a synthetic data generator, designed to be able to create data streams with controllable characteristics. The working principle of the generator is that ratings can be generated as the result of a stochastic process influenced by preference changes, and hence governed by evolving embedding matrices. The evolution essentially consists of the addition or removal of features.

The data generation process starts with a fixed number of users $N$ and items $M$, and relies on the initial non-negative embedding matrices $\mathbf{P} \in \mathbb{R}^{N \times K_0}$ and $\mathbf{Q} \in \mathbb{R}^{M \times K_0}$, where $K_0$ is the initial feature size. The latent features in the data generation process resemble latent semantic topics [16]: Given a user $u$, the embedding $\mathbf{p}_u$ encodes the leanings of $u$ for the given topic; in particular, $p_{u,k}$ is non-zero if there is a leaning of $u$ for topic (feature) $k$. Analogously, $q_{i,k}$ represents the relatedness of item $i$ to topic $k$. The generation process for $\mathbf{P}$ and $\mathbf{Q}$ proceeds in two steps. First, we build a tripartite graph $G = (V, E)$, where $V = (U, I, T)$, with $U$ representing the set of all users, $I$ the set of all items, and $T = \{1, \ldots, K_0\}$ the set of all features. Edges only connect topics to users or items and are generated through preferential attachment, by considering each topic in order. For a given topic $k$, we first sample the number of user (resp. item) neighbors $n_k$ from a Zipf distribution with parameter $\alpha$. Then, for each of these neighbors, we select a user (resp. an item) $x$ and add the edge $(x, k)$ to $E$. The element $x$ is stochastically selected from $U$ (resp. from $I$):

- with probability $p$, we randomly sample $x$ with uniform probability;
- otherwise (with probability $1-p$), we sample $x$ with probability proportional to its current degree.

The embeddings $\mathbf{P}$ and $\mathbf{Q}$ are extracted from the adjacency matrix $\mathbf{A}_G$ of $G$. By construction, $\mathbf{A}_G$ exhibits two main non-zero blocks, representing the edges from $U$ to $T$ and from $I$ to $T$, respectively. $\mathbf{P}$ corresponds to the block connecting users to topics, whereas $\mathbf{Q}$ corresponds to the block connecting items to topics. Figure 3 illustrates the process. In Fig. 3a, we see the tripartite graph connecting 7 users (blue) and items (red) to 4 features (green). The corresponding adjacency matrix is shown in Fig. 3b. The two main blocks of the matrix represent both the user and item embeddings. Notice that both embeddings are represented by binary matrices. Finally, a rating for a pair $(u, i)$ is generated by sampling from a Gaussian distribution with fixed variance and mean $\mu = (r_{max} - r_{min}) \cdot \mathbf{p}_u \mathbf{q}_i / \|\mathbf{q}_i\| + r_{min}$. The normalization of $\mathbf{q}_i$ guarantees that the maximal preference is only achieved when user and item completely overlap over the features.

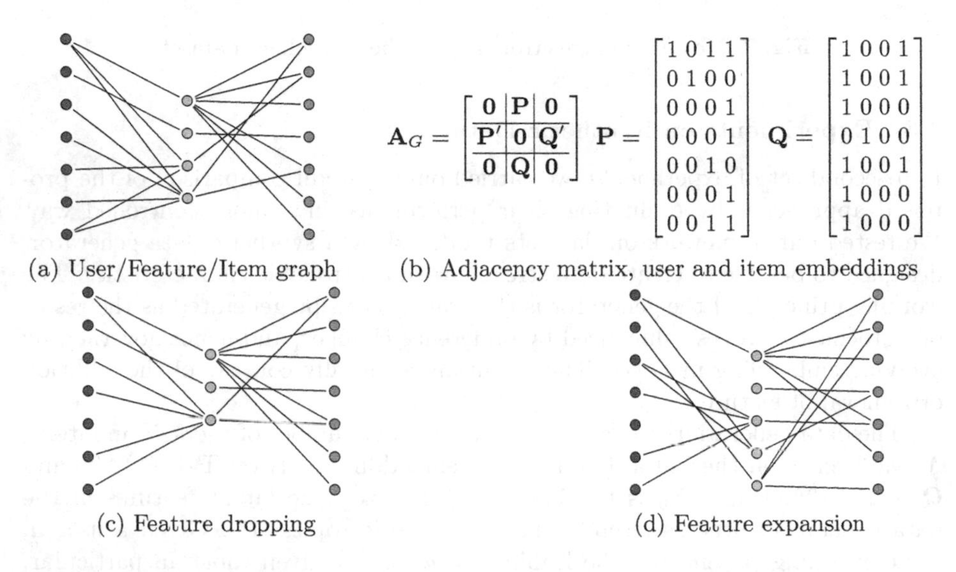


(a) User/Feature/Item graph (b) Adjacency matrix, user and item embeddings

(c) Feature dropping (d) Feature expansion

**Fig. 3.** An example illustrating the data generation process. (Color figure online)

Within the construction scheme outlined above, concept drift can be easily modeled, by evolving the underlying tripartite graph. New ratings can be generated accordingly from the updated embedding matrices. We consider two main drifting operations here: (1) elimination of a feature and (2) expansion with an additional feature. Elimination is simple and essentially consists of removing a feature node. The corresponding connections are rewired to an existing feature, according to a preferential attachment criterion (governed by parameters $p_d$ and $p_r$): Given an edge $(x, k)$ connected to an eliminated feature $k$, the edge is removed with probability $p_d$. If not removed, it is rewired to a random feature (either chosen uniformly with probability $p_r$, or with probability proportional to the feature degree with probability $1 - p_r$). Figure 3c shows an example where feature 3 from Fig. 3a is eliminated.

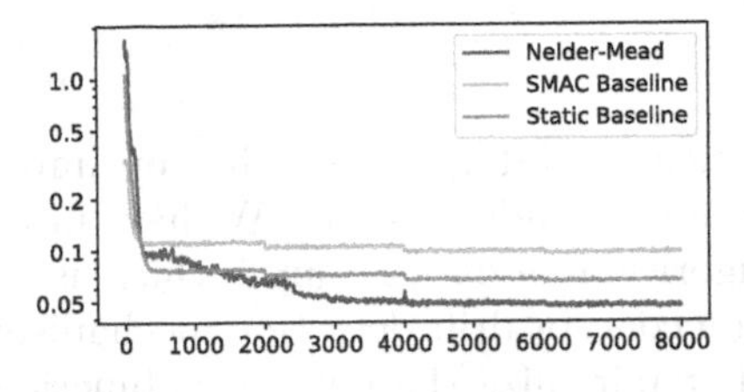


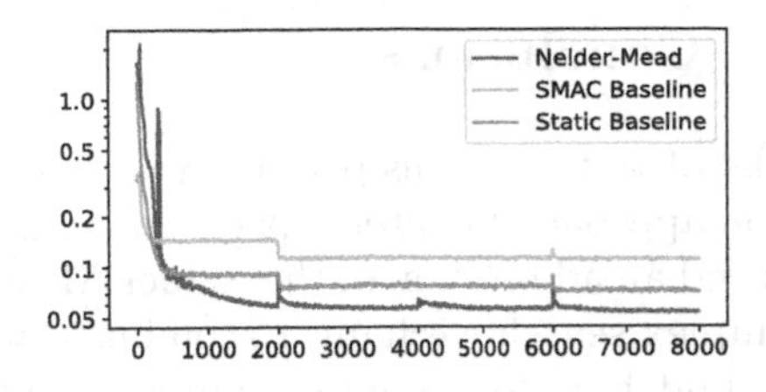


**Fig. 4.** Moving-average error rate for synthetic data (left: first experiment; right: second experiment; for details, see text).

The expansion follows a similar scheme: for each newly added feature, we fix the number of neighbors to connect, and select each such neighbor either randomly or by preferential attachment. The connection produces a new edge with probability $p_a$, and otherwise rewires one of the existing neighbor connections (with probability $1 - p_a$). Figure 3d illustrates how two new features are added and some edges are rewired.

We performed two experiments on a synthetic dataset with 10 000 users and 2 000 items. In both of these, we set $K_0 = 100$ and $\alpha = 1.1$. We generated a total of 8 000 chunks with 2 000 ratings each, with a drift every 2 000 chunks. In the first experiment, we set $p = 1$, and drift was controlled by $p_d = p_r = p_a = 1$. In the second experiment, we set $p = 0.9$, and drift was controlled by $p_d = p_r = p_a = 0.5$. In both cases, we experimented with expansion, by adding $K_0$ more features at each drift.

Figure 4 shows that our framework correctly detects the drift and restarts the hyper-parameter tuning process. More specifically, we observe that Nelder-Mead exhibits fast convergence and consistently adapts to all the injected occurrences of drift, which answers our second research question. Furthermore, we can notice a substantial difference between Nelder-Mead and SMAC baseline as well as the static matrix factorization model, resolving the third research question. Although the latter approach also adapts to changes, as a result of the SGD optimization, the automatic hyper-parameter optimization provided by the Nelder-Mead algorithm allows for a better adaptation to the changes triggered by the new features, and consequently yields a lower error rate. When using SMAC as an online configurator, we also find that model performance improves after each drift, indicating that the re-configuration procedure effectively accounts for the induced changes in the data. However, it does so less successfully than the Nelder-Mead approach and the static approach, which could be explained by the relatively small amount of data points SMAC is exposed to in the online setting. Possibly, the performance of this approach could be improved by increasing the chunk size $n$, thereby providing SMAC with a larger set of training data in the (re-)configuration procedure.

## 5 Conclusions

The objective of this research was to investigate and develop an online optimization approach for latent spaces in dynamic recommender systems. We have introduced an adaptation of the Nelder-Mead algorithm for data streams, which uses a simplex search mechanism, combined with a concept drift detection mechanism, to find hyper-parameter configurations that minimize the given loss function. Through experiments on real-world and artificial data sets, we have shown that the automatic selection of hyper-parameter settings has substantial impact on the outcomes of stream-based recommendations. In particular, we have demonstrated that i) our new approach achieves lower prediction error than a carefully tuned static matrix factorization model as well as the state-of-the-art configurator SMAC adapted to an online setting; ii) the concept drift detector is able to automatically trigger the search for a new optimal solution, which is an advantage when compared with static approaches.

The proposed approach can be adopted to other scenarios where an embedding model is used for predictive purposes. We plan to further explore the applicability of this optimization method in situations where the parameter space is more complex than the simple embedding size; these situations arise, for example, when using complex deep learning models requiring multiple components and specific tuning of the underlying components, such as convolutional architectures, recurrent layers or even attention models based on transformers.

**Acknowledgments.** The research reported in this work was partially supported by the EU H2020 ICT48 project "HumanE-AI-Net" under contract #952026. The support is gratefully acknowledged.

## References

1. Agarwal, D., Chen, B.C.: LDA: matrix factorization through latent Dirichlet allocation. In: ACM International Conference on Web Search and Data Mining, pp. 91–100 (2010)
2. Bahri, M., et al.: AutoML for stream k-nearest neighbors classification. In: IEEE International Conference on Big Data, pp. 597–602 (2020)
3. Barbieri, N., Manco, G.: An analysis of probabilistic methods for top-N recommendation in collaborative filtering. In: Joint European Conference on Machine Learning and Knowledge Discovery in Databases, pp. 172–187 (2011)
4. Bergstra, J., Bengio, Y.: Random search for hyper-parameter optimization. J. Mach. Learn. Res. **13**, 281–305 (2012)
5. Breiman, L.: Random forests. Mach. Learn. **45**(1), 5–32 (2001)
6. Carnein, M., et al.: confStream: automated algorithm selection and configuration of stream clustering algorithms. In: International Conference on Learning and Intelligent Optimization, pp. 80–95 (2020)
7. Carnein, M., et al.: Towards automated configuration of stream clustering algorithms. In: Joint European Conference on Machine Learning and Knowledge Discovery in Databases, pp. 137–143 (2019)

8. Celik, B., Vanschoren, J.: Adaptation strategies for automated machine learning on evolving data. In: arXiv preprint arXiv:2006.06480 (2020)
9. Feurer, M., Hutter, F.: Hyperparameter optimization. In: Automated Machine Learning, pp. 3–33 (2019)
10. Feurer, M., et al.: Auto-sklearn: efficient and robust automated machine learning. In: Automated Machine Learning, pp. 113–134 (2019)
11. Fitzgerald, T., et al.: Online search algorithm configuration. In: AAAI Conference on Artificial Intelligence, vol. 28 (2014)
12. Fitzgerald, T., et al.: ReACT: real-time algorithm configuration through tournaments. In: Annual Symposium on Combinatorial Search (2014)
13. Galuzzi, B.G., Giordani, I., Candelieri, A., Perego, R., Archetti, F.: Hyperparameter optimization for recommender systems through Bayesian optimization. CMS **17**(4), 495–515 (2020). https://doi.org/10.1007/s10287-020-00376-3
14. Gama, J., Sebastião, R., Rodrigues, P.P.: On evaluating stream learning algorithms. Mach. Learn. **90**(3), 317–346 (2013)
15. He, X., et al.: Neural collaborative filtering. In: International Conference on World Wide Web, pp. 173–182 (2017)
16. Hofmann, T.: Latent semantic models for collaborative filtering. ACM Trans. Inf. Syst. **22**(1), 89–115 (2004)
17. Hsu, C.-W., Chang, C.-C., Lin, C.-J.: A practical guide to support vector classification (2003)
18. Hutter, F., Hoos, H.H., Leyton-Brown, K.: Sequential model-based optimization for general algorithm configuration. In: International Conference on Learning and Intelligent Optimization, pp. 507–523 (2011)
19. Hutter, F., Kotthoff, L., Vanschoren, J.: Automated Machine Learning: Methods, Systems, Challenges. Springer Nature (2019). https://doi.org/10.1007/978-3-030-05318-5
20. Kabbur, S., Ning, X., Karypis, G.: FISM: factored item similarity models for top-N recommender systems. In: ACM SIGKDD International Conference on Knowledge Discovery and Data Mining, pp. 659–667 (2013)
21. Karimi, R., et al.: Non-myopic active learning for recommender systems based on matrix factorization. In: IEEE International Conference on Information Reuse Integration, pp. 299–303 (2011)
22. Kotthoff, L., et al.: Auto-WEKA: automatic model selection and hyperparameter optimization in WEKA. In: Automated Machine Learning, pp. 81–95 (2019)
23. Lawal, I.A., Abdulkarim, S.A.: Adaptive SVM for data stream classification. South Afr. Comput. J. **29**(1), 27–42 (2017)
24. Maclaurin, D., Duvenaud, D., Adams, R.: Gradient-based hyperparameter optimization through reversible learning. In: Proceedings of the 32nd International Conference on Machine Learning, vol. 37, pp. 2113–2122 (2015)
25. Madrid, J.G., et al.: Towards AutoML in the presence of drift: first results. In: arXiv preprint arXiv:1907.10772 (2019)
26. Nelder, J.A., Mead, R.: A simplex method for function minimization. Comput. J. **7**(4), 308–313 (1965)
27. Ning, X., Karypis, G.: SLIM: sparse linear methods for top-N recommender systems. In: IEEE International Conference on Data Mining, pp. 497–506 (2011)
28. Page, E.S.: Continuous inspection schemes. Biometrika **41**(1/2), 100–115 (1954)
29. Rendle, S., et al.: BPR: Bayesian personalized ranking from implicit feedback. In: Conference on Uncertainty in Artificial Intelligence, pp. 452–461 (2009)
30. Rendle, S., Zhang, L., Koren, Y.: On the difficulty of evaluating baselines: a study on recommender systems. In: arXiv preprint arXiv:2006.06480 (2019)

31. Rendle, S., et al.: Neural collaborative filtering vs. matrix factorization revisited. In: RecSys 2020. Virtual Event, Brazil: Association for Computing Machinery (2020)
32. van Rijn, J.N., Holmes, G., Pfahringer, B., Vanschoren, J.: Algorithm selection on data streams. In: Džeroski, S., Panov, P., Kocev, D., Todorovski, L. (eds.) DS 2014. LNCS (LNAI), vol. 8777, pp. 325–336. Springer, Cham (2014). https://doi.org/10.1007/978-3-319-11812-3_28
33. van Rijn, J.N., et al.: Having a blast: meta-learning and heterogeneous ensembles for data streams. In: IEEE International Conference on Data Mining. IEEE, pp. 1003–1008 (2015)
34. van Rijn, J.N., Holmes, G., Pfahringer, B., Vanschoren, J.: The online performance estimation framework: heterogeneous ensemble learning for data streams. Mach. Learn. **107**(1), 149–176 (2017). https://doi.org/10.1007/s10994-017-5686-9
35. Salakhutdinov, R., Mnih, A.: Probabilistic matrix factorization. In: Proceedings of the International Conference on Neural Information Processing Systems, pp. 1257–1264 (2008)
36. Veloso, B., Gama, J.: Self hyper-parameter tuning for stream classification algorithms. In: Gama, J., et al. (eds.) ITEM/IoT Streams -2020. CCIS, vol. 1325, pp. 3–13. Springer, Cham (2020). https://doi.org/10.1007/978-3-030-66770-2_1
37. Veloso, B., Gama, J., Malheiro, B.: Self hyper-parameter tuning for data streams. In: Soldatova, L., Vanschoren, J., Papadopoulos, G., Ceci, M. (eds.) DS 2018. LNCS (LNAI), vol. 11198, pp. 241–255. Springer, Cham (2018). https://doi.org/10.1007/978-3-030-01771-2_16
38. Veloso, B., et al.: Hyperparameter self-tuning for data streams. Inf. Fus. **76**, 75–86 (2021)
39. Veloso, B., Gama, J., Malheiro, B., Vinagre, J.: Self hyper-parameter tuning for stream recommendation algorithms. In: Monreale, A., et al. (eds.) ECML PKDD 2018. CCIS, vol. 967, pp. 91–102. Springer, Cham (2019). https://doi.org/10.1007/978-3-030-14880-5_8
40. Wang, C., Blei, D.: Collaborative topic modeling for recommending scientific articles. In: ACM SIGKDD International Conference on Knowledge Discovery and Data Mining, pp. 448–456 (2011)
41. Zhang, S., et al.: Deep learning based recommender system: a survey and new perspectives. In: ACM Comput. Surv. **52**(1), 1–38 (2019)

# Bayesian Optimization with a Prior for the Optimum

Artur Souza[1(✉)], Luigi Nardi[2,3], Leonardo B. Oliveira[1], Kunle Olukotun[3], Marius Lindauer[4], and Frank Hutter[5,6]

[1] Universidade Federal de Minas Gerais, Belo Horizonte, Brazil
{arturluis,leob}@dcc.ufmg.br
[2] Lund University, Lund, Sweden
luigi.nardi@cs.lth.se
[3] Stanford University, Stanford, USA
{lnardi,kunle}@stanford.edu
[4] Leibniz University Hannover, Hannover, Germany
lindauer@tnt.uni-hannover.de
[5] University of Freiburg, Freiburg im Breisgau, Germany
fh@cs.uni-freiburg.de
[6] Bosch Center for Artificial Intelligence, Renningen, Germany

**Abstract.** While Bayesian Optimization (BO) is a very popular method for optimizing expensive black-box functions, it fails to leverage the experience of domain experts. This causes BO to waste function evaluations on bad design choices (e.g., machine learning hyperparameters) that the expert already knows to work poorly. To address this issue, we introduce Bayesian Optimization with a Prior for the Optimum (BOPrO). BOPrO allows users to inject their knowledge into the optimization process in the form of priors about which parts of the input space will yield the best performance, rather than BO's standard priors over functions, which are much less intuitive for users. BOPrO then combines these priors with BO's standard probabilistic model to form a pseudo-posterior used to select which points to evaluate next. We show that BOPrO is around 6.67× faster than state-of-the-art methods on a common suite of benchmarks, and achieves a new state-of-the-art performance on a real-world hardware design application. We also show that BOPrO converges faster even if the priors for the optimum are not entirely accurate and that it robustly recovers from misleading priors.

## 1 Introduction

Bayesian Optimization (BO) is a data-efficient method for the joint optimization of design choices that has gained great popularity in recent years. It is impacting a wide range of areas, including hyperparameter optimization [10,41], AutoML [20], robotics [5], computer vision [30], Computer Go [6], hardware design [23,31], and many others. It promises greater automation so as to increase both product quality and human productivity. As a result, BO is also established in large tech companies, e.g., Google [13] and Facebook [1].


N. Oliver et al. (Eds.): ECML PKDD 2021, LNAI 12977, pp. 265–296, 2021.
https://doi.org/10.1007/978-3-030-86523-8_17

Nevertheless, domain experts often have substantial prior knowledge that standard BO cannot easily incorporate so far [44]. Users can incorporate prior knowledge by narrowing the search space; however, this type of hard prior can lead to poor performance by missing important regions. BO also supports a prior over functions $p(f)$, e.g., via a kernel function. However, this is not the prior domain experts have: they often know which ranges of hyperparameters tend to work best [36], and are able to specify a probability distribution $p_{\text{best}}(\boldsymbol{x})$ to quantify these priors; e.g., many users of the Adam optimizer [21] know that its best learning rate is often in the vicinity of $1 \times 10^{-3}$ (give or take an order of magnitude), yet may not know the accuracy they can achieve in a new application. Similarly, Clarke *et al.* [7] derived neural network hyperparameter priors for image datasets based on their experience. In these cases, users know potentially good values for a new application, but cannot be certain about them.

As a result, many competent users instead revert to manual search, which can fully incorporate their prior knowledge. A recent survey showed that most NeurIPS 2019 and ICLR 2020 papers reported having tuned hyperparameters used manual search, with only a very small fraction using BO [4]. In order for BO to be adopted widely, and accelerate progress in the ML community by tuning hyperparameters faster and better, it is, therefore, crucial to devise a method that fully incorporates expert knowledge about the location of the optimum areas into BO. In this paper, we introduce Bayesian Optimization with a Prior for the Optimum (BOPrO), a BO variant that combines priors for the optimum with a probabilistic model of the observations made. Our technical contributions are:

- We introduce *Bayesian Optimization with a Prior over the Optimum*, short *BOPrO*, which allows users to inject priors that were previously difficult to inject into BO, such as Gaussian, exponential, multimodal, and multivariate priors for the location of the optimum. To ensure robustness against misleading priors, BOPrO gives more importance to the data-driven model as iterations progress, gradually forgetting the prior.
- BOPrO's model bridges the gap between the well-established Tree-structured Parzen Estimator (TPE) methodology, which is based on Parzen kernel density estimators, and standard BO probabilistic models, such as Gaussian Processes (GPs). This is made possible by using the Probability of Improvement (PI) criterion to derive from BO's standard posterior over functions $p(f|(\boldsymbol{x}_i, y_i)_{i=1}^t)$ the probability of an input $\boldsymbol{x}$ leading to good function values.
- We demonstrate the effectiveness of BOPrO on a comprehensive set of synthetic benchmarks and real-world applications, showing that knowledge about the locality of an optimum helps BOPrO to achieve similar performance to current state-of-the-art on average 6.67× faster on synthetic benchmarks and 1.59× faster on a real-world application. BOPrO also achieves similar or better final performance on all benchmarks.

BOPrO is publicly available as part of the HyperMapper framework[1].

[1] https://github.com/luinardi/hypermapper/wiki/prior-injection.

# 2 Background

## 2.1 Bayesian Optimization

Bayesian Optimization (BO) is an approach for optimizing an unknown function $f : \mathcal{X} \rightarrow \mathbb{R}$ that is expensive to evaluate over an input space $\mathcal{X}$. In this paper, we aim to minimize $f$, i.e., find $\boldsymbol{x}^* \in \arg\min_{\boldsymbol{x} \in \mathcal{X}} f(\boldsymbol{x})$. BO approximates $\boldsymbol{x}^*$ with an optimal sequence of evaluations $\boldsymbol{x}_1, \boldsymbol{x}_2, \ldots \in \mathcal{X}$ that maximizes a utility metric, with each new $\boldsymbol{x}_{t+1}$ depending on the previous function values $y_1, y_2, \ldots, y_t$ at $\boldsymbol{x}_1, \ldots, \boldsymbol{x}_t$. BO achieves this by building a posterior on $f$ based on the set of evaluated points. At each iteration, a new point is selected and evaluated based on the posterior, and the posterior is updated to include the new point $(\boldsymbol{x}_{t+1}, y_{t+1})$.

The points explored by BO are dictated by the acquisition function, which attributes an utility to each $\boldsymbol{x} \in \mathcal{X}$ by balancing the predicted value and uncertainty of the prediction for each $\boldsymbol{x}$ [39]. In this work, as the acquisition function we choose Expected Improvement (EI) [29], which quantifies the expected improvement over the best function value found so far:

$$EI_{y_{inc}}(\boldsymbol{x}) := \int_{-\infty}^{\infty} \max(y_{inc} - y, 0) p(y|\boldsymbol{x}) dy, \tag{1}$$

where $y_{inc}$ is the incumbent function value, i.e., the best objective function value found so far, and $p(y|\boldsymbol{x})$ is given by a probabilistic model, e.g., a GP. Alternatives to EI would be Probability of Improvement (PI) [24], upper-confidence bounds (UCB) [42], entropy-based methods (e.g. [17]), and knowledge gradient [45].

## 2.2 Tree-Structured Parzen Estimator

The Tree-structured Parzen Estimator (TPE) method is a BO approach introduced by Bergstra *et al.* [3]. Whereas the standard probabilistic model in BO directly models $p(y|\boldsymbol{x})$, the TPE approach models $p(\boldsymbol{x}|y)$ and $p(y)$ instead.[2] This is done by constructing two parametric densities, $g(\boldsymbol{x})$ and $b(\boldsymbol{x})$, which are computed using configurations with function value below and above a given threshold $y^*$, respectively. The separating threshold $y^*$ is defined as a quantile $\gamma$ of the observed function values. Thus, TPE builds a density $g(\boldsymbol{x})$ using the best $\gamma$ configurations, and a density $b(\boldsymbol{x})$ using the remaining configurations. TPE uses the densities $g(\boldsymbol{x})$ and $b(\boldsymbol{x})$ to define $p(\boldsymbol{x}|y)$ as:

$$p(\boldsymbol{x}|y) = g(\boldsymbol{x}) I(y < y^*) + b(\boldsymbol{x})(1 - I(y < y^*)), \tag{2}$$

where $I(y < y^*)$ is 1 when $y < y^*$ and 0 otherwise. Bergstra *et al.* [3] show that the parametrization of the generative model $p(\boldsymbol{x}, y) = p(\boldsymbol{x}|y)p(y)$ facilitates the computation of EI as it leads to $EI_{y^*}(\boldsymbol{x}) \propto g(\boldsymbol{x})/b(\boldsymbol{x})$ and, thus, to maximize $EI$, they need to simply find the configurations that maximize the ratio $g(\boldsymbol{x})/b(\boldsymbol{x})$.

[2] Technically, the model does not parameterize $p(y)$, since it is computed based on the observed data points, which are heavily biased towards low values due to the optimization process. Instead, it parameterizes a dynamically changing $p_t(y)$, which helps to constantly challenge the model to yield better observations.

## 3 BO with a Prior for the Optimum

We now describe our BOPrO approach, which allows domain experts to inject prior knowledge about the locality of an optimum into the optimization. BOPrO combines this user-defined prior with a probabilistic model that captures the likelihood of the observed data $\mathcal{D}_t = (\boldsymbol{x}_i, y_i)_{i=1}^t$. BOPrO is independent of the probabilistic model being used; it can be freely combined with, e.g., Gaussian processes (GPs), random forests, or Bayesian NNs.

### 3.1 BOPrO Priors

BOPrO allows users to inject prior knowledge w.r.t. promising areas into BO. This is done via a prior distribution that informs where in the input space $\mathcal{X}$ we expect to find good $f(\boldsymbol{x})$ values. A point is considered "good" if it leads to low function values, and potentially to a global optimum. We denote the prior distribution $P_g(\boldsymbol{x})$, where $g$ denotes that this is a prior on good points and $\boldsymbol{x} \in \mathcal{X}$ is a given point. Examples of priors are shown in Figs. 2 and 3, additional examples of continuous and discrete priors are shown in Appendices A and D, respectively. Similarly, we define a prior on where in the input space we expect to have "bad" points. Although we could have a user-defined probability distribution $P_b(\boldsymbol{x})$, we aim to keep the decision-making load on users low and thus, for simplicity, only require the definition of $P_g(\boldsymbol{x})$ and compute $P_b(\boldsymbol{x}) = 1 - P_g(\boldsymbol{x})$.[3] $P_g(\boldsymbol{x})$ is normalized to $[0, 1]$ by min-max scaling before computing $P_b(\boldsymbol{x})$.

In practice, $\boldsymbol{x}$ contains several dimensions but it is difficult for domain experts to provide a joint prior distribution $P_g(\boldsymbol{x})$ for all of them. However, users can typically easily specify, e.g., sketch out, univariate or bivariate distributions for continuous dimensions or provide a list of probabilities for discrete dimensions. In BOPrO, users can define a complex multivariate distribution, but we expect the standard use case to be users choosing to specify univariate distributions, implicitly assuming a prior that factors as $P_g(\boldsymbol{x}) = \prod_{i=1}^{D} P_g(x_i)$, where $D$ is the number of dimensions in $\mathcal{X}$ and $x_i$ is the $i$-th input dimension of $\boldsymbol{x}$. To not assume unrealistically complex priors and to mimic our expected use case, we use factorized priors in our experiments; in Appendix E we show that these factorized priors can in fact lead to similar BO performance as multivariate priors.

### 3.2 Model

Whereas the standard probabilistic model in BO, e.g., a GP, quantifies $p(y|\boldsymbol{x})$ directly, that model is hard to combine with the prior $P_g(\boldsymbol{x})$. We therefore introduce a method to translate the standard probabilistic model $p(y|\boldsymbol{x})$ into a model that is easier to combine with this prior. Similar to the TPE work described

[3] We note that for continuous spaces, $P_b(\boldsymbol{x})$ is not a probability distribution as it does not integrate to 1 and therefore is only a pseudo-prior. For discrete spaces, we normalize $P_b(\boldsymbol{x})$ so that it sums to 1 and therefore is a proper distribution and prior.

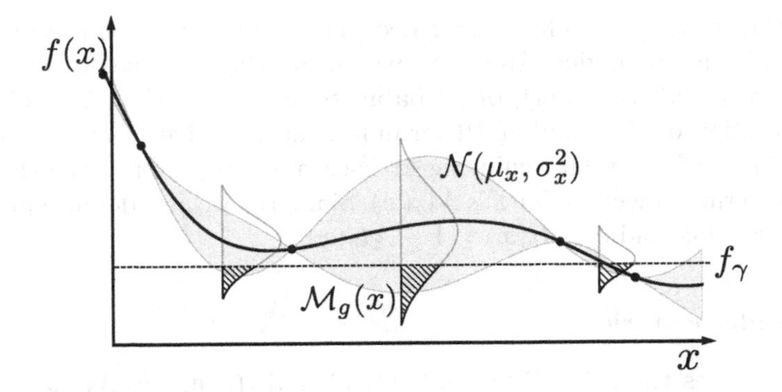


**Fig. 1.** Our model is composed by a probabilistic model and the probability of improving over the threshold $f_\gamma$, i.e., right tail of the Gaussian. The black curve is the probabilistic model's mean and the shaded area is the model's variance.

in Sect. 2.2, our generative model combines $p(\boldsymbol{x}|y)$ and $p(y)$ instead of directly modeling $p(y|\boldsymbol{x})$.

The computation we perform for this translation is to quantify the probability that a given input $\boldsymbol{x}$ is "good" under our standard probabilistic model $p(y|\boldsymbol{x})$. As in TPE, we define configurations as "good" if their observed $y$-value is below a certain quantile $\gamma$ of the observed function values (so that $p(y < f_\gamma) = \gamma$). We in addition exploit the fact that our standard probabilistic model $p(y|\boldsymbol{x})$ has a Gaussian form, and under this Gaussian prediction we can compute the probability $\mathcal{M}_g(\boldsymbol{x})$ of the function value lying below a certain quantile using the standard closed-form formula for PI [24]:

$$\mathcal{M}_g(\boldsymbol{x}) = p(f(\boldsymbol{x}) < f_\gamma|\boldsymbol{x}, \mathcal{D}_t) = \Phi\left(\frac{f_\gamma - \mu_{\boldsymbol{x}}}{\sigma_{\boldsymbol{x}}}\right), \tag{3}$$

where $\mathcal{D}_t = (\boldsymbol{x}_i, y_i)_{i=1}^t$ are the evaluated configurations, $\mu_{\boldsymbol{x}}$ and $\sigma_{\boldsymbol{x}}$ are the predictive mean and standard deviation of the probabilistic model at $\boldsymbol{x}$, and $\Phi$ is the standard normal CDF, see Fig. 1. Note that there are two probabilistic models here:

1. The standard probabilistic model of BO, with a structural prior over functions $p(f)$, updated by data $\mathcal{D}_t$ to yield a posterior over functions $p(f|\mathcal{D}_t)$, allowing us to quantify the probability $\mathcal{M}_g(\boldsymbol{x}) = p(f(x) < f_\gamma|\boldsymbol{x}, \mathcal{D}_t)$ in Eq. (3).[4]
2. The TPE-like generative model that combines $p(y)$ and $p(\boldsymbol{x}|y)$ instead of directly modeling $p(y|\boldsymbol{x})$.

Equation (3) bridges these two models by using the probability of improvement from BO's standard probabilistic model as the probability $\mathcal{M}_g(\boldsymbol{x})$ in TPE's

[4] We note that the structural prior $p(f)$ and the optimum-prior $P_g(\boldsymbol{x})$ provide orthogonal ways to input prior knowledge. $p(f)$ specifies our expectations about the structure and smoothness of the function, whereas $P_g(\boldsymbol{x})$ specifies knowledge about the location of the optimum..

model. Ultimately, this is a heuristic since there is no formal connection between the two probabilistic models. However, we believe that the use of BO's familiar, theoretically sound framework of probabilistic modeling of $p(y|x)$, followed by the computation of the familiar PI formula is an intuitive choice for obtaining the probability of an input achieving at least a given performance threshold – exactly the term we need for TPE's $\mathcal{M}_g(\boldsymbol{x})$. Similarly, we also define a probability $\mathcal{M}_b(\boldsymbol{x})$ of $\boldsymbol{x}$ being bad as $\mathcal{M}_b(\boldsymbol{x}) = 1 - \mathcal{M}_g(\boldsymbol{x})$.

### 3.3 Pseudo-posterior

BOPrO combines the prior $P_g(\boldsymbol{x})$ in Sect. 3.1 and the model $\mathcal{M}_g(\boldsymbol{x})$ in Eq. (3) into a pseudo-posterior on "good" points. This pseudo-posterior represents the updated beliefs on where we can find good points, based on the prior and data that has been observed. The pseudo-posterior is computed as the product:

$$g(\boldsymbol{x}) \propto P_g(\boldsymbol{x})\mathcal{M}_g(\boldsymbol{x})^{\frac{t}{\beta}}, \tag{4}$$

where $t$ is the current optimization iteration, $\beta$ is an optimization hyperparameter, $\mathcal{M}_g(\boldsymbol{x})$ is defined in Eq. (3), and $P_g(\boldsymbol{x})$ is the prior defined in Sect. 3.1, rescaled to [0, 1] using min-max scaling. We note that this pseudo-posterior is not normalized, but this suffices for BOPrO to determine the next $\boldsymbol{x}_t$ as the normalization constant cancels out (c.f. Sect. 3.5). Since $g(\boldsymbol{x})$ is not normalized and we include the exponent $t/\beta$ in Eq. (4), we refer to $g(\boldsymbol{x})$ as a pseudo-posterior, to emphasize that it is not a standard posterior probability distribution.

The $t/\beta$ fraction in Eq. (4) controls how much weight is given to $\mathcal{M}_g(\boldsymbol{x})$. As the optimization progresses, more weight is given to $\mathcal{M}_g(\boldsymbol{x})$ over $P_g(\boldsymbol{x})$. Intuitively, we put more emphasis on $\mathcal{M}_g(\boldsymbol{x})$ as it observes more data and becomes more accurate. We do this under the assumption that the model $\mathcal{M}_g(\boldsymbol{x})$ will eventually be better than the user at predicting where to find good points. This also allows to recover from misleading priors as we show in Sect. 4.1; similar to, and inspired by Bayesian models, the data ultimately washes out the prior. The $\beta$ hyperparameter defines the balance between prior and model, with higher $\beta$ values giving more importance to the prior.

We note that directly computing Eq. (4) can lead to numerical issues. Namely, the pseudo-posterior can reach extremely low values if the $P_g(\boldsymbol{x})$ and $\mathcal{M}_g(\boldsymbol{x})$ probabilities are low, especially as $t/\beta$ grows. To prevent this, in practice, BOPrO uses the logarithm of the pseudo-posterior instead:

$$\log(g(\boldsymbol{x})) \propto \log(P_g(\boldsymbol{x})) + \tfrac{t}{\beta} \cdot \log(\mathcal{M}_g(\boldsymbol{x})). \tag{5}$$

Once again, we also define an analogous pseudo-posterior distribution on bad $\boldsymbol{x}$: $b(\boldsymbol{x}) \propto P_b(\boldsymbol{x})\mathcal{M}_b(\boldsymbol{x})^{\frac{t}{\beta}}$. We then use these quantities to define $p(\boldsymbol{x}|y)$ as follows:

$$p(\boldsymbol{x}|y) \propto \begin{cases} g(\boldsymbol{x}) & \text{if } y < f_\gamma \\ b(\boldsymbol{x}) & \text{if } y \geq f_\gamma. \end{cases} \tag{6}$$

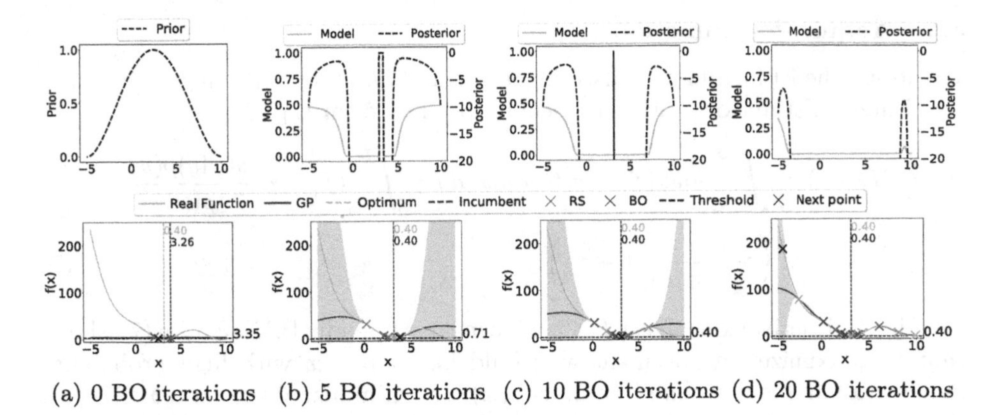


(a) 0 BO iterations (b) 5 BO iterations (c) 10 BO iterations (d) 20 BO iterations

**Fig. 2.** Breakdown of the prior $P_g(\boldsymbol{x}) = \mathcal{B}(3,3)$, the model $M_g(\boldsymbol{x})$, and the pseudo-posterior $g(\boldsymbol{x})$ (top row) on the 1D-Branin function (bottom row) and their evolution over the optimization iterations. In early iterations, the pseudo-posterior is high around the optimum, where both prior and model agree there are good points. In later iterations, it vanishes where the model is certain there will be no improvement and is high where there is uncertainty in the model. (Color figure online)

### 3.4 Model and Pseudo-posterior Visualization

We visualize the prior $P_g(\boldsymbol{x})$, the model $M_g(\boldsymbol{x})$, and the pseudo-posterior $g(\boldsymbol{x})$ and their evolution over the optimization iterations for a 1D-Branin function. We define the 1D-Branin by setting the second dimension of the function to the global optimum $x_2 = 2.275$ and optimizing the first dimension. We use a Beta distribution prior $P_g(\boldsymbol{x}) = \mathcal{B}(3,3)$, which resembles a truncated Gaussian centered close to the global optimum, and a GP as predictive model. We perform an initial design of $D + 1 = 2$ random points sampled from the prior and then run BOPrO for 20 iterations.

Figure 2 shows the optimization at different stages. Red crosses denote the initial design and blue/green crosses denote BOPrO samples, with green samples denoting later iterations. Figure 2a shows the initialization phase (bottom) and the Beta prior (top). After 5 iterations, in Fig. 2b (top), the pseudo-posterior is high near the global minimum, around $x = \pi$, where both the prior $P_g(\boldsymbol{x})$ and the model $\mathcal{M}_g(\boldsymbol{x})$ agree there are good points. After 10 iterations in Fig. 2c (top), there are three regions with high pseudo-posterior. The middle region, where BOPrO is exploiting until the optimum is found, and two regions to the right and left, which will lead to future exploration as shown in Fig. 2d (bottom) on the right and left of the global optimum in light green crosses. After 20 iterations, Fig. 2d (top), the pseudo-posterior vanishes where the model $\mathcal{M}_g(\boldsymbol{x})$ is certain there will be no improvement, but it is high where there is uncertainty in the GP.

### 3.5 Acquisition Function

We adopt the EI formulation used in Bergstra *et al.* [3] by replacing their Adaptive Parzen Estimators with our pseudo-posterior from Eq. (4), i.e.:

$$EI_{f_\gamma}(\boldsymbol{x}) := \int_{-\infty}^{\infty} \max(f_\gamma - y, 0) p(y|\boldsymbol{x}) dy = \int_{-\infty}^{f_\gamma} (f_\gamma - y) \frac{p(\boldsymbol{x}|y)p(y)}{p(\boldsymbol{x})} dy$$
$$\propto \left(\gamma + \frac{b(\boldsymbol{x})}{g(\boldsymbol{x})}(1-\gamma)\right)^{-1}. \quad (7)$$

The full derivation of Eq. (7) is shown in Appendix B. Equation (7) shows that to maximize improvement we would like points $\boldsymbol{x}$ with high probability under $g(\boldsymbol{x})$ and low probability under $b(\boldsymbol{x})$, i.e., minimizing the ratio $b(\boldsymbol{x})/g(\boldsymbol{x})$. We note that the point that minimizes the ratio for our unnormalized pseudo-posteriors will be the same that minimizes the ratio for the normalized pseudo-posterior and, thus, computing the normalized pseudo-posteriors is unnecessary.

The dynamics of the BOPrO algorithm can be understood in terms of the following proposition (proof in Appendix B):

**Proposition 1.** *Given $f_\gamma$, $P_g(\boldsymbol{x})$, $P_b(\boldsymbol{x})$, $\mathcal{M}_g(\boldsymbol{x})$, $\mathcal{M}_b(\boldsymbol{x})$, $g(\boldsymbol{x})$, $b(\boldsymbol{x})$, $p(\boldsymbol{x}|y)$, and $\beta$ as above, then*

$$\lim_{t\to\infty} \arg\max_{\boldsymbol{x}\in\mathcal{X}} EI_{f_\gamma}(\boldsymbol{x}) = \lim_{t\to\infty} \arg\max_{\boldsymbol{x}\in\mathcal{X}} \mathcal{M}_g(\boldsymbol{x}),$$

*where $EI_{f_\gamma}$ is the Expected Improvement acquisition function as defined in Eq. (7) and $\mathcal{M}_g(\boldsymbol{x})$ is as defined in Eq. (3).*

In early BO iterations the prior for the optimum will have a predominant role, but in later BO iterations the model will grow more important, and as Proposition 1 shows, if BOPrO is run long enough the prior washes out and BOPrO *only* trusts the model $\mathcal{M}_g(\boldsymbol{x})$ informed by the data. Since $\mathcal{M}_g(\boldsymbol{x})$ is the Probability of Improvement (PI) on the probabilistic model $p(y|\boldsymbol{x})$ then, in the limit, maximizing $EI_{f_\gamma}(\boldsymbol{x})$ is equivalent to maximizing the PI acquisition function on the probabilistic model $p(y|\boldsymbol{x})$. In other words, for high values of $t$, BOPrO converges to standard BO with a PI acquisition function.

### 3.6 Putting It All Together

Algorithm 1 shows the BOPrO algorithm. In Line 3, BOPrO starts with a design of experiments (DoE) phase, where it randomly samples a number of points from the user-defined prior $P_g(\boldsymbol{x})$. After initialization, the BO loop starts at Line 4. In each loop iteration, BOPrO fits the models $\mathcal{M}_g(\boldsymbol{x})$ and $\mathcal{M}_b(\boldsymbol{x})$ on the previously evaluated points (Lines 5 and 6) and computes the pseudo-posteriors $g(\boldsymbol{x})$ and $b(\boldsymbol{x})$ (Lines 7 and 8 respectively). The EI acquisition function is computed next, using the pseudo-posteriors, and the point that maximizes EI is selected as the next point to evaluate at Line 9. The black-box function evaluation is performed at Line 10. This BO loop is repeated for a predefined number of iterations, according to the user-defined budget $B$.

**Algorithm 1.** BOPrO Algorithm. $\mathcal{D}_t$ keeps track of all function evaluations so far: $(\boldsymbol{x}_i, y_i)_{i=1}^t$.

1: **Input:** Input space $\mathcal{X}$, user-defined prior distribution $P_g(\boldsymbol{x})$, quantile $\gamma$, initial model weight $\beta$, and BO budget $B$.
2: **Output:** Optimized point $\boldsymbol{x}_{inc}$.
3: $\mathcal{D}_1 \leftarrow initialize(\mathcal{X})$
4: **for** $t = 1$ **to** $B$ **do**
5: $\quad \mathcal{M}_g(\boldsymbol{x}) \leftarrow fit\_model\_good(\mathcal{D}_t)$ ▷ see Eq. (3)
6: $\quad \mathcal{M}_b(\boldsymbol{x}) \leftarrow fit\_model\_bad(\mathcal{D}_t)$
7: $\quad g(\boldsymbol{x}) \leftarrow P_g(\boldsymbol{x}) \cdot \mathcal{M}_g(\boldsymbol{x})^{\frac{t}{\beta}}$ ▷ see Eq. (4)
8: $\quad b(\boldsymbol{x}) \leftarrow P_b(\boldsymbol{x}) \cdot \mathcal{M}_b(\boldsymbol{x})^{\frac{t}{\beta}}$
9: $\quad \boldsymbol{x}_t \in \arg\max_{\boldsymbol{x} \in \mathcal{X}} EI_{f_\gamma}(\boldsymbol{x})$ ▷ see Eq. (5)
10: $\quad y_t \leftarrow f(\boldsymbol{x}_t)$
11: $\quad \mathcal{D}_{t+1} \leftarrow \mathcal{D}_t \cup (\boldsymbol{x}_t, y_t)$
12: **end for**
13: $\boldsymbol{x}_{inc} \leftarrow compute\_best(\mathcal{D}_{t+1})$
14: **return** $\boldsymbol{x}_{inc}$

## 4 Experiments

We implement both Gaussian processes (GPs) and random forests (RFs) as predictive models and use GPs in all experiments, except for our real-world experiments (Sect. 4.3), where we use RFs for a fair comparison. We set the model weight $\beta = 10$ and the model quantile to $\gamma = 0.05$, see our sensitivity hyperparameter study in Appendices I and J. We initialize BOPrO with $D + 1$ random samples from the prior. We optimize our EI acquisition function using a combination of multi-start local search [19] and CMA-ES [15]. We consider four synthetic benchmarks: Branin, SVM, FC-Net, and XGBoost, which are 2, 2, 6, and 8 dimensional, respectively. The last three are part of the Profet benchmarks [22], generated by a generative model built using performance data on OpenML or UCI datasets. See Appendix C for more details.

### 4.1 Prior Forgetting

We first show that BOPrO can recover from a misleading prior, thanks to our model $\mathcal{M}_g(\boldsymbol{x})$ and the $t/\beta$ parameter in Eq. (4). As BO progresses, $\mathcal{M}_g(\boldsymbol{x})$ becomes more accurate and receives more weight, guiding optimization away from the wrong prior and towards better function values.

Figure 3 shows BOPrO on the 1D Branin function with an exponential prior. Columns (b), (c), and (d) show BOPrO after $D + 1 = 2$ initial samples and 0, 10, 20 BO iterations, respectively. After initialization, as shown in Column (b), the pseudo-posterior is nearly identical to the exponential prior and guides BOPrO towards the region of the space on the right, which is towards the local optimum. This happens until the model $\mathcal{M}_g(\boldsymbol{x})$ becomes certain there will be no more improvement from sampling that region (Columns (c) and (d)). After that, $\mathcal{M}_g(\boldsymbol{x})$ guides the pseudo-posterior towards exploring regions with high

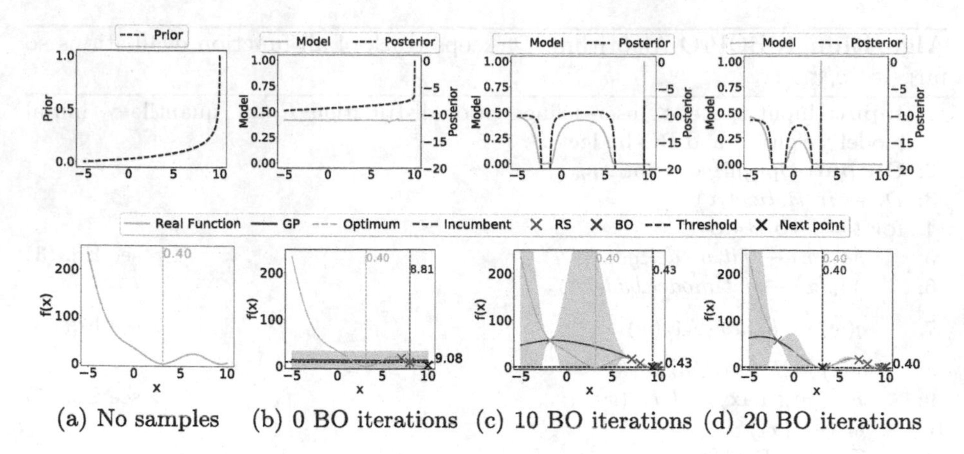


(a) No samples (b) 0 BO iterations (c) 10 BO iterations (d) 20 BO iterations

**Fig. 3.** BOPrO on the 1D Branin function. The leftmost column shows the exponential prior. The other columns show the model and the log pseudo-posterior after 0 (initialization only), 10, and 20 BO iterations. BOPrO forgets the wrong prior on the local optimum and converges to the global optimum.

uncertainty. Once the global minimum region is found, the pseudo-posterior starts balancing exploiting the global minimum and exploring regions with high uncertainty, as shown in Fig. 3d (bottom). Notably, the pseudo-posterior after $x > 4$ falls to 0 in Fig. 3d (top), as the model $\mathcal{M}_g(\boldsymbol{x})$ is certain there will be no improvement from sampling the region of the local optimum. We provide additional examples of forgetting in Appendix A, and a comparison of BOPrO with misleading priors, no prior, and correct priors in Appendix F.

## 4.2 Comparison Against Strong Baselines

We build two priors for the optimum in a controlled way and evaluate BOPrO's performance with these different prior strengths. We emphasize that in practice, manual priors would be based on the domain experts' expertise on their applications; here, we only use artificial priors to guarantee that our prior is not biased by our own expertise for the benchmarks we used. In practice, users will manually define these priors like in our real-world experiments (Sect. 4.3).

Our synthetic priors take the form of Gaussian distributions centered near the optimum. For each input $x \in \mathcal{X}$, we inject a prior of the form $\mathcal{N}(\mu_x, \sigma_x^2)$, where $\mu_x$ is sampled from a Gaussian centered at the optimum value $x_{opt}$[5] for that parameter $\mu_x \sim \mathcal{N}(x_{opt}, \sigma_x^2)$, and $\sigma_x$ is a hyperparameter of our experimental setup determining the prior's strength. For each run of BOPrO, we sample new $\mu_x$'s. This setup provides us with a synthetic prior that is close to the optimum, but not exactly centered at it, and allows us to control the strength of the prior by $\sigma_x$. We use two prior strengths in our experiments: a strong prior, computed with $\sigma_x = 0.01$, and a weak prior, computed with $\sigma_x = 0.1$.

[5] If the optimum for a benchmark is not known, we approximate it using the best value found during previous BO experiments.

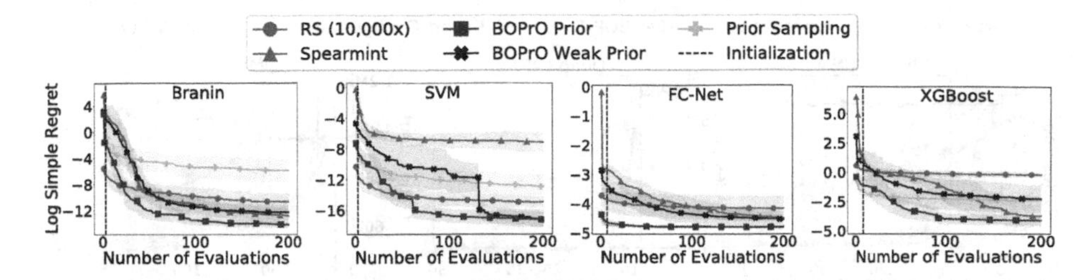


**Fig. 4.** Log regret comparison of BOPrO with weak and strong priors, sampling from the strong prior, 10,000× random search (RS), and Spearmint (mean +/− one std on 30 repetitions). We run each benchmark for 200 iterations.

Figure 4 compares BOPrO to other optimizers using the log simple regret on 30 runs on the synthetic benchmarks. We compare the results of BOPrO with weak and strong priors to 10,000× random search (RS, i.e., for each BO sample we draw 10,000 uniform random samples), sampling from the strong prior only, and Spearmint [41], a well-adopted BO approach using GPs and EI. In Appendix G, we also compare BOPrO to TPE, SMAC, and TuRBO [9,19,26]. Also, in Appendix H, we compare BOPrO to other baselines with the same prior initialization and show that the performance of the baselines remains similar.

BOPrO with a strong prior for the optimum beats 10,000× RS and BOPrO with a weak prior on all benchmarks. It also outperforms the performance of sampling from the strong prior; this is expected because the prior sampling cannot focus on the real location of the optimum. Both methods are identical in the initialization phase because they both sample from the prior in that phase.

BOPrO with a strong prior is also more sample efficient and finds better or similar results to Spearmint on all benchmarks. Importantly, in all our experiments, BOPrO with a good prior consistently shows tremendous speedups in the early phases of the optimization process, requiring on average only 15 iterations to reach the performance that Spearmint reaches after 100 iterations (6.67× faster). Thus, BOPrO makes use of the best of both worlds, leveraging prior knowledge and efficient optimization based on BO.

### 4.3 The Spatial Use-Case

We next apply BOPrO to the `Spatial` [23] real-world application. `Spatial` is a programming language and corresponding compiler for the design of application accelerators, i.e., FPGAs. We apply BOPrO to three `Spatial` benchmarks, namely, 7D shallow and deep CNNs, and a 10D molecular dynamics grid application (MD Grid). We compare the performance of BOPrO to RS, manual optimization, and HyperMapper [31], the current state-of-the-art BO solution for `Spatial`. For a fair comparison between BOPrO and HyperMapper, since HyperMapper uses RFs as its surrogate model, here, we also use RFs in BOPrO. The manual optimization and the prior for BOPrO were provided by an unbiased `Spatial` developer, who is not an author of this paper. The priors

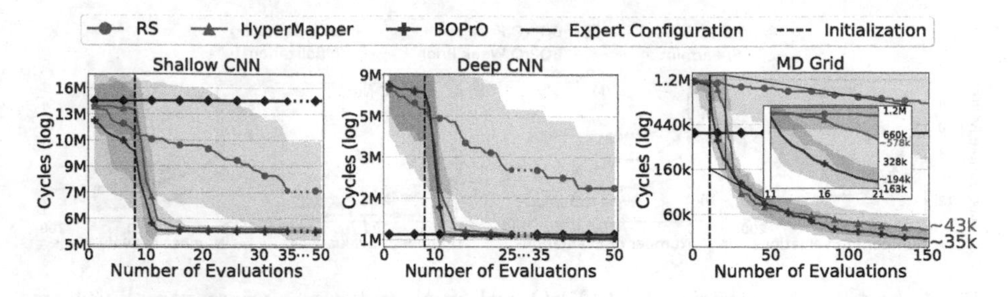


**Fig. 5.** Log regret comparison of random search (RS), HyperMapper, BOPrO, and manual optimization on Spatial. The line and shaded regions show mean and stdev after 30 repetitions. Vertical lines denote the end of the initialization phase.

were provided once and kept unchanged for the whole project. More details on the setup, including the priors used, are presented in Appendix D.

Figure 5 shows the log regret on the `Spatial` benchmarks. BOPrO vastly outperforms RS in all benchmarks. BOPrO is also able to leverage the expert's prior and outperforms the expert in all benchmarks (2.73×, 1.05×, and 10.41× speedup for shallow CNN, deep CNN, and MD Grid, respectively). In the MD Grid benchmark, BOPrO achieves better performance than HyperMapper in the early stages of optimization (2.98× speedup after the first 10 iterations, see the plot inset), and achieves better final performance (1.21× speedup). For context, this is a significant improvement in the FPGA field, where a 10% improvement could qualify for acceptance in a top-tier conference. In the CNN benchmarks, BOPrO converges to the minima regions faster than HyperMapper (1.18× and 2× faster for shallow and deep, respectively). Thus, BOPrO leverages both the expert's prior knowledge and BO to provide a new state of the art for `Spatial`.

# 5 Related Work

TPE by Bergstra *et al.* [3], the default optimizer in the HyperOpt package [2], supports limited hand-designed priors in the form of normal or log-normal distributions. We make three technical contributions that make BOPrO more flexible than TPE. First, BOPrO allow more flexible priors; second, BOPrO is agnostic to the probabilistic model used, allowing the use of more sample-efficient models (e.g., we use GPs and RFs in our experiments); and third, inspired by Bayesian models, BOPrO gives more importance to the data as iterations progress. We also show that BOPrO outperforms HyperOpt's TPE in Appendix G.

In parallel work, Li *et al.* [25] allow users to specify priors via a probability distribution. Their two-level approach samples a number of configurations by maximizing samples from a GP posterior and then chooses the configuration with the highest prior as the next to evaluate. In contrast, BOPrO leverages the information from the prior more directly; is agnostic to the probabilistic model used, which is important for applications with many discrete variables like our real-world application, where RFs outperform GPs; and provably recovers from misspecified priors, while in their approach the prior never gets washed out.

The work of Ramachandran *et al.* [37] also supports priors in the form of probability distributions. Their work uses the probability integral transform to warp the search space, stretching regions where the prior has high probability, and shrinking others. Once again, compared to their approach, BOPrO is agnostic to the probabilistic model used and directly controls the balance between prior and model via the $\beta$ hyperparameter. Additionally, BOPrO's probabilistic model is fitted independently from the prior, which ensures it is not biased by the prior, while their approach fits the model to a warped version of the space, transformed by the prior, making it difficult to recover from misleading priors.

The optimization tools SMAC [19] and iRace [28] also support simple hand-designed priors, e.g. log-transformations. However, these are not properly reflected in the predictive models and both cannot explicitly recover from bad priors.

Oh *et al.* [33] and Siivola *et al.* [40] propose structural priors for high-dimensional problems. They assume that users always center the search space at regions they expect to be good and then develop BO approaches that favor configurations near the center. However, this is a rigid assumption about optimum locality, which does not allow users to freely specify their priors. Similarly, Shahriari *et al.* [38] focus on unbounded search spaces. The priors in their work are not about good regions of the space, but rather a regularizer that penalizes configurations based on their distance to the user-defined search space. The priors are automatically derived from the search space and not provided by users.

Our work also relates to meta-learning for BO [43], where BO is applied to many similar optimization problems in a sequence such that knowledge about the general problem structure can be exploited in future optimization problems. In contrast to meta-learning, BOPrO allows human experts to explicitly specify their priors. Furthermore, BOPrO does not depend on meta-features [11] and incorporates the human's prior instead of information from different experiments [27].

## 6 Conclusions and Future Work

We have proposed a BO variant, BOPrO, that allows users to inject their expert knowledge into the optimization in the form of priors about which parts of the input space will yield the best performance. These are different than standard priors over functions which are much less intuitive for users. So far, BO failed to leverage the experience of domain experts, not only causing inefficiency but also driving users away from adopting BO because they could not exploit their years of knowledge in optimizing their black-box functions. BOPrO addresses this issue and we therefore expect it to facilitate the adoption of BO. We showed that BOPrO is 6.67× more sample efficient than strong BO baselines, and 10,000× faster than random search, on a common suite of benchmarks and achieves a new state-of-the-art performance on a real-world hardware design application. We also showed that BOPrO converges faster and robustly recovers from misleading priors. In future work, we will study how our approach can be used to leverage prior knowledge from meta-learning. Bringing these two worlds together will likely boost the performance of BO even further.

**Acknowledgments.** We thank Matthew Feldman for `Spatial` support. Luigi Nardi and Kunle Olukotun were supported in part by affiliate members and other supporters of the Stanford DAWN project—Ant Financial, Facebook, Google, Intel, Microsoft, NEC, SAP, Teradata, and VMware. Luigi Nardi was also partially supported by the Wallenberg AI, Autonomous Systems and Software Program (WASP) funded by the Knut and Alice Wallenberg Foundation. Artur Souza and Leonardo B. Oliveira were supported by CAPES, CNPq, and FAPEMIG. Frank Hutter acknowledges support by the European Research Council (ERC) under the European Union Horizon 2020 research and innovation programme through grant no. 716721. The computations were also enabled by resources provided by the Swedish National Infrastructure for Computing (SNIC) at LUNARC partially funded by the Swedish Research Council through grant agreement no. 2018-05973.

# A Prior Forgetting Supplementary Experiments

In this section, we show additional evidence that BOPrO can recover from wrongly defined priors so to complement Sect. 4.1. Figure 6 shows BOPrO on the 1D Branin function as in Fig. 3 but with a decay prior. Column (a) of Fig. 6 shows the decay prior and the 1D Branin function. This prior emphasizes the wrong belief that the optimum is likely located on the left side around x = −5 while the optimum is located at the orange dashed line. Columns (b), (c), and (d) of Fig. 6 show BOPrO on the 1D Branin after $D + 1 = 2$ initial samples and 0, 10, and 20 BO iterations, respectively. In the beginning of BO, as shown in column (b), the pseudo-posterior is nearly identical to the prior and guides

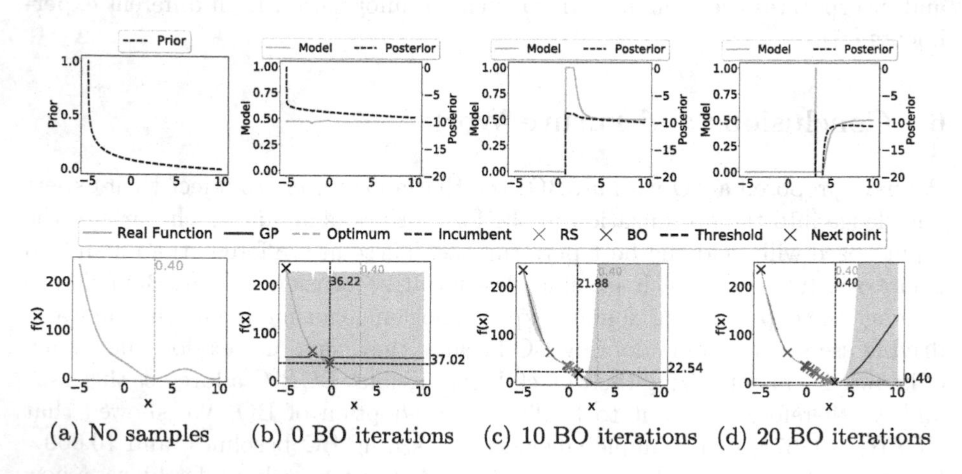


**Fig. 6.** BOPrO on the 1D Branin function with a decay prior. The leftmost column shows the log pseudo-posterior before any samples are evaluated, in this case, the pseudo-posterior is equal to the decay prior. The other columns show the model and pseudo-posterior after 0 (only random samples), 10, and 20 BO iterations. 2 random samples are used to initialize the GP model.

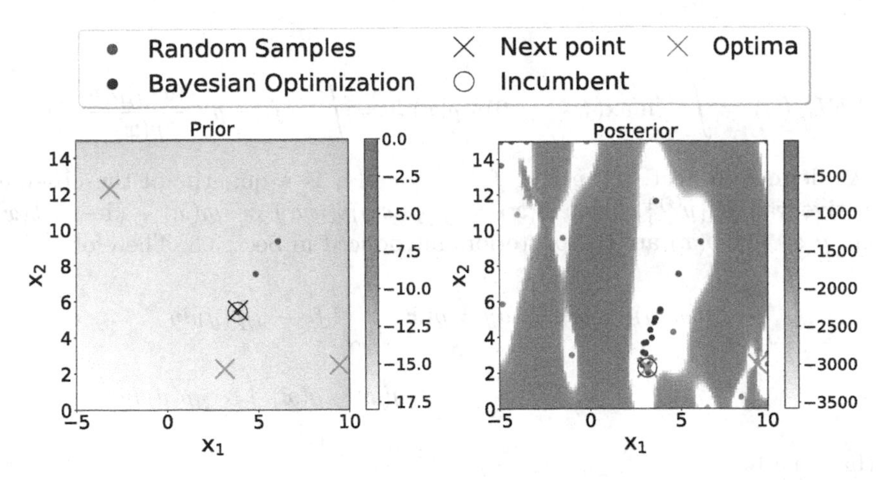


**Fig. 7.** BOPrO on the Branin function with exponential priors for both dimensions. (a) shows the log pseudo-posterior before any samples are evaluated, in this case, the pseudo-posterior is equal to the prior; the green crosses are the optima. (b) shows the result of optimization after 3 initialization samples drawn from the prior at random and 50 BO iterations. The dots in (b) show the points explored by BOPrO, with greener points denoting later iterations. The colored heatmap shows the log of the pseudo-posterior $g(\boldsymbol{x})$. (Color figure online)

BOPrO towards the left region of the space. As more points are sampled, the model becomes more accurate and starts guiding the pseudo-posterior away from the wrong prior (column (c)). Notably, the pseudo-posterior before x = 0 falls to 0, as the predictive model is certain there will be no improvement from sampling this region. After 20 iterations, BOPrO finds the optimum region, despite the poor start (column (d)). The peak in the pseudo-posterior in column (d) shows BOPrO will continue to exploit the optimum region as it is not certain if the exact optimum has been found. The pseudo-posterior is also high in the high uncertainty region after $x = 4$, showing BOPrO will explore that region after it finds the optimum.

Figure 7 shows BOPrO on the standard 2D Branin function. We use exponential priors for both dimensions, which guides optimization towards a region with only poor performing high function values. Figure 7a shows the prior and Fig. 7b shows optimization results after $D + 1 = 3$ initialization samples and 50 BO iterations. Note that, once again, optimization begins near the region incentivized by the prior, but moves away from the prior and towards the optima as BO progresses. After 50 BO iterations, BOPrO finds all three optima regions of the Branin.

# B Mathematical Derivations

## B.1 EI Derivation

Here, we provide a full derivation of Eq. (7):

$$EI_{f_\gamma}(\boldsymbol{x}) := \int_{-\infty}^{\infty} \max(f_\gamma - y, 0) p(y|\boldsymbol{x}) dy = \int_{-\infty}^{f_\gamma} (f_\gamma - y) \frac{p(\boldsymbol{x}|y)p(y)}{p(\boldsymbol{x})} dy.$$

As defined in Sect. 3.2, $p(y < f_\gamma) = \gamma$ and $\gamma$ is a quantile of the observed objective values $\{y^{(i)}\}$. Then $p(\boldsymbol{x}) = \int_{\mathbb{R}} p(\boldsymbol{x}|y)p(y)dy = \gamma g(\boldsymbol{x}) + (1-\gamma)b(\boldsymbol{x})$, where $g(\boldsymbol{x})$ and $b(\boldsymbol{x})$ are the posteriors introduced in Sect. 3.3. Therefore

$$\begin{aligned} \int_{-\infty}^{f_\gamma} (f_\gamma - y) p(\boldsymbol{x}|y) p(y) dy &= g(\boldsymbol{x}) \int_{-\infty}^{f_\gamma} (f_\gamma - y) p(y) dy \\ &= \gamma f_\gamma g(\boldsymbol{x}) - g(\boldsymbol{x}) \int_{-\infty}^{f_\gamma} y p(y) dy, \end{aligned} \tag{8}$$

so that finally

$$EI_{f_\gamma}(\boldsymbol{x}) = \frac{\gamma f_\gamma g(\boldsymbol{x}) - g(\boldsymbol{x}) \int_{-\infty}^{f_\gamma} y p(y) dy}{\gamma g(\boldsymbol{x}) + (1-\gamma) b(\boldsymbol{x})} \propto \left( \gamma + \frac{b(\boldsymbol{x})}{g(\boldsymbol{x})}(1-\gamma) \right)^{-1}. \tag{9}$$

### B.2 Proof of Proposition 1

Here, we provide the proof of Proposition 1:

$$\lim_{t\to\infty} \arg\max_{\boldsymbol{x}\in\mathcal{X}} EI_{f_\gamma}(\boldsymbol{x}) \tag{10}$$

$$= \lim_{t\to\infty} \arg\max_{\boldsymbol{x}\in\mathcal{X}} \int_{-\infty}^{f_\gamma} (f_\gamma - y) p(\boldsymbol{x}|y) p(y) dy \tag{11}$$

$$= \lim_{t\to\infty} \arg\max_{\boldsymbol{x}\in\mathcal{X}} g(\boldsymbol{x}) \int_{-\infty}^{f_\gamma} (f_\gamma - y) p(y) dy \tag{12}$$

$$= \lim_{t\to\infty} \arg\max_{\boldsymbol{x}\in\mathcal{X}} \left( \gamma f_\gamma g(\boldsymbol{x}) - g(\boldsymbol{x}) \int_{-\infty}^{f_\gamma} y p(y) dy \right) \tag{13}$$

$$= \lim_{t\to\infty} \arg\max_{\boldsymbol{x}\in\mathcal{X}} \frac{\gamma f_\gamma g(\boldsymbol{x}) - g(\boldsymbol{x}) \int_{-\infty}^{f_\gamma} y p(y) dy}{\gamma g(\boldsymbol{x}) + (1-\gamma) b(\boldsymbol{x})} \tag{14}$$

which, from Eq. (9), is equal to:

$$= \lim_{t\to\infty} \arg\max_{\boldsymbol{x}\in\mathcal{X}} \left( \gamma + \frac{b(\boldsymbol{x})}{g(\boldsymbol{x})}(1-\gamma) \right)^{-1} \tag{15}$$

we can take Eq. (15) to the power of $\frac{1}{t}$ without changing the expression, since the argument that maximizes EI does not change:

$$= \lim_{t\to\infty} \arg\max_{\boldsymbol{x}\in\mathcal{X}} \left( \gamma + \frac{b(\boldsymbol{x})}{g(\boldsymbol{x})}(1-\gamma) \right)^{-\frac{1}{t}} \tag{16}$$

substituting $g(x)$ and $b(x)$ using their definitions in Sect. 3.3:

$$= \lim_{t \to \infty} \arg\max_{\boldsymbol{x} \in \mathcal{X}} \left( \gamma + \frac{P_b(\boldsymbol{x}) \mathcal{M}_b(\boldsymbol{x})^{\frac{t}{\beta}}}{P_g(\boldsymbol{x}) \mathcal{M}_g(\boldsymbol{x})^{\frac{t}{\beta}}} (1-\gamma) \right)^{-\frac{1}{t}} \tag{17}$$

$$= \lim_{t \to \infty} \arg\max_{\boldsymbol{x} \in \mathcal{X}} \left( \frac{P_b(\boldsymbol{x}) \mathcal{M}_b(\boldsymbol{x})^{\frac{t}{\beta}}}{P_g(\boldsymbol{x}) \mathcal{M}_g(\boldsymbol{x})^{\frac{t}{\beta}}} (1-\gamma) \right)^{-\frac{1}{t}} \tag{18}$$

$$= \lim_{t \to \infty} \arg\max_{\boldsymbol{x} \in \mathcal{X}} \left( \frac{P_b(\boldsymbol{x})}{P_g(\boldsymbol{x})} \right)^{-\frac{1}{t}} \left( \frac{\mathcal{M}_b(\boldsymbol{x})^{\frac{t}{\beta}}}{\mathcal{M}_g(\boldsymbol{x})^{\frac{t}{\beta}}} \right)^{-\frac{1}{t}} (1-\gamma)^{-\frac{1}{t}} \tag{19}$$

$$= \lim_{t \to \infty} \arg\max_{\boldsymbol{x} \in \mathcal{X}} \left( \frac{P_b(\boldsymbol{x})}{P_g(\boldsymbol{x})} \right)^{-\frac{1}{t}} \left( \frac{\mathcal{M}_b(\boldsymbol{x})}{\mathcal{M}_g(\boldsymbol{x})} \right)^{-\frac{1}{\beta}} (1-\gamma)^{-\frac{1}{t}} \tag{20}$$

$$= \arg\max_{\boldsymbol{x} \in \mathcal{X}} \left( \frac{\mathcal{M}_b(\boldsymbol{x})}{\mathcal{M}_g(\boldsymbol{x})} \right)^{-\frac{1}{\beta}} \tag{21}$$

$$= \arg\max_{\boldsymbol{x} \in \mathcal{X}} \left( \frac{1 - \mathcal{M}_g(\boldsymbol{x})}{\mathcal{M}_g(\boldsymbol{x})} \right)^{-\frac{1}{\beta}} \tag{22}$$

$$= \arg\max_{\boldsymbol{x} \in \mathcal{X}} \left( \frac{1}{\mathcal{M}_g(\boldsymbol{x})} - 1 \right)^{-\frac{1}{\beta}} \tag{23}$$

$$= \arg\max_{\boldsymbol{x} \in \mathcal{X}} (\mathcal{M}_g(\boldsymbol{x}))^{\frac{1}{\beta}} \tag{24}$$

$$= \arg\max_{\boldsymbol{x} \in \mathcal{X}} \mathcal{M}_g(\boldsymbol{x}) \tag{25}$$

This shows that as iterations progress, the model grows more important. If BOPrO is run long enough, the prior washes out and BOPrO only trusts the probabilistic model. Since $\mathcal{M}_g(\boldsymbol{x})$ is the Probability of Improvement (PI) on the probabilistic model $p(y|\boldsymbol{x})$ then, in the limit, maximizing the acquisition function $EI_{f_\gamma}(\boldsymbol{x})$ is equivalent to maximizing the PI acquisition function on the probabilistic model $p(y|\boldsymbol{x})$. In other words, for high values of $t$, BOPrO converges to standard BO with a PI acquisition function.

## C Experimental Setup

We use a combination of publicly available implementations for our predictive models. For our Gaussian Process (GP) model, we use GPy's [14] GP implementation with the Matérn5/2 kernel. We use different length-scales for each input dimension, learned via Automatic Relevance Determination (ARD) [32]. For our Random Forests (RF), we use scikit-learn's RF implementation [35]. We set the fraction of features per split to 0.5, the minimum number of samples for a split to 5, and disable bagging. We also adapt our RF implementation to use the same split selection approach as Hutter *et al.* [18].

**Table 1.** Search spaces for our synthetic benchmarks. For the Profet benchmarks, we report the original ranges and whether or not a log scale was used.

| Benchmark | Parameter name | Parameter values | Log scale |
|---|---|---|---|
| Branin | $x_1$ | $[-5, 10]$ | – |
| | $x_2$ | $[0, 15]$ | – |
| SVM | C | $[e^{-10}, e^{10}]$ | ✓ |
| | $\gamma$ | $[e^{-10}, e^{10}]$ | ✓ |
| FCNet | Learning rate | $[10^{-6}, 10^{-1}]$ | ✓ |
| | Batch size | $[8, 128]$ | ✓ |
| | Units layer 1 | $[16, 512]$ | ✓ |
| | Units layer 2 | $[16, 512]$ | ✓ |
| | Dropout rate l1 | $[0.0, 0.99]$ | – |
| | Dropout rate l2 | $[0.0, 0.99]$ | – |
| XGBoost | Learning rate | $[10^{-6}, 10^{-1}]$ | ✓ |
| | Gamma | $[0, 2]$ | – |
| | L1 regularization | $[10^{-5}, 10^{3}]$ | ✓ |
| | L2 regularization | $[10^{-5}, 10^{3}]$ | ✓ |
| | Number of estimators | $[10, 500]$ | – |
| | Subsampling | $[0.1, 1]$ | – |
| | Maximum depth | $[1, 15]$ | – |
| | Minimum child weight | $[0, 20]$ | – |

For our constrained Bayesian Optimization (cBO) approach, we use scikit-learn's RF classifier, trained on previously explored configurations, to predict the probability of a configuration being feasible. We then weight our EI acquisition function by this probability of feasibility, as proposed by Gardner *et al.* [12]. We normalize our EI acquisition function before considering the probability of feasibility, to ensure both values are in the same range. This cBO implementation is used in the Spatial use-case as in Nardi *et al.* [31].

For all experiments, we set the model weight hyperparameter to $\beta = 10$ and the model quantile to $\gamma = 0.05$, see Appendices J and I. Before starting the main BO loop, BOPrO is initialized by random sampling $D + 1$ points from the prior, where $D$ is the number of input variables. We use the public implementation of Spearmint[6], which by default uses 2 random samples for initialization. We normalize our synthetic priors before computing the pseudo-posterior, to ensure they are in the same range as our model. We also implement interleaving which randomly samples a point to explore during BO with a 10% chance.

We optimize our EI acquisition function using a combination of a multi-start local search and CMA-ES [15]. Our multi-start local search is similar to the one used in SMAC [19]. Namely, we start local searches on the 10 best

[6] https://github.com/HIPS/Spearmint.

**Table 2.** Search space, priors, and expert configuration for the Shallow CNN application. The default value for each parameter is shown in bold.

| Parameter name | Type | Values | Expert | Prior |
|---|---|---|---|---|
| LP | Ordinal | [**1**, 4, 8, 16, 32] | 16 | [0.4, 0.065, 0.07, 0.065, 0.4] |
| P1 | Ordinal | [**1**, 2, 3, 4] | 1 | [0.1, 0.3, 0.3, 0.3] |
| SP | Ordinal | [**1**, 4, 8, 16, 32] | 16 | [0.4, 0.065, 0.07, 0.065, 0.4] |
| P2 | Ordinal | [**1**, 2, 3, 4] | 4 | [0.1, 0.3, 0.3, 0.3] |
| P3 | Ordinal | [**1**, 2, ..., 31, 32] | 1 | [0.1, 0.1, 0.033, 0.1, 0.021, 0.021, 0.021, 0.1, 0.021, 0.021, 0.021, 0.021, 0.021, 0.021, 0.021, 0.021, 0.021, 0.021, 0.021, 0.021, 0.021, 0.021, 0.021, 0.021, 0.021, 0.021, 0.021, 0.021, 0.021, 0.021, 0.021, 0.021] |
| P4 | Ordinal | [**1**, 2, ..., 47, 48] | 4 | [0.08, 0.0809, 0.0137, 0.1, 0.0137, 0.0137, 0.0137, 0.1, 0.0137, 0.0137, 0.0137, 0.05, 0.0137, 0.0137, 0.0137, 0.0137, 0.0137, 0.0137, 0.0137, 0.0137, 0.0137, 0.0137, 0.0137, 0.0137, 0.0137, 0.0137, 0.0137, 0.0137, 0.0137, 0.0137, 0.0137, 0.0137, 0.0137, 0.0137, 0.0137, 0.0137, 0.0137, 0.0137, 0.0137, 0.0137, 0.0137, 0.0137, 0.0137, 0.0137, 0.0137, 0.0137, 0.0137, 0.0137] |
| X276 | Categorical | [false, **true**] | true | [0.1, 0.9] |

points evaluated in previous BO iterations, on the 10 best performing points from a set of 10,000 random samples, on the 10 best performing points from 10,000 random samples drawn from the prior, and on the mode of the prior. To compute the neighbors of each of these 31 total points, we normalize the range of each parameter to $[0, 1]$ and randomly sample four neighbors from a truncated Gaussian centered at the original value and with standard deviation $\sigma = 0.1$. For CMA-ES, we use the public implementation of pycma [16]. We run pycma with two starting points, one at the incumbent and one at the mode of the prior. For both initializations we set $\sigma_0 = 0.2$. We only use CMA-ES for our continuous search space benchmarks.

We use four synthetic benchmarks in our experiments.

**Branin.** The Branin function is a well-known synthetic benchmark for optimization problems [8]. The Branin function has two input dimensions and three global minima.

**SVM.** A hyperparameter-optimization benchmark in 2D based on Profet [22]. This benchmark is generated by a generative meta-model built using a set of SVM classification models trained on 16 OpenML tasks. The benchmark has two input parameters, corresponding to SVM hyperparameters.

**FC-Net.** A hyperparameter and architecture optimization benchmark in 6D based on Profet. The FC-Net benchmark is generated by a generative meta-

**Table 3.** Search space, priors, and expert configuration for the Deep CNN application. The default value for each parameter is shown in bold.

| Parameter name | Type | Values | Expert | Prior |
|---|---|---|---|---|
| LP | Ordinal | [**1**, 4, 8, 16, 32] | 8 | [0.4, 0.065, 0.07, 0.065, 0.4] |
| P1 | Ordinal | [**1**, 2, 3, 4] | 1 | [0.4, 0.3, 0.2, 0.1] |
| SP | Ordinal | [**1**, 4, 8, 16, 32] | 8 | [0.4, 0.065, 0.07, 0.065, 0.4] |
| P2 | Ordinal | [**1**, 2, 3, 4] | 2 | [0.4, 0.3, 0.2, 0.1] |
| P3 | Ordinal | [**1**, 2, ..., 31, 32] | 1 | [0.04, 0.01, 0.01, 0.1, 0.01, 0.01, 0.01, 0.1, 0.01, 0.01, 0.01, 0.01, 0.01, 0.01, 0.01, 0.2, 0.01, 0.01, 0.01, 0.01, 0.01, 0.01, 0.01, 0.1, 0.01, 0.01, 0.01, 0.01, 0.01, 0.01, 0.01, 0.2] |
| P4 | Ordinal | [**1**, 2, ..., 47, 48] | 4 | [0.05, 0.005, 0.005, 0.005, 0.005, 0.005, 0.005, 0.13, 0.005, 0.005, 0.005, 0.005, 0.005, 0.005, 0.005, 0.2, 0.005, 0.005, 0.005, 0.005, 0.005, 0.005, 0.005, 0.11, 0.005, 0.005, 0.005, 0.005, 0.005, 0.005, 0.005, 0.2, 0.005, 0.005, 0.005, 0.005, 0.005, 0.005, 0.005, 0.005, 0.005, 0.005, 0.005, 0.005, 0.005, 0.005, 0.005, 0.1] |
| x276 | Categorical | [false, **true**] | true | [0.1, 0.9] |

model built using a set of feed-forward neural networks trained on the same 16 OpenML tasks as the SVM benchmark. The benchmark has six input parameters corresponding to network hyperparameters.

**XGBoost.** A hyperparameter-optimization benchmark in 8D based on Profet. The XGBoost benchmark is generated by a generative meta-model built using a set of XGBoost regression models in 11 UCI datasets. The benchmark has eight input parameters, corresponding to XGBoost hyperparameters.

The search spaces for each benchmark are summarized in Table 1. For the Profet benchmarks, we report the original ranges and whether or not a log scale was used. However, in practice, Profet's generative model transforms the range of all hyperparameters to a linear $[0, 1]$ range. We use Emukit's public implementation for these benchmarks [34].

## D Spatial Real-World Application

`Spatial` [23] is a programming language and corresponding compiler for the design of application accelerators on reconfigurable architectures, e.g. field-programmable gate arrays (FPGAs). These reconfigurable architectures are a type of logic chip that can be reconfigured via software to implement different applications. `Spatial` provides users with a high-level of abstraction for hardware design, so that they can easily design their own applications on FPGAs. It allows users to specify parameters that do not change the behavior of the

**Table 4.** Search space, priors, and expert configuration for the MD Grid application. The default value for each parameter is shown in bold.

| Parameter name | Type | Values | Expert | Prior |
|---|---|---|---|---|
| loop_grid0_z | Ordinal | [**1**, 2, ..., 15, 16] | 1 | [0.2, 0.1, 0.05, 0.05, 0.05, 0.05, 0.05, 0.05, 0.05, 0.05, 0.05, 0.05, 0.05, 0.05, 0.05, 0.05] |
| loop_q | Ordinal | [**1**, 2, ..., 31, 32] | 8 | [0.08, 0.08, 0.02, 0.1, 0.02, 0.02, 0.02, 0.1, 0.02, 0.02, 0.02, 0.02, 0.02, 0.02, 0.02, 0.1, 0.02, 0.02, 0.02, 0.02, 0.02, 0.02, 0.02, 0.02, 0.02, 0.02, 0.02, 0.02, 0.02, 0.02, 0.02, 0.02] |
| par_load | Ordinal | [**1**, 2, 4] | 1 | [0.45, 0.1, 0.45] |
| loop_p | Ordinal | [**1**, 2, ..., 31, 32] | 2 | [0.1, 0.1, 0.1, 0.1, 0.05, 0.03, 0.02, 0.02, 0.02, 0.02, 0.02, 0.02, 0.02, 0.02, 0.02, 0.02, 0.02, 0.02, 0.02, 0.02, 0.02, 0.02, 0.02, 0.02, 0.02, 0.02, 0.02, 0.02, 0.02, 0.02, 0.02, 0.02] |
| loop_grid0_x | Ordinal | [**1**, 2, ..., 15, 16] | 1 | [0.2, 0.1, 0.05, 0.05, 0.05, 0.05, 0.05, 0.05, 0.05, 0.05, 0.05, 0.05, 0.05, 0.05, 0.05, 0.05] |
| loop_grid1_z | Ordinal | [**1**, 2, ..., 15, 16] | 1 | [0.2, 0.2, 0.1, 0.1, 0.07, 0.03, 0.03, 0.03, 0.03, 0.03, 0.03, 0.03, 0.03, 0.03, 0.03, 0.03] |
| loop_grid0_y | Ordinal | [**1**, 2, ..., 15, 16] | 1 | [0.2, 0.1, 0.05, 0.05, 0.05, 0.05, 0.05, 0.05, 0.05, 0.05, 0.05, 0.05, 0.05, 0.05, 0.05, 0.05] |
| ATOM1LOOP | Categorical | [false, **true**] | true | [0.1, 0.9] |
| ATOM2LOOP | Categorical | [false, **true**] | true | [0.1, 0.9] |
| PLOOP | Categorical | [false, **true**] | true | [0.1, 0.9] |

application, but impact the runtime and resource-usage (e.g., logic units) of the final design. During compilation, the `Spatial` compiler estimates the ranges of these parameters and estimates the resource-usage and runtime of the application for different parameter values. These parameters can then be optimized during compilation in order to achieve the design with the fastest runtime. We fully integrate BOPrO as a pass in `Spatial`'s compiler, so that `Spatial` can automatically use BOPrO for the optimization during compilation. This enables `Spatial` to seamlessly call BOPrO during the compilation of any new application to guide the search towards the best design on an application-specific basis.

In our experiments, we introduce for the first time the automatic optimization of three `Spatial` real-world applications, namely, 7D shallow and deep CNNs, and a 10D molecular dynamics grid application. Previous work by Nardi *et al.* [31] had applied automatic optimization of `Spatial` parameters on a set of benchmarks but in our work we focus on real-world applications raising the bar of state-of-the-art automated hardware design optimization. BOPrO is used to optimize the parameters to find a design that leads to the fastest runtime. The search space for these three applications is based on ordinal and categorical parameters; to handle these discrete parameters in the best way we implement and use a Random Forests surrogate instead of a Gaussian Process one as explained in Appendix C. These parameters are application-specific and control how much of the FPGAs' resources we want to use to parallelize each step of the application's computation. The goal here is to find which steps are more important to parallelize in the final design, in order to achieve the fastest runtime. Some parameters also control whether we want to enable pipeline scheduling or not, which consumes resources but accelerates runtime, and others focus on memory management. We refer to Koeplinger *et al.* [23] and Nardi *et al.* [31] for more details on `Spatial`'s parameters.

The three `Spatial` benchmarks also have feasibility constraints in the search space, meaning that some parameter configurations are infeasible. A configuration is considered infeasible if the final design requires more logic resources than what the FPGA provides, i.e., it is not possible to perform FPGA synthesis because the design does not fit in the FPGA. To handle these constraints, we use our cBO implementation (Appendix C). Our goal is thus to find the design with the fastest runtime under the constraint that the design fits the FPGA resource budget.

The priors for these `Spatial` applications take the form of a list of probabilities, containing the probability of each ordinal or categorical value being good. Each benchmark also has a default configuration, which ensures all methods start with at least one feasible configuration. The priors and the default configuration for these benchmarks were provided once by an unbiased `Spatial` developer, who is not an author of this paper, and kept unchanged during the entire project. The search space, priors, and the expert configuration used in our experiments for each application are presented in Tables 2, 3, and 4.

## E Multivariate Prior Comparison

In this section we compare the performance of BOPrO with univariate and multivariate priors. For this, we construct synthetic univariate and multivariate priors using Kernel Density Estimation (KDE) with a Gaussian kernel. We build strong and weak versions of the KDE priors. The strong priors are computed using a KDE on the best $10D$ out of $10{,}000{,}000D$ uniformly sampled points, while the weak priors are computed using a KDE on the best $10D$ out of $1{,}000D$ uniformly sampled points. We use the same points for both univariate and multivariate priors. We use scipy's Gaussian KDE implementation, but adapt its Scott's Rule bandwidth to $100n^{-\frac{1}{d}}$, where $d$ is the number of variables in the KDE prior, to make our priors more peaked.

Figure 8 shows a log regret comparison of BOPrO with univariate and multivariate KDE priors. We note that in all cases BOPrO achieves similar performance with univariate and multivariate priors. For the Branin and SVM benchmarks, the weak multivariate prior leads to slightly better results than the weak univariate prior. However, we note that the difference is small, in the order of $10^{-4}$ and $10^{-6}$, respectively.

Surprisingly, for the XGBoost benchmark, the univariate version for both the weak and strong priors lead to better results than their respective multivariate counterparts, though, once again, the difference in performance is small, around

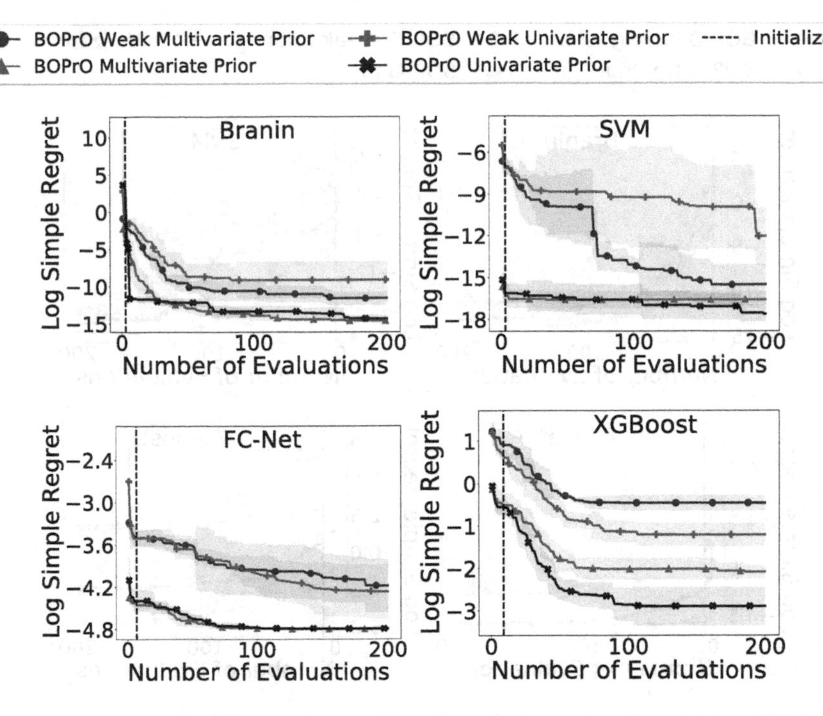


**Fig. 8.** Log regret comparison of BOPrO with multivariate and univariate KDE priors. The line and shaded regions show the mean and standard deviation of the log simple regret after 5 runs. All methods were initialized with $D+1$ random samples, where $D$ is the number of input dimensions, indicated by the vertical dashed line. We run the benchmarks for 200 iterations.

0.2 and 0.03 for the weak and strong prior, respectively, whereas the XGBoost benchmark can reach values as high as $f(\boldsymbol{x}) = 700$. Our hypothesis is that this difference comes from the bandwidth estimator ($100n^{-\frac{1}{d}}$), which leads to larger bandwidths, consequently, smoother priors, when a multivariate prior is constructed.

# F Misleading Prior Comparison

Figure 9 shows the effect of injecting a misleading prior in BOPrO. We compare BOPrO with a misleading prior, no prior, a weak prior, and a strong prior. For our misleading prior, we use a Gaussian centered at the worst point out of 10,000,000$D$ uniform random samples. Namely, for each parameter, we inject a prior of the form $\mathcal{N}(x_w, \sigma_w^2)$, where $x_w$ is the value of the parameter at the point with highest function value out of 10,000,000$D$ uniform random samples and $\sigma_w = 0.01$. For all benchmarks, we note that the misleading prior slows down convergence, as expected, since it pushes the optimization away from the optima in the initial phase. However, BOPrO is still able to forget the misleading prior and achieve similar regret to BOPrO without a prior.

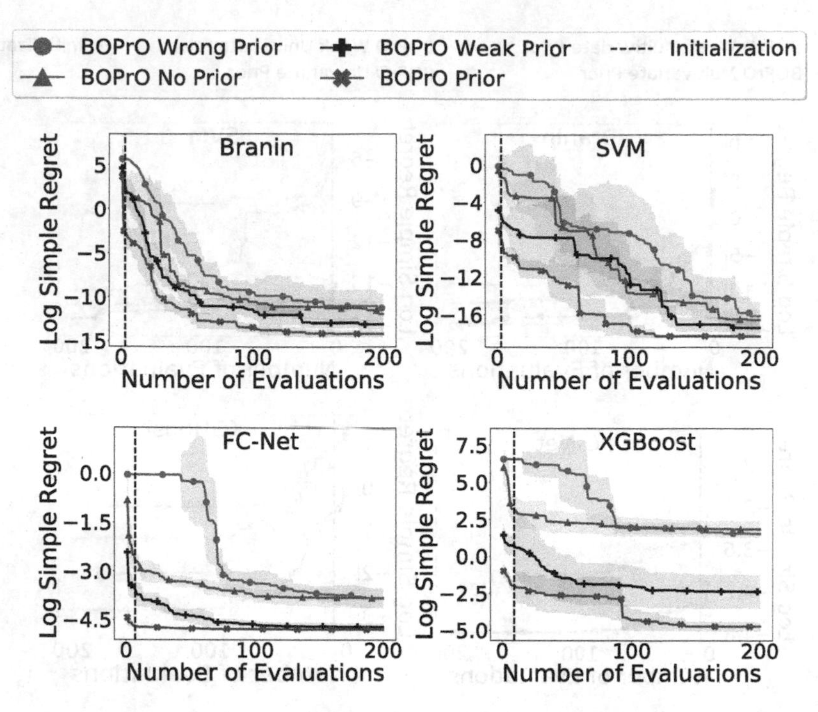


**Fig. 9.** Log regret comparison of BOPrO with varying prior quality. The line and shaded regions show the mean and standard deviation of the log simple regret after 5 runs. All methods were initialized with $D+1$ random samples, where $D$ is the number of input dimensions, indicated by the vertical dashed line. We run the benchmarks for 200 iterations.

## G Comparison to Other Baselines

We compare BOPrO to SMAC [19], TuRBO [9], and TPE [3] on our four synthetic benchmarks. We use Hyperopt's implementation[7] of TPE, the public implementation of TuRBO[8], and the SMAC3 Python implementation of SMAC[9]. Hyperopt defines priors as one of a list of supported distributions, including Uniform, Normal, and Lognormal distributions, while SMAC and TuRBO do not support priors on the locality of an optimum under the form of probability distributions.

For the three Profet benchmarks (SVM, FCNet, and XGBoost), we inject the strong priors defined in Sect. 4.2 into both Hyperopt and BOPrO. For Branin, we also inject the strong prior defined in Sect. 4.2 into BOPrO, however, we cannot inject this prior into Hyperopt. Our strong prior for Branin takes the form of a Gaussian mixture prior peaked at all three optima and Hyperopt

[7] https://github.com/hyperopt/hyperopt.
[8] https://github.com/uber-research/TuRBO.
[9] https://github.com/automl/SMAC3.

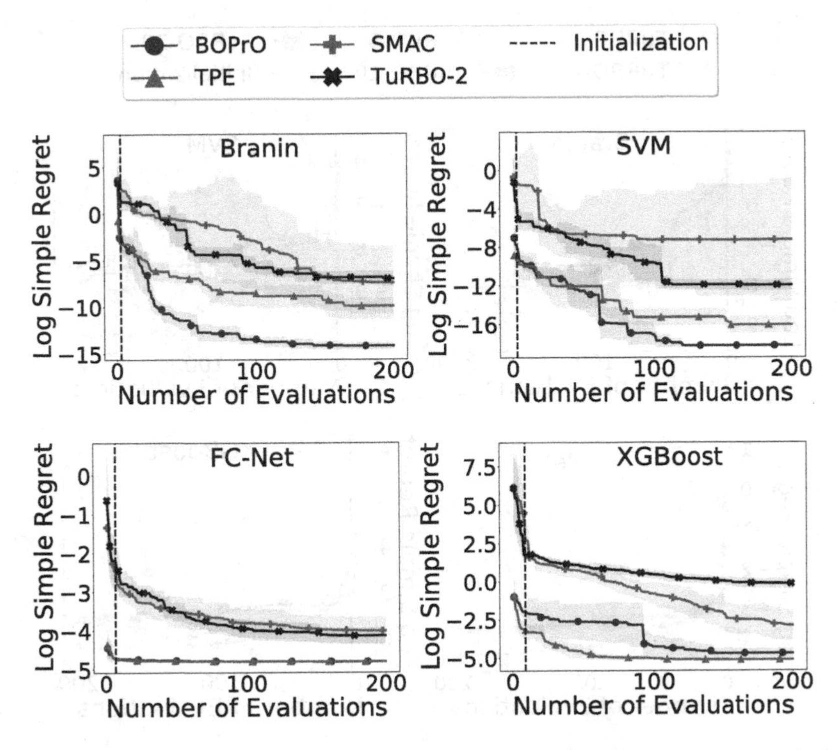


**Fig. 10.** Log regret comparison of BOPrO, SMAC, and TPE. The line and shaded regions show the mean and standard deviation of the log simple regret after 5 runs. BOPrO was initialized with $D + 1$ random samples, where $D$ is the number of input dimensions, indicated by the vertical dashed line. We run the benchmarks for 200 iterations.

does not support Gaussian mixture priors. Instead, for Hyperopt, we arbitrarily choose one of the optima $(\pi, 2.275)$ and use a Gaussian prior centered near that optimum. We note that since we compare all approaches based on the log simple regret, both priors are comparable in terms of prior strength, since finding one optimum or all three would lead to the same log regret. Also, we note that using Hyperopt's Gaussian priors leads to an unbounded search space, which sometimes leads TPE to suggest parameter configurations outside the allowed parameter range. To prevent these values from being evaluated, we convert values outside the parameter range to be equal to the upper or lower range limit, depending on which limit was exceeded. We do not inject any priors into SMAC and TuRBO, since these methods do not support priors about the locality of an optimum.

Figure 10 shows a log regret comparison between BOPrO, SMAC, TuRBO-2 and TPE on our four synthetic benchmarks. We use TuRBO-2 since it led to better performance overall compared to other variants, see Fig. 11. BOPrO achieves better performance than SMAC and TuRBO on all four benchmarks.

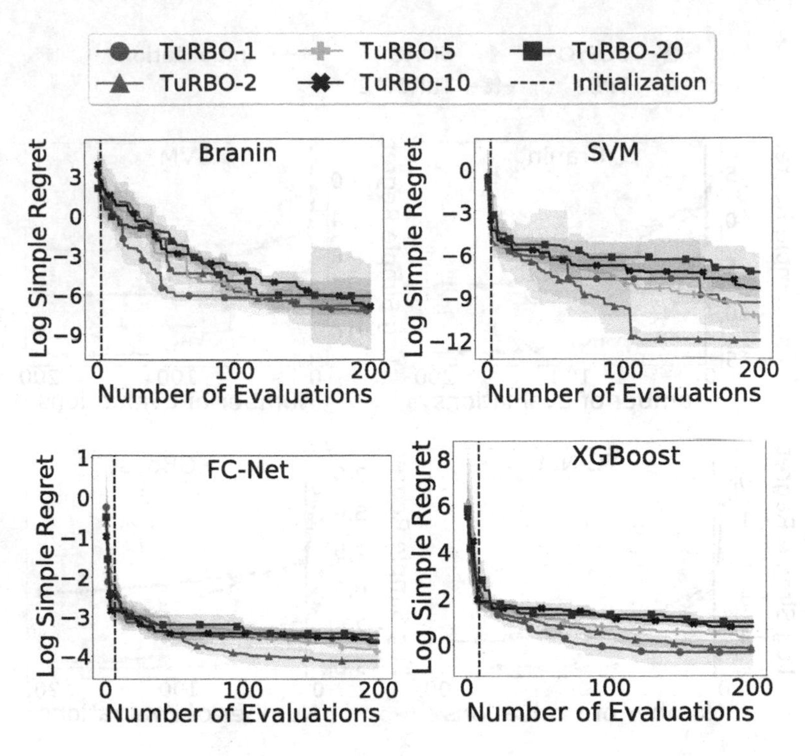


**Fig. 11.** Log regret comparison of TuRBO with different number of trust regions. TuRBO-$M$ denotes TuRBO with $M$ trust regions. The line and shaded regions show the mean and standard deviation of the log simple regret after 5 runs. TuRBO was initialized with $D+1$ uniform random samples, where $D$ is the number of input dimensions, indicated by the vertical dashed line. We run the benchmarks for 200 iterations.

Compared to TPE, BOPrO achieves similar or better performance on three of the four synthetic benchmarks, namely Branin, SVM, and FCNet, and slightly worse performance on XGBoost. We note, however, that the good performance of TPE on XGBoost may be an artifact of the approach of clipping values to its maximal or minimal values as mentioned above. In fact, the clipping nudges TPE towards promising configurations in this case, since XGBoost has low function values near the edges of the search space. Overall, the better performance of BOPrO is expected, since BOPrO is able to combine prior knowledge with more sample-efficient surrogates, which leads to better performance.

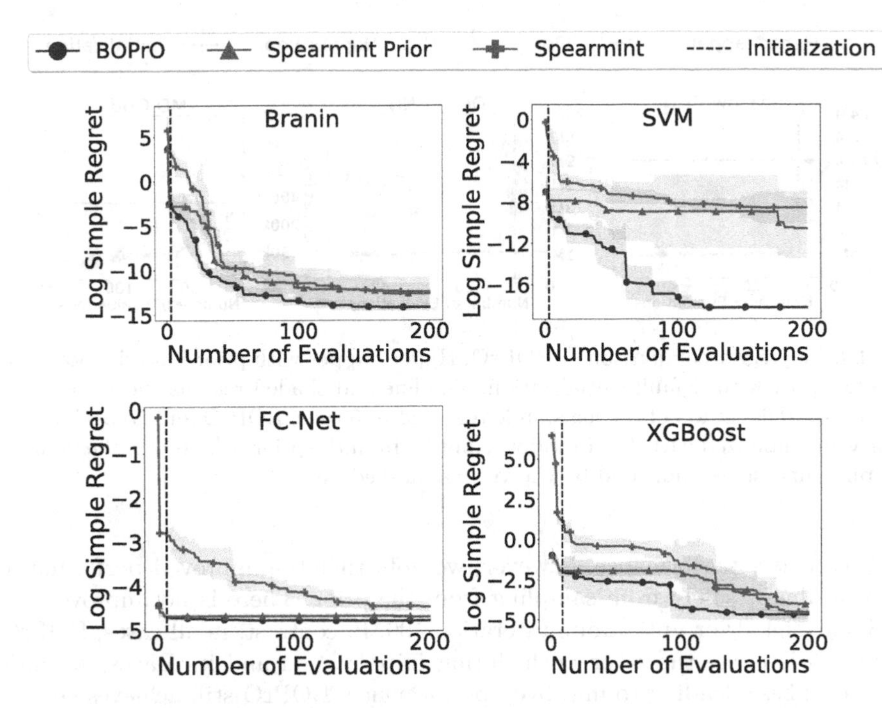


**Fig. 12.** Log regret comparison of BOPrO, Spearmint with prior initialization, and Spearmint with default initialization. The line and shaded regions show the mean and standard deviation of the log simple regret after 5 runs. BOPrO and Spearmint Prior were initialized with $D + 1$ random samples from the prior, where $D$ is the number of input dimensions, indicated by the vertical dashed line. We run the benchmarks for 200 iterations.

# H Prior Baselines Comparison

We show that simply initializing a BO method in the DoE phase by sampling from a prior on the locality of an optimum doesn't necessarily lead to better performance. Instead in BOPrO, it is the pseudo-posterior in Eq. (4) that drives its stronger performance by combining prior and new observations. To show that, we compare BOPrO with Spearmint and HyperMapper such as in section Sects. 4.2 and 4.3, respectively, but we initialize all three methods using the same approach. Namely, we initialize all methods with $D + 1$ samples from the prior. Our goal is to show that simply initializing Spearmint and HyperMapper with the prior will not lead to the same performance as BOPrO, because, unlike BOPrO, these baselines do not leverage the prior after the DoE initialization phase. We report results on both our synthetic and real-world benchmarks.

Figure 12 shows the comparison between BOPrO and Spearmint Prior. In most benchmarks, the prior initialization leads to similar final performance. In particular, for XGBoost, the prior leads to improvement in early iterations, but to worse final performance. We note that for FCNet, Spearmint Prior

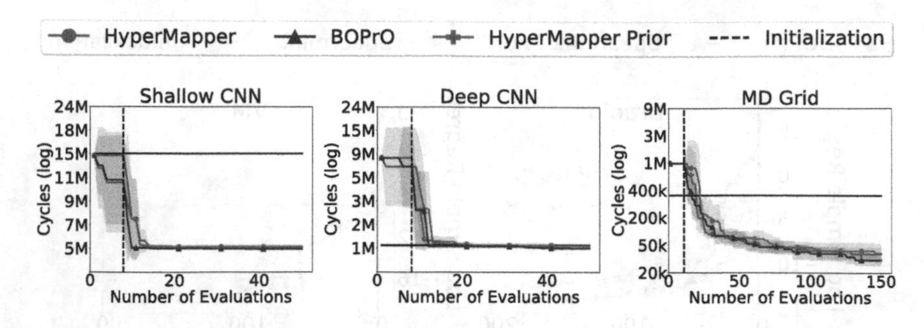


**Fig. 13.** Log regret comparison of BOPrO, HyperMapper with prior initialization, and HyperMapper with default initialization. The line and shaded regions show the mean and standard deviation of the log simple regret after 5 runs. BOPrO and HyperMapper Prior were initialized with $D+1$ random samples from the prior, where $D$ is the number of input dimensions, indicated by the vertical dashed line.

achieves better performance, however, we note that the improved performance is given almost solely from sampling from the prior. There is no improvement for Spearmint Prior until around iteration 190. In contrast, in all cases, BOPrO is able to leverage the prior both during initialization and its Bayesian Optimization phase, leading to improved performance. BOPrO still achieves similar or better performance than Spearmint Prior in all benchmarks.

Figure 13 shows similar results for our Spatial benchmarks. The prior does not lead HyperMapper to improved final performance. For the Shallow CNN benchmark, the prior leads HyperMapper to improved performance in early iterations, compared to HyperMapper with default initialization, but HyperMapper Prior is still outperformed by BOPrO. Additionally, the prior leads to degraded performance in the Deep CNN benchmark. These results confirm that BOPrO is able to leverage the prior in its pseudo-posterior during optimization, leading to improved performance in almost all benchmarks compared to state-of-the-art BO baselines.

# I $\gamma$-Sensitivity Study

We show the effect of the $\gamma$ hyperparameter introduced in Sect. 3.2 for the quantile identifying the points considered to be good. To show this, we compare the performance of BOPrO with our strong prior and different $\gamma$ values. For all experiments, we initialize BOPrO with $D+1$ random samples and then run BOPrO until it reaches $10D$ function evaluations. For each $\gamma$ value, we run BOPrO five times and report mean and standard deviation.

Figure 14 shows the results of our comparison. We first note that values near the lower and higher extremes lead to degraded performance, this is expected, since these values will lead to an excess of either exploitation or exploration. Further, we note that BOPrO achieves similar performance for all values of $\gamma$, however, values around $\gamma = 0.05$ consistently lead to better performance.

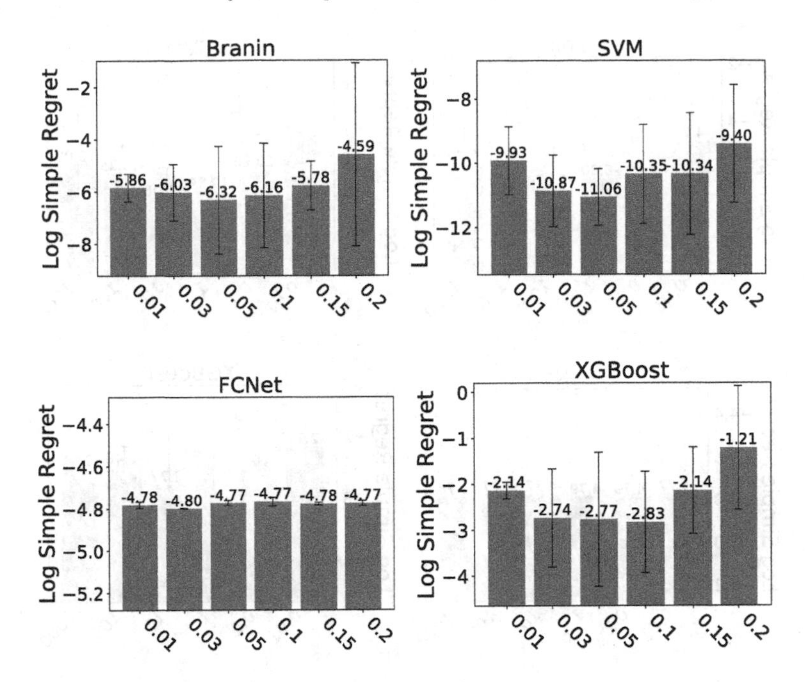


**Fig. 14.** Comparison of BOPrO with the strong prior and different values for the $\gamma$ hyperparameter on our four synthetic benchmarks. We run BOPrO with a budget of $10D$ function evaluations, including $D + 1$ randomly sampled DoE configurations.

## J $\beta$-Sensitivity Study

We show the effect of the $\beta$ hyperparameter introduced in Sect. 3.3 for controlling the influence of the prior over time. To show the effects of $\beta$, we compare the performance of BOPrO with our strong prior and different $\beta$ values on our four synthetic benchmarks. For all experiments, we initialize BOPrO with $D + 1$ random samples and then run BOPrO until it reaches $10D$ function evaluations. For each $\beta$ value, we run BOPrO five times and report mean and standard deviation.

Figure 15 shows the results of our comparison. We note that values of $\beta$ that are too low (near 0.01) or too high (near 1000) often lead to lower performance. This shows that putting too much emphasis on the model or the prior will lead to degraded performance, as expected. Further, we note that $\beta = 10$ lead to the best performance in three out of our four benchmarks. This result is reasonable, as $\beta = 10$ means BOPrO will put more emphasis on the prior in early iterations, when the model is still not accurate, and slowly shift towards putting more emphasis on the model as the model sees more data and becomes more accurate.

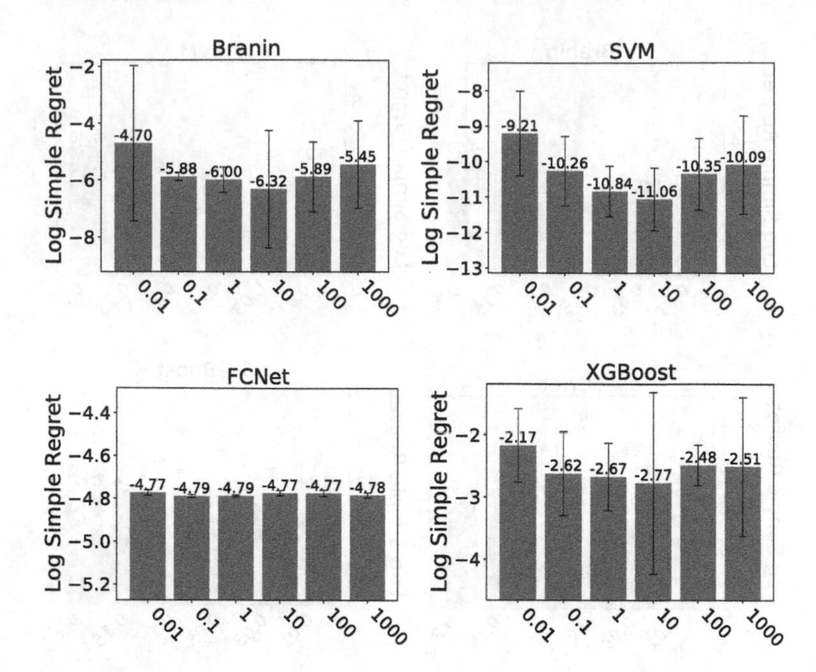


**Fig. 15.** Comparison of BOPrO with the strong prior and different values for the $\beta$ hyperparameter on our four synthetic benchmarks. We run BOPrO with a budget of $10D$ function evaluations, including $D+1$ randomly sampled DoE configurations.

## References


1. Balandat, M., et al.: BoTorch: a framework for efficient Monte-Carlo Bayesian optimization. In: Advances in Neural Information Processing Systems (2020)
2. Bergstra, J., Yamins, D., Cox, D.D.: Making a science of model search: hyperparameter optimization in hundreds of dimensions for vision architectures. In: International Conference on Machine Learning (2013)
3. Bergstra, J.S., Bardenet, R., Bengio, Y., Kégl, B.: Algorithms for hyper-parameter optimization. In: Advances in Neural Information Processing Systems (2011)
4. Bouthillier, X., Varoquaux, G.: Survey of machine-learning experimental methods at NeurIPS2019 and ICLR2020. Research report, Inria Saclay Ile de France (January 2020). https://hal.archives-ouvertes.fr/hal-02447823
5. Calandra, R., Seyfarth, A., Peters, J., Deisenroth, M.P.: Bayesian optimization for learning gaits under uncertainty. Ann. Math. Artif. Intell. **76**(1–2), 5–23 (2016)
6. Chen, Y., Huang, A., Wang, Z., Antonoglou, I., Schrittwieser, J., Silver, D., de Freitas, N.: Bayesian optimization in AlphaGo. CoRR abs/1812.06855 (2018)
7. Clarke, A., McMahon, B., Menon, P., Patel, K.: Optimizing hyperparams for image datasets in Fastai (2020). https://www.platform.ai/post/optimizing-hyperparams-for-image-datasets-in-fastai
8. Dixon, L.C.W.: The global optimization problem: an introduction. In: Toward Global Optimization 2, pp. 1–15 (1978)
9. Eriksson, D., Pearce, M., Gardner, J.R., Turner, R., Poloczek, M.: Scalable global optimization via local Bayesian optimization. In: Advances in Neural Information Processing Systems (2019)

10. Falkner, S., Klein, A., Hutter, F.: BOHB: robust and efficient hyperparameter optimization at scale. In: International Conference on Machine Learning (2018)
11. Feurer, M., Springenberg, J.T., Hutter, F.: Initializing bayesian hyperparameter optimization via meta-learning. In: AAAI Conference on Artificial Intelligence (2015)
12. Gardner, J.R., Kusner, M.J., Xu, Z.E., Weinberger, K.Q., Cunningham, J.P.: Bayesian optimization with inequality constraints. In: International Conference on Machine Learning (ICML) (2014)
13. Golovin, D., Solnik, B., Moitra, S., Kochanski, G., Karro, J., Sculley, D.: Google Vizier: a service for black-box optimization. In: SIGKDD International Conference on Knowledge Discovery and Data Mining (2017)
14. GPy: GPy: a Gaussian process framework in Python (since 2012). http://github.com/SheffieldML/GPy
15. Hansen, N., Ostermeier, A.: Adapting arbitrary normal mutation distributions in evolution strategies: the covariance matrix adaptation. In: Proceedings of IEEE International Conference on Evolutionary Computation (1996)
16. Hansen, N., Akimoto, Y., Baudis, P.: CMA-ES/pycma on GitHub
17. Hernández-Lobato, J.M., Hoffman, M.W., Ghahramani, Z.: Predictive entropy search for efficient global optimization of black-box functions. In: Advances in Neural Information Processing Systems (2014)
18. Hutter, F., Xu, L., Hoos, H., Leyton-Brown, K.: Algorithm runtime prediction: methods & evaluation. Artif. Intell. **206**, 79–111 (2014)
19. Hutter, F., Hoos, H.H., Leyton-Brown, K.: Sequential model-based optimization for general algorithm configuration. In: Learning and Intelligent Optimization Conference (2011)
20. Hutter, F., Kotthoff, L., Vanschoren, J. (eds.): Automated Machine Learning: Methods, Systems, Challenges. TSSCML, Springer, Cham (2019). https://doi.org/10.1007/978-3-030-05318-5
21. Kingma, D.P., Ba, J.: Adam: a method for stochastic optimization. In: International Conference on Learning Representations (2015)
22. Klein, A., Dai, Z., Hutter, F., Lawrence, N.D., Gonzalez, J.: Meta-surrogate benchmarking for hyperparameter optimization. In: Advances in Neural Information Processing Systems (2019)
23. Koeplinger, D., et al.: Spatial: a language and compiler for application accelerators. In: SIGPLAN Conference on Programming Language Design and Implementation (2018)
24. Kushner, H.J.: A new method of locating the maximum point of an arbitrary multipeak curve in the presence of noise. J. Basic Eng. **86**(1), 97–106 (1964)
25. Li, C., Gupta, S., Rana, S., Nguyen, V., Robles-Kelly, A., Venkatesh, S.: Incorporating expert prior knowledge into experimental design via posterior sampling. arXiv preprint arXiv:2002.11256 (2020)
26. Lindauer, M., Eggensperger, K., Feurer, M., Falkner, S., Biedenkapp, A., Hutter, F.: SMAC v3: algorithm configuration in Python (2017). https://github.com/automl/SMAC3
27. Lindauer, M., Hutter, F.: Warmstarting of model-based algorithm configuration. In: AAAI Conference on Artificial Intelligence (2018)
28. López-Ibáñez, M., Dubois-Lacoste, J., Pérez Cáceres, L., Stützle, T., Birattari, M.: The irace package: iterated racing for automatic algorithm configuration. Oper. Res. Perspect. **3**, 43–58 (2016)
29. Mockus, J., Tiesis, V., Zilinskas, A.: The application of Bayesian methods for seeking the extremum. In: Towards Global Optimization 2, pp. 117–129 (1978)

30. Nardi, L., Bodin, B., Saeedi, S., Vespa, E., Davison, A.J., Kelly, P.H.: Algorithmic performance-accuracy trade-off in 3d vision applications using hypermapper. In: International Parallel and Distributed Processing Symposium Workshops (2017)
31. Nardi, L., Koeplinger, D., Olukotun, K.: Practical design space exploration. In: International Symposium on Modeling, Analysis, and Simulation of Computer and Telecommunication Systems (2019)
32. Neal, R.M.: Bayesian Learning for Neural Networks, vol. 118. Springer, New York (1996). https://doi.org/10.1007/978-1-4612-0745-0
33. Oh, C., Gavves, E., Welling, M.: BOCK: Bayesian optimization with cylindrical kernels. In: International Conference on Machine Learning (2018)
34. Paleyes, A., Pullin, M., Mahsereci, M., Lawrence, N., González, J.: Emulation of physical processes with Emukit. In: Workshop on Machine Learning and the Physical Sciences, NeurIPS (2019)
35. Pedregosa, F., et al.: Scikit-learn: machine learning in Python. J. Mach. Learn. Res. **12**, 2825–2830 (2011)
36. Perrone, V., Shen, H., Seeger, M., Archambeau, C., Jenatton, R.: Learning search spaces for Bayesian optimization: another view of hyperparameter transfer learning. In: Advances in Neural Information Processing Systems (2019)
37. Ramachandran, A., Gupta, S., Rana, S., Li, C., Venkatesh, S.: Incorporating expert prior in Bayesian optimisation via space warping. Knowl. Based Syst. **195**, 105663 (2020)
38. Shahriari, B., Bouchard-Côté, A., Freitas, N.: Unbounded Bayesian optimization via regularization. In: Artificial Intelligence and Statistics. pp. 1168–1176 (2016)
39. Shahriari, B., Swersky, K., Wang, Z., Adams, R.P., De Freitas, N.: Taking the human out of the loop: a review of Bayesian optimization. Proc. IEEE **104**(1), 148–175 (2015)
40. Siivola, E., Vehtari, A., Vanhatalo, J., González, J., Andersen, M.R.: Correcting boundary over-exploration deficiencies in Bayesian optimization with virtual derivative sign observations. In: International Workshop on Machine Learning for Signal Processing (2018)
41. Snoek, J., Larochelle, H., Adams, R.P.: Practical Bayesian optimization of machine learning algorithms. In: Advances in Neural Information Processing Systems (2012)
42. Srinivas, N., Krause, A., Kakade, S.M., Seeger, M.W.: Gaussian process optimization in the bandit setting: no regret and experimental design. In: International Conference on Machine Learning (2010)
43. Hutter, F., Kotthoff, L., Vanschoren, J. (eds.): Automated Machine Learning. TSSCML, Springer, Cham (2019). https://doi.org/10.1007/978-3-030-05318-5
44. Wang, Q., et al.: ATMSeer: increasing transparency and controllability in automated machine learning. In: CHI Conference on Human Factors in Computing Systems (2019)
45. Wu, J., Poloczek, M., Wilson, A.G., Frazier, P.I.: Bayesian optimization with gradients. In: Advances in Neural Information Processing Systems 30 (2017)

# Rank Aggregation for Non-stationary Data Streams

Ekhine Irurozki[1,2(✉)], Aritz Perez[2(✉)], Jesus Lobo[3(✉)], and Javier Del Ser[3,4(✉)]

[1] Telecom Paris, Institut Polytechnique de Paris, Palaiseau, France
irurozki@telecom-paris.fr
[2] Basque Center for Applied Mathematics (BCAM), Bilbao, Spain
aperez@bcamath.org
[3] TECNALIA, Basque Research and Technology Alliance (BRTA), Bizkaia, Spain
{Jesus.Lopez,javier.delser}@tecnalia.com
[4] University of the Basque Country (UPV/EHU), Bilbao, Spain

**Abstract.** The problem of learning over non-stationary ranking streams arises naturally, particularly in recommender systems. The rankings represent the preferences of a population, and the non-stationarity means that the distribution of preferences changes over time. We propose an algorithm that learns the current distribution of ranking in an online manner. The bottleneck of this process is a rank aggregation problem.

We propose a generalization of the Borda algorithm for non-stationary ranking streams. As a main result, we bound the minimum number of samples required to output the ground truth with high probability. Besides, we show how the optimal parameters are set. Then, we generalize the whole family of weighted voting rules (the family to which Borda belongs) to situations in which some rankings are more *reliable* than others. We show that, under mild assumptions, this generalization can solve the problem of rank aggregation over non-stationary data streams.

**Keywords:** Preference learning · Rank aggregation · Borda · Evolving preferences · Voting · Concept drift

## 1 Introduction

In the rank aggregation problem, a population of voters cast their preferences over a set of candidates by providing each a ranking of these candidates. The goal is to summarize this collection of rankings $S$ with one single ranking $\pi$. A classical formulation of rank aggregation is to find the Kemeny ranking, the ranking that minimizes the number of discrepancies with $S$. Unfortunately, finding the Kemeny ranking is NP-hard.

**Electronic supplementary material** The online version of this chapter (https://doi.org/10.1007/978-3-030-86523-8_18) contains supplementary material, which is available to authorized users.


N. Oliver et al. (Eds.): ECML PKDD 2021, LNAI 12977, pp. 297–313, 2021.
https://doi.org/10.1007/978-3-030-86523-8_18

When the rankings are distributed according to the Mallows distribution Kemeny is known to be the maximum likelihood estimator of one model parameter and the Borda ranking a high-quality, low-cost approximation to the same parameter. Therefore, besides in Optimization [8], rank aggregation has been studied in Computational Social Choice [4] and Machine Learning [19,26] extensively, generally in a batch setting.

In this paper, we challenge the standard batch learning setting and consider stream learning [13,18], which refers to learning problems for which the data is continuously generated over time. This feature imposes severe computational complexity constraints on the proposed learning algorithms. For instance, in contrast to batch learning, in stream learning scenarios the data can no longer be completely stored, and the learning algorithms must be computationally efficient (linear or even sub-linear), to the extreme of operating close to real-time. These computational complexity constraints usually lead to models that are incrementally updated with the arrival of new data.

Besides, another major challenge when dealing with stream learning arises when the source producing stream data evolves over time, which yields to non-stationary data distributions. This phenomenon is known as concept drift (CD) [14]. In the stream learning scenarios with CD, the models have to be able to adapt to changes (drifts) in the data distribution (concept). We focus on this scenario that we call *evolving preferences.*

*Problem Statement.* In this paper, we assume that the data is given as a stream of rankings over $n$ items representing the evolving preferences of a population of voters. Rankings are i.i.d. according to a Mallows model (MM) -analogous to the Gaussian distribution defined over the permutation space-: MM are parametrized by (i) a modal ranking (representing the consensus of the population), and (ii) a parameter controlling the variance. Therefore, a stream of rankings with concept drift is naturally modelled by a sequence of MM with different modal rankings, i.e., an evolving MM. The goal is to develop an estimate of the current modal ranking of the evolving MM with low computational complexity and arbitrarily small error, that is able to deal with the CD phenomenon.

*Contributions.* There are three main contributions in this paper.

- We adapt the Borda algorithm to the stream learning scenario in which the distribution of the data changes, SL with CD. We denote this generalization of Borda unbalanced Borda (uBorda).
- We theoretically analyze uBorda and provide bounds on the number of samples required for recovering the last modal ranking of the evolving MM, in expectation and with arbitrary probability. Moreover, we show how to set an optimal learning parameter.

– We generalize whole families of voting rules[1] to manage evolving preferences. Moreover, we show that this is a particular case of the scenario in which some voters are more *trusted* than others. We denote this setting as *unbalanced voting* and show the empirical performance of this approach.

*Related Work.* Rank aggregation has been studied in Optimization [8], Computational Social Choice [4] and Machine Learning [19,26]. In the last decade, rank aggregation has been studied in on-line environments. In particular, [27] proposed a ranking reconstruction algorithm based on the pairwise comparisons of the items with near optimal relative loss bound. In [1], two efficient algorithms based on sorting procedures that predict the aggregated ranking with bounding maximal expected regret where introduced. In [6] an active learner was proposed that, querying items pair-wisely, returns the aggregate ranking with high probability.

Evolving preferences have been considered from an axiomatic perspective by modeling voters preferences with Markov decision processes [21], in fairness [12] and matching [15]. In [23], the authors analyze the behavior of the preferences of a population of voters under the plurality rule. In [22] the authors assume that each user preferences change over time. In contrast to the previous works, in this work we assume a probabilistic setting using a MM that changes over the time for modeling the evolving preferences of a population.

This paper is organized as follows: Sect. 2 includes the preliminaries of this paper. Section 3 presents the generalization of Borda and its theoretical guarantees. Section 4 introduces the adaptation of a whole family of voting rules to the case in which some voters are trusted more. Section 5 shows the empirical evaluation of the algorithms for the rank aggregation on stream learning with concept drift. Finally, Sect. 6 summarizes the main contributions of the work.

## 2 Preliminaries and Notation

Permutations are a bijection of the set $[n] = \{1, 2, \ldots, n\}$ onto itself and are represented as an ordered vector of $[n]$. The set of permutations of $[n]$ is denoted by $S_n$. Permutations will be denoted with the Greek letters $\sigma$ or $\pi$ and represent rankings, meaning that the ranking of element $i \in [n]$ in $\sigma$ is denoted by $\sigma(i)$. Rankings are used to represent preferences and we say that item $i$ is preferred to item $j$ when it has a lower ranking, $\sigma(i) < \sigma(j)$.

One of the most popular noisy models for permutations is the Mallows model (MM) [10,20]. The MM is an exponential location-scale model in which the location parameter (modal ranking) is a ranking denoted by $\pi$, and the scale

[1] Voting rules are functions that given a collections of rankings output the winner of the election and generalize the notion of rank aggregation. We generalize the Weighted Tournament solutions and Unweighted Tournament solutions families, from which Borda and Condorcet respectively are the best known members.

(or concentration parameter) is a non-negative real number denoted by $\theta$. In general, the probability of any ranking $\sigma \in S_n$ under the MM can be written as:

$$p(\sigma) = \frac{\exp(-\theta \cdot d(\sigma, \pi))}{\psi(\theta)}, \quad (1)$$

where $d(\cdot, \cdot)$ is a distance metric and $\psi(\theta)$ is a normalization factor that depends on the location parameter $\theta$. It has closed form expression for most interesting distances [16].

A random ranking $\sigma$ drawn from a MM centered at $\pi$ and with concentration parameter $\theta$ is denoted as $\sigma \sim MM(\pi, \theta)$. The expected rank of item $i$ for $\sigma \sim MM(\pi, \theta)$ is denoted by $\mathbb{E}_\pi[\sigma(i)]$, or simply by $\mathbb{E}[\sigma(i)]$ when it is clear from the context. To model preferences, the selected distance $d(\cdot, \cdot)$ for permutations is the Kendall's-$\tau$ distance (for other choice of distances see [16,25]). This distance counts the number of pairwise disagreements in two rankings,

$$d(\sigma, \pi) = \sum_{i<j} \mathbb{I}[(\sigma(i) - \sigma(j)) * (\pi(i) - \pi(j)) < 0], \quad (2)$$

where $\mathbb{I}$ is the indicator function.

Given a sample of rankings, the maximum likelihood estimation (MLE) of the parameters of a MM is usually computed in two steps. First, the MLE of the modal ranking $\pi$ is obtained. It corresponds to the Kemeny ranking [10] (see the following sections for further details). Second, given the MLE of the modal ranking, the MLE of the concentration parameter $\theta$ can be computed numerically.

### 2.1 Modeling Evolving Preferences: Evolving Mallows Model

In this paper, we assume that the data is a stream of preferences, i.e., a sequence of rankings and that our proposed algorithm has to update its estimate every time a new ranking is received. The rankings in the sequence may not all come from the same distribution, so in the next lines, we introduce a natural way of modeling preferences that evolve dynamically over time.

Let $\sigma_t$ for $t \geq 0$ be the $t-th$ ranking in the given sequence, being $\sigma_0$ the current ranking. We assume that the stream of rankings satisfy $\sigma_t \sim MM_t(\pi_t, \theta_t)$, for $t \geq 0$. This sequence of models is denoted the evolving Mallows model (EMM).

When two consecutive models, $MM_{t+1}(\pi_{t+1}, \theta_{t+1})$ and $MM_t(\pi_t, \theta_t)$, differ on their modal ranking, $\pi_{t+1} \neq \pi_t$, we say that at time $t$ there has been a drift in the modal ranking of the model. Under the Mallows models, the drifts in the modal ranking can be interpreted as a change in the consensus preference of a population of voters. When the concentration parameter changes over time, the drifts represent a change in the variability of the preferences of a population around the consensus ranking. For the rest of the paper, we focus on the scenario with drifts in the modal ranking.

Drifts can be classified as gradual and abrupt in terms of speed, being abrupt when a change happens suddenly between two concepts, and gradual when there is a smooth transition between both concepts. In this work, we have considered abrupt drifts in what refers to the speed of change (rankings generated from the old concept disappear suddenly and the new ones appear), and small-step/large-step (the Kendall's-$\tau$ distance between consecutive modal rankings is small/large).

## 3 Unbalanced Borda for Stream Ranking

In this section, we present a generalization of the well-known Borda count algorithm to rank aggregation for the context of stream data with concept drift. We want to point out that this approach is generalized in Sect. 4, for (1) the settings in which each ranking in the sample is not equally relevant (e.g., the agents have different reliabilities), and (2) for all the voting rules in the family of Weighted voting rules and Unweighted Voting rules (C1 and C2 families) [4].

The Borda algorithm [3] is one of the most popular methods for aggregating a sample of rankings, $S$, into one single ranking that best represents $S$. Borda ranks the items $[n]$ by their *Borda score* increasingly, where the Borda score $B(i)$ is the average ranking of each item $i$, for $i = 1, \ldots, n$:

$$B(i) = \frac{1}{|S|} \sum_{t \in S} \sigma_t(i).$$

Borda is the de-facto standard in the applied literature of rank aggregation. Firstly, it has a computational complexity of $\mathcal{O}(n \cdot (|S| + \log n))$. Secondly, it is guaranteed to be a good estimator of the modal ranking when the samples are i.i.d. according to a MM: it requires a polynomial number of samples with respect to $n$ to return the modal ranking of the MM with high probability [7]. In general, it is a 5-approximation to the Kemeny ranking [8], which, unfortunately, has been shown to be NP-hard to compute [9].

We propose unbalanced *Borda* (*uBorda*) a generalization of Borda for the rank aggregation problem in the context of streaming preferences. In uBorda, each ranking $\sigma_t$ has an associated weight $\rho^t$ that is proportional to the relevance of the ranking in the sample. The uBorda scores correspond to the weighted average of the given rankings. In other words, uBorda is equivalent to replicating each ranking $\sigma_t$ $\rho^t$ times and applying Borda.

In stream learning, votes arrive sequentially, at time-stamp $t$ rank $\sigma_t$ is received, being $t = 0$ the most recent ranking of a possible infinite sequence of votes. In order for uBorda to adapt to the concept drifts in the sequence, we propose to weight the ranking $\sigma_t$ by $\rho^t$ for a given parameter $\rho \in [0, 1]$. This choice leads to the following Borda score for each item $i$.

$$B(i) \propto \sum_{t \geq 0} \rho^t \cdot \sigma_t(i) \tag{3}$$

Intuitively, using these specific weights the voting rule *pays more attention* to recent rankings since $\rho^t$ exponentially decreases as the antiquity $t$ of the ranking $\sigma_t$ increases. The uBorda score can be incrementally computed as $B(i) \propto \sigma_0(i) + \rho \cdot B_1(i)$ where $B_1(i)$ denotes the previous uBorda score, computed using $\sigma_t$ for $t \geq 1$. Thus, in this streaming scenario, the uBorda scores can be updated incrementally in linear time $\mathcal{O}(n)$, and the Borda algorithm has a computational time complexity of $\mathcal{O}(n \log n)$, in the worst case.

By setting $\rho = 1$ we recover the classic Borda. When $0 \leq \rho < 1$, uBorda can be seen as a Borda algorithm that incorporates a forgetting mechanism to adapt the ranking aggregation process to drifts in streaming data.

In subsequent sections, we will theoretically analyze the properties of uBorda for ranking aggregation in streaming scenarios when the rankings are i.i.d. according to an EMM. First, next lemma provides some known intermediate result regarding the expected value of rankings obtained from a MM.

**Theorem 1** ([11]). *Given a sample $S$ of permutations i.i.d. according to a $MM(\pi, \theta)$, the Borda ranking is a consistent estimator of $\pi$.*

Following, we show a proof sketch to present the main ideas behind this result and introduce some new notation (see [11] for further details). The authors argue that, on average and for $\pi(i) < \pi(j)$, Borda[2] will rank $i$ above $j$ iff $\mathbb{E}_\pi[\sigma(i)] < \mathbb{E}_\pi[\sigma(j)]$. They reformulated the expression $\mathbb{E}_\pi[\sigma(j)] < \mathbb{E}_\pi[\sigma(i)]$ in a more convenient way:

$$\begin{aligned} \Delta_{ij} &= \mathbb{E}_\pi[\sigma(j)] - \mathbb{E}_\pi[\sigma(i)] \\ &= \sum_{\{\sigma:\sigma(i)<\sigma(j)\}} (\sigma(j) - \sigma(i))(p(\sigma) - p(\sigma\tau)), \end{aligned} \tag{4}$$

where $\tau$ is an inversion of positions $i$ and $j$. Clearly, the proof of Theorem 1 is equivalent to showing that $\Delta_{ij} > 0$ for every $i, j$ such that $\pi(i) < \pi(j)$. To see this, note that for the set of rankings $\{\sigma : \sigma(i) < \sigma(j)\}$ it holds that $(\sigma(j) - \sigma(i)) > 0$. When $p$ is a $MM(\pi, \theta)$, due to the strong unimodality of MMs, we can also state that $(p(\sigma) - p(\sigma\tau)) > 0$. Thus, given a MM with parameters $\pi$ and $\theta$ we have that $\Delta_{ij} > 0$ for every $i, j$ such that $\pi(i) < \pi(j)$.

In the following Lemma, we provide an intermediate result that relates the expected rankings of two MMs, $MM(\sigma, \theta)$ and $MM(\sigma\tau, \theta)$, where $\tau$ represents an inversion of positions $i$ and $j$.

**Lemma 2.** *Let $\sigma$ a ranking distributed according to $MM(\pi, \theta)$. Let $\tau$ be an inversion of $i$ and $j$ so that $d(\pi\tau, \pi) = 1$. Using the definition on Eq. (4), we can easily see that $\mathbb{E}_\pi[\sigma]$ and $\mathbb{E}_{\pi\tau}[\sigma]$ are related as follows:*

$$\begin{aligned} \mathbb{E}_{\pi\tau}[\sigma(i)] &= \mathbb{E}_\pi[\sigma(j)] = \mathbb{E}_\pi[\sigma(i)] + \Delta_{ij} \\ \mathbb{E}_{\pi\tau}[\sigma(j)] &= \mathbb{E}_\pi[\sigma(i)] \end{aligned} \tag{5}$$

[2] Note that, with a slight abuse in the notation, we are using $\sigma$ indistinctly for a ranking and for a random variable distributed according to a MM.

### 3.1 Sample Complexity for Returning $\pi_0$ on Average

In this section, we analyze theoretically uBorda for ranking aggregation using stream data distributed according to a EMM. We consider a possibly infinite sequence of rankings for which the current modal ranking is $\pi_0$ ant the last drift in the modal ranking has occurred $m$ batches before. We bound the number of batches since the last drift that uBorda needs to recover the current modal ranking $\pi_0$ on average.

**Theorem 3.** *Let $\pi(i) < \pi(j)$, and let $\tau$ be an inversion of $i$ and $j$ so that $d(\pi\tau, \pi) = 1$. Let $\sigma_t$ be a (possibly infinite) sequence of rankings generated by sampling an EMM, such that $\sigma_t \sim MM(\pi, \theta)$ for $t \geq m$ and $\sigma_t \sim MM(\pi\tau, \theta)$ for $m > t$. Then, uBorda returns the current ranking $\pi_0 = \pi\tau$ in expectation when*

$$m > \log_\rho 0.5.$$

*Proof.* The uBorda algorithm ranks the items in $k \in [n]$ w.r.t. the uBorda score $B(k) = \sum_{t\geq 0} \rho^t \sigma_t(k)$. Thus, uBorda recovers the current ranking $\pi_0$ if and only if the expression $\mathbb{E}[\sum_{t\geq 0} \rho^t \sigma_t(i)] < \mathbb{E}[\sum_{t\geq 0} \rho^t \sigma_t(j)]$ is satisfied for every pair $i, j$.

$$\mathbb{E}[\sum_{t\geq 0} \rho^t \sigma(i)] = \sum_{t\geq m} \rho^t \mathbb{E}_\pi[\sigma(i)] + \sum_{m>t} \rho^t \mathbb{E}_{\pi\tau}[\sigma(i)] = \sum_{t\geq 0} \rho^t \mathbb{E}_\pi[\sigma(i)] + \sum_{m>t} \rho^t \Delta_{ij}$$

$$\mathbb{E}[\sum_{t\geq 0} \rho^t \sigma(j)] = \sum_{t\geq m} \rho^t \mathbb{E}_\pi[\sigma(j)] + \sum_{m>t} \rho^t \mathbb{E}_{\pi\tau}[\sigma(j)] = \sum_{t\geq 0} \rho^t \mathbb{E}_\pi[\sigma(i)] + \sum_{t\geq m} \rho^t \Delta_{ij}$$

Therefore, the uBorda algorithm recovers the current ranking $\pi_0$ in expectation if and only if the next inequality holds:

$$\sum_{t<m} \rho^t < \sum_{t\geq m} \rho^t$$

Thus, given $m > 0$, we can satisfy the previous inequality by selecting $\rho$ according to $m > \log_\rho 0.5$, which concludes the proof.

Note that the right hand side expression decreases as $\rho$ decreases, for $\rho \in (0, 1]$, and in the limit, $\lim_{\rho\to 0} \log_\rho 0.5 = 0$. The intuitive conclusion is that as $\rho$ decreases uBorda is more reactive, that is, in expectation it needs less samples to accommodate to the drift. In other words, when $\rho$ decreases uBorda quickly forgets the old preferences and focuses on the most recently generated ones, so it can quickly adapt to recent drifts. One might feel tempted of lowering $\rho$ to guarantee that uBorda recovers the last modal ranking in expectation. However, as we decrease $\rho$ the variance on the uBorda score (see Eq. (3)) increases and, thus the variance of the ranking obtained with uBorda also increases. In the next section, we will show how the confidence of our estimated modal ranking decreases as $\rho$ decreases. Besides, we will provide a lower bound on the number of samples required by uBorda for recovering from a drift.

### 3.2 Sample Complexity for Returning $\pi_0$ with High Probability

In this section we consider a possibly infinite sequence of rankings and denote by $\pi_0$ the current consensus of the model, which has generated the last $m$ rankings. We bound the value of $m$ that uBorda needs to return $\pi_0$ with high probability in Theorem 5 and Corollary 6.

We present an intermediate result in Lemma 4, where we bound with high probability the difference between the uBorda score and its expected value for a given sub sequence of the stream of rankings. For this intermediate result we consider that there is no drift in the sample.

**Lemma 4.** *Let $\sigma_r, \sigma_{r+1}, \ldots, \sigma_s$ be $m = s - r + 1$ rankings i.i.d. distributed according to $MM(\pi, \theta)$. In the absence of drifts, the absolute difference between the uBorda score for item $i$, $(1-\rho) \cdot \sum_{t=r}^{s} \rho^t \sigma_t(i)$, and its expectation $\mathbb{E}[(1-\rho)\sum_{t=r}^{s} \rho^t \sigma_t(i)]$ is smaller than*

$$\epsilon_r^s = (n-1)(1-\rho) \cdot \sqrt{\frac{(\rho^{2r} - \rho^{2s})}{2 \cdot (1-\rho^2)} \cdot \log \frac{2}{\delta}} \tag{6}$$

*with, at least a probability of $1 - \delta$,*

$$P\left( \left| \sum_{t=r}^{s} \rho^t (\sigma_t(i) - \mathbb{E}[\sigma_t(i)]) \right| \leq \epsilon_r^s \right) \geq 1 - \delta$$

The proof, which is omitted, uses the Hoeffding's inequality and sums of geometric series.

The next result gives a lower bound on the number of samples required for recovering from a drift with high probability for a drift at distance 1. This result is generalized for distance $d$ drifts afterwards. The lower bound is given as a function of the concentration of the underlying distribution $\theta$ and parameter $\rho$.

**Theorem 5.** *Let $\pi(i) < \pi(j)$ and let $\tau$ be an inversion of $i$ and $j$ so that $d(\pi\tau, \pi) = 1$. Let $\sigma_t$ for $t \geq 0$ be a (possibly infinite) sequence of rankings generated by sampling an evolving MM, such that for $t \geq m$ then $\sigma_t \sim MM(\pi, \theta)$ and for $t < m$ then $\sigma_t \sim MM(\pi\tau, \theta)$. The number of samples that uBorda needs to return the current modal ranking $\pi_0 = \pi\tau$ in expectation with probability $1 - \delta$ is at least*

$$m > \log_\rho \left( \frac{-(1-\rho)^2}{\sqrt{1-\rho^2}} \frac{n\sqrt{0.5 \log \delta^{-1}}}{\Delta_{ij}} + 0.5 \right) \tag{7}$$

*where $\Delta_{ij}$ is defined in Eq. (4).*

*Proof.* Following the idea in Theorem 3, the algorithm uBorda returns the current ranking $\pi_0 = \pi\tau$ in expectation when the expected uBorda score (see Eq. (3)) of element $i$ is greater than of element $j$ as

$$\sum_{t=m}^{\infty} \rho^t \mathbb{E}_\pi[\sigma(j)] + \sum_{t=0}^{m-1} \rho^t \mathbb{E}_{\pi\tau}[\sigma(j)] < \sum_{t=m}^{\infty} \rho^t \mathbb{E}_\pi[\sigma(i)] + \sum_{t=0}^{m-1} \rho^t \mathbb{E}_{\pi\tau}[\sigma(i)], \tag{8}$$

which is based on Eq. (4). Equation (8) would be an accurate measure for the sample complexity if the expected value did not deviate at all from the sum. However, according to Lemma 4 we can upper bound the difference between the uBorda score as shown in Lemma 4. Next, we make use of the definition in Eq. (6) to define the deviations of the uBorda score before the drift, $\epsilon_m^\infty$, and after the drift, $\epsilon_0^{m-1}$ and let $\epsilon_0^\infty$. Note that the next inequality holds:

$$\epsilon_m^\infty + \epsilon_0^{m-1} = \epsilon_0^\infty \tag{9}$$

Therefore, we can state that with probability $1 - \delta$ we have recovered from a drift when the $m$ satisfies

$$\sum_{t=m}^{\infty} \rho^t \mathbb{E}_\pi[\sigma(j)] + \epsilon_m^\infty + \sum_{t=0}^{m-1} \rho^t \mathbb{E}_{\pi\tau}[\sigma(j)] + \epsilon_0^{m-1} <$$
$$\sum_{t=m}^{\infty} \rho^t \mathbb{E}_\pi[\sigma(i)] - \epsilon_m^\infty + \sum_{t=0}^{m-1} \rho^t \mathbb{E}_{\pi\tau}[\sigma(i)] - \epsilon_0^{m-1},$$

which implies that with probability $1 - \delta$ uBorda will recover $\pi\tau$ when the following expression holds:

$$\Delta_{ij} \sum_{t=0}^{m-1} \rho^t - 2\epsilon_0^{m-1} > \Delta_{ij} \sum_{t=m}^{\infty} \rho^t + 2\epsilon_m^\infty$$

After some algebra, and using the result in Eq. (9), we obtain the lower bound for $m$

$$\begin{aligned} \sum_{t=0}^{m-1} \rho^t \Delta_{ij} - \sum_{t=m}^{\infty} \rho^t \Delta_{ij} &> 2\epsilon_m^\infty + 2\epsilon_0^{m-1} > 2\epsilon_0^\infty \\ \frac{\rho^m - 1}{\rho - 1} - \frac{\rho^m}{1-\rho} &> \frac{\epsilon_0^\infty}{\Delta_{ij}} \\ \rho^m &> \frac{\epsilon_0^\infty(\rho - 1)}{\Delta_{ij}} + 0.5 \\ m > \log_\rho &\left( \frac{(\rho-1)(1-\rho)}{\sqrt{1-\rho^2}} \frac{n\sqrt{0.5 \log \delta^{-1}}}{\Delta_{ij}} + 0.5 \right) \end{aligned} \tag{10}$$

which concludes the proof.

The previous result gives a bound for drifts at distance 1 and it is generalized in the following Corollary for any drift.

**Corollary 6.** *For $\tau$ a drift at distance $d$, the maximum rank difference for item $i$, is $|\sigma(i) - \sigma\tau(i)| < d$. For $m$ as described in Theorem 5, after $dm$ observed samples, uBorda recovers the current central ranking with arbitrary high probability.*

### 3.3 Choosing $\rho$ Optimally

In this section how to compute the value of $\rho$ for recovering from a drift with high probability after receiving $m$ rankings.

**Theorem 7.** *Let $m$ be the number of rankings received after the last concept drift. uBorda recovers the true ranking with probability $(1-\delta)$ by using the forgetting parameter value $\rho* = \arg\max_\rho f(\rho, m)$ that can be found numerically, where*

$$f(\rho,m) = \frac{2\rho^m - 1}{\rho - 1}\left(\sqrt{\frac{\rho^{2m}}{1-\rho^2}\frac{1-\rho}{\rho^m}} + \sqrt{\frac{\rho^{2m}-1}{\rho^2-1}\frac{\rho-1}{\rho^m-1}}\right). \tag{11}$$

The previous result provides the optimal value of $\rho$ given the maximum delay allowed for determining a change in the consensus preference, i.e., maximize the probability of recovering the ground true ranking after $m$ rankings. Moreover, it has a probability of success as $1-\delta$. An more basic approach to include this kind of expert knowledge is to set a window of size $m$ and consider only the last $m$ rankings. In this case, the probability of success can be found with classic quality results for Borda [7]. However, our proposed uBorda has different advantages over traditional window approaches. First, uBorda allows handling and infinite sequence of permutations. This means that for every pair of items that have not suffered a drift there is an infinity number of samples available. Second, this analysis can be adapted to more general settings. In other words, this paper considers the particular situation in which recent rankings are more relevant than old ones but there are situations in which a subset of the rankings are more relevant than other for different reasons. For example, because some voters provide rankings that are more trustworthy than others. In this case, uBorda can be used as a rank aggregation procedure in which expert agents have a larger weight.

## 4 Generalizing Voting Rules

So far we have adapted the Borda algorithm to perform rank aggregation when the rankings lose importance over time. This section generalizes that setting in every possible way. First, we consider families of algorithms (which include Borda and Condorcet). Second, we consider that each ranking has different "importance" (rather than decreasing importance with time).[3]

We propose a general framework and skip the search for statistical properties for the voting rules, since it needs assumptions on the data generating process and is context-dependent. Instead, we validate empirically the Condorcet rule in the experimental section for the evolving preference setting, showing applicability and efficiency.

[3] We can say that some voters are more *trusted* than others.

In particular, we are given a set of rankings $S = \{\sigma_v : v \in [m]\}$ representing the preferences of a set of voters, where each voter $v < m$ has a weight $w_v \in \mathbb{R}^+$ representing our confidence in $\sigma_v$. There are similar contexts in crowd learning scenarios in which the weight of each voter is related to its reliability [17].

The most relevant voting rules are those in the Unweighted Tournament Solutions family (C1 voting rules) and the Weighted Tournament Solutions family (C2 voting rules [4][4]). We generalize these voting rules by generalizing the *frequency matrix* upon which all the rules in these families are defined. The weighted frequency matrix, $N$, is the following summary statistic of the sample of rankings $S$.

$$N_{ij} = \sum_{v<m} w_v \mathbb{I}[\sigma_v(i) < \sigma_v(j)], \tag{12}$$

where $w_v$ is the weight of the preference $\sigma_v$.[5] The family of C2 functions are given as a function of the *majority margin* matrix, wich is computed using the frequency matrix. The generalized majority matrix, $M$, is computed using $N$ as follows:

$$M_{ij} = N_{ij} - N_{ji} = 2N_{ij} - \sum_{v<m} w_v, \tag{13}$$

Intuitively, $M$ counts the difference on the weighting votes that prefer $i$ to $j$ and those that prefer $j$ to $i$. The Kemeny ranking can be formulated using the majority margins matrix in Eq. (13) as follows

$$\sigma_K = \arg\min_{\sigma \in S_n} \sum_{i,j \in [n]:\sigma(i)>\sigma(j)} M_{ij}. \tag{14}$$

The Kemeny ranking has been shown to maintain several interesting properties, such as being a median permutation of the sample of rankings, and the MLE of the modal ranking of the sample when the rankings are i.i.d. according to a Mallows model [10]. Unfortunately, the problem of computing the Kemeny ranking given a sample of rankings $S$ has been shown to be NP-hard [9]. Other interesting members of the family of the C2 family are a pairwise query algorithm that returns the Kemeny ranking with high probability [5], the ranked pairs method [24] and a 4/3-approximation of the Kemeny ranking [2].

The C1 family or Unweighted Voting Rules family uses the unweighted version of the graph defined by the majority margin matrix $M$. The most prominent member of this family is Condorcet: A Condorcet winner for a collection of rankings is an alternative $i$ that defeats every other alternative in the strict pairwise majority sense. By stating it as a function of Eq. (13), we can define Condorcet for balanced and unbalanced settings.

[4] Condorcet and Borda belong to C1 and C2 voting rules respectively.

[5] By setting $w_v = 1$ the standard frequency matrix is obtained.

**Definition 8** *(Condorcet). The Condorcet winner is the item $i$ for which $M_{ij} > 0$ for all $j$. For convenience, we define the number of Condorcet victories of item $i$ is $\{|j| : M_{ij} > 0\}$.*

The Condorcet winner does not always exist, and this holds for both balanced and unbalanced settings. This is known as the *Condorcet paradox*. We show strong experimental results in the following section.

# 5 Experiments

In this section, we provide empirical evidence of the strengths of uBorda to deal with the ranking aggregation problem in streaming scenarios. These experiments illustrate the theoretical results for the estimation of the ground true ranking after a drift (Sect. 3.2) and the tuning of the forgetting parameter $\rho$ (Sect. 3.3). Moreover, we show its applicability beyond the theoretical setting using different ranking models and several voting rules (see Sect. 4). For each parameter configuration, the results are run 10 times and the average is shown with a stroke line and the .95 confidence interval ($\lambda = 0.05$) shadowed.

## 5.1 Rank Aggregation for Dynamic Preferences with uBorda

This section analyses the performance of uBorda as a rank aggregation algorithm using streaming data of rankings. We consider a synthetic stream of 300 rankings. The larger $m$, the easier the learning problem from the statistical perspective and the sample size of $m = 20$ per drift is challenging even in the classical non-drift setting [7]. The stream of rankings has been sampled from an evolving **ranking model** that suffers concept drifts periodically. We have considered three types of evolving ranking models using the Mallows model, Plackett-Luce and Generalized Mallows model[6]. We consider two settings of each model, close to the uniform distribution or far from it.[7]

Between one subsequence and the next one, there has been a **drift** on the ground true ranking. The drift can be a small-step drift (drift at distance 1) or a large-step drift (the distance randomly increases in a number between 1 and $n$). In every case, the distance between the first and the last ground true ranking is maximal, $n * (n - 1)/2$.

For the visualization of the results, each ranking in the sequence lays in the x axis, and drifts are denoted with a grey vertical line in all plots. At each point of the sequence, the uBorda ranking $\hat{\sigma}_0$ estimates the ground true ranking $\sigma_0$ with the observed rankings so far, and therefore, the y axis measures the Kendall's-$\tau$ distance $d(\sigma_0, \hat{\sigma}_0)$.[8] Therefore, the smaller, the better.

[6] In the supplementary material.

[7] For the Mallows, the expected distance (which is a function of the spread parameter) is 0.1 and 0.4 of the maximum distance. For Plackett-Luce the weights are $w_i = n - i$ and $w_i = e^{n-i}$.

[8] We follow the classical test-then-train strategy.

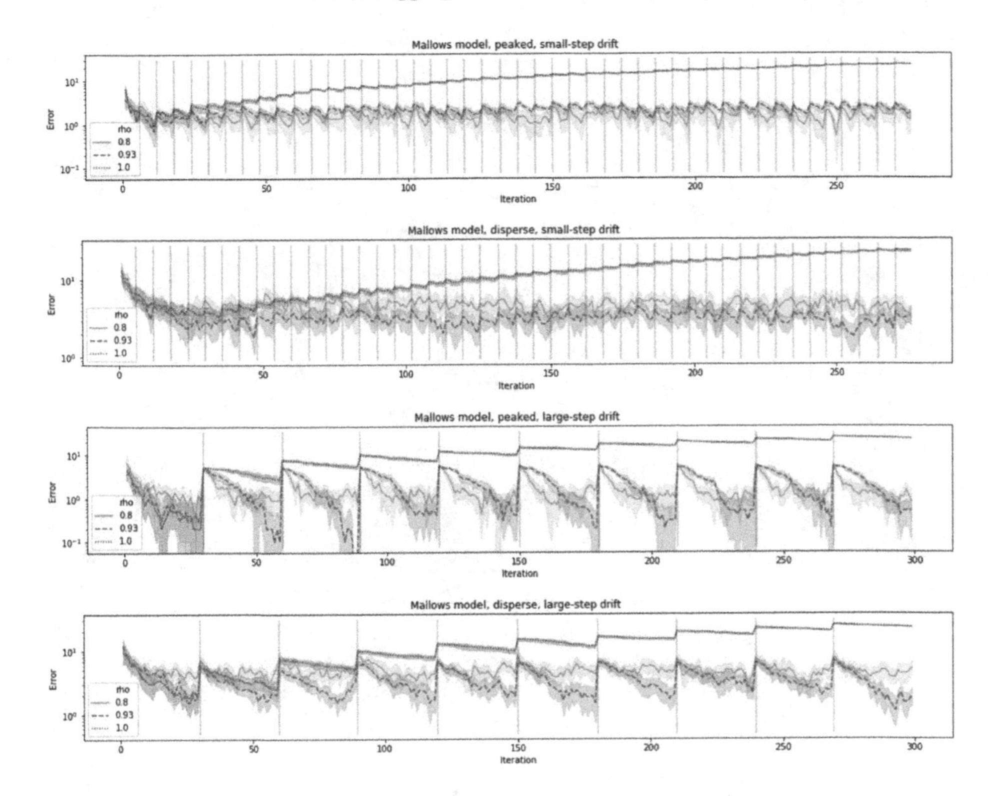


**Fig. 1.** uBorda algorithm for Mallows model with large and small step drifts

Simulating the inclusion of expert knowledge in the experiment, we expect the rank aggregation to recover from the drifts after $m = 20$ samples. Theorem 7 shows that finding the value for $\rho$ that minimizes the probability of error for a given number o samples is done by solving a convex optimization problem. For our choice of $m = 20$ the optimal $\rho$ is 0.93. For comparison, we use different fading factors $\rho \in \{0.8, 0.93, 1\}$, each corresponding to a different line in the plot.

Figure 1 shows the results of uBorda for the Mallows model and the particular configuration can be seen in the title. When a drift occurs (after each vertical grid-line), the error increases for every choice of $\rho$. However, differences in the value of $\rho$ cause critical differences in the behavior of uBorda: while $\rho = 1$ (the classic Borda algorithm) gets an increasing error with a larger sample, $\rho < 1$ can adapt to the drifts, consistently with the findings in the previous sections.

When $\rho = 0.93$ uBorda has the most accurate results. As shown in Theorem 7, this value maximizes the probability of recovering from a drift in 20 samples among the parameter values considered. Moreover, after these 20 samples, it recovers the ground true ranking with an error smaller than 0.1. Choosing $\rho = 0.8$ makes uBorda forget quicker the previous permutations and this can lead to a situation in which too few of the last permutations are considered to estimate the

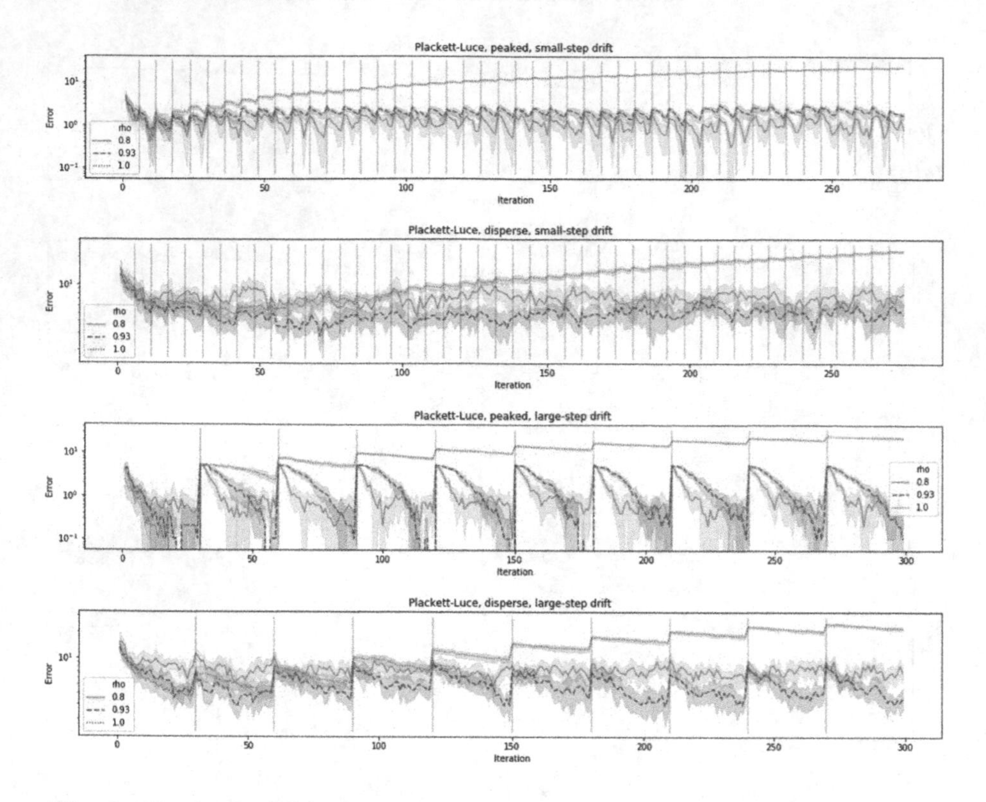


**Fig. 2.** uBorda algorithm for Plackett-Luce model with large and small step drifts

consensus. The more chaotic behavior of the smallest value of $\rho = 0.8$ (remind the logarithmic scale for the Y-axis) is related to this phenomenon in which few permutations are contributing to the estimation of the consensus, i.e., uBorda is aggregating a small number of rankings. Finally, for $\rho = 1$ uBorda is equivalent to Borda and does not forget the past rankings. That leads to an increasing error as new concept drifts occur because uBorda is not able to accommodate the changes in the consensus ranking of the evolving Mallows model.

Figure 2 show the results obtained with an evolving Plackett-Luce model. The conclusions are similar to those obtained for the evolving Mallows model. Since Plackett-Luce implies a transitivity pairwise matrix of the marginals, Borda is an unbiased estimator of the central ranking in a non-evolving setting [11]. For $\rho < 1$ uBorda can adapt to the changes in the ground true ranking, while the error of Borda (i.e., uBorda with $\rho = 1$) increases after each drift. The supplementary material shows similar results for other parameter setting and probability models in the probabilistic ranking literature.

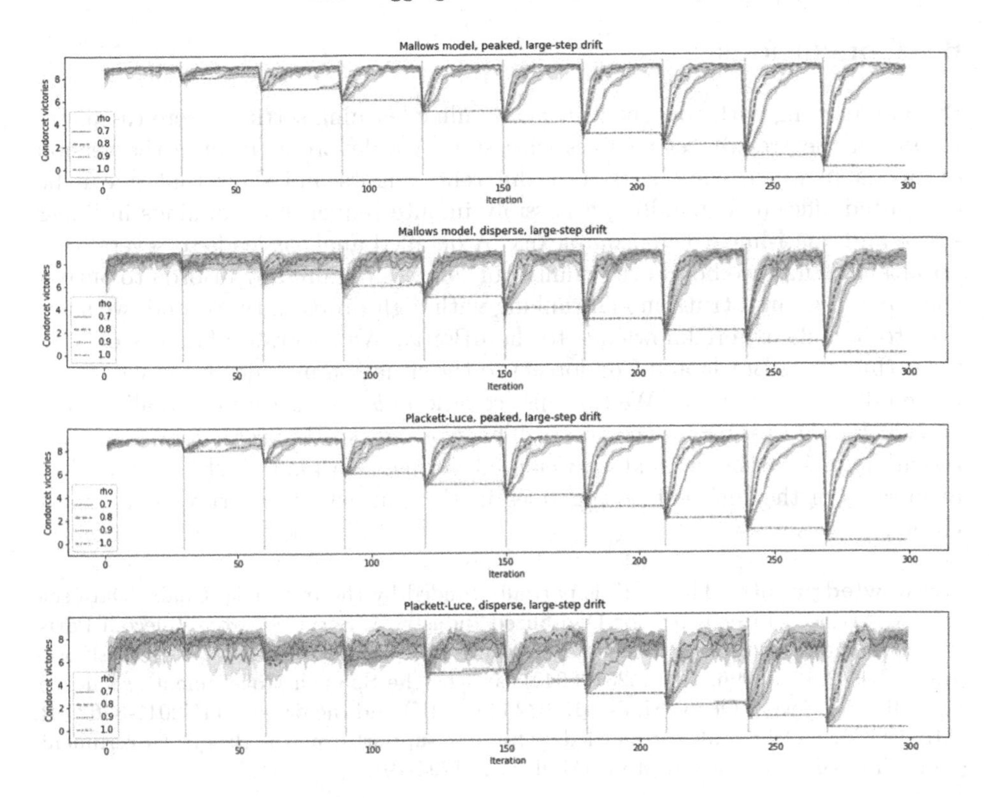


**Fig. 3.** Condorcet algorithm for Mallows and Plackett-Luce models

### 5.2 Condorcet Winner in Dynamic Preferences

Section 4 generalizes families of voting rules to the unbalanced voting (and evolving preferences) setting. In this section, we validate empirically the most relevant member of the C1 family, Condorcet (see Definition 8). In a similar setting to the previous experiments, we consider the Plackett-Luce and the Mallows models - since both possess a Condorcet winner. For each model, we consider the same peaked and disperse distributions, and the results are in Fig. 3. Drift $i$, represented as the $i$-th vertical line, is a swap between the first and the $i$-th item. A Condorcet winner will have $n-1=9$ Condorcet victories (y axis), therefore, the higher, the better. We show the results of 4 different settings of our algorithms, $\rho = [.7, .8, .9, 1]$.

In all the experiments, when $\rho = 1$ the Condorcet Victories of the winner item strictly decreases with each drift. When the forgetting parameter $\rho < 1$ the Condorcet Victories also decrease but recover after observing some rankings from the same distribution. This recovery is faster for smaller values of $\rho$, as we show theoretically for Borda in Sect. 3, and with higher variance also increases. Further experiments in the Supplementary material.

## 6 Conclusions

Stream learning with concept drift is an online learning setting where the distribution of the preferences changes with time. We denote as uBorda the version of Borda in which the voters have different weights and show that it can be computed efficiently, handling a possibly infinite sequence of rankings in linear space and quasi-linear time complexity. A detailed analysis leads to several contributions. First, we bound the number of samples required by uBorda to output the current ground truth modal ranking with high probability. Second, we show how to include expert knowledge to the uBorda. We generalize families of functions (including Borda and Condorcet) to the situation in which some voters are more reliable than others. We call this scenario *unbalanced voting* and allow arbitrary *reliability* values for the voters. This work, specially Sect. 4, opens natural research lines in the context of weighted voting. We plan to study statistical properties of the different voting rules in the context of expert vs non-expert voters.

**Acknowledgments.** This work is partially funded by the Industrial Chair "Data science & Artificial Intelligence for Digitalized Industry & Services" from Telecom Paris (France), the Basque Government through the BERC 2018–2021 and the Elkartek program (KK-2018/00096, KK-2020/00049), and by the Spanish Government excellence accreditation Severo Ochoa SEV-2013-0323 (MICIU) and the project TIN2017-82626-R (MINECO). J. Del Ser also acknowledges funding support from the Basque Government (Consolidated Research Gr. MATHMODE, IT1294-19).

## References

1. Ailon, N.: Improved bounds for online learning over the permutahedron and other ranking polytopes. In: Artificial Intelligence and Statistics, pp. 29–37 (2014)
2. Ailon, N., Charikar, M., Newman, A.: Aggregating inconsistent information: ranking and clustering. J. ACM **55**, 1–27 (2008)
3. Borda, J.: Memoire sur les elections au scrutin. Histoire de l'Academie Royal des Sciences **102**, 657–665 (1781)
4. Brandt, F., Conitzer, V., Endriss, U., Lang, J., Procaccia, A.D.: Handbook of Computational Social Choice (2016)
5. Braverman, M., Mossel, E.: Noisy sorting without resampling. In: Proceedings of the 19th Annual Symposium on Discrete Algorithms (2008)
6. Busa-Fekete, R., Hüllermeier, E., Szörényi, B.: Preference-based rank elicitation using statistical models: the case of Mallows. In: Proceedings of the 31th International Conference on Machine Learning (ICML), pp. 1071–1079 (2014)
7. Caragiannis, I., Procaccia, A.D., Shah, N.: When do noisy votes reveal the truth? In: Proceedings of the 14th ACM Conference on Electronic Commerce, pp. 143–160 (2013)
8. Coppersmith, D., Fleischer, L.K., Rurda, A.: Ordering by weighted number of wins gives a good ranking for weighted tournaments. ACM Trans. Algorithms **6**(3), 1–13 (2010)
9. Dwork, C., Kumar, R., Naor, M., Sivakumar, D.: Rank aggregation methods for the Web. In: International Conference on World Wide Web (2001)

10. Fligner, M.A., Verducci, J.S.: Distance based ranking models. J. Roy. Stat. Soc. **48**(3), 359–369 (1986)
11. Fligner, M.A., Verducci, J.S.: Multistage ranking models. J. Am. Stat. Assoc. **83**(403), 892–901 (1988)
12. Freeman, R., Zahedi, S.M., Conitzer, V.: Fair social choice in dynamic settings. In: Proceedings of the 26th International Joint Conference on Artificial Intelligence (IJCAI) (2017)
13. Gam, J.: Knowledge Discovery from Data Streams (2010)
14. Gama, J., Zliobaite, I., Bifet, A., Pechenizkiy, M., Bouchachia, A.: Survey on concept drift adaptation. ACM Comput. Surv. **46**(4), 44 (2014)
15. Hosseini, H., Larson, K., Cohen, R.: Matching with dynamic ordinal preferences. In: 29th AAAI Conference on Artificial Intelligence (2015)
16. Irurozki, E., Calvo, B., Lozano, J.A.: PerMallows: an R package for mallows and generalized mallows models. J. Stat. Softw. **71**, 1–30 (2019)
17. Karger, D.R., Oh, S., Shah, D.: Iterative learning for reliable crowdsourcing systems. In: Advances in Neural Information Processing Systems 24: 25th Annual Conference on Neural Information Processing Systems 2011, NIPS 2011 (2011)
18. Krempl, G., et al.: Open challenges for data stream mining research. ACM SIGKDD Explor. Newsl. **16**, 1–10 (2014)
19. Lu, T., Boutilier, C.: Learning Mallows models with pairwise preferences. In: Proceedings of the 28th International Conference on Machine Learning, ICML 2011 (2011)
20. Mallows, C.L.: Non-null ranking models. Biometrika **44**(1–2), 114–130 (1957)
21. Parkes, D.C., Procaccia, A.D.: Dynamic social choice with evolving preferences. In: 27th AAAI Conference on Artificial Intelligence (2013)
22. Siddiqui, Z.F., Tiakas, E., Symeonidis, P., Spiliopoulou, M., Manolopoulos, Y.: xStreams: recommending items to users with time-evolving preferences. In: Proceedings of the 4th International Conference on Web Intelligence, Mining and Semantics, WIMS 2014 (2014)
23. Tal, M., Meir, R., Gal, Y.: A study of human behavior in online voting. In: Proceedings of the 2015 International Conference on Autonomous Agents and Multiagent Systems (2015)
24. Tideman, T.N.: Independence of clones as a criterion for voting rules. Soc. Choice Welf. **4**, 185–206 (1987). https://doi.org/10.1007/BF00433944
25. Vitelli, V., Sørensen, Ø., Crispino, M., Frigessi, A., Arjas, E.: Probabilistic preference learning with the Mallows rank model. J. Mach. Learn. Res. **18**(1), 1–49 (2018)
26. Xia, L.: Learning and Decision-Making from Rank Data. Synthesis Lectures on Artificial Intelligence and Machine Learning, vol. 13, no. 1, pp. 1–159 (2019)
27. Yasutake, S., Hatano, K., Takimoto, E., Takeda, M., Hoi, S.C.H., Buntine, W.: Online rank aggregation. In: Asian Conference on Machine Learning (2012)

# Adaptive Optimizers with Sparse Group Lasso for Neural Networks in CTR Prediction

Yun Yue(✉), Yongchao Liu, Suo Tong, Minghao Li, Zhen Zhang, Chunyang Wen, Huanjun Bao, Lihong Gu, Jinjie Gu, and Yixiang Mu

Ant Group, No. 556 Xixi Road, Xihu District, Hangzhou, Zhejiang, China
{yueyun.yy,yongchao.ly,tongsuo.ts,chris.lmh,elliott.zz,chengfu.wcy, alex.bao,lihong.glh,jinjie.gujj,yixiang.myx}@antgroup.com

**Abstract.** We develop a novel framework that adds the regularizers of the sparse group lasso to a family of adaptive optimizers in deep learning, such as MOMENTUM, ADAGRAD, ADAM, AMSGRAD, ADAHESSIAN, and create a new class of optimizers, which are named GROUP MOMENTUM, GROUP ADAGRAD, GROUP ADAM, GROUP AMSGRAD and GROUP ADAHESSIAN, etc., accordingly. We establish theoretically proven convergence guarantees in the stochastic convex settings, based on primal-dual methods. We evaluate the regularized effect of our new optimizers on three large-scale real-world ad click datasets with state-of-the-art deep learning models. The experimental results reveal that compared with the original optimizers with the post-processing procedure which uses the magnitude pruning method, the performance of the models can be significantly improved on the same sparsity level. Furthermore, in comparison to the cases without magnitude pruning, our methods can achieve extremely high sparsity with significantly better or highly competitive performance.

**Keywords:** Adaptive optimizers · Sparse group lasso · DNN models · Online optimization

## 1 Introduction

With the development of deep learning, deep neural network (DNN) models have been widely used in various machine learning scenarios such as search, recommendation and advertisement, and achieved significant improvements. In the last decades, different kinds of optimization methods based on the variations of stochastic gradient descent (SGD) have been invented for training DNN models. However, most optimizers cannot directly produce sparsity which has been proven effective and efficient for saving computational resource and improving model performance especially in the scenarios of very high-dimensional data. Meanwhile, the simple rounding approach is very unreliable due to the inherent low accuracy of these optimizers.


N. Oliver et al. (Eds.): ECML PKDD 2021, LNAI 12977, pp. 314–329, 2021.
https://doi.org/10.1007/978-3-030-86523-8_19

In this paper, we develop a new class of optimization methods, that adds the regularizers especially sparse group lasso to prevalent adaptive optimizers, and retains the characteristics of the respective optimizers. Compared with the original optimizers with the post-processing procedure which use the magnitude pruning method, the performance of the models can be significantly improved on the same sparsity level. Furthermore, in comparison to the cases without magnitude pruning, the new optimizers can achieve extremely high sparsity with significantly better or highly competitive performance. In this section, we describe the two types of optimization methods, and explain the motivation of our work.

### 1.1 Adaptive Optimization Methods

Due to the simplicity and effectiveness, adaptive optimization methods [4,8,17,19,20,26,27] have become the de-facto standard algorithms used in deep learning. There are multiple variants, but they can be represented using the general update formula [19]:

$$x_{t+1} = x_t - \alpha_t m_t / \sqrt{V_t}, \tag{1}$$

where $\alpha_t$ is the step size, $m_t$ is the first moment term which is the weighted average of gradient $g_t$ and $V_t$ is the so called second moment term that adjusts updated velocity of variable $x_t$ in each direction. Here, $\sqrt{V_t} := V_t^{1/2}$, $m_t/\sqrt{V_t} := \sqrt{V_t}^{-1} \cdot m_t$. By setting different $m_t$, $V_t$ and $\alpha_t$ , we can derive different adaptive optimizers including Momentum [17], Adagrad [4], Adam [8], AMSGrad [19] and AdaHessian [26], etc. See Table 1.

**Table 1.** Adaptive optimizers with choosing different $m_t$, $V_t$ and $\alpha_t$.

| Optimizer | $m_t$ | $V_t$ | $\alpha_t$ |
|---|---|---|---|
| Sgd | $g_t$ | $\mathbb{I}$ | $\frac{\alpha}{\sqrt{t}}$ |
| Momentum | $\gamma m_{t-1} + g_t$ | $\mathbb{I}$ | $\alpha$ |
| Adagrad | $g_t$ | $\text{diag}(\sum_{i=1}^{t} g_i^2)/t$ | $\frac{\alpha}{\sqrt{t}}$ |
| Adam | $\beta_1 m_{t-1} + (1-\beta_1) g_t$ | $\beta_2 V_{t-1} + (1-\beta_2)\text{diag}(g_t^2)$ | $\frac{\alpha\sqrt{1-\beta_2^t}}{1-\beta_1^t}$ |
| AMSGrad | $\beta_1 m_{t-1} + (1-\beta_1) g_t$ | $\max(V_{t-1}, \beta_2 V_{t-1} + (1-\beta_2)\text{diag}(g_t^2))$ | $\frac{\alpha\sqrt{1-\beta_2^t}}{1-\beta_1^t}$ |
| AdaHessian | $\beta_1 m_{t-1} + (1-\beta_1) g_t$ | $\beta_2 V_{t-1} + (1-\beta_2) D_t^2$ * | $\frac{\alpha\sqrt{1-\beta_2^t}}{1-\beta_1^t}$ |

* $D_t = \text{diag}(H_t)$, where $H_t$ is the Hessian matrix.

### 1.2 Regularized Optimization Methods

Follow-the-regularized-leader (FTRL) [11,12] has been widely used in click-through rates (CTR) prediction problems, which adds $\ell_1$-regularization (lasso) to logistic regression and can effectively balance the performance of the model and the sparsity of features. The update formula [11] is:

$$x_{t+1} = \arg\min_{x} g_{1:t} \cdot x + \frac{1}{2}\sum_{s=1}^{t} \sigma_s \|x - x_s\|_2^2 + \lambda_1 \|x\|_1, \tag{2}$$

where $g_{1:t} = \sum_{s=1}^{t} g_s$, $\frac{1}{2}\sum_{s=1}^{t} \sigma_s \|x - x_s\|_2^2$ is the strong convex term that stabilizes the algorithm and $\lambda_1 \|x\|_1$ is the regularization term that produces sparsity. However, it doesn't work well in DNN models since one input feature can correspond to multiple weights and lasso only can make single weight zero hence can't effectively delete features.

To solve above problem, [16] adds the $\ell_{21}$-regularization (group lasso) to FTRL, which is named G-FTRL. [25] conducts the research on a group lasso method for online learning that adds $\ell_{21}$-regularization to the algorithm of Dual Averaging (DA) [15], which is named DA-GL. Even so, these two methods cannot be applied to other optimizers. Different scenarios are suitable for different optimizers in the deep learning fields. For example, MOMENTUM [17] is typically used in computer vision; ADAM [8] is used for training transformer models for natural language processing; and ADAGRAD [4] is used for recommendation systems. If we want to produce sparsity of the model in some scenario, we have to change optimizer which probably influence the performance of the model.

### 1.3 Motivation

Equation (1) can be rewritten into this form:

$$x_{t+1} = \arg\min_{x} m_t \cdot x + \frac{1}{2\alpha_t} \|\sqrt{V_t}^{\frac{1}{2}} (x - x_t)\|_2^2. \tag{3}$$

Furthermore, we can rewrite Eq. (3) into

$$x_{t+1} = \arg\min_{x} m_{1:t} \cdot x + \sum_{s=1}^{t} \frac{1}{2\alpha_s} \|Q_s^{\frac{1}{2}} (x - x_s)\|_2^2, \tag{4}$$

where $m_{1:t} = \sum_{s=1}^{t} m_s$, $\sum_{s=1}^{t} Q_s/\alpha_s = \sqrt{V_t}/\alpha_t$. It is easy to prove that Eq. (3) and Eq. (4) are equivalent using the method of induction. The matrices $Q_s$ can be interpreted as generalized learning rates. To our best knowledge, $V_t$ of Eq. (1) of all the adaptive optimization methods are diagonal for the computation simplicity. Therefore, we consider $Q_s$ as diagonal matrices throughout this paper.

We find that Eq. (4) is similar to Eq. (2) except for the regularization term. Therefore, we add the regularization term $\Psi(x)$ to Eq. (4), which is the sparse group lasso penalty also including $\ell_2$-regularization that can diffuse weights of neural networks. The concrete formula is:

$$\Psi_t(x) = \sum_{g=1}^{G} \left( \lambda_1 \|x^g\|_1 + \lambda_{21} \sqrt{d_{x^g}} \|A_t^{\frac{1}{2}} x^g\|_2 \right) + \lambda_2 \|x\|_2^2, \tag{5}$$

where $\lambda_1$, $\lambda_{21}$, $\lambda_2$ are regularization parameters of $\ell_1$, $\ell_{21}$, $\ell_2$ respectively, $G$ is the total number of groups of weights, $x^g$ is the weights of group $g$ and $d_{x^g}$ is the size of group $g$. In DNN models, each group is defined as the set of outgoing weights from a unit which can be an input feature, or a hidden neuron, or a bias unit (see, e.g., [22]). $A_t$ can be arbitrary positive matrix satisfying $A_{t+1} \succeq A_t$, e.g., $A_t = \mathbb{I}$. In Sect. 2.1, we let $A_t = (\sum_{s=1}^{t} \frac{Q_s^g}{2\alpha_s} + \lambda_2 \mathbb{I})$ just for solving the closed-form solution directly, where $Q_s^g$ is a diagonal matrix whose diagonal elements are part of $Q_s$ corresponding to $x_g$. The ultimate update formula is:

$$x_{t+1} = \arg\min_{x} m_{1:t} \cdot x + \sum_{s=1}^{t} \frac{1}{2\alpha_s} \|Q_s^{\frac{1}{2}}(x - x_s)\|_2^2 + \Psi_t(x). \quad (6)$$

### 1.4 Outline of Contents

The rest of the paper is organized as follows. In Sect. 1.5, we introduce the necessary notations and technical background.

In Sect. 2, we present the closed-form solution of Eq. (4) and the algorithm of general framework of adaptive optimization methods with sparse group lasso. We prove the algorithm is equivalent to adaptive optimization methods when regularization terms vanish. In the end, we give two concrete examples of the algorithm.[1]

In Sect. 3, we derive the regret bounds of the method and convergence rates.

In Sect. 4, we validate the performance of new optimizers in the public datasets.

In Sect. 5, we summarize the conclusion.

Appendices A–D [29] contain technical proofs of our main results and Appendix E [29] includes the additional details of the experiments of Sect. 4.

### 1.5 Notations and Technical Background

We use lowercase letters to denote scalars and vectors, and uppercase letters to denote matrices. We denote a sequence of vectors by subscripts, that is, $x_1, \ldots, x_t$, and entries of each vector by an additional subscript, e.g., $x_{t,i}$. We use the notation $g_{1:t}$ as a shorthand for $\sum_{s=1}^{t} g_s$. Similarly we write $m_{1:t}$ for a sum of the first moment $m_t$, and $f_{1:t}$ to denote the function $f_{1:t}(x) = \sum_{s=1}^{t} f_s(x)$. Let $M_t = [m_1 \cdots m_t]$ denote the matrix obtained by concatenating the vector sequence $\{m_t\}_{t \geq 1}$ and $M_{t,i}$ denote the $i$-th row of this matrix which amounts to the concatenation of the $i$-th component of each vector. The notation $A \succeq 0$ (resp. $A \succ 0$) for a matrix A means that A is symmetric and positive semidefinite (resp. definite). Similarly, the notations $A \succeq B$ and $A \succ B$ mean that $A - B \succeq 0$ and $A - B \succ 0$ respectively, and both tacitly assume that $A$ and $B$ are symmetric. Given $A \succeq 0$, we write $A^{\frac{1}{2}}$ for the square root of $A$, the unique $X \succeq 0$ such that $XX = A$ ([12], Sect. 1.4).

[1] The codes will be released if the paper is accepted.

Let $\mathcal{E}$ be a finite-dimension real vector space, endowed with the Mahalanobis norm $\|\cdot\|_A$ which is denoted by $\|\cdot\|_A = \sqrt{\langle \cdot, A \cdot \rangle}$ as induced by $A \succ 0$. Let $\mathcal{E}^*$ be the vector space of all linear functions on $\mathcal{E}$. The dual space $\mathcal{E}^*$ is endowed with the dual norm $\|\cdot\|_A^* = \sqrt{\langle \cdot, A^{-1} \cdot \rangle}$.

Let $\mathcal{Q}$ be a closed convex set in $\mathcal{E}$. A continuous function $h(x)$ is called *strongly convex* on $\mathcal{Q}$ with norm $\|\cdot\|_H$ if $\mathcal{Q} \subseteq \mathrm{dom}\, h$ and there exists a constant $\sigma > 0$ such that for all $x, y \in \mathcal{Q}$ and $\alpha \in [0, 1]$ we have

$$h(\alpha x + (1-\alpha)y) \leq \alpha h(x) + (1-\alpha)h(y) - \frac{1}{2}\sigma\alpha(1-\alpha)\|x - y\|_H^2.$$

The constant $\sigma$ is called the *convexity parameter* of $h(x)$, or the *modulus* of strong convexity. We also denote by $\|\cdot\|_h = \|\cdot\|_H$. Further, if $h$ is differentiable, we have

$$h(y) \geq h(x) + \langle \nabla h(x), y - x \rangle + \frac{\sigma}{2}\|x - y\|_h^2.$$

We use online convex optimization as our analysis framework. On each round $t = 1, \ldots, T$, a convex loss function $f_t : \mathcal{Q} \mapsto \mathbb{R}$ is chosen, and we pick a point $x_t \in \mathcal{Q}$ hence get loss $f_t(x_t)$. Our goal is minimizing the *regret* which is defined as the quantity

$$\mathcal{R}_T = \sum_{t=1}^{T} f_t(x_t) - \min_{x \in \mathcal{Q}} \sum_{t=1}^{T} f_t(x). \tag{7}$$

Online convex optimization can be seen as a generalization of stochastic convex optimization. Any regret minimizing algorithm can be converted to a stochastic optimization algorithm with convergence rate $O(\mathcal{R}_T/T)$ using an online-to-batch conversion technique [9].

In this paper, we assume $\mathcal{Q} \equiv \mathcal{E} = \mathbb{R}^n$, hence we have $\mathcal{E}^* = \mathbb{R}^n$. We write $s^T x$ or $s \cdot x$ for the standard inner product between $s, x \in \mathbb{R}^n$. For the standard Euclidean norm, $\|x\| = \|x\|_2 = \sqrt{\langle x, x \rangle}$ and $\|s\|_* = \|s\|_2$. We also use $\|x\|_1 = \sum_{i=1}^n |x^{(i)}|$ and $\|x\|_\infty = \max_i |x^{(i)}|$ to denote $\ell_1$-norm and $\ell_\infty$-norm respectively, where $x^{(i)}$ is the $i$-th element of $x$.

## 2 Algorithm

### 2.1 Closed-Form Solution

We will derive the closed-form solution of Eq. (6) with specific $A_t$ and Algorithm 1 with slight modification in this section. We have the following theorem.

**Theorem 1.** *Given $A_t = (\sum_{s=1}^t \frac{Q_s^g}{2\alpha_s} + \lambda_2 \mathbb{I})$ of Eq. (5), $z_t = z_{t-1} + m_t - \frac{Q_t}{\alpha_t} x_t$ at each iteration $t = 1, \ldots, T$ and $z_0 = \mathbf{0}$, the optimal solution of Eq. (6) is updated accordingly as follows:*

$$x_{t+1} = (\sum_{s=1}^{t} \frac{Q_s}{\alpha_s} + 2\lambda_2 \mathbb{I})^{-1} \max(1 - \frac{\sqrt{d_{x_t^g}}\lambda_{21}}{\|\tilde{s}_t\|_2}, 0) s_t \tag{8}$$

*where the i-th element of* $s_t$ *is defined as*

$$s_{t,i} = \begin{cases} 0 & \text{if } |z_{t,i}| \leq \lambda_1, \\ \text{sign}(z_{t,i})\lambda_1 - z_{t,i} & \text{otherwise,} \end{cases} \tag{9}$$

$\tilde{s}_t$ *is defined as*

$$\tilde{s}_t = (\sum_{s=1}^{t} \frac{Q_s}{2\alpha_s} + \lambda_2 \mathbb{I})^{-1} s_t \tag{10}$$

*and* $\sum_{s=1}^{t} \frac{Q_s}{\alpha_s}$ *is the diagonal and positive definite matrix.*

The proof of Theorem 1 is given in Appendix A [29]. Here $\tilde{s}_t$ can be considered as the weighted average of $s_t$. We slightly modify (8) where we replace $\tilde{s}_t$ with $s_t$ in practical algorithms. Our purpose is that the $\ell_{21}$-regularization does not depend on the second moment terms and other hyperparameters such as $\alpha_s$ and $\lambda_2$. The empirical experiment will also show that the algorithm using $s_t$ can improve accuracy over using $\tilde{s}_t$ in the same level of sparsity in Sect. 4.4. Therefore, we get Algorithm 1. Furthermore, we have the following theorem which shows the relationship between Algorithm 1 and adaptive optimization methods. The proof is given in Appendix B [29].

---

**Algorithm 1.** Generic framework of adaptive optimization methods with sparse group lasso

---

1: **Input:** parameters $\lambda_1, \lambda_{21}, \lambda_2$
   $x_1 \in \mathbb{R}^n$, step size $\{\alpha_t > 0\}_{t=0}^{T}$, sequence of functions $\{\phi_t, \psi_t\}_{t=1}^{T}$, initialize $z_0 = \mathbf{0}, V_0 = \mathbf{0}$
2: **for** $t = 1$ **to** $T$ **do**
3: $\quad g_t = \nabla f_t(x_t)$
4: $\quad m_t = \phi_t(g_1, \ldots, g_t)$ and $V_t = \psi_t(g_1, \ldots, g_t)$
5: $\quad \frac{Q_t}{\alpha_t} = \frac{\sqrt{V_t}}{\alpha_t} - \frac{\sqrt{V_{t-1}}}{\alpha_{t-1}}$
6: $\quad z_t \leftarrow z_{t-1} + m_t - \frac{Q_t}{\alpha_t} x_t$
7: $\quad$ **for** $i \in \{1, \ldots, n\}$ **do**
8: $\quad\quad s_{t,i} = \begin{cases} 0 & \text{if } |z_{t,i}| \leq \lambda_1 \\ \text{sign}(z_{t,i})\lambda_1 - z_{t,i} & \text{otherwise.} \end{cases}$
9: $\quad$ **end for**
10: $\quad x_{t+1} = (\frac{\sqrt{V_t}}{\alpha_t} + 2\lambda_2 \mathbb{I})^{-1} \max(1 - \frac{\sqrt{d_{x_t^g}}\lambda_{21}}{\|s_t\|_2}, 0) s_t$
11: **end for**

---

**Theorem 2.** *If regularization terms of Algorithm 1 vanish, Algorithm 1 is equivalent to Eq.* (1).

## 2.2 Concrete Examples

Using Algorithm 1, we can easily derive the new optimizers based on Adam [8], Adagrad [4] which we call Group Adam, Group Adagrad respectively.

**Algorithm 2.. Group Adam**

1: **Input:** parameters $\lambda_1, \lambda_{21}, \lambda_2, \beta_1, \beta_2, \epsilon$ $x_1 \in \mathbb{R}^n$, step size $\alpha$, initialize $z_0 = \mathbf{0}, \hat{m}_0 = \mathbf{0}, \hat{V}_0 = \mathbf{0}, V_0 = \mathbf{0}$
2: **for** $t = 1$ **to** $T$ **do**
3: $\quad g_t = \nabla f_t(x_t)$
4: $\quad \hat{m}_t \leftarrow \beta_1 \hat{m}_{t-1} + (1 - \beta_1) g_t$
5: $\quad m_t = \hat{m}_t / (1 - \beta_1^t)$
6: $\quad \hat{V}_t \leftarrow \beta_2 \hat{V}_{t-1} + (1 - \beta_2)\mathrm{diag}(g_t^2)$
7: $\quad V_t = \hat{V}_t / (1 - \beta_2^t)$
8: $\quad Q_t = \begin{cases} \sqrt{V_t} - \sqrt{V_{t-1}} + \epsilon\mathbb{I} & t = 1 \\ \sqrt{V_t} - \sqrt{V_{t-1}} & t > 1 \end{cases}$
9: $\quad z_t \leftarrow z_{t-1} + m_t - \frac{1}{\alpha} Q_t x_t$
10: $\quad$ **for** $i \in \{1, \ldots, n\}$ **do**
11: $\quad\quad s_{t,i} = -\,\mathrm{sign}(z_{t,i}) \max(|z_{t,i}| - \lambda_1, 0)$
12: $\quad$ **end for**
13: $\quad x_{t+1} = (\frac{\sqrt{V_t + \epsilon\mathbb{I}}}{\alpha} + 2\lambda_2 \mathbb{I})^{-1} \max(1 - \frac{\sqrt{d_{x_t^g}}\lambda_{21}}{\|s_t\|_2}, 0) s_t$
14: **end for**

**Algorithm 3.. Group Adagrad**

1: **Input:** parameters $\lambda_1, \lambda_{21}, \lambda_2, \epsilon$ $x_1 \in \mathbb{R}^n$, step size $\alpha$, initialize $z_0 = \mathbf{0}, V_0 = \mathbf{0}$
2: **for** $t = 1$ **to** $T$ **do**
3: $\quad g_t = \nabla f_t(x_t)$
4: $\quad m_t = g_t$
5: $\quad V_t = \begin{cases} V_{t-1} + \mathrm{diag}(g_t^2) + \epsilon\mathbb{I} & t = 1 \\ V_{t-1} + \mathrm{diag}(g_t^2) & t > 1 \end{cases}$
6: $\quad Q_t = \sqrt{V_t} - \sqrt{V_{t-1}}$
7: $\quad z_t \leftarrow z_{t-1} + m_t - \frac{1}{\alpha} Q_t x_t$
8: $\quad$ **for** $i \in \{1, \ldots, n\}$ **do**
9: $\quad\quad s_{t,i} = -\,\mathrm{sign}(z_{t,i}) \max(|z_{t,i}| - \lambda_1, 0)$
10: $\quad$ **end for**
11: $\quad x_{t+1} = (\frac{\sqrt{V_t}}{\alpha} + 2\lambda_2 \mathbb{I})^{-1} \max(1 - \frac{\sqrt{d_{x_t^g}}\lambda_{21}}{\|s_t\|_2}, 0) s_t$
12: **end for**

**Group Adam.** The detail of the algorithm is given in Algorithm 2. From Theorem 2, we know that when $\lambda_1$, $\lambda_2$, $\lambda_{21}$ are all zeros, Algorithm 2 is equivalent to ADAM [8].

**Group Adagrad.** The detail of the algorithm is given in Algorithm 3. Similarly, from Theorem 2, when $\lambda_1$, $\lambda_2$, $\lambda_{21}$ are all zeros, Algorithm 3 is equivalent to ADAGRAD [4]. Furthermore, we can find that when $\lambda_{21} = 0$, Algorithm 3 is equivalent to FTRL [11]. Therefore, GROUP ADAGRAD can also be called GROUP FTRL from the research of [16].

Similarly, GROUP MOMENTUM, GROUP AMSGRAD, GROUP ADAHESSIAN, etc., can be derived from MOMENTUM [17], AMSGRAD [19], ADAHESSIAN [26], etc., with the same framework and we will not list the details.

## 3 Convergence and Regret Analysis

Using the framework developed in [4,15,24], we have the following theorem providing the bound of the regret.

**Theorem 3.** *Let the sequence* $\{x_t\}$ *be defined by the update* (6) *and*

$$x_1 = \arg\min_{x \in \mathcal{Q}} \frac{1}{2}\|x - c\|_2^2, \tag{11}$$

*where* $c$ *is an arbitrary constant vector. Suppose* $f_t(x)$ *is convex for any* $t \geq 1$ *and there exists an optimal solution* $x^*$ *of* $\sum_{t=1}^T f_t(x)$, *i.e.,* $x^* = \arg\min_{x \in \mathcal{Q}} \sum_{t=1}^T f_t(x)$, *which satisfies the condition*

$$\langle m_{t-1}, x_t - x^* \rangle \geq 0, \quad t \in [T], \tag{12}$$

*where* $m_t$ *is the weighted average of the gradient* $f_t(x_t)$ *and* $[T] = \{1, \dots, T\}$ *for simplicity. Without loss of generality, we assume*

$$m_t = \gamma m_{t-1} + g_t, \tag{13}$$

*where* $\gamma < 1$ *and* $m_0 = 0$*. Then*

$$\mathcal{R}_T \leq \Psi_T(x^*) + \sum_{t=1}^{T} \frac{1}{2\alpha_t} \|Q_t^{\frac{1}{2}}(x^* - x_t)\|_2^2 + \frac{1}{2}\sum_{t=1}^{T} \|m_t\|_{h_{t-1}^*}^2, \tag{14}$$

*where* $\|\cdot\|_{h_t^*}$ *is the dual norm of* $\|\cdot\|_{h_t}$. $h_t$ *is* 1*-strongly convex with respect to* $\|\cdot\|_{\sqrt{V_t}/\alpha_t}$ *for* $t \in [T]$ *and* $h_0$ *is* 1*-strongly convex with respect to* $\|\cdot\|_2$.

The proof of Theorem 3 is given in Appendix C [29]. Since in most of adaptive optimizers, $V_t$ is the weighted average of $\text{diag}(g_t^2)$, without loss of generality, we assume $\alpha_t = \alpha$ and

$$V_t = \eta V_{t-1} + \text{diag}(g_t^2), \quad t \geq 1, \tag{15}$$

where $V_0 = 0$ and $\eta \leq 1$. Hence, we have the following lemma whose proof is given in Appendix D.1 [29].

**Lemma 1.** *Suppose* $V_t$ *is the weighted average of the square of the gradient which is defined by* (15), $\alpha_t = \alpha$, $m_t$ *is defined by* (13) *and one of the following conditions:*

1. $\eta = 1$,
2. $\eta < 1$, $\eta \geq \gamma$ *and* $\kappa V_t \succeq V_{t-1}$ *for all* $t \geq 1$ *where* $\kappa < 1$.

*is satisfied. Then we have*

$$\sum_{t=1}^{T} \|m_t\|^2_{(\frac{\sqrt{V_t}}{\alpha_t})^{-1}} < \frac{2\alpha}{1-\nu} \sum_{i=1}^{d} \|M_{T,i}\|_2, \tag{16}$$

*where* $\nu = \max(\gamma, \kappa)$ *and* $d$ *is the dimension of* $x_t$.

We can always add $\delta^2\mathbb{I}$ to $V_t$ at each step to ensure $V_t \succ 0$. Therefore, $h_t(x)$ is 1-strongly convex with respect to $\|\cdot\|_{\sqrt{\delta^2\mathbb{I}+V_t}/\alpha_t}$. Let $\delta \geq \max_{t\in[T]} \|g_t\|_\infty$, for $t > 1$, we have

$$\begin{aligned}
\|m_t\|^2_{h_{t-1}^*} &= \left\langle m_t, \alpha_t(\delta^2\mathbb{I} + V_{t-1})^{-\frac{1}{2}} m_t \right\rangle \leq \left\langle m_t, \alpha_t \left(\text{diag}(g_t^2) + \eta V_{t-1}\right)^{-\frac{1}{2}} m_t \right\rangle \\
&= \left\langle m_t, \alpha_t V_t^{-\frac{1}{2}} m_t \right\rangle = \|m_t\|^2_{(\frac{\sqrt{V_t}}{\alpha_t})^{-1}}.
\end{aligned} \tag{17}$$

For $t = 1$, we have

$$\begin{aligned}
\|m_1\|^2_{h_0^*} &= \left\langle m_1, \alpha_1(\delta^2\mathbb{I} + \mathbb{I})^{-\frac{1}{2}} m_1 \right\rangle \leq \left\langle m_1, \alpha_1 \left(\text{diag}^{-\frac{1}{2}}(g_1^2)\right) m_1 \right\rangle \\
&= \left\langle m_1, \alpha_1 V_1^{-\frac{1}{2}} m_1 \right\rangle = \|m_1\|^2_{(\frac{\sqrt{V_1}}{\alpha_1})^{-1}}.
\end{aligned} \tag{18}$$

From (17), (18) and Lemma 1, we have

**Lemma 2.** *Suppose $V_t$, $m_t$, $\alpha_t$, $\nu$, $d$ are defined the same as Lemma 1, $\max_{t\in[T]} \|g_t\|_\infty \leq \delta$, $\|\cdot\|^2_{h^*_t} = \left\langle \cdot, \alpha_t(\delta^2\mathbb{I} + V_t)^{-\frac{1}{2}} \cdot \right\rangle$ for $t \geq 1$ and $\|\cdot\|^2_{h^*_0} = \left\langle \cdot, \alpha_1\left((\delta^2+1)\mathbb{I}\right)^{-\frac{1}{2}} \cdot \right\rangle$. Then*

$$\sum_{t=1}^{T} \|m_t\|^2_{h^*_{t-1}} < \frac{2\alpha}{1-\nu} \sum_{i=1}^{d} \|M_{T,i}\|_2. \tag{19}$$

Therefore, from Theorem 3 and Lemma 2, we have

**Corollary 1.** *Suppose $V_t$, $m_t$, $\alpha_t$, $h^*_t$, $\nu$, $d$ are defined the same as Lemma 2, there exist constants $G$, $D_1$, $D_2$ such that $\max_{t\in[T]} \|g_t\|_\infty \leq G \leq \delta$, $\|x^*\|_\infty \leq D_1$ and $\max_{t\in[T]} \|x_t - x^*\|_\infty \leq D_2$. Then*

$$\mathcal{R}_T < dD_1 \left( \lambda_1 + \lambda_{21}(\frac{\sqrt{T}G}{2\alpha} + \lambda_2)^{\frac{1}{2}} + \lambda_2 D_1 \right) + dG\left(\frac{D_2^2}{2\alpha} + \frac{\alpha}{(1-\nu)^2}\right)\sqrt{T}. \tag{20}$$

The proof of Corollary 1 is given in Appendix D.2 [29]. Furthermore, from Corollary 1, we have

**Corollary 2.** *Suppose $m_t$ is defined as (13), $\alpha_t = \alpha$ and satisfies the condition (19). There exist constants $G$, $D_1$, $D_2$ such that $tG^2\mathbb{I} \succeq V_t$, $\max_{t\in[T]} \|g_t\|_\infty \leq G$, $\|x^*\|_\infty \leq D_1$ and $\max_{t\in[T]} \|x_t - x^*\|_\infty \leq D_2$. Then*

$$\mathcal{R}_T < dD_1 \left( \lambda_1 + \lambda_{21}(\frac{\sqrt{T}G}{2\alpha} + \lambda_2)^{\frac{1}{2}} + \lambda_2 D_1 \right) + dG\left(\frac{D_2^2}{2\alpha} + \frac{\alpha}{(1-\nu)^2}\right)\sqrt{T}. \tag{21}$$

Therefore, we know that the regret of the update (6) is $O(\sqrt{T})$ and can achieve the optimal convergence rate $O(1/\sqrt{T})$ under the conditions of Corollary 1 or Corollary 2.

## 4 Experiments

### 4.1 Experiment Setup

We test the algorithms on three different large-scale real-world datasets with different neural network structures. These datasets are various display ads logs for the purpose of predicting ads CTR. The details are as follows.

a) The Avazu CTR dataset [2] contains approximately 40M samples and 22 categorical features over 10 days. In order to handle categorical data, we use the one-hot-encoding based embedding technique (see, e.g., [23], Sect.2.1 or [13], Sect. 2.1.1) and get 9.4M features in total. For this dataset, the samples from the first 9 days (containing 8.7M one-hot features) are used for training, while the rest is for testing. Our DNN model follows the basic structure of most deep CTR models. Specifically, the model comprises one embedding layer, which maps each one-hot feature into 16-dimensional embeddings, and four fully connected layers (with output dimension of 64, 32, 16 and 1, respectively) in sequence.
b) The iPinYou dataset[2] [7] is another real-world dataset for ad click logs over 21 days. The dataset contains 16 categorical features[3]. After one-hot encoding, we get a dataset containing 19.5M instances with 1033.1K input dimensions. We keep the original train/test splitting scheme, where the training set contains 15.4M samples with 937.7K one-hot features. We use Outer Product-based Neural Network (OPNN) [18], and follow the standard settings of [18], i.e., one embedding layer with the embedding dimension of 10, one product layer and three hidden layers of size 512, 256, 128 respectively where we set dropout rate at 0.5.
c) The third dataset is the Criteo Display Ads dataset [3] which contains approximately 46M samples over 7 days. There are 13 integer features and 26 categorical features. After one-hot encoding of categorical features, we have a total of 33.8M features. We split the dataset into 7 partitions in chronological order and select the earliest 6 parts for training which contains 29.6M features and the rest for testing though the dataset has no timestamp. We use Deep & Cross Network (DCN) [23] and choose the following settings[4]: one embedding layer with embedding dimension 8, two deep layers of size 64 each, and two cross layers.

For the convenience of discussion, we use MLP, OPNN and DCN to represent the aforementioned three datasets coupled with their corresponding models. It is obvious that the embedding layer has most of parameters of the neural networks when the features have very high dimension, therefore we just add the regularization terms to the embedding layer. Furthermore, each embedding vector is considered as a group, and a visual comparison between $\ell_1$, $\ell_{21}$ and mixed regularization effect is given in Fig. 2 of [22].

We treat the training set as the streaming data, hence we train 1 epoch with a batch size of 512 and do the validation. The experiments are conducted with 4–9 workers and 2–3 parameter servers in the TensorFlow framework [1], which depends on the different sizes of the datasets. According to [5], area under the receiver-operator curve (AUC) is a good measurement in CTR estimation and AUC is widely adopted as the evaluation criterion in classification problems.

[2] We only use the data from season 2 and 3 because of the same data schema.
[3] See https://github.com/Atomu2014/Ads-RecSys-Datasets/ for details.
[4] Limited by training resources available, we don't use the optimal hyperparameter settings of [23].

Thus we choose AUC as our evaluation criterion. We explore 5 learning rates from 1e−5 to 1e−1 with increments of 10× and choose the one with the best AUC for each new optimizer in the case of no regularization terms (It is equivalent to the original optimizer according to Theorem 2). The details are listed in Table 5 of Appendix E [29]. All the experiments are run 5 times repeatedly and tested statistical significance using t-test. Without loss of generality, we choose two new optimizers to validate the performance, which are Group Adam and Group AdaGrad.

### 4.2 Adam vs. Group Adam

First, we compare the performance of the two optimizers on the same sparsity level. We set $\lambda_1 = \lambda_2 = 0$ and choose different values of $\lambda_{21}$ of Algorithm 2, i.e., Group Adam, and achieve the same sparsity with Adam that uses the magnitude pruning method. Since we should delete the entire embedding vector which the feature corresponds to, not a single weight, and the amount of the features will dynamically increase as the training goes on, our method is different from the commonly used method [28]. Concretely, our method works in three steps. The first step sorts the norm of embedding vector from largest to smallest, and keeps top N embedding vectors which depend on the sparsity when finishing the first phase of training. In the second step we fine-tune the model. Since some new or deleted features will appear in the model after training with new data, in the last step we need to prune the model again to ensure that the desired sparsity is reached. We use the schedule of keeping 0%, 10%, 20%, 30% training samples to fine tune, and choose the best one. The details are listed in Table 8 of Appendix E [29].

Table 2 reports the average results of the two optimizers in the three datasets. Note that Group Adam significantly outperforms Adam on the AUC metric in the same sparsity level for most experiments especially under extreme sparsity. (60% experiments show statistically significant with 90% confidence level, and 87.5% experiments show statistically significant with 90% confidence level when sparsity level is less than 5%). Furthermore, as shown in Fig. 1, the same $\ell_{21}$-regularization strength $\lambda_{21}$ has different effects of sparsity and accuracy on different datasets. The best choice of $\lambda_{21}$ depends on the dataset as well as the application (For example, if the memory of serving resource is limited, sparsity might be relatively more important). One can trade off accuracy to get more sparsity by increasing the value of $\lambda_{21}$.

Next, we compare the performance of Adam without post-processing procedure, i.e., no magnitude pruning, and Group Adam under extremely high sparsity. We search regularization terms according to AUC and the values are listed in Table 6 of Appendix E [29]. In general, good default settings of $\lambda_2$ is 1e−5. The results are shown in Table 3. Note that compared with Adam, Group Adam with appropriate regularization terms can achieve significantly better or highly competitive performance with producing extremely high sparsity.

**Table 2.** AUC for the two optimizers and sparsity (feature rate) in parentheses. The best AUC for each dataset on each sparsity level is bolded. The p-value of the t-test of AUC is also listed.

| $\lambda_{21}$ | MLP | | | OPNN | | | DCN | | |
|---|---|---|---|---|---|---|---|---|---|
| Group Adam | Adam | Group Adam | P-Value | Adam | Group Adam | P-Value | Adam | Group Adam | P-Value |
| 1e−4 | 0.7457 (0.974) | **0.7461** (0.974) | 0.470 | 0.7551 (0.078) | **0.7595** (0.078) | 0.086 | 0.8018 (0.518) | **0.8022** (0.518) | 0.105 |
| 5e−4 | 0.7464 (0.864) | **0.7468** (0.864) | 0.466 | 0.7491 (0.039) | **0.7573** (0.039) | 0.091 | 0.8017 (0.062) | **0.8019** (0.062) | 0.487 |
| 1e−3 | 0.7452 (0.701) | **0.7468** (0.701) | 0.058 | 0.7465 (0.032) | **0.7595** (0.032) | 0.014 | 0.8017 (0.018) | 0.8017 (0.018) | 0.943 |
| 5e−3 | 0.7457 (0.132) | **0.7464** (0.132) | 0.335 | 0.7509 (0.018) | **0.7561** (0.018) | 0.041 | 0.7995 (4.2e−3) | **0.8007** (4.2e−3) | 9.11e−3 |
| 1e−2 | 0.7444 (0.038) | **0.7466** (0.038) | 0.014 | 0.7396 (9.2e−3) | **0.7493** (9.2e−3) | 0.031 | 0.7972 (2.5e−3) | **0.7999** (2.5e−3) | 5.97e−7 |

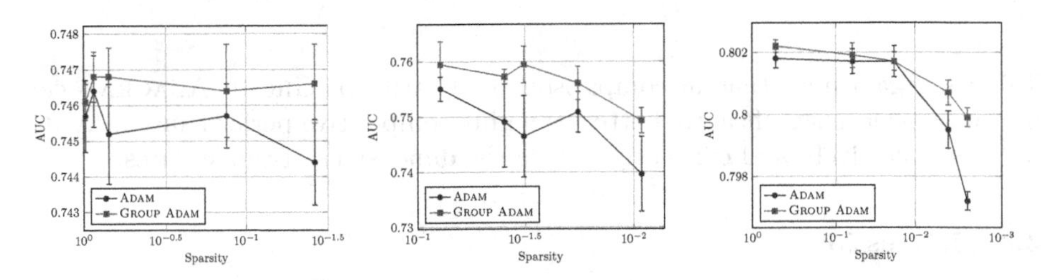


**Fig. 1.** AUC across different sparsity on two optimizers for the three datasets. MLP, OPNN and DCN are in left, middle, right column respectively. The x-axis is sparsity (number of non-zero features whose embedding vectors are not equal to **0** divided by the total number of features present in the training data). The y-axis is AUC. Error bars represent one standard deviation.

**Table 3.** AUC for three datasets and sparsity (feature rate) in parentheses. The best value for each dataset is bolded. The p-value of t-test is also listed.

| Dataset | Adam | Group Adam | P-Value | Adagrad | Group Adagrad | P-Value |
|---|---|---|---|---|---|---|
| MLP | 0.7458 (1.000) | **0.7486** (**0.018**) | 1.10e−3 (2.69e−11) | 0.7453 (1.000) | **0.7469** (**0.063**) | 0.106 (1.51e−9) |
| OPNN | 0.7588 (0.827) | **0.7617** (**0.130**) | 0.289 (6.20e−11) | 0.7556 (0.827) | **0.7595** (**0.016**) | 0.026 (<2.2e−16) |
| DCN | **0.8021** (1.000) | 0.8019 (**0.030**) | 0.422 (1.44e−11) | 0.7975 (1.000) | **0.7978** (**0.040**) | 0.198 (3.94e−11) |

### 4.3 Adagrad vs. Group Adagrad

We compare the performance of Adagrad without magnitude pruning and Group Adagrad under extremely high sparsity. The regularization terms we choose are listed in Table 7 of Appendix E [29]. The results are also shown in

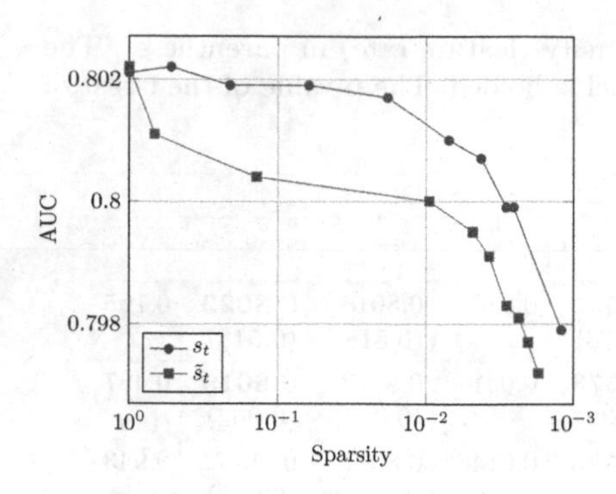


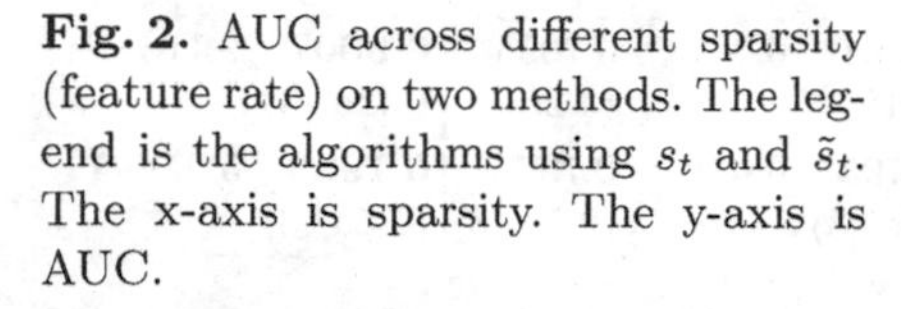
**Fig. 2.** AUC across different sparsity (feature rate) on two methods. The legend is the algorithms using $s_t$ and $\tilde{s}_t$. The x-axis is sparsity. The y-axis is AUC.

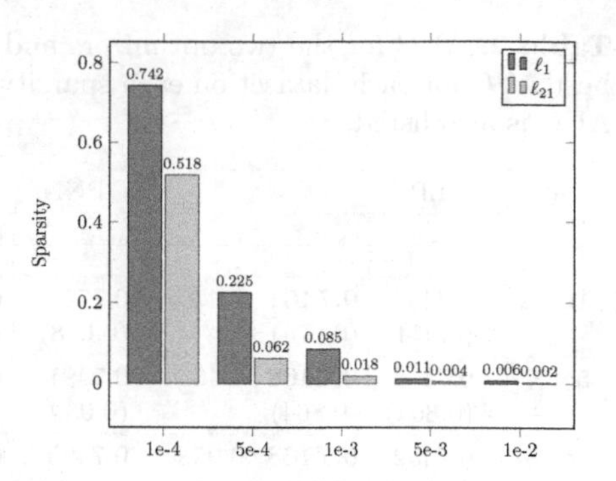


**Fig. 3.** The sparsity (feature rate) across different values of regularized terms. The legend is the regularized terms. The x-axis is the values of regularized terms. The y-axis is sparsity.

Table 3. Again note that in comparison to ADAGRAD, GROUP ADAGRAD can not only achieve significantly better or highly competitive performance of AUC, but also effectively and efficiently reduce the dimensions of the features.

### 4.4 Discussion

In this section we will compare the performance of $s_t$ with $\tilde{s}_t$ discussed in Sect. 2.1, i.e., using $\tilde{s}_t$ means that replacing $\|\tilde{s}_t\|$ with $\|s_t\|$ in line 10 of Algorithm 1. Furthermore, we will discuss the hyperparameters of $\ell_1$-regularization, $\ell_{21}$-regularization and embedding dimension to show how these hyperparameters affect the effects of regularization. Without loss of generality, all experiments are conducted on DCN using GROUP ADAM. The default settings of regularization terms are all zeros, unless otherwise specified.

**$s_t$ *vs.* $\tilde{s}_t$.** We choose $\ell_{21}$-regularization of $s_t$ and $\tilde{s}_t$ from 10 points in different sparsity levels. The details are listed in Table 9 of Appendix E [29]. As shown in Fig. 2, the algorithm using $s_t$ outperforms the one using $\tilde{s}_t$ in the same level of sparsity.

**$\ell_1$ *vs.* $\ell_{21}$.** From lines 8 and 10 of Algorithm 1, we know that if $z_t$ has the same elements, the values of $\ell_1$ and $\ell_{21}$, i.e., $\lambda_1$ and $\lambda_{21}$, have the same regularization effects. However, this situation almost cannot happen in reality. We compare the regularization performance with the same values of $\lambda_1$ and $\lambda_{21}$. The results are shown in Fig. 3. It is obvious that $\ell_{21}$-regularization is much more effective than $\ell_1$-regularization in producing sparsity. Therefore, if we just need to produce a sparse model, tuning $\lambda_{21}$ while keeping $\lambda_1 = 0$ is usually a simple but effective choice.

*Embedding Dimension.* Table 4 reports the average results of different embedding dimensions, whose regularization terms are same to DCN of Table 6 of Appendix E [29]. Note that the sparsity increases with the growth of the embedding dimension. The reason is that the square root of the embedding dimension is the multiplier of $\ell_{21}$-regularization.

**Table 4.** The sparsity (feature rate) for different embedding dimensions and AUC in parentheses. The best results are bolded.

| Embedding dimension | Group Adam |
|---|---|
| 4 | 0.074 (0.8008) |
| 8 | 0.030 (0.8019) |
| 16 | 0.012 (**0.8020**) |
| 32 | **0.008** (0.8011) |

## 5 Conclusion

In this paper, we propose a novel framework that adds the regularization terms to a family of adaptive optimizers for producing sparsity of DNN models. We apply this framework to create a new class of optimizers. We provide closed-form solutions and algorithms with slight modification. We built the relation between new and original optimizers, i.e., our new optimizers become equivalent with the corresponding original ones, once the regularization terms vanish. We theoretically prove the convergence rate of the regret and also conduct empirical evaluation on the proposed optimizers in comparison to the original optimizers with and without magnitude pruning. The results clearly demonstrate the advantages of our proposed optimizers in both getting significantly better performance and producing sparsity. Finally, it would be interesting in the future to investigate the convergence in non-convex settings and evaluate our optimizers on more applications from fields such as compute vision, natural language processing and etc.

## References

1. Abadi, M., et al.: TensorFlow: a system for large-scale machine learning. In: Keeton, K., Roscoe, T. (eds.) 12th USENIX Symposium on Operating Systems Design and Implementation, OSDI 2016, Savannah, GA, USA, 2–4 November 2016, pp. 265–283. USENIX Association (2016). https://www.usenix.org/conference/osdi16/technical-sessions/presentation/abadi
2. Avazu: Avazu click-through rate prediction (2015). https://www.kaggle.com/c/avazu-ctr-prediction/data
3. Criteo: Criteo display ad challenge (2014). http://labs.criteo.com/2014/02/kaggle-display-advertising-challenge-dataset

4. Duchi, J., Hazan, E., Singer, Y.: Adaptive subgradient methods for online learning and stochastic optimization. J. Mach. Learn. Res. **12**, 2121–2159 (2011). https://doi.org/10.5555/1953048.2021068
5. Graepel, T., Candela, J.Q., Borchert, T., Herbrich, R.: Web-scale Bayesian click-through rate prediction for sponsored search advertising in Microsoft's Bing search engine. In: Fürnkranz, J., Joachims, T. (eds.) Proceedings of the 27th International Conference on Machine Learning, ICML 2010, Haifa, Israel, 21–24 June 2010, pp. 13–20. Omnipress (2010). https://icml.cc/Conferences/2010/papers/901.pdf
6. Gupta, V., Koren, T., Singer, Y.: Shampoo: preconditioned stochastic tensor optimization. In: Dy, J.G., Krause, A. (eds.) Proceedings of the 35th International Conference on Machine Learning, ICML 2018, Stockholmsmässan, Stockholm, Sweden, 10–15 July 2018, vol. 80, pp. 1837–1845. PMLR (2018). http://proceedings.mlr.press/v80/gupta18a.html
7. Liao, H., Peng, L., Liu, Z., Shen, X.: IPinYou global RTB bidding algorithm competition (2013). https://www.kaggle.com/lastsummer/ipinyou
8. Kingma, D.P., Ba, J.L.: Adam: a method for stochastic optimization. In: Proceedings of the 3rd International Conference on Learning Representations, ICLR 2015, San Diego, CA, USA (2015)
9. Littlestone, N.: From on-line to batch learning. In: Rivest, R.L., Haussler, D., Warmuth, M.K. (eds.) Proceedings of the 2nd Annual Workshop on Computational Learning Theory, COLT 1989, Santa Cruz, CA, USA, 31 July–2 August 1989, pp. 269–284. Morgan Kaufmann (1989). http://dl.acm.org/citation.cfm?id=93365
10. McMahan, H.B.: Follow-the-regularized-leader and mirror descent: equivalence theorems and L1 regularization. In: Proceedings of the 14th International Conference on Artificial Intelligence and Statistics, AISTATS 2011, Fort Lauderdale, FL, USA, vol. 15, pp. 525–533. PMLR (2011)
11. McMahan, H.B., et al.: Ad click prediction: a view from the trenches. In: Proceedings of the 19th ACM SIGKDD International Conference on Knowledge Discovery and Data Mining, KDD 2013, Chicago, Illinois, USA, pp. 1222–1230. ACM (2013)
12. McMahan, H.B., Streeter, M.J.: Adaptive bound optimization for online convex optimization. In: The 23rd Conference on Learning Theory, COLT 2010, Haifa, Israel, 27–29 June 2010, pp. 244–256. Omnipress (2010). http://colt2010.haifa.il.ibm.com/papers/COLT2010proceedings.pdf#page=252
13. Naumov, M., et al.: Deep learning recommendation model for personalization and recommendation systems. CoRR abs/1906.00091 (2019). http://arxiv.org/abs/1906.00091
14. Nesterov, Y.E.: Smooth minimization of non-smooth functions. Math. Program. **103**, 127–152 (2005)
15. Nesterov, Y.E.: Primal-dual subgradient methods for convex problems. Math. Program. **120**(1), 221–259 (2009). https://doi.org/10.1007/s10107-007-0149-x
16. Ni, X., et al.: Feature selection for Facebook feed ranking system via a group-sparsity-regularized training algorithm. In: Proceedings of the 28th ACM International Conference on Information and Knowledge Management, CIKM 2019, Beijing, China, pp. 2085–2088. ACM (2019)
17. Polyak, B.T.: Some methods of speeding up the convergence of iteration methods. USSR Comput. Math. Math. Phys. **4**(5), 1–17 (1964). https://doi.org/10.1016/0041-5553(64)90137-5

18. Qu, Y., et al.: Product-based neural networks for user response prediction. In: Bonchi, F., Domingo-Ferrer, J., Baeza-Yates, R., Zhou, Z., Wu, X. (eds.) IEEE 16th International Conference on Data Mining, ICDM 2016, Barcelona, Spain, 12–15 December 2016, pp. 1149–1154. IEEE Computer Society (2016). https://doi.org/10.1109/ICDM.2016.0151
19. Reddi, S.J., Kale, S., Kumar, S.: On the convergence of Adam and beyond. In: Proceedings of the 6th International Conference on Learning Representations, ICLR 2018, Vancouver, BC, Canada. OpenReview.net (2018)
20. Robbins, H., Monro, S.: A stochastic approximation method. Ann. Math. Statist. **22**(3), 400–407 (1951)
21. Rockafellar, R.T.: Convex Analysis (Princeton Landmarks in Mathematics and Physics). Princeton University Press (1970)
22. Scardapane, S., Comminiello, D., Hussain, A., Uncini, A.: Group sparse regularization for deep neural networks. Neurocomputing **241**, 43–52 (2016). https://doi.org/10.1016/j.neucom.2017.02.029
23. Wang, R., Fu, B., Fu, G., Wang, M.: Deep & cross network for ad click predictions. In: Proceedings of the ADKDD 2017, Halifax, NS, Canada, 13–17 August 2017, pp. 12:1–12:7. ACM (2017). https://doi.org/10.1145/3124749.3124754
24. Xiao, L.: Dual averaging method for regularized stochastic learning and online optimization. J. Mach. Learn. Res. **11**, 2543–2596 (2010). https://doi.org/10.5555/1756006.1953017
25. Yang, H., Xu, Z., King, I., Lyu, M.R.: Online learning for group lasso. In: Proceedings of the 27th International Conference on Machine Learning, ICML 2010, Haifa, Israel, pp. 1191–1198. Omnipress (2010)
26. Yao, Z., Gholami, A., Shen, S., Keutzer, K., Mahoney, M.W.: ADAHESSIAN: an adaptive second order optimizer for machine learning. CoRR abs/2006.00719 (2020). https://arxiv.org/abs/2006.00719
27. Zeiler, M.D.: ADADELTA: an adaptive learning rate method. CoRR abs/1212.5701 (2012). https://arxiv.org/abs/1212.5701
28. Zhu, M., Gupta, S.: To prune, or not to prune: exploring the efficacy of pruning for model compression. In: 6th International Conference on Learning Representations, ICLR 2018, Workshop Track Proceedings Vancouver, BC, Canada, 30 April–3 May 2018. OpenReview.net (2018). https://openreview.net/forum?id=Sy1iIDkPM
29. Appendix. https://github.com/yadandan/adaptive_optimizers_with_sparse_group_lasso/blob/master/appendix.pdf

# Fast Conditional Network Compression Using Bayesian HyperNetworks

Phuoc Nguyen(✉), Truyen Tran, Ky Le, Sunil Gupta, Santu Rana, Dang Nguyen, Trong Nguyen, Shannon Ryan, and Svetha Venkatesh

A2I2, Deakin University, Geelong, Australia
{phuoc.nguyen,truyen.tran,k.le,sunil.gupta,santu.rana, d.nguyen,trong.nguyen,shannon.ryan,svetha.venkatesh}@deakin.edu.au

**Abstract.** We introduce a conditional compression problem and propose a fast framework for tackling it. The problem is how to quickly compress a pretrained large neural network into optimal smaller networks given target contexts, e.g., a context involving only a subset of classes or a context where only limited compute resource is available. To solve this, we propose an efficient Bayesian framework to compress a given large network into much smaller size tailored to meet each contextual requirement. We employ a hypernetwork to parameterize the posterior distribution of weights given conditional inputs and minimize a variational objective of this Bayesian neural network. To further reduce the network sizes, we propose a new input-output group sparsity factorization of weights to encourage more sparseness in the generated weights. Our methods can quickly generate compressed networks with significantly smaller sizes than baseline methods.

**Keywords:** Bayesian deep learning · Meta-learning · Network compression · Bayesian compression · Hyper networks

## 1 Introduction

Modern deep neural networks used in computer vision and natural language processing are very large function approximators with millions of trainable parameters. Their flexibility has led to undesirable properties such as overparameterization [6] and memorization of random patterns [1]. The large size makes it difficult to run the networks on edge devices, and wastes computational resources when run on servers. While it is sometimes possible to handcraft or deliberately search for more compact networks, it is often easier for practitioners to take large proven networks and compress them to fit the hardware and time constraints.

Recent efforts in network compression [10,19–21] have achieved significant progress in reducing large neural networks down to smaller sizes and cutting down run-time cost. The most popular approaches are to prune hidden units and filters of dense layers and convolutional filters or to quantize weights into lower bit precision. While the former helps slim down a network, thus reducing its size


N. Oliver et al. (Eds.): ECML PKDD 2021, LNAI 12977, pp. 330–345, 2021.
https://doi.org/10.1007/978-3-030-86523-8_20

and total floating point operations (FLOPS), the latter can make these networks run on limited compute hardware with integer arithmetic [14]. However, there is a compression-accuracy trade-off that prevent compressing large networks further and hinder their deployment in certain contexts with very low compute resources. It is largely unknown how to determine *a priori* if a certain degree of compression is achievable while satisfying context-specific constraints.

Addressing this limitation we introduce *the problem of contextual compression where a large network is conditionally compressed given operating constraints.* For instance, we need to deploy a large object recognition network trained on ImageNet with 1,000 classes in scenarios requiring only a subset of the classes. Using a single compressed network is clearly wasteful, and compressed networks are likely to bias minority classes [12]. Thus, there is an urgent need to optimally compress networks for different contexts.

Using only a partial network at inference time has been studied in a recent line of work known as conditional computation [2,3,24]. The strategy here is to use gating units to turn off unnecessary filters conditioned on inputs, therefore saving computation. While this reduces FLOPS at inference time, we still need to store and load the whole architecture onto run-time memory, making it unsuitable for edge devices.

To this end we propose a novel Bayesian framework for conditional compression given contextual run-time constraints. Key to the framework is a hypernetwork to parameterize the posterior distribution of weights of the target network given conditions. Given a pretrained large network, our hypernetwork generates compressed network weights for *on-the-fly* for each context. The conditional compression is implemented during training phase by minimizing a variational objective of the generated Bayesian neural network under different contexts. By using a sparsity inducing priors such as the scale mixture prior family [19] we can force the hypernetwork to generate very sparse hidden units networks, thus implementing context-based pruning. To encourage sparsifying the input and output neurons of layers jointly, we propose a novel input-output group sparsity factorization of the weight posterior distribution, in which the (input) group sparsity factorization of [19] is a special case. The framework is dubbed Bayesian Hypernetwork Compression (BHC).

We investigate the power of BHC on three general conditioning contexts: (a) given the subset of input features, (b) given the subset of output classes, and (c) constraints on expected model size. The model architectures include the popular LeNet-300-100 and LeNet-5-Caffe on MNIST dataset, and the VGG on CIFAR 10. Through a suite of experiments we demonstrate that BHC can quickly compress networks with sparse connectivity and low bit precision, resulting in significantly smaller memory footprints and computation than competing methods.

## 2 Related Work

Closest to our work are conditional computation methods [2,3,5,7] which can predict and turn off unnecessary filters before running, thus saving computation at inference time. However, these methods do not work for small compute devices. It also takes an extra step for the computation of which filters to include or remove. In [5] the masks for activation of each hidden layers are computed in parallel and for each sample independently, whereas in [27] the convolution kernels are parameterized using a condition from each input sample. In batch-shaping [2], large networks are slimmed down by using a residual network to predict masks. This is in contrast to our contextual compression proposal where a compressed network is generated once per context for run-time inference of any amount of data.

Also related to ours are meta-learning based compression methods. In [17], a hypernetwork is used to generate pruned convolution filters from sparse embedding vectors. In [18], a similar hypernetwork is used to generate various weights for a target network, then an evolutionary procedure is used to search for good-performing pruned networks. In contrast, we introduce a fast contextual compression problem where compressed networks is immediately generated given input conditions. We develop a principled Bayesian framework to model sparse posterior of weights which facilitates both pruning and bit precision reduction.

# 3 Preliminaries

## 3.1 Bayesian Neural Networks

A neural network is fully specified its computational graph whose nodes represent neural units equipped with activation functions and connectivity is specified by weight matrices. Let $W$ be the model weights. Bayesian neural network [22] provides a principled way of modeling uncertainties over weights $W$. By choosing a suitable prior distribution $p(W)$, we can use Bayes formula to update the posterior distribution of weights given data $\mathcal{D}$ as $p(W \mid \mathcal{D}) = \frac{p(\mathcal{D}|W)p(W)}{p(\mathcal{D})}$. This direct Bayesian inference is a long standing problem where the normalization term in the denominator requires marginalizing over the entire parameter space and is intractable to compute. Many recent advances in approximate inference for neural networks can be classified into sampling based or variational [4]. While the former can give better posterior approximation, recent literature focuses on the latter for neural networks due to its scalability [15,16,23]. Variational inference for the weight distribution often uses a simple posterior distribution such as normal distribution, $W \sim q(W) = \mathcal{N}(\mu_W, \Sigma_W)$ [19,22]. Inference is then amounted to optimizing its parameters,

$$\max_{\mu_W, \Sigma_w} \mathbb{E}_{W \sim q(W|\mu_W, \Sigma_W)} p(\mathcal{D}|W) - D_{KL}\left(q(W|\mu_W, \Sigma_W) \| p(W)\right)$$

An emerging problem for training Bayesian deep networks with many layers is in mini-batch training to optimize its posterior distribution. While reparameterization trick [16,23] can be used to sample one set of weights per update, it requires independent weights for independent inputs in each data mini-batch. Authors in [15] observed that by factorizing weights and noise and grouping noise by input neurons, they can "apply" the noise into the input instead of weights.

Then the (preactivation) output of the layer will be a normal distribution, linear transformation of the normal input vector [15],

$$y = x(Z \odot W) = (x \odot z)W$$
$$y_j \sim \mathcal{N}(\sum_i w_{ij} z_i x_i, \sum_i w_{ij}^2 z_i^2 x_i^2)$$

where the noise vector $z$ is reparameterized from the noise matrix $Z$ by grouping by input neurons, $x$ is the input vector, and $y = (y_j)$ is the output vector of the layer. This interpretation is used in variational dropout [21] for sparsifying networks.

### 3.2 Bayesian Compression

In [19], the authors explicitly describe the sparsity encouraging prior for neural network parameters by using a scale mixture prior from Bayesian statistics. Thereby, minimizing the KL divergence from this sparse prior to the weight posterior distribution will force weights to be sparse. This Bayesian compression framework for deep learning can be stated as follows.

Let us consider a particular weight matrix $[w_{ij}]_{i \in \overline{1,I}; j \in \overline{1,J}}$, where $w_{ij}$ is the connection from the $i$-th input unit to the $j$-th output unit, for $I$ input units and $J$ output units. We define a scale mixtures of normals for the prior over parameter $w$,

$$w_{ij} \sim \mathcal{N}(w_{ij}; 0, z_{ij}^2)$$

where $z \sim p(z)$ is some sparse prior. Marginalizing over $z$ gives heavier tails as well as peaks at zero, thus resulting in a sparse network. Instead of using one scale for each weight, it is beneficial to use one scale $z_i$ for all connections from neuron $i$. This has the effect of pruning (dropping) the entire neuron. The prior distribution of weight and scale variables are designed as:

$$p(z, W) = \prod_{i=1}^{I} p(z_i) \prod_{i,j}^{I,J} N(w_{ij} | 0, z_i^2) \tag{1}$$

A popular choice for the scale variable is the normal-Jeffreys prior $p(z) = \frac{1}{|z|}$. This improper prior (not normalizable) is non-informative and non parametric which make it suitable to model the scale for weights since it allows both small scales (weights are peak at zeros) and large scales (weights can spread out without being shrunk). Due to its irregularity, numerical approximation need to be made when computing KL divergences [15,21]. We suggest the references in [8,19] for more prior choices.

The posterior of weights and scales are approximated by normal distributions and they are factorized in the same way as the prior:

$$q_\phi(z, W) = \prod_{i=1}^{I} N\left(z_i \mid \mu_{z_i}, \sigma_z^2\right) \prod_{i,j}^{I,J} N\left(w_{ij} \mid z_i \mu_{ij}, z_i^2 \sigma_{ij}^2\right)$$

Note that $\sigma_z^2$ was reparameterized as $\mu_{z_i}^2 \alpha_i$ in [15,26] for the Gaussian dropout interpretation, where $\alpha_i$ is the dropout rate of the neuron $i$. This dropout rate $\alpha_i$ can be later compared to a threshold to decide a dropout mask for the input neurons.

The optimization of the variational parameters $\phi = (\mu_{ij}, \sigma_{ij}^2, \mu_{z_i}, \sigma_{z_i}^2)$ is minimizing a variational lower bound of the data log-likelihood $\log p(\mathcal{D})$:

$$L(\phi) = \mathbb{E}_{q_\phi(z)q_\phi(W|z)} \log p(\mathcal{D} \mid W) - \mathbb{E}_{q_\phi(z)} D_{KL}\left(q_\phi(W \mid z) \| p(W \mid z)\right) \\ - D_{KL}\left(q_\phi(z) \| p(z)\right)$$

At test time, one can recover a deterministic network by replacing the distribution of $W$ by its posterior mean times a dropout mask.

## 4 Conditional Compression

While the Bayesian compression framework of [19], explained in Sect. 3.2, provides a principled way for unconditional compression, it is not clear how to tune it to work in practical contexts with specific resource constraints. Thus we formulate a *new problem of conditional compression*, where we infer the network sparsity and bit precision *on-the-fly* conditioned on a specified context.

The compression context can be ideally arbitrary. In this work, we focus on three general types of conditioning context: (a) the subset of input features, (b) the subset of output classes, and (c) the expected model size. The input feature condition means that the feature or data distribution is limited to some subspace or input region, for example a subset of image channels or a subset of vocabulary. The output classes condition means that the classification output is limited to some sub-classes. The model size constraint reflects resource specification such as memory footprint, and this translates to reaching an equivalent compression rate for a given pre-trained network.

We then propose a new Bayesian framework to solve the problem, employing neural hypernetworks [9] to generate the distribution of compressed networks given the conditions. We term this new framework BHC, which stands for **B**ayesian **H**ypernetwork **C**ompression.

### 4.1 Conditional Bayesian Hypernetworks

Let $W_0$ be the pretrained network weights trained on the dataset $\mathcal{D}_0 = \{(x_0^n, y_0^n)\}_{n=1}^N$. We aim to quickly generate a compressed network $W_t$ from $W_0$ in some context $t$ with a constraint on the data domain or the compute resource, denoted by the condition $c_t$, i.e., $W_t = f_\gamma(c_t, W_0)$ for some hypernetwork $f_\gamma$ parameterized by $\gamma$. In theory we can train $f$ by running external compression algorithms with varying conditions to collect training data, but this is extremely expensive to cover the large condition space.

Inspired by the Bayesian compression framework of [19] (see Sect. 3.2), we introduce a new efficient Bayesian framework, in that $W_t$ is probabilistically

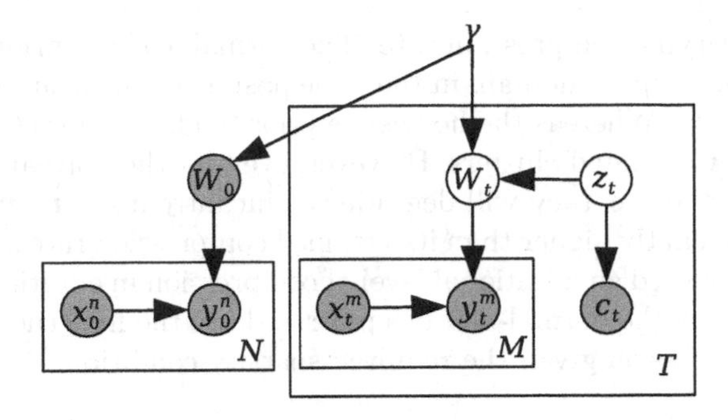


**Fig. 1.** The graphical model of our conditional compression framework. $W_0$ is the pre-trained network weights trained on the dataset $\mathcal{D}_0 = \{(x_0^n, y_0^n)\}_{n=1}^N$. $W_0$ is used to initialize the hypernetwork $\gamma$ to generate the initial means and variances of weights $W_t$. For each context $t$, the hypernetwork $\gamma$ generates a compressed $W_t$ conditioned on $z_t$ (and $c_t$ implicitly). Then it is used for the classification context on $\mathcal{D}_t = \{(x_t^m, y_t^m)\}_{m=1}^M$. When$W_0$ is absent, the framework can also be trained from scratch to generate $W_t$ for each context $t$.

generated by the hypernetwork at training time and the posterior of $W_t$ from the condition $c_t$ is estimated at inference time. Figure 1 shows the graphical model of the proposed method.

In this framework, we are given a set of $T$ classification scenarios (or contexts) $\mathcal{D} = \{(\mathcal{D}_t, c_t)\}_{t=1}^T$, each containing classification data $\mathcal{D}_t = \{(x_t^m, y_t^m)\}_{m=1}^M$ and constraint $c_t$. Marginalizing over all condition $c_t$'s then gives the posterior distribution of $W_t$, $q(W) = \int q(W_t|c_t)p(c_t)dc_t$, which is a continuous mixture distribution with mixing density $p(c_t)$. The compression is done by minimizing the KL distance from a normal-Jeffreys group sparsity prior to the weight posterior [19]. At test time, the hypernetwork generates a mask and the posterior mean of $W_t$ to produce a compressed deterministic network for classification on $\mathcal{D}_t$.

We solve this problem by modeling the joint distribution between the parameter $W$, the dropout mask $z$, and the condition $c$. First, we assume each mask is a latent variable associated with a given condition, and their joint distribution factors as $p(c, z) = p(c|z)p(z)$. We aims at learning an optimal variational posterior distribution $q_\phi(z|c_t)$ given this condition such that the weight $W$ is sparse and still performs well on the (conditioned) dataset $\mathcal{D}$. That is we maximize the variational objective:

$$\max_{\phi} \mathbb{E}_{z\sim q_\phi(z|c)} [\log p(\mathcal{D} \mid W, z) + \log p(c \mid z)] \\ - \mathbb{E}_{q_\phi(z)} [D_{KL} (q_\phi(W|z) \| p(W|z))] - D_{KL} (q_\phi(z|c) \| p(z)) \quad (2)$$

## 4.2 Compression Contexts

For the sparsity condition that directly limits the network capacity, we observe that varying the cutoff threshold on the logarithm of dropout rates can produce networks with varying compression rate. The normal-Jeffreys prior give rise to a continuum of the dropout domain in the scale posterior which supports compression at various rate. Whereas the horseshoe prior tends to separate the compression rates into two isolated clusters. However, even for the normal-Jeffreys prior, we observe that the accuracy will degrade significantly if the required compression rate is significantly higher than its attained compression rate at convergence. Therefore we proposed an additional level of compression in addition to the group sparsity induced by the normal-Jeffreys prior and use the hypernetwork to select the level of compression given the required sparsity condition.

### 4.3 Group Sparsity and Input-Output Group Sparsity Reparameterization

In [19], the author proposed coupling the scales of weights belonging to the same input neuron by sharing the scale variable $z$ in the joint prior as well as in the joint posterior,

$$p(W, z) \propto \prod_i^A \frac{1}{|z|} \prod_{i,j}^{A,B} \mathcal{N}(w_{ij}|0, z_i^2)$$

$$q_\phi(W, z) = \prod_i^A \mathcal{N}(z_i|\mu_{z_i}, \sigma_{z_i}^2) \prod_{i,j}^{A,B} \mathcal{N}(w_{ij}|z_i\mu_{ij}, z_i^2\sigma_{ij}^2)$$

This group sparsity distribution $p(z)$ is applied to the input of a layer with weight $W$. For deep networks with multiple layers $\{W_l\}_1^L$, we observe that the input sparsity $z^l$ of a layer $W_l$ should directly affect the input sparsity of the next layers. Therefore we introduce and additional scale variable $s$ for the output and propose to use a joint input-output group sparsity as follow,

$$p(W, z, s) \propto \prod_i^A \frac{1}{|z_i|} \prod_j^B \frac{1}{|s_j|} \prod_{i,j}^{A,B} \mathcal{N}(w_{ij}|0, z_i^2 s_j^2)$$

$$q_\phi(W, z, s) = \prod_i^A \mathcal{N}(z_i|\mu_{z_i}, \sigma_{z_i}^2) \prod_j^B \mathcal{N}(s_j|\mu_{s_j}, \sigma_{s_j}^2) \prod_{i,j}^{A,B} \mathcal{N}(w_{ij}|z_i s_j \mu_{ij}, z_i^2 s_j^2 \sigma_{ij}^2)$$

Note that this method applies to activation functions which have a zero fixed point, such as tanh, RELU, and ELU families, in order for the output group sparsity to carry on to the next layer input[1]. This creates a "pairwise" sparsity coupling one layer and the next layer since the output feature map of one layer will be the input to the next layer, where its sparsity pattern will be taken into account. This coupling of the input-output sparsity directly help increase the sparsity level of the network, thus can be used to achieve higher compression

[1] We assume the bias, if exist, can be included into the weight matrix and the input is augmented by a constant one.

level. In the forward step the weights $W$ is sampled conditioned on the sparsity of both the input, via $z$, and the output, via $s$. In the backward step, the gradient of $W$ is passed directly to update the input and output scales in pair.

The variational lower bound under this new prior and posterior is:

$$\mathcal{L}(\phi) = \mathbb{E}_{q_\phi(z,s)q_\phi(W|z,s)}[\log p(\mathcal{D}|W)] + \\ - \lambda_{KL}\left(\mathbb{E}_{q_\phi(z,s)}[D_{KL}(q_\phi(W|z,s)\|p(W|z,s))] + D_{KL}(q_\phi(z,s)\|p(z,s))\right) \tag{3}$$

where $\lambda_{KL}$ is the weighting factor to control the level of compression, which we will utilize in the experiments for different level of compression requirements. The negative KL divergence from the joint normal-Jeffreys scale prior to the joint Gaussian posterior in this new parameterization is as follow:

$$-D_{KL}(q_\phi(z,s)\|p(z,s)) \approx \sum_i^A (k_1\sigma(k_2 + k_3 \log\alpha_i) - 0.5m(-\log\alpha_i) - k_1) \\ + \sum_j^B (k_1\sigma(k_2 + k_3 \log\beta_j) - 0.5m(-\log\beta_j) - k_1),$$

where $\sigma(.)$, $m(.)$ are the sigmoid and softplus functions and $k_1 = 0.63576$, $k_2 = 1.87320$, $k_3 = 1.48695$, and $\log\alpha_i$ and $\log\beta_j$ are the dropout-rate vectors of the input and output respectively.

At test time, we use the posterior mean of $W$ and apply the masks $Z = m_z m_s^T$ to output the deterministic weight matrix,

$$\hat{W} = Z \odot \mathbb{E}_{q_\phi(z,s)q_\phi(\tilde{W})}\left[\text{diag}(z)\tilde{W}\text{diag}(s)\right] = \text{diag}(m_z \odot \mu_z) M_w \text{diag}(m_s \odot \mu_s) \tag{4}$$

where the mask $Z = m_z m_s^T$ is the binary mask matrix determined using the input mask $m_z$ and output mask $m_s$ determined by thresholding $z$ and $s$, and $M_w = (\mu_{ij})$ is the matrix of weight means. We use the following variational posterior marginal variance to determine the bit precision of the weights:

$$\mathbb{V}(w_{ij}) = \mathbb{V}(z_i \tilde{w}_{ij} s_j) = (\sigma_z^2 + \mu_z^2)(\sigma_{ij}^2 + \mu_{ij}^2)(\sigma_s^2 + \mu_s^2) - \mu_z^2 \mu_{ij}^2 \mu_s^2$$

### 4.4 Training Methods

We train the Bayesian compression models then quickly adapt to each condition using our framework. Therefore, when there is no condition, our method generates the compressed network which matches the performance of existing Bayesian compression networks. However when there are extra context information, we can input into our hypernetwork to generate a tailored, highly compressed network for that specific context.

**Algorithm 1:** BHC feedforward pass for fully connected layers.

**Require:** input $H$, hypernetwork $f_\phi$, condition $c \sim p(c)$

1 $M_w, \Sigma_w, \mu_z, \sigma_z, \mu_s, \sigma_s = f_\phi(c)$
2 $E_z \sim \mathcal{N}(0,1)$
3 $Z = \mu_z + \sigma_z \odot E_z$
4 $E_s \sim \mathcal{N}(0,1)$
5 $S = \mu_s + \sigma_s \odot E_s$
6 $M_h = ((H \odot Z)M_w) \odot S$
7 $V_h = ((H \odot Z)^2 \Sigma_w) \odot S^2$
8 $E \sim \mathcal{N}(0,1)$
9 return $M_h + \sqrt{V_h} \odot E$

**Algorithm 2:** BHC feedforward pass for convolution layers.

**Require:** input $H$, hypernetwork $f_\phi$, condition $c \sim p(c)$

1 $M_w, \Sigma_w, \mu_z, \sigma_z, \mu_s, \sigma_s = f_\phi(c)$
2 $E_z \sim \mathcal{N}(0,1)$
3 $Z = \mu_z + \sigma_z \odot E_z$
4 $E_s \sim \mathcal{N}(0,1)$
5 $S = \text{reshape}(\mu_s + \sigma_s \odot E_s, [1, 1, F, N])$
6 $M_h = ((H \odot Z) * M_w) \odot S$
7 $V_h = ((H \odot Z)^2 * \Sigma_w) \odot S^2$
8 $E \sim \mathcal{N}(0,1)$
9 return $M_h + \sqrt{V_h} \odot E$

Algorithms 1 and 2 show the forward pass of BHC networks for fully connected and convolution layers respectively. In each algorithm, $M_w = (\mu_{ij})$ and $\Sigma_w = (\sigma_{ij}^2)$ are the generated means and variances of each layers, $\mu_z$, $\sigma_z$ are the means and variances of the input group scale variables, $\mu_s$, $\sigma_s$ are the means and variances of the input group scale variables, $H$ is the input to the current layer, $N$ is the batch size, $F$ is the number of convolutional filters, and $*$ is the convolutional operator. For fully connected layers, we use one input (output) scale variable for each input (output) neuron to group weights connected to that neuron. For convolutional layers, we use only one input scale variable for all filters and one output scale variable for each filter output. We use the [width, height, filters, batch] dimension ordering. Local reparameterization [15] is used to efficiently sample the input, output group scale variables, and the activations. Both algorithms work as follow. In step 1, the mean and variance parameters of weights, input and output masks posterior distributions are generated given the condition $c$. In step 2–3, an input noise matrix $Z$ is sampled for the input matrix $H$. In step 4–5, an output noise matrix/tensor $S$ is sampled for the output matrix/tensor. In step 6 and 7, the output mean $M_h$ and variance $V_h$ is calculated and masked accordingly. Finally, an output sample is return by step 8–9.

# 5 Experiments

## 5.1 Settings

We design the experiments with varying conditions and test whether our Bayesian Hypernetwork Compression (BHC) method can compress and speedup a target network tailored to each condition. We examine three classes of conditions: *conditions on the input domain, conditions on the output classes,* and *conditions on the network capacity.*

**Network Architectures:** We investigate our method on the well-known neural network architectures of LeNet-300-100 and LeNet-5-Caffe on MNIST dataset, similarly with [21], and VGG on CIFAR 10, similarly with [19]. The input (output) groups of weight parameters are designed by coupling the scale variable for each input (pre-activation output) neuron for fully connected layers. For convolutional layers, we use a single scale variable for the input and coupling the scale variable for each filter. We simply use the same threshold for all layers for pruning and vary this threshold to determine the desired compression rate for each context[2]. We simply ignore the reconstruction loss of the condition, $\log p(c \mid z)$, in the optimization objective, Eq. 2, and consider the conditions in the train and test phases come from similar distributions to simplify the optimization problem. We leave this investigation for future work[3].

**Data Preparation:** In each experiment, we divide the dataset $\mathcal{D}_{\text{train}}$ test set $\mathcal{D}_{\text{test}}$ into tasks (contexts), add in a condition in each task, and finally create conditional datasets $\mathcal{D}_{\text{train}} = \{(X_t, Y_t), c_t)\}_{t=1}^{T}$ and $\mathcal{D}_{\text{test}} = \{(X_t, Y_t), c_t)\}_{t=1}^{T'}$. Our proposed networks are initialized with the given target network weights, then trained in minibatchs where each batch is a different task drawn randomly from $\mathcal{D}_{\text{train}}$. In the test phase, the hyper-network will generate the posterior mean over weights and masks given the condition. A threshold can then be varied and chosen to set the binary mask for weights such that a final compression rate is achieved for the final deterministic weights.

**Baselines:** We use the following baselines: sparse variational dropout (SparseVD) [21], Bayesian compression with group normal-Jeffreys (BC-GNJ) and with group horseshoe (BC-GHS) [20]. We also tested the input-output group sparsity method separately on the Lenet-300-100 architecture (denoted BC-IO-GNJ) and demonstrated that it has better compression rate than its BC-GNJ counter part[4]. We will use the proposed input-output group sparsity method for our BHC networks.

[2] Better thresholds can be chosen by visually inspection on the log dropout rates.

[3] This involves adding a generative network for the condition posterior given all layer-masks then maximizing its log-likelihood.

[4] We also tested BC-IO-GNJ on Lenet-5-Caffe and got better performance.

### 5.2 Implementation

Our method was implemented in the Flux framework [13]. We used Adam optimizer with default parameters and train our models for 300 epochs, and found it converges better when clipping the variances of the first layers of the networks as in [19]. Hence the first layer variance of LeNet-300-100 was clipped to $0.2^2$, the first layer variance of LeNet-5-Caffe to $0.5^2$, and the 64 and 128 feature maps layer in VGG to $0.1^2$ and the 256 feature maps layers to $0.2^2$.

For *subclass conditioning* we use a 2-layer MLP embedding network, each with hidden size 100 to embed the subclasses condition vector. For *input domain conditioning* we use a 2 layers convolutional neural network each with 20 filters, kernel width 3, and zero padding to embed the input domain condition vector.

The hypernetwork is a set of linear layers to map the embedding vectors into the posterior weights parameters: the posterior means, log variances, the input and output group scales. We initialize these hypernetwork weights using Kaiming uniform function [11] with gain 0.5. We initialize the hypernetworks such that it generates the pretrained weight $W_0$ with noise initially as follows. We set the bias of the mean-generating hypernetwork to $W_0$, the bias of log-variance-generating hypernetwork to a small number drawn from $\mathcal{N}(-9, 1e{-}4)$. For the scales, we set the bias of the scale-mean-generating hypernetwork to $\mathcal{N}(1, 1e{-}8)$, and the bias of its log-variance-generating hypernetwork to a small number drawn from $\mathcal{N}(-9, 1e{-}4)$.

### 5.3 Conditioning on Output Subclasses and Input Domain

For the condition on output subclasses, we take the vector of Bernoulli probabilities of the subclasses as condition, thus each dimension contains a probability representing the appearance of a class. We assume that the number of practical conditions is only a subset of all possible subsets, the power set $2^C$ where $C$ is the number of classes. In this experiments we limit our investigation to conditions that are the two consecutive random classes between 1 and $C$. In training phase, $c_y$ is the Bernoulli vector of length $C$ where each component represents the probability of appearance of that class. In test phase, the $c_y$ is the binary vector representing the appearance of the classes.

For the condition on the input domain, we use minibatchs as tasks and use features from the minibatchs as condition. We take the sample mean of the input distribution as the condition $c_x$, i.e., the arithmetic mean of the input tensor along the batch dimension. In practice, this can be the sample mean of the distribution of a representative dataset for the testing condition.

**Results on Compression Rates.** We present the weight compression rates of our methods to the baseline methods in Table 1. We use (i) compression by pruning input and output neurons and (ii) compression by reducing bit precision in addition to pruning the input and output neurons. These scenarios are practical since pruned weights can be used to create slimmer network, where as reduced precision weights can be exploited in future hardware [25]. We can observe that our method can achieve higher compression rates at lower error rates than the baselines. When condition on the subclasses, the compression rates is highest. When exploiting the weight variance to reduce bit precision in combination with pruning, a much higher compression rate is achieved, 216× for the LeNet-300-100 architecture, 1004× for the LeNet-5-Caffe architecture, and 73× for VGG architecture in this conditional setting.

**Table 1.** Compression results of Bayesian-hypernet compression (BHC) compared to sparse variational dropout (SparseVD), group normal-Jeffreys (BC-GNJ) and group horseshoe (BC-GHS). Results marked with * are from [19]. BHC method conditionally compresses the target network based on the domain input condition $c_x$ or the subclass condition $c_y$ in each context. We report the mean compression rate across tasks and the overall error on the testset.

| Model original error | Method | Pruning (error %) | Pruning and bit reduction (error %) |
|---|---|---|---|
| LeNet-300-100 1.6 | SparseVD* | 21 (1.8) | 84 (1.8) |
| | BC-GNJ* | 9 (1.8) | 36 (1.8) |
| | BC-GHS* | 9 (1.8) | 23 (1.9) |
| | BC-IO-GNJ | 12 (1.8) | 40 (1.9) |
| | BHC-$c_x$ | 26 (1.9) | 169 (1.9) |
| | BHC-$c_y$ | 54 (1.8) | 216 (1.9) |
| LeNet-5-Caffe 0.9 | SparseVD* | 63 (1.0) | 228 (1.0) |
| | BC-GNJ* | 108 (1.0) | 361 (1.0) |
| | BC-GHS* | 156 (1.0) | 419 (1.0) |
| | BHC-$c_x$ | 187 (0.2) | 608 (0.4) |
| | BHC-$c_y$ | 286 (0.4) | 1004 (0.5) |
| VGG 8.4 | BC-GNJ* | 14 (8.6) | 56 (8.8) |
| | BC-GHS* | 18 (9.0) | 59 (9.0) |
| | BHC-$c_x$ | 20 (7.0) | 66 (7.4) |
| | BHC-$c_y$ | 22 (2.0) | 73 (2.8) |

**Compressed Models.** We present the neuron group sparsity enforcing capabilities of our methods compared to baselines in Table 2. We show the pruned architecture and bit-precision per layer. We observe that our methods can generate very sparse input and output neurons of the target architecture. Especially, the input layer and the first hidden layers have significantly sparser connection and only need as low as 5 bit precision compared to the baselines. For Lenet-5-Caffe and VGG architecture, the first and last layers require a much smaller number of neurons. For Lenet-5-Caffe architecture, as low as 10 bit precision can be achieved. For VGG, the number of convolutional filters is also reduced significantly compared to sparse variational dropout, group normal-Jeffreys, and group horseshoe.

## 5.4 Conditioning on the Model Compression Rate

In this section, we investigate the Bayesian-hypernet compression method given different sparsity constraints. We compare BHC to Bayesian compression methods using normal-Jeffreys and horseshoe priors. We implement BHC using two broad levels of compression as condition: (1) high accuracy compression and (2) high compression rate with moderate accuracy drop. Finer compression rates can be chosen by varying the threshold in each level. Therefore, we use a binary condition input to the hypernetwork, $c_w = 0$ for high accuracy compression and $c_w = 1$ for high compression rate. In this case, the condition is discrete and BHC becomes a finite mixture model:

$$p(W) = \int p(W|c)p(c)dc = p(W|c)p(c=0) + p(W|c)p(c=1)$$

with mixing probabilities $p(c=0) = p(c=1) = \frac{1}{2}$. First, the user need to inputs $c_w \in \{0,1\}$ into BHC hypernetwork and get as output the posterior mean of weights $M_w = (\mu_{ij})$ and a dropout-rate vector $\log \alpha$. Then the user varies a threshold $\tau$ to create an input mask $m_z = I(\log \alpha < \tau)$ and output mask $m_s = I(\log \beta < \tau)$ for $z$ and $s$ and calculates the deterministic weight as $\hat{W} = \text{diag}(m_z) \odot M_w \odot \text{diag}(m_s)$, as in Eq. 4. By checking the compression rate and the accuracy of $\hat{W}$ on a validation set if it is available, the user can decide a final threshold $\tau$ to use.

We train BHC for $c_w = 0$ condition using a lower KL weighting, $\lambda_{KL} = 0.1$ in the training objective (Eq. 3), and we use a higher KL weighting, $\lambda_{KL} = 1$ for $c_w = 1$ condition. To speed up convergence, we choose a simple training strategy with this binary condition as follows[5]. We train the BHC hypernetwork with $c_w = 0$ condition for 100 epochs with $\lambda_{KL} = 0.1$ and freeze this generated weights in the hypernetwork bias output, by setting the trainable flag of the bias to false. Next, we train the BHC hypernetwork with $c_w = 1$ condition for another 200 epochs with $\lambda_{KL} = 1$.

**Results.** Figure 2 compares our Bayesian-hypernet compression (BHC) with Bayesian compression with normal Jeffreys and horseshoe priors by varying the threshold to output deterministic weights. As can be observed, our methods can generate models with high accuracy at low compression rate and with only moderate accuracy drop at high compression rate.

[5] In general, the condition can be sampled randomly from Bernoulli($\frac{1}{2}$) during training.

**Table 2.** Pruned architectures and bit precision of Bayesian-hypernet compression (BHC) compared to sparse variational dropout (SparseVD) [21], group normal-Jeffreys and group horseshoe [20]. Results marked with * are from [19].

| Network & size | Method | Pruned architecture | Bit-precision |
|---|---|---|---|
| Lenet-300-100 784-300-100 | SparseVD* | 512-114-72 | 8-11-14 |
| | BC-GNJ* | 278-98-13 | 8-9-14 |
| | BC-GHS* | 311-86-14 | 13-11-10 |
| | BC-IO-GNJ | 255-90-15 | 7-9-15 |
| | BHC-$c_x$ | 183-50-18 | 5-8-20 |
| | BHC-$c_y$ | 141-30-18 | 6-11-20 |
| Lenet-5-Caffe 20-50-800-500 | SparseVD* | 14-19-242-131 | 13-10-8-12 |
| | BC-GNJ* | 8-13-88-13 | 18-10-7-9 |
| | BC-GHS* | 5-10-76-16 | 10-10-14-13 |
| | BHC-$c_x$ | 4-4-64-2 | 12-10-11-11 |
| | BHC-$c_y$ | 4-4-64-3 | 12-9-11-9 |
| VGG $(2 \times 64)$- $(2 \times 128)$- -$(3 \times 256)$- $(8 \times 512)$ | BC-GNJ* | 63-64-128-128-245-155-63- -26-24-20-14-12-11-11-15 | 10-10-10-10-8-8-8- -5-5-5-5-5-6-7-11 |
| | BC-GHS* | 51-62-125-128-228-129-38- -13-9-6-5-6-6-6-20 | 11-12-9-14-10-8-5- -5-6-6-6-8-11-17-10 |
| | BHC-$c_x$ | 52-63-126-128-161-71- -22-19-11-10-6-7-8-8-9-13 | 12-7-5-7-6-8-9-8- -11-12-12-12-14-14-26-20 |
| | BHC-$c_y$ | 42-63-125-123-110-44-15- -12-8-6-5-5-6-6-6-8 | 12-11-8-13-6-9-10- -8-9-9-9-10-14-18-27-24 |

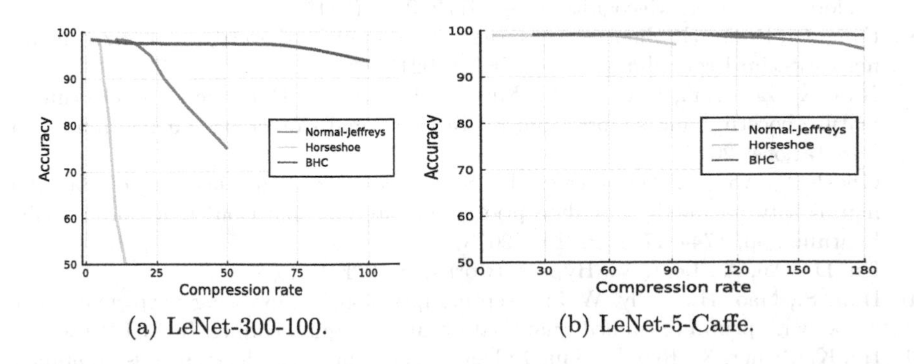


(a) LeNet-300-100. (b) LeNet-5-Caffe.

**Fig. 2.** Conditions on compression rate.

## 6 Conclusion

We introduced a new problem concerning contextual compression of deep neural networks. Given a large uncompressed pre-trained network, we compress it *on-the-fly* to satisfy resource constraints at run time, e.g., compression rate, memory, input size or class subset. Our work provides the first principled Bayesian hypernetwork compression framework towards contextual compression for edge devices. Testing on several popular architectures on MNIST and CIFAR 10, we demonstrated that our methods can quickly generate compressed networks with significantly smaller memory footprints and computation than competing methods. Future challenges would be scaling up the Bayesian hypernetwork compression method to recent large architectures with hundred of millions of parameters and applying the method in real world contextual compression tasks and other data domains such as texts and sounds.

**Acknowledgments.** This research was a collaboration between the Commonwealth Australia (represented by Department of Defence) and Deakin University, through a Defence Science Partnerships agreement.

## References

1. Arpit, D., et al.: A closer look at memorization in deep networks. In: International Conference on Machine Learning, pp. 233–242. PMLR (2017)
2. Bejnordi, B.E., Blankevoort, T., Welling, M.: Batch-shaping for learning conditional channel gated networks. In: International Conference on Learning Representations (2019)
3. Bengio, Y., Léonard, N., Courville, A.: Estimating or propagating gradients through stochastic neurons for conditional computation. arXiv preprint arXiv:1308.3432 (2013)
4. Blei, D.M., Kucukelbir, A., McAuliffe, J.D.: Variational inference: a review for statisticians. J. Am. Stat. Assoc. **112**(518), 859–877 (2017)
5. Chen, Z., Li, Y., Bengio, S., Si, S.: You look twice: GaterNet for dynamic filter selection in CNNs. In: Proceedings of the IEEE/CVF Conference on Computer Vision and Pattern Recognition, pp. 9172–9180 (2019)
6. Fang, C., Dong, H., Zhang, T.: Mathematical models of overparameterized neural networks. In: Proceedings of the IEEE (2021)
7. Gao, X., Zhao, Y., Dudziak, Ł., Mullins, R., Xu, C.: Dynamic channel pruning: feature boosting and suppression. In: International Conference on Learning Representations (2018)
8. Ghosh, S., Yao, J., Doshi-Velez, F.: Structured variational learning of Bayesian neural networks with horseshoe priors. In: International Conference on Machine Learning, pp. 1744–1753. PMLR (2018)
9. Ha, D., Dai, A., Le, Q.V.: Hypernetworks. In: ICLR (2017)
10. Han, S., Mao, H., Dally, W.J.: Deep compression: compressing deep neural networks with pruning, trained quantization and Huffman coding. In: ICLR (2016)
11. He, K., Zhang, X., Ren, S., Sun, J.: Delving deep into rectifiers: surpassing human-level performance on ImageNet classification. In: Proceedings of the IEEE International Conference on Computer Vision, pp. 1026–1034 (2015)

12. Hooker, S., Moorosi, N., Clark, G., Bengio, S., Denton, E.: Characterising bias in compressed models. In: ICML Workshop on Human Interpretability in Machine Learning (2020)
13. Innes, M., et al.: Fashionable modelling with flux. CoRR, abs/1811.01457 (2018)
14. Jacob, B., et al.: Quantization and training of neural networks for efficient integer-arithmetic-only inference. In: Proceedings of the IEEE Conference on Computer Vision and Pattern Recognition, pp. 2704–2713 (2018)
15. Kingma, D.P., Salimans, T., Welling, M.: Variational dropout and the local reparameterization trick. In: Proceedings of the 28th International Conference on Neural Information Processing Systems, vol. 2, pp. 2575–2583 (2015)
16. Kingma, D.P., Welling, M.: Auto-encoding variational Bayes. In: ICLR (2014)
17. Li, Y., Gu, S., Zhang, K., Van Gool, L., Timofte, R.: DHP: differentiable meta pruning via hypernetworks. arXiv preprint arXiv:2003.13683 (2020)
18. Liu, Z., et al.: MetaPruning: meta learning for automatic neural network channel pruning. In: Proceedings of the IEEE International Conference on Computer Vision, pp. 3296–3305 (2019)
19. Louizos, C., Ullrich, K., Welling, M.: Bayesian compression for deep learning. In: Advances in Neural Information Processing Systems, pp. 3288–3298 (2017)
20. Louizos, C., Welling, M., Kingma, D.P.: Learning sparse neural networks through l_0 regularization. In: International Conference on Learning Representations (2018)
21. Molchanov, D., Ashukha, A., Vetrov, D.: Variational dropout sparsifies deep neural networks. In: International Conference on Machine Learning, pp. 2498–2507. PMLR (2017)
22. Neal, R.M.: Bayesian Learning for Neural Networks, vol. 118. Springer, New York (2012). https://doi.org/10.1007/978-1-4612-0745-0
23. Rezende, D.J., Mohamed, S., Wierstra, D.: Stochastic backpropagation and approximate inference in deep generative models. In: International Conference on Machine Learning, pp. 1278–1286. PMLR (2014)
24. Shazeer, N., et al.: Outrageously large neural networks: the sparsely-gated mixture-of-experts layer. arXiv preprint arXiv:1701.06538 (2017)
25. Wang, K., Liu, Z., Lin, Y., Lin, J., Han, S.: HAQ: hardware-aware automated quantization with mixed precision. In: Proceedings of the IEEE Conference on Computer Vision and Pattern Recognition, pp. 8612–8620 (2019)
26. Wang, S., Manning, C.: Fast dropout training. In: International Conference on Machine Learning, pp. 118–126. PMLR (2013)
27. Yang, B., Bender, G., Le, Q.V., Ngiam, J.: CondConv: conditionally parameterized convolutions for efficient inference. In: Advances in Neural Information Processing Systems, pp. 1307–1318 (2019)

# Active Learning in Gaussian Process State Space Model

Hon Sum Alec Yu[1,2](✉), Dingling Yao[1], Christoph Zimmer[1], Marc Toussaint[2], and Duy Nguyen-Tuong[1]

[1] Bosch Center for Artificial Intelligence, 71272 Renningen, Germany
honsumalec.yu@de.bosch.com

[2] Learning and Intelligent System Lab, Technische Universität Berlin, Berlin, Germany

**Abstract.** We investigate active learning in Gaussian Process state-space models (GPSSM). Our problem is to actively steer the system through latent states by determining its inputs such that the underlying dynamics can be optimally learned by a GPSSM. In order that the most informative inputs are selected, we employ mutual information as our active learning criterion. In particular, we present two approaches for the approximation of mutual information for the GPSSM given latent states. The proposed approaches are evaluated in several physical systems where we actively learn the underlying non-linear dynamics represented by the state-space model.

**Keywords:** Active learning · Gaussian Process State Space Model · Mutual information

## 1 Introduction

State-space models (SSMs) are a compact representation of dynamical systems as a set of input, output and state variables where the transition could be deterministic or stochastic. SSMs are widely used in practice, where applications range from aerospace-related work (e.g. [16]), medicine (e.g. [1]) to economics and finance (e.g. [52]). The two important aspects of SSMs are *learning* and *control* and they have been a fruitful research field in classical engineering (e.g. [17,29]). Learning in SSM is about finding optimal hyperparameters so that the dynamical system is accurately captured whereas control refers to optimising a specific objective function based on inputs and states of a given SSM. The latter is often referred as *optimal control.* During the last decade, both aspects of SSMs also attracted the attention from the machine learning community (e.g. [18,41]). To capture the uncertainty of system dynamics, Bayesian models are

**Electronic supplementary material** The online version of this chapter (https://doi.org/10.1007/978-3-030-86523-8_21) contains supplementary material, which is available to authorized users.


N. Oliver et al. (Eds.): ECML PKDD 2021, LNAI 12977, pp. 346–361, 2021.
https://doi.org/10.1007/978-3-030-86523-8_21

often employed in the context of SSM. For example, a Gaussian Process (GP) prior [42] is mostly used to describe the transition between states, which leads to family of Gaussian Process State-Space Models (GPSSMs) with an impressive amount of progress has been made (e.g. [12,13,19,20,27,28,31,38,49], and more). Learning and inference of GPSSM are an active research topic in machine learning community and this paper focuses on *learning* of GPSSM.

While different approaches of GPSSM mainly address the modelling issues, the question about data-efficient learning of GPSSM remains largely unanswered. Such questions go naturally to the field of Active Learning (AL), which is the process of strategically selecting and generating new data for supervised learning. The purpose of AL is to learn the proposed model with the least amount of data possible. Though the majority of recent papers in this field focused on classification problems (e.g. [3,46]) and rather few were for regression (e.g. [7, 44]), Schreiter et al. [43] applied AL in GP regression and Zimmer et al. [53] extended the scope to learning time-series models. Both papers employed the entropy criterion for exploration, while using a greedy selection based on the maximum predictive variance of the GP. In context of state-space models for predictive control, Buisson et al. [6] proposed an AL approach for learning the transition dynamics via optimising the trajectory. For AL in GPSSM, Capone et al. [8] was the first to introduce the concept here. However, in both cases, Buisson et al. [6] and Capone et al. [8] simplified the settings by assuming the states are observable and measurable. By doing so, the resulting state-space models can be actively learned in the same way as standard GP regression models in a supervised manner.

In contrast to previous work, we present an AL strategy on GPSSM without relying on the assumption that the states are observed. That is, we actively steer the system through the latent state-space by determining its inputs such that the underlying dynamics can be optimally learned by a GPSSM. For the information criterion, we employ an approximated measure of the mutual information. We propose and discuss two different approaches for the approximation of mutual information in the context of unmeasurable, latent state-space.

This paper is on *learning* a stochastic nonlinear dynamical system actively from noisy data via probabilistic SSM. Its main contributions include:

- We derive tractable approximate mutual information estimates for GPSSM with latent states based on the state-of-the-art learning scheme and the technique of approximate Gaussian integral.
- We propose an AL strategy for GPSSM with latent states based on inputs.
- We conduct experiments for different examples to demonstrate the usefulness of our strategy in physical systems, which is a typical application for SSM.

The reminder of this paper is organised as follow: We give a brief overview of backgrounds in Sect. 2 and introduce GPSSM in Sect. 3. Section 4 shows our active learning strategies and Sect. 5 evaluates our concepts on several test scenarios. A conclusion is given in Sect. 6. For brevity, we leave detailed proof of propositions, experimental details, partial code and few further remarks in supplementary materials, which will be uploaded in due course.

## 2 Background

AL is itself a broad topic and here we refer readers to, for example, Settles [45] and Dasgupta [10] for an in-depth survey of the basic algorithmic and theoretical ideas. There are many different paradigms in AL. For example, *Bayesian Active Learning* incorporates the Bayesian framework and is often referred to AL with GP models (e.g. [26,50]), because GPs naturally carry the uncertainty measures. Specifically, Houlsby et al. [26] applied AL in GP classification problems, where they defined an acquisition function which estimates the quantity of mutual information between model predictions and model parameters. Other variants include *Batch Active Learning* and *Deep Bayesian Active Learning*. While Batch AL refers to making multiple queries in parallel (e.g. [26,30]), Deep Bayesian AL explores techniques for actively learning deep Bayesian models, e.g. Bayesian convolutional neural network [22]. For AL with regression models, Seo et al. [44] proposed a test point rejection strategy based on the posterior variance. Krause et al. [32] proposed an optimal exploration strategy for GP regression models, while providing bounds on the advantage of using AL. Srinivas et al. [47] further showed convergence rate for AL in GPs under specific kernels. Schreiter et al. [43] extended AL in GP regression by introducing constraints and Zimmer et al. [53] extended constrained AL in GP regression to time-series. They proposed using the determinant of the predictive covariance matrix defined in the GP as optimality criterion along with theoretical guarantees.

In dynamics modelling, SSM has been a long-lasting topic with enormous literature across different disciplines (e.g. [17,29,52]). In machine learning, the combination between SSM and GP began arguably from Wang et al. [51], where they learned the latent state of GPSSM via maximum a posteriori. Turner et al. [49] extended it to learning the transition function. Frigola et al. [20] presented a Bayesian treatment based on particle Markov-Chain-Monte-Carlo and Frigola et al. [19] further introduced a variational inference scheme to overcome the computational complexity of Monte-Carlo based methods. This becomes the state-of-the-art and due to the fact that this approach depends heavily on the approximated function, most recent work introduced their own variational inference scheme with their own advantages and disadvantages (e.g. [12,27,28,38]). For example, Doerr et al. [12] and Ialongo et al. [28] introduced their specific variational inference scheme to account for the dependence between transition function and latent states, which were treated as independent in earlier work.

To the best of our knowledge, AL and GPSSM have not been discussed together until Buisson et al. [6] first brought the topic of AL in learning the transition function using GP. They stated that the dynamical problem, i.e. learning the transition function, is fundamentally different from a static AL in GP problem, because we need to steer the system to certain states through the unknown transition function with a sequence of actions, which is the only component we can control. For each round of exploration, the next input is actively picked so that the model error can be maximally reduced. They proposed their AL strategy using the maximum entropy. Another recent work on AL in GPSSM is by Capone et al. [8]. They proposed taking the most informative data point while employing the mutual information between the most informative point and reference points. The states, in the existing work on AL in GPSSM, are hitherto assumed to be observable such that the model can be trained in a fashion similar to GP regression. Namely, in Capone et al. [8], the fact that states are observable simplify the estimation of the mutual information, resulting in two optimisation problems to solve in parallel.

Our paper differs from the previous work by presenting an AL strategy on GPSSM with states that are *latent*, resulting in a different estimation of the mutual information as exploration criterion while extending the scope of AL for GPSSM. We leave further quantitative discussion in Sect. 4.

## 3 Gaussian Process State-Space Model with Inputs

We consider a discrete-time sequence of $T$ observations $\boldsymbol{y}_{1:T} \equiv \{\boldsymbol{y}_t\}_{t=1}^T$, where each observed point $\boldsymbol{y}_t \in \mathcal{Y} \subseteq \mathbb{R}^{d_y}$ is generated by a corresponding latent variable $\boldsymbol{x}_t \in \mathcal{X} \subseteq \mathbb{R}^{d_x}$ and the previous step control $\boldsymbol{c}_{t-1} \in \mathcal{C} \subseteq \mathbb{R}^{d_c}$. The collection of latent states and controls are denoted by $\boldsymbol{x}_{0:T} \equiv \{\boldsymbol{x}_t\}_{t=0}^T$ and $\boldsymbol{c}_{0:T} \equiv \{\boldsymbol{c}_t\}_{t=0}^T$, respectively. These latent variables and controls are assumed to satisfy the Markov property, meaning that any $\boldsymbol{x}_{t+1}$ can be generated by only conditioning on $\boldsymbol{x}_t$, $\boldsymbol{c}_t$ and the transition function $f : \mathbb{R}^{d_x} \times \mathbb{R}^{d_c} \rightarrow \mathbb{R}^{d_x}$. To align with previous work, we use Gaussian distribution for both transition and observation density function. We also simplify the mean of the observation function by using linear mapping and tackle multivariate latent states by placing a GP prior on each dimension separately, as in, for instance, Ialongo et al. [28].

GPSSM is defined to be a probabilistic SSM with a GP prior over the transition function $f$, specified by

$$f \sim \mathcal{GP}(m(\cdot), k(\cdot, \cdot)), \tag{1}$$

$$\boldsymbol{x}_0 \sim p(\boldsymbol{x}_0), \tag{2}$$

$$\boldsymbol{x}_t | f(\boldsymbol{x}_{t-1}, \boldsymbol{c}_{t-1}) \sim \mathcal{N}(\boldsymbol{x}_t | f(\boldsymbol{x}_{t-1}, \boldsymbol{c}_{t-1}), \boldsymbol{Q}), \tag{3}$$

$$\boldsymbol{y}_t | \boldsymbol{x}_t \sim \mathcal{N}(\boldsymbol{y}_t | \boldsymbol{C}\boldsymbol{x}_t + \boldsymbol{d}, \boldsymbol{R}), \tag{4}$$

where in Eq. 1, $f$ is a GP governed by a given mean function $m(\cdot)$ and positive definite covariance function $k(\cdot, \cdot)$. The initial state $p(\boldsymbol{x}_0) = \mathcal{N}(\boldsymbol{x}_0 | \boldsymbol{\mu}_0, \boldsymbol{\Sigma}_0)$ in Eq. 2 is assumed known. $\boldsymbol{C}$ and $\boldsymbol{d}$ are parameters in the linear mapping, whereas $\boldsymbol{Q}$ and $\boldsymbol{R}$ are covariance matrices which capture process and observation noise, respectively. The advantages of linear mapping are that linear mean won't limit the range of systems that can be modelled and this reduces the non-identifiabilities between transitions and emissions. As a convention from previous work, we denote $\boldsymbol{f}_t \equiv f(\boldsymbol{x}_{t-1}, \boldsymbol{c}_{t-1})$ and Eq. 3 becomes

$$\boldsymbol{x}_t | \boldsymbol{f}_t \sim \mathcal{N}(\boldsymbol{x}_t | \boldsymbol{f}_t, \boldsymbol{Q}). \tag{5}$$

For brevity, we write $\tilde{\boldsymbol{x}}_* = (\boldsymbol{x}_*, \boldsymbol{c}_*)$, where $*$ could be a specific index or a collection of indices. Thus, the matrix of covariance functions are

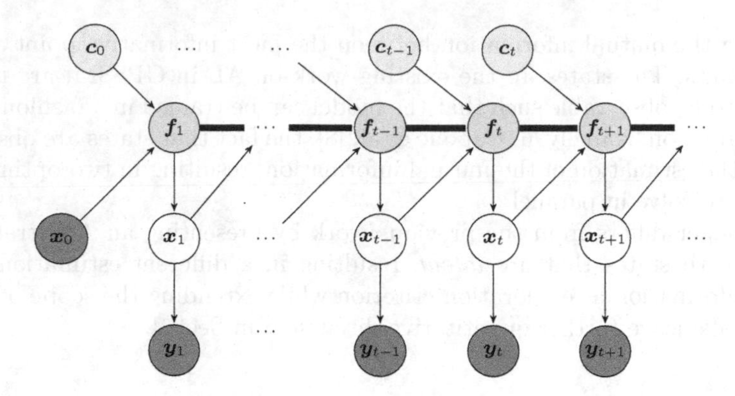

**Fig. 1.** Graphical model of GPSSM with control inputs as defined in Eq. 1–4. The observations $\boldsymbol{y}_t$ and initial state $\boldsymbol{x}_0$ are assumed known (filled in green). Control inputs $\boldsymbol{c}_t$ are independent variables (filled in yellow) whereas latent variables are denoted by $\boldsymbol{x}_t$. We use a thick, straight line to connect $\boldsymbol{f}_t \equiv f(\boldsymbol{x}_{t-1}, \boldsymbol{c}_{t-1})$ (filled in gray) to show that all variables are fully connected. (Color figure online)

denoted by $K_{i:j} := (k(\tilde{\boldsymbol{x}}_s, \tilde{\boldsymbol{x}}_t))_{s,t=i}^{j}$. For convenience, we write $k(\tilde{\boldsymbol{x}}_{i:j}, \tilde{\boldsymbol{x}}_k) \equiv (k(\tilde{\boldsymbol{x}}_i, \tilde{\boldsymbol{x}}_k), \cdots, k(\tilde{\boldsymbol{x}}_j, \tilde{\boldsymbol{x}}_k))$ as a vector collection of kernel evaluations. It holds that $k(\tilde{\boldsymbol{x}}_{i:j}, \tilde{\boldsymbol{x}}_k)^T = k(\tilde{\boldsymbol{x}}_k, \tilde{\boldsymbol{x}}_{i:j})$.

The control inputs $\boldsymbol{c}_t$ can be viewed as an augmented latent state. Since $\boldsymbol{c}_t$ is not a random variable, we always condition on this quantity. Most previous papers related to this model were either studied without controls or, if a control is involved, by assuming that this quantity was randomly distributed [36]. We, however, construct an AL strategy based on $\boldsymbol{c}_t$ to learn the GPSSM efficiently. Hence, we purposely keep this term in our definitions. Figure 1 shows a graphical model of such a setting. We would like to readdress here that $\boldsymbol{c}_t$ is a known quantity where our task is to optimise the value to select. This should not be confused with control in other contexts such as that in RL.

To complete the structure of GPSSM, let us define $\boldsymbol{f}_{1:T} \equiv \{\boldsymbol{f}_t\}_{t=1}^{T}$ and we can write the joint density as follows (e.g. Chap. 3 in Frigola [18]),

$$p(\boldsymbol{y}_{1:T}, \boldsymbol{x}_{0:T}, \boldsymbol{f}_{1:T}) = p(\boldsymbol{x}_0) \prod_{t=1}^{T} p(\boldsymbol{y}_t|\boldsymbol{x}_t) p(\boldsymbol{x}_t|\boldsymbol{f}_t) p(\boldsymbol{f}_t|\boldsymbol{x}_{0:t-1}, \boldsymbol{f}_{1:t-1}). \quad (6)$$

In particular, from Eq. 1, the last factor in Eq. 6 is derived in the same manner as the posterior in GP regression, given by

$$p(\boldsymbol{f}_t|\boldsymbol{x}_{0:t-1}, \boldsymbol{f}_{1:t-1}) = \mathcal{N}(\boldsymbol{f}_t|\mathcal{M}_{t-1}, \mathcal{K}_{t-1}), \quad (7)$$

where

$$\begin{aligned}
\mathcal{M}_{t-1} &= m(\tilde{\boldsymbol{x}}_{t-1}) + k(\tilde{\boldsymbol{x}}_{t-1}, \tilde{\boldsymbol{x}}_{0:t-2}) K_{0:t-2}^{-1} (\boldsymbol{f}_{1:t-1} - m(\tilde{\boldsymbol{x}}_{0:t-2}))^T, \\
\mathcal{K}_{t-1} &= k(\tilde{\boldsymbol{x}}_{t-1}, \tilde{\boldsymbol{x}}_{t-1}) - k(\tilde{\boldsymbol{x}}_{t-1}, \tilde{\boldsymbol{x}}_{0:t-2}) K_{0:t-2}^{-1} k(\tilde{\boldsymbol{x}}_{0:t-2}, \tilde{\boldsymbol{x}}_{t-1}),
\end{aligned}$$

and the term $\boldsymbol{f}_{1:t-1} - m(\tilde{\boldsymbol{x}}_{0:t-2})$ is written as

$$\boldsymbol{f}_{1:t-1} - m(\tilde{\boldsymbol{x}}_{0:t-2}) \equiv (\boldsymbol{f}_1 - m(\tilde{\boldsymbol{x}}_0), \cdots, \boldsymbol{f}_{t-1} - m(\tilde{\boldsymbol{x}}_{t-2})).$$

### 3.1 Learning and Prediction in GPSSM

In learning the posterior, our target is to compute

$$p(\boldsymbol{x}_{0:t}, \boldsymbol{f}_{1:t}|\boldsymbol{y}_{1:t}) = p(\boldsymbol{x}_{0:t}, \boldsymbol{f}_{1:t}, \boldsymbol{y}_{1:t})/p(\boldsymbol{y}_{1:t}).$$

The major challenge is the computation of $p(\boldsymbol{y}_{1:t})$. Throughout this paper, we are going to apply *variational inference* to train GPSSM, which is a technique based on making assumptions about the posterior over latent variables that leads to a tractable lower bound, often called *evidence lower bound* (ELBO). There are two reasons for choosing this approach, which is currently state-of-the-art in training GPSSM. First, variational inference is computationally efficient. Second, this approach will lead to an approximation of the posterior, which can be carried to the AL strategy, as we will present further in Sect. 4.

The first step of using variational inference in GPSSM is to introduce $M$ ($M \ll T$) inducing points $\boldsymbol{u} = \boldsymbol{u}_{1:M} = \{\boldsymbol{u}_i\}_{i=1}^{M}$ with their corresponding inputs. This is referred as sparse GP technique (see e.g. [21,48]) and the joint density is

$$p(\boldsymbol{y}_{1:T}, \boldsymbol{x}_{0:T}, \boldsymbol{f}_{1:T}, \boldsymbol{u}) = p(\boldsymbol{x}_0)p(\boldsymbol{u}) \prod_{t=1}^{T} p(\boldsymbol{y}_t|\boldsymbol{x}_t)p(\boldsymbol{x}_t|\boldsymbol{f}_t)p(\boldsymbol{f}_t|\boldsymbol{x}_{0:t-1}, \boldsymbol{f}_{1:t-1}, \boldsymbol{u}). \tag{8}$$

Then, the ELBO $\mathcal{L}_t$ to the log marginal likelihood of $p(\boldsymbol{y}_{1:t})$ in Eq. 6, which is based on KL divergence (see e.g. chapter 8.5 in Cover [9]), is given by

$$\log(p(\boldsymbol{y}_{1:t})) = \mathcal{L}_t + \mathrm{KL}\left[q(\boldsymbol{x}_{0:t}, \boldsymbol{f}_{1:t}) \| p(\boldsymbol{x}_{0:t}, \boldsymbol{f}_{1:t}|\boldsymbol{y}_{1:t})\right]$$

for $t = 1, \cdots, T$, where $q(\cdot)$ is the distribution function to approximate $p(\cdot)$. With further derivations from variational inference methodology, $\mathcal{L}_t$ is given by

$$\mathcal{L}_t = \int q(\boldsymbol{x}_{0:t}, \boldsymbol{f}_{1:t}, \boldsymbol{u}) \log\left(\frac{p(\boldsymbol{y}_{1:t}, \boldsymbol{x}_{0:t}, \boldsymbol{f}_{1:t}, \boldsymbol{u})}{q(\boldsymbol{x}_{0:t}, \boldsymbol{f}_{1:t}, \boldsymbol{u})}\right) d\boldsymbol{x}_{0:t} d\boldsymbol{f}_{1:t} d\boldsymbol{u}. \tag{9}$$

Analogously to Ialongo et al. [28], we set

$$q(\boldsymbol{x}_{0:t}, \boldsymbol{f}_{1:t}, \boldsymbol{u}) = q(\boldsymbol{u})q(\boldsymbol{x}_0) \prod_{i=1}^{t} q(\boldsymbol{x}_i|\boldsymbol{f}_i)p(\boldsymbol{f}_i|\boldsymbol{f}_{1:i-1}, \boldsymbol{x}_{0:i-1}, \boldsymbol{u}) \tag{10}$$

which leads to Eq. 17 of [28]

$$\begin{aligned}\mathcal{L}_T = &\int \sum_{t=1}^{T} q(\boldsymbol{x}_{0:T}) \log\left(p(\boldsymbol{y}_t|\boldsymbol{x}_t)\right) d\boldsymbol{x}_{0:T} - \sum_{t=1}^{T} \int q(\boldsymbol{f}_t)\mathrm{KL}\left[q(\boldsymbol{x}_t|\boldsymbol{f}_t) \| p(\boldsymbol{x}_t|\boldsymbol{f}_t)\right] d\boldsymbol{f}_t \\ &- \mathrm{KL}\left[q(\boldsymbol{x}_0) \| p(\boldsymbol{x}_0)\right] - \mathrm{KL}\left[q(\boldsymbol{u}) \| p(\boldsymbol{u})\right].\end{aligned} \tag{11}$$

Under sparse GP approximation, we can specify a free Gaussian density on the function values $\boldsymbol{u}$ giving $q(\boldsymbol{u}) = \mathcal{N}(\boldsymbol{u}|\boldsymbol{\mu_u}, \boldsymbol{\Sigma_u})$. The only term that requires further specification is $q(\boldsymbol{x}_i|\boldsymbol{f}_i)$ and [28] presented $q(\boldsymbol{x}_i|\boldsymbol{f}_i) = \mathcal{N}(\boldsymbol{x}_i|\boldsymbol{A}_{i-1}\tilde{\boldsymbol{f}}_{i-1} + \boldsymbol{b}_{i-1}, \boldsymbol{S}_{i-1})$, where $\tilde{\boldsymbol{f}}_i$ is a free variational parameter. $\boldsymbol{A}_i$, $\boldsymbol{b}_i$ and $\boldsymbol{S}_i$ are other free variational parameters depending on the choice of $\tilde{\boldsymbol{f}}_i$. This generalises a number of previous work. For example, by setting both $\tilde{\boldsymbol{f}}_i = \boldsymbol{x}_i$ in $q(\cdot|\cdot)$ and $q(\boldsymbol{x}) = q(\boldsymbol{x}_0)$, we recover the expression from [19] and the optimal $q(\boldsymbol{x})$ can be solved by calculus of variations. If we set $\tilde{\boldsymbol{f}}_i = \boldsymbol{x}_i$ only, then we have the Gaussian factorised approximation of [38]. In [28], they proposed setting $\tilde{\boldsymbol{f}}_i = k(\boldsymbol{x}_i, \boldsymbol{u})K_{\boldsymbol{uu}}^{-1}\boldsymbol{u}$ and the training turned out to be marginally better if a sufficiently large amount of iterations are allowed.

We choose to set only $\tilde{\boldsymbol{f}}_i = \boldsymbol{x}_i$ throughout this paper, which is the approach by [38]. This is because their approach has shown to have a stable improvement in model accuracy during the training phase. Also, its linear factorising nature, given a fixed and often small number of allowed iterations, is also beneficial for training time. Such setting leads to $q(\boldsymbol{x}_i|\boldsymbol{f}_i) = \mathcal{N}(\boldsymbol{x}_i|\boldsymbol{A}_{i-1}\boldsymbol{x}_{i-1} + \boldsymbol{b}_{i-1}, \boldsymbol{S}_{i-1})$, with the free parameters $\boldsymbol{S}_{i-1} = (\boldsymbol{Q}^{-1} + \boldsymbol{C}^T\boldsymbol{R}^{-1}\boldsymbol{C})^{-1}$, $\boldsymbol{A}_{i-1} = \boldsymbol{S}_{i-1}\boldsymbol{Q}^{-1}$ and $\boldsymbol{b}_{i-1} = \boldsymbol{S}_{i-1}\boldsymbol{C}^T\boldsymbol{R}^{-1}(\boldsymbol{y}_t - \boldsymbol{d})$.

The prediction with GPSSM can be done rather cheaply. Suppose we would like to predict the new observation $\boldsymbol{y}^\star$ based on a new control input $\boldsymbol{c}^\star$, by defining all training data as $\mathcal{D} \equiv \{\boldsymbol{y}_{1:T}, \boldsymbol{x}_{0:T}\}$, we can show easily [19] that the predictive distribution $p(\boldsymbol{y}^\star|\boldsymbol{c}^\star, \mathcal{D})$ is given by

$$p(\boldsymbol{y}^\star|\boldsymbol{c}^\star, \mathcal{D}) = \mathcal{N}(\boldsymbol{y}^\star|\boldsymbol{C}f(\boldsymbol{x}_T, \boldsymbol{c}^\star) + \boldsymbol{d}, \boldsymbol{R} + \boldsymbol{C}\boldsymbol{Q}\boldsymbol{C}^T). \tag{12}$$

In practice, we sample $f(\boldsymbol{x}_T, \boldsymbol{c}^\star)$, where the latent states $\boldsymbol{x}_{1:T}$ are already estimated via training. Then, we predict the new observation based on drawing samples from the distribution. This is the approach used by [28].

## 4 Active Learning Strategies

The previous section presents the modelling of GPSSM and in this section, we are interested in acquiring a strategy to actively learn the GPSSM by steering the system through the latent state-space. We employ AL as a query strategy to pick the most informative new control input $\boldsymbol{c}_t^*$ while observing $\boldsymbol{y}_t$, and to learn the model as we explore. Contrary to the two previous work [6,8], we treat the states as latent and unknown and $\boldsymbol{c}_t$ is the only controllable variable. For the exploration criterion, we use approximated mutual information. The usage of mutual information for AL has been well motivated by, for example, Krause et al. [33]. They pointed out that the mutual information might lead to a more accurate model than the differential entropy. This quantity was also shown to be the same as minimising the expected uncertainty of the model [15].

First, let us recall from Chap. 8.5 of Cover [9] that the mutual information between any two sets of random variables $\boldsymbol{Y}$ and $\boldsymbol{F}$ with joint density $p(\boldsymbol{Y}, \boldsymbol{F})$ is defined as

$$I(\boldsymbol{Y}; \boldsymbol{F}) = \int p(\boldsymbol{Y}, \boldsymbol{F}) \log\left(\frac{p(\boldsymbol{Y}, \boldsymbol{F})}{p(\boldsymbol{Y})p(\boldsymbol{F})}\right) d\boldsymbol{Y} d\boldsymbol{F}$$

and a well-known relationship between mutual information and differential entropy $h(\cdot)$ is given by $I(\boldsymbol{Y};\boldsymbol{F})=h(\boldsymbol{Y})-h(\boldsymbol{Y}|\boldsymbol{F})$.

For steering the system through latent state-space while gathering information for efficiently learning the model, a sensible quantity for exploration is the mutual information between the latest observations $\boldsymbol{y}_{t+1}$ and latest predicted transition functions $\boldsymbol{f}_{t+1}$. This quantity is an extension to the active strategy from Buisson et al. [6] by using mutual information as a criterion. That is,

$$\boldsymbol{c}_t^* = \underset{\boldsymbol{c}_t \in \mathcal{C}}{\operatorname{argmax}}\, I(\hat{\boldsymbol{y}}_{t+1}; \boldsymbol{f}_{t+1}), \tag{13}$$

where $\boldsymbol{c}_t$ is placed within the term $\boldsymbol{f}_{t+1} \equiv f(\boldsymbol{x}_t, \boldsymbol{c}_t)$. We use $\hat{\boldsymbol{y}}_{t+1}$ because this term is provided from predicting GPSSM with a given new control $\boldsymbol{c}_t$. Altogether, $\boldsymbol{c}_t$ is the only independent variable.

This expression is intractable and there have been enormous efforts in estimating it (e.g. [2,25,34,39,40]). However, most previous work assumed that we have little or no information on both entropies in the mutual information expression and this comes down to the fact that we do not know the corresponding probability density function (pdf). Their approach includes deriving bounds but extra functions such as *critic* or *variational parameters* are often required (e.g. Sect. 2 in [40]). However, should we have some information about this quantity, it is much better to make use of this known information instead. Therefore, we derive our estimate for the mutual information by directly approximating the pdf. The main advantages of our method versus others are: (1) As discussed in the previous paragraph, other methods often require extra functions called *critic* or *variational parameters*, or some free parameters to set. Finding the best settings of these extra functions and parameters could be complicated at times. Mcallester et al. [37] also showed that there are statistical limitations in various bounds of mutual information and a better approach is to measure this quantity as a difference in entropies. (2) Our mathematical derivation is purely via inequalities and numerical approximations, resulting an approximated closed form. Hence, the final expression derived can be easily implemented.

The quantity $I(\boldsymbol{y}_t; \boldsymbol{f}_t)$ is often studied in SSM (e.g. [15]), motivated by the assumption of Markov property. However, if the Markov assumption is not completely valid, as there are (weak) dependencies between the states, the mutual information between all observations $\boldsymbol{y}_{1:t+1}$ and all predictions from the transition function $\boldsymbol{f}_{1:t+1}$ within a time-horizon might present a better quantity. To the best of our knowledge, we are the first to propose using mutual information between all variables in time, and the AL strategy presented in Eq. 13 can be viewed as a special case of this general formulation. Thus, another more general AL strategy is

$$\boldsymbol{c}_t^* = \underset{\boldsymbol{c}_t \in \mathcal{C}}{\operatorname{argmax}}\, I(\boldsymbol{y}_{1:t}, \hat{\boldsymbol{y}}_{t+1}; \boldsymbol{f}_{1:t+1}). \tag{14}$$

Again, the term $\hat{\boldsymbol{y}}_{t+1}$ is provided from predicting GPSSM. This also leads to the third advantage of our method: (3) To the best of our knowledge, we are the first to estimate the total mutual information by leveraging the inference scheme of the model. This turns out to be computationally efficient as well.

**input :** Initial $T$ observations $\boldsymbol{y}_{1:T}$ and control inputs $\boldsymbol{c}_{0:T}$. Initial state $\boldsymbol{x}_0$.
**output:** An optimised GPSSM after taking $N$ exploration steps, the final data set including the explored observations $\boldsymbol{y}_{1:T+N}$ and the corresponding controls $\boldsymbol{c}_{0:T+N}$.

*Train the initial GPSSM* **for** $t = T, T+1, \cdots, T+N$ **do**

1. Solve $\boldsymbol{c}_t^*$ via equation (14) (or (13));
2. Evaluate $\boldsymbol{y}_{t+1}$ from the system;
3. Update the training data set;
4. Retrain GPSSM to update hyperparameters;
5. $t = t + 1$;

**end**

**Algorithm 1:** *Active Learning in GPSSM* on control through approximation of mutual information

For clarity, we refer $I(\hat{\boldsymbol{y}}_{t+1}; \boldsymbol{f}_{t+1})$ as *latest* mutual information (latMI) and $I(\boldsymbol{y}_{1:t}, \hat{\boldsymbol{y}}_{t+1}; \boldsymbol{f}_{1:t+1})$ as *total* mutual information (totMI). In the remainder of this section, we present how the two quantities can be approximated for GPSSM with latent states. We first present an estimation scheme for the *latest* mutual information to align with previous work. Then, we proceed to the estimation of our proposed *total* mutual information. Given the estimates of mutual information, we can formulate the AL strategy for GPSSM, as summarised in Algorithm 1.

### 4.1 Computation of Latest Mutual Information $I(\boldsymbol{y}_t; \boldsymbol{f}_t)$

To estimate the *latest* mutual information $I(\boldsymbol{y}_t; \boldsymbol{f}_t)$, we employ the Gaussian approximate integral derived by Girard [23]. Here, for two random variables $\boldsymbol{x}$ and $\boldsymbol{y}$ with $p(\boldsymbol{y}|\boldsymbol{x}) = \mathcal{N}(\boldsymbol{y}|\mu(\boldsymbol{x}), \sigma^2(\boldsymbol{x}))$ and $p(\boldsymbol{x}|\boldsymbol{u}, \boldsymbol{\Sigma}_{\boldsymbol{x}}) = \mathcal{N}(\boldsymbol{x}|\boldsymbol{u}, \boldsymbol{\Sigma}_{\boldsymbol{x}})$ for some mean $\boldsymbol{u}$ and variance $\Sigma_{\boldsymbol{x}}$, the Gaussian approximation yields

$$\int p(\boldsymbol{y}|\boldsymbol{x})p(\boldsymbol{x}|\boldsymbol{u}, \boldsymbol{\Sigma}_{\boldsymbol{x}})d\boldsymbol{x} \approx \mathcal{N}(M(\boldsymbol{u}, \boldsymbol{\Sigma}_{\boldsymbol{x}}), V(\boldsymbol{u}, \boldsymbol{\Sigma}_{\boldsymbol{x}})),$$

where $M(\boldsymbol{u}, \boldsymbol{\Sigma}_{\boldsymbol{x}})$ and $V(\boldsymbol{u}, \boldsymbol{\Sigma}_{\boldsymbol{x}})$ are integral functions to be evaluated (see supplementary materials for details). Based on this approximation, $I(\boldsymbol{y}_t; \boldsymbol{f}_t)$ can be approximately computed as in the following proposition.

**Proposition 1.** *Given the definition of GPSSM from Eq. 1–4, as well as the notation of $\mathcal{M}_t$ and $\mathcal{K}_t$ defined in Eq. 7. We define the approximation of the following integrals as*

$$\mathcal{N}(\boldsymbol{f}_1|M_1, V_1) := \mathcal{N}(\boldsymbol{f}_1|M(\boldsymbol{\mu}_0, \boldsymbol{\Sigma}_0), V(\boldsymbol{\mu}_0, \boldsymbol{\Sigma}_0))$$
$$\approx \int \mathcal{N}(\boldsymbol{f}_1|\mathcal{M}_0, \mathcal{K}_0)\mathcal{N}(\boldsymbol{x}_0|\boldsymbol{\mu}_0, \boldsymbol{\Sigma}_0)d\boldsymbol{x}_0$$

*and, recursively, for all $t = T, T+1, \cdots, T+N$,*

$$\mathcal{N}(\boldsymbol{f}_t|M_t, V_t) \approx \int \mathcal{N}(\boldsymbol{f}_t|\mathcal{M}_{t-1}, \mathcal{K}_{t-1})\mathcal{N}(\boldsymbol{x}_{t-1}|M_{t-1}, V_{t-1} + \boldsymbol{Q})d\boldsymbol{x}_{t-1}.$$

*Then, the latest mutual information is approximately*

$$I(\boldsymbol{y}_t; \boldsymbol{f}_t) \approx \frac{1}{2} \log \left( \frac{\det(\boldsymbol{R} + \boldsymbol{C}(V_t + \boldsymbol{Q})\boldsymbol{C}^T)}{\det(\boldsymbol{R} + \boldsymbol{C}\boldsymbol{Q}\boldsymbol{C}^T)} \right). \quad (15)$$

*Proof (sketch).* Based on the definition of mutual information, the numerator in Eq. 15 relies on the approximation via *moment matching*, whereas the denominator in Eq. 15 can be computed directly.

### 4.2 Computation of Total Mutual Information $I(\boldsymbol{y}_{1:t}; \boldsymbol{f}_{1:t})$

The ELBO $\mathcal{L}_t$ given by Eq. 11 can be utilised to estimate the *total* mutual information. Since this quantity is estimated via samples, this quantity is sample driven and the following proposition presents our computational approach.

**Proposition 2.** *Given the definition of GPSSM from Eq. 1–4, as well as the expression of the ELBO $\mathcal{L}_t$ from Eq. 11, for $t = T, T+1, \cdots, T+N$, if $S$ samples are drawn, the $s$-th sample estimate ($s = 1, \cdots, S$) of the mutual information between observations $\boldsymbol{y}_{1:t}$ and the prediction $\boldsymbol{f}_{1:t}$, denoted by $i_s$, is bounded by*

$$i_s \leq \sum_{i=1}^{t} \log(\mathcal{N}_s(\boldsymbol{y}_i | \boldsymbol{C}\boldsymbol{f}_i + \boldsymbol{d}, \boldsymbol{R} + \boldsymbol{C}\boldsymbol{Q}\boldsymbol{C}^T)) - \mathcal{L}_{t,s}, \quad (16)$$

*where $\mathcal{N}_s(\cdot)$ and $\mathcal{L}_{t,s}$ are the $s$-th sample estimate of the normal distribution and ELBO $\mathcal{L}_t$, respectively. The total mutual information is then approximated by $I(\boldsymbol{y}_{1:t}; \boldsymbol{f}_{1:t}) \approx \frac{1}{S} \sum_{s=1}^{S} i_s$.*

*Proof (sketch).* The proof employs mainly the definition of mutual information. The intractable part is bounded via ELBO, and the other term can be computed directly. Since it is a sample based approach, the bound from a sample becomes an approximation after taking all samples.

*Remark 3.* In computing Eq. 16, the sampling nature comes from the specific value of $\boldsymbol{x}_0$ drawn, and this value will be different between samples. Therefore, the evaluated $i_s$ will also be different.

The efficiency of this approach heavily relies on the inference of GPSSM and it is possible to use another form of ELBO by defining $q(\cdot)$ differently or even other inference techniques such as a MCMC based approaches. While different approaches will incur their corresponding computational effort, it is better to keep the same approach in training GPSSM and computing total mutual information in order to avoid extra computational cost, as we reuse the posterior with trained hyperparameters in estimating the total mutual information.

## 5 Experiments

In this section, our experiments demonstrate how our proposed AL strategies help in learning GPSSM. From our discussion during previous sections, the two

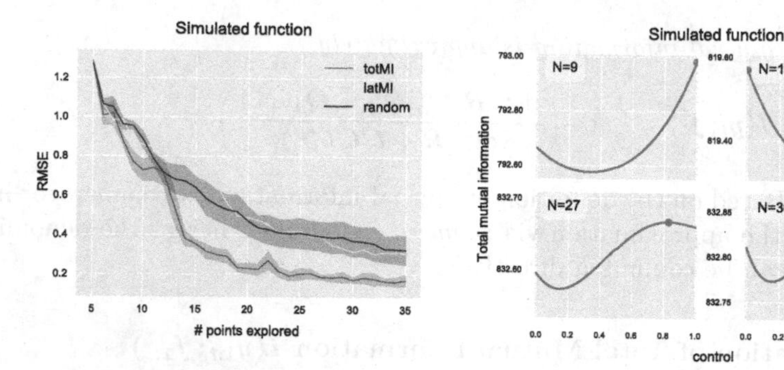


**Fig. 2.** The *left* diagram shows the results of the simulated function: Mean and standard deviation of RMSE for 30 exploration steps (starting with 5 initial points) are collected from independent trials. latMI is presented in yellow, totMI in red, and random exploration in blue. The right figures shows where the maximum *total* mutual information is attained when 9, 18, 27, 36 points are explored in one particular trial. (Color figure online)

latest work from Buisson et al. [6] and Capone et al. [8] assumed known states, so we cannot compare against them in the setting with unobservable states we focus on. Therefore, the reasonable benchmark we can compare to is the usual setting where models are learned with randomly distributed control inputs (e.g. [36]). We first perform an experiment on a simulated function, then on few non-linear physical problems.[1] The overall goal is to obtain a high accuracy for the GPSSM with as few selected points as possible.

## 5.1 Simulated Function

We first consider a modified kink function based on [28]. Our goal is to learn the dynamics of the system within a range of $\mathcal{X} = [-3.0, 1.1]$ for the latent states and $\mathcal{C} = [0, 1]$ for the control. We begin with 5 given points and explore 30 steps.

The left diagram in Fig. 2 depicts the RMSE of training sessions applying random selection, *latest* mutual information (latMI) and *total* mutual information (totMI) strategy, respectively. Each session consists of independent runs. Soon after collecting the first control input, both AL show results not worse than random but latMI is not significant, while totMI is more profound.

To illustrate why we need our AL strategies, especially totMI, we also present a snapshot of the position of the maximum totMI as we explore in the right of Fig. 2 in a trial. The selected diagram indicates that there is no trivial most informative control where we can attain maximum totMI. Therefore, optimisation in Eq. 14 (or 13) is necessary. These diagrams could change as we run different trials, and such evaluation will be more complicated as we increase the dimension of controls or the dynamics become more complex.

---

[1] The ground base of GPSSM code is based on Ialongo et al. [28], re-engineered in `GPflow` 2.1 [35].

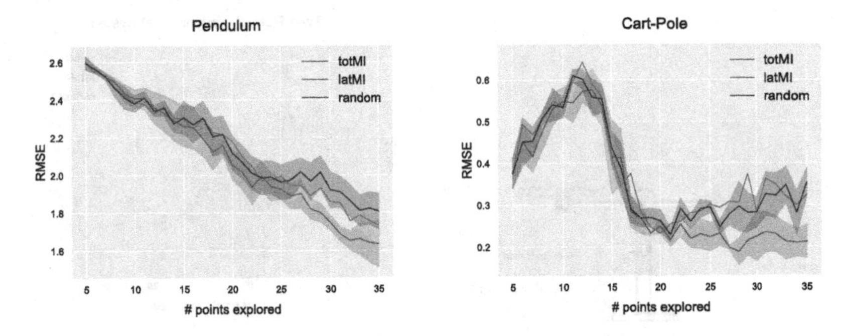


**Fig. 3.** The left diagram shows the results of the pendulum: Mean and standard deviation of RMSE for 30 exploration steps (starting with 5 initial points) are collected from independent trials. latMI is presented in yellow, totMI in red, and random exploration in blue. The right diagram shows the results of the cart-pole: Again, mean and standard deviation of RMSE for 30 exploration steps (starting with 5 initial points) are collected from independent trials. (Color figure online)

## 5.2 Pendulum and Cart-Pole

In these experiments, our aim is to learn the physics of the pendulum and cart-pole by actively controlling the input torque and the force of the cart, respectively. We evaluate the state-space model's accuracy via the angular position obtained from the model against the ground-truth, which is calculated via the equation of motions.

Figure 3 depicts the RMSE of training sessions applying random selection, *latest* mutual information and *total* mutual information strategy for pendulum (left) and cart-pole (right), respectively. Both experiments consist of independent runs. For the pendulum, both totMI and latMI strategy show advantages against the random exploration. However, totMI shows even better results as the exploration proceeds. For cart-pole, all strategies have, in the initial exploration phase, a raise in RMSE, which is subsequently reduced, as the model becomes more stable. latMI is indeed unable to attain noticeable improvement but totMI strategy still shows favourable performance via a consistently lower RMSE.

Since the physics of both systems are common enough that we can actually measure the states, we also compare our work directly with the AL strategy by Capone et al. [8], even though the fairness of such a comparison is debatable.

## 5.3 Twin-Rotor Aerodynamical System

As a more realistic setting, we test our AL strategy on a Twin-Rotor Aerodynamical System (TRAS). This is a typical design for control experiments, and its behaviour resembles that of a helicopter. There are two rotors – one horizontal and another vertical – joined by a beam. There is also a counter-weight, which determines a stable equilibrium position. When the system is switched off, the main rotor is lowered. The motion of the system is controlled by the two

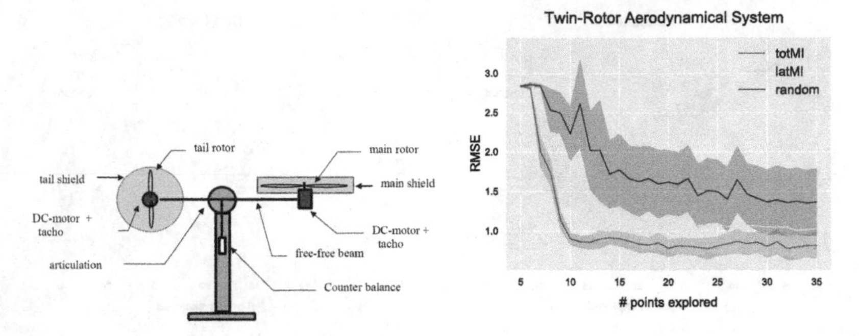


**Fig. 4.** The left diagram shows an example of a Twin-Rotor Aerodynamical System, taken from [4]. The right diagram shows the results of the Twin-Rotor Aerodynamical System: Mean and standard deviation of RMSE for 30 exploration steps (starting with 5 initial points) are collected from independent trials. latMI is presented in yellow; totMI in red and random exploration in blue. (Color figure online)

motor supply voltages to each motor. We demonstrate a simple example of such a system in the left diagram of Fig. 4.

Our aim is to learn the physics of the TRAS by actively control the two voltages to the system. We evaluate the model's accuracy via the angular position measured from the model against which is calculated via the equation of motions on both rotors. The physics of the TRAS is based on the description by [4]. The right diagram of Fig. 4 depicts the RMSE of training sessions applying random selection, latMI and totMI strategy, respectively. Each session consists of independent runs. Both latMI and totMI strategies show advantage against random exploration in presence of more sophisticated dynamics. However, GPSSM with totMI reaches an acceptable accuracy with much lower number of data points.

## 6 Discussion

We study AL in GPSSM by proposing a tractable mutual information estimates in order to select the most informative control inputs to maximise our model accuracy as we explore. Both *latest* and *total* mutual information are proposed. The empirical results show that *total* mutual information strategy outperforms the other approach and random exploration. There is a promising potential of our approach towards more complex industrial settings. Since we leverage ELBO to derive our estimates, we also expect our propositions to be applicable to future work on inference of GPSSM and other probabilistic models with latent space.

Another direction of extension is to look into the use of maximal mutual information in reinforcement learning (RL) community, as such strategy has been used in planning and control. For instance, Ding et al. [11] applied such strategy in Partially Observable Markov Decision Process (POMDP) framework in order to efficiently learn the RL model. However, there are key differences between the two themes. First, the word *control* is interpreted very differently.

In GPSSM, this refers to an augmented latent state and each input is simply a deterministic real number whereas in POMDP, this is a mapping between an action taken in certain state to maximise the expected return or reward. Second, AL in GPSSM has a very specific problem formulation whereas efficient learning in POMDP is a general framework with a large room of freedom to specify different components, leading to various research topics (e.g. [5,14,24]).

## References


1. Alaa, A.M., van der Schaar, M.: Attentive state-space modeling of disease progression. In: Advances in Neural Information Processing Systems, pp. 11338–11348 (2019)
2. Alemi, A., Poole, B., Fischer, I., Dillon, J., Saurous, R.A., Murphy, K.: Fixing a broken ELBO. In: International Conference on Machine Learning, pp. 159–168. PMLR (2018)
3. Amin, K., Cortes, C., DeSalvo, G., Rostamizadeh, A.: Understanding the effects of batching in online active learning. In: International Conference on Artificial Intelligence and Statistics, pp. 3482–3492 (2020)
4. Anonymous: Two rotor aero-dynamical system. user's manual. Technical report, Inteco (2006). www.inteco.com.pl
5. Berkenkamp, F., Nurchetta, M., Schoellig, A.P., Krause, A.: Safe model-based reinforcement learning with stability guarantees. In: Advances in Neural Information Processing Systems (2017)
6. Buisson-Fenet, M., Solowjow, F., Trimpe, S.: Actively learning Gaussian process dynamics. In: Learning for Dynamics and Control, pp. 5–15. PMLR (2020)
7. Cai, W., Zhang, M., Zhang, Y.: Batch mode active learning for regression with expected model change. IEEE Trans. Neural Netw. Learn. Syst. **28**, 1668–1681 (2016)
8. Capone, A., Noske, G., Umlauft, J., Beckers, T., Lederer, A., Hirche, S.: Localized active learning of Gaussian process state space models. In: Learning for Dynamics and Control, pp. 490–499. PMLR (2020)
9. Cover, T.M., Thomas, J.A.: Elements of Information Theory, 2nd edn. Wiley, Hoboken (2006)
10. Dasgupta, S.: Two faces of active learning. Theor. Comput. Sci. **412**(19), 1767–1781 (2011)
11. Ding, Y., Clavera, I., Abbeel, P.: Mutual information maximization for robust plannable representations. arXiv preprint arXiv:2005.08114 (2020)
12. Doerr, A., et al.: Probabilistic recurrent state-space models. In: International Conference on Machine Learning, pp. 1280–1289 (2018)
13. Eleftheriadis, S., Nicholson, T.F., Deisenroth, M.P., Hensman, J.: Identification of Gaussian process state space models. In: Advances in Neural Information Processing Systems, pp. 5309–5319 (2017)
14. Engel, Y., Mannor, S., Meir, R.: Reinforcement learning with Gaussian processes. In: Proceedings of the 22nd International Conference on Machine Learning, pp. 201–208 (2005)
15. Ertin, E., Fisher, J.W., Potter, L.C.: Maximum mutual information principle for dynamic sensor query problems. In: Zhao, F., Guibas, L. (eds.) IPSN 2003. LNCS, vol. 2634, pp. 405–416. Springer, Heidelberg (2003). https://doi.org/10.1007/3-540-36978-3_27

16. Faruqi, F.A.: State space model for autopilot design of aerospace vehicles. Technical report, Weapons Systems Division, Defence Science and Technology Organisation (DSTO), Edinburgh (Australia) (2007)
17. Friedland, B.: Control system design: an introduction to state-space methods. Courier Corporation (2012)
18. Frigola, R.: Bayesian time series learning with Gaussian processes. Ph.D. thesis, University of Cambridge (2015)
19. Frigola, R., Chen, Y., Rasmussen, C.E.: Variational Gaussian process state-space models. In: Advances in Neural Information Processing Systems, pp. 3680–3688 (2014)
20. Frigola, R., Lindstenk, F., Schön, T.B., Rasmussen, C.E.: Bayesian inference and learning in Gaussian process state-space models with particle MCMC. In: Advances in Neural Information Processing Systems, pp. 3156–3164 (2013)
21. G. de Matthews, A.G., Hensman, J., Turner, R., Ghahramani, Z.: On sparse variational methods and the Kullback-Leibler divergence between stochastic processes. In: Proceedings of the 19th International Conference on Artificial Intelligence and Statistics, pp. 231–239 (2016)
22. Gal, Y., Islam, R., Ghahramani, Z.: Deep Bayesian active learning with image data. In: International Conference on Machine Learning, pp. 1183–1192 (2017)
23. Girard, A.: Approximate methods for propagation of uncertainty with Gaussian process models. Ph.D. thesis, Citeseer (2004)
24. Grande, R., Walsh, T., How, J.: Sample efficient reinforcement learning with Gaussian processes. In: International Conference on Machine Learning, pp. 1332–1340. PMLR (2014)
25. Hoffman, M.D., Johnson, M.J.: ELBO surgery: yet another way to carve up the variational evidence lower bound. In: Workshop in Advances in Approximate Bayesian Inference, vol. 1, p. 2. NIPS (2016)
26. Houlsby, N., Huszár, F., Ghahramani, Z., Lengyel, M.: Bayesian active learning for classification and preference learning. arXiv preprint arXiv:1112.5745 (2011)
27. Ialongo, A.D., van der Wilk, M., Rasmussen, C.E.: Closed-form inference and prediction in Gaussian process state-space models. In: Advances in Neural Information Processing Systems Workshop (2017)
28. Ialongo, A.D., Wilk, M.V.D., Hensman, J., Rasmussen, C.E.: Overcoming mean-field approximations in recurrent Gaussian process models. In: International Conference on Machine Learning, pp. 2931–2940 (2019)
29. Kirk, D.E.: Optimal control theory: an introduction. Courier Corporation (2004)
30. Kirsch, A., van Amersfoort, J., Gal, Y.: BatchBald: efficient and diverse batch acquisition for deep Bayesian active learning. In: Advances in Neural Information Processing Systems, pp. 7024–7035 (2019)
31. Ko, J., Fox, D.: GP-BayesFilters: Bayesian filtering using Gaussian process prediction and observation models. Auton. Robot. **27**(1), 75–90 (2009)
32. Krause, A., Guestrin, C.: Nonmyopic active learning of Gaussian processes: an exploration-exploitation approach. In: Proceedings of the 24th International Conference on Machine Learning, pp. 449–456 (2007)
33. Krause, A., Singh, A., Guestrin, C.: Near-optimal sensor placements in Gaussian processes: theory, efficient algorithms and empirical studies. J. Mach. Learn. Res. **9**, 235–284 (2008)
34. Lombardi, D., Pant, S.: Nonparametric k-nearest-neighbor entropy estimator. Phys. Rev. E **93**(1), 013310 (2016)
35. Matthews, A.G.D.G., et al.: GPflow: a Gaussian process library using TensorFlow. J. Mach. Learn. Res. **18**(1), 1299–1304 (2017)

36. Mattos, C.L.C., Damianou, A., Barreto, G.A., Lawrence, N.D.: Latent autoregressive Gaussian processes models for robust system identification. IFAC-PapersOnLine **49**(7), 1121–1126 (2016)
37. McAllester, D., Stratos, K.: Formal limitations on the measurement of mutual information. In: International Conference on Artificial Intelligence and Statistics, pp. 875–884 (2020)
38. McHutchon, A.J.: Nonlinear modelling and control using Gaussian processes. Ph.D. thesis, University of Cambridge (2015)
39. Pérez-Cruz, F.: Estimation of information theoretic measures for continuous random variables. In: Advances in Neural Information Processing Systems, pp. 1257–1264 (2009)
40. Poole, B., Ozair, S., Oord, A.V.D., Alemi, A.A., Tucker, G.: On variational bounds of mutual information. In: International Conference on Machine Learning, pp. 5171–5180 (2019)
41. Rangapura, S.S., Seeger, M.W., Gasthaus, J., Stella, L., Wang, Y., Januschowski, T.: Deep state space models for time series forecasting. In: Advances in Neural Information Processing Systems, pp. 7785–7794 (2018)
42. Rasmussen, C.E., Williams, C.K.: Gaussian Processes for Machine Learning. The MIT Press, Cambridge (2006)
43. Schreiter, J., Nguyen-Tuong, D., Eberts, M., Bischoff, B., Markert, H., Toussaint, M.: Safe exploration for active learning with Gaussian processes. In: Bifet, A., et al. (eds.) ECML PKDD 2015. LNCS (LNAI), vol. 9286, pp. 133–149. Springer, Cham (2015). https://doi.org/10.1007/978-3-319-23461-8_9
44. Seo, S., Wallat, M., Graepel, T., Obermayer, K.: Gaussian process regression: active data selection and test point rejection. In: Sommer, G., Krüger, N., Perwass, C. (eds.) Mustererkennung 2000. Informatik aktuell. Springer, Heidelberg (2000). https://doi.org/10.1007/978-3-642-59802-9_4
45. Settles, B.: Active learning literature survey. Computer Sciences Technical report 1648, University of Wisconsin-Madison (2009)
46. Shui, C., Zhou, F., Gagné, C., Wang, B.: Deep active learning: unified and principled method for query and training. In: International Conference on Artificial Intelligence and Statistics, pp. 1308–1318. PMLR (2020)
47. Srinivas, N., Krause, A., Kakade, S.M., Seeger, M.W.: Information-theoretic regret bounds for Gaussian process optimization in the bandit setting. IEEE Trans. Inf. Theor. **58**(5), 3250–3265 (2012)
48. Titsias, M.: Variational learning of inducing variables in sparse Gaussian processes. In: International Conference on Artificial Intelligence and Statistics, pp. 567–574 (2009)
49. Turner, R., Deisenroth, M., Rasmussen, C.: State-space inference and learning with Gaussian processes. In: Proceedings of the 13th International Conference on Artificial Intelligence and Statistics, pp. 868–875 (2010)
50. Twomey, N., Diethe, T., Flach, P.: Bayesian active learning with evidence-based instance selection. In: Workshop on Learning over Multiple Contexts, European Conference on Machine Learning, ECML 2015 (2015)
51. Wang, J., Hertzmann, A., Fleet, D.J.: Gaussian process dynamical models. In: Advances in Neural Information Processing Systems, pp. 1441–1448 (2006)
52. Zeng, Y., Wu, S.: State-Space Models: Applications in Economics and Finance, vol. 1. Springer, Heidelberg (2013). https://doi.org/10.1007/978-1-4614-7789-1
53. Zimmer, C., Meister, M., Nguyen-Tuong, D.: Safe active learning for time-series modeling with Gaussian processes. In: Advances in Neural Information Processing Systems, pp. 2730–2739 (2018)

# Ensembling Shift Detectors: An Extensive Empirical Evaluation

Simona Maggio([✉]) and Léo Dreyfus-Schmidt

Dataiku Lab, Paris, France
{simona.maggio,leo.dreyfus-schmidt}@dataiku.com

**Abstract.** The term *dataset shift* refers to the situation where the data used to train a machine learning model is different from where the model operates. While several types of shifts naturally occur, existing shift detectors are usually designed to address only a specific type of shift. We propose a simple yet powerful technique to ensemble complementary shift detectors, while tuning the significance level of each detector's statistical test to the dataset. This enables a more robust shift detection, capable of addressing all different types of shift, which is essential in real-life settings where the precise shift type is often unknown. This approach is validated by a large-scale statistically sound benchmark study over various synthetic shifts applied to real-world structured datasets.

**Keywords:** Dataset shift · Shift detector · Ensemble · Hypothesis testing

## 1 Introduction

It is crucial for ML practitioners to detect possibly harmful changes of the data consumed by ML models, thus identifying situations of *dataset shift*, where the joint distribution $\mathbb{P}(X, Y)$ of the input features $X$ and output $Y$ in the source domain $S$, used to train the model, is different from the distribution in the target domain $T$, where the model operates.

There are many reasons why target domain data is different in practice from carefully collected source data. For instance, by sample selection bias, when some samples are less likely to be included in the source datasets because of a specific selection process implicitly dependent on the output variable. This is often the case for data collected through web surveys, where self-selection and under-representation of people with limited internet access prevent reliable inference [5]. Another example of dataset shift affecting the model validity is when the target variable distribution is not stationary. This is the case for pneumonia diagnosis models, which are sensitive to seasonal outbreaks [14].

Dataset shifts are usually categorised as follows: *(i)* prior shift indicates that the only changing factor is the prior distribution $\mathbb{P}_S(Y) \neq \mathbb{P}_T(Y)$ *(ii)* covariate shift indicates that the only changing factor is the covariates distribution

S. Maggio and L. Dreyfus-Schmidt–contributed equally.


N. Oliver et al. (Eds.): ECML PKDD 2021, LNAI 12977, pp. 362–377, 2021.
https://doi.org/10.1007/978-3-030-86523-8_22

$\mathbb{P}_S(X) \neq \mathbb{P}_T(X)$ *(iii)* sample selection bias impacts both prior and covariate distributions. Another cause for dataset change is concept shift, where the dependence of the target variable on the features changes from source to target domain. In this work we focus on unsupervised approaches, while concept shift detection requires target labels and so is not covered in this study.

To protect deployed models from dataset shift, we need to compare source and target data. The most challenging aspect of shift detection stems from the absence of ground truth for the target variable, preventing the use of simple model metric monitoring techniques. Thus, in order to detect changes in the output distribution, it is essential to find robust proxies.

There are two dominant approaches for generic shift detection, one comparing feature distributions, either directly or through a domain-discriminating classifier trained on the features from the two domains [21], the second comparing the distributions of the model predictions as proxies for the true prior distributions, called Black-Box Shift Detector (BBSD) [14]. The feature-based approaches are more sensitive to shifts impacting the feature distribution, while the BBSD has been specifically designed to deal with prior shift. In both cases, hypothesis testing is employed to seek statistically significant differences. Although these two approaches have different designs and purposes, they are usually compared on generic shift detection tasks mainly on datasets for visual classification [21].

The key idea of this paper is that ensembling shift detectors and adapting the significance level of their statistical tests is a more robust approach, especially suited for real-life settings where the precise shift type is unknown. This result is supported by the following contributions:

- We propose various ensemble schemes to combine feature- and prediction-based shift detectors and validate detectors ensembling as a more robust solution.
- Motivated by the observation that the statistical tests needed for drift detection have a different behaviour across datasets, we propose the adaptation of the significance level required for detection to the specific dataset under study.
- We perform a statistically sound benchmark study of base and ensembled shift detectors on structured datasets (21 OpenML[1] and Kaggle[2] classification datasets, largely differing in number of features and class proportions). Indeed the vast majority of deployed models in production consume structured data, highlighting the need for studies on model and data monitoring approaches for this modality. The shift detectors are evaluated against various types of synthetic drifts, including selection bias drift, which is less explored in the existing shift detection benchmarks although very frequent in real-world scenarios.
- Last but not least, we also investigate both theoretically and empirically the degradation of the BBSD detector, a core component of the detectors ensembles, as the predictive power of the underlying classifier drops.

[1] www.openml.org.
[2] www.kaggle.com.

After presenting related work, Sect. 3 describes the shift detection approaches in more details, as well as promoting the benefit of adapting the statistical tests to the dataset. Section 4 presents the experimental setting with a focus on synthetic drift generation, while Sect. 5 discusses the main outcomes and observations from the experiments.

## 2 Related Work

Dataset shift [16,20] has been addressed extensively in the context of **domain adaptation**, by exploring techniques aiming to align the model representations of source and target data, especially without target labels [13,22].

Label shift correction is a sub-field of domain adaptation that recently contributed to novel shift detection approaches. In particular, [14] proposed a shift detection approach based on the Black Box Shift Learning technique to correct label shift. [21] presented a general framework for shift detection and evaluated the BBSD and other state-of-the-art shift detection techniques on image datasets. A similar approach to correct prior shift is the Regularized Learning under Label Shifts [3], also relying on classifiers predictions. An alternative maximum likelihood approach to estimate target label distributions highlights the benefits of calibration under prior shift [1,11].

Best practices for responsible deployment of ML models include **data monitoring**, by inspecting incoming data for any signs of deviation from the expected scenario [2,12]. While dataset shift detection is an important part of ML monitoring system, recent work [10,24] has focused on directly predicting models performance drop for specific task-relevant drifts.

Besides monitoring possible distributional changes, the analysis of individual samples against the source dataset introduces the **anomaly detection** problem, extensively treated in the machine learning literature [7,15,23].

Another related field is **out-of-distribution detection** [25], which together with shift detection task has been used recently to evaluate ML models **predictive uncertainty** [18]. The study of predictive uncertainty offers a unifying view of the robustness a ML model should be equipped with to address any type of unexpected abnormality in input data. Finally **hypothesis testing** [26] is a core part of all shift detection techniques.

## 3 Shift Detectors

### 3.1 Notations and Problem Setup

Let $\mathcal{X}$ be an input space and $\mathcal{Y}$ be a label space where $|\mathcal{Y}|$ is finite. Let $X \in \mathcal{X}$ and $Y \in \mathcal{Y}$ be random variables. We denote by $f : X \rightarrow Y$ a predictor of the classification task at hand and we will refer to it as the *primary model*.

### 3.2 Feature-Based Detection

The simplest feature-based detection is univariate hypothesis testing on individual features. For $n$ features, this requires $n$ univariate hypotheses tests to separately compare the distributions of features between the source and the target domain, and eventually aggregate the $n$ $p$-values through Bonferroni correction.

However it can quickly become cumbersome when the feature space is large. For a more scalable solution, a more compact data representation can be obtained through standard dimensionality reduction techniques such as Principal Component Analysis (PCA) or Sparse Random Projection (SRP) [21].

In this work we do not consider detectors based on multivariate hypothesis testing as they were found to offer comparable performance to aggregated univariate tests [21].

A different solution to detect feature distributional changes in high dimensional datasets relies on domain-discriminating models [4]. The *domain classifier* is a model trained to discriminate source and target domains, using the very same features employed by the primary model. By design a domain classifier is meant to identify covariate shift and sample selection bias, but not prior shift which assumes unchanged feature distribution.

High accuracy of the classifier is a symptom of dataset shift, therefore a Binomial test with a null hypothesis of 0.5 accuracy for indistinguishable domains and balanced datasets is used to make sure the observed difference is statistically significant. One limitation of the domain classifier is the need of retraining for any new incoming target dataset.

### 3.3 Prediction-Based Detection

BBSD exploits the primary model to measure the distributions of predictions on source and target features, which are used as proxies for the true source and target label distributions. There are two variants of BBSD, one using the probability outputs (BBSDs), and the other using the hard-thresholded predictions (BBSDh).

For BBSDh, a $\chi^2$ test is used to compare source and target predicted class distributions, while for BBSDs the Kolmogorov-Smirnov (KS) test is employed to compare per-class probability distributions. The primary model predictions are $k$-dimensional with $k = |\mathcal{Y}|$ the number of classes in the primary task, thus the BBSDs shift detection requires $k$ univariate hypothesis tests. As in the previous section, Bonferroni correction [21] can be used to aggregate the $k$ KS tests.

Complementarily to the domain classifier, the BBSD has been specifically designed to address prior shift situations, but the distribution of predictions can also serve as proxy for the distribution of covariates [14].

### 3.4 Limitations of Shift Detectors

*BBSD Degrades as Primary Performances Drop.* We present two limitations of the above shift detection method. Firstly, the statistical power of BBSD is limited by the predictive power of the primary model. We theoretically prove in the Appendix[3] that there exist prior shifts such that the smaller the predictive power of the model is, the higher the $p$-values of the BBSD test are, reducing the sensitivity of the test. This phenomenon also occurs on more general type of shifts as validated experimentally (cf. Sect. 4).

*Detection Power Reduced by Bonferroni Correction.* Another limitation stems from the use of Bonferroni correction. Indeed, when the $p$-values of $k$ hypotheses are aggregated by Bonferroni correction, a significance level of $\alpha/k$ is required for each hypothesis to ensure that the Type-I error rate stays below $\alpha$. However if the hypotheses are not independent, this correction is too conservative and the family-wise Type-I error is controlled at a level strictly lower than $\alpha$, resulting in a loss of power [17,19]. This is the case for the BBSDs test as the underlying tests are done on the distribution of predicted probabilities of each class. For binary classification, the null hypothesis for one class and its complementary class are the same so that the Type-I error rate is actually bounded by $\alpha/2$ (instead of the expected $\alpha$), reducing the test's power. This is also the case for the univariate hypothesis testing on individual features, where the Bonferroni correction is conservative because of features correlation. We propose to address the latter limitation by adapting the significance level to the dataset.

### 3.5 Dataset Adaptive Significance Level

To improve the BBSDs test's power, keeping the Type-I error below a desired rate (i.e. 5%), we set the significance level to the 5% quantile of the empirical $p$-value distribution under the null hypothesis. We achieve this by performing 100 runs of shift detection comparing different random splits of the source dataset only.

A dataset-specific significance level is not only beneficial for shift detectors requiring aggregation of multiple univariate tests, but also when dealing with small datasets (i.e. from 10 to 100 samples). Indeed, performing statistical tests on small datasets yields $p$-values that can only assume few values. This quantization leads to non-uniform distributions of $p$-values under the null hypothesis so that a dataset-specific significance level is also beneficial.

The drawback of this adaptation represents the initial setup cost of running detection experiments on the source data, but the selected significance level can be used for all subsequent shift detection tasks for the given dataset.

Finally, for very imbalanced datasets (*creditcard*, *pc2* and *mc1* where the minority class represents less than 0.7%), the distribution of the BBSDs $p$-values under $H_0$ is extremely skewed towards 1. The study of the impact of imbalanced datasets on prediction-based detectors is left for future research.

As a natural extension of base shift detectors, the detectors ensembles also benefit from the adaptation of the significance level.

[3] See the complete version of this paper on arxiv.

### 3.6 Detectors Ensembles

We evaluate two ensembling strategies combining feature and prediction-based shift detectors, in order to exploit their complementary detection abilities:

- domain classifier trained on both features and primary model predictions;
- pair of statistical tests from complementary shift detectors with Bonferroni correction.

Ensembling strategies are naturally robust to various shift types, and represent a better fit to real-life settings where the precise shift type is often unknown. We confirm experimentally that ensembling retains the advantages of feature- and prediction-based shift detectors (cf. Sect. 4), providing a single practical tool for reliable drift monitoring.

## 4 Experimental Setup

The purpose of our experiments is to compare the different detection approaches on various types of simulated drift over a large collection of structured datasets. Specifically we aim at highlighting differences in the approaches related to both the detection power and efficiency, the latter expressed in terms of minimum number of samples required to detect a shift. All the details of the experiments setup and reproducibility of this work are available in the Appendix (see Footnote 3).

### 4.1 Datasets and Shift Simulation

In our experiments we use 21 datasets for classification tasks from OpenML and Kaggle as listed in Table 7 of the Appendix (see Footnote 4).

In order to simulate dataset drift we synthetically apply different types of shift to a split of the original dataset. The 10 simulated shifts belong to the following categories and are detailed in Table 1:

- Prior shift: generated by changing the fraction of samples belonging to a class.
- Covariate shift due to Gaussian noise: generated by adding Gaussian noise to some numeric features of a fraction of samples.
- Covariate shift due to Adversarial noise: generated by applying Adversarial noise to numeric features. It's a more subtle kind of noise, slightly changing the features but inducing the primary model to switch its predicted class.
- Selection bias: generated by selecting samples with a probability dependent on the sample features and implicitly dependent on the class, possibly oversampling by interpolation of existing observations.

The severity of those shifts is controlled by the parameters shown in Table 1. Our experiments explore the effect of changing those parameters values generating a total of 19 drifts, then averaging the outcome from the same shift type to build the results for the 10 shifts types in Table 1.

The combination of different shift types is also relevant as it can occur in real-life settings, for this reason the study of the detectors response to composite shifts will be considered in future work.

**Table 1.** Simulated shift types.

| Category | Shift type | Parameters | | Description |
|---|---|---|---|---|
| Prior shift | Knock-out | $s = 25\%$ | | Remove $s\%$ of majority class |
| | | $s = 40\%$ | | |
| | Only-One | | | Select only-one minority class |
| Covariate shift | Small Gaussian | $s = 50\%$ | $f = 50\%$ | Small amount of Gaussian Noise applied on $s\%$ of samples and $f\%$ of features |
| | | $s = 50\%$ | $f = 100\%$ | |
| | | $s = 100\%$ | $f = 50\%$ | |
| | | $s = 100\%$ | $f = 100\%$ | |
| | Medium Gaussian | $s = 50\%$ | $f = 50\%$ | Medium amount of Gaussian Noise applied on $s\%$ of samples and $f\%$ of features |
| | | $s = 50\%$ | $f = 100\%$ | |
| | | $s = 100\%$ | $f = 50\%$ | |
| | | $s = 100\%$ | $f = 100\%$ | |
| | Adversarial ZOO | $s = 25\%$ | | Zeroth-order optimization black box adversarial attack on $s\%$ of samples [8] |
| | | $s = 50\%$ | | |
| | Adversarial boundary | $s = 25\%$ | | Boundary black box adversarial attack on $s\%$ of samples [6] |
| | | $s = 50\%$ | | |
| Selection bias | Joint subsampling | | | Keeps an observation with probability decreasing as points are away from the samples mean |
| | Subsampling | | $f = 100\%$ | Subsample with low probability samples with low feature values separately for $f\%$ features |
| | Under-sampling | $s = 50\%$ | | Keep $s\%$ of samples, selecting samples close to the minority class (NearMiss3 heuristics) |
| | Over-sampling | $s = 50\%$ | | Replace $s\%$ of samples with samples interpolated from the remaining part |

### 4.2 Experiments

For each run of drift experiment, we test all the mentioned shift detection approaches on subsets of the source and target datasets with number of samples in [10, 100, 500, 1000, 2000]. The different detection approaches are reported using the following labels:

- *BBSDs*: soft version of BBSD with Random Forest primary model and KS test.
- *BBSDh*: hard version of BBSD with Random Forest primary model and $\chi^2$ test.
- *Test_X*: KS test on input features.
- *Test_PCA*: KS test on PCA-projected features.
- *Test_SRP*: KS test on SRP-projected features.
- *DC*: Random Forest domain classifier with Binomial test.

The ensembling strategies are reported using the following labels:

- *BBSDs + X*: *BBSDs* and *Test_X* with Bonferroni correction.
- *BBSDs + DC*: *BBSDs* and *DC* with Bonferroni correction.
- *DC**: *DC* trained on both features and primary predictions.

For any detector, its adaptive variant using a significance level tailored to the dataset (cf. Subsect. 3.5) is referred to with the suffix *(adapt)*, i.e. *BBSDs (adapt)*.

Overall, we have collected $p$-values from shift detection experiments for 19 different drifts, run 5 times, for 5 different sizes, for each of the 21 datasets. Furthermore the results over the different runs are averaged, as well as the results from the same types of drift with different drift parameters to yield 210 (21 datasets $\times 10$ shift types) detection results per dataset size per detector.

### 4.3 Metrics

The $p$-values of the detectors are averaged over multiple runs and when their value is less than the significance level, we consider that a drift has been detected. The significance level used in the non-adaptive version of the statistical test is 0.05.

Considering shift detectors as binary classifiers where one sample is one comparison of source and target datasets, we can measure its accuracy and true positive rate (TPR). When the accuracy is reported the evaluated shift scenarios include a negative (no shift) situation per dataset.

To measure the data efficiency of a shift detector, we define an **efficiency score** as the minimum size level required to detect the shift. As the detectors are evaluated at different datasets sizes $[10, 100, 500, 1000, 2000]$, they are assigned a score from 5 to 1. A detector failing to find the shift at any size is assigned a score of 0.

### 4.4 Statistical Comparison

In order to fairly compare the detection approaches, we use a Friedman test to statistically assess whether the detectors have different performances. When the conclusion is positive, a Nemenyi post-hoc test is used to determine pairwise equivalence of the methods [9].

Based on the efficiency score defined in Sect. 4.3, the average ranks for each detector are computed across all datasets. The Friedman test then checks whether the detectors average ranks are significantly different from the mean rank expected under the null-hypothesis. When the former test rejects the null-hypothesis, the Nemenyi test is used to seek for pairwise differences between ranks.

In order to have independent observations from different datasets, we perform these statistical tests separately for each of the 10 shift types. We use the library *autorank*[4] to perform this statistical comparison.

[4] https://github.com/sherbold/autorank.

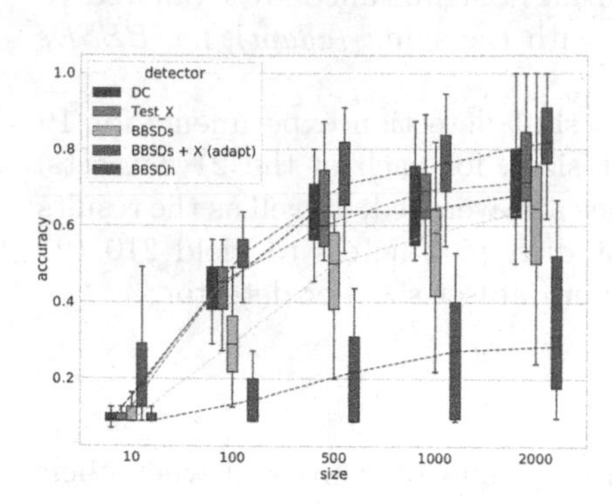


**Fig. 1.** Shift detectors accuracy by dataset size.

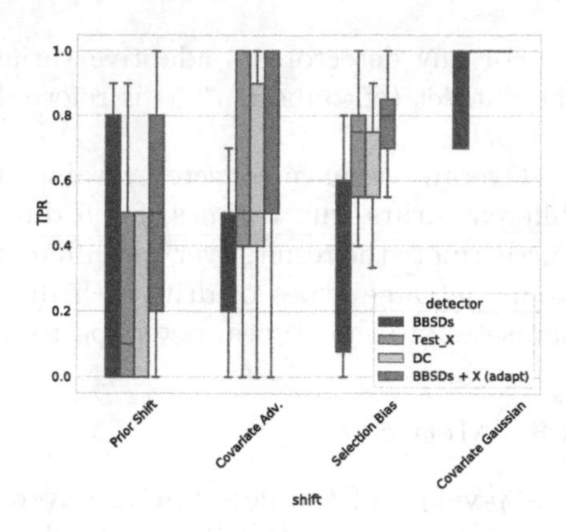


**Fig. 2.** Shift detectors true positive rate (TPR) by type of shift.

## 5 Results

We first give the experiment results comparing ensembled and base shift detectors over all datasets and by shift types. We then highlight the complementarity of feature- and prediction-based shift detectors motivating the ensembling strategy. Finally, we present ablation studies comparing base detectors and ensembles to their adaptive counterparts.

### 5.1 Ensembling Shift Detectors

When observing the average detection accuracy across datasets and shift types for different dataset sizes (Fig. 1) some simple detection approaches stand out (*Test_X* and *DC*), although slightly less accurate than the adaptive ensemble technique (*BBSDs+X (adapt)*). However the true positive rate for different shift types for a fixed size of 1000 (Fig. 2) reveals a more complex landscape, where the performance of base shift detectors is not uniform. For an easier reading of Figs. 1 and 2, some detectors have been omitted and the various shifts have been grouped by type (cf. Table 1), but all the detailed TPR and accuracy results for individual shift detectors and shift types can be found in Tables 8 and 9 of the Appendix (see Footnote 3).

In Tables 2 and 3 we report the mean accuracy by dataset size and mean TPR by type of shift for the base detectors and the detectors ensembles with adaptation of the significance level to the dataset. The ensemble *BBSDs+X (adapt)* comes out as the most accurate shift detector overall.

In addition to be sensitive at lower sizes *BBSDs+X (adapt)* is also able to capture alone most types of shifts. This approach represents a more robust solution able to detect the shift situations missed by either of the base shift detectors. It is interesting to notice that simpler ensembles *BBSDs+X (adapt)* and *BBSDs+DC (adapt)* perform very similarly to *DC**, while not requiring an additional training. Ensembling feature- and prediction-based shift detectors is a promising strategy for an effective and robust drift monitoring.

**Table 2.** Accuracy across all datasets by size (mean ± std). Best values and all values within 95% confidence interval are in bold.

| Size | BBSDs | Test_X | DC | Ensembles (adapt) | | |
|---|---|---|---|---|---|---|
| | | | | DC* | BBSDs + DC | BBSDs + X |
| 10 | 0.11 ± 0.03 | 0.10 ± 0.02 | 0.10 ± 0.02 | **0.20** ± 0.24 | 0.14 ± 0.03 | **0.22** ± 0.11 |
| 100 | 0.30 ± 0.10 | 0.45 ± 0.10 | 0.44 ± 0.09 | 0.46 ± 0.17 | **0.50** ± 0.10 | **0.54** ± 0.10 |
| 500 | 0.46 ± 0.14 | 0.65 ± 0.12 | 0.64 ± 0.10 | 0.63 ± 0.14 | 0.67 ± 0.12 | **0.73** ± 0.11 |
| 1000 | 0.54 ± 0.16 | 0.69 ± 0.11 | 0.66 ± 0.11 | 0.63 ± 0.13 | **0.72** ± 0.12 | **0.75** ± 0.12 |
| 2000 | 0.63 ± 0.19 | 0.75 ± 0.13 | 0.71 ± 0.14 | 0.67 ± 0.15 | **0.79** ± 0.12 | **0.83** ± 0.12 |

**Table 3.** TPR across all datasets by shift type at dataset size of 1000 (mean ± std). Best values and all values within 95% confidence interval are in bold.

| Shift type | BBSDs | Test_X | DC | Ensembles (adapt) | | |
|---|---|---|---|---|---|---|
| | | | | DC* | BBSDs + DC | BBSDs + X |
| Knock-out | **0.23** ± 0.30 | 0.11 ± 0.24 | 0.00 ± 0.00 | 0.01 ± 0.05 | **0.25** ± 0.30 | **0.34** ± 0.40 |
| Only-one | **0.60** ± 0.46 | 0.55 ± 0.49 | 0.46 ± 0.47 | 0.51 ± 0.48 | **0.70** ± 0.44 | **0.73** ± 0.41 |
| Small Gaussian | 0.74 ± 0.37 | 0.96 ± 0.14 | **1.00** ± 0.00 | 0.95 ± 0.12 | **1.00** ± 0.00 | 0.99 ± 0.04 |
| Medium Gaussian | 0.87 ± 0.32 | 0.98 ± 0.06 | **1.00** ± 0.00 | **1.00** ± 0.00 | 0.99 ± 0.04 | 0.99 ± 0.04 |
| Adv. ZOO | 0.21 ± 0.40 | **0.43** ± 0.51 | **0.42** ± 0.50 | **0.47** ± 0.50 | **0.43** ± 0.51 | **0.45** ± 0.49 |
| Adv. boundary | 0.72 ± 0.42 | 0.77 ± 0.35 | 0.70 ± 0.41 | 0.72 ± 0.40 | 0.74 ± 0.40 | **0.98** ± 0.06 |
| Subsampling joint | 0.04 ± 0.08 | 0.20 ± 0.30 | 0.10 ± 0.19 | **0.30** ± 0.35 | 0.16 ± 0.26 | **0.34** ± 0.30 |
| Subsampling | 0.42 ± 0.41 | **0.71** ± 0.45 | **0.70** ± 0.44 | 0.17 ± 0.29 | **0.72** ± 0.43 | **0.72** ± 0.45 |
| Under-sampling | 0.44 ± 0.44 | **1.00** ± 0.00 | **1.00** ± 0.00 | **1.00** ± 0.00 | **1.00** ± 0.00 | **1.00** ± 0.00 |
| Over-sampling | 0.57 ± 0.40 | **1.00** ± 0.00 | **1.00** ± 0.00 | **1.00** ± 0.00 | **1.00** ± 0.00 | **1.00** ± 0.00 |

### 5.2 Comparison of Base Shift Detectors

In order to have a global comparison of the performance of base detectors, we aggregate the results from different dataset sizes with the efficiency score and show the global average ranks by type of shift in Table 4. A first observation is that overall the feature-based detectors have lower rank, indicating that they are more effective than other approaches, requiring less data to detect a drift situation.

Indeed for all the types of shift the Friedman test at $p = 0.05$ reported that the approaches are not statistically equivalent.

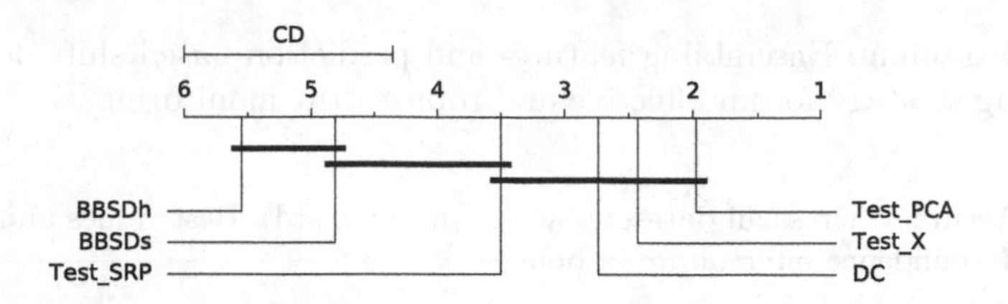


**Fig. 3.** Pairwise comparison of all detectors with the Nemenyi test for *Under-sampling* shift.

**Table 4.** Average ranks of 6 shift detectors based on efficiency score.

| Shift type | BBSDs | BBSDh | Test_X | Test_PCA | Test_SRP | DC |
|---|---|---|---|---|---|---|
| Knock-out | **2.71** | 3.02 | 3.74 | 3.67 | 3.62 | 4.24 |
| Only-one | **2.64** | 3.07 | 3.45 | 3.90 | 3.90 | 4.02 |
| Small Gaussian | 2.93 | 5.43 | 1.95 | 4.93 | 4.26 | **1.50** |
| Medium Gaussian | 2.36 | 5.52 | 2.26 | 4.79 | 4.14 | **1.93** |
| Adversarial ZOO | 3.81 | 4.21 | **2.60** | 3.93 | 3.40 | 3.05 |
| Adversarial boundary | 2.95 | 4.55 | **2.31** | 4.64 | 3.55 | 3.00 |
| Joint subsampling | 3.98 | 3.98 | 3.26 | **2.52** | 3.29 | 3.98 |
| Subsampling | 4.12 | 4.86 | **2.69** | 3.10 | 3.00 | 3.24 |
| Under-sampling | 4.81 | 5.55 | 2.43 | **1.98** | 3.50 | 2.74 |
| Over-sampling | 4.90 | 5.69 | 2.26 | **2.14** | 3.52 | 2.48 |
| All | 3.54 | 4.63 | **2.63** | 3.58 | 3.65 | 2.96 |

We thus look at the results of the post-hoc Nemenyi test, represented in Fig. 3 for the case of under-sampling shift. The Critical Distance (CD) between the average ranks reveals two groups of different detectors, with feature-based detectors outperforming the BBSD approaches. The *Test_PCA*, *Test_X*, *DC* detectors have statistically higher efficiency than *BBSDh* and *BBSDs*. The Nemenyi test at $p = 0.05$ is not powerful enough to draw conclusions about the other detectors.

Overall the Nemenyi tests (Fig. 4) highlight the following differences:

- *Prior shift*: prediction-based detectors perform better than other approaches when the label distribution is affected, as we could expect by design.
- *Gaussian noise*: domain classifier and direct input features testings are more effective than other approaches.
- *Adversarial Boundary*: BBSDs and direct feature testing perform slightly better than domain classifier and significantly better than BBSDh. This perturbation is too subtle to be spotted by the domain classifier, although its important impact on class distributions is easily detected by both BBSDs and Test_X.
- *Subsampling, Under-sampling, Over-sampling*: features-based detectors perform significantly better than prediction-based detectors on perturbations simulating selection bias.

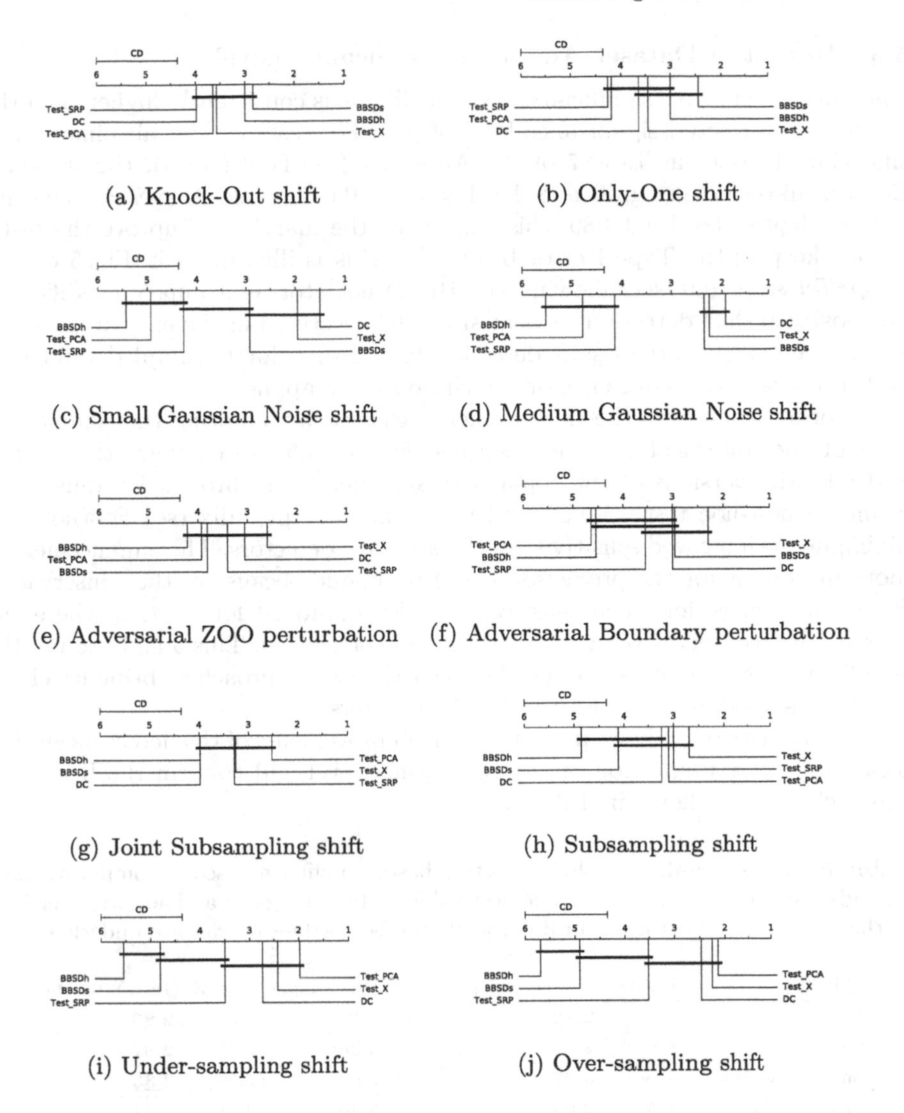


(a) Knock-Out shift

(b) Only-One shift

(c) Small Gaussian Noise shift

(d) Medium Gaussian Noise shift

(e) Adversarial ZOO perturbation

(f) Adversarial Boundary perturbation

(g) Joint Subsampling shift

(h) Subsampling shift

(i) Under-sampling shift

(j) Over-sampling shift

**Fig. 4.** Nemenyi post-hoc tests for the different types of drift.

The previous findings highlight the complementarity of feature- and prediction-based drift detectors on different drift scenarios, motivating the proposed ensembling strategy (cf. Sect. 3.6).

The individual results from each drift scenario per dataset are available in the paper repository[5] for full inspection.

[5] https://github.com/dataiku-research/drift_detectors_benchmark.

### 5.3 Impact of Dataset-Adaptive Significance Level

The dataset-adaptive significance level in BBSDs is consistently higher than the Bonferroni-corrected significance level of $\alpha/k$. For instance, for all binary classification datasets in Table 7 of the Appendix (see Footnote 3), the standard Bonferroni-corrected significance level is $\alpha/2 = 0.025$, while on average the computed adaptive level is 0.089: this gap shows the margin to improve the test's power, keeping the Type-I error below 5%. This is illustrated in Fig. 5 on the *MagicTelescope* dataset, showing the BBSD detector as a binary classifier on all (positive) shift detection tests at size 2000 described in the experiments and additional 100 (negative) shift detection tests on randomly sampled validation and test sets of the same size, on which no shift is applied.

In order to analyze the improvement achieved by adapting the significance level of the statistical tests as described in Sect. 3.5, we compare the regular and adaptive versions of the top-3 base shift detectors through Friedman and Nemenyi post-hoc test. Table 5 and Fig. 7 (in the Appendix (see Footnote 8)) highlight the improved sensitivity of the adaptive detectors. The enhancement is more important for the prediction-based technique because of the conservative Bonferroni correction. The sensitivity is also improved for *Test_X*, where the correction is conservative because of features correlation. This adaptation of the significance level reduces the gap between the two approaches, bringing closer prediction-based and feature-based shift detectors.

As an ablation study we investigate the improvements of the detectors ensembles due to the adaptation of the significance level and confirm the benefit of this technique as shown in Table 6.

**Table 5.** Average ranks of 3 shift detectors based on efficiency score, comparing fixed and adaptive significance levels. The best values between regular and adaptive versions of the same shift detector are in bold, while the best values overall are underlined.

| Shift type | BBSDs | BBSDs (adapt) | DC | DC (adapt) | Test_X | Test_X (adapt) |
|---|---|---|---|---|---|---|
| Knock-out | 3.17 | **<u>2.12</u>** | 4.43 | 4.43 | 4.02 | **2.83** |
| Only-one | 3.10 | **<u>2.29</u>** | 4.50 | **4.36** | 3.95 | **2.81** |
| Small Gaussian | 5.29 | **3.67** | 3.10 | **2.64** | 3.98 | **<u>2.33</u>** |
| Medium Gaussian | 4.38 | **2.93** | 3.76 | **3.40** | 4.31 | **<u>2.21</u>** |
| Adv. ZOO | 4.40 | **3.64** | 3.76 | **3.64** | 3.10 | **<u>2.45</u>** |
| Adv. boundary | 4.02 | **2.83** | 4.38 | **4.24** | 3.64 | **<u>1.88</u>** |
| Joint subsampling | 4.02 | **3.50** | 4.02 | **3.86** | 3.29 | **<u>2.31</u>** |
| Subsampling | 4.60 | **3.98** | 3.81 | **3.57** | 3.19 | **<u>1.86</u>** |
| Under-sampling | 5.38 | **4.50** | 3.17 | 3.17 | 2.81 | **<u>1.98</u>** |
| Over-sampling | 5.67 | **4.60** | 3.02 | **2.69** | 2.81 | **<u>2.21</u>** |

### 5.4 Impact of Model Quality on BBSDs

To confirm the limitations of BBSDs on the primary model performance (cf. Sect. 3), we apply a random perturbation to the primary model with probability

**Table 6.** Average ranks of 3 ensemble shift detectors based on efficiency score, comparing fixed and adaptive significance levels. The best values between regular and adaptive versions of the same shift detector are in bold, while the best values overall are underlined.

| Shift type | BBSDs + X | BBSDs + X (adapt) | BBSDs + DC | BBSDs + DC (adapt) | DC* | DC* (adapt) |
|---|---|---|---|---|---|---|
| Knock-out | 3.55 | **1.95** | 3.43 | **2.93** | 4.57 | 4.57 |
| Only-one | 3.55 | **2.62** | 3.40 | **2.83** | 4.36 | **4.24** |
| Small Gaussian | 4.29 | **2.26** | 3.40 | **2.62** | 4.36 | **4.07** |
| Medium Gaussian | 4.19 | **2.21** | 3.98 | **2.24** | 4.50 | **3.88** |
| Adv. ZOO | 3.24 | **2.57** | 3.81 | **3.52** | 3.85 | 3.85 |
| Adv. boundary | 3.21 | **1.88** | 3.86 | **3.17** | 4.62 | **4.15** |
| Joint subsampling | 3.24 | **2.43** | 3.79 | **3.55** | 4.00 | 4.00 |
| Subsampling | 2.67 | **2.19** | 3.31 | **2.79** | 5.02 | 5.02 |
| Under-sampling | 3.43 | **2.57** | 4.00 | **3.57** | 3.71 | 3.71 |
| Over-sampling | 3.60 | **2.88** | 3.88 | **3.17** | 3.65 | 3.65 |

$p$, yielding a primary model with quality $1-p$ in $[0.5, 1.0]$ and evaluate the power of the BBSDs test. As illustrated in Fig. 6, showing the average TPR across 21 datasets with size of 1000 samples, the power decreases with the primary model quality, along with the variance of the results across the datasets. More difficult shifts (such as low intensity or a small amount of drifted samples) require a better primary model for the BBSDs test power to hold (shift-specific details in Table 10 of the Appendix (see Footnote 3)). Regardless of the model quality, BBSDs is also less adequate in detecting some types of selection bias, as observed in Fig. 2.

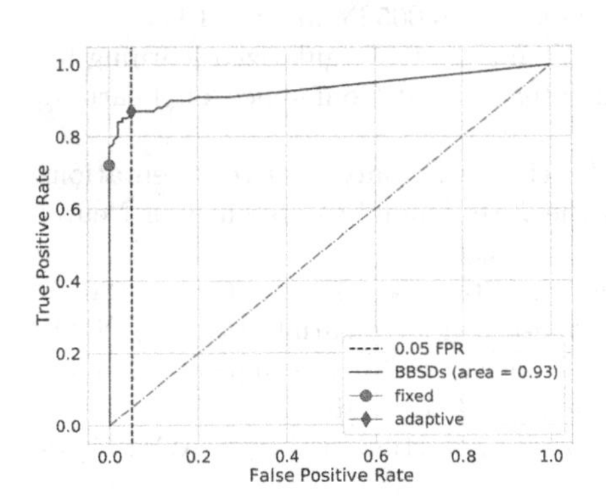


**Fig. 5.** ROC curve of BBSDs as a binary classifier on all positive and negative drift experiments for the *MagicTelescope* dataset.

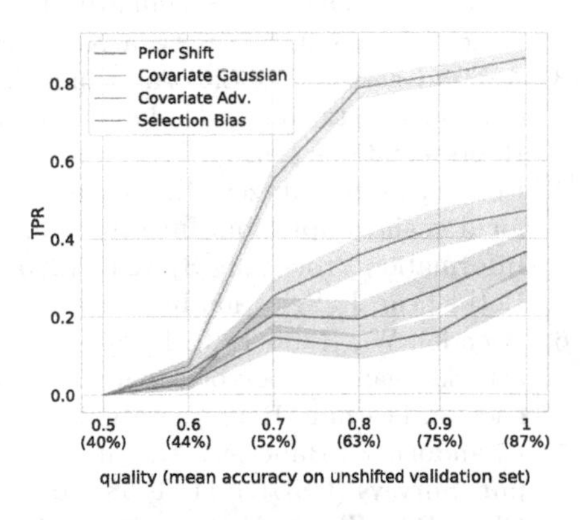


**Fig. 6.** Average TPR with standard deviation from the BBSDs test with perturbed primary model.

## 6 Conclusion

In this paper we propose a shift detectors ensembling technique capable of addressing all different types of studied drift scenarios. The key components of the proposed approach are the combination of complementary base detectors, designed to address different types of shift, and the adaptation of the significance level of the detectors statistical tests to the specific dataset under study.

The improved robustness of our approach is validated by a large-scale benchmark study comparing it to state-of-the-art shift detectors on 21 structured real-world datasets. For this purpose we simulate drifts with 10 different types of perturbations, including shift types not studied in previous works for tabular datasets, such as adversarial noise and sample selection bias. Our benchmark study highlights the complementarity of base drift detectors on different drift scenarios, motivating our ensembling approach. This benchmark also includes ablation studies showing that the dataset-adaptive significance level provides both base and ensembled shift detectors with higher detection power, while preserving the desired false positive rate.

Throughout our experiments, we observe that adaptive shift detectors ensembling represents the strongest strategy, robust to the various shift types, making this approach a natural choice for monitoring models in production in real-life settings, where the possible drift scenario is unknown.

## References

1. Alexandari, A., Kundaje, A., Shrikumar, A.: Maximum Likelihood with Bias-Corrected Calibration is Hard-To-Beat at Label Shift Adaptation. arXiv e-prints arXiv:1901.06852, January 2019
2. Amodei, D., Olah, C., Steinhardt, J., Christiano, P., Schulman, J., Mané, D.: Concrete problems in AI safety. arXiv e-prints arXiv:1606.06565, June 2016
3. Azizzadenesheli, K., Liu, A., Yang, F., Anandkumar, A.: Regularized learning for domain adaptation under label shifts. In: International Conference on Learning Representations (2019)
4. Ben-David, S., Blitzer, J., Crammer, K., Pereira, F.: Analysis of representations for domain adaptation. In: Proceedings of the 20th Annual Conference on Neural Information Processing Systems (2006)
5. Bethlehem, J.: Selection bias in web surveys. Int. Stat. Rev. **78**(2), 161–188 (2010)
6. Brendel, W., Rauber, J., Bethge, M.: Decision-based adversarial attacks: reliable attacks against black-box machine learning models. In: 6th International Conference on Learning Representations, ICLR 2018, pp. 1–12 (2018)
7. Chandola, V., Banerjee, A., Kumar, V.: Anomaly detection: a survey. ACM Comput. Surveys (CSUR) **41**, 1–58 (2009)
8. Chen, P.Y., Zhang, H., Sharma, Y., Yi, J., Hsieh, C.J.: ZOO: zeroth order optimization based black-box attacks to deep neural networks without training substitute models. In: AISec 2017 - Proceedings of the 10th ACM Workshop on Artificial Intelligence and Security, pp. 15–26 (2017)
9. Demšar, J.: Statistical comparisons of classifiers over multiple data sets. J. Mach. Learn. Res. **7**, 1–30 (2006)

10. Elsahar, H., Gallé, M.: To annotate or Not? Predicting performance drop under domain shift. In: Proceedings of the 2019 Conference on Empirical Methods in Natural Language Processing and the 9th International Joint Conference on Natural Language Processing (EMNLP-IJCNLP), pp. 2163–2173. Association for Computational Linguistics, Hong Kong, China, November 2019
11. Garg, S., Wu, Y., Balakrishnan, S., Lipton, Z.C.: A unified view of label shift estimation. arXiv e-prints arXiv:2003.07554, March 2020
12. Klaise, J., Van Looveren, A., Cox, C., Vacanti, G., Coca, A.: Monitoring and explainability of models in production. arXiv e-prints arXiv:2007.06299, July 2020
13. Kouw, W.M., Loog, M.: A review of domain adaptation without target labels. arXiv e-prints arXiv:1901.05335, January 2019
14. Lipton, Z.C., Wang, Y.X., Smola, A.: Detecting and correcting for label shift with black box predictors. arXiv e-prints arXiv:1802.03916, February 2018
15. Markou, M., Singh, S.: Novelty detection: a review: Part 1: statistical approaches. Signal Process. **83**, 2481–2497 (2003)
16. Moreno-Torres, J.G., Raeder, T., Alaiz-Rodríguez, R., Chawla, N.V., Herrera, F.: A unifying view on dataset shift in classification. Pattern Recogn. **45**(1), 521–530 (2012)
17. Nakagawa, S.: A farewell to Bonferroni: the problems of low statistical power and publication bias. Behav. Ecol. **15**(6), 1044–1045 (2004)
18. Ovadia, Y., et al.: Can you trust your model's uncertainty? Evaluating predictive uncertainty under dataset shift. arXiv e-prints arXiv:1906.02530, June 2019
19. Perneger, T.V.: What's wrong with Bonferroni adjustments. BMJ (Clin. Res. ed.) **316**(7139), 1236–1238 (1998)
20. Quinonero-Candela, J., Sugiyama, M., Schwaighofer, A., Lawrence, N.D.: Dataset Shift in Machine Learning (2009)
21. Rabanser, S., Günnemann, S., Lipton, Z.C.: Failing loudly: an empirical study of methods for detecting dataset shift. arXiv e-prints arXiv:1810.11953, October 2018
22. Redko, I., Morvant, E., Habrard, A., Sebban, M., Bennani, Y.: A survey on domain adaptation theory: learning bounds and theoretical guarantees. arXiv e-prints arXiv:2004.11829, April 2020
23. Ruff, L., et al.: A unifying review of deep and shallow anomaly detection. arXiv e-prints arXiv:2009.11732, September 2020
24. Schelter, S., Rukat, T., Biessmann, F.: Learning to validate the predictions of black box classifiers on unseen data. In: Proceedings of the 2020 ACM SIGMOD International Conference on Management of Data, pp. 1289–1299. SIGMOD 2020. Association for Computing Machinery, New York, NY, USA (2020)
25. Shafaei, A., Schmidt, M., Little, J.J.: A less biased evaluation of out-of-distribution sample detectors. In: 30th British Machine Vision Conference 2019 (2018)
26. Steinebach, J., Lehmann, E.L., Romano, J.P.: Testing statistical hypotheses. Metrika **64**(2), 255–256 (2006)

# Supervised Learning

# Adaptive Learning Rate and Momentum for Training Deep Neural Networks

Zhiyong Hao(✉), Yixuan Jiang, Huihua Yu, and Hsiao-Dong Chiang

Cornell University, Ithaca, NY 14850, USA
{zh272,yj373,hy437,hc63}@cornell.edu

**Abstract.** Recent progress on deep learning relies heavily on the quality and efficiency of training algorithms. In this paper, we develop a fast training method motivated by the nonlinear Conjugate Gradient (CG) framework. We propose the Conjugate Gradient with Quadratic line-search (CGQ) method. On the one hand, a quadratic line-search determines the step size according to current loss landscape. On the other hand, the momentum factor is dynamically updated in computing the conjugate gradient parameter (like Polak-Ribiere). Theoretical results to ensure the convergence of our method in strong convex settings is developed. And experiments in image classification datasets show that our method yields faster convergence than other local solvers and has better generalization capability (test set accuracy). One major advantage of the paper method is that tedious hand tuning of hyperparameters like the learning rate and momentum is avoided.

**Keywords:** Optimization algorithm · Line search · Deep learning

## 1 Introduction

We consider the minimization problem commonly used in the machine learning setting with the following finite-sum structure:

$$\min_{\theta} f(\theta) = \frac{1}{n} \sum_{i=1}^{n} f_i(\theta) \tag{1}$$

where $f : D \subset \mathbb{R}^n \to \mathbb{R}$ is continuously differentiable and $D$ is the search space. A point $\theta^* \in D$ is called a local minimum if $f(\theta^*) \leq f(\theta)$ for all $\theta \in D$ with $\|\theta - \theta^*\| < \sigma$ for $\sigma > 0$. A general training procedure is in Algorithm 1, where key steps are determining the update direction $p_t$ and step size $\alpha_t$ in each iteration.

Several methods propose various choices of search directions. The Gradient Descent (GD) method directly takes the negative gradient $-\nabla_\theta f(\theta)$. Newton's method selects $p_t$ as $-(\nabla_\theta^2 f(\theta))^{-1}\nabla_\theta f(\theta)$ (where $\nabla_\theta^2 f(\theta)$ is the Hessian matrix). On the other hand, momentum [19,23] is used to stabilize $p_t$ in stochastic optimization. Adam [12] formalizes $p_t$ as a moving average of first moments divided


N. Oliver et al. (Eds.): ECML PKDD 2021, LNAI 12977, pp. 381–396, 2021.
https://doi.org/10.1007/978-3-030-86523-8_23

**Algorithm 1.** Iterative Approach for Optimizing (1)

```
1: procedure OPTIMIZE(f(θ), maxiter)
2:     Initialize θ_0
3:     for t ← 1 to maxiter do
4:         Choose a search direction p_t;
5:         Determine a update step size α_t;
6:         θ_t ← θ_{t−1} + α_t p_t;
7:     return θ_t
```

by that of second moments. Other methods focus on the choice of $\alpha_t$. Various classic line-search methods were proposed, like the Newton-Raphson, Armijo, and Wolfe line search. Recently, [31] applied the Armijo rule in SGD. In this paper, we propose the Conjugate Gradient with Quadratic line-search method that jointly optimizes both $p_t$ and $\alpha_t$ for fast convergence.

The performance of SGD is well-known to be sensitive to step size $\alpha_t$ (or the learning rate called in machine learning). Algorithms, such as AdaGrad [7] and Adam [12], ease this limitation by tuning the learning rate dynamically using past gradient information. Line-search strategy is another option to adaptively tune step size $\alpha_t$. Exact line-search requires solving a simpler sub-optimization problem, such as applying quadratic interpolation [17]. An inexact line-search method can also be applied, such as the Armijo rule [1], to guarantee a sufficient decrease at each step. A stochastic variant of Armijo rule was proposed in [31] to set the step size for SGD.

Conjugate Gradient method (CG) is also popular for solving nonlinear optimization problems. In CG, $p_t$ is chosen as $(-\nabla f(\theta) + \beta_t p_{t-1})$, where $\beta_t$ is the conjugate parameter computed according to various formulas. $\alpha_t$ is determined by a line search $(\arg\min_\alpha f(\theta_t + \alpha p_t))$. In fact, CG can be understood as a Gradient Descent with an adaptive step size and dynamically updated momentum. For the classic CG method, step size is determined by the Newton-Raphson method or the Secant method, both of which need to exactly compute or approximate the Hessian matrix. This makes it less appealing to apply classic CG in deep learning.

In this paper, we propose the Conjugate Gradient with the Quadratic line-search method (CGQ), which uses quadratic interpolation line-search to choose the step size and adaptively tunes momentum term according to the Conjugate Gradient formula. CGQ requires no Hessian information and can be viewed as a variant of SGD with adaptive learning rate and momentum. For quadratic line-search, we propose two variants, *i.e.*, the 2-point method and the least squares method, which are introduced in detail in Sect. 3.2. And we illustrate how we dynamically determine the momentum term during training in Sect. 3.3. Our analyses indicate that SGD with 2-point quadratic line-search converges under convex and smooth settings. sCGQ is further proposed to improve efficiency and generalization on large models. Through various experiments, CGQ exhibits fast convergence, and is free from manually tuning the learning rate and momentum term.

Our major contributions are summarized as follows:

- A quadratic line-search method is proposed to dynamically adjust the learning rate of SGD.
- An adaptive scheme for computing the effective momentum factor of SGD is designed.
- Theoretical results are developed to prove the convergence of SGD with quadratic line-search.
- Line search on subset of batches is proposed to improve efficiency and generalization capability.

## 2 Related Work

In modern machine learning, SGD and its variants [7,12,16,24], are the most prevalent optimization methods used to train large-scale neural networks despite their simplicity. Theoretically, the performance of SGD with a constant learning rate has been analyzed that under some assumptions, such as, the Strong Growth Condition, has linear convergence if the learning rate is small enough [26]. However, there are still two non-negligible limitations: (1) the theoretical convergence rate is not as satisfying as the full gradient descent method; and (2) the performance is very sensitive to hyperparameters, especially the learning rate.

In SGD, random sampling introduces randomness and degrades the theoretical convergence of SGD. One direction of research focuses on reducing the variance of random sampling so that larger learning rates are allowed. Based on this idea, a family of SGD variants called Variance Reduction methods were proposed [6,11,25,27]. These methods have been proved to possess impressive theoretical convergence on both convex and non-convex optimization problems. However, the applications of these methods in machine learning are limited to optimization of simple logistic regression models because the additional memory and computational cost of these methods are unacceptable when training large-scale deep neural networks.

On the other hand, significant efforts seek to ameliorate SGD by designing an effective method to adaptively tune the learning rate as training. [21] divides the optimization process of SGD into transient phase and stationary phase, while the learning rate will be reduced after entering the stationary phase. In [29], scheduling learning rate cyclically for each iteration empirically boosts the convergence of SGD. Meta-learning [2] techniques that optimize the learning rate according to the hyper-gradient also have an outstanding performance in experiments. However, none of these techniques gives solid theoretical proof of fast convergence. In pursuit of an optimal learning rate schedule with a theoretical convergence guarantee, Vaswani *et al.* hybrid SGD with classic optimization techniques like Armijo line-search [31] and Polyak step size [3,15], and the developed methods not only have sound theory foundation but also empirically show much faster convergence than many widely-used optimization methods. Our work is inspired by Quadratic Interpolation (QI) technique used in line-search [30]. We developed two versions of quadratic line-search method to assist SGD to automatically determine an ideal learning rate schedule.

To accelerate convergence of SGD, heavy ball algorithm (SGD with fixed momentum) [23] is proposed. In practical neural network training, Nesterov accelerated momentum [19] is often used. As discussed in [4], heavy ball method is highly related to conjugate gradient (CG) method, and many efforts have been put into improving SGD by injecting the spirit of CG method. CG method presents promising performance in training an auto-encoder comparing with SGD and LBFGS [14]. A CG-based ADAM [13] was proposed and showed comparative performance to Adam, AdaGrad in solving text and image classification problems, but no generalization performance was provided. Jin *et al.* [10] proposed a stochastic CG with Wolfe line-search that was tested on low-dimensional datasets. All of these related works motivate us to build a bridge between SGD and CG method and create a dynamic acceleration scheme for SGD. Instead of tuning momentum term manually, we dynamically adjust it during training according to Polak-Ribiere formula [22] which shows the best based on our experiments.

Most recently, [18] independently proposed a line search method based on parabolic approximation (PAL). Our major differences are: 1). CGQ introduces feedback to check the estimation quality that improves reliability. 2). CGQ eliminates the momentum hyperparameter, as discussed in Sect. 3.3; 3). For stochastic settings, CGQ does line search on a subset of batches to improve the runtime and generalization; 4). Convergence analysis of CGQ is conducted with no quadratic objective assumption.

## 3 The CGQ Method

### 3.1 Overview

The Conjugate Gradient with Quadratic Line-Search (CGQ) method follows the iterative training framework in Algorithm 1, and eliminates hand-tuning two major hyperparameters in SGD: the learning rate and momentum factor.

In each training iteration, CGQ optimizes the learning rate $\alpha_t$ by estimating the one-dimensional loss landscape as a parabola. Two variants, the 2-point quadratic interpolation and the Least Squares estimation, are illustrated in Sect. 3.2. For direction $p_t$, methods from Conjugate Gradient are adopted to dynamically adjust the momentum factor in SGD instead of using a hand-tuned fixed momentum, which is described in Sect. 3.3.

### 3.2 Dynamically Adjusting the Learning Rate

**Two-Point Quadratic Interpolation.** Given a twice differenciable loss function $f(\theta)$ ($\theta \in \mathbb{R}^m$ is the parameter vector), its second order Taylor expansion around $\theta_0$ is expressed as:

$$\begin{aligned} f(\theta) = f(\theta_0) &+ \nabla f(\theta_0)^T(\theta - \theta_0) \\ &+ \frac{1}{2}(\theta - \theta_0)^T \nabla^2 f(\theta_0)(\theta - \theta_0) \\ &+ O((\theta - \theta_0)^2). \end{aligned} \tag{2}$$

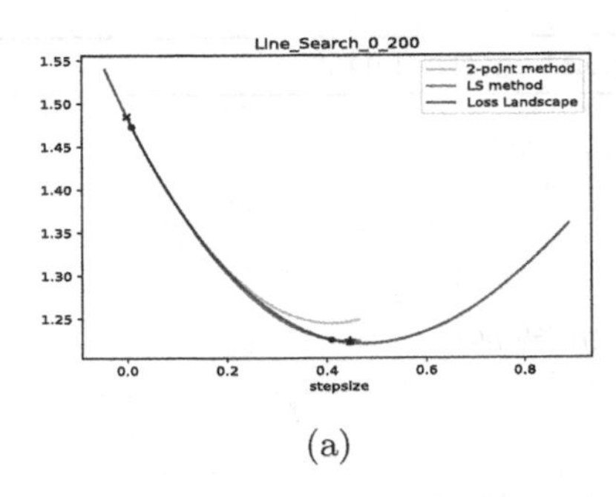


(a)

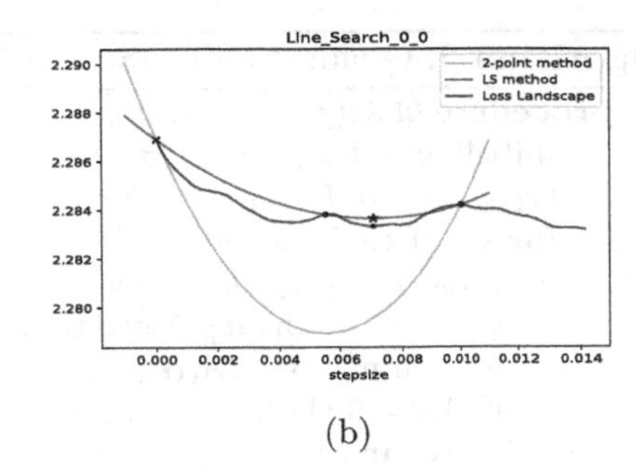


(b)

**Fig. 1.** Comparison of the two quadratic estimation algorithms (2-point method and least squares method). (a) When the landscape is smooth, both estimate well. (b) When the landscape is rough, least squares tend to capture more details. It is also observed in experiments that roughness only occurs at the beginning of a training, after which the landscapes are almost always like (a).

Along direction $p_t \in \mathbb{R}^m$, the 1D loss landscape $\phi(\alpha)$ w.r.t. the step size $\alpha \in \mathbb{R}$ can be described as:

$$\begin{aligned}\phi(\alpha) &= f(\theta - \alpha p_t) \\ &= f(\theta_0) - \alpha \cdot \ \nabla f(\theta_0)^T p_t \\ &+ \frac{1}{2}\alpha^2 \cdot p_t^T \nabla^2 f(\theta_0) p_t + O(\alpha^2).\end{aligned} \quad (3)$$

When the residual term $O(\alpha^2)$ is small enough, the line-search function can be expressed as:

$$\phi(\alpha) \approx q(\alpha) \triangleq \ A\alpha^2 + B\alpha + C \quad (4)$$

where $A = \frac{1}{2} p_t^T \nabla^2 f(\theta_0) p_t$, $B = -\nabla f(\theta_0)^T p_t$, and $C = f(\theta_0)$. When $A > 0$, $q(\alpha)$ reaches minimum at $\alpha^* = -B \ / \ (2A)$.

To get $A, B, C$ at each training iteration without evaluating the Hessian matrix $\nabla^2 f(\theta_0)$, we propose to apply the 2-point quadratic interpolation method. Given two sample points $(0, q(0))$, $(\alpha_1, q(\alpha_1))$ and the slope $q'(0)$, parabola parameters can be calculated as:

$$\begin{aligned}A &= \frac{q(\alpha_1) - q(0) - q'(0) \cdot \alpha_1}{\alpha_1^2} \\ B &= q'(0) \\ C &= q(0)\end{aligned} \quad (5)$$

The stopping criteria and details are shown in Algorithm 2.

One of the major advantages of our method is that when moving closer to a local optimum as training proceeds, $\alpha^*$ will decay automatically, due to the nature of a convex parabola. This is also verified and visualized in our experiment section. Another issue is when the initial $\gamma$ is too large compared to the estimated $\alpha^*$. In such cases, $q(\alpha)$ is not a good surrogate to represent the local landscape

**Algorithm 2.** Quadratic Line-Search (2-point interpolation)

```
1: procedure QLS2(f(·), θ_t, p_t, slope)
2:     Initialize: γ, K_max, α_max, σ
3:     (x_0, y_0) ← (0, f(θ_t))
4:     for k ← 1 to K_max do
5:         (x_1, y_1) ← (γ, f(θ_t + γ · p_t))
6:         q_{A,B,C}(·) ← Interpolate(x_0, y_0, x_1, y_1, slope)
7:         α* = min(−B / 2A, α_max)
8:         if A <= 0 then
9:             return γ
10:        else if q_ABC(α*) > 0 & f(θ_t + α* p_t) < f(θ_t) then
11:            return α*
12:        else
13:            γ ← α*
14:    return α*
```

of $\phi(\alpha)$ near the origin. This is handled by decaying the initial value of $\gamma$ when it is consistently larger than the recent average of $\alpha^*$. Through experiments, we observe that $A$ is usually estimated positive in training neural networks, while the $A < 0$ case is also properly dealt with in Algorithm 2.

**Convergence Analysis.** Our method is majorly motivated by empirical results. To give some insights on the theoretical side, we provide a convergence analysis of quadratic line-search under SGD update rule. To start with, we use the following assumptions on the loss function $f(\theta) = \sum_i f_i(\theta)$, which are commonly used for convergence analysis in deep learning [15,31]. (a) $f$ is *lower-bounded* by a finite value $f^*$; (b) *interpolation*: This requires that $\nabla f(\theta^*) = 0$ implies $\nabla f_i(\theta^*) = 0$ for all components $f_i$; (c) $f$ is $L$-*smooth, i.e.*, the gradient $\nabla f$ is $L$-Lipschitz continuous; and (d) $f$ is $\mu$ *strong-convex.* Proofs are given in https://arxiv.org/abs/2106.11548.

The following lemma provides the bound of the step size returned by quadratic line-search at each iteration $k$.

**Lemma 1.** *Assuming (1) $f_i$'s are $L_i$-smooth, (2) $\mu_i$ strong-convexity of $f_i$, the step size $\alpha_k$ returned by quadratic line-search in Algorithm 2 that is constrained within $(0, \alpha_{max}]$ satisfies:*

$$\alpha_k \in [\min\{\frac{1}{L_{ik}}, \alpha_{max}\}, \ \min\{\frac{1}{\mu_{ik}}, \alpha_{max}\}]. \tag{6}$$

Next, we provide the convergence of quadratic line-search under strong-convex and smooth assumptions.

**Theorem 2.** *Assuming (1) interpolation, (2) $f_i$'s are $L_i$-smooth, and (3) $\mu_i$ strong-convexity of $f_i$'s. When setting $\alpha_{max} \in (0, \ \min_i\{\frac{1}{L_i} + \frac{\mu_i}{L_i^2}\})$, the parame-*

*ters of a neural network trained with quadratic line-search satisfies:*

$$\mathbb{E}[\|\theta_k - \theta^*\|^2] \leq M^k \|\theta_0 - \theta^*\|^2 \tag{7}$$

*where* $M = \mathbb{E}_i[\max\{g_i(\alpha_{max}), 1 - \frac{\mu_i}{L_i}\}] \in (0,1)$ *is the convergence rate, and* $g_i(\alpha_k) = 1 - (\mu_i + L_i)\alpha_k + L_i^2\alpha_k^2$.

**Least Squares Estimation.** Notice that Theorem 2 assumes the loss landscape to be strongly convex and $L$-smooth. In reality, we find that such conditions are satisfied easier in the later than earlier phase of training. In other words, the loss landscape is more rough at the beginning, and the 2-point method sometimes counters difficulties getting a good estimation $q(\alpha)$. For instance, in rough surfaces like Fig. 1b, the 2-point method may focus too much on the local information at the origin (both zero's order and first order moments.), which may lead to poor estimation. This motivates us to further develop a more robust version of the quadratic line-search using least squares estimation.

Instead of only using two samples (plus gradient), the Least Squares estimator takes more samples. And instead of finding a perfect match on the samples, Least Squares minimizes the mean squared error among all samples. In our case, Least Squares minimizes the following objective:

$$\min_{\omega} \|X \cdot \omega - Y\|_2^2 = \frac{1}{n} \sum_i (A\alpha_i^2 + B\alpha_i + C - \phi(\alpha_i))^2 \tag{8}$$

where

$$X = \begin{bmatrix} \alpha_1^2 \; \alpha_1 \; 1 \\ \alpha_2^2 \; \alpha_2 \; 1 \\ \cdots \end{bmatrix} \in \mathbb{R}^{n \times 3} \tag{9}$$

is the sample matrix formulated by each step $\alpha_i$, $\omega = [A, B, C]^T$ are the coefficients to be optimized, and $Y = [\phi(\alpha_1), \phi(\alpha_2), ...]^T \in \mathbb{R}^n$ is the observation vector. The Least Squares estimator is: $\omega = (X^T X)^{-1} X^T Y$, where $(X^T X)^{-1} X^T$ is known as the Moore-Penrose inverse.

The algorithm is described in Algorithm 3. When the loss landscape $\phi(\alpha)$ is rough, as shown in Fig. 1b, the slope of one point is not useful in order to capture the landscape trend. By using least squares estimation on multiple sample points without using slope, the quadratic interpolation becomes insensitive to gradient information near the origin, and seeks to accommodate more global information. Through experiments in Sect. 4, we show that using least squares is more robust.

### 3.3 Dynamically Adjusting Momentum

In this section, we propose an automatic computation of the momentum factor. In SGD or its variants, this is manually set and fixed in value. We build a bridge between the momentum in SGD and the conjugate gradient parameter $\beta_t$. (Recall that in the Nonlinear Conjugate Gradient (CG) method, the conjugate direction

**Algorithm 3.** Quadratic line-search (Least Squares)

```
1: procedure QLS2(f(·), θ_t, p_t, K_max, α_max, slope)
2:     Initialize: γ(sample distance); α_0;
3:     X = {(0,1,0), (0,0,1)}; Y = {slope, f(θ_t)}
4:     for k ← 1 to K_max do
5:         x_k ← (γ^2, γ, 1)
6:         X ← X ⋃ {x_k}
7:         Y ← Y ⋃ {f(θ_t + γ · p_t)}
8:         q_{A,B,C}(·) ← LeastSquare(X, Y)
9:         if A <= 0 then
10:            return γ
11:        else if q_ABC(α*) > 0 & f(θ_t + α* p_t) < f(θ_t) then
12:            return α*
13:        else
14:            γ ← α*
15:    return α*
```

$s_t$ at iteration $t$ is updated as $s_t = -\nabla f + \beta_t \cdot s_{t-1}$, and parameters are updated by $x_{t+1} = x_t + \alpha_t \cdot s_t$.) $\beta_t$ has a deterministic optimal value (s.t. $s_t$ is conjugate to $s_{t-1}$) using the Hessian matrix in a quadratic objective. But for a general nonlinear case, various heuristic formulas are proposed [5,8,9,22,28]. $\alpha_t$ can be determined using either the Newton–Raphson method or a line-search such as the Armijo rule [1] and Wolfe conditions [32,33]. The Nonlinear CG method generalizes various gradient-based optimization methods. For instance, when $\beta_t = 0$ and $\alpha_t$ is constant, CG degenerates as the Gradient Descent,and when $\beta_t$ is a non-zero constant, it instantiates the Gradient Descent with momentum, which is also discussed in [4]. Moreover, the CG framework can also be used in the stochastic settings, where gradients $\nabla_\theta c(\theta)$ are replaced by partial gradients $\nabla_\theta\, c_{batch}(\theta)$ in each iteration.

With our proposed Quadratic line-search, the local solver is able to adaptively change the learning rate for performance. We next show that using the CG framework, the momentum term in SGD ($\beta_t$ in CG), which is usually constant in a SGD solver, can also be adaptively adjusted.

For a quadratic objective of the following form:

$$\min f(\theta) = \frac{1}{2}\theta^T Q\theta + b^T\theta \tag{10}$$

where $\theta \in \mathbb{R}^n$ is the model parameters, and $Q \in \mathbb{R}^{n\times n}$ and $b \in \mathbb{R}^n$ are problem-dependent matrices. The optimal $\beta_t$ can be chosen as

$$\beta_t = \frac{\nabla f(\theta_t)^T \cdot Q \cdot p_{t-1}}{p_{t-1}^T Q p_{t-1}} \tag{11}$$

where $p_{t-1}$ is the update direction at time $t-1$. Then the next update direction $p_t = -\nabla f(\theta_t) + \beta_t \cdot p_{t-1}$ is guaranteed to be $Q$-conjugate to $p_{t-1}$, which accelerates training by eliminating influences on previous progress. For general nonlinear objective functions with no explicit $Q$ and the Hessian is difficult to compute, different heuristics were proposed in [5,8,9,22,28]. Since the classic Gradient Descent with momentum method is a stationary version of the Conjugate Gradient [4], a dynamic choice of $\beta_t$, instead of fixed, can potentially speed up the training process. It is also notable that to compute $\beta_t$ using the aforementioned formula, no Hessian is required. For example, the Polak-Ribiere [22] formula:

$$\beta_t^{PR} = \frac{\nabla f(\theta_t)^T (\nabla f(\theta_t) - \nabla f(\theta_{t-1}))}{\nabla f(\theta_{t-1})^T \nabla f(\theta_{t-1})} \tag{12}$$

only requires the recent two gradients.

A bounded version of the Polak-Ribiere momentum is used, *i.e.*, to compute the momentum as $\beta'_t = \min\{\max\{0, \beta_t^{PR}\}, \beta_{max}\}$. Through experiments, it is shown that by applying such dynamic momentum $\beta'_t$, the performance is comparable to or even better than the best manual momentum. The major advantage is that another crucial hyperparameter is eliminated from hand tuning.

### 3.4 Optimizations for Large Datasets

Conceptually, CGQ performs a line search at every iteration. This unsurprisingly introduces more computation. To alleviate this issue on large systems, it is reasonable to use partially observed curvature (from batches) to estimate that of the whole dataset. This motivates us to implement the sCGQ that performs line search stochastically: For each training batch, the optimizer will perform a line search (that optimizes learning rate) with a certain probability $p$. For batches without line searches, the moving average of past line search results will be applied. Figure 2 show that this makes the run time of sCGQ with fixed momentum comparative to naive SGD, and sCGQ with dynamic momentum comparative to other line search methods. Experiments in Sect. 4.1 and the runtime comparison in Fig. 2 show that sCGQ benefits from both fast execution and good estimation of 1−D loss landscape (in terms of learning rate).

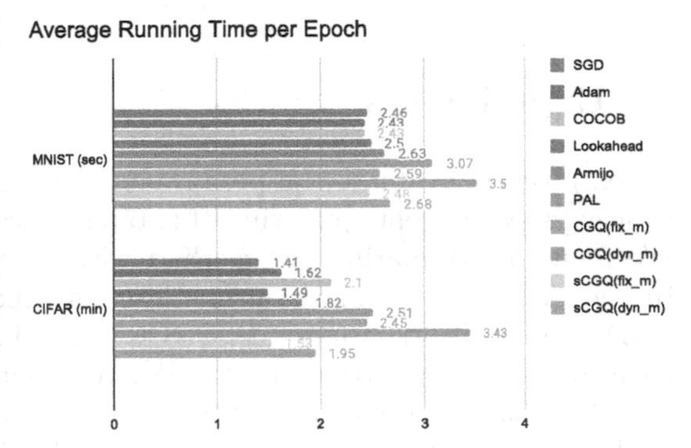


**Fig. 2.** Running time comparison. sCGQ: stochastic line search with probability 0.1. `fix_m`: fixed momentum; `dyn_m`: Polak-Ribiere momentum. 2-point interpolation and Least Squares takes similar time.

**Table 1.** Performance comparison of different optimization algorithms on the MNIST and SVHN datasets. Training budget: 20 epochs (MNIST)/40 epochs (SVHN). "-" means not converging.

| Method | Train loss/Test accuracy | | | |
|---|---|---|---|---|
| | MNIST | SVHN | | |
| | MLP | VGG-16 | ResNet-110 | DenseNet-100 |
| SGD | .076/97.24 | .080/93.16 | .217/92.68 | .158/94.34 |
| Adam [12] | .014/98.02 | .026/93.14 | .134/95.34 | .093/95.35 |
| Lookahead [34] | .131/95.99 | .039/93.95 | .157/93.14 | .112/94.12 |
| COCOB [20] | .267/92.72 | .036/91.81 | .069/90.34 | .035/91.30 |
| Armijo [31] | .023/97.85 | .008/92.42 | - | - |
| PAL [18] | .021/97.60 | - | .016/94.73 | .017/95.10 |
| CGQ (2pt) | **.007/98.24** | .011/94.70 | **.009**/95.31 | **.004/96.11** |
| CGQ (LS) | **.007/98.24** | **.006**/94.79 | .012/95.81 | .008/95.99 |
| sCGQ (2pt) | .008/**98.24** | .020/93.89 | .043/**95.84** | .029/94.23 |
| sCGQ (LS) | .008/98.12 | .016/**94.93** | .039/95.63 | .032/95.43 |

## 4 Experiments

To empirically evaluate the performance of the proposed method, we design a thorough experiment consisting of two parts described in the following sections. We focus on comparing our method with different configurations in Sect. 4.1. In Sect. 4.2, we benchmark our method with six popular optimization methods: SGD + momentum, Adam [12], COCOB [20], Lookahead optimizer [34], SGD + Armijo Line-Search [31], and PAL [18]. We manually tune their hyperparameters and the results summarized in the following sections are their best performances in the experiments to our best knowledge. Specifically for SGD and ADAM, the learning rates are fixed (SGD: 0.02 for VGG, 0.01 on MNIST, and 0.1 elsewhere; Adam: 0.001). There is no learning rate for COCOB. For our own method, we tested four varients: CGQ(2pt)/CGQ(LS): complete line search with quadratic interpolation or least squares estimation; sCGQ(2pt)/sCGQ(LS): stochastic line search with quadratic interpolation/least squares estimation. We showcase the fast convergence of our method on training multi-class classifiers for widely used benchmark image datasets including MNIST, SVHN, CIFAR-10, and CIFAR-100. For fairness, model parameters are initialized the same among all solvers for each test case, and we run multiple times and show the average performance for each case. All experiments are conducted on a NVIDIA GTX-1080Ti GPU, and the program is developed under PyTorch 1.7.0 framework. Code is available in https://github.com/zh272/CGQ-solver.

### 4.1 Ablation Test

**Dynamical Momentum.** Applying SGD with momentum to reach faster convergence has become common in deep neural network training. Heavy ball [23] and Nesterov momentum [19] are two classic static acceleration scheme that consider the momentum term as a fixed hyperparameter. In this section, we compare the performances of SGD with quadratic line-search plus (1) Heavy ball momentum[23] with coefficients from 0.1 to 0.9, (2) Fletcher-Reeves momentum [8], (3) Polak-Ribiere momentum [22], (4) Hestenes-Stiefel momentum [9], and (5) Dai-Yuan momentum [5].

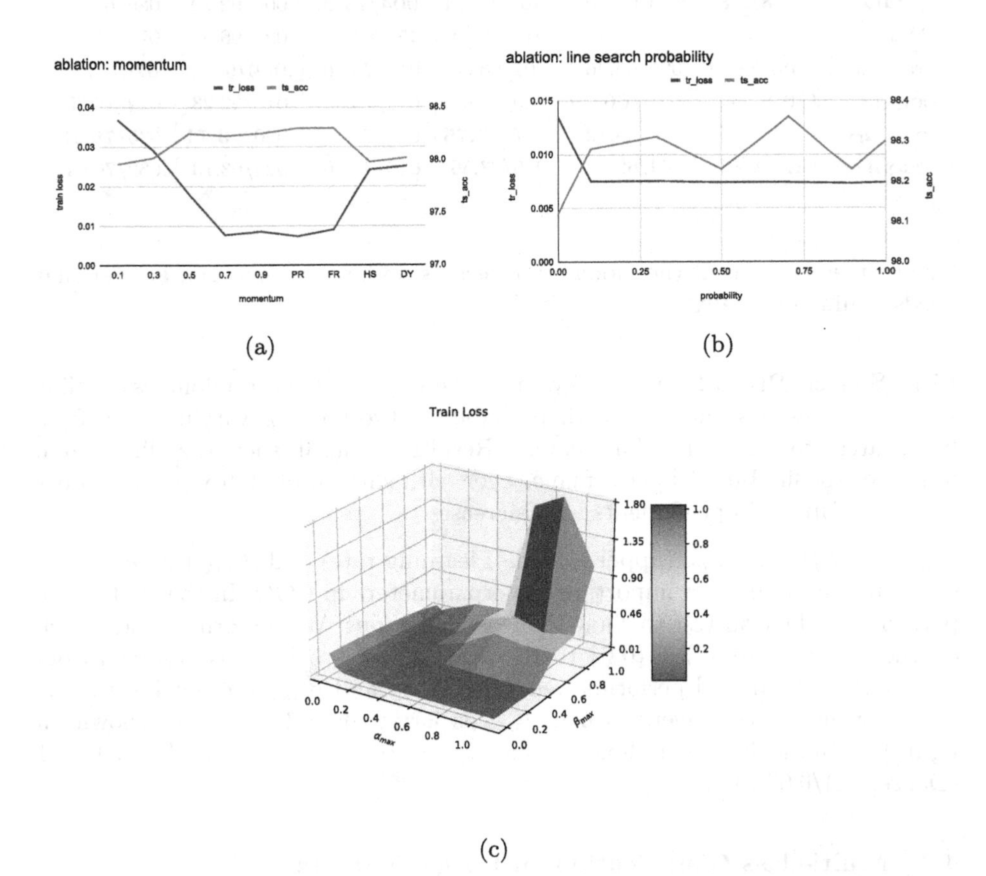


**Fig. 3.** Training loss (One-layer MLP trained on MNIST dataset) w.r.t.: (a) Different momentum settings. (b) Line search probability. "0": naive SGD with no line search; "1": line search on every mini-batch. (c) $\alpha_{max}$ and $\beta_{max}$.

The benchmark experiment is run on the MNIST dataset. The comparison results are summarized in Fig. 3a. On the basis of the results of the ablation test, it is promising to use Polak-Ribiere momentum as a dynamic acceleration scheme

**Table 2.** Performance comparison for the CIFAR-10 and CIFAR-100 datasets. Training budget: 200 epochs. "-" means the method does not converge.

| Method | Train loss/Test accuracy | | | | | |
|---|---|---|---|---|---|---|
| | VGG-16 | | ResNet-164 | | DenseNet-100 | |
| | CIFAR-10 | CIFAR-100 | CIFAR-10 | CIFAR-100 | CIFAR-10 | CIFAR-100 |
| SGD | .116/**89.64** | .685/64.37 | .271/86.06 | .860/63.55 | .255/86.71 | .901/62.06 |
| Adam | .044/88.06 | .408/55.21 | .144/90.06 | .419/68.07 | .127/90.42 | .364/68.23 |
| Lookahead | .069/89.21 | .369/60.97 | .155/88.33 | .526/59.21 | .153/88.50 | .594/61.62 |
| COCOB | .087/84.78 | .772/44.83 | .330/78.96 | 1.450/48.00 | .205/80.63 | 1.103/51.56 |
| Armijo | .298/81.18 | .025/62.50 | **.003**/93.64 | **.004**/73.26 | .009/92.72 | .086/65.55 |
| PAL | - | - | .017/92.04 | .452/64.01 | .023/90.99 | .072/65.90 |
| CGQ (2pt) | **.008**/ 88.21 | .025/65.46 | .007/93.04 | .014/75.26 | **.009**/91.49 | **.020**/75.01 |
| CGQ (LS) | .010/88.64 | **.019/66.49** | .010/93.52 | .019/73.43 | .010/92.23 | .029/74.35 |
| sCGQ (2pt) | .035/89.16 | .038/63.12 | .017/**93.75** | .016/**75.94** | .020/**93.71** | .035/**76.15** |
| sCGQ (LS) | .041/88.88 | .044/62.93 | .016/93.56 | .016/75.60 | .028/93.13 | .032/76.04 |

for our method so that the momentum term is no longer a hyperparameter that needs tuning manually.

**Line Search Probability.** To evaluate the impact of the randomness on line search, we test our method with probability thresholds $p$ varying from 0(no line search) to 1(complete line search). Results in Fig. 3b show that line search improves the final model performance even with small probability $p$, and it does not yield further improvements as $p$ increases.

**$\alpha_{max}$ and $\beta_{max}$** $\alpha_{max}$ (upper bound of learning rate) and $\beta_{max}$ (upper bound of momentum), are two important hyperparameters in CGQ. In this section we perform an ablation test to show the feasible regions. We perform a grid search on the 2D space by $\alpha_{max}$ (0.01 to 1.1) and $\beta_{max}$ (0 to 1). It is observed from Fig. 3c that the model performs well when $\alpha_{max}$ and $\beta_{max}$ are not both large. And the model will diverge when $\beta_{max}$ is larger than 1 (thus not shown on figure). Empirically good choices are: $\beta_{max} = 0.8$; $\alpha_{max} = 0.3$ (ResNet)/0.1 (DenseNet)/0.05 (VGG).

### 4.2 Multi-class Classification on Image Datasets

Recent evidence has shown that deep convolutional neural networks lead to impressive breakthroughs in multi-class image classification. We mainly select the three popular families of architectures of deep convolutional neural networks: ResNet, DenseNet and VGG to benchmark our method with the other powerful optimizers. The comparison results will be presented in the following two sections, according to different datasets used in experiments.

**MNIST and SVHN.** The MNIST dataset contains 60,000 hand-written digit images for training, and 10,000 images for testing. Images in MNIST are grayscale images with size 28 × 28. SVHN dataset consists of 73,257 images in the training set, 26032 images in the testing set and another 531,131 additional images provided for training. All the images in SVHN are colored digit images with a size of 32 × 32. For MNIST, we construct a fully connected network with one hidden layer of 1000 neurons as a toy benchmark example. For SVHN, we experiment with three more expressive architectures: VGG-16, ResNet-110, and DenseNet-100.

The main results are summarized in Table 1. We observed that with the same training budget, CGQ presents better performances in terms of both training loss and test accuracy. The stochastic version of We omit SGD with Armijo-line-search in the SVHN experiment since it appeared to be unstable and easy to diverge.

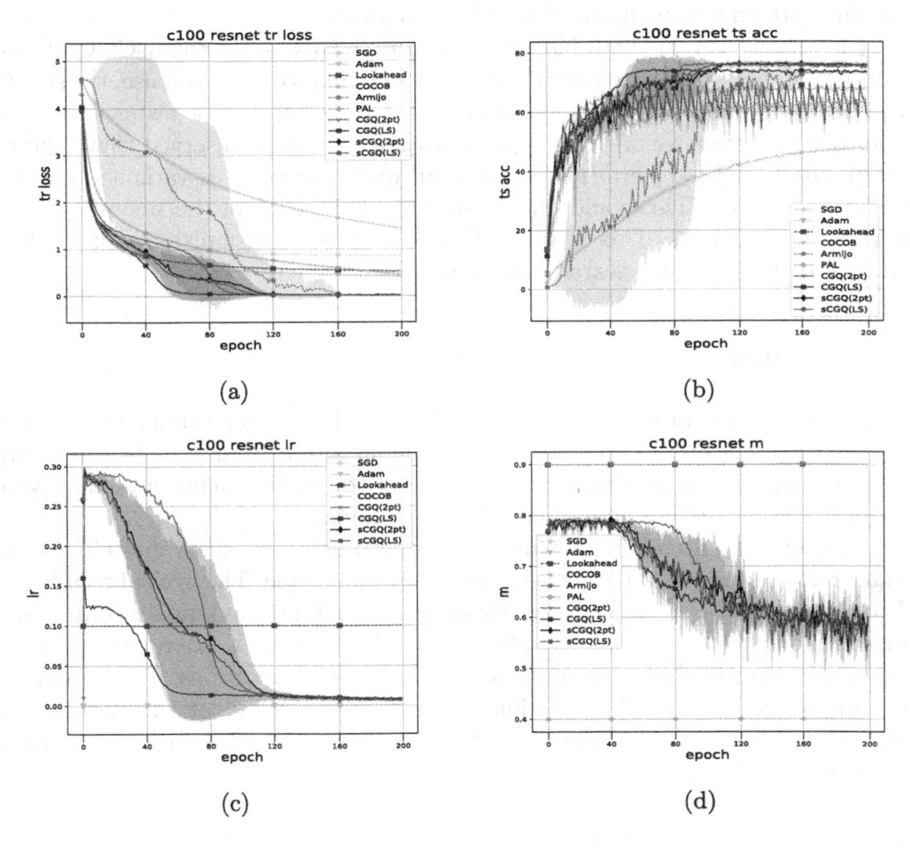


**Fig. 4.** Dynamics of training a ResNet-164 with various optimization methods on the CIFAR100. (a) Training Loss. (b) Test Accuracy. (c) Learning Rate. (d) Momentum

**CIFAR-10 and CIFAR-100.** The CIFAR-10 and CIFAR-100 datasets both consist of 50,000 images for training and 10,000 images for testing, while images are in 10 and 100 different classes, respectively. We run the experiments using VGG-16, ResNet-164, and DenseNet-100. Each of model-algorithm pair is trained for 200 epochs with same initialization points.

In Table 2, we record the training loss and test accuracy of each model and optimizer pair in the CIFAR-10 and CIFAR-100 experiments. In most of the cases, CGQ and sCGQ not only reach the best performance among all line search methods, but also converges to a local minimum faster. In the more challenging CIFAR-100 case, the margin is more significant. Figure 4 shows the dynamics for training ResNet on the CIFAR100 dataset. Each line plots the mean over multiple runs, while corresponding standard deviations are shown as shaded area around the mean. (a) and (b) tells that CGQ variants performs stably better in training loss and testing accuracy compared to other compared methods. Moreover, (c) and (d) demonstrates the automatic decaying effect on learning rate and momentum of the CGQ method.

It is observed from the tables that CGQ performs better than sCGQ on the training set, but sCGQ has better test time performance. In this sense, the sCGQ is preferable to CGQ for larger datasets, not only because of faster execution time, but also because of better generalization. It is also observed that 2-point interpolation method performs better than its Lease Squares variant on larger datasets (CIFAR). This could due to an underestimation of the smoothness for deep neural networks. This is also validated through a visualization of the loss landscape along the line search direction in Fig. 1.

## 5 Conclusion

In this paper, we propose CGQ method to replace hand tuning two crucial hyperparameters in SGD, *i.e.*, learning rate and momentum. The learning rate is determined by our quadratic line-search method, and momentum is adaptively computed using the bounded Polak-Ribiere formula.

Experiments on modern convolutional neural networks show that CGQ converges faster due to its adaptiveness to the loss landscape. Theoretical results are also provided towards the convergence of quadratic line search under convex and smooth assumptions. To improve efficiency on larger models, we further propose the sCGQ method that performs line search on fewer iterations. This modification improves not only the run time, but also the generalization capability in terms of test accuracy. In most cases, the CGQ method outperforms other local methods.

## References

1. Armijo, L.: Minimization of functions having Lipschitz continuous first partial derivatives. Pacific J. Math. **16**(1), 1–3 (1966)

2. Baydin, A.G., Cornish, R., Rubio, D.M., Schmidt, M., Wood, F.: Online learning rate adaptation with hypergradient descent. In: ICLR (2018)
3. Berrada, L., Zisserman, A., Kumar, M.P.: Training neural networks for and by interpolation. In: International Conference on Machine Learning (2020)
4. Bhaya, A., Kaszkurewicz, E.: Steepest descent with momentum for quadratic functions is a version of the conjugate gradient method. Neural Netw. **17**, 65–71 (2004). https://doi.org/10.1016/S0893-6080(03)00170-9
5. Dai, Y.H., Yuan, Y.X.: A nonlinear conjugate gradient method with a strong global convergence property. SIAM J. Optim. **10**(1), 177–182 (1999)
6. Defazio, A., Bach, F., Lacoste-Julien, S.: SAGA: a fast incremental gradient method with support for non-strongly convex composite objectives. In: Ghahramani, Z., Welling, M., Cortes, C., Lawrence, N., Weinberger, K.Q. (eds.) Advances in Neural Information Processing Systems, vol. 27, pp. 1646–1654. Curran Associates, Inc. (2014)
7. Duchi, J., Hazan, E., Singer, Y.: Adaptive subgradient methods for online learning and stochastic optimization. J. Mach. Learn. Res. **12**(61), 2121–2159 (2011)
8. Fletcher, R., Reeves, C.M.: Function minimization by conjugate gradients. Comput. J. **7**(2), 149–154 (1964). https://doi.org/10.1093/comjnl/7.2.149
9. Hestenes, M.R., Stiefel, E.: Methods of conjugate gradients for solving linear systems. J. Res. Natl. Bur. Stand. **49**, 409–436 (1952)
10. Jin, X.B., Zhang, X.Y., Huang, K., Geng, G.G.: Stochastic conjugate gradient algorithm with variance reduction. IEEE Trans. Neural Netw. Learn. Syst. **30**(5), 1360–1369 (2019). https://doi.org/10.1109/TNNLS.2018.2868835
11. Johnson, R., Zhang, T.: Accelerating stochastic gradient descent using predictive variance reduction. In: Burges, C.J.C., Bottou, L., Welling, M., Ghahramani, Z., Weinberger, K.Q. (eds.) Advances in Neural Information Processing Systems, vol. 26, pp. 315–323. Curran Associates, Inc. (2013)
12. Kingma, D.P., Ba, J.L.: Adam: a method for stochastic optimization. In: International Conference on Learning Representations (ICLR) (2015)
13. Kobayashi, Y., Iiduka, H.: Conjugate-gradient-based Adam for stochastic optimization and its application to deep learning (2020). http://arxiv.org/abs/2003.00231
14. Le, Q.V., Ngiam, J., Coates, A., Lahiri, A., Prochnow, B., Ng, A.Y.: On optimization methods for deep learning. In: ICML (2011)
15. Loizou, N., Vaswani, S., Laradji, I., Lacoste-Julien, S.: Stochastic Polyak step-size for SGD: an adaptive learning rate for fast convergence. In: Proceedings of the 24th International Conference on Artificial Intelligence and Statistics (AISTATS), pp. 1–33 (2021)
16. Loshchilov, I., Hutter, F.: Fixing weight decay regularization in Adam. arXiv:1711.05101 (2017)
17. Powell, M.J.D.: An efficient method for finding the minimum of a function of several variables without calculating derivatives. Comput. J. **7**(2), 155–162 (1964)
18. Mutschler, M., Zell, A.: Parabolic approximation line search for DNNs. In: NeurIPS (2020)
19. Nesterov, Y.: A method of solving a convex programming problem with convergence rate $o(1/k^2)$. Soviet Mathe. Doklady **27**, 372–376 (1983)
20. Orabona, F., Pal, D.: Coin betting and parameter-free online learning. In: Lee, D., Sugiyama, M., Luxburg, U., Guyon, I., Garnett, R. (eds.) Advances in Neural Information Processing Systems, vol. 29, pp. 577–585. Curran Associates, Inc. (2016)

21. Pesme, S., Dieuleveut, A., Flammarion, N.: On convergence-diagnostic based step sizes for stochastic gradient descent. In: Proceedings of the International Conference on Machine Learning 1 Pre-proceedings (ICML 2020) (2020)
22. Polak, E., Ribiere, G.: Note sur la convergence de méthodes de directions conjuguées. ESAIM: Math. Model. Numer. Anal. - Modélisation Mathématique et Analyse Numérique **3**(R1), 35–43 (1969)
23. Polyak, B.T.: Some methods of speeding up the convergence of iteration methods. USSR Comput. Math. Math. Phys. **4**(5), 1–17 (1964)
24. Reddi, S.J., Kale, S., Kumar, S.: On the convergence of Adam and beyond. In: The 35th International Conference on Machine Learning (ICML) (2018)
25. Schmidt, M., Le Roux, N., Bach, F.: Minimizing finite sums with the stochastic average gradient. Math. Program. **162**(1), 83–112 (2017)
26. Schmidt, M., Roux, N.L.: Fast convergence of stochastic gradient descent under a strong growth condition (2013)
27. Shalev-Shwartz, S., Zhang, T.: Stochastic dual coordinate ascent methods for regularized loss minimization. J. Mach. Learn. Res. **14**(1), 567599 (2013)
28. Shewchuk, J.R.: An introduction to the conjugate gradient method without the agonizing pain, August 1994. http://www.cs.cmu.edu/~quake-papers/painless-conjugate-gradient.pdf
29. Smith, L.N.: Cyclical learning rates for training neural networks. In: IEEE Winter Conference on Applications of Computer Vision (WACV) (2017)
30. Vandebogert, K.: Method of quadratic interpolation, September 2017. https://people.math.sc.edu/kellerlv/Quadratic_Interpolation.pdf
31. Vaswani, S., Mishkin, A., Laradji, I., Schmidt, M., Gidel, G., Lacoste-Julien, S.: Painless stochastic gradient: interpolation, line-search, and convergence rates. In: Advances in Neural Information Processing Systems, pp. 3727–3740 (2019)
32. Wolfe, P.: Convergence conditions for ascent methods. SIAM Rev. **11**(2), 226–000 (1969)
33. Wolfe, P.: Convergence conditions for ascent methods. II: some corrections. SIAM Rev. **13**(2), 185–000 (1969)
34. Zhang, M., Lucas, J., Ba, J., Hinton, G.E.: Lookahead optimizer: K steps forward, 1 step back. In: Wallach, H., Larochelle, H., Beygelzimer, A., d'Alché-Buc, F., Fox, E., Garnett, R. (eds.) Advances in Neural Information Processing Systems 32, pp. 9597–9608 (2019)

# Attack Transferability Characterization for Adversarially Robust Multi-label Classification

Zhuo Yang[1(✉)], Yufei Han[2], and Xiangliang Zhang[1]

[1] King Abdullah University of Science and Technology, Thuwal, Saudi Arabia
{zhuo.yang,xiangliang.zhang}@kaust.edu.sa
[2] CIDRE Team, Inria, France

**Abstract.** Despite of the pervasive existence of multi-label evasion attack, it is an open yet essential problem to characterize the origin of the adversarial vulnerability of a multi-label learning system and assess its attackability. In this study, we focus on non-targeted evasion attack against multi-label classifiers. The goal of the threat is to cause misclassification with respect to as many labels as possible, with the same input perturbation. Our work gains in-depth understanding about the multi-label adversarial attack by first characterizing the transferability of the attack based on the functional properties of the multi-label classifier. We unveil how the transferability level of the attack determines the attackability of the classifier via establishing an information-theoretic analysis of the adversarial risk. Furthermore, we propose a transferability-centered attackability assessment, named Soft Attackability Estimator (SAE), to evaluate the intrinsic vulnerability level of the targeted multi-label classifier. This estimator is then integrated as a transferability-tuning regularization term into the multi-label learning paradigm to achieve adversarially robust classification. The experimental study on real-world data echoes the theoretical analysis and verify the validity of the transferability-regularized multi-label learning method.

**Keywords:** Attackability of multi-label models · Attack transferability · Adversarial risk analysis · Robust training

## 1 Introduction

Adversarial evasion attack against real-world multi-label learning systems cannot only harm the system utility, but also facilitate advanced downstreaming cyber menaces [13]. For example, hackers embed toxic contents into images while hiding the malicious labels from the detection [8]. Stealthy harassment applications, such as phone call dictation and photo extraction, carefully shape the

**Electronic supplementary material** The online version of this chapter (https://doi.org/10.1007/978-3-030-86523-8_24) contains supplementary material, which is available to authorized users.


N. Oliver et al. (Eds.): ECML PKDD 2021, LNAI 12977, pp. 397–413, 2021.
https://doi.org/10.1007/978-3-030-86523-8_24

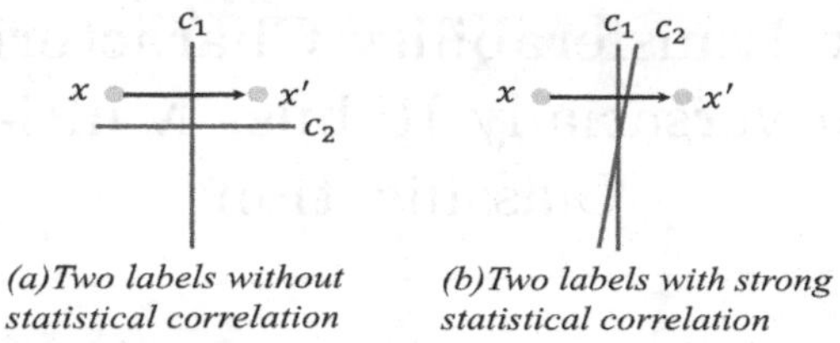

*(a) Two labels without statistical correlation*

*(b) Two labels with strong statistical correlation*

**Fig. 1.** A toy example of multi-label evasion attack.

app function descriptions to evade from the sanitary check of app stores [6,12]. Despite of the threatening impact, it remains an open problem to characterize key factors determining the attackability of a multi-label learning system. Compared to in-depth adversarial vulnerability study of single-label learning problems [5,16,18,20,23], this is a rarely explored, yet fundamental problem for trustworthy multi-label classification.

We focus on the non-targeted evasion attack against multi-label classifiers. In contrast to the single-label learning problem, the goal of the adversarial threat is to kill multiple birds with one stone: it aims at *changing as many label-wise outputs as possible simultaneously, with the same input.*

Figure 1 demonstrates a toy example of the threat scenario with two labels $l_1$ and $l_2$, with decision hyper-planes $c_1$ and $c_2$, respectively. Figure 1 (a) assumes no statistical correlation between the two labels. $c_1$ and $c_2$ are orthogonal therefore. In contrast, the two boundaries are well aligned in Fig. 1 (b), implying a strong correlation between $l_1$ and $l_2$. The injected evasion noise in both scenarios has the same magnitude to change $x$ to be $x'$, indicating the same attack strength. As we can see, the evasion attack can flip simultaneously the classifier's output with respect to both labels in Fig. 1 (b), due to the alignment between the decision boundaries of $l_1$ and $l_2$. However, in Fig. 1 (a), the evasion perturbation can only bring impacts to the decision output of $l_1$. As shown in the toy example, *whether the attack can transfer across different labels depends on the alignment of the decision hyper-planes, which is determined intrinsically by the correlation between the labels.* On the closely correlated labels, the multi-label classifier tends to produce the consistently same or converse decisions. The adversarial noise that successfully perturbs the decision over one label is likely to cause misclassification on the other labels. Bearing the goal of the non-targeted attack in mind, the transferability of the attack is closely related to the adversarial vulnerability of the targeted multi-label learning system. With a more transferable attack noise, the multi-label learning system is more attackable.

Given a multi-label learning task, our study aims at gaining in-depth understanding about the theoretical link between the transferability of the evasion attack across different labels and the adversarial vulnerability of the classifier. More specifically, we focus on characterizing the role of attack transferability in determining the adversarial risk of a multi-label classifier. Furthermore, we pursue a qualitative assessment of attack transferability based on the intrinsic functional properties of the classifier. It is beneficial to not only evaluate the attackability of the classifier, but also design a transferability-suppression regularization term to enhance the adversarial robustness of the classifier. In the community of multi-label learning, it is a well-known fact that capturing the correlation between labels helps to train accurate multi-label classifiers. However, our analysis unveils the other side of the story: encoding the label correlation can also make the classifier vulnerable in the evasion attack scenarios. Our contribution can be summarized as in the followings:

- We unveil the three key factors determining the adversarial risk of a multi-label classifier by establishing an information-theoretic upper bound of the adversarial risk. They are i) the conditional mutual information (CMI) between the training data and the learnt classifier [14]; ii) the transferability level of the attack; and iii) the strength of the evasion attack. Theoretical discussions over the first two factors unveil a dilemma: Encoding label correlation in the training data is the key-to-success in accurate adversary-free multi-label classification. However, it also increases the transferability of the attack, which makes the classifier vulnerable (with a higher adversarial risk).
- We propose an attackability assessment in Sect. 4 based on the unveiled link between the attack transferability and the adversarial risk. This attackability assessment is then integrated into the multi-label learning paradigm as a regularization term to suppress the attack transfer and enhance the adversarial robustness of the derived multi-label classifier.
- Our empirical study with real-world multi-label learning applications instantiates the theoretical study with both linear and deep multi-label models. The results confirm the trade-off between the utility of the classifier and its adversarial robustness by controlling the attack transferability.

## 2 Related Work

Bounding adversarial risk has been studied extensively in single-label learning scenarios [4,5,7,9,9,11,16–18,20,22,23]. They focus on identifying the upper bound of adversarial noise, which guarantees the stability of the targeted classifier's output, a.k.a. adversarial sphere. Notably, [4,9,16,22] study the association between adversarial robustness and the curvature of the learnt decision boundary. Strengthened further by [11,16,22], the expected classification risk under adversarial perturbation can be bounded by the model's Rademacher complexity of the targeted classifier. [21] extends the model complexity-dependent analysis to the multi-label learning problems and associates the Rademacher complexity with the nuclear norm of the model parameters.

Distinguished from single-label learning scenarios, the key-to-success of training an accurate multi-label classifier is to capture the correlation between the labels. More specifically, the alignment between the decision hyper-planes of the correlated labels helps to predict the occurrence of the labels. However, as revealed in [13,21], the evasion attack perturbation can transfer across the correlated labels: *the same input perturbation can affect the decision output of these labels.* It implies that the label correlation can be potentially beneficial to adversaries at the same time. Nevertheless, the relation between the transferability of the input perturbation and the adversarial vulnerability of the victim classifier cannot be characterized or measured by the Rademacher-complexity-based analysis conducted on the single-label case and [21]. Our work thus focuses on addressing the essential yet open problem from two perspectives. First, we target on establishing a theoretical link between the transferability measurement of the attack noise across multiple labels and the vulnerability of the classifier. Second, we conduct an information-theoretic analysis of the adversarial risk, which is an attack-strength-independent vulnerability assessment. This assessment can be used to guide proactive model hardening, e.g., robust model training, to improve the adversarial robustness of the classifier.

## 3 Vulnerability Assessment of Multi-label Classifiers

**Notations.** We use $z = (x, y)$ as a multi-label instance, with feature vector $x \in \mathbb{R}^d$ and label vector $y = \{-1, 1\}^m$, where $d$ and $m$ denote the feature dimension and the number of labels, respectively. Specially, we use $x_i$ and $y^i$ to denote the feature vector and the label vector of instance $z_i$ respectively and use $y_j$ to denote the $j$-th element of label vector $y$. Let $\mathcal{D}$ be the underlying distribution of $z$ and $z^n$ be a data set including $n$ instances. Let $h$ denote the multi-label classifier to learn from the data instances sampled from $\mathcal{D}$. The learning paradigm (possibly randomized) is thus noted as $\mathcal{A} : z^n \rightarrow h$. The probability distribution of the learning paradigm is $\mathcal{P}_{\mathcal{A}}$. The corresponding loss function of $\mathcal{A}$ is $\ell : h \times z \rightarrow \mathbb{R}$. $\|x\|_p$ $(p \geq 1)$ denotes the $L_p$ norm of a vector $x$. Without loss of generality, we choose $p = 2$ hereafter.

**Attackability of a Multi-label Classifier.** The attackability of $h$ is defined as the expected maximum number of flipped decision outputs by injecting the perturbation $r$ to $x$ within an attack budget $\varepsilon$:

$$\begin{aligned} &C^*(\mathcal{D}) = \mathop{\mathbb{E}}_{z\sim\mathcal{D}} \left( \max_{T, \|r^*\| \leq \varepsilon} \sum_{j=1}^{m} \mathbb{1}(y_j \neq sgn(h_j(x + r^*))) \right), \\ &\text{where } r^* = \mathop{argmin}_{r} \|r\|_p, \\ &s.t.\ y_j h_j(x + r^*) \leq 0\ (j \in T),\ \ y_j h_j(x + r^*) > 0\, (j \notin T). \end{aligned} \tag{1}$$

$T$ denotes the set of the attacked labels. $h_j(x + r)$ denotes the decision score of the label $j$ of the adversarial input. $\mathbb{1}(\cdot)$ is the indicator function. It is valued as 1 if the attack flips the decision of the label $j$ and 0 otherwise. With the same input $x$ and the same attack strength $\|r\|_p$, one multi-label classifier $h$ is more vulnerable to the evasion attack than the other $h'$, if $C_h^* > C_{h'}^*$.

### 3.1 Information-Theoretic Adversarial Risk Bound

Solving Eq. (1) directly for a given data instance $z$ reduces to an integer programming problem, as [21] did. Nevertheless, our goal is beyond solely empirically assessing the attackability of $h$ on a given set of instances. We are interested in **1)** establishing an upper bound of the expected misclassification risk of $h$ with the presence of adversary. It is helpful for characterizing the key factors deciding the adversarial risk of $h$; **2)** understanding the role of the transferability of the input perturbation across different labels in shaping the adversarial threat.

For a multi-label classifier $h$, $n$ legal instances $z^n = \{z_i\}$ $(i=1,2,\ldots,n$, $z_i = (x_i, y^i) \sim \mathcal{D})$ and the attack budget $\varepsilon$, we can estimate the expected adversarial risk of $h$ by evaluating the worst-case classification risk over the neighborhood $N(z_i) = \left\{(x'_i, y^i) \middle| \|x'_i - x_i\|_p \leq \varepsilon \right\}$. The expected and empirical adversarial risk $R_{\mathcal{D}}(h,\varepsilon)$ and $R^{emp}_{\mathcal{D}}(h,\varepsilon)$ give:

$$\begin{aligned} R_{\mathcal{D}}(h,\varepsilon) &= E_{\mathcal{A},z^n\sim\mathcal{D}^n}[E_{z\sim\mathcal{D}}[\max_{(x',y)\in N(z)} \ell(h(x'),y)]],\ \ h=\mathcal{A}(z^n), \\ R^{emp}_{\mathcal{D}}(h,\varepsilon) &= E_{\mathcal{A},z^n\sim\mathcal{D}^n}[\frac{1}{n}\sum_{i=1}^{n}[\max_{(x'_i,y^i)\in N(z_i)} \ell(h(x'_i),y^i)]],\ \ h=\mathcal{A}(z^n). \end{aligned} \tag{2}$$

The expectation in Eq. (2) is taken with respect to the joint distribution $\mathcal{D}^{\otimes n} \otimes \mathcal{P}_{\mathcal{A}}$ and $\mathcal{D}^n$ denotes the data distribution with $n$ instances. The expected adversarial risk $R_{\mathcal{D}}(h,\varepsilon)$ reflects the vulnerability level of the trained classifier $h$. Intuitively, a higher $R_{\mathcal{D}}(h,\varepsilon)$ indicates that the classifier $h$ trained with the learning paradigm $\mathcal{A}$ is easier to attack (more attackable). $R^{emp}_{\mathcal{D}}(h,\varepsilon)$ is the empirical evaluation of the attackability level. By definition, if $\mathcal{A}$ is deterministic and the binary 0-1 loss is adopted, $\sum_{i=1}^{n} C^*_h(z_i)$ gives $R^{emp}_{\mathcal{D}}(h,\varepsilon)$.

Theorem 1 establishes the upper bound of the adversarial risk $R_{\mathcal{D}}(h,\varepsilon)$ based on the conditional mutual information $CMI_{\mathcal{D},\mathcal{A}}$ between the legal data and the learning paradigm. Without loss of generality, the hinge loss is adopted to compute the misclassification risk of each $z$, i.e., $\ell(h, z=(x,y)) = \sum_{j=1}^{m} \max\{0, 1 - y_j h_j(x)\}$. We consider one of the most popularly used structures of multi-label classifiers, i.e., $h(x) = \mathbf{W}Rep(x)$, where $\mathbf{W} \in R^{m*d'}$ is the weight of a linear layer and $Rep(x) \in R^{d'}$ is a $d'$-dimensional representation vector of $x \in R^d$, e.g., from a non-linear network architecture. In Theorem 1, we assume a linear hypothesis $h$, i.e., $Rep(x) = x$ for the convenience of analysis. The conclusion holds for more advanced architectures, such as feed-forward neural networks.

**Theorem 1.** *Let $h = \mathbf{W}x$ be a linear multi-label classifier. We further denote $\mathcal{D} = (\mathcal{D}_1, \cdots, \mathcal{D}_m)$ and $\mathbf{W} = (\mathbf{w}_1, \cdots, \mathbf{w}_m)$, where $\mathcal{D}_j$ is the data distribution w.r.t. each label $j$ and $\mathbf{w}_j$ is the weight vector of the classifier of label $j$.*

$$\begin{aligned} R_{\mathcal{D}}(h,\varepsilon) \leq\ & R^{emp}_{\mathcal{D}}(h,\varepsilon) + \\ & \left(\tfrac{2}{n} CMI_{\mathcal{D},A} \mathop{\mathbb{E}}_{z=(x,y)\sim\mathcal{D}} \left[\sup_{\mathbf{W}\in\mathcal{W}_{\mathcal{A}}} (l(\mathbf{W},z) + C_{\mathbf{W},z}\varepsilon)^2\right]\right)^{1/2}, \end{aligned} \tag{3}$$

*where $\mathcal{W}_{\mathcal{A}}$ is the set including all possible weight vectors learned by $\mathcal{A}$ using the data set $z^n$ sampled from $\mathcal{D}^n$. $C_{\mathbf{W},z} = \max_{\{b_1,\cdots,b_m\}} \left\|\sum_{j=1}^{m} b_j y_j \mathbf{w}_j\right\|_2, b_j = \{0,1\}$. The empirical adversarial risk $R_{Z^n}(A,\varepsilon)$ has the upper bound:*

$$R^{emp}_{\mathcal{D}}(h,\varepsilon) \leq R^{emp}_{\mathcal{D}}(h,0) + \mathop{\mathbb{E}}_{z^n\sim\mathcal{D}^n,\mathcal{A}} \left[\sup_{\mathbf{W}\in\mathcal{W}_{\mathcal{A}}} \mathop{\mathbb{E}}_{z\in z^n} (C_{\mathbf{W},z}\varepsilon)\right], \tag{4}$$

*where* $R_{\mathcal{D}}^{emp}(h, 0)$ *denotes the empirical and adversarial-free classification risk.*
*We further provide the upper bound of* $CMI_{\mathcal{D},\mathcal{A}}$ *as:*

$$CMI_{\mathcal{D},\mathcal{A}} \leq ent(\mathbf{w}_1, \cdots, \mathbf{w}_m) + ent(\mathcal{D}_1, \cdots, \mathcal{D}_m) \tag{5}$$

*where* $ent(\cdot)$ *denotes the entropy of the concerned random variables.*

**Key Factors of Attackability.** The three key factors determining the adversarial risk (thus the attackability level) of the targeted multi-label classifier are: 1) $CMI_{\mathcal{D},\mathcal{A}}$; 2) $\mathbb{E}_{z} C_{\mathbf{W},z}$ ($\mathbb{E}_{z \leftarrow \mathcal{D}} C_{\mathbf{W},z}$ in Eq. (3) and $\mathbb{E}_{z \in z^n} C_{\mathbf{W},z}$ in Eq. (4)); and 3) the attack budget $\varepsilon$.

The last factor of the attack budget $\varepsilon$ is easy to understand. The targeted classifier is intuitively attackable if the adversary has more attack budget. The larger $\varepsilon$ is, the stronger the attack becomes and the adversarial risk rises accordingly. We then analyze the first factor $CMI_{\mathcal{D},\mathcal{A}}$. For a multi-label classifier $h$ accurately capturing the label correlation in the training data, the output from $h_j$ and $h_k$ are closely aligned w.r.t. the positively or negatively correlated labels $j$ and $k$. Specifically, in the linear case, the alignment between $h_j$ and $h_k$ can be presented by $s(h_j, h_k) = \max\{cos\langle \mathbf{w}_j, \mathbf{w}_k \rangle, cos\langle -\mathbf{w}_j, \mathbf{w}_k \rangle\}$, where $cos\langle *, * \rangle$ denotes the cosine similarity. As shown in Eq. (5), the alignment of the decision hyper-planes of the correlated labels reduce the uncertainty of $\mathbf{W} = \mathcal{A}(\mathcal{D})$. Correspondingly, the conditional mutual information $CMI_{\mathcal{D},\mathcal{A}}$ decreases if the label correlation is strong and the classifier perfectly encodes the correlation into the alignment of the label-wise decision hyper-planes. According to Eq. (3), it is consistent with the well recognized fact of adversary-free multi-label learning: encoding the label correlation in the classifier helps to achieve an accurate adversary-free multi-label classification.

**Lemma 1.** $\mathbb{E}_{z} C_{\mathbf{W},z}$ *reaches the maximum value, if for each pair of labels* $j$ *and* $k$, $\mathbb{E}_{z}\{cos\langle y_j\mathbf{w}_j, y_k\mathbf{w}_k \rangle\} = 1$.

The second factor $\mathbb{E}_{z} C_{\mathbf{W},z}$ *measures the transferability of the attack noise and demonstrates the impact of the transferability level on the attackability of the classifier.* With Lemma 1, we make the following analysis. **First**, for two labels $j$ and $k$ with strong positive or negative correlation in the training data, a large value of $\mathbb{E}_{z}\{cos\langle y_j\mathbf{w}_j, y_k\mathbf{w}_k \rangle\}$ indicates a high intensity of $s(h_j, h_k) = \max\{cos\langle \mathbf{w}_j, \mathbf{w}_k \rangle, cos\langle -\mathbf{w}_j, \mathbf{w}_k \rangle\}$. It represents that the decision hyper-planes $\mathbf{w}_j$ and $\mathbf{w}_k$ of the classifier $h$ are consistently aligned. Therefore, with the same attack strength encoded by $\|r\|_2 \leq \varepsilon$, the adversarial sample $x' = x + r$ tends to cause misclassification on both $h_j(x')$ and $h_k(x')$. Therefore, the attack perturbation's impact is easy to transfer between the correlated labels. Otherwise, $\mathbb{E}_{z}\{cos\langle y_j\mathbf{w}_j, y_k\mathbf{w}_k \rangle\} = 0$ indicates an orthogonal pair of $\mathbf{w}_j$ and $\mathbf{w}_k$. The adversarial perturbation $r$ may cause misclassification on one of the labels, but induce little bias to the decision output of the other. The attack cannot be transferred between the labels. Therefore, a higher/lower $\mathbb{E}_{z} C_{\mathbf{W},z}$ denotes higher/lower transferability of the attack perturbation. **Second**, according to Eq. (3) and Eq. (4), with an increasingly higher $\mathop{\mathbb{E}}_{z}\{cos\langle y_j\mathbf{w}_j, y_k\mathbf{w}_k\rangle\}$, the adversarial risk of the targeted classifier $h$ rises given a fixed attack budget $\varepsilon$. In summary, the alignment between the classifier's decision hyper-planes of different labels captures the label correlation. The alignment facilitates the attack to transfer across the labels. A multi-label classifier is more attackable if the attack is more transferable across the labels, as the attack can impact the decision of more labels at the same time.

*Remark 1.* **Trade-off between the generalization capability of the classifier on clean data and its adversarial robustness.**

Capturing the label correlation in the learnt multi-label classifier can be a double-edged sword. **On one hand**, encouraging alignment between the decision hyper-planes of the correlated labels reduces $CMI_{\mathcal{D},\mathcal{A}}$ under the adversary-free scenario ($\varepsilon = 0$ in Eq. (3)), thus reduces the expected misclassification risk. **On the other hand**, the alignment between the decision hyper-planes increases the transferability of the attack, which makes the classifier more vulnerable. *Controlling the alignment between the decision outputs of different labels can tune the trade-off between the utility and the adversarial robustness of the classifier.*

## 4 Transferrability Regularization for Adversarially Robust Multi-label Classification

Following the above discussion, an intuitive solution to achieve adversarially robust multi-label classification is to regularize $\mathop{\mathbb{E}}_{z\in z^n} C_{\mathbf{W},z}$ empirically, while minimizing the multi-label classification loss over the training data set $z^n$. We denote this training paradigm as **ARM-Primal**:

$$h^* = \arg\min_h \frac{1}{n}\ell(h, z_i) + \frac{\lambda}{n}\sum_{i=1}^{n} C_{\mathbf{W},z_i} \tag{6}$$

where $\lambda$ is the penalty parameter, and $C_{\mathbf{W},z_i}$ is given as in Theorem 1. As discussed in Sect. 3, the magnitude of $C_{\mathbf{W},z_i}$ in Eq. (6) reflects the alignment between the classifier's parameters $\{\mathbf{w}_1, \ldots, \mathbf{w}_m\}$. Penalizing large $C_{\mathbf{W},z_i}$ thus reduces the transferability of the input attack manipulation among different labels, which makes the learnt classifier $h$ more robust against the adversarial perturbation. However, *ARM-Primal* only considers the alignment between the parameters of the linear layer $\mathbf{w}_j$ $(j = 1, \ldots, m)$. This setting limits the flexibility of the regularization scheme from two perspectives. First, whether $h$ is attackable given a bounded attack budget also depends on the magnitude of the classification margin of the input instance [2,19]. Second, the regularization is only enforced over the linear layer's parameters of $h$. However, it is possible that the other layers could be relevant with the transferability of the attack noise. Adjusting the parameters of these layers can also help to control the attackability.

As an echo, we address accordingly the limits of *ARM-Primal*: **First**, a soft attackability estimator (**SAE**) for the targeted multi-label classifier $h$ is proposed to relax the NP-hard attackability assessment in Eq. (1). We show that the proposed *SAE* assesses quantitatively the transferability level of the input attack noise by considering both the alignment of the decision boundaries and the classification margin of the input data instance. The attackability of the classifier is unveiled to be proportional to the transferability of the attack. **Second**, *SAE* is then introduced as a regularization term to achieve a tunable trade-off between transferability control and classification accuracy of the targeted classifier $h$. It thus reaches a customized balance between adversarial attackability and utility of $h$ for multi-label learning practices.

### 4.1 Soft Attackability Estimator (SAE)

We first introduce the concept of *SAE* with the single-label classification setting and then extend it to the multi-label case. Suppose $h$ is a binary classifier and instance $x$ is predicted as positive if $h(x) > 0$ and vice versa. Let the adversarial perturbation be decomposed as $r = c\tilde{r}$, where $c = \|r\|_p$ and $\|\tilde{r}\|_p = 1$, i.e., $\tilde{r}$ shows the direction of the attack noise and $c$ indicates the strength of the attack along this direction. For the perturbed input $x' = x + c\tilde{r}$, the first-order approximation of $h(x')$ is given as:

$$h(x + c\tilde{r}) = h(x) + c\tilde{r}^T \nabla h(x), \;\; s.t. \;\; \|\tilde{r}\|_p = 1, \;\; c \geq 0 \tag{7}$$

where $\nabla h(x)$ denotes the gradient of $h$ to $x$. To deliver the attack successfully, the magnitude of the attack noise follows:

$$c \geq \frac{-h(x)}{\tilde{r}^T \nabla h(x)}. \tag{8}$$

The attackability of $h$ on $x$ along the direction of $\tilde{r}$ is proportional to $\frac{1}{c}$. The smaller $c$ is, the more attackable the classifier $h$ becomes.

Extending the notions to the multi-label setting, we define the multi-label classifier $h$**'s attackability at** $x$ **along the direction of** $\tilde{r}$:

$$A_{h(x),\tilde{r}} = \sum_{j=1}^{m} \max\{\frac{-\tilde{r}^T \nabla h_j(x)}{h_j(x)}, 0\}. \tag{9}$$

Note that in the multi-label setting, the adversarial perturbation $\tilde{r}$ may cause misclassification of $x$ for some labels, while enhancing the correct classification confidence for other labels, i.e., $\frac{-\tilde{r}^T \nabla h_j(x)}{h_j(x)}$ can be negative for the labels with enhanced correct classification confidences. We set the corresponding attackability level to 0, as the attack perturbation fails to cause misclassification.

The intensity of $A_{h(x),\tilde{r}}$ is proportional to the number of the labels whose decision outputs are flipped by the perturbation $\tilde{r}$. Compared to the hard-count based attackability measurement $C_h^*$ in Eq. (1), $A_{h(x),\tilde{r}}$ is a soft score quantifying the impact of the attack perturbation over the outputs of the classifier. It is therefore regarded as a *soft attackability estimator*.

**Transferrability defines attackability.** For simplicity, we denote $\frac{-\nabla h_j(x)}{h_j(x)}$ as $\mathbf{a}_j$, and $A_{h(x),\tilde{r}}$ can be further described as

$$\begin{aligned} A_{h(x),\tilde{r}} &= \tilde{r}^T \sum_{j \in S, S=\{j;\text{sgn}(-y_j \tilde{r}^T \nabla h_j(x))>0\}} \mathbf{a}_j \\ &= \|\tilde{r}\|_2 \sqrt{\sum_{j \in S} \|\mathbf{a}_j\|_2^2 + 2 \sum_{j<k;j,k \in S} \|\mathbf{a}_j\|_2 \|\mathbf{a}_k\|_2 \cos\langle \mathbf{a}_j, \mathbf{a}_k \rangle} \cos\left\langle \tilde{r}, \sum_{j \in S} \mathbf{a}_j \right\rangle \end{aligned} \tag{10}$$

As shown in Eq. (10), the transferability of the attack noise $\tilde{r}$ is measured by the cosine similarity between $\mathbf{a}_j$ and $\mathbf{a}_k$. Each $\mathbf{a}_j$ aligns with the principal eigenvector of the Fisher Information Matrix (FIM) of $h_j$ at the input instance $x$ [25]. It depicts the local geometrical profile of the decision boundaries of different labels near $x$. A larger cosine similarity between $\mathbf{a}_j$ and $\mathbf{a}_k$ indicates a stronger alignment of the decision boundaries of label $j$ and $k$ within the neighborhood of $x$. The attack noise $\tilde{r}$ thus causes closer magnitude of perturbation over $h_j(x)$ and $h_k(x)$ according to Eq. (9). It confirms the association between the transferability and the attackability, as unveiled by Eq. (3) and Eq. (4). Besides, the magnitude of the gradient $\mathbf{a_k} = \nabla h_k(x)$ also shapes the attackability level. A larger norm $\|\nabla h_k(x)\|_2$ indicates a less stable classification output within the $L_p$-ball centered at $x$, i.e., a higher attackability level of the classifier. Integrating both factors, $A_{h(x),\tilde{r}}$ is thus adopted as an empirical attackability estimator of $h$.

It is worth noting that **the proposed *SAE* reflects the transferability of the attack, regardless of the setting of attack budget**. As shown by Eq. (9), *SAE* is evaluated only with the gradient information of the classifier, which is independent of the attack capability of the adversary. In contrast, *GASE* in [21] depends on the prior knowledge about the attack budget of the adversary. In practical applications, the attack budget is usually case-dependent, which limits the use of *GASE* as a generic adversarial robustness evaluation tool. As an attack-strength-independent assessment, *SAE* can help to evaluate the attackability level of a classifier, before it is compromised by any specific attack. It is therefore can be used as a predicative guide for choosing adversarially robust multi-label learning architectures. In the linear case where $h(x) = \mathbf{W}x$, the cosine similarity $cos\langle \mathbf{a}_j, \mathbf{a}_k \rangle$ produces a similar alignment metric as $s(h_j, h_k) = \max\{cos\langle \mathbf{w}_j, \mathbf{w}_k \rangle, cos\langle -\mathbf{w}_j, \mathbf{w}_k \rangle\}$. According to Eq. (3) and (10), the higher the cosine similarity score $cos\langle \mathbf{a}_j, \mathbf{a}_k \rangle$ is, the higher $C_{\mathbf{W},z}$ in Eq. (3) and $A_{h(x),\tilde{r}}$ in Eq. (10) becomes. We thus measure the **attackability of** $h$ **at** $x$ as the maximum $A_{h(x),\tilde{r}}$ as:

$$\phi_{h,x} = \max_{\tilde{r}} A_{h(x),\tilde{r}}, \; s.t. \; \|\tilde{r}\|_p = 1 \tag{11}$$

We inherit the constraint $\|\tilde{r}\|_p = 1$ from Eq. (7). The resultant $\tilde{r}$ denotes the directions of the adversarial noise vector along which the attack can be maximally transferred. With this setting, we separate the derived transferability measurement with the attack strength. With the primal-dual conversion, we can obtain the solution to Eq. (11) as:

$$\phi_{h,x} = \max_{\{b_1,b_2,\cdots,b_m\}} \left\| \sum_{j=1}^{m} \frac{-b_j \nabla h_j(x)}{h_j(x)} \right\|_q , \quad (12)$$
$$s.t. \ \frac{1}{p} + \frac{1}{q} = 1, \ b_j = \{0, 1\},$$

where $p$ denotes the $L_p$ norm of the perturbations. Without loss of generality, we only discuss $p = 2$ of the $l_p$-norm in Eq. (12). As the objective function of Eq. (12) enjoys the submodularity property [1], we employ a simple yet effective greedy-based algorithm to solve Eq. (12). Algorithm 1 describes the greedy-search based solution to compute the *SAE* score.

**Algorithm 1:** The Greedy Solution to Soft Attackability Estimation

1 **Input:** $\left\{\frac{-\nabla h_1(x)}{h_1(x)}, \cdots, \frac{-\nabla h_m(x)}{h_m(x)}\right\}$.
2 **Output:** The set of selected labels $S$.
3 Initialize $S$ as an empty set. Set $LB = 0$ and $CB = 0$, where $LB$ denotes the best result of last iteration and $CB$ denotes the best result of current iteration.
4 **while** $|S| < m$ **do**
5 $\quad LB = CB$;
6 $\quad CB = \max_{\{1,\cdots,m\}-S} \left( \sum_{i \in S} \frac{-\nabla h_i(\mathbf{x})}{h_i(\mathbf{x})} + \frac{-\nabla h_j(\mathbf{x})}{h_j(\mathbf{x})} \right)$;
7 $\quad$ if $CB < LB$, break;
8 $\quad S = S + j$
9 **end**

## 4.2 SAE Regularized Multi-label Learning

We propose to enhance the adversarial robustness of a multi-label classifier by enforcing the control over the *SAE* score of the classifier explicitly during training. While we suppose $x$ is correctly classified during the theoretical analysis of attackability, it doesn't necessarily hold during training. For an originally misclassified data instance $x$, it is possible that $A_{h(x),\tilde{r}}$ can be valued to 0. In this case, the attack perturbation $r$ can augment the confidence of the misclassification. However, with $A_{h(x),\tilde{r}} = 0$, bare penalization can be enforced to suppress the bias. It may encourage further negative impact in the learnt classifier. To mitigate this issue, we slightly modify the definition of $A_{h(x),\tilde{r}}$ and use it as the transferability regularization term of multi-label learning, which gives:

$$\hat{A}_{h(x),\tilde{r}} = \sum_{j=1}^{m} \max\{\frac{-\tilde{r}^T y_j \nabla h_j(x)}{\max(e^{y_j h_j(x)}, \alpha)}, 0\}, \ \alpha > 0 \quad (13)$$

where $\alpha$ is set to prevent over-weighing. For an originally correctly classified instance ($y_j h_j(x) > 0$), $\hat{A}_{h(x),\tilde{r}}$ penalizes the attack transferability as $A_{h(x),\tilde{r}}$. For a misclassified instance ($y_j h_j(x) \leq 0$), minimizing $\hat{A}_{h(x),\tilde{r}}$ helps to reduce the confidence of the misclassification output. Using the exponential function in

$\hat{A}_{h(x),\tilde{r}}$, the misclassified instance with stronger confidence (more biased decision output) is assigned with an exponentially stronger penalty. This setting strengthens the error-correction effect of $\hat{A}_{h(x),\tilde{r}}$.

Similarly as in Eq. (12), we can define $\hat{\phi}_{h,x} = \max_{\tilde{r}} \hat{A}_{h(x),\tilde{r}}$ in Eq. (14). The objective function of the SAE regularized multi-label learning (named hereafter as **ARM-SAE**) gives in Eq. (15):

$$\begin{aligned} \hat{\phi}_{h,x} &= \max_{\{b_1,b_2,\cdots,b_m\}} \left\| \sum_{j=1}^{m} \frac{-b_j y_j \nabla h_j(x)}{\max(e^{y_j h_j(x)},\alpha)} \right\|_q, \\ s.t.\ & \tfrac{1}{p} + \tfrac{1}{q} = 1,\ b_j = \{0,1\}, \end{aligned} \tag{14}$$

$$l = \frac{1}{n}\sum_{i=1}^{n} \ell(h, z_i) + \frac{\lambda}{n}\sum_{i=1}^{n} \hat{\phi}_{h,x_i}, \tag{15}$$

where $\lambda$ is the regularization weight. $\hat{\phi}_{h,x}$ can be calculated using the greedy search solution as $\phi_{h,x}$. If the classifier $h$ takes a linear form, we can find that *ARM-SAE* reweighs the linear layer parameters of the classifier $\{\mathbf{w}_1, \cdots, \mathbf{w}_m\}$ with the weight $\frac{1}{\max(e^{y_j h_j(x)},\alpha)}$. Compared to *ARM-Primal* (see Eq. (12) to $C_{\mathbf{W},z}$ in Theorem 1), *ARM-SAE* enforces more transferability penalty over the instances with smaller classification margins. As unveiled in [21], these instances are easier to be perturbed for the attack purpose. Instead of penalizing each instance with the same weight as in *ARM-Primal*, *ARM-SAE* can thus perform a more flexible instance-adapted regularization.

# 5 Experiments

## 5.1 Experimental Setup

**Datasets.** We empirically evaluate our theoretical study on three data sets collected from real-world multi-label cyber-security applications (*Creepware*), object recognition (*VOC2012*) [3] and environment science (*Planet*) [10]. The descriptions of the datasets are given can be found in the supplementary file due to the space limit. The data sets are summarized in Table 1.

**Performance Benchmark.** Given a fixed attack strength of $\varepsilon$, we compute *the number of flipped labels* $C_h^*(z)$ *on each testing instance according to Eq. (1)* and take the average of the derived $\{C_h^*(z)\}$ (noted as $C_a$) as an overall estimation of attackability on the testing data set. Due to the NP-hard intrinsic of the combinatorial optimization problem in Eq. (1), we use *GASE* [21] to estimate empirically $C_h^*(z)$ and $C_a$. Besides, we measure the multi-label classification performance on the clean and adversarially modified testing instances with *Micro-F1* and *Macro-F1* scores.

**Table 1.** Summary of the used real-world data sets. $N$ is the number of instances. $m$ is the total number of labels. $l_{avg}$ is the average number of labels per instance. The F1-scores of the targeted classifiers on different data sets are also reported.

| Data set | $N$ | m | $l_{avg}$ | Micro F1 | Macro F1 | Classifier$_{target}$ |
|---|---|---|---|---|---|---|
| Creepware | 966 | 16 | 2.07 | 0.76 | 0.66 | SVM |
| VOC2012 | 17,125 | 20 | 1.39 | 0.83 | 0.74 | Inception-V3 |
| Planet | 40,479 | 17 | 2.87 | 0.82 | 0.36 | Inception-V3 |

**Table 2.** Attackability estimation by *SAE*. $\lambda_{nuclear}$ denotes the strength of nuclear-norm based regularization. *CC* and *P* denote the Spearman coefficient and the p-value between *GASE* and *SAE* scores on the testing instances.

| Data set | | $\longrightarrow$ *robustness increase* | | | | | $CC, P$ (*Spearman*) |
|---|---|---|---|---|---|---|---|
| *Creepware* | $\lambda_{nuclear}$ | 0 | 0.00001 | 0.0001 | 0.001 | 0.01 | $CC = 1$ $P = 0$ |
| | GASE ($C_a, \varepsilon = 0.5$) | 13.5 | 11.4 | 10.8 | 6.9 | 4.3 | |
| | SAE | 31.5 | 19.16 | 18.06 | 14.55 | 11.22 | |
| *VOC2012* | $\lambda_{nuclear}$ | 0 | 0.0001 | 0.001 | 0.01 | 0.1 | $CC = 1$ $P = 0$ |
| | GASE ($C_a, \varepsilon = 10$) | 10.8 | 10.1 | 9.3 | 8.5 | 4.9 | |
| | SAE | 157.6 | 127.3 | 77.6 | 69.1 | 61.0 | |
| *Planet* | $\lambda_{nuclear}$ | 0 | 0.0001 | 0.001 | 0.01 | 0.1 | $CC = 1$ $P = 0$ |
| | GASE ($C_a, \varepsilon = 2$) | 13.1 | 12.2 | 11.6 | 10.5 | 7.1 | |
| | SAE | 267.1 | 221.5 | 186.3 | 158.2 | 102.0 | |
| | | $\longrightarrow$ *attackability decrease* | | | | | |

**Targeted Classifiers.** We instantiate the study empirically with linear Support Vector Machine (SVM) and Deep Neural Nets (DNN) based multi-label classifiers. Linear SVM is applied on *Creepware*. DNN model Inception-V3 [15] is used on *VOC2012* and *Planet*. On each data set, we randomly choose 50%, 30% and 20% data instances for training, validation and testing to build the targeted multi-label classifier. In Table 1, we show *Micro-F1* and *Macro-F1* scores measured on the clean testing data to evaluate the classification performance of multi-label classifiers [24]. Note that accurate adversary-free multi-label classification is beyond our scope and these classifiers are used to verify the theoretical analysis and the proposed *ARM-SAE* method.

**Input Normalization and Reproduction.** We normalize the adversarially perturbed data during the attack process. Due to the space limit, we provide the parameter settings and the reproduction details in the supplementary file.

### 5.2 Effectivity of Soft Attackability Estimator (SAE)

In Table 2, we demonstrate the validity of the proposed *SAE* by checking the consistency between the *SAE* and the *GASE*-based attackability measurement

[21]. We adopt the **nuclear-norm** regularized training [21] to obtain an adversarially robust multi-label classifier. On the same training set, we increase the nuclear-norm regularization strength gradually to derive more robust architectures against the evasion attack. For each regularization strength, we can compute the *SAE* score of the classifier on the unperturbed testing instances. Similarly, by freezing the attack budget $\varepsilon$ on each data set, we can generate the *GASE* score ($C_a$) corresponding to each regularization strength. Note that *only the ranking orders of the SAE and GASE score matters in the attackability measurement by definition.* We use the ranking relation of the scores to select adversarially robust models. Therefore, we adopt the Spearman rank correlation coefficient to measure the consistency between *SAE* and *GASE*.

We use the *GASE* score as a baseline of attackability assessment. The *SAE* and *GASE* score are strongly and positively correlated over all the datasets according to the correlation metric. Furthermore, with a stronger robustness regularization, the *SAE* score decreases accordingly. It confirms that the intensity of the proposed *SAE* score capture the attackability level of the targeted classifier. This observation further validates empirically the motivation of using *SAE* in adjusting the adversarial robustness of the classifier.

The experimental study also shows the attack-strength-independent merit of the *SAE* over *GASE*. *SAE* is computed without knowing the setting of the attack budget. It thus reflects the intrinsic property of the classifier determining its adversarial vulnerability. In practice, this attack-strength-independent assessment can help to evaluate the attackability level of the deployed classifier, before it is compromised by any specific attack.

### 5.3 Effectiveness Evaluation of ARM-SAE

We compare the proposed *ARM-SAE* method to the state-of-the-art techniques in improving adversarial robustness of multi-label learning: the $L_2$-norm and the nuclear-norm regularized multi-label training [21]. Besides, we conduct an ablation study to verify the effectiveness of *ARM-SAE*.

- **$L_2$ Norm and Nuclear Norm Regularized Training.** Enforcing the $L_2$ and nuclear norm constraint helps to reduce the model complexity and thus enhance the model's adversarial robustness [16,21].
- **ARM-Single.** This variant of *ARM-SAE* is built by enforcing the transferability regularization with respect to individual labels separately:

$$\phi_{H\_\text{single}} = \sum_{i=1}^{n} \sum\nolimits_{k=1}^{m} \left\| \frac{\nabla h_k(x_i)}{\max(e^{y_k^i h_k(x)}, \alpha)} \right\|_2 . \tag{16}$$

  We compare *ARM-SAE* with *ARM-Single* to show the merit of jointly measuring and regularizing the impact of the input attack noise over all the labels. *ARM-SAE* tunes the transferability of the attack jointly, while *ARM-Single* enforces the penalization with respect to each label individually.
- **ARM-Primal.** We compare this variant to *ARM-SAE* to demonstrate the merit of *ARM-SAE* by 1) introducing the flexibility of penalizing the whole

**Table 3.** Effectiveness evaluation of ARM-SAE. For convenience, *non*, $L_2$, *nl*, *sg*, *pm* and *SAE* are used to denote the absence of regularization, $L_2$ *norm*, *nuclear-norm*, *ARM-single*, *ARM-Primal* and *ARM-SAE* based methods respectively. The best results are in bold.

| Budget | *Creepware* : Micro F1 = 0.76, Macro F1 = 0.66 (on clean data) | | | | | | | | | | | |
|---|---|---|---|---|---|---|---|---|---|---|---|---|
| | $\varepsilon = 0.05$ | | | | | | $\varepsilon = 0.2$ | | | | | |
| Regularizers | *non* | $L_2$ | *nl* | *sg* | *pm* | *SAE* | *non* | $l_2$ | *nl* | *sg* | *pm* | *SAE* |
| Micro F1 | 0.34 | 0.40 | 0.45 | 0.44 | 0.43 | **0.53** | 0.10 | 0.13 | 0.15 | 0.15 | 0.16 | **0.22** |
| Macro F1 | 0.33 | 0.39 | **0.43** | 0.39 | **0.43** | 0.42 | 0.12 | 0.15 | 0.20 | 0.17 | 0.20 | **0.25** |
| Budget | *VOC*2012 : Micro F1 = 0.83, Macro F1 = 0.74 (on clean data) | | | | | | | | | | | |
| | $\varepsilon = 0.1$ | | | | | | $\varepsilon = 1$ | | | | | |
| Regularizers | *non* | $L_2$ | *nl* | *sg* | *pm* | *SAE* | *non* | $l_2$ | *nl* | *sg* | *pm* | *SAE* |
| Micro F1 | 0.49 | 0.53 | 0.56 | 0.54 | 0.57 | **0.61** | 0.20 | 0.22 | 0.27 | 0.26 | 0.26 | **0.30** |
| Macro F1 | 0.29 | 0.31 | 0.33 | 0.31 | 0.36 | **0.38** | 0.12 | 0.16 | 0.22 | 0.17 | 0.20 | **0.23** |
| Budget | *Planet* : Micro F1 = 0.82, Macro F1 = 0.36 (on clean data) | | | | | | | | | | | |
| | $\varepsilon = 0.1$ | | | | | | $\varepsilon = 1$ | | | | | |
| Regularizers | *non* | $L_2$ | *nl* | *sg* | *pm* | *SAE* | *non* | $l_2$ | *nl* | *sg* | *pm* | *SAE* |
| Micro F1 | 0.41 | 0.49 | 0.45 | 0.48 | 0.49 | **0.53** | 0.06 | 0.09 | 0.08 | 0.10 | 0.09 | **0.13** |
| Macro F1 | 0.13 | 0.22 | 0.17 | 0.20 | 0.18 | **0.24** | 0.03 | 0.04 | 0.04 | 0.06 | 0.06 | **0.08** |

**Table 4.** Trade-off between generalization performance on clean data and adversarial robustness on *Creepware*. The attack budget $\varepsilon = 0.05$.

| $\lambda$ | 0 | $10^{-7}$ | $10^{-6}$ | $10^{-5}$ | $10^{-4}$ |
|---|---|---|---|---|---|
| $\phi_{align}$ | 0.23 | 0.22 | 0.20 | 0.15 | 0.12 |
| Micro F1 (clean) | 0.76 | 0.76 | 0.75 | 0.72 | 0.70 |
| Macro F1 (clean) | 0.66 | 0.63 | 0.56 | 0.50 | 0.46 |
| MIcro F1 (pert) | 0.34 | 0.35 | 0.39 | 0.44 | 0.53 |
| Macro F1 (pert) | 0.33 | 0.33 | 0.35 | 0.40 | 0.42 |

model architecture, instead of only the linear layer; 2) taking the impact of classification margin on adversarial risk [19] into the consideration.

Two different attack budgets $\varepsilon$ on each data set are introduced denoting varied attack strength. With each fixed $\varepsilon$, we compute the *Micro-F1* and *Macro-F1* scores of the targeted classifiers after retraining with the techniques above. Table 3 lists the classification accuracy over the adversarial testing instances using different robust training methods. In Table 3, we also show the multi-label classification accuracy (measured by two F1 scores) on the clean testing instances as a baseline. Consistently observed on the three datasets, even a small attack budget can deteriorate the classification accuracy drastically, which shows the vulnerability of multi-label classifiers. Generally, all the regularization method can improve the classification accuracy on the adversarial input. Among all the methods, *ARM-SAE* achieves the highest accuracy on the adversarial samples. It confirms the merit of *SAE* in controlling explicitly the transferability and then suppressing the attackability effectively. In addition, by regularizing jointly the attack transfer and exploiting classification margin for the attackability measurement, *ARM-SAE* achieves superior robustness over the two variants.

### 5.4 Validation of Trade-Off Between Generalization Performance on Clean Data and Adversarial Robustness

We validate the trade-off described in Remark 1. Without loss of generality, we conduct the case study on *Creepware*. Tuning the alignment between decision boundaries of different labels is achieved by conducting the *ARM-SAE* training as described in Eq. 15. We freeze $\varepsilon$ as 0.05 and vary the regularization weight $\lambda$ in Eq. (15) from $10^{-7}$ to $10^{-4}$ to show increasingly stronger regularization effects enforced on the alignment between decision boundaries of different labels. For each regularization strength, we train a multi-label classifier $h$ and evaluate quantitatively the averaged alignment level $\phi_{align} = \frac{1}{m^2} \sum\limits_{j,k \in \{1, \cdots, m\}} |\cos \langle \mathbf{w}_j, \mathbf{w}_k \rangle |$ between the decision hyperplanes of different labels. Table 4 shows the variation of $\phi_{align}$ and the Micro-/Macro-F1 accuracy of the trained multi-label classifier $h$ over the clean and adversarially perturbed data instances (Micros/Macro F1 (clean/pert)). With increasingly stronger robustness regularization, the averaged alignment level $\phi_{align}$ between the label-wise decision hyper-planes decreases accordingly. Simultaneously, we witness the rise of the classification accuracy of $h$ on the adversarially perturbed testing instances. It indicates the classifier $h$ is more robust to the attack perturbation. However, the Macro- and Micro-F1 scores of $h$ on the clean testing data drop with stronger alignment regularization. This observation is consistent with the discussion in Remark 1.

## 6 Conclusion

In this paper, we establish an information-theoretical adversarial risk bound of multi-label classification models. Our study identifies that the transferability of evasion attack across different labels determines the adversarial vulnerability of the classifier. Though capturing the label correlation improves the accuracy of adversary-free multi-label classification, our work unveils that it can also encourage transferable attack, which increases the adversarial risk. We show that the trade-off between the utility of the classifier and its adversarial robustness can be achieved by explicitly regularizing the transferability level of evasion attack in the learning process of multi-label classification models. Our empirical study demonstrates the applicability of the proposed transferability-regularized robust multi-label learning paradigm for both linear and non-linear classifies.

## References

1. Elenberg, E.R., Khanna, R., Dimakis, A.G., Negahban, S.: Restricted strong convexity implies weak submodularity. Ann. Statist. **46**, 3539–3568 (2016)
2. Elsayed, G.F., Krishnan, D., Mobahi, H., Regan, K.: Large margin deep networks for classification. In: NeuIPS (2018)
3. Everingham, M., Gool, L.V., Williams, C., Winn, J., Zisserman, A.: The pascal visual object classes challenge 2012 (voc2012) results (2012). http://www.pascal-network.org/challenges/VOC/voc2012/workshop/index.html
4. Fawzi, A., Moosavi-Dezfooli, S., Frossard, P.: Robustness of classifiers: from adversarial to random noise. In: NIPS, pp. 1632–1640 (2016)
5. Fawzi, A., Fawzi, O., Frossard, P.: Analysis of classifiers' robustness to adversarial perturbations. Mach. Learn. **107**, 481–508 (2018)
6. Freed, D., Palmer, J., Minchala, D., Levy, K., Ristenpart, T., Dell, N.: "A stalker's paradise": how intimate partner abusers exploit technology. In: The 2018 CHI Conference, pp. 1–13 (2018)
7. Gilmer, J., et al.: Adversarial spheres. CoRR (2018). http://arxiv.org/abs/1801.02774
8. Gupta, A., Lamba, H., Kumaraguru, P., Joshi, A.: Faking sandy: characterizing and identifying fake images on Twitter during hurricane sandy. In: WWW, pp. 729–736 (2013)
9. Hein, M., Andriushchenko, M.: Formal guarantees on the robustness of a classifier against adversarial manipulation. In: NeuIPS, pp. 2266–2276 (2017)
10. Kaggle: Planet: understanding the Amazon from space (2017). https://www.kaggle.com/c/planet-understanding-the-amazon-from-space/overview
11. Khim, J., Loh, P.L.: Adversarial risk bounds for binary classification via function transformation. arXiv (2018)
12. Roundy, K.A., et al.: The many kinds of creepware used for interpersonal attacks. In: IEEE Symposium on Security and Privacy (SP), pp. 626–643 (May 2020)
13. Song, Q., Jin, H., Huang, X., Hu, X.: Multi-label adversarial perturbations. In: ICDM, pp. 1242–1247 (2018)
14. Steinke, T., Zakynthinou, L.: Reasoning about generalization via conditional mutual information. In: COLT (2020)
15. Szegedy, C., Vanhoucke, V., Ioffe, S., Shlens, J., Wojna, Z.: Rethinking the inception architecture for computer vision. arXiv (2015)
16. Tu, Z., Zhang, J., Tao, D.: Theoretical analysis of adversarial learning: a minimax approach. In: NeuIPS, pp. 12259–12269 (2019)
17. Wang, Y., Jha, S., Chaudhuri, K.: Analyzing the robustness of nearest neighbors to adversarial examples. In: ICML, pp. 5133–5142 (2018)
18. Wang, Y., et al.: Attackability characterization of adversarial evasion attack on discrete data. In: KDD, pp. 1415–1425 (2020)
19. Yang, Y., et al.: Boundary thickness and robustness in learning models. In: NeuIPS (2020)
20. Yang, Y., Rashtchian, C., Zhang, H.: A closer look at accuracy vs. robustness. In: NeuIPS (2020)
21. Yang, Z., Han, Y., Zhang, X.: Characterizing the evasion attackability of multi-label classifiers. In: AAAI (2021)
22. Yin, D., Ramchandran, K., Bartlett, P.L.: Rademacher complexity for adversarially robust generalization. In: ICML, pp. 7085–7094 (2019)

23. Zhang, H., Yu, Y., Jiao, J., Xing, E.P., Ghaoui, L.E., Jordan, M.I.: Theoretically principled trade-off between robustness and accuracy. In: ICML (2019)
24. Zhang, M., Zhou, Z.: A review on multi-label learning algorithms. TKDE **26**(8), 1819–1837 (2013)
25. Zhao, C., Fletcher, P., Yu, M., Peng, Y., Zhang, G., Shen, C.: The adversarial attack and detection under the fisher information metric. In: AAAI, pp. 5869–5876 (2019)

# Differentiable Feature Selection, A Reparameterization Approach

Jérémie Donà[1(✉)] and Patrick Gallinari[1,2]

[1] Sorbonne Université, CNRS, LIP6, 75005 Paris, France
{jeremie.dona,patrick.gallinari}@lip6.fr
[2] Criteo AI Labs, Paris, France

**Abstract.** We consider the task of feature selection for reconstruction which consists in choosing a small subset of features from which whole data instances can be reconstructed. This is of particular importance in several contexts involving for example costly physical measurements, sensor placement or information compression. To break the intrinsic combinatorial nature of this problem, we formulate the task as optimizing a binary mask distribution enabling an accurate reconstruction. We then face two main challenges. One concerns differentiability issues due to the binary distribution. The second one corresponds to the elimination of redundant information by selecting variables in a correlated fashion which requires modeling the covariance of the binary distribution. We address both issues by introducing a relaxation of the problem via a novel reparameterization of the logitNormal distribution. We demonstrate that the proposed method provides an effective exploration scheme and leads to efficient feature selection for reconstruction through evaluation on several high dimensional image benchmarks. We show that the method leverages the intrinsic geometry of the data, facilitating reconstruction. (We refer to https://arxiv.org/abs/2107.10030 for a complete version of the article including proofs, thorough experimental details and results.)

**Keywords:** Representation learning · Sparse methods

## 1 Introduction

Learning sparse representations of data finds essential real-world applications as in budget learning where the problem is limited by the number of features available or in embedded systems where the hardware imposes computational limitations. Feature selection serves similar objectives giving insights about variable dependencies and reducing over-fitting [5]. Combined with a reconstruction objective, feature selection is a sensible problem when collecting data is expensive which is often the case with physical processes. For example, consider optimal sensor placement. This task consists in optimizing the location of sensors measuring a scalar field over an area of interest (e.g. pressure, temperature) to enable truthful reconstruction of the signal on the whole area. It finds applications in climate science [10,11], where key locations are monitored to evaluate


N. Oliver et al. (Eds.): ECML PKDD 2021, LNAI 12977, pp. 414–429, 2021.
https://doi.org/10.1007/978-3-030-86523-8_25

the impact of climate change on snow melt and Monsoon. These examples illustrate how feature selection for reconstruction may be critically enabling for large scale problems where measurements are costly.

Common practices for feature selection involves a $\ell_1$-regularization over the parameters of a linear model to promote sparsity [15]. Initiated by [15], several refinements have been developed for feature selection. For example, [14] employs a $\ell_{2,1}$-norm in a linear auto-encoder. [6] impose a $\ell_1$-penalty on the first layer of a deep auto-encoder to select features from the original signal. Finally, Group-Lasso methods extended lasso by applying the sparse $\ell_1$-penalty over precomputed chunks of variables to take prior knowledge into account while selecting features. Theses approaches suffer from two main limitations: the design of the groups for Group-Lasso methods and the loss of the intrinsic structure of the data as both [6,14] treat the input signal as a vector. Moreover, non-linear $\ell_1$ based methods for feature selection and reconstruction are intrinsically ill posed. Like Group-Lasso methods, our proposition aims at selecting variables in a correlated fashion, to eliminate redundant information, while leveraging the structure of the data. We illustrate its efficiency on images but it can be adapted to exploit patterns in other types of structured data as graphs.

We propose a novel sparse embedding method that can tackle feature selection through an end-to-end-approach. To do so, we investigate the learning of binary masks sampled from a distribution over binary matrices of the size of the image, with 1 indicating a selected pixel. We alleviate differentiability issues of learning categorical variables by relying on a continuous relaxation of the problem. The learned latent binary distribution is optimized via a stochastic exploration scheme. We consider the dependency between the selected pixels and we propose to sample the pixels in the mask in a correlated fashion to perform feature selection efficiently. Accordingly, we learn a correlated logitNormal distribution via the reparameterization trick allowing for an efficient exploration of the masks space while preserving structural information for reconstruction. Finally, sparsity in the embedding is enforced via a relaxation of the $\ell_0$-norm. To summarize, we aim at learning a binary mask for selecting pixels from a distribution of input signals $x$, with $x \in \mathbb{R}^{n \times n}$ for images, enabling an accurate reconstruction. We formulate our problem as learning jointly a parametric sampling operator $S$ which takes as input a random variable $z \in \mathcal{Z} \subseteq \mathbb{R}^d$ and outputs binary masks, i.e. $S : \mathcal{Z} \rightarrow \{0,1\}^{n \times n}$. We introduce two ways to learn the sampling operator $S$. For reconstruction, an additional operator denoted $G$ learns to reconstruct the data $x$ from the sparse measurements $s \odot x$. Our proposed approach is fully differentiable and can be optimized directly via back-propagation. Our main contributions are:

- We introduce a correlated logitNormal law to learn sparse binary masks, optimized thanks to the reparameterization trick. This reparameterization is motivated statistically. Sparsity is enforced via a relaxed $\ell_0$-norm.
- We formulate the feature selection task for 2-D data as the joint learning of a binary mask and a reconstruction operator and propose a novel approach to learn the parameters of the considered logitNormal law.

- We evidence the efficiency of our approach on several datasets: Mnist, CelebA and a complex geophysical dataset.

## 2 Related Work

Our objective of learning binary mask lies in between a few major domains: density modeling, feature selection and compressed sensing.

**Density Modeling via Reparameterization:** Sampling being not differentiable, different solutions have been developed in order to estimate the gradients of the parameters of a sampling operator. Gradient estimates through score functions [3, 7] usually suffer from high variance or bias. Reparameterization [22] provides an elegant way to solve the problem. It consists in sampling from a fixed distribution serving as input to a parametric transformation in order to obtain both the desired distribution and the gradient with respect to the parameters of interest. However, the learning of categorical variables remains tricky as optimizing on a discrete set lacks differentiability. Continuous relaxation of discrete variables enables parameters optimization through the reparameterization trick. Exploiting this idea, [34, 41] developed the concrete law as a reparameterization of the Gumbel max variable for sampling categorical variables [29]. Alternative distributions, defining relaxations of categorical variables can be learned by reparameterization such as the Dirichlet or logitNormal distribution [24, 26]. Nonetheless, most previous approaches learn factorized distribution, thus selecting variables independently when applied to a feature selection task. In contrast, we rely on the logitNormal distribution to propose a reparameterization scheme enabling us to sample the mask pixels jointly, taking into account dependencies between them and exploiting the patterns present in 2-D data.

**Feature Selection:** Wrapper methods, [5, 8, 17] select features for a downstream task whereas filter methods [18, 19, 25] rank the features according to tailored statistics. Our work belongs to the category of *embedded* methods, that address selection as part of the modeling process. $\ell_1$-penalization over parameters, as for instance in Lasso and in Group Lasso variants [9, 23, 31], is a prototypical embedded method. $\ell_1$-penalty was used for feature selection for example in [14, 33] learning a linear encoding with a $\ell_{2,1}$-constraint for a reconstruction objective. Auto-encoders [20] robustness to noise and sparsity is also exploited for feature selection [13, 16, 35]. For example, AEFS [6] extends Lasso with non linear auto-encoders, generalizing [33]. Another line of work learns embeddings preserving local properties of the data and then find the best variables in the original space to explain the learned embedding, using either $\ell_1$ or $\ell_{2,1}$ constraints [4, 12]. Closer to our work, [1] learn a matrix of weights $m$, where each row follow a concrete distribution [34]. That way each row of matrix $m$ samples one feature in x. The obtained linear projection $m.x$ is decoded by a neural network, and $m$ is trained to minimize the $\ell_2$-loss between reconstructions and targets. Because $x$ is treated as a vector, here too, the structure of the data is ignored and lost in the encoding process. Compared to these works, we leverage the dependencies between variables in the 2-D pixel distribution, by sampling binary masks via an adaptation of the logitNormal distribution.

**Compressed Sensing:** Our work is also related to compressed sensing (CS) where the objective is to reconstruct a signal from limited (linear) measurements [38]. Deep learning based compressed sensing algorithms have been developed recently: [37] use a pre-trained generative model and optimize the latent code to match generated measurements to the true ones; The measurement process can be optimized along with the reconstruction network as in [39]. Finally, [32] use a CS inspired method based on the pivots of a QR decomposition over the principal components matrix to optimize the placement of sensors for reconstruction, but scales poorly for large datasets. Our approach differs from CS. Indeed, for CS, measurements are linear combinations of the signal sources, whereas we consider pixels from the original image. Thus, when CS aims at reconstructing from linear measurements, our goal is to preserve the data structural information to select a minimum number of variables for reconstruction.

## 3 Method

We now detail our framework to learn correlated sparse binary masks for feature selection and reconstruction through an end-to-end approach. The choice of the logitNormal distribution, instead of the concrete distribution [34], is motivated by the simplicity to obtain correlated variables thanks to the stability of independent Gaussian law by addition as detailed below. We experimentally show in Sect. 4 that taking into account such correlations helps the feature selection task. This section is organised as follows: we first introduce in Sect. 3.1 some properties of the logitNormal distribution and sampling method for this distribution. We detail in Sect. 3.2 our parameterization for the learning of the masks distribution. Finally, in Sect. 3.3 we show how to enforce sparsity in our learned distribution before detailing our reconstruction objective in Sect. 3.4.

### 3.1 Preliminaries: logitNormal Law on $[0, 1]$

Our goal is to sample a categorical variable in a differentiable way. We propose to parameterize the sampling on the simplex by the logitNormal law, introduced in [2]. We detail this reparameterization scheme for the unidimensional case since we aim at learning binary encodings. It can be generalized to learn k-dimensional one-hot vector. Let $z \sim \mathcal{N}(\mu, \sigma)$, and $Y$ defined as:

$$Y = \text{sigmoid}(z) \tag{1}$$

Then $Y$ is said to follow a logitNormal law. This distribution defines a probability over $[0, 1]$, admits a density and its cumulative distribution function has an analytical expression used to enforce sparsity in Sect. 3.3.

This distribution can take various forms as shown in Fig. 1 and be flat as well as bi-modal. By introducing a temperature in the sigmoid so that we have, $\text{sigmoid}_\lambda(z) = \frac{1}{1+\exp^{-z/\lambda}}$, we can polarize the logitNormal distribution. In Proposition 1 we evidence the link between the 0-temperature logitNormal distribution and Bernoulli distribution:

**Proposition 1 (Limit Distribution).** *Let $W \in \mathbb{R}^n$ be a vector and $b \in \mathbb{R}$ a scalar. Let $Y = \text{sigmoid}_\lambda(W.z^T + b)$, where $z \sim \mathcal{N}(0, I_n)$, when $\lambda$ decrease towards 0, $Y$ converges in law towards a Bernoulli distribution and we have:*

$$\lim_{\lambda \to 0} \mathbb{P}(Y = 1) = 1 - \Phi(\frac{-b}{\sqrt{\sum_i w_i^2}}) \tag{2}$$

$$\lim_{\lambda \to 0} \mathbb{P}(Y = 0) = \Phi(\frac{-b}{\sqrt{\sum_i w_i^2}}) \tag{3}$$

*Where $\Phi$ is the cumulative distribution function of the Normal law $\mathcal{N}(0, 1)$,*

Proposition 1 characterizes the limit distribution as the temperature goes down to 0, and $Y$ defines a differentiable relaxation of a Bernoulli variable. This proposition is used to remove randomness in the learned distribution, see Sect. 4.

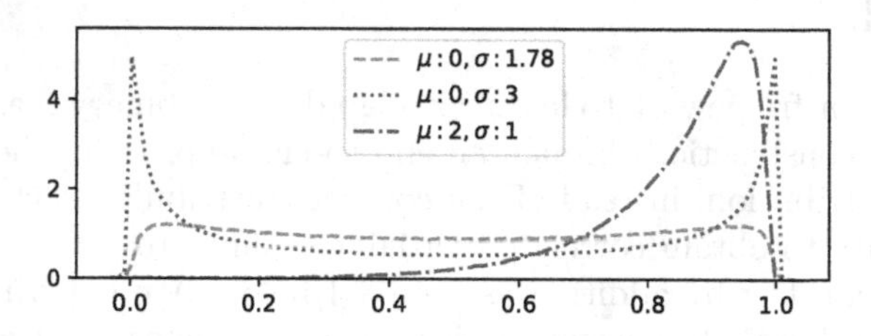


**Fig. 1.** Density of the logitNormal law for various couple $(\mu, \sigma)$: $(\mu = 0, \sigma = 1.78)$ (dashed line), $(\mu = 0, \sigma = 3)$ (dotted line) $(\mu = 2, \sigma = 1)$ (dotted and dashed).

We relax the objective of learning of a binary mask in $\{0, 1\}$ by learning in $[0, 1]$ using the logitNormal law. Let $m \in \mathbb{N}$, be the dimension of the desired logitNormal variable $Y$. A simple solution for learning the logitNormal distribution of the masks is via independent sampling.

**Independent Sampling:** A common assumption is that the logitNormal samples originate from a factorized Normal distribution [24]. Thus, the learned parameters of the distribution are: the average $\mu \in \mathbb{R}^m$ and the diagonal coefficients of the covariance matrix $\sigma \in \mathbb{R}^m$, according to:

$$Y = \text{sigmoid}_\lambda(\mu + z \odot \sigma) \tag{4}$$

where $\odot$ is the element-wise product and $Y \in \mathbb{R}^m$. Note that, for feature selection on images, one aims at learning a binary mask and thus the latent space has the same dimension as the images, i.e. $m = n \times n$, then $z \in \mathbb{R}^{n \times n}$.

This sampling method has two main drawbacks. First, the coordinates of $z$ are independent and so are the coordinates of $Y$, therefore such sampling scheme does not take correlations into account. Also, the dimension of the sampling space $\mathcal{Z}$ is the same as $Y$ which might be prohibitive for large images.

We address both limitations in the following section, by considering the relations between the pixel values. In that perspective, Group-Lasso selects variables among previously designed group of variables [31], reflecting different aspects of the data. Similarly, we want to select variables evidencing different facets of the signal to be observed. Indeed, finding the best subset of variables for the reconstruction implies to eliminate the redundancy in the signal and to explore the space of possible masks. We propose to do so by selecting the variables in a correlated fashion, avoiding the selection of redundant information.

**Correlated Sampling:** To palliate the limitations of independent sampling, we model the covariance between latent variables by learning linear combinations between the variables in the prior space $\mathcal{Z}$. Besides, considering dependencies between latent variables, this mechanism reduces the dimension of the sampling space $\mathcal{Z}$, allowing for a better exploration of the latent space. In order to generate correlated variables from a lower dimensional space, we investigate the following transformation: let $z \sim \mathcal{N}_d(0, I_d) \in \mathcal{Z} = \mathbb{R}^d$ with $d << m$, $W \in \mathcal{M}_{m,d}(\mathbb{R})$ a weight matrix of size $m \times d$ and $b \in \mathbb{R}^m$ a real vector, then

$$Y = \text{sigmoid}_\lambda(Wz + b) \tag{5}$$

represents $m$-one dimension logitNormal laws due to the stability of independent Gaussian laws by addition. However, the Normal law induced by $Wz + b$ has now a full covariance matrix and not only diagonal coefficient as in Eq. (4). This reparameterization provides a simple way to sample correlated (quasi)-binary variables, even for high dimension latent space, i.e. with $m$ large. Compared to [1], our proposition offers a significant advantage for feature selection in images. Indeed, let $G$ be the neural network aiming to reconstruct data $x$ from the selected variable. With our proposition $G$ can access a sparse version of the original signal $Y \odot x$ and can thus leverage both the pixel values and their position in the image for reconstruction. In [1] only the selected feature values without structural information are available for the reconstruction.

### 3.2 Parameterizing logitNormal Variables for Feature Selection

Now we have established how to compute correlated logitNormal variables following Eq. (5), we detail our parameterization for learning. Let $S : \mathcal{Z} \to [0,1]^{n \times n}$ be our sampling operator that generates a binary mask from a random sample $z$. We consider two approaches to parameterize $S$ so that it follows a logitNormal law. Our first proposition denoted vanilla parameterization directly optimizes $W$ and $b$ from Eq. (5), while our second approach proposes to explore and optimize the spaces of linear combinations $W$ and biases $b$.

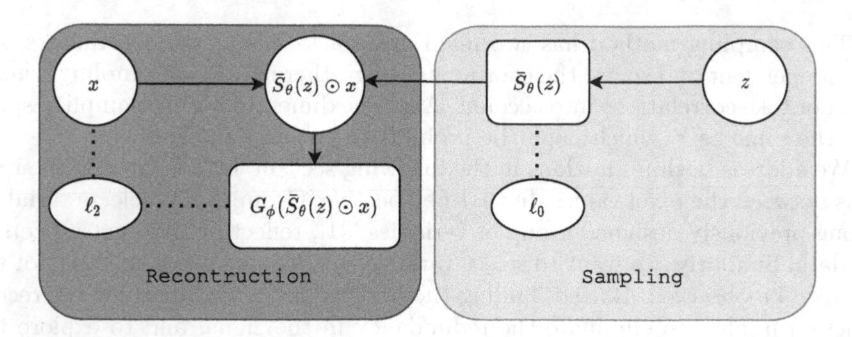


**Fig. 2.** Algorithmic flow of our framework for feature selection for reconstruction. $S_\theta(z)$ has a correlated logitNormal distribution. We sample $z \sim \mathcal{N}(0, 1)$. $\bar{S}_\theta(z)$ defines the binary masks and $G_\phi$ estimate $x$ from $x^{obs} = \bar{S}_\theta(z) \odot x$.

**Vanilla Parameterization:** A simple approach is to parameterize $S$ as $S_\theta$ according to Eq. (5). Then, the optimized parameters are: $\theta = (W, b)$ with $W \in \mathcal{M}_{n\times n,d}(\mathbb{R})$ and $b \in \mathbb{R}^{n\times n}$. This sampling process can be summarized by Eq. (6):

$$\begin{cases} \text{Initialize } W \in \mathcal{M}_{n\times n,d}(\mathbb{R}), b \in \mathbb{R}^{n\times n} \\ z \sim \mathcal{N}(0, I_d) & \text{(6a)} \\ S_\theta(z) = \text{sigmoid}(W.z + b) & \text{(6b)} \end{cases}$$

In that case each variable in $S_\theta(z)$ follows a logitNormal law. The selected variables are indicated for $S_\theta = 1$. The optimization process allows two degrees of freedom ($b$ and $W$) for the control of the variance, of the covariance and of the average of the variables of the masks. Note that, this parameterization corresponds to a linear layer followed by a sigmoid activation so that besides tractability for the distribution of $Y$, it presents the advantage of a simple implementation. Unlike [1], our proposition preserves the structure of the data.

**HyperNetworks Parameterization:** Aiming to learn a matrix $W$ and a bias vector $b$ that fully characterizes our logitNormal law as Eq. (5), we leveraged in Eq. (6) the stability of independent Gaussian law by addition. However, the space of the linear combinations to be learned is high dimensional and structured, hence hard to learn. Also, the optimization of the parameterization as Eq. (6) is highly dependent on the initialization, as we optimize $W$ and $b$ from a (randomly) chosen start point. Therefore, we want to be able to reach a wider space of parameters $W, b$. To do so, we build on [42] that successfully leverages latent code pre-processing with neural network in the context of adversarial learning for image generation, and [43] where a neural network generates the weights of another neural network to facilitate learning. Therefore, instead of learning directly $W, b$ as in the vanilla approach we propose to learn to sample on the space of linear combination $W$ and biases $b$. The core idea is to leverage neural networks expressivity to enrich the space of reachable matrices $W$ and vectors

$b$ compared to the vanilla approach. To do so we use the random sample $z$ to extract a representation vector $r \in \mathbb{R}^k$. This representation $r$ serves as input to neural networks $F_b$, $F_W$ providing estimates of $W$ and $b$. To sum up, in the HyperNetwork approach we learn a logitNormal law according to:

$$\begin{cases} z \sim \mathcal{N}(0, I_d),\ r = F_{rep}(z) \in \mathbb{R}^k, & \text{(7a)} \\ W = F_W(r) \in \mathcal{M}_{n\times n,d}(\mathbb{R}), & \text{(7b)} \\ b = F_b(r) \in \mathbb{R}^{n\times n}, \text{ and finally:} & \text{(7c)} \\ S_\theta(z) = \text{sigmoid}(F_b(r) + F_W(r).z), & \text{(7d)} \end{cases}$$

Note that as desired $S_\theta(z)$ follows a logitNormal law $\mathcal{LN}(F_b(r), F_W(r)^T.F_W(r))$.

This proposition presents several advantages. First, in Eq. (6) $W$ is a randomly initialized weight matrix, then we only explore one trajectory of optimization from this (randomly chosen) starting point. Also, instead of learning a distribution of masks, this parameterization learns a distribution of transport matrices and biases. Therefore, both $F_W$ and $F_b$ stochastically explore a direction for each sample of $z$, providing more feedback with respect to the objective of feature selection for reconstruction. This parameterization of $W$ and $b$ offers a way to explore efficiently the space of biases and linear combinations. Also, because it rely on matrix multiplication, this procedure is computationally barely less efficient than the naive one when $F_W$ and $F_b$ are small neural networks.

We show experimentally the superiority of this approach in Sect. 4.

### 3.3 Sparsity Constraint: $\ell_0$-Relaxation

We detail our approach promoting sparsity. Frequently, sparsity in regression settings is enforced thanks to a $\ell_1$ penalty on the parameters. However, $\ell_1$ approaches may suffer from a shrinking effect due to ill-posedness. Consequently, we introduce an alternative approximation of the $\ell_0$-formulation better suited to our feature selection application: we minimize the expected $\ell_0$-norm, i.e. the probability of each variable in our binary mask to be greater than 0. Thus, we need a non zero probability of sampling 0 which is not the case with the current scheme. Accordingly, we introduce a stretching scheme to obtain a non-zero mass at points 0 and 1 while maintaining differentiability.

**Stretched Distribution:** To create a mass at 0, we proceed as in [28]. Let $Y \in [0,1]^m$ be a logitNormal variable, $\gamma < 0$ and $\eta > 1$ and $HT$ be the hard-threshold function defined by $HT(Y) = \min(\max(Y, 0), 1)$, the stretching is defined as:

$$\bar{Y} = HT\{(\eta - \gamma)Y + \gamma\} \tag{8}$$

Thanks to this stretching of our distribution, we have a non zero probability to be zero, i.e. $\mathbb{P}(\bar{Y} = 0) > 0$ and also $\mathbb{P}(\bar{Y} = 1) > 0$. We can now derive a relaxed version of the $\ell_0$-norm penalizing the probability of the coordinates of $\bar{Y}$.

**Sparsity Constraint:** Let $L_0(\bar{Y})$ the expected $\ell_0$-norm of our stretched output $\bar{Y}$. Using the notation in Eq. 8, we have:

$$L_0(\bar{Y}) = \mathbb{E}[\ell_0(\bar{Y})] = \sum_{i=1}^{m} \mathbb{P}(\bar{Y}_i > 0) = \sum_{i=1}^{m} 1 - F_Y(\frac{-\gamma}{\eta - \gamma}), \tag{9}$$

where $F_Y$ denotes the cumulative distribution function (CDF) of $Y$. This loss constrains the random variable $Y$ to provide sparse outputs as long as we can estimate $F_Y$ in a differentiable way. In the case of the logitNormal law, we maintain tractability as $Y$ satisfies Eq. (5) or Eq. (4). Thus, for our $m$-dimensional logitNormal law defined as in Eq. (5), we have:

$$L_0(\bar{Y}) = \sum_i 1 - \Phi\Big(\frac{\log(\frac{-\gamma}{\eta}) - b}{\sqrt{\sum_j W_{j,i}^2}}\Big), \tag{10}$$

where $\Phi$ is the CDF of the unitary Normal law. Minimizing Eq. (10) promotes sparsity in the law of $Y$ by minimizing the expected true $\ell_0$-norm of the realisation of the random variable $Y$. We have developed a constraint that promotes sparsity in a differentiable way. Now we focus on how to learn efficiently the parameters of our correlated logitNormal law.

### 3.4 Reconstruction for Feature Selection

We have designed a sparsity cost function and detailed our parameterization to learn our sampling operator, we focus on the downstream task. Consider data $(x_i, y_i)_{i \in [1..N]}$, consisting in paired input $x$ and output $y$. Feature selection consists in selecting variables in $x$ with a mask $s$, so that the considered variables: $s \odot x$ explain at best $y$. Let $G$ be a prediction function and $\mathcal{L}$ a generic cost functional, feature selection writes as:

$$\min_{s,G} \mathbb{E}_{x,y}\ \mathcal{L}\big(G(s \odot x), y\big) \ \text{ s.t } ||s||_0 < \lambda, \tag{11}$$

In this work we focus on a reconstruction as final task, i.e. $y = x$. Besides the immediate application of such formulation to optimal sensors placement and data compression, reconstruction as downstream task requires no other source of data to perform feature selection. Naturally, this framework is adaptable to classification tasks. As a sparse auto-encoding technique, feature selection with a reconstruction objective aims at minimizing the reconstruction error while controlling the sparsity. In this case $G_\phi : \mathbb{R}^{n \times n} \to \mathbb{R}^{n \times n}$ is our reconstruction network (of parameter $\phi$) taking as inputs the sparse image. The feature selection task with an $\ell_2$-auto-encoding objective writes as:

$$\min_{\theta,\phi} \mathbb{E}_x ||G_\phi(\bar{S}_\theta(z) \odot x) - x||_2 + \lambda_{sparse} L_0(\bar{S}_\theta(z)) \tag{12}$$

A schematic view of our proposition, illustrating the sampling and the reconstruction component is available in Fig. 2.

# 4 Experiments

We provide experimental results on 3 datasets: MNIST, CelebA and a geophysical dataset resulting from complex climate simulations [30,40]. We use the traditional train-test split for MNIST and a 80-20 train-test split for the other datasets. The geophysical dataset is composed of surface temperatures anomalies (deviations between average temperature at each pixel for a reference period and observations) and contains 21000 samples (17000 for train). The data have both high (Gulf stream, circum-polar current ...) and low frequencies (higher temperature in the equatorial zone, difference between northern and southern hemispheres ...) that need to be treated accurately due to their influence on the Earth climate. Accuracy in the values of reconstructed pixel is then essential for the physical interpretation. These dense images represent complex dynamics and allow us to explore our method on data with crucial applications and characteristics very different from the digits and faces.

## 4.1 Experimental and Implementation Details

**Baselines:** Besides our models Vanilla logitNormal, denoted *VLN*, and its hyper-networks couterpart denoted *HNet-LN*, we consider as competing methods the following approaches:

1. Concrete-Autoencoder [1] denoted *CAE*.
2. To assess the relevance of our correlated proposition, we investigate a binary mask approach based on the independent logitNormal mask that corresponds to equation Eq. (4) denoted *ILN*,
3. Another independent binary mask method based on the concrete law [34], denoted *SCT*.

**Implementation Details:** For all binary mask based methods, we use a Resnet for $G_\phi$, [27] following the implementation of [21]. $F_{rep}$, $F_W$ and $F_b$ are two layers MLP with leaky relu activation. For CAE, because the structure of the data is lost in the encoding process, we train $G_\phi$ as a MLP for MNIST and a DcGAN for geophysical data and CelebA. The code is available at: https://github.com/JeremDona/feature_selection_public.

**Removing Randomness:** All masked based algorithms learn distributions of masks. To evaluate the feature selection capabilities, we evaluate the different algorithms using fixed masks. We rely on Proposition 1 to remove the randomness during test time. Let $S_\theta^0$ be the 0-temperature distribution of the estimated $S_\theta$. We first estimate the expected $\ell_0$-norm of the 0-temperature distribution: $L_0(S_\theta^0)$. We then estimate two masks selecting respectively the $10 \times \lfloor \frac{L_0(S_\theta^0)}{10} \rfloor$ and $10 \times \lceil \frac{L_0(S_\theta^0)}{10} \rceil$ most likely features (rounding $L_0(S_\theta^0)$ up and down to the nearest ten). This method has the advantage of implicitly fixing a threshold in the learned mask distribution to select or reject features.

We now illustrate the advantage of selecting features in a correlated fashion.

### 4.2 Independent vs Correlated Sampling Scheme

**Is a Covariance Matrix Learned?** Because we model the local dependencies in the sampling by learning linear mixing of latent variables $z$, we first verify the structure of the covariance matrix. Figure 3 reports the learned covariance matrix of the sampling for MNIST dataset using Eq. (6) method. Besides the diagonal, extra-diagonal structures emerge, revealing that local correlations are taken into account to learn the sampling distribution.

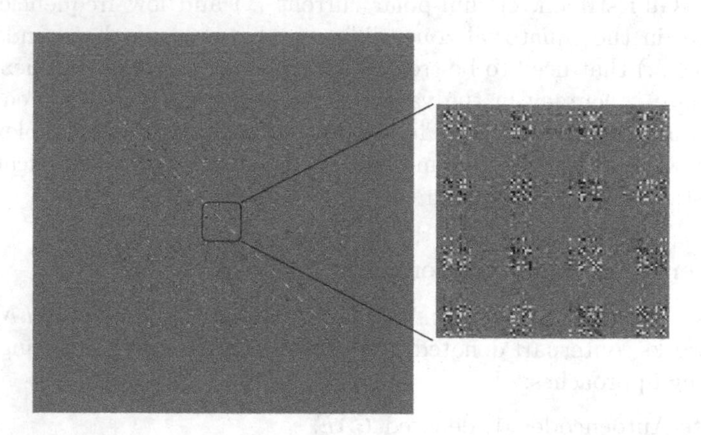

**Fig. 3.** Covariance matrix learned with Eq. (6), with $\approx$ 30 pixels selected. Yellow values indicates high positive covariance, blue ones low negative covariance (Color figure online)

**Independent Sampling Does Not Choose:** We show in Fig. 4 the empirical average of the sampled masks for each masked base competing algorithm where all algorithms were trained so that at $L_0(S^0_\theta) \approx 30$. Figure 4 clearly shows that concrete base algorithm (SCT) and in a lesser sense (ILN) do not select features, but rather put a uniformly low probability to sample pixels in the center of the image. This means that both algorithms struggle at discriminating important features from less relevant ones. On the other hand, our correlated propositions, Vanilla logitNormal (V-LN, Eq. (6)) and particularly the hyper-network

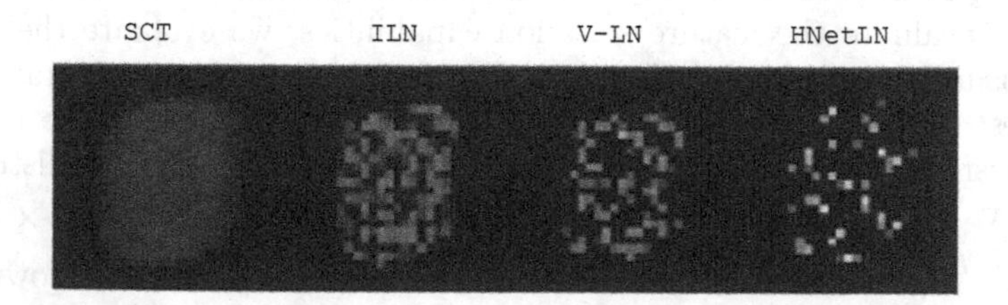


**Fig. 4.** Masks empirical distribution for competing binary masks algorithms on the MNIST datasets for about 30 features in the sampled mask

approach (HNetL, Eq. (7)) manage to sparsify the distribution prioritizing the selection of important pixels for reconstruction.

### 4.3 Feature Selection and Reconstruction

We now quantitatively estimate the impact of our choices on the reconstruction error on the various datasets. First, the mean squared error reconstruction results from Table 1 tells us that considering the spatial structure of the data enhances reconstruction performance. Indeed, mask based methods consistently over-perform CAE where the data structure is linearized in the encoding process. Furthermore for mask based method, correlated sampling (row V-LN and HNet-LN) also consistently improves over independent sampling based method (row ILN and SCT). Finally, our hyper-network HNet-Ln proposition also improves over the vanilla approach validating our proposition.

**Table 1.** Average Reconstruction Error (MSE) on MNIST, Climate and CelebA datasets for all considered baselines

| | MNIST | | | Climat | | | CelebA | | |
|---|---|---|---|---|---|---|---|---|---|
| # Features | 20 | 30 | 50 | 100 | 200 | 300 | 100 | 200 | 300 |
| CAE | 3.60 | 3.05 | 2.40 | 2.07 | 1.98 | 1.96 | 7.65 | 6.42 | 5.7 |
| ILN | 3.67 | 2.41 | 1.41 | 1.44 | 1.05 | 0.83 | 7.1 | 2.56 | 1.87 |
| SCT | 3.72 | 3.61 | 2.60 | 2.20 | 1.89 | 1.51 | 7.99 | 3.31 | 2.44 |
| VLN (Ours) | 3.22 | 2.19 | 1.33 | **1.11** | **0.93** | 0.79 | 3.11 | 1.96 | 1.50 |
| HNet-Ln (Ours) | **2.15** | **1.53** | **1.06** | 1.78 | 0.96 | **0.60** | **2.81** | **1.7** | **1.46** |

### 4.4 Quality of the Selected Features: MNIST Classification

We now assess the relevance of the selected features of our learned masks on another task. To do so, for each learned distribution we train a convolutional neural network, with a DcGAN architecture on MNIST classification task. Here also, the randomness in test set is removed. For each mask we run 5 experiments to account for the variability in the training. Classification results reported in Table 2 indicate that both our correlated logitNormal propositions consistently beat all considered baselines, validating our choices to learn a sampling scheme in a correlated fashion. Indeed, our propositions systematically reach the lowest minimum and average classification error.

**Table 2.** Classification error in percent for MNIST on test set for all considered baselines. Minimum and average are taken over 5 runs.

| # Features | 20 | | 30 | | 50 | |
|---|---|---|---|---|---|---|
| Metric | Min | Mean | Min | Mean | Min | Mean |
| CAE | 24.4 | 31.64 | 8.89 | 19.60 | 5.45 | 6.65 |
| ILN | 21.58 | 28.26 | 7.96 | 16.63 | 4.17 | 5.33 |
| SCT | 20.88 | 32.79 | 9.49 | 18.22 | 4.11 | 6.77 |
| VLN (Ours) | **12.15** | **24.74** | **6.38** | **15.07** | 3.32 | **4.67** |
| HNet-LN (Ours) | 19.23 | 25.07 | 7.24 | 17.80 | **2.84** | 6.45 |

### 4.5 Extension: cGAN

**Fig. 5.** Examples of masks (first row), reconstructions (second row) and true data (last row) for CelebA dataset using either a cGAN (4 first columns) or simple auto-encoding (4 last columns) for 200 selected features. Best viewed in color. (Color figure online)

We detailed in the previous experiments feature selection results obtained thanks to an $\ell_2$-auto-encoding approach. This choice was motivated because in physical measurement all points are equals: we don't want to favor the reconstruction of some part of the image while neglecting another. However, for images such as CelebA this assumption does not hold. Indeed, a realistically reconstructed face can be preferred to a truthful background. Moreover, $\ell_2$-auto-encoding suffers from blur in the reconstruction. In that perspective, we can leverage conditional generative adversarial networks (cGAN) approaches [21,36] that solves the blurriness occurring in $\ell_2$-decoding. We implement the cGAN approach of [21]. Figure 5 illustrates that despite both methods show good reconstruction, the cGAN approach on CelebA enables a stronger focus on faces facilitating realistic reconstruction.

## 5 Conclusion

In this work, we formulate the feature selection task as the learning of a binary mask. Aiming to select features in images for reconstruction, we developed a novel way to sample and learn a correlated discrete variable thanks to a reparameterization of the logitNormal distribution. The proposed learning framework also preserves the spatial structure of the data, enhancing reconstruction performance. We experimentally show that our proposition to explore the space of covariance matrices and average vectors as in Eq. (7) is efficient providing us with a sampling with lower variance. Finally, we experimentally evidenced the advantage of learning a correlated sampling scheme instead of independent ones.

## References


1. Balın, M.F., Abid, A., Zou, J.: Concrete autoencoders: differentiable feature selection and reconstruction. In: 36th International Conference on Machine Learning, in Proceedings of Machine Learning Research, vol. 97, pp. 444–453 (2019)
2. Aitchison, J., Shen, S.M.: Logistic-Normal Distributions: Some Properties and Uses. Biometrika 67, no. 2 (1980)
3. Bengio, Y., Nicholas, L., Courville, A.: Estimating or Propagating Gradients Through Stochastic Neurons for Conditional Computation. ArXiv abs/1308.3432 (2013)
4. Deng, C., Chiyuan, Z., Xiaofei, H.: Unsupervised feature selection for multi-cluster data. In: 16th ACM SIGKDD, International Conference on Knowledge Discovery and Data Mining (2010). https://doi.org/10.1145/1835804.1835848
5. Guyon, I., Elisseeff, A.: An introduction of variable and feature selection. J. Mach. Learn. Res. Special Issue Variable an Feature Selection **3**, 1157–1182 (2003). https://doi.org/10.1162/153244303322753616
6. Han, K., Wang, Y., Zhang, C., Li, C., Xu, C.: Autoencoder inspired unsupervised feature selection. In: IEEE International Conference on Acoustics, Speech and Signal Processing (ICASSP), pp. 2941–2945 (2018). https://doi.org/10.1109/ICASSP.2018.8462261
7. Williams, R.J: Simple statistical gradient-following algorithms for connectionist reinforcement learning. Mach. Learn. **8**, 229–256 (1992). https://doi.org/10.1007/BF00992696
8. Xing, E., Jordan, M., Karp, R.: Feature selection for high-dimensional genomic microarray data. In: Proceedings of the 8th International Conference on Machine Learning (2001)
9. Simon, N., Friedman, J., Hastie, T., Tibshirani, R.: A sparse-group lasso. J. Comput. Graph. Stat. **22**(2), 231–245 (2013)
10. McPhaden, M.J., Meyers, G., Ando, K., Masumoto, Y., Murty, V.S.N., Ravichandran, M., Syamsudin, F., Vialard, J., Yu, L., Yu, W.: RAMA: the research moored array for African-Asian-Australian monsoon analysis and prediction. Bull. Am. Meteor. Soc. **90**, 4 (2009)
11. Haeberli, W. Hoelzle, M. Paul, F. Zemp, M.: Integrated monitoring of mountain glaciers as key indicators of global climate change: the European Alps. Ann. Glaciol. **46**, 150–160. (2007)
12. Hou, C., Nie, F., Li, X., Yi, D., Wu, Y.: Joint embedding learning and sparse regression: a framework for unsupervised feature selection. IEEE Trans. Cybern. **44**(6), 793–804 (2014). https://doi.org/10.1109/TCYB.2013.2272642
13. Vincent, P., Larochelle, H., Bengio, Y., Manzagol, P.: Extracting and composing robust features with denoising autoencoders. In: Proceedings of the 25th International Conference on Machine Learning, pp. 1096–1103 (2008). https://doi.org/10.1145/1390156.1390294

14. Yang, S., Zhang, R., Nie, F., Li, X.: Unsupervised feature selection based on reconstruction error minimization. In: ICASSP - 2019 IEEE International Conference on Acoustics, Speech and Signal Processing (ICASSP), pp. 2107–2111 (2019). https://doi.org/10.1109/ICASSP.2019.8682731
15. Tibshirani, R.: Regression shrinkage and selection via the lasso. J. Royal Stat. Soc. Ser. B (Methodological), **58**(1), 267–288 (1996)
16. Han, K., Wang, Y., Zhang, C., Li, C.: Autoencoder Inspired Unsupervised Feature Selection, pp. 2941–2945 (2018). https://doi.org/10.1109/ICASSP.2018.8462261
17. Maldonado, S., Weber, R.: A wrapper method for feature selection using support vector machines. Inf. Sci. **179**(13), 2208–2217 (2009). https://doi.org/10.1016/j.ins.2009.02.014
18. Yu, L., Liu, H.: Feature selection for high-dimensional data: a fast correlation-based filter solution. In: 20th International Conference on Machine Learning, vol. 2, pp. 856–863 (2003)
19. Braunstein, S.L., Ghosh, S., Severini, S.: The Laplacian of a graph as a density matrix: a basic combinatorial approach to separability of mixed states. Ann. Comb. **10**, 291–317 (2006). https://doi.org/10.1007/s00026-006-0289-3
20. Hinton, G.E., Salakhutdinov, R.: Reducing the dimensionality of data with neural networks. Science **313**, 504–507 (2006)
21. Isola, P., Zhu, J., Zhou, T., Efros, A.: Image-to-image translation with conditional adversarial networks, pp. 5967–5976 (2017). https://doi.org/10.1109/CVPR.2017.632
22. Kingma, D. P., Welling, M.: Auto-encoding variational bayes. In: 2nd International Conference on Learning Representations, ICLR, Banff, AB, Canada (2014)
23. Zhou, Y., Jin, R., Hoi, S.C.: Exclusive lasso for multi-task feature selection. In: Proceedings of the Thirteenth International Conference on Artificial Intelligence and Statistics, in Proceedings of Machine Learning Research, vol. 9, pp. 988–995 (2010)
24. Kovciský, T., et al.: Semantic parsing with semi-supervised sequential autoencoders. In: Conference on Empirical Methods in Natural Language Processing (EMNLP) (2016)
25. Koller D., Sahami M.: Toward optimal feature selection. In: 13th International Conference on International Conference on Machine Learning (ICML). Morgan Kaufmann Publishers Inc., San Francisco, CA, USA (1996)
26. Figurnov, M., Mohamed, S., Mnih, A.: Implicit reparameterization gradients. In: 32nd International Conference on Neural Information Processing Systems (NIPS), pp. 439–450. Curran Associates Inc., Red Hook (2018)
27. He, K., Zhang, X., Ren, S., Sun, J.: Deep residual learning for image recognition. In: IEEE Conference on Computer Vision and Pattern Recognition (CVPR), pp. 770–778 (2016). https://doi.org/10.1109/CVPR.2016.90
28. Louizos, C., Welling, M., Kingma, D.: Learning sparse neural networks through $L_0$ regularization. In: International Conference on Learning Representations, ICLR (2017)
29. Luce, R.D.: Individual choice behavior. John Wiley (1959)
30. Sepulchre, P., et al.: IPSL-CM5A2 - an Earth system model designed for multi-millennial climate simulations. Geosci. Model Dev. **13**, 3011–3053 (2020). https://doi.org/10.5194/gmd-13-3011-2020
31. Yuan, M., Lin, Y.: Model selection and estimation in regression with grouped variables. J. Royal Stat. Soc. Series B **68**, 49–67 (2006). https://doi.org/10.1111/j.1467-9868.2005.00532.x

32. Manohar, K., Brunton, B., Kutz, J., Brunton, S.: Data-driven sparse sensor placement for reconstruction: demonstrating the benefits of exploiting known patterns. IEEE Control. Syst. **38**, 63–86 (2018)
33. Zhu, P., Zuo, W., Zhang, L., Hu, Q., Shiu, S.C.K.: Unsupervised feature selection by regularized self-representation. Pattern Recogn. **48**(2), 438–446 (2015). https://doi.org/10.1016/j.patcog.2014.08.006
34. Maddison, C., Mnih, A., Teh, Y.: The concrete distribution: a continuous relaxation of discrete random variables. In: International Conference on Learning Representations, ICLR (2017)
35. Makhzani, A., Frey, B.: k-sparse autoencoders. In: International Conference on Learning Representations, ICLR (2014)
36. Mirza, M., Osindero, S.: Conditional Generative Adversarial Nets (2014)
37. Bora, A., Jalal, A., Price, E., Dimakis, A.G.: Compressed sensing using generative models. In: 34th International Conference on Machine Learning (2017)
38. Donoho, D.: Compressed sensing. IEEE Trans. Inf. Theory **52**(4), 1289–1306 (2006). https://doi.org/10.1109/TIT.2006.871582
39. Wu, Y., Rosca, M., Lillicrap, T.: Deep compressed sensing. In: Proceedings of the 36th International Conference on Machine Learning, in Proceedings of Machine Learning Research 97, pp. 6850–6860 (2019)
40. IPSL: IPSL CMA5.2 simulation (2018). http://forge.ipsl.jussieu.fr/igcmg_doc/wiki/DocHconfigAipslcm5a2
41. Jang, E., Gu, S., Poole, B.: Categorical reparameterization with gumbel-softmax. In International Conference on Learning Representations, ICLR (2017)
42. Karras, T., Laine, S., Aila, T.: A style-based generator architecture for generative adversarial networks. In: IEEE/CVF Conference on Computer Vision and Pattern Recognition (CVPR), pp. 4396–4405 (2019)
43. Ha, D., Dai, A, Le, Q.: HyperNetworks. In: International Conference on Learning Representations, ICLR (2017)

# ATOM: Robustifying Out-of-Distribution Detection Using Outlier Mining

Jiefeng Chen[1(✉)], Yixuan Li[1], Xi Wu[2], Yingyu Liang[1], and Somesh Jha[1]

[1] Department of Computer Sciences, University of Wisconsin-Madison, 1210 W. Dayton Street, Madison, WI, USA
{jiefeng,sharonli,yliang,jha}@cs.wisc.edu
[2] Google, Madison, WI, USA
wuxi@google.com

**Abstract.** Detecting out-of-distribution (OOD) inputs is critical for safely deploying deep learning models in an open-world setting. However, existing OOD detection solutions can be brittle in the open world, facing various types of adversarial OOD inputs. While methods leveraging auxiliary OOD data have emerged, our analysis on illuminative examples reveals a key insight that the majority of auxiliary OOD examples may not meaningfully improve or even hurt the decision boundary of the OOD detector, which is also observed in empirical results on real data. In this paper, we provide a theoretically motivated method, *Adversarial Training with informative Outlier Mining* (ATOM), which improves the robustness of OOD detection. We show that, by mining informative auxiliary OOD data, one can significantly improve OOD detection performance, and somewhat surprisingly, generalize to unseen adversarial attacks. ATOM achieves **state-of-the-art** performance under a broad family of classic and adversarial OOD evaluation tasks. For example, on the CIFAR-10 in-distribution dataset, ATOM reduces the FPR (at TPR 95%) by up to 57.99% under adversarial OOD inputs, surpassing the previous best baseline by a large margin.

**Keywords:** Out-of-distribution detection · Outlier mining · Robustness

## 1 Introduction

Out-of-distribution (OOD) detection has become an indispensable part of building reliable open-world machine learning models [2]. An OOD detector determines whether an input is from the same distribution as the training data, or different distribution. As of recently a plethora of exciting literature has emerged to combat the problem of OOD detection [16,20,21,24,26–29,33].

---

The full version of this paper with a detailed appendix can be found at https://arxiv.org/pdf/2006.15207.pdf.


N. Oliver et al. (Eds.): ECML PKDD 2021, LNAI 12977, pp. 430–445, 2021.
https://doi.org/10.1007/978-3-030-86523-8_26

Despite the promise, previous methods primarily focused on clean OOD data, while largely underlooking the robustness aspect of OOD detection. Concerningly, recent works have shown the brittleness of OOD detection methods under adversarial perturbations [5,16,37]. As illustrated in Fig. 1, an OOD image (*e.g.*, mailbox) can be perturbed to be misclassified by the OOD detector as in-distribution (traffic sign data). Failing to detect such an *adversarial OOD example*[1] can be consequential in safety-critical applications such as autonomous driving [12]. Empirically on CIFAR-10, our analysis reveals that the false positive rate (FPR) of a competitive method Outlier Exposure [19] can increase from 3.66% to 99.94% under adversarial attack.

Motivated by this, we make an important step towards the robust OOD detection problem, and propose a novel training framework, ***A****dversarial* ***T****raining with informative* ***O****utlier* ***M****ining* (ATOM). Our key idea is to *selectively* utilize auxiliary outlier data for estimating a tight decision boundary between ID and OOD data, which leads to robust OOD detection performance. While recent methods [16,19,32,33] have leveraged auxiliary OOD data, we show that *randomly* selecting outlier samples for training yields a large portion of uninformative samples, which do not meaningfully improve the decision boundary between ID and OOD data (see Fig. 2). Our work demonstrates that by mining low OOD score data for training, one can significantly improve the robustness of an OOD detector, and somewhat surprisingly, generalize to unseen adversarial attacks.

We extensively evaluate ATOM on common OOD detection benchmarks, as well as a suite of adversarial OOD tasks, as illustrated in Fig. 1. ATOM achieves state-of-the-art performance, significantly outperforming competitive methods using standard training on random outliers [19,32,33], or using adversarial training on random outlier data [16]. On the classic OOD evaluation task (clean OOD data), ATOM achieves comparable and often better performance than current state-of-the-art methods. On $L_\infty$ OOD evaluation task, ATOM outperforms the best baseline ACET [16] by a large margin (e.g. **53.9%** false positive rate deduction on CIFAR-10). Moreover, our ablation study underlines the importance of having both adversarial training and outlier mining (ATOM) for achieving robust OOD detection.

Lastly, we provide theoretical analysis for ATOM, characterizing how outlier mining can better shape the decision boundary of the OOD detector. While hard negative mining has been explored in different domains of learning, e.g., object detection, deep metric learning [11,13,38], the vast literature of OOD detection has not explored this idea. Moreover, most uses of hard negative mining are on a heuristic basis, but in this paper, we derive precise formal guarantees with insights. Our **key contributions** are summarized as follows:

- We propose a novel training framework, adversarial training with outlier mining (ATOM), which facilitates efficient use of auxiliary outlier data to regularize the model for robust OOD detection.

[1] Adversarial OOD examples are constructed w.r.t the OOD detector, which is different from the standard notion of adversarial examples (constructed w.r.t the classification model).

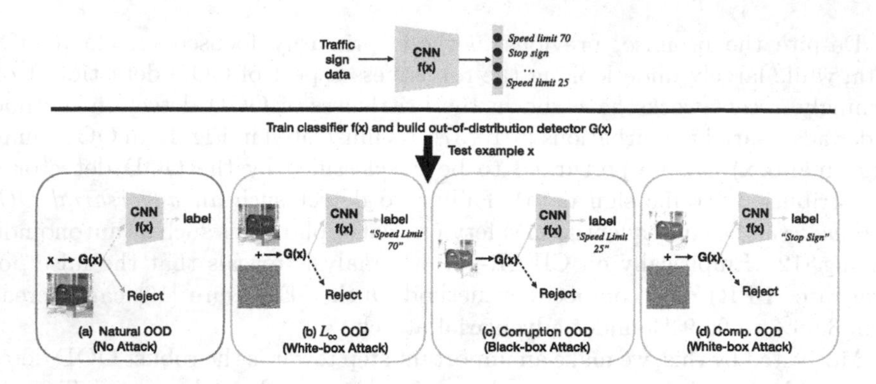


**Fig. 1. Robust out-of-distribution detection.** When deploying an image classification system (OOD detector $G(\mathbf{x})$ + image classifier $f(\mathbf{x})$) in an open world, there can be multiple types of OOD examples. We consider a broad family of OOD inputs, including (a) Natural OOD, (b) $L_\infty$ OOD, (c) corruption OOD, and (d) Compositional OOD. A detailed description of these OOD inputs can be found in Sect. 4.1. In (b-d), a perturbed OOD input (e.g., a perturbed mailbox image) can mislead the OOD detector to classify it as an in-distribution sample. This can trigger the downstream image classifier $f(\mathbf{x})$ to predict it as one of the in-distribution classes (e.g., speed limit 70). Through *adversarial training with informative outlier mining* (ATOM), our method can robustify the decision boundary of OOD detector $G(\mathbf{x})$, which leads to improved performance across all types of OOD inputs. Solid lines are actual computation flow.

- We perform extensive analysis and comparison with a diverse collection of OOD detection methods using: (1) pre-trained models, (2) models trained on randomly sampled outliers, (3) adversarial training. ATOM establishes **state-of-the-art** performance under a broad family of clean and adversarial OOD evaluation tasks.
- We contribute theoretical analysis formalizing the intuition of mining informative outliers for improving the robustness of OOD detection.
- Lastly, we provide a unified evaluation framework that allows future research examining the robustness of OOD detection algorithms under a broad family of OOD inputs. Our code and data are released to facilitate future research on robust OOD detection: https://github.com/jfc43/informative-outlier-mining.

## 2 Preliminaries

We consider the setting of multi-class classification. We consider a training dataset $\mathcal{D}_{\text{in}}^{\text{train}}$ drawn i.i.d. from a data distribution $P_{\boldsymbol{X},Y}$, where $\boldsymbol{X}$ is the sample space and $Y = \{1, 2, \cdots, K\}$ is the set of labels. In addition, we have an auxiliary outlier data $\mathcal{D}_{\text{out}}^{\text{auxiliary}}$ from distribution $U_{\mathbf{X}}$. The use of auxiliary outliers helps regularize the model for OOD detection, as shown in several recent works [16,25,29,32,33].

**Robust Out-of-Distribution Detection.** The goal is to learn a detector $G : \mathbf{x} \rightarrow \{-1, 1\}$, which outputs 1 for an in-distribution example $\mathbf{x}$ and output $-1$

for a clean or perturbed OOD example $\mathbf{x}$. Formally, let $\Omega(\mathbf{x})$ be a set of small perturbations on an OOD example $\mathbf{x}$. The detector is evaluated on $\mathbf{x}$ from $P_{\mathbf{X}}$ and on the worst-case input inside $\Omega(\mathbf{x})$ for an OOD example $\mathbf{x}$ from $Q_{\mathbf{X}}$. The false negative rate (FNR) and false positive rate (FPR) are defined as:

$$\mathrm{FNR}(G) = \mathbb{E}_{\mathbf{x}\sim P_{\mathbf{X}}}\mathbb{I}[G(\mathbf{x}) = -1], \quad \mathrm{FPR}(G; Q_{\mathbf{X}}, \Omega) = \mathbb{E}_{\mathbf{x}\sim Q_{\mathbf{X}}} \max_{\delta\in\Omega(\mathbf{x})} \mathbb{I}[G(\mathbf{x}+\delta) = 1].$$

**Remark.** Note that test-time OOD distribution $Q_{\mathbf{X}}$ is unknown, which can be different from $U_{\mathbf{X}}$. The difference between the auxiliary data $U_{\mathbf{X}}$ and test OOD data $Q_{\mathbf{X}}$ raises the fundamental question of how to effectively leverage $\mathcal{D}_{\text{out}}^{\text{auxiliary}}$ for improving learning the decision boundary between in- vs. OOD data. For terminology clarity, we refer to training OOD examples as *outliers*, and exclusively use *OOD data* to refer to test-time anomalous inputs.

# 3 Method

In this section, we introduce *Adversarial Training with informative Outlier Mining* (ATOM). We first present our method overview, and then describe details of the training objective with informative outlier mining.

**Method Overview: A Conceptual Example.** We use the terminology *outlier mining* to denote the process of selecting informative outlier training samples from the pool of auxiliary outlier data. We illustrate our idea with a toy example in Fig. 2, where in-distribution data consists of class-conditional Gaussians. Outlier training data is sampled from a uniform distribution from outside the support of in-distribution. Without outlier mining (*left*), we will almost sample those "easy" outliers and the decision boundary of the OOD detector learned can be loose. In contrast, with outlier mining (*right*), selective outliers close to the decision boundary between ID and OOD data, which improves OOD detection. This is particularly important for robust OOD detection where the boundary needs to have a margin from the OOD data so that even adversarial perturbation (red color) cannot move the OOD data points across the boundary. We proceed with describing the training mechanism that achieves our novel conceptual idea and will provide formal theoretical guarantees in Sect. 5.

## 3.1 ATOM: Adversarial Training with Informative Outlier Mining

**Training Objective.** The classification involves using a mixture of ID data and outlier samples. Specifically, we consider a $(K+1)$-way classifier network $f$, where the $(K+1)$-th class label indicates out-of-distribution class. Denote by $F_\theta(\mathbf{x})$ the softmax output of $f$ on $\mathbf{x}$. The robust training objective is given by

$$\underset{\theta}{\text{minimize}} \quad \mathbb{E}_{(\mathbf{x},y)\sim\mathcal{D}_{\text{in}}^{\text{train}}}[\ell(\mathbf{x}, y; F_\theta)] + \lambda \cdot \mathbb{E}_{\mathbf{x}\sim\mathcal{D}_{\text{out}}^{\text{train}}} \max_{\mathbf{x}'\in\Omega_{\infty,\epsilon}(\mathbf{x})} [\ell(\mathbf{x}', K+1; F_\theta)] \tag{1}$$

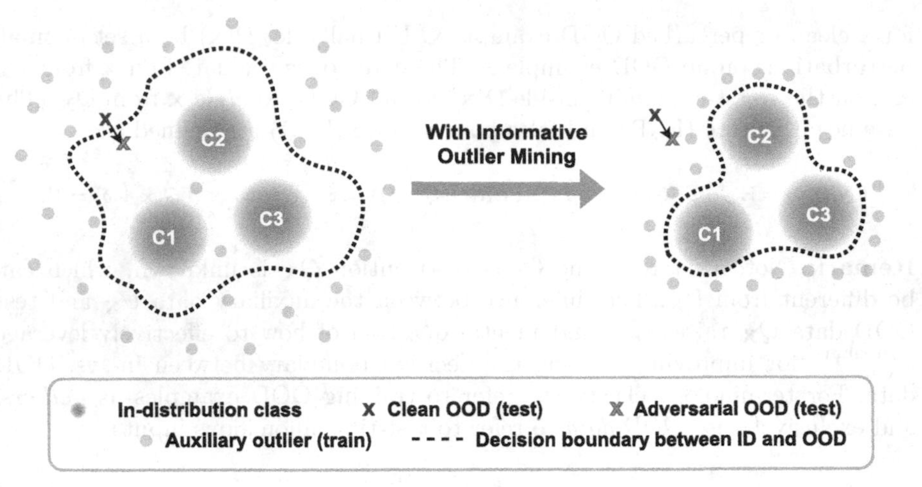


**Fig. 2.** A toy example in 2D space for illustration of informative outlier mining. With informative outlier mining, we can tighten the decision boundary and build a robust OOD detector.

where $\ell$ is the cross entropy loss, and $\mathcal{D}_{\text{out}}^{\text{train}}$ is the OOD training dataset. We use Projected Gradient Descent (PGD) [30] to solve the inner max of the objective, and apply it to half of a minibatch while keeping the other half clean to ensure performance on both clean and perturbed data.

Once trained, the OOD detector $G(\mathbf{x})$ can be constructed by:

$$G(\mathbf{x}) = \begin{cases} -1 & \text{if } F(\mathbf{x})_{K+1} \geq \gamma, \\ 1 & \text{if } F(\mathbf{x})_{K+1} < \gamma, \end{cases} \tag{2}$$

where $\gamma$ is the threshold, and in practice can be chosen on the in-distribution data so that a high fraction of the test examples are correctly classified by $G$. We call $F(\mathbf{x})_{K+1}$ the *OOD score* of $\mathbf{x}$. For an input labeled as in-distribution by $G$, one can obtain its semantic label using $\hat{F}(\mathbf{x})$:

$$\hat{F}(\mathbf{x}) = \underset{y \in \{1,2,\cdots,K\}}{\arg\max} \; F(\mathbf{x})_y \tag{3}$$

**Informative Outlier Mining.** We propose to adaptively choose OOD training examples where the detector is uncertain about. Specifically, during each training epoch, we randomly sample $N$ data points from the auxiliary OOD dataset $\mathcal{D}_{\text{out}}^{\text{auxiliary}}$, and use the current model to infer the OOD scores[2]. Next, we sort the data points according to the OOD scores and select a subset of $n < N$ data points, starting with the $qN^{\text{th}}$ data in the sorted list. We then use the selected samples as OOD training data $\mathcal{D}_{\text{out}}^{\text{train}}$ for the next epoch of training.

[2] Since the inference stage can be fully parallel, outlier mining can be applied with relatively low overhead.

**Algorithm 1:** ATOM: Adv. Training with informative Outlier Mining

**Input**: $\mathcal{D}_{\text{in}}^{\text{train}}$, $\mathcal{D}_{\text{out}}^{\text{auxiliary}}$, $F_\theta$, $m$, $N$, $n$, $q$
**Output**: $\hat{F}$, $G$
1 **for** $t = 1, 2, \cdots, m$ **do**
2 Randomly sample $N$ data points from $\mathcal{D}_{\text{out}}^{\text{auxiliary}}$ to get a candidate set $\mathcal{S}$;
3 Compute OOD scores on $\mathcal{S}$ using current model $F_\theta$ to get set $V = \{F(\mathbf{x})_{K+1} \mid \mathbf{x} \in \mathcal{S}\}$. Sort scores in $V$ from the lowest to the highest;
4 $\mathcal{D}_{\text{out}}^{\text{train}} \leftarrow V[qN : qN + n]$ ; /* $q \in [0, 1 - n/N]$ */
5 Train $F_\theta$ for one epoch using the training objective of (1);
6 **end**
7 Build $G$ and $\hat{F}$ using (2) and (3) respectively;

Intuitively, $q$ determines the *informativeness* of the sampled points w.r.t the OOD detector. The larger $q$ is, the less informative those sampled examples become. Note that informative outlier mining is performed on (non-adversarial) auxiliary OOD data. Selected examples are then used in the robust training objective (1).

We provide the complete training algorithm using informative outlier mining in Algorithm 1. Importantly, the use of informative outlier mining highlights the key difference between ATOM and previous work using ***randomly*** sampled outliers [16,19,32,33].

## 4 Experiments

In this section, we describe our experimental setup and show that ATOM can substantially improve OOD detection performance on both clean OOD data and adversarially perturbed OOD inputs. We also conducted extensive ablation analysis to explore different aspects of our algorithm.

### 4.1 Setup

**In-Distribution Datasets.** We use CIFAR-10, and CIFAR-100 [22] datasets as in-distribution datasets. We also show results on SVHN in Appendix B.8.

**Auxiliary OOD Datasets.** By default, we use 80 Million Tiny Images (TinyImages) [45] as $\mathcal{D}_{\text{out}}^{\text{auxiliary}}$, which is a common setting in prior works. We also use ImageNet-RC, a variant of ImageNet [7] as an alternative auxiliary OOD dataset.

**Out-of-Distribution Datasets.** For OOD test dataset, we follow common setup in literature and use six diverse datasets: `SVHN`, `Textures` [8], `Places365` [53], `LSUN (crop)`, `LSUN (resize)` [50], and `iSUN` [49].

**Hyperparameters.** The hyperparameter $q$ is chosen on a separate validation set from TinyImages, which is different from test-time OOD data (see Appendix B.9). Based on the validation, we set $q = 0.125$ for CIFAR-10 and $q = 0.5$ for

CIFAR-100. For all experiments, we set $\lambda = 1$. For CIFAR-10 and CIFAR-100, we set $N = 400,000$, and $n = 100,000$. More details about experimental set up are in Appendix B.1.

**Robust OOD Evaluation Tasks.** We consider the following family of OOD inputs, for which we provide details and visualizations in Appendix B.5:

- **Natural OOD:** This is equivalent to the classic OOD evaluation with clean OOD input $\mathbf{x}$, and $\Omega = \emptyset$.
- $L_\infty$ **attacked OOD (white-box):** We consider small $L_\infty$-norm bounded perturbations on an OOD input $\mathbf{x}$ [1,30], which induce the model to produce a high confidence score (or a low OOD score) for $\mathbf{x}$. We denote the adversarial perturbations by $\Omega_{\infty,\epsilon}(\mathbf{x})$, where $\epsilon$ is the adversarial budget. We provide attack algorithms for all eight OOD detection methods in Appendix B.4.
- **Corruption attacked OOD (black-box):** We consider a more realistic type of attack based on common corruptions [17], which could appear naturally in the physical world. For each OOD image, we generate 75 corrupted images (15 corruption types $\times$ 5 severity levels), and then select the one with the lowest OOD score.
- **Compositionally attacked OOD (white-box):** Lastly, we consider applying $L_\infty$-norm bounded attack and corruption attack jointly to an OOD input $\mathbf{x}$, as considered in [23].

**Evaluation Metrics.** We measure the following metrics: the false positive rate (FPR) at 5% false negative rate (FNR), and the area under the receiver operating characteristic curve (AUROC).

### 4.2 Results

**ATOM vs. Existing Methods.** We show in Table 1 that ATOM outperforms competitive OOD detection methods on both classic and adversarial OOD evaluation tasks. There are several salient observations. **First**, on classic OOD evaluation task (clean OOD data), ATOM achieves comparable or often even better performance than the current state-of-the-art methods. **Second**, on the existing adversarial OOD evaluation task, $L_\infty$ OOD, ATOM outperforms current state-of-the-art method ACET [16] by a large margin (e.g. on CIFAR-10, our method outperforms ACET by **53.9%** measured by FPR). **Third**, while ACET is somewhat brittle under the new Corruption OOD evaluation task, our method can generalize surprisingly well to the unknown corruption attacked OOD inputs, outperforming the best baseline by a large margin (e.g. on CIFAR-10, by up to **30.99%** measured by FPR). **Finally**, while almost every method fails under the hardest compositional OOD evaluation task, our method still achieves impressive results (e.g. on CIFAR-10, reduces the FPR by **57.99%**). The performance is noteworthy since our method is not trained explicitly on corrupted OOD inputs. Our training method leads to improved OOD detection while preserving classification performance on in-distribution data (see Appendix B.14). Consistent performance improvement is observed on *alternative in-distribution*

**Table 1.** Comparison with competitive OOD detection methods. We use DenseNet as network architecture for all methods. We evaluate on four types of OOD inputs: (1) natural OOD, (2) corruption attacked OOD, (3) $L_\infty$ attacked OOD, and (4) compositionally attacked OOD inputs. The description of these OOD inputs can be found in Sect. 4.1. ↑ indicates larger value is better, and ↓ indicates lower value is better. All values are percentages and are averaged over six different OOD test datasets described in Sect. 4.1. **Bold** numbers are superior results. Results on additional in-distribution dataset SVHN are provided in Appendix B.8. Results on a different architecture, WideResNet, are provided in Appendix B.12.

| $\mathcal{D}_{\text{in}}^{\text{test}}$ | Method | FPR (5% FNR) ↓ | AUROC ↑ | FPR (5% FNR) ↓ | AUROC ↑ | FPR (5% FNR) ↓ | AUROC ↑ | FPR (5% FNR) ↓ | AUROC ↑ |
|---|---|---|---|---|---|---|---|---|---|
| | | Natural OOD | | Corruption OOD | | $L_\infty$ OOD | | Comp. OOD | |
| CIFAR-10 | MSP [18] | 50.54 | 91.79 | 100.00 | 58.35 | 100.00 | 13.82 | 100.00 | 13.67 |
| | ODIN [27] | 21.65 | 94.66 | 99.37 | 51.44 | 99.99 | 0.18 | 100.00 | 0.01 |
| | Mahalanobis [26] | 26.95 | 90.30 | 91.92 | 43.94 | 95.07 | 12.47 | 99.88 | 1.58 |
| | SOFL [33] | 2.78 | 99.04 | 62.07 | 88.65 | 99.98 | 1.01 | 100.00 | 0.76 |
| | OE [19] | 3.66 | 98.82 | 56.25 | 90.66 | 99.94 | 0.34 | 99.99 | 0.16 |
| | ACET [16] | 12.28 | 97.67 | 66.93 | 88.43 | 74.45 | 78.05 | 96.88 | 53.71 |
| | CCU [32] | 3.39 | 98.92 | 56.76 | 89.38 | 99.91 | 0.35 | 99.97 | 0.21 |
| | ROWL [37] | 25.03 | 86.96 | 94.34 | 52.31 | 99.98 | 49.49 | 100.00 | 49.48 |
| | **ATOM** (ours) | **1.69** | **99.20** | **25.26** | **95.29** | **20.55** | **88.94** | **38.89** | **86.71** |
| CIFAR-100 | MSP [18] | 78.05 | 76.11 | 100.00 | 30.04 | 100.00 | 2.25 | 100.00 | 2.06 |
| | ODIN [27] | 56.77 | 83.62 | 100.00 | 36.95 | 100.00 | 0.14 | 100.00 | 0.00 |
| | Mahalanobis [26] | 42.63 | 87.86 | 95.92 | 42.96 | 95.44 | 15.87 | 99.86 | 2.08 |
| | SOFL [33] | 43.36 | 91.21 | 99.93 | 45.23 | 100.00 | 0.35 | 100.00 | 0.27 |
| | OE [19] | 49.21 | 88.05 | 99.96 | 45.01 | 100.00 | 0.94 | 100.00 | 0.59 |
| | ACET [16] | 50.93 | 89.29 | 99.53 | 54.19 | 76.27 | 59.45 | 99.71 | 38.63 |
| | CCU [32] | 43.04 | 90.95 | 99.90 | 48.34 | 100.00 | 0.75 | 100.00 | 0.48 |
| | ROWL [37] | 93.35 | 53.02 | 100.00 | 49.69 | 100.00 | 49.69 | 100.00 | 49.69 |
| | **ATOM** (ours) | **32.30** | **93.06** | **93.15** | **71.96** | **38.72** | **88.03** | **93.44** | **69.15** |

*datasets* (SVHN and CIFAR-100), *alternative network architecture* (WideResNet, Appendix B.12), and with *alternative auxiliary dataset* (ImageNet-RC, see Appendix B.11).

**Adversarial Training Alone is not Able to Achieve Strong OOD Robustness.** We perform an ablation study that isolates the effect of outlier mining. In particular, we use the same training objective as in Eq. (1), but with randomly sampled outliers. The results in Table 2 show AT (no outlier mining) is in general less robust. For example, under $L_\infty$ OOD, AT displays 23.76% and 31.61% reduction in FPR on CIFAR-10 and CIFAR-100 respectively. This validates the importance of outlier mining for robust OOD detection, which provably improves the decision boundary as we will show in Sect. 5.

**Effect of adversarial Training.** We perform an ablation study that isolates the effect of adversarial training. In particular, we consider the following objective without adversarial training:

$$\underset{\theta}{\text{minimize}} \quad \mathbb{E}_{(\mathbf{x},y)\sim\mathcal{D}_{\text{in}}^{\text{train}}}[\ell(\mathbf{x},y;\hat{F}_\theta)] + \lambda \cdot \mathbb{E}_{\mathbf{x}\sim\mathcal{D}_{\text{out}}^{\text{train}}}[\ell(\mathbf{x},K+1;\hat{F}_\theta)], \qquad (4)$$

which we name *Natural Training with informative Outlier Mining* (NTOM). In Table 2, we show that NTOM achieves comparable performance as ATOM on

**Table 2. Ablation** on ATOM training objective. We use DenseNet as network architecture. ↑ indicates larger value is better, and ↓ indicates lower value is better. All values are percentages and are averaged over six different OOD test datasets described in Sect. 4.1.

| $\mathcal{D}_{in}^{test}$ | Method | FPR (5% FNR) ↓ | AUROC ↑ | FPR (5% FNR) ↓ | AUROC ↑ | FPR (5% FNR) ↓ | AUROC ↑ | FPR (5% FNR) ↓ | AUROC ↑ |
|---|---|---|---|---|---|---|---|---|---|
| | | Natural OOD | | Corruption OOD | | $L_\infty$ OOD | | Comp. OOD | |
| CIFAR-10 | AT (no outlier mining) | 2.65 | 99.11 | 42.28 | 91.94 | 44.31 | 68.64 | 65.17 | 72.62 |
| | NTOM (no adversarial training) | 1.87 | **99.28** | 30.58 | 94.67 | 99.90 | 1.22 | 99.99 | 0.45 |
| | ATOM (ours) | **1.69** | 99.20 | **25.26** | **95.29** | **20.55** | **88.94** | **38.89** | **86.71** |
| CIFAR-100 | AT (no outlier mining) | 51.50 | 89.62 | 99.70 | 58.61 | 70.33 | 58.84 | 99.80 | 34.98 |
| | NTOM (no adversarial training) | 36.94 | 92.61 | 98.17 | 65.70 | 99.97 | 0.76 | 100.00 | 0.16 |
| | ATOM (ours) | **32.30** | **93.06** | **93.15** | **71.96** | **38.72** | **88.03** | **93.44** | **69.15** |

natural OOD and corruption OOD. However, NTOM is less robust under $L_\infty$ OOD (with 79.35% reduction in FPR on CIFAR-10) and compositional OOD inputs. This underlies the importance of having both adversarial training and outlier mining (ATOM) for overall good performance, particularly for robust OOD evaluation tasks.

**Effect of Sampling Parameter** $q$**.** Table 3 shows the performance with different sampling parameter $q$. For all three datasets, training on auxiliary outliers with large OOD scores (*i.e.*, too easy examples with $q = 0.75$) worsens the performance, which suggests the necessity to include examples on which the OOD detector is uncertain. Interestingly, in the setting where the in-distribution data and auxiliary OOD data are disjoint (*e.g.*, SVHN/TinyImages), $q = 0$ is optimal, which suggests that the hardest outliers are mostly useful for training. However, in a more realistic setting, the auxiliary OOD data can almost always contain data similar to in-distribution data (*e.g.*, CIFAR/TinyImages). Even without removing near-duplicates exhaustively, ATOM can adaptively avoid training on those near-duplicates of in-distribution data (e.g. using $q = 0.125$ for CIFAR-10 and $q = 0.5$ for CIFAR-100).

**Ablation on a Different Auxiliary Dataset.** To see the effect of the auxiliary dataset, we additionally experiment with ImageNet-RC as an alternative. We observe a consistent improvement of ATOM, and in many cases with performance better than using TinyImages. For example, on CIFAR-100, the FPR under natural OOD inputs is reduced from 32.30% (w/TinyImages) to 15.49% (w/ImageNet-RC). Interestingly, in all three datasets, using $q = 0$ (hardest outliers) yields the optimal performance since there are substantially fewer near-duplicates between ImageNet-RC and in-distribution data. This ablation suggests that ATOM's success does not depend on a particular auxiliary dataset. Full results are provided in Appendix B.11.

## 5 Theoretical Analysis

In this section, we provide theoretical insight on mining informative outliers for robust OOD detection. We proceed with a brief summary of our key results.

**Table 3.** Ablation study on $q$. We use DenseNet as network architecture. ↑ indicates larger value is better, and ↓ indicates lower value is better. All values are percentages and are averaged over six natural OOD test datasets mentioned in Sect. 4.1. Note: the hyperparameter $q$ is chosen on a separate validation set, which is different from test-time OOD data. See Appendix B.9 for details.

| $\mathcal{D}_{in}^{test}$ | Model | FPR (5% FNR) ↓ | AUROC ↑ | FPR (5% FNR) ↓ | AUROC ↑ | FPR (5% FNR) ↓ | AUROC ↑ | FPR (5% FNR) ↓ | AUROC ↑ |
|---|---|---|---|---|---|---|---|---|---|
| | | Natural OOD | | Corruption OOD | | $L_\infty$ OOD | | Comp. OOD | |
| SVHN | ATOM (q = 0.0) | 0.07 | 99.97 | 5.47 | 98.52 | 7.02 | 98.00 | 96.33 | 49.52 |
| | ATOM (q = 0.125) | 1.30 | 99.63 | 34.97 | 94.97 | 39.61 | 82.92 | 99.92 | 6.30 |
| | ATOM (q = 0.25) | 1.36 | 99.60 | 41.98 | 94.30 | 52.39 | 71.34 | 99.97 | 1.35 |
| | ATOM (q = 0.5) | 2.11 | 99.46 | 44.85 | 93.84 | 59.72 | 65.59 | 99.97 | 3.15 |
| | ATOM (q = 0.75) | 2.91 | 99.26 | 51.33 | 93.07 | 66.20 | 57.16 | 99.96 | 2.04 |
| CIFAR-10 | ATOM (q = 0.0) | 2.24 | 99.20 | 40.46 | 92.86 | 36.80 | 73.11 | 66.15 | 73.93 |
| | ATOM (q = 0.125) | 1.69 | 99.20 | 25.26 | 95.29 | 20.55 | 88.94 | 38.89 | 86.71 |
| | ATOM (q = 0.25) | 2.34 | 99.12 | 22.71 | 95.29 | 24.93 | 94.83 | 41.58 | 91.56 |
| | ATOM (q = 0.5) | 4.03 | 98.97 | 33.93 | 93.51 | 22.39 | 95.16 | 45.11 | 90.56 |
| | ATOM (q = 0.75) | 5.35 | 98.77 | 41.02 | 92.78 | 21.87 | 93.37 | 43.64 | 91.98 |
| CIFAR-100 | ATOM (q = 0.0) | 44.38 | 91.92 | 99.76 | 60.12 | 68.32 | 65.75 | 99.80 | 49.85 |
| | ATOM (q = 0.125) | 26.91 | 94.97 | 98.35 | 71.53 | 34.66 | 87.54 | 98.42 | 68.52 |
| | ATOM (q = 0.25) | 32.43 | 93.93 | 97.71 | 72.61 | 40.37 | 82.68 | 97.87 | 65.19 |
| | ATOM (q = 0.5) | 32.30 | 93.06 | 93.15 | 71.96 | 38.72 | 88.03 | 93.44 | 69.15 |
| | ATOM (q = 0.75) | 38.56 | 91.20 | 97.59 | 58.53 | 62.66 | 78.70 | 97.97 | 54.89 |

**Results Overview.** At a high level, our analysis provides two important insights. **First**, we show that with informative auxiliary OOD data, *less* in-distribution data is needed to build a robust OOD detector. **Second**, we show using outlier mining achieves a robust OOD detector in a more *realistic* case when the auxiliary OOD data contains many outliers that are far from the decision boundary (and thus non-informative), and may contain some in-distribution data. The above two insights are important for building a robust OOD detector in practice, particularly because labeled in-distribution data is expensive to obtain while auxiliary outlier data is relatively cheap to collect. *By performing outlier mining, one can effectively reduce the sample complexity while achieving strong robustness.* We provide the main results and intuition here and refer readers to Appendix A for the details and the proofs.

### 5.1 Setup

**Data Model.** To establish formal guarantees, we use a Gaussian $\mathcal{N}(\mu, \sigma^2 I)$ to model the in-distribution $P_{\mathbf{X}}$ and the test OOD distribution can be any distribution largely supported outside a ball around $\mu$. We consider robust OOD detection under adversarial perturbation with bounded $\ell_\infty$ norm, i.e., the perturbation $\|\delta\|_\infty \leq \epsilon$. Given $\mu \in \mathbb{R}^d, \sigma > 0, \gamma \in (0, \sqrt{d}), \epsilon_\tau > 0$, we consider the following data model:

- $P_{\mathbf{X}}$ **(in-distribution data)** is $\mathcal{N}(\mu, \sigma^2 I)$. The in-distribution data $\{\mathbf{x}_i\}_{i=1}^n$ is drawn from $P_{\mathbf{X}}$.
- $Q_{\mathbf{X}}$ **(out-of-distribution data)** can be any distribution from the family $\mathcal{Q} = \{Q_{\mathbf{X}} : \Pr_{\mathbf{x} \sim Q_{\mathbf{X}}}[\|\mathbf{x} - \mu\|_2 \leq \tau] \leq \epsilon_\tau\}$, where $\tau = \sigma\sqrt{d} + \sigma\gamma + \epsilon\sqrt{d}$.

- **Hypothesis class of OOD detector:** $\mathcal{G} = \{G_{u,r}(\mathbf{x}) : G_{u,r}(\mathbf{x}) = 2 \cdot \mathbb{I}[\|\mathbf{x} - u\|_2 \leq r] - 1, u \in \mathbb{R}^d, r \in \mathbb{R}_+\}$.

Here, $\gamma$ is a parameter indicating the margin between the in-distribution and OOD data, and $\epsilon_\tau$ is a small number bounding the probability mass the OOD distribution can have close to the in-distribution.

**Metrics.** For a detector $G$, we are interested in the False Negative Rate $\text{FNR}(G)$ and the worst False Positive Rate $\sup_{Q_\mathbf{X} \in \mathcal{Q}} \text{FPR}(G; Q_\mathbf{X}, \Omega_{\infty,\epsilon}(\mathbf{x}))$ over all the test OOD distributions $\mathcal{Q}$ under $\ell_\infty$ perturbations of magnitude $\epsilon$. For simplicity, we denote them as $\text{FNR}(G)$ and $\text{FPR}(G; \mathcal{Q})$.

While the Gaussian data model may be simpler than the practical data, its simplicity is desirable for our purpose of demonstrating our insights. Finally, the analysis can be generalized to mixtures of Gaussians which better models real-world data.

### 5.2 Learning with Informative Auxiliary Data

We show that informative auxiliary outliers can reduce the sample complexity for in-distribution data. Note that learning a robust detector requires to estimate $\mu$ to distance $\gamma\sigma$, which needs $\tilde{\Theta}(d/\gamma^2)$ in-distribution data, for example, one can compute a robust detector by:

$$u = \bar{\mathbf{x}} = \frac{1}{n}\sum_{i=1}^{n} \mathbf{x}_i, \quad r = (1 + \gamma/4\sqrt{d})\hat{\sigma}, \tag{5}$$

where $\hat{\sigma}^2 = \frac{1}{n}\sum_{i=1}^{n} \|\mathbf{x}_i - \bar{\mathbf{x}}\|_2^2$. Then we show that with informative auxiliary data, we need much less in-distribution data for learning. We model the auxiliary data $U_\mathbf{X}$ as a distribution over the sphere $\{\mathbf{x} : \|\mathbf{x} - \mu\|_2^2 = \sigma_o^2 d\}$ for $\sigma_o > \sigma$, and assume its density is at least $\eta$ times that of the uniform distribution on the sphere for some constant $\eta > 0$, i.e., it's surrounding the boundary of $P_\mathbf{X}$. Given $\{\mathbf{x}_i\}_{i=1}^{n}$ from $P_\mathbf{X}$ and $\{\tilde{\mathbf{x}}_i\}_{i=1}^{n'}$ from $U_\mathbf{X}$, a natural idea is to compute $\bar{\mathbf{x}}$ and $r$ as above as an intermediate solution, and refine it to have small errors on the auxiliary data under perturbation, i.e., find $u$ by minimizing a natural "margin loss":

$$u = \mathop{\arg\min}_{p: \|p - \bar{\mathbf{x}}\|_2 \leq s} \frac{1}{n'} \sum_{i=1}^{n'} \max_{\|\delta\|_\infty \leq \epsilon} \mathbb{I}\left[\|\tilde{\mathbf{x}}_i + \delta - p\|_2 < t\right] \tag{6}$$

where $s, t$ are hyper-parameters to be chosen. We show that with $\tilde{O}(d/\gamma^4)$ in-distribution data and sufficient auxiliary data can give a robust detector. See proof in Appendix A.2.

### 5.3 Learning with Informative Outlier Mining

In this subsection, we consider a more realistic data distribution where the auxiliary data can contain non-informative outliers (far away from the boundary), and in some cases mixed with in-distribution data. The non-informative outliers may not provide useful information to distinguish a good OOD detector statistically, which motivates the need for outlier mining.

**Uninformative Outliers can Lead to Bad Detectors.** To formalize, we model the non-informative ("easy" outlier) data as $Q_q = \mathcal{N}(0, \sigma_q^2 I)$, where $\sigma_q$ is large to ensure they are obvious outliers. The auxiliary data distribution $U_{\text{mix}}$ is then a mixture of $U_{\mathbf{X}}$, $Q_q$ and $P_{\mathbf{X}}$, where $Q_q$ has a large weight. Formally, $U_{\text{mix}} = \nu U_{\mathbf{X}} + (1 - 2\nu)Q_q + \nu P_{\mathbf{X}}$ for a small $\nu \in (0, 1)$. Then we see that the previous learning rule cannot work: those robust detectors (with $u$ of distance $O(\sigma\gamma)$ to $\mu$) and those bad ones (with $u$ far away from $\mu$) cannot be distinguished. There is only a small fraction of auxiliary data from $U_{\mathbf{X}}$ for distinguishing the good and bad detectors, while the majority (those from $Q_q$) do not differentiate them and some (those from $P_{\mathbf{X}}$) can even penalize the good ones and favor the bad ones.

**Informative Outlier Mining Improves the Detector with Reduced Sample Complexity.** The above failure case suggests that a more sophisticated method is needed. Below we show that outlier mining can help to identify informative data and improve the learning performance. It can remove most data outside $U_{\mathbf{X}}$, and keep the data from $U_{\mathbf{X}}$, and the previous method can work after outlier mining. We first use in-distribution data to get an intermediate solution $\bar{\mathbf{x}}$ and $r$ by Eqs. (5). Then, we use a simple thresholding mechanism to only pick points close to the decision boundary of the intermediate solution, which removes *non-informative outliers*. Specifically, we only select outliers with mild "confidence scores" w.r.t. the intermediate solution, i.e., the distances to $\bar{\mathbf{x}}$ fall in some interval $[a, b]$:

$$S := \{i : \|\tilde{\mathbf{x}}_i - \bar{\mathbf{x}}\|_2 \in [a, b], 1 \le i \le n'\} \tag{7}$$

The final solution $u_{\text{om}}$ is obtained by solving Eq. (6) on only $S$ instead of all auxiliary data. We can prove:

**Proposition 1. (Error bound with outlier mining).** *Suppose $\sigma^2\gamma^2 \ge C\epsilon\sigma_o d$ and $\sigma\sqrt{d} + C\sigma\gamma^2 < \sigma_o\sqrt{d} < C\sigma\sqrt{d}$ for a sufficiently large constant $C$, and $\sigma_q\sqrt{d} > 2(\sigma_o\sqrt{d} + \|\mu\|_2)$. For some absolute constant $c$ and any $\alpha \in (0, 1)$, if the number of in-distribution data $n \ge \frac{Cd}{\gamma^4}\log\frac{1}{\alpha}$ and the number of auxiliary data $n' \ge \frac{\exp(C\gamma^4)}{\nu^2\eta^2}\log\frac{d\sigma}{\alpha}$, then there exist parameter values $s, t, a, b$ such that with probability $\ge 1 - \alpha$, the detector $G_{u_{\text{om}},r}$ computed above satisfies:*

$$\text{FNR}(G_{u_{\text{om}},r}) \le \exp(-c\gamma^2), \quad \text{FPR}(G_{u_{\text{om}},r}; \mathcal{Q}) \le \epsilon_\tau.$$

This means that even in the presence of a large amount of uninformative or even harmful auxiliary data, we can successfully learn a good detector. Furthermore, this can reduce the sample size $n$ by a factor of $\gamma^2$. For example, when $\gamma = \Theta(d^{1/8})$, we only need $n = \tilde{\Theta}(\sqrt{d})$, while in the case without auxiliary data, we need $n = \tilde{\Theta}(d^{3/4})$.

**Remark.** We note that when $U_{\mathbf{X}}$ is as ideal as the uniform distribution over the sphere (i.e., $\eta = 1$), then we can let $u$ be the average of points in $S$ after mining, which will require $n' = \tilde{\Theta}(d/(\nu^2\gamma^2))$ auxiliary data, much less than that for more general $\eta$. We also note that our analysis and the result also hold for many other auxiliary data distributions $U_{\text{mix}}$, and the particular $U_{\text{mix}}$ used here is for the ease of explanation; see Appendix A for more discussions.

## 6 Related Work

**OOD Detection.** [18] introduced a baseline for OOD detection using the maximum softmax probability from a pre-trained network. Subsequent works improve the OOD uncertainty estimation by using deep ensembles [24], the calibrated softmax score [27], the Mahalanobis distance-based confidence score [26], as well as the energy score [29]. Some methods regularize the model with auxiliary anomalous data that were either realistic [19,33,35] or artificially generated by GANs [25]. Several other works [3,31,41] also explored regularizing the model to produce lower confidence for anomalous examples. Recent works have also studied the computational efficiency aspect of OOD detection [28] and large-scale OOD detection on ImageNet [21].

**Robustness of OOD Detection.** Worst-case aspects of OOD detection have been studied in [16,37]. However, these papers are primarily concerned with $L_\infty$ norm bounded adversarial attacks, while our evaluation also includes common image corruption attacks. Besides, [16,32] only evaluate adversarial robustness of OOD detection on random noise images, while we also evaluate it on natural OOD images. [32] has shown the first provable guarantees for worst-case OOD detection on some balls around uniform noise, and [5] studied the provable guarantees for worst-case OOD detection not only for noise but also for images from related but different image classification tasks. Our paper proposes ATOM which achieves state-of-the-art performance on a broader family of clean and perturbed OOD inputs. The key difference compared to prior work is introducing the informative outlier mining technique, which can significantly improve the generalization and robustness of OOD detection.

**Adversarial Robustness.** Adversarial examples [4,14,36,44] have received considerable attention in recent years. Many defense methods have been proposed to mitigate this problem. One of the most effective methods is adversarial training [30], which uses robust optimization techniques to render deep learning models resistant to adversarial attacks. [6,34,46,52] showed that unlabeled data could improve adversarial robustness for classification.

**Hard Example Mining.** Hard example mining was introduced in [43] for training face detection models, where they gradually grew the set of background examples by selecting those examples for which the detector triggered a false alarm. The idea has been used extensively for object detection literature [11, 13,38]. It also has been used extensively in deep metric learning [9,15,39,42,47] and deep embedding learning [10,40,48,51]. Although hard example mining has been used in various learning domains, to the best of our knowledge, we are the first to explore it to improve the robustness of out-of-distribution detection.

## 7 Conclusion

In this paper, we propose Adversarial Training with informative Outlier Mining (ATOM), a method that enhances the robustness of the OOD detector. We show the merit of adaptively selecting the OOD training examples which the OOD detector is uncertain about. Extensive experiments show ATOM can significantly improve the decision boundary of the OOD detector, achieving state-of-the-art performance under a broad family of *clean and perturbed* OOD evaluation tasks. We also provide a theoretical analysis that justifies the benefits of outlier mining. Further, our unified evaluation framework allows future research to examine the robustness of the OOD detector. We hope our research can raise more attention to a broader view of robustness in out-of-distribution detection.

**Acknowledgments.** The work is partially supported by Air Force Grant FA9550-18-1-0166, the National Science Foundation (NSF) Grants CCF-FMitF-1836978, IIS-2008559, SaTC-Frontiers-1804648 and CCF-1652140, and ARO grant number W911NF-17-1-0405. Jiefeng Chen and Somesh Jha are partially supported by the DARPA-GARD problem under agreement number 885000.

## References

1. Athalye, A., Carlini, N., Wagner, D.: Obfuscated gradients give a false sense of security: circumventing defenses to adversarial examples. In: ICML, pp. 274–283. PMLR (2018)
2. Bendale, A., Boult, T.: Towards open world recognition. In: CVPR. pp. 1893–1902 (2015)
3. Bevandić, P., Krešo, I., Oršić, M., Šegvić, S.: Discriminative out-of-distribution detection for semantic segmentation. arXiv preprint arXiv:1808.07703 (2018)
4. Biggio, B., et al.: Evasion attacks against machine learning at test time. In: Blockeel, H., Kersting, K., Nijssen, S., Železný, F. (eds.) ECML PKDD 2013. LNCS (LNAI), vol. 8190, pp. 387–402. Springer, Heidelberg (2013). https://doi.org/10.1007/978-3-642-40994-3_25
5. Bitterwolf, J., Meinke, A., Hein, M.: Certifiably adversarially robust detection of out-of-distribution data. In: NeurIPS 33 (2020)
6. Carmon, Y., Raghunathan, A., Schmidt, L., Duchi, J.C., Liang, P.S.: Unlabeled data improves adversarial robustness. In: NeurIPS, pp. 11190–11201 (2019)
7. Chrabaszcz, P., Loshchilov, I., Hutter, F.: A downsampled variant of imagenet as an alternative to the cifar datasets. arXiv preprint arXiv:1707.08819 (2017)
8. Cimpoi, M., Maji, S., Kokkinos, I., Mohamed, S., Vedaldi, A.: Describing textures in the wild. In: CVPR (2014)
9. Cui, Y., Zhou, F., Lin, Y., Belongie, S.: Fine-grained categorization and dataset bootstrapping using deep metric learning with humans in the loop. In: CVPR, pp. 1153–1162 (2016)

10. Duan, Y., Chen, L., Lu, J., Zhou, J.: Deep embedding learning with discriminative sampling policy. In: CVPR, pp. 4964–4973 (2019)
11. Felzenszwalb, P.F., Girshick, R.B., McAllester, D., Ramanan, D.: Object detection with discriminatively trained part-based models. IEEE Trans. Pattern Anal. Mach. Intell. **32**(9), 1627–1645 (2009)
12. Filos, A., Tigkas, P., McAllister, R., Rhinehart, N., Levine, S., Gal, Y.: Can autonomous vehicles identify, recover from, and adapt to distribution shifts? In: ICML, pp. 3145–3153. PMLR (2020)
13. Gidaris, S., Komodakis, N.: Object detection via a multi-region and semantic segmentation-aware CNN model. In: ICCV, pp. 1134–1142 (2015)
14. Goodfellow, I.J., Shlens, J., Szegedy, C.: Explaining and harnessing adversarial examples. ICLR (2015)
15. Harwood, B., Kumar BG, V., Carneiro, G., Reid, I., Drummond, T.: Smart mining for deep metric learning. In: ICCV, pp. 2821–2829 (2017)
16. Hein, M., Andriushchenko, M., Bitterwolf, J.: Why relu networks yield high-confidence predictions far away from the training data and how to mitigate the problem. In: CVPR, pp. 41–50 (2019)
17. Hendrycks, D., Dietterich, T.: Benchmarking neural network robustness to common corruptions and perturbations. In: ICLR (2019)
18. Hendrycks, D., Gimpel, K.: A baseline for detecting misclassified and out-of-distribution examples in neural networks. In: ICLR (2017)
19. Hendrycks, D., Mazeika, M., Dietterich, T.: Deep anomaly detection with outlier exposure. In: ICLR (2019)
20. Hsu, Y.C., Shen, Y., Jin, H., Kira, Z.: Generalized odin: detecting out-of-distribution image without learning from out-of-distribution data. In: CVPR (2020)
21. Huang, R., Li, Y.: Towards scaling out-of-distribution detection for large semantic space. In: CVPR (2021)
22. Krizhevsky, A., Hinton, G., et al.: Learning multiple layers of features from tiny images (2009)
23. Laidlaw, C., Feizi, S.: Functional adversarial attacks. In: NeurIPS, pp. 10408–10418 (2019)
24. Lakshminarayanan, B., Pritzel, A., Blundell, C.: Simple and scalable predictive uncertainty estimation using deep ensembles. In: NeurIPS, pp. 6402–6413 (2017)
25. Lee, K., Lee, H., Lee, K., Shin, J.: Training confidence-calibrated classifiers for detecting out-of-distribution samples. In: ICLR (2018)
26. Lee, K., Lee, K., Lee, H., Shin, J.: A simple unified framework for detecting out-of-distribution samples and adversarial attacks. In: NeurIPS, pp. 7167–7177 (2018)
27. Liang, S., Li, Y., Srikant, R.: Enhancing the reliability of out-of-distribution image detection in neural networks. In: ICLR (2018)
28. Lin, Z., Dutta, S., Li, Y.: Mood: Multi-level out-of-distribution detection. In: CVPR (2021)
29. Liu, W., Wang, X., Owens, J., Li, Y.: Energy-based out-of-distribution detection. In: NeurIPS (2020)
30. Madry, A., Makelov, A., Schmidt, L., Tsipras, D., Vladu, A.: Towards deep learning models resistant to adversarial attacks. In: ICLR (2018)
31. Malinin, A., Gales, M.: Predictive uncertainty estimation via prior networks. In: NeurIPS, pp. 7047–7058 (2018)
32. Meinke, A., Hein, M.: Towards neural networks that provably know when they don't know. In: ICLR (2020)

33. Mohseni, S., Pitale, M., Yadawa, J., Wang, Z.: Self-supervised learning for generalizable out-of-distribution detection. AAAI **34**, 5216–5223 (2020)
34. Najafi, A., Maeda, S.I., Koyama, M., Miyato, T.: Robustness to adversarial perturbations in learning from incomplete data. In: NeurIPS, pp. 5541–5551 (2019)
35. Papadopoulos, A., Rajati, M.R., Shaikh, N., Wang, J.: Outlier exposure with confidence control for out-of-distribution detection. Neurocomputing **441**, 138–150 (2021)
36. Papernot, N., McDaniel, P., Jha, S., Fredrikson, M., Celik, Z.B., Swami, A.: The limitations of deep learning in adversarial settings. In: 2016 IEEE European Symposium on Security and Privacy (EuroS&P), pp. 372–387. IEEE (2016)
37. Sehwag, V., Bhagoji, A.N., Song, L., Sitawarin, C., Cullina, D., Chiang, M., Mittal, P.: Analyzing the robustness of open-world machine learning. In: Proceedings of the 12th ACM Workshop on Artificial Intelligence and Security, pp. 105–116 (2019)
38. Shrivastava, A., Gupta, A., Girshick, R.: Training region-based object detectors with online hard example mining. In: CVPR, pp. 761–769 (2016)
39. Simo-Serra, E., Trulls, E., Ferraz, L., Kokkinos, I., Fua, P., Moreno-Noguer, F.: Discriminative learning of deep convolutional feature point descriptors. In: ICCV, pp. 118–126 (2015)
40. Smirnov, E., Melnikov, A., Oleinik, A., Ivanova, E., Kalinovskiy, I., Luckyanets, E.: Hard example mining with auxiliary embeddings. In: CVPR Workshops, pp. 37–46 (2018)
41. Subramanya, A., Srinivas, S., Babu, R.V.: Confidence estimation in deep neural networks via density modelling. arXiv preprint arXiv:1707.07013 (2017)
42. Suh, Y., Han, B., Kim, W., Lee, K.M.: Stochastic class-based hard example mining for deep metric learning. In: CVPR, pp. 7251–7259 (2019)
43. Sung, K.K.: Learning and example selection for object and pattern detection. Ph.D. thesis, Massachusetts Institute of Technology, Cambridge, MA, USA (1995)
44. Szegedy, C., et al.: Intriguing properties of neural networks. In: ICLR (2014)
45. Torralba, A., Fergus, R., Freeman, W.T.: 80 million tiny images: a large data set for nonparametric object and scene recognition. IEEE Trans. Pattern Anal. Mach. Intell. **30**(11), 1958–1970 (2008)
46. Uesato, J., Alayrac, J.B., Huang, P.S., Stanforth, R., Fawzi, A., Kohli, P.: Are labels required for improving adversarial robustness? NeurIPS (2019)
47. Wang, X., Gupta, A.: Unsupervised learning of visual representations using videos. In: ICCV, pp. 2794–2802 (2015)
48. Wu, C.Y., Manmatha, R., Smola, A.J., Krahenbuhl, P.: Sampling matters in deep embedding learning. In: ICCV, pp. 2840–2848 (2017)
49. Xu, P., Ehinger, K.A., Zhang, Y., Finkelstein, A., Kulkarni, S.R., Xiao, J.: Turkergaze: Crowdsourcing saliency with webcam based eye tracking. arXiv preprint arXiv:1504.06755 (2015)
50. Yu, F., Seff, A., Zhang, Y., Song, S., Funkhouser, T., Xiao, J.: Lsun: construction of a large-scale image dataset using deep learning with humans in the loop. arXiv preprint arXiv:1506.03365 (2015)
51. Yuan, Y., Yang, K., Zhang, C.: Hard-aware deeply cascaded embedding. In: ICCV, pp. 814–823 (2017)
52. Zhai, R., et al.: Adversarially robust generalization just requires more unlabeled data. arXiv preprint arXiv:1906.00555 (2019)
53. Zhou, B., Lapedriza, A., Khosla, A., Oliva, A., Torralba, A.: Places: a 10 million image database for scene recognition. IEEE Trans. Pattern Anal. Mach. Intell. **40**(6), 1452–1464 (2017)

# Robust Selection Stability Estimation in Correlated Spaces

Victor Hamer(✉) and Pierre Dupont

UCLouvain - ICTEAM/INGI/Machine Learning Group,
Place Sainte-Barbe 2, 1348 Louvain-la-Neuve, Belgium
{victor.hamer,pierre.dupont}@uclouvain.be

**Abstract.** The stability of feature selection refers to the variability of the selected feature sets induced by small changes of data sampling or analysis pipeline. Instability may strongly limit a sound interpretation of the selected features by domain experts. This work addresses the problem of assessing stability in the presence of correlated features. Correctly measuring selection stability in this context amounts to estimate to which extent several correlated variables contribute to predictive models, and how such contributions may change with the data sampling. We propose here a novel stability index taking into account such multivariate contributions. The shared contributions of several variables to predictive models do not only depend on the possible correlations between them. Computing this stability index therefore requires to solve a weighted bipartite matching problem to discover which variables actually share such contributions. We demonstrate that our novel approach provides more robust stability estimates than current measures, including existing ones taking into account feature correlations. The benefits of the proposed approach are demonstrated on simulated and real data, including microarray and mass spectrometry datasets. The code and datasets used in this paper are publicly available: https://github.com/hamerv/ecml21.

## 1 Introduction

*Feature selection*, *i.e.* the selection of a small subset of relevant features to be included in a predictive model, has already been studied in depth [3,8,10]. It has become compulsory for a wide variety of applications due to the appearance of very high dimensional data sets, notably in the biomedical domain [8].

Assessing feature selection has two distinct objectives: 1) a measure of the predictive performance of the models built on the selected features and 2) a measure of the stability of the selected features. Possible additional quality criteria are minimal model size or sparsity. Instability arises when the selected features drastically change after marginal modifications of the data sampling or processing pipeline. It prevents a correct and sound interpretation of the selected features and of the models built from them. One could even prefer a more stable modeling even if slightly less accurate [3,7].

V. Hamer—Research Fellow of the Fonds de la Recherche Scientifique - FNRS.


N. Oliver et al. (Eds.): ECML PKDD 2021, LNAI 12977, pp. 446–461, 2021.
https://doi.org/10.1007/978-3-030-86523-8_27

A typical protocol to assess selection stability amounts to run a feature selection algorithm over marginally modified training sets (*e.g.* through sub-sampling or bootstrapping) and to compare the selected feature sets across runs. Adequately evaluating the observed differences between these feature sets is the question we address here. A common measure is the Kuncheva index [4] which computes the proportion of common features across a pair of runs and reports the average over all pairs of runs, after correcting these proportions for chance. Such a measure and additional variants (briefly revisited in Sect. 2) only focus on the identities of selected features in each run but plainly ignore the possible correlations between such features. This could lead to a pessimistic estimation of the stability when some features are selected in one run and other features, distinct from but highly correlated to the initial ones, are selected in another run. Previous works specifically address this issue by proposing stability indices taking into account feature correlations (or, more generally, feature similarity values) [9,12].

In this work, we argue that correctly assessing feature selection stability should go beyond considering correlations between features. Indeed, the selected features are the input variables of predictive models. Measuring stability should also assess to which extent the selected features jointly contribute to a multivariate model, and how such contributions vary across selection runs. In the simplest case, the importance of a specific feature in a (generalized) linear model is directly proportional to its absolute weight value in such a model. Section 3 extends to non-linear models this notion of feature importance. Section 4 describes typical situations in which current stability measures are questionable. This analysis motivates our *maximum shared importance stability measure*, formally defined in Sect. 5. Computing this stability index requires to solve a weighted bipartite matching problem to discover which variables actually share such importance values across selection runs. Practical experiments on simulated and real data, including microarray and mass spectrometry datasets, are reported in Sect. 6 to illustrate the benefits of the proposed approach.

## 2 Related Work

Let $\mathcal{F}_{1\leq i\leq M}$ denote the subsets of features, among the $d$ original features, produced by a feature selection algorithm run on $M$ (*e.g.* bootstrap) samples of a training set. Let $k_i$ denote $|\mathcal{F}_i|$, the number of selected features in run $i$. The Kuncheva index [4] quantifies the stability across the $M$ selected subsets of features, whenever the number of selected features is constant across runs. Nogueira [6] generalizes the Kuncheva index to handle a varying number $k_i$ of selected features:

$$\phi = 1 - \frac{\frac{1}{d}\sum_{f=1}^{d} s_f^2}{\frac{\bar{k}}{d} * (1 - \frac{\bar{k}}{d})} \tag{1}$$

with $\bar{k}$ the mean number of selected features over all runs, and $s_f^2 = \frac{M}{M-1}\hat{p}_f(1 - \hat{p}_f)$ the estimator of the variance of the selection of feature $f$ over the $M$ selected subsets, where $\hat{p}_f$ is the fraction of times $f$ has been selected among them.

The Kuncheva index (KI) and the $\phi$ measure plainly ignore possible correlations between features. Inspired from Sechidis [9], we consider a scenario where a selection algorithm toggles, between pairs of correlated variables, across runs. More specifically, let us represent the selection matrix $\mathcal{Z}$ below with one row per selection run, $z_{i,f}$ indicating the selection of feature $f$ in run $i$

$$\mathcal{Z} = \begin{pmatrix} 1\,0\,1\,0\,0\,0\,...\,0 \\ 0\,1\,1\,0\,0\,0\,...\,0 \\ 1\,0\,0\,1\,0\,0\,...\,0 \\ 0\,1\,0\,1\,0\,0\,...\,0 \end{pmatrix} \begin{matrix} \mathbf{z}_{1,\cdot}, \text{ selected features in run 1} \\ \mathbf{z}_{2,\cdot}, \text{ selected features in run 2} \\ \cdots \\ \phantom{x} \end{matrix} \tag{2}$$

In this limit case scenario, the selection algorithm toggles between features (1,2) and between features (3,4), while no other feature is ever selected. Let us assume that each pair of these features is strongly correlated. Since the selection algorithm always picks one variable in each correlated group, the information captured is essentially the same in each run and the selection should be considered perfectly stable. Yet, the Kuncheva Index and $\phi$ would tend to $\frac{1}{3}$ as $d$ tends to infinity in such a scenario.

Sechidis [9] generalizes $\phi$ in order to accurately measure the selection stability in the presence of highly correlated variables:

$$\phi_{\mathcal{S}} = 1 - \frac{\mathrm{tr}(\mathcal{S}K_{\mathcal{Z}\mathcal{Z}})}{\mathrm{tr}(\mathcal{S}\Sigma^0)} \tag{3}$$

where the elements $s_{f,f'} \geq 0$ of the symmetric matrix $\mathcal{S}$ represent the correlations or, more generally, a similarity measure between features $f$ and $f'$, the matrix $K_{\mathcal{Z}\mathcal{Z}}$ is the variance-covariance matrix of $\mathcal{Z}$ and $\Sigma^0$ a normalization matrix. In the limit case scenario presented above and assuming perfect correlation between features 1 and 2, and, 3 and 4, respectively, $\phi_{\mathcal{S}} = 1$. This stability index $\phi_{\mathcal{S}}$ thus considers the feature correlations and only reduces to $\phi$ whenever $\mathcal{S}$ is the identity matrix. Yet, we show in Sects. 4 and 6.1 that this measure is not lower bounded and can tend to $-\infty$ in seemingly stable situations.

Similarly to our proposal (detailed in Sect. 5), Yu et al. [11] define a stability measure as the objective value of a maximum weighted bipartite matching problem. The two sets of vertices represent the selected features of two selection runs while the edge weights are the correlations between these features. The authors also propose a variant of their measure where each vertex can represent a correlated feature group as a whole. This measure is however not *fully defined*, as it requires the number of vertices (individually selected features or feature groups) to be constant across selection runs. This restriction is hardly met by all selection algorithms that could be considered in practice. In contrast, we show in Sect. 5 that our novel measure is fully defined and actually generalizes the measure proposed in [11].

The POGR index is another existing stability measure defined to handle feature correlations [12]:

$$\mathrm{POGR} = \frac{1}{M(M-1)} \sum_{i=1}^{M} \sum_{j=1, j\neq i}^{M} \frac{|\mathcal{F}_i \cap \mathcal{F}_j| + O_{i,j}}{k_i}, \tag{4}$$

with $|\mathcal{F}_i \cap \mathcal{F}_j|$ the number of features selected in both runs $i$ and $j$, and $O_{i,j}$ the number of selected features in run $i$ which are not selected in run $j$ but are significantly correlated to at least one feature selected in run $j$. In our limit case scenario (2), POGR=1 as the algorithm always selects a feature from each of the two correlated pairs. Yet, we also illustrate its limitations in Sect. 4 and 6.1.

We also consider a popular stability measure which estimates the stability of a *feature weighting*. It computes the average pairwise correlation between feature weights of different selection runs:

$$\phi_{\text{pears}} = \frac{2}{M(M-1)} \sum_{i=1}^{M-1} \sum_{j=i+1}^{M} \frac{\sum_{f=1}^{d} (w_{i,f} - \mu_i)(w_{j,f} - \mu_j)}{\sqrt{\sum_{f=1}^{d} (w_{i,f} - \mu_i)^2} * \sqrt{\sum_{f=1}^{d} (w_{j,f} - \mu_j)^2}}, \quad (5)$$

with $w_{i,f}$ the weight, or score, associated to feature $f$ in selection run $i$ and $\mu_i$ the average feature weight in this run. Whenever these weights are either 0 or 1 (indicating the selection status of the feature) and if the number of non-zero weights is constant across selection runs, $\phi_{\text{pears}}$ becomes equivalent to KI and $\phi$ [5]. In this work, we use as feature weights the feature importance values, formally defined in the next section.

## 3 Feature Importance

For each selection run $i$, the selected features $\mathcal{F}_i$ are assumed to be the input variables used to estimate a predictive model. We refer to $I_{i,f}$ as the importance value of the selected feature $f$ in the predictive model $\mathcal{P}_i$ of run $i$, with $I_{i,f} = 0$ if $f$ is not selected in run $i$. The binary matrix $\mathcal{Z}$ (introduced in (2)) is now replaced by a real matrix $I$ made of these positive importance values.

We define the importance of a feature $f$ in the predictive model $\mathcal{P}_i$, assumed here to be a classifier, as the inverse of the smallest noise applied to $f$ necessary to flip the decision of model $\mathcal{P}_i$, averaged over the $n$ learning examples. Formally,

$$I_{i,f} \triangleq \frac{1}{n} \sum_{l=1}^{n} \frac{k_i \times I_{i,f,\mathbf{x}_l}}{\sum_{f' \in \mathcal{F}_i}^{d} I_{i,f',\mathbf{x}_l}}, \quad I_{i,f,\mathbf{x}_l} = \frac{\sigma_f}{\delta_{\mathbf{x}_l,i,f}} \quad (6)$$

where $\delta_{\mathbf{x}_l,i,f}$ is the smallest additive change (in absolute value) required to feature $f$ such that the decision of the predictive model $\mathcal{P}_i$ on example $\mathbf{x}_l$ changes, and $\sigma_f$ the standard deviation of feature $f$. Intuitively, if one can change feature $f$ by large amounts without perturbing the decisions of the classifier, then $f$ is not important in its decisions. To the contrary, if a small change to $f$ causes a lot of decision switches, then the model is highly sensitive to it. Such a definition is highly reminiscent of the permutation test introduced by Breiman to quantify the importance of a feature in a Random Forest. Yet, our formulation has been preferred due to its high interpretability for linear models.

Indeed, for a linear model with weights $\mathbf{w}_i$, built on features normalized to unit variance, this formulation can be shown to be equivalent to

$$I_{i,f} = ||\mathbf{w}_i||_0 \times \frac{|w_{i,f}|}{||\mathbf{w}_i||_1}. \quad (7)$$

The importance of a feature in a linear model is proportional to the absolute value of its weight in such a model. We further normalize feature importance such that each row of the importance matrix $I$ sums up to $\bar{k}$, the average number of selected features.

## 4 Limits of Existing Stability Measures

We describe here 2 scenarios of feature selection in correlated feature spaces. We show that the existing stability measures exhibit undesirable behaviors in these scenarios, which will further motivates our novel stability measure introduced in Sect. 5.

Figures 1 and 2 illustrate the similarity between two selection runs. These figures represent feature importance in a more intuitive way than the importance matrix $I$. Each feature (or each group of highly correlated features) is identified by a unique color and the width of the rectangles correspond to the relative importance value of a feature (group) in the predictive model for this run. We assume for simplicity that the total number $d$ of features tends to infinity and that the only non-negligible feature correlations are the ones illustrated by explicit links in the figures. This implies that the denominator of $\phi_{\mathcal{S}}$ in Eq. (3) simply becomes $tr(\mathcal{S}\Sigma^0) = \bar{k}$, the average number of selected features.

In both scenarios, we consider a group of $q$ perfectly correlated features and we study what happens when $q$, starting from 1, gradually increases. The importance values of the features inside a correlated group can either be concentrated on a single feature in the group or, to the contrary, be divided, possibly unevenly, among several or all correlated features. No matter the case, we assume that the cumulative importance of all features within a correlated group roughly stay constant across selection runs. This assumption is actually confirmed in our later practical experiments reported in Fig. 4. Our limit case scenarios are designed in such a way that the stability value should be equal to $\frac{1}{2}$ and stay constant no matter the value of $q$. We show that no existing stability index satisfies this property.

In the first scenario (Fig. 1), a group of $q$ perfectly correlated features (orange) is selected with a *cumulative* importance of $\frac{\bar{k}}{4}$ in both selection runs, together with another feature (green). The two additional selected features differ in each run. Arguably, whether a large group of correlated features or a single feature from this group is selected should not impact stability, as long as the global contribution of the group to the predictive models is unchanged. In such a scenario we argue that the stability should be equal to $\frac{1}{2}$, independently of $q$, as half of the selected information is in common between both runs. In this scenario, $\phi_{\mathcal{S}}$, $\phi$, $\phi_{\text{pears}}$ and POGR correctly start at $\frac{1}{2}$ when $q = 1$ but $\phi_{\mathcal{S}}$, $\phi$ (both are equivalent in this example) and POGR increase and tend to 1 when $q$ tends to $\infty$ while $\phi_{\text{pears}}$ decreases with $q$. The precise dependencies of these measures on $q$ are summarized in Table 1.

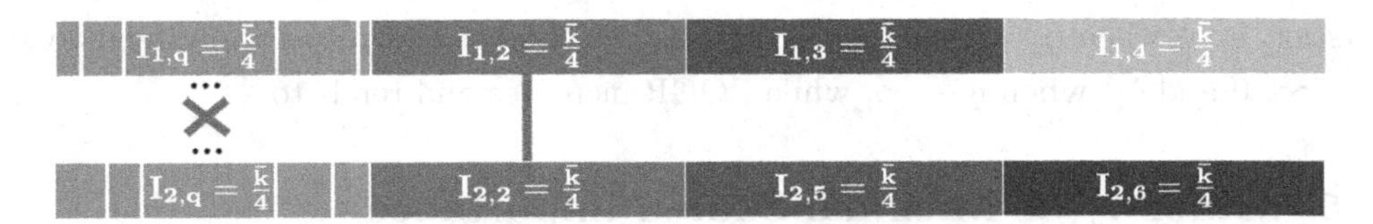


**Fig. 1.** Scenario 1. A group of $q$ perfectly correlated features (orange) is selected in both selection runs. Their specific importance in predictive models may vary but their *cumulative* importance is assumed constant across runs and equal to $\frac{k}{4}$. The stability should be equal to $\frac{1}{2}$ in such a scenario, independently of the $q$ value, because the information captured by predictive models is essentially the same in both runs. (Color figure online)

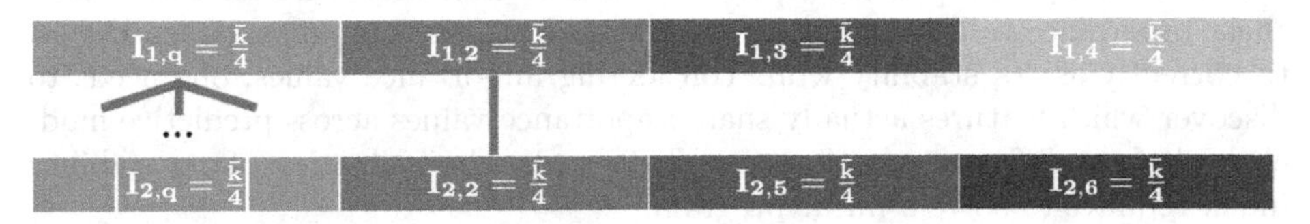


**Fig. 2.** Scenario 2. In the second selection run (below), a group of $q$ perfectly correlated features is selected with a cumulative importance of $\frac{k}{4}$. In the first run (top), a single feature from this group is selected and captures the whole group importance in the predictive model. The stability should be equal to $\frac{1}{2}$, independently of the $q$ value.

**Table 1.** Dependencies of $\phi_S$, $\phi$, $\phi_{\text{pears}}$ and POGR on the size $q$ of the correlated group in the scenarios represented in Figs. 1 (top) and 2 (bottom). In order to get a closed formula for $\phi_{\text{pears}}$, importance values are assumed to be evenly distributed within a correlated group.

| Measure | Value | $q = 1$ | $\lim_{q\to\infty}$ |
|---|---|---|---|
| $\phi_S$ | $\frac{q+1}{q+3}$ | $\frac{1}{2}$ | 1 |
| $\phi$ | $\frac{q+1}{q+3}$ | $\frac{1}{2}$ | 1 |
| $\phi_{\text{pears}}$ | $\frac{q+1}{3q+1}$ | $\frac{1}{2}$ | $\frac{1}{3}$ |
| POGR | $\frac{q+1}{q+3}$ | $\frac{1}{2}$ | 1 |

| Measure | Value | $q = 1$ | $\lim_{q\to\infty}$ |
|---|---|---|---|
| $\phi_S$ | $1 - \frac{q^2-2q+5}{q+7}$ | $\frac{1}{2}$ | $-\infty$ |
| $\phi$ | $1 - \frac{q+3}{q+7}$ | $\frac{1}{2}$ | 0 |
| $\phi_{\text{pears}}$ | $\frac{\frac{1}{4q}+\frac{1}{4}}{\sqrt{\frac{1}{4q}+\frac{3}{4}}}$ | $\frac{1}{2}$ | $\frac{\sqrt{3}}{6}$ |
| POGR | $\frac{1}{2}(\frac{1}{2} + \frac{q+1}{q+3})$ | $\frac{1}{2}$ | $\frac{3}{4}$ |

The second scenario (Fig. 2) is nearly identical to the first one, except that a single feature of the correlated group is selected in the first selection run. This feature captures the whole group importance. Again, stability should be equal to $\frac{1}{2}$, independently of $q$. In this second scenario, $\phi_S$, $\phi$, $\phi_{\text{pears}}$ and POGR correctly start at $\frac{1}{2}$ when $q = 1$ but $\phi_S$, $\phi$ and $\phi_{\text{pears}}$ decrease and respectively tend to $-\infty$, 0 and $\frac{\sqrt{3}}{6}$ when $q \to \infty$, while POGR increases and tends to $\frac{3}{4}$.

## 5 Stability as Maximal Shared Importance

In this section, we introduce a novel stability measure and show that it behaves adequately in the two scenarios discussed in Sect. 4. We further demonstrate that it generalizes previous work and prove several of its properties.

Unlike the limit cases presented in Sect. 4, actual correlations between features need not be restricted to 0 or 1 values. Moreover, observing some correlation between *e.g.* a pair of features does not guarantee that they will necessarily share their importance values in models taking these features as input variables. Hence, to correctly assess stability while considering importance values, one needs to discover which features actually share importance values across predictive models built from different selection runs. This is why the evaluation of our stability index requires to solve a linear program.

Let $\mathcal{S}$ be a symmetric similarity matrix with $s_{f,g}$ the *similarity value* between feature $f$ and feature $g$. We assume that such a similarity value falls in the $[0, 1]$ interval. It is supposed to be *a priori* defined or estimated over the whole training set for any pair of features. Typical choices include the absolute values of the Pearson's or Spearman's correlations, or mutual information normalized in the $[0, 1]$ interval over the whole training set.

For each pair $(i, j)$ of feature selection runs, one looks for the matching between features that maximizes a (similarity weighted) shared importance. Formally, $S(i, j)$ is the optimal objective value of the following constrained optimization problem:

$$S(i,j) = \max_{\mathbf{x}} \frac{\sum_{f,g} s_{f,g} \times x_{f,g}}{\bar{k}} \tag{8}$$

$$\text{subject to} \qquad \sum_{g \in \mathcal{F}_j} x_{f,g} \leq I_{i,f}, \quad \forall f \in \mathcal{F}_i \tag{9}$$

$$\sum_{f \in \mathcal{F}_i} x_{f,g} \leq I_{j,g}, \quad \forall g \in \mathcal{F}_j \tag{10}$$

$$x_{f,g} \geq 0, \quad \forall (f,g) \in (\mathcal{F}_i, \mathcal{F}_j) \tag{11}$$

The variables $x_{f,g}$ represent the *latent* amount of shared importance by feature $f$, selected in run $i$, and feature $g$, selected in run $j$. This shared importance is multiplied by the similarity $s_{f,g}$ between the two features. The feature $f$ can share its importance with several features of run $j$, but its total shared importance cannot exceed its own importance $I_{i,f}$ according to the constraint (9) (and its reciprocal (10)). Feature importance values are normalized such that $\sum_{f \in \mathcal{F}_i} I_{i,f} = \bar{k}$, the average number of selected features.[1] Since the objective (8) and all constraints are linear with respect to the variables $x_{f,g}$, this optimization problem can be efficiently solved by linear programming. The stability over $M$ feature selection runs is defined as the average pairwise optimal $S(i,j)$ values:

$$\phi_{\text{msi}} = \frac{2}{M(M-1)} \sum_{i=1}^{M} \sum_{j=i+1}^{M} S(i,j) \tag{12}$$

An example of optimal solution is depicted in Fig. 3. Features 1, 2, 3 and 4 are selected in run $i$ while features 2, 5, 6 and 7 are selected in run $j$. Feature 1 is heavily correlated to feature 6 ($s_{1,6} = 0.8$) and relatively well correlated to feature 5 ($s_{1,5} = 0.6$). Feature 6 is also somewhat correlated to feature 3 ($s_{3,6} = 0.4$). To maximize the objective, the link $x_{1,5}$ between features 1 and 5 is set to the maximum possible value, 0.7. The remaining importance of feature 1 is shared with feature 6 ($x^*_{1,6} = 0.6$) and the link between feature 3 and 6 is set to the maximum remaining importance of feature 6 ($x^*_{3,6} = 0.8$).

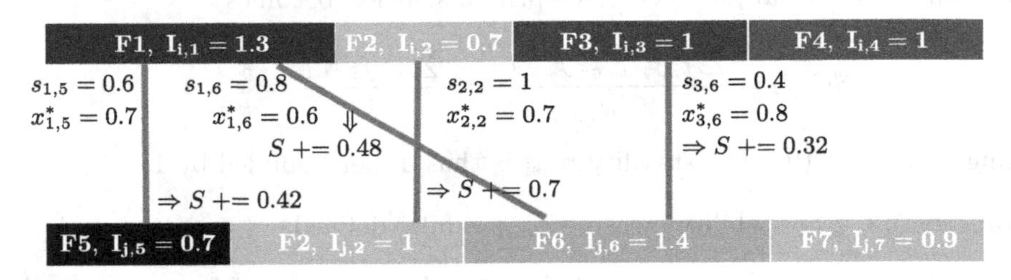


**Fig. 3.** Example of optimal solution with $S(i,j) = \frac{0.42+0.48+0.7+0.32}{4} = 0.48$. One can verify that $x^*_{1,5} + x^*_{1,6} \leq I_{i,1} = 1.3$ and $x^*_{1,6} + x^*_{3,6} \leq I_{j,6} = 1.4$.

The stability $\phi_{\text{msi}}$ behaves correctly in the two scenarios presented in Sect. 4. Considering the first scenario (Fig. 1), the constant cumulative importance of the correlated group, independently of its size $q$, guarantees that the stability is also constant with $q$. Indeed, the optimal solution verifies $\sum_{q',q'' \in [1,q]} x_{q',q''} = \frac{\bar{k}}{4}$ which implies

$$S(i,j) = \frac{\sum_{q',q'' \in [1,q]} s_{q',q''} x_{q',q''} + \frac{\bar{k}}{4}}{\bar{k}} = \frac{\frac{\bar{k}}{4} + \frac{\bar{k}}{4}}{\bar{k}} = \frac{1}{2}, \forall q \geq 1,$$

as all the similarities $s_{q,q'}$ are equal to 1. In scenario 2 (Fig. 2), the importance of feature 1 in the first run is shared among the $q$ correlated features in the second run: $x_{1,q'} = I_{2,q'}$, for all $1 \leq q' \leq q$, the optimal objective value is then

[1] Since this normalization is undefined for a run $i$ with $k_i = 0$ (a limit case with no feature selected in this run), we pose $S(i,j) = 0$ if $k_i = 0 \oplus k_j = 0$, and $S(i,j) = 1$ if $k_i = k_j = 0$, with $\oplus$ the XOR operator.

$$S(i,j) = \frac{\sum_{1 \le q' \le q} s_{1,q'} x_{1,q'} + \frac{\bar{k}}{4}}{\bar{k}} = \frac{\sum_{1 \le q' \le q} I_{2,q'} + \frac{\bar{k}}{4}}{\bar{k}} = \frac{\frac{\bar{k}}{4} + \frac{\bar{k}}{4}}{\bar{k}} = \frac{1}{2}, \forall q \ge 1.$$

Section 6.1 further illustrates on simulated data that our measure $\phi_{\text{msi}}$ does not suffer from the limitations of current measures.

We show below that $\phi_{\text{msi}}$ is bounded, which is necessary for a sound interpretation of the stability value, and fully defined (Propertys 5.1 and 5.2). We also demonstrate its maximality conditions in Property 5.3.

*Property 5.1.* The stability measure $\phi_{\text{msi}}$ is bounded in $[0, 1]$.

*Proof.* As every variable $x_{f,g}$ and every entry of the similarity matrix $\mathcal{S}$ are positive, the objective (8) is positive as well. The measure is thus lower-bounded by 0.

If one assumes maximally similar features ($s_{f,g} = 1, \forall f, g$) instead of their actual similarity values, the resulting optimization problem has an optimal objective value at least equal to the optimal objective value of the initial problem. The set of feasible solutions is the same as the constraints do not depend on $s_{f,g}$, and every solution has a larger or equal objective value than the corresponding solution of the initial problem. The optimal solution becomes

$$S(i,j) = \frac{\sum_{f \in \mathcal{F}_i} \sum_{g \in \mathcal{F}_j} x_{f,g}}{\bar{k}} \le \frac{\sum_{f \in \mathcal{F}_i} I_{i,f}}{\bar{k}} = \frac{\bar{k}}{\bar{k}} = 1, \tag{13}$$

using constraint (9). The stability $\phi_{\text{msi}}$ is thus upper-bounded by 1.

*Property 5.2.* The stability measure $\phi_{\text{msi}}$ is fully defined.

*Proof.* We have posed $S(i,j) = 0$ if $k_i = 0 \oplus k_j = 0$ and 1 if $k_i = k_j = 0$, with $\oplus$ the XOR operator. It remains to show that the optimization problem always admits a feasible solution when $k_i$ and $k_j$ are both non-zero. Since $I_{i,f} \ge 0, \forall f \in \mathcal{F}_i, \forall i$, the trivial assignation $x_{f,g} = 0, \forall f, g$ is a solution.

*Property 5.3.* The stability measure $\phi_{\text{msi}}$ is maximal (= 1) iff, for all pairs of runs $i$ and $j$, each feature importance in run $i$ can be fully shared with the importance of one or several perfectly correlated features in run $j$. In other words, there exists no link $x_{f,g} > 0$ with $s_{f,g} < 1$, and all constraints from the set (9) are active: $\sum_{g \in \mathcal{F}_j} x_{f,g} = I_{i,f}, \forall f \in \mathcal{F}_i$.

*Proof.* Suppose there exists a variable $x_{f,g} > 0$ with $s_{f,g} < 1$. Then increasing $s_{f,g}$ strictly increases the objective value. The previous objective value was thus not maximal. Suppose that $s_{f,g} = 1$ for every variable $x_{f,g} > 0$. Then, similarly to Eq. (13),

$$S(i,j) = \frac{\sum_{f \in \mathcal{F}_i} \sum_{g \in \mathcal{F}_j} x_{f,g}}{\bar{k}} \tag{14}$$

which is maximal (=1) iff $\sum_{g \in \mathcal{F}_j} x_{f,g} = I_{i,f}, \forall f \in \mathcal{F}_i$. By reciprocity, the set of constraints (9) is active iff the set of constraints (10) is active.

Theorem 5.1 shows that solving the optimization problem (8) is equivalent to solving a maximum weighted bi-partite matching problem whenever the importance of all selected features is evenly distributed between them in any given run and the number of selected features is constant across the $M$ runs. In this specific case, $\phi_{\mathrm{msi}}$ reduces to the measure proposed by Yu et al. [11].

**Theorem 5.1.** *Whenever the importance of all selected features is evenly distributed between them in any given run and the number of selected features is constant across the M runs, the constrained optimization problem (8) is a maximum weighted bi-partite matching problem.*

*Proof.* The assumptions imply $I_{i,f} = 1, \forall f \in \mathcal{F}_i, \forall i$. Problem (8) has the form of $\{\max cx | Ax \leq b,\ x \geq 0\}$, with $A$ the matrix of constraints (9) and (10). We first show that $A$ is totally submodular, *i.e.* every of its square submatrix has determinant $0, +1$ or $-1$. The following four conditions are sufficient for a matrix to be totally submodular [1]:

1. Every entry in $A$ is $0$, $+1$, or $-1$;
2. Every column of $A$ contains at most two non-zero (i.e., $+1$ or $-1$) entries;
3. If two non-zero entries in a column of $A$ have the same sign, then the row of one is in $B$ and the other in $C$, with $B$ and $C$ two disjoint sets of rows of $A$;
4. If two non-zero entries in a column of $A$ have opposite signs, then the rows of both are in $B$ or both in $C$, with $B$ and $C$ two disjoint sets of rows of $A$.

Condition 1 is satisfied as the coefficients multiplying $x_{f,g}$ in the set of constraints (9) and (10) are 0 or 1. Condition 2 holds because each variable $x_{f,g}$ has a non-zero coefficient in two constraints, one from the set of constraints (9) and one from (10). Condition 3 holds as we let $B$ be the rows of $A$ representing the set of constraints (9) and $C$ be the rows of $A$ corresponding to the set of constraints (10). If two non-zero entries in a column of $A$ have the same sign, then one row represents a constraint in set (9) (and is thus in $B$) while the other represents a constraint in set (10) (and is thus in $C$). Condition 4 is trivially satisfied as two non-zero entries of $A$ never have opposite signs.

If $A$ is totally submodular and $b$ is integral (which is here the case as $I_{i,f} = 1, \forall f \in \mathcal{F}_i, \forall i$), then linear programs of the forms $\{\max cx | Ax \leq b,\ x \geq 0\}$ have integral optimal solutions. In our case, the variables $x_{f,g}$ can only belong to $\{0, 1\}$ which makes the original optimization problem equivalent to maximum weighted bipartite matching.

Computing $S(i, j)$ requires to solve a linear programming problem with $k_i k_j$ variables, which can currently be done in $O((k_i k_j)^{2.055})$ time [2]. The overall time complexity to compute $\phi_{\mathrm{msi}}$ is then $O(M^2 \overline{(k_i k_j)^{2.055}}) \approx O(M^2 \overline{k}^{\,4.11})$. Even though this is a somewhat high computational cost, it does not depend on the typically very high number $d$ of features, unlike $\phi_{\mathcal{S}}$ which requires $O(d^2)$ time.

## 6 Experiments

In this section, we illustrate on simulated data that our proposed measure $\phi_{\mathrm{msi}}$ improves the behavior of current measures in the presence of highly correlated feature groups (Sect. 6.1). We further show on microarray and mass spectrometry data that the (accuracy, stability) Pareto fronts change when the stability measure includes feature correlation and feature importance values (Sect. 6.2).

Classification accuracy and stability of the feature selection are estimated through bootstrapping. Feature selection is applied on each bootstrap sample and the stability is computed across $M$ runs. Classification accuracy is evaluated on the out-of-bag examples of each run and its average value is reported.

### 6.1 Simulated Data

We use an artificially generated data set with $N = 5$ groups of variables. Each group contains $q$ features that are highly correlated to each other (average correlation of $\rho^g \geq 0.8$). In addition to these correlated groups, the data set contains $l = 1000$ variables. Feature values are sampled from two multivariate normal distributions using the `mvrnorm` R package. Positive examples ($n_+ = 100$) are sampled from a first distribution, centered on $\boldsymbol{\mu}_+$, a vector with $\mu_{+,f} = \mu_+^g$ if feature $f$ belongs to one of the $N = 5$ correlated groups, $\mu_+^{\neg g}$ otherwise. Negative examples ($n_- = 100$) are sampled from a second distribution, centered on $\boldsymbol{\mu}_- = -\boldsymbol{\mu}_+$. Both distributions have unit variance. We consider three scenarios with different values of $\mu_+^g$, $\mu_+^{\neg g}$ and $\rho^g$, specified in Table 2. In all scenarios, features inside a correlated group are very relevant to the binary prediction task, while features outside such groups are less but still marginally relevant. The group LASSO is used as feature selection method (scenarios 1 and 2), regularized such as to select all features inside a group or none of them. The standard LASSO, which tends to select only a few features inside each correlated group, is also evaluated (scenario 3). The regularization parameter $\lambda$ of the LASSO and group LASSO is chosen so as to select approximately 40 features when $q = 1$ (each correlated group is reduced to a single feature). For larger $q$ values, the $N = 5$ correlated groups are expected to be selected in most of the $M = 30$ selection runs while the selection of the additional features is likely to be unstable. This experiment is repeated 10 times using different generative seeds for the data sets and the mean stability values are reported on Fig. 5 as a function of $q$, the size of the correlated groups.

**Table 2.** Experimental settings for the 3 scenarios. The relevance of features inside one of the $N = 5$ correlated groups is related to $\mu_+^g$ while the relevance of features outside any group ($\sim \mu_+^{\neg g}$) is constant across scenarios. The average intra-group correlation is $\rho^g$ and inter-group correlation is negligible.

| Scenario | $\mu_+^g$ | $\mu_+^{\neg g}$ | $\rho^g$ | Method |
|---|---|---|---|---|
| 1 | 0.35 | 0.05 | 0.8 | Group LASSO |
| 2 | 0.5 | 0.05 | 0.8 | Group LASSO |
| 3 | 0.5 | 0.05 | 0.95 | LASSO |

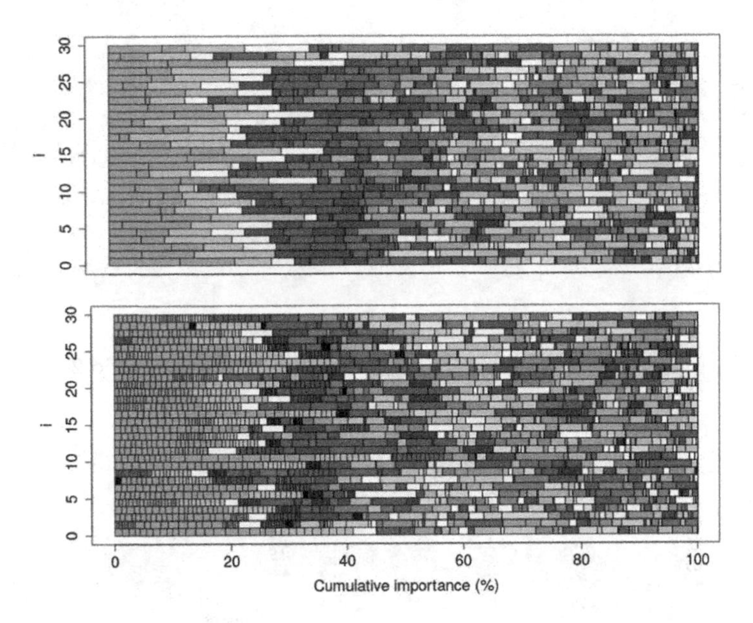


**Fig. 4.** Feature stability maps of the group LASSO (scenario 1) when the size of the correlated groups $q$ is equal to 1 (top) and 10 (bottom). As the cumulative importance of each group is approximately constant in both feature stability maps, their stability should be similar.

Figure 4 represents the cumulative importance of the features that are selected by the group LASSO in scenario 1 when $q = 1$ (top) and $q = 10$ (bottom). We use here a similar representation as in Figs. 1, 2 and 3, extended to $M = 30$ runs. We refer to such a representation as a *feature stability map.* Figure 4 illustrates that the group LASSO gives more importance to the features of the 5 "groups" when $q = 1$ as they are more relevant (by design) to the classification. When $q = 10$, the *cumulative* importance of each correlated group is approximately the same as in the $q = 1$ case, with the individual feature importance proportionally reduced. This result supports the assumptions made when defining the limit case represented in Fig. 2.

In this controlled experiment, the similarity matrix $\mathcal{S}$ is estimated as the absolute values of the pairwise Spearman's $\rho$ correlation: $s_{f,f'} = |\rho_{f,f'}|$. Such similarities are used when computing $\phi_{\text{msi}}$ and $\phi_{\mathcal{S}}$. Figure 5 compares $\phi$, $\phi_{\text{pears}}$ (with the importance values $I_{i,f}$ as feature weights), $\phi_{\text{msi}}$, $\phi_{\mathcal{S}}$ and POGR when the size $q$ of the correlated groups increases. When the group LASSO is used (Figs. 5a and 5b), $\phi$ and POGR increases with $q$ while $\phi_{\text{pears}}$ decreases with $q$.

The evolution of $\phi_{\mathcal{S}}$ depends on the scenario considered. The experiments reported in Fig. 5b are such that the correlated groups are sufficiently relevant for the classification so as to be selected in nearly all selection runs. In such a situation, $\phi_{\mathcal{S}}$ tends to increase with $q$. Whenever some correlated groups are

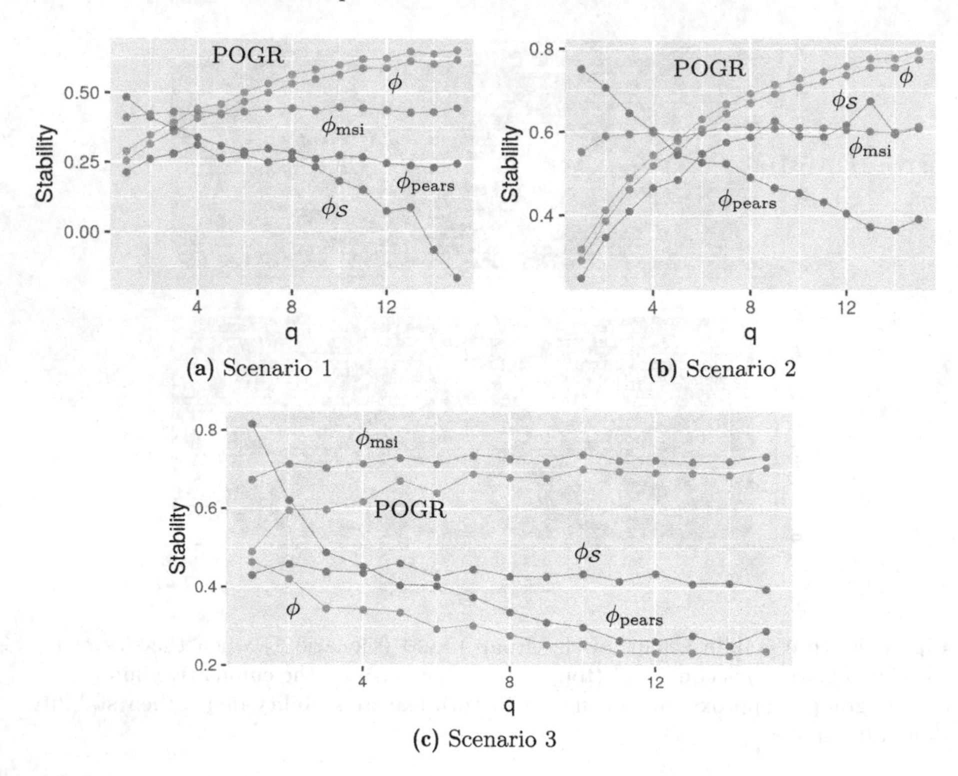


(a) Scenario 1

(b) Scenario 2

(c) Scenario 3

**Fig. 5.** Stability values of $\phi$, $\phi_S$, $\phi_{\text{msi}}$, $\phi_{\text{pears}}$ and POGR in the presence of highly correlated feature groups, in function of $q$, the size of such groups. The group LASSO is used for feature selection in (a) (scenario 1) and (b) (scenario 2), the LASSO in (c) (scenario 3). Given the design of these experiments, the stability value should not depend on $q$, which is only the case for $\phi_{\text{msi}}$ in the 3 scenarios.

regularly not selected, as in scenario 1 reported in Fig. 5a, $\phi_S$ actually tends to decrease with $q$ and decays to $-\infty$.

Figure 5c is obtained with the pure LASSO selection which selects a few features inside correlated groups instead of the whole groups. In such a scenario, $\phi$ and $\phi_{\text{pears}}$ decreases with $q$ as the selection of a few features inside each group becomes more and more unstable and because these measures do not take feature correlations into account. POGR somewhat increases with $q$. This is the only scenario for which $\phi_S$ is approximately constant, with a lower value than $\phi_{\text{msi}}$. The novel stability measure $\phi_{\text{msi}}$ is the only one to be approximately constant with $q$, in all 3 scenarios.

### 6.2 Stability of Standard Selection Methods

In this section, we report (accuracy, stability) Pareto fronts on real data sets, including microarray and mass spectrometry data. We compare here the standard stability $\phi$, which ignores feature correlations, and $\phi_{\text{msi}}$. The similarity matrix $\mathcal{S}$ is estimated as the absolute values of the pairwise Spearman's $\rho$ correlation: $s_{f,f'} = |\rho_{f,f'}|$. We consider the following feature selection methods and report performances over $M = 100$ selection runs.

- Random forests with 1000 trees. A first forest is learned on the original $d$ features. The 20 features whose removal would cause the greatest accuracy decrease on the out-of-bag examples are selected. A second forest of 1000 trees built on those 20 features is used for prediction.
- Logistic regression with a LASSO and ELASTIC-NET penalty. For the ELASTIC-NET $\lambda_1(\lambda_2 \text{L1} + (1 - \lambda_2)\text{L2})$, the parameter $\lambda_2$, which dictates the balance between L1 and L2 norms, is set to 0.8. The parameter $\lambda_1$ is set to select, on average, $\bar{k} = 20$ features over the $M$ runs.
- The RELIEF algorithm with 5 neighbors with equal weights to select 20 features. Predictive models are 5-NN classifiers.
- The logistic or hinge loss RFE. Predictive models are obtained by logistic regression or by fitting a linear SVM on the 20 selected features.
- The t-test, Mutual Information Maximization (MIM) (`infotheo` R package) and minimum Redundancy Maximum Relevance (mRMR) (`mRMRe` R package) information theoretic methods. For these three last approaches, the final model used for prediction is logistic regression estimated from the 20 selected features.

We report experiments on 5 microarray data sets (`alon`, `singh`, `chiaretti`, `gravier` and `borovecki`) from the `datamicroarray` R package and one mass-spectrometric data set, `arcene`, from the UCI machine learning repository. They have a high number of features (from 2,000 for `alon` to 22,283 for `borovecki`) with respect to the number of training examples (from 31 for `borovecki` to 198 for `arcene`), which generally leads to instability. In each selection run, the feature space is pre-filtered by keeping the 5,000 features with highest variance before running a specific selection algorithm.

Figure 6 represents the (accuracy, stability) Pareto front across all feature selection methods on two representative datasets (`chiaretti` (top) and `singh` (bottom)), for the two stability measures $\phi$ (left) and $\phi_{\text{msi}}$ (right). Results on all 6 datasets are summarized in Table 3. These results show that the choice of best performing feature selection methods highly depends on the stability measure used. Figure 6 and Table 3 further show that the stability $\phi_{\text{msi}}$ is much higher than $\phi$ for all selection methods. This phenomenon is the most pronounced on `borovecki` where $0.06 < \phi < 0.24$ and $0.7 < \phi_{\text{msi}} < 0.88$. This indicates that the observed instability is largely due to the high correlation between the input features. Even though the selection is unstable at the level of the input features, selected features from different selection runs tend to be highly correlated.

Correcting for the correlation between features, as done by $\phi_{\mathcal{S}}$, is not enough however. Stability results according to $\phi_{\mathcal{S}}$ are detailed in appendix, with even lower values than those of $\phi$. The proposed measure $\phi_{\text{msi}}$ behaves better because it does not only consider feature correlations but also feature importance values, which are matched between predictive models from several selection runs.

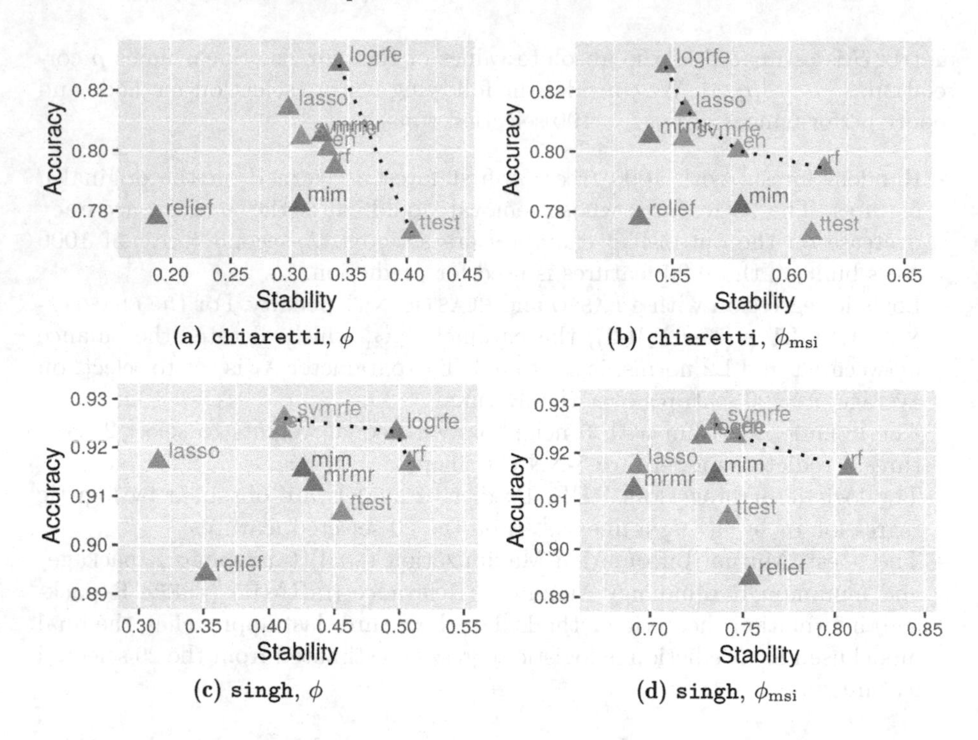


(a) **chiaretti**, $\phi$ (b) **chiaretti**, $\phi_{\text{msi}}$

(c) **singh**, $\phi$ (d) **singh**, $\phi_{\text{msi}}$

**Fig. 6.** Pareto fronts across selection methods obtained when $\phi$ (a,c) or $\phi_{\text{msi}}$ (b,d) estimates stability, on two representative datasets: **chiaretti** and **singh**.

**Table 3.** Stability ranges and Pareto fronts for $\phi$ and $\phi_{\text{msi}}$ on all datasets.

| Data | Range $\phi$ | Pareto front $\phi$ | Range $\phi_{\text{msi}}$ | Pareto front $\phi_{\text{msi}}$ | Range ac. |
|---|---|---|---|---|---|
| chiar. | 0.19–0.41 | logrfe/ttest | 0.54–0.62 | logrfe/lasso/en/rf | 0.77–0.83 |
| singh | 0.32–0.51 | svmrfe/logrfe/rf | 0.69–0.81 | svmrfe/en/rf | 0.89–0.93 |
| alon | 0.21–0.47 | logrfe/svmrfe | 0.64–0.76 | logrfe/en/ relief/svmrfe | 0.77–0.82 |
| grav. | 0.11–0.27 | logrfe/ttest | 0.36–0.47 | logrfe/ttest/en | 0.68–0.75 |
| arcene | 0.11–0.3 | rf/relief/ logrfe/en | 0.41–0.7 | rf | 0.68–0.75 |
| borov. | 0.06–0.24 | ttest | 0.7–0.88 | ttest/rf | 0.88–0.97 |

# 7 Conclusion

Current feature selection methods, especially applied to highly dimensional data, tend to suffer from instability since marginal modifications in data sampling may result in largely distinct selected feature sets. Such instability may strongly limits a sound interpretation of the selected variables.

In this work, we focus on estimating stability in feature spaces with strong feature correlations. We pose stability as the optimal objective value of a constrained optimization problem, which can be efficiently solved by linear programming. This objective depends on a similarity measure between features and on their relative importance values in predictive models. We demonstrate on handcrafted examples and on simulated data that our approach provides more relevant stability estimates than existing stability measures. Experimental results on microarray and mass spectrometry data also illustrate that a sound stability estimation may strongly affect the choice of selection method when picking an optimal trade-off between feature selection stability and predictive performance.

## References


1. Heller, I., Tompkins, C.: An extension of a theorem of dantzig's. Linear Inequalities Related Syst. **38**, 247–254 (1956)
2. Jiang, S., Song, Z., Weinstein, O., Zhang, H.: Faster dynamic matrix inverse for faster lps. in arxiv preprint (2020)
3. Kalousis, A., Prados, J., Hilario, M.: Stability of feature selection algorithms: a study on high-dimensional spaces. Knowl. Inf. Syst. **12**(1), 95–116 (2007)
4. Kuncheva, L.I.: A stability index for feature selection. In: Artificial Intelligence and Applications, pp. 421–427 (2007)
5. Nogueira, S., Brown, G.: Measuring the stability of feature selection. In: Frasconi, P., Landwehr, N., Manco, G., Vreeken, J. (eds.) ECML PKDD 2016. LNCS (LNAI), vol. 9852, pp. 442–457. Springer, Cham (2016). https://doi.org/10.1007/978-3-319-46227-1_28
6. Nogueira, S., Sechidis, K., Brown, G.: On the stability of feature selection algorithms. J. Mach. Learn. Res. **18**(1), 6345–6398 (2017)
7. Saeys, Y., Abeel, T., Van de Peer, Y.: Robust feature selection using ensemble feature selection techniques. In: Daelemans, W., Goethals, B., Morik, K. (eds.) ECML PKDD 2008. LNCS (LNAI), vol. 5212, pp. 313–325. Springer, Heidelberg (2008). https://doi.org/10.1007/978-3-540-87481-2_21
8. Saeys, Y., Inza, I., Larrañaga, P.: A review of feature selection techniques in bioinformatics. Bioinformatics **23**(19), 2507–2517 (2007)
9. Sechidis, K., Papangelou, K., Nogueira, S., Weatherall, J., Brown, G.: On the stability of feature selection in the presence of feature correlations. In: Proceedings of the 2019 European Conference on Machine Learning and Principles and Practice of Knowledge Discovery in Databases (2019)
10. Tang, J., Alelyani, S., Liu, H.: Feature selection for classification: a review. Data classification: Algorithms and applications, p. 37 (2014)
11. Yu, L., Ding, C., Loscalzo, S.: Stable feature selection via dense feature groups. In: Proceedings of the 14th ACM SIGKDD International Conference on Knowledge Discovery and Data Mining, pp. 803–811 (2008)
12. Zhang, M., Zhang, L., Zou, J., Yao, C., Xiao, H., Liu, Q., Wang, J., Wang, D., Wang, C., Guo, Z.: Evaluating reproducibility of differential expression discoveries in microarray studies by considering correlated molecular changes. Bioinformatics **25**(13), 1662–1668 (2009)

# Gradient-Based Label Binning in Multi-label Classification

Michael Rapp[1](✉), Eneldo Loza Mencía[1], Johannes Fürnkranz[2], and Eyke Hüllermeier[3]

[1] Knowledge Engineering Group, TU Darmstadt, Darmstadt, Germany
mrapp@ke.tu-darmstadt.de, research@eneldo.net

[2] Computational Data Analysis Group, JKU Linz, Linz, Austria
juffi@faw.jku.at

[3] Heinz Nixdorf Institute, Paderborn University, Paderborn, Germany
eyke@upb.de

**Abstract.** In multi-label classification, where a single example may be associated with several class labels at the same time, the ability to model dependencies between labels is considered crucial to effectively optimize non-decomposable evaluation measures, such as the Subset 0/1 loss. The gradient boosting framework provides a well-studied foundation for learning models that are specifically tailored to such a loss function and recent research attests the ability to achieve high predictive accuracy in the multi-label setting. The utilization of second-order derivatives, as used by many recent boosting approaches, helps to guide the minimization of non-decomposable losses, due to the information about pairs of labels it incorporates into the optimization process. On the downside, this comes with high computational costs, even if the number of labels is small. In this work, we address the computational bottleneck of such approach—the need to solve a system of linear equations—by integrating a novel approximation technique into the boosting procedure. Based on the derivatives computed during training, we dynamically group the labels into a predefined number of bins to impose an upper bound on the dimensionality of the linear system. Our experiments, using an existing rule-based algorithm, suggest that this may boost the speed of training, without any significant loss in predictive performance.

**Keywords:** Multi-label classification · Gradient boosting · Rule learning

## 1 Introduction

Due to its diverse applications, e.g., the annotation of text documents or images, *multi-label classification* (MLC) has become an established topic of research in the machine learning community (see, e.g., [24] or [8] for an overview). Unlike in traditional classification settings, like binary or multi-class classification, when dealing with multi-label data, a single example may correspond to several class


N. Oliver et al. (Eds.): ECML PKDD 2021, LNAI 12977, pp. 462–477, 2021.
https://doi.org/10.1007/978-3-030-86523-8_28

labels at the same time. As the labels that are assigned by a predictive model may partially match the true labeling, rather than being correct or incorrect as a whole, the quality of such predictions can be assessed in various ways. Due to this ambiguity, several meaningful evaluation measures with different characteristics have been proposed in the past (see, e.g., [22]). Usually, a single model cannot provide optimal predictions in terms of all of these measures. Moreover, empirical and theoretical results suggest that many measures benefit from the ability to model dependencies between the labels, if such patterns exist in the data [6]. As this is often the case in real-world scenarios, research on MLC is heavily driven by the motivation to capture correlations in the label space.

To account for the different properties of commonly used multi-label measures, the ability to tailor the training of predictive models to a certain target measure is a desirable property of MLC approaches. Methods based on *gradient boosting*, which guide the construction of an ensemble of weak learners towards the minimization of a given loss function, appear to be appealing with regard to this requirement. In fact, several boosting-based approaches for MLC have been proposed in the literature. Though many of these methods are restricted to the use of label-wise decomposable loss functions (e.g., [18,25] or [10]), including methods that focus on ranking losses (e.g., [5,11] or [17]), gradient boosting has also been used to minimize non-decomposable losses (e.g., [15] or [1]).

To be able to take dependencies between labels into account, problem transformation methods, such as *Label Powerset* [22], *RAKEL* [23] or *(Probabilistic) Classifier Chains* [4,16], transform the original learning task into several sub-problems that can be solved by the means of binary classification algorithms. Compared to binary relevance, where each label is considered in isolation, these approaches come with high computational demands. To compensate for this, methods like *HOMER* [21], *Compressed Sensing* [26], *Canonical Correlation Analysis* [19], *Principal Label Space Transformation* [20] or *Label Embeddings* [2,9,13] aim to reduce the complexity of the label space to be dealt with by multi-label classifiers. Notwithstanding that such a reduction in complexity is indispensable in cases where thousands or even millions of labels must be handled, it often remains unclear what measure such methods aim to optimize. In this work, we explicitly focus on the minimization of non-decomposable loss functions in cases where the original problem is tractable. We therefore aim at real-world problems with up to a few hundred labels, where such metrics, especially the Subset 0/1 loss, are considered as important quality measures.

As our contribution, we propose a novel method to be integrated into the gradient boosting framework. Based on the derivatives that guide the optimization process, it maps the labels to a predefined number of bins. If the loss function is non-decomposable, this reduction in dimensionality limits the computational efforts needed to evaluate potential weak learners. Unlike the reduction methods mentioned above, our approach dynamically adjusts to different regions in input space for which a learner may predict. Due to the exploitation of the derivatives, the impact of the approximation is kept at a minimum. We investigate the effects on training time and predictive performance using *BOOMER* [15], a boosting algorithm for learning multi-label rules. In general, the proposed method is not limited to rules and can easily be extended to gradient boosted decision trees.

## 2 Preliminaries

In this section, we briefly recapitulate the multi-label classification setting and introduce the notation used in this work. We also discuss the methodology used to tackle multi-label problems by utilizing the gradient boosting framework.

### 2.1 Multi-label Classification

We deal with multi-label classification as a supervised learning problem, where a model is fit to labeled training data $\mathcal{D} = \{(\boldsymbol{x}_1, \boldsymbol{y}_1), \ldots, (\boldsymbol{x}_N, \boldsymbol{y}_N)\} \subset \mathcal{X} \times \mathcal{Y}$. Each example in such data set is characterized by a vector $\boldsymbol{x} = (x_1, \ldots, x_L) \in \mathcal{X}$ that assigns constant values to numerical or nominal attributes $A_1, \ldots, A_L$. In addition, an example may be associated with an arbitrary number of labels out of a predefined label set $\mathcal{L} = \{\lambda_1, \ldots, \lambda_K\}$. The information, whether individual labels are relevant (1) or irrelevant ($-1$) to a training example, is specified in the form of a binary label vector $\boldsymbol{y} = (y_1, \ldots, y_K) \in \mathcal{Y}$. The goal is to learn a model $f : \mathcal{X} \rightarrow \mathcal{Y}$ that maps any given example to a predicted label vector $\hat{\boldsymbol{y}} = (\hat{y}_1, \ldots, \hat{y}_K) \in \mathcal{Y}$. It should generalize beyond the given training examples such that it can be used to obtain predictions for unseen data.

Ideally, the training process can be tailored to a certain loss function such that the predictions minimize the expected risk with respect to that particular loss. In multi-label classification several meaningful loss functions with different characteristics exist. In the literature, one does usually distinguish between label-wise *decomposable* loss functions, such as the Hamming loss (see, e.g., [22] for a definition of this particular loss function), and *non-decomposable* losses. The latter are considered to be particularly difficult to minimize, as it is necessary to take interactions between the labels into account [6]. Among this kind of loss functions is the *Subset 0/1 loss*, which we focus on in this work. Given true and predicted label vectors, it is defined as

$$\ell_{\text{Subs.}}(\boldsymbol{y}_n, \hat{\boldsymbol{y}}_n) := [\![\boldsymbol{y}_n \neq \hat{\boldsymbol{y}}_n]\!], \tag{1}$$

where $[\![x]\!]$ evaluates to 1 or 0, if the predicate $x$ is true or false, respectively. As a wrong prediction for a single label is penalized as much as predicting incorrectly for several labels, the minimization of the Subset 0/1 loss is very challenging. Due to its interesting properties and its prominent role in the literature, we consider it as an important representative of non-decomposable loss functions.

### 2.2 Multivariate Gradient Boosting

We build on a recently proposed extension to the popular gradient boosting framework that enables to minimize decomposable, as well as non-decomposable, loss functions in a multi-label setting [15]. Said approach aims at learning ensembles $F_T = \{f_1, \ldots, f_T\}$ that consist of several weak learners. In multi-label classification, each ensemble member can be considered as a predictive function that returns a vector of real-valued confidence scores

$$\hat{\boldsymbol{p}}_n^t = f_t(\boldsymbol{x}_n) = \left(\hat{p}_{n1}^t, \ldots, \hat{p}_{nK}^t\right) \in \mathbb{R}^K \tag{2}$$

for any given example. Each confidence score $\hat{p}_{nk}$ expresses a preference towards predicting the corresponding label $\lambda_k$ as relevant or irrelevant, depending on whether the score is positive or negative. To compute an ensemble's overall prediction, the vectors that are provided by its members are aggregated by calculating the element-wise sum

$$\hat{\boldsymbol{p}}_n = F_T\left(\boldsymbol{x}_n\right) = \hat{\boldsymbol{p}}_n^1 + \cdots + \hat{\boldsymbol{p}}_n^T \in \mathbb{R}^K, \tag{3}$$

which can be discretized in a second step to obtain a binary label vector.

An advantage of gradient boosting is the capability to tailor the training process to a certain (surrogate) loss function. Given a loss $\ell$, an ensemble should be trained such that the global objective

$$\mathcal{R}\left(F_T\right) = \sum_{n=1}^{N} \ell\left(\boldsymbol{y}_n, \hat{\boldsymbol{p}}_n\right) + \sum_{t=1}^{T} \Omega\left(f_t\right) \tag{4}$$

is minimized. The use of a suitable *regularization term* $\Omega$ may help to avoid overfitting and to converge towards a global optimum, if the loss function is not convex.

Gradient boosting is based on constructing an ensemble of additive functions following an iterative procedure, where new ensemble members are added step by step. To direct the step-wise training process towards a model that optimizes the global objective in the limit, (4) is rewritten based on the derivatives of the loss function. Like many recent boosting-based approaches (e.g., [3,12] or [25]), we rely on the second-order Taylor approximation. Given a loss function that is twice differentiable, this results in the stagewise objective function

$$\widetilde{\mathcal{R}}\left(f_t\right) = \sum_{n=1}^{N} \left(\boldsymbol{g}_n \hat{\boldsymbol{p}}_n^t + \frac{1}{2} \hat{\boldsymbol{p}}_n^t \boldsymbol{H}_n \hat{\boldsymbol{p}}_n^t\right) + \Omega\left(f_t\right), \tag{5}$$

which should be minimized by the ensemble member that is added at the $t$-th training iteration. The gradient vector $\boldsymbol{g}_n = \left(g_{ni}\right)_{1 \leq i \leq K}$ consist of the first-order partial derivatives of $\ell$ with respect to predictions of the current model for an example $\boldsymbol{x}_n$ and labels $\lambda_1, \ldots, \lambda_K$. Accordingly, the second-order partial derivatives form the Hessian matrix $\boldsymbol{H}_n = \left(h_{nij}\right)_{1 \leq i,j \leq K}$. The individual gradients and Hessians are formally defined as

$$g_i^n = \frac{\partial \ell}{\partial \hat{p}_{ni}}\left(\boldsymbol{y}_n, F_{t-1}\left(\boldsymbol{x}_n\right)\right) \quad \text{and} \quad h_{ij}^n = \frac{\partial \ell}{\partial \hat{p}_{ni} \partial \hat{p}_{nj}}\left(\boldsymbol{y}_n, F_{t-1}\left(\boldsymbol{x}_n\right)\right). \tag{6}$$

The confidence scores that are predicted by an ensemble member $f_t$ for individual labels must be chosen such that the stagewise objective in (5) is minimized. To derive a formula for calculating the predicted scores, the partial derivative of (5) with respect to the prediction for individual labels must be equated to zero. In case of a decomposable loss function, this results in a closed form solution that enables to compute the prediction for each label independently. In the general case, i.e., when the loss function is non-decomposable and the prediction should not be restricted to a single label, one obtains a system of $K$ linear equations

$$(H + R)\,\hat{\boldsymbol{p}} = -\boldsymbol{g}. \tag{7}$$

Whereas the elements of the Hessian matrix $H$ and the gradient vector $\boldsymbol{g}$ can be considered as coefficients and ordinates, the vector $\hat{\boldsymbol{p}}$ consists of the unknowns to be determined. The matrix $R$ is used to take the regularization into account. In this work, we use the $L_2$ regularization term

$$\Omega_{\text{L2}}\left(f_t\right) = \frac{1}{2}\omega \left\|\hat{\boldsymbol{p}}^t\right\|_2^2, \tag{8}$$

where $\|\boldsymbol{x}\|_2$ is the Euclidean norm and $\omega \geq 0$ controls the weight of the regularization. In this particular case, the regularization matrix $R = \text{diag}\,(\omega)$ is a diagonal matrix with the value $\omega$ on the diagonal.

As the target function to be minimized, we use the logistic loss function

$$\ell_{\text{ex.w.-log}}\left(\boldsymbol{y}_n, \hat{\boldsymbol{p}}_n\right) := \log\left(1 + \sum_{k=1}^{K} \exp\left(-y_{nk}\hat{p}_{nk}\right)\right), \tag{9}$$

which has previously been used with the BOOMER algorithm as a surrogate for the Subset 0/1 loss and was originally proposed by Amit et al. [1].

### 2.3 Ensembles of Multi-label Rules

We rely on multi-label rules as the individual building blocks of ensembles that are trained according to the methodology in Sect. 2.2. In accordance with (2), each rule can be considered as a function

$$f\left(\boldsymbol{x}\right) = b\left(\boldsymbol{x}\right)\hat{\boldsymbol{p}} \tag{10}$$

that predicts a vector of confidence scores for a given example.

The *body* of a rule $b : \mathcal{X} \to \{0, 1\}$ is a conjunction of one or several conditions that compare a given example's value for a particular attribute $A_l$ to a constant using a relational operator like $\leq$ and $>$, if the attribute is numerical, or $=$ and $\neq$, if it is nominal. If an example satisfies all conditions in the body, i.e., if $b\left(\boldsymbol{x}\right) = 1$, it is *covered* by the respective rule. In such case, the scores that are contained in the rule's *head* $\hat{\boldsymbol{p}} \in \mathbb{R}^K$ are returned. It assigns a positive or negative confidence score to each label, depending on whether the respective label is expected to be mostly relevant or irrelevant to the examples that belong to the region of the input space $\mathcal{X}$ that is covered by the rule. If an example is not covered, i.e., if $b\left(\boldsymbol{x}\right) = 0$, a null vector is returned. In such case, the rule does not have any effect on the overall prediction, as can be seen in (3).

If individual elements in a rule's head are set to zero, the rule does not provide a prediction for the corresponding labels. The experimental results reported by Rapp et al. [15] suggest that *single-label rules*, which only provide a non-zero prediction for a single label, tend to work well for minimizing decomposable loss functions. Compared to *multi-label rules*, which jointly predict for several labels, the induction of such rules is computationally less demanding, as a closed form solution for determining the predicted scores exists. However, the ability of multi-label rules to express local correlations between several labels, which hold for the examples they cover, has been shown to be crucial when it comes to non-decomposable losses.

To construct a rule that minimizes (5), we conduct a top-down greedy search, as it is commonly used in inductive rule learning (see, e.g., [7] for an overview on the topic). Initially, the search starts with an empty body that does not contain any conditions and therefore is satisfied by all examples. By adding new conditions to the body, the rule is successively specialized, resulting in less examples being covered in the process. For each candidate rule that results from adding a condition, the confidence scores to be predicted for the covered examples are calculated by solving (7). By substituting the calculated scores into (5), an estimate of the rule's quality is obtained. Among all possible refinements, the one that results in the greatest improvement in terms of quality is chosen. The search stops as soon as the rule cannot be improved by adding a condition.

Rules are closely related to the more commonly used decision trees, as each tree can be viewed as a set of non-overlapping rules. At each training iteration, a rule-based boosting algorithm focuses on a single region of the input space for which the model can be improved the most. In contrast, gradient boosted decision trees do always provide predictions for the entire input space. Due to their conceptual similarities, the ideas presented in this paper are not exclusive to rules, but can also be applied to decision trees.

## 3 Gradient-Based Label Binning

In this section, we present *Gradient-based Label Binning* (GBLB), a novel method that aims at reducing the computational costs of the multivariate boosting algorithm discussed in Sect. 2.2. Although the method can be used with any loss function, it is intended for use cases where a non-decomposable loss should be minimized. This is, because it explicitly addresses the computational bottleneck of such training procedure—the need to solve the linear system in (7)—which reduces to an operation with linear complexity in the decomposable case.

### 3.1 Complexity Analysis

The objective function in (5), each training iteration aims to minimize, depends on gradient vectors and Hessian matrices that correspond to individual training examples. Given $K$ labels, the former consist of $K$ elements, whereas the latter are symmetric matrices with $K\,(K+1)\,/\,2$ non-zero elements, one for each label, as well as for each pair of labels. The induction of a new rule, using a search algorithm as described in Sect. 2.3, requires to sum up the gradient vectors and Hessian matrices of the covered examples to form the linear system in (7). Instead of computing the sums for each candidate rule individually, the

**Algorithm 1:** Candidate evaluation without (left) / with GBLB (right)

**input** : Gradient vector $\boldsymbol{g}$, Hessian matrix $H$, $L_2$ regularization weight $\omega$
**output**: Predictions $\hat{\boldsymbol{y}}$, quality score $s$, mapping $\boldsymbol{m}$ (if GBLB is used)

| Without GBLB | With GBLB |
|---|---|
| 1 Regularization matrix $R = \text{diag}(\omega)$ | Mapping $\boldsymbol{m} = \text{MAP\_TO\_BINS}(\boldsymbol{g}, H, \omega)$<br>$\boldsymbol{g}, H, R = \text{AGGREGATE}(\boldsymbol{m}, \boldsymbol{g}, H, \omega)$ |
| **2** $\hat{\boldsymbol{y}} = \text{DSYSV}(-\boldsymbol{g}, H + R)$ ▷ cf. (7) | |
| **3** $s = \text{DDOT}(\hat{\boldsymbol{y}}, \boldsymbol{g}) +$<br>$(0.5 \cdot \text{DDOT}(\hat{\boldsymbol{y}}, \text{DSPMV}(\hat{\boldsymbol{y}}, H)))$ ▷ cf. (5) | *same as left* |
| **4 return** $\hat{\boldsymbol{y}}, s$ | **return** $\hat{\boldsymbol{y}}, s, \boldsymbol{m}$ |

candidates are processed in a predetermined order, such that each one covers one or several additional examples compared to its predecessor (see, e.g., [14] for an early description of this idea). As a result, an update of the sums with complexity $\mathcal{O}(K^2)$ must be performed for each example and attribute that is considered for making up new candidates.

Algorithm 1 shows the steps that are necessary to compute the confidence scores to be predicted by an individual candidate rule, as well as a score that assesses its quality, if the loss function is non-decomposable. The modifications that are necessary to implement GBLB are shown to the right of the original lines of code. Originally, the given gradient vector and Hessian matrix, which result from summation over the covered examples, are used as a basis to solve the linear system in (7) using the Lapack routine DSYSV (cf. Algorithm 1, line 2). The computation of a corresponding quality score by substituting the calculated scores into (5), involves invocations of the Blas operations DDOT for vector-vector multiplication, as well as DSPMV for vector-matrix multiplication (cf. Algorithm 1, line 3). Whereas the operation DDOT comes with linear costs, the DSPMV and DSYSV routines have quadratic and cubic complexity, i.e., $\mathcal{O}(K^2)$ and $\mathcal{O}(K^3)$, respectively[1]. As Algorithm 1 must be executed for each candidate rule, it is the computationally most expensive operation that takes part in a multivariate boosting algorithm aimed at the minimization of a non-decomposable loss function.

GBLB addresses the computational complexity of Algorithm 1 by mapping the available labels to a predefined number of bins $B$ and aggregating the elements of the gradient vector and Hessian matrix accordingly. If $B \ll K$, this significantly reduces their dimensionality and hence limits the costs of the Blas and Lapack routines. As a result, given that the overhead introduced by the mapping and aggregation functions is small, we expect an overall reduction in training time.

In this work, we do not address the computational costs of summing up the gradients that correspond to individual examples. However, the proposed method has been designed such that it can be combined with methods that are dedicated to this aspect. Albeit restricting themselves to decomposable losses, Si et al. [18]

[1] Information on the complexity of the Blas and Lapack routines used in this work can be found at http://www.netlib.org/lapack/lawnspdf/lawn41.pdf.

have proposed a promising method that ensures that many gradients evaluate to zero. This approach, which was partly adopted by Zhang and Jung [25], restricts the labels that must be considered to those with non-zero gradients. However, to maintain sparsity among the gradients, strict requirements must be fulfilled by the loss function. Among many others, the logistic loss function in (9) does not meet these requirements. The approach that is investigated in this work does not impose any restrictions on the loss function.

### 3.2 Mapping Labels to Bins

GBLB evolves around the idea of assigning the available labels $\lambda_1, \dots, \lambda_K$ to a predefined number of bins $\mathcal{B}_1, \dots, \mathcal{B}_B$ whenever a potential ensemble member is evaluated during training (cf. MAP_TO_BINS in Algorithm 1). To obtain the index of the bin, a particular label $\lambda_k$ should be assigned to, we use a mapping function $m : \mathbb{R} \rightarrow \mathbb{N}^+$ that depends on a given criterion $c_k \in \mathbb{R}$. In this work, we use the criterion

$$c_k = -\frac{g_k}{h_{kk} + \omega}, \tag{11}$$

which takes the gradient and Hessian for the respective label, as well as the $L_2$ regularization weight, into account. It corresponds to the optimal prediction when considering the label in isolation, i.e., when assuming that the predictions for other labels will be zero. As the criterion can be obtained for each label individually, the computational overhead is kept at a minimum.

Based on the assignments that are provided by a mapping function $m$, we denote the set of label indices that belong to the $b$-th bin as

$$\mathcal{B}_b = \{k \in \{1, \dots, K\} \mid m(c_k) = b\}. \tag{12}$$

Labels should be assigned to the same bin if the corresponding confidence scores, which will be presumably be predicted by an ensemble member, are close to each other. If the optimal scores to be predicted for certain labels are very different in absolute size or even differ in their sign, the respective labels should be mapped to different bins. Based on this premise, we limit the number of distinct scores, an ensemble member may predict, by enforcing the restriction

$$\hat{p}_i = \hat{p}_j, \forall i, j \in \mathcal{B}_b. \tag{13}$$

It requires that a single score is predicted for all labels that have been assigned to the same bin. Given that the mentioned prerequisites are met, we expect the difference between the scores that are predicted for a bin and those that are optimal with respect to its individual labels to be reasonably small.

### 3.3 Equal-Width Label Binning

Principally, different approaches to implement the mapping function $m$ are conceivable. We use *equal-width* binning, as this well-known method provides two advantages: First, unlike other methods, such as equal-frequency binning, it does not involve sorting and can therefore be applied in linear time. Second, the boundaries of the bins are chosen such that the absolute difference between the smallest and largest value in a bin, referred to as the *width*, is the same for all bins. As argued in Sect. 3.2, this is a desirable property in our particular use case. Furthermore, we want to prevent labels, for which the predicted score should be negative, from being assigned to the same bin as labels, for which the prediction should be positive. Otherwise, the predictions would be suboptimal for some of these labels. We therefore strictly separate between *negative* and *positive bins.* Given $B_\ominus$ negative and $B_\oplus$ positive bins, the width calculates as

$$w_\ominus = \frac{\max_\ominus - \min_\ominus}{B_\ominus} \quad \text{and} \quad w_\oplus = \frac{\max_\oplus - \min_\oplus}{B_\oplus}, \tag{14}$$

for the positive and negative bins, respectively. By $\max_\ominus$ and $\max_\oplus$ we denote the largest value in $\{c_1, \dots, c_K\}$ with negative and positive sign, respectively. Accordingly, $\min_\ominus$ and $\min_\oplus$ correspond to the smallest value with the respective sign. Labels for which $c_k = 0$, i.e., labels with zero gradients, can be ignored. As no improvement in terms of the loss function can be expected, we explicitly set the prediction to zero in such case.

Once the width of the negative and positive bins has been determined, the mapping from individual labels to one of the $B = B_\ominus + B_\oplus$ bins can be obtained via the function

$$m_{\text{eq.-width}}(c_k) = \begin{cases} \min\left(\lfloor \frac{c_k - \min_\ominus}{w_\ominus} \rfloor + 1, B_\ominus\right), & \text{if } c_k < 0 \\ \min\left(\lfloor \frac{c_k - \min_\oplus}{w_\oplus} \rfloor + 1, B_\oplus\right) + B_\ominus, & \text{if } c_k > 0. \end{cases} \tag{15}$$

### 3.4 Aggregation of Gradients and Hessians

By exploiting the restriction introduced in (13), the gradients and Hessians that correspond to labels in the same bin can be aggregated to obtain a gradient vector and a Hessian matrix with reduced dimensions (cf. AGGREGATE in Algorithm 1). To derive a formal description of this aggregation, we first rewrite the objective function (5) in terms of sums instead of using vector and matrix multiplications. This results in the formula

$$\widetilde{\mathcal{R}}(f_t) = \sum_{n=1}^{N} \sum_{i=1}^{K} \left( g_i^n \hat{p}_i + \frac{1}{2} \hat{p}_i \left( h_{ii}^n \hat{p}_i + \sum_{\substack{j=1, \\ j \neq i}}^{K} h_{ij}^n \hat{p}_j \right) \right) + \Omega(f_t). \tag{16}$$

Based on the constraint given in (13) and due to the distribution property of the multiplication, the equality

$$\sum_{i=1}^{K} x_i \hat{p}_i = \sum_{j=1}^{B} \left( \hat{p}_j \sum_{i \in \mathcal{B}_j} x_i \right), \tag{17}$$

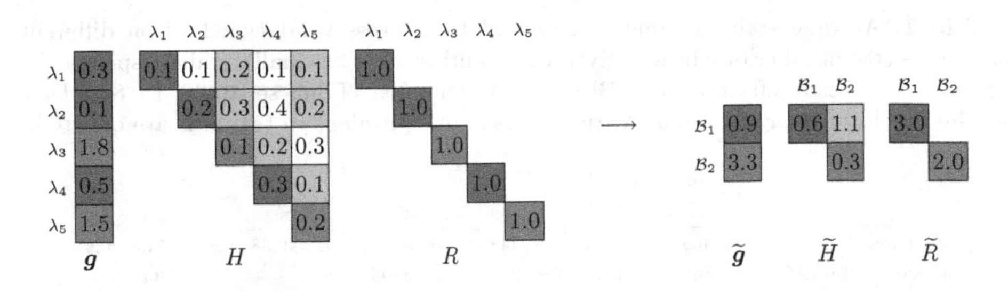


**Fig. 1.** Illustration of how a gradient vector, a Hessian matrix and a regularization matrix for five labels $\lambda_1, \dots, \lambda_5$ are aggregated with respect to two bins $\mathcal{B}_1 = \{1, 2, 4\}$ and $\mathcal{B}_2 = \{3, 5\}$ when using $L_2$ regularization with $\omega = 1$. Elements with the same color are added up for aggregation.

where $x_i$ is any term dependent on $i$, holds. It can be used to rewrite (16) in terms of sums over the bins, instead of sums over the individual labels. For brevity, we denote the sum of the gradients, as well as the sum of the elements on the diagonal of the Hessian matrix, that correspond to the labels in bin $\mathcal{B}_b$ as

$$\widetilde{g}_b = \sum_{i \in \mathcal{B}_b} g_i \quad \text{and} \quad \widetilde{h}_{bb} = \sum_{i \in \mathcal{B}_b} h_{ii}. \tag{18}$$

To abbreviate the sum of Hessians that correspond to a pair of labels that have been assigned to different bins $\mathcal{B}_b$ and $\mathcal{B}_q$, we use the short-hand notation

$$\widetilde{h}_{bq} = \sum_{i \in \mathcal{B}_b} \sum_{j \in \mathcal{B}_q} h_{ij}. \tag{19}$$

By exploiting (17) and using the abbreviations introduced above, the objective function in (16) can be rewritten as

$$\widetilde{\mathcal{R}}(f_t) = \sum_{n=1}^{N} \sum_{b=1}^{B} \left( \hat{p}_b \widetilde{g}_b^n + \frac{1}{2} \hat{p}_b \left( \hat{p}_b \widetilde{h}_b^n + \sum_{\substack{q=1,\\ q \neq b}}^{B} \hat{p}_q \widetilde{h}_{bq}^n \right) \right) + \Omega(f_t), \tag{20}$$

which can afterwards be turned into the original notation based on vector and matrix multiplications. The resulting formula

$$\widetilde{\mathcal{R}}(f_t) = \sum_{n=1}^{N} \left( \widetilde{\boldsymbol{g}}_n \hat{\boldsymbol{p}}_n^t + \frac{1}{2} \hat{\boldsymbol{p}}_n^t \widetilde{H}_n \hat{\boldsymbol{p}}_n^t \right) + \Omega(f_t) \tag{21}$$

has the same structure as originally shown in (5). However, the gradient vector $\boldsymbol{g}$ and the Hessian matrix $H$ have been replaced by $\widetilde{\boldsymbol{g}}$ and $\widetilde{H}$, respectively. Consequently, when calculating the scores to be predicted by an ensemble member

**Table 1.** Average training times (in seconds) per cross validation fold on different data sets (the number of labels is given in parentheses). The small numbers specify the speedup that results from using GBLB with the number of bins set to 32, 16, 8 and 4% of the labels, or using two bins. Variants that are equivalent to two bins are omitted.

| | | No | GBLB | | | | | | | | | |
|---|---|---|---|---|---|---|---|---|---|---|---|---|
| | | GBLB | 32% | | 16% | | 8% | | 4% | | 2 bins | |
| Eurlex-sm | (201) | 46947 | 54985 | 0.85 | 44872 | 1.05 | 38222 | 1.23 | 33658 | 1.39 | **21703** | 2.16 |
| EukaryotePseAAC | (22) | 16033 | 3593 | 4.46 | 2492 | 6.43 | 2195 | 7.30 | — | | **1534** | 10.45 |
| Reuters-K500 | (103) | 12093 | 6930 | 1.75 | 4197 | 2.88 | 3353 | 3.61 | 2803 | 4.31 | **2743** | 4.41 |
| Bibtex | (159) | 2507 | 2599 | 0.96 | 2765 | 0.91 | 2649 | 0.95 | 2456 | 1.02 | **2125** | 1.18 |
| Yeast | (14) | 2338 | 998 | 2.34 | 761 | 3.07 | 525 | 4.45 | — | | **521** | 4.49 |
| Birds | (19) | 2027 | 701 | 2.89 | 505 | 4.01 | 337 | 6.01 | — | | **336** | 6.03 |
| Yahoo-Social | (39) | 1193 | 261 | 4.57 | 217 | 5.50 | 192 | 6.21 | **139** | 8.58 | 175 | 6.82 |
| Yahoo-Computers | (33) | 874 | 172 | 5.08 | 134 | 6.52 | 126 | 6.94 | **101** | 8.65 | 123 | 7.11 |
| Yahoo-Science | (40) | 735 | 200 | 3.67 | 160 | 4.59 | 135 | 5.44 | **106** | 6.93 | 136 | 5.40 |
| Yahoo-Reference | (33) | 571 | 174 | 3.28 | 141 | 4.05 | 129 | 4.43 | **110** | 5.19 | 137 | 4.17 |
| Slashdot | (20) | 518 | 154 | 3.36 | 117 | 4.43 | **86** | 6.02 | — | | 119 | 4.35 |
| EukaryoteGO | (22) | 191 | 79 | 2.42 | 74 | 2.58 | **60** | 3.18 | — | | 64 | 2.98 |
| Enron | (53) | 181 | 69 | 2.62 | 52 | 3.48 | 48 | 3.77 | 47 | 3.85 | **44** | 4.11 |
| Medical | (45) | 170 | 60 | 2.83 | 57 | 2.98 | 55 | 3.09 | **50** | 3.40 | 51 | 3.33 |
| Langlog | (75) | 132 | 126 | 1.05 | 112 | 1.18 | 105 | 1.26 | **101** | 1.31 | 102 | 1.29 |
| Avg. Speedup | | | | 2.81 | | 3.58 | | 4.61 | | **4.86** | | 4.00 |

by solving (7), the coefficients and ordinates that take part in the linear system do not correspond to individual labels, but result from the sums in (18) and (19). As a result, number of linear equations has been reduced from the number of labels $K$ to the number of non-empty bins, which is at most $B$.

An example that illustrates the aggregation of a gradient vector and a Hessian matrix is given in Fig. 1. It also takes into account how the regularization matrix $R$ is affected. When dealing with bins instead of individual labels, the $L_2$ regularization term in (8) becomes

$$\Omega_{\mathrm{L2}}\left(f_t\right) = \frac{1}{2}\omega \sum_{b=1}^{B} \left( \left|\mathcal{B}_b\right| \hat{p}_b^2 \right), \tag{22}$$

where $|\mathcal{B}_b|$ denotes the number of labels that belong to a particular bin. As a consequence, the regularization matrix becomes $\widetilde{R} = \operatorname{diag}\left(\omega\left|\mathcal{B}_1\right|, \ldots, \omega\left|\mathcal{B}_B\right|\right)$.

## 4 Evaluation

To investigate in isolation the effects GBLB has on predictive performance and training time, we chose a single configuration of the BOOMER algorithm as the basis for our experiments. We used 10-fold cross validation to train models that are aimed at the minimization of the Subset 0/1 loss on commonly used benchmark data sets[2]. Each model consists of 5.000 rules that have been learned on

[2] All data sets are available at https://www.uco.es/kdis/mllresources.

**Table 2.** Predictive performance of different approaches in terms of the Subset 0/1 loss and the Hamming loss (smaller values are better).

| | | Label-wise | No GBLB | GBLB 32% | 16% | 8% | 4% | 2 bins |
|---|---|---|---|---|---|---|---|---|
| Subset 0/1 loss | Eurlex-sm | 61.63 | 69.53 | 45.07 | **45.03** | 45.30 | 45.08 | 47.32 |
| | EukaryotePseAAC | 85.09 | 65.43 | 65.37 | **65.28** | 65.68 | — | 65.52 |
| | Reuters-K500 | 71.37 | 71.07 | 53.70 | 53.22 | 53.07 | **52.90** | 53.40 |
| | Bibtex | 85.99 | 81.31 | **77.28** | 77.44 | 77.55 | 77.32 | 78.95 |
| | Yeast | 84.94 | 76.54 | 76.91 | 76.42 | 76.87 | — | **76.21** |
| | Birds | 45.29 | 45.30 | **45.14** | 45.45 | 45.60 | — | 46.53 |
| | Yahoo-Social | 50.65 | 64.30 | 34.49 | **34.40** | 34.84 | 35.47 | 35.37 |
| | Yahoo-Computers | 58.04 | **46.26** | 46.65 | 47.09 | 46.94 | 47.70 | 47.42 |
| | Yahoo-Science | 74.00 | 85.80 | **50.89** | 51.20 | 52.07 | 52.79 | 52.04 |
| | Yahoo-Reference | 58.19 | 74.14 | **39.82** | 40.16 | 40.73 | 40.51 | 40.48 |
| | Slashdot | 63.88 | **46.62** | 46.64 | 46.64 | 47.73 | — | 47.22 |
| | EukaryoteGO | 30.63 | 28.35 | 28.39 | 28.24 | **28.10** | — | 28.55 |
| | Enron | 88.19 | 83.14 | 83.32 | 83.32 | 83.38 | **82.91** | 82.97 |
| | Medical | 28.25 | 28.82 | 23.13 | **22.62** | 23.08 | 23.23 | 22.77 |
| | Langlog | 79.59 | 78.84 | 79.11 | 79.25 | **78.63** | 79.45 | 79.45 |
| Hamming loss | Eurlex-sm | 0.55 | 0.91 | 0.40 | **0.39** | 0.40 | 0.40 | 0.42 |
| | EukaryotePseAAC | **5.02** | 5.65 | 5.64 | 5.63 | 5.67 | — | 5.66 |
| | Reuters-K500 | 1.11 | 1.71 | 1.11 | **1.09** | **1.09** | **1.09** | 1.10 |
| | Bibtex | **1.25** | 1.45 | 1.27 | 1.27 | 1.27 | 1.28 | 1.31 |
| | Yeast | 19.75 | 19.01 | 18.87 | 19.08 | 19.01 | — | **18.80** |
| | Birds | 3.91 | 3.79 | 3.80 | 3.79 | **3.73** | — | 3.87 |
| | Yahoo-Social | 1.90 | 3.81 | **1.79** | 1.80 | 1.83 | 1.87 | 1.87 |
| | Yahoo-Computers | 3.10 | **2.97** | 3.00 | 3.02 | 3.03 | 3.08 | 3.06 |
| | Yahoo-Science | 2.83 | 5.85 | **2.74** | 2.75 | 2.81 | 2.84 | 2.79 |
| | Yahoo-Reference | 2.30 | 4.95 | **2.28** | 2.30 | 2.34 | 2.33 | 2.32 |
| | Slashdot | **4.02** | 4.24 | 4.24 | 4.25 | 4.37 | — | 4.30 |
| | EukaryoteGO | **1.89** | 1.95 | 1.95 | 1.94 | 1.92 | — | 1.98 |
| | Enron | **4.53** | 4.72 | 4.77 | 4.77 | 4.72 | 4.72 | 4.73 |
| | Medical | 0.84 | 1.05 | 0.80 | **0.77** | 0.79 | 0.81 | 0.79 |
| | Langlog | 1.52 | 1.52 | **1.50** | 1.51 | **1.50** | 1.52 | 1.52 |

varying subsets of the training examples, drawn with replacement. The refinement of rules has been restricted to random subsets of the available attributes. As the learning rate and the $L_2$ regularization weight, we used the default values 0.3 and 1.0, respectively. Besides the original algorithm proposed in [15], we tested an implementation that makes use of GBLB[3]. For a broad analysis, we set the maximum number of bins to 32, 16, 8, and 4% of the available labels. In addition, we investigated an extreme setting with two bins, where all labels with positive and negative criteria are assigned to the same bin, respectively.

Table 1 shows the average time per cross validation fold that is needed by the considered approaches for training. Compared to the baseline that does not use GBLB, the training time can always be reduced by utilizing GBLB with a

[3] An implementation is available at https://www.github.com/mrapp-ke/Boomer.

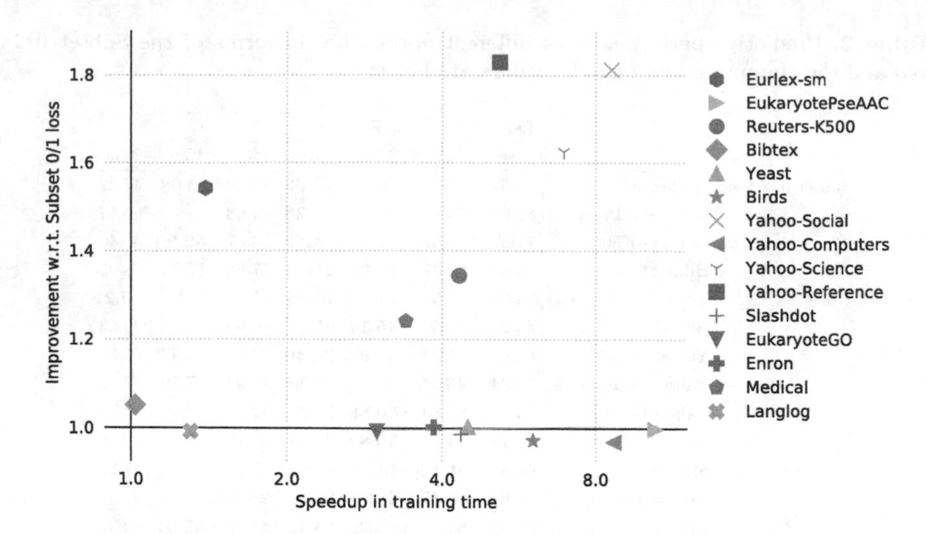


**Fig. 2.** Relative difference in training time and Subset 0/1 loss (both calculated as the baseline's value divided by the value of the respective approach) per cross validation fold that results from using GBLB with the number of bins set to 4% of the labels.

suitable number of bins. Using fewer bins tends to speed up the training process, although approaches that use the fewest bins are not always the fastest ones. On average, limiting the number of bins to 4% of the labels results in the greatest speedup (by factor 5). However, the possible speedup depends on the data set at hand. E.g., on the data set "EukaryotePseAac" the average training time is reduced by factor 10, whereas no significant speedup is achieved for "Bibtex".

To be useful in practice, the speedup that results from GBLB should not come with a significant deterioration in terms of the target loss. We therefore report the predictive performance of the considered approaches in Table 2. Besides the Subset 0/1 loss, which we aim to minimize in this work, we also include the Hamming loss as a commonly used representative of decomposable loss functions. When focusing on the Subset 0/1 loss, we observe that the baseline algorithm without GBLB exhibits subpar performance on some data sets, namely "Eurlex-sm", "Reuters-K500", "Bibtex", "Yahoo-Social", "Yahoo-Science", "Yahoo-Reference" and "Medical". This becomes especially evident when compared to an instantiation of the algorithm that targets the Hamming loss via minimization of a label-wise decomposable logistic loss function (cf. [15], Eq. 6). In said cases, the latter approach performs better even though it is not tailored to the Subset 0/1 loss. Although the baseline performance could most probably be improved by tuning the regularization weight, we decided against parameter tuning, as it exposes an interesting property of GBLB. On the mentioned data sets, approaches that use GBLB appear to be less prone to converge towards local minima. Regardless of the number of bins, they clearly outperform the baseline. According to the Friedman test, these differences are significant with $\alpha = 0.01$. The Nemenyi post-hoc test yields critical distances for each of

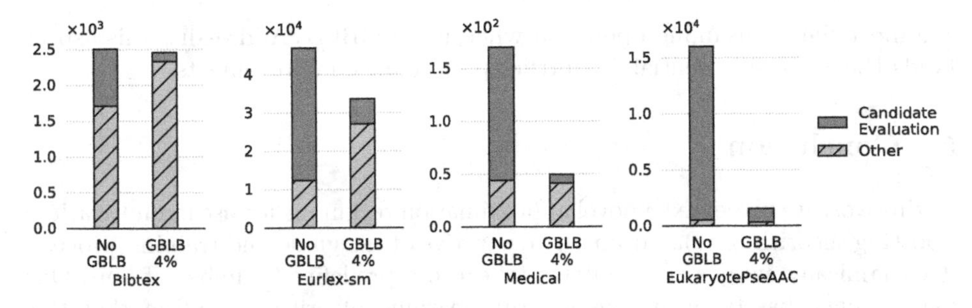


**Fig. 3.** Average proportion of training time per cross validation fold that is used for the evaluation of candidate rules without GBLB and when using GBLB with the number of bins set to 4% of the labels.

the GBLB-based approaches, when compared to the baseline. On the remaining data sets, where the baseline without GBLB already performs well, the use of GBLB produces competitive results. In these cases, the Friedman test confirms the null hypothesis with $\alpha = 0.1$. An overview of how the training time and the predictive performance in terms of the Subset 0/1 loss is affected, when restricting the number of bins to 4% of the labels, is given in Fig. 2.

To better understand the differences in speedups that may be achieved by using GBLB, a detailed analysis is given in the following for four data sets with varying characteristics. In Fig. 3, we depict the training time that is needed by the baseline approach, as well as by a GBLB-based approach with the number of bins set to 4% of the labels. Besides the total training time, we also show the amount of time spent on the evaluation of candidate rules (cf. Algorithm 1), which is the algorithmic aspect addressed by GBLB. For all given scenarios, it can be seen that the time needed for candidate evaluation could successfully be reduced. Nevertheless, the effects on the overall training time vastly differ. On the data sets "Bibtex" and "Eurlex-sm", the time spent on parts of the algorithm other than the candidate evaluation increased when using GBLB, which is a result of more specific rules being learned. On the one hand, this required more candidates to be evaluated and therefore hindered the overall speedup. On the other hand, the resulting rules clearly outperformed the baseline according to Table 2. On the data set "Bibtex", even without GBLB, the candidate evaluation was not the most expensive aspect of training. Due to its binary attributes, the number of potential candidates is small compared to the large number of examples. As a result, most of the computation time is spent on summing up the gradients and Hessians of individual examples (cf. Sect. 3.1). The impact of speeding up the candidate evaluation is therefore limited. On the data sets "Medical" and "EukaryotePseAAC", where the candidate evaluation was the most expensive aspect to begin with, a significant reduction of training time could be achieved by making that particular operation more efficient. The time spent on other parts of the algorithm remained mostly unaffected in these cases. As mentioned earlier, this includes the summation of gradients and Hessians, which becomes the most time consuming operation when using GBLB. Addressing this aspect holds the greatest potential for further performance improvements.

## 5 Conclusion

In this work we presented a novel approximation technique for use in multivariate boosting algorithms. Based on the derivatives that guide the training process, it dynamically assigns the available labels to a predefined number of bins. Our experiments, based on an existing rule learning algorithm, confirm that this reduction in dimensionality successfully reduces the training time that is needed for minimizing non-decomposable loss functions, such as the Subset 0/1 loss. According to our results, this speedup does not come with any significant loss in predictive performance. In several cases the proposed method even outperforms the baseline by a large extend due to its ability to overcome local minima without the necessity for extensive parameter tuning.

Despite our promising results, the use of non-decomposable loss functions in the boosting framework remains computationally challenging. Based on the analysis in this paper, we plan to extend our methodology with the ability to exploit sparsity in the label space. When combined with additional measures, the proposed method could become an integral part of more efficient algorithms that are capable of natively minimizing non-decomposable loss functions.

**Acknowledgments.** This work was supported by the German Research Foundation (DFG) under grant number 400845550. Computations were conducted on the Lichtenberg high performance computer of the TU Darmstadt.

## References

1. Amit, Y., Dekel, O., Singer, Y.: A boosting algorithm for label covering in multilabel problems. In: Proceedings of 11th International Conference on AI and Statistics (AISTATS), pp. 27–34 (2007)
2. Bhatia, K., Jain, H., Kar, P., Varma, M., Jain, P.: Sparse local embeddings for extreme multi-label classification. In: Proceedings of 28th International Conference on Neural Information Processing Systems (NIPS), pp. 730–738 (2015)
3. Chen, T., Guestrin, C.: XGBoost: a scalable tree boosting system. In: Proceedings of 22nd International Conference on Knowledge Discovery and Data Mining (KDD), pp. 785–794 (2016)
4. Cheng, W., Hüllermeier, E., Dembczyński, K.: Bayes optimal multilabel classification via probabilistic classifier chains. In: Proceedings of 27th International Conference on Machine Learning (ICML), pp. 279–286 (2010)
5. Dembczyński, K., Kotłowski, W., Hüllermeier, E.: Consistent multilabel ranking through univariate losses. In: Proceedings of 29th International Conference on Machine Learning (ICML), pp. 1319–1326 (2012)
6. Dembczyński, K., Waegeman, W., Cheng, W., Hüllermeier, E.: On label dependence and loss minimization in multi-label classification. Mach. Learn. **88**(1-2), 5–45 (2012)
7. Fürnkranz, J., Gamberger, D., Lavrač, N.: Foundations of Rule Learning. Springer Science & Business Media (2012)

8. Gibaja, E., Ventura, S.: Multi-label learning: a review of the state of the art and ongoing research. Wiley Interdisciplinary Rev. Data Mining Knowl. Discovery **4**(6), 411–444 (2014)
9. Huang, K.H., Lin, H.T.: Cost-sensitive label embedding for multi-label classification. Mach. Learn. **106**(9), 1725–1746 (2017)
10. Johnson, M., Cipolla, R.: Improved image annotation and labelling through multi-label boosting. In: Proceedings of British Machine Vision Conference (BMVC) (2005)
11. Jung, Y.H., Tewari, A.: Online boosting algorithms for multi-label ranking. In: Proceedings of 21st International Conference on AI and Statistics (AISTATS), pp. 279–287 (2018)
12. Ke, G., et al.: LightGBM: a highly efficient gradient boosting decision tree. In: Proceedings of 31st International Conference on Neural Information Processing Systems (NIPS) (2017)
13. Kumar, V., Pujari, A.K., Padmanabhan, V., Kagita, V.R.: Group preserving label embedding for multi-label classification. Pattern Recogn. **90**, 23–34 (2019)
14. Mehta, M., Agrawal, R., Rissanen, J.: SLIQ: a fast scalable classifier for data mining. In: Proceedings of International Conference on Extending Database Technology, pp. 18–32 (1996)
15. Rapp, M., Loza Mencía, E., Fürnkranz, J., Nguyen, V.L., Hüllermeier, E.: Learning gradient boosted multi-label classification rules. In: Proceedings of European Conference on Machine Learning and Knowledge Discovery in Databases (ECML-PKDD), pp. 124–140 (2020)
16. Read, J., Pfahringer, B., Holmes, G., Frank, E.: Classifier chains for multi-label classification. In: Proceedings of European Conference on Machine Learning and Knowledge Discovery in Databases (ECML-PKDD), pp. 254–269 (2009)
17. Schapire, R.E., Singer, Y.: BoosTexter: a boosting-based system for text categorization. Mach. Learn. **39**(2), 135–168 (2000)
18. Si, S., Zhang, H., Keerthi, S.S., Mahajan, D., Dhillon, I.S., Hsieh, C.J.: Gradient boosted decision trees for high dimensional sparse output. In: Proceedings of 34th International Conference on Machine Learning (ICML) pp. 3182–3190 (2017)
19. Sun, L., Ji, S., Ye, J.: Canonical correlation analysis for multilabel classification: a least-squares formulation, extensions, and analysis. IEEE Trans. Pattern Anal. Mach. Intell. **33**(1), 194–200 (2010)
20. Tai, F., Lin, H.T.: Multilabel classification with principal label space transformation. Neural Comput. **24**(9), 2508–2542 (2012)
21. Tsoumakas, G., Katakis, I., Vlahavas, I.: Effective and efficient multilabel classification in domains with large number of labels. In: Proceedings of ECML-PKDD 2008 Workshop on Mining Multidimensional Data, pp. 53–59 (2008)
22. Tsoumakas, G., Katakis, I., Vlahavas, I.: Mining multi-label data. In: Maimon, O., Rokach, L. (eds.) Data Mining and Knowledge Discovery Handbook, pp. 667–685. Springer, Boston (2010). https://doi.org/10.1007/978-0-387-09823-4_34
23. Tsoumakas, G., Vlahavas, I.: Random k-labelsets: an ensemble method for multi-label classification. In: Proceedings of European Conference on Machine Learning (ECML), pp. 406–417 (2007)
24. Zhang, M.L., Zhou, Z.H.: A review on multi-label learning algorithms. IEEE Trans. Knowl. Data Eng. **26**(8), 1819–1837 (2014)
25. Zhang, Z., Jung, C.: GBDT-MO: gradient-boosted decision trees for multiple outputs. IEEE Trans. Neural Networks Learn. Syst. (2020)
26. Zhou, T., Tao, D., Wu, X.: Compressed labeling on distilled labelsets for multi-label learning. Mach. Learn. **88**(1–2), 69–126 (2012)

# Joint Geometric and Topological Analysis of Hierarchical Datasets

Lior Aloni(✉), Omer Bobrowski, and Ronen Talmon

Technion, Israel Institute of Technology, Haifa, Israel
lioral@campus.technion.ac.il, {omer,ronen}@ee.technion.ac.il

**Abstract.** In a world abundant with diverse data arising from complex acquisition techniques, there is a growing need for new data analysis methods. In this paper we focus on high-dimensional data that are organized into several hierarchical datasets. We assume that each dataset consists of complex samples, and every sample has a distinct irregular structure modeled by a graph. The main novelty in this work lies in the combination of two complementing powerful data-analytic approaches: topological data analysis (TDA) and geometric manifold learning. Geometry primarily contains local information, while topology inherently provides global descriptors. Based on this combination, we present a method for building an informative representation of hierarchical datasets. At the finer (sample) level, we devise a new metric between samples based on manifold learning that facilitates quantitative structural analysis. At the coarser (dataset) level, we employ TDA to extract qualitative structural information from the datasets. We showcase the applicability and advantages of our method on simulated data and on a corpus of hyper-spectral images. We show that an ensemble of hyper-spectral images exhibits a hierarchical structure that fits well the considered setting. In addition, we show that our new method gives rise to superior classification results compared to state-of-the-art methods.

**Keywords:** Manifold learning · Diffusion maps · Topological data analysis · Persistent homology · Geometric learning

## 1 Introduction

Modern datasets often describe complex processes and convey a mixture of a large number of natural and man-made systems. Extracting the essential information underlying such datasets poses a significant challenge, as these are often high-dimensional, multimodal, and without a definitive ground truth. Moreover, the analysis of such data is highly sensitive to measurement noise and other experimental factors, such as sensor calibration and deployment. In order to cope with

**Electronic supplementary material** The online version of this chapter (https://doi.org/10.1007/978-3-030-86523-8_29) contains supplementary material, which is available to authorized users.


N. Oliver et al. (Eds.): ECML PKDD 2021, LNAI 12977, pp. 478–493, 2021.
https://doi.org/10.1007/978-3-030-86523-8_29

such an abundance, various data-analytic approaches have been developed, aimed at capturing the structure of the data. These approaches are often unsupervised, and are designed specifically to address the "curse of dimensionality" in data.

In this paper we consider two complementing approaches for such "structural" data analysis. Both of these approaches are based on the assumption that in high-dimensional real-world data, most of the information is concentrated around an intrinsic low-dimensional structure. Recovering the simplified underlying structure may reveal the true degrees of freedom of the data, reduce measurement noise, and facilitate efficient subsequent processing and analysis. The first approach we consider focuses on the *geometry* of the data, and is called *manifold learning* [2,8,24,27]. The second approach focuses on the *topology* of the data, and in known as *topological data analysis* (TDA) [5,28]. The key difference between these approaches is the distinction between local and precise phenomena (captured by geometry), and global qualitative phenomena (captured by topology). Briefly, the goal in manifold learning is to obtain an accurate geometric representation of the manifold that best describes the data. This is commonly accomplished by approximating the Laplace-Beltrami operator of the manifold. On the other end, TDA promotes the analysis of shapes and networks using qualitative topological features that are coordinate-free and robust under various types of deformations (e.g. the existence of holes). Our goal here is to take advantage of the strengths of each of these approaches, and combine them into a powerful geometric-topological framework. Conceptually, the common thread between manifold learning and TDA is the premise that the true information underlying the data is encapsulated in the "network" of associations within the data. Here lies another key difference between these two approaches. Manifold learning methods traditionally represent such networks as graphs (i.e. nodes and edges). While graphs serve as a powerful model for various applications, this approach is limited since it can only capture *pairwise* relationships between nodes. However, it is highly conceivable that complex data and networks consist of much more intricate interactions, involving more than just two nodes at a time. The methods developed in TDA focus on hypergraphs (simplicial complexes) that allow for high-order associations to be incorporated into the model [10,14,19].

In this work, we propose to combine manifold learning and TDA in order to provide informative representations of high-dimensional data, under the assumption that they can be arranged into several *hierarchical datasets* as follows. We assume that we have a collection of datasets, each consists of several complex samples, where each individual sample has a distinct irregular structure that can be captured by a weighted graph. Such datasets arise in many applications from a broad range of fields such as cytometry and gene expression in bioinformatics [12,13], social and computer network analysis [18,22], medical imaging [16], and geophysical tomography [1]. Following this hierarchy, our proposed method operates at two separate scales. At the finer (sample) scale, we use an operator-theoretic approach to attach operators to individual samples (graphs), pairs of samples, triplets, quadruplets, etc. These operators quantitatively describe the structure of each sample separately, as well as the common structure across samples. Specifically, we use the norm of these operators as a measure of similarity between samples, facilitating a transition from operator-theoretic analysis to affinity-based analysis. At the coarser (dataset) scale, we employ TDA to extract qualitative information from the datasets. Concretely, we use *persistent homology* [11,29] as a topological signature for each dataset. Persistent homology is a topological-algebraic tool that captures information about connectivity and holes at various scales. It is computed over the ensemble of samples contained in each dataset, which we model as a weighted simplicial complex, where the weights are derived from the geometric operators computed at the finer scale. The signature provided by persistent homology comes with a natural metric (the Wasserstein distance [7]), allowing us at the final stage to compare the structure of different datasets.

To demonstrate the advantages of our method, we apply it to Hyper-Spectral Imaging (HSI) [6]. HSI is a sensing technique aimed to obtain the electromagnetic spectrum at each pixel within an image, with the purpose of finding objects, identifying materials, or detecting processes. We test our unsupervised method on categorical hyper-spectral images [1] and show that it accurately distinguishes between the different categories. In addition, we show that in a (supervised) classification task based on the attained (unsupervised) representation and metric, our method outperforms a competing method based on deep learning.

The main contributions of this work are: (i) We introduce a powerful combination between geometry and topology, taking advantage of both local and global information contained in data. (ii) We propose a method for analyzing hierarchical datasets, that is data-driven and "model-free" (i.e. does not require prior knowledge or a rigid model). (iii) We introduce a new notion of affinity between manifolds, quantifying their commonality.

## 2 Problem Formulation

In this work, we consider the following hierarchical structure. At the top level, we have a collection of $N_D$ *datasets*

$$\mathcal{D} = \{D_1, \dots, D_{N_D}\}.$$

These datasets may vary in size, shape and origin. However, we assume that they all have the same prototypical structure, as follows. Each dataset $D$ consists of a collection of $N$ *samples*

$$D = \{S_1, \dots, S_N\},$$

and each sample is a collection of $L$ *observations*

$$S_i = \{x_{i,1} \dots, x_{i,L}\}, \; 1 \le i \le N.$$

Note that $N$ and $L$ may vary between datasets (for simplicity we omit the dataset index), but all the samples within a single dataset are of the same size $L$.

Next, we describe the structure of a single dataset $D$. Let $\{\mathcal{M}_\ell\}_{\ell=1}^{M}$ be a set of latent manifolds, and let $\Pi$ be their product

$$\Pi = \mathcal{M}_1 \times \cdots \times \mathcal{M}_M. \tag{1}$$

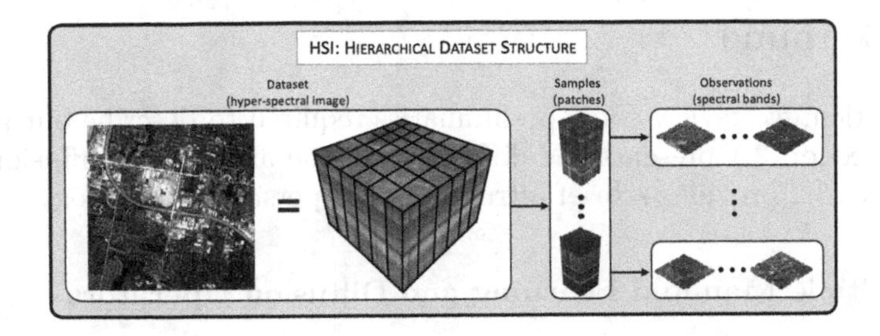


**Fig. 1.** The specification of the considered hierarchical structure for HSI.

We use $\Pi$ as a model for the common hidden space underlying the dataset $D$. Let $\mathcal{X} = \{x_1, \ldots, x_L\}$ be a set of points sampled from $\Pi$, where each point can be written as a tuple $x_j = (x_j^{(1)}, \ldots, x_j^{(M)})$ and $x_j^{(\ell)} \in \mathcal{M}_\ell$ for $\ell = 1, \ldots, M$.

Our main assumption here is that all samples $S_i$ are generated by the same set $\mathcal{X}$, while each sample contains information only about a subset of the manifolds in the product (1). The subset of manifolds corresponding to a sample $S_i$ is represented by a tuple of indices $I_i = (\ell_1^{(i)}, \ldots, \ell_m^{(i)})$ $(m \leq M)$, and a product manifold

$$\Pi_{I_i} = \mathcal{M}_{\ell_1^{(i)}} \times \cdots \times \mathcal{M}_{\ell_m^{(i)}}. \tag{2}$$

For convenience, for each $x \in \Pi$, we define the projection

$$x_{I_i} = (x^{(\ell_1^{(i)})}, \ldots, x^{(\ell_m^{(i)})}) \in \Pi_{I_i}.$$

Next, for each sample $S_i$ (and a corresponding subset $I_i$) we assume there is a function $g_i : \Pi_{I_i} \to \mathcal{O}_i$, for some target metric space $\mathcal{O}_i$. We define the observation function $f_i : \Pi \to \mathcal{O}_i$ as

$$f_i(x) = g_i(x_{I_i}) + \xi_i,$$

where $\xi_i \in \mathcal{O}_i$ denotes a random independent observation noise. Finally, the sample $S_i$ is defined as

$$S_i = \{x_{i,1}, \ldots, x_{i,L}\} = \{f_i(x_1), \ldots, f_i(x_L)\}.$$

In other words, each sample $S_i$ in the dataset $D$ reveals partial and noisy information about $\Pi$. As stated earlier, we assume that each of the datasets $D_k \in \mathcal{D}$ is generated by the model described above. However, the manifold $\Pi$, the functions $f_i, g_i$, and the parameters $L, M, N, I_i$ and $\xi_i$ may differ between datasets. Note that within a single dataset there is a correspondence between all samples $S_1, \ldots, S_N$, as they are generated by the same set of realizations $\mathcal{X}$.

In the context of HSI, the hierarchical structure described above is as follows. Each hyper-spectral image is viewed as a single dataset, the full spectrum of a single patch as a sample, and different spectral bands within a patch as observations (see Fig. 1 and Additional Materials for more details).

## 3 Background

In this section, we review some preliminaries required to describe our proposed method. Section 3.1 presents the diffusion and the alternating diffusion operators. Section 3.2 provides a brief introduction to persistent homology.

### 3.1 Multiple Manifold Learning and Diffusion Operators

Manifold learning is a class of unsupervised nonlinear data-driven methods for discovering the geometric structure underlying high dimensional data [2,8,24,27]. The main assumption in manifold learning is that high-dimensional data lie on a hidden lower-dimensional manifold.

One of the notable approaches in manifold learning is *diffusion maps* [8], in which diffusion operators built from data are shown to approximate the Laplace-Beltrami operator. This differential operator contains all the geometric information on the manifold [4,15], and thus its approximation provides means to incorporate geometric concepts such as metrics and embedding into data analysis tasks. In [17,26], an extension of diffusion maps for multiple datasets termed 'alternating diffusion' was introduced. This extension, which is based on the product of diffusion operators, was shown to recover the manifold structure common to multiple datasets. In this work, we utilize a variant of alternating diffusion, proposed in [25], which is briefly described in the remainder of this subsection.

Consider two diffeomorphic compact Riemannian manifolds without a boundary, denoted by $(\mathcal{M}_1, g_1)$ and $(\mathcal{M}_2, g_2)$, and a diffeomorphism $\phi : \mathcal{M}_1 \to \mathcal{M}_2$. For each manifold $\mathcal{M}_\ell$ ($\ell = \{1, 2\}$) and a pair of samples $x, x' \in \mathcal{M}_\ell$, let $k_\ell(x, x')$ be a Gaussian kernel based on the distance induced by the metric $g_\ell$ with a kernel scale $\epsilon_\ell > 0$. Define $d_\ell(x) = \int_{\mathcal{M}_\ell} k_\ell(x, x')\mu_\ell(x')dv_\ell(x')$, where $v_\ell(x)$ is the volume measure, and $\mu_\ell(x')$ is the density function of the samples on $\mathcal{M}_\ell$. Consider two kernel normalizations: $a_\ell(x, x') = \frac{k_\ell(x,x')}{d_\ell(x)}$ and $b_\ell(x, x') = \frac{k_\ell(x,x')}{d_\ell(x')}$. Based on the normalized kernel $a_\ell(x, x')$, define the *forward diffusion operator* by $A_\ell f(x) = \int_{\mathcal{M}_\ell} a_\ell(x, x')f(x')\mu_\ell(x')dv_\ell(x')$ for any function $f \in C^\infty(\mathcal{M}_\ell)$. Similarly, based on $b_\ell(x, x')$, define the *backward diffusion operator* by $B_\ell f(x) = \int_{\mathcal{M}_\ell} b_\ell(x, x')f(x')\mu_\ell(x')dv_\ell(x')$. In a single manifold setting, it is shown that when $\epsilon_\ell \to 0$, $A_\ell$ converges to a differential operator of an isotropic diffusion process on a space with non-uniform density $\mu_\ell(x)$, and $B_\ell$ converges to the backward Fokker-Planck operator, which coincides with the Laplace-Beltrami operator when the density $\mu_\ell(x)$ is uniform [21].

Next, for two manifolds, consider the following $C^\infty(\mathcal{M}_1) \to C^\infty(\mathcal{M}_1)$ composite operators,

$$Gf(x) = \phi^* A_2 (\phi^*)^{-1} B_1 f(x),$$

and

$$Hf(x) = A_1 \phi^* B_2 (\phi^*)^{-1} f(x)$$

for any function $f \in C^\infty(\mathcal{M}_1)$, where $\phi^* : C^\infty(\mathcal{M}_2) \to C^\infty(\mathcal{M}_1)$ denotes the pullback operator from $\mathcal{M}_2$ to $\mathcal{M}_1$ and $(\phi^*)^{-1}$ denotes the push-forward from $\mathcal{M}_2$ to $\mathcal{M}_1$, both corresponding to the diffeomorphism $\phi$.

In [17,26], it was shown that the two composite operators $G$ and $H$ recover the common structure between $\mathcal{M}_1$ and $\mathcal{M}_2$ and attenuate non-common structures that are often associated with noise and interference. In [25], the following symmetric alternating diffusion operator was introduced

$$S_{1,2}f(x) = \frac{1}{2}\Big(Gf(x) + Hf(x)\Big),$$

which, in addition to revealing the common structure as $G$ and $H$, has a real spectrum – a convenient property that allows to define spectral embeddings and spectral distances.

In practice, the operators defined above are approximated by matrices constructed from finite sets of data samples. Let $\{(x_i^{(1)}, x_i^{(2)})\}_{i=1}^L$ be a set of $L$ pairs of samples from $\mathcal{M}_1 \times \mathcal{M}_2$, such that $x_i^{(2)} = \phi(x_i^{(1)})$. For $\ell = \{1, 2\}$, let $\boldsymbol{W}_\ell$ be an $L \times L$ matrix, whose $(i, i')$-th element is given by $W_\ell(i, i') = k_\ell(x_i^{(\ell)}, x_{i'}^{(\ell)})$. Let $\boldsymbol{Q}_\ell = \text{diag}(\boldsymbol{W}_\ell \mathbf{1})$ be a diagonal matrix, where $\mathbf{1}$ is a column vector of all ones. The discrete counterparts of the operators $A_\ell$ and $B_\ell$ are then given by the matrices $\boldsymbol{K}_\ell = (\boldsymbol{Q}_\ell)^{-1}\boldsymbol{W}_\ell$ and $\boldsymbol{K}_\ell^T$, respectively, where $(\cdot)^T$ is the transpose operator. Consequently, the discrete counterparts of the operators $G$ and $H$ are $\boldsymbol{G} = \boldsymbol{K}_2\boldsymbol{K}_1^T$ and $\boldsymbol{H} = \boldsymbol{K}_1\boldsymbol{K}_2^T$, respectively. The symmetric matrix corresponding to the operator $S_{1,2}$ is then

$$\boldsymbol{S}_{1,2} = \boldsymbol{G} + \boldsymbol{H}.$$

### 3.2 Simplicial Complexes and Persistent Homology

At the heart of the topological layer in our proposed method, we will use (abstract) simplicial complexes to represent the structure of a dataset. Briefly, a simplicial complex is a discrete structure that contains vertices, edges, triangles and higher dimensional simplexes (i.e. it is a type of hypergraph). This collection has to be closed under inclusion – for every simplex we must also include all the faces on its boundary. See Fig. 2 in the Additional Materials for an example.

One of the many uses of simplicial complexes is in network modeling. While graphs take into account pairwise interactions between the nodes, simplicial complexes allow us to include information about the joint interaction of triplets, quadruplets, etc. We will use this property later, when studying the structure of samples within a dataset.

**Homology** is an algebraic structure that describes the shape of a topological space. Loosely speaking, for every topological space (e.g. a simplicial complex) we can define a sequence of vector spaces $H_0, H_1, H_2, \ldots$ where $H_0$ provides information about connected components, $H_1$ about closed loops surrounding holes, $H_2$ about closed surfaces enclosing cavities. Generally, we say that $H_k$ provides information about *k-dimensional cycles*, which can be thought of as

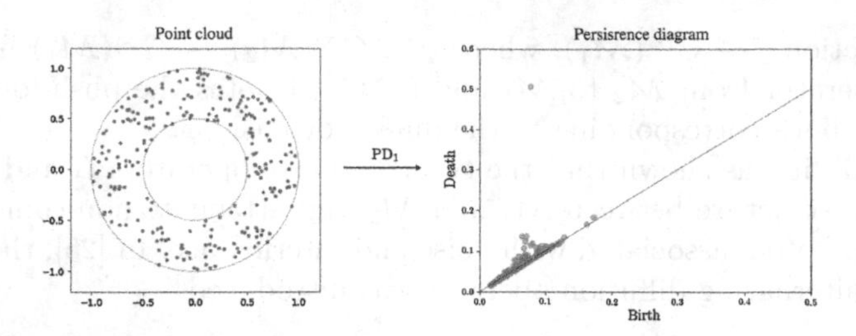


**Fig. 2.** Persistence diagram. Left: a point cloud generated in an annulus with a single hole (1-cycle). The filtration used is the union of balls around the points, for an increasing radius. Right: persistence diagram for 1-cycles. The birth/death axes represent radius values. The single feature away from the diagonal represents the hole of the annulus, while other cycles are considered "noise".

$k$-dimensional surfaces that are "empty" from within. We describe homology in more detail in Sect. 2.2 of the Additional Materials. In this work we will mainly use $H_0$ and $H_1$, i.e. information about connectivity and holes. However, the framework we develop can be used with any dimension of homology.

**Persistent homology** is one of the most heavily used tools in TDA [11, 29]. It can be viewed as a *multi-scale* version of homology, where instead of considering the structure of a single space, we track the evolution of cycles for a nested sequence of spaces, known as a *filtration*. As the spaces in a filtration grow, cycles of various dimensions may form (born) and later get filled in (die). The $k$-th persistent homology, denoted $\mathrm{PH}_k$, keeps a record of the birth-death process of $k$-cycles. Commonly, the information contained in $\mathrm{PH}_k$ is summarized using a *persistence diagram* – a set of points in $\mathbb{R}^2$ representing all the (birth,death) pairs for cycles in dimension $k$, and denoted $\mathrm{PD}_k$ (see Fig. 2). The motivation for using persistent homology is that it allows us to consider cycles at various scales, and identify those that seem to be prominent features of the data. In order to compare between persistence diagrams, we will employ the Wasserstein distance [7] defined as follows. Suppose that PD and PD$'$ are two persistence diagrams, then the $p$-Wasserstein distance is defined as

$$d_{W_p}(\mathrm{PD}, \mathrm{PD}') = \inf_{\phi:\widetilde{\mathrm{PD}} \to \widetilde{\mathrm{PD}}'} \Big( \sum_{\alpha \in \widetilde{\mathrm{PD}}} \|\alpha - \phi(\alpha)\|^p \Big)^{\frac{1}{p}}, \tag{3}$$

where $\widetilde{\mathrm{PD}}$ is an augmented version of PD that also includes the diagonal line $x = y$ (and the same goes for $\widetilde{\mathrm{PD}}'$). This augmentation is taken in order to allow cases where the $|\mathrm{PD}| \neq |\mathrm{PD}'|$. In other words, the Wasserstein distance is based on an optimal matching between features in PD and PD$'$.

## 4 Proposed Method

Recall the hierarchical dataset structure presented in Sect. 2. The processing method we propose for such datasets is hierarchical as well. At the fine level,

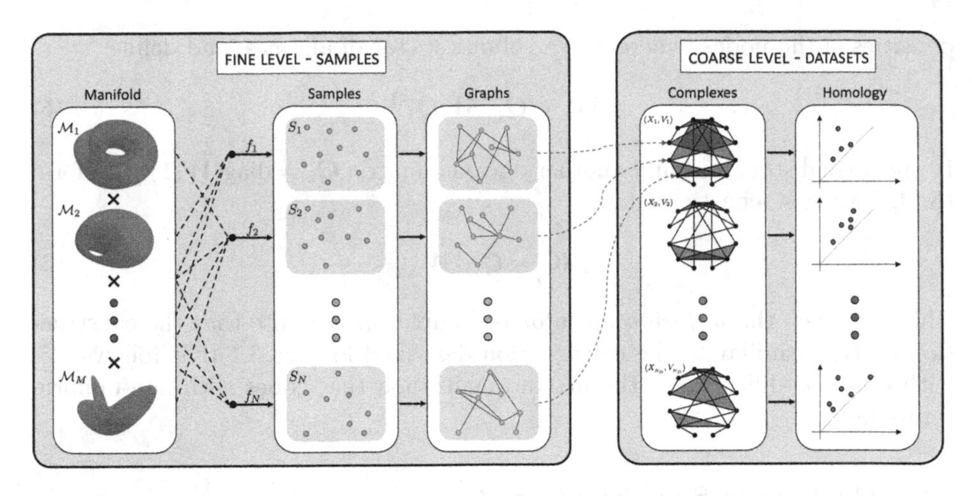


**Fig. 3.** Method outline. Each dataset (at the coarse level) is represented as a weighted simplicial complex, where the weights are calculated using the alternating diffusion operator between the sample graphs (see (8), (10)). The output is a set of persistence diagrams in the rightmost column, where each diagram summarizes a single dataset. We can compare the datasets using the Wasserstein distance (12).

each sample $S_i$ is treated as a weighted graph, which we analyze geometrically using a diffusion operator. At the coarse level, each dataset $D$ is considered as a weighted simplicial complex, from which we extract its persistent homology, enabling us to compare between different datasets using the Wasserstein distance. Figure 3 summarizes this pipline. The motivation for this analysis is the following. At the fine level, we use geometry in order to capture the detailed structure of a sample. Since all the samples within a dataset $D$ are assumed to be generated by the same set of realizations $\mathcal{X} \subset \Pi$, their geometry provides a solid measure of inter-sample similarity. Conversely, at the coarse level, the geometry of different datasets can be vastly different. Thus, in order to compare datasets, we propose to use topology as an informative representation of the global qualitative structure, rather than geometry.

### 4.1 The Sample Diffusion Operator

We treat each sample $S_i$ as a weighted graph, with weights calculated using a Gaussian kernel, forming an affinity matrix $\boldsymbol{W}_i \in \mathbb{R}^{L \times L}$, whose $(j_1, j_2)$-th element is given by

$$W_i(j_1, j_2) = \exp\left(-\frac{d_i^2(x_{i,j_1}, x_{i,j_2})}{\epsilon}\right), \tag{4}$$

where $d_i$ is a distance suitable for the observation space $\mathcal{O}_i$.

Next, following [8], we apply a two-step normalization. The first step is designed to handle a possibly non-uniform density of data points on the manifold. Let $\boldsymbol{Q}_i = \text{diag}(\boldsymbol{W}_i \mathbf{1})$ be a diagonal matrix that approximates the local densities of the nodes, where $\mathbf{1}$ is a column vector of all ones, and define

$$\widetilde{\boldsymbol{W}}_i = \boldsymbol{Q}_i^{-1} \boldsymbol{W}_i \boldsymbol{Q}_i^{-1}. \tag{5}$$

In the second step, we build another diagonal matrix $\widetilde{\boldsymbol{Q}}_i = \text{diag}(\widetilde{\boldsymbol{W}}_i \mathbf{1})$, and form the following stochastic matrix

$$\boldsymbol{K}_i = \widetilde{\boldsymbol{Q}}_i^{-1} \widetilde{\boldsymbol{W}}_i, \tag{6}$$

which is called the *diffusion operator* of sample $S_i$. We note that the construction of $\boldsymbol{K}_i$ is similar to the construction described in Sect. 3.1 that follows [25] with only one difference – the first normalization that copes with non-uniform sampling.

## 4.2 The Dataset Simplicial Complex

We construct a weighted simplicial complex for every dataset $D$, whose vertex set consists of the samples $\{S_i\}_{i=1}^N$. We assume that the simplicial complex is given (and depends on the problem at hand but not on the data), and we only need to determine the weights on the simplexes.

Considering our model in (1) and (2), we propose weights that are inversely correlated with the number of common hidden variables between the samples. Denote by $V$ the weight function for the simplexes representing dataset $D$. Ideally, for any $d$-dimensional simplex $\sigma = [i_1, \ldots, i_{d+1}]$ we want to have

$$V(\sigma) = U(|I_{i_1} \cap \cdots \cap I_{i_{d+1}}|), \tag{7}$$

where $I_i$ is the tuple of indexes corresponding to sample $S_i$ (see Sect. 2) and $U$ is a decreasing function. While this condition cannot hold in a strict sense (mainly due to observation noise), the method we devise below provides a close approximation.

We start with the edges. Let $\boldsymbol{K}_{i_1}, \boldsymbol{K}_{i_2}$ be a pair of diffusion operators for samples $S_{i_1}, S_{i_2} \in D$ (see Sect. 4.1). As described in Sect. 3.1, the work in [25], based on the notion of alternating diffusion [17], showed that one can reveal the common manifold structure between $S_{i_1}$ and $S_{i_2}$ by considering the symmetric alternating diffusion operator

$$\boldsymbol{S}_{i_1,i_2} = \boldsymbol{K}_{i_1} \boldsymbol{K}_{i_2}^T + \boldsymbol{K}_{i_2} \boldsymbol{K}_{i_1}^T. \tag{8}$$

As a heuristic, we propose to set the weight function $V$ to be the inverse of the Frobenius norm, i.e.

$$V([i_1, i_2]) = \|\boldsymbol{S}_{i_1,i_2}\|_F^{-1}. \tag{9}$$

The rationale behind this heuristic stems from the common practice in kernel methods. Typically, the eigenvalues of the kernel are used for evaluating the dominance of the component represented by the corresponding eigenvectors. Indeed, using the spectral distance was proposed in [23] in a setting where the samples form individual graphs, as in the current work. Here, we follow the same practice but with a kernel that captures only the common components. We will show empirically in Sect. 5 that indeed $V([i_1, i_2])$ inversely correlates with $|I_{i_1} \cap I_{i_2}|$, as desired.

Next, we consider triangles in our complex. In a similar spirit to (8), we define the three-way symmetric alternating diffusion operator by

$$\boldsymbol{S}_{i_1,i_2,i_3} = \boldsymbol{S}_{i_1,i_2}\boldsymbol{K}_{i_3}^T + \boldsymbol{K}_{i_3}\boldsymbol{S}_{i_1,i_2} + \boldsymbol{S}_{i_2,i_3}\boldsymbol{K}_{i_1}^T + \boldsymbol{K}_{i_1}\boldsymbol{S}_{i_2,i_3} + \boldsymbol{S}_{i_1,i_3}\boldsymbol{K}_{i_2}^T + \boldsymbol{K}_{i_2}\boldsymbol{S}_{i_1,i_3}. \tag{10}$$

The weight function of the corresponding triangle is then set as

$$V([i_1, i_2, i_3]) = \|\boldsymbol{S}_{i_1,i_2,i_3}\|_F^{-1}. \tag{11}$$

In Sect. 5 we also show empirically that $V([i_1, i_2, i_3])$ inversely correlates with $|I_{i_1} \cap I_{i_2} \cap I_{i_3}|$. In particular, we have $V([i_1, i_2]) \leq V([i_1, i_2, i_3])$ for all $i_1, i_2, i_3$, which is required in order to have a filtered complex.

In a similar spirit, one can define $V$ for simplexes of any dimension. However, for the simulation and application we consider here, edges and triangles suffice.

### 4.3 Topological Distance Between Datasets

The proposed pipeline concludes with a numerical measure of structural similarity between two datasets $D$ and $D'$. Recall that the output of the previous section are weighted simplicial complexes, denoted by the pairs $(X, V)$ and $(X', V')$, where $X$ and $X'$ are complexes and $V$ and $V'$ are the weight functions. We use each weight function to generate a filtration that in turn serves as the input to the persistent homology computation (see Sect. 3.2). The filtration we take is the *sublevel* set filtration $\{X_v\}_{v \in \mathbb{R}}$, where $X_v = \{\sigma : V(\sigma) \leq v\}$. Considering the weights constructed in (9) and (11) , this implies that simplexes that represent groups of samples that share more structure in common will appear earlier in the filtration.

Let $\mathrm{PD}_k$ and $\mathrm{PD}'_k$ be the $k$-th persistence diagrams of $(X, V)$ and $(X', V')$, respectively. We can then compare the topology of two datasets by calculating the Wasserstein distance

$$d_{\mathrm{dataset}}(D, D') = d_{W_p}(\mathrm{PD}_k, \mathrm{PD}'_k). \tag{12}$$

The choice of $p$ and $k$ depends on the application at hand. The entire pipeline is summarized in Algorithm 1. Note that it is currently described for $k = \{0, 1\}$, but once the weight function $V$ in Subsect. 4.2 is extended beyond edges and triangles to higher orders, the algorithm can be extended for $k \geq 2$ as well.

## 5 Simulation Study

In this section, we test the proposed framework on a toy problem, where we can manipulate and examine all the ingredients of our model and method.

**Algorithm 1:** A geometric-topological distance between two datasets

**Input** : Two hierarchical datasets: $D$ and $D'$
**Output** : Distance between the datasets: $d_{dataset}(D, D')$
**Parameters**: $k = \{0, 1\}$ (homology degree), $p$ (Wasserstein distance order), $\epsilon$ (kernel scale)

1. Construct a simplicial complex $X$ for each dataset as follows:
   (a) For each sample $S_i$, $i = 1, \ldots, N$ compute the diffusion operator $\boldsymbol{K}_i$, with scale parameter $\epsilon$ according to (4)–(6)
   (b) For all edges $(i_1, i_2)$:
      i. Compute the symmetric alternating diffusion operator $\boldsymbol{S}_{i_1,i_2}$ according to (8)
      ii. Set the weight function: $V([i_1, i_2]) = \|\boldsymbol{S}_{i_1,i_2}\|_F^{-1}$
   (c) For all triangles $(i_1, i_2, i_3)$:
      i. Compute the three-way symmetric alternating diffusion operator $\boldsymbol{S}_{i_1,i_2,i_3}$ according to (10)
      ii. Set the weights of $V([i_1, i_2, i_3]) = \|\boldsymbol{S}_{i_1,i_2,i_3}\|_F^{-1}$
2. Compute the $k$-th persistence diagram of the weighted complexes $(X, V)$ and $(X', V')$, corresponding to $D$ and $D'$, respectively.
3. Compute the distance between the persistence diagrams: $d_{W_p}(\text{PD}_k, \text{PD}'_k)$

We start with the description of a single dataset $D$. Revisiting the notation in Sect. 2, we assume that the latent manifold for each dataset is of the form $\Pi = \mathcal{M}_1 \times \cdots \times \mathcal{M}_M$, where each $\mathcal{M}_\ell$ is a circle of the form

$$\mathcal{M}_\ell = \{(\cos(\theta_\ell), \sin(\theta_\ell)) \,|\, 0 \leq \theta_\ell < 2\pi\}. \tag{13}$$

In other words, $\Pi$ is an $M$-dimensional torus. In this case, the latent realization set $\mathcal{X} = \{x_1, \ldots, x_L\}$ is a subset of $\mathbb{R}^{2M}$. We generate $\mathcal{X}$ by taking a sample of iid variables $\theta_\ell \sim U[0, 2\pi)$ for $\ell = 1, \ldots, M$. For all the samples $S_i$ $(1 \leq i \leq N)$ we take $|I_i| = 3$, and the observation space is then $\mathcal{O}_i = \mathbb{R}^6$. For every $x \in \Pi$, we define

$$g_i(x_{I_i}) = (R_1\cos(\theta_{\ell_1}), R_1\sin(\theta_{\ell_1}), R_2\cos(\theta_{\ell_2}), R_2\sin(\theta_{\ell_2}), R_3\cos(\theta_{\ell_3}), R_3\sin(\theta_{\ell_3}))$$

where $I_i = (\ell_1, \ell_2, \ell_3)$ are the indexes of the subset of manifolds viewed by sample $S_i$. The radii $R_1, R_2, R_3$ are generated uniformly at random in the interval $[1, R_{\max}]$, for each $i$ independently. The indexes $\ell_1, \ell_2, \ell_3$ are also chosen at random. Finally, the sample observation function is given by $f_i(x) = g_i(x_{I_i}) + \xi_i$, where $\xi_i \sim \mathcal{N}(0, \sigma_i^2 I)$ (independent between observations).

For the generation of all datasets, we use $N = 40$, and $L = 200$. The torus dimension $M$ varies between 3 and 30 across the datasets. As $M$ increases, the chances that the pair of samples $(S_i, S_j)$ has underlying circles in common decreases. Subsequently, the connectivity of the simplicial complex of the respective dataset decreases. Thus, $M$ strongly affects the affinity between the datasets.

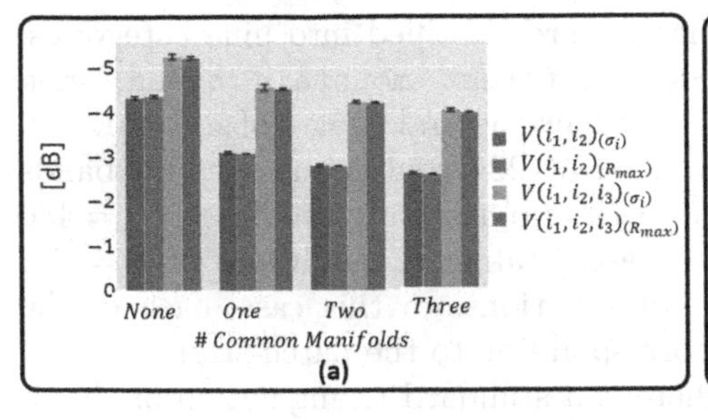


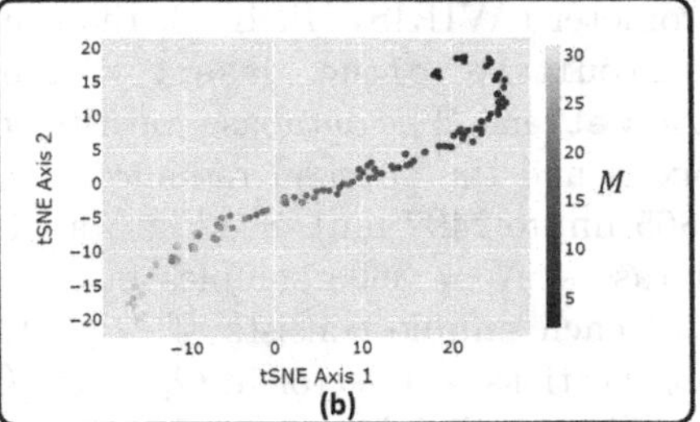


**Fig. 4.** (a) The mean and standard deviation of the weight function $V$ (for edges and triangles). We take 20 random realizations, as well as varying $R_{\max}$ between $1 - 15$ and $\sigma_i$ between $0.001 - 1000$. Note that we plot the value $\log(1 - V)$, and the $y$ axis is flipped, so this is indeed a monotone-decreasing behavior. (b) t-SNE embedding based on the Wasserstein distance for $H_1$ (holes), where the color indicates the size of the pool of underlying manifolds $M$.

In Sect. 4 we argued that the weight function $V$ defined in (9) and (11) is roughly decreasing in the number of common variables (7). Here, we provide an experimental evidence for that heuristic. In Fig. 4(a), we plot the mean and standard deviation of the weight function (in dB) as a function of the number of common indices (manifolds/circles). We calculate $V$ across 20 realizations, and across various choices of $R_{\max}$ $(1 - 15)$ and $\sigma_i$ $(0.001 - 1000)$. The results clearly indicate a monotone decreasing relationship between $V$ and the number of common manifolds. In addition, robustness to noise ($\sigma_i$) and to the particular observation space ($R_{\max}$) is demonstrated.

For each $3 \leq M \leq 30$ we generate 5 datasets with $R_{\max} = 15$ and $\sigma_i = 0.1$, so that overall there are $N_D = 140$ datasets. For each dataset, we follow Algorithm 1 and calculate the topological distance $d_{\text{dataset}}$ (12) using $k = 1$ (holes) and $p = 2$. Figure 4(b) presents the t-SNE [20] embedding based on the obtained distance matrix between all datasets. The color of each dataset indicates the value of $M$. Indeed, we observe that the datasets are organized according this value. In other words, our hierarchical geometric-topological analysis provides a metric between datasets that well-captures the similarity in terms of the global structure of the datasets.

## 6 Application to HSI

In this section we demonstrate the performance of our new geometric-topological framework on Hyper-Spectral Imaging (HSI). The structure of hyper-spectral images fits well with the considered hierarchical dataset model - a dataset here is a single image, and a sample $S_i$ within the dataset is a single square patch. The observations in each patch correspond to the content of the patch at separate spectral bands (see Fig. 1).

The HSI database contains images of various terrain patterns, taken from the NASA Jet Propulsion Laboratory's Airborne Visible InfraRed Imaging Spectrometer (AVIRIS) [1]. In [3] these images were classified into nine categories: `agriculture`, `cloud`, `desert`, `dense-urban`, `forest`, `mountain`, `ocean`, `snow` and `wetland`. The database consists of 486 hyper-spectral images of size $300 \times 300$ pixels and the spectral radiance is sampled at 224 contiguous spectral bands (365 nm to 2497 nm). See Fig. 5(a). In terms of our setting, we have $N_D = 486$ datasets, $N = 3600$ samples in each dataset (taking a patch size of $n = 5$), and each sample consists of $L = 224$ observations. In this case, each of the observations is a vector in $\mathcal{O}_i = \mathbb{R}^{25}$ (corresponding to the patch-size).

The simplicial complex $X$ we use here is a standard triangulation of the 2-dimensional grid of patches. This way the spatial organization of patches in the image is taken into account in the computation of the persistent homology.

We apply Algorithm 1 to the images (datasets) and obtain their pairwise distances. Figure 5(b) demonstrates how our new topological distance arranges the images in space. Specifically, we plot the t-SNE embedding [20] of the images (datasets) based on $d_{\text{dataset}}$ for $k = 1, p = 2$. Each point in the figure represents a single hyper-spectral image, colored by category. Importantly, the category information was not accessible to the (unsupervised) algorithm, and was added to the figure in order to evaluate the results. We observe that most images are grouped by category. In addition, the embedding also conveys the similarity between different categories, implying that this information is captured by $d_{\text{dataset}}$. For example, `agriculture` images (blue points) and `dense-urban` images (purple crosses) are embedded in adjacent locations, and indeed they share common patterns (e.g., grass areas). Conversely, `snow` images form their own separate cluster, as most of the snow instances do not have any common structure with the other categories.

For an objective evaluation of the results, we train an SVM classifier [9]. Prior to computing the SVM, we embed the images (datasets) into a Euclidean space using diffusion maps [8], and apply the classifier to the embedded images. For diffusion maps, the distance obtained by Algorithm 1 $d_{\text{dataset}}$ is used as input, and we generate embedding with 20 dimensions (see Sect. 2.1 in Additional Materials). For the classification, we divide the datasets into a train set and a test set with 10-fold cross validation; the reported results are the average over all folds. We use mean Average Precision (mAP) as the evaluation score.

We compare our results to the results reported in [3], where a deep learning approach was used for the classification of the images. To the best of our knowledge, the results in [3] are considered the state of the art for the ICONES dataset. In Table 1, we present the obtained classification results. In order to make a fair comparison with the reported results in [3], we show the mAP obtained on the train sets. We observe that our method achieves superior results. In addition, we report that our method obtains 0.81 mAP on the test sets.

In order to test the sensitivity of the proposed algorithm to the choice of hyper-parameters, in Fig. 5(c) we present the train scores (top) and test scores (bottom) as a function of the two key hyper-parameters – the patch size $n$ and and kernel scale $\epsilon$ (normalized by the median of the distances in the affinity matrix (4)). The correspondence of the colors between the two figures, as well as

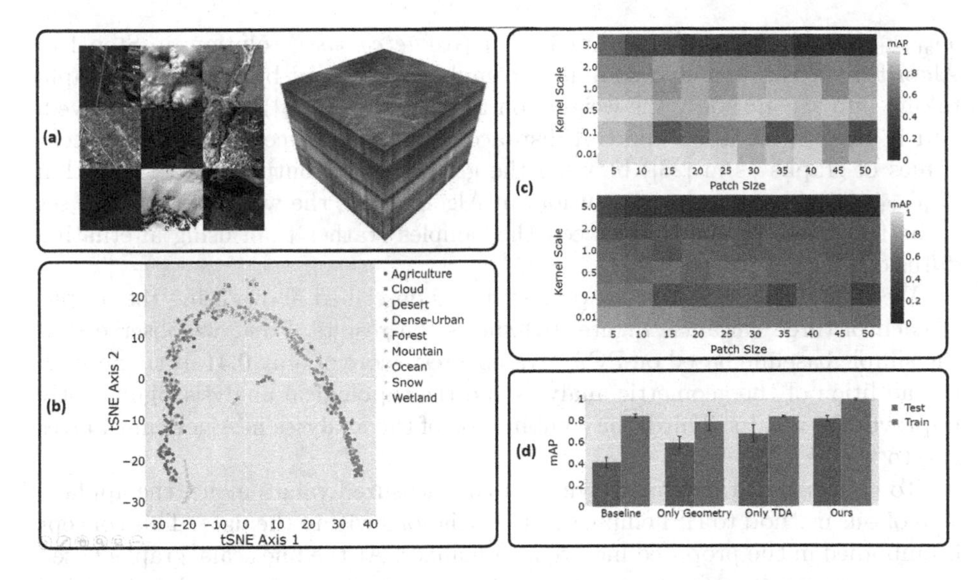


**Fig. 5.** (a) Left: examples of the terrains of hyper-spectral images from different categories (RGB). Right: hyper-spectral image example, stack of 224 spectral bands. (b) t-SNE embedding based on the Wasserstein distance for $H_1$ (holes) between the hyper-spectral images. Each category is denoted by a different color and marker. (c) Classification results (mAP score) as a function of two hyper-parameters: kernel scale and patch size. Top: train score. Bottom: test score. (d) Numerical ablation study – classification results (mAP score) evaluating the contribution of each component in our method.

the apparent smoothness of the color gradient within each image imply robustness to hyper-parameter tuning. Further, we can optimize the hyper-parameters using the train set without leading to an overfit.

Next, we perform an ablation study to evaluate the contribution of the geometric and topological analyses separately. In order to do so, we consider three variants of the algorithm. (i) A *'baseline'* solution: here we replace both the geometric and the topological components with the following implementation which was inspired by [3]. We split the $L = 224$ spectral bands into 5 contiguous ranges. For each range, we apply Principal Component Analysis (PCA) and keep only the principal component in order to reduce the clutter and to get the essence of the spectral information. Next, the pairwise Euclidean distance between the principal components is considered as the counterpart of

**Table 1.** Classification Results (mAP).

| | agric. | cloud | desert | dense-urban | forest | mountain | ocean | snow | wetland | All |
|---|---|---|---|---|---|---|---|---|---|---|
| [3] | 0.48 | 0.66 | 0.5 | 0.86 | 0.57 | 0.64 | 0.83 | 0.57 | 0.23 | 0.59 |
| Ours | 1.0 | 0.95 | 1.0 | 1.0 | 0.99 | 1.0 | 1.0 | 1.0 | 0.82 | 0.98 |

$d_{\text{dataset}}$ (the output of Algorithm 1). (ii) *Geometry-based* solution: in Step 1 of Algorithm 1, the weighted simplicial complex is replaced by a weighted graph, taking into account only the weights on the edges $V([i_1, i_2])$. Step 2 is removed, and in Step 3, we use a spectral distance (the $L_2$ distance between the eigenvalues of graphs as in [23]) between the graphs as the output of the algorithm $d_{\text{dataset}}$. (iii) *Topology-based* solution: in Algorithm 1, the weight function is set to be the cross-correlation between the samples (rather than using alternating diffusion).

We repeat the use of an SVM classifier as described above using the output of each of three variants. Figure 5(d) shows the results. First, we observe that the simple baseline based on PCA attains a test score of only 0.41 mAP. Second, the addition of the geometric analysis and the topological analysis significantly improves the results. Third, the combination of the analyses in Algorithm 1 gives rise to the best results.

To conclude, rather than relying on the measured values alone, the application of our method to HSI emphasizes associations within the data. This concept is embodied in the proposed hierarchical manner. At the fine scale, graphs based on local spectral associations are constructed. At the coarse scale, simplicial complexes based on global spatial associations are formed. The combination of the structures at the two scales, involving both spectral an spatial information, is shown to be beneficial and gives rise to an informative and useful representation of the images.

**Acknowledgement.** OB was supported in part by the Israel Science Foundation, Grant 1965/19. LA and RT were supported by the European Union's Horizon 2020 research and innovation programme under grant agreement No. 802735-ERC-DIFFOP. We would like to thank the anonymous referees for useful comments about our manuscript.

## References

1. NASA jet propulsion laboratory's airborne visible infrared imaging spectrometer (AVIRIS). https://aviris.jpl.nasa.gov/
2. Belkin, M., Niyogi, P.: Laplacian eigenmaps for dimensionality reduction and data representation. Neural Comput. **15**(6), 1373–1396 (2003)
3. Ben-Ahmed, O., Urruty, T., Richard, N., Fernandez-Maloigne, C.: Toward content-based hyperspectral remote sensing image retrieval (cb-hrsir): a preliminary study based on spectral sensitivity functions. Remote Sens. **11**(5), 600 (2019)
4. Bérard, P., Besson, G., Gallot, S.: Embedding Riemannian manifolds by their heat kernel. Geom. Funct. Anal. **4**(4), 373–398 (1994)
5. Carlsson, G.: Topology and data. Bull. Am. Math. Soc. **46**(2), 255–308 (2009)
6. Chang, C.I.: Hyperspectral Imaging: Techniques for Spectral Detection and Classification, vol. 1. Springer, New York (2003)
7. Cohen-Steiner, D., Edelsbrunner, H., Harer, J., Mileyko, Y.: Lipschitz functions have L p-stable persistence. Found. Comput. Math. **10**(2), 127–139 (2010)
8. Coifman, R.R., Lafon, S.: Diffusion maps. Appl. Comput. Harmon. Anal. **21**(1), 5–30 (2006)

9. Cortes, C., Vapnik, V.: Support-vector networks. Mach. Learn. **20**(3), 273–297 (1995)
10. Dabaghian, Y., Mémoli, F., Frank, L., Carlsson, G.: A topological paradigm for hippocampal spatial map formation using persistent homology. PLoS Comput. Biol. **8**(8), e1002581 (2012)
11. Edelsbrunner, H., Harer, J.: Persistent homology-a survey. Contemp. Math. **453**, 257–282 (2008)
12. Edgar, R., Domrachev, M.E.A.: Gene expression omnibus: NCBI gene expression and hybridization array data repository. Nucleic Acids Res. **30**(1), 207–210 (2002)
13. Giesen, C., Wang, Hao AO, E.A.: Highly multiplexed imaging of tumor tissues with subcellular resolution by mass cytometry. Nat. Methods **11**(4), 417–422 (2014)
14. Giusti, C., Pastalkova, E., Curto, C.: Clique topology reveals intrinsic geometric structure in neural correlations. Proc. Natl. Acad. Sci. **112**(44), 13455–13460 (2015)
15. Jones, P.W., Maggioni, M., Schul, R.: Manifold parametrizations by eigenfunctions of the laplacian and heat kernels. Proc. Natl. Acad. Sci. **105**(6), 1803–1808 (2008)
16. LaMontagne, P.J., Benzinger, L.: Oasis-3: longitudinal neuroimaging, clinical, and cognitive dataset for normal aging and alzheimer disease. MedRxiv (2019)
17. Lederman, R.R., Talmon, R.: Learning the geometry of common latent variables using alternating-diffusion. Appl. Comput. Harmon. Anal. **44**(3), 509–536 (2018)
18. Leskovec, J., Krevl, A.: SNAP datasets: Stanford large network dataset collection. http://snap.stanford.edu/data (2014)
19. Lum, P.Y., Singh, G., Lehman, A., Ishkanov, T., Vejdemo-Johansson, M., Alagappan, M., Carlsson, J., Carlsson, G.: Extracting insights from the shape of complex data using topology. Sci. Rep. **3**(1), 1–8 (2013)
20. Van der Maaten, L., Hinton, G.: Visualizing data using t-SNE. J. Mach. Learn. Res. **9**(11), 2579–2605 (2008)
21. Nadler, B., Lafon, S., Coifman, R.R., Kevrekidis, I.G.: Diffusion maps, spectral clustering and reaction coordinates of dynamical systems. Appl. Comput. Harmon. Anal. **21**(1), 113–127 (2006)
22. Naitzat, G., Zhitnikov, A., Lim, L.H.: Topology of deep neural networks. J. Mach. Learn. Res. **21**(184), 1–40 (2020)
23. Rajendran, K., Kattis, A., Holiday, A., Kondor, R., Kevrekidis, I.G.: Data mining when each data point is a network. In: Gurevich, P., Hell, J., Sandstede, B., Scheel, A. (eds.) PaDy 2016. SPMS, vol. 205, pp. 289–317. Springer, Cham (2017). https://doi.org/10.1007/978-3-319-64173-7_17
24. Roweis, S.T., Saul, L.K.: Nonlinear dimensionality reduction by locally linear embedding. Sci. **290**(5500), 2323–2326 (2000)
25. Shnitzer, T., Ben-Chen, M., Guibas, L., Talmon, R., Wu, H.T.: Recovering hidden components in multimodal data with composite diffusion operators. SIAM J. Math. Data Sci. **1**(3), 588–616 (2019)
26. Talmon, R., Wu, H.T.: Latent common manifold learning with alternating diffusion: analysis and applications. Appl. Comput. Harmon. Anal. **47**(3), 848–892 (2019)
27. Tenenbaum, J.B., De Silva, V., Langford, J.C.: A global geometric framework for nonlinear dimensionality reduction. Sci. **290**(5500), 2319–2323 (2000)
28. Wasserman, L.: Topological data analysis. Annu. Rev. Stat. Appl. **5**, 501–532 (2018)
29. Zomorodian, A., Carlsson, G.: Computing persistent homology. Discrete Comput. Geom. **33**(2), 249–274 (2005)

# Reparameterized Sampling for Generative Adversarial Networks

Yifei Wang[1], Yisen Wang[2](✉), Jiansheng Yang[1], and Zhouchen Lin[2]

[1] School of Mathematical Sciences, Peking University, Beijing 100871, China
yifei_wang@pku.edu.cn, yjs@math.pku.edu.cn
[2] Key Laboratory of Machine Perception (MoE), School of EECS, Peking University, Beijing 100871, China
{yisen.wang,zlin}@pku.edu.cn

**Abstract.** Recently, sampling methods have been successfully applied to enhance the sample quality of Generative Adversarial Networks (GANs). However, in practice, they typically have poor sample efficiency because of the independent proposal sampling from the generator. In this work, we propose REP-GAN, a novel sampling method that allows general dependent proposals by REParameterizing the Markov chains into the latent space of the generator. Theoretically, we show that our reparameterized proposal admits a closed-form Metropolis-Hastings acceptance ratio. Empirically, extensive experiments on synthetic and real datasets demonstrate that our REP-GAN largely improves the sample efficiency and obtains better sample quality simultaneously.

**Keywords:** Generative Adversarial Networks · Sampling · Markov Chain Monte Carlo · Reparameterization

## 1 Introduction

Generative Adversarial Networks (GANs) [9] have achieved a great success on generating realistic images in recent years [4,12]. Unlike previous models that explicitly parameterize the data distribution, GANs rely on an alternative optimization between a generator and a discriminator to learn the data distribution implicitly. However, in practice, samples generated by GANs still suffer from problems such as mode collapse and bad artifacts.

Recently, sampling methods have shown promising results on enhancing the sample quality of GANs by making use of the information in the discriminator. In the alternative training scheme of GANs, the generator only performs a few updates for the inner loop and has not fully utilized the density ratio information estimated by the discriminator. Thus, after GAN training, the sampling methods propose to further utilize this information to bridge the gap between the

**Electronic supplementary material** The online version of this chapter (https://doi.org/10.1007/978-3-030-86523-8_30) contains supplementary material, which is available to authorized users.


N. Oliver et al. (Eds.): ECML PKDD 2021, LNAI 12977, pp. 494–509, 2021.
https://doi.org/10.1007/978-3-030-86523-8_30

generative distribution and the data distribution in a fine-grained manner. For example, DRS [2] applies rejection sampling, and MH-GAN [27] adopts Markov chain Monte Carlo (MCMC) sampling for the improved sample quality of GANs. Nevertheless, these methods still suffer a lot from the sample efficiency problem. For example, as will be shown in Sect. 5, MH-GAN's average acceptance ratio on CIFAR10 can be lower than 5%, which makes the Markov chains slow to mix. As MH-GAN adopts an *independent* proposal $q$, i.e., $q(\mathbf{x}'|\mathbf{x}) = q(\mathbf{x}')$, the difference between samples can be so large that the proposal gets rejected easily.

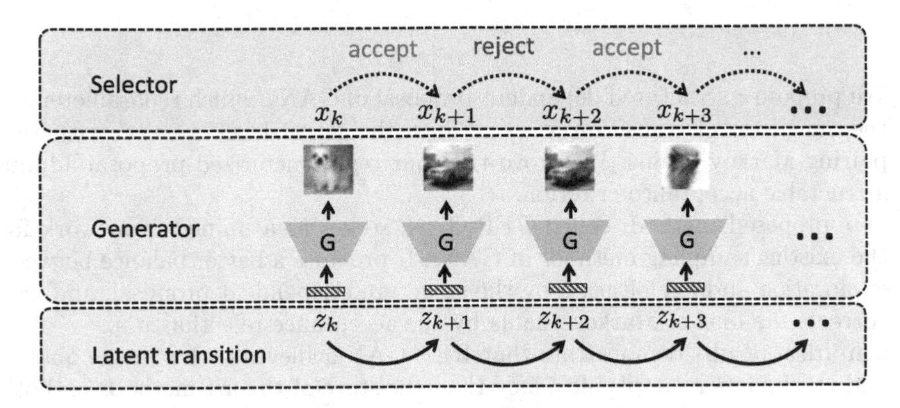


**Fig. 1.** Illustration of REP-GAN's reparameterized proposal with two pairing Markov chains, one in the latent space $\mathcal{Z}$, and the other in the sample space $\mathcal{X}$.

To address this limitation, we propose to generalize the independent proposal to a general *dependent* proposal $q(\mathbf{x}'|\mathbf{x})$. To the end, the proposed sample can be a refinement of the previous one, which leads to a higher acceptance ratio and better sample quality. We can also balance between the exploration and exploitation of the Markov chains by tuning the step size. However, it is hard to design a proper dependent proposal in the high dimensional sample space $\mathcal{X}$ because the energy landscape could be very complex [19].

Nevertheless, we notice that the generative distribution $p_g(\mathbf{x})$ of GANs is implicitly defined as the push-forward of the latent prior distribution $p_0(\mathbf{z})$, and designing proposals in the low dimensional latent space is generally much easier. Hence, GAN's latent variable structure motivates us to design a *structured dependent proposal* with two pairing Markov chains, one in the sample space $\mathcal{X}$ and the other in the latent space $\mathcal{Z}$. As shown in Fig. 1, given the current pairing samples $(\mathbf{z}_k, \mathbf{x}_k)$, we draw the next proposal $\mathbf{x}'$ in a bottom-to-up way: 1) drawing a latent proposal $\mathbf{z}'$ following $q(\mathbf{z}'|\mathbf{z}_k)$; 2) pushing it forward through the generator and getting the sample proposal $\mathbf{x}' = G(\mathbf{z}')$; 3) assigning $\mathbf{x}_{k+1} = \mathbf{x}'$ if the proposal $\mathbf{x}'$ is accepted, otherwise $\mathbf{x}_{k+1} = \mathbf{x}_k$ if rejected. By utilizing the underlying structure of GANs, the proposed reparameterized sampler becomes more efficient in the low-dimensional latent space. We summarize our main contributions as follows:

**Table 1.** Comparison of sampling methods for GANs in terms of three effective sampling mechanisms.

| Method | Rejection step | Markov chain | Latent gradient proposal |
|---|---|---|---|
| GAN | ✗ | ✗ | ✗ |
| DRS [2] | ✓ | ✗ | ✗ |
| MH-GAN [27] | ✓ | ✓ | ✗ |
| DDLS [5] | ✗ | ✓ | ✓ |
| REP-GAN (ours) | ✓ | ✓ | ✓ |

- We propose a structured dependent proposal of GANs, which reparameterizes the sample-level transition $\mathbf{x} \to \mathbf{x}'$ into the latent-level $\mathbf{z} \to \mathbf{z}'$ with two pairing Markov chains. We prove that our reparameterized proposal admits a tractable acceptance criterion.
- Our proposed method, called REP-GAN, serves as a unified framework for the existing sampling methods of GANs. It provides a better balance between exploration and exploitation by the structured dependent proposal, and also corrects the bias of Markov chains by the acceptance-rejection step.
- Empirical results demonstrate that REP-GAN achieves better image quality and much higher sample efficiency than the state-of-the-art methods on both synthetic and real datasets.

## 2 Related Work

Although GANs are able to synthesize high-quality images, the minimax nature of GANs makes it quite unstable, which usually results in degraded sample quality. A vast literature has been developed to fix the problems of GANs ever since, including network modules [18], training mechanisms [17] and objectives [1].

Moreover, there is another line of work using sampling methods to improve the sample quality of GANs. DRS [2] firstly proposes to use rejection sampling. MH-GAN [27] instead uses the Metropolis-Hasting (MH) algorithm with an independent proposal. DDLS [5] and DCD [24] apply gradient-based proposals by viewing GAN as an energy-based model. Tanaka et al. [25] proposes a similar gradient-based method named DOT from the perspective of optimal transport.

Different from them, our REP-GAN introduces a structured dependent proposal through latent reparameterization, and includes all three effective sampling mechanisms, the Markov Chain Monte Carlo method, the acceptance-rejection step, and the latent gradient-based proposal, to further improve the sample efficiency. As shown in Table 1, many existing works are special cases of our REP-GAN.

Our method also belongs to the part of the literature that combine MCMC and neural networks for better sample quality. Previously, some works combine variational autoencoders [13] and MCMC to bridge the amorization gap [11,15, 22], while others directly learn a neural proposal function for MCMC [14,23,28]. Our work instead reparameterizes the high-dimensional sample-level transition into a simpler low-dimensional latent space via the learned generator network.

## 3 Background

GANs model the data distribution $p_d(\mathbf{x})$ implicitly with a generator $G: \mathcal{Z} \to \mathcal{X}$ mapping from a low-dimensional latent space $\mathcal{Z}$ to a high-dimensional sample space $\mathcal{X}$,

$$\mathbf{x} = G(\mathbf{z}), \quad \mathbf{z} \sim p_0(\mathbf{z}), \tag{1}$$

where the sample $\mathbf{x}$ follows the generative distribution $p_g(\mathbf{x})$ and the latent variable $\mathbf{z}$ follows the prior distribution $p_0(\mathbf{z})$, e.g., a standard normal distribution $\mathcal{N}(\mathbf{0}, \mathbf{I})$. In GANs, a discriminator $D: \mathcal{X} \to [0,1]$ is learned to distinguish samples from $p_d(\mathbf{x})$ and $p_g(\mathbf{x})$ in an adversarial way

$$\min_G \max_D \mathbb{E}_{\mathbf{x} \sim p_d(\mathbf{x})} \log(D(\mathbf{x})) + \mathbb{E}_{\mathbf{z} \sim p_0(\mathbf{z})} \log(1 - D(G(\mathbf{z}))). \tag{2}$$

[9] point out that an optimal discriminator $D$ implies the density ratio between the data and generative distributions

$$D(\mathbf{x}) = \frac{p_d(\mathbf{x})}{p_d(\mathbf{x}) + p_g(\mathbf{x})} \quad \Rightarrow \quad \frac{p_d(\mathbf{x})}{p_g(\mathbf{x})} = \frac{1}{D(\mathbf{x})^{-1} - 1}. \tag{3}$$

Markov Chain Monte Carlo (MCMC) refers to a kind of sampling methods that draw a chain of samples $\mathbf{x}_{1:K} \in \mathcal{X}^K$ from a target distribution $p_t(\mathbf{x})$. We denote the initial distribution as $p_0^x(\mathbf{x})$ and the proposal distribution as $q(\mathbf{x}'|\mathbf{x}_k)$. With the Metropolis-Hastings (MH) algorithm, we accept the proposal $\mathbf{x}' \sim q(\mathbf{x}'|\mathbf{x}_k)$ with probability

$$\alpha(\mathbf{x}', \mathbf{x}_k) = \min\left(1, \frac{p_t(\mathbf{x}')\, q(\mathbf{x}_k|\mathbf{x}')}{p_t(\mathbf{x}_k)\, q(\mathbf{x}'|\mathbf{x}_k)}\right) \in [0,1]. \tag{4}$$

If $\mathbf{x}'$ is accepted, $\mathbf{x}_{k+1} = \mathbf{x}'$, otherwise $\mathbf{x}_{k+1} = \mathbf{x}_k$. Under mild assumptions, the Markov chain is guaranteed to converge to $p_t(\mathbf{x})$ as $K \to \infty$. In practice, the sample efficiency of MCMC crucially depends on the proposal distribution to trade off between exploration and exploitation.

## 4 The Proposed REP-GAN

In this section, we first review MH-GAN and point out the limitations. We then propose our structured dependent proposal to overcome these obstacles, and finally discuss its theoretical properties as well as practical implementations.

### 4.1 From Independent Proposal to Dependent Proposal

MH-GAN [27] first proposes to improve GAN sampling with MCMC. Specifically, given a perfect discriminator $D$ and a decent (but imperfect) generator $G$ after training, they take the data distribution $p_d(\mathbf{x})$ as the target distribution and use the generator distribution $p_g(\mathbf{x})$ as an independent proposal

$$\mathbf{x}' \sim q\left(\mathbf{x}'|\mathbf{x}_k\right) = q\left(\mathbf{x}'\right) = p_g(\mathbf{x}'). \tag{5}$$

With the MH criterion (Eq. (4)) and the density ratio (Eq. (3)), we should accept $\mathbf{x}'$ with probability

$$\alpha_{\text{MH}}\left(\mathbf{x}', \mathbf{x}_k\right) = \min\left(1, \frac{p_d\left(\mathbf{x}'\right) q\left(\mathbf{x}_k\right)}{p_d\left(\mathbf{x}_k\right) q\left(\mathbf{x}'\right)}\right) = \min\left(1, \frac{D\left(\mathbf{x}_k\right)^{-1} - 1}{D\left(\mathbf{x}'\right)^{-1} - 1}\right). \tag{6}$$

However, to achieve tractability, MH-GAN adopts an independent proposal $q(\mathbf{x}')$ with poor sample efficiency. As the proposed sample $\mathbf{x}'$ is independent of the current sample $\mathbf{x}_k$, the difference between the two samples can be so large that it results in a very low acceptance probability. Consequently, samples can be trapped in the same place for a long time, leading to a very slow mixing of the chain.

A natural solution is to take a *dependent* proposal $q(\mathbf{x}'|\mathbf{x}_k)$ that will propose a sample $\mathbf{x}'$ close to the current one $\mathbf{x}_k$, which is more likely to be accepted. Nevertheless, the problem of such a dependent proposal is that its MH acceptance criterion

$$\alpha_{\text{DEP}}\left(\mathbf{x}', \mathbf{x}_k\right) = \min\left(1, \frac{p_d\left(\mathbf{x}'\right) q\left(\mathbf{x}_k|\mathbf{x}'\right)}{p_d\left(\mathbf{x}_k\right) q\left(\mathbf{x}'|\mathbf{x}_k\right)}\right), \tag{7}$$

is generally intractable because the data density $p_d(\mathbf{x})$ is unknown. Besides, it is hard to design a proper dependent proposal $q(\mathbf{x}'|\mathbf{x}_k)$ in the high dimensional sample space $\mathcal{X}$ with complex landscape. These obstacles prevent us from adopting a dependent proposal that is more suitable for MCMC.

### 4.2 A Tractable Structured Dependent Proposal with Reparameterized Markov Chains

As discussed above, the major difficulty of a general dependent proposal $q(\mathbf{x}'|\mathbf{x}_k)$ is to compute the MH criterion. We show that it can be made tractable by considering an additional pairing Markov chain in the latent space.

As we know, samples of GANs lie in a low-dimensional manifold induced by the push-forward of the latent variable [1]. Suppose that at the $k$-th step of the Markov chain, we have a GAN sample $\mathbf{x}_k$ with latent $\mathbf{z}_k$. Instead of drawing a sample $\mathbf{x}'$ directly from a sample-level proposal distribution $q(\mathbf{x}'|\mathbf{x}_k)$, we first draw a latent proposal $\mathbf{z}'$ from a dependent latent proposal distribution $q(\mathbf{z}'|\mathbf{z}_k)$. Afterward, we push the latent $\mathbf{z}'$ forward through the generator and get the output $\mathbf{x}'$ as our sample proposal.

As illustrated in Fig. 1, our bottom-to-up proposal relies on the transition reparameterization with two pairing Markov chains in the sample space $\mathcal{X}$ and the latent space $\mathcal{Z}$. Hence we call it a REP (reparameterized) proposal. Through a learned generator, we transport the transition $\mathbf{x}_k \to \mathbf{x}'$ in the high dimensional space $\mathcal{X}$ into the low dimensional space $\mathcal{Z}$, $\mathbf{z}_k \to \mathbf{z}'$, which enjoys a much better landscape and makes it easier to design proposals in MCMC algorithms. For example, the latent target distribution is nearly standard normal when the generator is nearly perfect. In fact, under mild conditions, the REP proposal distribution $q_{\text{REP}}(\mathbf{x}'|\mathbf{x}_k)$ and the latent proposal distribution $q(\mathbf{z}'|\mathbf{z}_k)$ are tied with the following change of variables [3,7]

$$\log q_{\text{REP}}(\mathbf{x}'|\mathbf{x}_k) = \log q(\mathbf{x}'|\mathbf{z}_k) = \log q(\mathbf{z}'|\mathbf{z}_k) - \frac{1}{2}\log\det J_{\mathbf{z}'}^\top J_{\mathbf{z}'}, \tag{8}$$

where $J_{\mathbf{z}}$ denotes the Jacobian matrix of the push-forward $G$ at $\mathbf{z}$, i.e., $[J_{\mathbf{z}}]_{ij} = \partial\,\mathbf{x}_i/\partial\,\mathbf{z}_j, \mathbf{x} = G(\mathbf{z})$.

Nevertheless, it remains unclear whether we can perform the MH test to decide the acceptance of the proposal $\mathbf{x}'$. Note that a general dependent proposal distribution does not meet a tractable MH acceptance criterion (Eq. (7)). Perhaps surprisingly, it can be shown that with our structured REP proposal, the MH acceptance criterion is tractable for general latent proposals $q(\mathbf{z}'|\mathbf{z}_k)$.

**Theorem 1.** *Consider a Markov chain of GAN samples $\mathbf{x}_{1:K}$ with initial distribution $p_g(\mathbf{x})$. For step $k+1$, we accept our REP proposal $\mathbf{x}' \sim q_{\text{REP}}(\mathbf{x}'|\mathbf{x}_k)$ with probability*

$$\alpha_{\text{REP}}\left(\mathbf{x}', \mathbf{x}_k\right) = \min\left(1, \frac{p_0(\mathbf{z}')q(\mathbf{z}_k|\mathbf{z}')}{p_0(\mathbf{z}_k)q(\mathbf{z}'|\mathbf{z}_k)} \cdot \frac{D(\mathbf{x}_k)^{-1}-1}{D(\mathbf{x}')^{-1}-1}\right), \tag{9}$$

*i.e. let $\mathbf{x}_{k+1} = \mathbf{x}'$ if $\mathbf{x}'$ is accepted and $\mathbf{x}_{k+1} = \mathbf{x}_k$ otherwise. Further assume the chain is irreducible, aperiodic and not transient. Then, according to the Metropolis-Hastings algorithm, the stationary distribution of this Markov chain is the data distribution $p_d(\mathbf{x})$ [6].*

*Proof.* Note that similar to Eq. (8), we also have the change of variables between $p_g(\mathbf{x})$ and $p_0(\mathbf{z})$,

$$\log p_g(\mathbf{x})|_{\mathbf{x}=G(\mathbf{z})} = \log p_0(\mathbf{z}) - \frac{1}{2}\log\det J_{\mathbf{z}}^\top J_{\mathbf{z}}. \tag{10}$$

According to [6], the assumptions that the chain is irreducible, aperiodic, and not transient make sure that the chain has a unique stationary distribution, and the MH algorithm ensures that this stationary distribution equals to the target distribution $p_d(\mathbf{x})$. Thus we only need to show that the MH criterion in Eq. (9) holds. Together with Eq. (3), (7) and (8), we have

$$\begin{aligned}\alpha_{\mathrm{REP}}(\mathbf{x}',\mathbf{x}_k) &= \frac{p_d(\mathbf{x}')\,q(\mathbf{x}_k|\mathbf{x}')}{p_d(\mathbf{x}_k)\,q(\mathbf{x}'|\mathbf{x}_k)} = \frac{p_d(\mathbf{x}')q(\mathbf{z}_k|\mathbf{z}')\left(\det J_{\mathbf{z}_k}^\top J_{\mathbf{z}_k}\right)^{-\frac{1}{2}} p_g(\mathbf{x}_k)p_g(\mathbf{x}')}{p_d(\mathbf{x}_k)q(\mathbf{z}'|\mathbf{z}_k)\left(\det J_{\mathbf{z}'}^\top J_{\mathbf{z}'}\right)^{-\frac{1}{2}} p_g(\mathbf{x}')p_g(\mathbf{x}_k)} \\ &= \frac{q(\mathbf{z}_k|\mathbf{z}')\left(\det J_{\mathbf{z}_k}^\top J_{\mathbf{z}_k}\right)^{-\frac{1}{2}} p_0(\mathbf{z}')\left(\det J_{\mathbf{z}'}^\top J_{\mathbf{z}'}\right)^{-\frac{1}{2}}(D(\mathbf{x}_k)^{-1}-1)}{q(\mathbf{z}'|\mathbf{z}_k)\left(\det J_{\mathbf{z}'}^\top J_{\mathbf{z}'}\right)^{-\frac{1}{2}} p_0(\mathbf{z}_k)\left(\det J_{\mathbf{z}_k}^\top J_{\mathbf{z}_k}\right)^{-\frac{1}{2}}(D(\mathbf{x}')^{-1}-1)} \\ &= \frac{p_0(\mathbf{z}')q(\mathbf{z}_k|\mathbf{z}')(D(\mathbf{x}_k)^{-1}-1)}{p_0(\mathbf{z}_k)q(\mathbf{z}'|\mathbf{z}_k)(D(\mathbf{x}')^{-1}-1)},\end{aligned} \tag{11}$$

which is the acceptance ratio as desired. Q.E.D.

The theorem above demonstrates the following favorable properties of our method:

- The discriminator score ratio is the same as $\alpha_{\mathrm{MH}}(\mathbf{x}',\mathbf{x}_k)$, but MH-GAN is restricted to a specific independent proposal. Our method instead works for any latent proposal $q(\mathbf{z}'|\mathbf{z}_k)$. When we take $q(\mathbf{z}'|\mathbf{z}_k) = p_0(\mathbf{z}')$, our method reduces to MH-GAN.
- Compared to $\alpha_{\mathrm{DEP}}(\mathbf{x}',\mathbf{x}_k)$ of a general dependent proposal (Eq. (7)), the unknown data distributions terms are successfully cancelled in the reparameterized acceptance criterion.
- The reparameterized MH acceptance criterion becomes tractable as it only involves the latent priors, the latent proposal distributions, and the discriminator scores.

Combining the REP proposal $q_{\mathrm{REP}}(\mathbf{x}'|\mathbf{x}_k)$ and its tractable MH criterion $\alpha_{\mathrm{REP}}(\mathbf{x}',\mathbf{x}_k)$, we have developed a novel sampling method for GANs, coined as REP-GAN. See Appendix 1 for a detailed description. Moreover, our method can serve as a general approximate inference technique for Bayesian models by bridging MCMC and GANs. Previous works [10,16,26] also propose to avoid the bad geometry of a complex probability measure by reparameterizing the Markov transitions into a simpler measure. However, these methods are limited to explicit invertible mappings without dimensionality reduction. With this work, we are the first to show that it is also tractable to conduct such model-based reparameterization with implicit models like GANs.

### 4.3 A Practical Implementation

REP-GAN enables us to utilize the vast literature of existing MCMC algorithms [19] to design dependent proposals for GANs. We take Langevin Monte Carlo (LMC) as an example. As an Euler-Maruyama discretization of the Langevin dynamics, LMC updates the Markov chain with

$$\mathbf{x}_{k+1} = \mathbf{x}_k + \frac{\tau}{2}\nabla_{\mathbf{x}} \log p_t(\mathbf{x}_k) + \sqrt{\tau}\cdot\boldsymbol{\varepsilon}, \quad \boldsymbol{\varepsilon} \sim \mathcal{N}(\mathbf{0},\mathbf{I}), \tag{12}$$

for a target distribution $p_t(\mathbf{x})$. Compared to MH-GAN, LMC utilizes the gradient information to explore the energy landscape more efficiently. However, if we directly take the (unknown) data distribution $p_d(\mathbf{x})$ as the target distribution $p_t(\mathbf{x})$, LMC does not meet a tractable update rule.

As discussed above, the reparameterization of REP-GAN makes it easier to design transitions in the low-dimensional latent space. Hence, we instead propose to use LMC for the latent Markov chain. We assume that the data distribution also lies in the low-dimensional manifold induced by the generator, i.e., $\mathrm{Supp}\,(p_d) \subset \mathrm{Im}(G)$. This implies that the data distribution $p_d(\mathbf{x})$ also has a pairing distribution in the latent space, denoted as $p_t(\mathbf{z})$. They are tied with the change of variables

$$\log p_d(\mathbf{x})|_{\mathbf{x}=G(\mathbf{z})} = \log p_t(\mathbf{z}) - \frac{1}{2}\log\det J_{\mathbf{z}}^\top J_{\mathbf{z}}, \tag{13}$$

Taking $p_t(\mathbf{z})$ as the (unknown) target distribution of the latent Markov chain, we have the following Latent LMC (L2MC) proposal

$$\begin{aligned}
\mathbf{z}' &= \mathbf{z}_k + \frac{\tau}{2}\nabla_{\mathbf{z}}\log p_t(\mathbf{z}_k) + \sqrt{\tau}\cdot\varepsilon \\
&= \mathbf{z}_k + \frac{\tau}{2}\nabla_{\mathbf{z}}\log\frac{p_t(\mathbf{z}_k)\left(\det J_{\mathbf{z}_k}^\top J_{\mathbf{z}_k}\right)^{-\frac{1}{2}}}{p_0(\mathbf{z}_k)\left(\det J_{\mathbf{z}_k}^\top J_{\mathbf{z}_k}\right)^{-\frac{1}{2}}} + \frac{\tau}{2}\nabla_{\mathbf{z}}\log p_0(\mathbf{z}_k) + \sqrt{\tau}\cdot\varepsilon \\
&= \mathbf{z}_k + \frac{\tau}{2}\nabla_{\mathbf{z}}\log\frac{p_d(\mathbf{x}_k)}{p_g(\mathbf{x}_k)} + \frac{\tau}{2}\nabla_{\mathbf{z}}\log p_0(\mathbf{z}_k) + \sqrt{\tau}\cdot\varepsilon \\
&= \mathbf{z}_k - \frac{\tau}{2}\nabla_{\mathbf{z}}\log(D^{-1}(\mathbf{x}_k) - 1) + \frac{\tau}{2}\nabla_{\mathbf{z}}\log p_0(\mathbf{z}_k) + \sqrt{\tau}\cdot\varepsilon, \quad \varepsilon\sim\mathcal{N}(\mathbf{0},\mathbf{I}),
\end{aligned} \tag{14}$$

where $\mathbf{x}_k = G(\mathbf{z}_k)$. As we can see, L2MC is made tractable by our structured dependent proposal with pairing Markov chains. DDLS [5] proposes a similar Langevin proposal by formalizing GANs as an implicit energy-based model, while here we provide a straightforward derivation through reparameterization. Our major difference to DDLS is that REP-GAN also includes a tractable MH correction step (Eq. (9)), which accounts for the numerical errors introduced by the discretization in Eq. (12) and ensures that detailed balance holds.

We give a detailed description of the algorithm procedure of our REP-GAN in Algorithm 1.

### 4.4 Extension to WGAN

Our method can also be extended to other kinds of GAN, like Wasserstein GAN (WGAN) [1]. The WGAN objective is

$$\min_G \max_D \mathbb{E}_{\mathbf{x}\sim p_d(\mathbf{x})}[D(\mathbf{x})] - \mathbb{E}_{\mathbf{x}\sim p_g(\mathbf{x})}[D(\mathbf{x})], \tag{15}$$

where $D: \mathcal{X} \to \mathbb{R}$ is restricted to be a Lipschitz function. Under certain conditions, WGAN also implies an approximate estimation of the density ratio [5],

$$D(\mathbf{x}) \approx \log\frac{p_d(\mathbf{x})}{p_g(\mathbf{x})} + \text{const} \quad\Rightarrow\quad \frac{p_d(\mathbf{x})}{p_g(\mathbf{x})} \approx \exp(D(\mathbf{x}))\cdot\text{const}. \tag{16}$$

**Algorithm 1.** GAN sampling with Reparameterized Markov chains (REP-GAN)

**Input:** trained GAN with (calibrated) discriminator $D$ and generator $G$, Markov chain length $K$, latent prior distribution $p_0(\mathbf{z})$, latent proposal distribution $q(\mathbf{z}'|\mathbf{z}_k)$;
**Output:** an improved GAN sample $\mathbf{x}_K$;

Draw an initial sample $\mathbf{x}_1$: 1) draw initial latent $\mathbf{z}_1 \sim p_0(\mathbf{z})$ and 2) push forward $\mathbf{x}_1 = G(\mathbf{z}_1)$;
**for** each step $k \in [1, K-1]$ **do**
    Draw a REP proposal $\mathbf{x}' \sim q_{\text{REP}}(\mathbf{x}'|\mathbf{x}_k)$: 1) draw a latent proposal $\mathbf{z}' \sim q(\mathbf{z}'|\mathbf{z}_k)$, and 2) push forward $\mathbf{x}' = G(\mathbf{z}')$;
    Calculate the MH acceptance criterion $\alpha_{\text{REP}}(\mathbf{x}_k, \mathbf{x}')$ following Eq. (9);
    Decide the acceptance of $\mathbf{x}'$ with probability $\alpha_{\text{REP}}(\mathbf{x}_k, \mathbf{x}')$;
    **if** $\mathbf{x}'$ is accepted **then**
        Let $\mathbf{x}_{k+1} = \mathbf{x}', \mathbf{z}_{k+1} = \mathbf{z}'$
    **else**
        Let $\mathbf{x}_{k+1} = \mathbf{x}_k, \mathbf{z}_{k+1} = \mathbf{z}_k$
    **end if**
**end for**

Following the same derivations as in Eqs. (11) and (14), we will have the WGAN version of REP-GAN. Specifically, with $\mathbf{x}_k = G(\mathbf{z}_k)$, the L2MC proposal follows

$$\mathbf{z}' = \mathbf{z}_k + \frac{\tau}{2}\nabla_{\mathbf{z}} D(\mathbf{x}_k) + \frac{\tau}{2}\nabla_{\mathbf{z}} \log p_0(\mathbf{z}_k) + \sqrt{\tau} \cdot \varepsilon, \quad \varepsilon \sim \mathcal{N}(\mathbf{0}, \mathbf{I}), \tag{17}$$

and the MH acceptance criterion is

$$\alpha_{REP-W}(\mathbf{x}', \mathbf{x}_k) = \min\left(1, \frac{q(\mathbf{z}_k|\mathbf{z}')p_0(\mathbf{z}')}{q(\mathbf{z}'|\mathbf{z}_k)p_0(\mathbf{z}_k)} \cdot \frac{\exp(D(\mathbf{x}'))}{\exp(D(\mathbf{x}_k))}\right). \tag{18}$$

# 5 Experiments

We evaluate our method on two synthetic datasets and two real-world image datasets as follows.

## 5.1 Manifold Dataset

Following DOT [25] and DDLS [5], we apply REP-GAN to the Swiss Roll dataset, where data samples lie on a Swiss roll manifold in the two-dimensional space. We construct the dataset by scikit-learn with 100,000 samples, and train a WGAN with the same architecture as DOT and DDLS, where both the generator and discriminator are fully connected neural networks with leaky ReLU nonlinearities. We optimize the model using the Adam optimizer, with learning rate 0.0001. After training, we draw 1,000 samples with different sampling methods. Following previous practice, we initialize a Markov chain with a GAN sample, run it for $K = 100$ steps, and collect the last example for evaluation.

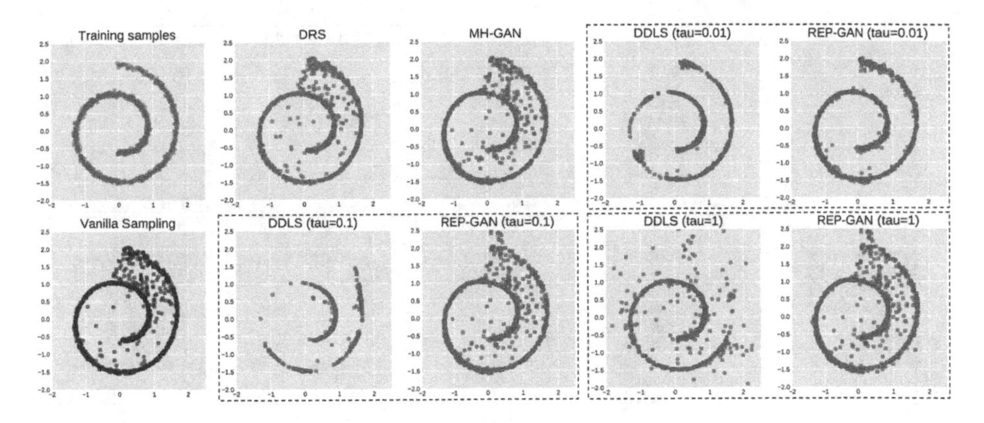


**Fig. 2.** Visualization of samples with different sampling methods on the Swiss Roll dataset. Here tau denotes the Langevin step size in Eq. (17).

As shown in Fig. 2, with appropriate step size ($\tau = 0.01$), the gradient-based methods (DDLS and REP-GAN) outperform independent proposals (DRS and MH-GAN) by a large margin, while DDLS is more discontinuous on shape compared to REP-GAN. In DDLS, when the step size becomes too large ($\tau = 0.1, 1$), the numerical error of the Langevin dynamics becomes so large that the chain either collapses or diverges. In contrast, those bad proposals are rejected by the MH correction steps of REP-GAN, which prevents the misbehavior of the Markov chain.

### 5.2 Multi-modal Dataset

As GANs are known to suffer from the mode collapse problem [8], we also compare different GAN sampling methods in terms of modeling multi-modal distributions. Specifically, we consider the 25-Gaussians dataset that is widely discussed in previous work [2,5,27]. The dataset is generated by a mixture of twenty-five two-dimensional isotropic Gaussian distributions with variance 0.01, and means separated by 1, arranged in a grid. We train a small GAN with the standard WGAN-GP objective following the setup in [25]. After training, we draw 1,000 samples with different sampling methods.

As shown in Fig. 3, compared to MH-GAN, the gradient-based methods (DDLS and ours) produce much better samples close to the data distribution with proper step size ($\tau = 0.01$). Comparing DDLS and our REP-GAN, we can notice that DDLS tends to concentrate so much on the mode centers that its standard deviation can be even smaller than the data distribution. Instead, our method preserves more sample diversity while concentrating on the mode centers. This difference becomes more obvious as the step size $\tau$ becomes larger. When $\tau = 0.1$, as marked with blue circles, DDLS samples become so concentrated that some modes are even missed. When $\tau = 1$, DDLS samples diverge

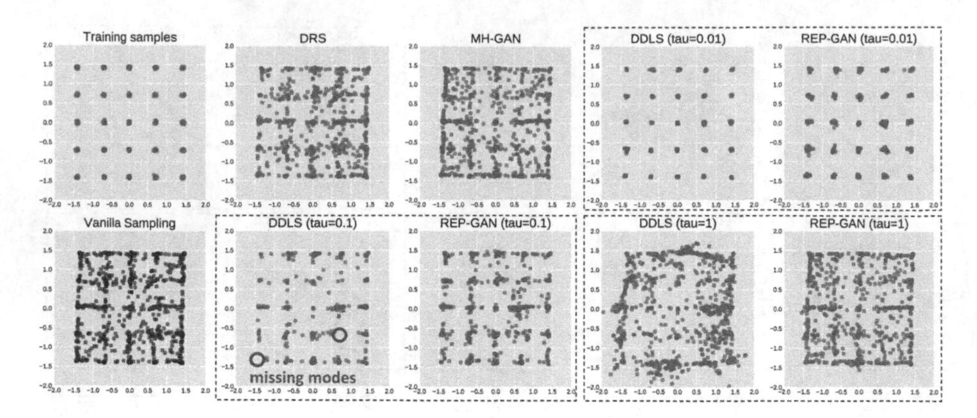


**Fig. 3.** Visualization of samples with different sampling methods on the 25-Gaussians dataset. Here $\tau$ denotes the Langevin step size in Eq. (17).

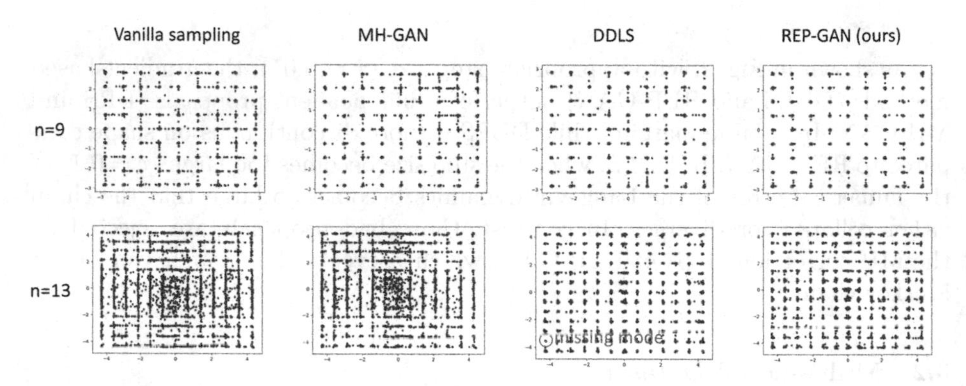


**Fig. 4.** Visualization of the mixture-of-Gaussian experiments with $9 \times 9$ (1st row) and $13 \times 13$ (2nd row) modes with proper step size $\tau = 0.01$. True data points are shown in grey (in background), and generated points are shown in blue. (Color figure online)

far beyond the $5 \times 5$ grid. In comparison, our REP-GAN is more stable because the MH correction steps account for the numerical errors caused by large $\tau$.

These distinctions also become even more obvious when we scale to more modes. As shown in Fig. 4, we also compare them w.r.t. mixture of Gaussians with $9 \times 9$ and $13 \times 13$ modes, respectively. Under the more challenging scenarios, we can see that the gradient-based methods still consistently outperforms MH-GAN. Besides, our REP-GAN has a more clear advantage over DDLS. Specifically, for $9 \times 9$ modes, our REP-GAN produces samples that are less noisy, while preserving all the modes. For $13 \times 13$ modes, DDLS makes a critical mistake that it drops one of the modes. As discussed above, we believe this is because DDLS has a bias towards regions with high probability, while ignoring the diversity of the distribution. In comparison, REP-GAN effectively prevents such bias by the MH correction steps.

**Table 2.** Inception scores of different sampling methods on CIFAR-10 and CelebA, with the DCGAN and WGAN backbones.

| Method | CIFAR-10 | | CelebA | |
|---|---|---|---|---|
| | DCGAN | WGAN | DCGAN | WGAN |
| GAN | 3.219 | 3.740 | 2.332 | 2.788 |
| DRS [2] | 3.073 | 3.137 | 2.869 | 2.861 |
| MH-GAN [27] | 3.225 | 3.851 | **3.106** | 2.889 |
| DDLS [5] | 3.152 | 3.547 | 2.534 | 2.862 |
| REP-GAN (ours) | **3.541** | **4.035** | 2.686 | **2.943** |

### 5.3 Real-World Image Dataset

Following MH-GAN [27], we conduct experiments on two real-world image datasets, CIFAR-10 and CelebA, for two models, DCGAN [20] and WGAN [1]. We adopt the DCGAN generator and discriminator networks as our backbone networks. Following the conventional evaluation protocol, we initialize each Markov chain with a GAN sample, run it for 640 steps, and take the last sample for evaluation. We collect 50,000 samples to evaluate the Inception Score[1] [21]. The step size $\tau$ of our L2MC proposal is 0.01 on CIFAR-10 and 0.1 on CelebA. We calibrate the discriminator with Logistic Regression as in [27].

From Table 2, we can see our method outperforms the state-of-the-art sampling methods in most cases. In Table 3, we also present the average Inception Score and acceptance ratio during the training process. As shown in Table 3a, our REP-GAN can still outperform previous sampling methods consistently and significantly. Besides, in Table 3b, we find that the average acceptance ratio of MH-GAN is lower than 0.05 in most cases, which is extremely low. While with our reparameterized dependent proposal, REP-GAN achieves an acceptance ratio between 0.2 and 0.5, which is known to be a relatively good tradeoff for MCMC algorithms.

### 5.4 Algorithmic Analysis

***Ablation Study.*** We conduct an ablation study of the proposed sampling algorithm, REP-GAN, and the results are shown in Table 4. We can see that without our proposed reparameterized (REP) proposal, the acceptance ratio is very small (with an independent proposal instead). Consequently, the sample quality degrades significantly. Also, we can find that the MH correction step also matters a lot, without which the sample quality of Langevin sampling becomes even worse than the independent proposal. The ablation study shows the necessity of both REP proposal and MH rejection steps in the design of our REP-GAN.

[1] For fair comparison, our training and evaluation follows the the official code of MH-GAN [27]: https://github.com/uber-research/metropolis-hastings-gans.

**Table 3.** Average Inception Score (a) and acceptance ratio (b) vs. training epochs with DCGAN on CIFAR-10.

(a) Inception Score (mean ± std)

| Epoch | 20 | 21 | 22 | 23 | 24 |
|---|---|---|---|---|---|
| GAN | 2.482 ± 0.027 | 3.836 ± 0.046 | 3.154 ± 0.014 | 3.383 ± 0.046 | 3.219 ± 0.036 |
| MH-GAN | 2.356 ± 0.023 | 3.891 ± 0.040 | 3.278 ± 0.033 | 3.458 ± 0.029 | 3.225 ± 0.029 |
| DDLS | 2.419 ± 0.021 | 3.332 ± 0.025 | 2.996 ± 0.035 | 3.255 ± 0.045 | 3.152 ± 0.028 |
| REP-GAN | **2.487** ± 0.019 | **3.954** ± 0.046 | **3.294** ± 0.030 | **3.534** ± 0.035 | **3.541** ± 0.038 |

(b) Average Acceptance Ratio (mean ± std)

| Epoch | 20 | 21 | 22 | 23 | 24 |
|---|---|---|---|---|---|
| MH-GAN | 0.028 ± 0.143 | 0.053 ± 0.188 | 0.060 ± 0.199 | 0.021 ± 0.126 | 0.027 ± 0.141 |
| REP-GAN | **0.435** ± 0.384 | **0.350** ± 0.380 | **0.287** ± 0.365 | **0.208** ± 0.335 | **0.471** ± 0.384 |

**Table 4.** Ablation study of our REP-GAN with Inception Scores (IS) and acceptance ratios on CIFAR-10 with two backbone models, DCGAN and WGAN.

| Method | DCGAN | | WGAN | |
|---|---|---|---|---|
| | Accept ratio | IS | Accept Ratio | IS |
| REP-GAN | **0.447** ± 0.384 | **3.541** ± 0.038 | **0.205** ± 0.330 | **4.035** ± 0.036 |
| REP-GAN w/o REP proposal | 0.027 ± 0.141 | 3.225 ± 0.029 | 0.027 ± 0.141 | 3.851 ± 0.044 |
| REP-GAN w/o MH rejection | – | 3.152 ± 0.028 | – | 3.547 ± 0.029 |

***Markov Chain Visualization.*** In Fig. 5, we demonstrate two Markov chains sampled with different methods. We can see that MH-GAN is often trapped in the same place because of the independent proposals. DDLS and REP-GAN instead gradually refine the samples with gradient steps. In addition, compared the gradient-based methods, we can see that the MH rejection steps of REP-GAN help avoid some bad artifacts in the images. For example, in the camel-like images marked in red, the body of the camel is separated in the sample of DDLS (middle) while it is not in the sample of REP-GAN (bottom). Note that, the evaluation protocol only needs the last step of the chain, thus we prefer a small step size that finetunes the initial samples for better sample quality. As shown in Fig. 6, our REP proposal can also produce very diverse images with a large step size.

***Computation Overhead.*** We also compare the computation cost of the gradient-based sampling methods, DDLS and REP-GAN. They take 88.94 s and 88.85 s, respectively, hence the difference is negligible. Without the MH-step, our method takes 87.62 s, meaning that the additional MH-step only costs 1.4% computation overhead, which is also negligible, but it brings a significant improvement of sample quality as shown in Table 4.

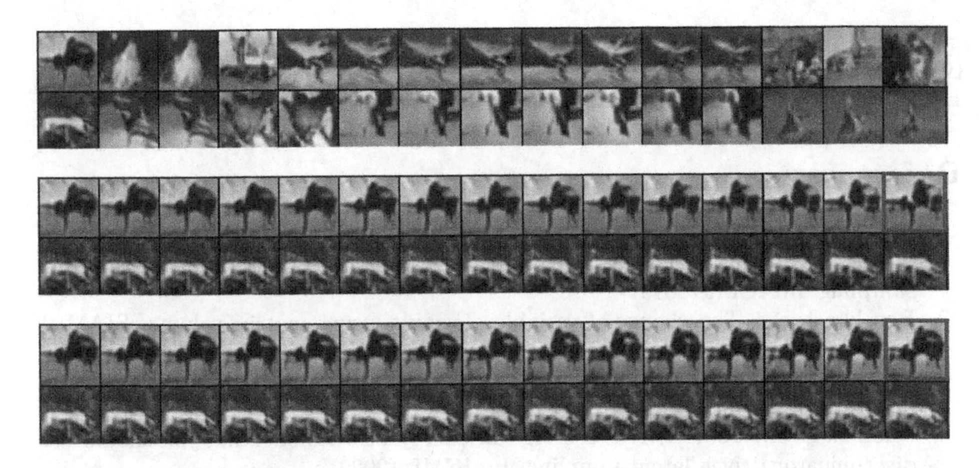

**Fig. 5.** The first 15 steps of two Markov chains with the same initial samples, generated by MH-GAN (top), DDLS (middle), and REP-GAN (bottom). (Color figure online)

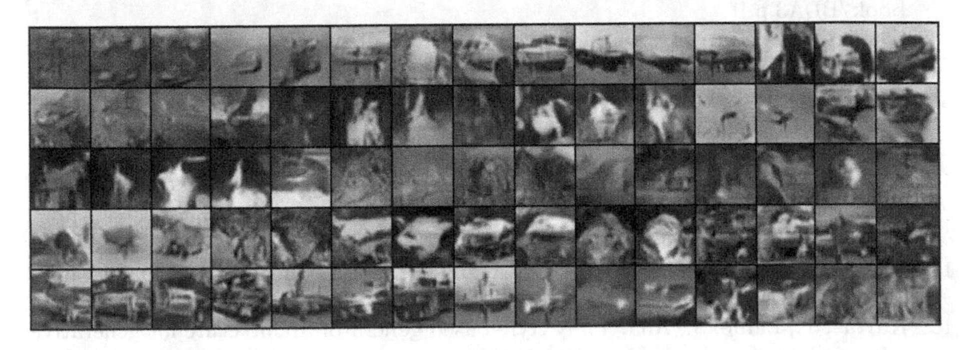

**Fig. 6.** Visualization of 5 Markov chains of our REP proposals (i.e., REP-GAN without the MH rejection steps) with a large step size ($\tau = 1$).

## 6 Conclusion

In this paper, we have proposed a novel method, REP-GAN, to improve the sampling of GAN. We devise a structured dependent proposal that reparameterizes the sample-level transition of GAN into the latent-level transition. More importantly, we first prove that this general proposal admits a tractable MH criterion. Experiments show our method does not only improve sample efficiency but also demonstrate state-of-the-art sample quality on benchmark datasets over existing sampling methods.

**Acknowledgement.** Yisen Wang is supported by the National Natural Science Foundation of China under Grant No. 62006153 and Project 2020BD006 supported by PKU-Baidu Fund. Jiansheng Yang is supported by the National Science Foundation of China under Grant No. 11961141007. Zhouchen Lin is supported by the National Natural Science Foundation of China (Grant No. s 61625301 and 61731018), Project 2020BD006 supported by PKU-Baidu Fund, Major Scientific Research Project of Zhejiang Lab (Grant No. s 2019KB0AC01 and 2019KB0AB02), and Beijing Academy of Artificial Intelligence.

## References

1. Arjovsky, M., Chintala, S., Bottou, L.: Wasserstein GAN. In: ICML (2017)
2. Azadi, S., Olsson, C., Darrell, T., Goodfellow, I., Odena, A.: Discriminator rejection sampling. In: ICLR (2019)
3. Ben-Israel, A.: The change-of-variables formula using matrix volume. SIAM J. Matrix Anal. Appl. **21**(1), 300–312 (1999)
4. Brock, A., Donahue, J., Simonyan, K.: Large scale GAN training for high fidelity natural image synthesis. In: ICLR (2019)
5. Che, T., et al.: Your GAN is secretly an energy-based model and you should use discriminator driven latent sampling. In: ICML (2020)
6. Gelman, A., Carlin, J.B., Stern, H.S., Dunson, D.B., Vehtari, A., Rubin, D.B.: Bayesian data. Analysis 278–280 (2013). http://www.stat.columbia.edu/~gelman/book/BDA3.pdf
7. Gemici, M.C., Rezende, D., Mohamed, S.: Normalizing flows on Riemannian manifolds. arXiv preprint arXiv:1611.02304 (2016)
8. Goodfellow, I.: NIPS 2016 tutorial: Generative adversarial networks. arXiv preprint arXiv:1701.00160 (2016)
9. Goodfellow, I., et al.: Generative adversarial nets. In: NeurIPS (2014)
10. Hoffman, M., Sountsov, P., Dillon, J.V., Langmore, I., Tran, D., Vasudevan, S.: Neutra-lizing bad geometry in Hamiltonian Monte Carlo using neural transport. arXiv preprint arXiv:1903.03704 (2019)
11. Hoffman, M.D.: Learning deep latent Gaussian models with Markov chain Monte Carlo. In: ICML (2017)
12. Karras, T., Laine, S., Aila, T.: A style-based generator architecture for generative adversarial networks. In: NeurIPS (2019)
13. Kingma, D.P., Welling, M.: Auto-encoding variational bayes. In: ICLR (2014)
14. Levy, D., Hoffman, M.D., Sohl-Dickstein, J.: Generalizing Hamiltonian Monte Carlo with neural networks. In: ICLR (2018)
15. Li, Y., Turner, R.E., Liu, Q.: Approximate inference with amortised MCMC. arXiv preprint arXiv:1702.08343 (2017)
16. Marzouk, Y., Moselhy, T., Parno, M., Spantini, A.: An introduction to sampling via measure transport. arXiv preprint arXiv:1602.05023 (2016)
17. Metz, L., Poole, B., Pfau, D., Sohl-Dickstein, J.: Unrolled generative adversarial networks. In: ICLR (2017)
18. Miyato, T., Kataoka, T., Koyama, M., Yoshida, Y.: Spectral normalization for generative adversarial networks. In: ICLR (2018)
19. Neal, R.M., et al.: MCMC using Hamiltonian dynamics. Handb. Markov Chain Monte Carlo **54**, 113–162 (2010)
20. Radford, A., Metz, L., Chintala, S.: Unsupervised representation learning with deep convolutional generative adversarial networks. arXiv preprint arXiv:1511.06434 (2015)
21. Salimans, T., Goodfellow, I., Zaremba, W., Cheung, V., Radford, A., Chen, X.: Improved techniques for training GANs. In: NeurIPS (2016)
22. Salimans, T., Kingma, D., Welling, M.: Markov chain Monte Carlo and variational inference: Bridging the gap. In: ICML (2015)

23. Song, J., Zhao, S., Ermon, S.: A-NICE-MC: Adversarial training for MCMC. In: NeurIPS (2017)
24. Song, Y., Ye, Q., Xu, M., Liu, T.Y.: Discriminator contrastive divergence: semi-amortized generative modeling by exploring energy of the discriminator. arXiv preprint arXiv:2004.01704 (2020)
25. Tanaka, A.: Discriminator optimal transport. In: NeurIPS (2019)
26. Titsias, M.K.: Learning model reparametrizations: implicit variational inference by fitting MCMC distributions. arXiv preprint arXiv:1708.01529 (2017)
27. Turner, R., Hung, J., Saatci, Y., Yosinski, J.: Metropolis-Hastings generative adversarial networks. In: ICML (2019)
28. Wang, T., Wu, Y., Moore, D., Russell, S.J.: Meta-learning MCMC proposals. In: NeurIPS (2018)

# Asymptotic Statistical Analysis of Sparse Group LASSO via Approximate Message Passing

Kan Chen(✉), Zhiqi Bu, and Shiyun Xu

Graduate Group of Applied Mathematics and Computational Science, University of Pennsylvania, Philadelphia, PA 19104, USA
{kanchen,zbu,shiyunxu}@upenn.edu

**Abstract.** Sparse Group LASSO (SGL) is a regularized model for high-dimensional linear regression problems with grouped covariates. SGL applies $l_1$ and $l_2$ penalties on the individual predictors and group predictors, respectively, to guarantee sparse effects both on the inter-group and within-group levels. In this paper, we apply the approximate message passing (AMP) algorithm to efficiently solve the SGL problem under Gaussian random designs. We further use the recently developed state evolution analysis of AMP to derive an asymptotically exact characterization of SGL solution. This allows us to conduct multiple fine-grained statistical analyses of SGL, through which we investigate the effects of the group information and $\gamma$ (proportion of $\ell_1$ penalty). With the lens of various performance measures, we show that SGL with small $\gamma$ benefits significantly from the group information and can outperform other SGL (including LASSO) or regularized models which does not exploit the group information, in terms of the recovery rate of signal, false discovery rate and mean squared error.

## 1 Introduction

Suppose we observe an $n \times p$ design matrix $\mathbf{X}$, and the response $\mathbf{y} \in \mathbb{R}^n$ which is modeled by

$$\mathbf{y} = \mathbf{X}\boldsymbol{\beta} + \boldsymbol{w} \tag{1}$$

in which $\boldsymbol{w} \in \mathbb{R}^n$ is a noise vector.

In many real life applications, we encounter the $p \gg n$ case in which standard linear regression fails. To address this issue, [35] introduced the LASSO by adding the $\ell_1$ penalty and minimizing

$$\frac{1}{2}\|\mathbf{y} - \mathbf{X}\boldsymbol{\beta}\|_2^2 + \lambda\|\boldsymbol{\beta}\|_1. \tag{2}$$

**Electronic supplementary material** The online version of this chapter (https://doi.org/10.1007/978-3-030-86523-8_31) contains supplementary material, which is available to authorized users.


N. Oliver et al. (Eds.): ECML PKDD 2021, LNAI 12977, pp. 510–526, 2021.
https://doi.org/10.1007/978-3-030-86523-8_31

Suppose further that the predictors are divided into multiple groups. One could select a few of the groups rather than a few components of $\boldsymbol{\beta}$. An analogous procedure for group selection is Group LASSO [40]:

$$\min_{\boldsymbol{\beta}} \frac{1}{2}\|\mathbf{y} - \sum_{l=1}^{L} \mathbf{X}_l \boldsymbol{\beta}_l\|_2^2 + \lambda \sum_{l=1}^{L} \sqrt{p_l}\|\boldsymbol{\beta}_l\|_2 \tag{3}$$

where $\mathbf{X}_l$ represents the predictors corresponding to the $l$-th group, with corresponding coefficient vector $\boldsymbol{\beta}_l$. Here the group information $\boldsymbol{g} \in \mathbb{R}^p$ is implicit in the definition of $\mathbf{X}_l$, indicating that the $i$-th predictor belongs to the $g_i$-th group. We assume that $p$ predictors are divided into $L$ groups and denote the size of the $l$-th group as $p_l$.

However, for a group selected by Group LASSO, all entries in the group are nonzero. To yield both sparsity of groups and sparsity within each group, [33] introduced the Sparse Group LASSO problem as follows:

$$\min_{\boldsymbol{\beta} \in \mathbb{R}^p} \frac{1}{2}\|\mathbf{y} - \sum_{l=1}^{L} \mathbf{X}_l \boldsymbol{\beta}_l\|_2^2 + (1-\gamma)\lambda \sum_{l=1}^{L} \sqrt{p_l}\|\boldsymbol{\beta}_l\|_2 + \gamma\lambda\|\boldsymbol{\beta}\|_1 \tag{4}$$

where $\gamma \in [0, 1]$ refers to the proportion of the LASSO fit in the overall penalty. If $\gamma = 1$, SGL is purely LASSO, while if $\gamma = 0$, SGL reduces to Group LASSO. We denote the solution to the SGL problem as $\hat{\boldsymbol{\beta}}$.

As a convex optimization problem, SGL can be solved by many existing methods, including ISTA (Iterative Shrinkage-Thresholding Algorithm) [10] and FISTA (Fast Iterative Shrinkage-Thresholding Algorithm) [5]. Both methods rely on the derivation of the proximal operator of the SGL problem, which is non-trivial due to the *non-separability* of the penalty when $\gamma \in [0, 1)$. In previous literature, [33] used the blockwise gradient descent algorithm with backtracking, instead of the proximal approach, to solve this problem. Other method like fast block gradient descent [19] can also be implemented.

In this paper, we derive the proximal operator for SGL and establish *approximate message passing* (AMP) [1,3,4,11,13,25] for SGL from this new approach. We then analyze the algorithmic aspects of SGL via AMP. In general, AMP is a class of computationally efficient gradient-based algorithms originating from graphical models and extensively studied for many compressed sensing problems [22,29].

We derive, for fixed $\gamma$, the SGL AMP as follows: set $\boldsymbol{\beta}^0 = \mathbf{0}, \boldsymbol{z}^0 = y$ and for $t > 0$,

$$\boldsymbol{\beta}^{t+1} = \eta_\gamma(\mathbf{X}^\top \mathbf{z}^t + \boldsymbol{\beta}^t, \theta_t) \tag{5}$$

$$\mathbf{z}^{t+1} = \mathbf{y} - \mathbf{X}\boldsymbol{\beta}^{t+1} + \frac{1}{\delta}\mathbf{z}^t \langle \eta'_\gamma(\mathbf{X}^\top \mathbf{z}^t + \boldsymbol{\beta}^t, \theta_t) \rangle. \tag{6}$$

Here the threshold $\theta_t$ is carefully designed and can be found in [4]. $\langle \boldsymbol{v} \rangle := \sum_{i=1}^{p} v_i/p$ is the average of vector $\boldsymbol{v}$. Furthermore, $\eta_\gamma$ is the proximal operator

$$\eta_\gamma(\boldsymbol{s},\lambda) := \underset{\boldsymbol{b}}{\operatorname{argmin}} \frac{1}{2}\|\boldsymbol{s}-\boldsymbol{b}\|^2 + (1-\gamma)\lambda\sum_{l=1}^{L}\sqrt{p_l}\|\boldsymbol{b}_l\|_2 + \gamma\lambda\|\boldsymbol{b}\|_1$$

and $\eta'_\gamma := \nabla \circ \eta_\gamma$ is the diagonal of the Jacobian matrix of the proximal operator with respect to its first argument $\boldsymbol{s}$, with $\circ$ being the Hadamard product.

*Empirically*, the simulation results in Fig. 1 and Table 1 demonstrate the supremacy of AMP convergence speed over the two most well-known proximal gradient descent methods, ISTA and FISTA. We also compare these methods to the Nesterov-accelerated blockwise descent in [33] and in R package SGL. We note that the Nesterov-accelerated ISTA (i.e. FISTA) outperforms the accelerated blockwise descent in terms of both the number of iterations and the wall-clock time (see Fig. 11(a) and the detailed analysis in Appendix B). This observation suggests that using the proximal operator not only requires fewer iterations but also reduces the complexity of computation at each iteration. We pause to emphasize that, in general, the cost function $\mathcal{C}_{\mathbf{X},\boldsymbol{y}}(\boldsymbol{\beta}) := \frac{1}{2}\|\mathbf{y}-\sum_{l=1}^{L}\mathbf{X}_l\boldsymbol{\beta}_l\|_2^2 + (1-\gamma)\lambda\sum_{l=1}^{L}\sqrt{p_l}\|\boldsymbol{\beta}_l\|_2 + \gamma\lambda\|\boldsymbol{\beta}\|_1$ is not strictly convex. We choose the optimization error (mean squared error, or MSE, between $\boldsymbol{\beta}^t$ and $\boldsymbol{\beta}$) as the measure of convergence, as there may exist $\hat{\boldsymbol{\beta}}$ far from $\boldsymbol{\beta}$ for which $\mathcal{C}(\hat{\boldsymbol{\beta}})$ is close to $\mathcal{C}(\boldsymbol{\beta})$.

**Table 1.** Same settings as in Fig. 1 except $\lambda = 1$ and the prior $\beta_0$ is $5\times$ Bernoulli (0.1).

| Number of Iterations | | | | |
|---|---|---|---|---|
| **MSE** | $10^{-2}$ | $10^{-3}$ | $10^{-4}$ | $10^{-5}$ |
| ISTA | 309 | 629 | 988 | 1367 |
| FISTA | 42 | 81 | 158 | 230 |
| AMP | 4 | 6 | 14 | 35 |

In [4,26], it has been rigorously proved that applying AMP to LASSO shows nice statistical properties of the LASSO solution. However, applying AMP to a non-separable regularizer is more challenging. Along this line of research, recently, SLOPE AMP was rigorously developed in [7] and [8] further analyzed a class of non-separable penalties which become asymptotically separable. Nevertheless, the question of whether AMP provably solves and characterizes SGL remains open for researchers.

Our contributions are as follows. We first derive a proximal operator of SGL on which the SGL AMP is based. We prove that the algorithm solves the SGL problem *asymptotically exactly* under i.i.d. Gaussian designs. The proof leverages the recent state evolution analysis [6] for non-separable penalties and shows that the state evolution characterizes the asymptotically exact behaviors of $\hat{\boldsymbol{\beta}}$.

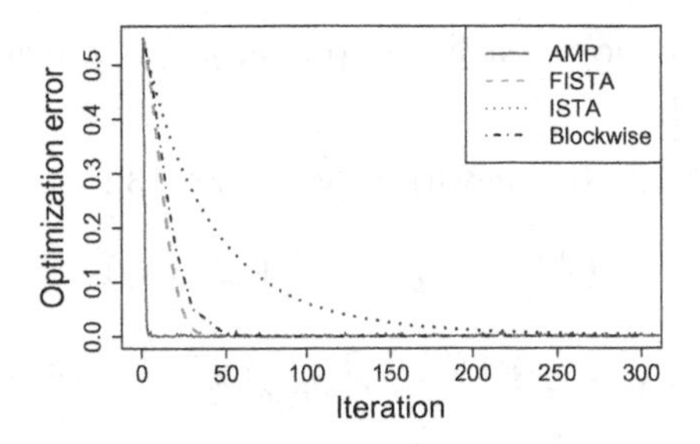


**Fig. 1.** $p = 4000, n = 2000, \gamma = 0.5, \boldsymbol{g} = (1, \cdots, 1)$, the entries of $\mathbf{X}$ are i.i.d. $\mathcal{N}(0, 1/n)$, $\lambda = 2$, and the prior $\beta$ is $5\times$Bernoulli (0.2).

Specifically, the distribution of SGL solution is completely specified by a few parameters that are the solution to a certain fixed-point equation asymptotically. As a consequence, we can use the characterization of the SGL solution to analyze the behaviors of the $\hat{\boldsymbol{\beta}}$ precisely. In particular, we investigate the effects of group information and $\gamma$ on the model performance. Empirically, we find that SGL can benefit significantly from good group information in the sense of MSE and False Discovery Rate/Proportion.

The rest of this paper is divided into four sections. In Sect. 2, we give some preliminary background of the AMP algorithm. In Sect. 3, we state our main theorems about the convergence and the characterization. In Sect. 4, we show some simulation results. In Sect. 5, we conclude our paper and list some possible extensions of future work.

# 2 Algorithm

## 2.1 Approximate Message Passing

*Assumptions for AMP*

- **(A1)** The measurement matrix $\mathbf{X}$ has independent entries following $\mathcal{N}(0, \frac{1}{n})$.
- **(A2)** The elements of signal $\boldsymbol{\beta}$ are i.i.d. copies of a random variable $\Pi$ with $\mathbb{E}(\Pi^2 \max\{0, \log(\Pi)\}) < \infty$. We use $\mathbf{\Pi} \in \mathbb{R}^p$ to denote random vector with each component following i.i.d. $\Pi$.
- **(A3)** The elements of noise $\boldsymbol{w}$ are i.i.d. $W$ with $\sigma_w^2 := \mathbb{E}(W^2) < \infty$.
- **(A4)** The ratio of sample size to feature size $\frac{n}{p}$ approaches a constant $\delta \in (0, \infty)$ as $n, p \to \infty$.

We note that the assumptions are the same as in [7] and the second-moment assumptions **(A2)** and **(A3)** can be relaxed. For example, we can instead assume that $\boldsymbol{w}$ has an empirical distribution that converges weakly to $W$, with $\|\boldsymbol{w}\|^2/p \to \mathbb{E}(W^2) < \infty$. In general, we may extend assumptions **(A1)** and **(A2)** to a much broader range of cases, as discussed in Sect. 4.3. Additionally, we need one extra assumption for the group information as follows.

- (**A5**) The relative ratio of each group size, $p_l/p$, converges to $r_l \in (0,1)$ as $p \to \infty$.

Now we write the SGL AMP algorithm based on [13]:

$$\boldsymbol{\beta}^{t+1} = \eta_{\gamma,\boldsymbol{g}}(\mathbf{X}^\top \mathbf{z}^t + \boldsymbol{\beta}^t, \alpha\tau_t) \tag{7}$$

$$\mathbf{z}^{t+1} = \mathbf{y} - \mathbf{X}\boldsymbol{\beta}^{t+1} + \frac{1}{\delta}\mathbf{z}^t \langle \eta'_{\gamma,\boldsymbol{g}}(\mathbf{X}^\top \mathbf{z}^t + \boldsymbol{\beta}^t, \alpha\tau_t)\rangle \tag{8}$$

$$\tau_{t+1}^2 = \sigma_w^2 + \lim_{p\to\infty} \frac{1}{\delta p}\mathbb{E}\|\eta_{\gamma,\boldsymbol{g}}(\boldsymbol{\Pi} + \tau_t \boldsymbol{Z}, \alpha\tau_t) - \boldsymbol{\Pi}\|_2^2 \tag{9}$$

where $\boldsymbol{Z}$ is the standard Gaussian $\mathcal{N}(0, \mathcal{I}_p)$ and the expectation is taken with respect to both $\boldsymbol{\Pi}$ and $\boldsymbol{Z}$. We denote $\eta_{\gamma,\boldsymbol{g}}(\boldsymbol{s},\lambda) : \mathbb{R}^p \times \mathbb{R} \to \mathbb{R}^p$ as the proximal operator for SGL, which we will derive in Sect. 2.2. We notice that, comparing AMP to the standard proximal gradient descent, the thresholds are related to $(\alpha, \tau_t)$ instead of to $\lambda$. On one hand, $\tau_t$ is derived from Eq. (9), known as the **state evolution**, which relies on $\alpha$. On the other hand, $\alpha$ corresponds uniquely to $\lambda$ via equation (10) which is so called **calibration**: with $\tau_* := \lim_{t\to\infty} \tau_t$,

$$\lambda = \alpha\tau_* \left(1 - \lim_{p\to\infty} \frac{1}{\delta}\langle \eta'_{\gamma,\boldsymbol{g}}(\boldsymbol{\Pi} + \tau_* \boldsymbol{Z}, \alpha\tau_*)\rangle\right). \tag{10}$$

### 2.2 Proximal Operator and Derivative

In this section we derive the proximal operator [10] for SGL. In comparison to [15,16,28,33], which all used subgradient conditions to solve SGL, our proximal-based methods can be much more efficient in terms of convergence speed and accuracy (for details of comparison, see Appendix B).

Now we derive the proximal operator for SGL. Denote $\beta^{(j)}$ as the $j$-th component of $\boldsymbol{\beta}$ and the cost function $\mathcal{C}_{\lambda,\gamma,\boldsymbol{g}}$ as

$$\mathcal{C}_{\lambda,\gamma,\boldsymbol{g}}(\boldsymbol{s},\boldsymbol{\beta}) = \frac{1}{2}\|\boldsymbol{s} - \boldsymbol{\beta}\|_2^2 + (1-\gamma)\lambda \sum_{l=1}^{L} \sqrt{p_l}\|\boldsymbol{\beta}_l\|_2 + \gamma\lambda\|\boldsymbol{\beta}\|_1.$$

We sometimes ignore the dependence on the subscripts when there is no confusion. When $g_j = l$ and $\|\boldsymbol{\beta}_l\|_2 \neq 0$, we set $\frac{\partial \mathcal{C}}{\partial \beta^{(j)}} = 0$ and denote $l_j := \{i : g_i = g_j\}$. Then the explicit formula is

$$\eta_\gamma(\boldsymbol{s},\lambda)^{(j)} = \eta_{\text{soft}}(s^{(j)}, \gamma\lambda)\left(1 - \frac{(1-\gamma)\lambda\sqrt{p_{l_j}}}{\|\eta_{\text{soft}}(\boldsymbol{s}_{l_j}, \gamma\lambda)\|_2}\right) \tag{11}$$

when $\eta_{\text{soft}}(\boldsymbol{s}_{l_j}, \gamma\lambda) \in \mathbb{R}^{p_{l_j}}$ has a non-zero norm. Here

$$\eta_{\text{soft}}(x; b) = \begin{cases} x - b & x > b \\ x + b & x < -b \\ 0 & \text{otherwise} \end{cases} \tag{12}$$

is the soft-thresholding operator.

We emphasize that (11) is incomplete due the non-differentiability of $\|\boldsymbol{\beta}_l\|_2$ at $\boldsymbol{\beta}_l = 0$. Denoting the right hand side of equation (11) as $\hat{\eta}_\gamma^{(j)}$, the full formula of the proximal operator is

$$\eta_\gamma(\boldsymbol{s}, \lambda)^{(j)} = \begin{cases} \hat{\eta}_\gamma^{(j)} & \text{if } \|\eta_{\text{soft}}(\boldsymbol{s}_{l_j}, \gamma\lambda)\|_2 > (1-\gamma)\lambda\sqrt{p_{l_j}} \\ 0 & \text{otherwise} \end{cases} \tag{13}$$

The details of derivation are left to Appendix A.

Similarly, the derivative of the proximal operator w.r.t. $s$ also has two forms, split by the same conditions as above. For simplicity we only describe the non-zero form here:

$$\eta'_\gamma(\boldsymbol{s}, \lambda)^{(j)} = \mathbf{1}\{|s^{(j)}| > \gamma\lambda\} \cdot \left[1 - \frac{(1-\gamma)\lambda\sqrt{p_{l_j}}}{\|\eta_{\text{soft}}(\boldsymbol{s}_{l_j}, \gamma\lambda)\|_2}\left(1 - \frac{\eta^2_{\text{soft}}(s^{(j)}, \gamma\lambda)}{\|\eta_{\text{soft}}(\boldsymbol{s}_{l_j}, \gamma\lambda)\|_2^2}\right)\right] \tag{14}$$

We note that in [33], the SGL problem is divided into $L$ sub-problems according to the group information. Then subgradient conditions are used to construct majorization-minimization problems. These problems are solved cyclically, via accelerated gradient descent, in a blockwise or groupwise manner. Along this line of research, there have been other blockwise descent methods designed for Group LASSO [24,39]. In contrast, our proximal operator is unified, as can be seen by comparing Algorithm 1 and Algorithm 2 in Appendix B. We note that our proximal operator also has a groupwise flavor, but all groups are updated independently and simultaneously, thus improving the convergence speed. Since SGL AMP is built on the proximal operator, we will validate the proximal approach by introducing ISTA and FISTA for SGL, with detailed complexity analysis and convergence analysis in Appendix B.

# 3 Main Result

## 3.1 State Evolution and Calibration

Notice that in SGL AMP, we use $\theta_t$ as the threshold, whose design requires state evolution and calibration. Thus we start with some properties of state evolution recursion (9). To simplify the analysis, we consider the finite approximation of state evolution and present precise conditions which guarantee that the state evolution converges efficiently.

**Proposition 1.** *Let* $\mathbf{F}_\gamma(\tau_t^2, \alpha\tau_t) = \sigma_w^2 + \frac{1}{\delta p}\mathbb{E}\|\eta_\gamma(\boldsymbol{\Pi} + \tau_t \boldsymbol{Z}, \alpha\tau_t) - \boldsymbol{\Pi}\|_2^2$ *and define* $\mathcal{A}(\gamma) = \{\alpha : \delta \geq 2T(\gamma\alpha) - 2(1-\gamma)\alpha\sqrt{2T(\gamma\alpha)} + (1-\gamma)^2\alpha^2\}$ *with* $T(z) = (1+z^2)\Phi(-z) - z\phi(z)$, $\phi(z)$ *being the standard Gaussian density and* $\Phi(z) = \int_{-\infty}^z \phi(x)dx$. *For any* $\sigma_w^2 > 0$, $\alpha \in \mathcal{A}(\gamma)$ *, the fixed point equation* $\tau^2 = \mathbf{F}_\gamma(\tau^2, \alpha\tau)$ *admits a unique solution. Denoting the solution as* $\tau_* = \tau_*(\alpha)$, *we have* $\lim_{t\to\infty}\tau_t \to \tau_*(\alpha)$, *where the convergence is monotone under any initial condition. Finally* $\left|\frac{d\mathbf{F}_\gamma}{d\tau^2}\right| < 1$ *at* $\tau = \tau_*$

A demonstration of $\mathcal{A}$ is given in Fig. 2. We note that for all $\gamma < 1$, $\mathcal{A}$ has upper and lower bounds; however, when $\gamma = 1$, i.e. for LASSO, there is no upper bound. We provide the proof of this statement in Appendix D.

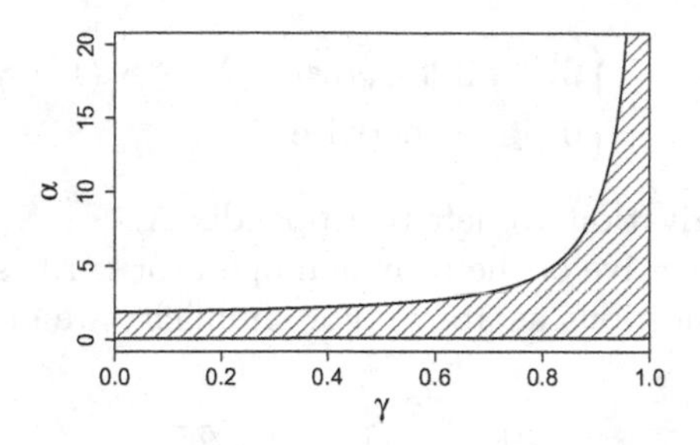


**Fig. 2.** $\mathcal{A}$ when $\delta = 0.2$ is represented in the red shaded region. (Color figure online)

*Remark 1.* When $\alpha$ is outside the set $\mathcal{A}$ in Proposition 1, we must have $\alpha > \mathcal{A}_{\max}$, since we use a non-negative $\lambda$ as penalty which guarantees $\alpha > \mathcal{A}_{\min}$. To see this, consider $\alpha \notin \mathcal{A}$, we have $\tau = \infty$ and hence the dominant term in $\pi + \tau \boldsymbol{Z}$ is $\tau \boldsymbol{Z}$. We can view $\pi$ as if vanishing and easily derive $\alpha > \mathcal{A}_{\min}$ from the state evolution. For $\alpha > \mathcal{A}_{\max}$, the state evolution in fact still converges.[1]

Before we employ the finite approximation of state evolution to describe the calibration (10), we explain the necessity of calibration by the following lemma.

**Lemma 1.** *For fixed $\gamma$, a stationary point $\hat{\boldsymbol{\beta}}$ with corresponding $\hat{\mathbf{z}}$ of the AMP iteration (7), (8) with $\theta_t = \theta_*$ is a minimizer of the SGL cost function in* (4) *with* $\lambda = \theta_* \left(1 - \frac{1}{\delta}\langle \eta'_\gamma(\mathbf{X}^\top \hat{\mathbf{z}} + \hat{\boldsymbol{\beta}}, \theta_*)\rangle\right)$.

Setting $\theta_* = \alpha\tau_*$, we are in the position to define the finite approximation of calibration between $\alpha$ and $\lambda$ by

$$\lambda = \alpha\tau_* \left(1 - \frac{1}{\delta}\langle \eta'_\gamma(\boldsymbol{\Pi} + \tau_* \boldsymbol{Z}, \alpha\tau_*)\rangle\right) \tag{15}$$

In practice, we need to invert (15) to input $\lambda$ and recover

$$\alpha(\lambda) \in \{a \in \mathcal{A} : \lambda(a) = \lambda\} \tag{16}$$

The next proposition and corollary imply that the mapping of $\lambda \to \alpha(\lambda)$ is well-defined and easy to compute.

[1] We note that $\mathcal{A}$ is a sufficient but not necessary condition for the state evolution to converge. The reason that we split the analysis at $\alpha = \mathcal{A}_{\max}$ is because, for $\alpha > \mathcal{A}_{\max}$, the SGL estimator is 0. Besides, we note that the set $\mathcal{A}$ only affects the state evolution. Hence when $\alpha > \mathcal{A}_{\max}$, the calibration is still valid and the mapping between $\alpha$ and $\lambda$ is monotone.

**Proposition 2.** *The function* $\alpha \to \lambda(\alpha)$ *is continuous on* $\mathcal{A}(\gamma)$ *with* $\lambda(\min \mathcal{A}) = -\infty$ *and* $\lambda(\max \mathcal{A}) = \lambda_{\max}$ *for some constant* $\lambda_{\max}$ *depending on* $\Pi$ *and* $\gamma$. *Therefore, the function* $\lambda \to \alpha(\lambda)$ *satisfying* $\alpha(\lambda) \in \{\alpha \in \mathcal{A}(\gamma) : \lambda(\alpha) = \lambda\}$ *exists where* $\lambda \in (-\infty, \lambda_{\max})$.

Given $\lambda$, Proposition 2 claims that $\alpha$ exists and the following result guarantees its uniqueness.

**Corollary 1.** *For* $\lambda < \lambda_{\max}, \sigma_{\boldsymbol{w}}^2 > 0$, *there exists a unique* $\alpha \in \mathcal{A}(\gamma)$ *such that* $\lambda(\alpha) = \lambda$ *as defined in* (15). *Hence the function* $\lambda \to \alpha(\lambda)$ *is continuous and non-decreasing with* $\alpha((-\infty, \lambda_{\max})) = \mathcal{A}(\gamma)$.

The proofs of these statements are left in Appendix E.

### 3.2 AMP Characterizes SGL Estimate

Having described the state evolution, we now state our main theoretical results. We establish an asymptotic equality between $\hat{\boldsymbol{\beta}}$ and $\eta_\gamma$ in pseudo-Lipschitz norm, which allows the fine-grained statistical analysis of the SGL minimizer.

**Definition 1** [6]: *For* $k \in \mathbb{N}_+$, *a function* $\phi : \mathbb{R}^d \to \mathbb{R}$ *is* ***pseudo-Lipschitz*** *of order* $k$, *if there exists a constant* $L$ *such that for* $\mathbf{a}, \mathbf{b} \in \mathbb{R}^d$,

$$|\phi(\boldsymbol{a}) - \phi(\boldsymbol{b})| \leq L\Big(1 + \Big(\frac{\|\boldsymbol{a}\|}{\sqrt{d}}\Big)^{k-1} + \Big(\frac{\|\boldsymbol{b}\|}{\sqrt{d}}\Big)^{k-1}\Big)\Big(\frac{\|\boldsymbol{a} - \boldsymbol{b}\|}{\sqrt{d}}\Big). \tag{17}$$

*A sequence (in* $p$*) of pseudo-Lipschitz functions* $\{\phi_p\}_{p \in \mathbb{N}_+}$ *is* ***uniformly pseudo-Lipschitz*** *of order* $k$ *if, denoting by* $L_p$ *the pseudo-Lipschitz constant of* $\phi_p$, $L_p < \infty$ *for each* $p$ *and* $\limsup_{p \to \infty} L_p < \infty$.

**Theorem 1.** *Under the assumptions (**A1**)–(**A5**), for any uniformly pseudo-Lipschitz sequence of function* $\varphi_p : \mathbb{R}^p \times \mathbb{R}^p \to \mathbb{R}$ *and for* $\mathbf{Z} \sim \mathcal{N}(0, \mathcal{I}_p), \boldsymbol{\Pi} \sim p_\Pi$,

$$\lim_{p \to \infty} \varphi_p(\hat{\boldsymbol{\beta}}, \boldsymbol{\beta}) = \lim_t \lim_p \mathbb{E}[\varphi_p(\eta_\gamma(\boldsymbol{\Pi} + \tau_t \boldsymbol{Z}; \alpha \tau_t), \boldsymbol{\Pi})].$$

The proof of this theorem is left in Appendix C. Essentially, up to a uniformly pseudo-Lipschitz loss, we can replace $\hat{\boldsymbol{\beta}}$ by $\eta_\gamma$ in the large system limit. The distribution of $\eta_\gamma$ is explicit, thus allowing the analysis of certain quantities. Specifically, if we use $\varphi_p(\boldsymbol{a}, \boldsymbol{b}) = \frac{1}{p}\|\boldsymbol{a} - \boldsymbol{b}\|_2^2$, the MSE between $\hat{\boldsymbol{\beta}}$ and $\boldsymbol{\beta}$ can be characterized by $\tau$.

**Corollary 2.** *Under the assumptions (**A1**)–(**A5**), then almost surely*

$$\lim_{p \to \infty} \frac{1}{p}\|\hat{\boldsymbol{\beta}} - \boldsymbol{\beta}\|_2^2 = \delta(\tau_*^2 - \sigma_{\boldsymbol{w}}^2)$$

*Proof (Proof of Corollary 2).* Applying Theorem 1 to the pseudo-Lipschitz loss function and letting $\varphi_p(\boldsymbol{a}, \boldsymbol{b}) = \|\boldsymbol{a} - \boldsymbol{b}\|_2^2$, we obtain

$$\lim_{p\to\infty} \|\hat{\boldsymbol{\beta}} - \boldsymbol{\beta}\|_2^2 = \lim_{t\to\infty} \mathbb{E}\left[\varphi_p(\eta_\gamma(\boldsymbol{\Pi} + \tau_t \boldsymbol{Z}; \alpha\tau_t), \boldsymbol{\Pi})\right].$$

The result follows from the state evolution (9) since

$$\lim_{t\to\infty} \mathbb{E}\left[\varphi_p(\eta_\gamma(\Pi + \tau_t Z; \alpha\tau_t), \Pi)\right] = \delta(\tau_*^2 - \sigma_w^2).$$

Now that we have demonstrated the usefulness of our main theoretical result, we prove Theorem 1 at a high level. We first show the convergence of $\boldsymbol{\beta}^t$ to $\hat{\boldsymbol{\beta}}$, i.e. the AMP iterates converge to the true minimizer.

**Theorem 2.** *Under assumptions **(A1)**-**(A5)**, for the output of the AMP algorithm in (7) and the Sparse Group LASSO estimator given by the solution of (4),*

$$\lim_{p\to\infty} \frac{1}{p} \|\hat{\boldsymbol{\beta}} - \boldsymbol{\beta}^t\|_2^2 = k_t, \quad \text{where } \lim_{t\to\infty} k_t = 0$$

The proof is similar to the proof of [4] Theorem 1.8. The difference is incurred by the existence of the $\ell_2$ norm which imposes the group structure. We leave the proof in Appendix C.

In addition to Theorem 2, we borrow the state evolution analysis from [6] Theorem 14 to complete the proof of Theorem 1.

**Lemma 2** [6]. *Under assumptions **(A1)** - **(A5)**, given that **(S1)** and **(S2)** are satisfied, consider the recursion* (7) *and* (8). *For any uniformly pseudo-Lipschitz sequence of functions $\phi_n : \mathbb{R}^n \times \mathbb{R}^n \to \mathbb{R}$ and $\varphi_p : \mathbb{R}^p \times \mathbb{R}^p \to \mathbb{R}$,*

$$\phi_n(\mathbf{z}^t, \mathbf{w}) \xrightarrow{\mathbb{P}} \mathbb{E}\left[\phi_n(\mathbf{w} + \sqrt{\tau_t^2 - \sigma_w^2}\mathbf{Z}', \mathbf{w})\right] \tag{18}$$

$$\varphi_p(\boldsymbol{\beta}^t + \mathbf{X}^\top \mathbf{z}^t, \Pi) \xrightarrow{\mathbb{P}} \mathbb{E}\left[\varphi_p(\Pi + \tau_t \mathbf{Z}, \Pi)\right] \tag{19}$$

*where $\tau_t$ is defined in* (9), $\mathbf{Z}' \sim \mathcal{N}(\mathbf{0}, \mathcal{I}_n)$ *and* $\mathbf{Z} \sim \mathcal{N}(\mathbf{0}, \mathcal{I}_p)$.

To see that Theorem 1 holds, we obtain that $\boldsymbol{\beta}^t + \boldsymbol{X}^\top \boldsymbol{z}^t \approx \boldsymbol{\Pi} + \tau_t \boldsymbol{Z}$ from Lemma 2. Together with $\boldsymbol{\beta}^{t+1} = \eta_\gamma(\mathbf{X}^\top \mathbf{z}^t + \boldsymbol{\beta}^t, \alpha\tau_t)$ from (7), we have $\boldsymbol{\beta}^t \approx \eta_\gamma(\boldsymbol{\Pi} + \tau_t \boldsymbol{Z}, \alpha\tau_t)$. Finally Theorem 2 and to obtain that $\hat{\boldsymbol{\beta}} = \eta_\gamma(\boldsymbol{\beta}^t + \mathbf{X}^\top \mathbf{z}^t, \alpha\tau_t)$ within uniformly pseudo-Lipschitz loss, in large system limit.

### 3.3 TPP and FDP Trade-Off of SGL

We now introduce the definition of '**SGL path**' to study statistical quantities: for a fixed $\gamma$, the SGL path is the space of all SGL solutions $\hat{\boldsymbol{\beta}}(\lambda)$ as $\lambda$ varies in $(0, \infty)$. From now on, our simulations will focus on the MSE, power and FDR estimated by AMP instead of the empirical values.

We focus on the single group TPP-FDP trade-off of SGL, i.e. with mixed group information. With the characterization by AMP, we can analytically represent the trade-off curve and extend to the multi-group version analogously. We state our asymptotic result which describes FDP given TPP.

**Theorem 3.** *For any fixed* $\delta, \epsilon, \gamma \in (0, 1)$ *and define*

$$q(u; \delta, \epsilon, \gamma) = \frac{2(1-\epsilon)\Phi(-\gamma t^*(u))}{2(1-\epsilon)\Phi(-\gamma t^*(u)) + \epsilon u} \tag{20}$$

*where* $t^*(u)$ *is the largest positive root of the equation (39) (see Appendix F). Under assumptions for AMP and single group information, FDP*($\lambda$) *of SGL solution satisfies* $\mathbb{P}\Big(q(TPP(\lambda); \delta, \epsilon, \gamma) - c \leq FDP(\lambda) \leq 1 - \epsilon + c\Big) \overset{n\to\infty}{\longrightarrow} 1$ *for any* $\lambda$ *and arbitrarily small* $c > 0$.

We leave the proof in Appendix F in which we consider a special class of priors: the infinity-or-nothing signals. These priors are known to achieve the optimal trade-off in LASSO [34], i.e. with properly tuned $\lambda$, LASSO attains its lowest FDP with such priors. In Fig. 3, we visualize the single group TPP-FDP trade-off in Theorem 1.

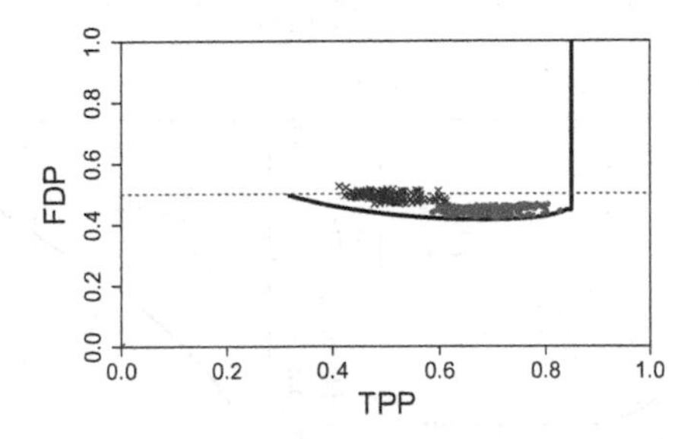


**Fig. 3.** Two priors approximating the single group FDP lower and upper bounds. $\gamma = 0.5$, $\epsilon = 0.5, \delta = 0.4, \mathbf{X} \in \mathbb{R}^{400\times1000}$ is i.i.d. generated from $\mathcal{N}(0, 1/20)$. Blue: $\Pi_1$ takes values in $\{100, 0.1, 0\}$ with probability $(0.25, 0.25, 0.5)$. Red: $\Pi_2$ takes values in $\{1000, 1, 0\}$ with probability $(0.001, 0.499, 0.5)$

We further compare single group SGL TPP-FDP trade-off with different $\gamma$ and $\delta$. In Fig. 4, we observe that with larger $\gamma$, both the supreme TPP and the inferior TPP become smaller. In particular, when $\gamma = 0$ (Group LASSO), the trade-off curve shrinks to a single point. Only when $\gamma = 1$ (LASSO), the trade-off is monotone and allows zero FDP to be achievable. Additionally, as the dimension of data becomes higher, the trade-off shrinks away from 100% TPP.

From the viewpoint of recovery rate, which is defined as the upper bound of TPP, $\sup\{u : q(u) < 1-\epsilon\}$, Group LASSO is optimal with perfect group information and significantly outperforms other SGL including LASSO. Figure 5 clearly shows that, for fixed $\delta > 0$, the recovery rate of SGL and especially LASSO decays with less sparse signals. For fixed $\epsilon > 0$, the recovery rate increases as we move from low dimension to higher ones. It is interesting to observe that Group LASSO remains 100% recovery rate regardless of $\epsilon$ and $\delta$, while LASSO is much more sensitive.

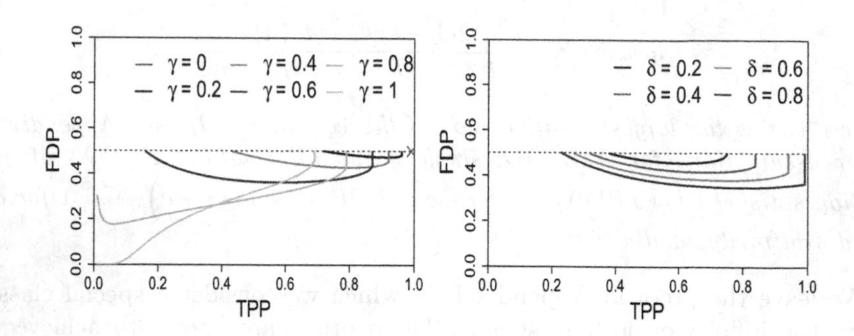


**Fig. 4.** TPP and FDP trade-off curve. Left: $\delta = 0.5, \epsilon = 0.5$. Right: $\gamma = 0.5, \epsilon = 0.5$.

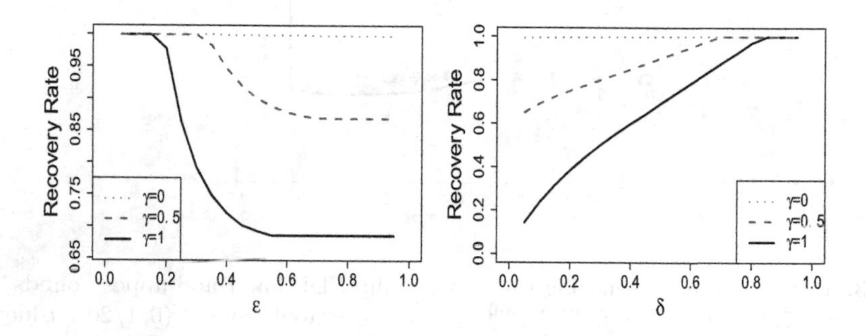


**Fig. 5.** Recovery rate with perfect group information. Left: $\delta = 0.5$. Right: $\epsilon = 0.5$.

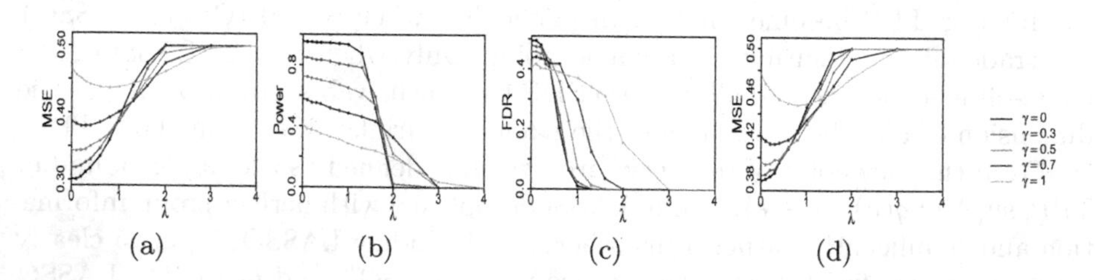


**Fig. 6.** (a) (b) (c) MSE/Power/FDR on SGL path given perfect group information; (d) MSE on SGL path given mixed group information.

## 4 Simulation

Throughout this section, unless specified otherwise, the default setting is as follows: $\delta := n/p = 0.25$, $\boldsymbol{X}$ has i.i.d. entries from $\mathcal{N}(0, 1/n)$, $\gamma = 0.5$, $\sigma_w = 0$ and the prior is Bernoulli(0.5). We consider at most two groups and set the perfect group information as default. The group information $\boldsymbol{g}$ is perfect if all the true signals are classified into one group, while the other group contains all the null signals. In contrast, the mixed group information is the case when only one group exists, so that the true and null signals are mixed.

### 4.1 State Evolution Characterization

First we demonstrate that AMP state evolution indeed characterizes the solution $\hat{\boldsymbol{\beta}}$ by $\eta_\gamma(\boldsymbol{\beta}+\tau \boldsymbol{Z}, \alpha\tau)$ asymptotically accurately. Figure 7(a) clearly visualizes and confirms Theorem 1 by the distributional similarity. In Fig. 7(b), the empirical MSE $\|\hat{\boldsymbol{\beta}} - \boldsymbol{\beta}\|^2/p$ has a mean close to the AMP estimate $\delta(\tau^2 - \sigma_w^2)$ in Corollary 2 and the variance decreases as the dimension increases.

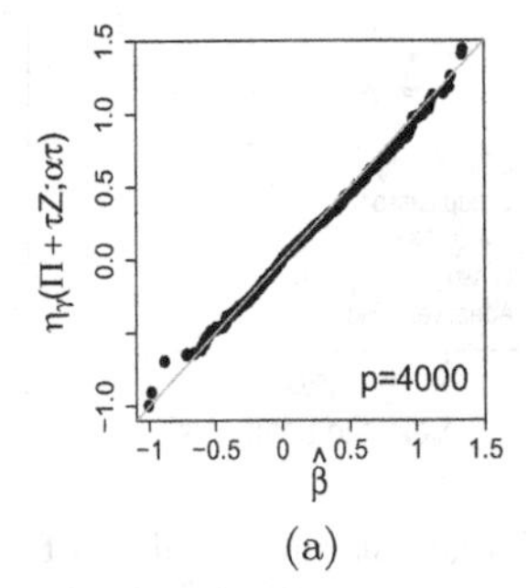


(a)

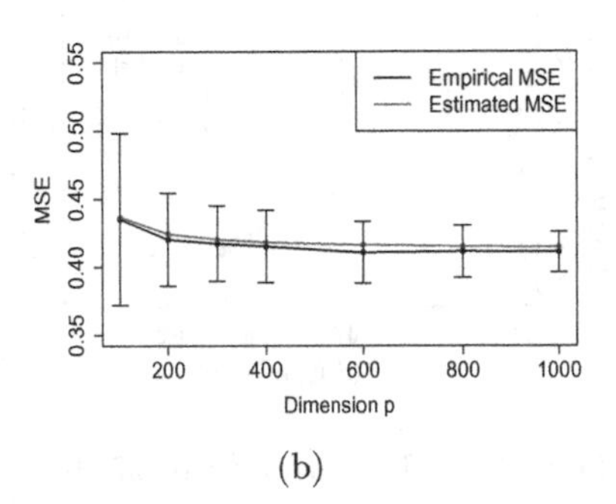


(b)

**Fig. 7.** Illustration of Theorem 1: AMP characterizes SGL solution. Notice $\epsilon = 0.5, \delta = 0.25, \sigma_w = 1, \gamma = 0.5$ and group information is perfect. (a) Quantile-quantile plot of $\hat{\boldsymbol{\beta}}$ and AMP solution with $\lambda = 1$; (b) Empirical and estimated MSE by AMP over 100 independent runs with $\alpha = 1$ ($\lambda = 0.32$). The error bars correspond to one standard deviation.

### 4.2 Benefits of Groups

Now we investigate the benefits of groups under different scenarios. We set the dimension $p = 400$ and the noise $\sigma_w = 0$. Figure 6(a) plots MSE against $\lambda$ given perfect group information, while Fig. 6(d) demonstrates the case with the mixed group information. Figure 6(b) and 6(c) plot the power and FDR against $\lambda$ given the perfect group information.

We observe that, fixing models, better group information helps the models achieve better performances, especially when $\gamma$ is small, i.e. when SGL is closer to the Group LASSO. By comparing Fig. 6(a) and Fig. 6(d), we see an increase of

MSE when signals are mixed by the group information. Somewhat surprisingly, even SGL with the mixed group information may achieve better MSE than LASSO, which does not use any group information.

On the other hand, for fixed group information, models with different $\gamma$ enjoy the benefit of group information differently: in Fig. 6(a), we notice that the performance depends on the penalty $\lambda$: if $\lambda$ is small enough, then SGL with smaller $\gamma$ performs better; if $\lambda$ is sufficiently large, then SGL with larger $\gamma$ may be favored.

We compare SGL with other $\ell_1$-regularized models, namely the LASSO, Group LASSO, adaptive LASSO [41] and elasitc net [42] in Fig. 8, given perfect group information. In the type I/II tradeoff, Group LASSO and SGL with $\gamma = 0.5$ demonstrates dominating performance over other models. However, in terms of the estimation MSE (between $\hat{\boldsymbol{\beta}}$ and $\boldsymbol{\beta}$), SGL allows to achieve smaller MSE but selects more features. In both figures, SGL shows a piecewise continuous pattern and it would be interesting to derive the explicit form with AMP in the future.

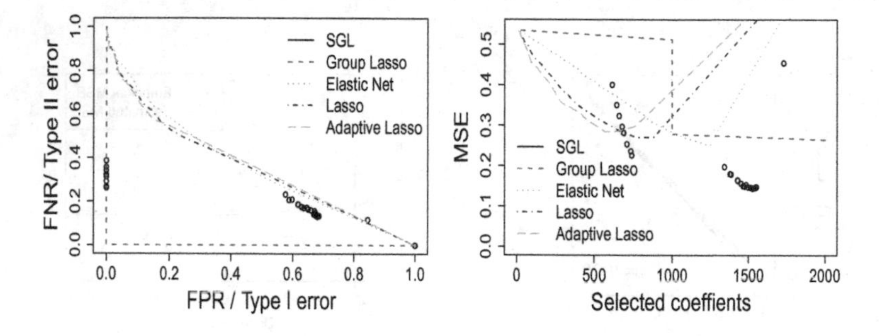


**Fig. 8.** Left: Type I/II error tradeoff. Right: MSE against the number of selected coefficients with perfect group information. Here $\sigma_w = 0, \boldsymbol{X} \in \mathbb{R}^{1000\times2000}$ is i.i.d. Gaussian and the prior follows a Bernoulli-Gaussian with 0.5 probability being standard Gaussian.

We further compare SGL AMP to the MMSE AMP (or Bayes-AMP) [12], which by design finds the minimum MSE over a wide class of convex regularized estimators [9]. In Fig. 9, we plot SGL with different $\gamma$ and carefully tune the penalty of each model to achieve its minimum MSE. We summarize that, empirically, SGL AMP with good group information is very competitive to MMSE AMP and smaller $\gamma$ leads to better performance. Nevertheless, this observed pattern may break if the group information is less correct.

### 4.3 Extensions of SGL AMP

The theoretical result of vanilla AMP assumes that the design matrix $\boldsymbol{X}$ is i.i.d. Gaussian (or sub-Gaussian) [2,3]. The convergence of AMP may be difficult if not impossible on the real-world data. Nevertheless, empirical results suggest that

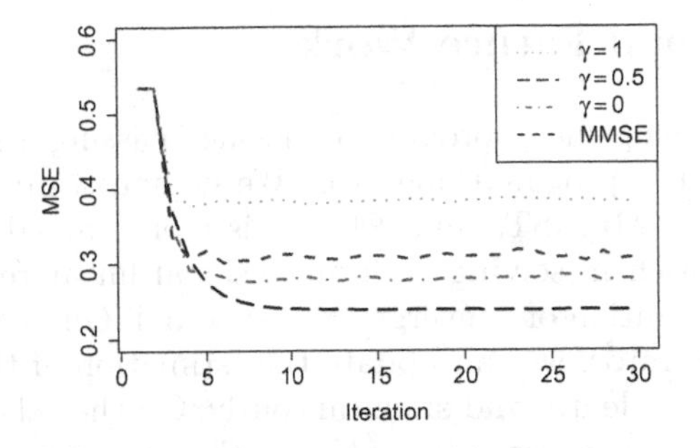


**Fig. 9.** MSE of AMP algorithms with perfect group information. Here $n = 1000, p = 2000, \sigma_w = 0, \boldsymbol{X}$ is i.i.d. Gaussian and the prior follows a Bernoulli-Gaussian with 0.5 probability being standard Gaussian.

AMP works on a much broader class of data matrices even without theoretical guarantees. In our experiments, we observe that the performance of AMP is very similar to Fig. 1 for i.i.d. non-Gaussian data matrices (c.f. Appendix G).

On the other hand, we may relax the assumption of 'i.i.d.' by leveraging a variant of AMP, called vector-AMP or VAMP [30]. It has been rigorously shown that VAMP works on a larger class of data matrices, i.e. the right rotationally-invariant matrices. We emphasize that applying VAMP to non-separable penalties is in general an open problem, though there has been some progress for certain specific type of non-separability [23]. In Appendix G, we substitute the soft-thresholding function with the SGL proximal operator to extend the LASSO VAMP to the SGL VAMP. We implement our SGL VAMP (Algorithm 3) on the Adult Data Set [21], which contains 32,561 samples and 124 features to predict an adult's income influenced by the individual's education level, age, gender, occupation, and etc. We observe that on this specific dataset, SGL VAMP converges in one single iteration and shows its potential to work on other real datasets.

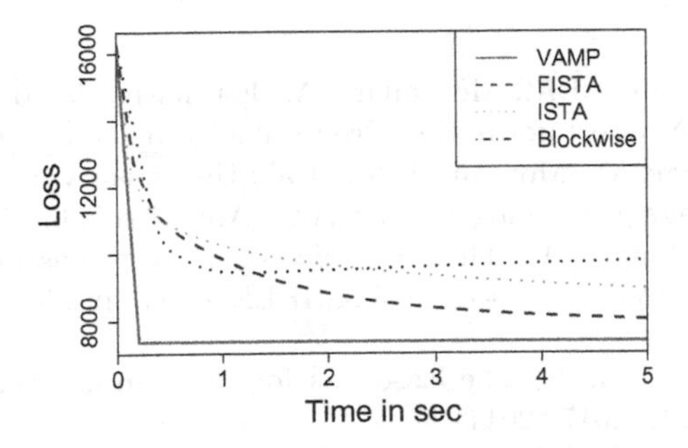


**Fig. 10.** VAMP algorithm on the Adult income dataset with $\gamma = 0.5, \boldsymbol{g} = (1, \cdots, 1)$, $\lambda = 5$ and damping constant $\mathcal{D} = 0.1$.

## 5 Discussion and Future Work

In this work, we develop the approximate message passing algorithm for Sparse Group LASSO via the proximal operator. We demonstrate that the proximal approach, including AMP, ISTA and FISTA, is more unified and efficient than the blockwise approach in solving high-dimensional linear regression problems. The key to the acceleration of convergence is two-fold. On one hand, by employing the proximal operator, we can update the estimation of the SGL minimizer for each group independently and simultaneously. On the other hand, AMP has an extra 'Onsager reaction term', $\langle\eta'_{\gamma,\boldsymbol{g}}(\mathbf{X}^*\mathbf{z}^t+\boldsymbol{\beta}^t,\alpha\tau_t)\rangle$, which corrects the algorithm at each step non-trivially.

Our analysis of SGL AMP reveals some important results on the state evolution and the calibration. For example, the state evolution of SGL AMP only works on a bounded domain of $\alpha$, whereas in the LASSO case, the $\alpha$ domain is not bounded above and makes the penalty tuning more difficult. We then prove that SGL AMP converges to the true minimizer and characterizes the solution exactly, in an asymptotic sense. We highlight that such characterization is empirically accurate in the finite sample scenario and allows us to analyze certain statistical quantities of the SGL solution closely, such as $\ell_2$ risk, type I/II errors as well as the effect of the group information on these measures.

Our work suggests several possible future research. In one direction, it is promising to extend the proximal algorithms (especially AMP) to a broader class of models with structured sparsity, such as the sparse linear regression with overlapping groups, Group SLOPE and the sparse group logistic regression. On a different road, although AMP is robust in distributional assumptions in the sense of fast convergence under i.i.d. non-Gaussian measurements, multiple variants of AMP may be applied to adapt to real-world data. To name a few, one may look into SURE-AMP [17], EM-AMP [36,38] and VAMP [30] to relax the known signal assumption and non-i.i.d. measurement assumption.

## References

1. Bayati, M., Erdogdu, M.A., Montanari, A.: Estimating lasso risk and noise level. In: Advances in Neural Information Processing Systems, pp. 944–952 (2013)
2. Bayati, M., Lelarge, M., Montanari, A., et al.: Universality in polytope phase transitions and message passing algorithms. Ann. Appl. Probab. **25**(2), 753–822 (2015)
3. Bayati, M., Montanari, A.: The dynamics of message passing on dense graphs, with applications to compressed sensing. IEEE Trans. Inf. Theory **57**(2), 764–785 (2011)
4. Bayati, M., Montanari, A.: The lasso risk for gaussian matrices. IEEE Trans. Inf. Theory **58**(4), 1997–2017 (2011)
5. Beck, A., Teboulle, M.: A fast iterative shrinkage-thresholding algorithm for linear inverse problems. SIAM J. Imag. Sci. **2**(1), 183–202 (2009)
6. Berthier, R., Montanari, A., Nguyen, P.M.: State evolution for approximate message passing with non-separable functions. arXiv preprint arXiv:1708.03950 (2017)

7. Bu, Z., Klusowski, J., Rush, C., Su, W.: Algorithmic analysis and statistical estimation of slope via approximate message passing. arXiv preprint arXiv:1907.07502 (2019)
8. Celentano, M.: Approximate separability of symmetrically penalized least squares in high dimensions: characterization and consequences. arXiv preprint arXiv:1906.10319 (2019)
9. Celentano, M., Montanari, A.: Fundamental barriers to high-dimensional regression with convex penalties. arXiv preprint arXiv:1903.10603 (2019)
10. Daubechies, I., Defrise, M., De Mol, C.: An iterative thresholding algorithm for linear inverse problems with a sparsity constraint. Commun. Pure Appl. Math. J. Issued Courant Inst. Math. Sci. **57**(11), 1413–1457 (2004)
11. Donoho, D.L., Maleki, A., Montanari, A.: Message-passing algorithms for compressed sensing. Proc. Natl. Acad. Sci. **106**(45), 18914–18919 (2009)
12. Donoho, D.L., Maleki, A., Montanari, A.: Message passing algorithms for compressed sensing: I. motivation and construction. In: 2010 IEEE Information Theory Workshop on Information Theory (ITW 2010, Cairo), pp. 1–5. IEEE (2010)
13. Donoho, D.L., Maleki, A., Montanari, A.: How to design message passing algorithms for compressed sensing. preprint (2011)
14. Donoho, D.L., Maleki, A., Montanari, A.: The noise-sensitivity phase transition in compressed sensing. IEEE Trans. Inf. Theory **57**(10), 6920–6941 (2011)
15. Foygel, R., Drton, M.: Exact block-wise optimization in group lasso and sparse group lasso for linear regression. arXiv preprint arXiv:1010.3320 (2010)
16. Friedman, J., Hastie, T., Tibshirani, R.: A note on the group lasso and a sparse group lasso. arXiv preprint arXiv:1001.0736 (2010)
17. Guo, C., Davies, M.E.: Near optimal compressed sensing without priors: parametric sure approximate message passing. IEEE Trans. Signal Process. **63**(8), 2130–2141 (2015)
18. Hu, H., Lu, Y.M.: Asymptotics and optimal designs of slope for sparse linear regression. In: 2019 IEEE International Symposium on Information Theory (ISIT), pp. 375–379. IEEE (2019)
19. Ida, Y., Fujiwara, Y., Kashima, H.: Fast sparse group lasso (2019)
20. Doob, J.L.: Stochastic Processes, vol. 101. Wiley, New York (1953)
21. Kohavi, R.: Scaling up the accuracy of naive-bayes classifiers: a decision-tree hybrid. KDD **96**, 202–207 (1996)
22. Krzakala, F., Mézard, M., Sausset, F., Sun, Y., Zdeborová, L.: Probabilistic reconstruction in compressed sensing: algorithms, phase diagrams, and threshold achieving matrices. J. Stat. Mech: Theory Exp. **2012**(08), P08009 (2012)
23. Manoel, A., Krzakala, F., Varoquaux, G., Thirion, B., Zdeborová, L.: Approximate message-passing for convex optimization with non-separable penalties. arXiv preprint arXiv:1809.06304 (2018)
24. Meier, L., Van De Geer, S., Bühlmann, P.: The group lasso for logistic regression. J. Roy. Stat. Soc. Ser. B (Statistical Methodology) **70**(1), 53–71 (2008)
25. Montanari, A., Eldar, Y., Kutyniok, G.: Graphical models concepts in compressed sensing. Compressed Sensing: Theory and Applications, pp. 394–438 (2012)
26. Mousavi, A., Maleki, A., Baraniuk, R.G., et al.: Consistent parameter estimation for lasso and approximate message passing. Ann. Stat. **46**(1), 119–148 (2018)
27. Parikh, N., Boyd, S., et al.: Proximal algorithms. Foundations Trends Optimization **1**(3), 127–239 (2014)
28. Puig, A.T., Wiesel, A., Hero, A.O.: A multidimensional shrinkage-thresholding operator. In: 2009 IEEE/SP 15th Workshop on Statistical Signal Processing, pp. 113–116. IEEE (2009)

29. Rangan, S.: Generalized approximate message passing for estimation with random linear mixing. In: 2011 IEEE International Symposium on Information Theory Proceedings, pp. 2168–2172. IEEE (2011)
30. Rangan, S., Schniter, P., Fletcher, A.K.: Vector approximate message passing. IEEE Trans. Inf. Theory **65**, 6664–6684 (2019)
31. Rangana, S., Schniterb, P., Fletcherc, A.K., Sarkar, S.: On the convergence of approximate message passing with arbitrary matrices. IEEE Trans. Inf. Theory **65**(9), 5339–5351 (2019)
32. Shi, H.J.M., Tu, S., Xu, Y., Yin, W.: A primer on coordinate descent algorithms. arXiv preprint arXiv:1610.00040 (2016)
33. Simon, N., Friedman, J., Hastie, T., Tibshirani, R.: A sparse-group lasso. J. Comput. Graph. Stat. **22**(2), 231–245 (2013)
34. Su, W., Bogdan, M., Candes, E., et al.: False discoveries occur early on the lasso path. Ann. Stat. **45**(5), 2133–2150 (2017)
35. Tibshirani, R.: Regression shrinkage and selection via the lasso. J. Roy. Stat. Soc.: Ser. B (Methodol.) **58**(1), 267–288 (1996)
36. Vila, J., Schniter, P.: Expectation-maximization bernoulli-gaussian approximate message passing. In: 2011 Conference Record of the Forty Fifth Asilomar Conference on Signals, Systems and Computers (ASILOMAR), pp. 799–803. IEEE (2011)
37. Vila, J., Schniter, P., Rangan, S., Krzakala, F., Zdeborová, L.: Adaptive damping and mean removal for the generalized approximate message passing algorithm. In: 2015 IEEE International Conference on Acoustics, Speech and Signal Processing (ICASSP), pp. 2021–2025. IEEE (2015)
38. Vila, J.P., Schniter, P.: Expectation-maximization gaussian-mixture approximate message passing. IEEE Trans. Signal Process. **61**(19), 4658–4672 (2013)
39. Yang, Y., Zou, H.: A fast unified algorithm for solving group-lasso penalize learning problems. Stat. Comput. **25**(6), 1129–1141 (2014). https://doi.org/10.1007/s11222-014-9498-5
40. Yuan, M., Lin, Y.: Model selection and estimation in regression with grouped variables. J. Roy. Stat. Soc. Ser. B (Statistical Methodology) **68**(1), 49–67 (2006)
41. Zou, H.: The adaptive lasso and its oracle properties. J. Am. Stat. Assoc. **101**(476), 1418–1429 (2006)
42. Zou, H., Hastie, T.: Regularization and variable selection via the elastic net. J. Roy. Stat. Soc. Ser. B (statistical methodology) **67**(2), 301–320 (2005)

# Sparse Information Filter for Fast Gaussian Process Regression

Lucas Kania[1(✉)], Manuel Schürch[1,2], Dario Azzimonti[2], and Alessio Benavoli[3]

[1] Università della Svizzera italiana (USI), Via Buffi 13, Lugano, Switzerland
lucas.kania@usi.ch
[2] Istituto Dalle Molle di Studi sull'Intelligenza Artificiale (IDSIA), Via la Santa 1, Lugano, Switzerland
{manuel.schuerch,dario.azzimonti}@idsia.ch
[3] School of Computer Science and Statistics, Trinity College, Dublin, Ireland
alessio.benavoli@tcd.ie

**Abstract.** Gaussian processes (GPs) are an important tool in machine learning and applied mathematics with applications ranging from Bayesian optimization to calibration of computer experiments. They constitute a powerful kernelized non-parametric method with well-calibrated uncertainty estimates, however, off-the-shelf GP inference procedures are limited to datasets with a few thousand data points because of their cubic computational complexity. For this reason, many sparse GPs techniques were developed over the past years. In this paper, we focus on GP regression tasks and propose a new algorithm to train variational sparse GP models. An analytical posterior update expression based on the Information Filter is derived for the variational sparse GP model. We benchmark our method on several real datasets with millions of data points against the state-of-the-art Stochastic Variational GP (SVGP) and sparse orthogonal variational inference for Gaussian Processes (SOLVEGP). Our method achieves comparable performances to SVGP and SOLVEGP while providing considerable speed-ups. Specifically, it is consistently four times faster than SVGP and on average 2.5 times faster than SOLVEGP.

**Keywords:** Gaussian process regression · Sparse variational method · Information filter variational bound

## 1 Introduction

Gaussian processes (GPs) are an important machine learning tool [14] widely used for regression and classification tasks. The well-calibrated uncertainty quantification provided by GPs is important in a wide range of applications such as

**Electronic supplementary material** The online version of this chapter (https://doi.org/10.1007/978-3-030-86523-8_32) contains supplementary material, which is available to authorized users.


N. Oliver et al. (Eds.): ECML PKDD 2021, LNAI 12977, pp. 527–542, 2021.
https://doi.org/10.1007/978-3-030-86523-8_32

Bayesian optimization [20], visualization [10], and analysis of computer experiments [16]. The main drawback of GPs is that they scale poorly with the training size.

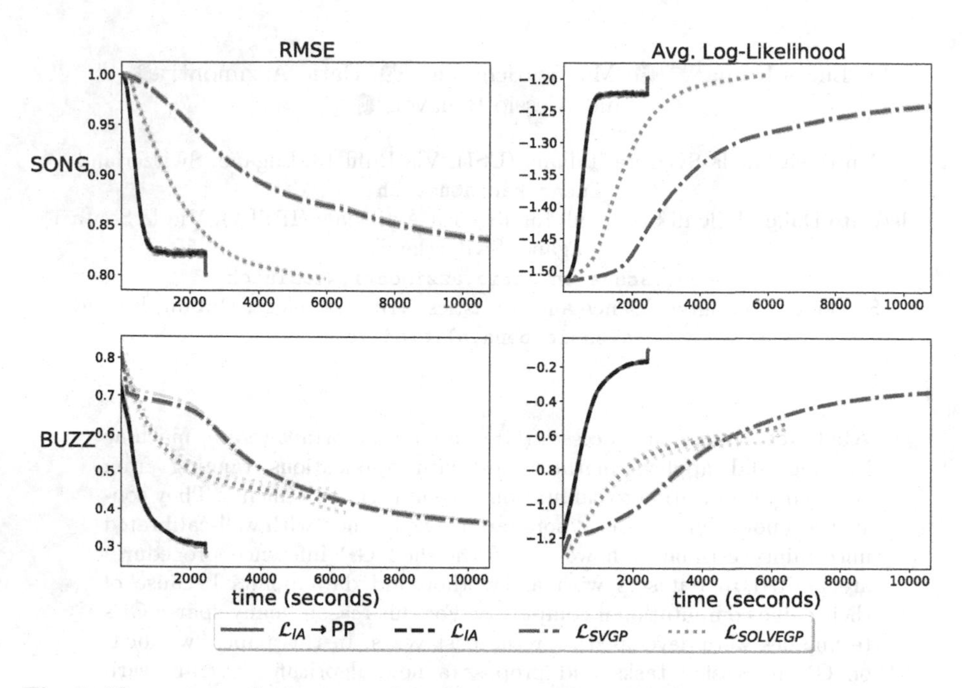


**Fig. 1.** The main method presented in this paper, $\mathcal{L}_{\mathrm{IA}} \rightarrow$ PP, compared against $\mathcal{L}_{\mathrm{SVGP}}$ [7] and $\mathcal{L}_{\mathrm{SOLVEGP}}$ [21] on the high dimensional UCI datasets SONG (89 dimensions) and BUZZ (77 dimensions). All methods run for the same number of iteration, plotted against wall-clock time. This behaviour is also observed in the other experiments, see Sect. 4.3.

Given $N$ data points, GP training has a computational complexity of $O(N^3)$ due to the inversion of a $N \times N$ covariance matrix. In the last few decades, different families of methods were introduced to address this issue. Aggregation or statistical consensus methods [5,8,11,24] solve the issue by training smaller GP models and then aggregating them. Other techniques exploit numerical linear algebra approximations to efficiently solve the inversion problem [26] or state-space representations [17,18].

In this work, we focus on the family of sparse inducing points approximations [13]. Such methods employ $M \ll N$ inducing points to summarize the whole training data and reduce the computational complexity to $O(NM^2)$. In such models, the key parameters are the positions of the inducing points which need to be optimized. Initial attempts did not guarantee a convergence to full GP [4,22], however, ref. [23] introduced a variational lower bound which links the sparse GP approximation to a full GP. This method (Variational Free Energy, VFE) approximates the posterior with a variational distribution chosen by minimizing the Kullback-Leibler divergence between the variational distribution and the exact posterior.

The method proposed by [23] allows for large training data sizes, however, it is still not appropriate for big data since the optimization of the lower bound, required to choose the positions of the inducing points, cannot be split into mini-batches. In order to address this issue, [7] proposed a stochastic gradient variational method (SVGP) employing an uncollapsed version of the lower bound which splits into a sum over mini-batches and allows for stochastic gradient descent. In a regression setting, VFE [23] provides analytical updates for the variational distribution while SVGP [7] requires the optimization of all the variational distribution's parameters. This increases the size of the parameter space and can lead to unstable results. Natural gradients [15] are often used to speed up the convergence of this optimization problem. Nonetheless, analytical updates for regression tasks could lead to higher speed-ups.

Reference [19] proposed a recursive collapsed lower bound for regression that exploits analytical updates for the posterior distribution and splits into a sum over mini-batches. Consequently, the method can be scaled to millions of data points by stochastic gradient descent. This recursive approach provides a performance competitive with SVGP both in terms of accuracy and computational time, however, it requires storing past gradients in memory. When the input space dimension or the number of inducing points is very large, this method becomes problematic memory-wise as the past Jacobian matrices are cumbersome to store.

In this paper, we address this issue and propose a simple and cheap training method that efficiently achieves state-of-the-art performance in practice. Figure 1 shows an example of the behaviour of our method, denoted $\mathcal{L}_{\text{IA}} \to \text{PP}$, compared against SVGP [7] and SOLVEGP [21] on two high dimensional large UCI [6] datasets. The plots show root mean squared error (RMSE) and average log-likelihood as a function of computational time. All methods are run for a fixed number of iterations; note the computational speed-ups achieved by $\mathcal{L}_{\text{IA}} \to \text{PP}$.

We develop our method with a straightforward approach: first, we stochastically train the model parameters on independent mini-batches and then we compute the full approximate posterior on the whole dataset with analytical updates. In particular we

1. formulate the VFE model with an Information Filter (IF) approach allowing for analytical posterior updates in natural parameters. We would like to stress that IF is used to reformulate posterior updates and not to obtain alternative state-space representations;
2. describe a training algorithm, $\mathcal{L}_{\text{IA}} \to \text{PP}$, that employs the IF formulation on independent mini-batches as a warm-up phase and recovers the previously ignored data dependencies by employing analytic posterior updates in the final phase of the optimization;
3. show on real datasets that our training method achieves comparable performances with respect to state-of-the-art techniques (SVGP [7], SOLVEGP [21]) in a fraction of their runtime as shown in Fig. 1 and Sect. 4.3.

## 2 Gaussian Process Regression

### 2.1 Full Gaussian Processes

Consider a dataset $\mathcal{D} = (x_i, y_i)_{i=1}^N$ of input points $X = (x_i)_{i=1}^N$, $x_i \in \mathbb{R}^D$ and observations $y_i \in \mathbb{R}$, where the $i$th observation $y_i$ is the sum of an unknown function $f : \mathbb{R}^D \to \mathbb{R}$ evaluated at $x_i$ and independent Gaussian noise, i.e.

$$y_i = f(x_i) + \epsilon_i \overset{iid}{\sim} \mathcal{N}\left(f(x_i), \sigma_n^2\right). \tag{1}$$

We model $f$ by using a Gaussian Process [14] with mean function $m$ and a covariance function $k$. A GP is a stochastic process such that the joint distribution of any finite collection of evaluations of $f$ is distributed as a multivariate Gaussian

$$\begin{bmatrix} f(x_1) \\ \vdots \\ f(x_n) \end{bmatrix} \sim \mathcal{N}\left( \begin{bmatrix} m(x_1) \\ \vdots \\ m(x_n) \end{bmatrix}, \begin{bmatrix} k(x_1, x_1) & \cdots & k(x_1, x_n) \\ \vdots & & \vdots \\ k(x_n, x_1) & \cdots & k(x_n, x_n) \end{bmatrix} \right),$$

where $m : \mathbb{R}^D \to \mathbb{R}$ is an arbitrary function and $k : \mathbb{R}^D \times \mathbb{R}^D \to \mathbb{R}$ is a positive definite kernel. Both $m$ and $k$ could depend on a vector of parameters $\theta \in \Theta$; in this paper we assume that $m \equiv 0$ and $k$ is a kernel from a parametric family such as the radial basis functions (RBF) or the Matérn family, see [14, Chapter 4].

Let $\mathbf{f} = [f(x_1), \ldots, f(x_N)]$, $\mathbf{y} = [y_1, \ldots, y_N]$ and consider test points $X_* = (x_j^*)_{j=1}^A$ with the respective function values $\mathbf{f}_* = f(X_*)$. The joint distribution of $(\mathbf{y}, \mathbf{f}_*)$ is normally distributed due to Eq (1). Thus, the predictive distribution can be obtained by conditioning $\mathbf{f}_*|\mathbf{y} \sim \mathcal{N}(\mathbf{f}_*|\mu_{\mathbf{y}}, \Sigma_{\mathbf{y}})$ [14, Chapter 2] where

$$\mu_{\mathbf{y}} = \mathrm{K}_{\mathrm{X}_*\mathrm{X}}(\mathrm{K}_{\mathrm{XX}} + \sigma_n^2 \mathbb{I})^{-1}\mathbf{y};$$
$$\Sigma_{\mathbf{y}} = \mathrm{K}_{\mathrm{X}_*\mathrm{X}_*} - \mathrm{K}_{\mathrm{X}_*\mathrm{X}}(\mathrm{K}_{\mathrm{XX}} + \sigma_n^2 \mathbb{I})^{-1}\mathrm{K}_{\mathrm{XX}_*}.$$

where $\mathbb{I}$ is the identity matrix and $K_{VW} := [k(v_i, w_j)]_{i,j}$ is the kernel matrix obtained for two sets of points: $V = (v_i)_{i=1}^{N_V} \subseteq \mathbb{R}^D$ and $W = (w_j)_{j=1}^{N_W} \subseteq \mathbb{R}^D$. Given the prior $\mathbf{f} \sim \mathcal{N}(0, \mathrm{K}_{\mathrm{XX}})$, we can compute the log marginal likelihood $\log p(\mathbf{y}|\theta) = \mathcal{N}\left(\mathbf{y}|0, \mathrm{K}_{\mathrm{XX}} + \sigma_n^2 \mathbb{I}\right)$ which is a function of $\theta$, the parameters associated to the kernel $k$, and of the Gaussian noise variance $\sigma_n^2$. We can estimate those parameters by maximizing $\log p(\mathbf{y}|\theta)$. Note that both training and marginal likelihood computation require the inversion of a $N \times N$ matrix which makes this method infeasible for large datasets due to the $O(N^3)$ time complexity of the operation.

### 2.2 Sparse Inducing Points Gaussian Processes

The cubic time complexity required by GP regression motivated the development of sparse methods for GP regression. The key idea is to find a set of $M$ so-called inducing points, $R = (r_j)_{j=1}^M$ where $r_j \in \mathbb{R}^D$, such that $\mathbf{f}_{\mathrm{R}} := (f(r_j))_{j=1}^M \in \mathbb{R}^M$

are an approximate sufficient statistic of the whole dataset [13]. Thereby, the key challenge is to find the location of these inducing points. Early attempts [4,22] used the marginal likelihood to select the inducing points, however, they are prone to overfitting. Reference [23] instead proposed to perform approximate variational inference, leading to the Variational Free Energy (VFE) method which converges to full GP as $M$ increases. The inducing points and the parameters are selected by maximizing the following VFE lower bound

$$\mathcal{L}_{\text{VFE}} = \log \mathcal{N}\left(\mathbf{y}|\mathbb{0}, \mathrm{Q_{XX}} + \sigma_n^2 \mathbb{I}\right) - \frac{Tr(\mathrm{K_{XX}} - \mathrm{Q_{XX}})}{2\sigma_n^2}, \tag{2}$$

where $\mathrm{Q_{XX}} = \mathrm{H_X K_{RR} H_X^T}$ and $\mathrm{H_X} = \mathrm{K_{XR} K_{RR}^{-1}}$, see [23] for details. In this method, only $\mathrm{K_{RR}}$ needs to be inverted, therefore training with $\mathcal{L}_{\text{VFE}}$ has time complexity $O(M^3 + NM)$. This allows training (sparse) GPs with tens of thousands of points; the method, however, becomes infeasible for larger $N$ because the lower bound in Eq. (2) cannot be split into mini-batches and optimized stochastically.

### 2.3 Stochastic Variational Gaussian Processes

Stochastic Variational Gaussian Processes (SVGP) [7] avoid the above-mentioned problem by splitting the data $\mathcal{D}$ into $K$ mini-batches of $B$ points, i.e. $(\mathbf{y}_k, X_k) \in \mathbb{R}^B \times \mathbb{R}^{B\times D}$ for $k = 1, \ldots, K$, and by stochastically optimizing the following uncollapsed lower bound

$$\mathcal{L}_{\text{SVGP}} = \left(\sum_{k=1}^{K} \mathbb{E}_{\mathbf{f}_k \sim q(\cdot)}[\log p(\mathbf{y}_k|\mathbf{f}_k, \theta)]\right) - \mathbb{KL}[q(\mathbf{f}_{\mathrm{R}})||p(\mathbf{f}_{\mathrm{R}}|\theta)]. \tag{3}$$

This lower bound uses a normal variational distribution $q(\mathbf{f}_{\mathrm{R}})$ where the variational parameters, i.e. mean and covariance, need to be optimized numerically in addition to the inducing points and the kernel parameters. Natural gradients [15] for the variational parameters speed up the task. Nonetheless, all entries in the inducing point's posterior mean vector and posterior covariance matrix have to be estimated numerically. The lack of analytical posterior updates introduces $O(M^2)$ additional variational parameters, whose optimization becomes cumbersome when using a large number of inducing points.

### 2.4 Recursively Estimated Sparse Gaussian Processes

Reference [19] obtained analytical updates of the inducing points' posterior by recursively applying the technique introduced by [23]. That is, the distribution $p(\mathbf{f}_R|\mathbf{y}_{1:k})$ is approximated by a moment parameterized distribution $q_k(\mathbf{f}_{\mathrm{R}}) = \mathcal{N}(\mathbf{f}_{\mathrm{R}}|\mu_k, \Sigma_k)$. Recursively performing variational inference leads to the bound

$$\mathcal{L}_{\text{REC}} = \sum_{k=1}^{K} \log \mathcal{N}\left(\mathbf{y}_k|\mathrm{H}_{\mathrm{X_k}}\mu_{k-1}, S_k\right) - \frac{Tr(\mathrm{K_{X_K X_K}} - \mathrm{Q_{X_K X_K}})}{2\sigma_n^2} \tag{4}$$

where $S_k = \mathrm{H}_{\mathrm{X_k}} \Sigma_{k-1} \mathrm{H}_{\mathrm{X_k}}^{\mathrm{T}} + \sigma_n^2 \mathbb{I}$ and $\mathrm{H}_{\mathrm{X_k}} = \mathrm{K}_{\mathrm{X_k R}} \mathrm{K}_{\mathrm{RR}}^{-1}$. The posterior can be analytically computed by the moment parameterized posterior propagation

$$\mu_k = \Sigma_k \left( \frac{\mathrm{H}_{\mathrm{X_k}}^{\mathrm{T}} \mathbf{y}_k}{\sigma_n^2} + \Sigma_{k-1}^{-1} \mu_{k-1} \right), \tag{5}$$

$$\Sigma_k = \left( \Sigma_{k-1}^{-1} + \frac{\mathrm{H}_{\mathrm{X_k}}^{\mathrm{T}} \mathrm{H}_{\mathrm{X_k}}}{\sigma_n^2} \right)^{-1}. \tag{6}$$

By keeping the same parameters for the whole epoch, the posterior equals that of its batch counterpart VFE. Similarly, the sum of $\mathcal{L}_{\mathrm{REC}}$ over all mini-batches is equal to the batch lower bound $\mathcal{L}_{\mathrm{VFE}}$. However, the authors proposed to stochastically approximate its gradient by computing the gradient w.r.t. one mini-batch and plugging-in the derivative of the variational distribution parameters from the previous iteration in order to speed up the optimization. This method has two main drawbacks. Firstly, it is constrained to use small learning rates for the approximation to be valid and stable in practice. Second, the gradients of the last iteration must be stored. Such storage becomes problematic when the input space dimension or the number of inducing points is very large.

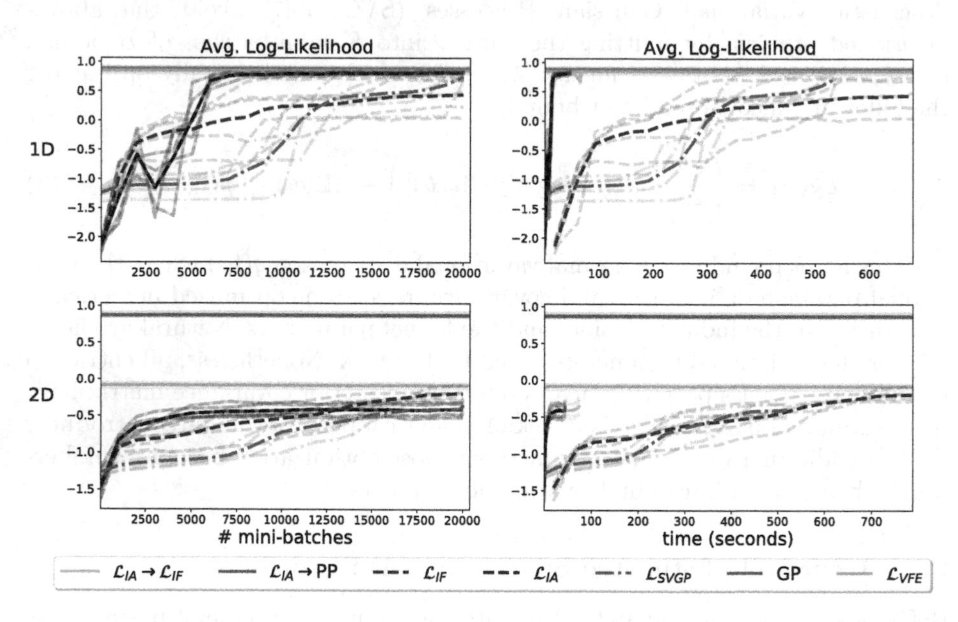


**Fig. 2.** Toy experiments: 1D and 2D generated GP data. Performances of GP and $\mathcal{L}_{\mathrm{VFE}}$ in horizontal lines as their optimization does not use mini-batches. On the left comparison in number of mini-batches, on the right runtime comparison.

# 3 Information Filter for Sparse Gaussian Processes

In order to overcome the previous issues but keep the advantage of analytical posterior updates, we propose an Information Filter update for the posterior, a very efficient and easy to interpret method for estimation in sparse GP methods.

Alternatively to the moment parameterized propagation in Eqs. (5) and (6), the posterior over the inducing points can be more efficiently propagated using a natural parameterization $\mathcal{N}^{-1}(\mathbf{f}_R|\eta_k, \Lambda_k)$ with $\eta_k = \Sigma_k^{-1}\mu_k$ and $\Lambda_k = \Sigma_k^{-1}$. The advantage of this Information Filter formulation are the compact updates

$$\eta_k = \eta_{k-1} + \frac{1}{\sigma_n^2}\mathrm{H}_{\mathrm{X_k}}^{\mathrm{T}}\mathbf{y}_k \quad \text{and} \quad \eta_0 = \mathbb{0}; \tag{7}$$

$$\Lambda_k = \Lambda_{k-1} + \frac{1}{\sigma_n^2}\mathrm{H}_{\mathrm{X_k}}^{\mathrm{T}}\mathrm{H}_{\mathrm{X_k}} \quad \text{and} \quad \Lambda_0 = \mathrm{K}_{\mathrm{RR}}^{-1}, \tag{8}$$

where $\mathrm{H}_{\mathrm{X_k}} = \mathrm{K}_{\mathrm{X_kR}}\mathrm{K}_{\mathrm{RR}}^{-1}$. Further computational efficiency is gained by using the equivalent rotated parameterization: $\eta_{\mathrm{k}}^{\mathrm{R}} = \mathrm{K}_{\mathrm{RR}}\eta_k$ and $\Lambda_{\mathrm{k}}^{\mathrm{R}} = \mathrm{K}_{\mathrm{RR}}\Lambda_k\mathrm{K}_{\mathrm{RR}}$ with updates

$$\eta_{\mathrm{k}}^{\mathrm{R}} = \eta_{k-1}^{R} + \frac{1}{\sigma_n^2}\mathrm{K}_{\mathrm{RX_k}}\mathbf{y}_k \quad \text{and} \quad \eta_0^R = \mathbb{0}; \tag{9}$$

$$\Lambda_{\mathrm{k}}^{\mathrm{R}} = \Lambda_{k-1}^{R} + \frac{1}{\sigma_n^2}\mathrm{K}_{\mathrm{RX_k}}\mathrm{K}_{\mathrm{X_kR}} \quad \text{and} \quad \Lambda_0^R = \mathrm{K}_{\mathrm{RR}}. \tag{10}$$

This parameterization constitutes a computational shortcut since several matrix computations in each step can be avoided. The overall computational complexity of the method remains the same as in $\mathcal{L}_{\mathrm{REC}}$, however, it allows for smaller constant terms, thus reducing computational time in practice. Particularly, this Information Filter propagation is more efficient than the Kalman Filter formulation used by [19] when the mini-batch size is greater than the number of inducing points, which is usually the case in practice. The corresponding lower bound is

$$\mathcal{L}_{\mathrm{IF}} = \sum_{k=1}^{K} \log \mathcal{N}^{-1}\left(r_k|0, S_k^{-1}\right) - \frac{Tr(\mathrm{K_{X_KX_K}} - \mathrm{Q_{X_KX_K}})}{2\sigma_n^2}, \tag{11}$$

where $r_k = \mathbf{y}_k - \text{Cholesky}(\Lambda_{\mathrm{k}}^{\mathrm{R}})^{-1}\mathrm{K}_{\mathrm{RX_k}}$ and $S_k^{-1} = \frac{\mathbb{I}}{\sigma_n^2} - \frac{1}{\sigma_n^4}\mathrm{K}_{\mathrm{X_kR}}(\Lambda_{\mathrm{k}}^{\mathrm{R}})^{-1}\mathrm{K}_{\mathrm{RX_k}}$.

The bound consist of two parts: the first term corresponds to the log marginal likelihood of each mini-batch, and the second term is the correction term resulting from the variational optimization. We refer the reader to the supplementary material for the derivation and efficient computation of the posterior propagation and the lower bound in terms of the rotated posterior mean and covariance.

Computing $\mathcal{L}_{\mathrm{IF}}$ for an epoch using the rotated parameterization in Eqs. (9) and (10) while keeping the same parameters for all mini-batches produces a posterior that equals its batch counterpart VFE.

### 3.1 Stochastic Hyperparameter Optimization

The additive structure of the $\mathcal{L}_{\mathrm{IF}}$, together with the efficient IF posterior propagation allows for more frequent parameter updates. Analogously to the approach in [19], $\mathcal{L}_{\mathrm{IF}}$ can be stochastically optimized w.r.t. to all the kernel parameters and the inducing points' locations. Let $\mathcal{L}_{\mathrm{IF}k}$ be the k-term of the $\mathcal{L}_{\mathrm{IF}}$ sum

$$\mathcal{L}_{\mathrm{IF}k}(\theta_k, \Lambda^R_{k-1}, \eta^R_{k-1}) = l_k(\theta_k, \Lambda^R_{k-1}, \eta^R_{k-1}) - a_k(\theta_k),$$

where $l_k = \log \mathcal{N}^{-1}\left(r_k|0, S_k^{-1}\right)$ and $a_k = \frac{Tr\left(\mathrm{K_{X_KX_K}} - \mathrm{Q_{X_KX_K}}\right)}{2\sigma_n^2}$. The gradient of $\mathcal{L}_{IF}k$ w.r.t. the parameters at iteration $t$, denoted $\theta_t$, can be approximated using Jacobian matrices of the previous iteration

$$\frac{\partial \mathcal{L}_{\mathrm{IF}k}}{\partial \theta_t} \approx \frac{\partial l_k}{\partial \theta_t} + \frac{\partial l_k}{\partial \Lambda^R_{k-1}} \frac{\partial \Lambda^R_{k-1}}{\partial \theta_{t-1}} + \frac{\partial l_k}{\partial \eta^R_{k-1}} \frac{\partial \eta^R_{k-1}}{\partial \theta_{t-1}} - \frac{\partial a_k}{\partial \theta_t}. \tag{12}$$

This recursive propagation of the gradients of the rotated posterior mean and covariance indirectly takes into account all the past gradients via $\frac{\partial \Lambda^R_{k-1}}{\partial \theta_{t-1}}$ and $\frac{\partial \eta^R_{k-1}}{\partial \theta_{t-1}}$. It would be exact and optimal if the derivatives with respect to the current parameters were available. However, when changing the parameters too fast between iterations, e.g. due to a large learning rate, this approximation is too rough and leads to unstable optimization results in practice. Furthermore, the performance gain provided by the recursive gradient propagation does not compensate for the additional storage requirements on the order of $O(M^2U)$ where $U$ is the number of hyperparameters. For instance, if 500 inducing points are used in a 10-dimensional problem, only storing the Jacobian matrices requires 10 GB, under double float precision, i.e. around the memory limit of the GPUs used for this work. Additionally, using the approximated posterior mean and covariance in the initial stages of the optimization slows down the convergence due to the lasting effect of the parameters' random initialization. This effect is noticeable in the test cases presented in Fig. 2: compare $\mathcal{L}_{\mathrm{IF}}$ with the more stable method introduce below called $\mathcal{L}_{\mathrm{IA}}$.

*Independence Assumption for Optimization.* In order to circumvent these instabilities in the optimization part, we propose a fast and efficient method that ignores the (approximated) correlations between the mini-batches in the stochastic gradient computation in the beginning. Specifically, the mini-batches are assumed to be mutually independent, which simplifies the lower-bound of the marginal likelihood to

$$\mathcal{L}_{\mathrm{IA}} = \sum_{k=1}^{K} \log \mathcal{N}^{-1}\left(\mathbf{y}_k|0, \frac{\mathbb{I}}{\sigma_n^2} - \frac{1}{\sigma_n^4}\mathrm{Q_{X_KX_K}}\right) - \frac{Tr(\mathrm{K_{X_KX_K}} - \mathrm{Q_{X_KX_K}})}{2\sigma_n^2}. \tag{13}$$

**Algorithm 1:** $\mathcal{L}_{\text{IA}} \rightarrow$ PP Algorithm

```
1 Choose number of epochs E_IA for L_IA ;
2 Split D into K mini-batches of B points ;
3 for k = 1 → K · E_IA do
4 |   compute gradients on the kth mini-batch, eq. (13);
5 |   update the hyperparameters ;
6 end
7 for k = 1 → K do                                  // i.e. for one epoch
8 |   compute posterior propagation with eqs. (9),(10) ;
9 end
```

*Recovery of the Full Posterior.* In order to incorporate all the dependencies once reasonable hyperparameters are achieved, we could switch to optimize the bound $\mathcal{L}_{\text{IF}}$ during the last few epochs. Eventually, all the ignored data dependencies are taken into account with this method, denoted $\mathcal{L}_{\text{IA}} \rightarrow \mathcal{L}_{\text{IF}}$.

A further computational shortcut is to propagate the inducing points posterior according to Eqs. (9), (10), and not perform any optimization in the last epoch. This method, denoted $\mathcal{L}_{\text{IA}} \rightarrow$ PP, constitutes a practical alternative when the computation and storage of the Jacobian matrices is costly, for instance in high dimensional problems where a large number of inducing points are needed. For clarity, a pseudo-code of the method is provided in Algorithm 1. Note that the full sparse posterior distribution including all data is achieved. The independence assumption influences only the optimization of the hyperparameters. No approximation is used when computing the posterior distribution.

*Distributed Posterior Propagation.* An advantage of the Information Filter formulation is that it allows computing the posterior in parallel. Note that Eqs. (9), (10) can be rewritten as

$$\eta_K^R = \sum_{k=1}^{K} \frac{1}{\sigma_n^2} \mathrm{K}_{\mathrm{RX_k}} \mathbf{y}_k \text{ and } \Lambda_K^R = \Lambda_0^R + \sum_{k=1}^{K} \frac{1}{\sigma_n^2} \mathrm{K}_{\mathrm{RX_k}} \mathrm{K}_{\mathrm{X_kR}}.$$

Since the sums terms are functionally independent, the computation of the posterior distribution can be easily distributed via a map-reduce scheme.

*Prediction.* Given a new $X_* \in \mathbb{R}^{A \times D}$, the predictive distribution after seeing $\mathbf{y}_{1:k}$ of the sparse GP methods can be computed by

$$p\left(\mathbf{f}_* \mid \mathbf{y}_{1:k}\right) = \int p\left(\mathbf{f}_* \mid \mathbf{f}_{\mathrm{R}}\right) p\left(\mathbf{f}_{\mathrm{R}} \mid \mathbf{y}_{1:k}\right) \mathrm{d}\mathbf{f}_{\mathrm{R}} = \mathcal{N}\left(\mathbf{f}_* | \mu_*, \Sigma_*\right)$$

$$\text{where } \mu_* = \mathrm{K}_{\mathrm{X_*R}} (\Lambda_{\mathrm{k}}^{\mathrm{R}})^{-1} \eta_{\mathrm{k}}^{\mathrm{R}};$$

$$\Sigma_* = \mathrm{K}_{\mathrm{X_*X_*}} - \mathrm{Q}_{\mathrm{X_*X_*}} + \mathrm{K}_{\mathrm{X_*R}} (\Lambda_{\mathrm{k}}^{\mathrm{R}})^{-1} \mathrm{K}_{\mathrm{RX_*}}.$$

The predictions for $\mathbf{y}_*$ are obtained by adding $\sigma_n^2 \mathbb{I}$ to the covariance of $\mathbf{f}_* | \mathbf{y}_{1:k}$. A detailed derivation is provided in the supplementary material.

## 4 Experiments

We repeated every experiment presented in this Sect. 5 times, using different random splits for each dataset. We always use the same hardware equipped with a GeForce RTX 2080 Ti and an Intel(R) Xeon(R) Gold 5217 (3 GHz).

Furthermore, we switch from optimizing $\mathcal{L}_{\text{IA}}$ to the posterior propagation, denoted $\mathcal{L}_{\text{IA}} \rightarrow$ PP, in the last 5 epochs of training for all the synthetic and real datasets. Although only one epoch is needed to incorporate all the information of a dataset, we do 5 epochs so the effect is noticeable in the figures. Alternatively, an adaptive switch could be used. For simplicity, we briefly discuss it in the supplementary material. Moreover, we did not exploit the parallelization of the posterior propagation in order to be able to compare all training algorithms on the number of iterations.

For full GP, VFE and SVGP, we used the GPflow 2.0 library [12,25], which is based on Tensorflow 2.0 [1]. We further compare with the recently introduced SOLVEGP [21], implemented in GPflow 1.5.1 and Tensorflow 1.15.4. Our algorithms, namely the training of $\mathcal{L}_{\text{IF}}$, $\mathcal{L}_{\text{IA}} \rightarrow \mathcal{L}_{\text{IF}}$ and $\mathcal{L}_{\text{IA}} \rightarrow$ PP were implemented in Tensorflow 2.0[1]. All stochastic methods were optimized with the ADAM optimizer [9] using the default settings ($\beta_1 = 0.9, \beta_2 = 0.999, \epsilon = 1e^{-7}$); we used the recommended settings ($\gamma = 0.1$) for the stochastic optimization of natural gradients in SVGP [15].

We do not compare the methods against optimizing $\mathcal{L}_{\text{REC}}$ [19] because their code does not run on GPU, making the computational times not comparable. Another candidate method that is not considered is Exact Gaussian Processes (ExactGP) [26]. The aim of approximated sparse GPs (SGPs, which includes our method, SVGP and SOLVEGP) and ExactGPs are different, the latter propose a computationally expensive procedure to compute a full GP while the former propose a computationally cheap approximation of a SGP. Consequently, an ExactGP would yield higher accuracy than an approximated SGP. The accuracy of approximated SGP is upper-bounded by VFE's accuracy which is itself upper-bounded by full GP's accuracy. Moreover, for high dimensional datasets such as BUZZ or SONG used later, ExactGP requires an infrastructure with many GPUs in parallel, e.g. 8 in their experiments, which makes the method hard to replicate in practice.

We compare each method in terms of average log-likelihood, root mean squared error (RMSE) and computational time. The comparison in computational time is fair because all methods were run on the same hardware which ran exclusively the training procedure.

### 4.1 Toy Data

We start by showcasing the methods on a simple dataset generated by a SGP with 2000 random inducing points, and a RBF kernel with variance 1, Gaussian noise 0.01 and lengthscales 0.1 and 0.2 for 1 and 2 dimensions correspondingly.

[1] The code is available at https://github.com/lkania/Sparse-IF-for-Fast-GP.

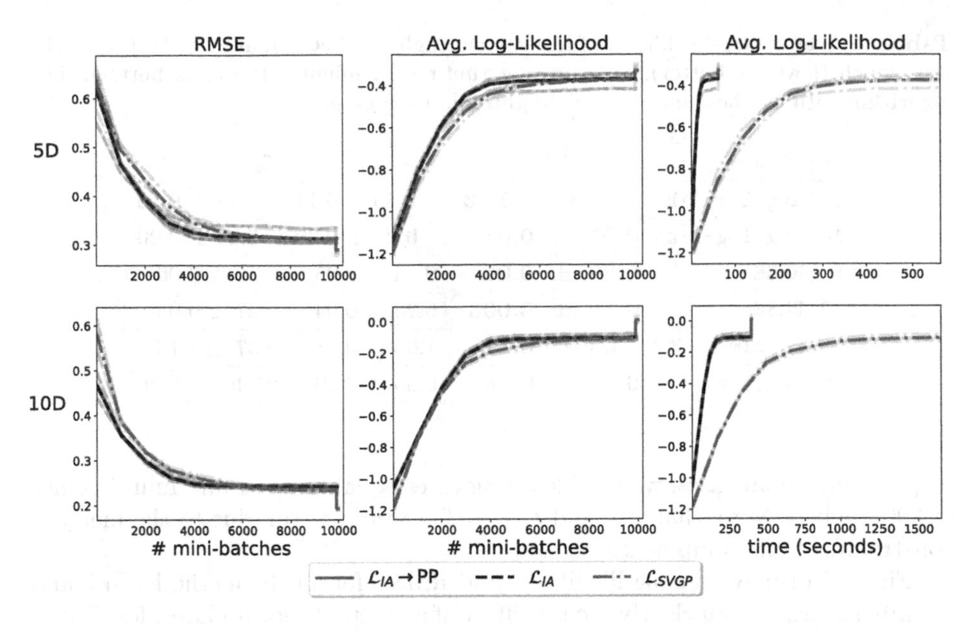


**Fig. 3.** Synthetic experiments: 5D and 10D data generated from GPs. RMSE vs number of mini-batches (left), average log-likelihood vs number of mini-batches (center), and average log-likelihood vs runtime (right).

We consider $N = 5000$ training points and 500 test points. In all experiments, the initial parameters for the RBF kernel were 1 for the lengthscales and Gaussian noise, and 2 for the variance. Moreover, 20 inducing points were randomly selected from the training data. In this example, the switch for $\mathcal{L}_{\text{IA}} \to \mathcal{L}_{\text{IF}}$ and $\mathcal{L}_{\text{IA}} \to$ PP was done in the last 200 epochs to make the difference between them visually noticeable. All methods were run for 20K iterations with a learning rate of 0.001 and mini-batches of 500 data points.

Figure 2 shows that in 1D all sparse training methods converge to the $\mathcal{L}_{\text{VFE}}$ solution, which itself converges to the GP solution. However, this is not the case in the 2D example. $\mathcal{L}_{\text{IF}}$ comes closer to the VFE solution, followed by $\mathcal{L}_{\text{IA}} \to \mathcal{L}_{\text{IF}}$. Note that the convergence of $\mathcal{L}_{\text{IA}} \to \mathcal{L}_{\text{IF}}$ is expected since using $\mathcal{L}_{\text{IA}}$ at the beginning is just a way of *warm-starting* the parameters for $\mathcal{L}_{\text{IF}}$.

## 4.2 Synthetic Data

Using the same generating process used in the previous experiments, we produced two 100K points datasets for 5 and 10 dimensions using an RBF kernel with variance 1, Gaussian noise 0.01, and lengthscales 0.5 and 1 respectively. The SGPs models, with a RBF kernel, were initialized with Gaussian noise 1, kernel variance equal to 2 and the lengthscales all equal to 1 in the 5-dimensional case, and 2 in the 10-dimensional problem. All methods run for 10K iterations in all the datasets using 100 and 500 inducing points for 5 and 10 dimensions

**Table 1.** Average log-likelihood of last epoch (higher is better), average RMSE of the last epoch (lower is better), and average runtime in minutes (lower is better). The algorithm with the best metric is highlighted in each case.

| D | Metric | $\mathcal{L}_{\text{IA}} \to$ PP | $\mathcal{L}_{\text{IA}}$ | $\mathcal{L}_{\text{SVGP}}$ |
|---|---|---|---|---|
| 5 | avg log-lik | **−0.33 ± 0.03** | −0.45 ± 0.14 | −0.48 ± 0.15 |
| 10 | avg log-lik | **0.01 ± 0.01** | −0.21 ± 0.20 | −0.24 ± 0.20 |
| 5 | RMSE | **0.29 ± 0.01** | 0.34 ± 0.05 | 0.36 ± 0.06 |
| 10 | RMSE | **0.19 ± 0.003** | 0.26 ± 0.04 | 0.27 ± 0.05 |
| 5 | runtime | **1.03 ± 0.11** | 1.05 ± 0.11 | 9.37 ± 1.16 |
| 10 | runtime | **6.59 ± 0.06** | 6.65 ± 0.09 | 27.40 ± 0.79 |

respectively. In all cases, we used learning rates equal to 0.001 and mini-batches of 5000 points. Note, that $\mathcal{L}_{\text{IF}}$ and $\mathcal{L}_{\text{IA}} \to \mathcal{L}_{\text{IF}}$ were not run due to the memory constraints of our equipment.

Figure 3 displays the log-likelihood and RMSE for all the methods. In 5 and 10 dimensions, we can clearly see the effect of a few posterior updates, for $\mathcal{L}_{\text{IA}} \to$ PP, after fixing the parameters with $\mathcal{L}_{\text{IA}}$. Additionally, Table 1 displays the average runtimes for each method. Note that $\mathcal{L}_{\text{IA}} \to$ PP offers a speed-up of 9× and 4× over $\mathcal{L}_{\text{SVGP}}$ for the 5- and 10-dimensional datasets respectively.

### 4.3 Real Data

**Table 2.** Training dataset size, test dataset size and dimension of each one of the benchmarked datasets. The last two columns show the learning rate and the number of iterations for all training methods run in each dataset.

| Name | Train | Test | D | Learning rate | Iterations |
|---|---|---|---|---|---|
| AIRLINES | 1.6M | 100K | 8 | 0.0001 | 60000 |
| GAS | 1.4M | 100K | 17 | 0.0001 | 40000 |
| BUZZ | 0.46M | 100K | 77 | 0.0001 | 60000 |
| SONG | 0.41M | 100K | 89 | 0.0001 | 60000 |
| SGEMM | 0.19M | 48K | 14 | 0.001 | 15000 |
| PROTEIN | 0.036M | 9K | 9 | 0.001 | 5000 |
| BIKE | 0.013M | 3.4K | 12 | 0.0001 | 40000 |

We trained SGPs, whose RBF kernels had all their hyperparameters initialized to 1, using $\mathcal{L}_{\text{SVGP}}$, $\mathcal{L}_{\text{SOLVEGP}}$, $\mathcal{L}_{\text{IA}}$ and $\mathcal{L}_{\text{IA}} \to$ PP in several real-world datasets, shown in Table 2, using mini-batches of 5000 points. All datasets were downloaded from the UCI repository [6] except for AIRLINES, where we follow the original example in [7].

For each training algorithm, we present average log-likelihood, RMSE and runtime for the same number of iterations. Figure 1 shows the results corresponding to the datasets SONG and BUZZ. Tables 3, 4 and 5 report the average log-likelihood, RMSE and runtime of all the algorithms. Note that training with $\mathcal{L}_{\text{IA}} \to \text{PP}$ always provides comparable results to the state-of-the-art. However, such results are achieved, on average, approximately 4 times faster than $\mathcal{L}_{\text{SVGP}}$ and 2.5 times faster than $\mathcal{L}_{\text{SOLVEGP}}$, see the last two columns of Table 5. While $\mathcal{L}_{\text{IA}} \to \text{PP}$ is uniformly around 4 times faster than $\mathcal{L}_{\text{SVGP}}$, its performance with respect to $\mathcal{L}_{\text{SOLVEGP}}$ strongly depends on the dataset. In general, $\mathcal{L}_{\text{SOLVEGP}}$ is faster on smaller and simpler datasets, while it becomes computationally costly on noisy datasets like `AIRLINE` or `BUZZ`.

The results show that in high dimensional, large problems a simple method such as $\mathcal{L}_{\text{IA}} \to \text{PP}$ guarantees fast results and should be used in practice. Note that, compared to $\mathcal{L}_{\text{IA}}$, the inducing point posterior propagation step added in $\mathcal{L}_{\text{IA}} \to \text{PP}$ always increases performances (lower RMSE, higher log-likelihood) and it is not noticeable in terms of computational time.

**Table 3.** Average Log-Likelihood of the last epoch, higher is better. The algorithm with the highest average mean log-likelihood is highlighted for each dataset.

| Dataset | $\mathcal{L}_{\text{IA}} \to \text{PP}$ | $\mathcal{L}_{\text{IA}}$ | $\mathcal{L}_{\text{SVGP}}$ | $\mathcal{L}_{\text{SOLVEGP}}$ |
|---|---|---|---|---|
| AIRLINES | **−1.31 ± 0.01** | −1.33 ± 0.01 | −1.33 ± 0.01 | −1.32 ± 0.01 |
| GAS | −0.13 ± 0.02 | −0.24 ± 0.02 | −0.25 ± 0.01 | **−0.05 ± 0.05** |
| BUZZ | **−0.10 ± 0.01** | −0.17 ± 0.01 | −0.36 ± 0.01 | −0.55 ± 0.04 |
| SONG | **−1.20 ± 0.001** | −1.22 ± 0.002 | −1.25 ± 0.004 | **−1.20 ± 0.002** |
| SGEMM | 0.47 ± 0.002 | 0.33 ± 0.02 | 0.28 ± 0.05 | **0.63 ± 0.17** |
| PROTEIN | **−1.04 ± 0.01** | −1.13 ± 0.05 | −1.14 ± 0.08 | −1.06 ± 0.02 |
| BIKE | **0.08 ± 0.02** | 0.03 ± 0.02 | 0.04 ± 0.02 | 0.01 ± 0.04 |

**Table 4.** Average RMSE of the last epoch, lower is better. The algorithm with the lowest average mean RMSE is highlighted for each dataset.

| Dataset | $\mathcal{L}_{\text{IA}} \to \text{PP}$ | $\mathcal{L}_{\text{IA}}$ | $\mathcal{L}_{\text{SVGP}}$ | $\mathcal{L}_{\text{SOLVEGP}}$ |
|---|---|---|---|---|
| AIRLINES | **0.89 ± 0.01** | 0.92 ± 0.01 | 0.92 ± 0.01 | 0.90 ± 0.01 |
| GAS | 0.28 ± 0.01 | 0.32 ± 0.01 | 0.32 ± 0.01 | **0.25 ± 0.01** |
| BUZZ | **0.28 ± 0.01** | 0.31 ± 0.004 | 0.37 ± 0.01 | 0.42 ± 0.02 |
| SONG | **0.80 ± 0.001** | 0.82 ± 0.002 | 0.84 ± 0.003 | **0.80 ± 0.002** |
| SGEMM | 0.14 ± 0.001 | 0.17 ± 0.004 | 0.17 ± 0.01 | **0.13 ± 0.03** |
| PROTEIN | **0.68 ± 0.005** | 0.74 ± 0.01 | 0.74 ± 0.04 | 0.69 ± 0.01 |
| BIKE | **0.22 ± 0.004** | 0.24 ± 0.005 | 0.23 ± 0.004 | 0.23 ± 0.01 |

**Table 5.** Average runtimes in hours, lower is better. The algorithm with the lowest average mean runtime is highlighted for each dataset. $\Delta_{\text{SV}}$ is the ratio between the average runtimes of $\mathcal{L}_{\text{SVGP}}$ and $\mathcal{L}_{\text{IA}} \to \text{PP}$. $\Delta_{\text{SOLVE}}$ is the ratio between the average runtimes of $\mathcal{L}_{\text{SOLVEGP}}$ and $\mathcal{L}_{\text{IA}} \to \text{PP}$.

| Dataset | $\mathcal{L}_{\text{IA}} \to \text{PP}$ | $\mathcal{L}_{\text{IA}}$ | $\mathcal{L}_{\text{SVGP}}$ | $\mathcal{L}_{\text{SOLVEGP}}$ | $\Delta_{\text{SV}}$ | $\Delta_{\text{SOLVE}}$ |
|---|---|---|---|---|---|---|
| AIRLINES | **0.65 ± 0.004** | 0.66 ± 0.004 | 2.78 ± 0.07 | 3.72 ± 0.16 | 4.29 | 5.75 |
| GAS | **0.43 ± 0.001** | 0.44 ± 0.004 | 1.84 ± 0.04 | 2.20 ± 0.11 | 4.28 | 5.12 |
| BUZZ | **0.67 ± 0.004** | 0.68 ± 0.005 | 2.95 ± 0.04 | 1.79 ± 0.08 | 4.38 | 2.66 |
| SONG | **0.68 ± 0.01** | 0.69 ± 0.01 | 3.00 ± 0.11 | 1.65 ± 0.02 | 4.41 | 2.43 |
| SGEMM | **0.16 ± 0.001** | 0.16 ± 0.001 | 0.69 ± 0.02 | 0.30 ± 0.01 | 4.22 | 1.84 |
| PROTEIN | **0.05 ± 0.0002** | 0.05 ± 0.0001 | 0.23 ± 0.004 | 0.07 ± 0.001 | 4.20 | 1.33 |
| BIKE | **0.44 ± 0.005** | 0.44 ± 0.003 | 1.88 ± 0.07 | 0.56 ± 0.02 | 4.27 | 1.28 |

# 5 Conclusion

In this paper, we presented $\mathcal{L}_{\text{IA}} \to \text{PP}$, a fast method to train sparse inducing points GP models. Our method is based on

(A) optimizing a simple lower bound under the assumption of independence between mini-batches, until the convergence of hyperparameters;
(B) recovering a posteriori the dependencies between mini-batches by exactly computing the posterior distribution.

We focused on VFE-like models and provided a method that achieves a performance comparable to state-of-the-art techniques with considerable speed-ups.

The method could be adapted to the power EP model [3]. Additionally, our lower bound exploits an Information Filter formulation that could potentially be distributed. We outlined how to distribute the posterior propagation, which could be exploited by future work to distribute the computation of gradients.

Our technique is based on analytical updates of the posterior that are in general not available for classic GP classification models. Recently, SkewGP for classification [2] proposed a full classification model with analytical updates for the posterior. Our work could be exploited to provide fast sparse training for such models.

**Acknowledgments.** Lucas Kania thankfully acknowledges the support of the Swiss National Science Foundation (grant number 200021_188534). Manuel Schürch and Dario Azzimonti gratefully acknowledge the support of the Swiss National Research Programme 75 "Big Data" (grant number 407540_167199/1). All authors would like to thank the IDSIA Robotics Lab for granting access to their computational facilities.

# References

1. Abadi, M., et al.: TensorFlow: large-scale machine learning on heterogeneous systems (2015). https://www.tensorflow.org/. Software available from tensorflow.org

2. Benavoli, A., Azzimonti, D., Piga, D.: Skew gaussian processes for classification. In: Machine Learning and Knowledge Discovery in Databases - European Conference, ECML PKDD 2020. LNCS. Springer, Heidelberg (2020)
3. Bui, T.D., Yan, J., Turner, R.E.: A unifying framework for sparse Gaussian process approximation using power expectation propagation. J. Mach. Learn. Res. **18**, 1–72 (2017)
4. Csató, L., Opper, M.: Sparse on-line Gaussian processes. Neural Comput. **14**(3), 641–668 (2002)
5. Deisenroth, M.P., Ng, J.W.: Distributed Gaussian processes. In: 32nd International Conference on Machine Learning, ICML 2015.,vol. 37, pp. 1481–1490 (2015)
6. Dua, D., Graff, C.: UCI machine learning repository (2017). http://archive.ics.uci.edu/ml
7. Hensman, J., Fusi, N., Lawrence, N.D.: Gaussian processes for big data. In: Proceedings of the Twenty-Ninth Conference on Uncertainty in Artificial Intelligence (2013)
8. Hinton, G.E.: Training products of experts by minimizing contrastive divergence. Neural Comput. **14**(8), 1771–1800 (2002)
9. Kingma, D.P., Ba, J.: Adam: a method for stochastic optimization. arXiv preprint arXiv:1412.6980 (2014)
10. Lawrence, N.: Probabilistic non-linear principal component analysis with Gaussian process latent variable models. J. Mach. Learn. Res. **6**, 1783–1816 (2005)
11. Liu, H., Cai, J., Wang, Y., Ong, Y.S.: Generalized robust Bayesian committee machine for large-scale Gaussian process regression. In: 35th International Conference on Machine Learning, Stockholm, Sweden. ICML 2018, vol. 7, pp. 4898–4910 (2018)
12. Matthews, A.G.D.G., et al.: GPflow: a Gaussian process library using TensorFlow. J. Mach. Learn. Res. **18**(40), 1–6 (2017)
13. Quiñonero-Candela, J., Rasmussen, C.E.: A unifying view of sparse approximate Gaussian process regression. J. Mach. Learn. Res. **6**, 1939–1959 (2005)
14. Rasmussen, C.E., Williams, C.K.I.: Gaussian Processes for Machine Learning. MIT Press (2006)
15. Salimbeni, H., Eleftheriadis, S., Hensman, J.: Natural gradients in practice: non-conjugate variational inference in Gaussian process models. In: International Conference on Artificial Intelligence and Statistics, AISTATS, vol. 2018, pp. 689–697 (2018)
16. Santner, T.J., Williams, B.J., Notz, W.I.: The Design and Analysis of Computer Experiments. SSS, Springer, New York (2018). https://doi.org/10.1007/978-1-4757-3799-8
17. Särkkä, S., Hartikainen, J.: Infinite-dimensional Kalman filtering approach to spatio-temporal Gaussian process regression. J. Mach. Learn. Res. **22**, 993–1001 (2012)
18. Särkkä, S., Solin, A.: Applied Stochastic Differential Equations. Cambridge University Press, Cambridge (2019)
19. Schürch, M., Azzimonti, D., Benavoli, A., Zaffalon, M.: Recursive estimation for sparse Gaussian process regression. Automatica **120**, 109127 (2020)
20. Shahriari, B., Swersky, K., Wang, Z., Adams, R.P., de Freitas, N.: Taking the human out of the loop: a review of Bayesian optimization. Proc. IEEE **104**(1), 148–175 (2016)
21. Shi, J., Titsias, M.K., Mnih, A.: Sparse orthogonal variational inference for Gaussian processes. In: Proceedings of the 23rd International Conference on Artificial Intelligence and Statistics (AISTATS), Palermo, Italy, vol. 108 (2020)

22. Snelson, E., Ghahramani, Z.: Sparse Gaussian processes using pseudo-inputs Edward. In: Weiss, Y., Schölkopf, B., Platt, C., J. (eds.) Advances in Neural Information Processing Systems, vol. 18. pp. 1257–1264. MIT Press (2006)
23. Titsias, M.K.: Variational learning of inducing variables in sparse Gaussian processes. In: Proceedings of the 12th International Conference on Artificial Intelligence and Statistics (AISTATS), vol. 5, pp. 567–574 (2009)
24. Tresp, V.: A Bayesian committee machine. Neural Computation **12**, 2719–2741 (2000)
25. van der Wilk, M., Dutordoir, V., John, S., Artemev, A., Adam, V., Hensman, J.: A framework for interdomain and multioutput Gaussian processes. arXiv:2003.01115 (2020)
26. Wang, K., Pleiss, G., Gardner, J., Tyree, S., Weinberger, K.Q., Wilson, A.G.: Exact Gaussian processes on a million data points. In: Wallach, H., Larochelle, H., Beygelzimer, A., d'Alché Buc, F., Fox, E., Garnett, R. (eds.) Advances in Neural Information Processing Systems, vol. 32. Curran Associates, Inc. (2019)

# Bayesian Crowdsourcing with Constraints

Panagiotis A. Traganitis(✉) and Georgios B. Giannakis

Electrical and Computer Engineering Department, and Digital Technology Center, University of Minnesota, Minneapolis, MN 55414, USA
{traga003,georgios}@umn.edu

**Abstract.** Crowdsourcing has emerged as a powerful paradigm for efficiently labeling large datasets and performing various learning tasks, by leveraging crowds of human annotators. When additional information is available about the data, constrained or semi-supervised crowdsourcing approaches that enhance the aggregation of labels from human annotators are well motivated. This work deals with constrained crowdsourced classification with instance-level constraints, that capture relationships between pairs of data. A Bayesian algorithm based on variational inference is developed, and its quantifiably improved performance, compared to unsupervised crowdsourcing, is analytically and empirically validated on several crowdsourcing datasets.

**Keywords:** Crowdsourcing · Variational inference · Bayesian · Semi-supervised · Ensemble learning

## 1 Introduction

Crowdsourcing, as the name suggests, harnesses crowds of human annotators, using services such as Amazon's Mechanical Turk [16], to perform various learning tasks such as labeling, image tagging and natural language annotations, among others [11]. Even though crowdsourcing can be efficient and relatively inexpensive, inference of true labels from the noisy responses provided by multiple annotators of unknown expertise can be challenging, especially in the typical *unsupervised* scenario, where no ground-truth data is available.

It is thus practical to look for side information that can be beneficial to the crowdsourcing task, either from experts or from physical constraints associated with the task. For example, queries with known answers may be injected in crowdsourcing tasks in Amazon's Mechanical Turk, in order to assist with the evaluation of annotator reliability. Methods that leverage such information fall under the constrained or semi-supervised learning paradigm [5]. Seeking improved performance in the crowdsourcing task, we focus on constrained crowdsourcing, and investigate instance-level or pairwise constraints, such as must- and cannot-link constraints, which show up as side information in unsupervised tasks, such as clustering [3]. Compared to label constraints, instance-level ones

Work in this paper was supported by NSF grant 1901134.


N. Oliver et al. (Eds.): ECML PKDD 2021, LNAI 12977, pp. 543–559, 2021.
https://doi.org/10.1007/978-3-030-86523-8_33

provide 'weaker information,' as they describe relationships between pairs of data, instead of anchoring the label of a datum. Semi-supervised learning with few label constraints is typically employed when acquiring ground-truth labels is time consuming or expensive, e.g. annotation of biomedical images from a medical professional. Instance-level constraints on the other hand are used when domain knowledge is easier to encode between pairs of data, e.g. road lane identification from GPS data.

To accommodate such available information, the present work capitalizes on recent advances in Bayesian inference. Relative to deterministic methods, the Bayesian approach allows for seamless integration of prior information for the crowdsourcing task, such as annotator performance from previous tasks, and at the same time enables uncertainty quantification of the fused label estimates and parameters of interest. Our major contributions can be summarized as follows: (i) We develop a Bayesian algorithm for crowdsourcing with pairwise constraints; (ii) We derive novel error bounds for the unsupervised variational Bayes crowdsourcing algorithm, in Theorem 1. These error bounds are extended for the proposed constrained algorithm in Theorem 2; and (iii) Guided by the aforementioned theoretical analysis, we provide a constraint selection scheme. In addition to the error bounds, the performance of the proposed algorithm is evaluated with extensive numerical tests. Those corroborate that there are classification performance gains to be harnessed, even when using weaker side information, such as the aforementioned must- and cannot-link constraints.

**Notation:** Unless otherwise noted, lowercase bold fonts, $\boldsymbol{x}$, denote vectors, uppercase ones, $\mathbf{X}$, represent matrices, and calligraphic uppercase, $\mathcal{X}$, stand for sets. The $(i, j)$th entry of matrix $\mathbf{X}$ is denoted by $[\mathbf{X}]_{ij}$. Pr or $p$ denotes probability, or the probability mass function; $\sim$ denotes "distributed as," $|\mathcal{X}|$ is the cardinality of set $\mathcal{X}$, $\mathbb{E}[\cdot]$ denotes expectation, and $\mathbb{1}(\mathcal{A})$ is the indicator function for the event $\mathcal{A}$, that takes value 1, when $\mathcal{A}$ occurs, and is 0 otherwise.

## 2 Problem Formulation and Preliminaries

Consider a dataset consisting of $N$ data $\{x_n\}_{n=1}^N$, with each datum belonging to one of $K$ classes with corresponding labels $\{y_n\}_{n=1}^N$; that is, $y_n = k$ if $x_n$ belongs to class $k$. Consider now $M$ annotators that observe $\{x_n\}_{n=1}^N$, and provide estimates of labels. Let $\check{y}_n^{(m)} \in \{1, \ldots, K\}$ be the label estimate of $x_n$ assigned by the $m$-th annotator. If an annotator has not provided a response for datum $n$ we set $\check{y}_n^{(m)} = 0$. Let $\check{\mathbf{Y}}$ be the $M \times N$ matrix of annotator responses with entries $[\check{\mathbf{Y}}]_{mn} = \check{y}_n^{(m)}$, and $\mathbf{y} := [y_1, \ldots, y_N]^\top$ the $N \times 1$ vector of ground truth labels. The task of *crowdsourced classification* is: Given only the annotator responses in $\check{\mathbf{Y}}$, the goal is find the ground-truth label estimates $\{\hat{y}_n\}_{n=1}^N$.

Per datum $x_n$, the true label $y_n$ is assumed drawn from a categorical distribution with parameters $\boldsymbol{\pi} := [\pi_1, \ldots, \pi_K]^\top$, where $\pi_k := \Pr(y_n = k)$. Further, consider that each learner has a fixed probability of deciding that a datum belongs to class $k'$, when presented with a datum of class $k$; thus, annotator behavior is presumed invariant across the dataset. The performance of an annotator $m$ is then characterized by the so-called *confusion* matrix $\mathbf{\Gamma}^{(m)}$, whose $(k, k')$-th entry is $[\mathbf{\Gamma}^{(m)}]_{k,k'} := \gamma^{(m)}_{k,k'} = \Pr\left(\check{y}_n^{(m)} = k'|y_n = k\right)$. The $K \times K$ confusion matrix showcases the statistical behavior of an annotator, as each row provides the annotator's probability of deciding the correct class, when presented with a datum from each class. Collect all annotator confusion matrices in $\mathbf{\Gamma} := [\mathbf{\Gamma}^{(1)}, \ldots, \mathbf{\Gamma}^{(m)}]$. Responses of different annotators per datum $n$ are presumed conditionally independent, given the ground-truth label $y_n$; that is, $p\left(\check{y}_n^{(1)} = k_1, \ldots, \check{y}_n^{(M)} = k_M|y_n = k\right) = \prod_{m=1}^{M} p\left(\check{y}_n^{(m)} = k_m|y_n = k\right)$. The latter is a standard assumption that is commonly employed in crowdsourcing works [7,13,37]. Finally, most annotators are assumed to be better than random.

## 2.1 Prior Works

Arguably the simplest approach to fusing crowdsourced labels is majority voting, where the estimated label of a datum is the one most annotators agree upon. This presumes that all annotators are equally "reliable," which may be unrealistic. Aiming at high-performance label fusion, several approaches estimate annotator parameters, meaning the confusion matrices as well as class priors. A popular approach is joint maximum likelihood (ML) estimation of the unknown labels $\mathbf{y}$, and the aforementioned confusion matrices using the expectation-maximization (EM) algorithm [7]. As EM guarantees convergence to a locally optimal solution, alternative estimation methods have been recently advocated. Spectral methods invoke second- and third-order moments of annotator responses to infer the unknown annotator parameters [12,13,31,37].

A Bayesian treatment of crowdsourced learning, termed *Bayesian Classifier Combination* was introduced in [15]. This approach used Gibbs sampling to estimate the parameters of interest, while [28] introduced a variational Bayes EM (VBEM) method for the same task. Other Bayesian approaches infer communities of annotators, enhancing the quality of aggregated labels [19,35]. For sequential or networked data, [20,23,32,33] advocated variational inference and EM-based alternatives. All aforementioned approaches utilize only $\check{\mathbf{Y}}$. When features $\{x_n\}_{n=1}^N$ are also available, parametric models [22], approaches based on Gaussian Processes [25,26], or deep learning [1,2,24,27] can be employed to classify the data, and simultaneously learn a classifier.

Current constrained or semi-supervised approaches to crowdsourcing extend the EM algorithm of Dawid and Skene [30], by including a few ground-truth labels. The model in [36] includes a graph between datapoints to enable label prediction for data that have not been annotated, while [17] puts forth a max-margin majority vote method when label constraints are available. When features $\{x_n\}_{n=1}^N$ are available, [14] relies on a parametric model to learn a binary classifier from crowd responses and expert labels.

The present work develops a Bayesian *constrained* approach to crowdsourced classification, by adapting the popular variational inference framework [4].

In addition, it provides novel performance analysis for both Bayesian unsupervised and semi-supervised approaches. The proposed method does not require features $\{x_n\}_{n=1}^N$, but relies only on annotator labels – a first attempt at incorporating instance-level constraints for the crowdsourced classification task.

## 3 Variational Inference for Crowdsourcing

Before presenting our constrained approach, this section will recap a Bayesian treatment of the crowdsourcing problem. First, variational Bayes is presented in Sect. 3.1, followed by an inference algorithm for crowdsourcing [28] in Sect. 3.2. A novel performance analysis for the VBEM algorithm is also presented in Sect. 3.2.

### 3.1 Variational Bayes

Consider a set of observed data collected in $\mathbf{X}$, and a set of latent variables and parameters in $\mathbf{Z}$ that depend on $\mathbf{X}$. Variational Bayes seeks $\mathbf{Z}$ that maximize the marginal of $\mathbf{X}$, by treating both latent variables and parameters as random [4]. EM in contrast treats $\mathbf{Z}$ as deterministic and provides point estimates. The log-marginal of $\mathbf{X}$ can be written as

$$\ln p(\mathbf{X}) = \int q(\mathbf{Z}) \ln \frac{p(\mathbf{X},\mathbf{Z})}{q(\mathbf{Z})} d\mathbf{Z} - \int q(\mathbf{Z}) \ln \frac{p(\mathbf{Z}|\mathbf{X})}{q(\mathbf{Z})} d\mathbf{Z} = L(q) + \mathrm{KL}(q||p) \quad (1)$$

where $L(q) := \int q(\mathbf{Z}) \ln \frac{p(\mathbf{X},\mathbf{Z})}{q(\mathbf{Z})} d\mathbf{Z}$, and $\mathrm{KL}(q||p) = -\int q(\mathbf{Z}) \ln \frac{p(\mathbf{Z}|\mathbf{X})}{q(\mathbf{Z})} d\mathbf{Z}$ denotes the Kullback-Leibler divergence between pdfs $q$ and $p$ [6]. This expression is maximized for $q(\mathbf{Z}) = p(\mathbf{Z}|\mathbf{X})$, however when $p(\mathbf{Z}|\mathbf{X})$ is an intractable pdf, one may seek distributions $q$ from a prescribed tractable family $\mathcal{Q}$, such that $\mathrm{KL}(q||p)$ is minimized. One such family is the family of factorized distributions, which also go by the name of mean field distributions. Under the mean field paradigm, the variational distribution $q(\mathbf{Z})$ is decomposed into a product of single variable factors, $q(\mathbf{Z}) = \prod_i q(\mathbf{Z}_i)$, with $q(\mathbf{Z}_i)$ denoting the variational distribution corresponding to the variable $\mathbf{Z}_i$.

It can be shown that the optimal updates for each factor are given by [4]

$$\ln q^*(\mathbf{Z}_i) = \mathbb{E}_{-\mathbf{Z}_i}\left[\ln p(\mathbf{X},\mathbf{Z})\right] + c \quad (2)$$

where $c$ is an appropriate constant, and the $-\mathbf{Z}_i$ subscript denotes that the expectation is taken w.r.t. the terms in $q$ that do not involve $\mathbf{Z}_i$, that is $\prod_{j\neq i} q(\mathbf{Z}_j)$. These optimal factors can be estimated iteratively, and such a procedure is guaranteed to converge at least to a local maximum of (1).

### 3.2 Variational EM for Crowdsourcing

Next, we will outline how variational inference can be used to derive an iterative algorithm for crowdsourced classification [28]. The Bayesian treatment of the crowdsourcing problem dictates the use of prior distributions on the parameters of interest, namely $\boldsymbol{\pi}$, and $\boldsymbol{\Gamma}$. The probabilities $\boldsymbol{\pi}$ are assigned a Dirichlet distribution prior with parameters $\boldsymbol{\alpha}_0 := [\alpha_{0,1}, \ldots, \alpha_{0,K}]^\top$, that is $\boldsymbol{\pi} \sim \text{Dir}(\boldsymbol{\pi}; \boldsymbol{\alpha}_0)$, whereas for an annotator $m$, the columns $\{\boldsymbol{\gamma}_k^{(m)}\}_{k=1}^K$ of its confusion matrix are considered independent, and $\boldsymbol{\gamma}_k^{(m)}$ is assigned a Dirichlet distribution prior with parameters $\boldsymbol{\beta}_{0,k}^{(m)} := [\beta_{0,k,1}^{(m)}, \ldots, \beta_{0,k,K}^{(m)}]^\top$, respectively. These priors on the parameters of interest are especially useful when only few data have been annotated, and can also capture annotator behavior from previous crowdsourcing tasks. The joint distribution of $\mathbf{y}, \check{\mathbf{Y}}, \boldsymbol{\pi}$ and $\boldsymbol{\Gamma}$ is

$$p(\mathbf{y}, \boldsymbol{\pi}, \check{\mathbf{Y}}, \boldsymbol{\Gamma}; \boldsymbol{\alpha}_0, \mathbf{B}_0) = \prod_{n=1}^{N} \pi_{y_n} \prod_{m=1}^{M} \prod_{k'=1}^{K} (\gamma_{y_n,k'}^{(m)})^{\delta_{n,k'}^{(m)}} p(\boldsymbol{\pi}; \boldsymbol{\alpha}_0) p(\boldsymbol{\Gamma}; \mathbf{B}_0) \quad (3)$$

with $\mathbf{B}_0$ collecting all prior parameters $\boldsymbol{\beta}_{0,k}^{(m)}$ for $k = 1, \ldots, K$ and $m = 1, \ldots, M$, and $\delta_{n,k}^{(m)} := \mathbb{1}(\check{y}_n^{(m)} = k)$. The parametrization on $\boldsymbol{\alpha}_0, \mathbf{B}_0$ will be henceforth implicit for brevity.

With the goal of estimating the unknown variables and parameters of interest, we will use the approach outlined in the previous subsection, to approximate $p(\mathbf{y}, \boldsymbol{\pi}, \boldsymbol{\Gamma}|\check{\mathbf{Y}})$ using a variational distribution $q(\mathbf{y}, \boldsymbol{\pi}, \boldsymbol{\Gamma})$. Under the mean-field class, this variational distribution factors across the unknown variables as $q(\mathbf{y}, \boldsymbol{\pi}, \boldsymbol{\Gamma}) = q(\mathbf{y})q(\boldsymbol{\pi})q(\boldsymbol{\Gamma})$. Since the data are assumed i.i.d. [cf. Sect. 2], $q(\mathbf{y})$ is further decomposed into $q(\mathbf{y}) = \prod_{n=1}^{N} q(y_n)$. In addition, since annotators are assumed independent and the columns of their confusion matrices are also independent, we have $q(\boldsymbol{\Gamma}) = \prod_{m=1}^{M} \prod_{k=1}^{K} q(\boldsymbol{\gamma}_k^{(m)})$. The variational Bayes EM algorithm is an iterative algorithm, with each iteration consisting of two steps: the variational E-step, where the latent variables $\mathbf{y}$ are estimated; and the variational M-step, where the distribution of the parameters of interest $\boldsymbol{\pi}, \boldsymbol{\Gamma}$ are estimated.

At iteration $t+1$, the variational distribution for the unknown label $y_n$ is given by

$$\ln q_{t+1}(y_n) = \mathbb{E}_{-y_n} \left[\ln p(\mathbf{y}, \boldsymbol{\pi}, \check{\mathbf{Y}}, \boldsymbol{\Gamma})\right] + c. \quad (4)$$

The subscript of $q$ denotes the iteration index. The expectation in (4) is taken w.r.t. the terms in $q_t$ that do not involve $y_n$, that is $\prod_{n' \neq n} q_t(y_{n'}) q_t(\boldsymbol{\pi}) q_t(\boldsymbol{\Gamma})$. Upon expanding $p(\boldsymbol{y}, \boldsymbol{\pi}, \check{\mathbf{Y}}, \boldsymbol{\Gamma})$, (4) becomes

$$\ln q_{t+1}(y_n) = \mathbb{E}_{\boldsymbol{\pi}} \left[\ln \pi_{y_n}\right] + \mathbb{E}_{\boldsymbol{\Gamma}} \left[\sum_{m=1}^{M} \sum_{k'=1}^{K} \delta_{n,k'}^{(m)} \ln \gamma_{y_n,k'}^{(m)}\right] + c. \quad (5)$$

Accordingly, the update for the class priors in $\boldsymbol{\pi}$ is

$$\ln q_{t+1}(\boldsymbol{\pi}) = \mathbb{E}_{-\boldsymbol{\pi}} \left[\ln p(\mathbf{y}, \boldsymbol{\pi}, \check{\mathbf{Y}}, \boldsymbol{\Gamma})\right] + c \quad (6)$$

where the expectation is taken w.r.t. $q_{t+1}(\mathbf{y})q_t(\boldsymbol{\Gamma})$. Based on (6), it can be shown [28] that

$$q_{t+1}(\boldsymbol{\pi}) \propto \text{Dir}(\boldsymbol{\pi}; \boldsymbol{\alpha}_{t+1}) \quad (7)$$

where $\boldsymbol{\alpha}_{t+1} := [\alpha_{t+1,1}, \ldots, \alpha_{t+1,K}]^\top$ with $\alpha_{t+1,k} = N_{t+1,k} + \alpha_{0,k}$, $N_{t+1,k} := \sum_{n=1}^{N} q_{t+1}(y_n = k)$. As a direct consequence of this, the term involving $\boldsymbol{\pi}$ in (5) is given by $\mathbb{E}_{\boldsymbol{\pi}}[\ln \pi_k] = \psi(\alpha_{t,k}) - \psi(\bar{\alpha}_t)$, with $\psi$ denoting the digamma function, and $\bar{\alpha}_t := \sum_{k=1}^{K} \alpha_{t,k}$.

For the $k$-th column of $\boldsymbol{\Gamma}^{(m)}$ the update takes the form

$$\ln q_{t+1}(\boldsymbol{\gamma}_k^{(m)}) = \mathbb{E}_{-\boldsymbol{\gamma}_k^{(m)}}\left[\ln p(\mathbf{y}, \boldsymbol{\pi}, \check{\mathbf{Y}}, \boldsymbol{\Gamma})\right] + c \tag{8}$$

Using steps similar to the update of $q(\boldsymbol{\pi})$, one can show that

$$q_{t+1}(\boldsymbol{\gamma}_k^{(m)}) \propto \text{Dir}(\boldsymbol{\gamma}_k^{(m)}; \boldsymbol{\beta}_{t+1,k}^{(m)}) \tag{9}$$

with $\boldsymbol{\beta}_{t+1,k}^{(m)} := [\beta_{t+1,k,1}^{(m)}, \ldots, \beta_{t+1,k,1}^{(m)}]$, $\beta_{t+1,k,k'}^{(m)} := N_{t+1,k,k'}^{(m)} + \beta_{0,k,k'}^{(m)}$ and $N_{t+1,k,k'}^{(m)} := \sum_{n=1}^{N} q_{t+1}(y_n = k)\delta_{n,k'}^{(m)}$. Consequently at iteration $t+1$, and upon defining $\bar{\beta}_{t+1,k}^{(m)} := \sum_{\ell}^{K} \beta_{t+1,k,\ell}^{(m)}$, we have $\mathbb{E}_{\boldsymbol{\Gamma}}[\ln \gamma_{k,k'}^{(m)}] = \psi\left(\beta_{t+1,k,k'}^{(m)}\right) - \psi\left(\bar{\beta}_{t+1,k}^{(m)}\right)$. Given initial values for $\mathbb{E}_{\boldsymbol{\pi}}[\ln \pi_{y_n}]$ and $\mathbb{E}_{\boldsymbol{\Gamma}}\left[\sum_{m=1}^{M} \ln \gamma_{y_n, \check{y}_n^{(m)}}^{(m)}\right]$ per variational E-step, first the variational distribution for each datum $q(y_n)$ is computed using (4). Using the recently computed $q(\mathbf{y})$ at the variational M-step, the variational distribution for class priors $q(\boldsymbol{\pi})$ is updated via (6) and the variational distributions for each column of the confusion matrices $q(\boldsymbol{\gamma}_k^{(m)})$ are updated via (8). The E- and M-steps are repeated until convergence. Finally, the fused labels $\hat{\mathbf{y}}$ are recovered at the last iteration $T$ as

$$\hat{y}_n = \underset{k}{\text{argmax}}\ q_T(y_n = k). \tag{10}$$

The overall computational complexity of this algorithm is $\mathcal{O}(NKMT)$. The updates of this VBEM algorithm are very similar to the corresponding updates of the EM algorithm for crowdsourcing [7].

**Performance Analysis.** Let $\boldsymbol{\pi}^*$ comprise the optimal class priors, $\boldsymbol{\Gamma}^*$ the optimal confusion matrices, and $\mu_m := \Pr(\check{y}_n^{(m)} \neq 0)$ the probability an annotator will provide a response. The following theorem establishes that when VBEM is properly initialized, the estimation errors for $\boldsymbol{\pi}, \boldsymbol{\Gamma}$ and the labels $\mathbf{y}$ are bounded. Such proper initializations can be achieved, for instance, using the spectral approaches of [12,31,37].

**Theorem 1.** *Suppose that $\gamma_{k,k'}^{*(m)} \geq \rho_\gamma$ for all $k, k', m$, $\pi_k^* \geq \rho_\pi$ for all $k$, and that VBEM is initialized so that $|\mathbb{E}[\pi_k] - \pi_k^*| \leq \varepsilon_{\pi,0}$ for all $k$ and $|\mathbb{E}[\gamma_{k,k'}^{(m)}] - \gamma_{k,k'}^{*(m)}| \leq \varepsilon_{\gamma,0}$ for all $k, k', m$. It then holds w.p. at least $1-\nu$ that for iterations $t = 1, \ldots, T$, update* (4) *yields*

$$\max_k |q_t(y_n = k) - \mathbb{1}(y_n = k)| \leq \varepsilon_{q,t} := K \exp(-U)$$
$$U := D + f_\pi(\varepsilon_{\pi,t-1}) + M f_\gamma(\varepsilon_{\gamma,t-1}),$$

*where $D$ is a quantity related to properties of the dataset and the annotators, while $f_\pi, f_\gamma$ are decreasing functions. Consequently for $k, k' = 1, \ldots, K$, $m = 1, \ldots, M$ updates* (6) *and* (8) *yield respectively*

$$|\mathbb{E}[\pi_k] - \pi_k^*| \leq \epsilon_{\pi,k,t} := \frac{N(\varepsilon_{q,t} + g_\pi(\nu)) + \alpha_{0,k} + \rho_\pi \bar{\alpha}_0}{N + \bar{\alpha}_0} \tag{11}$$

$$|\mathbb{E}[\gamma_{k,k'}^{(m)}] - \gamma_{k,k'}^{*(m)}| \leq \epsilon_{\gamma,k,k',t}^{(m)} := \frac{2N(g_\gamma(\nu) + \varepsilon_{q,t}) + \beta_{0,k,k'}^{(m)} + \bar{\beta}_{0,k}^{(m)}}{N\mu_m \pi_k^* - N\frac{g_\gamma(\nu)}{\gamma_{k,k'}^{*(m)}} - N\varepsilon_{q,t} + \bar{\beta}_k^{(m)}} \tag{12}$$

*with $g_\pi(\nu)$ and $g_\gamma(\nu)$ being decreasing functions of $\nu$ and $\varepsilon_{\pi,t} := \max_k \epsilon_{\pi,k,t}, \varepsilon_{\gamma,t} := \max_{k,k',m} \epsilon_{\gamma,k,k',t}^{(m)}$.*

Detailed theorem proofs are deferred to Appendix A of the supplementary material [34]. Theorem 1 shows that lowering the upper bound on label errors $\varepsilon_q$, will reduce the estimation error upper bounds of $\{\gamma_{k,k'}^{(m)}\}$ and $\boldsymbol{\pi}$. This in turn can further reduce $\varepsilon_{q,t}$, as it is proportional to $\varepsilon_\pi$, and $\varepsilon_\gamma$. With a VBEM algorithm for crowdsourcing and its performance analysis at hand, the ensuing section will introduce our constrained variational Bayes algorithm that can incorporate additional information to enhance label aggregation.

## 4 Constrained Crowdsourcing

This section deals with a constrained (or semi-supervised) variational Bayes approach to crowdsourcing. Here, additional information to the crowdsourcing task is available in the form of pairwise constraints, that indicate relationships between pairs of data. Throughout this section, $N_C$ will denote the number of available constraints that are collected in the set $\mathcal{C}$.

First, we note that when label constraints are available, the aforementioned VBEM algorithm can readily handle them, in a manner similar to [30]. Let $\mathcal{C}$ denote the set of indices for data with label constraints $\{y_n\}_{n \in \mathcal{C}}$ available. These constraints can then be incorporated by fixing the values of $\{q(y_n)\}_{n \in \mathcal{C}}$ to 1 for all iterations. The variational distributions for data $n \notin \mathcal{C}$ are updated according to (5), while confusion matrices and prior probabilities according to (8) and (6), respectively.

Next, we consider the case of instance-level constraints, which are the main focus of this work. Such information may be easier to obtain than the label constraints of the previous subsection, as pairwise constraints encapsulate relationships between pairs of data and not "hard" label information. The pairwise constraints considered here take the form of must-link and cannot-link relationships, and are collected respectively in the sets $\mathcal{C}_{\text{ML}}$ and $\mathcal{C}_{\text{CL}}$, and $\mathcal{C} = \mathcal{C}_{\text{ML}} \cup \mathcal{C}_{\text{CL}}$, $N_{\text{ML}} = |\mathcal{C}_{\text{ML}}|, N_{\text{CL}} = |\mathcal{C}_{\text{CL}}|$. All constraints consist of tuples $(i, j) \in \{1, \ldots, N\} \times \{1, \ldots, N\}$. A must-link constraint $(i, j), i \neq j$ indicates that two data points, $\boldsymbol{x}_i$ and $\boldsymbol{x}_j$ must belong to the same class, i.e. $y_i = y_j$, whereas a cannot-link constraint $(i', j')$ indicates that two points $\boldsymbol{x}_{i'}$ and $\boldsymbol{x}_{j'}$ are not in the same class; $y_{i'} \neq y_{j'}$. Note that instance level constraints naturally describe a graph $\mathcal{G}$, whose $N$ nodes correspond to the data $\{x_n\}_{n=1}^N$, and whose (weighted) edges are the available constraints. This realization is the cornerstone of the proposed algorithm, suggesting that instance-level constraints can be incorporated in the crowdsourcing task by means of a probabilistic graphical model.

Specifically, we will encode constraints in the marginal pmf of the unknown labels $p(\mathbf{y})$ using a Markov Random Field (MRF), which implies that for all $n = 1, \ldots, N$, the local Markov property holds, that is $p(y_n|\mathbf{y}_{-n}) = p(y_n|\mathbf{y}_{\mathcal{C}_n})$, where $\mathbf{y}_{-n}$ denotes a vector containing all labels except $y_n$, while $\mathbf{y}_{\mathcal{C}_n}$ a vector containing labels for $y_{n'}$, where $n' \in \mathcal{C}_n$, and $\mathcal{C}_n$ is a set containing indices such that $n' : (n, n') \in \mathcal{C}$. By the Hammersley-Clifford theorem [10], the marginal pmf of the unknown labels is given by

$$p(\mathbf{y}) = \frac{1}{Z} \exp\left( \sum_{n=1}^{N} \ln \pi_{y_n} + \eta \sum_{n' \in \mathcal{C}_n} V(y_n, y_{n'}) \right) \tag{13}$$

where $Z$ is the (typically intractable) normalization constant, and $\eta > 0$ is a tunable parameter that specifies how much weight we assign to the constraints. Here, we define the $V$ function as $V(y_n, y_{n'}) = w_{n,n'} \mathbb{1}(y_n = y_{n'})$, where

$$w_{n,n'} = \mathbb{1}\left((n, n') \in \mathcal{C}_{\mathrm{ML}}\right) - \mathbb{1}\left((n, n') \in \mathcal{C}_{\mathrm{CL}}\right) \tag{14}$$

The weights $w_{n,n'}$ can also take other real values, indicating the confidence one has per constraint $(n, n')$, but here we will focus on $-1, 0, 1$ values. This $V(y_n, y_{n'})$ term promotes the same label for data with must-link constraints, whereas it penalizes similar labels for data with cannot-link constraints. Note that more sophisticated choices for $V$ can also be used. Nevertheless, this particular choice leads to a simple algorithm with quantifiable performance, as will be seen in the ensuing sections. Since the number of constraints is typically small, for the majority of data the term $\sum V(y_n, y_{n'})$ will be 0. Thus, the $\ln \pi$ term in the exponent of (13) acts similarly to the prior probabilities of Sect. 3.2. Again, adopting a Dirichlet prior $\mathrm{Dir}(\boldsymbol{\pi}; \boldsymbol{\alpha}_0)$, and $\gamma_k^{(m)} \sim \mathrm{Dir}(\gamma_k^{(m)}; \boldsymbol{\beta}_{0,k}^{(m)})$ for all $m, k$, and using mean-field VB as before, the variational update for the $n$-th label becomes

$$\begin{aligned} \ln q_{t+1}(y_n = k) &= \mathbb{E}_{\pi}\left[\ln \pi_k\right] + c \\ &+ \mathbb{E}_{\boldsymbol{\Gamma}}\left[\sum_{m=1}^{M} \sum_{k'=1}^{K} \delta_{n,k'}^{(m)} \ln \gamma_{k,k'}^{(m)}\right] + \eta \sum_{n' \in \mathcal{C}_n} w_{n,n'} q_t(y_{n'} = k). \end{aligned} \tag{15}$$

where we have used $\mathbb{E}_{y_{n'}}[\mathbb{1}(y_{n'} = k)] = q_t(y_{n'} = k)$. The variational update for label $y_n$ is similar to the one in (5), with the addition of the $\eta \sum_{n' \in \mathcal{C}_n} w_{n,n'} q_t(y_{n'} = k)$ term that captures the labels of the data that are related (through the instance constraints) to the $n$-th datum. Updating label distributions via (15), updates for $\boldsymbol{\pi}$ and $\boldsymbol{\Gamma}$ remain identical to those in VBEM

**Algorithm 1.** Crowdsourcing with instance-level constraints

**Input:** Annotator responses $\check{\mathbf{Y}}$, initial $\boldsymbol{\pi}$, $\boldsymbol{\Gamma}$, $\{q_0(y_n = k\}_{n,k=1}^{N,K}$, constraints $\mathcal{C}$, parameters $\boldsymbol{\alpha}_0$, $\mathbf{B}_0$, $\eta$
**Output:** Estimates $\mathbb{E}[\boldsymbol{\pi}]$, $\mathbb{E}[\boldsymbol{\Gamma}]$, $\hat{\mathbf{y}}$.
**while** not converged **do**
  Update $q_{t+1}(y_n)$ via (15)
  Update $q_{t+1}(\boldsymbol{\pi})$ using $q_{t+1}(y_n)$, (6)
  Update $q_{t+1}(\gamma_k^{(m)})$ $\forall m, k$, using $q_{t+1}(y_n)$, (8).
  $t \leftarrow t+1$
**end while**
Estimate fused data labels using (10).

[cf. Sect. 3.2]. The instance-level semi-supervised crowdsourcing algorithm is outlined in Algorithm 1. As with its plain vanilla counterpart of Sect. 3.2, Algorithm 1 maintains the asymptotic complexity of $\mathcal{O}(NMKT)$.

### 4.1 Performance Analysis

Let $\tilde{C}$ with cardinality $|\tilde{C}| = \tilde{N}_C$ be a set comprising indices of data that take part in at least one constraint, and let $\tilde{C}^c$ with $|\tilde{C}^c| = \bar{N}_C$ denote its complement. The next theorem quantifies the performance of Algorithm 1.

**Theorem 2.** *Consider the same setup as in Theorem 1, instance level constraints collected in* $\mathcal{C}$, *and the initalization of Algorithm 1 satisfying* $\max_k |q_0(y_n = k) - \mathbb{1}(y_n = k)| \leq \varepsilon_{q,0}$ *for all* $k, n$. *Then the following hold w.p. at least* $1 - \nu$: *For iterations* $t = 1, \ldots, T$, *update* (15) *yields for* $n \in \tilde{C}$

$$\max_k |q_t(y_n = k) - \mathbb{1}(y_n = k)| \leq \tilde{\varepsilon}_{q,t} \tag{16}$$

$$\tilde{\varepsilon}_{q,t} := \max_n K \exp\left(-U - \eta W_n\right) = \max_n \varepsilon_{q,t} \exp(-\eta W_n)$$

$$W_n := N_{\mathrm{ML,n}}(1 - 2\varepsilon_{q,t-1}) - 2N_{\mathrm{CL,n}}\varepsilon_{q,t-1} + N_{\mathrm{CL,n,min}}$$

*with* $N_{\mathrm{ML,n}}, N_{\mathrm{CL,n}}$ *denoting the number of must- and cannot-link constraints for datum* $n$ *respectively, and* $N_{\mathrm{CL,n,min}} := \min_k N_{\mathrm{CL,n,k}}$, *where* $N_{\mathrm{CL,n,k}}$ *is the number of cannot-link constraints of datum* $n$, *that belong to class* $k$. *For data without constraints, that is* $n \in \tilde{C}^c$

$$\max_k |q_t(y_n = k) - \mathbb{1}(y_n = k)| \leq \varepsilon_{q,t} = K \exp(-U),$$

*with* $U$ *as defined in Theorem 1. For* $k, k' = 1, \ldots, K$, $m = 1, \ldots, M$ *updates* (6) *and* (8) *yield respectively*

$$|\mathbb{E}[\pi_k] - \pi_k^*| \leq \epsilon_{\pi,k,t} := \frac{\tilde{N}_C\tilde{\varepsilon}_{q,t} + \bar{N}_C\varepsilon_{q,t} + Ng_\pi(\nu) + \alpha_{0,k} + \rho_\pi\bar{\alpha}_0}{N + \bar{\alpha}_0} \tag{17}$$

$$|\mathbb{E}[\gamma_{k,k'}^{(m)}] - \gamma_{k,k'}^{*(m)}| \leq \epsilon_{\gamma,k,k',t}^{(m)} := \frac{2Ng_\gamma(\nu) + 2\tilde{N}_C\tilde{\varepsilon}_{q,t} + 2\bar{N}_C\varepsilon_{q,t} + \beta_{0,k,k'}^{(m)} + \bar{\beta}_{0,k}^{(m)}}{N\mu_m\pi_k^* - N\frac{g_\gamma(\nu)}{\gamma_{k,k'}^{*(m)}} - \tilde{N}_C\tilde{\varepsilon}_{q,t} - \bar{N}_C\varepsilon_{q,t} + \bar{\beta}_k^{(m)}} \tag{18}$$

*and* $\varepsilon_{\pi,t} := \max_k \epsilon_{\pi,k,t}$, $\varepsilon_{\gamma,t} := \max_{k,k',m} \epsilon_{\gamma,k,k',t}^{(m)}$.

Comparing Theorem 2 to Theorem 1, the error bounds for the labels involved in constraints introduced by Algorithm 1 will be smaller than the corresponding bounds by VBEM, as long as $\max_n W_n > 0$. This in turn reduces the error bounds for the parameters of interest, thus justifying the improved performance of Algorithm 1 when provided with good initialization. Such an initialization can be achieved by using spectral methods, or, by utilizing the output of the aforementioned unconstrained VBEM algorithm.

### 4.2 Choosing $\eta$.

Proper choice of $\eta$ is critical for the performance of the proposed algorithm, as it represents the weight assigned to the available constraints. Here, using the number of violated constraints $N_{\mathrm{V}} := \sum_{(n,n')\in\mathcal{C}_{\mathrm{ML}}} \mathbb{1}(y_n \neq y_{n'}) + \sum_{(n,n')\in\mathcal{C}_{\mathrm{CL}}} \mathbb{1}(y_n = y_{n'})$ as a proxy for performance, grid search can be used to select $\eta$ from a (ideally small) set of possible values $\mathcal{H}$.

### 4.3 Selecting Instance-Level Constraints

In some cases, acquiring pairwise constraints, can be costly and time consuming. This motivates judicious selection of data we would prefer to query constraints for. Here, we outline an approach for selecting constraints using only the annotator responses $\check{\mathbf{Y}}$. In addition to providing error bounds for VBEM under instance level constraints, Theorem 2 reveals how to select the $N_C$ constraints in $\mathcal{C}$. Equations (17) and (18) show that in order to minimize the error bounds of the priors and confusion matrices, the data with the largest label errors $|q(y_n = k) - \mathbb{1}(y_n = k)|$ should be included in $\mathcal{C}$. The smaller error bounds of (17) and (18) then result in smaller error bounds for the labels, thus improving the overall classification performance. Let $\tilde{\mathcal{C}}_u$ denote the set of data for which we wish to reduce the label error. In order to minimize the error $\tilde{\varepsilon}_q$ per element of $\tilde{\mathcal{C}}_u$, the term $W_n$ in the exponent of (16) should be maximized. To this end, data in $\tilde{\mathcal{C}}_u$ should be connected with instance-level constraints to data that exhibit small errors $\varepsilon_q$. Data connected to each $n \in \tilde{\mathcal{C}}_u$, are collected in the set $\tilde{\mathcal{C}}_c(n)$.

Since the true label errors are not available, one has to resort to surrogates for them. One class of such surrogates, typically used in active learning [9], are the so-called uncertainty measures. These quantities estimate how uncertain the classifier (in this case the crowd) is about each datum. Intuitively, data for which the crowd is highly uncertain are more likely to be misclassified. In our setup, such uncertainty measures can be obtained using the results provided by VBEM, and specifically the label posteriors $\{q_T(y_n)\}_{n=1}^N$. As an alternative, approximate posteriors can be found without using the VBEM algorithm, by taking a histogram of annotator responses per datum. Using these posteriors, one can measure the uncertainty of the crowd for each datum. Here, we opted for the so-called best-versus-second-best uncertainty measure, which per datum $n$ is given by

$$H(\hat{y}_n) = \max_k q_T(y_n = k) - \max_{k' \neq k} q_T(y_n = k'). \tag{19}$$

**Table 1.** Dataset properties

| Dataset | $N$ | $M$ | $K$ | $\tilde{\delta}$ |
|---|---|---|---|---|
| RTE | 800 | 164 | 2 | 48.78 |
| Sentence polarity | 5,000 | 203 | 2 | 136.68 |
| Dog | 807 | 109 | 5 | 74.03 |
| Web | 2,665 | 177 | 5 | 87.94 |

This quantity shows how close the two largest posteriors for each datum are; larger values imply that the crowd is highly certain for a datum, whereas smaller ones indicate uncertainty.

Using the results of VBEM and the uncertainty measure in (19), $\lfloor N_C/K \rfloor$ are randomly selected without replacement, to be included in $\tilde{\mathcal{C}}_u$, with probabilities $\lambda_n \propto 1 - H(\hat{y}_n)$. For each $n \in \tilde{\mathcal{C}}_u$, $K$ data are randomly selected, with probabilities $\lambda_n^\dagger \propto H(\hat{y}_n)$, and are included in $\tilde{\mathcal{C}}_c(n)$. Using this procedure, data in $\tilde{\mathcal{C}}_c(n), n \in \tilde{\mathcal{C}}_u$ are likely the ones that the crowd is certain of. Finally, the links $\{(n, n'), n \in \tilde{\mathcal{C}}_u, n' \in \tilde{\mathcal{C}}_c(n)\}$ are queried and included in the constraint set $\mathcal{C}$. Note here that each uncertain data point is connected to $K$ certain ones, to increase the likelihood that its label error will decrease.

*Remark 3.* The principles outlined in this subsection can be leveraged to develop active semi-supervised crowdsourcing algorithms. We defer such approaches to future work.

## 5 Numerical Tests

Performance of the proposed semi-supervised approach is evaluated here on several popular crowdsourcing datasets. Variational Bayes with instance-level constraints (abbreviated as *VB - ILC*) [cf. Sect. 4, Algorithm 1] is compared to majority voting (abbreviated as *MV*), the variational Bayes method of Sect. 3.2 that does not include side-information (abbreviated as *VB*), and the EM algorithm of [7] (abbreviated as *DS*) that also does not utilize any side information. Here, simple baselines are chosen to showcase the importance of including constraints in the crowdsourcing task. To further show the effect of pairwise constraints *VB - ILC* is compared to Variational Bayes using label constraints (abbreviated as *VB - LC*), as outlined at the beginning of Sect. 4.

As figure of merit, the macro-averaged F-score [21] is adopted to indicate the per-class performance of an algorithm. The datasets considered here are the RTE [29], Sentence Polarity [25], Dog [8], and Web [38]; see Table 1 for their properties, where $\tilde{\delta}$ denotes the average number of responses per annotator. For the datasets presented here, results indicating the micro-averaged F-score, alongside dataset descriptions and results with 6 additional datasets are included in Appendix B of the supplementary material [34]. MATLAB [18] was used throughout, and all results represent the average over 20 Monte Carlo runs. In all experiments *VB*

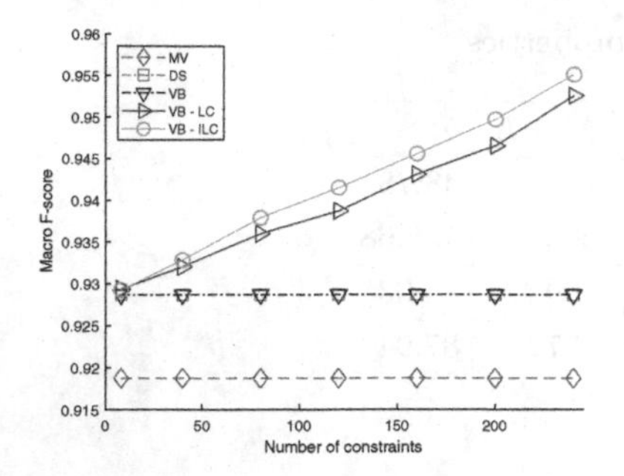


Fig. 1. Macro F-score for the RTE dataset, with the same number of constraints for *VB-LC* and *VB - ILC*

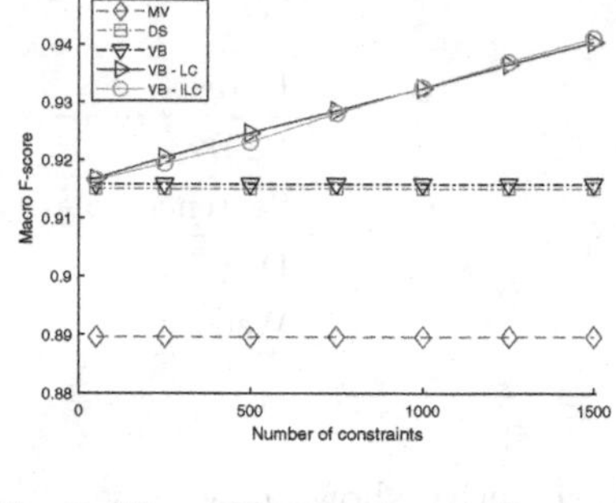


Fig. 2. Macro F-score for the Sentence Polarity dataset, with the same number of constraints for *VB-LC* and *VB - ILC*

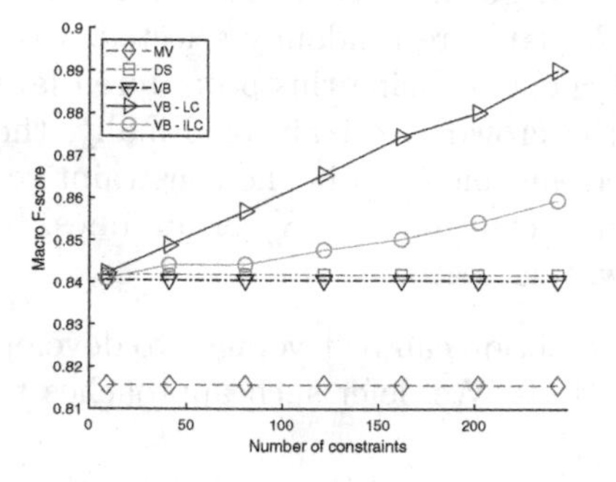


Fig. 3. Macro F-score for the Dog dataset, with the same number of constraints for *VB-LC* and *VB - ILC*

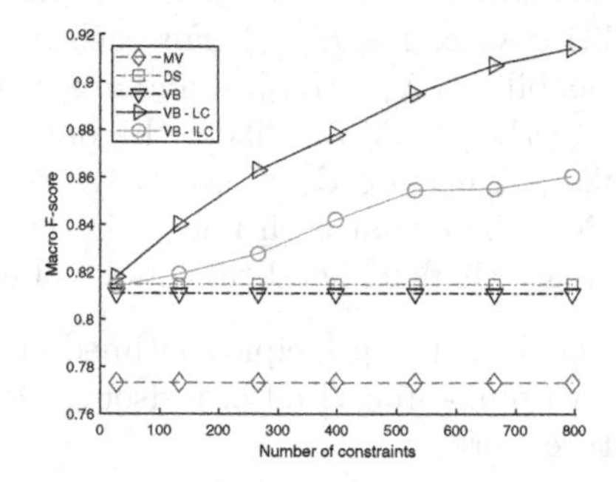


Fig. 4. Macro F-score for the Web dataset, with the same number of constraints for *VB-LC* and *VB - ILC*

and *DS* are initialized using majority voting, while *VB - LC* and *VB - ILC* are initialized using the results of *VB*, since, as shown in Theorem 2 *VB-ILC* requires good initialization. For the RTE, Dog, and Sentence Polarity datasets, the prior parameters are set as $\boldsymbol{\alpha}_0 = \mathbf{1}$, where $\mathbf{1}$ denotes the all-ones vector, and $\boldsymbol{\beta}_{0,k}^{(m)}$ is a vector with $K$ at its $k$-th entry and ones everywhere else, for all $k, m$. For the Web dataset, all priors are set to uniform as this provides the best performance for the VB based algorithms. The values of $\eta$ for Algorithm 1 are chosen as described in Sect. 4.2 from the set $\mathcal{H} = \{0.01, 0.05, 0.1, 0.2, 0.5, 1, 2, 5, 10, 20, 100, 500\}$. When instance-level constraints in $\mathcal{C}$ are provided to *VB-ILC*, constraints that can be logically derived from the ones in $\mathcal{C}$ are also included. For example, if $(i, j) \in \mathcal{C}_{\text{ML}}$ and $(j, k) \in \mathcal{C}_{\text{ML}}$, a new must-link constraint $(i, k)$ will be added. Similarly, if $(i, j) \in \mathcal{C}_{\text{ML}}$ and $(j, k) \in \mathcal{C}_{\text{CL}}$, then a cannot-link constraint $(i, k)$ will be added.

Figs. 1, 2, 3, 4 show classification performance when *VB-LC* and *VB-ILC* are provided with $N_C$ randomly selected constraints, that is $N_C$ label constraints for *VB-LC* and $N_C$ instance-level constraints for *VB-ILC*. *VB - LC* and *VB -*

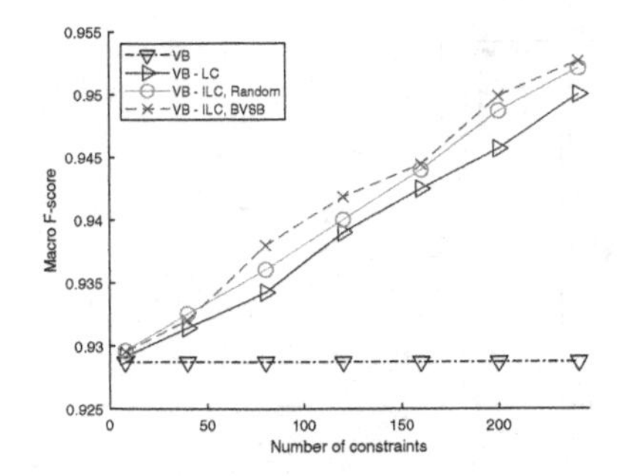


**Fig. 5.** Macro F-score for the RTE dataset, with random and uncertainty based constraint selection for *VB-ILC*

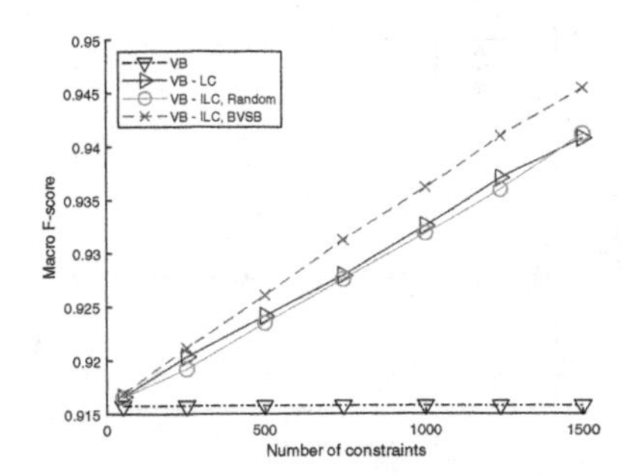


**Fig. 6.** Macro F-score for the Sentence polarity dataset, with random and uncertainty based constraint selection for *VB-ILC*

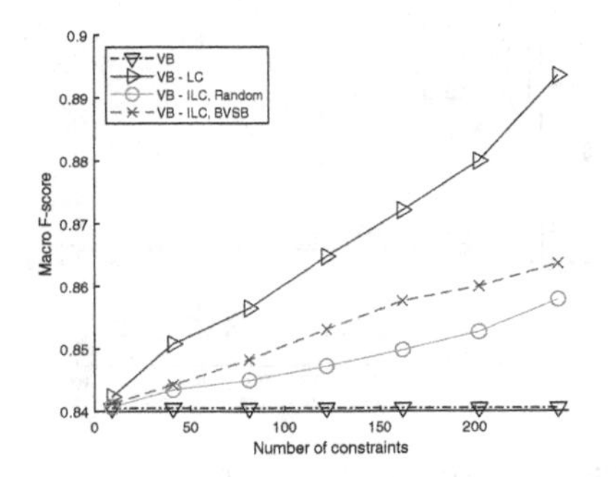


**Fig. 7.** Macro F-score for the Dog dataset, with random and uncertainty based constraint selection for *VB-ILC*

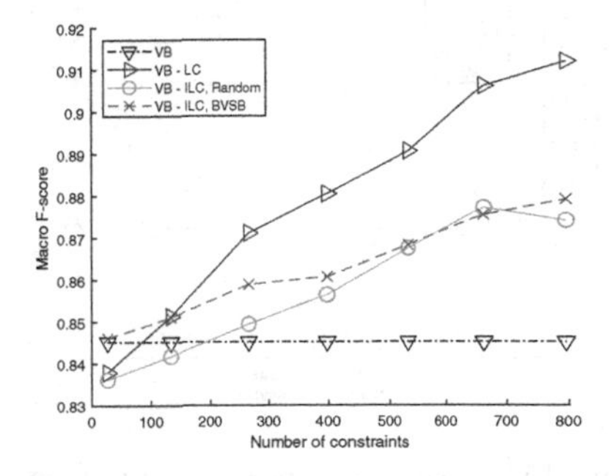


**Fig. 8.** Macro F-score for the Web dataset, with random and uncertainty based constraint selection for *VB-ILC*

*ILC* exhibit improved performance as the number of constraints increases. As expected, *VB - LC* outperforms *VB - ILC*, when $K > 2$, since label information is stronger than instance-level information. Interestingly, for $K = 2$, *VB-ILC* exhibits comparable performance to *VB-LC* in the Sentence polarity dataset, and outperforms *VB-LC* in the RTE dataset. Nevertheless, *VB-ILC* outperforms its unsupervised counterparts as $N_C$ increases, corroborating that even relatively weak constraints are useful. The performance of the constraint selection scheme of Sect. 4.3 is evaluated in Figs. 5, 6, 7, 8. The uncertainty sampling based variant of *VB - ILC* is denoted as *VB - ILC, BVSB*. Uncertainty sampling for selecting label constraints is clearly beneficial, as it provides performance gains compared to randomly selecting constraints in all considered datasets. Figures 9, 10, 11, 12 show the effect of providing *VB-ILC* with constraints derived from $N_C$ label constraints. When provided with $N_C$ label constraints from $K$ classes, and with $z_k$ denoting the proportion of constraints from class $k$, the resulting number of must-link constraints is $N_{\mathrm{ML}} = \sum_{k=1}^{K} \binom{N_C z_k}{2}$, since for each class every two points must be connected. The number of cannot-link constraints is

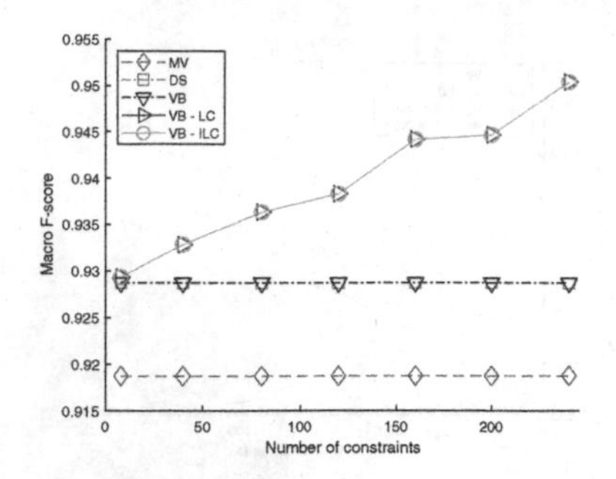


**Fig. 9.** Macro F-score for the RTE dataset, with label derived constraints for *VB-ILC*

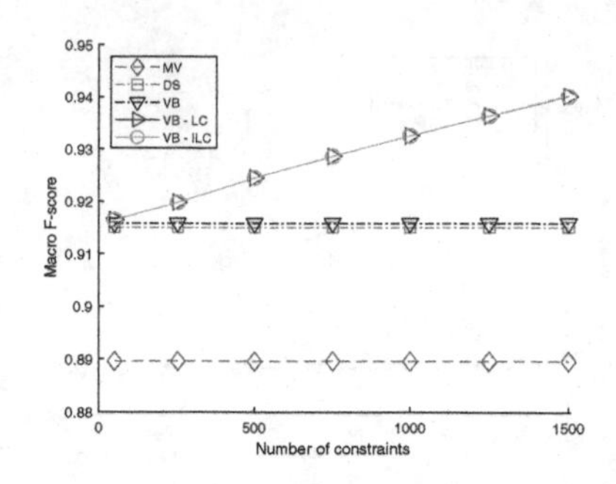


**Fig. 10.** Macro F-score for the Sentence Polarity dataset, with label derived constraints for *VB-ILC*

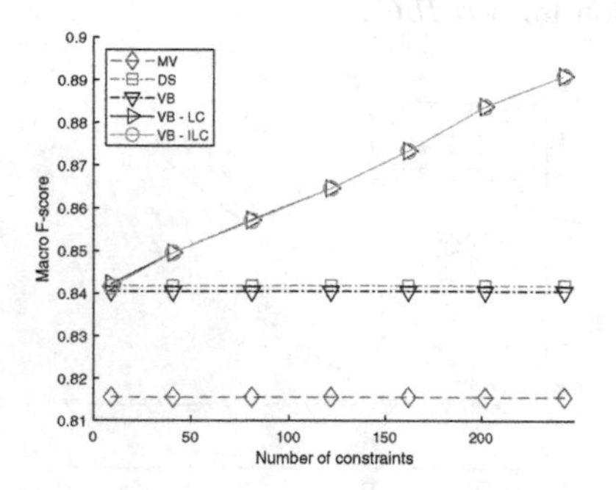


**Fig. 11.** Macro F-score for the Dog dataset, with label derived constraints for *VB-ILC*

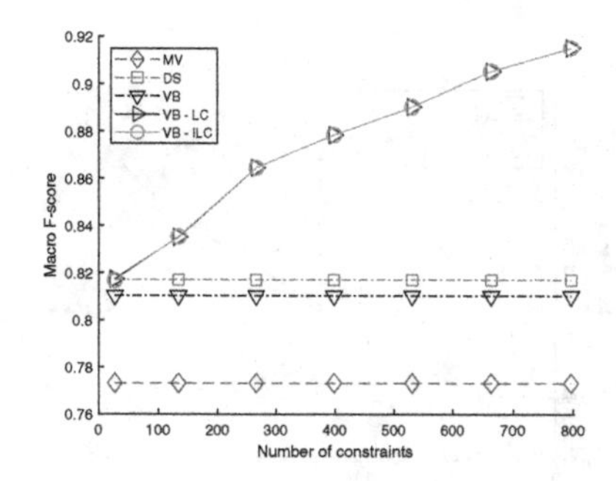


**Fig. 12.** Macro F-score for the Web dataset, with label derived constraints for *VB-ILC*

$N_{\text{CL}} = \sum_{k=1}^{K} \sum_{k'=1}^{K-1} N_C^2 z_k z_{k'+1}$, as for each class every point must be connected to all points belonging to other classes. In this case, *VB - ILC* almost matches the performance of *VB - LC*. This indicates that when provided with an adequate number of instance-level constraints, *VB - ILC* performs as well as if label constraints had been provided.

## 6 Conclusions

This paper investigated constrained crowdsourcing with pairwise constraints, that encode relationships between data. The performance of the proposed algorithm was analytically and empirically evaluated on popular real crowdsourcing datasets. Future research will involve distributed and online implementations of the proposed algorithm, and other types of constraints alongside semi-supervised crowdsourcing with dependent annotators, and dependent data.

## References


1. Albarqouni, S., Baur, C., Achilles, F., Belagiannis, V., Demirci, S., Navab, N.: Aggnet: deep learning from crowds for mitosis detection in breast cancer histology images. IEEE Trans. Med. Imaging **35**(5), 1313–1321 (2016). https://doi.org/10.1109/TMI.2016.2528120
2. Atarashi, K., Oyama, S., Kurihara, M.: Semi-supervised learning from crowds using deep generative models. In: Proceedings of the 32nd AAAI Conference on Artificial Intelligence (2018)
3. Basu, S., Davidson, I., Wagstaff, K.: Constrained Clustering: Advances in Algorithms, Theory, and Applications, 1 edn, Chapman & Hall/CRC, Boca Raton (2008)
4. Beal, M.J.: Variational Algorithms for Approximate Bayesian Inference. Ph.D. thesis, University of London, University College London (United Kingdom (2003)
5. Chapelle, O., Schlkopf, B., Zien, A.: Semi-Supervised Learning, 1st edn, The MIT Press, Cambridge (2010)
6. Cover, T.M., Thomas, J.A.: Elements of Information Theory. Wiley, New Delhi (2012)
7. Dawid, A.P., Skene, A.M.: Maximum likelihood estimation of observer error-rates using the EM algorithm. Appl. Stat. **28**(1), 20–28 (1979)
8. Deng, J., Dong, W., Socher, R., Li, L.J., Li, K., Fei-Fei, L: Imagenet: a large-scale hierarchical image database. In: Proceedings of IEEE Conference on Computer Vision and Pattern Recognition, pp. 248–255 (2009)
9. Fu, Y., Zhu, X., Li, B.: A survey on instance selection for active learning. J. Knowl. Inf. Syst. **35**(2), 249–283 (2013)
10. Hammersley, J.M., Clifford, P.E.: Markov random fields on finite graphs and lattices. Unpublished manuscript (1971)
11. Howe, J.: The rise of crowdsourcing. Wired Mag. **14**(6), 1–4 (2006)
12. Ibrahim, S., Fu, X., Kargas, N., Huang, K.: Crowdsourcing via pairwise co-occurrences: identifiability and algorithms. In: Advances in Neural Information Processing Systems, vol. 32, pp. 7847–7857. Curran Associates, Inc. (2019)
13. Jaffe, A., Nadler, B., Kluger, Y.: Estimating the accuracies of multiple classifiers without labeled data. In: AISTATS. vol. 2, p. 4. San Diego, CA (2015)
14. Kajino, H., Tsuboi, Y., Sato, I., Kashima, H.: Learning from crowds and experts. In: Workshops at the Twenty-Sixth AAAI Conference on Artificial Intelligence (2012)
15. Kim, H.C., Ghahramani, Z.: Bayesian classifier combination. In: Proceedings of Machine Learning Research, vol. 22, pp. 619–627. PMLR, La Palma, Canary Islands (21–23 April 2012)
16. Kittur, A., Chi, E.H., Suh, B.: Crowdsourcing user studies with mechanical turk. In: Proceedings of SIGCHI Conference on Human Factors in Computing Systems, pp. 453–456. ACM, Florence, Italy (2008)
17. Liu, M., Jiang, L., Liu, J., Wang, X., Zhu, J., Liu, S.: Improving learning-from-crowds through expert validation. In: Proceedings of the Twenty-Sixth International Joint Conference on Artificial Intelligence, IJCAI-17, pp. 2329–2336 (2017). https://doi.org/10.24963/ijcai.2017/324
18. MATLAB: version 9.7.0 (R2019b). The MathWorks Inc., Natick, Massachusetts (2019)
19. Moreno, P.G., Artés-Rodríguez, A., Teh, Y.W., Perez-Cruz, F.: Bayesian nonparametric crowdsourcing. J. Mach. Learn. Res. **16**(1), 1607–1627 (2015)

20. Nguyen, A.T., Wallace, B.C., Li, J.J., Nenkova, A., Lease, M.: Aggregating and predicting sequence labels from crowd annotations. Proc. Conf. Assoc. Comput. Linguist. Meet **2017**, 299–309 (2017)
21. Powers, D.M.W.: Evaluation: from precision, recall and f-measure to roc, informedness, markedness and correlation. Intl. J. Mach. Learn. Technol. **2**, 37–63 (2011)
22. Raykar, V.C., Yu, S., Zhao, L.H., Valadez, G.H., Florin, C., Bogoni, L., Moy, L.: Learning from crowds. J. Mach. Learn. Res. **11**, 1297–1322 (2010)
23. Rodrigues, F., Pereira, F., Ribeiro, B.: Sequence labeling with multiple annotators. Mach. Learn. **95**(2), 165–181 (2014)
24. Rodrigues, F., Pereira, F.: Deep learning from crowds. In: Proceedings of the AAAI Conference on Artificial Intelligence, vol. 32, no. 1 (April 2018)
25. Rodrigues, F., Pereira, F., Ribeiro, B.: Gaussian process classification and active learning with multiple annotators. In: Proceedings of the 31st International Conference on Machine Learning. Proceedings of Machine Learning Research, vol. 32, pp. 433–441. PMLR, Bejing, China (22–24 June 2014)
26. Ruiz, P., Morales-Álvarez, P., Molina, R., Katsaggelos, A.: Learning from crowds with variational gaussian processes. Pattern Recogn. **88**, 298–311 (2019)
27. Shi, W., Sheng, V.S., Li, X., Gu, B.: Semi-supervised multi-label learning from crowds via deep sequential generative model. In: Proceedings of the 26th ACM SIGKDD International Conference on Knowledge Discovery & Data Mining, KDD 2020, pp. 1141–1149. Association for Computing Machinery, New York (2020). https://doi.org/10.1145/3394486.3403167
28. Simpson, E., Roberts, S., Psorakis, I., Smith, A.: Dynamic bayesian combination of multiple imperfect classifiers. In: Guy, T., Karny, M., Wolpert, D. (eds.) Decision Making and Imperfection. Studies in Computational Intelligence, vol. 474, pp. 1–35. Springer, Berlin (2013). https://doi.org/10.1007/978-3-642-36406-8_1
29. Snow, R., O'Connor, B., Jurafsky, D., Ng, A.Y.: Cheap and fast–but is it good? evaluating non-expert annotations for natural language tasks. In: Proceedings of the Conference on Empirical Methods in Natural Language Processing, EMNLP 2008, pp. 254–263. Association for Computational Linguistics, USA (2008)
30. Tang, W., Lease, M.: Semi-supervised consensus labeling for crowdsourcing. In: ACM SIGIR Workshop on Crowdsourcing for Information Retrieval (CIR), 2011 (2011)
31. Traganitis, P.A., Pagès-Zamora, A., Giannakis, G.B.: Blind multiclass ensemble classification. IEEE Trans. Signal Process. **66**(18), 4737–4752 (2018). https://doi.org/10.1109/TSP.2018.2860562
32. Traganitis, P.A.: Blind ensemble classification of sequential data. In: IEEE Data Science Workshop, Minneapolis, MN (June 2019)
33. Traganitis, P.A., Giannakis, G.B.: Unsupervised ensemble classification with sequential and networked data. IEEE Trans. Knowl. Data Eng. (2020). https://doi.org/10.1109/TKDE.2020.3046645
34. Traganitis, P.A., Giannakis, G.B.: Bayesian Crowdsourcing with Constraints. CoRR abs/2012.11048 (2021). https://arxiv.org/abs/2012.11048
35. Venanzi, M., Guiver, J., Kazai, G., Kohli, P., Shokouhi, M.: Community-based bayesian aggregation models for crowdsourcing. In: Proceedings of 23rd International Conference on World Wide Web, pp. 155–164. ACM (April 2014)
36. Yan, Y., Rosales, R., Fung, G., Dy, J.: Modeling multiple annotator expertise in the semi-supervised learning scenario. In: Proceedings of the Twenty-Sixth Conference on Uncertainty in Artificial Intelligence, UAI 2010, pp. 674–682. AUAI Press, Arlington, Virginia, USA (2010)

37. Zhang, Y., Chen, X., Zhou, D., Jordan, M.I.: Spectral methods meet EM: a provably optimal algorithm for crowdsourcing. In: Advances in Neural Information Processing Systems, pp. 1260–1268 (2014)
38. Zhou, D., Basu, S., Mao, Y., Platt, J.C.: Learning from the wisdom of crowds by minimax entropy. In: Advances in Neural Information Processing Systems 25, pp. 2195–2203. Curran Associates, Inc. (2012)

# Text Mining and Natural Language Processing

# VOGUE: Answer Verbalization Through Multi-Task Learning

Endri Kacupaj[1(✉)], Shyamnath Premnadh[1], Kuldeep Singh[2], Jens Lehmann[1,3], and Maria Maleshkova[1]

[1] University of Bonn, Bonn, Germany
{kacupaj,jens.lehmann,maleshkova}@cs.uni-bonn.de, s6shprem@uni-bonn.de
[2] Zerotha Research and Cerence GmbH, Aachen, Germany
kuldeep.singh1@cerence.com
[3] Fraunhofer IAIS, Dresden, Germany
jens.lehmann@iais.fraunhofer.de

**Abstract.** In recent years, there have been significant developments in Question Answering over Knowledge Graphs (KGQA). Despite all the notable advancements, current KGQA systems only focus on answer generation techniques and not on answer verbalization. However, in real-world scenarios (e.g., voice assistants such as Alexa, Siri, etc.), users prefer verbalized answers instead of a generated response. This paper addresses the task of answer verbalization for (complex) question answering over knowledge graphs. In this context, we propose a multi-task-based answer verbalization framework: VOGUE (**V**erbalization thr**O**u**G**h m**U**lti-task l**E**arning). The VOGUE framework attempts to generate a verbalized answer using a hybrid approach through a multi-task learning paradigm. Our framework can generate results based on using questions and queries as inputs concurrently. VOGUE comprises four modules that are trained simultaneously through multi-task learning. We evaluate our framework on existing datasets for answer verbalization, and it outperforms all current baselines on both BLEU and METEOR scores.

**Keywords:** Answer verbalization · Question answering · Knowledge graphs · Multi-task learning · Natural language generation

## 1 Introduction

In recent years, publicly available knowledge graphs (KG) (e.g., DBpedia [18], Wikidata [31]) have been broadly adopted as a source of knowledge in several tasks such as entity linking, relation extraction, and question answering [14]. Question answering (QA) over knowledge graphs, in particular, is an essential task that maps a user's utterance to a query over a KG to retrieve the correct answer [26]. The initial knowledge graph question answering systems (KGQA) were mostly template- or rule-based systems with limited learnable modules [29].


N. Oliver et al. (Eds.): ECML PKDD 2021, LNAI 12977, pp. 563–579, 2021.
https://doi.org/10.1007/978-3-030-86523-8_34

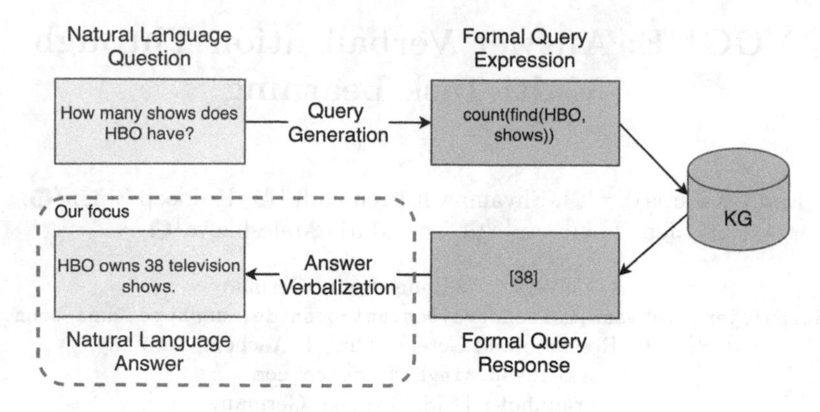


**Fig. 1.** A QA pipeline with integrated answer verbalization module. Our focus is the answer verbalization task as we assume logical form is generated by a QA system using the input question.

With the increasing popularity of intelligent personal assistants (e.g., Alexa, Siri), the research focus has been shifted to conversational question answering over KGs (ConvQA) that involve single-turn/multi-turn dialogues [13].

Existing open-source KGQA systems are restricted to only generating or producing answers without verbalizing them in natural language [10]. The lack of verbalization makes the interaction with the user not natural in contrast to voice assistants such as Siri and Alexa. Figure 1 depicts an ideal integration of a QA pipeline with answer verbalization. For instance, assuming that the answer to the exemplary question, "How many shows does HBO have?" is not known by the user. Suppose the QA system only responds with a number (e.g., 38) as an answer (similar to open-source KGQA systems), with no further explanation. In that case, the user might need to refer to an external data source to verify the answer. In an attempt to enable the users to verify the answer provided by a QA system, researchers employed techniques such as (i) revealing the generated formal query [9], (ii) graphical visualizations of the formal query [35] and (iii) verbalizing the formal query [8]. Understanding the necessity of verbalized answers in the KGQA domain, recently, several datasets have been proposed [4,15]. For the answer verbalization task, the system has to verbalize the answer to convey not only the information requested by the user but also additional characteristics that indicate how the answer was determined. In our exemplary question (from [13] dataset), a verbalized response would look like, "HBO owns 38 television shows." or "There are 38 TV shows whose owner is HBO.". Both answers allow the user to verify that the system retrieved the total number of TV shows owned by HBO. In the literature, there exist empirical results showing that answer verbalization quantitatively and qualitatively improves the ability to understand the answer [13]. However, it remains an open question – How can we verbalize an answer, given a logical form and an input question. With our work we address precisely this open and highly relevant research question.

In this paper, we propose VOGUE, the first approach dedicated to verbalize answers for KGQA. Our idea is to employ the question (user utterance) and the QA system-generated query as inputs. We refer to this strategy as "hybrid", since the final verbalized answer is produced using both the question and query concurrently. This work argues that leveraging content from both sources allows the model for better convergence and provides new, improved results. Furthermore, we complement our hypothesis with multi-task learning paradigms, since multi-task learning has been quite efficient for different system architectures [6], including question answering systems [14,25]. Our framework can receive two (e.g., question & query) or even one (e.g., question) input. It consists of four modules that are trained simultaneously to generate the verbalized answer. The first module employs a dual transformer-based encoder architecture for encoding the inputs. The second module determines whether the encoded inputs are relevant and decides if both will be used for verbalization. The third module consists of a cross-attention network that performs question and query matching by jointly modeling the relationships of question words and query actions. Finally, the last module employs a transformer decoder that is used to generate the final verbalization. Our work makes the following key contributions:

- We introduce the first multi-task-based hybrid answer verbalization framework that consists of four modules trained simultaneously.
- We propose a similarity threshold and cross attention modules to determine the relevance between the inputs and fuse information to employ a hybrid strategy.
- We provide an extensive evaluation and ablation study of the proposed framework on three QA datasets with answer verbalization. Our evaluation results establish a new baseline for answer verbalization task, which we believe will drive future research in a newly studied problem.

To facilitate reproducibility and reuse, our framework implementation is publicly available[1]. The structure of the paper is as follows: Sect. 2 summarizes the related work. Section 3 provides the task definition. Section 4 presents the proposed framework. Section 5 describes the experiments, results, ablation study and error analysis. We conclude in Sect. 6.

## 2 Related Work

As part of the related work we describe previous efforts and refer to different approaches from research fields, including task-oriented dialog systems, WebNLG, and KGQA systems.

A task-oriented dialogue system aims to help the user complete certain tasks in a specific domain (e.g. restaurant booking, weather query, or flight booking), making it valuable for real-world business. Typically, task-oriented dialogue systems are built on top of a structured ontology, which defines the tasks' domain

[1] https://github.com/endrikacupaj/VOGUE.

knowledge. Bordes et al. [5] formalized the task-oriented dialogue as a reading comprehension task regarding the dialogue history as context, user utterance as the question, and system response as the answer. In their work, authors utilized end-to-end memory networks for multi-turn inference. In [19], authors proposed a two-step seq2seq generation model, which bypassed the structured dialogue act representation and only retain the dialogue state representation. Kassawat et al. [16] proposed RNN-based end-to-end encoder-decoder architecture, which employs joint embeddings of the knowledge graph and the corpus as input. The model provides an additional integration of user intent and text generation, trained through a multi-task learning paradigm.

The WebNLG is a challenge that consists of mapping structured data to a textual representation. The dataset [11] contains data/text pairs where the data is a set of triples extracted from DBpedia, and the text is the verbalization of these triples. The dataset has been promoted for the development of 1) RDF verbalizers and 2) microplanners to handle a wide range of linguistic constructions. In our case, we only focus on related work pertaining to RDF verbalizers. Zhao et al. [34] propose DualEnc, a dual encoding model that can incorporate the graph structure and cater to the linear structure of the output text. Song et al. [27] proposes a graph-to-text approach that leverages richer training signals to guide the model for preserving input information. They introduce two types of autoencoding losses, each individually focusing on different aspects of input graphs. The losses are then back-propagated to calibrate the model via multi-task training. Liu et al. [20] propose an attention-based model, which mainly contains an entity extraction module and a relation detection module. The model devises a supervised multi-head self-attention mechanism as the relation detection module to learn the token-level correlation for each relation type separately.

The fields of task-oriented dialogue systems and WebNLG contain various approaches for generating text; nevertheless, none of them can be applied directly to solve answer verbalization for KGQA systems. Most task-oriented dialogue systems are designed and implemented to fit their corresponding task, and therefore they would not be suitable for open-domain knowledge graphs (e.g. Wikidata, DBpedia). Regarding WebNLG, the task only considers triples or graph structure data as input. In answer verbalization, the model input can be the question and/or the query. While the query can be translated into a graph structure, there is no support for textual information such as the question.

## 3 Task Definition

In this work, we target the problem of answer verbalization for KGQA. A semantic parsing-based QA system maps the question into executable logical forms and then executes it on a KG to produce the answer. For our task, given the question, the generated logical form, and the extracted answer, we aim to generate a natural language sentence, with the requirements that it is grammatically sound and correctly represents all the information in the question, logical form, and answer. Formally, let $X, Y$ denote the source-target pair. $X$ contains the set of

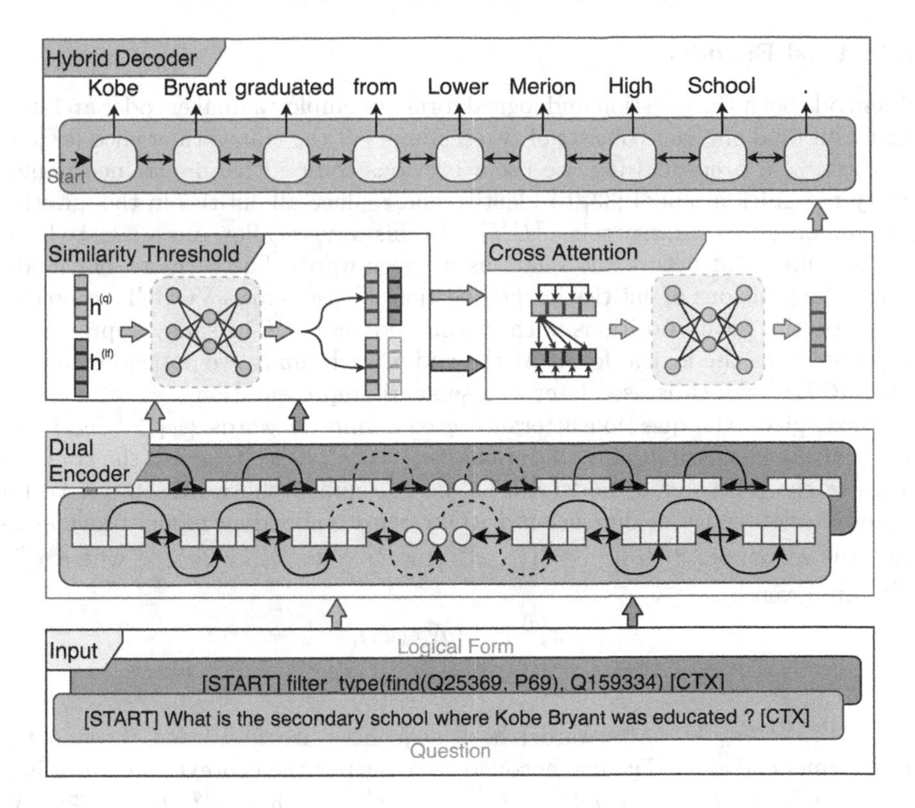


**Fig. 2.** VOGUE's architecture. It consists of four modules: 1) A dual encoder that is responsible to encode both inputs (question, logical form). 2) A similarity threshold module that determines whether the encoded inputs are relevant and determines if both will be used for verbalization. 3) A cross-attention module that performs question and query matching by jointly modeling the relationships of question words and query actions. 4) A hybrid decoder that generates the verbalized answer using the information of both question and logical form representations from the cross-attention module.

questions, logical forms, answers, and $Y$ corresponds to $y_1, y_2, ..., y_m$, which is the verbalized answer of $X$. The goal of the answer verbalization is to learn a distribution $p(Y|X)$ to generate natural language text describing the answer automatically.

## 4 Approach

In question answering, the input data consists of question $u$ and its answer $a$, extracted from the knowledge graph. The QA system will map the question to a logical form $z$ depending on the context. For answer verbalization, VOGUE maps the question, logical form, and answer to natural language sentence $s$. Figure 2 shows the architecture of VOGUE.

### 4.1 Dual Encoder

To encode both the question and logical form, we employ a dual encoder architecture. Our dual encoder consists of two instances of the transformer encoder [30].

First, as a preprocessing, we use a previous competitive pre-trained named entity recognition model [33] to identify and replace all entities in the question with a more general entity token $[ENT]$. In this way, we allow our model to focus on the sentence structure and relations between words. Furthermore, our model learns the positions of entities in the question. It also allows VOGUE to predict the respective entity positions in the verbalized answer. The same preprocessing step applies to the logical form. At the end of each input, we append a context token $[CTX]$, which is used later as a semantic representation.

Next, given the question utterance $q$ containing $n$ words $\{w_1, \ldots, w_n\}$ and the logical form $l$ containing $m$ actions $\{a_1, \ldots, a_m\}$, we tokenize the contexts and use the pre-trained model GloVe [24] to embed the words into a vector representation space of dimension $d^2$. Our word embedding model provides us with the sequences $x^{(q)} = \{x_1^{(q)}, \ldots, x_n^{(q)}\}$, $x^{(lf)} = \{x_1^{(lf)}, \ldots, x_m^{(lf)}\}$ where $x_i^{(q)}$, $x_i^{(lf)}$ are given by,

$$\begin{aligned} x_i^{(q)} &= GloVe(w_i), \\ x_i^{(lf)} &= GloVe(a_i), \end{aligned} \tag{1}$$

and $x_i^{(q)}, x_i^{(lf)} \in \mathbb{R}^d$. Afterwards, both sequences are forwarded through the transformer encoders. The two encoders here output the contextual embeddings $h^{(q)} = \{h_1^{(q)}, \ldots, h_n^{(q)}\}$ and $h^{(lf)} = \{h_1^{(lf)}, \ldots, h_m^{(lf)}\}$, where $h_i^{(q)}, h_i^{(lf)} \in \mathbb{R}^d$. We define this as:

$$\begin{aligned} h^{(q)} &= encoder_q(x_q; \theta^{(enc_q)}), \\ h^{(lf)} &= encoder_{lf}(x_{lf}; \theta^{(enc_{lf})}), \end{aligned} \tag{2}$$

where $\theta^{(enc_q)}$, $\theta^{(enc_{lf})}$ are the encoders trainable parameters.

### 4.2 Similarity Threshold

Given the encoded question utterance and logical form, VOGUE's second module is responsible for learning the relevance between the inputs and determining whether we will employ both for verbalization. This module is necessary when we want to utilize our framework alongside a question answering system. If we assume that the QA system is perfect and always produces correct logical forms, this module can be skipped. However, in a real-world scenario, QA systems are far from perfect. Therefore, we employ this module, which intends to identify the threshold for determining if two inputs are similar or not. The input here is the concatenation of the hidden states of the encoded question utterance $h^{(q)}$ and logical form $h^{(lf)}$. The module will perform binary classification on the vocabulary $V^{(st)} = \{0, 1\}$, where 0 indicates that there is no high relevance between the

[2] We employ the same dimension $d$ for all the representations, unless it is explicitly mentioned.

inputs, and only the question will be used for verbalization. While 1 allows us to use both and continue with the next module. Overall, our similarity threshold module is implemented using two linear layers, a Leaky ReLU activation function and a softmax for the predictions. Formally we define the module as:

$$\begin{aligned} h^{(st)} &= LeakyReLU(\boldsymbol{W}^{(st_1)}[h^{(q)}; h^{(lf)}]), \\ p^{(st)} &= softmax(\boldsymbol{W}^{(st_2)} h^{(st)}), \end{aligned} \tag{3}$$

where $\boldsymbol{W}^{(st_1)} \in \mathbb{R}^{d \times 2d}$ are the weights of the first linear layer and $h^{(st)}$ is the hidden state of the module. $\boldsymbol{W}^{(st_2)} \in \mathbb{R}^{|V^{(st)}| \times d}$ are the weights of the second linear layer, $|V^{(st)}|$ is the size of the vocabulary and $p^{(st)}$ denotes the probability distribution over the vocabulary indices.

### 4.3 Cross Attention

Inspired by recent computer vision research [22,32], we employ a cross-attention module that exploits relationships between the inputs and fuses information. The module here performs question and logical form matching by jointly modeling the relationships of question words and logical form actions. Our cross-attention approach is a variation of the self-attention mechanism [30]. In the self-attention mechanism the output is determined by a query and a set of key-value pairs. Given the stacked encoded question and logical form, $h^{(qlf)} = \binom{h^{(q)}}{h^{(lf)}} = \{h_1^{(q)}, \ldots, h_n^{(q)}; h_1^{(lf)}, \ldots, h_m^{(lf)}\}$, where $h^{(qlf)} \in \mathbb{R}^{2 \times d}$ we calculate the query and key-value pairs using three linear projections:

$$\begin{aligned} \boldsymbol{Q}^{(qlf)} &= \boldsymbol{W}^{(Q)} h^{(qlf)} = \begin{pmatrix} \boldsymbol{W}^{(Q)} h^{(q)} \\ \boldsymbol{W}^{(Q)} h^{(lf)} \end{pmatrix} = \begin{pmatrix} \boldsymbol{Q}^{(q)} \\ \boldsymbol{Q}^{(lf)} \end{pmatrix}, \\ \boldsymbol{K}^{(qlf)} &= \boldsymbol{W}^{(K)} h^{(qlf)} = \begin{pmatrix} \boldsymbol{W}^{(K)} h^{(q)} \\ \boldsymbol{W}^{(K)} h^{(lf)} \end{pmatrix} = \begin{pmatrix} \boldsymbol{K}^{(q)} \\ \boldsymbol{K}^{(lf)} \end{pmatrix}, \\ \boldsymbol{V}^{(qlf)} &= \boldsymbol{W}^{(V)} h^{(qlf)} = \begin{pmatrix} \boldsymbol{W}^{(V)} h^{(q)} \\ \boldsymbol{W}^{(V)} h^{(lf)} \end{pmatrix} = \begin{pmatrix} \boldsymbol{V}^{(q)} \\ \boldsymbol{V}^{(lf)} \end{pmatrix}, \end{aligned} \tag{4}$$

where $\boldsymbol{W}^{(Q)}, \boldsymbol{W}^{(K)}, \boldsymbol{W}^{(V)} \in \mathbb{R}^{d \times d}$ are the weights of the linear layers and $\boldsymbol{Q}^{(qlf)}, \boldsymbol{K}^{(qlf)}, \boldsymbol{V}^{(qlf)}$ are the query, key and value of the stacked question and logical form. Next, for calculating the cross-attention we simplify the "*Scaled Dot-Product Attention*" [30] step by removing the scaling factor and softmax.

We end-up calculating the attention of our input as:

$$\begin{aligned}
Attention(\boldsymbol{Q}^{(qlf)}, \boldsymbol{K}^{(qlf)}, \boldsymbol{V}^{(qlf)}) &= \boldsymbol{Q}^{(qlf)} \boldsymbol{K}^{(qlf)T} \cdot \boldsymbol{V}^{(qlf)} \\
&= \begin{pmatrix} \boldsymbol{Q}^{(q)} \\ \boldsymbol{Q}^{(lf)} \end{pmatrix} (\boldsymbol{K}^{(q)T} \boldsymbol{K}^{(lf)T}) \cdot \begin{pmatrix} \boldsymbol{V}^{(q)} \\ \boldsymbol{V}^{(lf)} \end{pmatrix} \\
&= \begin{pmatrix} \boldsymbol{Q}^{(q)} \boldsymbol{K}^{(q)T} & \boldsymbol{Q}^{(q)} \boldsymbol{K}^{(lf)T} \\ \boldsymbol{Q}^{(lf)} \boldsymbol{K}^{(q)T} & \boldsymbol{Q}^{(lf)} \boldsymbol{K}^{(lf)T} \end{pmatrix} \cdot \begin{pmatrix} \boldsymbol{V}^{(q)} \\ \boldsymbol{V}^{(lf)} \end{pmatrix} \\
&= \begin{pmatrix} \boldsymbol{Q}^{(q)} \boldsymbol{K}^{(q)T} \boldsymbol{V}^{(q)} + \boldsymbol{Q}^{(q)} \boldsymbol{K}^{(lf)T} \boldsymbol{V}^{(lf)} \\ \boldsymbol{Q}^{(lf)} \boldsymbol{K}^{(lf)T} \boldsymbol{V}^{(lf)} + \boldsymbol{Q}^{(lf)} \boldsymbol{K}^{(q)T} \boldsymbol{V}^{(q)} \end{pmatrix}.
\end{aligned} \tag{5}$$

When we calculate the cross-attention for the question, we also use the key-value pair from the logical form $(\boldsymbol{K}^{(lf)}, \boldsymbol{V}^{(lf)})$, the same applies when calculating the cross-attention for the logical form. After calculating the cross-attentions, we use the same steps as in the transformer to produce the new representations for our inputs. Finally, considering $h^{(qca)}, h^{(lfca)}$ the output representations of the cross-attention module for the question and logical form respectively, we concatenate them and forward them to the hybrid decoder module.

### 4.4 Hybrid Decoder

To translate the input question and logical form into a sequence of words (verbalized answer), we utilize a transformer decoder architecture [30], which employs the multi-head attention mechanism. The decoder will generate the final natural language answer. The output here is dependent on the cross-attention embedding $h^{(ca)}$. Here we define the decoder vocabulary as

$$V^{(dec)} = V^{(vt)} \cup \{ [START], [END], [ENT], [ANS] \}, \tag{6}$$

where $V^{(vt)}$ is the vocabulary with all the distinct tokens from our verbalizations. As we can see, the decoder vocabulary contains four additional helper tokes, where two of them ($[START]$, $[END]$) indicate when the decoding process starts and ends, while the other two ($[ENT]$, $[ANS]$) are used to specify the position of the entities and the answer on the final verbalized sequence. On top of the decoder stack, we employ a linear layer alongside a softmax to calculate each token's probability scores in the vocabulary. We define the decoder stack output as follows:

$$\begin{aligned}
h^{(dec)} &= decoder(h^{(ca)}; \theta^{(dec)}), \\
p_t^{(dec)} &= softmax(\boldsymbol{W}^{(dec)} h_t^{(dec)}),
\end{aligned} \tag{7}$$

where $h_t^{(dec)}$ is the hidden state in time step $t$, $\theta^{(dec)}$ are the decoder trainable parameters, $\boldsymbol{W}^{(dec)} \in \mathbb{R}^{|V^{(dec)}| \times 2d}$ are the linear layer weights, and $p_t^{(dec)} \in \mathbb{R}^{|V^{(dec)}|}$ is the probability distribution over the decoder vocabulary in time step $t$. The $|V^{(dec)}|$ denotes the decoder's vocabulary size.

**Table 1.** Dataset statistics, including the (average) number of tokens per question sentence, the (average) number of tokens per answer sentence and the vocabulary list size.

| Dataset | Train | Test | Question len. | Answer len. | Vocabulary |
|---|---|---|---|---|---|
| VQuAnDa | 4000 | 1000 | 12.27 | 16.95 | 10431 |
| ParaQA | 12637 | 3177 | 12.27 | 17.06 | 12755 |
| VANiLLa | 85729 | 21433 | 8.96 | 8.98 | 50505 |

### 4.5 Learning

The framework consists of four trainable modules. However, we apply a loss function only on two of them (similarity threshold and hybrid decoder). The dual encoder and cross-attention modules are trained based on the similarity threshold and hybrid decoder's signal. To account for multi-tasking, we perform a weighted average of all the single losses:

$$L = \lambda_1 L^{st} + \lambda_2 L^{dec}, \tag{8}$$

where $\lambda_1, \lambda_2$ are the relative weights learned during training considering the difference in magnitude between losses by consolidating the log standard deviation [1,6]. $L^{st}$ and $L^{dec}$ are the respective negative log-likelihood losses of the similarity threshold and hybrid decoder modules. These losses are defined as:

$$\begin{aligned} L^{st} &= -\sum_{j=1}^{2d} logp(y_j^{(st)}|x), \\ L^{dec} &= -\sum_{k=1}^{m} logp(y_k^{(dec)}|x), \end{aligned} \tag{9}$$

where $m$ is the length of the gold logical form. $y_j^{(st)} \in V^{(st)}$ are the gold labels for the similarity threshold and $y_k^{(dec)} \in V^{(dec)}$ are the gold labels for the decoder. The model benefits from each module's supervision signals, which improves the performance in the given task.

## 5 Experiments

### 5.1 Experimental Setup

**Datasets.** We perform experiments on three answer verbalization datasets (cf., Table 1). Below we provide a brief description of these:

- VQuAnDa [15] is the first QA dataset, which provides the verbalization of the answer in natural language. It contains 5000 "*complex*" questions with their SPARQL queries and answers verbalization. The dataset consists of 5042 entities and 615 relations.

- ParaQA [13] is a QA dataset with multiple paraphrased responses. The dataset was created using a semi-automated framework for generating diverse paraphrasing of the answers using techniques such as back-translation. It contains 5000 "*complex*" question-answer pairs with a minimum of two and a maximum of eight unique paraphrased responses for each question.
- VANiLLa [4] is a QA dataset that offers answers in natural language sentences. The answer sentences in this dataset are syntactically and semantically closer to the question than the triple fact. The dataset consists of over $100k$ "*simple*" questions.

**Model Configuration.** For simplicity, to represent the logical forms, we employ the same grammar as in [14]. Our approach can be used with any other grammar or even directly with SPARQL queries. However, we believe it is better to employ semantic grammar from a state-of-the-art QA model. To properly train the similarity threshold module, we had to introduce negative logical forms for each question. We did that by corrupting the gold logical forms, either by replacing a random action or finding another "*similar*" logical form from the dataset based on the Levenshtein distance. For all the modules in our framework, we employ an embedding dimension of 300. A transformer encoder and decoder having two layers and six heads for the multi-head attention model is used. We apply dropout [28] with a probability 0.1. For the optimization, we use the Noam optimizer proposed by [30]. The number of training parameters for VQuAnDa, ParaQA, and VANiLLa datasets are 12.9M, 14.9M, and 46.8M respectively.

**Model for Comparison.** We compare our framework with the four baselines that have been evaluated on the considered datasets. All baselines consist of sequence to sequence architectures, a family of machine learning approaches used for language processing and often used for natural language generation tasks. The first model consists of an RNN [21] based architecture, the second uses a convolutional network [12], the third employs a transformer network [30], while the last one uses pre-trained BERT [7] model. For a fair comparison with our framework, we report the baselines' results using the question and the logical form as separate inputs considering that baselines are limited to accept both inputs together.

**Evaluation Metrics.** We use the same metrics as employed by the authors of the three existing datasets [4,13,15] on the previously mentioned baselines. The BLEU score, as defined by [23], analyzes the co-occurrences of n-grams in the reference and the proposed responses. It computes the n-gram precision for the whole dataset, which is then multiplied by a brevity penalty to penalize short translations. We report results for BLEU-4. The METEOR score introduced

**Table 2.** Results on answer verbalization. VOGUE outperforms all existing baselines and achieves the new state of the art for both the BLEU and METEOR scores. The baseline experiment results are reported with two inputs: Question (**Q**) and gold Logical Form (**LF**), while VOGUE employs a Hybrid (**H**) approach.

| | BLEU | | | METEOR | | |
|---|---|---|---|---|---|---|
| Models | VQuAnDa | ParaQA | VANiLLa | VQuAnDa | ParaQA | VANiLLa |
| RNN [21] (**Q**) | 15.43 | 22.45 | 16.66 | 53.15 | 58.41 | 58.67 |
| RNN [21] (**LF**) | 20.19 | 26.36 | 16.45 | 57.06 | 61.87 | 55.34 |
| Convolutional [12] (**Q**) | 21.32 | 25.94 | 15.42 | 57.54 | 60.82 | 61.14 |
| Convolutional [12] (**LF**) | 26.02 | 31.89 | 16.89 | 64.30 | 65.85 | 58.72 |
| Transformer [30] (**Q**) | 18.37 | 23.61 | 30.80 | 56.83 | 59.63 | 62.16 |
| Transformer [30] (**LF**) | 23.18 | 28.01 | 28.12 | 60.17 | 63.75 | 59.01 |
| BERT [7] (**Q**) | 22.78 | 26.12 | 31.32 | 59.28 | 62.59 | 62.96 |
| BERT [7] (**LF**) | 26.48 | 30.31 | 30.11 | 65.92 | 65.92 | 59.27 |
| VOGUE (Ours) (**H**) | **28.76** | **32.05** | **35.46** | **67.21** | **68.85** | **65.04** |

by [2] is based on the harmonic mean of uni-gram precision and recall, with recall weighted higher than precision. Both metrics can be in the range of 0.0 and 100, with 100 being the best score.

### 5.2 Results

Table 2 summarizes the results comparing the VOGUE framework to the previous baselines for answer verbalization. VOGUE significantly outperforms the earlier baselines for both the BLEU and METEOR scores. While for the other baselines, we perform experiments with two different inputs (Question, gold Logical Form), VOGUE is the only one that directly uses both inputs (Hybrid). As we can see, for both datasets VQuAnDa and ParaQA, all baselines perform slightly worse when receiving the question as input compared to the gold logical form. This is due to the constant input pattern templates that the logical forms have. However, this does not apply to the VANiLLa dataset since it only contains simple questions. VOGUE achieves a BLEU score of 28.76 on VQuAnDa, which is 2 points higher than the second-best BERT (LF). The same applies to the METEOR score. Regarding ParaQA, VOGUE performs slightly better than the second Convolutional (LF) on BLEU score, while on METEOR score, the margin increases to 3 points. Finally, for the VANiLLa dataset, VOGUE performs considerably better compared to other baselines.

### 5.3 Ablation Study

***Integration with Semantic Parsing based QA system***
The logical forms used for the results in Table 2 are the gold ones, and therefore the performance of all baselines, including our framework, is boosted. In our first

**Table 3.** Results of the answer verbalization with a semantic parsing QA system. VOGUE still outperforms all baselines. For the baselines we employ only the question as input, while our framework employs the similarity threshold module to determine whether a hybrid verbalization can be performed.

| | BLEU | | | METEOR | | |
|---|---|---|---|---|---|---|
| Models | VQuAnDa | ParaQA | VANiLLa | VQuAnDa | ParaQA | VANiLLa |
| RNN [21] | 15.43 | 22.45 | 16.66 | 53.15 | 58.41 | 58.67 |
| Convolutional [12] | 21.32 | 25.94 | 15.42 | 57.54 | 60.82 | 61.14 |
| Transformer [30] | 18.37 | 23.61 | 30.80 | 56.83 | 59.63 | 62.16 |
| BERT [7] | 22.78 | 26.12 | 31.32 | 59.28 | 62.59 | 62.96 |
| VOGUE (Ours) | **25.76** | **28.42** | **33.14** | **64.61** | **67.52** | **63.69** |

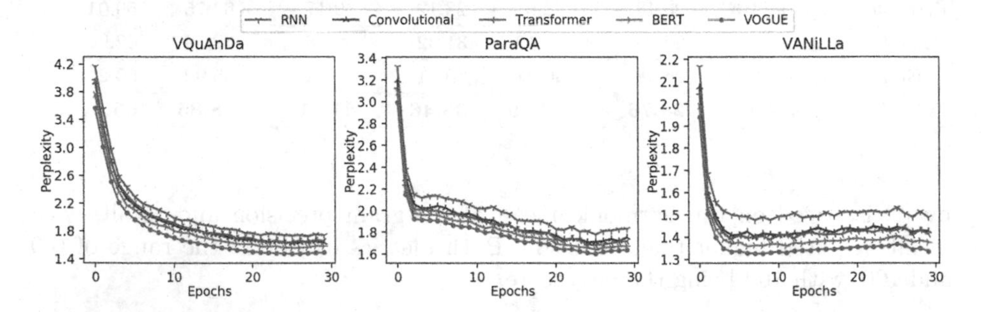


**Fig. 3.** Perplexity curves for all three answer verbalization datasets.

ablation study, we want to perform experiments in an end-to-end manner with a semantic parsing QA system, alongside the models, to understand our framework's superior performance. In this experiment, we train a simple, sequence-to-sequence-based semantic parser system to generate the logical forms by using the questions. As expected, the generated logical forms are not all correct, and therefore this affects the verbalization results. However, in Table 3, we can see that VOGUE still outperforms all baselines in this setting. An important role here plays the similarity threshold module, enabling a hybrid approach even in a real-world scenario. We can only use the question as input for the baselines since we do not have the gold logical forms. Here, it is also interesting that in two of the datasets, our framework outperforms the baselines with a more significant margin than before (c.f. Table 3, METEOR-VQuAnDa, METEOR-ParaQA). Finally, Fig. 3 illustrates the perplexity results, which show how well a probability distribution predicts a sample. A low perplexity indicates the probability distribution is good at predicting the sample. As we can see, our framework achieves the lowest perplexity values on all three datasets compared to other the baselines.

***Impact of Cross Attention and Multi-Task Learning***
Our second ablation experiment demonstrates the vitality of the cross-attention module and multi-task learning strategy. We first remove the cross-attention

**Table 4.** Ablation study results that indicate the effectiveness of cross attention and multi-task learning. The first row contains the results of the VOGUE framework when training all four modules with multi-task learning. The second and third rows selectively remove the cross attention and the multi-task learning from VOGUE. Best values in bold.

| | BLEU | | | METEOR | | |
|---|---|---|---|---|---|---|
| Ablation | VQuAnDa | ParaQA | VANiLLa | VQuAnDa | ParaQA | VANiLLa |
| Ours | **28.76** | **32.05** | **35.46** | **67.21** | **68.85** | **65.04** |
| w/o Cross Attention | 26.24 | 30.59 | 30.94 | 64.93 | 66.16 | 62.12 |
| w/o Multi-Task Learning | 25.74 | 28.15 | 29.07 | 62.31 | 63.84 | 61.49 |

**Table 5.** Similarity threshold f1-score results for each dataset.

| | F1-score | | |
|---|---|---|---|
| Module | VQuAnDa | ParaQA | VANiLLa |
| Similarity threshold | 64.73 | 58.55 | 98.76 |

module from our framework. Instead, we only concatenate the question and logical form to generate the verbalization. As observed in Table 4, we obtain worse results compared to the original configuration of VOGUE. A simple concatenation does not interchange any information between the question and the logical form, and therefore the results are expected to be lower. The cross-attention module is intentionally built to determine relevance between inputs by jointly modeling the relationship between the question words and logical form actions. Next, we train all modules independently and join them on inference to understand multi-task learning efficacy. As observed, our results have a negative impact when a multi-task learning strategy is not employed.

***Similarity Threshold Task Analysis***
Table 5 illustrates the performance of similarity threshold module. We observe that the module performs fairly well on VQuAnDa and ParaQA with f1 scores of 64.73 and 58.55, respectively. Both datasets contain complex questions. Hence, predicting the similarity between the question and the logical form is not easy. However, as long as the module's score is beyond 50, we are confident that using the similarity threshold module can improve our frameworks' answer verbalization results. For the VANiLLa dataset, the performance is incredibly high, with a score of 98.76. This is because the dataset contains only simple questions. Consequently, a single template pattern is employed for this dataset, and the module here has to predict if the logical form contains the correct triple relation. The task is much easier to perform compared to complex questions. Overall, the module results are pretty accurate and encourage us to apply them in our task.

### 5.4 Error Analysis

For the error analysis, we randomly sampled 100 incorrect predictions for human evaluation. We detail the reasons for two types of errors observed in the analysis:

**Words Mischoose.** A common error of VOGUE is mischoosing a word in the answer verbalization sentence. For instance, for the question "*Give me a count of everything owned by the network whose sister name is The CW?*" our framework generated the answer "*There are 156 television shows whose network's sister station is The CW.*". However, the gold reference here is "*There are 156 things whose network's sister name is The CW.*" As we can see, our framework misselected words in two parts of the sentence. The first one is the word "*things*", where it predicted "*television shows*". The second one is the word "*name*", where our model predicted "*station*". Both model predictions ("*television shows*", "*station*") are correlated, since they belong to the same context. Such errors do not heavily penalize the overall performance. For the example mentioned above, the BLEU and METEOR score is positive, with values 35.74 and 81.52, respectively.

**Factual Errors.** Another type of error of VOGUE is when it misses the semantic meaning and produces irrelevant results. It contributes to a major chunk of overall errors. There are two cases that can cause observed errors. The first one is the lack of reasoning for similar context data. When facing examples with the limited context in the dataset, the model would most definitely fail to reproduce the same context in the answer sentence. One can solve the issue by enriching the training data with other examples containing similar contexts. The second reason for having factual errors is when similarity threshold module fails to determine the inputs' relevance. As illustrated before, using the similarity threshold allows to successfully adopt a hybrid approach in a real-world scenario (*QA + Answer Verbalization*) and exceed any previous baseline performance.

## 6 Conclusions

The considered hypothesis in the paper was to study the impact of jointly utilizing the question and logical form on the answer verbalization task. We empirically observed that the proposed "hybrid" approach implemented in the VOGUE framework provides a flexibility to be deployed in a real world scenario where a QA system not always will produce the correct logical form. We systematically studied the impact of our choices in the proposed architecture. For instance, the ablation study demonstrates the effectiveness of multi-task learning and cross-attention module. Albeit effective, VOGUE is the first step towards a more extensive research agenda. Based on our observations, we submit the following open research questions in this domain: 1) KGs have been recently used as a source of external knowledge in the tasks such as entity and relation linking, which are also prominent for question answering [3]. It is yet to be studied if external knowledge from KGs may impact the answer verbalization. 2) There are empirical evaluations that for AI systems, the explanations regarding the retrieved answers improve trustworthiness, especially in wrong prediction [17]. Hence, how an answer verbalization can be explained remains an important open research direction. 3) In our work, we focused on English as an underlying language, and a multi-lingual approach is the next viable step.

## References


1. Armitage, J., Kacupaj, E., Tahmasebzadeh, G., Maleshkova, M.S., Ewerth, R., Lehmann, J.: Mlm: a benchmark dataset for multitask learning with multiple languages and modalities. In: 29th ACM CIKM. ACM (2020)
2. Banerjee, S., Lavie, A.: METEOR: an automatic metric for MT evaluation with improved correlation with human judgments. In: Proceedings of the ACL Workshop on Intrinsic and Extrinsic Evaluation Measures for Machine Translation and/or Summarization. ACL (2005)
3. Bastos, A., et al.: Recon: Relation extraction using knowledge graph context in a graph neural network. In: Proceedings of The Web Conference (WWW), p. N/A (2021)
4. Biswas, D., Dubey, M., Rashad Al Hasan Rony, M., Lehmann, J.: VANiLLa: Verbalized Answers in Natural Language at Large Scale. arXiv e-prints arXiv:2105.11407 (2021)
5. Bordes, A., Boureau, Y.L., Weston, J.: Learning end-to-end goal-oriented dialog. In: 5th ICLR, 2017 (2017)
6. Cipolla, R., Gal, Y., Kendall, A.: Multi-task learning using uncertainty to weigh losses for scene geometry and semantics. In: CVPR (2018)
7. Devlin, J., Chang, M.W., Lee, K., Toutanova, K.: BERT: pre-training of deep bidirectional transformers for language understanding. In: NAACL. ACL (2019)
8. Ell, B., Harth, A., Simperl, E.: Sparql query verbalization for explaining semantic search engine queries. In: ESWC (2014)
9. Ferré, S.: Sparklis: an expressive query builder for sparql endpoints with guidance in natural language. In: Semantic Web (2017)
10. Fu, B., Qiu, Y., Tang, C., Li, Y., Yu, H., Sun, J.: A survey on complex question answering over knowledge base: Recent advances and challenges. arXiv preprint arXiv:2007.13069 (2020)
11. Gardent, C., Shimorina, A., Narayan, S., Perez-Beltrachini, L.: Creating training corpora for NLG micro-planners. In: 55th ACL. ACL (2017)
12. Gehring, J., Auli, M., Grangier, D., Yarats, D., Dauphin, Y.N.: Convolutional sequence to sequence learning. In: 34th ICML (2017)
13. Kacupaj, E., Banerjee, B., Singh, K., Lehmann, J.: Paraqa: a question answering dataset with paraphrase responses for single-turn conversation. In: Eighteenth ESWC (2021)
14. Kacupaj, E., Plepi, J., Singh, K., Thakkar, H., Lehmann, J., Maleshkova, M.: Conversational question answering over knowledge graphs with transformer and graph attention networks. In: The 16th Conference of the European Chapter of the Association for Computational Linguistics (2021)

15. Kacupaj, E., Zafar, H., Lehmann, J., Maleshkova, M.: VQuAnDa: verbalization question answering dataset. In: Harth, A., Kirrane, S., Ngonga Ngomo, A.-C., Paulheim, H., Rula, A., Gentile, A.L., Haase, P., Cochez, M. (eds.) ESWC 2020. LNCS, vol. 12123, pp. 531–547. Springer, Cham (2020). https://doi.org/10.1007/978-3-030-49461-2_31
16. Kassawat, F., Chaudhuri, D., Lehmann, J.: Incorporating joint embeddings into goal-oriented dialogues with multi-task learning. In: Hitzler, P., Fernández, M., Janowicz, K., Zaveri, A., Gray, A.J.G., Lopez, V., Haller, A., Hammar, K. (eds.) ESWC 2019. LNCS, vol. 11503, pp. 225–239. Springer, Cham (2019). https://doi.org/10.1007/978-3-030-21348-0_15
17. Kouki, P., Schaffer, J., Pujara, J., O'Donovan, J., Getoor, L.: User preferences for hybrid explanations. In: Proceedings of the Eleventh ACM Conference on Recommender Systems, pp. 84–88 (2017)
18. Lehmann, J., et al.: Dbpedia - a large-scale, multilingual knowledge base extracted from wikipedia. In: Semantic Web (2015)
19. Lei, W., Jin, X., Kan, M.Y., Ren, Z., He, X., Yin, D.: Sequicity: simplifying task-oriented dialogue systems with single sequence-to-sequence architectures. In: 56th ACL. ACL (2018)
20. Liu, J., Chen, S., Wang, B., Zhang, J., Li, N., Xu, T.: Attention as relation: learning supervised multi-head self-attention for relation extraction. In: IJCAI-20. IJCAI (2020)
21. Luong, T., Pham, H., Manning, C.D.: Effective approaches to attention-based neural machine translation. In: EMNLP. ACL (2015)
22. Mohla, S., Pande, S., Banerjee, B., Chaudhuri, S.: Fusatnet: dual attention based spectrospatial multimodal fusion network for hyperspectral and lidar classification. In: 2020 IEEE/CVF Conference on CVPRW (2020)
23. Papineni, K., Roukos, S., Ward, T., Zhu, W.J.: Bleu: a method for automatic evaluation of machine translation. In: 40th ACL (2002)
24. Pennington, J., Socher, R., Manning, C.: Glove: global vectors for word representation. In: EMNLP. ACL (2014)
25. Plepi, J., Kacupaj, E., Singh, K., Thakkar, H., Lehmann, J.: Context transformer with stacked pointer networks for conversational question answering over knowledge graphs. In: Eighteenth ESWC (2021)
26. Singh, K., et al.: Why reinvent the wheel: let's build question answering systems together. In: Proceedings of the 2018 World Wide Web Conference (2018)
27. Song, L., et al.: Structural information preserving for graph-to-text generation. In: 58th ACL. ACL (2020)
28. Srivastava, N., Hinton, G., Krizhevsky, A., Sutskever, I., Salakhutdinov, R.: Dropout: a simple way to prevent neural networks from overfitting. J. Mach. Learn. Res. **15**(1), 1929–1958 (2014)
29. Unger, C., Bühmann, L., Lehmann, J., Ngonga Ngomo, A.C., Gerber, D., Cimiano, P.: Template-based question answering over rdf data. In: Proceedings of the 21st International Conference on World Wide Web (2012)
30. Vaswani, A., et al.: Attention is all you need. In: NIPS (2017)
31. Vrandečić, D., Krötzsch, M.: Wikidata: A free collaborative knowledgebase. ACM, Commun (2014)
32. Wei, X., Zhang, T., Li, Y., Zhang, Y., Wu, F.: Multi-modality cross attention network for image and sentence matching. In: 2020 IEEE/CVF Conference on CVPR (2020)

33. Yamada, I., Asai, A., Shindo, H., Takeda, H., Matsumoto, Y.: LUKE: deep contextualized entity representations with entity-aware self-attention. In: EMNLP. ACL (2020)
34. Zhao, C., Walker, M., Chaturvedi, S.: Bridging the structural gap between encoding and decoding for data-to-text generation. In: 58th ACL. ACL (2020)
35. Zheng, W., Cheng, H., Zou, L., Yu, J.X., Zhao, K.: Natural language question/answering: let users talk with the knowledge graph. In: 2017 ACM CIKM (2017)

# NA-Aware Machine Reading Comprehension for Document-Level Relation Extraction

Zhenyu Zhang[1,2], Bowen Yu[1,2], Xiaobo Shu[1,2], and Tingwen Liu[1,2](✉)

[1] Institute of Information Engineering, Chinese Academy of Sciences, Beijing, China
{zhangzhenyu1996,yubowen,shuxiaobo,liutingwen}@iie.ac.cn
[2] School of Cyber Security, University of Chinese Academy of Sciences, Beijing, China

**Abstract.** Document-level relation extraction aims to identify semantic relations between target entities from the document.Most of the existing work roughly treats the document as a long sequence and produces target-agnostic representation for relation prediction, limiting the model's ability to focus on the relevant context of target entities. In this paper, we reformulate the document-level relation extraction task and propose a NA-aware machine Reading Comprehension (NARC) model to tackle this problem. Specifically, the input sequence formulated as the concatenation of a head entity and a document is fed into the encoder to obtain comprehensive target-aware representations for each entity. In this way, the relation extraction task is converted into a reading comprehension problem by taking all the tail entities as candidate answers. Then, we add an artificial answer `NO-ANSWER` (NA) for each query and dynamically generate a NA score based on the decomposition and composition of all candidate tail entity features, which finally weighs the prediction results to alleviate the negative effect of having too many no-answer instances after task reformulation. Experimental results on DocRED with extensive analysis demonstrate the effectiveness of NARC.

**Keywords:** Document-level relation extraction · Machine reading comprehension · No-answer query

## 1 Introduction

Reading text to identify and extract relational facts in the form of (*head entity, relation, tail entity*) is one of the fundamental tasks in data mining and natural language processing. For quite some time, researchers mainly focus on extracting facts from a sentence, i.e., sentence-level relation extraction [8,34,35], However, such an ideal setting makes it powerless to handle a large number of inter-sentence relational triples in reality. To move relation extraction forward from sentence-level to document-level, the DocRED dataset is proposed recently [31], in which each document is annotated with a set of named entities and relations.


N. Oliver et al. (Eds.): ECML PKDD 2021, LNAI 12977, pp. 580–595, 2021.
https://doi.org/10.1007/978-3-030-86523-8_35

> **One Love (Blue album)**
> [1] *One Love* is the second studio album by English boy band *Blue*, released on *4 November 2002* in the *United Kingdom* and on *21 October 2003* in the *United States*. [2] The album peaked at number one on the *UK Albums Chart*, where it stayed for one week. On *20 December 2003* it was certified *4 ×Platinum* in the UK. ... [4] Three singles were released from the album: "*One Love*", which peaked at number three, "*Sorry Seems to Be the Hardest Word*", featuring *Elton John*, which peaked at number one, and "*U Make Me Wanna*", which peaked at number four.

**Subject:** *One Love, Sorry Seems to Be the Hardest Word, U Make Me Wanna*
**Object:** *4 November 2002, 21 October 2003*
**Relation:** *publication date*

**Subject:** *One Love, Sorry Seems to Be the Hardest Word, U Make Me Wanna*
**Object:** *Blue*
**Relation:** *performer*

**Fig. 1.** An example from DocRED. Word spans with the same color indicate the same named entity, and the key clues for relation inference are underlined.

In Fig. 1, we show an example in DocRED development set to illustrate the challenging yet practical extension: for the extraction of relational fact (`U Make Me Wanna`, `performer`, `Blue`), one has to first identify the fact that `U Make Me Wanna` is a music single in `One Love` from sentence 4, then identify the facts `One Love` is an album by `Blue` from sentence 1, and finally infer from these facts that the `performer` of `U Make Me Wanna` is `Blue`.

In recent times, there are considerable efforts devoted to document-level relation extraction. Some popular techniques in sentence-level relation extraction (e.g., attention mechanism, graph neural networks, and pre-trained language models) are introduced and make remarkable improvements [17,24,32]. Specifically, most of them take the document as a long sequence and generate target-agnostic representations, then perform relation classification for each entity pair. Despite the great success, we argue that learning general representation is sub-optimal for extracting relations between specific target entities from the long document, since some target-irrelevant words could introduce noise and cause confusion to the relation prediction.

Inspired by the current trend of formalizing NLP problems as machine reading comprehension (MRC) style tasks [7,13,30], we propose NARC, a NA-aware MRC model, to address this issue. Instead of treating document-level relation extraction as a simple entity pair classification problem, NARC first formulates it as a MRC task by taking the given head entity as query and all tail entities in the document as candidate answers, then performing relation classification for each candidate tail entity. Specifically, the input sequence of NARC is organized as the concatenation of the head entity and the document in the form of "`[CLS]+ Head Entity + [SEP] + DOCUMENT + [SEP]`", and then fed into the query-context encoder, which is made up of a pre-trained language model followed by a simple entity graph. The former serves the target-aware context encoding, while the later is constructed to perform multi-hop reasoning.

However, one barrier in such task formulation is the troublesome No-Answer (NA) problem. Considering a document with $n$ entities, MRC-style formulation requires $n$ enumerations for a complete extraction, in which many queries have no correct answer (65.5% in DocRED) since there are a great number of

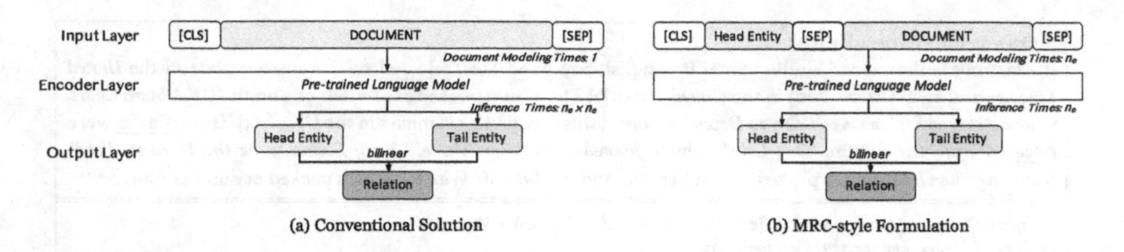


**Fig. 2.** Diagrams of the (a) conventional and (b) MRC-style paradigms for document-level relation extraction with pre-trained language models.

entity pairs in a document that do not hold pre-defined relations. To fill this gap, we append a special candidate `NO-ANSWER` for each query. As a result, the number of queries pointing to `NO-ANSWER` often exceeds that of other queries with valid answers, causing extremely imbalanced data distribution. To mitigate this adverse effect, we introduce a novel answer vector assembler module after task reformulation, which firstly integrates features from different layers of the encoder as the final representation of each entity, then vectorizes the human-made candidate `NO-ANSWER` with a decomposition-composition strategy, where each candidate tail entity vector is first decomposed into the relevant and irrelevant components with respect to the head entity, and then composed to a query-specific NA vector. Finally, this vector is projected into a NA score, which weighs the predicted relation scores to take the probability of `NO-ANSWER` into account (Fig. 2).

Experiments conducted on DocRED, the largest public document-level relation extraction dataset, show that the proposed NARC model achieves superior performance over previous competing approaches. Extensive validation studies demonstrate the effectiveness of our MRC-style task formulation and the NA-aware learning strategy.

## 2 Task Formulation

In this section, we first briefly recall some basic concepts and classic baselines for document-level relation extraction, and then describe the task transformation.

Formally, given a document $\mathcal{D} = \{w_i\}_{i=1}^{n_w}$ and its entity set $\mathcal{E} = \{e_i\}_{i=1}^{n_e}$, where $e_i$ is the $i$-th entity with $n_m^i$ mentions $\mathcal{M}_i = \{m_j\}_{j=1}^{n_m^i}$. The goal of document-level relation extraction is to predict all the relations $\mathcal{R}' \in \mathcal{R} = \{r^i\}_{i=1}^{n_r}$ between every possible entity pair. Named entity mentions corresponding to the same entity have been assigned with the same entity id in the annotation. Considering that many relational facts express in multiple sentences, the document-level task is more complicated than the traditional sentence-level task. The model is expected to have a powerful ability to extract relational evidence from the long text and eliminate the interference of noise information.

**Conventional Solution.** Previous work usually takes the document as a long sequence, and converts it into hidden states with kinds of encoders. Typically, Wang et al. [26] packed the input sequence to "`[CLS] + DOCUMENT + [SEP]`"

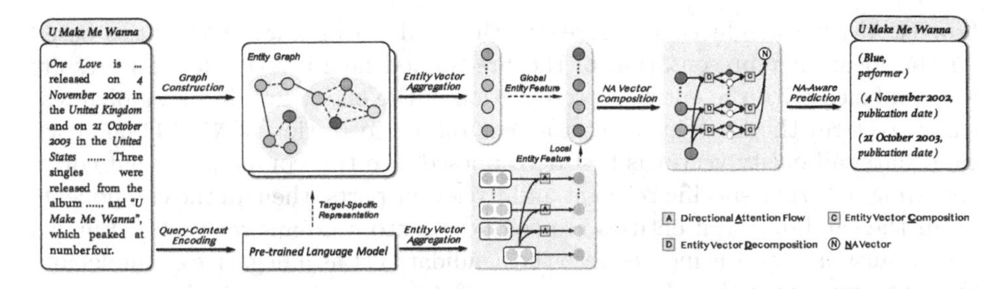


**Fig. 3.** Overview of the NARC model. The model receives a document and a target *head entity* at a time, and outputs all the related (*tail entity, relation*) pairs. Here we take *U Make Me Wanna* (symbols with green color) as an example. (Color figure online)

and employed BERT [6] as the encoder. Then, the document-level relation extraction task is treated as a multi-label classification problem. For each target entity pair, it gathers entity representations from the hidden states and employs the Sigmoid function to compute the probability of each relation. Obviously, this practice encodes the document once to produce target-agnostic representations, and the classification enumerates all possible entity pairs by $n_e \times n_e$ times.

**MRC-style Formulation.** Different from the conventional solution, we propose that document-level relation extraction can be formulated as a MRC problem, in which the model is expected to answer: "*which entities in the document have what relations with the target head entity?*". Under such formulation, the input sequence is modified to "`[CLS] + Head Entity + [SEP] + DOCUMENT + [SEP]`", then pre-trained language models is able to output target-specific representations, which have benefits in filtering irrelevant information of the target entity pair as revealed in the experiments. In this paradigm, the times of document modeling and classification are both $n_e$. However, the enumeration of head entities inevitably introduces a number of No-Answer (NA) queries since many entity pairs in the document do not hold pre-defined relations, which act as negative samples and will damage the model performance due to the data imbalance. To be compatible with the unforeseen situation, we add a special candidate `NO-ANSWER` for all instances. In the end, how to solve the no-answer problem becomes the key barrier of applying MRC-style formulation into the document-level relation extraction task.

# 3 NA-aware MRC (NARC)

This section provides NARC in detail. It formulates document-level relation extraction as a machine reading comprehension problem based on a query-context encoder, and solves the no-answer (NA) issue with the answer vector assembler and NA-aware predictor. As shown in Fig. 3, we first feed (*head entity, document*) pairs into a pre-trained language model, then the vectorized document tokens pass through a stacked entity graph to derive semantic evidence from the document and enable multi-hop reasoning. Next, the directional attention flow (DAF) is introduced to aggregate the local features for tail entities based on the mention representation of the pre-trained language model. The results are combined with the entity representation in the entity graph (i.e., global features) to form the final entity vector. For the vectorization of NO-ANSWER, each candidate tail entity vector is first decomposed into two components that corresponding to target-specific relevant and irrelevant parts, then all the components of all the candidate tail entities are composed into a no-answer vector. Finally, the no-answer vector is merged into the candidate list as a negative example, and the no-answer score is calculated based on the vector to weigh the prediction. In this way, the model could induce low confidence when there is no valid tail entity due to the dominance of irrelevant components in the no-answer vector.

### 3.1 Query-Context Encoder

Following the MRC-style formulation, the document (*context*) is concatenated with the head entity (*query*) and fed into a pre-trained language model. By introducing such a packed sequence, advanced pre-trained language models such as BERT [6] can encode the document in a query-aware manner owning to the sufficiently deep self-attention architectures. Beyond that, the great success of integrating graph neural networks with pre-trained language models makes it a popular document encoding structure in natural language processing. Here we directly borrow the representative model Entity Graph [5] from multi-hop MRC to achieve global entity features, where mentions of entities are regarded as nodes in the graph while edges encode relations between different mentions (e.g., within- and cross-sentence coreference links or simply co-occurrence in a sentence). Then, the relational graph convolutional networks (R-GCN [20]) are applied to the graph and trained to perform multi-hop relation reasoning[1].

### 3.2 Answer Vector Assembler

It is intuitive that the global entity features obtained from Entity Graph could be treated as the final representations for relation prediction. However, it may fail to effectively exploit the local contextual interaction between mentions. To assemble comprehensive representations vectors for entities and the man-made option NO-ANSWER, we propose the entity vector aggregation and NA vector composition modules in this section.

**Entity Vector Aggregation.** In a document, one entity could be mentioned multiple times, and these mentions are the exact elements involved in relation expression and reasoning. To capture such local features, we extract all mention-level representations for each entity from the output of pre-trained language models. Apparently, the importance varies among different tail entity mentions for the target head entity. Thus we introduce directional attention flow (DAF), a variety of BiDAF [21], to measure the difference and compress the mention features into an embedding for each candidate tail entity.

[1] For more details about the construction process of Entity Graph, we recommend readers to reference the original paper [5].

Given the head entity $e_h$ and a candidate tail entity $e_t$, the similarity matrix $\mathbf{S}_{ht} \in \mathbb{R}^{n_m^h \times n_m^t}$ is first calculated by

$$\mathbf{S} = avg_{-1}\mathcal{F}_s([\mathbf{M}_h; \mathbf{M}_t; \mathbf{M}_h \odot \mathbf{M}_t]), \tag{1}$$

where $\mathbf{M}_h \in \mathbb{R}^{n_m^h \times d}$ and $\mathbf{M}_t \in \mathbb{R}^{n_m^t \times d}$ are the mention feature matrixes for these two entities, in which each mention feature is generated by the mean-pooling over corresponding word embeddings. $\mathcal{F}_s$ is a linear transformation, $avg_{-1}$ stands for the average operation in the last dimension. Next, we design the head-to-tail attention matrix $\mathbf{M}_{h2t} \in \mathbb{R}^{n_m^t \times d}$, which signifies the tail mentions that are most related to each mention in the head entity, via

$$\mathbf{M}_{h2t} = dup(softmax(max_{col}(\mathbf{S})))^\top \mathbf{M}_t, \tag{2}$$

where $max_{col}$ is the maximum function applied on across column of a matrix, which transforms $\mathbf{S}_{ht}$ into $\mathbb{R}^{1 \times n_m^t}$. Then the *dup* function duplicates it for $n_m^h$ times into shape $\mathbb{R}^{n_m^h \times n_m^t}$.

The output of DAF is the head mention feature matrix $\mathbf{M}_h$ and head-to-tail attention matrix $\mathbf{M}_{h2t}$. Finally, we utilize mean-pooling to obtain local entity features and concatenate them with the global entity features $\mathbf{e}^G \in \mathbb{R}^d$ generated by the entity graph in query-context encoder:

$$\mathbf{e}_h = [mean(\mathbf{M}_h); \mathbf{e}_h^G], \quad \mathbf{e}_t = [mean(\mathbf{M}_{h2t}); \mathbf{e}_t^G]. \tag{3}$$

**NA Vector Composition.** Different from other candidate answers that point to specific entities, "NO-ANSWER" (NA) is a man-made option without corresponding representation. To meet this challenge, we assume that each candidate entity vector could be decomposed into relevant and irrelevant parts with respect to the target head entity and later composited to derive the NA vector based on all candidate tail entities. In other words, every candidate tail entity contributes to the vectorization of NO-ANSWER. The key intuition behind this is that NO-ANSWER could be regarded as an option similar to the none-of-the-above in multiple-choice questions, only after comprehensively considering all other candidate answers can one make such a choice.

Formally, based on the final representation of given head entity $\mathbf{e}_h \in \mathbb{R}^{2d}$, each candidate tail entity vector $\mathbf{e}_t \in \mathbb{R}^{2d}$ is expected to be decomposed into a relevant part $\mathbf{e}_t^+ \in \mathbb{R}^{2d}$ and an irrelevant part $\mathbf{e}_t^- \in \mathbb{R}^{2d}$. Here we adapt the linear decomposition strategy proposed in sentence similarity learning [28] to meet this demand:

$$\mathbf{e}_t^+ = \frac{\mathbf{e}_h^\top \mathbf{e}_t}{\mathbf{e}_t^\top \mathbf{e}_t}\mathbf{e}_t, \quad \mathbf{e}_t^- = \mathbf{e}_t - \mathbf{e}_t^+. \tag{4}$$

The motivation here is that the more similar between $\mathbf{e}_h$ and $\mathbf{e}_t$, the higher the correlation between the head entity and the candidate tail entity, thus the higher proportion of $\mathbf{e}_t$ should be assigned to the similar component. In the composition step, we extract features from both the relevant matrix and the irrelevant matrix for each tail entity as follows:

$$\mathbf{e}_t^n = tanh(\mathbf{W}_{cr}\mathbf{e}_t^+ + \mathbf{W}_{ci}\mathbf{e}_t^- + b_c). \tag{5}$$

where $\mathbf{W}_{cr/ci} \in \mathbb{R}^{2d\times 2d}$ and $b_c \in \mathbb{R}^{2d}$ are trainable weight matrix and bias vector respectively. Afterwards, we apply a max-pooling over all candidate tail entities to obtain the representation of NO-ANSWER:

$$\mathbf{n} = max\{\mathbf{e}_t^n\}_{t=1}^{n_e}. \tag{6}$$

### 3.3 NA-Aware Predictor

In the prediction stage, we hope that the model has a preliminary perception about whether there is a valid answer to the query, then give the final relation prediction based on the perception. Moreover, NO-ANSWER is regarded as a special candidate entity, which takes $\mathbf{n} \in \mathbb{R}^{2d}$ as representation, thus introducing additional negative examples to guide the model optimization.

Specifically, we pass the NA vector through a linear transformation $\mathcal{F}_n$ followed by a sigmoid function $\delta$ to obtain a score that points to NO-ANSWER for the given query: $s_n = \delta(\mathcal{F}_n(\mathbf{n}))$. Next, the NA score is combined with the output logits as an auxiliary weight to achieve the NA-aware prediction:

$$\mathbf{r}_{ht} = \begin{cases} (1 - s_n) \cdot bili(\mathbf{e}_h, \mathbf{e}_t), \text{ if } e_t \in \mathcal{E}, \\ s_n \cdot bili(\mathbf{e}_h, \mathbf{n}), \text{ if } e_t \text{ is NO-ANSWER}. \end{cases} \tag{7}$$

where *bili* denotes the bilinear layer.

**Training and Inference.** Considering that there are multiple relations between an entity pair ($e_h$, $e_t$), we take the relation prediction as a multiple binary classification problem, and choose the binary cross-entropy loss between the prediction and ground truth as the optimization objective:

$$\mathcal{L} = -\sum_{i=1}^{n_r} \left(y_{ht}^i \cdot \log(r_{ht}^i) + (1 - y_{ht}^i) \cdot \log(1 - r_{ht}^i)\right) \tag{8}$$

where $r_{ht}^i \in (0, 1)$ is the $i$-th dimension of $\mathbf{r}_{ht}$, indicating the prediction possibility of $i$-th relation, and $y_{ht}^i \in \{0, 1\}$ is the corresponding ground truth label. Specially, $y_{ht}^i$ is always 0 if $e_t$ is NO-ANSWER.

Following previous work [31], we determine a thresholds $\theta$ based on the micro F1 on the development set. With the threshold, we classify a triplet $(e_h, r_{ht}^i, e_t)$ as positive result if $r_{ht}^i > \theta$ or negative result otherwise in the test period. It is worth noting that we omit the relational triples whose tail entity is"NO-ANSWER" in inference. Finally, We combine the predictions from every sequence generated from the same document and with different queries, in order to obtain all relational facts over the document.

## 4 Experiments

### 4.1 Dataset

We evaluate our model on the public benchmark dataset, DocRED [31]. It is constructed from Wikipedia and Wikidata, covers a broad range of categories with

**Table 1.** Statistics of the DocRED dataset.

| | # Doc | # Fact | # Pos. Pair | # Neg. Pair | # Rel |
|---|---|---|---|---|---|
| Train | 3,053 | 34,715 | 38,269 | 1,160,470 | 96 |
| Dev | 1,000 | 11,790 | 12,332 | 384,467 | 96 |
| Test | 1,000 | 12,101 | 12,842 | 379,316 | 96 |

96 relation types, and is the largest human-annotated dataset for general domain document-level relation extraction. Documents in DocRED contain about 9 sentences and 20 entities on average, and more than 40.7% relation facts can only be extracted from multiple sentences. Moreover, 61.1% relation instances require various inference skills such as multi-hop reasoning. We follow the official partition of the dataset (i.e., 3053 documents for training, 1000 for development, and 1000 for test) and show the statistics in Table 1.

### 4.2 Implementation Details

We implement NARC with PyTorch 1.4.0 and *bert-base-uncased* model. The concatenated sequence in the input layer is trimmed to a maximum length of 512. The embedding size of BERT is 768, a linear-transformation layer is utilized to project the BERT embedding into a low-dimensional space with the same size of the hidden state, which is set to 200 (chosen from $[100, 150, 200, 250]$). Besides, the layer number of Entity Graph is set to 2 (chosen from $[1, 2, 3, 4]$), the batch size is set to 10 (chosen from $[5, 8, 10, 12]$), the learning rate is set to $1e^{-5}$ (chosen from $1e^{-4}$ to $1e^{-6}$). We optimize our model with Adam and run it on one 16G Tesla V100 GPU for 50 epochs. All hyper-parameters are tuned on the development set. Evaluation on the test set is done through CodaLab[2]. Following popular choices and previous work, we choose micro F1 and micro Ign F1 as evaluation metrics. Ign F1 denotes F1 excluding relational facts that appear in both the training set and the development or test set.

### 4.3 Baselines

We compare our NARC model with the following two types of baselines.

**Baselines w/o BERT:** On this track, we select 5 representative classic models without BERT. (1–3) CNN/BiLSTM/ContextAware [31]: These models leverage different neural architectures to encode the document, which are all text-based models and official baselines released by the authors of DocRED. (4) AGGCN [8]: It is the state-of-the-art sentence-level relation extraction model, which takes full dependency trees as inputs and constructs latent structure by self-attention. (5) EoG [4]: It constructs an edge-oriented graph and uses an iterative algorithm

[2] https://competitions.codalab.org/competitions/20717.

**Table 2.** Main results on DocRED, bold marks highest number among all the models. BERT-MRC indicates the vanilla MRC-style formulation without NA-related module, and $\text{NARC}_{\text{w/o EG}}$ indicates the NARC model without entity graph.

| (year) Model | Dev | Test |
|---|---|---|
| | Ign F1 / F1 | Ign F1 / F1 |
| (2019) CNN [31] | 41,58/43.45 | 40.33/42.26 |
| (2019) LSTM [31] | 48.44/50.68 | 47.71/50.07 |
| (2019) BiLSTM [31] | 48.87/50.94 | 48.78/51.06 |
| (2019) ContextAware [31] | 48.94/51.09 | 48.40/50.70 |
| (2019) AGGNN [8] | 46.29/52.47 | 48.89/51.45 |
| (2019) EoG [4] | 45.94 2.15 | 49.48/51.82 |
| (2019) BERT-RE [26] | 52.04/54.18 | 51.44/53.60 |
| (2020) BERT-HIN [24] | 54.29/56.31 | 53.70/55.60 |
| (2020) BERT-Coref [32] | 55.32 / 57.51 | 54.54/56.96 |
| (2020) BERT-GLRE [25] | –/ – | 55.40/57.40 |
| (2020) BERT-LSR [17] | 52.43/59.00 | **56.97**/59.05 |
| (ours) BERT-MRC | 55.49/57.59 | 54.86/57.13 |
| (ours) $\text{NARC}_{\text{w/o EG}}$ | 56.94/59.05 | 55.99/58.33 |
| (ours) NARC | **57.73/59.84** | 56.71/**59.17** |

over the graph edges, which is a recent state-of-the-art model in biomedical domain document-level relation extraction.

**Baselines w/ BERT:** On this track, we select 5 recent methods that adopt bert-base as the basic encoder. (1) BERT-RE [26]: It is the standard form of using BERT for relation extraction described in Sect. 2. (2) BERT-HIN [24]: It aggregates inference information from entity, sentence, and document levels with a hierarchical inference network to predict relation. (3) BERT-Coref [32]: It proposes an auxiliary training task to enhance the reasoning ability of BERT by capturing the co-refer relations between noun phrases. (4) BERT-GLRE [25]: It is a graph-based model by encoding the document information in terms of entity global and local features as well as relation features. (5) BERT-LSR [17]: It dynamically constructs a latent document-level graph for information aggregation in the entire document with an iterative refinement strategy.

### 4.4 Performance Comparison

Comparing the performance of different models in Table 2, the first conclusion we draw is that *NARC outperforms all baseline models in almost all the evaluation matrices*, which demonstrates the effectiveness of our NA-aware MRC solution, as well as the motivation of formulating document-level relation extraction as a machine reading comprehension problem. Secondly, *BERT-MRC outperforms*

**Table 3.** Ablation study on DocRED development set to investigate the influence of different modules in NARC. † indicates that we also remove the NA-Aware Prediction module, because it relies on the NA Vector Composition.

| | Ign F1 | F1 |
|---|---|---|
| NARC | 57.73 | 59.84 |
| – Entity Vector Aggregation | 56.36 | 58.58 |
| – NA-Aware Prediction | 56.86 | 59.01 |
| – NA Vector Composition† | 56.84 | 58.98 |

*BERT-RE by a significant margin.* We consider that the MRC-style model captures the interaction between the head entity and the document based on the deep self-attention structure, which helps to extract establish target-centric representations and extract information from relevant tokens from the document. Thirdly, *$NARC_{\text{w/o EG}}$ improves BERT-MRC by about 1.5% in F1 score.* We attribute the performance gain to the composition of NA vector. As the training set contains queries with and without valid answers, the vectorization process associated with all entities allows the model to automatically learn when to pool relevant and irrelevant portions to construct the NA vector. In optimization, the NA vector is used to increase or decrease the confidence of prediction, and thus makes the model aware of no-answer queries and alleviates its harmful effects. Lastly, *NARC exhibits a remarkable gain compared with $NARC_{\text{w/o EG}}$*, which demonstrates that the graph structure can exploit useful reasoning information among entities to capture rich non-local dependencies.

The effectiveness of each module in NARC is investigated in Table 3. From these ablations, we observe that: (1) The operation of Entity Vector Aggregation is indispensable since the ablation hurts the final result by 1.26% F1. It verifies the effectiveness of integrating the global and local features for an entity, as well as introducing the directional attention flow to take into account the fine-grained interaction between mention pairs. (2) NA-Aware Prediction is also a necessary component that contributes 0.83% gain of F1 to the ultimate performance. This is strong evidence that the NA score associated with all candidate tail entities is capable of providing powerful guidance for the final prediction. (3) When further removing NA Vector Composition, there are only slight fluctuations in performance. In other words, there is no remarkable improvement when only composing the NA vector as negative samples but not using the NA-aware prediction. The principle behind this phenomenon is that merely adding negative samples is not an effective way to boost performance, even if the generated negative samples are incredibly informative.

### 4.5 Performance Analysis

To further analyze the performance of NARC, we split the DocRED development set into several subsets based on different analytical strategies and report the performance of different models in each subset.

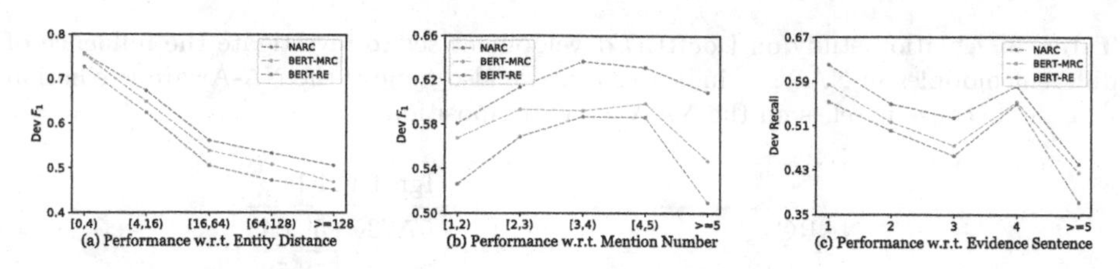


**Fig. 4.** Performance analysis on (a) detecting long-distance relations, (b) aggregating multiple mention information, and (c) reasoning multi-hop relations. We report the *F1* score for the first two analyses, while report *Recall* for the last one.

**Performance on Various Distances between Entities.** In this part, we examine the model performance in terms of entity distance, which is defined as the relative distances of the first mentions of the two entities in the document. As shown in Fig. 4(a), the F1 score suffers a quick and pronounced drop with the increase of entity distance, which is accordant with human intuition that detecting long-distance relations is still a challenging problem for current relation extraction models. Nevertheless, BERT-MRC consistently outperform BERT-RE by a sizable margin, due to the strong power of our MRC-style in reserving the relevant content of the target entities. Moreover, NARC outperforms all other baselines as the entity distance increases. This is because NARC breaks the limitation of sequence modeling and effectively captures long-distance interactions of semantic information by introducing the graph structure.

**Performance on Various Amounts of Entity Mentions.** To explore the capacity of different models in aggregating information from multiple mentions, we measure the performance in terms of the average mention number for each entity pair and report the F1 score in Fig. 4(b). Interestingly, all the models do not achieve their best performance when the mention number is small. We explain that the relevant information carried by a single mention is quite limited, making relations harder to be predicted, especially when the extraction scope is enlarged to the document level, the long-distance between two mentions making relations harder to be predicted. When the number of mentions is large, the performance of BERT-RE and NARC is devastating once again. This is because not every entity mention is involved in the relational facts, and aggregating information indiscriminately may introduce a large amount of noisy context, which will confuse the classifier. On the contrary, the directional attention flow measures the tail entity mentions and selects the most important one for each head entity mention, so that the proposed NARC maintains a relatively high performance when there are many mentions of the entity pair.

**Performance on Various Amounts of Evidence Sentences.** To assess the model's ability in multi-hop reasoning, we plot the preference curve in Fig. 4(c) when different amounts of evidence sentences are available for the relational facts. Unlike the previous two statistical features, the evidence sentence number is a semantic feature and can only be counted for positive labels, thus we report

**Table 4.** F1 score w.r.t. NA query ratio. Both *0x* and *All* indicate no negative sampling process, the former does not use NA query, while the latter uses all NA queries.

| | 0× | 1× | 2× | 3× | All |
|---|---|---|---|---|---|
| BERT-MRC | 67.32 | 63.33 | 59.74 | 57.66 | 57.59 |
| NARC | 68.06 | 64.41 | 61.29 | 59.70 | 59.84 |

**Table 5.** Computational cost analysis on DocRED dev set. For the test time, we execute 5 independent runs and report the average value for each model.

| | BERT-RE | BERT-MRC | NARC |
|---|---|---|---|
| Para. Num | 114.3M | 114.3M | 127.4M |
| Test Time | 134.3 s | 408.8 s | 416.2 s |

the Recall score for evaluation. Again, NARC outperforms all methods. Furthermore, the results indicate that the performance gap between NARC and BERT-RE/MRC reaches the maximum when the number of evidence sentences is 3. It is because a 2-layer entity graph is constructed in NARC. Typically, considering there are three entities (*head entity*, *tail entity*, and a *relay entity*) distributed in three sentences, the reasoning chain *head-relay-tail* could be exactly achieved by two times message propagation. From this viewpoint, it is a natural phenomenon that the gap gradually decreases with the further increase of evidence sentence numbers.

**NA Influence Analysis.** NA query is an unexpected problem that arises after formulating document-level relation extraction as machine reading comprehension, and we assume that it drags down the model performance. In this experiment, we conduct random negative sampling for NA queries based on the number of non-NA queries in each document, and report the results conditioned on different negative ratios on the development set, as summarized in Table 4. We observe that NARC achieves relatively similar performance with BERT-MRC when there is no negative instance (0x). With the increase of the proportion of negative samples, the performance gap increases gradually, demonstrating that NARC is effective in mitigating the negative effect of having too many NA queries.

### 4.6 Computational Cost Analysis

While BERT-RE runs document modeling only once to create general representations and extract all possible relational facts, NARC enumerates the document $n_e$ times to establish representations specific to each target head entity. This means NARC is more time-consuming than BERT-RE in theory ($\mathcal{O}(n_e)$ vs. $\mathcal{O}(1)$), To study the actual computational cost, we run them on the DocRED development set with the same setting and present the results in Table 5. The test time of BERT-MRC and NARC is very close, both about 3 times of BERT-RE. This is an acceptable result, because intuitively speaking, the time overhead of BERT-MRC seems to be 20 times that of BERT-RE (there is an average of 20 entities in a document). We assume the reason is that the inference complexity of BERT-MRC is one order less than that of BERT-RE ($\mathcal{O}(n_e)$ vs. $\mathcal{O}(n_e^2)$). Through further investigation, we find that BERT-RE needs to enumerate and preprocess all possible entity pairs for each input in the dataloader, which is an extraordinarily time-consuming process, accounting for 85% of the test time. If we only calculate the inference time without considering the data processing, BERT-MRC takes about 0.7s and BERT-RE takes about 1.2s for a batch. Moreover, we may also prune some queries to further accelerate in real application since some types of entities may not become the head entity. Taken altogether, NARC is not as time-consuming as expected. It sacrifices a little efficiency in exchange for a substantial performance improvement.

## 5 Related Work

This work builds on a rich line of recent efforts on relation extraction and machine reading comprehension models.

**Relation Extraction.** Relation extraction is always a research hotspot in the field of data mining and natural language processing. Early approaches mainly focus on the sentence-level relation extraction [33–36], which aims at predicting the relation label between two entities in a sentence. This kind of method does not consider interactions across mentions and ignores relations expressed across sentence boundaries. Afterward, many researchers show interest in the cross-sentence relation extraction problem [9,18,22], yet they restrict all relation candidates in a continuous sentence span with a fixed length. Meanwhile, there are also some efforts to expand the extraction scope to the entire document in the biomedical domain but only considering a few relations [3,4,37]. However, these idealized settings make their solutions not suitable for complex and diversified real-world scenarios. Recently, Yao et al. [31] propose DocRED, a large-scale document-level relation extraction dataset with 96 relations that constructed from Wikipedia and Wikidata. Nowadays, the document-level relation extraction task has attracted a lot of researchers' interest [17,24,26,32].

In the long history of relation extraction, how to fully capture the specific information related to the target entities is an eternal topic. Wang et al. [27] propose diagonal attention to depict the strength of connections between entity and context for sentence-level relation classification. He et al. [10] first utilize intra- and inter-sentence attentions to learn syntax-aware entity embedding, and then combine sentence and entity embedding for distantly supervised relation extraction. Li et al. [15] incorporate an entity-aware embedding module and a selective gate mechanism to integrate task-specific entity information into word embeddings. Beyond that, Jia et al. [12] propose an entity-centric, multi-scale representation learning on a different level for $n$-ary relation extraction. However, due to a large number of entity pairs in documents, there are few works to consider entity-specific text-modeling in document-level relation extraction.

**Machine Reading Comprehension.** Machine reading comprehension is a general and extensible task form, and many tasks in natural language processing can be framed as reading comprehension: Li et al. [13] propose a MRC-based unified framework to handle both flat and nested named entity recognition. Li et al. [14] formulate the entity-relation extraction task as a multi-turn question-answering problem. The most similar task to document-level relation extraction is multi-hop machine reading comprehension [29], which takes (*head entity, relation, ?*), not utterance, as query. The last few years have witnessed significant progress on this task: Typically, De Cao et al. [5] introduce Entity-GCN, which takes entity mentions as nodes and learns to answer questions with graph convolutional networks. On this basis, Cao et al. [2] apply bi-directional attention between graph nodes and queries to learn query-aware representation for reading comprehension. This success inspires us to pay more attention to the interaction between query and document, along with the reasoning process in multi-hop relations.

In the formulation of multi-hop machine reading comprehension, every query could retrieval an accurate answer from its candidate list, which is inconsistent with the scenario of document-level relation extraction. Recently, Rajpurkar et al. [19] release SQuAD 2.0 by augmenting the SQuAD dataset with unanswerable questions, which officially opens the curtain for solving unanswerable questions in span-based machine reading comprehension. Then some approaches for the challenging problem are proposed: Sun et al. [23] present a unified model with a no-answer pointer and answer verifier to predict whether the question is answerable. Hu et al. [11] introduce a read-then-verify system to check whether the extracted answer is legitimate or not. However, considering the technical gap between span-based and multi-hop reading comprehension (i.e., a sequence labeling problem vs. a classification problem), how to deal with numerous no-answer queries is still an open problem after we transform the document-level relation extraction into the paradigm of machine reading comprehension.

## 6 Conclusion

In this paper, we propose a NA-aware MRC model for document-level relation extraction, connecting the relation extraction problem to the well-studied machine reading comprehension field. The proposed approach facilitates the model focusing on the context related to each given head entity in the document, and yields significant improvements compared to the conventional solution. Interesting future work directions include employing other advanced pre-trained language models (e.g., DeFormer[1], Roberta [16]) to further improve the efficiency and performance, as well as adapting the proposed paradigm and model to other knowledge-guided tasks in information extraction (e.g., event extraction).

**Acknowledgments.** We would like to thank all reviewers for their insightful comments and suggestions. The work is supported by the Strategic Priority Research Program of Chinese Academy of Sciences (grant No.XDC02040400), and the Youth Innovation Promotion Association of Chinese Academy of Sciences (grant No.2021153).

## References

1. Cao, Q., Trivedi, H., Balasubramanian, A., Balasubramanian, N.: Deformer: Decomposing pre-trained transformers for faster question answering. In: Proceedings of ACL, pp. 4487–4497 (2020)
2. Cao, Y., Fang, M., Tao, D.: Bag: Bi-directional attention entity graph convolutional network for multi-hop reasoning question answering. In: Proceedings of NAACL, pp. 357–362 (2019)
3. Christopoulou, F., Miwa, M., Ananiadou, S.: A walk-based model on entity graphs for relation extraction. In: Proceedings of ACL, pp. 81–88 (2018)
4. Christopoulou, F., Miwa, M., Ananiadou, S.: Connecting the dots: document-level neural relation extraction with edge-oriented graphs. In: Proceedings of EMNLP, pp. 4927–4938 (2019)
5. De Cao, N., Aziz, W., Titov, I.: Question answering by reasoning across documents with graph convolutional networks. In: Proceedings of NAACL, pp. 2306–2317 (2019)
6. Devlin, J., Chang, M.W., Lee, K., Toutanova, K.: Bert: pre-training of deep bidirectional transformers for language understanding. In: Proceedings of NAACL, pp. 4171–4186 (2019)
7. Feng, R., Yuan, J., Zhang, C.: Probing and fine-tuning reading comprehension models for few-shot event extraction. arXiv preprint arXiv:2010.11325 (2020)
8. Guo, Z., Zhang, Y., Lu, W.: Attention guided graph convolutional networks for relation extraction. In: Proceedings of ACL, pp. 241–251 (2019)
9. Gupta, P., Rajaram, S., Schütze, H., Runkler, T.: Neural relation extraction within and across sentence boundaries. In: Proceedings of AAAI, pp. 6513–6520 (2019)
10. He, Z., Chen, W., Li, Z., Zhang, M., Zhang, W., Zhang, M.: See: syntax-aware entity embedding for neural relation extraction. In: Proceedings of AAAI, pp. 5795–5802 (2018)
11. Hu, M., Wei, F., Peng, Y., Huang, Z., Yang, N., Li, D.: Read+verify: machine reading comprehension with unanswerable questions. In: Proceedings of AAAI, pp. 6529–6537 (2019)
12. Jia, R., Wong, C., Poon, H.: Document-level n-ary relation extraction with multiscale representation learning. In: Proceedings of NAACL, pp. 3693–3704 (2019)
13. Li, X., Feng, J., Meng, Y., Han, Q., Wu, F., Li, J.: A unified mrc framework for named entity recognition. In: Proceedings of ACL, pp. 5849–5859 (2019)
14. Li, X., et al.: Entity-relation extraction as multi-turn question answering. In: Proceedings of ACL, pp. 1340–1350 (2019)
15. Li, Y., et al.: Self-attention enhanced selective gate with entity-aware embedding for distantly supervised relation extraction. In: Proceedings of AAAI, pp. 8269–8276 (2020)
16. Liu, Y., et al.: Roberta: a robustly optimized bert pretraining approach. arXiv preprint arXiv:1907.11692 (2019)
17. Nan, G., Guo, Z., Sekulić, I., Lu, W.: Reasoning with latent structure refinement for document-level relation extraction. In: Proceedings of ACL, pp. 1546–1557 (2020)

18. Peng, N., Poon, H., Quirk, C., Toutanova, K., Yih, W.T.: Cross-sentence n-ary relation extraction with graph lstms. TACL **5**, 101–115 (2017)
19. Rajpurkar, P., Jia, R., Liang, P.: Know what you don't know: unanswerable questions for squad. In: Proceedings of ACL, pp. 784–789 (2018)
20. Schlichtkrull, M., Kipf, T.N., Bloem, P., van den Berg, R., Titov, I., Welling, M.: Modeling relational data with graph convolutional networks. In: Proceedings of ESWC, pp. 593–607 (2018)
21. Seo, M., Kembhavi, A., Farhadi, A., Hajishirzi, H.: Bidirectional attention flow for machine comprehension. arXiv preprint arXiv:1611.01603 (2016)
22. Song, L., Zhang, Y., Wang, Z., Gildea, D.: N-ary relation extraction using graph-state lstm. In: Proceedings of EMNLP, pp. 2226–2235 (2018)
23. Sun, F., Li, L., Qiu, X., Liu, Y.: U-net: machine reading comprehension with unanswerable questions. arXiv preprint arXiv:1810.06638 (2018)
24. Tang, H., et al.: Hin: Hierarchical inference network for document-level relation extraction. In: Proceedings of PAKDD, pp. 197–209 (2020)
25. Wang, D., Hu, W., Cao, E., Sun, W.: Global-to-local neural networks for document-level relation extraction. In: Proceedings of EMNLP, pp. 3711–3721 (2020)
26. Wang, H., Focke, C., Sylvester, R., Mishra, N., Wang, W.: Fine-tune bert for docred with two-step process. arXiv preprint arXiv:1909.11898 (2019)
27. Wang, L., Cao, Z., De Melo, G., Liu, Z.: Relation classification via multi-level attention cnns. In: Proceedings of ACL, pp. 1298–1307 (2016)
28. Wang, Z., Mi, H., Ittycheriah, A.: Sentence similarity learning by lexical decomposition and composition. In: Proceedings of COLING, pp. 1340–1349 (2016)
29. Welbl, J., Stenetorp, P., Riedel, S.: Constructing datasets for multi-hop reading comprehension across documents. TACL **6**, 287–302 (2018)
30. Wu, W., Wang, F., Yuan, A., Wu, F., Li, J.: Coreference resolution as query-based span prediction. In: Proceedings of ACL, pp. 6953–6963 (2020)
31. Yao, Y., et al.: Docred: a large-scale document-level relation extraction dataset. In: Proceedings of ACL (2019)
32. Ye, D., Lin, Y., Du, J., Liu, Z., Sun, M., Liu, Z.: Coreferential reasoning learning for language representation. In: Proceedings of EMNLP (2020)
33. Zeng, D., Liu, K., Chen, Y., Zhao, J.: Distant supervision for relation extraction via piecewise convolutional neural networks. In: Proceedings of EMNLP, pp. 1753–1762 (2015)
34. Zeng, D., Liu, K., Lai, S., Zhou, G., Zhao, J.: Relation classification via convolutional deep neural network. In: Proceedings of COLING, pp. 2335–2344 (2014)
35. Zhang, Y., Zhong, V., Chen, D., Angeli, G., Manning, C.D.: Position-aware attention and supervised data improve slot filling. In: Proceedings of EMNLP, pp. 35–45 (2017)
36. Zhang, Z., Shu, X., Yu, B., Liu, T., Zhao, J., Li, Q., Guo, L.: Distilling knowledge from well-informed soft labels for neural relation extraction. In: Proceedings of AAAI, pp. 9620–9627 (2020)
37. Zhang, Z., Yu, B., Shu, X., Liu, T., Tang, H., Wang, Y., Guo, L.: Document-level relation extraction with dual-tier heterogeneous graph. In: Proceedings of COLING, pp. 1630–1641 (2020)

# Follow Your Path: A Progressive Method for Knowledge Distillation

Wenxian Shi, Yuxuan Song, Hao Zhou(✉), Bohan Li, and Lei Li(✉)

Bytedance AI Lab, Shanghai, China
{shiwenxian,songyuxuan,zhouhao.nlp,libohan.06,lileilab}@bytedance.com

**Abstract.** Deep neural networks often have huge number of parameters, which posts challenges in deployment in application scenarios with limited memory and computation capacity. Knowledge distillation is one approach to derive compact models from bigger ones. However, it has been observed that a converged heavy teacher model is strongly constrained for learning a compact student network and could make the optimization subject to poor local optima. In this paper, we propose ProKT, a new model-agnostic method by projecting the supervision signals of a teacher model into the student's parameter space. Such projection is implemented by decomposing the training objective into local intermediate targets with approximate mirror descent technique. The proposed method could be less sensitive with the quirks during optimization which could result in a better local optima. Experiments on both image and text datasets show that our proposed ProKT consistently achieves superior performance comparing to other existing knowledge distillation methods.

**Keywords:** Knowledge distillation · Curriculum learning · Deep learning · Image classification · Text classification · Model miniaturization

## 1 Introduction

Advanced deep learning models have shown impressive abilities in solving numerous machine learning tasks [5,8,23]. However, the advanced heavy models are not compatible with many real-world application scenarios due to the low inference efficiency and high energy consumption. Hence preserving the model capacity using fewer parameters has been an active research direction during recent years [10,22,35]. Knowledge distillation [10] is an essential way in the field which refers to a model-agnostic method where a model with fewer parameters (student) is optimized to minimize some statistical discrepancy between its predictions distribution and the predictions of a higher capacity model (teacher).

W. Shi and Y. Song—Equal contribution.


N. Oliver et al. (Eds.): ECML PKDD 2021, LNAI 12977, pp. 596–611, 2021.
https://doi.org/10.1007/978-3-030-86523-8_36

Recently, it has been observed that employing a static target as the distillation objective would leash the effectiveness of the knowledge distillation method [13,19] when the capacity gap between student and teacher model is large. The underlying reason lies in common sense that optimizing deep learning models with gradient descent is favorable to the target which is close to their model family [21]. To counter the above issues, designing the intermediate target has been a popular solution: Teacher-Assistant learning [13] shows that within the same architecture setting, gradually increasing the teacher size will promote the distillation performance; Route-Constrained Optimization (RCO) [19] uses the intermediate model during the teacher's training process as the anchor to constrain the optimization path of the student, which could close the performance gap between student and teacher model.

One reasonable explanation beyond the above facts could be derived from the perspective of curriculum learning [2]: the learning process will be boosted if the goal is set suitable to the underlying learning preference (bias). The most common arrangement for the tasks is to gradually increase the difficulties during the learning procedures such as pre-training [29]. Correspondingly, TA-learning views the model with more similar capacity/model-size as the easier tasks while RCO views the model with more similar performance as the easier tasks, etc.

In this paper, we argue that the utility of the teacher is not necessarily fully explored in previous approaches. First, the intermediate targets usually discretize the training process as several periods and the unsmoothness of target changes in optimization procedure will hurt the very property of introducing intermediate goals. Second, manual design of the learning procedure is needed which is hard to control and adapt among different tasks. Finally, the statistical dependency between the student and intermediate target is never explicitly constrained.

To counter the above obstacles, we propose ProKT, a new knowledge distillation method, which better leverages the supervision signal of the teacher to improve the optimization path of student. Our method is mainly inspired by the guided policy search in reinforcement learning [17], where the intermediate target constructed by the teacher should be approximately projected on the student parameter space. More intuitively, the key motivation is to make the teacher model aware of the optimization progress of student model hence the student could get the "hand-on" supervision to get out of the poor minimal or bypass the barrier in the optimization landscape.

The main contribution of this paper is that we propose a simple yet effective model-agnostic method for knowledge distillation, where intermediate targets are constructed by a model with the same architecture of teacher and trained by approximate mirror descent. We empirically evaluate our methods on a variety of challenging knowledge distillation setting on both image data and text data. We find that our method outperforms the vanilla knowledge distillation approach consistently with a large margin, which even leads to significant improvements compared to several strong baselines and achieves state-of-the-art on several knowledge distillation benchmark settings.

## 2 Related Work

In this section, we discuss several most related literature in model miniaturization and knowledge distillation.

**Model Miniaturization.** There has been a fruitful line of research dedicated to modifying the model structure to achieve fast inference during the test time. For instance, MobileNet [11] and ShuffleNet [38] modify the convolution operator to reduce the computational burden. And the method of model pruning tries to compress the large network by removing the redundant connection in the large networks. The connections are removed either based on the weight magnitude or the impact on the loss function. One important hyperparameter of the model pruning is the compression ratio of each layer. [9] proposes the automatical tuning strategy instead of setting the ratio manually which are proved to promote the performance.

**Knowledge Distillation.** Knowledge distillation focuses on boosting the performance while the small network architecture is fixed. [3,10] introduced the idea of distilling knowledge from a heavy model with a relatively smaller and faster model which could preserve the generalization power. To this end, [3] proposes to match the logits of the student and teacher model, and [10] tends to decrease the statistical dependency between the output probability distributions of the student model and the teacher model. And [39] proposes the deep mutual learning which demonstrates that bi-jective learning process could boost the distillation performance. Orthogonal to output matching, many works have been conducted on matching the student model and teacher by enforcing the alignment on the latent representation [12,28,36]. This branch of works typically involves prior knowledge towards the network architectures of student and teacher model which is more favorable to distill from the model with the same architecture. In the context of knowledge distillation, our method is mostly related to TA-learning [19] and the Route-Constraint Optimization(RCO) [13] which improved the optimization of student model by designing a sequence of intermediate targets to impose constraint on the optimization path. Both of the above methods could be well motivated in the context of curriculum learning, while the underlying assumption indeed varies: TA-learning views the increasing order of the model capacity implied a suitable learning trajectory; while RCO considers the increasing order of the model performance forms a favorable learning curriculum for student. However, there have been several limitations. For example, the sequence of learning targets that are set before the training process needs to be manually designed. Besides, targets are also independent of the states of the student which does not enjoy all the merits of curriculum learning.

**Connections to Other Fields.** Introducing a local target within the training procedure is a widely applied spirit in many fields of machine learning. [20] introduce the guided policy search where a local policy is then introduced to provide the local improved trajectory, which has been proved to be useful towards bypassing the bad local minima. [7] augmented the training trajectories by introducing the so called "coaching" distribution to ease the training burden and similarity. [16] introduce a family of smooth policy classes to reduce smooth imitation learning to a regression problem. [18] introduce an intermediate target so-called mediator during the training of the auto-regressive language model, while the information discrepancy between the intermediate target and the model is constrained through the Kullback-Leibler(KL) divergence. Moreover, [4] utilized the interpolation between the generator's output and target as the bridge to alleviate data sparsity and overfitting problems of MLE training. Expect from the distinct research communities and objectives, our method also differs from their methods in both the selection of intermediate targets, *i.e.* learned online versus designed by hands, and the theoretical motivation, *i.e.* the explicit constrain in mirror descent guarantee the good property on improvement.

# 3 Methodology

In this section, we first introduce the background of knowledge distillation and notations in Sect. 3.1. Then, in Sect. 3.2, we generalize and formalize the knowledge distillation methods with intermediate targets. In Sect. 3.3, we mainly introduce the details of our method ProKT.

## 3.1 Background on Knowledge Distillation

To start with, we introduce the necessary notations and backgrounds which are most related to our work. Taking an K-class classification task as an example, the inputs and label tuple is denoted as $(x, y) \in \mathcal{X} \times \mathcal{Y}$ and the label $y$ is usually in the format of a one-hot vector with dimension $K$. The objective in this setting is to learn a parameterized function approximator: $f(x;\theta) : \mathcal{X} \rightarrow \mathcal{Y}$. Typically, the function could be characterized as the deep neural networks. With the logits output as $u$, the output distribution $q$ of the neural network $f(x;\theta)$ could be acquired by applying the softmax function over the logits output $u$: $q_i = \frac{\exp(u_i/T)}{\sum_{j=1}^{K}\exp(u_j/T)}$, where $T$ corresponds to the temperature. The objective of knowledge distillation could be then written as:

$$\mathcal{L}_{\mathrm{KD}}(\theta) = (1-\alpha)H\left(y, q_s(\theta)\right) + \alpha T^2 H(p_t, q_s(\theta)). \quad (1)$$

Here $H$ denotes the cross entropy objective, *i.e.*, $H(p,q) = \sum_{i=1}^{K} -p_i \log q_i$ which is the KL divergence between $p$ and $q$ minus the entropy of $p$ (usually constant when $p = y$). $p_t$ is the output distribution of a given teacher model and $\alpha$ is the balanced weight between the standard cross entropy loss and the knowledge distillation loss from teacher. $T$ is the temperature. In the following formulations, we omit the $T$ by setting $T = 1$.

## 3.2 Knowledge Distillation with Dynamic Target

In this section, we generalize and formalize the knowledge distillation methods with intermediate targets. We propose that previous knowledge distillation methods, either with a static target (i.e., the vanilla KD) or with hand-crafted discrete targets (i.e., Route-Constraint Optimization (RCO) [13]), cannot make full use of the knowledge from teacher. Instead, a dynamic and continuous sequence of targets is a better choice, and then we propose our method in the next section.

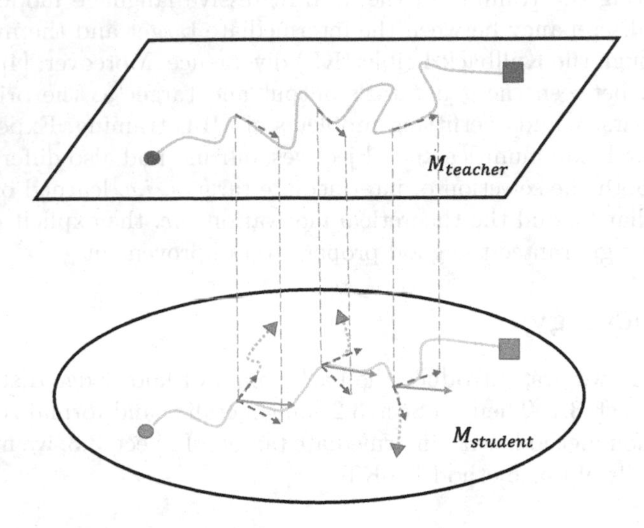


**Fig. 1.** $\mathcal{M}_{teacher}$ and $\mathcal{M}_{student}$ refer to the output manifolds of student model and teacher model. The lines between circles (●,●) to squares (■,■) imply the learning trajectories in the distribution level. The intuition of ProKT is to avoid bad local optimas (triangles (▲)) by conducting supervision signal projection.

Firstly, we generalize and formalize the knowledge distillation methods with intermediate targets, named as *sequential optimization knowledge distillation* (SOKD) methods. Instead of conducting a static teacher model in vanilla KD, the targets to the student model of SOKD methods are changed during the training time. Without loss of generality, we denote the sequence of intermediate target distributions as $P_t = [p_t^1, p_t^2, \cdots, p_t^m, \cdots]$. Starting from a random initialized parameters $\theta^0$, the student model is optimized by gradient descent methods to mimic its intermediate target $p_t^m$:

$$\theta^m = \theta^{m-1} - \beta \nabla_\theta \mathcal{L}^m(\theta^{m-1}), \tag{2}$$

$$\mathcal{L}^m(\theta) = (1-\alpha) H(y, q_s(\theta)) + \alpha H\left(p_t^m, q_s(\theta)\right). \tag{3}$$

One choice to organize the intermediate targets is to split the training process into intervals and adopt a fixed target in each intervals, named as discrete targets. For example, the Route-Constraint Optimization (RCO) [13] saves the un-convergent checkpoints of teacher during the teacher's training to construct the target sequence. The learning target of student is changed every few epochs.

However, the targets are changed discontinuously in the turning points between discrete intervals, which would incur negative effects on the dynamic knowledge distillation. Firstly, switching to a target that is too difficult for the student model would undermine the advantages of curriculum learning. If the target is changed sharply to a model with large complexity improvement, it is hard for student to learn. Besides, the ineligible gap between adjacent targets would make the training process unstable and hurt the convergence property during the optimization [40].

Therefore, we propose to replace the discrete target sequence with a continuous and dynamic one, whose targets are adjusted smoothly and dynamically according to the status of student model. In continuous target sequence, targets in each step are changed smoothly with ascending performance. In that case, if the student learns the target well in current step, the target of the next step is easier to learn because of the slight performance gap. The training process is stable as well, because the training targets are improved smoothly. Specifically, the optimization trajectories of the teacher model naturally offer continuous supervision signals for the student. In our work, we propose to conduct the optimization trajectories of teacher model as the continuous targets. Besides, to ensure that intermediate teachers are kept easy to learn for students, we introduce an explicit constraint in the objective of the teacher. This constraint dynamically adjusts the updating path of the teacher according to learning progress of the student. The key motivation of our method is illustrated in Fig. 1.

### 3.3 Progressive Knowledge Teaching

In this section, we firstly propose the SOKD adopting the optimization trajectories of teacher as the continuous targets. The learning process is that every time the teacher model updates one step towards the ground-truth, the student model updates one step towards the new teacher. Then based on this, we propose the *Progressive Knowledge Teaching* (ProKT), which modifies the updating objective of the teacher by explicitly constraining it in the neighbourhood of student model.

To construct the target sequence with continuous ascending target distributions, a natural selection is the gradient flow of the optimization procedure of the teacher distribution. With the student $q_{\theta_s}$ and teacher model $p_{\theta_t}$ initialized at the same starting point (e.g., $q_{\theta_s^0}(y|x) = p_{\theta_t^0}(y|x) = Uniform(1, K)$), we iteratively update the teacher model and the student model according to the following loss functions:

$$\theta_t^{m+1} = \theta_t^m - \eta_t \nabla \mathcal{L}_t(\theta_t^m), \quad \mathcal{L}_t(\theta_t) = H(y, p_{\theta_t}), \tag{4}$$

$$\theta_s^{m+1} = \theta_s^m - \eta_s \nabla \mathcal{L}_s(\theta_s, p_{\theta_t^{m+1}}), \quad \mathcal{L}_s(\theta_s) = H(p_{\theta_t}, q_{\theta_s}). \tag{5}$$

Here, the $\eta_t$ and $\eta_s$ are learning rates of student and teacher models, respectively. Starting with the same initialized distribution, the teacher model is updated firstly by running a step of stochastic gradient descent. Then, the student model learns from the updated teacher model. In this process, the student could learn from the optimization trajectories of the teacher model, which provides the knowledge of how the teacher model is optimized from a random classifier to a good approximator. Compared with the discrete case such as RCO, the targets are improved progressively and smoothly.

**Algorithm 1.** ProKT

1: **Input:** Initialized student model $q_{\theta_s}$ and teacher model $q_{\theta_t}$. Data set $\mathcal{D}$.
2: **while** not converged **do**
3: Sample a batch of input $(x, y)$ from the dataset $\mathcal{D}$.
4: update teacher by $\theta_t \leftarrow \theta_t - \eta_t \nabla_{\theta_t} \hat{\mathcal{L}_{\theta_t}}$.
5: update student by $\theta_s \leftarrow \theta_s - \eta_s \nabla_{\theta_s} \mathcal{L}(\theta_s)$.
6: **end while**

However, simply conducting iterative optimization following Eq. 4 with gradient descent could not guarantee the teacher would stay close to the student model even with a small update step. The gradient descent step of teacher in Eq. 4 is equivalent to solving the following formulation:

$$\theta_t^{m+1} = \arg\min_{\theta} \mathcal{L}\left(\theta_t^m\right) + \nabla_\theta \mathcal{L}(\theta)^\top \left(\theta - \theta_t^m\right) + \frac{1}{2}\eta_t \left\|\theta - \theta_t^m\right\|^2,$$

which only seeks the solution in the neighborhood of current parameter $\theta_t^m$ in terms of the Euclidean distance. Unfortunately, there is no explicit constraint that the target distribution $p_{\theta_t^{m+1}}(y|x)$ stays close to $p_{\theta_t^m}(y|x)$. Besides, because the learning process of teacher model is ignorant of how the student model has been trained, it is probably that the gap between student model and teacher model grows cumulatively.

Therefore, in order to constrain the target distribution to be easy-to-learn for the student, we modify the training objective of teacher model in Eq. 4 by explicitly bounding the KL divergence between the teacher distribution and student distribution:

$$\theta_t^{m+1} = \min_{\theta_t} H(y, p_{\theta_t}) \quad \text{s.t. } D_{\mathrm{KL}}(q_{\theta_s}^m, p_{\theta_t}) \leq \epsilon. \tag{6}$$

The $\epsilon$ controls the how close the teacher model for the next step to the student model. In this case, we make an approximation that if the KL divergence of target distribution and the current student distribution is small, this target is easy for student to learn. By optimizing the Eq. 6, the teacher is chosen as the best approximator of the teacher model's family in the neighbour of student distribution.

With slight variant of the Lagrangian formula of Eq. 6, the learning objective of teacher model in ProKT is

$$\hat{\mathcal{L}}_{\theta_t} = (1 - \lambda) H(y, p_{\theta_t}) + \lambda H(q_{\theta_s}, p_{\theta_t}), \tag{7}$$

in which the hyper-parameter $\lambda$ controls the difficulty of teacher model compared with student model. The overall algorithm is summarized in Algorithm 1. The proposed method also ensemble the spirit of mirror descent [1] which we provide a more detailed discussion in the Appendix.

## 4 Experiments

In this section, we empirically test the validity of our method in both image and text classification benchmarks. Results show that ProKT achieves significant improvement in a wide range of tasks and model architectures.

### 4.1 Setup

In order to evaluate the performance of ProKT under different knowledge distillation settings, we implement the ProKT in different tasks (image recognition and text classification), different network architectures, and different training objectives.

**Image Recognition.** The image classification experiments are conducted in CIFAR-100 [15] following [30].

**Settings.** Following the [30], we compare the performance of knowledge distillation methods under various architecture of teacher and student models. We use the following models as teacher or student models: vgg [26], MobileNetV2 [25] (with a width multiplier of 0.5), ShuffleNetV1 [38], ShuffleNetV2 [26], Wide Residual Network (WRN-*d*-*w*) [37] (with depth $d$ and width factor $w$) and ResNet [8]. To evaluate the ProKT under different distillation loss, we conduct the ProKT with standard KL divergence loss and contrastive representation distillation loss proposed by CRD [30].

**Baselines.** We compare our model with the following baselines: vanilla KD [10], CRD [30] and RCO [13]. Results of baselines are from the report of [30], except for the RCO [13], which is implemented by ourselves.

**Text Classification.** Text classification experiments are conducted following the setting of [31] and [12] on the GLUE [33] benchmark.

**Datasets.** We evaluate our method for sentiment classification on SST-2 [27], natural language inference on MNLI [34] and QNLI [24], and paraphrase similarity matching on MRPC [6] and QQP[1].

**Settings.** The teacher model is the BERT-base [5] fine-tuned in the training set, which is a 12-layer Transformers [32] with 768 hidden units. Following the setting of [31] and [12], a BERT of 6 layer Transformers and 786 hidden units is conducted as the student model. We use the pretrained 6 layer BERT model released by [31][2], and fine-tune it in the training set. For distillation between heterogeneous architectures, we use a single-layer bi-LSTM with 300 embedding size and 300 hidden size as student model. We did not pretrain the bi-LSTM models. We implement the basic ProKT with standard KL divergence loss, and

[1] https://data.quora.com/First-Quora-Dataset-Release-Question-Pairs.
[2] https://github.com/google-research/bert.

**Table 1.** Top-1 test *accuracy* (%) of student networks distilled from teacher with different network architectures on CIFAR100. Results except the RCO, ProKT and CRD+ProKT are from [30].

| Teacher<br>Student | vgg13<br>MobileNetV2 | ResNet50<br>MobileNetV2 | ResNet50<br>vgg8 | resnet 32 × 4<br>ShuffleNetV1 | resnet 32 × 4<br>ShuffleNetV2 | WRN-40-2<br>ShuffleNetV1 |
|---|---|---|---|---|---|---|
| Teacher | 74.64 | 79.34 | 79.34 | 79.42 | 79.42 | 75.61 |
| Student | 64.6 | 64.6 | 70.36 | 70.5 | 71.82 | 70.5 |
| KD* | 67.37 | 67.35 | 73.81 | 74.07 | 74.45 | 74.83 |
| RCO | 68.42 | 68.95 | 73.85 | 75.62 | **76.26** | 75.53 |
| ProKT | **68.79** | **69.32** | **73.88** | **75.79** | 75.59 | **76.02** |
| CRD | 69.73 | 69.11 | 74.30 | 75.11 | 75.65 | 76.05 |
| CRD+KD | **69.94** | 69.54 | 74.58 | 75.12 | 76.05 | 76.27 |
| CRD+ProKT | 69.59 | **69.93** | **75.14** | **76.0** | **76.86** | **76.76** |

combine our method with the TinyBERT [12] by replacing the second stage of fine-tuning TinyBERT with our ProKT. To fair comparison, we use the pre-trained TinyBERT released by [12] when combing our ProKT with TinyBERT. More experimental details are listed in the Appendix.

**Baselines.** We compare our method with following baselines: (1) BERT + Finetune, fine-tune the BERT student on training set; (2) BERT/bi-LSTM + KD, fine-tune the BERT student or train the bi-LSTM on training set using the vanilla knowledge distillation loss [10]; (3) Route Constrained Optimization (RCO) [13], use 4 un-convergent teacher checkpoints as intermediate training targets; (4) bi-LSTM: train bi-LSTM in training set; (5) TinyBERT [12]: match the attentions and representations of student model with teacher model on the first stage and then fine-tune by the vanilla KD loss on the second stage. For vanilla KD methods, we set the temperature as 1.0 and only use the KL divergence with teacher outputs as loss. We also compare our method with the results reported by [28] and [31].

### 4.2 Results

Results of image classification on CIFAR100 are shown in Table 1. The performance is evaluated by top-1 accuracy. Results of text classification are shown in Table 2. The accuracy or f1-score on test set are obtained by submitting to the GLUE [33] website. Results on both text and image classification tasks show that ProKT achieves the best performance under almost all model settings.

Results show that the continuous and dynamic targets are helpful to take advantage of the knowledge from the teacher. Although adopting discrete targets in RCO could improve the performance to vanilla KD, our ProKT with continuous and dynamic targets is more effective in teaching student. To further show the effectiveness of continuity and adaptiveness (i.e., the KL divergence term to student in the update of teacher) in ProKT respectively, we test the results of ProKT with $\lambda = 0$, in which the targets are improved smoothly but without the adjustment towards the student. As shown in Table 2, the continuous targets are better than discrete targets (i.e., RCO), while incorporating

**Table 2.** Test results of different knowledge distillation methods in GLUE.

| Model | SST-2 (acc) | MRPC (f1/acc) | QQP (f1/acc) | MNLI (acc m/mm) | QNLI (acc) |
|---|---|---|---|---|---|
| $BERT_{12}$ (teacher) | 93.4 | 88.0/83.2 | 71.4/89.2 | 84.3/83.4 | 91.1 |
| PF [31] | 91.8 | 86.8/81.7 | 70.4/88.9 | 82.8/82.2 | 88.9 |
| PKD [28] | 92.0 | 85.0/79.9 | 70.7/88.9 | 81.5/81.0 | 89.0 |
| $BERT_6$ + Finetune | 92.6 | 86.3/81.4 | 70.4/88.9 | 82.0/80.4 | 89.3 |
| $BERT_6$ + KD | 90.8 | 86.7/81.4 | 70.5/88.9 | 81.6/80.8 | 88.9 |
| $BERT_6$ + RCO | 92.6 | 86.8/81.4 | 70.4/88.7 | 82.3/81.2 | 89.3 |
| $BERT_6$ + ProKT ($\lambda = 0$) | 92.9 | **87.1/82.3** | 70.7/88.9 | 82.5/81.3 | 89.4 |
| $BERT_6$ + ProKT | **93.3** | 87.0/82.3 | **70.9/88.9** | **82.9/82.2** | **89.7** |
| $TinyBERT_6$ [12] | 93.1 | 87.3/82.6 | **71.6/89.1** | **84.6**/83.2 | 90.4 |
| $TinyBERT_6$ + ProKT | **93.6** | **88.1/83.8** | 71.2/89.2 | 84.2/**83.4** | **90.9** |
| bi-LSTM | 86.3 | 76.2/67.0 | 60.1/80.7 | 66.9/66.6 | 73.2 |
| bi-LSTM + KD | 86.4 | 77.7/68.1 | **60.7/81.2** | 68.1/67.6 | 72.7 |
| bi-LSTM + RCO | 86.7 | 76.0/67.3 | 60.1/80.4 | 66.9/67.6 | 72.5 |
| bi-LSTM + ProKT ($\lambda = 0$) | 86.2 | 80.1/71.8 | 59.7/79.7 | 68.4/68.3 | 73.5 |
| bi-LSTM + ProKT | **88.3** | **80.3/71.0** | 60.2/80.4 | **68.8/69.1** | **76.1** |

the constraint from student when updating teacher could further improve the performance.

ProKT is effective as well when it is combined with different objective of knowledge distillation. When combined with contrastive representation learning loss in CRD, as shown in Table 1, and combined with TinyBERT in Table 2, ProKT could further boost the performance and achieves the state-of-the-art results in almost all settings.

ProKT is especially effective when the student is of different structure with teacher. As shown in Table 2, when the student is bi-LSTM, directly distilling knowledge from a pre-trained BERT has a minor effect. ProKT could improve a larger margin for bi-LSTM than small BERT when distilled from BERT-base. Since learning from a heterogeneous teacher is more difficult, exposing teacher's training process to student could offer better guidance to the student.

### 4.3 Discussion

**Training Dynamics.** To visualize the training dynamics of teacher model and student model, we show the training loss of student model and the training accuracy of teacher model in Fig. 2. The training losses are calculated by the KL divergence between the student model and their intermediate targets. Figure 2a shows that the divergence between student and teacher in ProKT (i.e., the training loss for ProKT) is smooth and well bounded to a relative small value. For discrete targets in RCO, the divergence is bounded well in the beginning of training. However, at the target switching points, there are impulses in the training curve and then the loss is kept to a relative larger value.

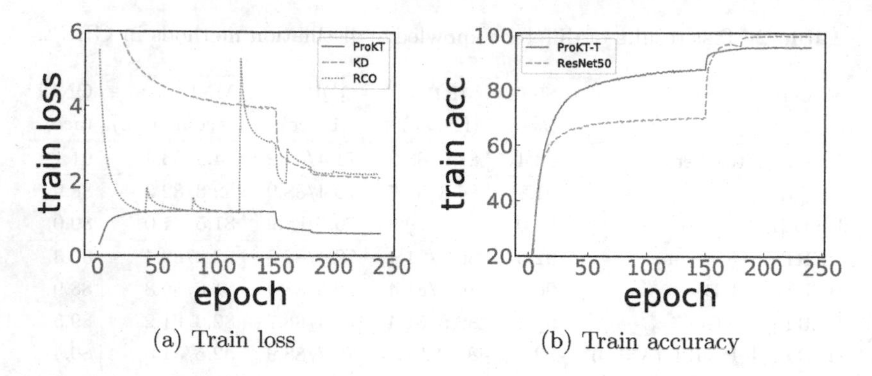


(a) Train loss (b) Train accuracy

**Fig. 2.** Training loss and accuracy for MobileNetV2 distilled from ResNet50 on CIFAR 100.

Then, we examine the performance of teacher model in Fig. 2b. ResNet50 refers to the teacher model which is trained by vanilla loss. While the ProKT-T denotes the teacher model which updated by the ProKT loss. It could be found that the performance of teacher model in ProKT deteriorates because of the "local" constraint from student. However, the lower training accuracy for teacher model does not affect the training performance of the student model as illustrated in the Table 1. These results show that a better teacher could not guarantee a better student model, which further justifies our intuition that involving local targets is beneficial for the learning of the student model.

**Ablation Study.** To test the impact of the constraint from student in Eq. 6, test and valid accuracy with respect to different $\lambda$ for image and text classification tasks are shown in Fig. 3. It is illustrated that the performance is improved in an appropriate range of $\lambda$, which means that the constraint term is helpful to provide appropriate targets. However, when the $\lambda$ is too large, the regularization from student will heavily damage the training of teacher and the performance of student will drop.

**Training Cost.** In our ProKT, the teacher model should be trained as well as the student models, which brings extra training cost compared with directly training the student models. Taking the distillation from $\text{BERT}_{12}$ to $\text{BERT}_6$ as an example, the time multiples of training by ProKT relative to training vanilla KD is listed in Table 3. On average, the training time for ProKT is about 2x to vanilla KD. However, the training time is not a bottleneck in practice. Because the model is trained once but runs unlimitedly, inference time is the main concern in the deployment of neural models. Our model has the same inference complexity as vanilla KD.

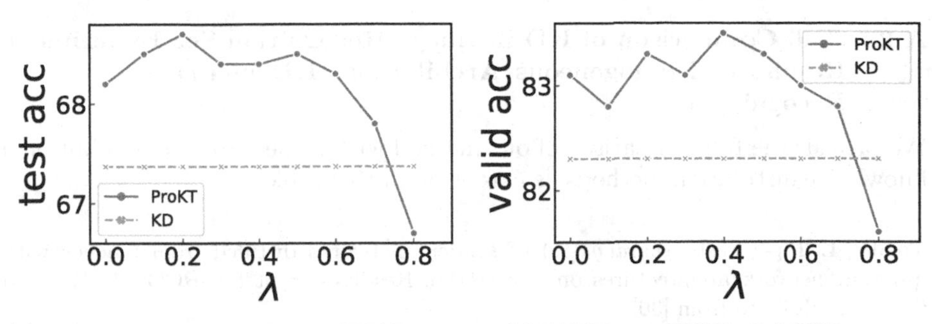


(a) VGG13 to MobileNetV2 in CIFAR 100. (b) $\text{BERT}_{12}$ to $\text{BERT}_6$ in MNLI-mm.

**Fig. 3.** Test/valid accuracy with different value of $\lambda$ for teacher and student model in ProKT.

**Table 3.** The time multiples of training by ProKT relative to training vanilla KD.

| Dataset | SST-2 | MRPC | QQP | MNLI | QNLI |
|---|---|---|---|---|---|
| Time cost of ProKT | 2.1x | 2.0x | 1.8x | 1.7x | 1.8x |

## 5 Conclusion

We propose a novel model agnostic knowledge distillation method, ProKT. The method projects the step-by-step supervision signal on the optimization procedure of student with an approximate mirror descent fashion, *i.e.*, student model learns from a dynamic teacher sequence while the progressive teacher is aware of the learning process of student. Experimental results show that ProKT achieves good performance in knowledge distillation for both image and text classification tasks.

**Acknowledgement.** We thanks the colleagues in MLNLC group for the helpful discussions and insightful comments.

## A Appendix

### A.1 Experimental Details for Text Classification

We use the pre-trained BERTs released by [31] except for TinyBERTs. For TinyBERTs, we use the pre-trained model released by [12][3]. We fine-tune 4 epoch for non-distillation training and 6 epoch for distillation training. Adam [14] optimizer with learning rate 0.001 is used for biLSTM and with a learning rate from {3e-5, 5e-5, 1e-4} is used for BERTs. The hyper-parameter of $\lambda$ in Eq. 6 is chosen according to the performance in the validation set. For ProKT in TinyBERT, we use the data argumentation following [12].

[3] https://github.com/huawei-noah/Pretrained-Language-Model/tree/master/TinyBERT.

### A.2 Full Comparison of KD in Image Recognition Sec Experiment Results of Homogeneous Architecture KD in Image Recognition

We provide the full comparison of our method with respect to several additional knowledge distillation methods as extension in the Table 4.

**Table 4.** Top-1 test *accuracy* (%) of student networks distilled from teacher with different network architectures on CIFAR100. Results except the RCO, ProKT and CRD+ProKT are from [30].

| Teacher<br>Student | vgg13<br>MobileNetV2 | ResNet50<br>MobileNetV2 | ResNet50<br>vgg8 | resnet 32 × 4<br>ShuffleNetV1 | resnet 32 × 4<br>ShuffleNetV2 | WRN-40-2<br>ShuffleNetV1 |
|---|---|---|---|---|---|---|
| Teacher | 74.64 | 79.34 | 79.34 | 79.42 | 79.42 | 75.61 |
| Student | 64.6 | 64.6 | 70.36 | 70.5 | 71.82 | 70.5 |
| KD* | 67.37 | 67.35 | 73.81 | 74.07 | 74.45 | 74.83 |
| FitNet* | 64.14 | 63.16 | 70.69 | 73.59 | 73.54 | 73.73 |
| AT | 59.40 | 58.58 | 71.84 | 71.73 | 72.73 | 73.32 |
| SP | 66.30 | 68.08 | 73.34 | 73.48 | 74.56 | 74.52 |
| CC | 64.86 | 65.43 | 70.25 | 71.14 | 71.29 | 71.38 |
| VID | 65.56 | 67.57 | 70.30 | 73.38 | 73.40 | 73.61 |
| RKD | 64.52 | 64.43 | 71.50 | 72.28 | 73.21 | 72.21 |
| PKT | 67.13 | 66.52 | 73.01 | 74.10 | 74.69 | 73.89 |
| AB | 66.06 | 67.20 | 70.65 | 73.55 | 74.31 | 73.34 |
| FT* | 61.78 | 60.99 | 70.29 | 71.75 | 72.50 | 72.03 |
| NST* | 58.16 | 64.96 | 71.28 | 74.12 | 74.68 | 74.89 |
| RCO | 68.42 | 68.95 | 73.85 | 75.62 | **76.26** | 75.53 |
| ProKT | **68.79** | **69.32** | **73.88** | **75.79** | 75.59 | **76.02** |
| CRD | 69.73 | 69.11 | 74.30 | 75.11 | 75.65 | 76.05 |
| CRD+KD | **69.94** | 69.54 | 74.58 | 75.12 | 76.05 | 76.27 |
| CRD+ProKT | 69.59 | **69.93** | **75.14** | **76.0** | **76.86** | **76.76** |

### A.3 ProKT as Approximate Mirror Descent

Following the assumption that supervised learning could globally solve a convex optimization problem, it could be shown the proposed method corresponds to a special case of mirror descent [1] with the objective as $H(y, q_{\theta_s})$. Note the optimization procedure is conducted on the output distribution space, the constraint is the solution must lie on the manifold of output distributions which could be characterized in the same way as the student model. We use $\mathcal{Q}_{\theta_s}$ to denote the possible output distribution family with the same parameterization as the student model.

**Proposition 1.** *The proposed ProKT solves the optimization problem:*

$$q_{\theta_s} \leftarrow \arg \min_{q_{\theta_s} \in \mathcal{Q}_{\theta_s}} H(y, q_{\theta_s})$$

*with mirror descent by iteratively conducting the following two step optimization at step m:*

$$q_{\theta_t}^m \leftarrow \arg\min_{q_{\theta_t}} H(y, q_{\theta_t}) \ s.t. \ D_{KL}\left(q_{\theta_s}^m, q_{\theta_t}^m\right) \leq \epsilon, \quad q_{\theta_s}^{m+1} \leftarrow \arg \min_{q_{\theta_s} \in \mathcal{Q}_{\theta_s}} D_{KL}\left(q_{\theta_t}^m, q_{\theta_s}\right) \quad (8)$$

The first step is to find a better output distribution which minimizes the classification task and is close to the previous student distribution $q_{\theta_s}^{m}$ under the KL divergence. While the second step projects the distribution in the distribution family $\mathcal{Q}_{\theta_s}$ in terms of the KL divergence. The monotonic property directly follows the monotonic improvement in mirror descent [1].

## References


1. Beck, A., Teboulle, M.: Mirror descent and nonlinear projected subgradient methods for convex optimization. Oper. Res. Lett. **31**(3), 167–175 (2003)
2. Bengio, Y., Louradour, J., Collobert, R., Weston, J.: Curriculum learning. In: Proceedings of the 26th Annual International Conference on Machine Learning, pp. 41–48 (2009)
3. Buciluč, C., Caruana, R., Niculescu-Mizil, A.: Model compression. In: Proceedings of the 12th ACM SIGKDD International Conference on Knowledge Discovery and Data Mining, pp. 535–541 (2006)
4. Chen, W., et al.: Generative bridging network in neural sequence prediction. arXiv preprint arXiv:1706.09152 (2017)
5. Devlin, J., Chang, M.W., Lee, K., Toutanova, K.: Bert: Pre-training of deep bidirectional transformers for language understanding. arXiv preprint arXiv:1810.04805 (2018)
6. Dolan, W.B., Brockett, C.: Automatically constructing a corpus of sentential paraphrases. In: Proceedings of the Third International Workshop on Paraphrasing (IWP2005) (2005)
7. He, H., Eisner, J., Daume, H.: Imitation learning by coaching. In: Advances in Neural Information Processing Systems, pp. 3149–3157 (2012)
8. He, K., Zhang, X., Ren, S., Sun, J.: Deep residual learning for image recognition. In: Proceedings of the IEEE Conference on Computer Vision and Pattern Recognition, pp. 770–778 (2016)
9. He, Y., Lin, J., Liu, Z., Wang, H., Li, L.J., Han, S.: Amc: automl for model compression and acceleration on mobile devices. In: Proceedings of the European Conference on Computer Vision (ECCV), pp. 784–800 (2018)
10. Hinton, G., Vinyals, O., Dean, J.: Distilling the knowledge in a neural network. arXiv preprint arXiv:1503.02531 (2015)
11. Howard, A.G., et al.: Mobilenets: Efficient convolutional neural networks for mobile vision applications. arXiv preprint arXiv:1704.04861 (2017)
12. Jiao, X., et al.: Tinybert: Distilling bert for natural language understanding. arXiv preprint arXiv:1909.10351 (2019)
13. Jin, X., et al.: Knowledge distillation via route constrained optimization. In: Proceedings of the IEEE International Conference on Computer Vision, pp. 1345–1354 (2019)
14. Kingma, D.P., Ba, J.: Adam: A method for stochastic optimization. arXiv preprint arXiv:1412.6980 (2014)
15. Krizhevsky, A., Hinton, G., et al.: Learning multiple layers of features from tiny images (2009)
16. Le, H.M., Kang, A., Yue, Y., Carr, P.: Smooth imitation learning for online sequence prediction. arXiv preprint arXiv:1606.00968 (2016)
17. Levine, S., Koltun, V.: Guided policy search. In: International Conference on Machine Learning, pp. 1–9 (2013)

18. Lu, S., Yu, L., Feng, S., Zhu, Y., Zhang, W., Yu, Y.: Cot: Cooperative training for generative modeling of discrete data. arXiv preprint arXiv:1804.03782 (2018)
19. Mirzadeh, S.I., Farajtabar, M., Li, A., Ghasemzadeh, H.: Improved knowledge distillation via teacher assistant: Bridging the gap between student and teacher. arXiv preprint arXiv:1902.03393 (2019)
20. Montgomery, W.H., Levine, S.: Guided policy search via approximate mirror descent. In: Advances in Neural Information Processing Systems, pp. 4008–4016 (2016)
21. Phuong, M., Lampert, C.: Towards understanding knowledge distillation. In: International Conference on Machine Learning, pp. 5142–5151 (2019)
22. Polino, A., Pascanu, R., Alistarh, D.: Model compression via distillation and quantization. arXiv preprint arXiv:1802.05668 (2018)
23. Radford, A., Narasimhan, K., Salimans, T., Sutskever, I.: Improving language understanding by generative pre-training. https://s3-us-west-2.amazonaws.com/openai-assets/researchcovers/languageunsupervised/languageunderstandingpaper.pdf (2018)
24. Rajpurkar, P., Zhang, J., Lopyrev, K., Liang, P.: Squad: 100,000+ questions for machine comprehension of text. arXiv preprint arXiv:1606.05250 (2016)
25. Sandler, M., Howard, A., Zhu, M., Zhmoginov, A., Chen, L.C.: Mobilenetv 2: inverted residuals and linear bottlenecks. In: Proceedings of the IEEE Conference on Computer Vision and Pattern Recognition, pp. 4510–4520 (2018)
26. Simonyan, K., Zisserman, A.: Very deep convolutional networks for large-scale image recognition. arXiv preprint arXiv:1409.1556 (2014)
27. Socher, R., et al.: Recursive deep models for semantic compositionality over a sentiment treebank. In: Proceedings of the 2013 Conference on Empirical Methods in Natural Language Processing, pp. 1631–1642 (2013)
28. Sun, S., Cheng, Y., Gan, Z., Liu, J.: Patient knowledge distillation for bert model compression. arXiv preprint arXiv:1908.09355 (2019)
29. Sutskever, I., Hinton, G.E., Taylor, G.W.: The recurrent temporal restricted boltzmann machine. In: Advances in Neural Information Processing Systems, pp. 1601–1608 (2009)
30. Tian, Y., Krishnan, D., Isola, P.: Contrastive representation distillation. arXiv preprint arXiv:1910.10699 (2019)
31. Turc, I., Chang, M.W., Lee, K., Toutanova, K.: Well-read students learn better: On the importance of pre-training compact models. arXiv preprint arXiv:1908.08962 (2019)
32. Vaswani, A., et al.: Attention is all you need. In: Advances in Neural Information Processing Systems, pp. 5998–6008 (2017)
33. Wang, A., Singh, A., Michael, J., Hill, F., Levy, O., Bowman, S.R.: Glue: a multi-task benchmark and analysis platform for natural language understanding. arXiv preprint arXiv:1804.07461 (2018)
34. Williams, A., Nangia, N., Bowman, S.R.: A broad-coverage challenge corpus for sentence understanding through inference. arXiv preprint arXiv:1704.05426 (2017)
35. Wu, J., Leng, C., Wang, Y., Hu, Q., Cheng, J.: Quantized convolutional neural networks for mobile devices. In: Proceedings of the IEEE Conference on Computer Vision and Pattern Recognition, pp. 4820–4828 (2016)
36. Yim, J., Joo, D., Bae, J., Kim, J.: A gift from knowledge distillation: fast optimization, network minimization and transfer learning. In: Proceedings of the IEEE Conference on Computer Vision and Pattern Recognition, pp. 4133–4141 (2017)
37. Zagoruyko, S., Komodakis, N.: Wide residual networks. arXiv preprint arXiv:1605.07146 (2016)

38. Zhang, X., Zhou, X., Lin, M., Sun, J.: Shufflenet: an extremely efficient convolutional neural network for mobile devices. In: Proceedings of the IEEE Conference on Computer Vision and Pattern Recognition, pp. 6848–6856 (2018)
39. Zhang, Y., Xiang, T., Hospedales, T.M., Lu, H.: Deep mutual learning. In: Proceedings of the IEEE Conference on Computer Vision and Pattern Recognition, pp. 4320–4328 (2018)
40. Zhou, Z., Zhang, Q., Lu, G., Wang, H., Zhang, W., Yu, Y.: Adashift: Decorrelation and convergence of adaptive learning rate methods. arXiv preprint arXiv:1810.00143 (2018)

# TaxoRef: Embeddings Evaluation for AI-driven Taxonomy Refinement

Lorenzo Malandri[1,3](✉), Fabio Mercorio[1,3](✉), Mario Mezzanzanica[1,3], and Navid Nobani[2]

[1] Department of Statistics and Quantitative Methods, University of Milano Bicocca, Milan, Italy
{lorenzo.malandri,fabio.mercorio,mario.mezzanzanica,navid.nobani}@unimib.it
[2] Department of Informatics, Systems and Communication, University of Milano Bicocca, Milan, Italy
[3] CRISP Research Centre, University of Milano Bicocca, Milan, Italy

**Abstract.** Taxonomies provide a structured representation of semantic relations between lexical terms. In the case of standard official taxonomies, the refinement task consists of maintaining them updated over time, while preserving their original structure. To date, most of the approaches for automated taxonomy refinement rely on word vector models. However, none of them considers to what extent those models encode the taxonomic similarity between words. Motivated by this, we propose and implement `TaxoRef`, a methodology that (i) synthesises the semantic similarity between taxonomic elements through a new metric, namely `HSS`, (ii) evaluates to what extent the embeddings generated from a text corpus preserve those similarity relations and (iii) uses the best embedding resulted from this evaluation to perform taxonomy refinement. `TaxoRef` is a part of the research activity of a 4-year EU project that collects and classifies millions of Online Job Ads for the 27+1 EU countries. It has been tested over 2M ICT job ads classified over ESCO, the European standard occupation and skill taxonomy.

Experimental results confirm (i) the `HSS` outperforms previous metrics for semantic similarity in taxonomies, and (ii) `TaxoRef` accurately encodes similarities among occupations, suggesting a refinement strategy.

**Keywords:** Taxonomy refinement · Semantic similarity · Word embeddings evaluation

## 1 Introduction and Motivation

Taxonomies are a natural way to represent and organise concepts in a hierarchical manner. They are pivotal for machine understanding, natural language processing, and decision-making tasks. However, taxonomies are domain-dependent, usually have low coverage and their manual creation and update are time-consuming requiring a domain-specific knowledge [16]. For these reasons, many


N. Oliver et al. (Eds.): ECML PKDD 2021, LNAI 12977, pp. 612–627, 2021.
https://doi.org/10.1007/978-3-030-86523-8_37

researchers have tried to automatically infer semantic information from domain-specific text corpora to build or update taxonomies. Unlike the automated construction of new taxonomies from scratch, which is a well-established research area [44], the refinement of existing hierarchies is gaining in importance. Due to the evolution of human languages and proliferation of online contents, it is often required to improve existing taxonomies while maintaining their structure. To date, the most adopted approaches to enrich or extend standard *de-jure* taxonomies lean on expert panels and surveys; while these approaches entirely rely on human knowledge and have no support from the AI-side, in this paper we resort to word embeddings, an outcome of the distributional semantics field. Word embeddings rely on the assumption that words occurring in the same context tend to have similar meanings. These methods are semi-supervised and knowledge-poor, thus suitable for large corpora and evolving scenarios.

As a contribution of the paper, we propose a method for the evaluation of word embeddings based on their ability to represent an existing hierarchy.

Evaluating the intrinsic quality of vector space models, as well as their impact when used as input of specific tasks (*aka*, extrinsic quality), has a very practical significance (see, e.g. [10]), as this affects the trustworthiness of the overall process or system in which they are used (see, e.g., [30,45]). We may argue that the well-known principle *"garbage-in, garbage-out"* of the data quality research field, also applies to word embedding, as *the lower the quality of the embeddings, the lower the effectiveness of the tasks based on them.*

This thought, along with the need of keeping updated current taxonomies, inspired this paper, that is framed within the research activities of an ongoing European tender[1] for the Cedefop EU Agency[2]. The project aims at realising a European system to collect and classify Online Job Vacancies (OJVs)[3] for the whole EU country members through machine learning [8]. OJVs are encoded within ESCO [4], an extensive taxonomy with 2,942 occupations and 13,485 skills serving as a *lingua franca* to compare labour market dynamics across borders. To fill the gap between the lexicon used in labour market demand (*de facto*), and the one used in standard taxonomies (*de jure*), we propose `TaxoRef`, a method for encoding the semantic similarity of taxonomic concepts into a vector representation synthesised from the data.

**Contribution.** The contributions of this work are:

1. First, we define a domain-independent metric, i.e., the *Hierarchical Semantic Similarity* (`HSS`) to measure the pairwise semantic similarity among words in a taxonomy;

[1] Real-time Labour Market information on Skill Requirements: Setting up the EUsystem for online vacancy analysis. https://goo.gl/5FZS3E (2016) .

[2] The European Center for the Development of Vocational Training https://www.cedefop.europa.eu/.

[3] An Online Job Vacancy (OJV, *aka*, job offers) is a document containing a *title* - that shortly summarises the job position - and a *full description*, that advertises the skills a candidate should hold.

[4] The European Taxonomy of Occupations, Skills & Qualifications https://ec.europa.eu/.

2. Second, we define and implement TaxoRef to select embeddings from a large text corpus that preserves the similarity relationships synthesised from a domain-specific taxonomy. Then, it uses the selected embedding to estimate to what extent the data adheres the taxonomy, identifying new relations between entities and concepts to allow the taxonomy to represent the data better.
3. Third, we apply TaxoRef to a real-life scenario, framed within an EU Project (See Footnote 1) in the context of Labour Market.

To date, `TaxoRef` is the first approach aimed at generating and evaluating word embeddings for labour market data.

The paper is organised as follows: we first present the related work, then we describe `TaxoRef` and we apply it to a real-life case in the field of Labour Market. Finally, some concluding remarks are drawn.

## 2 The Significance of Analysing Job Ads

In recent years, the European Labour demand conveyed through specialised Web portals and services has grown exponentially. This also contributed to introducing the term *"Labour Market Intelligence"* (LMI), which refers to the use and design of AI algorithms and frameworks to analyse Labour Market Data for supporting decision making (see, e.g., [20,42,47]).

Nowadays, the problem of monitoring, analysing, and understanding labour market changes (i) timely and (ii) at a very fine-grained geographical level, has become practically significant in our daily lives. Recently, machine learning has been applied to compute the effect of robotisation within occupations in the US labour market [15] as well as to analyse skill relevance in the US standard taxonomy O*Net [2], just to cite a few. In 2016 the EU Cedefop agency - aimed at supporting the development of European Vocational Education and Training - has launched a European tender for realising a machine-learning-based system able to collect and classify Web job vacancies from all 28 EU country members using the ESCO hierarchy for reasoning over the 32 languages of the Union (See Footnote 1). Preliminary results of this project have focused on analysing lexicon extracted from OJVs as well as to identify novel occupations and skills (see, e.g. [7,8,18,19,27]).

As one might note, the use of classified OJVs and skills, in turn, enables several third-party research studies to understand and explain complex labour market phenomena. Just to give a very recent few examples, in [12] authors used OJVs for estimating the impact of AI in job automation and measuring the impact of digital/soft skills within occupations; On May 2020, the EU Cedefop Agency has been started using those OJVs to build an index named Cov19R that identifies workers with a higher risk of COVID-19 exposure, who need greater social distancing, affecting their current and future job performance capacity[5].

[5] https://www.cedefop.europa.eu/en/news-and-press/news/cedefop-creates-cov19r-social-distancing-risk-index-which-eu-jobs-are-more-risk.

While on the one side ESCO allows comparing different labour markets, on the other side it does not encode the characteristics and peculiarities of such labour markets, in terms of skills advertised - that vary Country by Country - and the meaning of terms, that are used differently based on the level of maturity of local labour markets. In such a scenario, TaxoRef would allow encoding semantic similarities as they emerge from OJVs within ESCO.

## 3 Preliminaries and Related Work

In this section, we survey the state of the art on taxonomy refinement and embeddings evaluation, showing how these techniques are related to TaxoRef. Then, some background notions are introduced to explain how TaxoRef works.

*Taxonomy Refinement.* While the automatic extraction of taxonomies from text corpora has received considerable attention in past literature [44], the evaluation and refinement of existing taxonomies are growing in importance in all the scenarios where the user wants to update a taxonomy maintaining its structure rather than rebuilding it from scratch. Most of the existing methods are either domain-dependent [34] or related to lexical structures specific to the existing hierarchies [37]. Two recent TExEval tasks at SemEval-2016 [5,6] introduce a setting for the evaluation of taxonomies extracted from a test corpus, using standard precision, recall and F1 measures over gold benchmark relations from WordNet and other well-known resources, resorting to human evaluation for relations not included in the benchmark. Though interesting, this methodology relies on existing resources, which to some extent, could be inaccurate themselves. [3] employ Poincaré embeddings to find a child node that is assigned to a wrong parent. Although the goal of this research is quite similar to ours, in the detection of outliers they use Poincaré embeddings trained on the taxonomy, without considering any external corpus, thus information on the use of taxonomic terms. Though all these approaches are very relevant, none of them exploits in any way the structure of an existing hierarchy to preserve taxonomic relations in word embeddings.

*Evaluation Methods for Word Embeddings.* When it comes to classifying embedding evaluation methods, researchers almost univocally agree on *intrinsic* vs *extrinsic* division (see, e.g., [41]).

*Intrinsic metrics* attempt to increase the syntactic or semantic relationships between words either by directly getting human judgement (Comparative intrinsic evaluation [41]) or by comparing the aggregated results with the pre-constructed datasets (Absolute intrinsic evaluation). Some of the limitations of *intrinsic metrics* are: (i) suffering from word sense ambiguity (faced by a human tester) and subjectivity (see, e.g. [43]); (ii) facing difficulties in finding significant differences between models due to the small size and low inter-annotator agreement of existing datasets and (iii) need for constructing judgement datasets for each language [24].

*Extrinsic metrics* perform the evaluation by using embeddings as features for specific downstream tasks. For instance, question deduplication [25],

Part-of-Speech (POS) Tagging, Language Modelling [38] and Named Entity Recognition (NER) [17], just to cite a few. As for drawbacks, extrinsic metrics (i) are computationally heavy [43]; (ii) have high complexity of creating gold standard datasets for downstream tasks, and (iii) lack the performance consistency on downstream tasks [41]. The metric we propose in this paper are similar to the *intrinsic thesaurus-based evaluation metrics* as they use an expert-constructed taxonomy to evaluate the word vectors as an intrinsic metric [4]. These approaches typically evaluate embeddings through their correlation with manually crafted lexical resources, like expert rating of similarity or relatedness between hierarchical elements. However, those resources are usually limited and hard to create and maintain. Furthermore, human similarity judgments evaluate ex novo the semantic similarity between taxonomic elements, but the structure of the taxonomy already encodes information about the relations between its elements. In this research, instead, we exploit the information encoded in an existing taxonomy to build a benchmark for the evaluation of word embeddings. As far as we know, this is the first attempt to encode similarity relationships from a manually built semantic hierarchy into a distributed word representation, automatically generated from a text corpus.

# 4 The TaxoRef Approach

Below we present TaxoRef, depicted in Fig. 1. The method is composed of four tasks. In *Task 1* we define a measure of similarity within a semantic hierarchy, which will serve as a basis for the evaluation of the embeddings. In *Task 2* we generate embeddings from a domain-specific corpus, and we identify vector pairs which represent a pair of elements in the semantic hierarchy. In *Task 3* we evaluate the embeddings generated in Task 2 using three different criteria. Lastly, in *Task 4*, we evaluate the actual taxonomy suggesting entities which should be assigned to a different concept.

## 4.1 Task 1: Hierarchical Semantic Similarity (HSS)

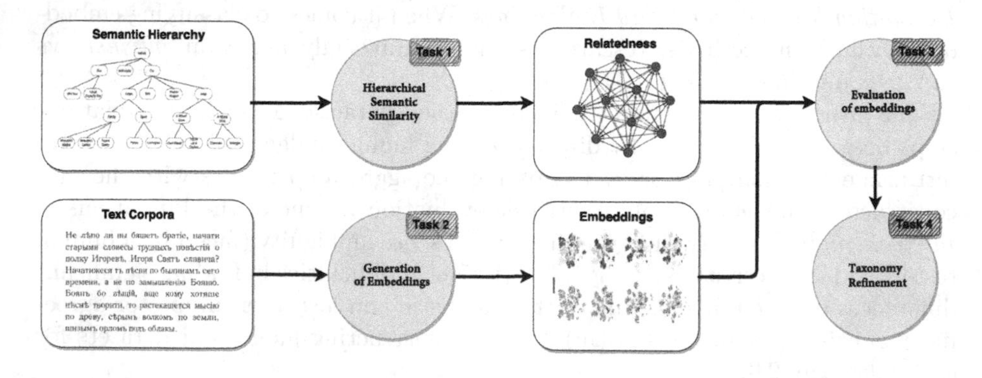


**Fig. 1.** A graphical overview of TaxoRef

Below we introduce a measure of similarity in a taxonomy, based on the concept of information content. Intuitively, the lower the rank of a concept $c \in \mathcal{C}$ which contains two entities, the higher the information content ($IC$) the two entities share. According to information theory, the $IC$ of a concept $c$ can be approximated by its negative log-likelihood: $IC(c) = -\log p(c)$ where $p(c)$ is the probability of encountering the concept $c$. Building on [40], we can supplement the taxonomy with a probability measure $p : \mathcal{C} \rightarrow [0, 1]$ such that for every concept $c \in \mathcal{C}$, $p(c)$ is the probability of encountering an instance of the concept $c$. It follows that $p$ (i) is monotonic and (ii) decreases with the rank of the taxonomy, i.e., if $c_1$ is a sub-concept of $c_2$, then $p(c_1) \leq p(c_2)$. To estimate the values of $p$, in [40] the author uses the frequency of the concepts in a large text corpus. Anyway, our purpose is to infer the similarity values intrinsic to the semantic hierarchy, since we want to extend a semantic hierarchy built by human experts. As a consequence, we use the frequencies of concepts and entities in the taxonomy to compute the value of $p$, that is $\hat{p}(c) = \frac{N_c}{N}$ where $N$ is the cardinality, i.e. the number of entities (words), of the taxonomy and $N_c$ the sum of the cardinality of the concept $c$ with the cardinality of all its hyponyms. Note that $\hat{p}(c)$ is monotonic and increases with granularity, thus respects our definition of $p$.

Now, given two words $w_1$ and $w_2$, [40] defines $c_1 \in s(w_1)$ and $c_2 \in s(w_2)$ all the concepts containing $w_1$ and $w_2$ respectively, i.e. the *senses* of $w_1$ and $w_2$. Therefore, there are $S_{w_1} \times S_{w_2}$ possible combinations of their word senses, where $S_{w_1}$ and $S_{w_2}$ are the cardinality of $s(w_1)$ and $s(w_2)$ respectively. We can now define $\mathcal{L}$ as the set of all the lowest common ancestor (LCA) for all the combinations of $c_1 \in s(w_1), c_2 \in s(w_2)$. The hierarchical semantic similarity between the words $w_1$ and $w_2$ can be defined as:

$$sim_{\mathtt{HSS}}(w_1, w_2) = \sum_{\ell \in \mathcal{L}} \hat{p}\left(\ell = L \mid w_1, w_2\right) \times IC(L) \quad (1)$$

where $\hat{p}\left(\ell = L \mid w_1, w_2\right)$ is the probability of $L$ being the LCA of $w_1, w_2$, and can be computed as follows applying the Bayes theorem:

$$\hat{p}\left(\ell = L \mid w_1, w_2\right) = \frac{\hat{p}\left(w_1, w_2 \mid \ell = L\right) \hat{p}\left(L\right)}{\hat{p}\left(w_1, w_2\right)} \quad (2)$$

We define $N_\ell$ as the cardinality of $\ell$ and all its descendants. Now we can rewrite the numerator of Eq. 2 as:

$$\hat{p}\left(w_1, w_2 \mid \ell = L\right) \hat{p}\left(L\right) = \frac{S_{<w_1,w_2>\in\ell}}{|descend(\ell)|^2} \times \frac{N_\ell}{N} \quad (3)$$

where the first leg of the *rhs* is the class conditional probability of the pair $< w_1, w_2 >$ and the second one is the marginal probability of class $\ell$. The term $|descend(\ell)|$ represents the number of subconcepts of $\ell$. Since we could have at most one word sense $w_i$ for each concept $c$, $|descend(\ell)|^2$ represents the maximum number of combinations of word senses $< w_1, w_2 >$ which have $\ell$ as LCA. $S_{<w_1,w_2>\in L}$ is the number of pairs of senses of word $w_1$ and $w_2$ which have

$L$ as LCA. The denominator can be written as:

$$\hat{p}(w_1, w_2) = \sum_{k \in \mathcal{L}} \frac{S_{<w_1, w_2> \in k}}{|descend(k)|^2} \tag{4}$$

### 4.2 Task 2: Generation of Embeddings

In this task, we generate several vector representation of the OJVs text corpus through fastText. FastText is an extension of word2vec for scalable word representation and classification which considers sub-word information by representing each word as the sum of its character $n$-gram vectors. This is important in corpora with short texts (like OJVs titles) and many words with the same root (e.g. engineer-engineering,developer-developing). Other embedding models have been evaluated along with fastText. Nevertheless, none of them fits our conditions. Classical embedding models [31,36], embeddings specifically designed to fit taxonomic data [14,23], and hyperbolic and spherical embeddings like HyperVec [33] or JoSe [29] don't consider subword information. Moreover, HyperVec uses hypernym-hyponym relationships for training, while we train our models on a OJVs corpus which has not such relationships. Context embeddings (see e.g. [13]) represent different word uses based on their context. It is not suitable for our case, since we aim at comparing words in a corpus and their specific sense within a taxonomy.

### 4.3 Task 3: Embeddings Evaluation

We describe three criteria to select the embeddings which better represents a semantic hierarchy. In essence, we want the relationship between vectors to reflect as more as possible the semantic similarity between words in the taxonomy.

**Cosine Similarity:** We maximise the cosine similarity between all the pairs of words $w$ belonging to the same concept $c \in \mathcal{C}$. We consider words belonging to the same concept to be semantically similar; thus we select the embedding which maximises the cosine Similarity between all the pairs of vectors of words belonging to the same concept.

**Silhouette Coefficient:** With cosine similarity, we maximise the similarity between words, but we do not discriminate for words that belong to different categories, as the silhouette coefficient does, taking into account the mean distance of word $i$ with all other words within the same concept, and $b(i)$ is the minimum mean distance of word $i$ to any word of another concept.

**Spearman's Rank Correlation Coefficient:** In *Task 1* we augmented the semantic hierarchy with a value of similarity inferred from the taxonomic relations between elements. We can compute the Spearman's rank correlation coefficient between the Cosine Similarity of the embeddings and the *hrs* measured above. The Spearman's Rank Correlation coefficient is a non-parametric measure of rank correlation often employed for intrinsic embeddings evaluation [4,41].

### 4.4 Task 4: Taxonomy Refinement

In this section, we present the criteria for moving a taxonomic entity to a different concept. We implement it applying the Bayes theorem. Thus, the probability that the word $w$, represented in the embedding space by the vector $\mathbf{v}$, belongs to the concept $c$ is given by:

$$p(c \mid \mathbf{v}) = \frac{p(\mathbf{v} \mid c)\, p(c)}{p(\mathbf{v})} \tag{5}$$

Thus the word $w$ is assigned to the class $c_i$ iff:

$$p(\mathbf{v} \mid c = c_i)\, p(c = c_i) \geq p(\mathbf{v} \mid c = c_j)\, p(c = c_j), c_j \in C \setminus c_i \tag{6}$$

Where the prior probability $p(c)$ is estimated through class frequency and the likelihood $p(\mathbf{v} \mid c)$, is: $p(\mathbf{v} \mid c) = p(v_1 \mid c) \times p(v_2 \mid c) \times ... \times p(v_D \mid c)$ where we assume conditional independence between the elements $v_1, v_2, ..., v_D$ of the vector $\mathbf{v}$, analogously to the Naive Bayes classifier. The probability $p(v_i \mid c)$ is estimated by a Gaussian density function for $\forall i \in 1, 2, ..., D$.

**Experimental Settings for Reproducibility.** All the experiments have been performed over an Intel i7 processor with Ubuntu-20, equipped with 32GB RAM. `TaxoRef` is implemented with Python 3.7. Tuning parameters are reported for each experiments while the source code of `TaxoRef` is provided on Github at https://github.com/Crisp-Unimib/TaxoRef.

### 4.5 Benchmarking HSS

Before employing the `HSS` for applications in the Labour Market, we test it against benchmark measures for semantic similarity in taxonomies: WUP [46], LC [26], Shortest Path, Resnik [40] and Jiang-Conrath [22]. See [21] for a survey. Following the state of the art, we evaluate the correlation of similarity scores obtained with `HSS` and benchmark metrics, with similarity generated by humans. We calculate the pairwise Pearson and Spearman correlation between measures of semantic similarity in taxonomies and three benchmark datasets with human-evaluated similarity: WSS [1], MT287 [39] and MEN [9]. For the sake of comparison and only for this task, we do not use ESCO to compute the `HSS`, but the WordNet taxonomy [32] and its benchmark metrics [35] used through the NLTK interface. Results are reported in Table 1. `HSS`, in general, outperforms the rest of the measures, both in terms of Pearson and Spearman correlations with the human-annotated datasets, with exception of the Pearson correlation with the human judjements in the MEN dataset, which ties with the measure of semantic similarity from [46]. These results confirm the performance superiority of `HSS` respect to the benchmark measures.

**Table 1.** Benchmarking HSS against other measures of semantic similarity in taxonomies: WUP, LC, Shortest Path, Resnik and Jiang-Conrath. A recent survey on those metrics can be found at [21]

| | HSS (ours) | | WUP | | LC | | Shortest path | | Resnik | | Jiang-Conrath | |
|---|---|---|---|---|---|---|---|---|---|---|---|---|
| | Pearson | Spearman | Pearson | Spearman | Pearson | Spearman | Pearson | Spearman | Pearson | Spearman | Pearson | Spearman |
| MEN | **0.41** | **0.33** | 0.36 | **0.33** | 0.14 | 0.05 | 0.07 | 0.03 | 0.05 | 0.03 | -0.05 | -0.04 |
| WSS | **0.68** | **0.65** | 0.58 | 0.59 | 0.36 | 0.23 | 0.16 | 0.1 | 0.02 | -0.03 | 0.04 | 0.06 |
| MT287 | **0.46** | **0.31** | 0.4 | 0.28 | 0.26 | 0.12 | 0.11 | 0.11 | 0.03 | 0.04 | 0.18 | 0.16 |

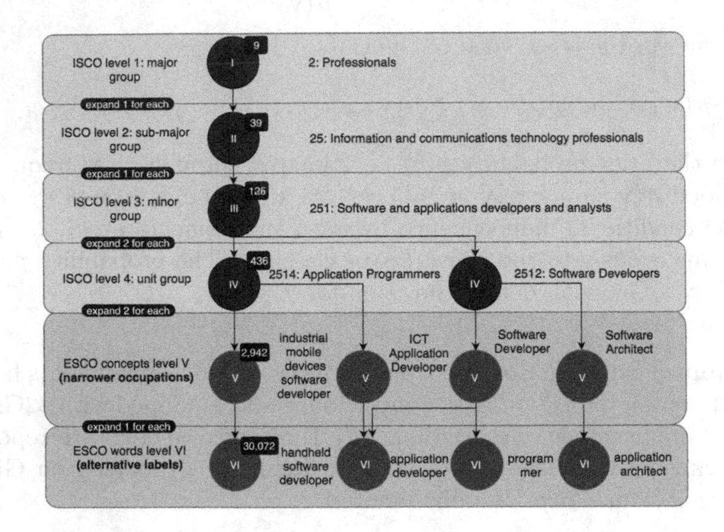


**Fig. 2.** The ESCO taxonomy built on top of ISCO

## 5 Experimental Results on 2M+ UK Online Job Ads

While on the one side, ESCO allows comparing different labour markets, on the other side it does not encode the characteristics and peculiarities of such labour markets, in terms of skills advertised - that vary country by country - and the meaning of terms, that are used differently based on the level of maturity of local labour markets. TaxoRef would allow encoding semantic similarities as they emerge from OJVs (i.e., the labour market demand) within ESCO, identifying relationships that might refine ESCO to fit the labour market demand better. In [18], we used the approach employed by TaxoRef to identify novel occupations to enrich the ESCO taxonomy with new emerging jobs.

**Data Overview.** Our experiments rely on the use of a large corpus of OJVs collected from online sources within the project (See Footnote 1)[6]. We selected the titles of all 2,119,493 online job vacancies referring to ICT jobs in the United Kingdom during the year 2018. Concepts belonging to the fifth and highest level of ESCO (*c5*) are called *narrower labels*, while all the hyponyms of the same

[6] Preliminary results at https://tinyurl.com/skillovate.

narrow label, called *alternative labels*, are co-hyponyms[7] and are different terms which express the same kind of occupation. As we can see in Fig. 2, the ISCO classification assigns a code only to the first four levels ($c4$). To evaluate the similarity between narrower and alternative labels, we assigned a new code to each narrower label. For instance, if the concept *2512*, at the fourth level of ESCO has two narrower labels as hyponyms, their codes will be 2512_01, and 2512_02 respectively.

**Generation of Embeddings.** We trained our vector model with the fastText library using both *skipgram* and *cbow*. We tested five values of the size of the embeddings (5, 25, 100, 300, 500), five for the number of epochs (10, 25, 100, 200, 300) and four for the learning rate (0.05, 0.1, 0.2, 0.5) for a total of 600 embeddings. All the subwords with 3 to 10 letters were considered. Average training times (with std) in seconds were $246 \pm 333$.

### 5.1 Embeddings Evaluation

In this section we evaluate how well the three embeddings selected by the criteria presented above preserve the semantic similarity between concepts in the ESCO taxonomy. The three embeddings which scored the best result for each criterion are shown in Table 2.

*Intrinsic Evaluation: The Mann-Whitney U-test.* [28] is a non-parametric test on whether a random variable is stochastically larger than another one. We want to assess to what extent the similarity of words in the same class (*co-hyponyms*) is different from the similarity of words in different classes. For this reason, we define *intra-class similarity* as the cosine similarity between all the pairs of occupations belonging to the same ESCO c5 class and *inter-class similarity* as the cosine similarity between all the pairs of occupations belonging to different ESCO c5 concepts. The *U-test* reports a p-value $\leq 0.0001$ for all models, and a statistic of 582, 194 for Model 1, 321, 016 for Model 2 and 647, 116 for Model 3.

**Table 2.** Selected embeddings and their evaluation criteria

| Model Name | Params Training mode\|dim\|epochs\|lr | Sim | Sil | SR ($\rho$, P-Value) |
|---|---|---|---|---|
| M1 | CBOW \|5\|25\|0.1 | **0.954** | 0.866 | (-0.023, $\leq$ 0.001) |
| M2 | CBOW \|25\|200\|0.05 | 0.919 | **0.875** | (-0.020, $\simeq$ 0.712) |
| M3 | CBOW \|100\|200\|0.1 | 0.866 | 0.843 | (**0.356**, $\leq$ 0.0001) |

**Table 3.** Classification results

| Model Name | Criterion | Precision | Recall | f1-score | Accuracy |
|---|---|---|---|---|---|
| M1 | Cos Sim | 0.522 | 0.688 | 0.513 | 0.877 |
| M2 | Silhouette | 0.663 | 0.715 | 0.685 | 0.982 |
| M3 | HSS | **0.964** | **0.801** | **0.865** | **0.994** |

[7] co-hyponyms refer to hierarchical concepts which share the same hypernym.

The *U-test* is positive and significant for each of the three criteria, which means that the cosine similarity between embeddings of occupations belong to the same class is stochastically larger than the embeddings similarities for occupations of different classes. This result holds for all the three selected embeddings, even though we have the larger statistics with the embedding chosen by HSS, and the smallest with the embedding chosen by silhouette.

*Extrinsic Evaluation: Classification.* We tested the selected embeddings through a binary classification task, predicting whether a pair of occupations belong to the same c5 concept (class 1) or not (class 0). In essence, given two $w_1$ and $w_2$, we predict if they are *co-hyponym.* We employ a Multilayer Perceptron (MLP) with a single hidden layer with 100 neurons and a dense output layer with ReLu activation. We test different parameters settings, with number of epochs in [10, 20, 50, 100, 200], batch size in [8,16,32,64] and *RMSprop*, *Adam* and *SGD* as optimisers. Average training time for each parameters combination (with std) in seconds was $253 \pm 374$. We divided the dataset in train and test, with a split of 70% train and 30% test. We select the best parameters based on the f1-score of 3-fold cross-validation on the training set. Since the dataset was highly imbalanced (309 samples of class 1 vs 24,444 samples of class 0) we over-sampled the training set using SMOTE [11]. The best parameters are batch size = 8, epochs = 200, Optimizer = SGD with f1-score of 0.69 for Model 1, batch size = 16, epochs = 200, Optimizer = RMSprop with f1-score of 0.52 for Model 2 and batch size = 32, epochs = 100, Optimizer = Adam with f1-score of 0.83 for Model 3. Classification scores on the test set are reported in the Table3. First column reports the criterion used for the selection of the embedding (see Table 2). Because the test set is highly imbalanced, we show the macro average of the four scores, to give more importance to the minority class.

**Table 4.** The fraction of ESCO V and VI level words to be assigned to each ESCO IV level concept according to the criterion in Task 4. The rows represent ESCO IV concepts. For each concept (row), the column *Accordance* reports the fraction of occupations terms which are assigned to the same concept by Eq. 6, while the column *Refinement* shows the fraction of occupation terms assigned to a different ESCO IV level concept, which is specified (note that only the main ones are presented, and rounded to the second decimal place, thus not all the rows sum to 1). Missing concepts do not need a refinement (i.e., Accordance = 1)

| ESCO concept | Accordance | Refinement (fraction) | | | | |
|---|---|---|---|---|---|---|
| 1330 ICT service managers | 1330 (0.5) | 2511 (0.18) | 2519 (0.12) | 2514 (0.06) | 2513 (0.06) | 3512 (0.06) |
| 2511 Systems analysts | 2511 (0.77) | 2512 (0.08) | 1330 (0.05) | 2521 (0.03) | 2513 (0.03) | 2522 (0.03) |
| 2512 SW developers | 2512 (0.82) | 2514 (0.11) | 2511 (0.05) | | | |
| 2513 Web and multimedia developers | 2513 (0.85) | 2512 (0.14) | | | | |
| 2519 SW & Application developers | 2519 (0.9) | 2514 (0.1) | | | | |
| 2521 DB designers and admin | 2521 (0.66) | 2511 (0.33) | | | | |
| 2529 DB and network professionals | 2529 (0.69) | 2511 (0.15) | 2522 (0.08) | 2519 (0.08) | | |

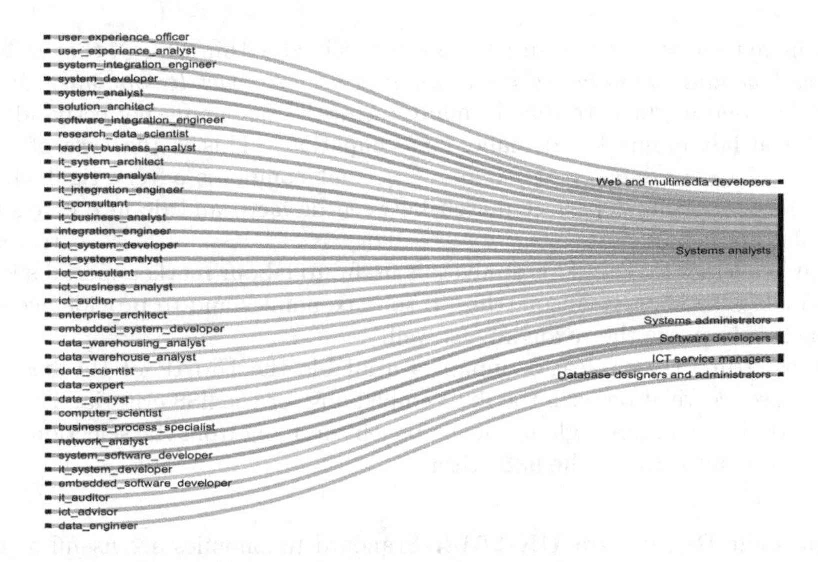


**Fig. 3.** Example of refinement for the ESCO concept *2511 Systems Analysts*

The vector representation selected is Table 3, that shows HSS outperforms the other two in classifying *co-hyponyms*. While all the three embeddings perform quite well on class 0 (non *co-hyponym* words), with and *f1-score* of the second and the third close to 1, only the third model shows good performances for class 1 (*co-hyponym* words).

### 5.2 Result Comments

**Comments on the Best Embedding.** As discussed in the introduction part of this paper, the evaluation and selection of the best model are mandatory activities that affect the trustworthiness of all the tasks that use such an embedding as input. To better clarify the matter, in Fig. 4 we provide a scatter plot produced over the best embedding model - as emerges from Table 3 - generated by means of UMAP. Each icon is assigned to one ISCO level 4 group, as in Fig. 2. The ESCO concepts and words belonging to each group are showed, distinguishing between narrower occupations (shallow shape) and alternative labels (filled shape). The embedding shown in Fig. 4 encodes the occupations as they emerge from the UK labour market (2M+ OJVs in 2018) within the ESCO taxonomy. This is beneficial for labour market specialists as a way to understand and monitor labour market dynamics. Specifically, one might observe that though a *data engineer* and a *data scientist* were designed to be co-hyponyms in ESCO, as they belong both to the *2511: Systems Analysts* ISCO group, their meaning is quite different in the real-labour market, as any computer scientist knows. The former indeed shares much more with *2521: Database designers and administrators* rather than its theoretical group. Still along this line, an *IT security technician* seems to be also co-hyponym with occupations in *2529: Database and network professionals*,

much more than with terms in its class as specified within ESCO, that is *3512: Information and communications technology user support technicians*. On the other side, one might note that in many cases, the taxonomy perfectly adheres to the real labour market demand for occupations. This is the case of *3521: Broadcasting and audio-visual technicians*, that composes a very tight cluster in the map, showing a perfect match between de-facto and de-jure categories. This also applies to ∗ *3513: Computer network and systems technicians*, even though in a lesser extent. This analysis is useful to labour market specialists and policymakers to identify mismatches in the taxonomies and to provide accurate feedback to improve the taxonomy as well.
At https://tinyurl.com/scatter-umap is available the UMAP of the best and worst (lower correlation with the HSS) embedding for the HSS criteria. The comparison of the two gives a glance of the benefit deriving from the selection of the best embedding through the HSS criterion.

**Refinement Results on UK-2018.** Standard taxonomies are useful as they represent a lingua franca for knowledge-sharing many domains. However, as they are built periodically by a panel of experts following a top-down approach, they quickly become obsolete, losing their ability to represent the domain. This is why there is a growing number of attempts to refine taxonomies following a data-driven paradigm. In this section, we employ the methodology described above to find the most suitable concept for each taxonomic entity.

Applying Eq. 6, we find that for the 83.4% of the words $w$ analysed, the concept $c_i$ with the highest probability to be its hypernym, is its current hypernym in ESCO. In Tab. 4, we present the evaluation for each ICT ESCO concept, excluding those with less than 10 entities. For instance, we can see that all the occupations of the group *3521: Broadcasting and audio-visual technicians* are tightly related among themselves, and none of them is moved to a different group. Conversely, only 50% of the occupations in group *1330: ICT Service Managers* are assigned to the group *1330* itself. For instance, the web project manager is assigned to the class ◀ *2513: Web and multimedia developers*. Note that these results depend on the corpus chosen for the embeddings generation. In this case, we choose ICT related OJVs posted in 2018 in the UK.

Figure 3 shows the refinement proposed by TaxoRef for the occupations belonging to the ESCO concept *2511 - Systems Analysts*. For the 7% (28 over 36) of the occupations the ESCO taxonomy and TaxoRef are in accordance, assigning them to the class *2511 - Systems Analysts*. For the remaining 8 occupations, TaxoRef suggests a different classification. The occupation *user experience officer*, for instance, is reassigned to the ESCO class *2513 - Web and Multimedia Developers*, the *data Engineer* to the class *2521 - DB Designers and Administrators* (as it was clear from Fig. 4) and so on. All the suggestions for the refinement are highly plausible and can constitute the basis for a discussion among experts on the accordance between the *de jure* taxonomy ESCO and the *de facto* labour market in a specific context as it emerges from labour market demand (OJVs).

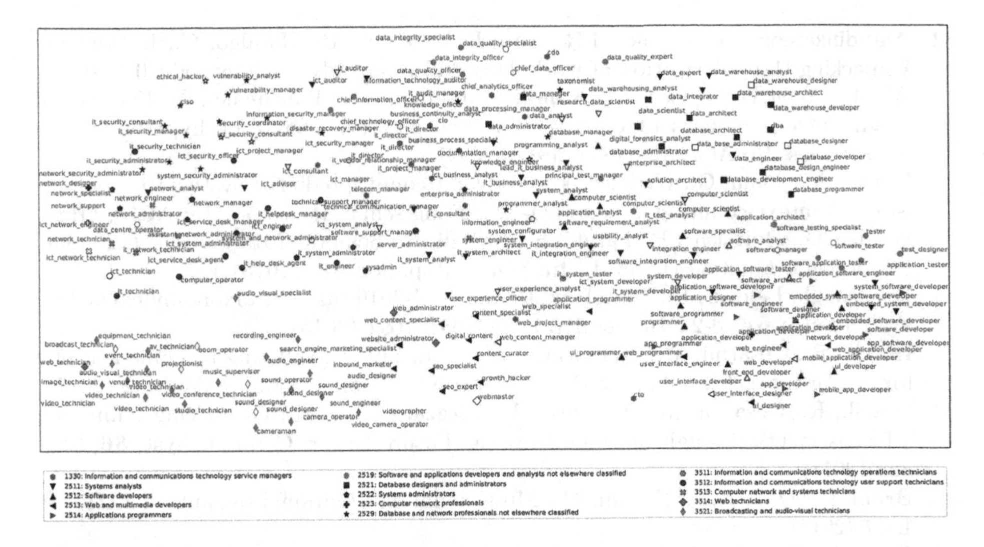


**Fig. 4.** UMAP plot of the **best** word-embedding model resulting from Tab. 2 (Model 3 with *CBOW dim=100, epochs=200 and learning rate = 0.1*). Each icon is assigned to one ISCO level 4 group, as in Fig. 2. The ESCO concepts and words belonging to each group are showed, distinguishing between narrower occupations (shallow shapes) and alternative labels (filled shapse). Available at higher resolution at https://tinyurl.com/scatter-umap

## 6 Conclusion and Future Work

In this paper we presented `TaxoRef`, an unsupervised method for taxonomy refinement via word embeddings. `TaxoRef`, unlike its predecessors, considers the embedding model that better encodes the similarity relationships from an existing taxonomy. To this end, we developed a measure of semantic similarity between taxonomic elements (i.e., the `HSS`). Finally, we used the selected embedding to refine an existing taxonomy using Bayesian theorem.

We applied `TaxoRef` to both (i) benchmark datasets and metrics, showing that `HSS` outperforms state-of-the-art metrics [21] and (ii) a real-word application framed within the research activities of a Labour Market Intelligence EU Project (See Footnote 1), processing more than 2M UK-2018 ICT job ads. The intrinsic and an extrinsic evaluations showed that `TaxoRef` can effectively encode similarities among ESCO concepts, enabling concept refinements to allow the taxonomy to adhere better to the real labour market demand.

## References

1. Agirre, E., Alfonseca, E., Hall, K., Kravalova, J., Pasca, M., Soroa, A.: A study on similarity and relatedness using distributional and wordnet-based approaches (2009)

2. Alabdulkareem, A., Frank, M.R., Sun, L., AlShebli, B., Hidalgo, C., Rahwan, I.: Unpacking the polarization of workplace skills. Sci. Adv. **4**(7), eaao6030 (2018)
3. Aly, R., Acharya, S., Ossa, A., Köhn, A., Biemann, C., Panchenko, A.: Every child should have parents: a taxonomy refinement algorithm based on hyperbolic term embeddings. In: ACL, pp. 4811–4817 (2019)
4. Baroni, M., Dinu, G., Kruszewski, G.: Don't count, predict! a systematic comparison of context-counting vs. context-predicting semantic vectors. In: ACL (2014)
5. Bordea, G., Buitelaar, P., Faralli, S., Navigli, R.: Semeval-2015 task 17: taxonomy extraction evaluation (texeval). In: SemEval, p. 902–910 (2015)
6. Bordea, G., Lefever, E., Buitelaar, P.: Semeval-2016 task 13: taxonomy extraction evaluation (texeval-2). In: SemEval, pp. 1081–1091 (2016)
7. Boselli, R., Cesarini, M., Mercorio, F., Mezzanzanica, M.: Using machine learning for labour market intelligence. In: ECML-PKDD, vol. 10536, pp. 330–342 (2017)
8. Boselli, R., Cesarini, M., Mercorio, F., Mezzanzanica, M.: Classifying online job advertisements through machine learning. Future Gener. Comput. Syst. **86**, 319–328 (2018)
9. Bruni, E., Tran, N.K., Baroni, M.: Multimodal distributional semantics. JAIR **49**, 1–47 (2014)
10. Camacho-Collados, J., Pilehvar, M.T.: From word to sense embeddings: a survey on vector representations of meaning. JAIR **63**, 743–788 (2018)
11. Chawla, N.V., Bowyer, K.W., Hall, L.O., Kegelmeyer, W.P.: Smote: synthetic minority over-sampling technique. JAIR **16**, 321–357 (2002)
12. Colombo, E., Mercorio, F., Mezzanzanica, M.: Ai meets labor market: exploring the link between automation and skills. Inf. Econ. Policy **47**, 27–37 (2019)
13. Devlin, J., Chang, M.W., Lee, K., Toutanova, K.: Bert: Pre-training of deep bidirectional transformers for language understanding. arXiv preprint (2018)
14. Faruqui, M., Dodge, J., Jauhar, S.K., Dyer, C., Hovy, E., Smith, N.A.: Retrofitting word vectors to semantic lexicons. arXiv preprint arXiv:1411.4166 (2014)
15. Frey, C.B., Osborne, M.A.: The future of employment: How susceptible are jobs to computerisation? technological forecasting and social change (2017)
16. Fu, R., Guo, J., Qin, B., Che, W., Wang, H., Liu, T.: Learning semantic hierarchies via word embeddings. In: ACL, pp. 1199–1209 (2014)
17. Ghannay, S., Favre, B., Esteve, Y., Camelin, N.: Word embedding evaluation and combination. In: LREC (2016)
18. Giabelli, A., Malandri, L., Mercorio, F., Mezzanzanica, M., Seveso, A.: NEO: a tool for taxonomy enrichment with new emerging occupations. In: ISWC (2020)
19. Giabelli, A., Malandri, L., Mercorio, F., Mezzanzanica, M., Seveso, A.: NEO: a system for identifying new emerging occupation from job ads. In: AAAI (2021)
20. Giabelli, A., Malandri, L., Mercorio, F., Mezzanzanica, M., Seveso, A.: Skills2job: a recommender system that encodes job offer embeddings on graph databases. Appl. Soft Comput. **101**, 107049 (2021). https://doi.org/10.1016/j.asoc.2020.107049
21. Jauhiainen, T., Lui, M., Zampieri, M., Baldwin, T., Lindén, K.: Automatic language identification in texts: a survey. JAIR **65**, 675–782 (2019)
22. Jiang, J.J., Conrath, D.W.: Semantic similarity based on corpus statistics and lexical taxonomy. arXiv preprint cmp-lg/9709008 (1997)
23. Kiela, D., Hill, F., Clark, S.: Specializing word embeddings for similarity or relatedness. In: EMNLP, pp. 2044–2048 (2015)
24. Köhn, A.: What's in an embedding? analyzing word embeddings through multilingual evaluation. In: EMNLP (2015)
25. Lau, J.H., Baldwin, T.: An empirical evaluation of doc2vec with practical insights into document embedding generation. arXiv preprint arXiv:1607.05368 (2016)

26. Leacock, C., Chodorow, M.: Combining local context and wordnet similarity for word sense identification. WordNet Electron. Lexical Database **49**(2), 265–283 (1998)
27. Malandri, L., Mercorio, F., Mezzanzanica, M., Nobani, N.: MEET-LM: a method for embeddings evaluation for taxonomic data in the labour market. Comput. Ind. **124** (2021). https://doi.org/10.1016/j.compind.2020.103341
28. Mann, H.B., Whitney, D.R.: On a test of whether one of two random variables is stochastically larger than the other. Ann. Math. Stat. (1947)
29. Meng, Y., et al.: Spherical text embedding. In: NIPS (2019)
30. Mezzanzanica, M., Boselli, R., Cesarini, M., Mercorio, F.: A model-based evaluation of data quality activities in KDD. Inf. Process. Manag. **51**(2), 144–166 (2015)
31. Mikolov, T., Sutskever, I., Chen, K., Corrado, G.S., Dean, J.: Distributed representations of words and phrases and their compositionality. In: NIPS (2013)
32. Miller, G.A.: Wordnet: a lexical database for English. Comm. ACM **38**, 39–41 (1995)
33. Nguyen, K.A., Köper, M., Walde, S.S.i., Vu, N.T.: Hierarchical embeddings for hypernymy detection and directionality. arXiv preprint arXiv:1707.07273 (2017)
34. O'Hara, T.D., Hugall, A.F., Thuy, B., Stöhr, S., Martynov, A.V.: Restructuring higher taxonomy using broad-scale phylogenomics: the living ophiuroidea. Molec. Phylogenet. Evol. **107**, 415–430 (2017)
35. Pedersen, T., Patwardhan, S., Michelizzi, J., et al.: Wordnet: Similarity-measuring the relatedness of concepts. In: AAAI, vol. 4, pp. 25–29 (2004)
36. Pennington, J., Socher, R., Manning, C.: Glove: global vectors for word representation. In: EMNLP, pp. 1532–1543 (2014)
37. Ponzetto, S.P., Navigli, R.: Large-scale taxonomy mapping for restructuring and integrating wikipedia. In: IJCAI (2009)
38. Press, O., Wolf, L.: Using the output embedding to improve language models. In: EACL, p. 157 (2017)
39. Radinsky, K., Agichtein, E., Gabrilovich, E., Markovitch, S.: A word at a time: computing word relatedness using temporal semantic analysis. In: WWW (2011)
40. Resnik, P.: Semantic similarity in a taxonomy: an information-based measure and its application to problems of ambiguity in natural language. JAIR **11**, 95–130 (1999)
41. Schnabel, T., Labutov, I., Mimno, D., Joachims, T.: Evaluation methods for unsupervised word embeddings. In: EMNLP (2015)
42. Vinel, M., Ryazanov, I., Botov, D., Nikolaev, I.: Experimental comparison of unsupervised approaches in the task of separating specializations within professions in job vacancies. In: Ustalov, D., Filchenkov, A., Pivovarova, L. (eds.) AINL 2019. CCIS, vol. 1119, pp. 99–112. Springer, Cham (2019). https://doi.org/10.1007/978-3-030-34518-1_7
43. Wang, B., Wang, A., Chen, F., Wang, Y., Kuo, C.C.J.: Evaluating word embedding models: methods and experimental results. In: APSIPA TSIP (2019)
44. Wang, C., He, X., Zhou, A.: A short survey on taxonomy learning from text corpora: issues, resources and recent advances. In: EMNLP (2017)
45. Wang, R.Y., Strong, D.M.: Beyond accuracy: what data quality means to data consumers. J. Manag. Inf. Syst. **12**(4), 5–33 (1996)
46. Wu, Z., Palmer, M.: Verbs semantics and lexical selection. In: ACL (1994)
47. Zhang, D., Liu, J., Zhu, H., Liu, Y., Wang, L., Xiong, H.: Job2vec: job title benchmarking with collective multi-view representation learning. In: CIKM (2019)

# MaxVA: Fast Adaptation of Step Sizes by Maximizing Observed Variance of Gradients

Chen Zhu[1(✉)], Yu Cheng[2], Zhe Gan[2], Furong Huang[1], Jingjing Liu[2,3], and Tom Goldstein[1]

[1] University of Maryland, College Park, USA
{chenzhu,furongh,tomg}@umd.edu
[2] Microsoft, Redmond, USA
{yu.cheng,zhe.gan,jingjl}@microsoft.com
[3] Tsinghua University, Beijing, USA
jjliu@air.tsinghua.edu.cn

**Abstract.** Adaptive gradient methods such as RMSPROP and ADAM use exponential moving estimate of the squared gradient to compute adaptive step sizes, achieving better convergence than SGD in face of noisy objectives. However, ADAM can have undesirable convergence behaviors due to unstable or extreme adaptive learning rates. Methods such as AMSGRAD and ADABOUND have been proposed to stabilize the adaptive learning rates of ADAM in the later stage of training, but they do not outperform ADAM in some practical tasks such as training Transformers [27]. In this paper, we propose an adaptive learning rate principle, in which the running mean of squared gradient in ADAM is replaced by a weighted mean, with weights chosen to maximize the estimated variance of each coordinate. This results in a faster adaptation to the local gradient variance, which leads to more desirable empirical convergence behaviors than ADAM. We prove the proposed algorithm converges under mild assumptions for nonconvex stochastic optimization problems, and demonstrate the improved efficacy of our adaptive averaging approach on machine translation, natural language understanding and large-batch pretraining of BERT. The code is available at https://github.com/zhuchen03/MaxVA.

## 1 Introduction

Stochastic Gradient Descent (SGD) and its variants are commonly used for training deep neural networks because of their effectiveness and efficiency. In their simplest form, gradient methods train a network by iteratively moving each parameter in the direction of the negative gradient (or the running average of gradients) of the loss function on a randomly sampled mini-batch of training data. A scalar learning rate is also applied to control the size of the update. In contrast, *adaptive* stochastic gradient methods use coordinate-specific learning rates, which are inversely proportional to the square root of the running mean of


N. Oliver et al. (Eds.): ECML PKDD 2021, LNAI 12977, pp. 628–643, 2021.
https://doi.org/10.1007/978-3-030-86523-8_38

squared gradients [8,12,26]. Such methods are proposed to improve the stability of SGD on non-stationary problems, and have achieved success in different fields across Speech, Computer Vision, and Natural Language Processing.

Large pretrained Transformer-based language models have achieved remarkable successes in various language tasks [3,7,13,17,22]. The original Transformer architecture (Post-LN Transformers) often demonstrates better performance than its Pre-LN variant [16], but its gradient has high variance during training. A warmup learning rate schedule or small initial adaptive learning rates [15] are required for its convergence. [36] shows that SGD fails to train Transformers without gradient clipping, and adaptivity is important for stabilizing optimization under the heavy-tailed noise in Transformer's gradients. This indicates that the strategy of AdaBound [19], which is to transition from Adam into SGD, may fail on Post-LN Transformers (see Appendix D for instance). However, the adaptive learning rate of Adam can be unstable in the later stage of training, and such instability sometimes leads to sub-optimal solutions or even non-convergent behavior on some simple problems [19,23]. AMSGrad [23] was proposed to deal with this issue by computing the adaptive learning rate with an update rule that guarantees monotonically decaying adaptive learning rates for each coordinate, but to our knowledge, it has not been widely deployed to enhance Adam for training Transformer-based language models.

In this work, we explore a different approach to improving the stability of adaptive learning rates. We propose *Maximum Variation Averaging* (MaxVA), which computes the running average of squared gradients using dynamic, rather than constant, coordinate-wise weights. These weights are chosen so that the estimated variance of gradients is maximized, to enable a faster adaptation to the changing variance of gradients. The MaxVA weights for maximizing this variance have a simple closed-form solution that requires little storage or computational cost. With MaxVA, the adaptive optimizer 1) takes a smaller step size when abnormally large gradient is present, to improve stability; 2) takes a larger step size when abnormally small gradient is prevent, to avoid spurious minima and achieve better generalization [14]; 3) takes a steady step size when gradients are stable and within estimated deviation, to ensure convergence [23]. In the large-batch setting of BERT pretraining, where the total number of iterations is sharply reduced and a faster adaptation in each step is more important, MaxVA achieves faster convergence and obtain models with better test performance on downstream tasks than both Adam and Lamb [32]. Extensive experiments on both synthetic and practical datasets demonstrate that MaxVA leads to an improved adaptability and stability for Adam, yielding better test set performance than Adam on a variety of tasks. We also prove MaxVA converges under mild assumptions in the nonconvex stochastic optimization setting.

## 2 Preliminary and Definitions

By default, all vector-vector operators are element-wise in the following sections. Let $\theta \in \mathbb{R}^d$ be the parameters of the network to be trained, $\ell(x;\theta)$ is the loss of the model with parameters $\theta$ evaluated at $x$. Our goal is to minimize the expected risk on the data distribution defined as:

$$f(\theta) = \mathbb{E}_{x \sim \mathcal{D}} \left[ \ell(x; \theta) \right]. \tag{1}$$

In most deep learning problems, only a finite number of potentially noisy samples can be used to approximate Eq. 1, and the gradients are computed on randomly sampled minibatches during training. Stochastic regularizations such as Dropout [25] are commonly used for training Transformer-based language models [27,38], which further adds to the randomness of the gradients. Thus, it is important to design optimizers that tolerate noisy gradients. ADAM [12] is an effective optimizer that adapts to such noisy gradients. It keeps exponential moving averages $m_t$ and $v_t$ of past gradients $g_1, ..., g_{t-1}$, defined as:

$$\tilde{m}_t = \alpha \tilde{m}_{t-1} + (1-\alpha) g_t, \quad m_t = \frac{\tilde{m}_t}{1-\alpha^{t+1}},$$

$$\tilde{v}_t = \beta \tilde{v}_{t-1} + (1-\beta) g_t^2, \quad v_t = \frac{\tilde{v}_t}{1-\beta^{t+1}},$$

where $\alpha, \beta \in [0,1]$, $g_t = \nabla_\theta \ell(x_t; \theta_t)$ is the gradient of the $t$-th minibatch $x_t$, $\tilde{m}_0 = \tilde{v}_0 = 0$, and $m_t, v_t$ corrects this zero-initialization bias of $\tilde{m}_t, \tilde{v}_t$ [12]. ADAM updates the parameters with the estimated moments as $\theta_{t+1} = \theta_t - \eta_t \frac{m_t}{\sqrt{v_t}+\epsilon}$, where $\epsilon > 0$ is a small constant for numerical stability.

If we assume that the distribution of the stochastic gradient is constant within the effective horizon of the running average, then $m_t$ and $v_t$ will be estimates of the first and second moments of the gradient $g_t$ [1]. Same as other adaptive methods such as ADAM and the recently proposed AdaBelief [39], we adopt this assumption throughout training. With this assumption, at time $t$, we assume $\mathbb{E}[m_t] \approx \nabla f_t$, $\mathbb{E}[v_t] \approx \nabla f_t^2 + \sigma_t^2$, where $\sigma_t^2$ is the variance of $g_t$. ADAM, RMSPROP and other variants that divide the update steps by $\sqrt{v_t}$ can be seen as adapting to the gradient variance under this assumption when $m_t$ is small. These adaptive methods take smaller step sizes when the estimated variance $\sigma_t^2 = v_t - m_t^2$ is high. Higher local gradient variance indicates higher local curvature, and vice versa. In certain quadratic approximations to the loss function, this variance is proportional to the curvature [24] (Eq. 12 of our paper). Therefore, like a diagonal approximation to Newton's method, such adaptative learning rates adapt to the curvature and can accelerate the convergence of first-order methods.

However, the adaptive learning rate $\eta_t/(\sqrt{v_t}+\epsilon)$ of ADAM and RMSPROP can take extreme values, causing convergence to undesirable solutions [6,29]. [23] gave one such counter example where gradients in the correct direction are large but occur at a low frequency, and ADAM converges to the solution of maximum regret. They solve this issue by keeping track of the maximum $v_t$ for each coordinate throughout training with a new variable $\hat{v}_t$, and replace the adaptive learning rate with $\eta_t/\sqrt{\hat{v}_t}$ to enforce monotonically descreasing learning rates. Extremely small adaptive learning rates can also cause undesirable convergence behavior, as demonstrated by a counter example from [19].

## 3 Maximizing the Variance of Running Estimations

**Motivation.** We propose to mitigate the undesirable convergence issue of ADAM by changing the constant running average coefficient $\beta$ for the second moment into an adaptive one. The idea is to allow $\beta_t$ to adopt the value that maximizes the estimated variance of the gradient at each iteration $t$. As a result, our method will assign a higher coefficient $(1-\beta_t)$ to $g_t$ when it deviates too much from the estimated mean, resulting in a smaller step size when the gradient is too large to avoid overshooting, and a larger step size when the gradient is abnormally small to avoid spurious local minima. By contrast, if the gradient is stable and close to the estimated mean, which often happens near a flat minimum, our method will assign minimum coefficient to $g_t$ to maintain the value of $v_t$ and take steady steps towards the minimum. Therefore, our method can use $\beta_t$ as the adaptive running average coefficient to take steps that are cautious enough to avoid instability and spurious minima but aggressive enough to make progress. An illustrative example is given in Fig. 1.

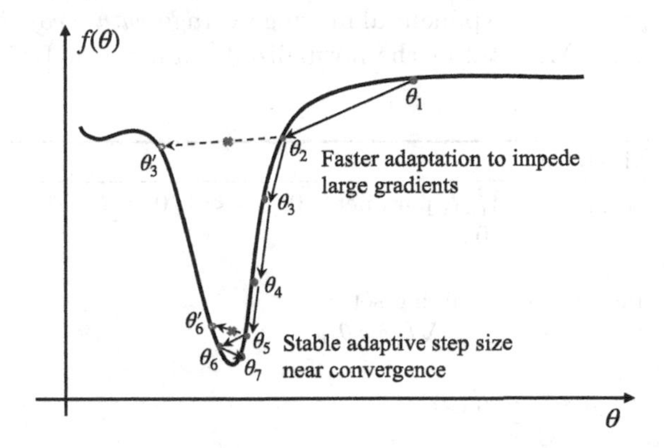


**Fig. 1.** An illustrative example of MaxVA. It adopts a smaller adaptive learning rate when an abnormally large gradient appears at $\theta_2$ by choosing a smaller $\beta$, to prevent overshooting to $\theta_3'$. When the gradient is more stable near convergence at $\theta_5$, it uses a larger $\beta$ for the slowest change in adaptive learning rate, to prevent overshooting to $\theta_6'$.

**Maximum Variation Averaging.** Formally, we estimate the variance of the gradient at each coordinate by keeping track of the zeroth, first, and second moments of the gradient as functions of the adaptive running average coefficient $\beta_t$, denoted as $w_t(\beta_t)$, $\tilde{u}_t(\beta_t)$ and $\tilde{v}_t(\beta_t)$, respectively:

$$w_t(\beta_t) = \beta_t w_{t-1}(\beta_{t-1}) + (1-\beta_t), \tag{2}$$

$$\tilde{u}_t(\beta_t) = \beta_t \tilde{u}_{t-1}(\beta_{t-1}) + (1-\beta_t) g_t, \tag{3}$$

$$\tilde{v}_t(\beta_t) = \beta_t \tilde{v}_{t-1}(\beta_{t-1}) + (1-\beta_t) g_t^2. \tag{4}$$

The zeroth moment $w_t(\beta_t)$ is used to normalize $\tilde{u}_t(\beta_t)$ and $\tilde{v}_t(\beta_t)$ to achieve bias-corrected estimates $u_t(\beta_t) = \tilde{u}_t(\beta_t)/w_t(\beta_t)$ and $v_t(\beta_t) = \tilde{v}_t(\beta_t)/w_t(\beta_t)$ for the first and second moments, so that the estimates are not biased towards zero ($\tilde{m}_0 = \tilde{v}_0 = 0$) [12].

Under our assumptions, the bias-corrected local estimate of the gradient variance is $\sigma_t^2 = \tilde{v}_t(\beta_t)/w_t(\beta_t) - [\tilde{u}_t(\beta_t)/w_t(\beta_t)]^2$. Taking the arg max for $\sigma_t^2$, we find the $\beta_t$ that achieves the maximal variance for each coordinate $i$:

$$\beta_{t,i} = \arg\max_{\beta} \sigma_{t,i}^2 = \arg\max_{\beta} v_{t,i}(\beta) - [u_{t,i}(\beta)]^2. \tag{5}$$

We call our approach to finding adaptive running average coefficient $\beta_t$ *Maximum Variation Averaging* (MaxVA). We plug MaxVA into ADAM and its variant LAPROP [40], which results in two novel algorithms, MADAM and LAMADAM, listed in Algorithm 1 and Algorithm 2 (in the Appendix). Different from ADAM, LAPROP uses $v_t$ to normalize the gradients before taking the running average, which results in higher empirical stability under various hyperparameters. Note, we only apply the adaptive $\beta_t$ to the *second* moment $u_t(\beta_t)$ used for scaling the learning rate; $m_t$ is still an exponential moving average *with a constant coefficient* $\alpha$ of the gradient for MADAM or the normalized gradient for LAMADAM.

**Algorithm 1.** MADAM

1: **Input:** Learning rate $\{\eta_t\}_{t=1}^T$, parameter $0 < \alpha < 1$, $0 < \underline{\beta} < \bar{\beta} < 1$, $\epsilon > 0$
2: Set $\tilde{m}_0 = \tilde{u}_0 = \tilde{v}_0 = w_0 = 0$
3: **for** $t = 1$ **to** $T$ **do**
4: Draw samples $S_t$ from training set
5: Compute $g_t = \frac{1}{|S_t|} \sum_{x_k \in S_t} \nabla \ell(x_k; \theta_t)$
6: $\tilde{m}_t = \alpha \tilde{m}_{t-1} + (1-\alpha) g_t$
7: $\tilde{\beta}_t = \arg\max_{\beta} v_t(\beta) - u_t^2(\beta)$ ▷ see Eq 6
8: $\beta_t = \max(\underline{\beta}, \min(\bar{\beta}, \tilde{\beta}_t))$
9: $\tilde{u}_t = \beta_t \tilde{u}_{t-1} + (1-\beta_t) g_t$
10: $\tilde{v}_t = \beta_t \tilde{v}_{t-1} + (1-\beta_t) g_t^2$
11: $w_t = \beta_t w_{t-1} + (1-\beta_t)$
12: $\theta_t = \theta_{t-1} - \eta_t \frac{\sqrt{w_t}}{1-\alpha^t} \frac{\tilde{m}_t}{\sqrt{\tilde{v}_t} + \epsilon}$

**Finding $\beta_t$ via a Closed-form Solution.** The maximization for $\beta_t$ in Eq. 5 is quadratic and has a relatively simple closed-form solution that produces maximal $\sigma_t^2$ for each coordinate:

$$\beta_t = \frac{\Delta g_t^2 + \sigma_{t-1}^2}{w_{t-1}(\Delta g_t^2 - \sigma_{t-1}^2) + \Delta g_t^2 + \sigma_{t-1}^2}, \tag{6}$$

where all variables are vectors and all the operations are elementwise, $\Delta g_t = (g_t - u_{t-1})$ is the deviation of the gradient $g_t$ from the estimated mean $u_{t-1}$, $\sigma_{t-1}^2 =$

$v_{t-1} - u_{t-1}^2$ is the estimated variance, and we have abbreviated $u_{t-1}(\beta_{t-1})$, $v_{t-1}(\beta_{t-1})$ and $w_{t-1}(\beta_{t-1})$ into $u_{t-1}, v_{t-1}$ and $w_{t-1}$. We use this abbreviation in the following sections, and defer the derivation of Eq. 6 to Appendix A.

**Implementation Notes.** We apply MaxVA in every step except for the first step, where the gradient variance one can observe is zero. So for Algorithm 1 and Algorithm 2 we define:

$$\tilde{u}_1 = (1-\beta_1)g_1, \tilde{v}_1 = (1-\beta_1)g_1^2, w_1 = 1-\beta_1. \tag{7}$$

The coefficient $\beta_1$ for $t = 1$ is set to a constant that is the same as typical values for ADAM. To obtain a valid running average, we clip $\beta_t$ so that $\underline{\beta} \leq \beta_t \leq \bar{\beta}$, where the typical values are $\underline{\beta} = 0.5, 0.98 \leq \bar{\beta} \leq 1$. For convenience, we set $\beta_1 = \bar{\beta}$ by default. For $t > 1$, since $0 < \beta_t \leq 1$, $w_t$ will monotonically increase from $(1-\beta_1)$ to 1. Before clipping, for any $g_t, u_{t-1}, v_{t-1}$ satisfying $v_{t-1} - u_{t-1}^2 > 0$ in Eq. 6, we have $\beta_t \in [1/(1+w_{t-1}), 1/(1-w_{t-1})]$. As a result, the lower bound that we use $(\underline{\beta} = 0.5)$ is tight and does not really change the value of $\beta_t$, and as $t \to \infty$, $w_t \to 1$ and $\beta_t \in [0.5, \infty]$. We have a special case at $t = 2$, where $\beta_t$ is a constant $1/(2-\beta_1)$.

In practice, we also add a small coefficient $\delta > 0$ to the denominator of Eq. 6 to prevent division by zero, which will have negligible effect on the value of $\beta_t$ and does not violate the maximum variation objective (Eq. 5). All the derivations for these conclusions are deferred to Appendix C.

**Effect of Maximum Variation Averaging.** By definition, we have $\sigma_{t-1}^2 \geq 0$, but in most cases $\sigma_{t-1}^2 > 0$. When $\sigma_{t-1}^2 > 0$, we define a new variable $R_t = \Delta g_t^2/\sigma_{t-1}^2$, which represents the degree of deviation of gradient $g_t$ from the current estimated average. Then, we can rewrite:

$$\beta_t = \frac{R_t + 1}{(1+w_t)R_t + 1 - w_t}. \tag{8}$$

From Eq. 8, we can see $\beta_t$ monotonically decreases from $1/(1-w_t)$ to $1/(1+w_t)$ as $R_t$ increases from 0 to $\infty$, and equals to 1 when $R_t = 1$. As a result, for each coordinate, if $R_t \gg 1$, $g_t$ deviates much more than $\sigma_{t-1}$ from $u_{t-1}$, and MaxVA will find a smaller $\beta_t$ and therefore a higher weight $(1-\beta_t)$ on $g_t^2$ to adapt to the change faster. This helps to avoid overshooting when abnormally large gradient is present (see Fig. 1), and avoids spurious sharp local minima where gradients are abnormally small. With a faster response to abnormal gradients, MaxVA is better at handling the heavy-tailed distribution of gradients in the process of training Transformers [36]. In practice, $v_t$ tends to be larger than ADAM/LAPROP using a constant $\bar{\beta}$, but as we will show in the experiments, using a larger learning rate counters such an effect and achieves better results.

On the other hand, if $R_t < 1$, or the deviation of the gradient $g_t$ from the current running mean $u_{t-1}$ is within the estimated standard deviation $\sigma_{t-1}$, we will use $\bar{\beta}$ to update $\tilde{v}_t$, which is the smallest change we allow for $\tilde{v}_t$. This tends to happen in the later phase of training, where the gradient variance decreases. MaxVA will adopt a steady step towards convergence by finding the slowest rate to update $\tilde{v}_t$. This allows large values of $\tilde{v}_t$ to last for a longer horizon even compared with setting $\beta_t$ to a constant $\bar{\beta}$ on the same sequence, since we have assigned more mass to large gradients, which can be seen as an adaptive version of AMSGRAD. Note that MaxVA and AMSGRAD can be complementary approaches if applied together, which we have found helpful for Image Classification on CIFAR10/100.

**Convergence Analysis.** We prove the convergence of MaxVA in the nonconvex stochastic optimization setting. For the sake of simplicity, we analyze the case where $\alpha = 0$, which is effectively applying MaxVA to RMSPROP. We leave the analysis for $\alpha \neq 0$ for future research. We assume the function $\ell$ is $L$-smooth in $\theta$, i.e., there exists a constant $L$ such that for all $\theta_1, \theta_2 \in \mathbb{R}^d, x \in \mathcal{X}$,

$$\|\nabla_\theta \ell(x;\theta_1) - \nabla_\theta \ell(x;\theta_2)\| \leq L\|\theta_1 - \theta_2\|. \tag{9}$$

This automatically implies that $f(\theta) = \mathbb{E}[\ell(x;\theta)]$ is $L$-smooth. Such a smoothness assumption holds for networks with smooth activation functions, e.g., Transformers that use the GELU activation [11]. We also need to assume function $\ell$ has bounded gradient, i.e., $\|\nabla_\theta \ell(x;\theta)\|_\infty \leq G$ for all $\theta \in \mathbb{R}^d, x \in \mathcal{X}$. As typically used in the analysis of stochastic first-order methods [9,33], we assume the stochastic gradient has bounded variance: $\mathbb{E}[[\nabla_\theta \ell(x;\theta)]_i - [\nabla_\theta f(\theta)]_i]^2 \leq \sigma^2$ for all $\theta \in \mathbb{R}^d$. Further, we assume the batch size increases with time as $b_t = t$, which is also adopted in the analysis of SIGNSGD [2], and holds in our large batch experiments. Theorem 1 gives a "worst-case" convergence rate of MaxVA to a stationary point under these assumptions, where the dependence of $\beta_t$ on $g_t$ is ignored and we only consider the worst-case of $\beta_t$ in each step. The proof is given in Appendix B.

**Theorem 1.** *Define* $w_0 = 1$. *Let* $\eta_t = \eta$ *and* $b_t = t$ *for all* $t \in [T]$. *Furthermore, we assume* $\epsilon, \underline{\beta}, \bar{\beta}, \eta$ *are chosen such that* $\eta \leq \frac{\epsilon}{2L}$, $1 - \underline{\beta} \leq \frac{\epsilon^2}{16G^2}$, *and* $\bar{\beta} \leq 2\underline{\beta}$. *Then for* $\theta_t$ *generated using* MADAM, *we have the following bound:*

$$\mathbb{E}\|\nabla f(\theta_a)\|^2 \leq O\left(\frac{f(\theta_1) - f(\theta^*)}{\eta T} + \frac{2\sigma d G}{\epsilon\sqrt{T}}\right), \tag{10}$$

*where* $\theta^*$ *is an optimal solution to minimize the objective in Eq. 1, and* $\theta_a$ *is an iterate uniformly randomly chosen from* $\{\theta_1, ..., \theta_T\}$.

## 4 Experiments on Synthetic Data

For a quantitative control of the stochasticity and data distribution, which affects the difficulty of the problem and the efficacy of the optimizers, we compare MADAM and the baselines in two sets of synthetic data, and demonstrate the efficacy of MaxVA with statistical significance on a large number of instances.

The first dataset simulates prevalent machine learning settings, where mini-batch stochastic gradient methods are applied on a finite set of samples, on which we show MADAM fixes the nonconvergence issue of ADAM and achieves faster convergence rate than AMSGRAD. The second dataset evaluates the algorithms under different curvatures and gradient noise levels, where we show MADAM achieves both lower loss and variance than fine-tuned ADAM at convergence (Fig. 2).

### 4.1 Convergence with Stochastic Gradients

Since MaxVA maximizes the variance and the gradient converges to zero in most cases, MADAM biases towards larger $v_t$ than ADAM but does not require $v_t$ to be monotonically increasing, which is like an adaptive version of AMSGRAD. To highlight the difference, we compare ADAM, MADAM and AMSGRAD on the synthetic dataset from [6] simulating training with stochastic mini batches on a finite set of samples. Formally, let $\mathbb{1}_{[\cdot]}$ be the indicator function. We consider the problem $\min_\theta f(\theta) = \sum_{i=1}^{11} \ell_i(\theta)$ where

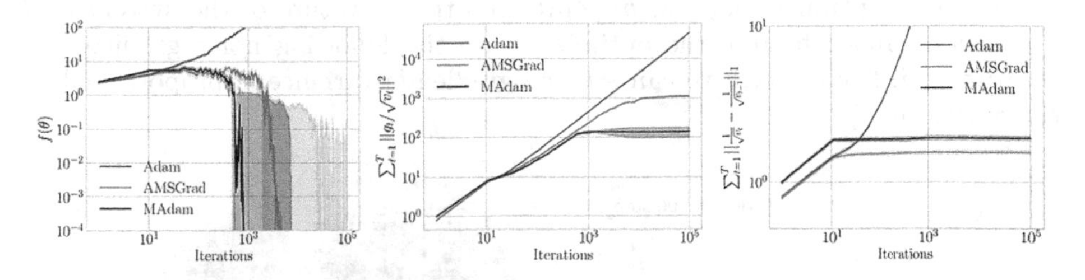


**Fig. 2.** Median and standard error (100 runs) of objective value ($f(\theta)$), accumulated update size ($\sum_{t=1}^{T} ||g_t/\sqrt{v_t}||^2$) and total change in adaptive learning rate ($\sum_{t=1}^{T} ||\frac{1}{\sqrt{v_t}} - \frac{1}{\sqrt{v_{t-1}}}||_1$) for ADAM, AMSGRAD, MADAM on the problem in Eq. 11.

$$\ell_i(\theta) = \begin{cases} \mathbb{1}_{i=1} 5.5\theta^2 + \mathbb{1}_{i\neq 1}(-0.5\theta^2), & \text{if } |\theta| \leq 1; \\ \mathbb{1}_{i=1}(11|\theta| - 5.5) + \mathbb{1}_{i\neq 1}(-|\theta| + 0.5), & \text{otherwise.} \end{cases} \tag{11}$$

At every step, a random index $i$ is sampled uniformly from $i \in [11]$, and the gradient $\nabla \ell_i(\theta)$ is used by the optimizer. The only stationary point where $\nabla f(\theta) = 0$ is $\theta = 0$. We set $\alpha = 0, \beta = 0.9$ for ADAM and AMSGRAD. For MADAM, we set $\alpha = 0, (\underline{\beta}, \bar{\beta}) = (0.5, 1)$. We select the best *constant* learning rates for the three algorithms, see Appendix E for details.

We plot the median and standard error of the objective ($f(\theta)$), accumulated update size ($S_1 = \sum_{t=1}^{T} ||g_t/\sqrt{v_t}||^2$), and total change in adaptive step size ($S_2 = \sum_{t=1}^{T} ||\frac{1}{\sqrt{v_t}} - \frac{1}{\sqrt{v_{t-1}}}||_1$) over 100 runs in Fig. 11. The optimal learning rates for these optimziers are different, so for fair comparisons, we have ignored the constant learning rate in $S_1$ and $S_2$. From the curves of $f(\theta)$, we can see ADAM diverges, and MADAM converges faster than AMSGRAD in the later stage. As shown by the $S_2$ curves, the adaptive step sizes of MADAM and AMSGRAD all converged to some constant values after about 10 steps, but MADAM converges faster on both $f(\theta)$ and $S_1$, indicating the adaptive step size found by MADAM fits the geometry of the problem better than AMSGRAD. This also shows $S_1 + S_2$ of MADAM has a smaller slope than AMSGRAD in the log-scale plots after 10 iterations, leading to a faster theoretical convergence rate in the bound given by [6]. The slightly larger variation in adaptive step sizes of MADAM at the beginning of training, shown by the larger $S_2$ values, demonstrates MADAM adapts faster to the changing gradients than AMSGRAD, achieved by dynamically selecting $\beta < 0.9$.

### 4.2 Convergence in the Noisy Quadratic Model

We analyze the ability of MADAM to adapt to curvature and gradient noise on the simple but illustrative Noisy Quadratic Model (NQM), which has been widely adopted for analyzing optimization dynamics [24,30,35,37]. The loss function is defined as $f(\theta) = \mathbb{E}_{x \sim \mathcal{N}(0,\sigma^2 I)} \left[\frac{1}{2}\sum_{i=1}^{d} h_i(\theta_i - x_i)^2\right]$, where $x$ is a noisy observation of the ground-truth parameter $\theta^* = 0$, simulating the gradient noise in stochastic optimization, and $h_i$ represents the curvature of the system in $d$ dimensions. In each step, the optimizers use the following noisy gradient for coordinate $i$, from which we can see the gradient's variance is proportional to the curvature $h_i^2$:

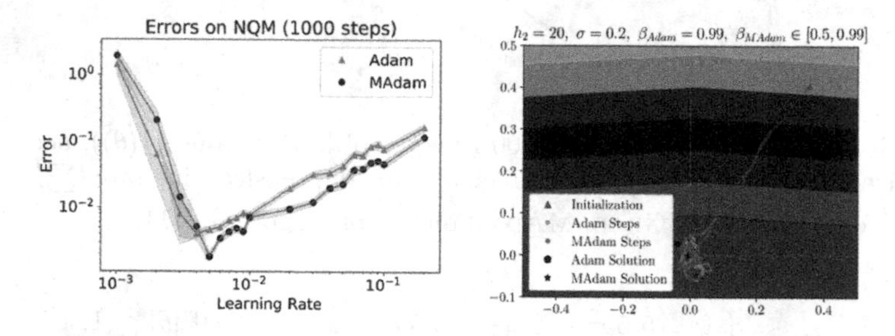


**Fig. 3.** Results on NQM. The left figure shows the mean and standard error of the loss under different learning rates $\eta$, computed over 100 runs at each point. We select the best $\beta$ for ADAM at each $\eta$. The best results (mean and variance) of ADAM and MADAM are 1.84e-3 (2.51e-4) and 4.05e-3 (4.84e-4) respectively. Figure on the right gives a qualitative example of the trajectories of two approaches.

$$\nabla_{\theta_i} \ell(\sigma\epsilon_i; \theta_i) = h_i(\theta_i - \sigma\epsilon_i), \epsilon_i \sim \mathcal{N}(0, 1). \tag{12}$$

To validate the effectiveness of MaxVA, we compare MADAM with ADAM under a variety of different curvatures $h$ and noise level $\sigma$ on an NQM with $d = 2$. For each setting of $h$ and $\sigma$, we test both algorithms on a variety of learning rates. For ADAM, we additionally choose the best $\beta$ and report the best results. See Appendix F for details. We run each setting 100 times to report the mean and standard error. MAdam consistently achieves 30–40% lower average loss with smaller standard error in all settings. Figure 3 shows the results for one of the settings, from which we find the best result of MAdam is better than Adam under any choice of $\beta$ and learning rate, confirming the advantage of MaxVA. From the qualitative example, MaxVA also demonstrates smaller variance near convergence, enabled by a quicker response to impede the noise with a smaller $\beta_t$. More experimental results under other settings are provided in Appendix F.

## 5 Experiments on Practical Datasets

In this section, we evaluate MAdam and LaMAdam on a variety of tasks against well-calibrated baselines: IWSLT'14 DE-EN/WMT'16 EN-DE for neural machine translation, the GLUE benchmark for natural language understanding, and pretraining the BERT-Base model. We also provide results on image classification. We use the decoupled weight decay [18] in all our experiments. Across all the plots in this section, we define the average step size at time $t$ as the average of $|\eta_t m_t/(\sqrt{v_t}+\epsilon)|$ for Adam/MAdam and $|\eta_t m_t|$ for LaProp/LaMAdam over all the entries.

### 5.1 Image Classification

To evaluate the effectiveness of MaxVA for image classification, we compare with SGD, Adam, LaProp [40] and AdaBelief [39] in training ResNet18 [10] on CIFAR10, CIFAR100 and ImageNet. On all the datasets, we perform a grid search for the learning rate and weight decay, and report the best results for each method in Table 1. For CIFAR10/100, we train ResNet18 with a batch size of 128 for 200 epochs. We also find AMSGrad [23] improves the classification accuracy of all adaptive methods evaluated on CIFAR10/100, so we apply AMSGrad in all experiments with adaptive methods. On ImageNet, we use the implementation

**Table 1.** Comparing adaptive methods with exhaustively fine-tuned SGD on CIFAR10/100 and ImageNet. CIFAR10/100 experiments are the median (standard error) over 4 runs. *: The results of AdaBelief are from their paper [39] with a ResNet34, while our results are with ResNet18.

| Model | CIFAR-10 | CIFAR-100 | ImageNet |
|---|---|---|---|
| SGD | 95.44 (.04) | 79.62 (.07) | 70.18 |
| Adam | 95.37 (.03) | 78.77 (.07) | 66.54 |
| LaProp | 95.34 (.03) | 78.36 (.07) | 70.02 |
| AdaBelief | 95.30* | 77.30* | 70.08 |
| MAdam (ours) | **95.51** (.09) | **79.32** (.08) | 69.96 |
| LaMAdam (ours) | 95.38 (.11) | 79.21 (.11) | **70.16** |

from torchvision and the default multi-step learning rate schedule. We do not use AMSGrad in this case. Further details are in Appendix G.

Despite achieving a marginal improvement on CIFAR10, adaptive methods often underperforms carefully tuned SGD on CIFAR100 and ImageNet when training popular architectures such as ResNet, as confirmed by [15,29,37]. Nevertheless, with the proposed MaxVA, we shrink the gap between adaptive methods and carefully tuned SGD on these image classification datasets, and achieve top-1 accuracy very close to SGD on ImageNet. Note our results with ResNet18 is better than the recent AdaBelief's results with ResNet34 on CIFAR10/CIFAR100 (95.51/79.32 vs. 95.30/77.30 approximately), as well as AdaBelief with ResNet18 on ImageNet (70.16 vs. 70.08) [39].

### 5.2 Neural Machine Translation

We train Transformers from scratch with LaProp and LaMAdam on IWSLT'14 German-to-English (DE-EN) translation [4] and WMT'16 English-to-German (EN-DE) translation, based on the implementation of fairseq.[1] We do not compare with SGD, since it is unstable for Transformers [36]. We also show in Appendix D that AdaBound cannot achieve any good result without degenerating into Adam. More details are in Appendix H.

**Table 2.** BLEU score for training transformers on machine translation datasets. We report the median and standard error for IWSLT'14 over 5 runs. Results of other meethods are from the AdaBelief paper [39].

| Method | IWSLT'14 DE-EN | WMT'16 EN-DE |
|---|---|---|
| RAdam | 35.51 | – |
| AdaBelief | 35.90 | – |
| LaProp(ours) | 35.98 (0.06) | 27.02 |
| LaMAdam(ours) | **36.09** (0.04) | **27.11** |

IWSLT'14 DE-EN has 160k training examples, on which we use a Transformer with 512-dimensional word embeddings and 1024 FFN dimensions. We train it for 60k iterations, with up to 4096 tokens in each minibatch. Results are listed in Table 2. Note the baseline's BLEU score is already 1.22 higher than the best results reported in [15] using the same model. As shown in Appendix H, LaMAdam uses much smaller update size than LaProp, and it is not able for LaProp to achieve better results even when we scale its learning rate to get similar update sizes as LaMAdam, indicating MaxVA helps to find a better minimum not achievable by using constant $\beta$.

WMT'16 EN-DE has 4.5M training examples, where same as [21], we use a larger Transformer with 1024-dimensional word embeddings and 4096 FFN

[1] https://github.com/pytorch/fairseq.

dimensions. Each batch has up to 480k tokens. We train for 32k iterations using the same inverse square root learning rate schedule as [27]. We evaluate the *single model* BLEU on newstest2013, unlike [15] where models in the last 20 epochs are averaged to get the results. As shown in Table 2, LAMADAM also achieves better results.

### 5.3 General Language Understanding Evaluation (GLUE)

**Table 3.** Results (median and variance) on the dev sets of GLUE based on finetuning the RoBERTa-base model ([17]), from 4 runs with the same hyperparameter but different random seeds.

| Method | MNLI | QNLI | QQP | RTE | SST-2 | MRPC | CoLA | STS-B |
|---|---|---|---|---|---|---|---|---|
| | (Acc) | (Acc) | (Acc) | (Acc) | (Acc) | (Acc) | (Mcc) | (Pearson) |
| Reported | 87.6 | 92.8 | 91.9 | 78.7 | 94.8 | 90.2 | 63.6 | 91.2 |
| ADAM | 87.70 (.03) | 92.85 (.06) | 91.80 (.03) | 79.25 (.71) | 94.75 (.08) | 88.50 (.24) | 61.92 (1.1) | 91.17 (.13) |
| LAPROP | 87.80 (.04) | 92.85 (.13) | 91.80 (.03) | 78.00 (.46) | 94.65 (.11) | 89.20 (.20) | 63.01 (.61) | 91.17 (.06) |
| MADAM | **87.90** (.08) | 92.95 (.07) | 91.85 (.03) | 79.60 (.66) | 94.85 (.12) | 89.70 (.17) | 63.33 (.60) | 91.28 (.03) |
| LAMADAM | 87.80 (.03) | **93.05** (.05) | **91.85** (.05) | **80.15** (.64) | **95.15** (.15) | **90.20** (.20) | **63.84** (.85) | **91.36** (.04) |

To evaluate MaxVA for transfer learning, we fine-tune pre-trained RoBERTa-base model [17] on 8 of the 9 tasks of the GLEU benchmark [28]. Following prevalent validation settings [7,13,22], we report the median and standard error for fine-tuning the RoBERTa-base model [17] over 4 runs where only the random seeds are changed. The results are in Table 3. MADAM and LAMADAM give better scores than the corresponding baselines in the 8 tasks. More experimental details are in Appendix I.

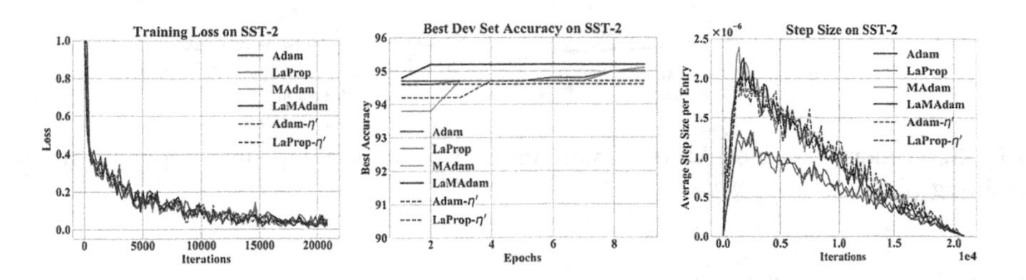


**Fig. 4.** Training loss, validation accuracy and step size of various optimization methods on SST-2. All optimizers here use $\lambda = 0.1$. ADAM and LAPROP use $(\eta, \beta)$=(1e-5, 0.98), MADAM and LAMADAM use $(\eta, \underline{\beta}, \bar{\beta})$=(4e-5, 0.5, 0.98), ADAM-$\eta'$ and LAPROP-$\eta'$ use $(\eta, \beta)$=(1.6e-5, 0.98).

To highlight the difference of the optimizers, we compare the training loss, dev set accuracy and the average step size on SST-2, as shown in Fig. 4. Different from Machine Translation experiments where we train the Transformers from scratch, the adaptive step size of MAdam/LaMAdam is higher in this transfer learning setting. The ratio of the learning rate and step size of MaxVA to non-MaxVA optimizers are 4 and 1.8 respectively on GLUE, while on IWSLT'14 the two ratios are 2 and (approximately) 0.875. Because we start from a pre-trained model, the heavy tail of the gradient is alleviated, just as the BERT model in the later stage of training as shown by [36], and the curvature of the loss landscape should be smaller. Therefore, MaxVA selects larger adaptive step sizes for better convergence. Same as in the Machine Translation experiments, the highest test accuracy of Adam/LaProp cannot reach the same value as MAdam/LaMAdam by simply scaling the base learning rate $\eta$ to reach similar step sizes as MAdam/LaMAdam.

### 5.4 Large-Batch Pretraining for BERT

We use the NVIDIA BERT pretraining repository to perform large-batch pre-training for BERT-Base model on the Wikipedia Corpus only.[2] Each run takes about 52 h on 8 V100 GPUs. Training is divided into two phases: the first phase uses a batch size of 64K with input sequence length 128 for 7,038 steps; the second phase uses a batch size 32K with input sequence length 512 for 1563 steps. The total of steps is significantly smaller than the 1,000,000 steps used in the small-batch training of [7]. Therefore, a faster adaptation to curvature in each step is more important.

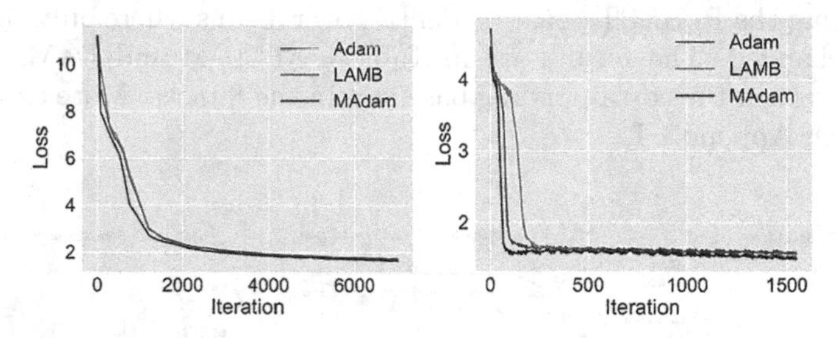


**Fig. 5.** Training losses of Adam, Lamb and MAdam on Wikipedia Corpus in the two training phases.

This point is validated by the faster convergence of MAdam in both phases, as shown in the training loss curves in Fig. 5. Contrary to the observation by [32], Adam even converges faster than Lamb in the earlier iterations. [32] only explored weight decay of up to 0.01 for Adam, but we find using larger weight decay of 0.1 together with gradient clipping ($\|g_t\|_2 \leq 1$, same as Lamb) stabilizes Adam. We inherit this setting for MAdam. For MAdam and Adam, we do a

[2] Note the results from the repository are for BERT-Large trained with additional data from BookCorpus.

grid search on the learning rate of phase 1 while keeping the ratios of learning rate in phase 1 and phase 2 to the same as LAMB. We use $\bar{\beta} = 0.999, \underline{\beta} = 0.5$ for MADAM. For LAMB, we use the default setting from the aforementioned repository.

The faster adaptation of MaxVA improves the stability, which enables MADAM to use a much larger learning rate to achieve faster convergence than ADAM. The best learning rate for MADAM is 3.4e–3. We tried learning rates in {7e–4, 8e–4, 9e–4, 1e–3} for ADAM, and find it always diverges when the learning rate is higher or equal to 9e–4. The best result of ADAM is achieved with learning rate 8e–4. MADAM achieves a training loss of 1.492, while LAMB achieves a training loss of 1.507, and ADAM has the worst training loss 1.568. The test scores of the models pretrained with MADAM/LAMB/ADAM are 88.53/87.60/88.07 (F1) and 82.10/81.40/80.78 (Accuracy) on SQuAD v1.1 and MNLI, respectively.

## 6 Related Work

Various adaptive methods have been proposed and broadly applied in deep learning [8,12,26,34]. [23] proposed to compute the adaptive learning rate with the coordinate-wise maximum value of $v_t$ so that the adaptive learning rate does not increase. ADABOUND [19] clips the adaptive learning rate of ADAM with a decreasing upper bound and an increasing lower bound. Lookahead [37] computes weight updates by looking ahead at the sequence of "fast weights" generated by another optimizer. Padam [5] improves the generalization of adaptive methods by choosing a proper exponent for the $v_t$ of AMSGRAD. LAPROP [40] uses local running estimation of the variance to normalize the gradients, resulting in higher empirical stability. RAdam [15] was recently invented to free ADAM from the warmup schedule for training Transformers. [20] found that using a linear warmup over $2 \cdot (1 - \beta_2)^{-1}$ iterations for ADAM achieves almost the same convergence as RAdam. [31] proposes Layer-wise Adaptive Rate Scaling (LARS), and scales the batch size to 16,384 for training ResNet50. LAMB [32] applies a similar layer-wise learning rate on ADAM to improve LARS on training BERT. Starting from a similar motivation of adapting to the curvature, the recent work AdaBelief [39] directly estimates the exponential running average of the gradient deviation to compute the adaptive step sizes. Our approach finds the averaging coefficients $\beta_t$ automatically by maximizing the estimated variance for a faster adaptation to the curvature, which could be complementary to all the aforementioned methods, and is the first to explore in this direction to our knowledge.

## 7 Conclusion

In this paper, we present Maximum Variation Averaging (MaxVA), a novel adaptive learning rate scheme that replaces the exponential running average of squared gradient with an adaptive weighted mean. In each step, MaxVA chooses the weight $\beta_t$ for each coordinate, such that the esimated gradient variance is maximized. This enables MaxVA to: (1) take smaller steps when large curvatures or abnormally large gradients are present, which leads to more desirable convergence behaviors in face of noisy gradients; (2) adapt faster to the geometry of the objective, achieving faster convergence in the large-batch setting. We illustrate how our method improves convergence by a better adaptation to variance, and demonstrate strong empirical results on a wide range of tasks. We prove MaxVA converges in the nonconvex stochastic optimization setting under mild assumptions.

## References

1. Balles, L., Hennig, P.: Dissecting adam: the sign, magnitude and variance of stochastic gradients. In: ICML, pp. 404–413 (2018)
2. Bernstein, J., Wang, Y.X., Azizzadenesheli, K., Anandkumar, A.: signsgd: compressed optimisation for non-convex problems. In: ICML, pp. 560–569 (2018)
3. Brown, T.B., et al.: Language models are few-shot learners. arXiv preprint arXiv:2005.14165 (2020)
4. Cettolo, M., Niehues, J., Stüker, S., Bentivogli, L., Federico, M.: Report on the 11th iwslt evaluation campaign, iwslt 2014. In: IWSLT, vol. 57 (2014)
5. Chen, J., Zhou, D., Tang, Y., Yang, Z., Gu, Q.: Closing the generalization gap of adaptive gradient methods in training deep neural networks. arXiv:1806.06763 (2018)
6. Chen, X., Liu, S., Sun, R., Hong, M.: On the convergence of a class of adam-type algorithms for non-convex optimization. In: ICLR (2019)
7. Devlin, J., Chang, M.W., Lee, K., Toutanova, K.: Bert: pre-training of deep bidirectional transformers for language understanding. In: NAACL, pp. 4171–4186 (2019)
8. Duchi, J., Hazan, E., Singer, Y.: Adaptive subgradient methods for online learning and stochastic optimization. JMLR (2011)
9. Ghadimi, S., Lan, G.: Stochastic first-and zeroth-order methods for nonconvex stochastic programming. SIAM J. Optim. **23**(4), 2341–2368 (2013)
10. He, K., Zhang, X., Ren, S., Sun, J.: Deep residual learning for image recognition. In: CVPR, pp. 770–778 (2016)
11. Hendrycks, D., Gimpel, K.: Gaussian error linear units (gelus). arXiv preprint arXiv:1606.08415 (2016)
12. Kingma, D.P., Ba, J.: Adam: a method for stochastic optimization. In: Bengio, Y., LeCun, Y. (eds.) ICLR (2015)
13. Lan, Z., Chen, M., Goodman, S., Gimpel, K., Sharma, P., Soricut, R.: Albert: a lite bert for self-supervised learning of language representations. In: ICLR (2020)
14. Li, H., Xu, Z., Taylor, G., Studer, C., Goldstein, T.: Visualizing the loss landscape of neural nets. In: Advances in Neural Information Processing Systems, pp. 6389–6399 (2018)
15. Liu, L., et al.: On the variance of the adaptive learning rate and beyond. In: ICLR (2020)
16. Liu, L., Liu, X., Gao, J., Chen, W., Han, J.: Understanding the difficulty of training transformers. arXiv:2004.08249 (2020)
17. Liu, Y., et al.: Roberta: a robustly optimized bert pretraining approach. arXiv:1907.11692 (2019)
18. Loshchilov, I., Hutter, F.: Decoupled weight decay regularization. In: ICLR (2018)

19. Luo, L., Xiong, Y., Liu, Y., Sun, X.: Adaptive gradient methods with dynamic bound of learning rate. In: ICLR (2019)
20. Ma, J., Yarats, D.: On the adequacy of untuned warmup for adaptive optimization. arXiv:1910.04209 (2019)
21. Ott, M., Edunov, S., Grangier, D., Auli, M.: Scaling neural machine translation. In: WMT, pp. 1–9 (2018)
22. Raffel, C., et al.: Exploring the limits of transfer learning with a unified text-to-text transformer. arXiv:1910.10683 (2019)
23. Reddi, S.J., Kale, S., Kumar, S.: On the convergence of adam and beyond. In: ICLR (2018)
24. Schaul, T., Zhang, S., LeCun, Y.: No more pesky learning rates. In: ICML, pp. 343–351 (2013)
25. Srivastava, N., Hinton, G., Krizhevsky, A., Sutskever, I., Salakhutdinov, R.: Dropout: a simple way to prevent neural networks from overfitting. JMLR (2014)
26. Tieleman, T., Hinton, G.: Lecture 6.5–RmsProp: divide the gradient by a running average of its recent magnitude. COURSERA (2012)
27. Vaswani, A., et al.: Attention is all you need. In: NeurIPS (2017)
28. Wang, A., Singh, A., Michael, J., Hill, F., Levy, O., Bowman, S.R.: Glue: a multi-task benchmark and analysis platform for natural language understanding. In: EMNLP (2018)
29. Wilson, A.C., Roelofs, R., Stern, M., Srebro, N., Recht, B.: The marginal value of adaptive gradient methods in machine learning. In: Neurips, pp. 4148–4158 (2017)
30. Wu, Y., Ren, M., Liao, R., Grosse, R.: Understanding short-horizon bias in stochastic meta-optimization. arXiv:1803.02021 (2018)
31. You, Y., Gitman, I., Ginsburg, B.: Scaling SGD batch size to 32k for imagenet training. CoRR abs/1708.03888 (2017)
32. You, Y., et al.: Large batch optimization for deep learning: training bert in 76 minutes. In: ICLR (2020)
33. Zaheer, M., Reddi, S., Sachan, D., Kale, S., Kumar, S.: Adaptive methods for nonconvex optimization. In: NeurIPS, pp. 9793–9803 (2018)
34. Zeiler, M.D.: ADADELTA: an adaptive learning rate method. CoRR (2012)
35. Zhang, G., et al.: Which algorithmic choices matter at which batch sizes? insights from a noisy quadratic model. In: NeurIPS, pp. 8194–8205 (2019)
36. Zhang, J., et al.: Why are adaptive methods good for attention models? In: NeurIPS 33 (2020)
37. Zhang, M.R., Lucas, J., Ba, J., Hinton, G.E.: Lookahead optimizer: k steps forward, 1 step back. In: NeurIPS (2019)
38. Zhu, C., Cheng, Y., Gan, Z., Sun, S., Goldstein, T., Liu, J.: Freelb: enhanced adversarial training for natural language understanding. In: ICLR (2020)
39. Zhuang, J., et al.: Adabelief optimizer: adapting stepsizes by the belief in observed gradients. In: NeurIPS (2020)
40. Ziyin, L., Wang, Z.T., Ueda, M.: Laprop: a better way to combine momentum with adaptive gradient. arXiv:2002.04839 (2020)

# Augmenting Open-Domain Event Detection with Synthetic Data from GPT-2

Amir Pouran Ben Veyseh[1(✉)], Minh Van Nguyen[1], Bonan Min[2], and Thien Huu Nguyen[1]

[1] Department of Computer and Information Science, University of Oregon, Eugene, OR, USA
{apouranb,minhnv,thien}@cs.uoregon.edu
[2] Raytheon BBN Technologies, Cambridge, USA
bonan.min@raytheon.com

**Abstract.** Open-domain event detection (ODED) aims to identify event mentions of all possible types in text. A challenge for ODED research is the lack of large training datasets. In this work, we explore a novel method to overcome this challenge by fine-tuning the powerful pre-trained language model GPT-2 on existing datasets to automatically generate new training data for ODED. To address the noises presented in the generated data, we propose a novel teacher-student architecture where the teacher model is used to capture anchor knowledge on sentence representations and data type difference. The student model is then trained on the combination of the original and generated data and regularized to be consistent with the anchor knowledge from the teacher. We introduce novel regularization mechanism based on mutual information and optimal transport to achieve the knowledge consistency between the student and the teacher. Moreover, we propose a dynamic sample weighting technique for the generated examples based on optimal transport and data clustering. Our experiments on three benchmark datasets demonstrate the effectiveness of the propped model, yielding state-of-the-art performance for such datasets.

**Keywords:** Event extraction · Natural language processing · Data augmentation · GPT-2 · Teacher-student architecture

## 1 Introduction

In Natural Language Processing (NLP), events are mentioned in text and refer to changes of states of real word entities [33]. Each event in text is associated with an event trigger, which is the main word (most often a single verb or nominalization) to evoke the event (e.g., *agree*, *death*). As such, Event Detection (ED) is one of the important tasks in Information Extraction of NLP whose goal is to identify trigger words of events in text. For instance, in the sentence "*Ames*


N. Oliver et al. (Eds.): ECML PKDD 2021, LNAI 12977, pp. 644–660, 2021.
https://doi.org/10.1007/978-3-030-86523-8_39

***recruited*** *her as an informant in 1983, then* ***married*** *her two years later*", an ED system should be able to recognize "*recruited*" and "*married*" as trigger words of two events of semantic types *Employment* and *Marriage* respectively (also called event mentions). Due to its ubiquity, detecting event mentions is an integral part of natural language understanding that can support question answering, text sumarization, and knowledge base population (among others).

There is a wealth of prior work on event detection [5,11,15,24,25,37]. However, most of prior works on ED assumes a predefined and small set of event types for some specific domain that prevents the extraction of open-domain events (i.e., of all possible types). This work focuses on open-domain ED (ODED) to fill this gap. One hurdle for developing effective ODED models is the lack of large training datasets. To obtain more training signals/data for ED, prior works have resorted to unsupervised [9,40], distantly supervised [3,12] or domain adaptation [22] methods. The common characteristics of these methods involves the requirement of large-scale (though unlabeled) texts replete with event mentions. In this work, we propose a novel method to generate additional training data for ODED solely based on the pre-trained language model GPT-2 [32]. In particular, given the training data in an existing ODED dataset, we fine-tune GPT-2 to automatically generate sentences that are annotated with open-domain event mentions. The generated data will then be combined with the original training data to train models for ODED. To our knowledge, this is the first work that leverages GPT-2 for data generation/augmentation for ED.

How can we efficiently combine the generated sentences with the original data? The answer for this question is important as the generated samples might be noisy/erroneous and simply concatenating them with the original training data to train models might even hurt the performance. To address this issue, prior work on automatic data generation with GPT-2 [2,39] has explored sample filtering techniques to remove noisy samples from the generated data. This is done once before the model training and the remaining generated data is added into the original data to train models. However, the noisiness criteria for sample selection in such prior work is solely determined by fixed heuristics that might not be optimal. To better handle the noisy data from GPT-2, we propose to instead keep all the generated samples and devise mechanisms to directly address noisy examples during the training process of the models for ODED. As such, we argue that noisy samples would cause the representations/knowledge learned via the combination of synthetic and original data to diverge significantly from those learned solely via the original training data. To minimize the effect of noisy samples, we can thus use the knowledge in the original data as an anchor and enforce the induced knowledge from the synthetic and original data to be consistent/close to those anchor. In this way, the models can still benefit from the new information presented in the synthetic data while avoiding its inherent noises via the consistency constraint.

To this end, the teacher-student architecture is employed to achieve the knowledge consistency. Here, the teacher model is first trained on the original training data for ODED to induce anchor knowledge. The student will be trained afterward from both generated and original data. Two learning principles are introduced to accommodate consistency between the knowledge from the student and the anchor knowledge: (1) Representations obtained by the teacher and the student on the same sentences should be compatible with each other. As such, we propose to increase the mutual information between the teacher and student networks in the learning process; (2) Both the teacher and the student should discern the same level of difference between the original and the generated data. For this principle, we propose two aspects to assess the distance between these two types of data for difference consistency enforcement, i.e., representations and event-containing likelihoods for sentences. In particular, instead of only relying on the similarity of the induced representations, our distance function for two sentence examples in ODED also consider the similar likelihood for the sentences to contain event mentions (i.e., event-containing sentences are more similar in ODED). Accordingly, our model leverages Optimal Transport, a method to find the cheapest transformation between two data distributions, as a natural solution to simultaneously incorporate the two aspects into the distance function for synthetic and original data for ODED.

Finally, to further regulate the synthetic data in the training process, we seek to assign weights to the generated samples in the training loss function to control their contribution for ODED. In particular, we compute the weight of each synthetic data point based on its diversity w.r.t. the other generated samples and its novelty w.r.t. the original training data. Our model features dynamic updates of the data point weights after each training epoch to better reflect the training process. Extensive experiments on three benchmark ODED datasets reveal the effectiveness of the proposed model and lead to state-of-the-art performance for all the datasets.

## 2 Model

We formulate ODED as a sequence labeling task. Given a sentence $S = [w_1, w_2, \ldots, w_N]$, the goal is to make a prediction for each word $w_i$, indicating whether it is triggering an event or not (i.e., a binary classification for each word). Our proposed approach for ODED consists of two major stages: (1) Data Generation and (2) Task Modeling.

### 2.1 Data Generation

Our approach is motivated by GPT-2, a transformer-based language model [32] that has shown impressive capability to generate fluent and meaningful text. As such, we aim to utilize GPT-2 to automatically generate training data for ODED[1] and combine it with existing datasets to train better ODED models. To this end, given a training dataset for ODED, we first attempt to fine-tune the pre-trained GPT-2 model on this dataset so the knowledge about open-domain

[1] We use the small version of GPT-2 in this work.

event mentions can be injected into the language model. The expectation is to encourage GPT-2 to produce synthetic sentences marked with event mentions afterward.

Formally, for each sentence $S = [w_1, w_2, \ldots, w_N]$ in the original training dataset $\mathcal{O}$, we first insert two special tokens $TRG_s$ and $TRG_e$ right before and after (respectively) each event trigger word in $S$. For instance, assuming $w_t$ is the only event trigger in $S$, the resulting sentence $S'$ after insertion would be: $S' = [w_1, \ldots, TRG_s, w_t, TRG_e, \ldots, w_N]$. Afterward, we fine-tune the pre-trained GPT-2 model on the newly generated sentences $S'$ in an auto-regressive fashion (i.e., predicting the next token in $S'$). Finally, the fine-tuned GPT-2 is employed to generate $|\mathcal{O}^+|$ sentences where $\mathcal{O}^+$ is the set of sentences with at least one event mention in $\mathcal{O}$ (i.e., positive samples). We denote the generated dataset by $\mathcal{G}$. Note that our generation process ensures all the sentences in $\mathcal{G}$ to involve at least one marked event trigger from GPT-2.

To evaluate the quality of the generated data, we manually assess 100 sentences that are generated by GPT-2 after being fine-tuned on the ODED dataset LitBank [33]. Among these samples, we find that 85% of the sentences are grammatically correct, semantically sound, and similar to the original sentences in LitBank. Also, 80% of the marked event triggers in the sentences are correct. Refer to Sect. 3.5 for examples of generated sentences, noises and more data analysis.

## 2.2 Task Modeling

Our generated data in $\mathcal{G}$ is noisy as the sentences might not be natural and the event triggers might be incorrectly marked. To effectively combine $\mathcal{G}$ with the original data $\mathcal{O}$, this section describes our proposed teacher-student framework to overcome the noise in the generated data to train ODED models.

**Base Model:** The student and teacher models in our framework employ the same base network architecture for ODED (with different parameters). As such, following the recent ED model in [36], our base architecture for ODED starts with the pre-trained BERT model [6] whose output representations are consumed by a Bi-directional LSTM (BiLSTM), and then followed by a feed-forward network to make event predictions for each word in $S$. In particular, we first feed the input sentence $S = [w_1, w_2, \ldots, w_N]$ into the pre-trained $\text{BERT}_{base}$ model. Afterward, the hidden vectors from the last layer of the BERT model are consumed by BiLSTM (i.e., the encoder) to generate the word representations $H = [h_1, h_2, \ldots, h_N]$. The representation vector $h_i$ for $w_i \in S$ is then sent to a two-layer feed-forward network with the sigmoid function in the end to obtain a probability distribution $P(\cdot|S, i)$ over possible event labels for $w_i$, i.e., *EVENT* or *NONE*. Finally, we use the negative log-likelihood as the loss function to train our models for ODED: $\mathcal{L}_{pred} = -\sum_{i=1}^{N} \log P(y_i|S, i)$ where $y_i$ is the golden event label for $w_i \in S$.

**Knowledge Consistency Enforcement:** As motivated in the introduction, the teacher model $\mathcal{T}$ in our framework is first trained on the original noiseless data $\mathcal{O}$ to capture the anchor knowledge (using the loss $\mathcal{L}_{pred}$). The data combination $\mathcal{O} \cup \mathcal{G}$ will then be utilized to train the student model $\mathcal{S}$, also serving as our final model for ODED. To mitigate the noise from $\mathcal{G}$, the student model $\mathcal{S}$ would be regularized to be consistent with the anchor knowledge from the teacher. As such, we employ two types of anchor knowledge for teacher-student consistency, i.e., the representations of the input sentences and the difference between the synthetic and original data as perceived by the teacher.

First, to enforce the teacher-student compatibility based on representations of input sentences, we first feed the input sentence $S$ into both the teacher and the student networks, and compute the representations of the sentence by max-pooling the representation vectors returned by the BiLSTM encoders for the words, i.e., $h^{\mathcal{T}} = MAX\_POOL[h_1^{\mathcal{T}}, \dots, h_N^{\mathcal{T}}]$ for the teacher and similarly $h^{\mathcal{S}}$ for the student. The goal of the representation compatibility is to encourage the similarity between $h^{\mathcal{T}}$ and $h^{\mathcal{S}}$ for the same $S$. As such, one simple method is to directly compute the distance between the two vectors and use this distance as an auxiliary training loss. However, this might restrict the learning capability of the student as less room for variation in $h^{\mathcal{S}}$ is granted. To allow more variation in $h^{\mathcal{S}}$ and still maintain essential similarity with $h^{\mathcal{T}}$, we propose to instead maximize the mutual information (MI) between $h^{\mathcal{T}}$ and $h^{\mathcal{S}}$. However, computing MI between $h^{\mathcal{T}}$ and $h^{\mathcal{S}}$ is intractable due their high dimensionality. To address this issue, following [8], we treat $h^{\mathcal{T}}$ and $h^{\mathcal{S}}$ as random variables and estimate their MI via the Jenson-Shanon divergence between the joint distribution, i.e., $P_{\mathcal{T},\mathcal{S}}(h^{\mathcal{T}}, h^{\mathcal{S}})$, and the product of the marginal distributions, i.e., $P_{\mathcal{T}}(h^{\mathcal{T}}) * P_{\mathcal{S}}(h^{\mathcal{S}})$. In particular, the Jenson-Shanon divergence computation can be fulfilled by a feed-forward discriminator, denoted by $D$, which distinguishes the samples of the joint distribution $P_{\mathcal{T},\mathcal{S}}(h^{\mathcal{T}}, h^{\mathcal{S}})$ from those of the product distribution $P_{\mathcal{T}}(h^{\mathcal{T}}) * P_{\mathcal{S}}(h^{\mathcal{S}})$. In this work, we sample from $P_{\mathcal{T},\mathcal{S}}(h^{\mathcal{T}}, h^{\mathcal{S}})$ by directly concatenating $h^{\mathcal{T}}$ and $h^{\mathcal{S}}$. The samples from $P_{\mathcal{T}}(h^{\mathcal{T}}) * P_{\mathcal{S}}(h^{\mathcal{S}})$ are obtained by concatenating $h^{\mathcal{T}}$ with the student-based representation vector for another sentence randomly chosen from the same mini-batch, i.e., $h'^{\mathcal{S}}$. Finally, the samples are sent to the discriminator $D$ to produce scores for whether the samples are from the joint distribution or not. The negative log-likelihoods of the samples are then included in the overall loss function to serve as a way to maximize the MI between $h^{\mathcal{S}}$ and $h^{\mathcal{T}}$:

$$\mathcal{L}_{MI} = -\log(D([h^{\mathcal{T}}, h^{\mathcal{S}})])) - \log(1 - D([h^{\mathcal{T}}, h'^{\mathcal{S}}])) \tag{1}$$

Our second teacher-student consistency for learning is realized by the difference between synthetic and original data in $\mathcal{G}$ and $\mathcal{O}$. As such, our principle is to treat the teacher-based distance between the sentences in $\mathcal{G}$ and $\mathcal{O}$ as an anchor. The student should then adhere to this anchor distance to avoid significant divergence from the teacher for noise mitigation. Given that, the major question is how to effectively estimate the distance/difference between $\mathcal{G}$ and $\mathcal{O}$ for this consistency regularization. To this end, as presented in the introduction, our intuition is to simultaneously consider the representations and the event-containing likelihoods of the sentences in $\mathcal{G}$ and $\mathcal{O}$ for this distance function. To implement this intuition, we first discuss how we compute the distance $D^{\mathcal{S}}_{\mathcal{G},\mathcal{O}}$ between $\mathcal{G}$ and $\mathcal{O}$ for the student model $\mathcal{S}$.

In particular, given the student network $\mathcal{S}$, we first obtain a representation vector $h^S$ for each sentence $S$ in $\mathcal{G} \cup \mathcal{O}$ as done in the MI estimation above. Afterward, to obtain an event-containing likelihood score $p^S$ for $S$, we compute the maximum of the probabilities for being an event trigger of the words $w_i \in S$: $p^S = \max_{i=1..N}(P^{\mathcal{S}}(EVENT|S,i))$ where $P^{\mathcal{S}}(\cdot|S,i)$ is the probability distribution over the possible event labels for $w_i$ based on the student model $\mathcal{S}$. To facilitate the distance estimation for $\mathcal{G}$ and $\mathcal{O}$, let $R^o = [h_1^{\mathcal{S},o}, \dots, h_n^{\mathcal{S},o}]$ and $P^o = [p_1^{\mathcal{S},o}, \dots, p_n^{\mathcal{S},o}]$ be the sets of representation vectors and event-containing likelihood scores (respectively) for the sentences in $\mathcal{O}$ ($n = |\mathcal{O}|$). Similarly, let $R^g = [h_1^{\mathcal{S},g}, \dots, h_m^{\mathcal{S},g}]$ and $P^g = [p_1^{\mathcal{S},g}, \dots, p_m^{\mathcal{S},g}]$ be the sets of representation vectors and event-containing likelihood scores (respectively) for the sentences in $\mathcal{G}$ ($m = |\mathcal{G}|$). Consequently, to simultaneously exploit $R^o, P^o, R^g$ and $P^g$ for distance estimation, we seek to find an optimal alignment between sentences in $\mathcal{G}$ and $\mathcal{O}$ such that two sentences with closest representations and event-containing likelihoods have better chance to be aligned to each other. This problem can be then solved naturally with optimal transport (OT) methods that enable the computation of the optimal mapping between two probability distributions.

Formally, given the probability distributions $p(x)$ and $q(y)$ over the domains $\mathcal{X}$ and $\mathcal{Y}$, and the cost function $C(x,y) : \mathcal{X} \times \mathcal{Y} \rightarrow \mathbb{R}_+$ for mapping $\mathcal{X}$ to $\mathcal{Y}$, OT finds the optimal joint distribution $\pi^*(x,y)$ (over $\mathcal{X} \times \mathcal{Y}$) with marginals $p(x)$ and $q(y)$, i.e., the cheapest transportation of $p(x)$ to $q(y)$, by solving the following problem:

$$\begin{aligned} \pi^*(x,y) = \min_{\pi \in \Pi(x,y)} \int_{\mathcal{Y}} \int_{\mathcal{X}} \pi(x,y) C(x,y) dx dy \\ \textbf{s.t. } x \sim p(x) \text{ and } y \sim q(y), \end{aligned} \tag{2}$$

where $\Pi(x,y)$ is the set of all joint distributions with marginals $p(x)$ and $q(y)$. Note that if the distributions $p(x)$ and $q(y)$ are discrete, the integrals in Eq. 2 are replaced with a sum and the joint distribution $\pi^*(x,y)$ is represented by a matrix whose entry $\pi(x_i, y_j)$ represents the probability of transforming the data point $x_i$ to $y_j$ to convert the distribution $p(x)$ to $q(y)$. To this end, our model defines the domains $\mathcal{X}$ and $\mathcal{Y}$ for OT via the representation spaces of the sets $R^g$ and $R^o$. As such, the cost function $C(x,y)$ is defined by the Euclidean distance between the representation vectors of the corresponding elements, i.e., $C(h_i^{\mathcal{S},g}, h_j^{\mathcal{S},o}) = \|h_i^{\mathcal{S},g} - h_j^{\mathcal{S},o}\|$. Also, the probability distributions $p(x)$ and $q(y)$ are defined over the normalized likelihood scores $P^g$ and $P^o$, i.e., $p(h_i^{\mathcal{S},g}) = softmax(P^g)$ and $p(h_i^{\mathcal{S},o}) = softmax(P^o)$. Based on these definitions, the optimal transport $\pi^*(h_i^{\mathcal{S},g}, h_j^{\mathcal{S},o})$, which is obtained by solving Eq. 2, could be used to compute the smallest transportation cost between $\mathcal{G}$ and $\mathcal{O}$ (i.e., Wasserstein distance), serving as the distance function $D^{\mathcal{S}}_{\mathcal{G},\mathcal{O}}$ in our model: $D^{\mathcal{S}}_{\mathcal{G},\mathcal{O}} = \Sigma_{i=1}^m \Sigma_{j=1}^n \pi^*(h_i^{\mathcal{S},g}, h_j^{\mathcal{S},o}) \|h_i^{\mathcal{S},g} - h_j^{\mathcal{S},o}\|$.

Note that as solving the OT problem in Eq. 2 is intractable, we employ the entropy-based approximation of OT and solve it with the Sinkhorn algorithm [30].

In the next step, we apply the same procedure obtain the distance $D^{\mathcal{T}}_{\mathcal{G},\mathcal{O}}$ between $\mathcal{G}$ for the teacher network. Finally, to realize the $\mathcal{G}$-$\mathcal{O}$ distance consistency between the student and the teacher, we introduce the difference $\mathcal{L}_{diff}$ between $D^{\mathcal{T}}_{\mathcal{G},\mathcal{O}}$ and $D^{\mathcal{S}}_{\mathcal{G},\mathcal{O}}$ into the overall loss function:

$$\mathcal{L}_{diff} = |D^{\mathcal{T}}_{\mathcal{G},\mathcal{O}} - D^{\mathcal{S}}_{\mathcal{G},\mathcal{O}}| \quad (3)$$

Note that in the actual implementation, we only compute $\mathcal{L}_{diff}$ for each mini-batch and add it to the loss function to enable efficient solving of the OT optimization.

**Dynamic Sample Weighting:** To further improve the representation learning of the student model for ODED, we seek to control the contribution of each generated sample in $\mathcal{G}$ for the overall training loss. In particular, we weight each generated sample in the loss function based on an estimated degree to which the sample can provide new training signals for the original data $\mathcal{O}$. In our model, the weight for each synthetic sample is determined based on its diversity score w.r.t. to other generated sentences in $\mathcal{G}$ and its novelty score w.r.t the original data. As such, we dynamically computes the weights during the training process where the weight and scores for each generated sample are updated immediately after each epoch to better capture the progress of the model.

Accordingly, to compute diversity scores for synthetic samples, we first cluster all generated sentences in $\mathcal{G}$ using their student-based representation vectors $h_i^{\mathcal{S},g}$ from the BiLSTM encoder and $K$-mean clustering[2]. This produces $K$ centroid vectors $c^g = [c_1^g, \ldots, c_K^g]$ for $K$ clusters $C^g = [C_1^g, \ldots, C_K^g]$ (respectively) in $\mathcal{G}$. Let $k(i)$ be the cluster index of the $i$-the sentence $S_i \in \mathcal{G}$ $(1 \le k(i) \le K)$. The diversity score $s_i^{div}$ for $S_i$ in our model is then computed based on the distance between its representation $h_i^{\mathcal{S},g}$ and the corresponding centroid vector $c_{k(i)}^g$ (i.e., the farther $S_i$ is toward the center, the more it contributes to the diversity in $\mathcal{G}$): $a_i = \|h_i^{\mathcal{S},g} - c_{k(i)}^g\|$, $s_i^{div} = e^{a_i} / \sum_{S_j \in C_{k(i)}^g} e^{a_j}$.

In the next step, we aim to compute novelty scores for the generated samples in $\mathcal{G}$ via their distances toward the original data $\mathcal{O}$ (i.e., the farther $S_i$ is toward $\mathcal{O}$, the more novel it is for the training data). As such, we reuse the idea of OT-based distance between $\mathcal{G}$ and $\mathcal{O}$ to estimate the novelty score for the synthetic samples in $\mathcal{G}$ (i.e., based on both the representation vectors and the event-containing likelihoods of sentences). In particular, to enable efficient OT computation, we also first cluster the sentences in the original training dataset $\mathcal{O}$ using their representation vectors $h_i^{\mathcal{S},o} \in R^o$ from BiLSTM and $K$-mean clustering. The outputs from this clustering involve $K$ centroid vectors $c^o = [c_1^o, \ldots, c_K^o]$ for the $K$ clusters $C^o = [C_1^o, \ldots, C_K^o]$ (respectively) in $\mathcal{O}$. Afterward, we attempt to compute the optimal transport between the cluster sets $C^g$ and $C^o$ for the

[2] $K = 10$ produces the best performance in our study.

synthetic and original data whose results are leveraged to obtain novelty scores for generated samples later.

To this end, we directly use the centroid vectors in $c^g$ and $c^o$ as the representations for the clusters in $C^g$ and $C^o$, serving as the domains $\mathcal{X}$ and $\mathcal{Y}$ for OT. Also, the event-containing likelihood score $p_i^t$ for the cluster $C_i^t$ ($t \in \{g, o\}$ to indicate synthetic or original data, $1 \leq i \leq K$) is estimated via the average of the student-based likelihood scores $p_j^{S,t}$ of the sentences in $C_i^t$: $p_i^t = \frac{1}{|C_i^t|} \sum_{S_j \in C_i^t} p_j^{S,t}$. Given the representations and event-containing likelihood scores, we follow the same described procedure to obtain the optimal transport $\pi_c^*(c_i^g, c_j^o)$ between the cluster sets $C^g$ and $C^o$ (i.e., by solving Eq. 2 and using the Euclidean distance for the cost function $C(x, y)$). In the next step, we align each cluster $C_i^g$ in $\mathcal{G}$ to cluster $C_{i*}^o$ in $\mathcal{O}$ that has the highest probability in the optimal transport, i.e., $i^* = \text{argmax}_j \, \pi_c^*(c_i^g, c_j^o)$. The distance $d_i$ between the cluster $C_i^g$ in $\mathcal{G}$ toward the original data $\mathcal{O}$ is then obtained via the distance between the centroid vectors of $C_i^g$ and $C_{i*}^o$: $b_i = \|c_i^g - c_{i*}^o\|, d_i = e^{b_i} / \sum_{j=1..K} e^{b_j}$.

As such, the novelty score $s_i^{nov}$ for a generated sample $S_i \in \mathcal{G}$ in our work will be set directly to the distance of its corresponding cluster $C_{k(i)}^g$ toward the original data: $s_i^{nov} = d_{k(i)}$.

Finally, to aggregate the diversity and novelty scores for each sentence $S_i \in \mathcal{G}$ to compute its weight $s_i^{comb}$ in the training loss, we set $s_i^{comb} = (1-\alpha)s_i^{nov} + \alpha s_i^{div}$ and normalize it over the generated samples ($\alpha$ is a trade-off parameter). For convenience, we consider the examples in the original data $\mathcal{O}$ as having the weight of 1. The final training loss function for the student network over an example $S \in \mathcal{G} \cup \mathcal{O}$ with the weight $s^{comb}$ is thus: $\mathcal{L} = s^{comb} * \mathcal{L}_{pred} + \beta * \mathcal{L}_{MI} + \gamma * \mathcal{L}_{diff}$ where $\beta$ and $\gamma$ are the trade-off parameters.

## 3 Experiments

### 3.1 Datasets and Hyper-parameters

We use the following three ODED datasets to evaluate the models[3]:

- **LitBank** [33]: This dataset provides event annotation for 100 English literary texts. We use the same data split (train/dev/test) as [33] for fair comparison.
- **TimeBank** [31]: This dataset annotates 183 English news articles with event mentions and temporal relations. To make fair comparison with prior work [22], we exclude event mentions in TimeBank that have not occurred (e.g., future, hypothetical or negated events) and apply the the same data split.
- **SW100** [3]: This corpus presents event annotations for 100 documents in 10 different domains from Simple Wikipedia. The original data split in [3] is inherited in our experiment.

[3] Note that we do not use the ACE 2005 dataset [35] as it only focuses on a small set of event types in the news domain, thus being not appropriate for our open-domain setting of event detection.

**Table 1.** Model performance (F1 scores) on the test data of the datasets. Rows with * indicate results taken from the corresponding papers.

| | LitBank | TimeBank | SW100 |
|---|---|---|---|
| BiLSTM-Base$_{\mathcal{O}}$ | 73.4 | 81.9 | 85.3 |
| BiLSTM-Base$_{\mathcal{G}}$ | 75.9 | 80.0 | 83.8 |
| BiLSTM-Base$_{\mathcal{O}+\mathcal{G}}$ | 74.0 | 82.3 | 84.3 |
| Confidence-Filtering [2] | 73.9 | 82.5 | 86.1 |
| Novelty-Filtering [39] | 74.9 | 81.4 | 86.8 |
| BiLSTM+BERT* [33] | 73.9 | – | – |
| GatedGCN* [15] | 74.5 | – | – |
| BERT-A* [22] | – | 82.6 | – |
| DS-BiLSTM* [3] | – | – | 72.4 |
| GPT-Augmented (proposed) | **77.4** | **85.0** | **89.7** |

In our experiments we use the small version of GPT-2 to generate data. All the hyperparameters for the proposed model are selected based on the F1 scores on the development set of LitBank. The same hyper-parameters from this fine-tuning are then applied for the other datasets for consistency. In the base model, we use the BERT$_{base}$ version, 200 dimensions in the hidden states of the BiLSTM, and 2 layers for the feed-forward neural network with 200 hidden dimensions to predict events. The discriminator $D$ of the MI component consists of an one-layer feed-forward neural network with 200 hidden dimensions. The trade-off parameters $\alpha$, $\beta$ and $\gamma$ are set to 0.6, 0.1, and 0.05, respectively. The learning rate is set to 0.3 for the Adam optimizer and the batch size of 50 is employed during training. Finally, note that we do not update the BERT model for word embeddings in this work due to its better performance on the development data of LitBank. We use a single GeForce RTX 2080 GPU with 11 GB memory to run the models in this work. PyTorch 1.1 is used to implement the models.

### 3.2 Baselines

We compare our model (called **GPT-Augmented**) with the following baselines:

- Base models: The baselines in this group all employ the base architecture in Sect. 2.2. Three versions of this base model are possible, i.e., **BiLSTM-Base**$_{\mathcal{O}}$, **BiLSTM-Base**$_{\mathcal{G}}$, and **BiLSTM-Base**$_{\mathcal{O}+\mathcal{G}}$, depending on whether it is trained on the original training data $\mathcal{O}$, the generated data $\mathcal{G}$, or the combination of $\mathcal{O}$ and $\mathcal{G}$ respectively.
- Noisy mitigation methods: We consider two recent methods that are proposed to mitigate the noise from the GPT-generated data to train models: (i) **Confidence-Filtering**: Inspired by LAMBADA [2], this baseline filters the

generated data based on the confidence of the trained base model BiLSTM-Base$_{\mathcal{O}}$ in predicting labels for the generated data. Specifically, BiLSTM-Base$_{\mathcal{O}}$ is first employed to make predictions on the generated data. The synthetic sentences in $\mathcal{G}$ are then sorted via the confidence of the model (i.e., the event-containing likelihoods with the event probabilities from BiLSTM-Base$_{\mathcal{O}}$). Only sentences with a confidence above a tuned threshold are kept and combined with the original data to retrain the base model; and (ii) **Novelty-Filtering**: Inspired by [39], this baseline also filters the generated samples in $\mathcal{G}$ as Confidence-Filtering; however, instead of using the confidence, it employs novelty scores. As such, the novelty score for one generated sentence $S \in \mathcal{G}$ is obtained via the average of the Euclidean distances between the representations of $S$ and each sentence in the original training data $\mathcal{O}$. Here, the representations of the examples are also computed by the trained BiLSTM-Base$_{\mathcal{O}}$ model.

- State-of-the-art models: We also compare GPT-Augmented with the existing systems that report the state-of-the-art performance on each dataset. Namely, on LitBank we compare with the **BiLSTM+BERT** model in [33] and the recent **GatedGCN** model in [15]. On TimeBank, we compare with the in-domain performance of the **BERT-A** model in [22] that takes an adversarial domain adaptation approach. Finally, on SW100 dataset, we compare with the **DS-BiLSTM** model in [3] that augments the training data with distant supervision data.

### 3.3 Results

The performance (i.e., F1 scores) of the models on three datasets is shown in Table 1. As can be seen, the proposed model significantly outperforms the baselines (with $p < 0.01$) over all the three datasets. Specifically, compared to BiLSTM-Base$_{\mathcal{O}}$, GPT-Augmented improves F1 scores by at least 3%[4]. We attribute this improvement to the new training signals from the GPT-generated data and the effectiveness of the proposed techniques for noise mitigation and generated sample weighting for ODED. Also, the necessity of the proposed knowledge consistency enforcement for noise mitigation and sample weighting can be better revealed by considering the baselines BiLSTM-Base$_{\mathcal{O}+\mathcal{G}}$, Confidence-Filtering, and Novelty-Filtering that all exploit the generated data $\mathcal{G}$, but considerably under-perform the proposed model.

### 3.4 Ablation Study

This section studies the contribution of different components of the proposed model. We analyze the performance of GPT-Augmented on the LitBank development set when its two major components are removed or altered, i.e., (1) Knowledge Consistency Enforcement (KCE) and (2) Dynamic Sample Weighting (DSW).

[4] In the experiments, we learn that augmenting the models with GPT-generated data is more helpful for recalls.

**Knowledge Consistency Enforcement:** First, for KCE, we examine the following baselines:

(i) Full-MI: This model excludes the representation consistency based on MI from GPT-Augmented;
(ii) Full-MI+Euclidean: This model replaces the MI-based loss $\mathcal{L}_{MI}$ with Euclidean loss in GPT-Augmented (i.e., $\|h^{\mathcal{S}} - h^{\mathcal{T}}\|$);
(iii) Full-OT: This model eliminates the OT-based loss $\mathcal{L}_{OT}$ for data difference compatibility from GPT-Augmented;
(iv) Full-OT$^{rep}$: This baseline still employs the OT-based loss $\mathcal{L}_{OT}$; however, instead of computing the cost function $C(h_i^{\mathcal{S},g}, h_j^{\mathcal{S},o})$ based on representation distances, this model uses a constant cost function, i.e., $C(h_i^{\mathcal{S},g}, h_j^{\mathcal{S},o}) = 1$. This baseline aims to show that sentence representations play an important role for distance estimation between synthetic and original data for ODED;
(v) Full-OT$^{event}$: Instead of using event-containing likelihoods of sentences to compute the distributions $p(h_i^{\mathcal{S},g})$ and $p(h_i^{\mathcal{S},o})$ in the OT-based loss for GPT-Augmented, this method utilizes uniform distributions. The goal of this baseline is to demonstrate that event-containing likelihoods of sentences are crucial for the distance estimation between synthetic and original data for ODED;
(vi) Full-OT+Euclidean: This baseline completely replaces the OT loss $\mathcal{L}_{diff}$ with the Euclidean distance between the original data $\mathcal{O}$ and the generated data $\mathcal{G}$ for data difference consistency in KCE, i.e., the distance $D^{\mathcal{S}}_{\mathcal{G},\mathcal{O}}$ for the student network $\mathcal{S}$ is computed by: $D^{\mathcal{S}}_{\mathcal{G},\mathcal{O}} = \frac{1}{mn}\Sigma_{i=1}^{m}\Sigma_{j=1}^{n}\|h_i^{\mathcal{S},g} - h_j^{\mathcal{S},o}\|$;
(vii) Full-OT-MI: This baseline completely removes the OT- and MT-based loss; it excludes the KCE component and the teacher $\mathcal{T}$.

**Table 2.** Performance on the LitBank development set.

| Model | P | R | F1 |
|---|---|---|---|
| GPT-Augmented (Full) | 78.9 | 82.0 | **80.4** |
| Full-MI | 74.1 | 84.6 | 79.1 |
| Full-MI+Euclidean | 74.4 | 84.4 | 79.1 |
| Full-OT | 74.0 | 84.4 | 78.9 |
| Full-OT$^{rep}$ | 75.9 | 82.6 | 79.1 |
| Full-OT$^{event}$ | 76.1 | 79.8 | 77.9 |
| Full-OT+Euclidean | 76.1 | 81.8 | 78.9 |
| Full-OT-MI | 74.3 | 82.2 | 78.1 |

Table 2 shows the performance of the baseline models for KCE. It is clear from the table that both MI- and OT-based losses for KCE are important for

GPT-Augmented to achieve its best performance. Comparing GPT-Augmented and Full-MI+Euclidean, we see that using MI to realize the representation-based KCE is significantly better than its alternatives (i.e., via Euclidean distance) for GPT-Augmented. Finally, comparing different methods to estimate the distance between the generated and the original data for KCE (i.e., Full-OT$^{rep}$, Full-OT$^{event}$, and Full-OT+Euclidean), the OT-based approach in GPT-Augmented clearly demonstrates its benefits with the highest performance. This testifies to the importance of both sentence representations and event-containing likelihoods for the computation of the distance between $\mathcal{G}$ and $\mathcal{O}$ for ODED.

**Dynamic Sample Weighting:** Second, for the DSW component, we study the contribution of the proposed techniques for the sample weights with diversity and novelty scores. To this end, we evaluate the following models:

(i) Full-Div: This baseline excludes the diversity score from the weight of each generated sentence in the training loss of GPT-Augment, i.e., $s_i^{comb} = s_i^{nov}$.;
(ii) Full-Nov: This baseline excludes the novelty score $s_i^{nov}$ from GPT-Augmented, i.e., $s_i^{comb} = s_i^{div}$.;
(iii) Full-Nov$^{rep}$: In the computation of novelty scores using OT, this baseline replaces the cost function based on the centroid vectors with a constant function, i.e., $C(c_i^g, c_j^o) = 1$.;
(iv) Full-Nov$^{event}$: This model employs uniform distributions for the probability distributions in the OT-based computation for novelty scores (i.e., ignoring the event-containing likelihoods).;
(v) Full-Cluster: This baseline aims to completely eliminate the data clustering in GPT-Augmented.;
(vi) Full-Div-Nov: This component entirely removes DSW, using the same weight for all the samples, i.e., $s_i^{comb} = 1$.

**Table 3.** Performance on the LitBank development set.

| Model | P | R | F1 |
|---|---|---|---|
| GPT-Augmented (Full) | 78.9 | 82.0 | **80.4** |
| Full-Div | 78.1 | 80.4 | 79.2 |
| Full-Nov | 77.9 | 80.1 | 79.0 |
| Full-Nov$^{rep}$ | 76.9 | 79.4 | 78.1 |
| Full-Nov$^{event}$ | 77.2 | 80.6 | 78.8 |
| Full-Cluster | 74.2 | 82.6 | 78.2 |
| Full-Div-Nov | 75.1 | 82.9 | 78.8 |

The performance of the baseline models for DSW are presented in Table 3. The Table shows that both novelty and diversity scores are important for the best performance of GPT-Augmented. In addition, computing these scores without clustering cannot reach the best performance, indicating the necessity of clustering the data before computing the scores. Finally, the superiority of GPT-Augmented over Full-Nov$^{rep}$ and Full-Nov$^{event}$, emphasizes the benefits of leveraging both representations and event-containing likelihoods of data clusters with OT to obtain the distance between original and GPT-generated data for ODED.

### 3.5 Analysis

To further evaluate the novelty/similarity of the event mention patterns captured in the generated data $\mathcal{G}$ w.r.t the original data $\mathcal{O}$, we train the base model in Sect. 2.2 only on the original data $\mathcal{O}$ (i.e., BiLSTM-Base$_{\mathcal{O}}$) and directly evaluate its performance on the generated data $\mathcal{G}$. The precision, recall and F1 scores for this experiment using the LitBank dataset are 94.9%, 87.1% and 90.8%, respectively. As such, the high precision of BiLSTM-Base$_{\mathcal{O}}$ on $\mathcal{G}$ shows that the generated data is highly compatible with the original data. More importantly, the lower recall indicates that the generated data does involve event mention patterns that are novel for the original data. Therefore, training ODED models on the combination of generated and original data could improve the models' performance, as demonstrated in our experiments.

**Table 4.** The performance of GPT-Augmented on the LitBank dev set with different sizes of the generated data $\mathcal{G}$.

| $\|\mathcal{G}\|$ | P | R | F1 |
|---|---|---|---|
| 0.5 * $\|\mathcal{O}^+\|$ | 77.6 | 80.6 | 79.1 |
| 1.0 * $\|\mathcal{O}^+\|$ | 78.9 | 82.0 | 80.4 |
| 2.0 * $\|\mathcal{O}^+\|$ | 75.0 | 81.6 | 78.2 |
| 3.0 * $\|\mathcal{O}^+\|$ | 76.2 | 76.0 | 76.1 |

To understand the effect of the generated data size for GPT-Augmented, Table 4 reports the performance of GPT-Augmented on the development data of LitBank when we vary the size of the generated data $\mathcal{G}$. On the one hand, we find that decreasing the size $|\mathcal{G}|$ of the generated data to half of positive samples $\mathcal{O}^+$ in the original data $\mathcal{O}$ cannot fully benefit from the generated data from GPT-2. On the other hand, increasing the size for $\mathcal{G}$ (i.e., to two or three times larger than $\mathcal{O}^+$) could introduce more noises and hurt the model performance.

Table 6 presents some noiseless examples generated from GPT-2 while Table 5 shows some noisy sentences along with their categories in the generated data.

**Table 5.** Samples of noisy generated sentences for the LitBank dataset from GPT-2. Event triggers are shown in boldface and surrounded with special tokens $TRG_s$ and $TRG_e$, generated by GPT-2. In the first example, i.e., "*Inconsistency*" error, the model changes the pronoun from *he* to *she* in the first and second clause in the sentence. For the error of "*Missing Triggers*", GPT-2 fails to mark the word *arrived* as an event trigger. Finally, in the example for the error of "*Incorrect Triggers*", there is no event trigger in the sentence and GPT-2 incorrectly marks "*revolution*" as an event trigger.

| Error | Sentence |
|---|---|
| Inconsistency | I $TRG_s$ **heard** $TRG_e$ that he was very fond of women and I had a good time at the dinner table; |
| | But in the mean time she $TRG_s$ **told** $TRG_e$ me in the end it was all about her! |
| Repetition | "That's my boy" $TRG_s$ **said** $TRG_e$ the little boy"and that's my boy" |
| Meaninglessness | He $TRG_s$ **looked** $TRG_e$ down at his watch and then at the clock in his shoes |
| Missing Triggers | He $TRG_s$ **left** $TRG_e$ London yesterday, at 2 P.M, and arrived at Paris two hours later; |
| Incorrect Triggers | There has been no $TRG_s$ **revolution** $TRG_e$ in this country since 1960 |

**Table 6.** Samples of generated sentences for each dataset. Event triggers are shown in boldface and surrounded with special tokens $TRG_s$ and $TRG_e$, generated by GPT-2.

| Dataset | Sentence |
|---|---|
| LitBank | I was excited to go with him, as we had $TRG_s$ **met** $TRG_e$ earlier before this |
| LitBank | We $TRG_s$ **went** $TRG_e$ out in groups and $TRG_s$ **came** $TRG_e$ after a long journey |
| TimeBank | Last year, Beijing companies $TRG_s$ **proposed** $TRG_e$ more $TRG_s$ **discounts** $TRG_e$ than western companies |
| TimeBank | Some experts $TRG_s$ **anticipate** $TRG_e$ that the long $TRG_s$ **recession** $TRG_e$ $TRG_s$ **resulted** $TRG_e$ in all absolute $TRG_s$ **loss** $TRG_e$ for all players |
| SW100 | Scientists have $TRG_s$ **found** $TRG_e$ that this plant is so small that no rat has been $TRG_s$ **fed** $TRG_e$ by it |
| SW100 | 40 years ago, the British government $TRG_s$ **entered** $TRG_e$ into an $TRG_s$ **agreement** $TRG_e$ with all groups in $TRG_s$ **rebellions** $TRG_e$ |

## 4 Related Work

The major approaches for ED involve early feature-based models [1,11,16–18, 20,21,37] and recent deep learning models [5,10,14,26,27,34,38,43,44]. Open domain event detection has attracted more attentions recently [3,22,33].

A challenge for ED in general and for ODED in particular is the scarcity of large training datasets. As such, prior work has attempted to address this issue via unsupervised [9,40], semi-supervised [7,10,17], distantly supervised [3,23,41] or domain adaptation [22] models. In this work, we take a novel approach by utilizing the pre-trained language model GPT-2 to generate new training data for ODED.

Using GPT-2 to generate training data has been studied very recently for other NLP tasks, including relation extraction [28], multi-label text classification [42], commonsense reasoning [39], event influence prediction [19], knowledge base completion [4], sentiment, intent and question classification [2,13], and spoken language understanding [29]. In order to deal with noisy generated data, these approaches have only heuristically and statically filtered out noisy data. Unlike such prior work, our model keeps all the generated data and introduce a novel teacher-student framework to allow models learn with noisy data, featuring MI- and OT-based KCE and DSW mechanisms.

## 5 Conclusion

We propose a novel approach to generate training data for ODED by fine-tuning the pre-trained language model GPT-2. To deal with the noises in the generated data, we propose a teacher-student framework in which the teacher model is used to capture anchor knowledge (i.e., sentence representations and synthetic-original data difference) to regularize the student model. We present novel mechanisms to encourage the knowledge consistency between the student and the teacher based on MI and OT. We also introduce a dynamic method to weight the generated sentences. In the future, we plan to apply the proposed method to other related NLP tasks.

**Acknowledgments.** This research has been supported by the Army Research Office (ARO) grant W911NF-21-1-0112 and the NSF grant CNS-1747798 to the IUCRC Center for Big Learning. This research is also based upon work supported by the Office of the Director of National Intelligence (ODNI), Intelligence Advanced Research Projects Activity (IARPA), via IARPA Contract No. 2019-19051600006 under the Better Extraction from Text Towards Enhanced Retrieval (BETTER) Program. The views and conclusions contained herein are those of the authors and should not be interpreted as necessarily representing the official policies, either expressed or implied, of ARO, ODNI, IARPA, the Department of Defense, or the U.S. Government.

## References

1. Ahn, D.: The stages of event extraction. In: Proceedings of the Workshop on Annotating and Reasoning about Time and Events (2006)
2. Anaby-Tavor, A., et al.: Do not have enough data? deep learning to the rescue! In: AAAI (2020)
3. Araki, J., Mitamura, T.: Open-domain event detection using distant supervision. In: COLING (2018)

4. Bosselut, A., Rashkin, H., Sap, M., Malaviya, C., Celikyilmaz, A., Choi, Y.: Comet: commonsense transformers for automatic knowledge graph construction. In: ACL (2019)
5. Chen, Y., Xu, L., Liu, K., Zeng, D., Zhao, J.: Event extraction via dynamic multi-pooling convolutional neural networks. In: ACL-IJCNLP (2015)
6. Devlin, J., Chang, M.W., Lee, K., Toutanova, K.: BERT: Pre-training of deep bidirectional transformers for language understanding. In: NAACL-HLT (2019)
7. Ferguson, J., Lockard, C., Weld, D.S., Hajishirzi, H.: Semi-supervised event extraction with paraphrase clusters. In: NAACL (2018)
8. Hjelm, R.D., et al.: Learning deep representations by mutual information estimation and maximization. In: ICLR (2019)
9. Huang, L., et al.: Liberal event extraction and event schema induction. In: ACL (2016)
10. Huang, R., Riloff, E.: Bootstrapped training of event extraction classifiers. In: EACL (2012)
11. Ji, H., Grishman, R.: Refining event extraction through cross-document inference. In: ACL (2008)
12. Keith, K., Handler, A., Pinkham, M., Magliozzi, C., McDuffie, J., O'Connor, B.: Identifying civilians killed by police with distantly supervised entity-event extraction. In: EMNLP (2017)
13. Kumar, V., Choudhary, A., Cho, E.: Data augmentation using pre-trained transformer models. arXiv preprint arXiv:2003.02245 (2020)
14. Lai, V.D., Dernoncourt, F., Nguyen, T.H.: Exploiting the matching information in the support set for few shot event classification. In: Proceedings of the 24th Pacific-Asia Conference on Knowledge Discovery and Data Mining (PAKDD) (2020)
15. Lai, V.D., Nguyen, T.N., Nguyen, T.H.: Event detection: gate diversity and syntactic importance scores for graph convolution neural networks. In: EMNLP (2020)
16. Li, Q., Ji, H., Huang, L.: Joint event extraction via structured prediction with global features. In: ACL (2013)
17. Liao, S., Grishman, R.: Filtered ranking for bootstrapping in event extraction. In: COLING (2010)
18. Liao, S., Grishman, R.: Using document level cross-event inference to improve event extraction. In: ACL (2010)
19. Madaan, A., Rajagopal, D., Yang, Y., Ravichander, A., Hovy, E., Prabhumoye, S.: Eigen: event influence generation using pre-trained language models. arXiv preprint arXiv:2010.11764 (2020)
20. McClosky, D., Surdeanu, M., Manning, C.: Event extraction as dependency parsing. In: BioNLP Shared Task Workshop (2011)
21. Miwa, M., Thompson, P., Korkontzelos, I., Ananiadou, S.: Comparable study of event extraction in newswire and biomedical domains. In: COLING (2014)
22. Naik, A., Rosé, C.: Towards open domain event trigger identification using adversarial domain adaptation. In: ACL (2020)
23. Nguyen, M., Nguyen, T.H.: Who is killed by police: introducing supervised attention for hierarchical lstms. In: COLING (2018)
24. Nguyen, T.H., Cho, K., Grishman, R.: Joint event extraction via recurrent neural networks. In: NAACL (2016a)
25. Nguyen, T.H., Grishman, R.: Event detection and domain adaptation with convolutional neural networks. In: ACL (2015b)
26. Nguyen, T.H., Grishman, R.: Graph convolutional networks with argument-aware pooling for event detection. In: AAAI (2018)

27. Nguyen, T.M., Nguyen, T.H.: One for all: neural joint modeling of entities and events. In: AAAI (2019)
28. Papanikolaou, Y., Pierleoni, A.: Dare: data augmented relation extraction with gpt-2. In: SciNLP workshop at AKBC (2020)
29. Peng, B., Zhu, C., Zeng, M., Gao, J.: Data augmentation for spoken language understanding via pretrained models. arXiv preprint arXiv:2004.13952 (2020)
30. Peyre, G., Cuturi, M.: Computational optimal transport: with applications to data science. In: Foundations and Trends in Machine Learning (2019)
31. Pustejovsky, J., et al.: The timebank corpus. In: Corpus linguistics (2003)
32. Radford, A., Wu, J., Child, R., Luan, D., Amodei, D., Sutskever, I.: Language models are unsupervised multitask learners. OpenAI blog **1**(8), 9 (2019)
33. Sims, M., Park, J.H., Bamman, D.: Literary event detection. In: ACL (2019)
34. Trong, H.M.D., Le, D.T., Veyseh, A.P.B., Nguyen, T., Nguyen, T.H.: Introducing a new dataset for event detection in cybersecurity texts. In: EMNLP (2020)
35. Walker, C., Strassel, S., Medero, J., Maeda, K.: Ace 2005 multilingual training corpus. In: LDC, Philadelphia (2006)
36. Wang, X., Han, X., Liu, Z., Sun, M., Li, P.: Adversarial training for weakly supervised event detection. In: NAACL-HLT (2019)
37. Yang, B., Mitchell, T.M.: Joint extraction of events and entities within a document context. In: NAACL-HLT (2016)
38. Yang, S., Feng, D., Qiao, L., Kan, Z., Li, D.: Exploring pre-trained language models for event extraction and generation. In: ACL (2019)
39. Yang, Y., et al.: Generative data augmentation for commonsense reasoning. In: Findings of EMNLP 2020 (2020)
40. Yuan, Q., et al.: Open-schema event profiling for massive news corpora. In: CIKM (2018)
41. Zeng, Y., et al.: Scale up event extraction learning via automatic training data generation. In: AAAI (2017)
42. Zhang, D., Li, T., Zhang, H., Yin, B.: On data augmentation for extreme multi-label classification. arXiv preprint arXiv:2009.10778 (2020)
43. Zhang, J., Qin, Y., Zhang, Y., Liu, M., Ji, D.: Extracting entities and events as a single task using a transition-based neural model. In: IJCAI (2019)
44. Zhang, Y., et al.: A question answering-based framework for one-step event argument extraction. IEEE Access **8**, 65420–65431 (2020)

# Enhancing Summarization with Text Classification via Topic Consistency

Jingzhou Liu(✉) and Yiming Yang

Carnegie Mellon University, Pittsburgh, PA, USA
{liujingzhou,yiming}@cs.cmu.edu

**Abstract.** The recent success of abstractive summarization is partly due to the availability of large-volume and high-quality human-produced summaries for training, which are extremely expensive to obtain. In this paper, we aim to improve state-of-the-art summarization models by utilizing less expensive text classification data. Specifically, we use an eXtreme Multi-label Text Classification (XMTC) classifier to predict relevant category labels for each input document, and impose topic consistency in the system-produced summary or in the document encoder shared by both the classifier and the summarization model. In other words, we use the classifier to distill the training of the summarization model with respect to topical consistency between the input document and the system-generated summary. Technically, we propose two novel formulations for this objective, namely a multi-task approach, and a policy gradient approach. Our experiments show that both approaches significantly improve a state-of-the-art BART summarization model on the CNNDM and XSum datasets. In addition, we propose a new evaluation metric, CON, that measures the topic consistency between the input document and the summary. We show that CON has high correlation with human judgements and is a good complementary metric to the commonly used ROUGE scores.

**Keywords:** Text summarization · Extreme multi-label text classification · Multi-task learning · Policy gradient

## 1 Introduction

Text summarization is the task of condensing a given document into a shorter piece of textual summary, which preserves the main contents of the input. Existing approaches can be characterized into two categories, namely extractive and abstractive. Extractive methods compose each summary by extracting a subset of sentences from the input document, and abstractive methods produce each summary based on an underlying generative model, where the output may include the words or phrases beyond the input text. Generally speaking, extractive summaries are more fluent and accurate, while abstractive summaries can be globally more coherent and versatile. This paper focuses on improving abstractive summarization.


N. Oliver et al. (Eds.): ECML PKDD 2021, LNAI 12977, pp. 661–676, 2021.
https://doi.org/10.1007/978-3-030-86523-8_40

Significant improvements have been made recently in abstractive summarization [4,7,8,10,25,29,30], thanks to the rapid development of neural sequence-to-sequence learning techniques [31], such as attention [1], copy mechanism [11,12], coverage [32] and reinforcement training [28]. However, many of these successful methods heavily rely on the availability of large-volume and high-quality of human-annotated summaries for training, which are extremely expensive to obtain. Later proposed models try to address this issue by first pre-training on massive unannotated corpora and then fine-tuning on supervised data [15,27,39]. A less explored direction is to leverage less expensive supervised data for other natural language processing tasks, such as those for classification, tokenization, textual entailment, etc. In this paper, we focus on the effective use of large-volume labeled documents in the area of eXtreme Multi-label Text Classification [17] (XMTC) for improving the state-of-art models in abstractive summarization.

The XMTC task is to find each document the most relevant subset of labels from an extremely large space of categories. For example, a news article reporting on the 2020 US election is likely to be labeled "US politics", "US election", "Donald Trump", "Joe Biden", etc., by an XMTC classifier trained with labeled Wikipedia articles. Those predicted category labels can be naturally viewed as a topical summary of the input article at a relatively high granularity, without including all the detailed information. In other words, a good summary of the input document should preserve the semantic information presented by those category labels at the desirable granularity. Then, the related question for research is: if we have sufficient labeled data to train a high-quality XMTC classifier, how can we use such a classifier to improve the training of a generative model for abstractive summarization?

We answer the above question by enforcing topic consistency between the classifier-predicted category labels and the system-produced summary for the input document. The reasons of focusing on XMTC are its multi-label predictions per document and the huge label spaces in general, ensuring adequate topic coverage and diversity for representing documents and their summaries. Specifically, in this paper, we utilize a convolutional neural net-based, efficient XMTC model, namely XML-CNN [17] trained on a Wikipedia dataset.

We propose two novel approaches to imposing topic consistency on summarization models: (1) a multi-tasking approach, and (2) a policy gradient approach. In the multi-tasking approach (as shown in Fig. 1), the pre-trained XMTC classifier is applied to the input document to predict its topic distribution on one hand, and an additional MLP added to the encoder of the summarization model also predicts a topic distribution based on the latent embedding of the document for summarization. The encoder and the MLP are jointly trained to produce similar distributions, in addition to the seq2seq generative summarization objective. Thus, the summarization model is guided by the XMTC classifier to capture the most essential information with respect to subject topics from the input document. In the policy gradient approach (as shown in Fig. 2), we apply the same pre-trained XMTC classifier to both the input document and

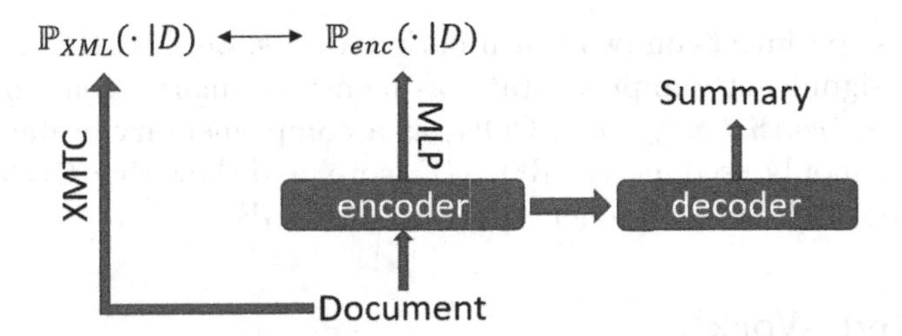


**Fig. 1.** Overview of the multi-tasking method. The XMTC classifier is applied to the input document to obtain document topic distribution, and the encoder of the summarization model is extended with an MLP as an induced classifier. The encoder induced classifier is trained to predict the document distribution, in addition to the normal sequence generation objective.

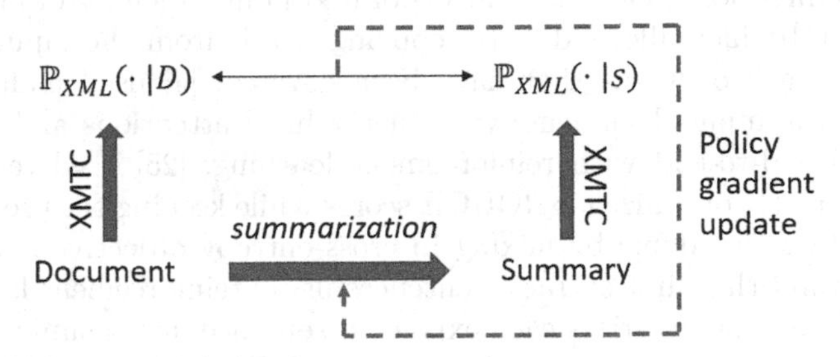


**Fig. 2.** Overview of the policy gradient method. The XMTC classifier is applied to both the input document and the system-produced summary, and then the predicted topic distributions are directly compared, sending training signal back to the summarization model via policy gradient.

the system-produced summary, and directly force these two distributions to be close to each other by minimizing a consistency reward and propagating back to the summarization model via policy gradient (Sect. 3). The two proposed approaches are generally applicable to any seq2seq-based summarization models. In this paper, we use a state-of-the-art summarization model, BART [15], as our base model, and we show that both of our proposed approaches produce significant improvements over the basic BART model (Sect. 4).

Additionally, for thorough evaluation, we compute topic consistency between an input document and a summary (topical information overlapping), and use this score as a complementary quality measure to the most commonly used n-gram overlapping-based ROUGE scores [16]. We show that CON scores have higher correlation with human evaluation scores than ROUGE, and therefore, should be informative when used as additional evaluation metric for methods comparison.

To summarize, the contributions of this paper are threefold:

1. We propose two novel approaches to imposing topic consistency to improve abstractive summarization models, by utilizing extreme multi-label text classification (XMTC).

2. Through experiments on two benchmark datasets, we show that our proposed methods significantly improve state-of-the-art summarization models.
3. We propose *consistency score*, CON, as a complementary evaluation metric to the commonly used metric, ROUGE score, and show that CON has higher correlation with human judgements than ROUGE.

## 2 Related Work

### 2.1 Abstractive Summarization

[29] and [8] were among the first to apply neural networks and attention mechanisms to abstractive summarization and since then substantial effort has been spent in this direction. [30] further incorporated pointer-generator and coverage mechanism [11] which allows directly copying words from the input document and keeping track of words that have been covered. [4] used multiple agents to represent the input documents with hierarchical attentions and trained the entire model end-to-end with reinforcement learning. [25] used reinforcement learning to directly optimize for ROUGE scores while keeping the produced summaries fluent and readable by mixing in cross-entropy objectives. [7] proposed a hybrid method that first extracts sentences using reinforcement learning, and then abstractively summarizes each extracted sentence into a summary.

As large-scale pre-training models such as BERT [9], roBERTa [19], and XLNet [37] were introduced, we have seen further improvements in abstractive summarization. [18] directly used pre-trained BERT encoders in summarization models. T5 [27], BART [15] and PEGASUS [39] pre-trained large-scale sequence-to-sequence models on massive unannotated corpora with specially designed tasks, and then further fine-tuned their models on downstream tasks (e.g. summarization) to achieve superior performances. [26] pre-trained a sequence-to-sequence model with future n-gram prediction to avoid overfitting on strong local correlations.

### 2.2 Combining Summarization and Text Classification

Summarization and text classification are two important and extensively studied tasks in natural language processing, and there are existing works trying to combine these two tasks. [3] proposed to utilize a multi-class (as opposed to multi-label) classifier to extract document representations, and to further use the predicted category to rank sentences in the document to compose extractive summaries. Other works combined sentiment classification with review summarization [5,20,36]: [36] proposed to jointly train a sentiment classifier and review summarizer by using semantic/sentiment/aspect attentions to better capture the review sentiment; [20] proposed an end-to-end model where a sentiment classifier is on top of the summarization model and jointly trained; [5] proposed a dual-view model where different sentiment classifiers are applied to reviews and summaries, for measuring sentiment inconsistency. All of these works focused on review summarization and preserving sentiment in the input reviews using (multi-class) sentiment classification, and they worked on datasets where both sentiment classification and review summarization supervision is available. To the best of our knowledge, this work is the first to utilize *external multi-label* text classifiers to capture *important topical information* to enhance abstractive summarization models.

### 2.3 Summarization Evaluation Metrics

Most commonly used metrics for evaluating text generation are based on counting n-gram overlapping between documents and summaries, such as ROUGE [16], BLEU [23] and METEOR [2]. However, it is known that these n-gram-based metrics have several deficiencies [14,34,40]. For example, n-gram-based metrics often fail to robustly match paraphrases due to string matching or heuristic matching, and n-gram-based metrics are usually insensitive to semantic errors and factual inconsistencies. To address these issues, several works have been proposed to improve or replace existing n-gram-based metrics. BERTScore [40] proposed to use contextual embeddings from BERT to compute similarity scores. [14] formulated the procedure of fact checking between input documents and summaries as a natural language inference problem. [34] further proposed to evaluate factual consistency in a question-answering framework: questions are generated based on the ground-truth summary, and then the QA system needs to answer the generated questions based on the system-produced summary. Our proposed consistency scores CON (Sect. 3.3), on the other hand, measures the topical information overlapping between a document and a summary, providing a cheap and efficient complementary metric to the most commonly used ROUGE scores.

## 3 Methodology

### 3.1 XMTC for Topic Distribution Prediction

Recall that extreme multi-label text classification (XMTC) is the problem of mapping each document to the subset of relevant categories in an extremely large space of predefined categories. Given document set $D$ and category set $L$, we train the classifier $f(d) \rightarrow \{0,1\}^{|L|}$ for each $d \in \mathcal{D}$, where the size of category set $L$ is usually extremely large, up to millions. As discussed in Sect. 1, the system-predicted categories for each document should capture important topical information about the input document, and provide useful guidance to the training of summarization models. To implement this idea, we utilize a representative convolutional neural net-based XMTC classifier, namely XML-CNN [17], which performs strongly on evaluation benchmarks and keeps a good balance among classification accuracy, model simplicity and computation scalability [6,38].

An overview of the XML-CNN model is shown in Fig. 3. It consists of a convolutional layer, a dynamic pooling layer, a bottleneck layer, and an output

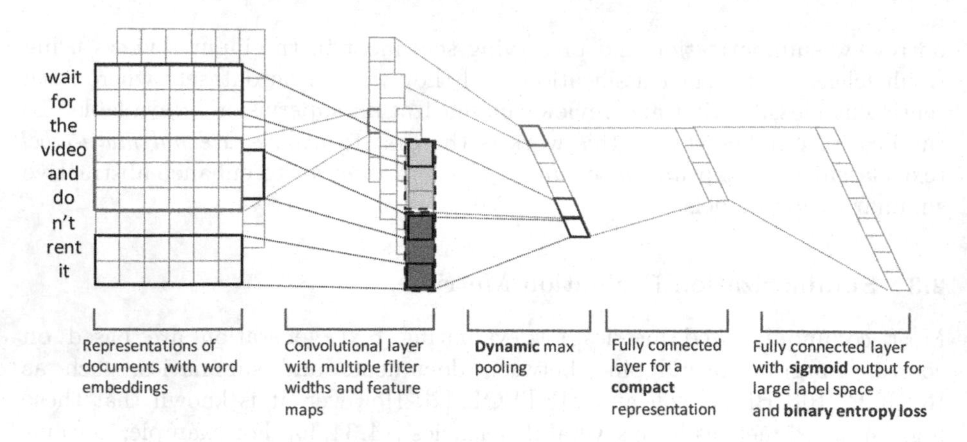


**Fig. 3.** Overview of XML-CNN, figure taken from [17].

layer that predicts a probabilistic score for each label in the category space. Sorting and thresholding on the predicted scores allow us to select the top-ranking categories (e.g., 5—20) for each input document and set the rest of the labels with zero scores. Then we normalize the topic scores to obtain a topic distribution conditioned on each input document. Let us denote the classifier by $C(\cdot)$, and then the document-conditioned topic scores for the document ($d$) and the system-produced summary ($s$) are $C(d) \in \mathcal{R}^{|L|}$ and $C(s) \in \mathcal{R}^{|L|}$, respectively. Our goal in this paper is to impose topic consistency between $C(d)$ and $C(s)$.

**Table 1.** Data Statistics of Wiki-30K: $L$ is the total number of class labels, $\bar{L}$ is the average number of label per document, $\tilde{L}$ is the average number of documents per label, $\bar{W}$ and $\hat{W}$ are the average number of words per document in the training/testing set.

| #Training | #Testing | #Vocab | $L$ | $\bar{L}$ | $\tilde{L}$ | $\bar{W}$ | $\hat{W}$ |
|---|---|---|---|---|---|---|---|
| 12,959 | 5,992 | 100,819 | 29,947 | 18.74 | 8.11 | 2,247 | 2,210 |

In the experiments of this paper, we train the XML-CNN classifier on a benchmark dataset named Wiki-30K [17], which is a collection of Wikipedia pages with human-curated category labels. Table 1 shows the dataset statistics. The label set consists of the 30K relatively popular category labels of Wikipedia articles. We choose this dataset to train our XML-CNN model because of its broad topic coverage and sufficient diversity, and hence rich enough for representing the topics in a wide range of documents and their summaries at various granularity levels. We also found the XML-CNN trained on this dataset can produce accurate predictions, which is also a desired property for the classifier to succeed in classification-enhanced summarization models.

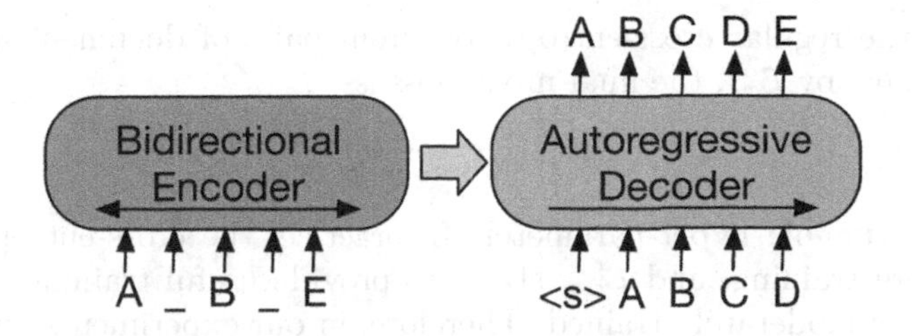


**Fig. 4.** Overview of BART, figure taken from [15].

### 3.2 Enhancing Summarization with Topic Consistency

Our base summarization model is BART [15], a state-of-the-art summarization model, as shown in Fig. 4. It is a sequence-to-sequence model, where the encoder and decoder are transformers [33]. BART is pre-trained on large corpora using a variety of transformation tasks, including token masking/deletion/infilling, etc. To perform summarization task, BART is further fine-tuned on summarization datasets [15]. We start from a pre-trained BART model (without fine-tuning), and propose two approaches to imposing topic consistency on the system-produced summaries.

**Multi-Tasking.** A natural idea is to formulate this as a multi-task learning problem, by enabling the summarization model also to be able to generate a topic distribution and making this generated topic distribution to be close to the one generated from the input document.

To achieve this, we add an additional MLP onto the encoder of the summarization model as an induced classifier. Specifically, the output of the encoder is a set of hidden states, which are averaged into a single vector, and then a two-layer MLP with separate sigmoid activation output is added on top of it. This two-layer MLP has the same shape as the last two layers of XML-CNN (Fig. 3), and its parameters are initialized from our trained XML-CNN classifier (later both the encoder-decoder and the addition MLP are trainable). Now the encoder of the summarization model and the additional MLP together can be viewed as an induced classifier, denoted by $C_{enc}(\cdot)$.

Then, given training data of (document, summary) pairs $\{(d_i, s_i)\}_1^n$, we can define the consistency loss as

$$\mathcal{L}_{con} = \frac{1}{n}\sum_{i=1}^{n} ||C(d_i)^{(k)} - C_{enc}(d_i)||_2^2, \tag{1}$$

where $C(d_i)^{(k)}$ means only preserving the $k$ largest entries in $C(d_i)$ and setting the rest to zeros. The reason we only focus on the top $k$ labels is that since the label set $L$ is huge (30K labels in Wiki30K), the predicted scores for most tail labels are quite noisy and unreliable, and that we only want the system-produced summary to capture the most salient and confident topics from the input document. The influence of different choices of $k$ is investigated in Sect. 4.3.

Denote the regular cross-entropy loss from pairs of document and ground-truth summary by $\mathcal{L}_{xe}$, the final mixed loss is

$$\mathcal{L}_{MT} = \mathcal{L}_{xe} + \alpha \mathcal{L}_{con}. \tag{2}$$

Here $\alpha$ is a tunable hyper-parameter. In practice, the cross-entropy objective requires more training, and $\mathcal{L}_{con}$ starts to provide useful training signals after the encoder is moderately trained. Therefore, in our experiments, we first train the model only using $\mathcal{L}_{xe}$ for 3 epochs, and then further train the entire model using the mixed loss in Eq. 2 until convergence.

**Policy Gradient.** Another approach is to directly impose the constraints on the system-produced summaries using policy gradient method. As shown in Fig. 2, a summary is generated from the summarization model without modification, and then the same, fixed XML-CNN classifier is applied to both the input document and the produced summary to obtained document topic distribution and summary topic distribution, denoted by $\mathbb{P}_{XML}(\cdot|d)$ and $\mathbb{P}_{XML}(\cdot|s)$, respectively. Based on these, a reward function measuring topic consistency between the input document and summary can be defined as

$$r_{con}(d, s) = -dist(\mathbb{P}_{XML}(\cdot|d), \mathbb{P}_{XML}(\cdot|s)), \tag{3}$$

where $dist(\cdot, \cdot)$ is a distance function. Topic distributions $\mathbb{P}_{XML}(\cdot|d)$ and $\mathbb{P}_{XML}(\cdot|s)$ come from the output of the XML-CNN model $C(d)$ and $C(s)$ (by normalizing the predicted scores), and we implement the reward function as

$$r_{con}(d, s) = -\frac{1}{k}||C(d)^{(k)} - C(s)^{(k)}||_2^2, \tag{4}$$

where the superscription $(k)$ indicates preserving only the $k$ largest entries and setting the rest to zeros.

The operation of feeding generated summaries to the XMTC classifier is non-differentiable, and so the training signals from the topic distribution mismatch cannot be back-propagated to the summarization model. To address this, we adopt the reinforcement learning-based training method. The summarization model can be viewed as a policy for generating summaries, denoted by $p_\theta$, where $\theta$ is the parameters of the summarization model. Using the REINFORCE [35] algorithm, we minimize the RL loss function

$$\min_\theta \mathcal{L}_{rl} = -\mathbb{E}_{d\sim D}\mathbb{E}_{s\sim p_\theta(d)}[r_{con}(d, s)], \tag{5}$$

and its one sample derivative approximation is

$$\nabla_\theta \mathcal{L}_{rl} = -(r_{con}(d, s^*) - b_e)\nabla_\theta \log p_\theta(d, s^*). \tag{6}$$

Here $s^*$ is the best summary generated by $p_\theta$ using beam search, and $b_e$ is a baseline estimator to reduce variance during training. Following [24] and [28], baseline $b_e$ is set to be $r_{con}(d, s^a)$, where $s^a$ is obtained by top-1 greedy decoding.

Following [24,25], we also optimize the cross-entropy generation loss in addition to the REINFORCE loss, to maintain the fluency and readability of the generated summaries. The cross-entropy loss is denoted by $\mathcal{L}_{xe}$, same as in Eq. 2, and the final mixed loss is

$$\mathcal{L}_{PG} = \mathcal{L}_{xe} + \beta\mathcal{L}_{rl}. \tag{7}$$

Here $\beta$ is a tunable hyper-parameter, and similar to the multi-tasking method, the summarization model is trained using only $\mathcal{L}_{xe}$ for 3 epochs first before using the mixed loss. The pseudo-code for policy gradient training is shown in Algorithm 1.

**Algorithm 1:** Pseudo-code for policy gradient approach training.

```
Data: Training samples T = {(d_i, s_i)}_1^n, XMTC classifier C
Result: Summarization model p_θ
Train p_θ with cross-entropy objective for t_1 epochs on T;
for t_2 epochs do
    for training sample (d, s) ∈ T do
        g_xe ← gradient from cross-entropy objective;
        generate summary s* ← p_θ(d) using beam search;
        generate baseline summary s^a ← p_θ(d) using greedy decoding;
        compute reward r_con(d, s*) and baseline b_e = r_con(d, s^a) (Eq 4);
        g_rl ← gradient from RL objective (Eq 6);
        update θ with g_xe + βg_rl;
return p_θ
```

### 3.3 Topic Consistency as an Additional Evaluation Metric

The idea of measuring the consistency between the input document and summary itself can be utilized to evaluate summary quality. Specifically, we define *consistency score* (CON) for a document-summary pair as

$$CON(d, s) = \frac{1}{k}||C(d)^{(k)} - C(s)^{(k)}||_2. \tag{8}$$

This formulation is similar to the consistency reward defined in Eq. 4. Only the largest $k$ entries in the document and summary topic distributions are preserved to reduce noise, and then $l_2$ distance is computed. Intuitively, lower consistency scores, i.e., more shared topics between the document and summary, indicate better summary quality.

Recently, the most commonly used n-gram overlapping-based summarization evaluation metrics (such as ROUGE score) have been criticized [14,25,34,40] for not being able to robustly match paraphrases, under-penalizing small word or ordering changes which lead to serious hallucinations (more details in Sect. 2.3), and several alternative metrics have been proposed [14,34,40]. Compared to these newly proposed metrics (which are based on question-answering, BERT, etc.), our proposed consistency score, CON, processes coarser information, but its underlying XML-CNN model is more efficient and more robust. It measures topical information overlapping between a document and a summary, and serves as a cheap and efficient complementary metric to the traditional n-gram overlapping-based ROUGE scores.

# 4 Experiments

## 4.1 Datasets

We conduct experiments on two benchmark datasets (statistics in Table 2).

**Table 2.** Dataset statistics.

| Dataset | # Sample split | # Words (doc) | # Sents (doc) | # Words (summary) | # Sents (summary) |
|---|---|---|---|---|---|
| CNNDM | 287,227/13,368/11,490 | 651.9 | 30.0 | 52.6 | 3.6 |
| XSum | 204,045/11,332/11,334 | 431.1 | 19.8 | 23.3 | 1 |

**CNN/Daily Mail.** [13] is a large-scale dataset consisting of news articles from the CNN and Daily Mail news publishers, and the reference summaries are human-generated highlights associated with these news articles. We do not anonymize named entities and follow the pre-processing procedure in [30]. It has been reported in previous literature that lead bias (important information of an article is covered in the first few sentences) is quite serious [30].

**XSum.** [21] is an extreme news summarization dataset consisting of BBC articles and accompanying single-sentence summaries, where the single-sentence summaries are professionally written introductory sentences. Compared to the CNN/Daily Mail dataset, XSum is less biased toward extractive methods in that gold summaries in XSum contain significantly more novel n-grams than in CNN/Daily Mail, and so XSum is a more abstractive dataset.

## 4.2 Comparing Methods

- **Refresh** [22] is a strong supervised extractive method. It treats extractive summarization task as a sentence ranking task, and globally optimizes ROUGE score through a reinforcement learning objective.
- **Pointer-Generator** [30] is a supervised abstractive method that is based on the sequence-to-sequence framework, and is able to directly copy phrases from the input document via copy attention mechanism.
- **BertSumExtAbs** [18] is a supervised hybrid method that uses pre-trained large BERT models as document encoder, and then further fine-tunes the model on supervised summarization datasets.

- **Pegasus** [39] pre-trains a transformer-based sequence-to-sequence model on large-scale unannotated news corpora with masked language model and gap sentences generation, which are designed for the summarization task. The pre-trained model is then fine-tuned on supervised summarization datasets. Pegasus has two versions: $\text{Pegasus}_{base}$, which has 12-layer transformer encoder/decoder (24 layers in total), and $\text{Pegasus}_{large}$, which has 16-layer transformer encoder/decoder (32 layers in total).
- **BART** [15] is also a pre-trained sequence-to-sequence model fine-tuned on summarization dataset (more details in Sect. 3.2). It has 12-layer transformer encoder/decoder (24 layers in total).

**Table 3.** Main results on CNNDM and XSum datasets. ROUGE scores and consistency scores (CON) are reported. 3 sentences are extracted for refresh, instead of 4 [41]. For XMTC-enhanced models (row 6&7) and computing CON scores, $k$ is set to 5. * indicates statistically significantly better than the best results in the same column through row 0–5.

| ID | Method | CNNDM | | | | XSum | | | |
|---|---|---|---|---|---|---|---|---|---|
| | | R-1 | R-2 | R-L | CON | R-1 | R-2 | R-L | CON |
| 0 | Lead | 40.5 | 17.7 | 36.7 | 0.25 | 16.3 | 1.6 | 12.0 | 0.48 |
| 1 | Refresh | 41.3 | 18.4 | 37.5 | 0.23 | – | – | – | – |
| 2 | Pointer-Generator | 39.5 | 17.3 | 36.4 | 0.27 | 29.7 | 9.2 | 23.2 | 0.36 |
| 3 | BertSumExtAbs | 42.1 | 19.6 | 39.2 | 0.23 | 38.8 | 16.5 | 31.3 | 0.34 |
| 4 | $\text{Pegasus}_{base}$ (24 layers) | 41.8 | 18.8 | 38.9 | 0.22 | 39.8 | 16.6 | 31.7 | 0.34 |
| 5 | BART (24 layers) | 44.2 | 21.3 | 40.9 | 0.20 | 45.1 | 22.3 | 37.3 | 0.31 |
| *Our Methods* | | | | | | | | | |
| 6 | BART-Enhanced-MT (24 layers) | 44.8* | 21.6 | 41.6* | 0.18* | 45.7* | 22.8* | 37.8* | 0.29* |
| 7 | BART-Enhanced-PG (24 layers) | **45.2*** | **22.0*** | **42.1*** | **0.14*** | **45.9*** | **23.0*** | **38.1*** | **0.25*** |

### 4.3 Main Results

Main results are shown in Table 3. Statistical significance tests are conducted (including later experiments, Table 5 and 6) using paired t-test on summary-level with p value at the 5% level. Lead (row 0) is an extractive baseline that simply takes the first three sentences in CNNDM and the first sentence in XSum as summaries. Row 6&7 are our proposed methods that enhance a pre-trained BART summarization model with XMTC predictions via multi-tasking and policy gradient, respectively. Comparing row 3–7 with row 0–2, it is clear that models utilizing large pre-trained models achieve superior performances. Pegasus (row 4) and BART (row 5) are two state-of-the-art methods. They have the same architecture, and the differences are their pre-training corpora and pre-training tasks. Comparing row 6&7 with BART (row 5), both of our two proposed methods surpass their base BART model. Specifically, policy gradient method (row 7) leads to larger improvements than multi-tasking method (row 6), showing that directly imposing topic consistency via policy gradient is more effective than indirect multi-tasking method. Moreover, improvement of our methods (row 6&7) over BART (row 5) is slightly larger in CNNDM than in XSum, probably due to the fact that summaries in CNNDM are longer than in XSum (52.6 vs 23.3) and so summaries in CNNDM may carry more topical information.

The larger version of Pegasus (row 4) contains 36 layers, and here we did not run our model using 36 layers due to computation resource limitations. The results of $Pegasus_{large}$ (36 layers) on CNNDM and XSum in ROUGE scores are (44.2,21.5,41.1) and (47.2,24.6,39.3), respectively [39]. Our proposed 24-layer PG model (row 7) is better than $Pegasus_{large}$ on CNNDM but worse on XSum. However, our proposed methods of incorporating XMTC are orthogonal directions of improving summarization to better pre-training techniques (e.g. BART, Pegasus). Our proposed methods can be generally applied to enhance any sequence-to-sequence-based summarization model.

**Table 4.** Experiments on CNNDM with different $k$.

| Method | | CNNDM | | |
|---|---|---|---|---|
| | | R-1 | R-2 | R-L |
| Bart | | 44.2 | 21.3 | 40.9 |
| Bart-Enhanced-PG | $k = 1$ | 44.10 | 21.35 | 40.96 |
| | $k = 5$ | 45.22 | 21.96 | 42.10 |
| | $k = 10$ | **45.26** | **22.04** | **42.13** |
| | $k = 20$ | 45.13 | 21.68 | 41.85 |

In previous experiments, $k$ (the number of preserved labels) is set to 5. To investigate the influence of $k$ on the final performances, additional experiments on CNNDM using BART-Enhanced-PG are conducted with different $k$'s. As shown in Table 4, when $k = 1$, the performance is much worse than larger $k$'s, and is close to BART, indicating that extracting only one topic label from the document and summary does not provide much useful training signal and lead to any improvement. When $k = 5, 10, 20$, the performances are close, with $k = 20$ being slightly worse. Based on these observations, our proposed approaches to enhancing summarization models are quite robust when $k$ is between 5 and 20 (i.e., adequate but not too noisy topic information is leveraged).

### 4.4 Human Evaluation

To further confirm the effectiveness of our proposed methods, we conduct human evaluations on summaries produced by BART, BART-Enhanced-MT and BART-Enhanced-PG. We sampled 200 documents from the test set, and AMT workers were asked to rate the system-produced and ground-truth summaries' quality given the input original document, on a 1–5 scale (higher is better). Each summary was shown to three different workers, whose ratings were averaged into a final rating. The results are shown in Table 5. We can see that the

**Table 5.** Human evaluation. Ratings are on a 1–5 scale, higher is better. * indicates statistically significantly better than ground-truth.

| Summary | CNNDM | XSum |
|---|---|---|
| Ground-Truth | 3.2 | 3.5 |
| Bart | 3.1 | 3.3 |
| Bart-Enhanced-MT | 3.2 | 3.3 |
| Bart-Enhanced-PG | **3.4*** | **3.4** |

human evaluation results roughly agree with the automatic evaluation results in Table 3. Notably, in CNNDM, the summaries produced by Bart-Enhanced-PG are statistically significantly better than ground-truth summaries.

**Table 6.** Absolute Pearson correlation coefficients between automatic metrics and human ratings. * indicates statistically significantly better than the second best.

| Metric | CNNDM | XSum |
|---|---|---|
| ROUGE-1 | 0.33 | 0.18 |
| ROUGE-2 | 0.21 | 0.12 |
| ROUGE-L | 0.30 | 0.11 |
| CON | **0.42*** | **0.21** |

**Topic Consistency.** To investigate the consistency scores CON (Sect. 3.3), we further compute Pearson correlation coefficients (per summary) between various automatic evaluation metrics in Table 3 and human ratings in Table 5. The results are shown in Tabel 6. Clearly in this experiment, CON scores match human judgements better than ROUGE scores (with statistical significance on CNNDM). It should be noted that CON scores only consider shared topical information between input documents and system-produced summaries, but not the summaries' readability, fluency, grammaticality, etc., and so conceptually CON is not suitable to be used alone, but rather as a complementary metric to n-gram overlapping based metrics such as ROUGE scores for comparing methods (e.g., when ROUGE scores of two methods are close).

## 5 Conclusions

In this paper, we propose to utilize the cheaper extreme multi-label text classification (XMTC) data to enhance abstractive summarization models, where it is much more expensive to obtain supervised data. We propose two methods to impose topic consistency on the input documents and system-produced summaries using an XML-CNN classifier trained on a Wikipedia dataset, namely a multi-tasking method and a policy gradient method. Both methods manage to significantly improve a state-of-the-art BART summarization model. We also propose consistency score, CON, for evaluating summary quality, and show that CON has higher correlation with human judgements than the most commonly used ROUGE scores. As for related future research directions, we would like to investigate the effective use of topic hierarchies and labeled documents to improve topic-conditioned summarization and multi-granularity summarization.

**Acknowledgments.** We thank all the reviewers for their helpful comments. This work is supported in part by National Science Foundation (NSF) under grant IIS-1546329.

## References

1. Bahdanau, D., Cho, K., Bengio, Y.: Neural machine translation by jointly learning to align and translate. arXiv preprint arXiv:1409.0473 (2014)
2. Banerjee, S., Lavie, A.: Meteor: An automatic metric for mt evaluation with improved correlation with human judgments. In: Proceedings of the acl Workshop on Intrinsic and Extrinsic Evaluation Measures for Machine Translation and/or Summarization, pp. 65–72 (2005)
3. Cao, Z., Li, W., Li, S., Wei, F.: Improving multi-document summarization via text classification. In: Proceedings of the AAAI Conference on Artificial Intelligence, vol. 31 (2017)
4. Celikyilmaz, A., Bosselut, A., He, X., Choi, Y.: Deep communicating agents for abstractive summarization. In: Proceedings of the 2018 Conference of the North American Chapter of the Association for Computational Linguistics: Human Language Technologies (vol. 1 Long Papers), pp. 1662–1675 (2018)
5. Chan, H.P., Chen, W., King, I.: A unified dual-view model for review summarization and sentiment classification with inconsistency loss. In: Proceedings of the 43rd International ACM SIGIR Conference on Research and Development in Information Retrieval, pp. 1191–1200 (2020)
6. Chang, W.C., Yu, H.F., Zhong, K., Yang, Y., Dhillon, I.S.: Taming pretrained transformers for extreme multi-label text classification. In: Proceedings of the 26th ACM SIGKDD International Conference on Knowledge Discovery & Data Mining, pp. 3163–3171 (2020)
7. Chen, Y.C., Bansal, M.: Fast abstractive summarization with reinforce-selected sentence rewriting. In: Proceedings of the 56th Annual Meeting of the Association for Computational Linguistics (vol. 1: Long Papers), pp. 675–686 (2018)
8. Chopra, S., Auli, M., Rush, A.M.: Abstractive sentence summarization with attentive recurrent neural networks. In: Proceedings of the 2016 Conference of the North American Chapter of the Association for Computational Linguistics: Human Language Technologies, pp. 93–98 (2016)
9. Devlin, J., Chang, M.W., Lee, K., Toutanova, K.: Bert: pre-training of deep bidirectional transformers for language understanding. In: Proceedings of the 2019 Conference of the North American Chapter of the Association for Computational Linguistics: Human Language Technologies (vol. 1 Long and Short Papers), pp. 4171–4186 (2019)

10. Gehrmann, S., Deng, Y., Rush, A.M.: Bottom-up abstractive summarization. In: Proceedings of the 2018 Conference on Empirical Methods in Natural Language Processing, pp. 4098–4109 (2018)
11. Gu, J., Lu, Z., Li, H., Li, V.O.: Incorporating copying mechanism in sequence-to-sequence learning. In: Proceedings of the 54th Annual Meeting of the Association for Computational Linguistics (vol. 1: Long Papers), pp. 1631–1640 (2016)
12. Gulcehre, C., Ahn, S., Nallapati, R., Zhou, B., Bengio, Y.: Pointing the unknown words. In: Proceedings of the 54th Annual Meeting of the Association for Computational Linguistics (vol. 1: Long Papers), pp. 140–149 (2016)
13. Hermann, K.M., et al.: Teaching machines to read and comprehend. In: Advances in Neural Information Processing Systems, pp. 1693–1701 (2015)
14. Kryscinski, W., McCann, B., Xiong, C., Socher, R.: Evaluating the factual consistency of abstractive text summarization. In: Proceedings of the 2020 Conference on Empirical Methods in Natural Language Processing (EMNLP), pp. 9332–9346 (2020)
15. Lewis, M., et al.: Bart: Denoising sequence-to-sequence pre-training for natural language generation, translation, and comprehension. In: Proceedings of the 58th Annual Meeting of the Association for Computational Linguistics, pp. 7871–7880 (2020)
16. Lin, C.Y., Hovy, E.: Automatic evaluation of summaries using n-gram co-occurrence statistics. In: Proceedings of the 2003 Human Language Technology Conference of the North American Chapter of the Association for Computational Linguistics, pp. 150–157 (2003)
17. Liu, J., Chang, W.C., Wu, Y., Yang, Y.: Deep learning for extreme multi-label text classification. In: Proceedings of the 40th International ACM SIGIR Conference on Research and Development in Information Retrieval, pp. 115–124 (2017)
18. Liu, Y., Lapata, M.: Text summarization with pretrained encoders. In: Proceedings of the 2019 Conference on Empirical Methods in Natural Language Processing and the 9th International Joint Conference on Natural Language Processing (EMNLP-IJCNLP), pp. 3730–3740 (2019)
19. Liu, Y., et al.: Roberta: A robustly optimized bert pretraining approach. arXiv preprint arXiv:1907.11692 (2019)
20. Ma, S., Sun, X., Lin, J., Ren, X.: A hierarchical end-to-end model for jointly improving text summarization and sentiment classification. In: Proceedings of the 27th International Joint Conference on Artificial Intelligence, pp. 4251–4257 (2018)
21. Narayan, S., Cohen, S.B., Lapata, M.: Don't give me the details, just the summary! topic-aware convolutional neural networks for extreme summarization. In: Proceedings of the 2018 Conference on Empirical Methods in Natural Language Processing, pp. 1797–1807 (2018)
22. Narayan, S., Cohen, S.B., Lapata, M.: Ranking sentences for extractive summarization with reinforcement learning. In: Proceedings of the 2018 Conference of the North American Chapter of the Association for Computational Linguistics: Human Language Technologies (vol. 1 Long Papers), pp. 1747–1759 (2018)
23. Papineni, K., Roukos, S., Ward, T., Zhu, W.J.: Bleu: a method for automatic evaluation of machine translation. In: Proceedings of the 40th Annual Meeting of the Association for Computational Linguistics, pp. 311–318 (2002)
24. Pasunuru, R., Bansal, M.: Multi-reward reinforced summarization with saliency and entailment. In: Proceedings of the 2018 Conference of the North American Chapter of the Association for Computational Linguistics: Human Language Technologies (vol. 2 Short Papers), pp. 646–653 (2018)

25. Paulus, R., Xiong, C., Socher, R.: A deep reinforced model for abstractive summarization. In: International Conference on Learning Representations (2018)
26. Qi, W., et al.: Prophetnet: Predicting future n-gram for sequence-to-sequence pre-training. In: Proceedings of the 2020 Conference on Empirical Methods in Natural Language Processing: Findings, pp. 2401–2410 (2020)
27. Raffel, C., et al.: Exploring the limits of transfer learning with a unified text-to-text transformer. J. Mach. Learn. Res. **21**, 1–67 (2020)
28. Rennie, S.J., Marcheret, E., Mroueh, Y., Ross, J., Goel, V.: Self-critical sequence training for image captioning. In: Proceedings of the IEEE Conference on Computer Vision and Pattern Recognition, pp. 7008–7024 (2017)
29. Rush, A.M., Chopra, S., Weston, J.: A neural attention model for abstractive sentence summarization. In: Proceedings of the 2015 Conference on Empirical Methods in Natural Language Processing, pp. 379–389 (2015)
30. See, A., Liu, P.J., Manning, C.D.: Get to the point: Summarization with pointer-generator networks. In: Proceedings of the 55th Annual Meeting of the Association for Computational Linguistics (vol. 1: Long Papers), pp. 1073–1083 (2017)
31. Sutskever, I., Vinyals, O., Le, Q.V.: Sequence to sequence learning with neural networks. In: Advances in Neural Information Processing Systems, pp. 3104–3112 (2014)
32. Tu, Z., Lu, Z., Liu, Y., Liu, X., Li, H.: Modeling coverage for neural machine translation. In: Proceedings of the 54th Annual Meeting of the Association for Computational Linguistics (vol. 1: Long Papers), pp. 76–85 (2016)
33. Vaswani, A., et al.: Attention is all you need. In: Advances in Neural Information Processing Systems, pp. 5998–6008 (2017)
34. Wang, A., Cho, K., Lewis, M.: Asking and answering questions to evaluate the factual consistency of summaries. In: Proceedings of the 58th Annual Meeting of the Association for Computational Linguistics, pp. 5008–5020 (2020)
35. Williams, R.J.: Simple statistical gradient-following algorithms for connectionist reinforcement learning. Mach. Learn. **8**(3–4), 229–256 (1992)
36. Yang, M., Qu, Q., Shen, Y., Liu, Q., Zhao, W., Zhu, J.: Aspect and sentiment aware abstractive review summarization. In: Proceedings of the 27th International Conference on Computational Linguistics, pp. 1110–1120 (2018)
37. Yang, Z., Dai, Z., Yang, Y., Carbonell, J., Salakhutdinov, R.R., Le, Q.V.: Xlnet: generalized autoregressive pretraining for language understanding. In: Advances in Neural Information Processing Systems, pp. 5753–5763 (2019)
38. You, R., Zhang, Z., Wang, Z., Dai, S., Mamitsuka, H., Zhu, S.: Attentionxml: label tree-based attention-aware deep model for high-performance extreme multi-label text classification. Adv. Neural Inf. Process. Syst. **32**, 5820–5830 (2019)
39. Zhang, J., Zhao, Y., Saleh, M., Liu, P.: Pegasus: pre-training with extracted gap-sentences for abstractive summarization. In: International Conference on Machine Learning, pp. 11328–11339. PMLR (2020)
40. Zhang, T., Kishore, V., Wu, F., Weinberger, K.Q., Artzi, Y.: Bertscore: evaluating text generation with bert. In: International Conference on Learning Representations (2019)
41. Zheng, H., Lapata, M.: Sentence centrality revisited for unsupervised summarization. In: Proceedings of the 57th Annual Meeting of the Association for Computational Linguistics, pp. 6236–6247 (2019)

# Transformers: "The End of History" for Natural Language Processing?

Anton Chernyavskiy[1(✉)], Dmitry Ilvovsky[1], and Preslav Nakov[2]

[1] HSE University, Moscow, Russian Federation
aschernyavskiy_1@edu.hse.ru, dilvovsky@hse.ru
[2] Qatar Computing Research Institute, HBKU, Doha, Qatar
pnakov@hbku.edu.qa

**Abstract.** Recent advances in neural architectures, such as the Transformer, coupled with the emergence of large-scale pre-trained models such as BERT, have revolutionized the field of Natural Language Processing (NLP), pushing the state of the art for a number of NLP tasks. A rich family of variations of these models has been proposed, such as RoBERTa, ALBERT, and XLNet, but fundamentally, they all remain limited in their ability to model certain kinds of information, and they cannot cope with certain information sources, which was easy for pre-existing models. Thus, here we aim to shed light on some important theoretical limitations of pre-trained BERT-style models that are inherent in the general Transformer architecture. First, we demonstrate in practice on two general types of tasks—segmentation and segment labeling—and on four datasets that these limitations are indeed harmful and that addressing them, even in some very simple and naïve ways, can yield sizable improvements over vanilla RoBERTa and XLNet models. Then, we offer a more general discussion on desiderata for future additions to the Transformer architecture that would increase its expressiveness, which we hope could help in the design of the next generation of deep NLP architectures.

**Keywords:** Transformers · Limitations · Segmentation · Sequence classification

## 1 Introduction

The history of Natural Language Processing (NLP) has seen several stages: first, rule-based, e.g., think of the expert systems of the 80s, then came the statistical revolution, and now along came the neural revolution. The latter was enabled by a combination of deep neural architectures, specialized hardware, and the existence of large volumes of data. Yet, the revolution was going slower in NLP compared to other fields such as Computer Vision, which were quickly and deeply transformed by the emergence of large-scale pre-trained models, which were in turn enabled by the emergence of large datasets such as ImageNet.


N. Oliver et al. (Eds.): ECML PKDD 2021, LNAI 12977, pp. 677–693, 2021.
https://doi.org/10.1007/978-3-030-86523-8_41

Things changed in 2018, when NLP finally got its "ImageNet moment" with the invention of BERT [9].[1] This was enabled by recent advances in neural architectures, such as the Transformer [31], followed by the emergence of large-scale pre-trained models such as BERT, which eventually revolutionized NLP and pushed the state of the art for a number of NLP tasks. A rich family of variations of these models have been proposed, such as RoBERTa [20], ALBERT [18], and XLNet [34]. For some researchers, it felt like this might very well be the "End of History" for NLP (à la Fukuyama[2]).

It was not too long before researchers started realizing that BERT and Transformer architectures in general, despite their phenomenal success, remained fundamentally limited in their ability to model certain kinds of information, which was natural and simple for the old-fashioned feature-based models. Although BERT does encode some syntax, semantic, and linguistic features, it may not use them in downstream tasks [16]. It ignores negation [11], and it might need to be combined with Conditional Random Fields (CRF) to improve its performance for some tasks and languages, most notably for sequence classification tasks [28]. There is a range of sequence tagging tasks where entities have different lengths (not 1–3 words as in the classical named entity recognition formulation), and sometimes their continuity is required, e.g., for tagging in court papers. Moreover, in some problem formulations, it is important to accurately process the boundaries of the spans (in particular, the punctuation symbols), which turns out to be something that Transformers are not particularly good at (as we will discuss below).

In many sequence classification tasks, some classes are described by specific features. Besides, a very large contextual window may be required for the correct classification, which is a problem for Transformers because of the quadratic complexity of calculating their attention weights.[3]

Is it possible to guarantee that BERT-style models will carefully analyze all these cases? This is what we aim to explore below. Our contributions can be summarized as follows:

- We explore some theoretical limitations of pre-trained BERT-style models when applied to sequence segmentation and labeling tasks. We argue that these limitations are not limitations of a specific model, but stem from the general Transformer architecture.
- We demonstrate in practice on two different tasks (one on segmentation, and one on segment labeling) and on four datasets that it is possible to improve over state-of-the-art models such as BERT, RoBERTa, XLNet, and this can be achieved with simple and naïve approaches, such as feature engineering and post-processing.

[1] A notable previous promising attempt was ELMo [21], but it became largely outdated in less than a year.

[2] http://en.wikipedia.org/wiki/The_End_of_History_and_the_Last_Man.

[3] Some solutions were proposed such as Longformer [3], Performer [4], Linformer [33], Linear Transformer [15], and Big Bird [35].

- Finally, we propose desiderata for attributes to add to the Transformer architecture in order to increase its expressiveness, which could guide the design of the next generation of deep NLP architectures.

The rest of our paper is structured as follows. Section 2 summarizes related prior research. Section 3 describes the tasks we address. Section 4 presents the models and the modifications thereof. Section 5 outlines the experimental setup. Section 6 describes the experiments and the evaluation results. Section 7 provides key points that lead to further general potential improvements of Transformers. Section 8 concludes and points to possible directions for future work.

## 2 Related Work

*Studies of What BERT Learns and What it Can Represent.* There is a large number of papers that study what kind of information can be learned with BERT-style models and how attention layers capture this information; a survey is presented in [26]. It was shown that BERT learns syntactic features [12,19], semantic roles and entities types [30], linguistic information and subject-verb agreement [13]. Note that the papers that explore what BERT-style models can encode do not indicate that they directly use such knowledge [16]. Instead, we focus on what is *not* modeled, and we explore some general limitations.

*Limitations of BERT/Transformer.* Indeed, Kovaleva et al. (2019) [16] revealed that vertical self-attention patterns generally come from pre-training tasks rather than from task-specific linguistic reasoning and the model is over-parametrized. Ettinger (2020) [11] demonstrated that BERT encodes some semantics, but is fully insensitive to negation. Sun et al. (2020) [29] showed that BERT-style models are erroneous in simple cases, e.g., they do not correctly process word sequences with misspellings. They also have bad representations of floating point numbers for the same tokenization reason [32]. Moreover, it is easy to attack them with adversarial examples [14]. Durrani et al. (2019) [10] showed that BERT subtoken-based representations are better for modeling syntax, while ELMo character-based representations are preferable for modeling morphology. It also should be noticed that hyper-parameter tuning is a very non-trivial task, not only for NLP engineers but also for advanced NLP researchers [23]. Most of these limitations are low-level and technical, or are related to a specific architecture (such as BERT). In contrast, we single out the general limitations of the Transformer at a higher level, but which can be technically confirmed, and provide desiderata for their elimination.

*Fixes of BERT/Transformer.* Many improvements of the original BERT model have been proposed: RoBERTa (changed the language model masking, the learning rate, the dataset size), DistilBERT [27] (distillation to significantly reduce the number of parameters), ALBERT (cross-layer parameter sharing, factorized embedding parametrization), Transformer-XL [8] (recurrence mechanism and relative positional encoding to improve sequence modeling), XLNet (permutation language modeling to better model bidirectional relations), BERT-CRF [1,28] (dependencies between the posteriors for structure prediction helped in some tasks and languages), KnowBERT [22] (incorporates external knowledge). Most of these models pay attention only to 1–2 concrete fixes, whereas our paper aims at more general Transformer limitations.

## 3 Tasks

In this section, we describe two tasks and four datasets that we used for experiments.

### 3.1 Propaganda Detection

We choose the task of Detecting Propaganda Techniques in News Articles (SemEval-2020 Task 11)[4] as the main for experiments. Generally, it is formulated as finding and classifying all propagandistic fragments in the text [6]. To do this, two subtasks are proposed: (*i*) *span identification (SI)*, i.e., selection of all propaganda spans within the article, (*ii*) *technique classification (TC)*, i.e., multi-label classification of each span into 14 classes. The corpus with a detailed description of propaganda techniques is presented in [7].

The motivation for choosing this task is triggered by several factors. First, two technically different problems are considered, which can be formulated at a general level (multi-label sequence classification and binary token labeling). Second, this task has specificity necessary for our research, unlike standard named entity recognition. Thus, traditional NLP methods can be applied over the set of hand-crafted features: sentiment, readability scores, length, etc. Here, length is a strong feature due to the data statistics [7]. Moreover, spans can be nested, while span borders vary widely and may include punctuation symbols. Moreover, sometimes Transformer-based models face the problem of limited input sequence length. In this task, such a problem appears with the classification of "Repetition" spans. By definition, this class includes spans that have an intentional repetition of the same information. This information can be repeated both in the same sentence and in very distant parts of the text.

### 3.2 Keyphrase Extraction

In order to demonstrate the transferability of the studied limitations between datasets, we further experimented with the task of Extracting Keyphrases and Relations from Scientific Publications, using the dataset from SemEval-2017 Task 10 [2]. We focus on the following two subtasks: (*i*) *keyphrase identification (KI)*, i.e., search of all keyphrases within the text, (*ii*) *keyphrase classification (KC)*, i.e., multi-class classification of given keyphrases into three classes.

[4] The official task webpage: http://propaganda.qcri.org/semeval2020-task11/.

According to the data statistics, the length of the phrases is a strong feature. Also, phrases can be nested inside one other, and many of them are repeated across different articles. So, these subtasks allow us to demonstrate a number of issues.

# 4 Method

Initially, we selected the most successful approach as the baseline from Transformer-based models (BERT, RoBERTa, ALBERT, and XLNet). In both propaganda detection subtasks, it turned out to be RoBERTa, which is an optimized version of the standard BERT with a modified pre-training procedure. Whereas in both keyphrase extraction subtasks, it turned out to be XLNet, which is a language model that aims to better study bidirectional links or relationships in a sequence of words. From a theoretical point of view, investigated results and researched problems should be typical for other Transformer-based models, such as BERT and DistilBERT. Nonetheless, we additionally conduct experiments with both XLNet and RoBERTa for both tasks for a better demonstration of the universality of our findings.

## 4.1 Token Classification

We reformulate the SI and the KI tasks as "named entity recognition" tasks. Specifically, in the SI task, for each span, all of its internal tokens are assigned to the "`PROP`" class and the rest to "`O`" (**O**utside). Thus, this is a binary token classification task. At the same time, various types of encoding formats are studied. Except for the above described Inside-Outside classification, we further consider BIO (**B**egin) and BIEOS (**B**egin, **E**nd, **S**ingle are added) tags encodings. Such markups theoretically can provide better processing for border tokens [25].

In order to ensure the sustainability of the trained models, we create an ensemble of three models trained with the same hyper-parameters, but using different random seeds. We merge the intersecting spans during the ensemble procedure (intervals union).

*End-to-End Training with CRFs.* Conditional Random Fields (CRF) [17] can qualitatively track the dependencies between the tags in the markup. Therefore, this approach has gained great popularity in solving the problem of extracting named entities with LSTMs or RNNs. Advanced Transformer-based models generally can model relationships between words at a good level due to the attention mechanism, but adding a CRF layer theoretically is not unnecessary. The idea is that we need to model the relationships not only between the input tokens but also between the output labels.

Our preliminary study showed that both RoBERTa and XLNet are make classification errors even when choosing tags in named entity recognition (NER) encodings with clear rules. For example, in the case of BIO, the "`I-PROP`" tag can only go after the "`B-PROP`" tag. However, RoBERTa produced results with a sequence of tags such as "O-PROP I-PROP O-PROP" for some inputs. Here, it is hard to determine where the error was, but the CRF handles such cases from a probabilistic point of view. We use the CRF layer instead of the standard model classification head to apply the end-to-end training. Here, we model connections only between neighboring subtokens since our main goal is the proper sequence analysis. Thus, the subtokens that are not placed at the beginning of words are ignored (i.e., of the format *##smth*).

*RoBERTa, XLNet, and Punctuation Symbols.* In the SI task, there is one more problem that even the CRF layer cannot always handle. It is the processing of punctuation and quotation marks at the span borders. Clark et al. (2019) [5] and Kovaleva et al. (2019) [16] showed that BERT generally has high token–token attention to the [SEP] token, to the periods, and to the commas, as they are the most frequent tokens. However, we found out that large attention weights to punctuation may still not be enough for some tasks.

In fact, a simple rule can be formulated to address this problem: a span cannot begin or end with a punctuation symbol, unless it is enclosed in quotation marks. With this observation in mind, we apply post-processing of the spans borders by adding missing quotation marks and also by filtering punctuation symbols in case they were absent.

### 4.2 Sequence Classification

We model the TC task in the same way as the KC task, that is, as a multi-class sequence classification problem. We create a fairly strong baseline to achieve better results. First, the context is used, since spans in both tasks can have various meanings in different contexts. Thus, we select the entire sentence that contains the span for this purpose. In this case, we consider two possible options: (*i*) highlight the span with the special limiting tokens and submit to the model only one input; (*ii*) make two inputs: one for the span and one for the context. Moreover, to provide a better initialization of the model and to share some additional knowledge from other data in the TC task, we apply the transfer learning strategy from the SI task.

Just like in the token classification problem, we compose an ensemble of models for the same architecture, but with three different random seed initializations. We do this in order to stabilize the model, and this is not a typical ensemble of different models.

*Input Length.* BERT does not have a mechanism to perform explicit character/word/subword counting. Exactly this problem and the lack of good consistency between the predicted tags may cause a problem with punctuation (quotation marks) in the sequence tagging task, since BERT theoretically cannot accurately account for the number of opening/closing quotation marks (as it cannot count).

In order to explicitly take into account the input sequence size in the model, we add a length feature to the [CLS] token embedding, as it should contain all

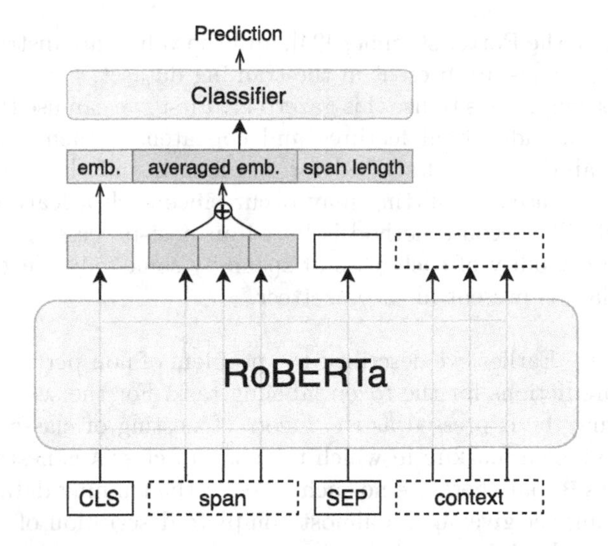


**Fig. 1.** The RoBERTa model takes an input span and the context (sentence with the span). It combines the embedding of the [CLS] token, the averaged embedding of all span tokens, and the span length as a feature.

the necessary information to solve the task (see Fig. 1). It may be also useful to pre-process the length feature through binning. In this case, it is possible to additionally create trainable embeddings associated with each bin or directly to add an external knowledge from a gazetteer containing relevant information about the dataset according to the given bin (we will consider gazetteers below).

In addition to the input length in characters (or in tokens), it may be useful to add other quantitative features such as the number of question or exclamation symbols.

*Span Embeddings.* In the TC task, we concatenate the [CLS] token representation with the span embedding obtained by averaging all token embeddings from the last layer to submit to the classifier (end-to-end training). Note that the added embedding contains information about the degree of propaganda in the classified span as an initialization, since we transfer a model from another task. Moreover, this model can reconfigure it to serve other features during the training process. Also, it may be useful to join embeddings obtained by the max-pool operation or taken from other layers.

*Training a Hand-Crafted Gazetteer.* Gazetteers can provide external relevant information about entities in NER tasks. As some propaganda techniques are often described by the same words, it might be a good idea to construct and to use a gazetteer of words for each technique. While in NER, gazetteers are externally constructed to provide additional knowledge, here we use the training data to construct our gazetteer. We create a hash map, where the keys are spans pre-processed by the Porter stemmer [24], and the values are distributions of the classes in which spans are present in the training dataset.

There are several ways to use this gazetteer. First, we can use these frequency representations as additional features and concatenate them with the [CLS] token in the same way as described for the length and the span embedding. However, in this case, over-fitting may occur since such a feature will contain a correct label. The second method is based on post-processing. The idea is to increase the probability of each class of spans by some value (e.g., +0.5) if the span of this class is present in the gazetteer.

*Class Insertions.* Earlier, we described the problem of non-perfect spatially consistent class predictions for the token labeling task. For the sequence classification task, it may be expressed as the incorrect nesting of classes. That is, the model can produce a markup in which the span of class A is nested in the span of another class B, but there are no such cases in the training data. If we believe that the training set gives us an almost complete description of the researched problem, such a classification obviously cannot be correct.

The simplest solution is again post-processing. One possibility is to choose a pair of spans that have maximal predicted probability and the correct nesting. Another option is to choose a pair of classes with a maximal probability $p(x)p(y)p(A)$. Here, $p(x)$ is the predicted probability that the span has the label $x$, and $p(A)$ is the estimated probability of the nesting case $A$, where a span of class $x$ is inside the span of class $y$. To estimate $p(A)$, we calculate the co-occurrence matrix of nesting classes in the training set, and we apply softmax with temperature $t$ over this matrix to obtain probabilities. The temperature parameter is adjusted for each model on validation. We use the first approach in the TC task. As there are only three classes and all class insertions are possible, we apply the second approach with $t = 0.26$ in the KC task.

*Specific Classes: "Repetition".* In some cases, the entire text of the input document might be needed as a context (rather than just the current sentence) in order for the model to be able to predict the correct propaganda technique for a given span. This is the case of the *repetition* technique.

As a solution, we apply a special post-processing step. Let $k$ be the number of occurrences of the considered span in the set of spans allocated for prediction within the article and $p$ be the probability of the *repetition* class predicted by the source model. We apply the following formula:

$$\hat{p} = \begin{cases} 1, & \text{if } k \geq 3 \text{ or } (k = 2 \text{ and } p \geq t_1) \\ 0, & \text{if } k = 1 \text{ and } p \leq t_2 \\ p, & \text{otherwise} \end{cases} \tag{1}$$

We use the following values for the probability thresholds: $t_1 = 0.001$ and $t_2 = 0.99$. Note that since the repetition may be contained in the span itself, it is incorrect to nullify the probabilities of the unique spans.

*Multi-label Classification.* If the same span can have multiple labels, it is necessary to apply supplementary post-processing of the predictions. Thus, if the same span is asked several times during the testing process (the span is determined by its coordinates in the text, and in the TC task, multiple labels are signalled by repeating the same span multiple times in the test set), then we assign different labels to the different instances of that span, namely the top among the most likely predictions.

## 5 Experimental Setup

Below, we describe the data we used and the parameter settings for our experiments for all the tasks.

### 5.1 Data

*Propaganda Detection.* The dataset provided for the SemEval-2020 task 11 contains 371 English articles for training, 75 for development, and 90 for testing. Together, the training and the testing sets contain 6,129 annotated spans. While there was an original partitioning of the data into training, development, and testing, the latter was only available via the task leaderboard, and was not released. Thus, we additionally randomly split the training data using a 80:20 ratio to obtain new training and validation sets. The evaluation measure for the SI task is the variant of the $F_1$ measure described in [7]: it penalizes for predicting too long or too short spans (compared to the gold span) and generally correlates with the standard $F_1$ score for tokens. Micro-averaged $F_1$ score is used for the TC task, which is equivalent to accuracy.

*Keyphrase Extraction.* The dataset provided for SemEval-2017 task 10 contains 350 English documents for training, 50 for development, and 100 for testing. In total, the training and the testing sets contain 9,945 annotated keyphrases. The evaluation measure for both sub-tasks is micro-averaged $F_1$ score.

### 5.2 Parameter Setting

We started with pre-trained model checkpoints and baselines as from the HuggingFace Transformers library,[5] and we implemented our modifications on top of them. We used RoBERTa-large and XLNet-large, as they performed better than their base versions in our preliminary experiments.

We selected hyper-parameters according to the recommendations in the original papers using our validation set and we made about 10–20 runs to find the best configuration. We used grid-search over {5e-6, 1e-5, 2e-5, 3e-5, 5e-5} for the optimal learning rate. Thus, we fix the following in the propaganda detection problem: learning rate of 2e-5 (3e-5 for XLNet in the TC task), batch size of 24,

[5] http://github.com/huggingface/transformers.

maximum sequence length of 128 (128 is fixed as it is long enough to encode the span; besides, there are very few long sentences in our datasets), Adam optimizer with a linear warm-up of 500 steps. The sequence length and the batch size are selected as the maximum possible for our GPU machine (3 GeForce GTX 1080 GPUs). We performed training for 30 epochs with savings every two epochs and we selected the best checkpoints on the validation set (typically, it was 10–20 epochs). We found that uncased models should be used for the SI task, whereas the cased model were better for the TC task.

**Table 1.** Analysis of RoBERTa and XLNet modifications for sequential classification tasks: span identification and keyphrase identification. *Overall* is the simultaneous application of two improvements.

| Task | Approach | F1 |
|---|---|---|
| SI | RoBERTa (BIO encoding) | 46.91 |
| | + CRF | 48.54 $\uparrow$**1.63** |
| | + punctuation post-processing | 47.54 $\uparrow$**0.63** |
| | *Overall* | 48.87 $\uparrow$**1.96** |
| | XLNet (BIO encoding) | 46.47 |
| | + CRF | 46.68 $\uparrow$**0.21** |
| | + punctuation post-processing | 46.76 $\uparrow$**0.29** |
| | *Overall* | 47.05 $\uparrow$**0.58** |
| **KI** | RoBERTa (BIO encoding) | 57.85 |
| | + CRF | 58.59 $\uparrow$**0.74** |
| | XLNet (BIO encoding) | 58.80 |
| | + CRF | 60.11 $\uparrow$**1.31** |

For keyphrase extraction, for the KI task, we used a learning rate of 2e-5 (and 3e-5 for RoBERTa-CRF), a batch size of 12, a maximum sequence length of 64, Adam optimizer with a linear warm-up of 60 steps. For the KC task, we used a learning rate of 2e-5 (1e-5 for XLNet-Length) and a head learning rate of 1e-4 (in cases with the *Length* feature), batch size of 20 (10 for XLNet-Length), maximum sequence length of 128, and the Adam optimizer with a linear warm-up of 200 steps. We performed training for 10 epochs, saving each epoch and selecting the best one on the validation set.

The training stage in a distributed setting takes approximately 2.38 min per epoch (+0.05 for the *avg. embedding* modification) for the TC task. For the SI task, it takes 6.55 min per epoch for RoBERTa (+1.27 for CRF), and 6.75 min per epoch for XLNet (+1.08 for CRF).

## 6 Experiments and Results

### 6.1 Token Classification

We experimented with BIOES, BIO, and IO encodings, and we found that BIO performed best, both when using CRF and without it. Thus, we used the BIO encoding in our experiments. We further observed a much better recall with minor loss in precision for our ensemble with span merging.

A comparison of the described approaches for the SI and the KI tasks is presented in Table 1. Although the sequential predictions of the models are generally consistent, adding a CRF layer on top improves the results. Manual analysis of the output for the SI task has revealed that about 3.5% of the predicted tags were illegal sequences, e.g., an "`I-PROP`" tag following an "`O`" tag.

**Table 2.** Analysis of the improvements using RoBERTa and XLNet for the TC task on the development set. Shown is micro-$F_1$ score.

| Technique classification | | |
|---|---|---|
| Approach | RoBERTa | XLNet |
| Baseline | 62.75 | 58.23 |
| + length | 63.50 $\uparrow_{0.75}$ | 59.64 $\uparrow_{1.41}$ |
| + averaged span embededding | 62.94 $\uparrow_{0.19}$ | 59.64 $\uparrow_{1.41}$ |
| + multi-label | 63.78 $\uparrow_{1.03}$ | 59.27 $\uparrow_{1.04}$ |
| + gazetteer post-processing | 62.84 $\uparrow_{0.10}$ | 58.33 $\uparrow_{0.10}$ |
| + *repetition* post-processing | 66.79 $\uparrow_{4.04}$ | 62.46 $\uparrow_{3.67}$ |
| + class insertions | 62.65 $\downarrow_{0.10}$ | 57.85 $\downarrow_{0.38}$ |

**Table 3.** Analysis of improvements for the KC task using RoBERTa and XLNet on the development set in the multi-label mode. Shown is micro-$F_1$ score.

| Keyphrase classification | | |
|---|---|---|
| Approach | RoBERTa | XLNet |
| Baseline | 77.18 | 78.50 |
| + length | 77.38 $\uparrow_{0.20}$ | 78.65 $\uparrow_{0.15}$ |
| + gazetteer post-processing | 77.43 $\uparrow_{0.25}$ | 78.69 $\uparrow_{0.19}$ |
| + class insertions | 77.82 $\uparrow_{0.64}$ | 78.69 $\uparrow_{0.19}$ |

Also, we figured out that neither XLNet nor RoBERTa could learn the described rule for quotes and punctuation symbols. Moreover, adding CRF also does not help solve the problem according to the better "overall" score in the table. We analyzed the source of these errors. Indeed, there were some annotation errors. However, the vast majority of the errors related to punctuation at the boundaries were actually model errors. E.g., in an example like `<"It is what it is.">`, where the entire text (including the quotation marks) had to be detected, the model would propose sequences like `<"It is what it is>` or `<It is what it is.>`. Thus, there is a common problem for all Transformer-based models—lack of consistency for sequential tag predictions.

### 6.2 Sequence Classification

We took models that use separate inputs (span and context) for all experiments, as they yielded better results on the validation set. The results for the customized models are shown in Tables 2 and 3 for the Technique Classification (TC) and the Keyphrase Classification (KC) tasks, respectively. We also studied the impact of the natural multi-label formulation of the TC task (see Table 4). We can see that all directions of quality changes were the same.

**Table 4.** Analysis of the improvements for the TC task using RoBERTa and XLNet on the development set in the multi-label mode. Shown is micro-$F_1$ score.

| Technique Classification | | |
|---|---|---|
| Approach | RoBERTa | XLNet |
| Baseline+multi-label | 63.78 | 59.27 |
| + length | 64.72 $_{\uparrow\mathbf{0.94}}$ | 60.68 $_{\uparrow\mathbf{1.41}}$ |
| + averaged span embededding | 64.25 $_{\uparrow\mathbf{0.47}}$ | 60.77 $_{\uparrow\mathbf{1.50}}$ |
| + gazetteer post-processing | 63.87 $_{\uparrow\mathbf{0.09}}$ | 59.36 $_{\uparrow\mathbf{0.09}}$ |
| + *repetition* post-processing | 67.54 $_{\uparrow\mathbf{3.76}}$ | 63.50 $_{\uparrow\mathbf{4.23}}$ |
| + class insertions | 63.69 $_{\downarrow 0.09}$ | 58.89 $_{\downarrow 0.38}$ |

**Table 5.** An incremental analysis of the proposed approach for the TC task on the development set.

| Technique Classification | |
|---|---|
| Approach | F1-score |
| RoBERTa | 62.08 |
| + length and averaged span embedding | 62.27 $_{\uparrow\mathbf{0.19}}$ |
| + multi-label correction | 63.50 $_{\uparrow\mathbf{1.23}}$ |
| + class insertions | 63.69 $_{\uparrow\mathbf{0.19}}$ |
| + *repetition* post-processing | 66.89 $_{\uparrow\mathbf{3.20}}$ |
| + gazetteer post-processing | 67.07 $_{\uparrow\mathbf{0.18}}$ |

Although positional embeddings are used in BERT-like models, our experiments showed that they are not enough to model the length of the span. Indeed, the results for systems that explicitly use length improved both for RoBERTa and for XLNet, for both tasks.

According to the source implementation of RoBERTa, XLNet, and other similar models, only the [CLS] token embedding is used for sequence classification. However, in the TC task, it turned out that the remaining tokens can also be useful, as in the averaging approach.

Moreover, the use of knowledge from the training set through post-processing with a gazetteer consistently improved the results for both models. Yet, it can also introduce errors since it ignores context. That is why we did not set 100% probabilities for the corrected classes.

As for the sequential consistency of labels, the systems produced output with unacceptable nesting of spans of incompatible classes. Thus, correcting such cases can also have a positive impact (see Table 3). However, a correct nesting does not guarantee correct final markup, since we only post-process predictions. Better results can be achieved if the model tries to learn this as part of training.

The tables show that the highest quality increase for the TC task was achieved by correcting the *repetition* class. This is because this class is very frequent, but it often requires considering a larger context.

We also examined the impact of each modification on RoBERTa for the TC task, applying an incremental analysis on the development set (Table 5). We can see that our proposed modifications are compatible and can be used together.

Finally, note that while better pre-training could make some of the discussed problems less severe, it is still true that certain limitations are more "theoretical" and that they would not be resolved by simple pre-training. For example, there is nothing in the Transformer architecture that would allow it to model the segment length, etc.

## 7 Discussion

Below we describe a desiderata to add to the Transformer in order to increase its expressiveness, which could guide the design of the next generation of general Transformer architectures.

*Length.* We have seen that length is important for the sequence labeling task. However, it would be important for a number of other NLP tasks, e.g., in seq2seq models. For example, in Neural Machine Translation, if we have an input sentence of length 20, it might be bad to generate a translation of length 2 or of length 200. Similarly, in abstractive neural text summarization, we might want to be able to inform the model about the expected target length of the summary: should it be 10 words long? 100-word long?

*External Knowledge.* Gazetteers are an important source of external knowledge, and it is important to have a mechanism to incorporate such knowledge. A promising idea in this direction is KnowBERT [22], which injects Wikipedia knowledge when pre-training BERT.

*Global Consistency.* For structure prediction tasks, such as sequence segmentation and labeling, e.g., named entity recognition, shallow parsing, and relation extraction, it is important to model the dependency between the output labels. This can be done by adding a CRF layer on top of BERT, but it would be nice to have this as part of the general model. More generally, for many text generation tasks, it is essential to encourage the global consistency of the output text, e.g., to avoid repetitions. This is important for machine translation, text summarization, chat bots, dialog systems, etc.

*Symbolic vs. Distributed Representation.* Transformers are inherently based on distributed representations for words and tokens. This can have limitations, e.g., we have seen that BERT cannot pay attention to specific symbols in the input such as specific punctuation symbols like quotation marks. Having a hybrid symbolic-distributed representation might help address these kinds of limitations. It might also make it easier to model external knowledge, e.g., in the form of gazetteers.

## 8 Conclusion and Future Work

We have shed light on some important theoretical limitations of pre-trained BERT-style models that are inherent in the general Transformer architecture. In particular, we demonstrated on two different tasks—one on segmentation, and one on segment labeling—and four datasets that these limitations are indeed harmful and that addressing them, even in some very simple and naïve ways, can yield sizable improvements over vanilla BERT, RoBERTa, and XLNet models. Then, we offered a more general discussion on desiderata for future additions to the Transformer architecture in order to increase its expressiveness, which we hope could help in the design of the next generation of deep NLP architectures.

In future work, we plan to analyze more BERT-style architectures, especially such requiring text generation, as here we did not touch the generation component of the Transformer. We further want to experiment with a pre-formulation of the task as span enumeration instead of sequence labeling with BIO tags. Moreover, we plan to explore a wider range of NLP problems, again with a focus on such involving text generation, e.g., machine translation, text summarization, and dialog systems.

**Acknowledgments.** Anton Chernyavskiy and Dmitry Ilvovsky performed this research within the framework of the HSE University Basic Research Program.

Preslav Nakov contributed as part of the Tanbih mega-project (http://tanbih.qcri.org/), which is developed at the Qatar Computing Research Institute, HBKU, and aims to limit the impact of "fake news," propaganda, and media bias by making users aware of what they are reading.

## References


1. Arkhipov, M., Trofimova, M., Kuratov, Y., Sorokin, A.: Tuning multilingual transformers for language-specific named entity recognition. In: Proceedings of the 7th Workshop on Balto-Slavic Natural Language Processing (BSNLP 2019), pp. 89–93. Florence, Italy (2019)
2. Augenstein, I., Das, M., Riedel, S., Vikraman, L., McCallum, A.: SemEval 2017 task 10: ScienceIE - extracting keyphrases and relations from scientific publications. In: Proceedings of the 11th International Workshop on Semantic Evaluation (SemEval 2017), pp. 546–555. Vancouver, Canada (2017)
3. Beltagy, I., Peters, M.E., Cohan, A.: Longformer: the long-document transformer. In: ArXiv (2020)
4. Choromanski, K., et al.: Rethinking attention with performers. In: Proceedings of the 9th International Conference on Learning Representations (ICLR 2021) (2021)
5. Clark, K., Khandelwal, U., Levy, O., Manning, C.D.: What does BERT look at? An analysis of BERT's attention, ArXiv (2019)
6. Da San Martino, G., Barrón-Cedeño, A., Wachsmuth, H., Petrov, R., Nakov, P.: SemEval-2020 task 11: detection of propaganda techniques in news articles. In: Proceedings of the 14th International Workshop on Semantic Evaluation (SemEval 2020), Barcelona, Spain (2020)
7. Da San Martino, G., Yu, S., Barrón-Cedeño, A., Petrov, R., Nakov, P.: Fine-grained analysis of propaganda in news article. In: Proceedings of the 2019 Conference on Empirical Methods in Natural Language Processing and the 9th International Joint Conference on Natural Language Processing (EMNLP-IJCNLP 2019), pp. 5636–5646. Hong Kong, China (2019)
8. Dai, Z., Yang, Z., Yang, Y., Carbonell, J., Le, Q., Salakhutdinov, R.: TransformerXL: attentive language models beyond a fixed-length context. In: Proceedings of the 57th Annual Meeting of the Association for Computational Linguistics (ACL 2019), pp. 2978–2988. Florence, Italy (2019)
9. Devlin, J., Chang, M.W., Lee, K., Toutanova, K.: BERT: pre-training of deep bidirectional transformers for language understanding. In: Proceedings of the 2019 Conference of the North American Chapter of the Association for Computational Linguistics: Human Language Technologies (NAACL-HLT 2019), pp. 4171–4186. Minneapolis, MN, USA (2019)
10. Durrani, N., Dalvi, F., Sajjad, H., Belinkov, Y., Nakov, P.: One size does not fit all: Comparing NMT representations of different granularities. In: Proceedings of the 2019 Conference of the North American Chapter of the Association for Computational Linguistics: Human Language Technologies (NAACL-HLT 2019), pp. 1504–1516. Minneapolis, MN, USA (2019)
11. Ettinger, A.: What BERT is not: lessons from a new suite of psycholinguistic diagnostics for language models. Trans. Assoc. Comput. Linguist. **8**, 34–48 (2020)
12. Goldberg, Y.: Assessing bert's syntactic abilities (2019)
13. Jawahar, G., Sagot, B., Seddah, D.: What does BERT learn about the structure of language? In: Proceedings of the 57th Annual Meeting of the Association for Computational Linguistics (ACL 2019), pp. 3651–3657. Florence, Italy (2019)
14. Jin, D., Jin, Z., Zhou, J.T., Szolovits, P.: Is BERT really robust? a strong baseline for natural language attack on text classification and entailment. In: Proceedings of the 34th Conference on Artificial Intelligence (AAAI 2020), pp. 8018–8025 (2019)
15. Katharopoulos, A., Vyas, A., Pappas, N., Fleuret, F.: Transformers are RNNs: fast autoregressive transformers with linear attention. In: Proceedings of the 37th International Conference on Machine Learning (ICML 2020), pp. 5156–5165 (2020)

16. Kovaleva, O., Romanov, A., Rogers, A., Rumshisky, A.: Revealing the dark secrets of BERT. In: Proceedings of the Conference on Empirical Methods in Natural Language Processing and the International Joint Conference on Natural Language Processing (EMNLP-IJCNLP 2019), pp. 4365–4374. Hong Kong, China (2019)
17. Lafferty, J.D., McCallum, A., Pereira, F.C.N.: Conditional random fields: probabilistic models for segmenting and labeling sequence data. In: Proceedings of the Eighteenth International Conference on Machine Learning (ICML 2001), pp. 282–289. Williamstown, MA, USA (2001)
18. Lan, Z., Chen, M., Goodman, S., Gimpel, K., Sharma, P., Soricut, R.: ALBERT: a lite BERT for self-supervised learning of language representations. In: ArXiv (2019)
19. Liu, N.F., Gardner, M., Belinkov, Y., Peters, M.E., Smith, N.A.: Linguistic knowledge and transferability of contextual representations. In: Proceedings of the 2019 Conference of the North American Chapter of the Association for Computational Linguistics: Human Language Technologies (NAACL-HLT 2019), pp. 1073–1094. Minneapolis, MN, USA (2019)
20. Liu, Y., et al.: RoBERTa: A robustly optimized BERT pretraining approach. In: ArXiv (2019)
21. Peters, M., et al.: Deep contextualized word representations. In: Proceedings of the 2018 Conference of the North American Chapter of the Association for Computational Linguistics: Human Language Technologies (NAACL-HLT 2018), pp. 2227–2237. New Orleans, LA, USA (2018)
22. Peters, M.E., et al.: Knowledge enhanced contextual word representations. In: Proceedings of the 2019 Conference on Empirical Methods in Natural Language Processing and the 9th International Joint Conference on Natural Language Processing (EMNLP-IJCNLP 2019), pp. 43–54. Hong Kong, China (2019)
23. Popel, M., Bojar, O.: Training tips for the transformer model. Prague Bull. Math. Linguist. **110**(1), 43–70 (2018)
24. Porter, M.F.: An algorithm for suffix stripping. Program **14**(3), 130–137 (1980)
25. Ratinov, L.A., Roth, D.: Design challenges and misconceptions in named entity recognition. In: Proceedings of the Thirteenth Conference on Computational Natural Language Learning (CoNLL 2009), pp. 147–155. Boulder, CO, USA (2009)
26. Rogers, A., Kovaleva, O., Rumshisky, A.: A primer in BERTology: what we know about how BERT works. Trans. Assoc. Comput. Linguist. **8**, 842–866 (2020)
27. Sanh, V., Debut, L., Chaumond, J., Wolf, T.: DistilBERT, a distilled version of BERT: smaller, faster, cheaper and lighter. In: ArXiv (2019)
28. Souza, F., Nogueira, R., Lotufo, R.: Portuguese named entity recognition using BERT-CRF. In: Arxiv (2019)
29. Sun, L., et al.: Adv-BERT: BERT is not robust on misspellings! generating nature adversarial samples on BERT. In: Arxiv (2020)
30. Tenney, I., et al.: What do you learn from context? Probing for sentence structure in contextualized word representations. In: Arxiv (2019)
31. Vaswani, A., et al.: Attention is all you need. In: Arxiv (2017)
32. Wallace, E., Wang, Y., Li, S., Singh, S., Gardner, M.: Do NLP models know numbers? probing numeracy in embeddings. In: Proceedings of the 2019 Conference on Empirical Methods in Natural Language Processing and the 9th International Joint Conference on Natural Language Processing (EMNLP-IJCNLP 2019), pp. 5307–5315. Hong Kong, China (2019)
33. Wang, S., Li, B.Z., Khabsa, M., Fang, H., Ma, H.: Linformer: self-attention with linear complexity. In: Arxiv (2020)

34. Yang, Z., Dai, Z., Yang, Y., Carbonell, J., Salakhutdinov, R.R., Le, Q.V.: XLNet: generalized autoregressive pretraining for language understanding. In: Proceedings of the Annual Conference on Neural Information Processing Systems (NeurIPS 2019), pp. 5753–5763 (2019)
35. Zaheer, M., et al.: Big bird: transformers for longer sequences. In: Proceedings of the Annual Conference on Neural Information Processing Systems (NeurIPS 2020) (2020)

# Image Processing, Computer Vision and Visual Analytics

# Subspace Clustering Based Analysis of Neural Networks

Uday Singh Saini[1(✉)], Pravallika Devineni[2], and Evangelos E. Papalexakis[1]

[1] University of California Riverside, 900 University Avenue, Riverside, CA, USA
usain001@ucr.edu, epapalex@cs.ucr.edu
[2] Oak Ridge National Laboratory, Oak Ridge, TN, USA
devinenip@ornl.gov

**Abstract.** Tools to analyze the latent space of deep neural networks provide a step towards better understanding them. In this work, we motivate sparse subspace clustering (SSC) with an aim to learn affinity graphs from the latent structure of a given neural network layer trained over a set of inputs. We then use tools from Community Detection to quantify structures present in the input. These experiments reveal that as we go deeper in a network, inputs tend to have an increasing affinity to other inputs of the same class. Subsequently, we utilise matrix similarity measures to perform layer-wise comparisons between affinity graphs. In doing so we first demonstrate that when comparing a given layer currently under training to its final state, the shallower the layer of the network, the quicker it is to converge than the deeper layers. When performing a pairwise analysis of the entire network architecture, we observe that, as the network increases in size, it reorganises from a state where each layer is moderately similar to its neighbours, to a state where layers within a block have high similarity than to layers in other blocks. Finally, we analyze the learned affinity graphs of the final convolutional layer of the network and demonstrate how an input's local neighbourhood affects its classification by the network.

## 1 Introduction

With the emergence of deep neural networks in a variety of domains, there is a need for understanding and characterising the inner workings and operations of these models. With critical applications such as autonomous vehicles, healthcare, and criminal justice relying on neural network-based models, model interpretability has become an important and necessary aspect of machine learning. In the domain of theoretical analysis of neural networks, previous works like [7,13], and [14] focus on the approximation capability of neural networks, while works like [12] and [19] focus on the interpretations of neural networks as kernels. In the domain of experimental analysis, works like [21,25] and [2] focus on visualising inputs, filters and neurons of a network.

Our work focuses on interpreting the behaviour of neural networks by analyzing subspaces of latent representations learned by the layers of a neural network.


N. Oliver et al. (Eds.): ECML PKDD 2021, LNAI 12977, pp. 697–712, 2021.
https://doi.org/10.1007/978-3-030-86523-8_42

Most closely related to our approach are works like SVCCA [20], PWCCA [15] and Linear-CKA [10,18]. SVCCA and PWCCA operate directly on the activations of a neural network for a given set of inputs and allows comparison between two such sources of activations over the same inputs, thereby facilitating a comparison between different layers of the same network, or even different networks. On the other hand, [10] and [18] take the same activations over a set of inputs and use it to construct a pairwise similarity matrix with the help of a kernel. Taking these pairwise similarity matrices obtained via the response of two different layers to a set of inputs, they then compare the two kernel matrices using Centered Kernel Alignment (CKA) [4]. This lets them directly compare any two layers from different sources. For most of their experiments, [10] and [18] motivate the theory and use of linear kernels, henceforth we call their method Linear-CKA. All these methods offer architecture agnostic ways to analyze neural networks.

We investigate Sparse Subspace Clustering (SSC) [5] as an alternative to Kernel-based pairwise Similarity matrices over inputs. Methods like SSC help us learn a connectivity graph over data points that are assumed to belong to an underlying union of subspaces embedded in a given space. Such methods are typically based on the **Self-Expressiveness** property in a union of subspaces, where each data point can be represented as a sparse linear combination of points from its own subspace. Learning such a graph helps impart transparency to the learning dynamics of the neural network and also presents us with a way to represent each layer of a neural network in an architecture agnostic manner. This facilitates the comparison of layers within a network and across various architectures.

We combine SSC with CKA to provide an alternate similarity measurement tool for latent representations learned by neural networks. It is akin to translating a layer's activations over a set of inputs into a connectivity graph and then using CKA to compute the similarity between graphs arising from two different neural network representations.

Our main contributions are the following:

1. We utilise SSC to learn a connectivity graph over inputs in the latent space. We then employ graph modularity [16] to characterise the community structure of each input's neighbourhood. We demonstrate that as we go deeper into the layers of the network, the community around each input becomes more homogeneous, i.e., a higher proportion of an input's neighbourhood in the Affinity Graph is of the same class as the input itself. Additionally, we also demonstrate the utility of SSC-CKA to capture network training dynamics. We observe that when compared to shallower layers, the deeper layers of the network tend to take longer to converge to their final state. This is a corroboration of similar observations made in SVCCA [20].
2. We demonstrate the ability of SSC-CKA to visualize and analyze entire network architectures by capitalising on its ability to perform pairwise comparison of two different layers of a network (or two different networks). These experiments set along the lines of Linear-CKA [10], demonstrate the effects of depth, width, epochs and training data size on the network architecture.

3. We demonstrate the ability of SSC to interpret the latent spaces of a given layer of the network. SSC allows us to represent each input as a weighted and sparse linear combination of other inputs in the same subspace. Here we demonstrate a strong correlation between classes of network's prediction for the input and the class having the highest weightage in the local neighbourhood of the input. This hints towards neural networks separating out classes into disjoint subspaces and presents interesting opportunities and directions going forward for creating tools to analyze neural networks.

## 2 Background and Method

In this section, we lay the background on the Sparse Subspace Clustering, its optimization through ADMM [3] and motivate its fitness for the purpose of interpreting subspaces learned by neural networks.

### 2.1 Sparse Spectral Clustering

Subspace clustering refers to the task of clustering the data into their original subspaces and uncovering the underlying structure of the data. Given some high-dimensional data, subspace clustering attempts to model the data as samples drawn from a union of multiple low-dimensional linear subspaces [5]. First, we present Sparse Spectral Clustering (SSC) for the case of uncorrupted data. As mentioned earlier, SSC typically works well when the underlying data follows Self-Expressive Property, i.e., each data point can be represented by a linear combination of points that belong to the same subspace.

Let $X = [\mathbf{x}_1, \ldots, \mathbf{x}_N]$, such that $X \in \mathbf{R}^{d \times N}$ represents the data matrix. Let $C = [\mathbf{c}_1, \ldots, \mathbf{c}_N]$, such that $C \in \mathbf{R}^{N \times N}$ and $\mathbf{c}_i \in \mathbf{R}^N$, where the entry $(i, j)$ of the matrix $C$, given by $c_{ij}$ represents the weight of data point $\mathbf{x}_j \in \mathbf{R}^d$ in the linear combination to reconstruct $\mathbf{x}_i \in \mathbf{R}^d$. Mathematically, a noiseless model for SSC is equivalent to Eq. 1. However, as shown in [1], this problem is NP-hard.

$$\min_{\mathbf{c}_i} ||\mathbf{c}_i||_0 \quad \text{s.t.} \quad \mathbf{x}_i = X\mathbf{c}_i,\ c_{ii} = 0 \quad \forall i \in \{1, \ldots, N\} \tag{1}$$

We therefore focus on Eq. 2, a convex relaxation of Eq. 1 which is also robust to noise and solve this optimization problem using Algorithm 1 as shown in [23].

$$\min_{C,Z} ||C||_1 + \frac{\tau}{2}||X - XZ||_F^2 \quad \text{s.t.} \quad Z = C - diag(C) \tag{2}$$

### 2.2 Centered Kernel Alignment

Centered Kernel Alignment [4], as defined in Eq. 3, between two similarity matrices $X$ and $Y$ is an isotropic invariant similarity index that relies upon Hilbert-Schmidt Independence Criterion (HSIC) [6] to determine statistical dependence between two sets of variables.

$$CKA(X,Y) = \frac{HSIC(X,Y)}{\sqrt{HSIC(X,X)HSIC(Y,Y)}} \tag{3}$$

where HSIC between a pair of $N \times N$ matrices is defined in Eq. 4

$$HSIC(X,Y) = \frac{trace(HXHHYH)}{(N-1)^2} \quad (4)$$

where H is a Centering matrix given by $H = I - \frac{1}{N}\mathbf{1}\mathbf{1}^T$.

### 2.3 Meta Algorithm

Our goal is to take a matrix of neural activations $X \in \mathbf{R}^{d_1 \times N}$, where $d_1$ is the number of neurons in the given layer and $N$ is the number of examples for which we obtain the activations, and construct an affinity matrix $C_X \in \mathbf{R}^{N \times N}$, via Sparse Subspace Clustering [5], where each non diagonal entry $(i,j)$ of $C_X$ denotes the affinity between input samples $i$ and $j$. This gives us the ability to compare 2 matrices of neural activations, say $X \in \mathbf{R}^{d_1 \times N}$ and $Y \in \mathbf{R}^{d_2 \times N}$ by representing them as 2 Affinity Matrices $\in \mathbf{R}^{N \times N}$, namely $C_X$ and $C_Y$ and then using Centered Kernel Alignment [4] to compare the similarity of the 2 Affinity Matrices. This procedure helps us compare 2 different layers of a network, or 2 different layers of 2 architecturally different neural networks. We would like to point out that $C_X = |C| + |C^T|$ where the matrix $C$ is obtained from Algorithm 1, where $|C|$ represents the absolute value function applied element-wise to $C$.

**Algorithm 1:** Matrix LASSO Minimization by ADMM

**Data**: Data Matrix $X$

**Result**: Sparse Representation $C$

1 **initialization:** $C^0 = \mathbf{0}$, $\Lambda_2^0 = \mathbf{0}$, $\mu_2 > 0$;

2 **while** *not converged* **do**

3 $\quad Z^{k+1} = (\tau X^T X + \mu_2 I)^{-1}(\tau X^T X + \mu_2(C^k - \frac{\Lambda_2^k}{\mu_2}))$;

4 $\quad C^{k+1} = \mathbf{S}_{\frac{1}{\mu_2}}(Z^{k+1} + \frac{\Lambda_2^k}{\mu_2})$; $\mathbf{S}$ : Shrinkage operator

$\quad C^{k+1} = C^{K+1} - diag(C^{k+1})$;

5 $\quad \Lambda_2^{k+1} = \Lambda_2^k + \mu_2(Z^{K+1} - C^{k+1})$;

6 **end**

## 3 Problem and Experimental Setup

In this paper, we experiment with VGGs [22], ResNets [8], Wide-ResNets [24] and DenseNets [9] trained on CIFAR-10 and CIFAR-100 datasets [11]. Our approach relies upon having access to activations of various hidden layers in the network for each input instance. As an example, for the first set of experiments in this study, we take outputs from a few pooling layers at the end of each block of a DenseNet to study the community structure of its subspace and also to learn the layer-wise dynamics of training progression. In another setting, we take focus on all the convolutional layers in the network and get their activations for an input.

### 3.1 Problem Formulation

We attempt to interpret and analyze neural networks by taking a matrix of its activations (layer-wise) $X \in \mathbf{R}^{d \times N}$, where $d$ is the number of neurons in the subject layer and $N$ is the number of inputs used for analysis. We first aim to learn a connectivity graph between these inputs based on their affinity scores obtained from SSC. Then we use this connectivity matrix to perform analysis pertaining to the community structure of the graph in Sect. 4, and in Sect. 6 we analyze instance based neighbourhoods of various inputs to better reconcile network predictions with the need to human oriented explanations

### 3.2 Experimental Details

All networks, unless otherwise stated, were trained with a learning rate of 0.1 and a weight decay of $5 \times 10^{-4}$ with a learning rate step multiplier of 0.2 applied after every 30 epochs. All networks were trained for 100 epochs since that was sufficient to achieve optimal performance, with the exception of VGG-29, which required 160 epochs. For SSC computations, both $\tau$ and $\mu$ were set to 10, with $\mu$ being adaptive based on Eq. 3.13 in [3]. We recommend choosing $\tau$ and $\mu$ between 10 and 100. Our implementation is publicly available on GitHub[1].

## 4 Analysis of Network Training Dynamics

In this section, we analyze behaviour of the network as its training progresses. To demonstrate the training dynamics, we train the following networks[2] - ResNet [8], Wide ResNet [24] and DenseNet [9] on the datasets CIFAR-10 and CIFAR-100 [11]. Both ResNets and DenseNets are constructed by stacking and joining different residual blocks with multiple convolution layers residing in each block. For brevity and scalability of this experiment, we use the output of residual blocks instead of every convolution layer and the final classification layers. For a given combination of the network and the dataset, we present three results - layer-wise modularity of the sparse subspace affinity graphs, SSC-CKA based layer-wise similarity to understand training dynamics, and analogous training dynamic analysis with Linear-CKA.

### 4.1 Community Structure via Graph Modularity

We analyze the community structure of the sparse subspace affinity graph for various layers at each epoch. We do this by calculating the modularity [16] of the learned subspace affinity graph. Modularity is a measure of the structure of a graph, measuring the density of connections within a module or community.

[1] URL - https://github.com/23Uday/Subspace-Clustering-based-analysis-of-Neural-Networks.

[2] Networks used from: https://github.com/kuangliu/pytorch-cifar and https://github.com/meliketoy/wide-resnet.pytorch.

High modularity in a graph indicates dense connectivity between nodes of the same type, while low modularity indicates dense connectivity between nodes of different types, where a node type is its class label. Mathematically, modularity of a graph can be represented as follows

$$Q = \frac{1}{2m} \sum_{ij} \left( A_{ij} - \frac{k_i k_j}{2m} \right) \delta(c_i, c_j) \tag{5}$$

where $m$ is the number of edges in the graph, $A_{ij}$ is the weight of the edge between nodes $i$ and $j$, $k_i$ is the degree of node $i$, $c_i$ is the class label of node $i$ and $\delta(c_i, c_j)$ is 1 if node $i$ and node $j$ are of the same class and 0 otherwise.

We present this analysis in Fig. 1a and Fig. 2a. We observe that as we go deeper in the network, the modularity of the learned sparse subspace affinity graph increases, implying that earlier layers of the network cluster examples of different classes together in the same subspace and deeper layers of the network tend to separate out classes into different disjoint subspaces. Another noteworthy observation is that the modularity scores of subspace affinity graph learned from earlier layers tends to saturate earlier in training when compared to the modularity of the subspace affinity graphs of deeper layers. This phenomenon is consistent across different architectures and different datasets. A similar observation regarding earlier saturation of shallower layers in training dynamics is also made in [20], albeit in context of representational similarity, which we address next.

### 4.2 Layer-Wise Training Dynamics and Convergence

For our second analysis, we compute the Centered Kernel Alignment (CKA) scores between the sparse subspace affinity graph of a layer at a given epoch and the same layer after the final epoch. This allows us to observe the rates at which layers converge to their final states. These results are shown in Fig. 1b and Fig. 2b. We observe a similar pattern of shallower layers converging to their final representation much earlier, when compared to deeper layers of the network. Another key observation that we make pertains to the behaviour of representations when the learning rate of the optimiser is reduced to improve convergence. For training of all the networks, we start with a learning rate of 0.1 and reduce it by a factor of 0.2 after every 30 epochs using SGD with learning rate decay. At each step size decay, in addition to an improvement in network accuracy, we also observe a jump in the CKA similarity scores of the epoch's subspace affinity graph towards the CKA scores of final epoch's subspace affinity graph. We note that this jump is less prominent in shallower layers of the network when compared to the deeper layers as shallower layers converge to their final representations much earlier in training when compared to deeper layers. This bottom-up convergence observation is similar to observations made in [15], and [20] as described earlier. This jump also signifies a marked deviation from the representations of the layer during it's previous state - before the step decay, and current state - after the step decay. These observations highlight the role of step decay in Stochastic Gradient Descent based training of Neural networks to escape saddle points and other spurious local minima.

### 4.3 Comparison with Linear-CKA

As a comparison with related works in the area in Fig. 1c and Fig. 2c we present the same training dynamics as previous paragraph, this time by analyzing the Linear Kernel-CKA [10]. We observe a much cleaner bottom-up convergence of layers when compared with our method SSC-CKA, but the rates of convergence to final state is much faster when compared with the rates from SSC-CKA.

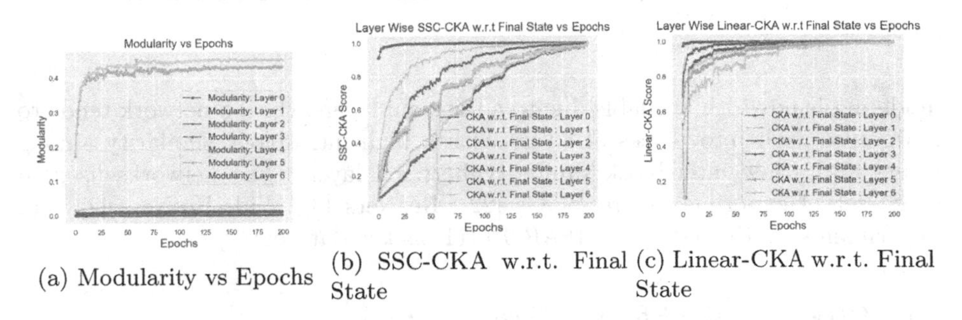


(a) Modularity vs Epochs (b) SSC-CKA w.r.t. Final State (c) Linear-CKA w.r.t. Final State

**Fig. 1.** Analysis of SSC affinity graphs using modularity and CKA and it's comparison with a linear kernel affinity graph on DenseNet-121 network using CIFAR-10 dataset.

## 5 Analysis of Network Architecture

In this section, we utilise SSC-CKA to visualize network architectures. We do this by using CKA to compare the Subspace Affinity Graphs obtained by applying SSC on the activations of two different layers of the network. We try to focus on various inter-layer dynamics in different scenarios so as to learn intrinsic behaviours of the network and the architecture class it belongs to. In pursuit of these goals, we evaluate and demonstrate the architectural behaviour of various networks as a function of network architectural depth, network width, training duration in terms of epochs and finally as a function of quantity of training data. The motivation behind these experiments is to decipher how the latent space structure evolves through different layers of the network and how does a given set of layers add to the modeling power of the network. In doing so, we observe that deeper networks, wider networks, prematurely trained networks (networks trained for fewer epochs than optimal) and malnourished networks (networks trained on less training data but trained till saturation of performance) tend to develop prominent and mostly non-overlapping block diagonal structures in their layer-wise SSC-CKA heatmaps. This indicates that as the network's modeling capacity increases, a given amount training data is unable to exhaust the spare

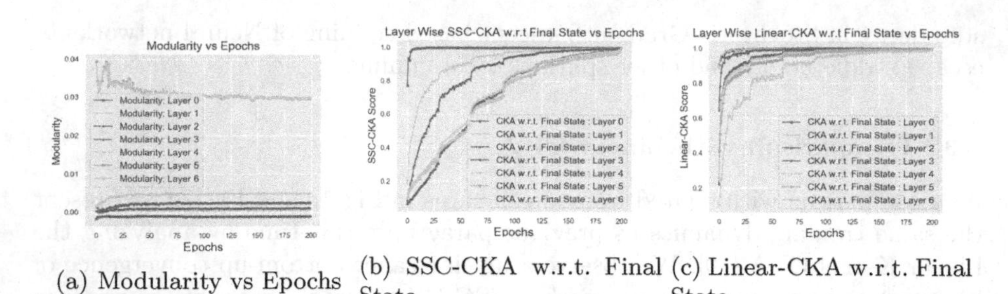


(a) Modularity vs Epochs (b) SSC-CKA w.r.t. Final State (c) Linear-CKA w.r.t. Final State

**Fig. 2.** Analysis of SSC affinity graphs using modularity and CKA and it's comparison with a linear kernel affinity graph on DenseNet-121 network using CIFAR-100 dataset.

modeling bandwidth available due to additional layers, thus, a network tends to reorganise itself into blocks of layers with a high intra-block similarity among layers and a low inter-block similarity between layers. The networks used in these set of experiments are VGGs [22], ResNets [8], Wide-ResNets [24], all experiments conducted with CIFAR-10 [11] as the dataset.

### 5.1 Observing the Effects of Depth

In this section we demonstrate the effect of depth on neural network architecture by analyzing the SSC-CKA maps of different CNNs with varying depths. Through Fig. 3a–Fig. 3d, we demonstrate the effects of increasing depths in VGG architecture based networks. In Fig. 3a, we train a VGG-11 and reach around 92% accuracy on the evaluation set. Applying SSC-CKA to this network, we observe a prominently diagonal similarity matrix indicating that most layers, especially the earlier ones learn unique representations and as we go deeper in the architecture, some inter layer similarity appears. Upon increasing the depth in the architecture to a VGG-16 as shown in Fig. 3b, we observe a slight improvement in accuracy 94%, but we begin to observe a diagonally dominant similarity matrix where shallower layers have some degree of similarity with their neighbours, but deeper layers, especially towards the end tend to form blocks of layers which are very similar to each other indicating that the network doesn't need to utilize the additional bandwidth to learn newer features in order to better learn and generalise. In Fig. 3c, we show a similar observation for VGG-24 that attained an accuracy of 93%. However, as we increase the number of convolutional layers to 29 (VGG-29) Fig. 3d, the performance of the network drops to around 62%, which is symptomatic of overfitting. We also observe a reinforcement in the prominence of the block diagonal structure signifying network over-parametrization relative to the data.

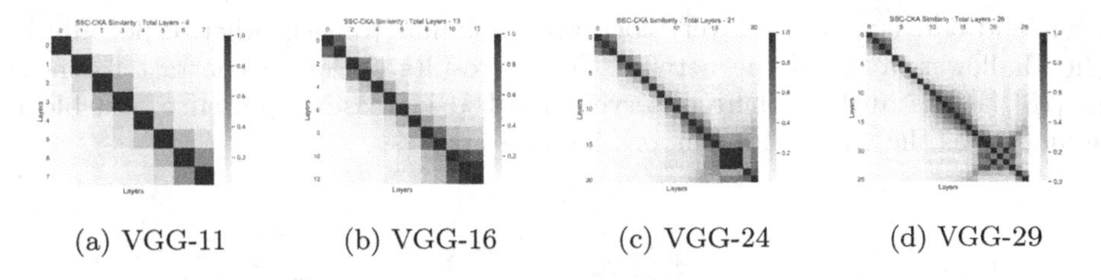

(a) VGG-11 (b) VGG-16 (c) VGG-24 (d) VGG-29

**Fig. 3.** Left to Right : Pairwise SSC-CKA between all convolution layers of VGG-11: (accuracy 92%), VGG-16: (accuracy 94%), VGG-24: (accuracy 93%) and VGG-29: (accuracy 62%) respectively on CIFAR-10 dataset.

Next we demonstrate a similar set of results for ResNets of various depths, Fig. 4a–Fig. 4c. In case of ResNets, in addition to the block diagonal structure we also notice a chess-board pattern inside the blocks themselves, where every layer is similar to every alternate layer in a block. This is a consequence of skip-connections present inside resnet blocks, which allow inputs from one layer to propagate much deeper into the network. The observations in this experiment are in-line with the ones made in [10] and [18].

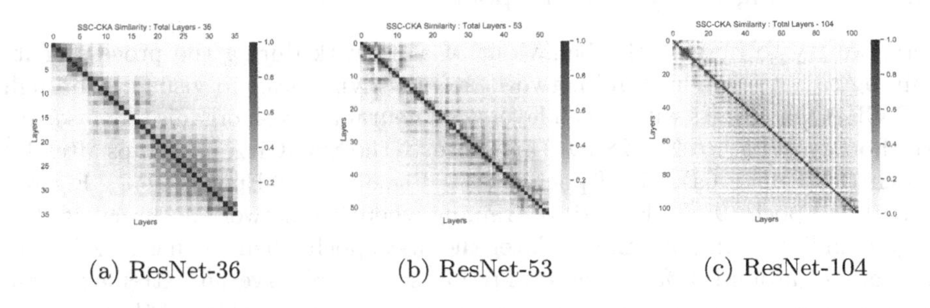


(a) ResNet-36 (b) ResNet-53 (c) ResNet-104

**Fig. 4.** Left to right : pairwise SSC-CKA between all convolution layers of ResNet-36: (accuracy 94%), ResNet-53: (accuracy 94%), ResNet-104: (accuracy 93%) respectively on CIFAR 10 dataset.

### 5.2 Observing the Effects of Width

In this experiment, we aim to study the effects of increasing the width of an architecture for a given depth. We use Wide-ResNet-64 network architecture with varying depth configurations (width 2x, 6x and 10x) and train them to their peak performance. The results of SSC-CKA are presented in Fig. 5a–Fig. 5c. In case of Wide-ResNets for a given depth, we observe that the architectural structure of various width configurations is very similar. The network can be divided into three distinct and disjoint blocks of layers, each having a high intra-block similarity and low inter-block similarity. Furthermore, we also notice the network wide presence of the chess-board pattern inherent to ResNets where even and odd layers in a layer block are dissimilar to each other, especially in the shallower layers of the network. These results present a contrast to those in [18], as the authors there observe a marked increase in presence of a block structure as the width of a network increases.

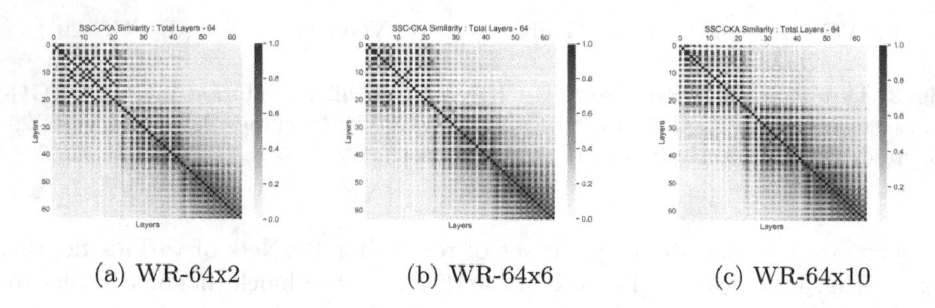


(a) WR-64x2 (b) WR-64x6 (c) WR-64x10

**Fig. 5.** Left to right : pairwise SSC-CKA between all convolution layers of wide ResNet-64 with width 2x(Accuracy-95%), 6x(Accuracy-95%) and 10x(Accuracy-96%), respectively on CIFAR-10 dataset.

### 5.3 Observing the Effects of Epochs

Here, we try to observe the behaviour of a network during the process of its training to better understand network training dynamics and visualize through SSC-CKA the process with which layers of a neural network organise themselves. We choose a Wide ResNet-28-by-2 and observe the SSC-CKA heatmaps after $1^{st}$ Epoch (35% - Fig. 6a), $31^{st}$ Epoch (84% - Fig. 6b), $61^{st}$ Epoch (91% - Fig. 6c) and $100^{th}$ Epoch (95% - Fig. 6d) of training where the network accuracy at that stage is indicated in parenthesis. After the first epoch when the network has an accuracy of around 35% on the evaluation set, we observe an extremely large block diagonal structure encapsulating about the first two-third of the network. Subsequently as the network trains and improves its performance, the large global structure present in the earlier layers makes way for a smaller and more contained local block diagonal structures. This observation into network training dynamics can provide us with alternate ways to determine the maturation of the training process, one that doesn't necessarily need any labeled data. This experiment can be seen as an extension of the procedure demonstrated in Fig. 1b and Fig. 2b, but instead of comparing a few layers of the network per epoch to their final state, we compare all convolutional layers after every few epochs.

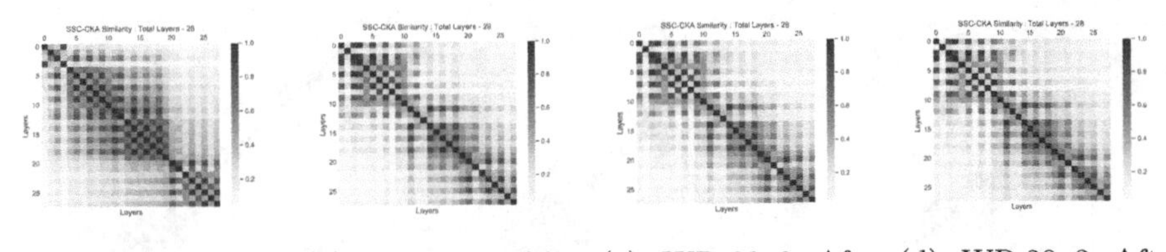

(a) WR-28x2 After $1^{st}$ Epoch. (b) WR-28x2 After $31^{st}$ Epoch. (c) WR-28x2 After $61^{st}$ Epoch. (d) WR-28x2 After $100^{th}$ Epoch.

**Fig. 6.** Left to right : pairwise SSC-CKA between all convolution layers of various wide ResNet-28x2 After $1^{st}$, $31^{st}$, $61^{st}$ and $100^{th}$ epoch in training on CIFAR-10 dataset.

### 5.4 Observing the Effects of Quantity of Training Data

Next, along the lines of [18] we setup an experiment to demonstrate the emergence of the block diagonal structure in SSC-CKA heatmaps in networks that are over-parameterized relative to the training data. To simulate network size relative to data we train the network on 3 different dataset configurations (of CIFAR-10), the first configuration uses only 5% of the original CIFAR-10 training set, the second uses 50% of the training set and the third uses the entire set. The network used in these sets of experiments is a ResNet-36 trained till saturation of performance on the evaluation set. The network achieves an accuracy of 56% when trained with only 5% of the training data - Fig. 7a, 92% when trained with half the training set - Fig. 7b and 94% when trained with the entire training data - Fig. 7c. Comparing the network that was trained on only 5% of the dataset to the other two configurations we observe a more pronounced block-diagonal structure consisting of two distinct and disjointed blocks comprising the first half of the network in the former configuration. As the network is less starved for data, the block diagonal structure that was prevalent in the first half starts to become less recognisable until it breaks down into much more localised structures as seen in Fig. 7c. These observations through SSC-CKA re-affirm the assertions made in Linear-CKA [10, 18] that over-parametrized networks are prone to developing a block diagonal structure as shown in their heat-maps for pairwise similarity. Please note that the second half of all the networks still retains a block diagonal structure. This is in accordance with the earlier observations made by increasing the depths of neural networks.

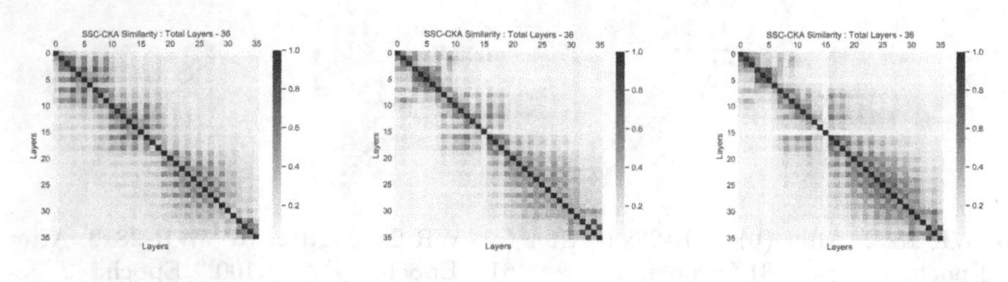


(a) ResNet-36 - 5% Training Data (b) ResNet-36 - 50% Training Data (c) ResNet-36 - 100% Training Data

**Fig. 7.** Left to right : pairwise SSC-CKA between all convolution layers of various ResNet-36 when trained on 5%, 50% and 100% training data on CIFAR-10 dataset.

# 6 Analysis of Inputs

In this section, we focus our analysis on individual inputs fed to the network as a part of the study. The goal is to highlight the ability of the model to interpret each input in terms of it's affinity to it's neighbours, and demonstrate experimentally that the neural network tries to separate out input of different classes into different mostly disjoint subspaces.

## 6.1 Layer-Wise Latent Space Visualisation

To begin the analysis, we first present a selective layer-by-layer two-dimensional embedding of all correctly classified inputs to the network. We obtain the embeddings by taking the top two eigenvectors obtained from the decomposition of the respective layer's affinity matrix (Normalised Laplacian Matrix can also be used), which we do along the lines of [17] and [5]. In Fig. 8a–Fig. 8f, we present the layer-wise analysis for six layers within the network, namely layer 1, 7, 14, 21, 28 and 36 respectively. In accordance with the observation made in Fig. 4 regarding the layer-by-layer modularity scores of the subspace affinity graph as we go deeper in the network, we observe a similar and perhaps a corroborating phenomenon, where inputs belonging to the same or similar classes get clustered closer to each other in a section of the latent space, and the inputs that belong to different classes get pushed into disjoint sections of the latent space, as we go deeper in the network.

## 6.2 Model Explanation by Instance Neighbourhood Visualisation

Here, we analyze the failure cases and focus on the inputs where the network failed to classify the input correctly and chose inputs which had the highest output softmax scores among the incorrectly classified inputs, i.e. the network was confidently wrong for those inputs. For the purpose of this subsequent analysis, we take the SSC-based affinity of the network's final convolution layer.

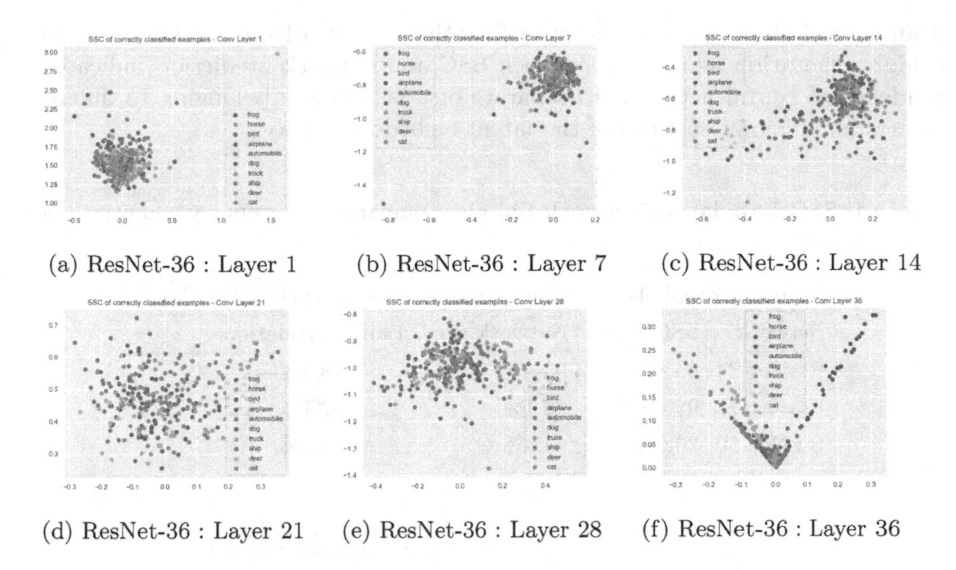


(a) ResNet-36 : Layer 1 (b) ResNet-36 : Layer 7 (c) ResNet-36 : Layer 14

(d) ResNet-36 : Layer 21 (e) ResNet-36 : Layer 28 (f) ResNet-36 : Layer 36

**Fig. 8.** Left to right : spectral embeddings of layer-wise affinity graphs learned via a ResNet-36 - data set : CIFAR 10

In Fig. 9a we show the input image of a cat that was incorrectly classified as a bird with a softmax score of 0.97. Figure 9b shows the normalised class-wise distribution of affinity scores for this input summed over all the images. This plot clearly shows the three strongest classes for which the given input has a strong affinity for, namely, bird, airplane and cat, in that order. At this juncture, we make a note that the predicted label, though incorrectly, of this input by the network seems to be the class for which this input has the highest affinity for as determined by Sparse Subspace Clustering Algorithm. Figure 9c–Fig. 9j show the top-8 images, in descending order, for which the given input of cat has the highest affinity for.

Continuing discussion on the previous observation, we further expand on that observation in Table 1, where we present the accuracy of the network on the testing set in two scenarios. The first scenario assigns an output label to a given input based on the highest aggregate class-wise affinity score for that input and the second scenario assigns the same output label to the input as the networks prediction through it's classification layers. The results of those are presented in columns 2 and 3 labeled 'SSC Label' and 'Network Prediction' respectively, in Table 1. We observe that both these 'accuracies' turn out to be 95.8%, Please be advised that the equality of values is a coincidence. Thus, given the activations of the final convolutional layer for the entire testing set we observe that the SSC based Affinity scoring assigns labels which are competitive with what the classification layers of the network could learn. This is further demonstrated in column 4 of Table 1 which shows a 98.3% agreement between one to one comparisons of network assigned labels with SSC derived labels. Row

3 and row 4 show the same metrics for other 2 networks, on CIFAR-10. Such a high annotation agreement between SSC and network prediction indicates a tendency of neural networks to separate out data points belonging to different classes into disjoint sections of the latent embedding space.

**Table 1. SSC Labels vs. Network Labels**: Comparison of accuracy and correlation on the test set

| Accuracy of labels when compared to ground truth (CIFAR-10) | | | |
|---|---|---|---|
| Network | SSC Label | Network prediction | Correlation |
| ResNet-34 | 95.8% | 95.8% | 98.3% |
| ResNet-18 | 95.4% | 96.1% | 97% |
| ResNet-50 | 90.6% | 94.8% | 93.5% |

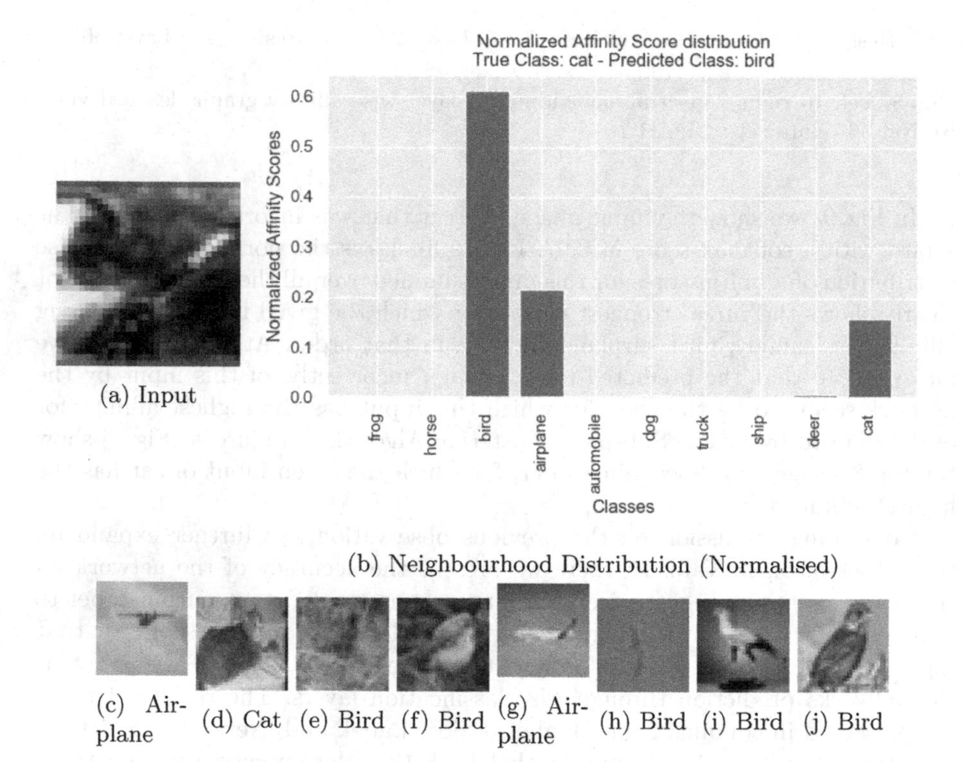


(a) Input

(b) Neighbourhood Distribution (Normalised)

(c) Airplane (d) Cat (e) Bird (f) Bird (g) Airplane (h) Bird (i) Bird (j) Bird

**Fig. 9.** 9(a): A Cat, classified as a bird with a softmax score of 0.97. 9(b): Distribution of it's neighbourhood affinity scores (Normalised). 9(c): Airplane. 9(d): Cat. 9(e): Bird. 9(f): Bird. 9(g): Airplane. 9(h): Bird. 9(i): Bird. 9(j): Bird. Network: ResNet-36 - Data Set : CIFAR 10

## 7 Conclusions

Our work proposes a two-step framework to analyze deep neural networks. This framework combines Sparse Subspace Clustering and Centered Kernel Alignment to provide the ability to analyze the network on a macro and micro level. The macro analysis helps in visualising network architectures as a function of network depth, width, training epochs and training data quantity and provides an insight into network architecture and training dynamics of the network. It also provides a framework to compare two architecturally different neural network based on a common set of inputs. The framework also provides the ability to micro analyze the network in the form of instance based interpretability by providing a measure of a degree of closeness each input has to decision boundaries for different classes in the loss landscape of a network.

**Acknowledgements.** This research was supported by National Science Foundation Grant No. 1901379, the U.S. Department of Energy, Office of Science, Office of Advanced Scientific Computing Research (ASCR), Scientific Discovery through Advanced Computing (SciDAC) program, specifically the RAPIDS-2 SciDAC institute. Any opinions, findings, and conclusions or recommendations expressed in this material are those of the author(s) and do not necessarily reflect the views of the funding parties. The authors would like to thank Tushar Nagarajan for helpful discussions and anonymous reviewers for critical feedback.

## References

1. Amaldi, E., Kann, V.: On the approximability of minimizing nonzero variables or unsatisfied relations in linear systems. Theor. Comput. Sci. **209**(1–2), 237–260 (Dec 1998)
2. Bau, D., Zhou, B., Khosla, A., Oliva, A., Torralba, A.: Network dissection: quantifying interpretability of deep visual representations (2017)
3. Boyd, S., Parikh, N., Chu, E., Peleato, B., Eckstein, J.: Distributed optimization and statistical learning via the alternating direction method of multipliers. Found. Trends Mach. Learn. **3**(1), 1–122 (2011)
4. Cortes, C., Mohri, M., Rostamizadeh, A.: Algorithms for learning kernels based on centered alignment. CoRR abs/1203.0550 (2012). http://arxiv.org/abs/1203.0550
5. Elhamifar, E., Vidal, R.: Sparse subspace clustering: algorithm, theory, and applications. IEEE Trans. Pattern Anal. Mach. Intell. **35**(11), 2765–2781 (2013)
6. Gretton, A., Bousquet, O., Smola, A., Schölkopf, B.: Measuring statistical dependence with hilbert-schmidt norms. In: Jain, S., Simon, H.U., Tomita, E. (eds.) ALT 2005. LNCS (LNAI), vol. 3734, pp. 63–77. Springer, Heidelberg (2005). https://doi.org/10.1007/11564089_7
7. Hanin, B., Sellke, M.: Approximating continuous functions by relu nets of minimal width (2018)
8. He, K., Zhang, X., Ren, S., Sun, J.: Deep residual learning for image recognition (2015)
9. Huang, G., Liu, Z., van der Maaten, L., Weinberger, K.Q.: Densely connected convolutional networks (2018)

10. Kornblith, S., Norouzi, M., Lee, H., Hinton, G.: Similarity of neural network representations revisited (2019)
11. Krizhevsky, A.: Learning multiple layers of features from tiny images. Technical Report (2009)
12. Lee, J., Bahri, Y., Novak, R., Schoenholz, S.S., Pennington, J., Sohl-Dickstein, J.: Deep neural networks as gaussian processes (2018)
13. Lin, H., Jegelka, S.: Resnet with one-neuron hidden layers is a universal approximator (2018)
14. Lu, Z., Pu, H., Wang, F., Hu, Z., Wang, L.: The expressive power of neural networks: a view from the width (2017)
15. Morcos, A.S., Raghu, M., Bengio, S.: Insights on representational similarity in neural networks with canonical correlation (2018)
16. Newman, M.E.J.: Modularity and community structure in networks. Proc. Nat. Acad. Sci. **103**(23), 8577–8582 (2006)
17. Ng, A., Jordan, M., Weiss, Y.: On spectral clustering: analysis and an algorithm. In: Dietterich, T., Becker, S., Ghahramani, Z. (eds.) Advances in Neural Information Processing Systems. vol. 14. MIT Press (2002). https://proceedings.neurips.cc/paper/2001/file/801272ee79cfde7fa5960571fee36b9b-Paper.pdf
18. Nguyen, T., Raghu, M., Kornblith, S.: Do wide and deep networks learn the same things? uncovering how neural network representations vary with width and depth (2020)
19. Novak, R., et al.: Bayesian deep convolutional networks with many channels are gaussian processes (2020)
20. Raghu, M., Gilmer, J., Yosinski, J., Sohl-Dickstein, J.: Svcca: singular vector canonical correlation analysis for deep learning dynamics and interpretability (2017)
21. Simonyan, K., Vedaldi, A., Zisserman, A.: Deep inside convolutional networks: visualising image classification models and saliency maps (2014)
22. Simonyan, K., Zisserman, A.: Very deep convolutional networks for large-scale image recognition (2015)
23. Vidal, R., Ma, Y., Sastry, S.S.: Generalized Principal Component Analysis, 1st edn. Springer, New York (2016)
24. Zagoruyko, S., Komodakis, N.: Wide residual networks (2017)
25. Zeiler, M.D., Fergus, R.: Visualizing and understanding convolutional networks (2013)

# Invertible Manifold Learning for Dimension Reduction

Siyuan Li[1,2], Haitao Lin[1,2], Zelin Zang[1,2], Lirong Wu[1,2], Jun Xia[1,2], and Stan Z. Li[1,2(✉)]

[1] AI Lab, School of Engineering, Westlake University, Hangzhou, Zhejiang, China
{lisiyuan,linhaitao,zangzelin,wulirong,xiajun,Stan.ZQ.Li}@westlake.edu.cn
[2] Institute of Advanced Technology, Westlake Institute for Advanced Study, Hangzhou, Zhejiang, China

**Abstract.** Dimension reduction (DR) aims to learn low-dimensional representations of high-dimensional data with the preservation of essential information. In the context of manifold learning, we define that the representation after information-lossless DR preserves the topological and geometric properties of data manifolds formally, and propose a novel two-stage DR method, called invertible manifold learning (*inv-ML*) to bridge the gap between theoretical information-lossless and practical DR. The first stage includes a homeomorphic *sparse coordinate transformation* to learn low-dimensional representations without destroying topology and a *local isometry* constraint to preserve local geometry. In the second stage, a *linear compression* is implemented for the trade-off between the target dimension and the incurred information loss in excessive DR scenarios. Experiments are conducted on seven datasets with a neural network implementation of *inv-ML*, called *i-ML-Enc*. Empirically, *i-ML-Enc* achieves invertible DR in comparison with typical existing methods as well as reveals the characteristics of the learned manifolds. Through latent space interpolation on real-world datasets, we find that the reliability of tangent space approximated by the local neighborhood is the key to the success of manifold-based DR algorithms.

**Keywords:** Dimension reduction · Manifold learning · Deep learning · Inverse problem

## 1 Introduction

In real-world scenarios, it is widely believed that the loss of data information is inevitable after dimension reduction (DR), though the goal of DR is to preserve as much data information as possible in the low-dimensional space. Most methods try to preserve some essential information of data after DR, e.g., geometric structure within the data, which is usually achieved by preserving the distance in high and low-dimensional space. In the case of linear DR, compressed sensing [5] breaks this common sense with practical sparse conditions of the given data. The lower bound of target dimension and the information loss for linear


N. Oliver et al. (Eds.): ECML PKDD 2021, LNAI 12977, pp. 713–728, 2021.
https://doi.org/10.1007/978-3-030-86523-8_43

DR are provided by Johnson–Lindenstrauss Theorem [10] with the pairwise distance. In the case of nonlinear dimension reduction (NLDR), however, it has not been thoroughly discussed, i.e., what structures within data are necessary to preserve, how to maintain these structures after NLDR, and how much information can be preserved under different cases? From the perspective of manifold learning, a popular *manifold assumption* is widely adopted that the given data has relatively low-dimensional intrinsic structures. Classical manifold-based DR methods [24,28] work well on synthetic manifold datasets, but usually fail to yield good results in the many practical cases. Therefore, there is still a gap between theoretical and real-world applications of manifold-based DR.

Here, we give a detailed discussion of these problems in the context of manifold learning and define that the representation after information-lossless DR should preserve the topology and geometry of input data. On the one hand, the representation should demonstrate some geometric properties after DR, or it will be meaningless. For example, if the distance between point $A$ and $B$ is larger than that between $A$ and $C$ on the data manifold, the low-dimension representation should preserve the order to revealing the simalarity of data points. On the other hand, the topological properties can be preserved if the DR transformation is a continuous bijective mapping, i.e., homeomorphism, leading to the information-lossless mapping.

To achieve the information-lossless DR, we propose an invertible NLDR process, called *inv-ML*, combining *sparse coordinate transformation* and *local isometry* constraint which preserve the property of topology and geometry respectively. In terms of the target dimension and information loss, we discuss different cases of NLDR in manifold learning. We instantiate *inv-ML* as a neural network called *i-ML-Enc* via a cascade of equidimensional layers and a linear transform layer. The proposed loss terms and network structures are explainable. Sufficient experiments are conducted to validate invertible NLDR abilities of *i-ML-Enc* and analyze learned representations to reveal inherent difficulties of classical manifold learning empirically.

We summarize our main contributions as follows:

- Introduce an invertible NLDR process *inv-ML* to fill the gap between theoretical information-lossless and real-world applications of NLDR.
- Verify the proposed *inv-ML* in different cases by designing an invertible neural network *i-ML-Enc* which produces explainable NLDR results and achieves state-of-the-art performance on benchmark datasets.
- Reveals characteristics of the learned low-dimensional representation by latent space interpolation.

## 2 Related Work

*Manifold Learning.* Most classical DR or NLDR methods aim to preserve the geometric properties of manifolds. The Isomap [26] based methods aim to preserve the global metric between every pair of sample points. For example, [17] can be regarded as such methods based on the push-forward Riemannian metric. For the other aspect, LLE [24] based methods try to preserve local geometry after DR, whose derivatives like LTSA [29], MLLE [28], etc. have been widely used but usually fail in the high-dimensional case. Recently, based on local properties of manifolds, MLDL [15] was proposed as a robust NLDR method implemented by a neural network. However, those methods ignore the retention of topology. In contrast, we take the preservation of both geometry and topology into consideration, trying to maintain these properties of manifolds even in cases of excessive dimension reduction when the target dimension $s'$ is smaller than $s$.

*Invertible Model.* From AutoEncoder (AE) [7], the fundamental neural network based model, having achieved DR and cut information loss by minimizing the reconstruction loss, some AE based generative models like VAE [13] and manifold-based NLDR models like TopoAE [19] and GRAE [6] have emerged. These methods cannot avoid information loss after NLDR, and thus, some invertible models consist of a series of equidimensional layers have been proposed, some of which aim to generate samples by density estimation through layers [1,3,4], and the other of which are established for other targets, e.g., validating the mutual information bottleneck [9]. Different from the methods mentioned above, our proposed *i-ML-Enc* is a neural network based encoder, with NLDR as well as maintaining structures of raw data points based on manifold assumption via a series of equidimensional layers.

# 3 Proposed Method

Firstly, we state the information-lossless DR problem in Sect. 3.1. Then, the proposed invertible NLDR process *inv-ML* is specifically discussed in Sect. 3.2 and Sect. 3.3. Finally, we instantiate the proposed *inv-ML* as *i-ML-Enc* in Sect. 3.4.

## 3.1 Problem Statement

To start, we first make out a theoretical definition of information-lossless DR of a data manifold. The structures of the manifold from which data points are sampled from include topology and geometry, if the transformed manifold preserves these two structures after a dimension reduction process, this DR process is defined as information-lossless.

*Topology Preservation.* The topological property is what is invariant under a homeomorphism, and thus what we want to achieve is to construct a homeomorphism for dimension reduction, removing the redundant dimensions while preserving invariant topology. To be more specific, $f : \mathcal{M}_0^d \to \mathbb{R}^m$ is a smooth mapping of a differential manifold into another, and if $f$ is a homeomorphism of $\mathcal{M}_0^d$ into $\mathcal{M}_1^d = f(\mathcal{M}_0^d) \subset \mathbb{R}^m$, we call $f$ is an embedding of $\mathcal{M}_0^d$ into $\mathbb{R}^m$. Assume that the data set $\mathcal{X} = \{\boldsymbol{x}_j | 1 \leq j \leq n\}$ sampled from the compact manifold $\mathcal{M}_1^d \subset \mathbb{R}^m$ which we call the data manifold and is homeomorphic to $\mathcal{M}_0^d$. For the sample points we get are represented in the coordinate after inclusion mapping $i_1$, we can only regard them as points from Euclidean space $\mathbb{R}^m$

without any prior knowledge, and learn to approximate the data manifold in the latent space $Z$. According to the Whitney Embedding Theorem [25], $\mathcal{M}_0^d$ can be embedded smoothly into $\mathbb{R}^{2d}$ by a homeomorphism $g$. Rather than to find the $f^{-1} : \mathcal{M}_1^d \rightarrow \mathcal{M}_0^d$, our goal is to seek a smooth map $h : \mathcal{M}_1^d \rightarrow \mathbb{R}^s \subset \mathbb{R}^{2d}$, where $h = g \circ f^{-1}$ is a homeomorphism of $\mathcal{M}_1^d$ into $\mathcal{M}_2^d = h(\mathcal{M}_1^d)$ and $d \leq s \leq 2d \ll m$, and thus the $dim(h(\mathcal{X})) = s$, which achieves the DR while preserving the topology. Owing to the homeomorphism $h$ we seek as a DR mapping, the data manifold $\mathcal{M}_1^d$ is reconstructible via $\mathcal{M}_1^d = h^{-1} \circ h(\mathcal{M}_1^d)$, by which we mean $h$ a topology preserving DR as well as information-lossless DR (Figs. 1 and 2).

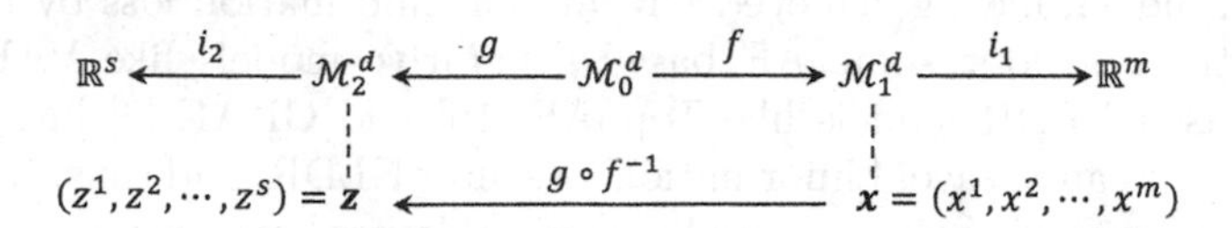


**Fig. 1.** Illustration of the process of NLDR. The dash line links $\mathcal{M}_1^d$ and $\boldsymbol{x}$ means $\boldsymbol{x}$ is sampled from $\mathcal{M}_1^d$, and it is represented in the Euclidean space $\mathbb{R}^m$ after an inclusion mapping $i_1$. We aim to approximate $\mathcal{M}_1^d$ from the observed sample $\boldsymbol{x}$. For the topology preserving dimension reduction methods, it aims to find a homeomorphism $g \circ f^{-1}$ to map $\boldsymbol{x}$ into $\boldsymbol{z}$ which is embedded in $\mathbb{R}^s$.

*Geometry Preservation.* While the topology of the data manifold $\mathcal{M}_1^d$ can be preserved by the homeomorphism $h$ discussed above, it may distort the geometry. To preserve the local geometry of the data manifold, we choose pairwise distance as the key geometric property, i.e. the DR mapping should be isometric on the tangent space $\mathcal{T}_p\mathcal{M}_1^d$ for every $p \in \mathcal{M}_1^d$, indicating that $d_{\mathcal{M}_1^d}(u, v) = d_{\mathcal{M}_2^d}(h(u), h(v)), \forall u, v \in \mathcal{T}_p\mathcal{M}_1^d$. By Nash's Embedding Theorem [20], any smooth manifold of class $C^k$ with $k \geq 3$ and dimension $d$ can be embedded isometrically in the Euclidean space $\mathbb{R}^s$ with $s$ polynomial in $d$.

*Noise Perturbation.* In the real-world scenarios, sample points are not lied on the ideal manifold strictly due to the limitation of sampling, e.g., non-uniform sampling noises. When the DR method is very robust to the noise, it is reasonable to ignore the effects of the noise and learn the representation $Z$ from the given data. Therefore, the intrinsic dimension of $\mathcal{X}$ is approximate to $d$, resulting in the lowest isometric embedding dimension is larger than $s$.

### 3.2 Methods for Structure Preservation

*Canonical Embedding for Homeomorphism.* To seek the smooth homeomorphism $h$, we turn to the theorem of local canonical form of immersion [18]. Let $f : \mathcal{M} \rightarrow \mathcal{N}$ an immersion, and for any $p \in \mathcal{M}$, there exist local coordinate systems $(U, \phi)$

around $p$ and $(V, \psi)$ around $f(p)$ such that $\psi \circ f \circ \phi^{-1} : \phi(U) \to \psi(V)$ is a canonical embedding, which reads

$$\psi \circ f \circ \phi^{-1}(x^1, x^2, \ldots, x^d) = (x^1, x^2, \ldots, x^d, 0, 0, \ldots, 0). \tag{1}$$

In our case, let $\mathcal{M} = \mathcal{M}_2^d$, and $\mathcal{N} = \mathcal{M}_1^d$, any point $\boldsymbol{z} = (z^1, z^2, \ldots, z^s) \in \mathcal{M}_1^d \subset \mathbb{R}^s$ can be mapped to a point in $\mathbb{R}^m$ by the canonical embedding

$$\psi \circ h^{-1}(z^1, z^2, \ldots, z^s) = (z^1, z^2, \ldots, z^s, 0, 0, \ldots, 0). \tag{2}$$

For the point $\boldsymbol{z}$ is regarded as a point in $\mathbb{R}^s$, $\phi = \mathbb{I}$ is an identity mapping, and for $h = g \circ f^{-1}$ is a homeomorphism, $h^{-1}$ is continuous. The Eq. (2) can be written as

$$\begin{aligned}(z^1, z^2, \ldots, z^s) &= h \circ \psi^{-1}(z^1, z^2, \ldots, z^s, 0, 0, \ldots, 0) \\ &= h(x^1, x^2, \ldots, x^m).\end{aligned} \tag{3}$$

Therefore, to reduce $dim(\mathcal{X}) = m$ to $s$, we can decompose $h$ into $\psi$ and $h \circ \psi^{-1}$, by firstly finding a homeomorphic coordinate transformation $\psi$ to map $\boldsymbol{x} = (x^1, x^2, \ldots, x^m)$ into $\psi(\boldsymbol{x}) = (z^1, z^2, \ldots, z^s, 0, 0, \ldots, 0)$, which is called a *sparse coordinate transformation*, and $h \circ \psi^{-1}$ can be easily obtained by Eq. (2). We denote $h \circ \psi^{-1}$ by $h_0$ and call it a *sparse compression*. The theorem holds for any manifold, while in our case, we aims to find the mapping of $\mathcal{X} \subset \mathbb{R}^m$ into $\mathbb{R}^s$, so the local coordinate systems can be extended to the whole space of $\mathbb{R}^m$.

*Local Isometry Constraint.* The prior local isometry constraint is applied under the manifold assumption, which aims to preserve distances (or some other metrics) locally so that $d_{\mathcal{M}_1^d}(u, v) = d_{\mathcal{M}_2^d}(h(u), h(v))$, $\forall u, v \in \mathcal{T}_p\mathcal{M}_1^d$.

### 3.3 Linear Compression

With the former discussed method, manifold-based NLDR can be achieved with topology and geometry preserved, i.e. $s$-sparse representation in $\mathbb{R}^m$. However, the target dimension $s'$ may be even less than $s$, further compression can be performed through the *linear compression* $h_0' : \mathbb{R}^m \to \mathbb{R}^{s'}$ instead of *sparse compression*, where $h_0'(\boldsymbol{z}) = W_{m \times s'}\boldsymbol{z}$, with minor information loss. In general, the *sparse compression* is a particular case of *linear compression* with $h_0(\boldsymbol{z}) = h_0'(\boldsymbol{z}) = \Lambda \boldsymbol{z}$, where $\Lambda = (\delta_{i,j})_{m \times s}$ and $\delta_{i,j}$ is the Kronecker delta. We discusses the information loss caused by a linear compression under different target dimensions $s'$ as following cases.

*Ideal Case.* In the case of $d \leq s \leq s'$, based on compressed sensing, we can reconstruct the raw input data after the NLDR process without loss of any information by solving the sparse optimization problem mentioned in Sect. 2 when the transformation matrix $W_{m \times s'}$ has the full rank of the column. In the case of $d \leq s' < s$, it is inevitable to drop the topological properties because the two spaces before and after NLDR are not homeomorphic. It is reduced to local geometry preservation by LIS constraint. However, in the case of $s' \leq d < s$, both topological and geometric information is lost to varying degrees. Therefore, we can only try to retain as much geometric structure as possible.

*Practical Case.* In real-world scenarios, the target dimension $s'$ is usually lower than $s$, even lower than $d$. Meanwhile, the data sampling rate is quite low, and the clustering effect is extremely significant, indicating that it is possible to approximate $\mathcal{M}_1$ by low-dimensional hyperplane in the Euclidean space. In the case of $s' < s$, we can retain the prior Euclidean topological structure as additional topological information of raw data points. It is reduced to replace the global topology with some relative structures between each cluster.

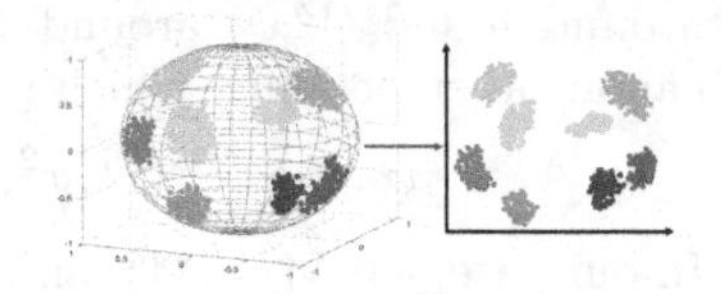

**Fig. 2.** Assuming data points are non-uniform sampled from a high-dimensional hypersphere, it is no need to maintain the global topology for the sparsity and clustering effect.

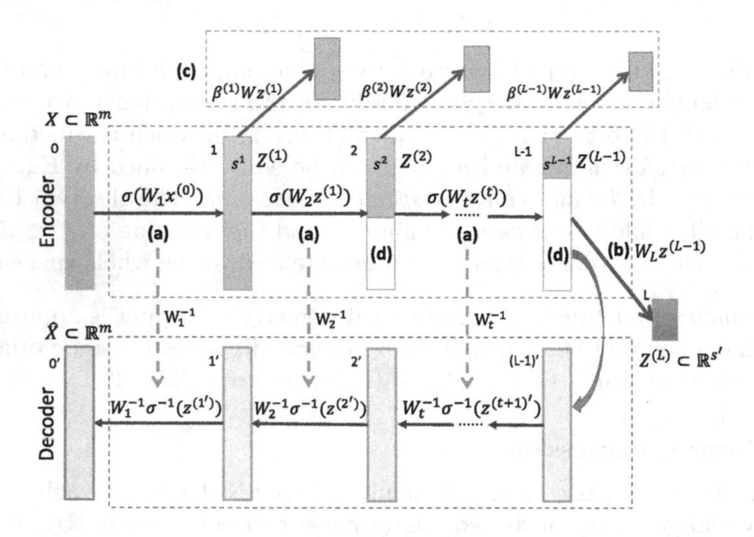


**Fig. 3.** Architecture of the proposed neural network implementation *i-ML-Enc*. The first $L-1$ layers equidimensional mapping in the green dash box are the first stage that achieves $s$-sparse, and they have an inverse process in the purple dash box. (a) denotes a layer of nonlinear homeomorphism transformation (red arrow). (b) linearly transforms (blue arrow) $s$-sparse representation in $\mathbb{R}^m$ into $\mathbb{R}^{s'}$ as the second stage. (c) represents a *extra head* by linear transformations, which will be removed after training. (d) indicates the padding zeros of the $l$-th layer to force $d^{(l)}$-sparse. (Color figure online)

### 3.4 Network Implementation

Based on Sect. 3.2 and Sect. 3.3, we propose a neural network *i-ML-Enc* which achieves two-stage NLDR preserving both topology and geometry, as shown in Fig. 3. In this section, we will introduce the function of proposed network structures and loss terms respectively, including the orthogonal loss, padding loss, and *extra heads* for the first stage, the LIS loss and push-away loss for the second stage.

*Cascade of Homeomorphisms.* Since the *sparse coordinate transformation* $\psi$ (and its inverse) can be highly nonlinear and complex, we decompose it into a cascade of $L-1$ isometric homeomorphisms $\psi = \psi^{(L-1)} \circ \ldots \circ \psi^{(2)} \circ \psi^{(1)}$, which can be achieved by $L-1$ equidimensional network layers. For each $\psi^{(l)}$, it is a *sparse coordinate transformation*, where $\psi^{l}(z^{1,(l)}, z^{2,(l)}, \ldots, z^{s_l,(l)}, 0, \ldots, 0) = (z^{1,(l+1)}, z^{2,(l+1)}, \ldots, z^{s_{l+1},(l+1)}, 0, \ldots, 0)$ with $s_{l+1} < s_l$ and $s_{L-1} = s$. The layer-wise transformation $Z^{(l+1)} = \psi^{(l)}(Z^{(l)})$ and its inverse can be written as

$$Z^{(l+1)} = \sigma(W_l X^{(l)}),\ Z^{(l)'} = W_l^{-1}(\sigma^{-1}(Z^{(l+1)'})), \tag{4}$$

in which $W_l$ is the $l$-th weight matrix of the neural network to be learned, and $\sigma(.)$ is a nonlinear activation. The bias term is removed here to facilitate its simple inverse structure.

*Orthogonal Loss.* Each layer-wise transformation is thought to be a homeomorphism between $Z^{(l)}$ and $Z^{(l+1)}$ in the first $L-1$ layers, and we want it to be a nearly isometric as

$$(1-\epsilon)\|\boldsymbol{x}_1 - \boldsymbol{x}_2\| \le \|W(\boldsymbol{x}_1 - \boldsymbol{x}_2)\| \le (1+\epsilon)\|\boldsymbol{x}_1 - \boldsymbol{x}_2\|, \tag{5}$$

where $\epsilon \in (0,1)$ is a rather small constant and $W$ is a linear measurement of signal $\boldsymbol{x}_1$ and $\boldsymbol{x}_2$. Because the activation function $\sigma(.)$ is monotonous, we can rewrite Eq.(5) as

$$L_{orth} = \sum_{l=1}^{L-1} \alpha^{(l)} \rho(W_l^T W_l - I), \tag{6}$$

where $\{\alpha^{(l)}\}$ are the loss weights. Notice that $\rho(W) = \sup_{\boldsymbol{z}\in\mathbb{R}^m, \boldsymbol{z}\neq\boldsymbol{0}} \frac{|W\boldsymbol{z}|}{|\boldsymbol{z}|}$ is the spectral norm of $W$, and the loss term can be written as $\rho(W_l^T W_l - I) = \sup_{\boldsymbol{z}\in\mathbb{R}^m, \boldsymbol{z}\neq\boldsymbol{0}} |\frac{|W\boldsymbol{z}|}{|\boldsymbol{z}|}|$ which is equivalent to force each $W_l$ to be an orthogonal matrix. The orthogonal constraint allows simple calculation of the inverse of $W_l$.

*Padding Loss.* To force sparsity from the second to $(L-1)$-th layers, we add a zero padding loss to each of these layers. For the $l$-th layer whose target dimension is $s_l$, pad the last $m - s_l$ elements of $\boldsymbol{z}^{(l+1)}$ with zeros and panish these elements with $L_1$ norm loss:

$$L_{pad} = \sum_{l=2}^{L-1} \beta^{(l)} \sum_{i=s^{(l)}}^{m} |\boldsymbol{z}_i^{(l+1)}|, \tag{7}$$

where $\{\beta^{(l)}\}$ are loss weights. The target dimension $s_l$ can be set heuristically.

*Linear Transformation Head.* We use the linear transformation head to achieve the linear compression step in our NLDR process, which is a transformation between the orthogonal basis of high dimension and lower dimension. Thus, we apply the row orthogonal constraint to $W_L$.

*LIS Loss.* Since the linear DR is applied at the end of the NLDR process, we apply *locally isometric smoothness* (LIS) constraint [15] to preserve the local geometric properties. Take the LIS loss in the $l$-th layer as an example:

$$L_{LIS} = \sum_{i=1}^{n} \sum_{j \in \mathcal{N}_i^k} \left\| d_X(\boldsymbol{x}_i, \boldsymbol{x}_j) - d_Z(\boldsymbol{z}_i^{(l)}, \boldsymbol{z}_j^{(l)}) \right\|, \tag{8}$$

where $\mathcal{N}_i^k$ is a set of $x_i$'s $k$-nearest neighborhood in the input space, and $d_X$ and $d_Z$ are the distance of the input and the latent space, which can be approximated by Euclidean distance in local open sets.

*Push-Away Loss.* In the real case discussed in Sect. 3.3, the latent space of the $(L-1)$-th layer can approximately be a hyperplane in Euclidean space, so that we introduce push-away loss to repel the non-adjacent sample points of each $x_i$ in its $B$-radius neighborhood in the latent space. It deflates the manifold locally when acting together with $L_{LIS}$ in the linear DR. Similarly, $L_{push}$ is applied after the linear transformation in the $l$-th layer:

$$L_{push} = -\sum_{i=1}^{n} \sum_{j \in \mathcal{N}_i^k} \mathbf{1}_{d_Z(\boldsymbol{z}_i^{(l)}, \boldsymbol{z}_j^{(l)}) < B} \log \left(1 + d_Z(\boldsymbol{z}_i^{(l)}, \boldsymbol{z}_j^{(l)})\right), \tag{9}$$

where $\mathbf{1}(.) \in \{0, 1\}$ is the indicator function for the bound of $B$.

*Extra Head.* In order to force the first $L-1$ layers of the network to achieve NLDR gradually, we introduce auxiliary DR branches, called *extra heads*, after the second layer to the $(L-1)$-th layer. The structure of each *extra head* is the same as the linear transformation head and will be discarded after training. $L_{extra}$ is written as

$$L_{extra} = \sum_{l=1}^{L-1} \gamma^{(l)} (L_{LIS} + \mu^{(l)} L_{push}), \tag{10}$$

where $\{\gamma^{(l)}\}$ and $\{\mu^{(l)}\}$ are loss weights which can be set based on $\{s_l\}$.

*Inverse Process.* The inverse process is the decoder directly obtained by the first $L-1$ layers of the encoder given by Eq. (4), which is not involved in the training process. When the target dimension $s'$ is equal to $s$, the inverse of the layer-$L$ can be solved by some existing methods such as compressed sensing or eigenvalue decomposition.

## 4 Experiment

In this section, we first evaluate the proposed *inv-ML* achieved by *i-ML-Enc* in Sect. 4.1, then investigate the property of data manifolds with *i-ML-Enc* in

Sect. 4.2. The properties of *i-ML-Enc* are further studied in Sect. 4.3. We carry out experiments on **seven datasets**: (i) Swiss roll [23], (ii) Spheres [19] and Half Spheres, (iii) USPS [8], (iv) MNIST [14], (v) KMNIST [2], (vi) FMNIST [27], (vii) COIL-20 [21]. The first two datasets are uniformly sampled on synthetic manifolds which can reflect mathematical properties of NLDR. The later five are real-world datasets where samples lie on circular manifolds (COIL-20) and cluster manifolds (MNIST, USPS, KMNIST, FMNIST). The following settings of *i-ML-Enc* are used for all datasets: LeakyReLU with $\alpha = 0.1$; Adam optimizer [12] with learning rate $lr = 0.001$ for 8000 epochs; the local neighborhood is determined by kNN with $k = 15$. The implementation is based on the PyTorch 1.3.0 library running on NVIDIA v100 GPU, and the source code is available at https://github.com/Westlake-AI/inv-ML.

**Table 1.** Comparison in NLDR, invertible and generalization qualities on MNIST and COIL-20.

| | Algorithm | RMSE | MNE | Trust | Cont | $l$-MSE | Acc |
|---|---|---|---|---|---|---|---|
| MNIST | MLLE | – | – | 0.6709 | 0.6573 | 36.80 | 0.8341 |
| | t-SNE | – | – | 0.9896 | 0.9886 | 48.07 | 0.9246 |
| | ML-Enc | – | – | 0.9862 | **0.9927** | 18.98 | 0.9326 |
| | VAE | 0.5263 | 33.17 | 0.9712 | 0.9703 | 22.79 | 0.8652 |
| | GRAE | 0.4324 | 17.32 | 0.9811 | 0.9796 | 20.45 | 0.8769 |
| | TopoAE | 0.5178 | 31.45 | **0.9915** | 0.9878 | 24.98 | 0.8993 |
| | ML-AE | 0.4012 | 16.84 | 0.9893 | 0.9926 | 19.05 | **0.9340** |
| | i-ML-Enc (L) | **0.0457** | **0.5085** | 0.9906 | 0.9912 | **18.16** | 0.9316 |
| | INN | 0.0615 | 0.5384 | 0.9851 | 0.9823 | 7.494 | 0.9176 |
| | i-RevNet | 0.0443 | **0.4679** | 0.9118 | 0.8785 | 6.958 | – |
| | i-ResNet | 0.0502 | 0.6422 | 0.9149 | 0.8922 | 10.78 | – |
| | i-ML-Enc(L-1) | **0.0407** | 0.5085 | **0.9986** | **0.9973** | **5.895** | **0.9580** |
| COIL-20 | t-SNE | – | – | 0.9911 | **0.9954** | 17.22 | 0.9039 |
| | ML-Enc | – | – | 0.9920 | 0.9889 | **9.961** | **0.9564** |
| | AE | 0.3507 | 24.09 | 0.9745 | 0.9413 | 11.45 | 0.8958 |
| | GRAE | 0.2685 | 23.57 | 0.9840 | 0.9705 | 25.36 | 0.8912 |
| | TopoAE | 0.4712 | 26.66 | 0.9768 | 0.9625 | 27.19 | 0.9043 |
| | ML-AE | 0.1220 | 16.86 | 0.9914 | 0.9885 | 10.34 | 0.9548 |
| | i-ML-Enc (L) | **0.0312** | **1.026** | **0.9921** | 0.9871 | 11.13 | 0.9386 |
| | INN | 0.0758 | 0.8075 | 0.9791 | 0.9681 | 8.595 | 0.9936 |
| | i-RevNet | 0.0508 | 0.7544 | 0.9316 | 0.9278 | 9.803 | – |
| | i-ResNet | 0.0544 | **0.7391** | 0.9258 | 0.9136 | 10.41 | – |
| | i-ML-Enc(L-1) | **0.0312** | 0.9263 | **0.9940** | **0.9937** | **7.539** | **1.000** |

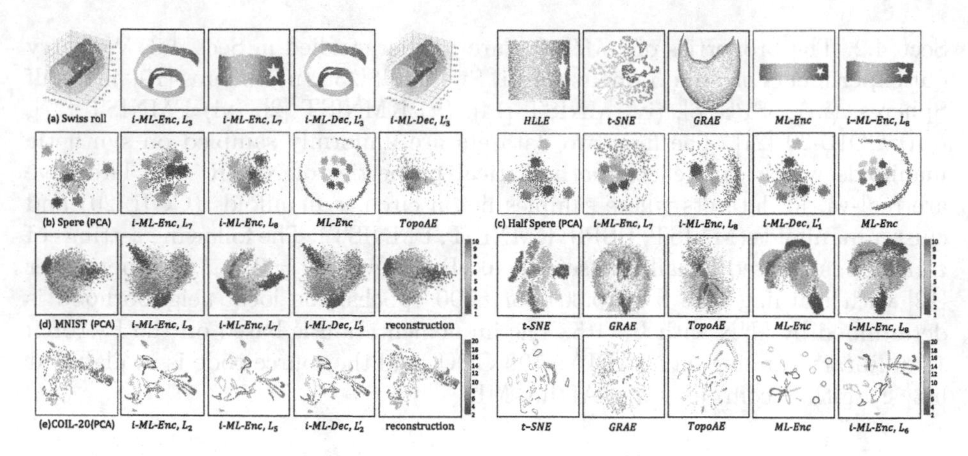


**Fig. 4.** Visualization of NLDR results of *i-ML-Enc* and relevant methods. All the high-dimensional results are visualized by PCA and the target dimension $s' = 2$. (a) shows NLDR and its inverse process of *i-ML-Enc* on the test set of Swiss roll in the case of $d = s = s'$. We show the cases of $s' < d \leq s$ and $s' = d \leq s$ by comparing (b)(c): (b) shows the failure case of reducing spheres $S^{100}$ sampled in $\mathbb{R}^{101}$ into 10-D, while (c) shows results of reducing half-spheres $S^{10}$ sampled in $\mathbb{R}^{101}$ into 10-D. In (b), TopoAE preserves the topological structure of those hyper-spheres. ML-Enc only maintains the geometric structure of circles but collapses into bad topological structures. In both cases, *i-ML-Enc* maintains the same topology as the input data in the first 7 layers, though it fails to achieve NLDR in (b). (d) and (e) show results of two sparse cases on MNIST and COIL-20: The left columns provide the invertible NLDR process of *i-ML-Enc* which are homeomorphic mappings. Because of the clustering effect, it is vital to focus on the local geometric structure while simply preserving the correct relationship between sub-manifolds. The results of ML-Enc and t-SNE show clear cluster structures and geometric structures of sub-manifolds. GRAE and TopoAE show more mixed results because of their over-reliance on topological structures. The results of *i-ML-Enc* provide similar local structures shapes as ML-Enc, but more connection between clusters.

### 4.1 Methods Comparison

To verify the invertible NLDR ability of *i-ML-Enc* and analyze different cases of NLDR, we compare it with eight typical methods in NLDR and inverse scenarios on both synthetic (Swiss roll, Spheres and Half Spheres) and real-world datasets (USPS, MNIST, FMNIST and COIL-20). **Eight methods for manifold learning**: Isomap [26], MLLE [28], t-SNE [16] and ML-Enc [15] are compared for NLDR; four AE-based methods VAE [13], GRAE [6], TopoAE [19] and ML-AE [15] are compared for reconstructible manifold learning. **Three methods for inverse models**: INN [22], i-RevNet [9], and i-ResNet [1] are compared for bijective inverse property. Among them, i-RevNet and i-ResNet are supervised algorithms while the rest are unsupervised. For a fair comparison in this experiment, we adopt 8 layers neural network for all the network-based methods except i-RevNet and i-ResNet. **Hyperparameter** values of *i-ML-Enc* and configurations of datasets are provided in Appendix A.2.

*Evalution Metrics.* We evaluate an invertible NLDR algorithm from three aspects: (i) Invertible property. Reconstruction MSE (**RMSE**) and maximum norm error (**MNE**) measure the difference between the input data and reconstruction results by norm-based errors. (ii) NLDR quality. Trustworthiness (**Trust**), Continuity (**Cont**) [11], and latent MSE (**l-MSE**) [15] are used to evaluate the quality of the low-dimensional representation. (iii) Generalization ability. Mean accuracy (**Acc**) of linear classification on the learned representation measures models' generalization ability to downstream tasks. Their exact definitions and purpose are given in Appendix A.1.

*Comparison and Conclusion.* Table 1 compares the *i-ML-Enc* with the relevant methods on MNIST and FMNIST datasets, more results and detailed analysis on other datasets are given in Appendix A.2. The process of invertible NLDR of *i-ML-Enc* and comparing results of typical methods are visualized in Fig. 4. We can conclude: (i) *i-ML-Enc* achieves invertible NLDR in the first stage with great NLDR and generalization qualities. The representation in the $L-1$-th layer of *i-ML-Enc* mostly outperforms all comparing methods for both invertible and NLDR metrics without losing information of the data, while other methods drop geometric and topological information to some extent. (ii) *i-ML-Enc* tries to keep more geometric and topological structure in the second stage in the case of $s^{'} < d \leq s$. The $L$-th layer of *i-ML-Enc* shows high consistency with its $L-1$-th layer and comparable NLDR performance in visualization results. (iii) *i-ML-Enc* provides more reliable and explainable representations of the data manifold because of its good mathematic properties.

### 4.2 Latent Space Interpolation

Since the first stage of *i-ML-Enc* is nearly homeomorphism, we perform linear interpolation experiments in both the input space and the $L-1$-th layer latent space to analyze the intrinsic continuous manifold and verify latent results by its inverse process. A good low-dimensional representation of the manifold should not only preserve the local properties, but also be flatter (with lower curvature) than the high-dimensional input space. Thus, we expect that local linear interpolation results in the latent space should be more reliable than in the input space.

*Interpolation Datasets.* The manifold learning difficulties of five datasets can be roughly analyzed in terms of **sampling ratio**, **image entropy**, **texture**, and performances on **classification tasks**: (i) Sampling ratio. The input dimension and sample number reflect the sampling ratio. In the case of sufficient sampling, the sample number nearly has an exponential relationship with the input dimension. Thus, the sampling ratio of USPS is higher than others. (ii) Image entropy. The Shannon entropy of the histogram measures the information content of images. It shows that USPS has richer grayscale than MNIST(256). The information content of MNIST(784), KMNIST, and FMNIST shows an increasing trend. (iii) Texture. The standard deviation (std) of the histogram reflects the texture information in images. (iv) Classification tasks. Performances of kNN classifier [23] on the input space reflect the credibility of the neighborhood system. The credibility decreases gradually from USPS, MNIST, KMNIST to FMNIST. In a nutshell, we can conclude that the complexity of data manifolds increases from USPS(256), MNIST(256), MNIST(784), KMNIST(784) to FMNIST(784).

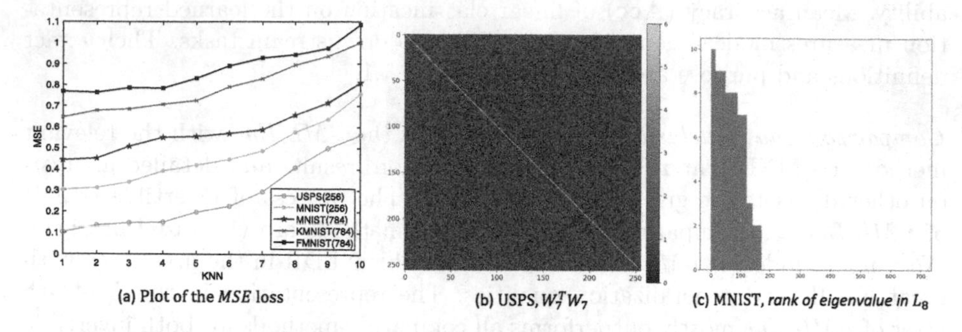


(a) Plot of the *MSE* loss (b) USPS, $W_7^T W_7$ (c) MNIST, *rank of eigenvalue in* $L_8$

**Fig. 5.** (a) shows the MSE loss of 1 to 10 nearest neighbors interpolation results on five datasets. It reflects the reliability of linear approximation in different low-dimensional representations. (b) shows the orthogonality of the weight matrix $W_7$ in *inv-ML-Enc* trained on USPS ($256 \times 256$) dataset. The elements are ranged from $10^0$ to $10^5$ after min-max normalization and rescaling, indicating that $W_7$ is nearly an orthogonal matrix. (c) shows the rank of eigenvalues (by SVD) of the 8-th layer output of *i-ML-Enc* on MNIST test set, which range from $10^0$ to $10^10$ by rescaling. The matrix rank of the output is 125, and the extra 46-D can be regarded as some machine errors when performs PCA.

*K-Nearest Neighbor Interpolation.* We verify the reliability of the low-dimensional representation in a small local system by kNN interpolation. Given a sample $\boldsymbol{x}_i$, randomly select $\boldsymbol{x}_j$ in $\boldsymbol{x}_i$'s k-nearest neighborhood in the latent space to form a sample pair $(\boldsymbol{x}_i, \boldsymbol{x}_j)$. Perform linear interpolation of the latent representation of the pair and get reconstruction results for evaluation as: $\hat{\boldsymbol{x}}_{i,j}^t = \psi^{-1}(t\psi(\boldsymbol{x}_i) + (1-t)\psi(\boldsymbol{x}_j))$, $t \in [0,1]$. The experiment is performed on *i-ML-Enc* with $L = 6$ and $K = 15$, training with 9298 samples for USPS and MNIST(256), 20000 sapmles for MNIST(784), KMNIST, FMNIST. We evaluate kNN interpolation from two aspects: (i) Calculate the MSE loss between reconstruction results of the latent interpolation $\hat{\boldsymbol{x}}_{i,j}^t$ and the corresponding input interpolation results $\boldsymbol{x}_{i,j}^t = t\boldsymbol{x}_i + (1-t)\boldsymbol{x}_j$. A larger MSE loss indicates the worse fitting to the data manifold. Notice that this MSE loss is only a rough measurement of kNN interpolation when $k$ is small. Figure 5 shows evaluation results with $k = 1, 2, ..., 10$. (ii) Visualize typical results of the input space and the latent space for comparison, as shown in Fig. 6. More results and analysis are given in Appendix A.3. We further employ *geodesic interpolation* between two distant samples pairs in the latent space to analyze topological structures. Given a sample pair $(x_i, x_j)$ from different clusters, we select the three intermediate sample pairs $(x_i, x_{i_1})$, $(x_{i_1}, x_{i_2})$, $(x_{i_2}, x_j)$ with $k \leq 20$ along the geodesic path in latent space. Visualization results are given in Appendix A.3. The latent results show no overlap of multiple submanifolds in the geodesic path.

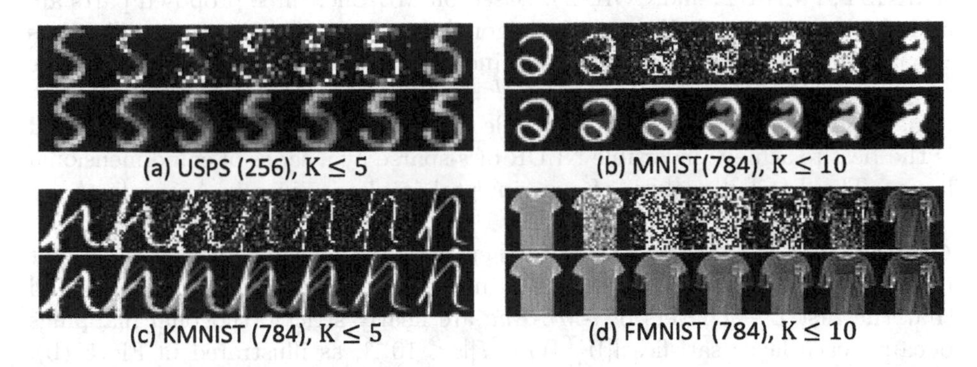


**Fig. 6.** Results of kNN interpolation. For each dataset, the upper and lower rows show latent space and input space results respectively. From a overall aspect, the latent results show more noise because *inv-ML-Enc* is not a AE-based or generative model which optimize reconstruction results explicitly. But the latent results are more reliable than the input. For example, left latent interpolation results are similar to the left sample which show less overlapping and pseudo-contour than the input results.

*Comparison and Conclusion.* Compared with results of the kNN and geodesic interpolation, we can conclude: (i) Because of the sparsity of the latent space, noises are inevitable on the latent results. Empirically, the reliability of the latent interpolation decreases with the expansion of the local neighborhood on the same dataset. (ii) The latent results of kNN interpolation get worse in the following cases: for similar manifolds, when the sampling rate is lower (indicated by USPS(256), MNIST(256) and MNIST(784)); with the same sampling rate, the manifold becomes more complex (indicated by MNIST(784), KMNIST to FMNIST). They indicate that the confidence of the tangent space estimated by local neighborhood decreases on more complex manifolds with sparse sampling. (iii) The interpolation between two samples in latent space is smoother than that in the input space, validating the flatness and density of the lower-dimensional representation learned by *i-ML-Enc*. Overall, we infer that the unreliable approximation of the local tangent space by the local neighborhood is the basic reason for the manifold learning fails in the real-world case, because the geometry should be preserved in the first place. To come up with this common situation, it is necessary to import other prior assumptions or knowledge when the sampling rate of data manifolds is relatively low, e.g., the Euclidean space assumption, semantic information of down-stream tasks.

### 4.3 Analysis

*Analysis on Loss Terms.* We perform an ablation study to evaluate the effects of the proposed network structure and loss terms in *i-ML-Enc* on MNIST, USPS, KMNIST, FMNIST, and COIL-20. Based on ML-Enc, three proposed parts are added: the *extra head* (**Ex**), the orthogonal loss $\mathcal{L}_{orth}$ (**Orth**), the padding loss $\mathcal{L}_{pad}$ (**Pad**). Besides the previous six indicators, we introduce the rank of the output matrix of the layer $L-1$ as $r(Z^{L-1})$, to measure the sparsity of the high-dimensional representation. We conclude that the combination **Ex+Orth+Pad** is the best to achieve invertible NLDR of $s$-sparse by a series of equidimensional layers. The detailed analysis of experimental results is given in Appendix A.4.

*Orthogonality and Sparsity.* We further discuss the orthogonality of weight matrices and $s$-sparse representations in the first stage of *i-ML-Enc.* We find that the first $L-1$ layers of *i-ML-Enc* are nearly strict orthogonal mappings because each layer satisfies $||W_l^T W_l - I|| < 10^{-5}$, as illustrated in Fig. 5 (b). Meanwhile, the $L-1$-th layer output of *i-ML-Enc* achieves sparsity. Taking the 8-th layer output of *i-ML-Enc* on MNIST test set as an example, as shown in Fig. 5 (c). We can construct a 125-D linear subspace with 125 orthogonal base vectors decomposed from the output matrix and reconstruct to the original space (784-D) without losing information by PCA [23] and its inverse transform. It indicates a low-dimensional constrain is learned by *inv-ML-Enc.* Thus, we conclude that an invertible NLDR of data manifolds can be learned by *i-ML-Enc* in the *sparse coordinate transformation.*

*Relationship Between s-Sparse and Intrinsic Dimension d.* We notice that the $s$-sparse achieved by the first stage of *i-ML-Enc* is higher than the approximate intrinsic dimension $d$ on each dataset, e.g. 116-sparse on USPS and 125-sparse on MNIST. We found the following reasons: (i) Because the data manifolds are usually quite complex but sampling sparsely, the lowest isometric embedding dimension is between $d$ to $2d$ according to Nash Embedding Theorem and the hyper-plane hypothesis. The $s$ obtained by *i-ML-Enc* on each dataset is nearly in the interval of $[d, 2d]$, which is not the true intrinsic dimension of the manifolds. (ii) The proposed *i-ML-Enc* is not optimized enough, which serves as a simple network implementation of inv-ML. We need to design a better implementation model if we want to approach the lower embedding dimension to preserve both geometry and topology (Table 2).

**Table 2.** Ablation study of proposed loss terms in *i-ML-Enc* on MNIST.

| | | RMSE | MNE | Trust | Cont | Acc | $r(Z^{L-1})$ |
|---|---|---|---|---|---|---|---|
| MNIST | ML-AE | 0.4012 | 16.84 | 0.9893 | 0.9926 | **0.9340** | 15 |
| | ML-Enc | – | – | 0.9862 | **0.9927** | 0.9326 | 14 |
| | +Ex | – | – | 0.9891 | 0.9812 | 0.9316 | **12** |
| | +Orth | **0.0056** | **0.1275** | 0.9652 | 0.9578 | 0.8807 | 716 |
| | +Ex+Orth | 0.0341 | 0.4255 | 0.9874 | **0.9927** | 0.9298 | 361 |
| | +Ex+Orth+Pad | 0.0457 | 0.5085 | **0.9906** | 0.9912 | 0.9316 | 125 |

## 5 Conclusion

To fill the gap between theoretical and real-world applications of manifold-based DR, we introduce a novel invertible NLDR process *inv-ML* and a neural network implementation *inv-ML-Enc* to verify the proposed process. Firstly, the *sparse coordinate transformation* is learned to find a flatter and denser low-dimensional representation with preservation of geometry and topology of data manifolds. Secondly, we discuss the condition of NLDR and information loss with different target dimensions in *linear compression.* Experiment results of *i-ML-Enc* on seven datasets validate the proposed invertible NLDR process and the sparsity of learned low-dimensional representations. Further, the interpolation experiments reveal that finding a reliable tangent space by the local neighborhood on real-world datasets is the inherent defect of manifold-based DR methods.

**Acknowledgments.** This work was done during the internship of Siyuan Li and Haitao Lin at Westlake University. We thank Di Wu for helpful insights on hyper-parameters tuning.

## References

1. Behrmann, J., Grathwohl, W., Chen, R.T.Q., Duvenaud, D., Jacobsen, J.: Invertible residual networks. In: International Conference on Machine Learning (ICML) (2019)
2. Clanuwat, T., Bober-Irizar, M., Kitamoto, A., Lamb, A., Yamamoto, K., Ha, D.: Deep learning for classical japanese literature. arXiv preprint arXiv:1812.01718 (2018)
3. Dinh, L., Krueger, D., Bengio, Y.: NICE: non-linear independent components estimation. In: International Conference on Learning Representations (ICLR) (2015)
4. Dinh, L., Sohl-Dickstein, J., Bengio, S.: Density estimation using real NVP. In: International Conference on Learning Representations (ICLR) (2017)
5. Donoho, D.L.: Compressed sensing. IEEE Trans. Inf. Theory **52**, 1289–1306 (2006)
6. Duque, A.F., Morin, S., Wolf, G., Moon, K.R.: Extendable and invertible manifold learning with geometry regularized autoencoders. arXiv preprint arXiv:2007.07142 (2020)
7. Hinton, G.E., Salakhutdinov, R.R.: Reducing the dimensionality of data with neural networks. Science **313**(5786), 504–507 (2006)

8. Hull, J.: Database for handwritten text recognition research. IEEE Trans. Pattern Anal. Mach. Intell. **16**, 550–554 (1994)
9. Jacobsen, J., Smeulders, A.W.M., Oyallon, E.: i-revnet: deep invertible networks. In: International Conference on Learning Representations (ICLR) (2018)
10. Johnson, W.B., Lindenstrauss, J.: Extensions of lipschitz maps into a hilbert space. Contemp. Math. **26**, 189–206 (1984)
11. Kaski, S., Venna, J.: Visualizing gene interaction graphs with local multidimensional scaling. In: European Symposium on Artificial Neural Networks, pp. 557–562 (2006)
12. Kingma, D.P., Ba, J.: Adam: A method for stochastic optimization. In: International Conference on Learning Representations (ICLR) (2015)
13. Kingma, D.P., Welling, M.: Auto-encoding variational bayes. In: International Conference on Learning Representations (ICLR) (2014)
14. LeCun, Y., Bottou, L., Haffner, P.: Gradient-based learning applied to document recognition. Proc. IEEE **86**(11), 2278–2324 (1998)
15. Li, S.Z., Zhang, Z., Wu, L.: Markov-lipschitz deep learning. arXiv preprint arXiv:2006.08256 (2020)
16. Maaten, L.V.D., Hinton, G.: Visualizing data using t-sne. J. Mach. Learn. Res. **9**, 2579–2605 (2008)
17. McQueen, J., Meila, M., Joncas, D.: Nearly isometric embedding by relaxation. In: Proceedings of the 29th Neural Information Processing Systems (NIPS), pp. 2631–2639 (2016)
18. Mei, J.: Introduction to Manifold and Geometry. Beijing Science Press, Beijing (2013)
19. Moor, M., Horn, M., Rieck, B., Borgwardt, K.: Topological autoencoders. In: International Conference on Machine Learning (ICML) (2020)
20. Nash, J.: The imbedding problem for riemannian manifolds. Ann. Math. **63**, 20–63 (1956)
21. Nene, S.A., Nayar, S.K., Murase, H.: Columbia object image library (coil-20). Technical Report, Columbia University (1996). https://www.cs.columbia.edu/CAVE/software/softlib/coil-20.php
22. Nguyen, T.-G.L., Ardizzone, L., Köthe, U.: Training Invertible Neural Networks as Autoencoders. In: Fink, G.A., Frintrop, S., Jiang, X. (eds.) DAGM GCPR 2019. LNCS, vol. 11824, pp. 442–455. Springer, Cham (2019). https://doi.org/10.1007/978-3-030-33676-9_31
23. Pedregosa, F., et al.: Édouard Duchesnay: Scikit-learn: machine learning in python. J. Mach. Learn. Res. **12**(85), 2825–2830 (2011)
24. Roweis, S.T., Saul, L.K.: Nonlinear dimensionality reduction by locally linear embedding. Science **290**, 2323–2326 (2000)
25. Seshadri, H., Verma, K.: The embedding theorems of Whitney and Nash. Resonance **21**(9), 815–826 (2016). https://doi.org/10.1007/s12045-016-0387-4
26. Tenenbaum, J.B., De Silva, V., Langford, J.C.: A global geometric framework for nonlinear dimensionality reduction. Science **290**(5500), 2319–2323 (2000)
27. Xiao, H., Rasul, K., Vollgraf, R.: Fashion-mnist: a novel image dataset for benchmarking machine learning algorithms. arXiv preprint arXiv:1708.07747 (2017)
28. Zhang, Z., Wang, J.: Mlle: modified locally linear embedding using multiple weights. In: Advances in Neural Information Processing systems, pp. 1593–1600 (2007)
29. Zhang, Z., Zha, H.: Principal manifolds and nonlinear dimensionality reduction via tangent space alignment. SIAM J. Sci. Comput. **26**(1), 313–338 (2004)

# Small-Vote Sample Selection for Label-Noise Learning

Youze Xu[1], Yan Yan[1(✉)], Jing-Hao Xue[2], Yang Lu[1], and Hanzi Wang[1]

[1] Fujian Key Laboratory of Sensing and Computing for Smart City, School of Informatics, Xiamen University, Xiamen, China
xuyouze@stu.xmu.edu.cn, {yanyan,luyang,hanzi.wang}@xmu.edu.cn
[2] Department of Statistical Science, University College London, London, UK
jinghao.xue@ucl.ac.uk

**Abstract.** The small-loss criterion is widely used in recent label-noise learning methods. However, such a criterion only considers the loss of each training sample in a mini-batch but *ignores* the loss distribution in the whole training set. Moreover, the selection of clean samples depends on a *heuristic* clean data rate. As a result, some noisy-labeled samples are easily identified as clean ones, and vice versa. In this paper, we propose a novel yet simple sample selection method, which mainly consists of a Hierarchical Voting Scheme (HVS) and an Adaptive Clean data rate Estimation Strategy (ACES), to accurately identify clean samples and noisy-labeled samples for robust learning. Specifically, we propose HVS to effectively combine the global vote and the local vote, so that both *epoch-level* and *batch-level* information is exploited to assign a hierarchical vote for each mini-batch sample. Based on HVS, we further develop ACES to *adaptively* estimate the clean data rate by leveraging a 1D Gaussian Mixture Model (GMM). Experimental results show that our proposed method consistently outperforms several state-of-the-art label-noise learning methods on both synthetic and real-world noisy benchmark datasets.

**Keywords:** Noisy labels · Label-noise learning · Sample selection

## 1 Introduction

In recent years, Deep Neural Networks (DNNs) based methods have achieved remarkable success in a variety of artificial intelligence-related tasks. Generally, these methods rely heavily on a large number of high-quality annotated training samples. Unfortunately, collecting large-scale samples with fully accurate annotations is labor-intensive and time-consuming, which unavoidably yields noisy labels [2,11,27]. Many DNNs-based methods easily overfit noisy-labeled samples, mainly due to the high learning capability of DNNs involving millions of parameters. A recent study [26] has shown that DNNs severely suffer from poor generalization capability when they are trained on the samples containing noisy labels.


N. Oliver et al. (Eds.): ECML PKDD 2021, LNAI 12977, pp. 729–744, 2021.
https://doi.org/10.1007/978-3-030-86523-8_44

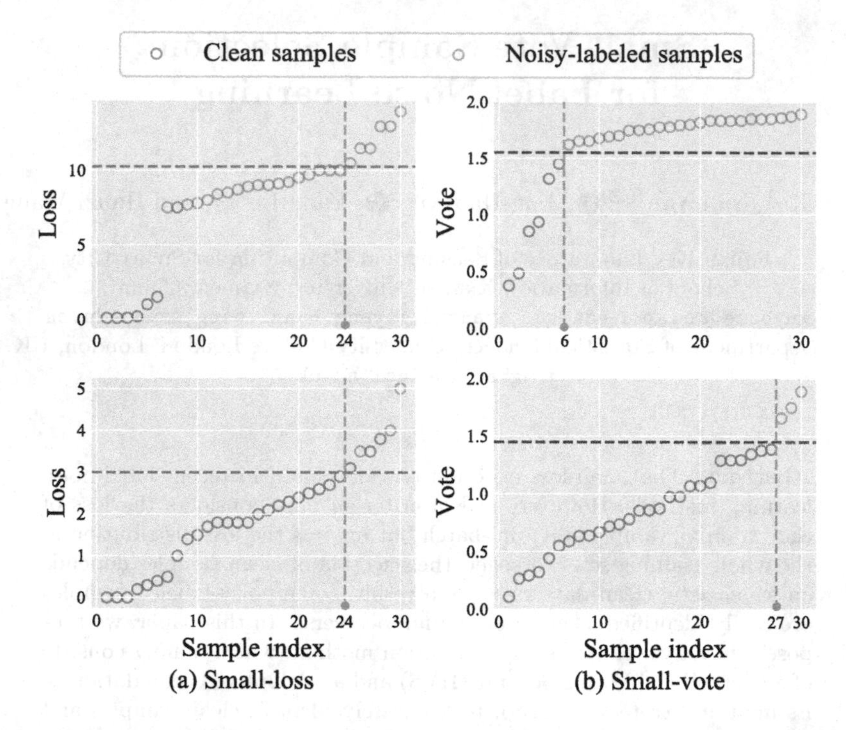


**Fig. 1.** Comparison between (a) the small-loss criterion and (b) our proposed small-vote sample selection method in the two randomly chosen mini-batches (the batch size is 30) during training. The purple dotted lines represent the thresholds obtained by the heuristic clean data rate (80% in two cases) and the green dotted ones represent the adaptive thresholds obtained by our method. The samples whose indices are smaller than or equal to the orange dotted lines are selected as clean samples. (Color figure online)

To alleviate the adverse effects of noisy labels, label-noise learning methods have been proposed to learn robust representations. Some methods [4,10,14,19,24] address the problem of noisy labels by selecting or weighting clean samples for each mini-batch during training. These methods usually take advantage of the small-loss criterion, which first identifies small-loss training samples as clean ones and then uses them for updating the network parameters [7,20]. Such a criterion is well justified by the memorization effect that DNNs are able to learn simple and general patterns from clean samples before fitting noisy-labeled samples [1]. However, the small-loss criterion only considers the loss of training samples in each single mini-batch but ignores the loss distribution in the whole training set. Moreover, a heuristic clean data rate is used to select clean samples, whereas the noisy label distribution varies in randomly chosen mini-batches. Therefore, the small-loss criterion may not accurately identify clean samples and noisy-labeled samples, as illustrated in Fig. 1(a). This raises the difficulty of learning robust models. Hence, how to accurately distinguish clean samples from noisy-labeled samples remains a great challenge.

To address the above challenge, in this paper, we propose a novel small-vote sample selection method to accurately select clean samples and noisy-labeled samples, and robustly train DNN simultaneously. Specifically, our proposed method mainly consists of a Hierarchical Voting Scheme (HVS) and an Adaptive Clean data rate Estimation Strategy (ACES). First, we develop HVS to assign a hierarchical vote for each mini-batch sample. Then, based on HVS, we introduce ACES to estimate the clean data rate by leveraging a 1D Gaussian Mixture Model (GMM). Some intermediate training results are given in Fig. 1(b). Obviously, our proposed method adaptively estimates clean data rates reflecting well the different proportions of clean samples in the two randomly chosen mini-batches.

The contributions of this paper are summarized as follows:

- We propose a novel yet simple small-vote sample selection method, which performs noisy label detection and learns from noisy data in an end-to-end manner. In particular, we develop HVS to effectively combine the global vote from previous epochs and the local vote from the current mini-batch. In this way, both epoch-level and batch-level information can be fully exploited for voting. Based on HVS, we design ACES to adaptively and accurately identify clean samples, guiding the learning of a robust model.
- We conduct extensive experiments on four benchmark datasets (including MNIST, CIFAR-10, CIFAR-100, and Clothing 1M) with synthetic and real-world noisy labels. Without bells and whistles, our proposed method achieves excellent performance in terms of both test accuracy and label F1-score in comparison with state-of-the-art label-noise learning methods. Moreover, we show the good generalization capability of our method by introducing the proposed sample selection method into several representative label-noise learning methods.

## 2 Related Work

Roughly speaking, existing label-noise learning methods can be classified into three categories, including label transition matrix estimation, robust regularization, and sample selection.

**Label Transition Matrix Estimation.** This category of methods is based on the estimation of the label transition matrix, which characterizes the label transition probabilities from a true class to an assigned one [22]. For example, Goldberger *et al.* [5] add an additional softmax layer in the neural network to model the label transition matrix. Patrini *et al.* [15] develop a two-step solution to heuristically estimate the label transition matrix. Yao *et al.* [23] introduce an intermediate class to decompose the original label transition matrix into the product of two easy-to-estimate transition matrices. Note that this category of methods cannot deal with a large number of labels and is fragile to a large ratio of noisy-labeled samples.

**Robust Regularization.** This type of methods leverages robust regularization techniques to avoid overfitting on noisy labels and thus improve the generalization ability of DNNs. Pereyra *et al.* [16] estimate the marginalized effect of noisy labels during training and prevent DNN from assigning a full probability to the noisy-labeled sample, thereby reducing overfitting. Zhang *et al.* [28] regularize the DNN to favor simple linear behaviors in-between training samples to address the overfitting problem. Although these methods have achieved promising performance, they usually depend on additional hyperparameters that are sensitive to the data type [13,17]. Moreover, some methods may completely memorize noisy-labeled samples with high capacity networks, since DNNs are often over-parameterized [6].

**Sample Selection.** Recently, sample selection methods, which aim to select clean samples from noisy training data, have attracted considerable attention in label-noise learning. The small-loss criterion that identifies samples with small training losses as clean samples has been widely used in recent methods. For example, MentorNet [10] develops a collaborative learning paradigm, where a mentor network is first pre-trained and then used to select clean samples based on the small-loss criterion for guiding the training of a student network. Co-teaching [7] also involves two networks, where small-loss samples are selected by each network and fed into the peer network to update the network parameters in each mini-batch. Different from Co-teaching, Co-teaching+ [24] only selects small-loss samples with different prediction results from two networks. JoCoR [20] calculates a joint loss with co-regularization for each sample based on two networks, and then chooses the small-loss samples to simultaneously update the parameters of two networks. Yao *et al.* [22] use an AutoML method to dynamically determine the noise rate during the training process.

Different from conventional small-loss criterion based sample selection methods that only take into account the loss of each training sample in the current mini-batch, we also exploit the loss distribution in the whole training set at previous training epochs. This enables our method to more accurately identify clean samples or noisy-labeled samples in each mini-batch during training.

## 3 Proposed Method

In this section, we develop a novel and effective sample selection method for label-noise learning. After introducing preliminaries, we present the key components of our proposed method in detail.

### 3.1 Preliminaries

Considering a $K$-class classification problem with the noisy training data $\mathcal{D} = \{\mathbf{x}_i, y_i\}_{i=1}^{N}$, where $\mathbf{x}_i$ denotes the $i$-th sample (e.g., an image) in the training set and $y_i \in \{1, 2, \ldots, K\}$ represents the label corresponding to the sample $\mathbf{x}_i$. A sample is noisy-labeled when the corresponding label mismatches its ground-truth label. A large number of methods [7,10,20,24] identify the samples that are likely to be clean ones by using the small-loss criterion, and thus a robust model can be trained with small-loss samples in each mini-batch. The details of the small-loss criterion are described as follows.

A mini-batch data $\mathcal{D}_b^t = \{\mathbf{x}_j^b, y_j^b\}_{j=1}^J$ at epoch $t$ is randomly drawn from the noisy training data $\mathcal{D}$, where $\mathbf{x}_j^b$ and $y_j^b$ represent the $j$-th training sample in mini-batch $b$ and the corresponding label, respectively. $J$ denotes the batch size. The loss of each sample can be obtained by feeding $\mathcal{D}_b^t$ into the model and then used to identify clean samples. The selected clean data $\tilde{\mathcal{D}}_b^t$ can be formulated as

$$\tilde{\mathcal{D}}_b^t = \mathop{\arg\min}_{\mathcal{D}':|\mathcal{D}'|\geq\lambda^t\cdot|\mathcal{D}_b^t|} L(f_{\mathbf{\Theta}}, \mathcal{D}'), \quad (1)$$

where $f_{\mathbf{\Theta}}$ denotes the network with the parameters $\mathbf{\Theta}$, $L$ represents the cross-entropy loss, and $\lambda^t$ denotes the clean data rate which controls how many small-loss samples should be selected into $\tilde{\mathcal{D}}_b^t$. $\lambda^t$ is heuristically defined as

$$\lambda^t = 1 - \tau \cdot \min(\frac{t}{T}, 1), \quad (2)$$

where $\tau$ is the estimated noise rate, which can be inferred using validation sets [12,25]. The value of $\lambda^t$ decreases quickly at the first $T$ epochs until reaching $1-\tau$. Then, the selected clean data $\tilde{D}_b^t$ are used to calculate the average loss for updating the network parameters $\mathbf{\Theta}$.

Despite its popularity, the small-loss criterion suffers from the following two limitations. First, this criterion only considers the losses of mini-batch samples but ignores the loss distribution of all the samples in the whole training set. Such a way is not globally optimal. For example, as shown in the top row of Fig. 1(a), when a mini-batch mainly contains noisy-labeled samples, the losses of this mini-batch may not be effectively used to indicate whether the label is noisy or clean. Second, the clean data rate $\lambda^t$ is critical to exploit the memorization effect [1]. But according to Eq. (2), the value of $\lambda^t$ is usually set without fully exploiting the knowledge on the data. In other words, $\lambda^t$ is heuristic and it is often difficult to manually determine $\lambda^t$ for each dataset. Therefore, some noisy-labeled samples may be improperly selected as clean ones, and vice versa (note that the mini-batch samples are randomly chosen from the whole training set). This clearly leads to a performance decrease.

The above limitations motivate us to formulate and design an effective small-vote sample selection method, which not only takes both the whole training set and the current mini-batch into account, but also adaptively selects clean samples. The proposed method mainly consists of a novel Hierarchical Voting Scheme (HVS) and an Adaptive Clean data rate Estimation Strategy (ACES).

### 3.2 Hierarchical Voting Scheme (HVS)

HVS combines the global vote (based on the loss distributions of all the samples at previous epochs) and the local vote (based on the losses of current mini-batch samples) to assign a hierarchical vote for each mini-batch sample. In general, we compute and sort the loss of each sample in the whole training set after each epoch, and combine the normalized rank indices of each mini-batch sample at previous epochs as the global vote. Similarly, we view the normalized rank index of each sample in the current mini-batch as the local vote. Then, a hierarchical vote is performed by combining the global vote and the local vote from the epoch-level and batch-level, respectively.

To be specific, at epoch $t$, we compute the losses of all the samples in the whole training data. Suppose that the cross-entropy losses for the noisy training data are denoted as $\mathcal{L}^t = \{l_1^t, \cdots, l_N^t\}$, where $l_i^t$ represents the loss of the $i$-th sample at epoch $t$. Then, we sort all the elements in $\mathcal{L}^t$ in the ascending order to obtain the sorted set $\mathcal{L}^t = \{l_{\mu_1}^t, \cdots, l_{\mu_N}^t\}$, where the permutation $\{\mu_1, \cdots, \mu_N\}$ is obtained such that $l_{\mu_1}^t \leq \cdots, \leq l_{\mu_N}^t$. Hence, the normalized rank index set $\mathcal{P}^t$ at epoch $t$ can be formulated as

$$\mathcal{P}^t = \{p_{\mathbf{x}_{\mu_j}}^t | p_{\mathbf{x}_{\mu_j}}^t = j/N, \forall\ \mathbf{x}_{\mu_j} \in \mathcal{D}\}, \quad (3)$$

where $p_{\mathbf{x}_{\mu_j}}^t \in [0, 1]$ represents the normalized rank index of the $\mu_j$-th sample.

For the global vote, we vote each mini-batch sample based on the normalized rank index set obtained at previous $C$ training epochs, which can be formulated as

$$\mathcal{G}_b^t = \{g_{\mathbf{x}_j^b}^t | g_{\mathbf{x}_j^b}^t = \frac{1}{C} \sum_{c=1}^{C} p_{\mathbf{x}_j^b}^{(t-c)}, \forall\ \mathbf{x}_j^b \in \mathcal{D}_b^t\}, \quad (4)$$

where $g_{\mathbf{x}_j^b}^t \in [0, 1]$ indicates the global vote of $\mathbf{x}_j^b$ for mini-batch $b$ at epoch $t$. $C$ is a hyper-parameter used to control how many previous epochs are used to perform the global vote, and $p_{\mathbf{x}_j^b}^{(t-c)}$ represents the normalized rank index of $\mathbf{x}_j^b$ at epoch $(t - c)$.

Similarly, the losses for mini-batch $b$ at epoch $t$ are denoted as $\mathcal{L}_b^t = \{l_{1b}^t, \cdots, l_{Jb}^t\}$, where $l_{jb}^t$ represents the loss of the $j$-th sample in the mini-batch. All the elements in $\mathcal{L}_b^t$ are sorted in the ascending order to obtain the sorted set $\mathcal{L}_b^t = \{l_{\nu_1 b}^t, \cdots, l_{\nu_J b}^t\}$, where the permutation $\{\nu_1, \cdots, \nu_J\}$ is obtained such that $l_{\nu_1 b}^t \leq \cdots, \leq l_{\nu_J b}^t$. The normalized rank index set $\hat{\mathcal{P}}_b^t$ in the mini-batch is

$$\hat{\mathcal{P}}_b^t = \{\hat{p}_{\mathbf{x}_{\nu_j}^b}^t | \hat{p}_{\mathbf{x}_{\nu_j}^b}^t = j/J, \forall\ \mathbf{x}_{\nu_j}^b \in \mathcal{D}_b^t\}, \quad (5)$$

where $\hat{p}_{\mathbf{x}_{\nu_j}^b}^t \in [0, 1]$ indicates the local vote of $\mathbf{x}_{\nu_j}^b$.

Finally, since both the global vote and the local vote have the same value range, we define the hierarchical votes of mini-batch samples at epoch $t$ by simply combining them, i.e.,

$$\mathcal{V}_b^t = \{v_{\mathbf{x}_j^b}^t | v_{\mathbf{x}_j^b}^t = g_{\mathbf{x}_j^b}^t + \hat{p}_{\mathbf{x}_j^b}^t, \forall\ \mathbf{x}_j^b \in \mathcal{D}_b^t\}, \quad (6)$$

where $v_{\mathbf{x}_j^b}^t \in [0, 2]$ represents the hierarchical vote of $\mathbf{x}_j^b$ for mini-batch $b$ at epoch $t$. It is worth noting that instead of directly relying on the loss of the sample, we

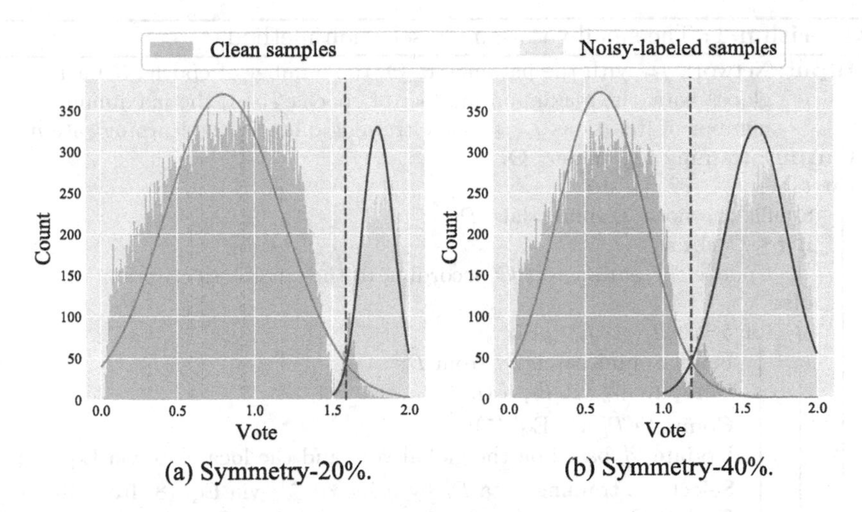


(a) Symmetry-20%. (b) Symmetry-40%.

**Fig. 2.** Histograms of hierarchical votes at an epoch on the training sets involving (a) symmetry-20% and (b) symmetry-40% noisy labels from the CIFAR-10 dataset. The ground-truth noisy-labeled samples and clean samples are marked with different colors. The red dotted lines denote the thresholds estimated by ACES. (Color figure online)

take advantage of the normalized rank index for both the global vote and the local vote. Such a manner is able to address the problem of different scales of loss distributions at epochs and mini-batches.

### 3.3 Adaptive Clean Data Rate Estimation Strategy (ACES)

Intuitively, a noisy-labeled sample tends to have a higher hierarchical vote than a clean sample (note that a sample will be assigned with a high hierarchical vote if the corresponding local and global votes show large normalized rank indices). In Fig. 2, we visualize the histograms of hierarchical votes at an epoch on the training sets involving two different levels of noisy labels from the CIFAR-10 dataset. We can find that the histogram of samples at an epoch shows two distinct modes which correspond to noisy-labeled samples and clean samples, respectively.

The $\lambda^t$ defined in Eq. (2) depends on a fixed noise rate $\tau$, which is heuristic. As a result, some noisy-labeled samples are incorrectly identified as clean ones, and vice versa. Such a manner leads to a performance decrease. Therefore, we develop ACES to adaptively estimate the clean data rate by taking advantage of a 1D Gaussian Mixture Model (GMM) to model the histogram with two modes.

More specifically, given the noisy training data $\mathcal{D}$ and the corresponding hierarchical votes $\mathcal{V}^{(t-1)}$ at epoch $(t-1)$, we fit these votes using a 1D GMM with two components:

$$F(\mathcal{V}^{(t-1)}) = \sum_{i=1,2} \pi_i^t \mathcal{N}\{\mathcal{V}^{(t-1)} | m_i^t, \gamma_i^t\}, \tag{7}$$

**Algorithm 1:** The small-vote sample selection method

**Input**: Network $f_{\Theta}$ with the parameters $\Theta$, the number of epochs $C$ for the global vote, the maximum number of epochs $T_{max}$, the maximum number of iterations $I_{max}$, noisy training data $\mathcal{D}$, the learning rate $\eta$;
**Output**: training parameters $\Theta$;

```
1  for t = 1, 2, ..., T_max do
2      Shuffle the noisy training data D;
3      if t ≤ C then
4          Update the parameters Θ according to the small-loss criterion;
5      else
6          for b = 1, 2, ..., I_max do
7              Fetch a mini-batch D_b^t from D;
8              Compute G_b^t via Eq. (4);
9              Compute P̂_b^t via Eq. (5);
10             Update V_b^t based on the global vote and the local vote via Eq. (6);
11             Select the training data D̂_b^t by using α^(t−1) via Eq. (8) from D_b^t;
12             Update the parameters Θ via gradient descent based on the selected
               training data;
13         end
14     end
15     Compute and store P^t via Eq. (3);
16     Compute and store the threshold α^t;
17     Update the learning rate η;
18 end
19 return Θ
```

where $\mathcal{N}$ denotes a Gaussian distribution, $m_i^t$ and $\gamma_i^t$ are the mean and standard deviation of the $i$-th component, respectively. $\pi_i^t$ represents the weight of the $i$-th component. The parameters of 1D GMM can be estimated by using the EM algorithm [3]. Then, the threshold $\alpha^{(t-1)}$ is determined by finding the intersection point of two Gaussians.

After obtaining the threshold $\alpha^{(t-1)}$, we can select the mini-batch samples whose votes are lower than the threshold as clean samples at epoch $t$. Mathematically, the selected training data $\hat{D}_b^t$ can be obtained as

$$\hat{D}_b^t = \{\mathbf{x}_j^b | v_{\mathbf{x}_j^b}^t \leq \alpha^{(t-1)}, \ \forall \ \mathbf{x}_j^b \in D_b^t\}, \tag{8}$$

where $v_{\mathbf{x}_j^b}^t$ is the hierarchical vote of $\mathbf{x}_j^b$ for mini-batch $b$ at epoch $t$. Hence, the clean data rate is estimated as $|\hat{D}_b^t|/J$.

The overall procedure of our proposed small-vote sample selection method is shown in Algorithm 1. It is worth noting that our proposed method can be viewed as an extension of the small-loss criterion. When only the local vote (defined in Eq. (5)) and the heuristic clean data rate (defined in Eq. (2)) are adopted, our proposed method degenerates to the small-loss criterion.

**Table 1.** The details of four benchmark datasets.

| Datasets | # Of train | # Of test | # Of class | Size |
|---|---|---|---|---|
| MNIST | $60K$ | $10K$ | 10 | $28 \times 28$ |
| CIFAR-10 | $50K$ | $10K$ | 10 | $32 \times 32$ |
| CIFAR-100 | $50K$ | $10K$ | 100 | $32 \times 32$ |
| Clothing1M | $1M$ | $10K$ | 14 | $256 \times 256$ |

## 4 Experiments

We conduct experiments on several commonly used benchmark datasets and compare the proposed method with several state-of-the-art label-noise learning methods.

**Datasets.** We show the effectiveness of our proposed method on four benchmark datasets, MNIST, CIFAR-10, CIFAR-100, and Clothing 1M [21]. These datasets are widely used for evaluating the performance of label-noise learning methods [7,20]. The details of the four benchmark datasets are shown in Table 1.

For MNIST, CIFAR-10, and CIFAR-100, we follow the common settings to add synthetic noise into the training sets, as done in the literature [7,9]. Specifically, the noisy labels are modified in the following two ways: (1) Symmetry flipping: a sample is assigned to a uniform random label rather than its true label with the probability $p_s$, where $p_s = 20\%$ or $40\%$ in our experiments, as done in [18]; (2) Pair flipping: a sample in one class is assigned to have the same label of another class [7]. The probability $p_k$ of sample mislabelling in a class is simply set to 40% in our experiments due to space limits. Similar results can be observed for other values of $p_s$ or $p_k$.

For Clothing 1M, we follow the same settings as [20]. We use $1M$ images with noisy labels as the training set and $10K$ clean images as the test set. For each image in the Clothing 1M dataset, we resize it to $256 \times 256$ and crop the middle $224 \times 224$ as the input of the model.

**Competing Methods.** We compare our proposed small-vote sample selection method (called as small-vote) with the following state-of-the-art label-noise learning methods, including Co-teaching [7], Co-teaching+ [24], O2U-Net [9], and JoCoR [20]. The baseline method that trains the standard Convolutional Neural Network (CNN) with the small-loss criterion is also used.

**Network Structure and Optimizer.** For a fair comparison, we re-implement all the methods based on the open-source codes by PyTorch and conduct all the experiments on a NIVIDIA 2080Ti GPU. For MNIST, CIFAR-10, and CIFAR-100, we adopt a 9-layer CNN [7]. For Clothing 1M, we use ResNet-18 [8]. All experiments are trained for 200 epochs with the Adam optimizer ($momentum =$ 0.9). The batch size is set to 128 for each dataset. The number of epochs $C$ for the global vote is set to 10. Similarly to O2U-Net [9], we use a linear decrease

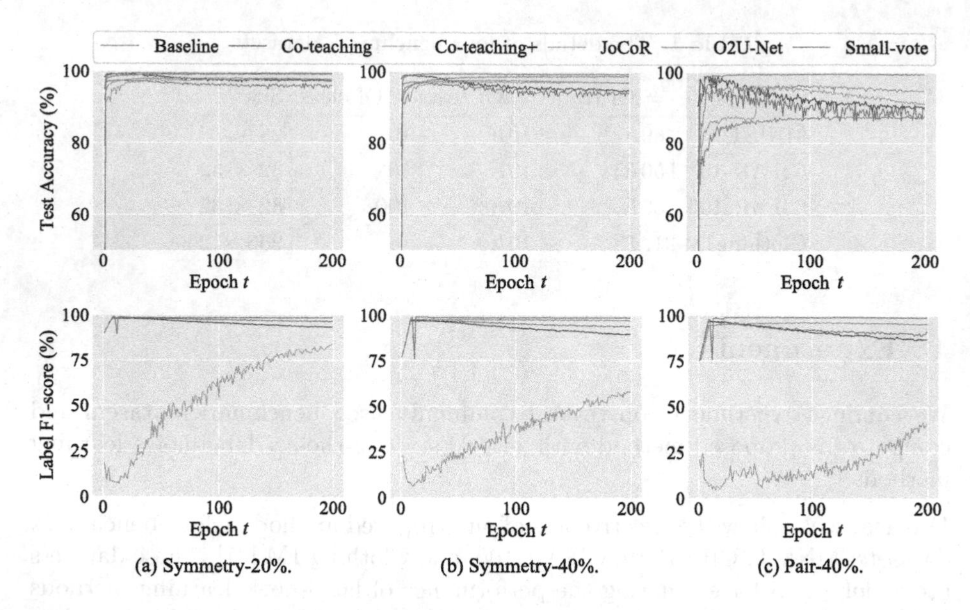


**Fig. 3.** Results on the MNIST dataset under different settings of noisy labels. Top: test accuracy (%), vs. epochs; bottom: label F1-score (%) vs. epochs.

function to cyclically adjust the learning rate, which linearly decreases from 0.1 to 0.001 in a cycle round. All our code will be released soon.

**Evaluation Metrics.** We use two commonly used evaluation metrics: test accuracy (i.e., *test accuracy = (# of correct predictions)/(# of test data)*) and label F1-score (i.e., *Label F1-score* $= \frac{2 \times P \times R}{P+R}$, where $P =$ *(# of clean labels)/(# of all selected labels)* is the label precision and $R =$ *(# of clean labels)/(# of all clean samples)* is the label recall).

### 4.1 Comparisons with State-of-the-Arts

We evaluate the performance obtained by all the competing methods on the synthetic noisy labels by using MNIST, CIFAR-10, and CIFAR-100. Figures 3, 4 and 5 show the results on the three datasets, respectively. We report the test accuracy vs. the number of epochs and the label F1-score vs. the number of epochs on the three datasets under different settings of noisy labels (including Symmetry-20%, Symmetry-40%, and Pair-40%).

Moreover, we also show the superiority of our proposed method on the real-world noisy labels by using the Clothing1M dataset. The comparison results are given in Table 2, where "best" and "last" respectively denote the trained models at the epoch (when the validation accuracy is optimal) and at the end of training epochs.

**Results on MNIST.** As shown in Fig. 3, our proposed small-vote method outperforms the other competing methods in terms of both test accuracy and label

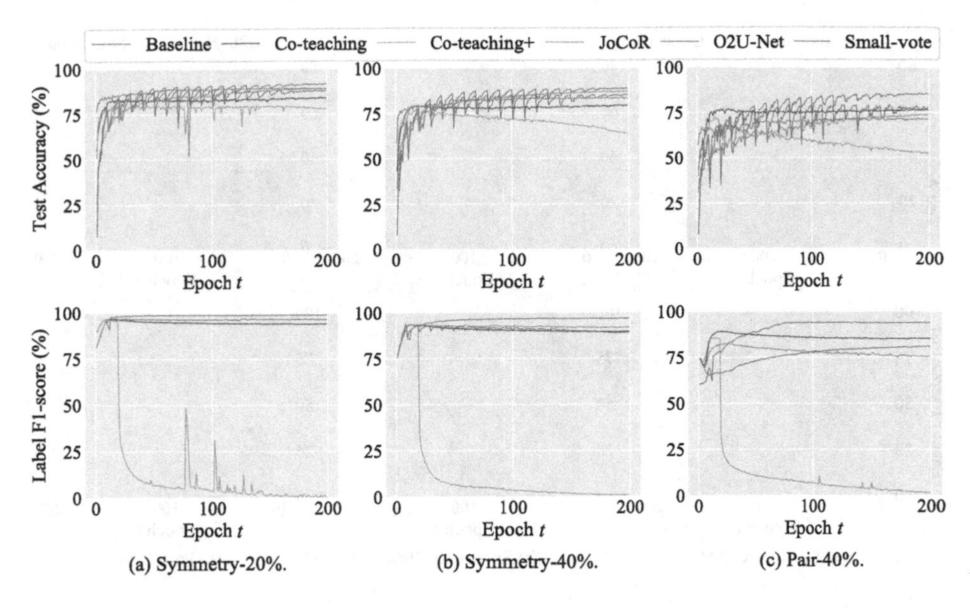


(a) Symmetry-20%. (b) Symmetry-40%. (c) Pair-40%.

**Fig. 4.** Results on the CIFAR-10 dataset under different settings of noisy labels. Top: test accuracy (%), vs. epochs; bottom: label F1-score (%) vs. epochs.

F1-score for different settings of noisy labels. This is because HVS effectively assigns a hierarchical vote for each mini-batch sample and ACES adaptively estimates the clean data rate, leading to accurate identification of both noisy-labeled samples and clean samples. O2U-Net involves two stages, where noisy labels are detected in the first stage and then a model is trained on clean data in the second stage. However, O2U-Net does not fully use the memorization effect to alleviate the negative impact of noisy-labeled samples for training. Hence, the test accuracy obtained by O2U-Net is inferior to our method. Note that O2U-Net is not shown in the second row of Fig. 3 since no sample selection is used in O2U-Net.

**Results on CIFAR-10.** As shown in the first row of Fig. 4, our proposed small-vote method outperforms the other competing methods with a large margin. The recent state-of-the-art JoCoR obtains much worse performance than our method in terms of average test accuracy over the last ten epochs (about 11.96%, 2.56%, and 2.49% decrease under three settings of noisy labels). From the second row of Fig. 4, the label F1-scores obtained by Co-teaching, Co-teaching+, and JoCoR gradually decline after several epochs. In contrast, our small-vote method not only achieves high label F1-scores under all the settings, but also shows better performance at larger epochs. This can be ascribed to the effectiveness of our method for discriminating noisy-labeled samples from clean samples, enabling the capability of learning a more robust model.

**Results on CIFAR-100.** CIFAR-100 is more challenging than MNIST and CIFAR-10. The overall test accuracy obtained by all the methods on CIFAR-100

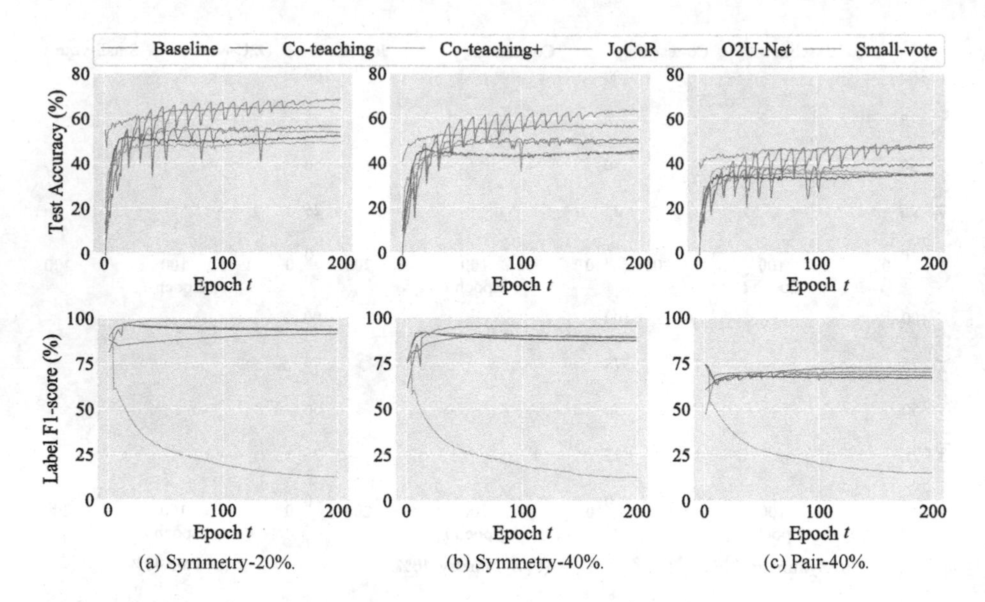


(a) Symmetry-20%. (b) Symmetry-40%. (c) Pair-40%.

**Fig. 5.** Results on the CIFAR-100 dataset under different settings of noisy labels. Top: test accuracy (%), vs. epochs; bottom: label F1-score (%) vs. epochs.

is much lower than that on CIFAR-10, since there are more classes in CIFAR-100. As shown in Fig. 5, the label F1-score obtained by other competing methods first increases and then gradually decreases during the training. This shows that the ability to identify noisy-labeled samples is limited for these methods. However, the label F1-score obtained by the proposed method keeps very stable after several epochs under different settings, leading to better test accuracy. Compared with the baseline method, our proposed method obtains much better performance, which shows the excellent classification ability of our method. In general, our proposed method performs favorably against the other competing methods.

**Results on Clothing 1M.** From Table 2, our proposed small-vote method obtains better results than the other competing methods on best. Moreover, small-vote achieves a significant improvement in accuracy of 11.53% over Co-teaching+, and an improvement of 0.53% over the JoCoR on last. Therefore, small-vote can effectively identify noisy-labeled samples and clean samples on the dataset containing real-world noisy labels. In summary, our proposed method achieves state-of-the-art results against several competing methods on four datasets with synthetic and real-world noisy labels.

### 4.2 Ablation Studies

In this subsection, we perform ablation studies to analyze the effectiveness of key components of small-vote and the influence of the key parameter on CIFAR-10.

**Table 2.** Results on the clothing 1M dataset. The test accuracy (%) is used for performance comparison. The best results are highlighted in bold.

| Methods | Best | Last |
|---|---|---|
| Co-teaching | 69.21 | 68.51 |
| Co-teaching+ | 59.32 | 58.79 |
| JoCoR | 70.30 | 69.79 |
| Baseline | 68.72 | 68.21 |
| Small-vote | **70.43** | **70.32** |

**Table 3.** Ablation study of key components of our small-vote on CIFAR-10. The average test accuracy (%) over the last ten epochs is used for performance comparison. The best results are highlighted in bold.

| Methods | Symmetry-20% | Symmetry-40% | Pair-40% |
|---|---|---|---|
| HVS-L | 89.67 | 86.54 | 80.46 |
| HVS-G | 90.12 | 87.53 | 83.12 |
| HVS | 90.99 | 88.27 | 84.51 |
| Small-vote | **91.56** | **89.59** | **85.70** |

Moreover, we show the generalization capability of our method by integrating small-vote with several state-of-the-art label-noise learning methods.

First, to evaluate the importance of two main components (i.e., HVS and ACES) in small-vote, we compare the following variants. The HVS-G and HVS-L methods denote our proposed method based on the global vote and the local vote, respectively, without using ACES (we use $\lambda^t$ defined in Eq. (2) to select clean samples instead). The HVS method denotes our proposed method based on HVS without using ACES. The results are given in Table 3. Note that HVS-L is equivalent to the baseline method.

Compared with HVS-L, HVS achieves a performance boost. HVS-L only relies on the local information from the current mini-batch and uses a heuristic clean date rate. On the contrary, HVS effectively combines the global information from previous epochs and the local information from the current mini-batch. HVS also obtains higher test accuracy than HVS-G. This shows that both the global vote and the local vote play an important role in sample selection. Small-vote outperforms HVS under different settings of noisy labels. This can be ascribed to the adoption of ACES. ACES is a data-dependent clean data rate estimation strategy, which can effectively address the problem of unknown noisy label distribution in the mini-batch (note that the mini-batch is randomly chosen from the whole training data). Therefore, small-vote is able to identify clean and noisy-labeled samples more accurately, thereby leading to a more robust DNN model.

**Table 4.** Influence of different values of $C$ on CIFAR-10. The average test accuracy (%) over the last ten epochs is used for performance comparison. The best results are highlighted in bold.

| $C$ | Symmetry-20% | Symmetry-40% | Pair-40% |
|---|---|---|---|
| 0 | 89.67 | 86.54 | 80.46 |
| 5 | 91.05 | 88.10 | 84.95 |
| 10 | **91.56** | **89.59** | **85.70** |
| 50 | 90.87 | 88.50 | 82.05 |

**Table 5.** Performance comparison between small-loss and small-vote in the frameworks of Co-teaching, Co-teaching+, and JoCoR on the CIFAR-10 dataset. The average test accuracy (%) over the last ten epochs is used for performance comparison. The best results are highlighted in bold.

| Methods | Symmetry-20% | Symmetry-40% | Pair-40% |
|---|---|---|---|
| Co-teaching (small-loss) | 83.92 | 79.25 | 73.93 |
| Co-teaching (small-vote) | **85.44** | **81.35** | **75.17** |
| Co-teaching+ (small-loss) | 78.09 | 64.27 | 52.79 |
| Co-teaching+ (small-vote) | **80.32** | **66.24** | **66.67** |
| JoCoR (small-loss) | 89.07 | 86.75 | 73.73 |
| JoCoR (small-vote) | **90.32** | **88.09** | **82.50** |

Second, we evaluate the influence of the key parameter $C$ defined in Eq. (4) on the final performance. We set the values of $C$ to 0, 5, 10, and 50. The results are given in Table 4. We can see that our small-vote achieves the best performance when the value of $C$ is set to 10. When the value of $C$ is set to 0, only the local vote is used. When the value of $C$ is large, too old historical information is exploited to perform the global vote. For these two extreme cases, the performance drops. Therefore, in all the experiments, we fix the value of $C$ to 10.

Finally, to demonstrate the generalization ability of our small-vote sample selection method, we replace the small-loss criterion used in several state-of-the-art label-noise learning methods (including Co-teaching, Co-teaching+, and JoCoR) with our proposed small-vote. Table 5 shows the test accuracy obtained by different methods on CIFAR-10. As we can see, our small-vote sample selection outperforms the small-loss criterion with a moderate margin for different types and levels of noisy labels in the frameworks of Co-teaching, Co-teaching+, and JoCoR. Therefore, our small-vote sample selection is able to distinguish clean samples from noisy-labeled samples more effectively than the small-loss criterion, leading to performance improvements. The above results further show the great generalization of small-vote for label-noise learning.

## 5 Conclusion

In this paper, we have proposed a simple yet effective small-vote sample selection method for label-noise learning. The proposed method is comprised of two main components, including a Hierarchical Voting Scheme (HVS) and an Adaptive Clean data rate Estimation Strategy (ACES). HVS effectively combines the global vote from previous epochs and the local vote from the current mini-batch to assign hierarchal votes for each mini-batch. Based on HVS, ACES adaptively estimates the clean data rate, so that clean samples and noisy-labeled samples can be accurately identified in the mini-batch. Experimental results on noisy-labeled data from four benchmark datasets including MNIST, CIFAR-10, CIFAR-100, and Clothing1M have shown the superiority of our proposed method over several state-of-the-art methods. Moreover, the good generalization capability of our method has been verified by incorporating our small-vote method into representative label-noise learning methods.

**Acknowledgment.** This work was supported by the Open Research Projects of Zhejiang Lab under Grant 2021KB0AB03, by the National Natural Science Foundation of China under Grants 62071404 and 61872307, by the Natural Science Foundation of Fujian Province under Grant 2020J01001, and by the Youth Innovation Foundation of Xiamen City under Grant 3502Z20206046.

## References

1. Arpit, D., et al.: A closer look at memorization in deep networks. In: ICML, pp. 233–242 (2017)
2. Bootkrajang, J., Kabán, A.: Label-noise robust logistic regression and its applications. In: ECML-PKDD, pp. 143–158 (2012)
3. Dempster, A.P., Laird, N.M., Rubin, D.B.: Maximum likelihood from incomplete data via the em algorithm. J. Roy. Stat. Soc. Ser. B (Methodological) **39**(1), 1–22 (1977)
4. Feng, L., Shu, S., Lin, Z., Lv, F., Li, L., An, B.: Can cross entropy loss be robust to label noise. In: IJCAI, pp. 2206–2212 (2020)
5. Goldberger, J., Ben-Reuven, E.: Training deep neural-networks using a noise adaptation layer. In: ICLR (2017)
6. Han, B., Niu, G., Yu, X., Yao, Q., Xu, M., Tsang, I., Sugiyama, M.: Sigua: forgetting may make learning with noisy labels more robust. In: ICML, pp. 4006–4016 (2020)
7. Han, B., et al.: Co-teaching: robust training of deep neural networks with extremely noisy labels. In: NeurIPS, pp. 8527–8537 (2018)
8. He, K., Zhang, X., Ren, S., Sun, J.: Deep residual learning for image recognition. In: CVPR, pp. 770–778 (2016)
9. Huang, J., Qu, L., Jia, R., Zhao, B.: O2u-net: a simple noisy label detection approach for deep neural networks. In: ICCV, pp. 3326–3334 (2019)
10. Jiang, L., Zhou, Z., Leung, T., Li, L.J., Fei-Fei, L.: Mentornet: learning data-driven curriculum for very deep neural networks on corrupted labels. In: ICML, pp. 2304–2313 (2018)

11. Kumar, A., Shah, A., Raj, B., Hauptmann, A.: Learning sound events from webly labeled data. In: IJCAI, pp. 2772–2778 (2019)
12. Liu, T., Tao, D.: Classification with noisy labels by importance reweighting. IEEE Trans. Pattern Anal. Mach. Intell. **38**(3), 447–461 (2015)
13. Luo, Y., Han, B., Gong, C.: A bi-level formulation for label noise learning with spectral cluster discovery. In: IJCAI, pp. 2605–2611 (2020)
14. Malach, E., Shalev-Shwartz, S.: Decoupling "when to update" from "how to update". In: NeurIPS, pp. 960–970 (2017)
15. Patrini, G., Rozza, A., Krishna Menon, A., Nock, R., Qu, L.: Making deep neural networks robust to label noise: a loss correction approach. In: CVPR, pp. 1944–1952 (2017)
16. Pereyra, G., Tucker, G., Chorowski, J., Kaiser, Ł., Hinton, G.: Regularizing neural networks by penalizing confident output distributions. In: ICLRW (2017)
17. Song, H., Kim, M., Park, D., Lee, J.G.: Learning from noisy labels with deep neural networks: A survey. arXiv preprint arXiv:2007.08199 (2020)
18. Van Rooyen, B., Menon, A., Williamson, R.C.: Learning with symmetric label noise: the importance of being unhinged. In: NeurIPS, pp. 10–18 (2015)
19. Vembu, S., Zilles, S.: Interactive learning from multiple noisy labels. In: ECML-PKDD, pp. 493–508 (2016)
20. Wei, H., Feng, L., Chen, X., An, B.: Combating noisy labels by agreement: a joint training method with co-regularization. In: CVPR, pp. 13726–13735 (2020)
21. Xiao, T., Xia, T., Yang, Y., Huang, C., Wang, X.: Learning from massive noisy labeled data for image classification. In: CVPR, pp. 2691–2699 (2015)
22. Yao, Q., Yang, H., Han, B., Niu, G., Kwok, J.T.Y.: Searching to exploit memorization effect in learning with noisy labels. In: ICML, pp. 10789–10798 (2020)
23. Yao, Y., et al.: Dual t: reducing estimation error for transition matrix in label-noise learning. In: NeurIPS (2020)
24. Yu, X., Han, B., Yao, J., Niu, G., Tsang, I.W., Sugiyama, M.: How does disagreement help generalization against label corruption? In: ICML, pp. 7164–7173 (2019)
25. Yu, X., Liu, T., Gong, M., Batmanghelich, K., Tao, D.: An efficient and provable approach for mixture proportion estimation using linear independence assumption. In: CVPR, pp. 4480–4489 (2018)
26. Zhang, C., Bengio, S., Hardt, M., Recht, B., Vinyals, O.: Understanding deep learning requires rethinking generalization. In: ICLR (2017)
27. Zhang, H., et al.: Learning with noise: improving distantly-supervised fine-grained entity typing via automatic relabeling. In: IJCAI, pp. 3808–3815 (2020)
28. Zhang, H., Cisse, M., Dauphin, Y.N., Lopez-Paz, D.: Mixup: beyond empirical risk minimization. In: ICLR (2018)

# Iterated Matrix Reordering

Gauthier Van Vracem and Siegfried Nijssen(✉)

UCLouvain - ICTEAM/INGI, Louvain-la-Neuve, Belgium
siegfried.nijssen@uclouvain.be

**Abstract.** Heatmaps are a popular data visualization technique that allows to visualize a matrix or table in its entirety. An important step in the creation of insightful heatmaps is the determination of a good order for the rows and the columns, that is, the use of appropriate matrix reordering or seriation techniques. Unfortunately, by using artificial data with known patterns, it can be shown that existing matrix ordering techniques often fail to identify good orderings in data in the presence of noise. In this paper, we propose a novel technique that addresses this weakness. Its key idea is to make an underlying base matrix ordering technique more robust to noise by embedding it into an iterated loop with image processing techniques. Experiments show that this iterative technique improves the quality of the matrix ordering found by all base ordering methods evaluated, both for artificial and real-world data, while still offering high levels of computational performance as well.

**Keywords:** Matrix reordering · Matrix seriation · Pattern mining · Clustering · Convolution

## 1 Introduction

*Heatmaps* are a popular data visualization technique that allows to easily visualize a matrix or table in its entirety. In the process of creating a heatmap an important step is however the determination of the order of the rows and columns used in the visualization; depending on the order chosen, the visualization may highlight different structures or patterns, such as clusters, that are of interest to understand the data being visualized. The challenge is that this specific order is not necessarily known and it is necessary to implement different *matrix reordering* algorithms that aim to find a permutation that best displays the data.

Many methods have been proposed in the literature that allow to permute a matrix in order to highlight a particular pattern (see [2] for a survey). However, in practice these methods are very sensitive to noise. This is illustrated in Fig. 1 for a banded Boolean matrix. On the right-hand side of Fig. 1(a) we show the output of a reordering algorithm, a TSP solver in this case, applied on a random permutation of a banded matrix without noise, shown on the left-hand side; the algorithm successfully identifies the banded structure. However, if we flip each


N. Oliver et al. (Eds.): ECML PKDD 2021, LNAI 12977, pp. 745–761, 2021.
https://doi.org/10.1007/978-3-030-86523-8_45

entry of the matrix with a 25% probability, the reordering method no longer recovers the banded structure, as shown in Fig. 1(b). Similar observations also hold for other matrices and ordering methods, as we will see in the experimental section of this work.

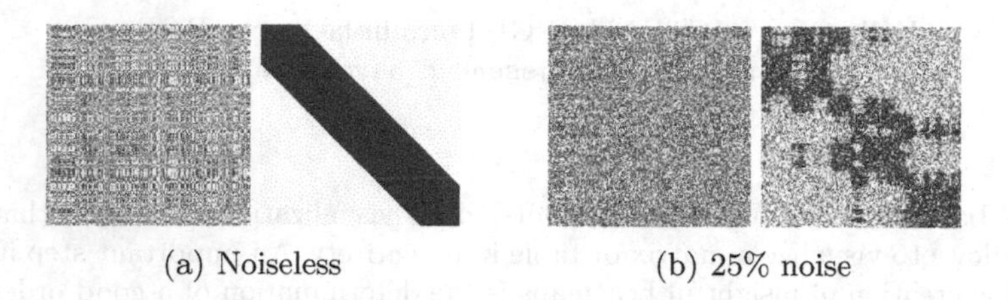

(a) Noiseless (b) 25% noise

**Fig. 1.** Impact of noise on the ordering of the same matrix. On the left, a random permutation. On the right, the matrix after applying the same ordering method

In this paper we address this problem. We propose a novel type of method that allows to reorder data such that structures are also recovered in the presence of higher levels of noise. The key idea of this novel type of method is to reuse less effective methods by integrating them into an iterative loop that abstracts from noise by using convolution. *Convolution* is a technique that can be used to blur an image in order to reduce noise. We use convolution for different purposes: on the one hand, to reduce the noise of the matrix in a thresholding process; on the other hand to asses the quality of the order produced. The key idea is therefore to temporarily transform the data by removing the noise and then to reorder this simpler noiseless matrix with conventional methods. The developed technique allows to quickly order sparse or dense binary data while eliminating the noisiness. The advantage of the framework is that it allows to use any existing method (as a black box), and, as we will show experimentally, enhances the quality of the outputs of these existing algorithms.

The paper is structured as follows: in Sect. 2, we discuss the state-of-the-art methods. In Sect. 3 we introduce our framework. In Sect. 4 we conduct different experiments comparing existing methods and the framework on both artificial and real data. Finally, in Sect. 5, we lay the foundations for possible future improvements.

## 2 Related Work

Even though our technique can also be used on non-binary data, in this work we focus on binary data, as many ordering techniques already exist for such data and hence we can evaluate our contributions well for such data. We will distinguish two types of binary matrices. The most generic form of binary data is a matrix of size $m \times n$, where $m$ and $n$ are not necessarily equal. Such *tabular* or *two-way two-mode* data can also be seen as a bipartite graph; the rows and columns of such matrices can be ordered independently from each other. A specific form of binary data are $n \times n$ matrices which represent *graphs* or *two-way one-mode* data (adjacency matrix). For such data the same order is typically used for both the rows and the columns.

To order rows and columns, we will distinguish three families of methods: *distance-based*, *structure-based* and *convolution-based* methods.

***Distance-based*** methods aim to put together rows and columns that are similar. To do this, the original matrix is transformed into a distance or *dissimilarity matrix* $\mathbf{D}$. Each entry $\mathbf{D}_{i,j}$ represents the distance between row $i$ and $j$ (Euclidean or Manhattan distance for example). For *network* matrices, only one dissimilarity matrix is created. For *tabular* data two matrices are created (one for the columns and one for the rows), which results in two ordering problems. Once these matrices are built, their elements are reordered. This problem is known as the *Seriation problem* and is a combinatorial task that aims at ordering a one-dimensional set of objects $\{O_1, ..., O_n\}$ in order to optimize a merit/loss function. The *Seriation library* [8] in R offers a wide range of methods and heuristics. Behrisch et al. [2] provide a survey detailing and comparing these methods.

The advantage of dissimilarity matrices is that they allow the problem to be viewed from different perspectives. One way of solving it is to solve a *TSP* (traveling salesman problem) on the dissimilarity matrix [4], where a path over all the rows (or columns) of the matrix is found that maximized the similarity between each pair of adjacent rows.

One of the most used methods is hierachical clustering, in which a dendogram is built over data; using various algorithms, from this dendogram a seriation can be created that respects the dendogram.

All the methods described above, being based on the distance between rows, are efficient when there is little noise in the matrix. They also have the advantage of being applicable on both binary and numerical data. However, the introduction of noise can strongly disturb this similarity to the point that these methods can no longer produce a relevant order. This issue is similar for sparse matrices.

***Structure-based*** methods aim to swap rows/columns in such a way that the resulting image looks most like a specific structure. Contrary to the previously described methods, the ordering of rows and columns are no longer two independent problems, but are solved in a combined process. Two basic methods aim at discovering different structures [3,7] and work exclusively on binary data:

*Nested* ordering consists in counting respectively for the columns and rows the number of 1s and then sorting them by these values. This allows to discover structures where there is a hierarchy between the different matrix entries.

The *barycenter* heuristic method aims at revealing a banded structure. For each row a barycentric index is computed which corresponds to the average position for all the 1. The rows are reordered according to this index, after which the process is iterated by alternating columns and rows until convergence.

These methods suffer from similar disadvantages as distance-based methods; they assume that there is a specific pattern to discover and are also very sensitive to noise.

The third type of method consists of methods based on ***convolution***, such as ConvoMap [1]. This method optimizes an evaluation criterion that measures the quality of an order by computing the difference between the image in this order and a blurred version of the same image. The intuition is that if applying a blurring filter does not change an image significantly, this indicates that there are not many black-white transitions in the image; there are clear black and white parts in the image that remain unmodified by the blur. It was shown that this criterion indeed scores best those orders in which a clear structure visible. The main weakness of the ConvoMap method is that its local search algorithm is inefficient; in earlier work execution times of multiple hours were reported [1]. In this paper, we propose to reuse this evaluation criterion, but we integrate it in a very different approach: a framework that aims to reuse any ordering method as a black box component and to boost it by taking advantage of convolution.

## 3 Iterative Ordering Using Convolution

We consider a binary matrix $\mathbf{M}$ of size $m \times n$ with $\mathbf{M}_{i,j} \in \{0, 1\}$, or an $n \times n$ matrix for networks, with $\mathbf{M}_{i,j} = \mathbf{M}_{j,i}$ and $\mathbf{M}_{i,i} = 0$.

The key idea of our iterative method is that we embed a *base ordering method* $\varphi_b$ in an iterative loop to obtain a better output ordering, similar in spirit to how boosting methods improve the classification accuracy of an underlying base learning algorithm. We make the following two assumptions on the underlying base ordering method:

1. on data with a perfect, noiseless structure (such as a banded structure, a nested structure, a block structure, and so on), the ordering method will recover this structure;
2. on noisy data the ordering method will recover some of the structure in the data, although in an imperfect manner.

We will refer to the imperfect, incomplete patterns identified by the base method as *embryonic patterns*. Table 2 shows that most existing methods in literature are able to generate these fragments of dense patterns.

An overview of our iterative method is provided in Fig. 2. Initially, the base method $\varphi_b$ is used to produce a preliminary order; we assume that this order is informative, but imperfect due to noise. Subsequently, we apply convolution and thresholding to remove this noise from the image. The resulting data is once more put in the base ordering method, exploiting the assumption that for data without noise the method will produce a better ordering. The resulting order is used to reorder the original data (renoising), after which the process is repeated, until a stopping criterion indicates the process has converged.

We optimize the process by including an additional smoothing step in which we use local search to locally improve the solution.

The details of our approach are provided below.

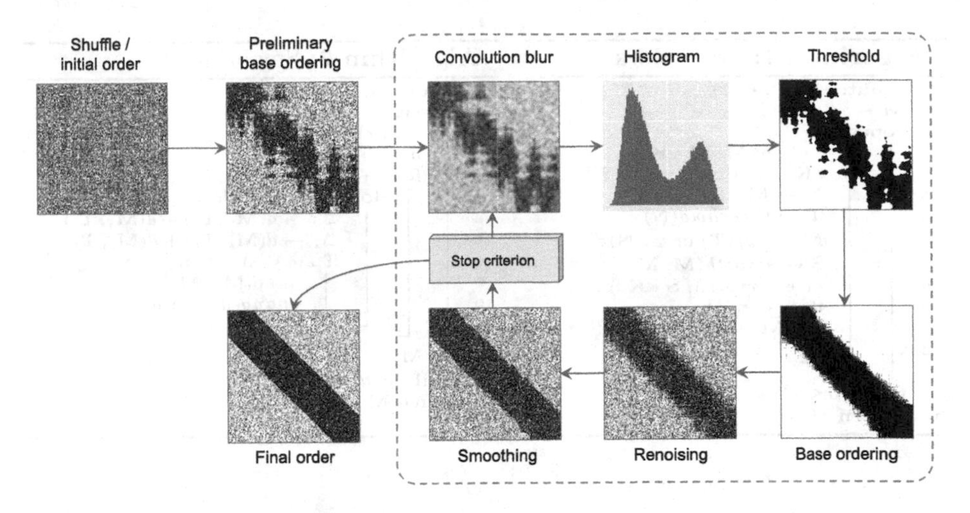


**Fig. 2.** Overview of the iterative framework with thresholding enabled

*Blur.* The objective of convolution and thresholding is to obtain a clearer image with less variation; this should make it easier to order the data by the base algorithm. To achieve this goal a *convolution* operation is performed on the matrix. *Convolution* is an image processing technique that consists in filtering an image by applying a *kernel* matrix $\mathbf{K}$ of size $k_m \times k_n$. Each pixel of the image is multiplied by the value of the surrounding pixels weighted by a specific weight. These weights are determined by the chosen kernel. We define the convolution operation as follows:

$$(\mathbf{M} * \mathbf{K})_{i,j} = \sum_{r=1}^{k_m} \sum_{c=1}^{k_n} \mathbf{K}_{r,c} \cdot \mathbf{M}_{i+r-\lceil k_m/2 \rceil, j+c-\lceil k_n/2 \rceil} \quad (1)$$

The effect of the convolution depends on the kernel used; some kernels have the effect of sharpening the image, others blur the picture. In this paper we only consider kernels that fall in this second category. The larger the kernel is, the stronger the removal of noise will be. In our algorithm we allow for several predefined kernels, but we will limit ourselves to one type of kernel in this paper: the linear kernel of size $k \times k$, defined as follows: $\mathbf{K}_{r,c} = k_{mid} - max(|k_{mid} - r - 1|, |k_{mid} - c - 1|) + 1$ with $k_{mid} = \lfloor k/2 \rfloor$.

*Threshold (optional).* After applying the blur convolution, the initial matrix is transformed from a binary one into a continuous one $\mathbf{N}$ with $\mathbf{N}_{i,j} \in [0, 1]$. Two strategies are therefore possible at this point. The first one consists in directly reordering the convoluted matrix with the base method. This is only possible for *distance-based* methods. In this case the thresholding step is **optional**. The second option consists in reconverting the blurred matrix into a binary one by

**Algorithm 1:** Framework

```
   input: M, K_c, κ, φ_b
1  M ← φ_b(M)
2  repeat
3      e_0 ← error(M, M * K_c)
4      for ∀K_i ∈ κ do
5          N ← M * K_i
6          T ← threshold(N)
7          M' ← φ_b(T) or φ_b(N)
8          S ← smooth(M, M')
9          e_1 ← error(S, S * K_c)
10         if e_1 < e_0 then
11             M ← S
12             break
13 until e0 ≤ e1;
14 return M
```

**Algorithm 2:** Smoothing

```
   input: M, T
1  repeat
2      changed ← false
3      h ← height(M)
4      for i ← 1 to h − 1 do
5          for j ← 2 to h do
6              Δ_0 ← d(M_i, T_j) + d(M_j, T_i)
7              Δ_1 ← d(M_i, T_i) + d(M_j, T_j)
8              if Δ_0 < Δ_1 then
9                  swap(M_i, M_j)
10                 changed ← true
11     M ← M^t, T ← T^t
12 until ¬changed;
13 return M
```

applying a *thresholding* operation. The *thresholded* version of a Matrix $\mathbf{N}$ is defined by:

$$threshold(\mathbf{N}, t)_{i,j} = \begin{cases} 1 & \text{if } \mathbf{N}_{i,j} > t \\ 0 & \text{if } \mathbf{N}_{i,j} \leq t \end{cases} \tag{2}$$

This filter consists in choosing a threshold $t \in [0, 1]$, which is used to partition the values of the matrix into two classes. To achieve this, the $OTSU$ method is used. This is a popular algorithm [6] that constructs a histogram of the intensities of the matrix values and then partitions it into two classes in order to minimize their intra-class variance. This method is based on the shape of the histogram and makes the assumption that the distribution of the values is modeled by a bi-modal distribution. This assumption is partially satisfied if the original image contains a well defined pattern divided into different dense areas.

*Smoothing Refinement.* Once the convolution and optionally the threshold are applied, a noiseless binary matrix is obtained, which contains only the shape of the *embryonic pattern*. It is then easier to reorder this new simpler matrix. The embedded base algorithm is then executed and the produced order of rows and columns is applied to the initial matrix. This operation is called *renoising*.

Nevertheless some defects can remain, notably for the quality of the edges as shown in Table 5. To address this issue, we apply a refinement operation that we refer to as *smoothing*. The pseudo-code for this operation is provided in Algorithm 2. This technique consists in permuting the columns and rows of the matrix in order to minimize the difference between the renoised matrix $\mathbf{M}$ and its thresholded version $\mathbf{T}$ computed previously. We define the function $d(\mathbf{M}_i, \mathbf{T}_j)$ as the Hamming distance between row $i$ of the matrix $\mathbf{M}$ and row $j$ of the matrix $\mathbf{T}$. Essentially, we locally optimize the order of the renoised data such that this data more closely resembles the thresholded version.

*Stopping Criterion.* The operations described above are repeated iteratively. For each iteration, a new order is produced but the challenge is to measure and quantify the quality of this produced image and inferring an evaluation or stopping criterion. Here we reuse the scoring function used in the ConvoMap algorithm [1], where the goal is to ensure that the image looks as similar as possible to its blurred version. The scoring function used as criterion is defined as the difference between a matrix **M** and its convoluted version after applying a blurring kernel **K**: $error(\mathbf{M}, \mathbf{M} * \mathbf{K})$. The difference between two matrix is defined as follows:

$$error(\mathbf{M}, \mathbf{M}') = \sum_{r=1}^{m} \sum_{c=1}^{n} |\mathbf{M}_{r,c} - \mathbf{M}'_{r,c}| \tag{3}$$

*Framework Overview.* Pseudo-code for the framework is provided in Algorithm 1. Within the algorithm, we use a unique kernel $\mathbf{K}_c$ as evaluation criterion and a sequence of kernels $\kappa$ for the convolution/threshold process. In the most simple setting, $|\kappa| = 1$. *Thresholding* and *smoothing* are steps that can be enabled or disabled. At the end of each iteration, the quality of the new ordered matrix is evaluated using $\mathbf{K}_c$. If the order of this candidate is improved compared to the previous one, the execution continues with it. If no improvement is found, we can continue the process with the next kernel in $\kappa$; this is useful for instance to repeat the process with a more pronounced blur. The execution ends eventually if all kernels fail to improve the last solution. The order of $\kappa$ is not modified during the execution for the sake of simplicity and to make the results more predictable.

## 4 Experiments

In this section, we will evaluate the performance of the framework as well as the quality of the results produced for different types of datasets. The experiments will be guided by the following questions:

***Q1*** How does the iterative framework improve existing methods in terms of quality, for tabular data (Q1a) and for sparse network data (Q1b)?
***Q2*** How does the iterative framework compare to existing methods with respect to execution time?
***Q3*** What is the impact of noise on the ordering?
***Q4*** How well does the smoothing refinement improve the results?
***Q5*** For which type of data is thresholding necessary?

### 4.1 Experimental Setup

The framework is configured as follows. We have defined a sequence of kernels ($\kappa$) containing linear kernels of size $n \times n$ with respectively $n \in \{3, 5, 7, 9, 15, 25\}$. A linear kernel of size $49 \times 49$ is considered as evaluation criterion $K_c$. We have set the maximum number of iterations at 50. Except if specified otherwise, *smoothing* is enabled for all our experiments, *thresholding* is active for tabular data and inactive for network data.

Any matrix ordering method can be embedded into the framework. We made the choice to benchmark different methods belonging to the *R seriation package* [8] as distance-based methods. We have selected the following methods: TSP, OLO, GW, MDS, Spectral, QAP, HC. The barycentric method [7] will also be evaluated as well as the ConvoMap algorithm [1]. These methods are integrated into the framework in order to evaluate how our approach improves the quality of the order.

### 4.2 Datasets

The experiments were conducted on 2 types of data:
**Tabular Data:** We consider 4 dense artificial datasets, each featuring a different pattern: *Pareto*, *Banded*, *Blocks* and *Triangles*. These datasets are generated for a size of $300 \times 300$. Noise is then added to these matrices by flipping each bit with a probability of $p$. We also evaluate the framework on real world data. *Mammals* [5] is a dataset that records the presence of mammals species (124 columns) at locations across Europe (2200 rows). A cell in the table is positive if a particular species has been seen at a specific location.

**Networks:** here we consider 4 artificial and 4 real world datasets that represents sparse networks. We consider 4 types of popular structures as artificial data: *Assortative*, *Core-periphery*, *Ordered* and *Disassortative* as presented by [9]. The specificity of these matrices is that they are partitioned in clusters or communities. This ground-truth information will be used for the evaluation of the different methods. In addition, we will evaluate the framework on real world data: (1) *political blog networks* for the US 2004 elections where the communities (ground-truth) are known [10]. In this work, we treated these graphs as undirected graphs, and removed nodes with less than 5 connections. (2) *A mail network* [11] between email addresses of a large European institution. The dataset originally contains 42 departments, but has been preprocessed to remove nodes with less than 5 connections and keep the 6 biggest communities. (3) *Indian village data* [12] representing an Indian village. The relationships inside the matrix represent the interaction links between different households. The caste of these households is known and is used as ground-truth information. (4) Finally we use *news data* [13] corresponding to the adjacency matrix of news documents divided into 3 categories. The links model the similarity between these documents. We applied a threshold value of 0.05. In all these cases, we simplified the data, as the visualization would otherwise be too large.

### 4.3 Results

We answer ***Q1a*** by comparing the order produced by the conventional methods (*basic*) and its embedded version in our *iterative* framework.

We benchmark all the methods listed above for the four dense datasets with a noise level $p = 0.2$. Before reordering the matrices, the columns and rows are shuffled. Evaluating the quality of an order is not straightforward, as a number of different orders could yield images of similar quality as an image created for the original order of the data (including, for instance, flipping the image horizontally or vertically). As a proxy of the quality of an image, we choose to use the ConvoMap evaluation criterion, as this was shown in previous work to be a reasonably proxy for orders of good quality [1]. The obtained scores are summarized in Table 1, where we also show the score of the original, unshuffled matrix as a reference to facilitate the comparison. Table 2 shows the produced images for some of these methods, allowing to also visually verify that the evaluation score used is a reasonable proxy for the quality of an image.

**Table 1.** Error scores comparison. The scores better than the original order are highlighted. The best score for a given dataset is indicated in **bold**

| Method | Pareto 20% | | Banded 20% | | Blocks 20% | | Triangles 20% | |
|---|---|---|---|---|---|---|---|---|
| | basic | iterative | basic | iterative | basic | iterative | basic | iterative |
| original | 30,123 | | 30,839 | | 31,732 | | 31,213 | |
| ConvoMap | 30,174 | - | **30,759** | - | 31,151 | - | 31,259 | - |
| barycentric | 36,087 | 30,168 | 33,549 | 30,819 | 33,851 | 31,057 | 32,908 | 31,711 |
| TSP | 32,469 | 30,155 | 34,076 | 30,831 | 32,814 | 31,040 | 32,313 | 30,650 |
| HC | 32,823 | 31,403 | 34,636 | 33,309 | 32,563 | 31,775 | 32,926 | 31,900 |
| GW | 31,258 | 30,115 | 32,686 | 30,822 | 31,758 | 31,145 | 32,272 | 30,753 |
| MDS | 30,690 | 30,118 | 35,698 | 35,698 | 34,475 | 34,465 | 32,721 | 32,079 |
| MDS angle | 30,898 | 30,870 | 31,079 | 30,817 | 31,673 | **31,006** | 33,509 | 31,523 |
| Spectral | 30,344 | 30,091 | 34,765 | 32,502 | 35,932 | 33,437 | 35,587 | 31,437 |
| QAP 2SUM | 30,298 | 30,116 | 34,553 | 31,386 | 34,338 | 32,533 | 35,834 | 33,057 |
| QAP BAR | 30,162 | **30,077** | 31,511 | 30,775 | 32,029 | 31,691 | 30,735 | **30,500** |
| OLO | 31,275 | 30,115 | 32,081 | 30,820 | 31,439 | 31,420 | 31,951 | 30,743 |

Since the framework uses the *basic* solution as a starting point/initial solution in order to enhance it, it is expected that the score of the *iterative* version is systematically better. We observe that the *basic* method is unable to produce a good order (only pieces of *embryonic patterns* that are difficult to interpret). However, *iterative* is able to recover the original pattern for almost all methods and datasets; the ConvoMap score of the solution that is found by the iterative method is in fact slightly better than the score of the original data in more than 50% of the cases. It should be noted that a method is only able to recover the original pattern if it is able to do so on noiseless matrices. The hierachical clustering (HC) approach is not able to structure a linear pattern like in the banded dataset, so it cannot be enhanced by the framework.

We also compare the results of the framework with the ConvoMap algorithm in Table 3. The produced ordered matrices are similar. This is not surprising since both rely on convolution. However, our framework produces these results in a much shorter time, several orders of magnitude shorter. This is explained by the fact that ConvoMap uses the evaluation criterion as a scoring function through a local search algorithm that tries to optimize this criterion through random moves. In our framework we use faster methods, i.e. the base methods, repeatedly on simplified noise-free matrices.

**Table 2.** Comparative results between the original algorithm and its embedded version into the iterative approach

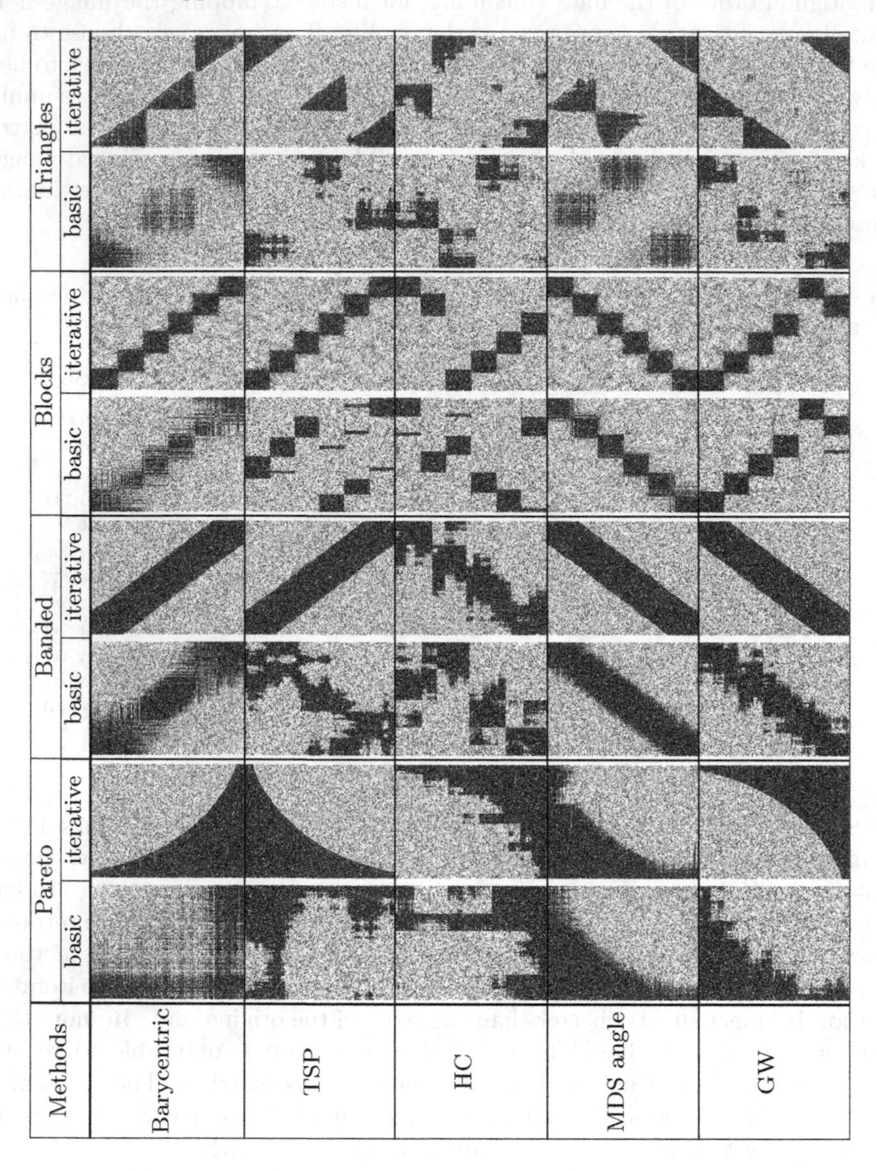


We show the results for the mammals dataset in Fig. 3. We use two *structure-based* algorithms (nested and barycentric). As expected the nested method gives the best results: it is known that in warmer areas the diversity of mammals increases compared to colder areas. This nested pattern is however more clearly visible in the output of the iterative method. Moreover, in the output of the iterative method we also notice a cluster of species outside the nested struc-

**Table 3.** Comparative results for ConvoMap and iterated (TSP) algorithm

| Pareto | | | Banded | | |
|---|---|---|---|---|---|
| Original | ConvoMap | Iterative | Original | ConvoMap | Iterative |
|  |  |  |  | 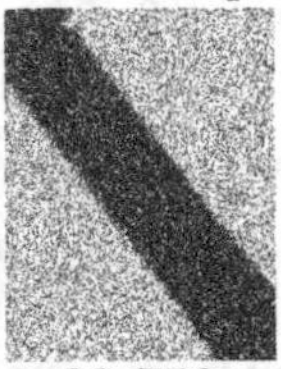 |  |
| **30 123** | 30 174 | 30 155 | 30 839 | **30 759** | 30 831 |
| | 20 841 s | 98 s | | 22 754 s | 109 s |
| Blocks | | | Triangles | | |
| Original | ConvoMap | Iterative | Original | ConvoMap | Iterative |
|  | 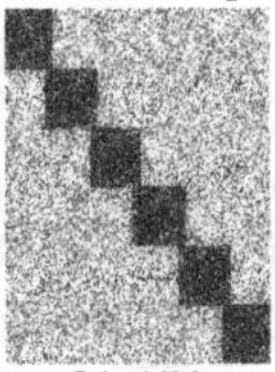 |  |  | 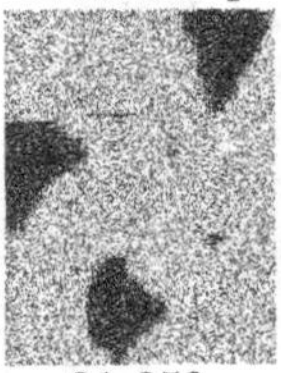 |  |
| 31 732 | 31 151 | **31 040** | 31 213 | 31 259 | **30 650** |
| | 40 489 s | 75 s | | 18 551 s | 148 s |

ture, occurring in areas where the number of other species is relatively small. This cluster corresponds to species that are almost exclusively located in northern Europe (Scandinavia, blue label) (*Gulo-Gulo*, *Alces-Alces*, ...). This useful information is not visible in the output of the original methods.

We address ***Q1b*** by applying the same experimental process to the network data described above. The quality of the ordering will be judged (1) subjectively by looking at the image produced (2) with an accuracy score. Each row/node of the adjacency matrix is associated with a label corresponding to the community. The better the ordering, the more likely these labels will be ordered together. To calculate the accuracy, we determine for each position the most common label among the *10-nearest* rows, if this majority label corresponds to the label of the row it is a *match*. The ratio $\frac{maches}{nodes}$ corresponds to the accuracy score.

Table 4 summarizes the accuracy scores for the different communities by comparing the basic and iterated version. We observe that almost all methods are improved with respect to the accuracy score on both synthetic and real data. Figure 4a to 4d show the produced matrices for synthetic networks with the labels of the different communities. We use HC (hierarchical clustering) as base method but other methods produce visually comparable results. The experiment show that that the *basic* version is not effective, but the iterative method identifies the clusters accurately.

Figure 4e to 4h show the results for real data. The base methods used are those that give the best results for their iterated version. Once more the iterated method correctly partitions the matrix into the corresponding clusters, in con-

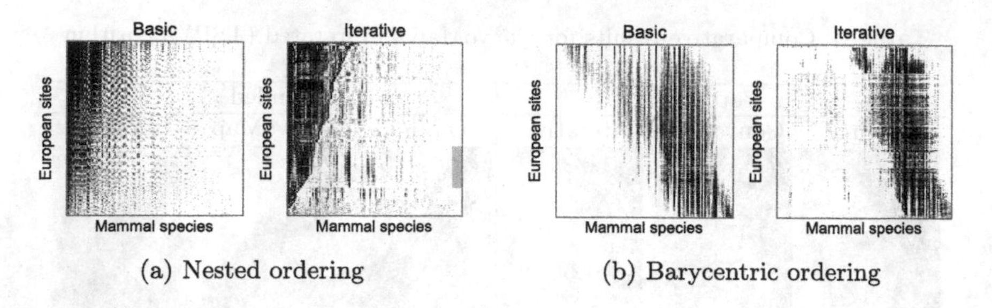


(a) Nested ordering (b) Barycentric ordering

**Fig. 3.** European mammals dataset: comparison between basic and iterative

tract to the base method. Another feature is that the granularity of the noise is significantly smoothed.

**Table 4.** Experiments for synthetic and real network reordering: community labels ($k$) accuracy assessment between the basic ($b$) and iterated version ($i$)

| Dataset | $n$ | $k$ | HC | | OLO | | GW | | MDS | | TSP | |
|---|---|---|---|---|---|---|---|---|---|---|---|---|
| | | | $b$ | $i$ | $b$ | $i$ | $b$ | $i$ | $b$ | $i$ | $b$ | $i$ |
| *a) Artificial data* | | | | | | | | | | | | |
| Assortative | 500 | 5 | 81.2 | **93.8** | 84.6 | **96.2** | 50.2 | **95.0** | 64.2 | **93.4** | 87.0 | **94.4** |
| Core-periphery | 400 | 4 | 86.2 | **95.2** | 86.5 | **94.5** | 93.2 | **96.7** | 95.2 | **95.7** | 65.7 | **89.5** |
| Dissasortative | 400 | 4 | 89.0 | **96.5** | 90.7 | **97.7** | 81.2 | **97.5** | 85.0 | **96.5** | 87.5 | **96.5** |
| Ordered | 350 | 5 | 90.8 | **95.1** | 93.1 | **98.8** | 90.0 | **98.0** | 88.2 | **96.2** | 84.6 | **94.5** |
| *b) Real data* | | | | | | | | | | | | |
| Political blogs [10] | 852 | 2 | 94.1 | **95.5** | 90.2 | **97.2** | 92.2 | **96.1** | **97.3** | 96.5 | 95.8 | **96.4** |
| EU-emails [11] | 329 | 6 | 78.7 | **87.2** | 89.0 | **92.7** | 82.7 | **84.0** | 63.2 | **91.7** | 90.9 | **94.2** |
| Village castes [12] | 356 | 3 | 81.7 | **85.1** | 82.8 | **86.2** | 74.4 | **87.1** | **87.9** | 87.6 | 82.0 | **87.9** |
| News [13] | 600 | 3 | 83.7 | **87.2** | 85.7 | **88.7** | 79.8 | **88.2** | 81.3 | **88.3** | **89.0** | 87.1 |

To address ***Q2***, we already show in Table 3 the computational time for different datasets. Experiments show that the type of data (sparse, noise) has little impact and that it is essentially (*1*) the base method used (*2*) the size and number of kernels used for the convolution that are the most determining factors. The execution time is intrinsically linked to the number of iterations (Fig. 5) and to the thresholding and smoothing. The methods that have been benchmarked (from the R seriation library) are optimized and have more or less similar times so that the average execution time of the framework for data of medium size ($300 \times 300$) rarely exceeds 3 min of computation time. It should be noted that the more the execution progresses, the smaller the improvement gain will be. Approaches for early stopping could potentially provide shorter execution times.

Table 5 compares the same dataset (band) for different levels of noise ($p = 0.1, 0.2, 0.3, 0.35$). The objective of this experiment is to answer questions ***Q3*** and ***Q4***. The method studied is the *barycentric* one. The first row of the table corresponds to the results produced by the *basic* version. The second row is the iterative method but without the smoothing enable. As expected, the stronger

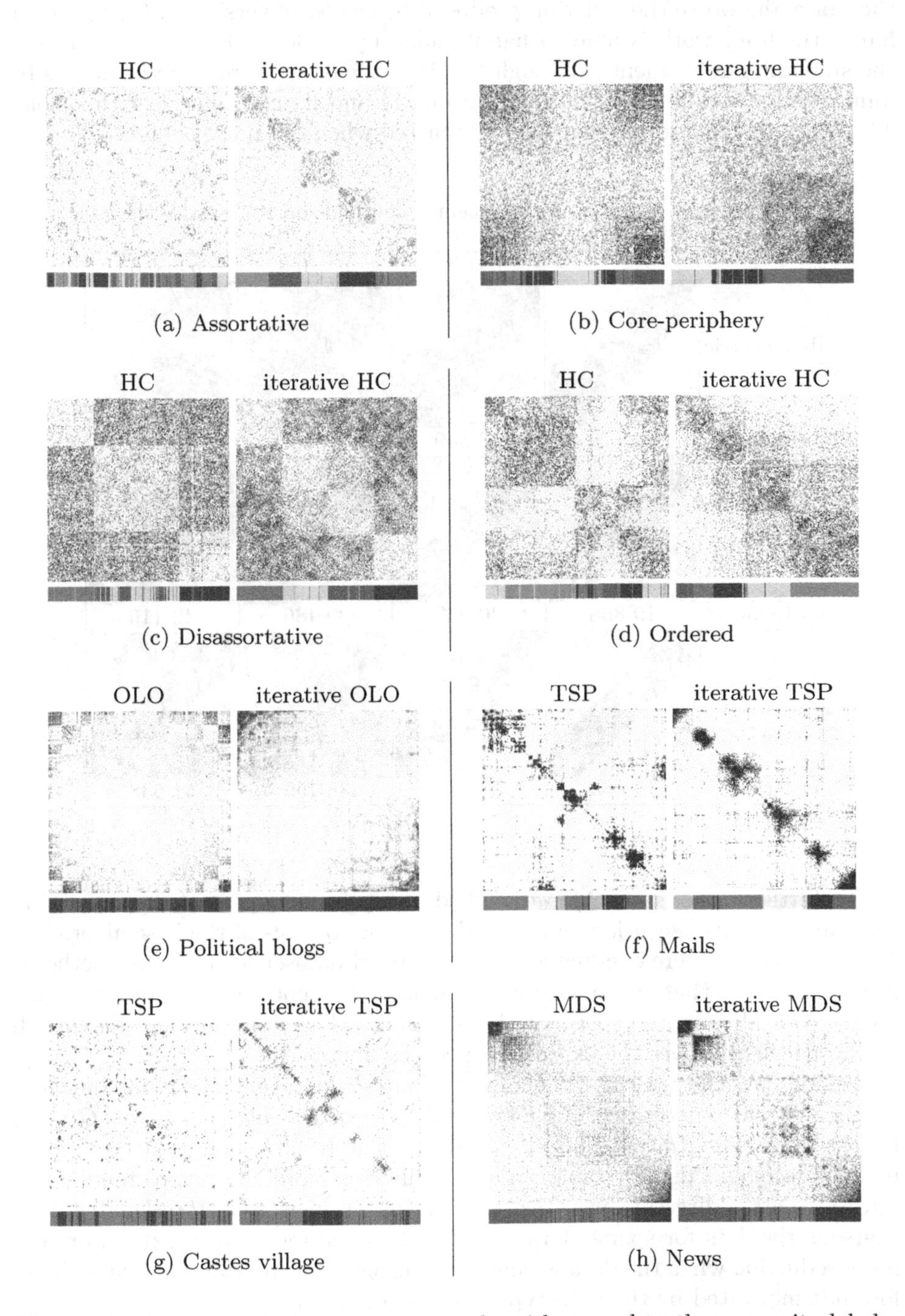


**Fig. 4.** Results for synthetic and real networks with ground truth community labels

the noise, the worse the ordering produced by the *basic* version is. On the other hand, the framework is able to handle noise up to 30–35%. The advantage of the smoothing refinement is to slightly adjust the order or rows and columns by minimizing the difference between the matrix and its convoluted or threshold. The impact of this refinement is more marked when the noise is high.

**Table 5.** Noise variation for barycentric methods on the banded dataset

| | $p = 0.1$ | $p = 0.2$ | $p = 0.3$ | $p = 0.35$ |
|---|---|---|---|---|
| barycentric | 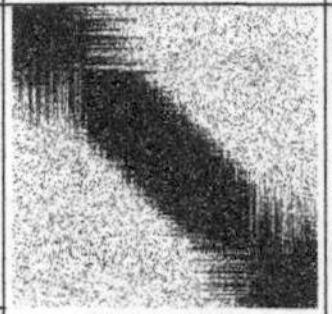 | 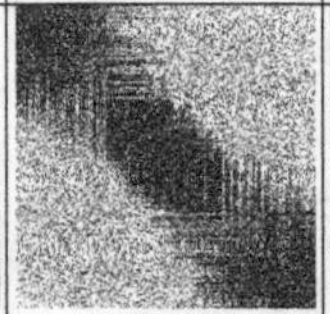 |  |  |
| criterion | 22 179 | 33 549 | 40 510 | 42 506 |
| iterative simple |  |  | 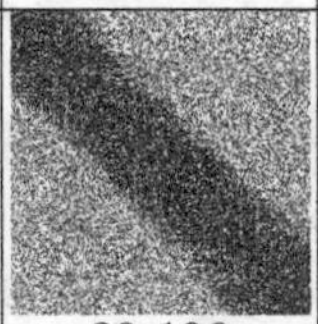 | 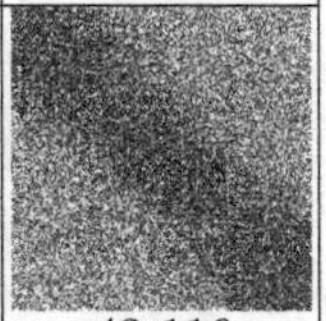 |
| criterion | 19 868 | 30 933 | 39 186 | 42 116 |
| iterative smooth |  |  |  |  |
| criterion | 19 863 | 30 819 | 38 759 | 41 538 |

To further investigate ***Q4***, we studied in Fig. 5 the impact of the smoothing refinement on the execution time and the number of runs of the base algorithm. The experiments were conducted on a 25% band dataset for different methods. A key finding is that the smoothing refinement has not only an impact on the quality but also on the speed of the algorithm. The solution converges much faster, which improves the execution speed by about 50%.

Finally, we study in Fig. 6 the impact of *thresholding* and *smoothing* on dense tables and sparse networks. The interest of this experiment is to answer ***Q5*** and to determine when *thresholding* is necessary. The results show that this refinement is recommended for tabular data, while it is more efficient to reorder the blurred matrix directly for networks. The intuition is that *thresholding* only coarsens the dots for sparse data while it allows better pattern extraction and noise reduction when the data is dense. The choice of using *thresholding* is therefore not motivated by the data type, but by the sparsity.

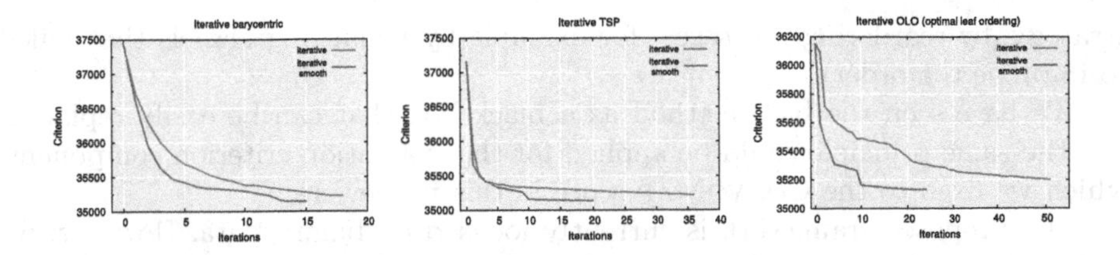


**Fig. 5.** Smoothing impact on number of iterations for a band dataset ($p = 0.25$)

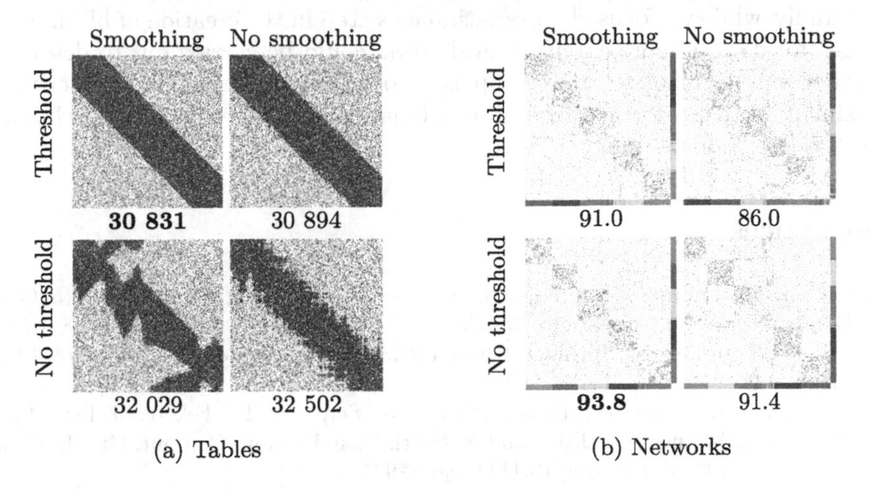


**Fig. 6.** Impact of thresholding and smoothing. TSP method used on (a) with criterion and hierarchical clustering on (b) with label accuracy

## 5 Conclusions and Future Work

We have introduced a new framework for boosting the quality of the output produced by matrix reordering or seriation methods, for a large variety of datasets, and in particular those that are noisy and sparse. The key idea is to blur the finer structure in a dataset away, hence allowing the reordering to focus on the high level structure; the blurring is progressively reduced in successive iterations. Our results on a wide range of datasets show that the method outperforms existing methods both in terms of the quality of the output and in terms of the time necessary to find the ordering.

Many adjustments and improvements can nevertheless be added to the framework. For example, we have chosen to use a single base embedded method, but a variant would be to consider a set of different methods that would be applied in parallel in order to diversify the order produced; we could also apply a different method for the initial ordering.

We choose a fixed the set of kernels used in this work. An alternative would be to generate the kernels automatically by analyzing the noisiness or sparsity of the data. We can also consider a more refined method of kernel replacement strategy, by reordering the $\kappa$ set (for example by penalizing kernels that failed to improve the order).

We have seen the base method as a black box that can be easily replaced, but the same principle could be applied for the evaluation criterion component, which we fixed to the ConvoMap score in this framework.

The proposed framework is currently focused on binary data. However, we believe it is robust enough to be easily adapted to numerical data (biological datasets, gene expression) or a mix of categorical and numerical data.

Finally, while we focused on seriation as a step in the creation of black/white images for data, a more extensive evaluation could be carried out to determine whether this form of iterative loop is also useful when solving clustering or classification tasks. for instance, in combination with deep learning techniques also based on convolution.

## References

1. Bollen, T., Leurquin, G., Nijssen, S.: ConvoMap: Using Convolution to Order Boolean Data. In: Duivesteijn, W., Siebes, A., Ukkonen, A. (eds.) IDA 2018. LNCS, vol. 11191, pp. 62–74. Springer, Cham (2018). https://doi.org/10.1007/978-3-030-01768-2_6
2. Behrisch, M., Bach, B., Henry Riche, N., Schreck, T., Fekete, J.-D.: Matrix Reordering Methods for Table and Network Visualization. Comput. Graph. Forum **35**, (2016). https://doi.org/10.1111/cgf.12935
3. Garriga, G., Junttila, E., Mannila, H.: Banded structure in binary matrices. Knowl. Inf. Syst. **28**, 197–226 (2008)
4. Hubert, L.J.: Some applications of graph theory and related nonmetric techniques to problems of approximate seriation: The case of symmetric proximity measures. Br. J. Math. Stat. Psychol. **27**, 133–153 (1974)
5. Mitchell-Jones, A. J., et al.: The atlas of European mammals, vol. 3. Academic Press, London (1999)
6. Sezgin, M., Sankur, B.: Survey over image thresholding techniques and quantitative performance evaluation. J. Electron. Imaging **13**(1), 146–165 (2004)
7. Makinen, E., Siirtola, H.: The barycenter heuristic and the reorderable matrix. Informatica (Slovenia) **29**(3), 357–364 (2005)
8. Hahsler, M., Buchta, C., Hornik, K.: Seriation: Infrastructure for Ordering Objects Using Seriation. R package version 1.2-9 (2020)
9. Faskowitz, J., Yan, X., Zuo, X.N., et al.: Weighted stochastic block models of the human connectome across the life span. Sci Rep **8**, 12997 (2018)
10. Adamic, L.A. Glance, N. The political Blogosphere and the 2004 U.S. election: divided they blog. In: Proceedings of the 3rd international workshop on Link discovery, LinkKDD 2005, New York, pp. 36–43. ACM (2005)
11. Yin, H., Benson, A.R., Leskovec, J., Gleich, D.F. Local higher-order graph clustering. In: Proceedings of the 23rd ACM SIGKDD International Conference on Knowledge Discovery and Data Mining (2017)

12. Banerjee, A., Chandrasekhar, A.G., Duflo, E., Jackson, M.O.: The Diffusion of Microfinance (2013). https://doi.org/10.7910/DVN/U3BIHX,HarvardDataverse, V9
13. Ding, C.H., He, X., Zha, H., Gu, M., Simon, H.D. A min-max cut algorithm for graph partitioning and data clustering. In: Proceedings IEEE International Conference on Data Mining (ICDM), 2001, pp. 107–114. IEEE (2001)

# Semi-supervised Semantic Visualization for Networked Documents

Delvin Ce Zhang(✉) and Hady W. Lauw

School of Computing and Information Systems, Singapore Management University, Singapore, Singapore
{cezhang.2018,hadywlauw}@smu.edu.sg

**Abstract.** Semantic interpretability and visual expressivity are important objectives in exploratory analysis of text. On the one hand, while some documents may have explicit categories, we could develop a better understanding of a corpus by studying its finer-grained structures, which may be latent. By inferring latent topics and discovering keywords associated with each topic, one obtains a semantic interpretation of the corpus. One the other hand, by visualizing documents, latent topics, and category labels on the same plot, one gains a bird's eye view of the relationships among documents, topics, and various categories. Semantic visualization is a class of methods that unify both topic modeling and visualization. In this paper, we propose a novel semantic visualization model for networked documents that incorporates partial labels. We introduce coordinate-based label distribution and label-dependent topic distribution to visualize documents, topics, and labels in a semi-supervised way. We further derive three variants for singly-labeled, multi-labeled, and hierarchically-labeled documents. The focus on semi-supervision that employs variants of labeling structures is particularly novel. Experiments verify the efficacy of our model against baselines.

**Keywords:** Semantic visualization · Topic modeling · Dimensionality reduction · Generative models

## 1 Introduction

While text documents are mainly expressed in words, in many cases they are interconnected in a network, e.g., Web page hyperlinks or paper citations. When exploring such a corpus, we seek a comprehensive understanding in terms of both latent semantics and document proximity. On one hand, *topic modeling* excels at latent semantics. It represents each document by a topic distribution, and a topic is described by a group of keywords. Lacking visual interpretation, it requires cognitive efforts to summarize. *Visualization*, on the other hand, provides another view to understand the corpus by projecting high-dimensional documents to a low-dimensional space (2D or 3D), so similar documents could be found in spatial proximities. But it offers no lexical nor semantic interpretability. Given such tradeoffs, a promising direction is to pursue the 'joint' avenue of *semantic visualization*, which conducts topic modeling and visualization simultaneously, and visualizes documents and topics in the same scatterplot.


N. Oliver et al. (Eds.): ECML PKDD 2021, LNAI 12977, pp. 762–778, 2021.
https://doi.org/10.1007/978-3-030-86523-8_46

Existing semantic visualization models are mainly *unsupervised*. They do not take advantage of the fact that many corpora are partially labeled. Documents may be partitioned into categories, such as primary areas of academic publications. From this observation, we draw three critical insights. First, visualizing category labels in addition to documents and topics would better flesh out the corpus structure, as labels summarize a group of topics, and topics characterize documents. Second, by exploiting label structure, we could improve topic modeling, as documents within the same category would share topics or neighbors. Third, even partially available labels would be useful, if the modeling could induce probabilistic labels in a semi-supervised way. Note that labels are different from topics. The former capture a category of documents and are explicit and observed, the latter are completely latent. A document usually has more latent topics than observed labels. Documents of the same label may still vary in topics.

Of particular interest is the existence of several label structures. Single-label would be the most common, each document is assigned only one label. Alternatively, documents may also be tagged, giving rise to a multi-label structure, e.g., news articles with multiple tags. In some scenarios, the categorization may even be hierarchical, e.g., academic papers from the same area further fall into different sub-areas. We seek to design a semantic visualization model capable of accommodating different label structures.

The proposed model is called SemiVN, a **Semi**-supervised topic model for semantic **V**isualization of **N**etworked documents. The *first* key design is to introduce coordinate-based label distribution and label-dependent topic distribution to visualize documents, topics, and labels on the same scatterplot. One can infer how documents relate to topics and how topics relate to labels by visually sensing relative distances. *Second*, to support multiple label structures, we further enrich label-dependent topic distribution and derive three variants for single, multiple, and hierarchical labeling, respectively. *Third*, by deterministically supervising observed labels and probabilistically modeling unobserved labels, SemiVN benefits from partially available labels in a semi-supervised manner.

To demonstrate one of SemiVN's use cases, Fig. 1 is a screenshot of an interactive interface of SemiVN's output of the Coronavirus news corpus[1]. Effectively understanding newsstream in terms of their main topics could help in selecting articles of interest to readers efficiently. SemiVN generates topic-word and label-word distributions for semantic interpretability, as seen by the example word clouds. A label is rendered as a black triangle. Right-clicking would reveal its word cloud, which represents a summary of documents. Topics further split a label into sub-concepts. A topic is rendered as a white circle. For instance, the word cloud of the label around the bottom gray area reveals *health and hospital*. In turn, its surrounding topic further focuses on *health situation of Boris Johnson*. To see the content of a document, one can left-click on one of the colored circles (the color reflects the category of the document), revealing the content in a separate window below the scatterplot. The placement of documents on the plot reveals the coherence within each category, as well as the potential semantic relations across labels and topics. For instance, *Economy, business, and finance* category (blue) lies in the center, suggesting that economy is associated with diverse industries and influenced by Coronavirus from many aspects. SemiVN unifies semantic interpretability and visual expressivity and provides a holistic understanding of the corpus.

[1] https://aylien.com/blog/free-coronavirus-news-dataset.

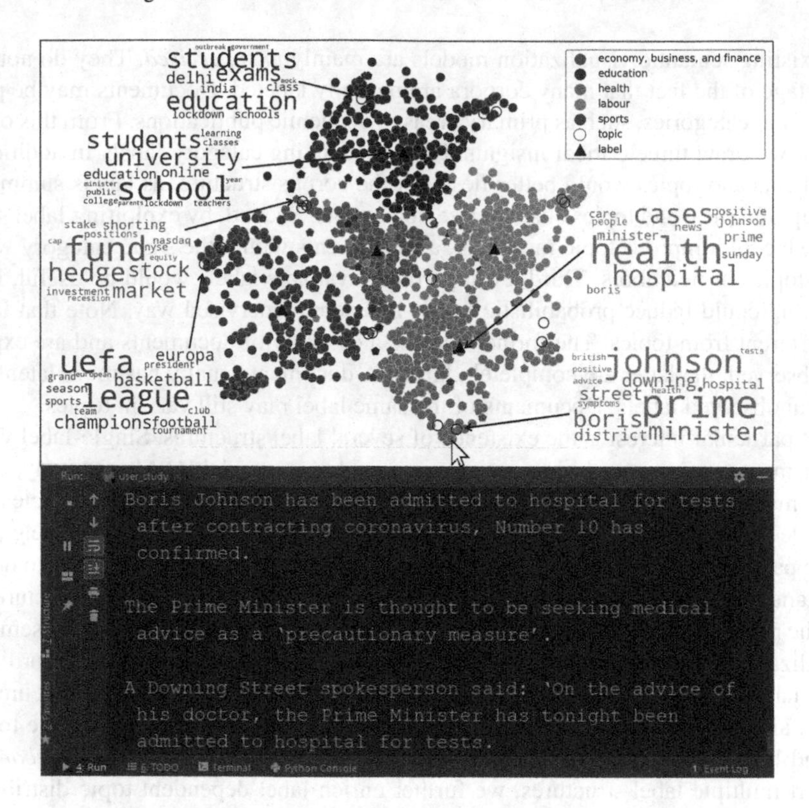


**Fig. 1.** Semantic visualization of Coronavirus news corpus with 20 topics. (Color figure online)

The joining of visualization and topic modeling, within a semi-supervised framework, lends SemiVN new capabilities not existing in prior models. Pure visualization tools, such as the widely used t-SNE [18] does not model topics, and cannot express main topics in the scatterplot. This motivates the development of semantic visualization. Prior works in semantic visualization are mostly unsupervised, except ContraVis [14], which requires full supervision and accommodates only single labels. SemiVN is designed in a semi-supervised manner to leverage a proportion of labeled documents to visualize all and extends beyond single labeling to multiple and hierarchical labeling.

**Problem.** Let $\mathcal{G} = \{\mathcal{D}, \mathcal{E}, \mathcal{L}\}$ be a document network with labels. $\mathcal{D} = \{\mathbf{d}_i\}_{i=1}^{N}$ is document set. Each document $\mathbf{d} \in \mathbb{R}^{|\mathcal{V}|}$ is a vector in the vocabulary space $\mathcal{V}$. We use *tf-idf* to represent $\mathbf{d}$. $\mathcal{E} \subseteq \mathcal{D} \times \mathcal{D}$ contains edges, where $e_{ij} \in \mathcal{E}$ if there is an edge between document $i$ and $j$. Here we model an undirected network, $e_{ij} = e_{ji}$. We will use *edge* and *link* interchangeably. Document $i$'s neighbors $\mathcal{N}(i)$ are those directly linked to $i$. As in [28], when no appropriate links are observed in a corpus, one could alternatively induce similarity-based $k$NN document network based on *tf-idf* cosine similarity. The set $\mathcal{L}$ has observed labels where $\ell_i \subseteq \mathcal{L}$ if we observe document $i$'s label(s) $\ell_i$.

Given a partially labeled document network $\mathcal{G}$ as input, the goal is to find visual coordinates for *i)* $N$ documents $\{x_i\}_{i=1}^{N}$, *ii)* $T$ topics $\{\phi_t\}_{t=1}^{T}$, and *iii)* $L$ labels $\{\psi_l\}_{l=1}^{L}$, where the Euclidean distances among coordinates reflect distributions of document-topic, document-label, and label-topic pair.

**Contributions.** *First*, we introduce coordinate-based label distribution and label-dependent topic distribution, and propose a novel semi-supervised topic model for networked documents that unifies semantic and visual expressivity. *Second*, we extend our model for singly-labeled, multi-labeled and hierarchically-labeled documents. *Third*, our model outperforms baselines quantitatively and qualitatively on public datasets[2].

## 2 Related Work

**Semantic Visualization.** Incorporating topic modeling into data visualization is referred to as *semantic visualization*. One pioneering model is PLSV [8]. PLANE [15] is the first attempt on semantic visualization of networked documents. However, these models learn coordinates in an unsupervised way, and do not embed labels to reflect corpus hierarchy. While [14] incorporates labels for contrastive visualization, it requires labels for all the documents in the corpus, which precludes the use of unlabeled documents or class prediction. On the contrary, SemiVN unifies networked documents, topics, and labels into the same visualization scatterplot in a semi-supervised way. There are also models [5,16] visualizing documents, but without the notion of labels as well.

**Document Network Embedding.** Previously, topic models for document networks are based on graphical models, e.g., RTM [4] leverages topics of two documents to predict the link. More recent models are based on neural approaches. NRTM [1] extends VAE [10] by introducing a multi-layer perception [2] for link prediction. Adjacent-Encoder [31] models network structure by neighboring document reconstruction. These embed networked documents into topic space only, without any visualization. For the latter, one needs a post-hoc embedding using dimensionality reduction (e.g., t-SNE [18]). In contrast, SemiVN systematically incorporates topic modeling and visualization as a joint approach without the necessity for post-hoc embedding. There are other models that learn node embeddings on attributed graphs [12,29], but they are not topic nor visualization models. They do not generate topic-word matrix, and the learned embeddings are not topics. Their learning process does not offer visualization.

**(Semi-) Supervised Topic Modeling.** Supervised and semi-supervised topic models are those methods that embed both textual content and document labels and produce label-dependent topic distributions. Graphical models include sLDA [20] and DiscLDA [13] for single labeling, LLDA [24] for multi-labeling, and PLDA [25] for partially labeling documents. SemiVAE [11] and MVAE [30] are based on Auto-Encoder, a neural topic model. Similarly, these models do not have an in-built visualization aspect, thus need a post-hoc technique for visual comparison. We also distinguish SemiVN from hierarchical topic modeling, such as nCRP [7], which learns hierarchical topics unsupervisedly. SemiVN's topics are not hierarchical and are semi-supervised.

[2] Source code and datasets are available at https://github.com/cezhang01/semivn.

# 3 Model Architecture and Analysis

In this section, we describe the technical details of proposed generative approach, whose graphical models are given by Fig. 2. See Table 1 for the summary of notations.

**Table 1.** Summary of notations.

| Notation | Description |
|---|---|
| $\mathcal{G}$ | Document network |
| $\mathcal{D}$ | Document set |
| $\mathcal{E}$ | Edge set |
| $\mathcal{L}$ | Label set |
| $\mathcal{V}$ | Vocabulary |
| $\mathbf{d}_i$ | Document $i$'s $tf - idf$ representation in the vocabulary space, $\mathbf{d}_i \in \mathbb{R}^{\lvert\mathcal{V}\rvert}$ |
| $\mathcal{N}(i)$ | Document $i$'s neighbor set |
| $\ell_i$ | Document $i$'s observed label(s) |
| $N$ | Number of documents, $N = \lvert\mathcal{D}\rvert$ |
| $T$ | Number of topics |
| $L$ | Number of labels |
| $x_i$ | Visualization coordinate of document $i$ |
| $\phi_t$ | Visualization coordinate of topic $t$ |
| $\psi_l$ | Visualization coordinate of label $l$ |
| $\hat{\mathbf{y}}_i$ | Document $i$'s estimated label distribution, $\hat{\mathbf{y}}_i \in \mathbb{R}^L$, or $\hat{\mathbf{y}}_i \in \mathbb{R}^D$ (hierarchical variant only) |
| $\mathbf{y}_i$ | Document $i$'s ground-truth label distribution, $\mathbf{y}_i \in \mathbb{R}^L$, or $\mathbf{y}_i \in \mathbb{R}^D$ (hierarchical variant only) |
| $D$ | Depth of the hierarchical softmax tree |
| $H$ | Number of different paths on the tree |
| $h$ | A path on the tree |
| $M$ | Number of negative samples |
| $\mathbf{t}_i$ | Document $i$'s topic distribution, $\mathbf{t}_i \in \mathbb{R}^T$ |
| $\hat{\mathbf{d}}_i$ | Document $i$'s generated content, $\hat{\mathbf{d}}_i \in \mathbb{R}^{\lvert\mathcal{V}\rvert}$ |
| $N_l$ | Number of documents with observed label $l$ |
| $z$ | Dimension of visualization coordinates (2 or 3 in general) |

## 3.1 Coordinate-Based Distribution

To tightly couple topic modeling and visualization, we devise a model whose parameters are visualization coordinates that give rise to the probability distributions that underlie a topic model. We define coordinate-based label distribution and label-dependent topic distribution, then discuss three modelings: labels $\mathcal{L}$, links $\mathcal{E}$, and text $\mathcal{D}$.

Labels represent main categories of a corpus and summarize a group of topics; topics in turn characterize documents. We preserve corpus structure with a nested approach. First, we introduce a label distribution $p(l|i)$ for document $i$. The generation of each link $e_{ij}$ can be characterized as follows. For document $i$, we draw its label $l \sim p(l|i)$, representing $i$'s main category. Its linked neighbor $j$ is then generated based on $i$ and its label $l$ by $j \sim p(j|i,l)$. Formally,

$$p(e_{ij}) = p(j|i)p(i) \propto \sum_{l} p(j|i,l)p(l|i) \tag{1}$$

where we assume $p(i) = \frac{1}{N}$. Since label is a general description of corpus, and groups a set of topics, given document $i$ and its label assignment, we factorize $p(j|i,l)$ into topic distributions $\sum_t p(j|t)p(t|i,l)$. Equation 1 can be rewritten as

$$p(e_{ij}) \propto \textstyle\sum_l p(j|i,l)p(l|i) = \sum_l \sum_t p(j|t)p(t|i,l)p(l|i) = \sum_t p(j|t) \sum_l p(t|i,l)p(l|i). \tag{2}$$

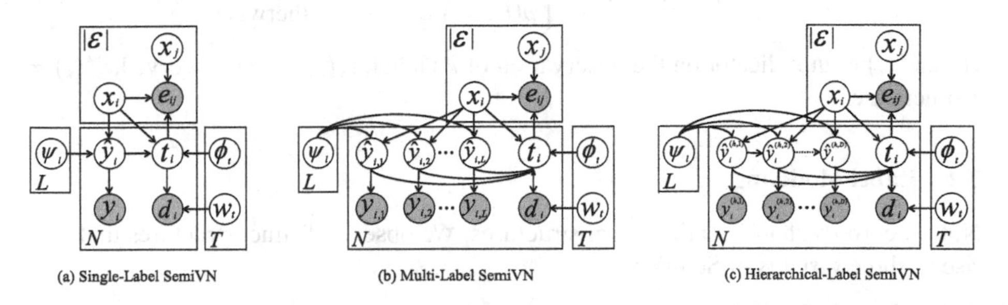


**Fig. 2.** Graphical models of (a) Single-Label SemiVN, (b) Multi-Label SemiVN, and (c) Hierarchical-Label SemiVN.

We interpret Eq. 2 as follows. Each document $i$ is represented by its label distribution $p(l|i)$, wherein each label is decomposed into topic distribution $p(t|i,l)$. In turn, each topic generates neighboring document $j$ by $p(j|t)$. Each link is generated in a nested process, and corpus structure is preserved.

Since we are interested in modeling visualization coordinates, we define label distribution $p(l|i)$ as

$$p(l|i) = \frac{\exp(-\frac{1}{2}||x_i - \psi_l||^2)}{\sum_{l'} \exp(-\frac{1}{2}||x_i - \psi_{l'}||^2)}. \tag{3}$$

This is expressed in terms of the Euclidean distances between a document $i$'s coordinate $x_i$ and those of different labels $\psi_{l'}$. The closer is $x_i$ to a specific $\psi_l$, the higher is the probability $p(l|i)$, which aligns with the objective of semantic visualization. In turn, for each document and label, we introduce a coordinate-based label-dependent topic distribution $p(t|i,l)$

$$p(t|i,l) = \frac{\exp(-\frac{1}{2}||x_i - \phi_t||^2)\exp(-\frac{1}{2}||\psi_l - \phi_t||^2)}{\sum_{t'} \exp(-\frac{1}{2}||x_i - \phi_{t'}||^2)\exp(-\frac{1}{2}||\psi_l - \phi_{t'}||^2)}. \tag{4}$$

The topic distribution is jointly determined by both document $i$'s coordinate $x_i$ and its label coordinate $\psi_l$. Document $i$ has a high topic probability $p(t|i,l)$ when it is close to topic $\phi_t$, plus $\phi_t$ is a nearby topic of label $\psi_l$. Finally, $p(j|t)$ is similarly defined.

$$p(j|t) = \frac{\exp(-\frac{1}{2}||\phi_t - x_j||^2)}{\sum_{j'} \exp(-\frac{1}{2}||\phi_t - x_{j'}||^2)}. \tag{5}$$

So far, we have assumed no label has been observed, and we model such uncertainty in a probabilistic way. Since documents are partially labeled, if we observe document $i$'s label $\ell_i$, its deterministic label distribution is $p(\ell_i|i) = 1$ and $p(l \neq \ell_i|i) = 0$. We substitute it into topic distribution ($\sum_l p(t|i,l)p(l|i)$ at Eq. 2), and obtain $p(t|i,\ell_i)$, instead of a summation over all possible labels. We rewrite Eq. 2 below.

$$\begin{aligned} p(e_{ij}|\mathbb{I}(\ell_i)) &\propto \sum_t p(j|t)p(t|i,\mathbb{I}(\ell_i)), \\ \text{where } p(t|i,\mathbb{I}(\ell_i)) &= \begin{cases} \sum_l p(t|i,l)p(l|i) & \text{if } \mathbb{I}(\ell_i)=\emptyset, \\ p(t|i,\ell_i) & \text{otherwise.} \end{cases} \end{aligned} \tag{6}$$

Here $\mathbb{I}(\ell_i)$ is an indicator on the observation of $i$'s label, $\mathbb{I}(\ell_i) = \ell_i$ if observed, $\mathbb{I}(\ell_i) = \emptyset$ otherwise.

### 3.2 Label Modeling

Not all corpora share identical label structures. We observe distinct structures that give rise to three variants of SemiVN.

**Single-Label.** Each document has onc label. Still, we observe the labels of only a proportion of documents in the corpus. With coordinate-based label distribution $\hat{\mathbf{y}}_i = p(l|i) = [\hat{y}_{i,1}, \hat{y}_{i,2}, ..., \hat{y}_{i,L}]^T$ estimated by Eq. 3, we maximize the following log-likelihood for document $i$'s observed label $\ell_i$.

$$\mathcal{J}_{label} = \log p(\mathbf{y}_i|i) = \sum_{l=1}^{L} y_{i,l} \log \hat{y}_{i,l}. \tag{7}$$

Here $\mathbf{y}_i = [y_{i,1}, y_{i,2}, ..., y_{i,L}]^T$ is the ground-truth label distribution with $y_{i,l=\ell_i} = 1$ and $y_{i,l\neq\ell_i} = 0$.

**Multi-label.** SemiVN can be extended to model multi-labeled documents. In this case, document $i$'s observed label set contains more than one label, $|\ell_i| > 1$. Coordinate-based label distribution at Eq. 3 $\hat{\mathbf{y}}_i = [\hat{y}_{i,1}, \hat{y}_{i,2}, ..., \hat{y}_{i,L}]^T$ is no longer softmax. Each single label probability is modified to

$$\hat{y}_{i,l} = \sigma(-\frac{1}{2}||x_i - \psi_l||^2), \quad (l = 1, 2, ..., L), \tag{8}$$

$$\mathcal{J}_{label} = \log p(\mathbf{y}_i|i) = \sum_{l=1}^{L} y_{i,l} \log \hat{y}_{i,l} + (1 - y_{i,l}) \log(1 - \hat{y}_{i,l}). \tag{9}$$

Again $\mathbf{y}_i$ is the ground-truth label distribution. Coordinate-based label-dependent topic distribution Eq. 4 is extended to

$$p(t|i,\ell_i) = \frac{\exp(-\frac{1}{2}||x_i - \phi_t||^2)\prod_{l\in\ell_i}\exp(-\frac{1}{2}||\psi_l - \phi_t||^2)}{\sum_{t'}\exp(-\frac{1}{2}||x_i - \phi_{t'}||^2)\prod_{l\in\ell_i}\exp(-\frac{1}{2}||\psi_l - \phi_{t'}||^2)}. \quad (10)$$

**Hierarchical-Label.** In contrast to independent labels in the multi-label scenario, hierarchical-label relies on label dependency in a $D$-level tree (with labels as nodes). Document $i$'s label is thus a path on the tree $\ell_i = \{\ell_{i,d}\}_{d=1}^{D}$. See Fig. 3 for illustration of NET dataset. Motivated by hierarchical softmax [23], we modify coordinate-based label distribution Eq. 3 to $\hat{\mathbf{y}}_i = [\hat{y}_i^{(1)}, \hat{y}_i^{(2)}, ..., \hat{y}_i^{(H)}]^T$, where $H$ is the number of different paths on the tree. The probability of each path is $\hat{y}_i^{(h)} = \prod_{d=1}^{D}\hat{y}_i^{(h,d)}$, and each single-label probability $\hat{y}_i^{(h,d)}$ of $d^{th}$ label on path $h$ is Eq. 8. The log-likelihood function is similar to Eq. 9, except that the summation is in terms of $H$, rather than $L$. Finally, its label-dependent topic distribution aligns with Eq. 10.

### 3.3 Link and Content Modeling

**Link Modeling.** To model all the links in a given network, we maximize the log-likelihood of each observed link, $\log p(e_{ij}|\mathbb{I}(\ell_i))$ at Eq. 6. Directly maximizing this objective is intractable for large networks, we instead maximize its lower bound below.

$$\mathcal{J}_{link} = E_{t\sim q(t|i,j,\mathbb{I}(\ell_i))}[\log p(j|t)] - KL[q(t|i,j,\mathbb{I}(\ell_i))||p(t|i,\mathbb{I}(\ell_i))]. \quad (11)$$

Here $q(t|i,j,\mathbb{I}(l_i))$ is a variational distribution that approximates the true posterior $p(t|i,j,\mathbb{I}(l_i))$, and KL divergence $KL(\cdot||\cdot)$ measures the difference between two distributions. We similarly define variational distribution $q(t|i,j,\mathbb{I}(l_i))$ as

$$q(t|i,j,\mathbb{I}(\ell_i)) = \begin{cases} \sum_l q(t|i,j,l)q(l|i,j) & \text{if } \mathbb{I}(\ell_i) = \emptyset, \\ q(t|i,j,\ell_i) & \text{otherwise.} \end{cases} \quad (12)$$

We parameterize $q(l|i,j)$ and $q(t|i,j,\ell_i)$ using coordinates.

$$q(l|i,j) = \frac{\exp(-\frac{1}{2}||x_i \oplus x_j - \psi_l||^2)}{\sum_{l'}\exp(-\frac{1}{2}||x_i \oplus x_j - \psi_{l'}||^2)} \quad (13)$$

$$q(t|i,j,l) = \frac{\exp(-\frac{1}{2}||x_i \oplus x_j - \phi_t||^2)\exp(-\frac{1}{2}||\psi_l - \phi_t||^2)}{\sum_{t'}\exp(-\frac{1}{2}||x_i \oplus x_j - \phi_{t'}||^2)\exp(-\frac{1}{2}||\psi_l - \phi_{t'}||^2)}. \quad (14)$$

We use $\oplus$ to denote element-wise average operation. For each link $e_{ij}$, we utilize Eq. 12 to evaluate topic distribution, and adopt gumbel-softmax reparameterization [9,19] to sample a topic. Evaluating $\log p(j|t)$ in Eq. 11 is computationally expensive on large networks, since it requires summation over all the documents. Inspired by negative sampling [22], we replace $\log p(j|t)$ with

$$\log\sigma(-\frac{1}{2}||\phi_t - x_j||^2) + \sum_{m=1}^{M} E_{v\sim P_n(v)}[\log(1-\sigma(-\frac{1}{2}||\phi_t - x_v||^2))], \quad (15)$$

$\sigma(x) = \frac{1}{1+\exp(-x)}$ is sigmoid, $P_n(v)$ is a noise distribution over documents, and $M$ is the number of negative samples.

**Content Modeling.** Another important objective is topic modeling by learning topic-word associations. Following previous neural topic models [27,31], with coordinate-based label-dependent topic distribution $\mathbf{t}_i = p(t|i, \mathbb{I}(\ell_i))$ at Eq. 6, we generate its observed plain text, and parameterize this decoder using a fully connected neural network by $\hat{\mathbf{d}}_i = p(\mathbf{d}_i|\mathbf{t}_i) = \sigma(\mathbf{W}\mathbf{t}_i + \mathbf{b})$. Here $\sigma(x)$ is sigmoid function, $\mathbf{W} \in \mathbb{R}^{|\mathcal{V}| \times T}$ represents topic-word associations, and $\mathbf{b} \in \mathbb{R}^{|\mathcal{V}|}$ is bias. The log-likelihood of the observed textual content $\log p(\mathbf{d}_i|\mathbf{t}_i)$ is

$$\mathcal{J}_{content} = \sum_{w=1}^{|\mathcal{V}|} d_{i,w} \log \hat{d}_{i,w} + (1 - d_{i,w}) \log(1 - \hat{d}_{i,w}). \tag{16}$$

**Table 2.** Dataset statistics.

| Name | #Documents | #Links | Vocabulary | #Labels | Labeling |
|---|---|---|---|---|---|
| DS | 570 | 1,336 | 3,085 | 9 | Single |
| ML | 1,980 | 5,748 | 4,431 | 7 | Single |
| COVID | 1,500 | 6,418 | 2,226 | 5 | Single |
| NET | 1,278 | 4,610 | 6,832 | 6 | Hierarchical |
| DBLP | 14,036 | 40,269 | 8,600 | 4 | Multiple |

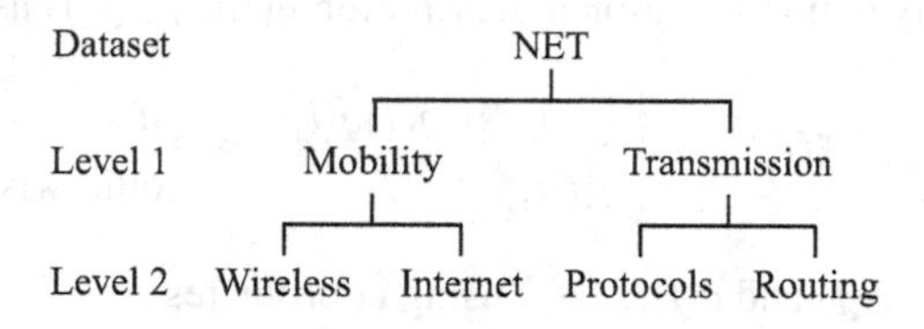


**Fig. 3.** Label hierarchy of NET.

### 3.4 The Complete Model

Given a document network $\mathcal{G}$ with links $\mathcal{E}$, document content $\mathcal{D}$, and a proportion of labels $\mathcal{L}$, putting the three components together, we obtain $\mathcal{J} = \mathcal{J}_{link} + \mathcal{J}_{content} + \mathcal{J}_{label}$ as the overall log-likelihood. We intuit that two linked documents are similar if both share many common neighbors. Thus, we add a label smoothness regularizer to objective function, which helps to distinguish different neighbors and encourages strongly connected neighbors to have similar label distributions. Specifically, the regularizer is

$$\mathcal{J}_{reg} = \sum_{e_{ij}} \alpha_{ij} d(p(l|i), p(l|j)), \quad \alpha_{ij} = \frac{|\mathcal{N}(i) \cap \mathcal{N}(j)|}{|\mathcal{N}(i) \cup \mathcal{N}(j)|}. \tag{17}$$

$\mathcal{N}(i)$ is $i$'s neighbor set. As in [26], $\alpha_{ij}$ is a similarity measure based on common neighbors. $d(\cdot,\cdot)$ measures the difference between two distributions. We use KL divergence for single-label and squared difference for multi- and hierarchical-label. Although KL is asymmetric, i.e., $KL(p(l|i), p(l|j)) \neq KL(p(l|j), p(l|i))$, in this paper we model undirected links and consider both $e_{ij}$ and $e_{ji}$, which removes the effect of asymmetry. Finally, the ultimate loss is (we take negative for log-likelihood for minimization)

$$\mathcal{J} = -\mathcal{J}_{link} - \mathcal{J}_{content} - \mathcal{J}_{label} + \lambda\mathcal{J}_{reg}. \tag{18}$$

$\lambda$ is a balancing hyperparameter.

**Inference.** After convergence, in addition to the visualization coordinates, we obtain topic-word association matrix $\mathbf{W}$ in the content decoder. To infer label-word association, we have $p(\mathbf{d}|l) = \sum_t p(\mathbf{d}|t) \sum_{i\in\mathcal{D}} p(t|i,l)p(i|l)$. Label-dependent topic distribution $p(t|i,l)$ is Eq. 4. $p(i|l) = \frac{1}{N_l}$ if $\ell_i = l$, 0 otherwise. $N_l$ is number of documents with label $l$. $p(\mathbf{d}|t)$ is content decoder. The keywords (word cloud) of each topic $t$ and label $l$ are those with highest value at $p(\mathbf{d}|t)$ and $p(\mathbf{d}|l)$, respectively. Every topic and label has its own word cloud, but for clarity, we only show some topics and labels. As in previous semantic visualization works, word cloud is not our design, our focus is to extract latent semantics and learn coordinates to visualize documents, topics, and labels. Similarly, as with previous works, including t-SNE, SemiVN is transductive. Our emphasis is using SemiVN to explore existing documents in a corpus for visual and semantic understanding.

**Complexity.** Single- and multi-labeling is $\mathcal{O}(\sum_l N_l Lz)$, hierarchical-labeling is $\mathcal{O}(\sum_l N_l DHz)$. $z$ is the dimension of visual coordinates (typically 2 or 3). Note that $\sum_l N_l = N$ only if all the documents are labeled, $\sum_l N_l < N$ otherwise. For simplicity, we use $F$ to denote $L$ and $DH$. Link modeling is $\mathcal{O}(|\mathcal{E}|(zT|\ell_i|_{max} + zM))$. Content modeling is $\mathcal{O}(NT|\mathcal{V}|)$. Putting all three together, we obtain $\mathcal{O}(|\mathcal{E}|(zdT|\ell_i|_{max} + zM) + NT|\mathcal{V}| + \sum_l N_l Fz)$. SemiVN converges in one hour on DBLP dataset (see Table 2), while some baseline, PLANE [15], even did not converge in 48 h. Evaluations were conducted on a machine with Intel Xeon E5-2650v4 2.20 GHz CPU and 256 GB RAM.

## 4 Experiments

Experimental objective is to investigate the quality of visual coordinates. Evaluating visualization is indeed not an easy task. After reviewing many previous visualization works, we summarize some standard experiments, including coordinate classification, link prediction, and topic interpretability as quantitative tasks. In addition, we further conduct user study, involving both static and interactive study.

**Datasets.** Cora [21] is a public collection of papers with abstract as content and citations as links. Two papers are linked by an undirected link if one cites the other. We extracted three independent datasets, Data Structure (DS), Machine Learning (ML), and Networking (NET). DS and ML contain singly labeled documents, NET is organized into hierarchical labels (Fig. 3). Besides Cora, we created a co-authorship network

DBLP. Each author is represented by the aggregation of her publications. Two authors are linked if they have collaboration. If an author publishes at least three papers on one type of conference, we consider her having the corresponding label. Around 11% authors have more than one label. We also created a Coronavirus news corpus. Each article belongs to one category. Since no appropriate links are observed, we generate $k$NN ($k = 5$) network using $tf - idf$ cosine similarity. Table 2 shows the statistics.

**Baselines.** We compare to several categories of baselines. *i*) **Topic modeling on networked documents**, including RTM, NRTM, and Adjacent-Encoder. *ii*) **Semantic visualization**. PLSV visualizes documents individually, and PLANE visualizes networked documents. Neither has labels. *iii*) **Semi-supervised topic model**, including PLDA, SemiVAE, and MVAE. In addition, recently there are models for attributed graph embedding. Strictly speaking, they are not topic models, nor baselines. For completeness, we still compare to *iv*) GraphTSNE [17] with GCN [12] and t-SNE as a joint model. Models in category *i*) and *iii*) extract topics only, we pipeline their topics by t-SNE to obtain coordinates. By comparison to these disjoint models, we showcase the advantage of jointly modeling topics and visualization. The comparison to joint models (PLSV, PLANE, GraphTSNE) shows the importance of modeling labels. Each result is obtained by 5 independent runs.

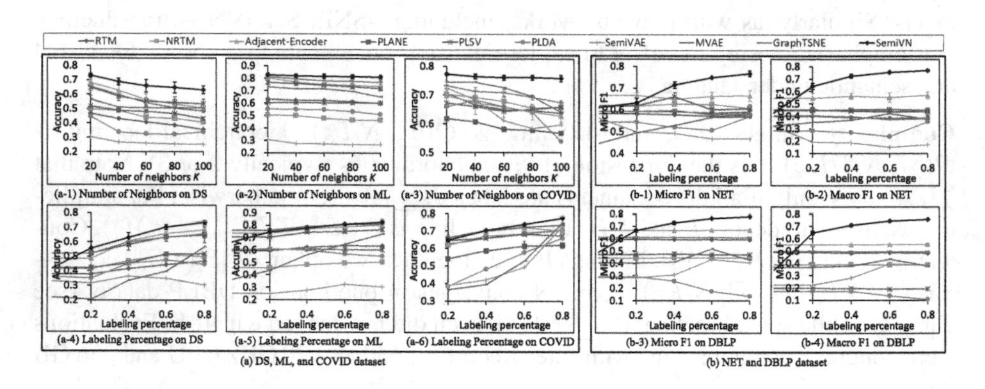


**Fig. 4.** Coordinate classification on five datasets.

Hyperparameters are chosen based on validation set. We set 2, 0.01, and 0.01 as Dirichlet prior for RTM, PLANE, and PLSV, respectively. We input texts, adjacency matrix, and labels for MVAE. Other baselines use default settings. For SemiVN, we set the number of negative samples $M$ to 5. Regularizer $\lambda$ is searched in $[0.1, 0.2, 0.5, 1, 2, 5]$ and set to 1. $tf - idf$ is generated by sklearn (https://scikit-learn.org).

## 4.1 Quantitative Evaluation

**Coordinate Classification.** A good visualization is expected to group coordinates from the same category closely, and separate different categories. For DS, ML, COVID, we adopt $K$-nearest neighbors as classifier. Our goal is to use labeled coordinates to predict labels for the unlabeled coordinates. We report classification accuracy at $T = 30$ in Fig. 4(a). For clarity, we only report std.dev. of SemiVN and best-performing baselines. We first fix 80% labeling percentage (we further split 10% among them for validation), and vary $K$ for classification. Figure 4(a-1-3) summarizes the results. Although SemiVN performs similarly with GraphTSNE at $K = 20$ on ML, as $K$ increases, SemiVN stays stable, but GraphTSNE deteriorates its results. This verifies that SemiVN benefits from modeling labels to better separate different groups of coordinates. We then fix $K = 20$ for $K$NN and vary the percentage of labeled coordinates for training. Figure 4(a-4-6) reveals that as labeling increases, most models improve results. SemiVN significantly outperforms baselines on DS. It is competitive with GraphT-SNE on ML and COVID, but still outperforms the best topic model, Adjacent-Encoder, showcasing SemiVN's advantage of jointly modeling topics and visualization.

NET and DBLP represent a multi-label classification task, thus we train a one-vs-the-rest logistic regression as a multi-label classifier. We report Micro and Macro F1 scores. We exclude PLANE on DBLP, since it did not converge in 48 h. Figure 4(b) presents the results at $T = 30$ when varying labeling percentage. Overall, semi-supervised models tend to improve results with increasing labeling.

**Coordinate-Based Link Prediction.** Following previous work in semantic visualization [15], we could use coordinates of two documents to predict a link. Following [18], the link probability is $p(e_{ij}) \propto \frac{1}{1+||x_i - x_j||^2}$. For documents with more than three links, we randomly hold out one. In total 15%–17% links are hidden. We sample the same number of disconnected pairs as negative instances. The remaining network is used for training. Our goal is to predict the held-out links. We report F1 score and AUC in Table 3. As mentioned, we pipeline some baselines with t-SNE. These disjoint models would increment errors from two separate components, thus achieving worse results than SemiVN. SemiVN outperforms PLSV and PLANE, indicating that incorporating labels can indeed help to group related coordinates together and predict links better. We do not report the results on COVID dataset, since this dataset does not explicitly observe links. It contains texts and labels only, and we induce $k$NN network based on

**Table 3.** F1 score and AUC of link prediction at $T = 30$ (results are in percentage).

| Model | DS | | ML | | NET | | DBLP | |
|---|---|---|---|---|---|---|---|---|
| | F1 | AUC | F1 | AUC | F1 | AUC | F1 | AUC |
| RTM | 52.4 ± 0.5 | 80.1 ± 0.7 | 46.8 ± 0.4 | 75.0 ± 0.4 | 47.9 ± 0.5 | 72.6 ± 0.3 | 53.5 ± 0.3 | 81.1 ± 0.2 |
| NRTM | 55.7 ± 0.4 | 83.6 ± 0.8 | 32.9 ± 0.7 | 65.6 ± 0.5 | 47.6 ± 0.7 | 72.2 ± 0.6 | 57.6 ± 0.2 | 59.3 ± 0.1 |
| Adjacent-encoder | 72.8 ± 0.3 | 94.2 ± 0.4 | 65.3 ± 0.2 | 90.3 ± 0.2 | 67.5 ± 0.1 | 86.9 ± 0.3 | 74.0 ± 0.1 | **90.3 ± 0.1** |
| PLANE | 75.4 ± 0.3 | 95.0 ± 0.4 | 53.3 ± 0.4 | 90.6 ± 0.4 | 61.7 ± 0.4 | 92.6 ± 0.3 | – | – |
| PLSV | 59.1 ± 0.6 | 55.4 ± 1.6 | 64.6 ± 0.2 | 50.1 ± 0.2 | 42.8 ± 0.2 | 76.6 ± 0.4 | 45.0 ± 0.5 | 84.0 ± 0.1 |
| PLDA | 59.3 ± 0.9 | 79.2 ± 1.0 | 50.6 ± 0.2 | 79.9 ± 0.4 | 48.2 ± 0.5 | 74.7 ± 0.3 | 53.6 ± 0.4 | 82.2 ± 0.1 |
| SemiVAE | 34.3 ± 0.6 | 61.1 ± 1.6 | 38.8 ± 0.3 | 48.7 ± 1.1 | 60.6 ± 0.5 | 55.1 ± 0.6 | 33.9 ± 0.0 | 61.8 ± 0.2 |
| MVAE | 60.8 ± 0.3 | 86.6 ± 0.8 | 42.1 ± 0.8 | 82.2 ± 0.3 | 67.3 ± 0.4 | 71.1 ± 0.3 | 14.9 ± 0.2 | 63.0 ± 0.1 |
| GraphTSNE | 77.7 ± 0.6 | 95.7 ± 0.4 | 65.9 ± 0.4 | 90.8 ± 0.3 | 70.2 ± 0.5 | 90.1 ± 0.2 | 61.4 ± 0.1 | 87.9 ± 0.1 |
| SemiVN | **80.5 ± 0.8** | **96.0 ± 0.3** | **79.0 ± 0.9** | **92.9 ± 0.6** | **78.6 ± 0.4** | **92.9 ± 0.1** | **78.3 ± 0.4** | 88.2 ± 0.4 |

**Table 4.** Topic coherence NPMI (in percentage) at $T = 30$. GraphTSNE is not a topic model, thus is not included.

| Model | NPMI | | | | |
|---|---|---|---|---|---|
| | DS | ML | COVID | NET | DBLP |
| RTM | $8.5 \pm 0.4$ | $7.4 \pm 0.3$ | $22.8 \pm 0.4$ | $14.5 \pm 0.7$ | $2.8 \pm 0.2$ |
| NRTM | $6.7 \pm 0.3$ | $7.6 \pm 0.2$ | $19.5 \pm 1.5$ | $12.4 \pm 0.2$ | $6.0 \pm 0.9$ |
| Adjacent-encoder | $5.6 \pm 0.5$ | $8.6 \pm 0.9$ | $9.9 \pm 1.6$ | $11.2 \pm 1.2$ | $7.5 \pm 1.3$ |
| PLANE | $8.2 \pm 0.1$ | $9.0 \pm 0.2$ | $21.1 \pm 0.8$ | $14.5 \pm 0.6$ | – |
| PLSV | $8.6 \pm 0.2$ | **$10.1 \pm 0.3$** | **$25.9 \pm 0.7$** | $15.4 \pm 0.2$ | $6.9 \pm 0.6$ |
| PLDA | $4.0 \pm 0.5$ | $2.8 \pm 0.3$ | $9.1 \pm 0.4$ | $5.0 \pm 0.6$ | $4.8 \pm 0.2$ |
| SemiVAE | $3.5 \pm 0.8$ | $7.1 \pm 1.0$ | $9.1 \pm 0.6$ | $8.4 \pm 2.1$ | $1.2 \pm 1.0$ |
| MVAE | $4.4 \pm 0.6$ | $6.3 \pm 0.5$ | $7.9 \pm 0.3$ | $11.0 \pm 0.9$ | $1.8 \pm 0.7$ |
| SemiVN (topic) | **$9.8 \pm 0.3$** | **$9.9 \pm 0.4$** | **$25.0 \pm 1.6$** | **$17.6 \pm 0.7$** | **$8.4 \pm 0.7$** |
| SemiVN (label) | $9.5 \pm 0.4$ | $9.7 \pm 0.2$ | $24.2 \pm 1.1$ | $17.0 \pm 0.9$ (level 1)<br>$17.2 \pm 1.3$ (level 2) | $8.1 \pm 0.2$ |

$tf-idf$ cosine similarity. Unlike other datasets where we predict citations or coauthorship, predicting links on COVID does not make any sense in real-word scenarios.

**Topic Interpretability.** We use normalized PMI (NPMI) [3] to evaluate the coherence of top 10 words of each topic. *Google Web 1T 5-gram Version 1* [6] is the external corpus for evaluation. GraphTSNE is not a topic model, thus is excluded. Table 4 shows that SemiVN (topic) generates more coherent words and interpretable topics than others. This supports the importance of labels to improve the quality of topic model.

In addition to those of topics, we also evaluate word coherence of labels. No baseline extracts label-word association. SemiVN (label) is consistently lower than topics', since labels' associated keywords are overly general and capture multiple aspects, resulting in fewer co-occurrences.

**Analysis.** To evaluate the label smoothness regularizer, we compare to two variants. *i)* SemiVN$-reg$ removes the regularizer. *ii)* SemiVN$-\alpha$ maintains it, but uses the same $\alpha_{ij}$ in Eq. 17, and neighbors are equally important. Table 5 shows that *i)* regularizer is helpful to embed neighbors closely, thus achieves better results; *ii)* modeling neighbors differently is necessary, since disregarding it leads to worse performance.

**Table 5.** Classification accuracy (in percentage) of variants at $T = 30$, $K = 20$, 80% labeling.

| Model | DS | ML | COVID |
|---|---|---|---|
| SemiVN$-reg$ | $70.6 \pm 1.6$ | $78.7 \pm 2.5$ | $70.9 \pm 1.0$ |
| SemiVN$-\alpha$ | $72.2 \pm 2.0$ | $81.2 \pm 2.2$ | $74.3 \pm 0.7$ |
| SemiVN | **$73.1 \pm 1.3$** | **$82.6 \pm 1.3$** | **$77.3 \pm 0.3$** |

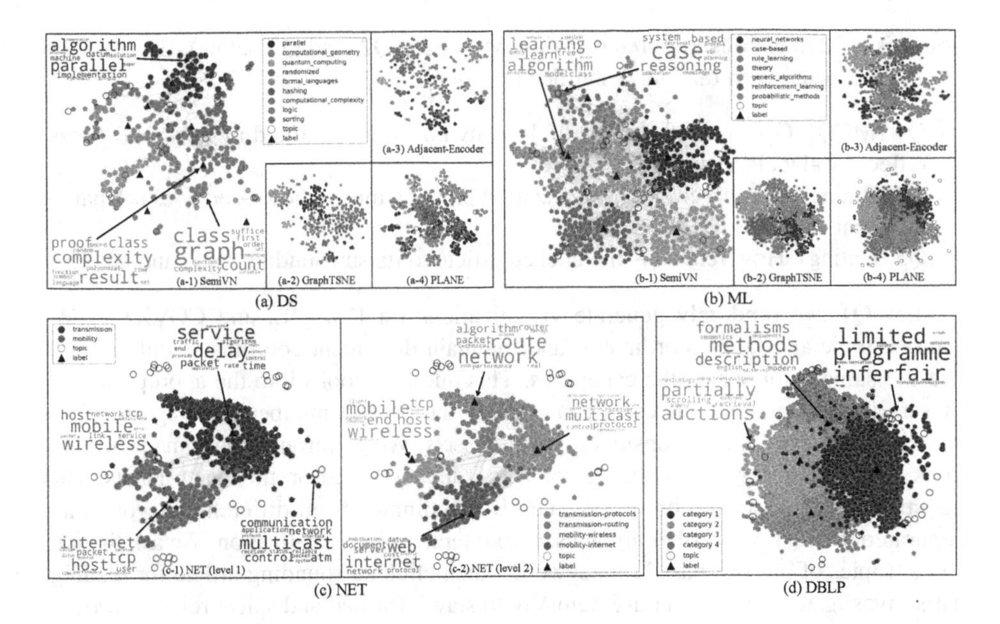


**Fig. 5.** Semantic visualization with $T = 30$ topics and 80% labeling (best seen in color). (Color figure online)

## 4.2 Visualization

To sense how SemiVN embeds networked documents, topics, and labels into the same scatterplot, we present visualizations in Fig. 5. See Fig. 1 for COVID. The similarity of topics and labels can be revealed by the relative distance among coordinates. Similar topics and labels tend to group together, distinct ones are separate. Labels and topics split visual space into different semantic subspaces. NET's labels are hierarchical at Fig. 5(c). Label coordinates in Fig. 5(c-1) center within two categories, while in Fig. 5(c-2) the second-tier labels separate into four subspaces. Overall, SemiVN produces clearer separation than baselines.

**Table 6.** Results of user study.

| Question | Adjacent-encoder | PLANE | GraphTSNE | SemiVN |
|---|---|---|---|---|
| Q1 | 10.0% | 40.0% | 2.5% | **47.5%** |
| Q2 | 2.67/5 | 3.08/5 | 2.55/5 | **4.17**/5 |
| Q3 | 2.19/5 | 3.08/5 | 1.75/5 | **4.69**/5 |

## 4.3 User Study

We conduct a user study to test the effectiveness of visualization from human perspectives. We design a survey involving 20 participants who are not authors. The survey comprises 22 questions, of three question types (Q1, Q2, and Q3 below). Each participant is presented with randomly shuffled questions.

- Q1 (MCQ): Given masked plots of 4 anonymized and shuffled models, which best reflects <#labels> clusters?
- Q2 (Rating): Given a colored visualization plot of a model, how good does it separate different categories?
- Q3 (Rating): How related is the clicked article to its surrounding topic and label?

For Q1, we randomly generate visualizations on DS, ML, and COVID. $i$) We remove topic and label coordinates, and maintain document coordinates only. $ii$) We use the same color for all the categories. This question looks into the appropriateness of semi-supervision, i.e., if users can identify the correct number of categories from masked plots. For Q2, we consider all five datasets. We go through the same procedure $i$) as above. Different from $ii$), we color coordinates based on their own labels. This question tests the separation quality, i.e., if coordinates from different categories are separated. For Q3, we further allow users to interact with visualization. We ask them to select topics of interest based on keywords, then click surrounding article for reading. This investigates if users can use SemiVN to stay informed and select relevant articles in practice, and if the article is visualized at the correct place w.r.t. topics and labels.

We compare SemiVN to three representative baselines: Adjacent-Encoder, PLANE, and GraphTSNE. Q1 contains multiple-choice questions, we report the percentage of participants who favor each model. Q2 and Q3 involve ratings from 1 (terrible) to 5 (excellent), we report the average rating of each model. Table 6 shows that SemiVN outperforms baselines on three types of questions, indicating that users are more satisfied with its ability to visualize documents, topics, and labels. PLANE is the best among baselines, verifying the advantage of modeling topics and visualization jointly.

## 5 Conclusion

We propose SemiVN, a semi-supervised semantic visualization model that embeds networked documents, topics, and labels into the same visualization scatterplot. Its versatility accommodates different labeling structures: single-label, multi-label, and hierarchical-label variants. Extensive experiments verify the effectiveness of SemiVN in both semantic interpretability and visual expressivity. Future work includes extending SemiVN to inductive scenarios so as to generalize to unseen documents.

**Acknowledgments.** This research is supported by the National Research Foundation, Prime Minister's Office, Singapore under its NRF Fellowship Programme (Award No. NRF-NRFF2016-07).

## References

1. Bai, H., Chen, Z., Lyu, M.R., King, I., Xu, Z.: Neural relational topic models for scientific article analysis. In: Proceedings of the 27th ACM International Conference on Information and Knowledge Management, pp. 27–36 (2018)

2. Bishop, C. M.: Pattern Recognition and Machine Learning, Springer, Heidelberg (2006). ISBN 978-0-387-31073-2
3. Bouma, G.: Normalized (pointwise) mutual information in collocation extraction. In: Proceedings of GSCL, pp. 31–40 (2009)
4. Chang J., Blei D.: Relational topic models for document networks. In: Artificial Intelligence and Statistics, pp. 81–88 (2009)
5. Choo, J., Lee, C., Reddy, C.K., Park, H.: Utopian: user-driven topic modeling based on interactive nonnegative matrix factorization. IEEE Trans. Visual Comput. Graph. **19**(12), 1992–2001 (2013)
6. Evert, S.: Google web 1t 5-grams made easy (but not for the computer). In: Proceedings of the NAACL HLT 2010 Sixth Web as Corpus Workshop, pp. 32–40 (2010)
7. Blei, D.M., Griffiths, T.L., Jordan, M.I., Tenenbaum, J.B.: Hierarchical topic models and the nested Chinese restaurant process. In: Advances in Neural Information Processing Systems, pp. 17–24 (2004)
8. Iwata, T., Yamada, T., Ueda, N.: Probabilistic latent semantic visualization: topic model for visualizing documents. In: Proceedings of the 14th ACM SIGKDD International Conference on Knowledge Discovery and Data Mining, pp. 363–371 (2008)
9. Jang, E., Gu, S., Poole, B.: Categorical reparameterization with Gumbel-Softmax. In: Proceedings of International Conference on Learning Representations (2017)
10. Kingma, D.P., Welling, M.: Auto-encoding variational Bayes. arXiv preprint arXiv:1312.6114 (2013)
11. Kingma, D.P., Rezende, D.J., Mohamed, S., Welling, M.: Semi-supervised learning with deep generative models. In: Proceedings of Advances in Neural Information Processing Systems, pp. 3581–3589 (2014)
12. Kipf, T.N., Welling, M.: Semi-supervised classification with graph convolutional networks. arXiv preprint arXiv:1609.02907 (2016)
13. Lacoste-Julien, S., Sha, F., Jordan, M.I.: DiscLDA: discriminative learning for dimensionality reduction and classification. In: Proceedings of Advances in Neural Information Processing Systems, pp. 897–904. (2009)
14. Le, T., Akoglu, L.: ContraVis: contrastive and visual topic modeling for comparing document collections. In: Proceedings of The World Wide Web Conference, pp. 928–938 (2019)
15. Le, T.M., Lauw, H.W.: Probabilistic latent document network embedding. In: 2014 IEEE International Conference on Data Mining, pp. 270–279 (2014)
16. Lee, H., Kihm, J., Choo, J., Stasko, J., Park, H.: iVisClustering: an interactive visual document clustering via topic modeling. In: Computer graphics forum, vol. 31, No. 3pt3, pp. 1155–1164. Oxford, UK: Blackwell Publishing Ltd. (2012)
17. Leow, Y.Y., Laurent, T., Bresson, X.: GraphTSNE: a visualization technique for graph-structured data. arXiv preprint arXiv:1904.06915 (2019)
18. Van der Maaten, L., Hinton, G.: Visualizing data using t-SNE. J. Mach. Learn. Res. **9**, 2579–2605 (2008)
19. Maddison, C.J., Mnih, A., Teh, Y.W.: The concrete distribution: a continuous relaxation of discrete random variables. In: Proceedings of International Conference on Learning Representations (2017)
20. Blei, D.M., McAuliffe, J.D.: Supervised topic models. In: Proceedings of Advances in Neural Information Processing Systems, pp. 121–128 (2008)
21. McCallum, A.K., Nigam, K., Rennie, J., Seymore, K.: Automating the construction of internet portals with machine learning. Inf. Retrieval **3**(2), 127–163 (2000)
22. Mikolov, T., Sutskever, I., Chen, K., Corrado, G.S., Dean, J.: Distributed representations of words and phrases and their compositionality. In: Advances in Neural Information Processing Systems, pp. 3111–3119 (2013)

23. Morin, F., Bengio, Y.: Hierarchical probabilistic neural network language model. In: International Workshop on Artificial Intelligence and Statistics, pp. 246–252 (2005)
24. Ramage, D., Hall, D., Nallapati, R., Manning, C.D.: Labeled LDA: a supervised topic model for credit attribution in multi-labeled corpora. In: Proceedings of the 2009 Conference on Empirical Methods in Natural Language Processing, pp. 248–256 (2009)
25. Ramage, D., Manning, C.D., Dumais, S.: Partially labeled topic models for interpretable text mining. In: Proceedings of the 17th ACM SIGKDD International Conference on Knowledge Discovery and Data Mining, pp. 457–465 (2011)
26. Rozemberczki, B., Davies, R., Sarkar, R., Sutton, C.: GEMSEC: graph embedding with self clustering. In: Proceedings of the 2019 IEEE/ACM International Conference on Advances in Social Networks Analysis and Mining, pp. 65–72 (2019)
27. Srivastava, A., Sutton, C.: Autoencoding variational inference for topic models. arXiv preprint arXiv:1703.01488 (2017)
28. Tang, J., Liu, J., Zhang, M., Mei, Q.: Visualizing large-scale and high-dimensional data. In: Proceedings of the 25th International Conference on World Wide Web, pp. 287–297 (2016)
29. Veličković, P., Cucurull, G., Casanova, A., Romero, A., Lio, P., Bengio, Y.: Graph attention networks. In: Proceedings of International Conference on Learning Representations (2018)
30. Wu, M., Goodman, N.: Multimodal generative models for scalable weakly-supervised learning. In: Proceedings of Advances in Neural Information Processing Systems, pp. 5575–5585 (2018)
31. Zhang, C., Lauw, H.W.: Topic modeling on document networks with adjacent-encoder. In: Proceedings of the AAAI Conference on Artificial Intelligence, vol. 34(04), pp. 6737–6745 (2020)

# Self-supervised Multi-task Representation Learning for Sequential Medical Images

Nanqing Dong[1(✉)], Michael Kampffmeyer[2], and Irina Voiculescu[1]

[1] Department of Computer Science, University of Oxford, Oxford OX1 3QD, UK
nanqing.dong@cs.ox.ac.uk

[2] Department of Physics and Technology, UiT The Arctic University of Norway, 9019 Tromsø, Norway

**Abstract.** Self-supervised representation learning has achieved promising results for downstream visual tasks in natural images. However, its use in the medical domain, where there is an underlying anatomical structural similarity, remains underexplored. To address this shortcoming, we propose a self-supervised multi-task representation learning framework for sequential 2D medical images, which explicitly aims to exploit the underlying structures via multiple pretext tasks. Unlike the current state-of-the-art methods, which are designed to only pre-train the encoder for instance discrimination tasks, the proposed framework can pre-train the encoder and the decoder at the same time for dense prediction tasks. We evaluate the representations extracted by the proposed framework on two public whole heart segmentation datasets from different domains. The experimental results show that our proposed framework outperforms MoCo V2, a strong representation learning baseline. Given only a small amount of labeled data, the segmentation networks pre-trained by the proposed framework on unlabeled data can achieve better results than their counterparts trained by standard supervised approaches.

**Keywords:** Self-supervised learning · Multi-task learning · Medical image segmentation

## 1 Introduction

Fueled by the recent success of convolutional neural networks (CNN), deep learning (DL) has led to many breakthroughs in computer vision tasks, benefiting from large-scale training data. However, under standard supervised learning (SL), preparing a large training dataset requires extensive and costly hand annotation, especially in the medical domain, where the annotation requires further domain expertise from the clinicians. To mitigate this data scarcity challenge in SL, there is a renaissance of research on self-supervised learning (SSL) [34]. SSL aims to learn meaningful representations from the unlabeled data in SSL and then transfer the extracted representations for the downstream task with small-scale labeled data. For simplicity, we use the term self-supervised learning and the term self-supervised representation learning interchangeably in this


N. Oliver et al. (Eds.): ECML PKDD 2021, LNAI 12977, pp. 779–794, 2021.
https://doi.org/10.1007/978-3-030-86523-8_47

work. The state-of-the-art (SOTA) SSL methods [6,9,20,27,36] have demonstrated that, on various downstream tasks, a model trained with only unlabeled data plus a small amount of labeled data can achieve comparable performance to the same model trained with large amounts of labeled data.

Although similar SSL techniques [7] have been applied in medical tasks and have achieved promising performance, several questions remain unsolved. First, SOTA SSL methods leverage instance-wise differences among the unlabeled data by *contrastive learning*. For example, the models are pre-trained on ImageNet [11] and fine-tuned with PASCAL VOC [15] and MS COCO [24] for the downstream tasks. As shown in Fig. 1, compared with natural images from ImageNet [11] which belong to particular categories, the medical images show clear differences: (1) a natural image usually has a single object of interest centered in the image, while a medical image usually contains more than one semantic class; (2) the background (BG) in medical images usually contains more supportive semantic information than natural images; (3) for a particular class, the instance-wise difference is more obvious in natural images than medical images. For medical images such as computerized tomography (CT) and magnetic resonance imaging (MRI), a scan contains a series of slices for the same patient. The difference between neighbored slices is commonly negligible by human eyes (see Fig. 2 for example). Further, to utilize instance discrimination as the *pretext* task, SOTA SSL methods can only pre-train the encoder for image classification tasks, due to the nature of the instance discrimination task. In contrast to image classification, where an input image is mapped to a single label, dense prediction tasks are expected to learn a pixel-wise mapping between the input and the output. For downstream tasks such as depth estimation, edge detection, and surface normal estimation, contrastive learning cannot learn representations for the decoder. In addition, medical datasets are usually much smaller than general-purpose datasets such as ImageNet. This would limit the performance of contrastive learning methods, which rely on large-scale training data to catch the instance-wise difference [6,20]. Last but not least, directly applying SSL methods on medical images does not utilize domain-specific knowledge that is particular to the medical domain. Unlike general objects, medical objects such as human organs or human structures usually share statistical similarities in terms of the location, shape, and size among different patients [14]. There have been studies of utilizing such an anatomical structural similarity, as a free lunch, in medical image analysis [10,13,14,30]. See Fig. 2 for the intuition of anatomical structural similarity.

To bridge the methodological gaps discussed above, we propose a novel SSL framework for medical images such as CT scans or MRI scans. We propose two pretext tasks that can be formulated as two SSL problems by utilizing the characteristics of the medical data. Concretely, for a single slice in a series of medical images in sequential order, we try to reconstruct two slices that have a fixed distance to it on both sides. Given a CNN with an end-to-end pixel-wise mapping

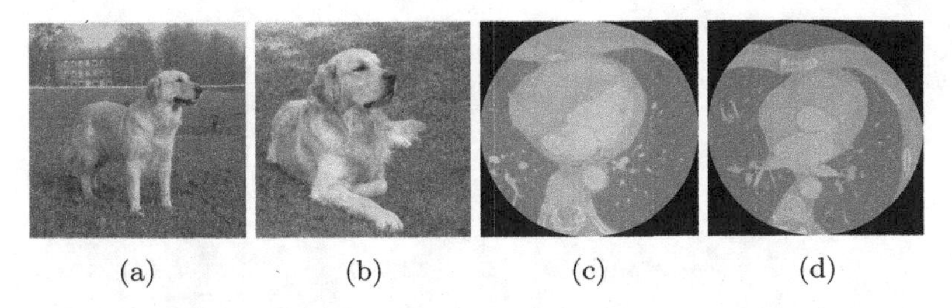

(a) (b) (c) (d)

**Fig. 1.** Visual comparison between instance-wise difference. (a) and (b) are two Golden Retrievers sampled from the ImageNet dataset [11]. (c) and (d) are two slices sampled from two patients' CT scans, which are collected in the CT-WHS dataset [43].

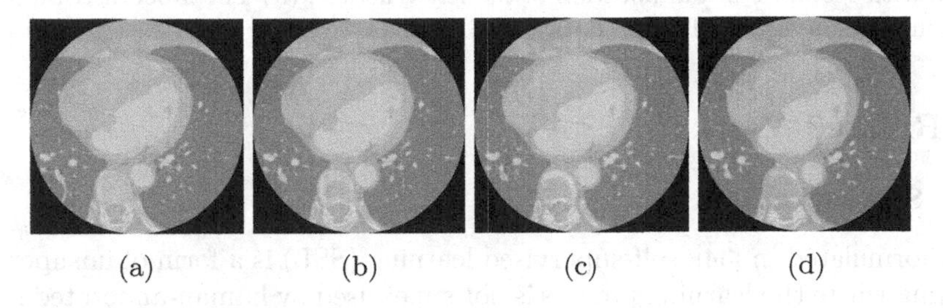

(a) (b) (c) (d)

**Fig. 2.** A piece of sequential slices of a patient's CT scan. The CT scan is collected in the CT-WHS dataset [43].

(e.g. U-Net [33]), two pretext tasks are trained jointly in a multi-task learning (MTL) formulation to learn domain-specific knowledge for the CNN. Given limited data, we use MTL to improve the generalization of the CNN. If a CNN can be decomposed into an encoder and decoder separately (e.g. FCN [25] where the encoder can be viewed as a standard feature extractor for image classification), we extend the proposed framework to integrate the instance discrimination task for the encoder as the third pretext task. We evaluate the proposed framework in medical image segmentation tasks where we pre-train a segmentation network on unlabeled data first and then fine-tune the model with small-scale labeled data. The segmentation network pre-trained by the proposed framework can outperform the same network pre-trained by MoCo [20], the SOTA SSL method, in the whole heart segmentation tasks.

Our main contributions can be summarized as follows: 1) We propose a simple SSL framework for dense prediction tasks on medical images with inherent sequential order. 2) We are the first to exploit anatomical structural similarity to integrate MTL with SSL. 3) We extend the proposed framework to incorporate the concept of contrastive learning.

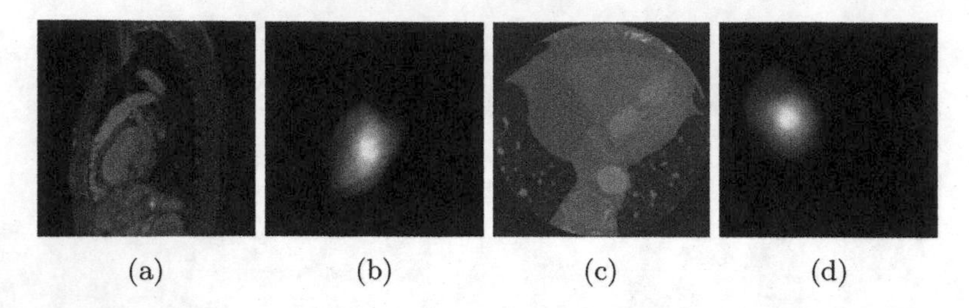


**Fig. 3.** Illustration of anatomical structural similarity: (a) An axial CT image with ground truth annotation of the right atrium. (b) The label distribution (normalized density heatmap) of right atriums in the CT-WHS dataset [42]. (c) A sagittal MRI image with ground truth annotation of the left ventricle. (d) The label distributions of left ventricles in the MRI-WHS dataset [43].

## 2 Related Works

### 2.1 Self-supervised Learning

First formulated in [34], self-supervised learning (SSL) is a form of unsupervised learning where the learning process is not supervised by human-annotated labels. The concept of SSL originated from cognitive science by observing how babies interact with new environments [18]. Meanwhile, representation learning aims to extract useful representations from data [2] in the context of DL. An important application of SSL is to learn transferable representations for downstream tasks, e.g. common downstream tasks for visual understanding include image recognition [21,35], object detection [23,32], and semantic segmentation [5,25]. With this definition, SSL can also be understood as transfer learning from unlabeled data [31].

**Pretext Tasks.** In the recent renaissance of SSL in visual understanding tasks, the role of SSL in each target task is associated with the corresponding *pretext* task. By solving a pretext task, the model extracts meaningful representations for the target task. Thus, the model is pre-trained on the pretext task for the target task. For example, [12] utilizes two CNNs with shared weights to predict the relative positions of two patches randomly cropped from the same image. Similarly, for an image divided into a $3 \times 3$ grid, [28] permutes the order of 9 patches and predicted the index of the chosen permutation, just like solving jigsaw puzzles.[40] colorizes the grayscale images by using the lightness channel $L$ as input to predict the corresponding $a$ and $b$ color channels of the image in the CIE *Lab* colorspace. [17] randomly rotates the images by multiples of 90 degrees and predicts the rotation. Designing good pretext tasks requires extra effort and is challenging, as *a good self-supervised task is neither simple nor ambiguous* [28].

**Contrastive Learning.** Contrastive learning was first developed as a learning paradigm for neural networks to identify what makes two objects similar or different [1]. In the literature of SSL, contrastive learning utilizes the instance-wise difference and systematically defines the pretext tasks as a simple instance discrimination task [20]. Recently, SOTA contrastive learning methods [6,20,27,36] have been proposed based on InfoNCE [29], a contrastive loss motivated by *noise-contrastive estimation* (NCE) [19]. By minimizing InfoNCE, the model is expected to learn the invariant features shared by a positive pair [6,27,36], where a positive pair is usually defined as two stochastically augmented views from the same instance. Note, SOTA contrastive learning methods usually rely on large-scale datasets, which are often unavailable in the medical domain.

### 2.2 Multi-task Learning

Multi-task learning (MTL) [4] is a learning paradigm inspired by human learning activities where the knowledge learned from previous tasks can help learn a new task. MTL aims to improve the generalization performance of all the tasks by leveraging useful information contained in multiple related tasks [38]. In the era of DL, we use a model to map the input to the output, given a specific task. In contrast to single-task learning, where each task is handled by an independent model, MTL can reduce the memory footprint, increase overall inference speed, and improve the model performance. Moreover, when the associated tasks contain complementary information, MTL can regularize each single task. For dense prediction tasks, a good example is semantic segmentation, where we always assume that the classes of interest are mutually exclusive. Depending on the data modality of the input and the task affinity [37] between tasks, there are various types of MTL. We depict the workflows for the situations that the tasks share the same input in Fig. 4. Given the same input, pixel-level tasks in visual understanding often have similar characteristics, which can be potentially used to boost the performance by MTL [41].

## 3 Method

Let $X$ be an unlabeled set consisting of sequences of medical images for $N$ patients, i.e. $X = \{\boldsymbol{x}_i\}_{i=1}^{N}$. Each example consists of a sequence of medical images $\boldsymbol{x}_i = \{x_i^j\}_{j=1}^{n_i}$ for the patient $i$. For example, the sequences of medical images could be CT scans or MRI scans. The goal is to learn meaningful representations from $X$ for a downstream dense prediction task, such as semantic segmentation on medical images.

### 3.1 Self-supervised Multi-tasking Learning

Based on the empirical observation of anatomical structural similarity, we assume that each medical image $x_i^j$ follows an unknown distribution $\mathcal{X}$. We aim to utilize this similarity. Given a neural network backbone for a dense prediction task, we create two pretext tasks which can be formulated in a MTL setting. Concretely, given a sequence $\boldsymbol{x}_i$, we use an anchor image $x_i^j$ to predict the image $t$ steps before $x_i^{j-t}$ and the image $t$ steps behind $x_i^{j+t}$, where $t$ is an integer. Formally, let the neural network backbone we are interested (including the encoder and

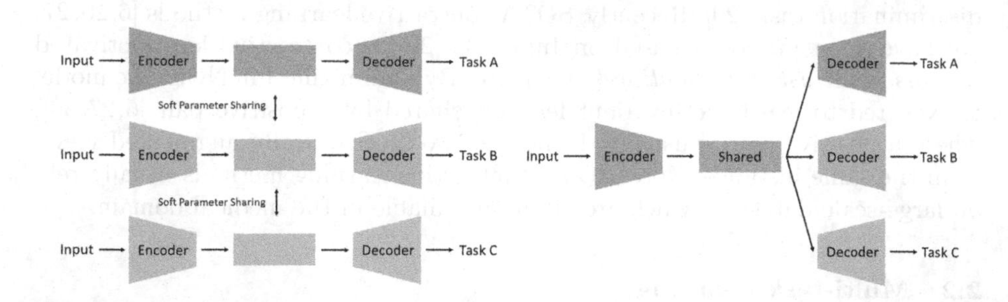


(a) Soft Parameter Sharing (b) Hard Parameter Sharing

**Fig. 4.** Two common MTL workflows for dense prediction tasks given the same input. (a) Soft parameter sharing: The different tasks have separate models (in different colors), where the parameters are communicated between models. (b) Hard parameter sharing: The different tasks share the same encoder and network backbone (in purple) but independent decoders. (Color figure online)

the decoder) be $f_\theta$ and the auxiliary task decoders be $g_{\phi^-}$ for $x_i^{j-t}$ and $g_{\phi^+}$ for $x_i^{j+t}$ respectively. The loss function is

$$\mathcal{L}_{pretext} = \sum_i ||g_{\phi^-}(f_\theta(x_i^j)) - x_i^{j-t}|| + ||g_{\phi^+}(f_\theta(x_i^j)) - x_i^{j+t}||, \quad (1)$$

where $||\cdot||$ denotes a distance measure in Euclidean space. For simplicity, we use a standard Euclidean distance (i.e. mean squared error). The overall learning framework is illustrated in Fig. 5.

Intuitively, when we are learning the mappings from $x_i^j$ to $x_i^{j-t}$ and $x_i^{j+t}$, the anatomical structural similarity (e.g. the relative location, shape, and size of the organs and structures) is extracted by the neural network backbone. For the downstream tasks such as medical image segmentation, the extracted knowledge should play an important role as there is an overlap of the semantic information shared between the pretext tasks and the downstream tasks. Note, without the regularization of MTL, i.e. if there is only one pretext task, the neural network backbone could only memorize information for just one direction, which might be an easy pretext task to learn and make the learned representations less meaningful for the downstream tasks. Theoretically, we can have $2|\mathcal{T}|$ pretext tasks for $t \in \mathcal{T}$. In this work, we show that two pretext tasks are sufficient to learn meaningful representations for the downstream tasks.

### 3.2 Integration with Instance Discrimination

As discussed in Sect. 2.1, contrastive learning can be viewed as a generalized pretext task in SSL. As contrastive learning has shown promising performance in SSL for the encoder of image classification tasks, it is natural to consider

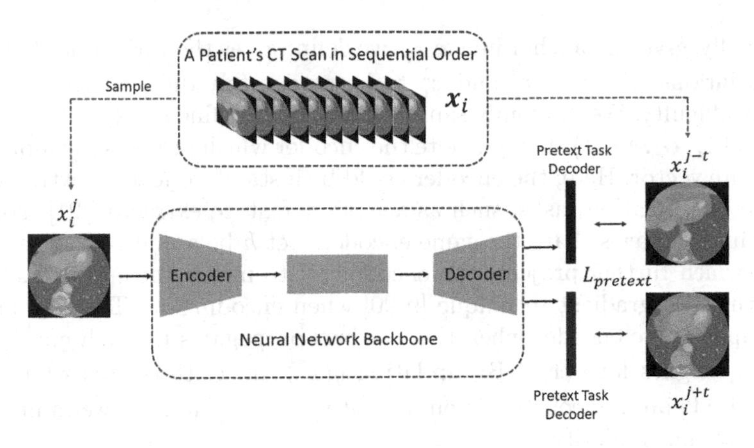


**Fig. 5.** Illustration of the proposed self-supervised multi-task learning framework. Given a sequence of medical images $\boldsymbol{x}_i$ (e.g. a CT scan of a patient), randomly sample $x_i^j$ from the sequence and get corresponding $x_i^{j-t}$ and $x_i^{j+t}$ if possible. The architecture in purple is the neural network backbone that we are interested in (i.e. $f_\theta$) and the two architectures in red and blue are two decoders for two pretext tasks respectively (i.e. $g_{\phi^-}$ and $g_{\phi^+}$). (Color figure online)

including instance discrimination as the third pretext task in the MTL formulation when the neural network backbone can be perfectly decomposed as an encoder and a decoder. Here, we require that the encoder and the decoder can be trained independently (although we train them jointly).

For STOA contrastive learning methods, a positive pair is defined as two augmented views from the same instance and a negative pair is defined as two augmented views from two different instances. The common data augmentation policies include the combinations of cropping, resizing, flipping, color distortion by grayscale conversion, color distortion by jittering, cutout, Gaussian noise, Gaussian blurring, rotation, and Sobel filtering [6]. However, most of these data augmentation policies can not be applied to medical images directly for three reasons. First, medical images tend to be grayscale images or can only be transformed to grayscale images. Second, medical images are sensitive to local texture, which may be changed by the data augmentation. Third, unlike the random crops that contrastive learning methods usually work on, dense prediction tasks for medical images usually require the whole image as the input.

As discussed in Sect. 1, medical images share more similarities in the objects of interest than general objects in natural images due to the anatomical structural similarity. Instead of defining negative pairs, we only define positive pairs, inspired by Siamese networks [3,9]. Based on the anatomical structural similarity, we propose to define two slices from the same scan as a positive pair. Intuitively, we are using two *natural* variants of the same instance rather than *synthetic* variants (i.e. augmented views) of the same instance. For example, Fig. 2(a) and Fig. 2(d) can be viewed as a natural positive pair.

Formally, given an anchor image $x_i^j$, we define $x_i^k$ as the variant of $x_i^j$. To avoid trivial solutions, we want $x_i^j$ and $x_i^k$ to be moderately different to increase the learning difficulty. We randomly sample $k$ where we define $k \in \{j-2t, \cdots, j-t-1\} \cup \{j+t+1, \cdots, j+2t\}$. Let $q$ denote the encoder which projects the input image to a feature vector. Here, the encoder could be a standard feature extractor used in image classification tasks, such as a ResNet feature extractor [21]. Following [9], two input images share the same encoder. Let $h$ be a multi-layer perceptron (MLP), which further projects the encoded $x_i^k$ to match the encoded $x_i^j$. Note, we use the stop gradient technique [9,20] when encoding $x_i^k$. That is to say, we do not update the encoder when the loss backpropagates through $q(x_i^k)$, i.e. we use fixed weights for $q(x_i^k)$. By updating $q(x_i^j)$ alone, three pretext tasks can be optimized simultaneously. Given a positive pair $x_i^j$ and $x_i^k$, we minimize the negative cosine similarity

$$\mathcal{L}_{sim} = -\frac{q(x_i^k)}{||q(x_i^k)||_2} \cdot \frac{h(q(x_i^j))}{||h(q(x_i^j))||_2}, \tag{2}$$

where each encoded feature vector is normalized by its $l_2$ norm. The motivation here is to learn the invariance between two images. Given two slices from the same patient, the invariance shared between two images is the general knowledge of human structures for the region of interest.

The final optimization object for the self-supervised MTL is to minimize the total loss of the three pretext tasks. We have

$$\mathcal{L}_{self} = \mathcal{L}_{pretext} + \lambda \mathcal{L}_{sim}, \tag{3}$$

where $\lambda$ is the hyperparameter to control the weight of $\mathcal{L}_{sim}$. In this work, to balance the weights among three pretext tasks, we use set $\lambda = 1$[1]. The complete learning framework is presented in Fig. 6. Note, for Sect. 3.1, there is no assumption of the architecture of the neural network, i.e. the learning framework should apply to any dense prediction task. However, for Sect. 3.2, we assume that the encoder should be the standard feature extractor for image classification tasks. We will empirically evaluate both frameworks in Sect. 4.

# 4 Experiments

We evaluate the proposed SSL frameworks on the whole heart segmentation (WHS) task. Unlike general semantic segmentation tasks that commonly take standard RGB images as input, WHS could have different source domains, namely CT scans and MRI scans. CT scans and MRI scans show variations in data modalities, which can be viewed in Fig. 3. The purposes of the experiments are twofold. First, we want to validate the theoretical advantages of the proposed framework. Second, we want to show that the proposed framework can extract meaningful representations for different types of data.

[1] Task balancing is a topic of active research in MTL, which is beyond the scope of discussion in this work. We refer the interested readers to [38] for details.

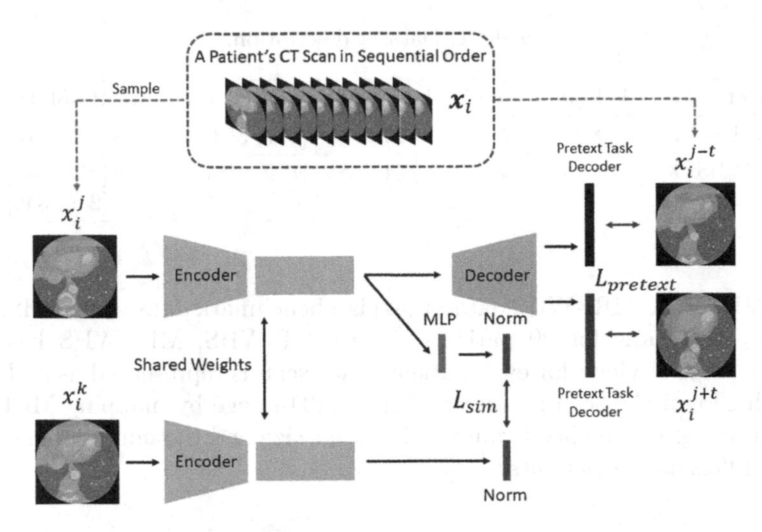


**Fig. 6.** Illustration of the proposed self-supervised multi-task learning framework with the additional instance discrimination pretext task. In addition to Fig. 5, we minimize the similarity between the encoded $x_i^j$ and the encoded $x_i^k$, where two images share the same encoder. Note, we only update $q(x_i^j)$, the branch with the MLP in the MTL formulation. The neural network backbone that we are interested in includes an encoder (the architecture in yellow) and a decoder (the architecture in purple). The auxiliary MLP (the architecture in green) will not be used in the downstream tasks. (Color figure online)

### 4.1 Datasets

We use two public benchmark datasets for WHS.[2] See Table 1 for the statistics of the datasets. Each dataset contains manual segmentation masks of 7 substructures of the heart for 20 patients: the left ventricle blood cavity, the right ventricle blood cavity, the left atrium blood cavity, the right atrium blood cavity, the myocardium of the left ventricle, the ascending aorta, and the pulmonary artery. See Table 1 for the description of the datasets.

**CT-WHS.** The CT-WHS dataset [43] is a benchmark dataset in WHS, which contains CT scans for 20 patients. CT-WHS has axial views for all the patients. Each CT scan is represented as a 3D array and each slice of the scan is converted to a 2D image by mapping Hounsfield units to grayscale pixel values. The number of slices per patient differs across patients. Each slice has a fixed resolution of $512 \times 512$.

[2] http://www.sdspeople.fudan.edu.cn/zhuangxiahai/0/mmwhs/.

**Table 1.** Dataset description.

| Dataset | Axial | Frontal | Sagittal | # scans | # slices per scan | Resolution |
|---|---|---|---|---|---|---|
| CT-WHS | ✓ | ✗ | ✗ | 20 | [177, 363] | $512 \times 512$ |
| MRI-WHS | ✗ | ✗ | ✓ | 14 | [120, 180] | $[256, 340] \times [256, 340]$ |

**MRI-WHS.** The MRI-WHS dataset [42] is a benchmark dataset in WHS, which contains MRI scans for 20 patients. Unlike CT-WHS, MRI-WHS has either frontal or sagittal views for each patient. Each scan is represented as a 3D array and each slice of the scan is converted into a 2D image by mapping MRI intensity values to grayscale pixel values. The image size and the number of slices per patient differs across patients.

### 4.2 Experimental Setup

**Implementation.** For Sect. 3.1, we use a standard U-Net [33] as the neural network backbone. For simplicity, we replace the last convolutional layer of U-Net as two 1 convolutional layers with 64 channels for the input and 1 channel for the output. That is to say, we maximally shared the neural network backbone in MTL. For Sect. 3.2, as we need to decompose the neural network backbone into an encoder and a decoder, we choose FCN [25] as the neural network backbone. We use the ResNet50 [21] as the encoder. More precisely, the ResNet50 in this work denotes the ResNet50 architecture without the final fully-connected layer. The MLP consists of 2 fully-connected layers, whose number input channels and output channels are $2048 \mapsto 512$ and $512 \mapsto 2048$. The decoder is trained as stated in Sect. 3.1. There is limited literature for utilizing anatomical structural similarity in self-supervised multi-task learning for dense prediction tasks on medical images. For the baseline model, we use the SOTA contrastive learning framework MoCo V2 [8][3]. For a fair comparison, we also use ResNet50 as the encoder. When fine-tuning with the downstream task, the decoder is randomly initialized. All models are implemented by PyTorch on an NVIDIA Tesla V100 GPU.

**Hyperparameters.** For a fair comparison, we use the same set of hyperparameters for all models and all networks (encoder and decoder) are initialized with the same random seed. We use an Adam optimizer [22] with a fixed learning rate of $10^{-3}$ across all experiments. The batch size is 16. Note, we do not use any data augmentation in the proposed framework. For MoCo V2, we use the default hyperparameters except for $K$ and the stochastic data augmentation policy. We set $K$ to 1024 because we have much smaller datasets than ImageNet[4]. As discussed in Sect. 3.2, data augmentation plays an important role in

[3] https://github.com/facebookresearch/moco.

[4] $K$ is originally 65532.

**Table 2.** Proxy evaluation of self-supervised multi-task learning on axial CT scans with U-Net as the neural network backbone.

| Method | LV | RV | LA | RA | mIOU |
|---|---|---|---|---|---|
| w/o pre-training | 0.445 | 0.378 | 0.427 | 0.429 | 0.420 |
| SSMTL (t = 5) | **0.656** | 0.496 | **0.657** | 0.537 | **0.586** |
| SSMTL (t = 10) | 0.571 | 0.478 | 0.545 | **0.558** | 0.537 |
| SSMTL (t = 15) | 0.590 | **0.499** | 0.541 | 0.479 | 0.527 |

SOTA contrastive learning frameworks. The original data augmentation policy is designed for RGB images instead of grayscale images. Here, we therefore use random flipping and stochastic Gaussian blurring as proposed in MoCo V2 [8].

**Training and Evaluation.** For consistency among the datasets, all images and corresponding annotations are resized to 256×256. The resizing is also important for the baseline MoCo V2. The encoder for MoCo V2 is designed for standard ImageNet image size i.e. 224 × 224. Each image is pre-processed by instance normalization. For the cardiac segmentation tasks on CT scans, we use all slices of the 20 patients for the self-supervised pre-training. All models are pre-trained for the same number of epochs for a fair comparison. In terms of evaluation, the benchmark *linear classification protocol* [6,8,20,27,39] only applies to image classification tasks. Instead, we use the performance of supervised semantic segmentation as a *proxy* measurement for the quality of the learned representations. We choose four classes of interest in the cardiac segmentation task: the left ventricle blood cavity (LV), the right ventricle blood cavity (RV), the left atrium blood cavity (LA), and the right atrium blood cavity (RA). We split the 20 patients into a training set of 5 CT scans and a test set of 15 CT scans. As in practical situations, the clinical annotators will only annotate a small number of slices for each scan. We randomly sampled 20 annotated slices from each scan, resulting in a total of 100 slices as the training data. We use such as small training data to simulate the challenging data scarcity situation and also demonstrate the efficiency of the proposed framework. Given the self-supervised pre-trained neural network backbone, we fine-tune the model with the small training set and report the Intersection-Over-Union (IOU) of each class of interest and the mean IOU (mIOU) on the test set. The same training strategy applies to the MRI scans in the set. However, we only use 14 sagittal MRIs as the self-supervised pre-training data. For the evaluation, we split the 14 patients into a training set of 4 MRI scans and a test set of 10 MRIs. Similarly, we randomly sampled 25 annotated slices from each scan as the training images.

### 4.3 Results

We first evaluate the proposed framework in Sect. 3.1 on CT and MRI scans. We pre-trained the U-Net on CT scans for 100 epochs and on MRI scans for 400 epochs because CT scans have around 3 times more slices than MRI scans.

**Table 3.** Proxy evaluation of self-supervised multi-task learning on sagittal MRI scans with U-Net as the neural network backbone.

| Method | LV | RV | LA | RA | mIOU |
|---|---|---|---|---|---|
| w/o pre-training | 0.447 | 0.491 | 0.260 | 0.267 | 0.354 |
| SSMTL (t = 5) | 0.552 | 0.489 | **0.297** | **0.371** | **0.427** |
| SSMTL (t = 10) | 0.600 | **0.509** | 0.257 | 0.271 | 0.409 |
| SSMTL (t = 15) | **0.640** | **0.509** | 0.283 | 0.234 | 0.417 |

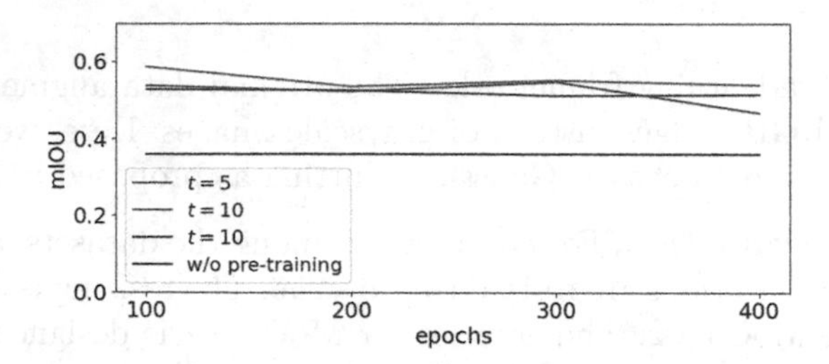


**Fig. 7.** The learning goal for pretext tasks might not be well-aligned with the learning goal for downstream tasks: more pre-training epochs do not always help the downstream tasks.

As discussed in Sect. 4.2, we use the downstream task cardiac segmentation as a proxy evaluation. For CT scans, the models are trained with 100 labeled CT slices until convergence and tested with 4080 CT slices. For MRI scans, the number of training and test slices are 100 and 1480, respectively. The results of the segmentation performance are reported in Table 2 and Table 3. We denote the proposed framework as SSMTL. The U-Net pre-trained with SSMTL outperforms the U-Net without pre-training by a large margin on both datasets. In fact, this large margin is caused by the data scarcity. With insufficient labeled images, which is quite common in the medical domain, traditional supervised approaches could easily fail. The proposed framework is an efficient alternative. The overall best performance on both datasets is achieved for $t = 5$, but in practice, the choice of $t$ depends on the slice thickness. It is worth mentioning that the length of the sequence and the shape of the structure would influence the performance of representation learning. As shown in Table 2 and Table 3, the pre-training leads to more performance gain for scans with more sequential slices and larger structures. We also perform an ablation study for the number of pre-training epochs in Fig. 7. More pre-training epochs might not always help because the model could overfit to the pretext tasks. This also leads to an interesting research question about how to measure the task affinity between pretext tasks and downstream tasks, which is left as future work.

We repeat the previous experiment to evaluate the extended framework proposed in Sect. 3.2. We denote this extension of SSMTL as SSMTL+. We use $t = 5$. This time, we use a FCN with a ResNet50 encoder. The results are

**Table 4.** Proxy evaluation of self-supervised multi-task learning on CT scans with ResNet-FCN as the neural network backbone.

| Method | LV | RV | LA | RA | mIOU |
|---|---|---|---|---|---|
| w/o pre-training | 0.548 | 0.475 | 0.491 | 0.421 | 0.483 |
| MoCo V2 | 0.586 | 0.491 | 0.512 | 0.424 | 0.503 |
| SSMTL | 0.573 | 0.487 | 0.451 | 0.398 | 0.477 |
| SSMTL+ | 0.607 | 0.487 | 0.523 | 0.434 | 0.513 |
| SSMTL (U-Net) | **0.656** | **0.496** | **0.657** | **0.537** | **0.586** |

**Table 5.** Proxy evaluation of self-supervised multi-task learning on sagittal MRI scans with ResNet-FCN as the neural network backbone.

| Method | LV | RV | LA | RA | mIOU |
|---|---|---|---|---|---|
| w/o pre-training | 0.445 | 0.415 | 0.186 | 0.254 | 0.325 |
| MoCo V2 | 0.481 | 0.443 | 0.226 | 0.276 | 0.357 |
| SSMTL | 0.454 | 0.399 | 0.145 | 0.243 | 0.310 |
| SSMTL+ | 0.501 | 0.473 | 0.211 | 0.257 | 0.361 |
| SSMTL (U-Net) | **0.552** | **0.489** | **0.297** | **0.371** | **0.427** |

presented in Table 4 and Table 5. Surprisingly, although a ResNet-FCN without pre-training shows much better results than its U-Net counterpart, U-Net pre-trained with SSMTL outperforms ResNet-FCN pre-trained with SSMTL+. Another interesting phenomenon is that ResNet-FCN pre-trained with SSMTL has a decreased performance. Both phenomena can be explained by the relationship between the network architecture and the target task. Note, there is a anatomical difference between U-Net and FCN, where U-Net has a balanced architecture between the encoder and the decoder but FCN puts more weight on the encoder. This enables ResNet-FCN to be more sensitive to semantic information, as ResNet50 is a seminal feature extractor, but also weakens its learning ability for dense prediction tasks with less semantic contents (i.e. no semantic labels). The pretext tasks proposed in Sect. 3.1 are not designed to extract semantic information as a segmentation task. So FCN might not be the suitable neural network backbone. SSMTL+ actually mitigates the issue with the additional instance discrimination task and shows slightly better performance than MoCo V2. Compared with benchmark SSL pre-training datasets such as ImageNet-1$M$ [11] and Instagram-1$B$ [26], the data scarcity in medical tasks will impair the performance of data-driven SSL models such as MoCo V2.

**Denoising.** We have another ablation study to validate our hypothesis of the relationship between the architecture and the target task. Here, we examine the proposed framework with a simple downstream task, denoising. Denoising is a dense prediction task with pixel-to-pixel mapping. The pretext tasks are highly correlated with denoising as they are both reconstructing images. Moreover,

**Table 6.** Proxy evaluation of self-supervised multi-task learning with denoising on CT scans.

| Method | Network | PSNR |
|---|---|---|
| w/o pre-training | U-Net | 36.04 |
| SSMTL | U-Net | 37.22 |
| w/o pre-training | ResNet-FCN | 34.85 |
| MoCo V2 | ResNet-FCN | 32.85 |
| SSMTL | ResNet-FCN | 35.38 |
| SSMTL+ | ResNet-FCN | 34.66 |

there are no semantic labels involved. We use the same training and test split for CT scans. We add synthetic noise to the original images and use them as training/test images. We treat the original images as the ground truth. Following [16], we implement the noise model as a zero-mean Gaussian distribution. We utilize the models pre-trained from previous experiments. The training for the denoising downstream task is performed by minimizing the L1 loss between the noisy input and the clean original images. We report the peak signal-to-noise ratio (PSNR) in Table 6, where U-Net outperforms FCN by a large margin.

**Limitations.** Finally, we want to clarify that the experiments in this section are only used to validate the theoretical discussion in a simplified scenario. The proposed framework is designed for sequential medical images only. In practice, the medical tasks could have more complex problem settings and data challenges, which require further consideration. In addition, we conclude that the choice of the neural network backbone should be dependent on the downstream dense prediction tasks. A possible future research direction could be using *neural architecture search* [14] to find the optimal network.

## 5 Conclusion

In this work, we propose a self-supervised representation learning framework for dense prediction tasks on sequential medical images. The proposed framework utilizes anatomical structural similarity between humans to integrate MTL and SSL. The theoretical discussion and empirical analysis show that, on label-efficient medical image analysis, the proposed framework has several advantages over SOTA SSL methods, which are originally designed for natural images. Limited by space, we only investigate a few downstream tasks on medical images. In the future, we will generalize the proposed framework for more dense prediction tasks in the medical domain and study the task affinity between the pretext tasks and the downstream tasks.

**Acknowledgements.** We would like to thank Huawei Technologies Co., Ltd. for providing GPU computing service for this study. The work was partially funded by the Research Council of Norway grants no. 315029, 309439, and 303514.

## References


1. Baldi, P., Pineda, F.: Contrastive learning and neural oscillations. Neural Comput. **3**, 526–545 (1991)
2. Bengio, Y., Courville, A., Vincent, P.: Representation learning: a review and new perspectives. IEEE TPAMI **35**, 1798–1828 (2013)
3. Bromley, J., Guyon, I., LeCun, Y., Säckinger, E., Shah, R.: Signature verification using a "Siamese" time delay neural network. In: NeurIPS, pp. 737–744 (1993)
4. Caruana, R.: Multitask learning. Mach. Learn. **28**, 41–75 (1997)
5. Chen, L., Papandreou, G., Kokkinos, I., Murphy, K., Yuille, A.: DeepLab: semantic image segmentation with deep convolutional nets, Atrous convolution, and fully connected CRFs. IEEE TPAMI **40**, 834–848 (2017)
6. Chen, T., Kornblith, S., Norouzi, M., Hinton, G.: A simple framework for contrastive learning of visual representations. In: ICML (2020)
7. Chen, X., Yao, L., Zhou, T., Dong, J., Zhang, Y.: Momentum contrastive learning for few-shot COVID-19 diagnosis from chest CT images. Patt. Recogn. **113**, 107826 (2021)
8. Chen, X., Fan, H., Girshick, R., He, K.: Improved baselines with momentum contrastive learning. ArXiv Preprint ArXiv:2003.04297 (2020)
9. Chen, X., He, K.: Exploring simple Siamese representation learning. In: CVPR (2021)
10. Dai, W., Dong, N., Wang, Z., Liang, X., Zhang, H., Xing, E.P.: SCAN: structure correcting adversarial network for organ segmentation in chest X-Rays. In: Stoyanov, D., et al. (eds.) DLMIA/ML-CDS -2018. LNCS, vol. 11045, pp. 263–273. Springer, Cham (2018). https://doi.org/10.1007/978-3-030-00889-5_30
11. Deng, J., Dong, W., Socher, R., Li, L., Li, K., Li, F.: ImageNet: a large-scale hierarchical image database. In: CVPR, pp. 248–255 (2009)
12. Doersch, C., Gupta, A., Efros, A.: Unsupervised visual representation learning by context prediction. In: ICCV, pp. 1422–1430 (2015)
13. Dong, N., Kampffmeyer, M., Liang, X., Wang, Z., Dai, W., Xing, E.: Unsupervised domain adaptation for automatic estimation of cardiothoracic ratio. In: MICCAI, pp. 544–552 (2018)
14. Dong, N., Xu, M., Liang, X., Jiang, Y., Dai, W., Xing, E.: neural architecture search for adversarial medical image segmentation. In: MICCAI, pp. 828–836 (2019)
15. Everingham, M., Van Gool, L., Williams, C., Winn, J., Zisserman, A.: The Pascal visual object classes (VOC) challenge. IJCV **88**, 303–338 (2010)
16. Foi, A., Trimeche, M., Katkovnik, V., Egiazarian, K.: Practical Poissonian-Gaussian noise modeling and fitting for single-image raw-data. IEEE TIP **17**, 1737–1754 (2008)
17. Gidaris, S., Singh, P., Komodakis, N.: Unsupervised representation learning by predicting image rotations. In: ICLR (2018)
18. Gopnik, A., Meltzoff, A., Kuhl, P.: The Scientist in the Crib: Minds, Brains, and How Children Learn. William Morrow and Co, New York (1999)
19. Gutmann, M., Hyvärinen, A.: Noise-contrastive estimation: a new estimation principle for unnormalized statistical models. In: AISTATS, pp. 297–304 (2010)
20. He, K., Fan, H., Wu, Y., Xie, S., Girshick, R.: Momentum contrast for unsupervised visual representation learning. In: CVPR, pp. 9729–9738 (2020)
21. He, K., Zhang, X., Ren, S., Sun, J.: Deep residual learning for image recognition. In: CVPR, pp. 770–778 (2016)
22. Kingma, D., Ba, J.: A method for stochastic optimization. In: ICLR, Adam (2015)

23. Lin, T., Dollár, P., Girshick, R., He, K., Hariharan, B., Belongie, S.: Feature pyramid networks for object detection. In: CVPR, pp. 2117–2125 (2017)
24. Lin, T., et al.: Microsoft coco: common objects in context. In: ECCV, pp. 740–755 (2014)
25. Long, J., Shelhamer, E., Darrell, T.: Fully convolutional networks for semantic segmentation. In: CVPR, pp. 3431–3440 (2015)
26. Mahajan, D., et al.: Exploring the limits of weakly supervised pretraining. In: ECCV, pp. 181–196 (2018)
27. Misra, I., Maaten, L.: Self-supervised learning of pretext-invariant representations. In: CVPR, pp. 6707–6717 (2020)
28. Noroozi, M., Favaro, P.: Unsupervised learning of visual representations by solving Jigsaw puzzles. In: ECCV, pp. 69–84 (2016)
29. Oord, A., Li, Y., Vinyals, O.: Representation learning with contrastive predictive coding. ArXiv Preprint ArXiv:1807.03748 (2018)
30. Ouyang, C., Biffi, C., Chen, C., Kart, T., Qiu, H., Rueckert, D.: Self-supervision with superpixels: training few-shot medical image segmentation without annotation. In: ECCV, pp. 762–780 (2020)
31. Raina, R., Battle, A., Lee, H., Packer, B., Ng, A.: Self-taught learning: transfer learning from unlabeled data. In: ICML, pp. 759–766 (2007)
32. Ren, S., He, K., Girshick, R., Sun, J.: Faster R-CNN: towards real-time object detection with region proposal networks. IEEE TPAMI **39**, 1137–1149 (2016)
33. Ronneberger, O., Fischer, P., Brox, T.: U-Net: convolutional networks for biomedical image segmentation. In: MICCAI, pp. 234–241 (2015)
34. de Sa, V.: Learning classification with unlabeled data. In: NeurIPS, pp. 112–119 (1994)
35. Simonyan, K., Zisserman, A.: Very deep convolutional networks for large-scale image recognition. In: ICLR (2015)
36. Tian, Y., Sun, C., Poole, B., Krishnan, D., Schmid, C., Isola, P.: What makes for good views for contrastive learning. NeurIPS **33**, 6827–6839 (2020)
37. Vandenhende, S., Georgoulis, S., De Brabandere, B., Van Gool, L.: Branched multi-task networks: deciding what layers to share. In: BMVC (2020)
38. Vandenhende, S., Georgoulis, S., Van Gansbeke, W., Proesmans, M., Dai, D., Van Gool, L.: A survey. IEEE TPAMI, Multi-task learning for dense prediction tasks (2021)
39. Wu, Z., Xiong, Y., Yu, S., Lin, D.: Unsupervised feature learning via non-parametric instance discrimination. In: CVPR, pp. 3733–3742 (2018)
40. Zhang, R., Isola, P., Efros, A.: Colorful image colorization. In: ECCV, pp. 649–666 (2016)
41. Zhang, Z., Cui, Z., Xu, C., Yan, Y., Sebe, N., Yang, J.: Pattern-affinitive propagation across depth, surface normal and semantic segmentation. In: CVPR, pp. 4106–4115 (2019)
42. Zhuang, X., et al.: Multiatlas whole heart segmentation of CT data using conditional entropy for atlas ranking and selection. Med. Phys. **42**, 3822–3833 (2015)
43. Zhuang, X., Rhode, K., Razavi, R., Hawkes, D., Ourselin, S.: A registration-based propagation framework for automatic whole heart segmentation of cardiac MRI. IEEE TMI **29**, 1612–1625 (2010)

# Label-Assisted Memory Autoencoder for Unsupervised Out-of-Distribution Detection

Shuyi Zhang[1,2,3], Chao Pan[1,2], Liyan Song[1,2], Xiaoyu Wu[4], Zheng Hu[4], Ke Pei[5], Peter Tino[3], and Xin Yao[1,2,3](✉)

[1] Research Institute of Trustworthy Autonomous Systems, Southern University of Science and Technology (SUSTech), Shenzhen, China
xiny@sustc.edu.cn
[2] Guangdong Provincial Key Laboratory of Brain-inspired Intelligent Computation, Department of Computer Science and Engineering, Southern University of Science and Technology (SUSTech), Shenzhen, China
[3] CERCIA, School of Computer Science, University of Birmingham, Birmingham, UK
[4] RAMS Reliability Technology Laboratory, Huawei Technology Co. Ltd., Shenzhen, China
[5] TTE-DE RAMS Laboratory, Huawei Technology Co. Ltd., Munich, Germany

**Abstract.** Out-of-Distribution (OoD) detectors based on AutoEncoder (AE) rely on an underlying assumption that an AE network cannot reconstruct OoD data as good as in-distribution (ID) data when it is constructed based on ID data only. However, this assumption may be violated in practice, resulting in a degradation in detection performance. Therefore, alleviating the factors violating this assumption can potentially improve the robustness of OoD performance. Our empirical studies also show that image complexity can be another factor hindering detection performance for AE-based detectors. To cater for these issues, we propose two OoD detectors LAMAE and LAMAE+. Both can be trained without the availability of any OoD-related data. The key idea is to regularize the AE network architecture with a classifier and a label-assisted memory to confine the reconstruction of OoD data while retaining the reconstruction ability for ID data. We also adjust the reconstruction error by taking image complexity into consideration. Experimental studies show that the proposed OoD detectors can perform well on a wider range of OoD scenarios.

## 1 Introduction

Deep neural networks have been playing an increasingly important role in many applications, such as autonomous driving [37] and surveillance tracking [38]. When deploying neural networks in real-world applications, there is often very little control over the distribution of test data. The existence of test examples belonging to a different distribution from the training one, also known as Out-of-Distribution (OoD), may cause conventional classification models no longer


N. Oliver et al. (Eds.): ECML PKDD 2021, LNAI 12977, pp. 795–810, 2021.
https://doi.org/10.1007/978-3-030-86523-8_48

suitable to be used [10]. Therefore, it is of crucial importance to identify OoD examples in order to maintain the reliability of classification models.

Many OoD detectors have been developed lately [10,17–19,25]. But a large body of them requires the availability of OoD data to tune the hyperparameters of the deep networks, being less applicable as OoD data are not typically accessible in reality. Generative models such as deep AutoEncoder (AE) [27] are exempted from this problem when used for OoD detection since they rely on the assumption that when trained with ID data only, an AE network produces higher reconstruction error for unseen OoD data than ID data.

Many AE-based OoD detectors have been proposed based on this assumption [9,33]. However, this assumption may be violated in some scenarios. Observations have demonstrated that its validity depends on the specific characteristics of OoD examples. Sometimes AE-based OoD detector can "generalize" so well that it can also reconstruct OoD data with low reconstruction error, causing unsatisfactory detection performance [7,9]. When the training dataset contains multiple classes instead of only one, which is of more practical use in the real-world, empirical studies of state-of-the-art (SOTA) AE-based OoD detectors reveal an even larger deterioration of detection performance, showing a need for dealing with such learning scenario.

In addition, there has been no systematic study characterizing different types of OoD aiming for analysing the cause of performance degradation of AE-based detectors. The only categorization of OoD is discussed in [11] based on the semantics of ID and OoD examples. The study concluded that detection difficulty would increase when OoD examples possess the same semantic meaning as the ID examples, which coincides with our findings. Nonetheless, based on our preliminary investigation on the detection performance of various types of OoD examples, we noticed that inherent image complexity may be another factor causing OoD performance degradation. As an extreme example, a constant image (i.e., with same-valued pixels) that is of low complexity can always be reconstructed very well. We further noticed that most AE-based methods suffer in such a scenario. Therefore, a more thorough OoD characterization is preferred, which can not only allow us to scrutinize the reason behind the performance variation, but also help researchers to provide more targeted solutions.

To address the above issues, we propose two OoD detectors, namely LAMAE (Label-Assisted Memory AutoEncoder) and LAMAE+, as well as a new criterion to characterize OoD scenarios. Both detectors can be trained without the availability of any information from OoD data. The key idea is to leverage the information of the class-labels of ID data so that the reconstruction of OoD data is constrained while the reconstruction capability for ID data can be retained. Hence, differentiation between ID and OoD examples can be promoted. Furthermore, we provide a finer characterization based on image complexity to investigate the reason for performance degradation of some particular types of OoD. To mitigate the bias induced by inherent image complexity, we propose an entropy-based metric, namely Complexity Normalizer (CN), to adjust the reconstruction error, and incorporate CN metric in the OoD model, forming LAMAE+.

The contributions of this paper are as follows:

1. We propose a new unsupervised OoD detector (LAMAE) that does not require OoD examples for training, neither do we make any assumptions.
2. We provide a finer characterization of OoD scenarios and discuss their relationship to detection performance.
3. Based on the proposed OoD characterization, we further propose a new metric to adjust the reconstruction error so that the refined OoD detector (LAMAE+) performs well on a wider range of different OoD types.

The rest of this paper is organised as follows. Section 2 discusses related work for OoD detection in the literature. Section 3 explains when and why existing OoD detectors may fail, based on which we propose two detectors LAMAE and LAMAE+. The effectiveness of the proposed OoD detectors are evaluated experimentally in Sect. 4. The paper is concluded in Sect. 5.

## 2 Background

This paper considers detecting OoD samples in the context of image classification. When training a classification model, we have a training dataset $D_{in} = \{(\mathbf{x}_i, y_i)\}_{i=1}^{T}$ where $\mathbf{x}_i \in \mathcal{X} = \mathcal{R}^d$ is a d-dimensional feature vector representation of an image data and $y_i = 1, \cdots, S$ is the class label. All training samples are in-distributed as $p_{in}(\mathbf{x}, y)$. The purpose of OoD detection is to identify input examples $\mathbf{x} \sim p_{out}$ where $p_{out} \neq p_{in}$. According to [11], OoDs can be categorized into two types: semantic and non-semantic. Semantic OoDs include data from a distribution $p_{out}(\mathbf{x}, \overline{y})$ with $\{\overline{y}\} \cap \{y\} = \emptyset$. Non-semantic OoDs include data from $p_{out}(\mathbf{x}, y)$, that is, data from the same object class but presented with different styles. It was also concluded that OoD datasets with both types of distribution shifts are the easiest to detect, followed by non-semantic OoD. Semantic OoD turns out to be the hardest one to detect [11].

Many algorithms have been proposed to detect OoD examples [1,10,17–19,21,25,29,36,39]. However, most of them require the aid of some kind of genuine or synthetic OoD examples in the training stage. This is an unrealistic requirement since in reality, it is often hard, if not impossible, to gain any information regarding OoD a priori. Therefore, we review only OoD detectors that do not require OoD examples for training in this section.

### 2.1 AE-based Detectors

OoD detectors based on generative models such as AEs naturally possess a characteristic of being able to detect OoD in an unsupervised manner [7]. There are two ways of using AEs for OoD detection [28]. Firstly, an AE can be used to learn a low-dimensional representation of the input data, then distance-based metrics can be applied to assess the discrepancy between newly arrived test examples and the ID dataset [3,28,35,40]. Secondly, the reconstruction error or probability of the test example is calculated directly and used for detection. This work follows the latter strategy.

The reconstruction of AEs has been used extensively for OoD detection to tackle various issues that may exist. For instance, Zhou et al. proposed a robust AE that is capable of detecting anomalies when no clean, noise-free data is available during training [39]. Chen et al. addresses the same issue with AE ensemble by randomly varying the connectivity architecture of the base AE [6]. In contrast, this work focuses on the setting where only clean ID data are available for training.

Ana and Cho adopts a variational autoencoder (VAE) [14] as the base model and utilizes reconstruction probability in a similar manner as reconstruction error to detect OoD [2]. OoDs are expected to have low probability density. SSVAE [4] is a more advanced VAE for semi-supervised learning. The authors supplement the classification loss with the VAE loss so that the performance for both classification and OoD detection can be improved. However, VAEs have their own limitations such as the Gaussian prior assumption on the latent space. Furthermore, many recent work challenge the use of reconstruction likelihood of flow-based generative models such as VAE for OoD detection because extensive experiments have shown that it is not a reliable metric as expected [12,23].

On the other hand, the reconstruction error of vanilla AEs is a more straightforward metric to use. However, there are also other issues with them. Denouden et al. [7] noticed that AEs can sometimes reconstruct the semantic OoD examples with less error than ID examples. To solve this issue, they adjusted the reconstruction-based detection criterion by adding the Mahalanobis distance between the test sample and the training set mean within the AE latent space. MemAE [9] is another recently proposed method aiming to improve detection for this type of OoD. It incorporates within the training stage a memory to store prototypical elements of the ID data. Hence, the reconstruction of any test examples will be forced to be more similar to the most representative ID examples. Thus, the reconstruction error will be strengthened for OoD examples. This particular issue raised by the methods above is in accordance with the findings in [11]. That is, semantic OoDs are more difficult to detect. The abovementioned attempts are only tested on one-class ID training data only. However, the difficulty in identifying semantic OoD examples does not only exists in the one-class setting. In fact, the challenge becomes more problematic when there are multiple classes within the ID data, which is a more practical scenario. In this case, characterizing OoD examples based on their semantic meaning may be inadequate. This is explained in more detail in Sect. 3.2, where we also provide a novel methodology to effectively solve this problem.

Our OoD detectors are built upon MemAE [9], which is explained in more details as follows. MemAE endorses a memory component into the traditional AE architecture as shown in Fig. 1. The encoder $f_e(\cdot)$ maps an input image $\mathbf{x} \in \mathcal{X}$ to a latent space $\mathcal{Z} = \mathcal{R}^C$ via $\mathbf{z} = f_e(\mathbf{x}; \theta_e)$, where $\theta_e$ represents the encoder-specific model parameter. Before the latent vector $\mathbf{z}$ is forwarded to the

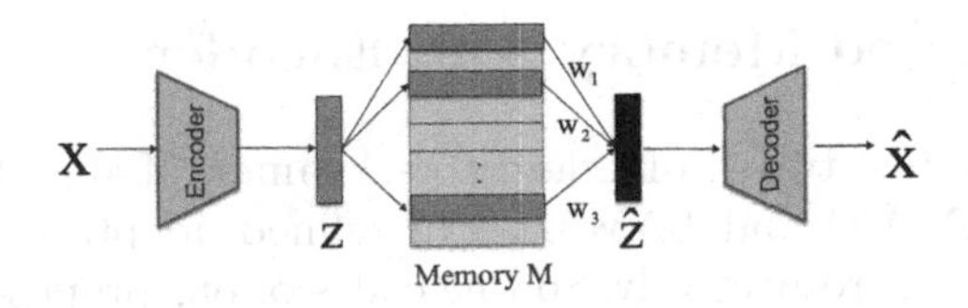


**Fig. 1.** Framework of MemAE [9]

decoder, the memory module $\mathbf{M} \in \mathcal{R}^{N \times C}$ containing $N$ prototypical vectors $\mathbf{m}_i$, each of dimension $1 \times C$, is put in place, where $N$ is a predefined parameter for the memory size. $\mathbf{M}$ is designed to record the prototypical normal patterns of ID data $D_{in}$, which is updated at each epoch in the training phase. Once a new training example is received, cosine similarity between the encoded vector $\mathbf{z}$ and each memory item $\mathbf{m}_i$ is calculated as $d(\mathbf{z}, \mathbf{m_i}) = \frac{\mathbf{z}\mathbf{m_i}^T}{||\mathbf{z}|| \cdot ||\mathbf{m_i}||}$ for $\forall\, i = \{1, \cdots, N\}$. The weight vector $\mathbf{w} = [w_i, \cdots, w_N]$ is calculated via a softmax operation $w_i = \frac{exp(d(\mathbf{z},\mathbf{m_i}))}{\sum_{j=1}^{N} exp(d(\mathbf{z},\mathbf{m_j}))}$ with $\sum_{i=1}^{N} w_i = 1$. To further limit the reconstruction ability for OoD examples, MemAE applied a hard shrinkage technique on $\mathbf{w}$, promoting the sparsity of model parameters.

The latent representation fed to the decoder is then $\hat{\mathbf{z}} = \mathbf{w}\mathbf{M} = \sum_{i=1}^{N} w_i \mathbf{m_i}$. The reconstructed image is $\hat{\mathbf{x}} = f_d(\hat{\mathbf{z}}; \theta_d)$ where $\theta_d$ represents the decoder-specific model parameter. The MemAE loss function considers two terms: the reconstruction loss and an entropy for promoting the sparsity of $\mathbf{w}$. The memory is fixed after the training stage. In the testing phase, all examples are forced to be constructed with prototypical components of the ID data, resulting in significant reconstruction errors for OoDs.

### 2.2 Non-AE-Based Detectors

One-class classification are popularly used for OoD detection [24,26]. Nonetheless, when the number of data dimensions is high, which is a typical issue of image data, these approaches can suffer from the curse of dimensionality.

Other types of OoD detectors also exist. For instance, Shalev et al. [30] utilize extra supervision by training networks to predict word embedding of class labels. It needs to combine the outputs of several similar networks to detect OoD examples. GODIN [11] is a very recently proposed OoD detector. It is an improvement of one of the benchmark detectors, ODIN [19], which utilizes class posterior probability produced by a softmax classifier for detection. Unlike ODIN, it decomposes the class posterior probability using the rule of conditional probability during training and uses only the numerator, i.e., the joint class-domain probability for detection. GODIN frees the algorithm from explicit parameter-tuning with respect to specific OoD datasets.

# 3 Label-Assisted Memory AutoEncoder

This section explains two OoD detectors, namely Label-Assisted Memory AutoEncoder (LAMAE) and LAMAE+ (a refined adaptation of LAMAE) in Sect. 3.1 and Sect. 3.2, respectively. Source codes of our proposed algorithms are available at https://github.com/fzjcdt/LAMAE.

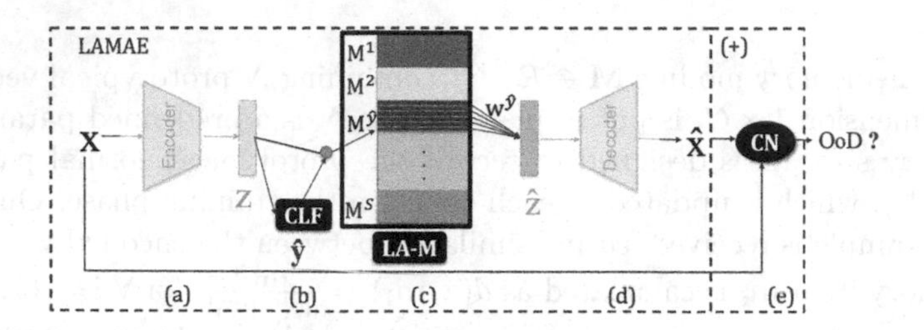


**Fig. 2.** Framework of our proposed LAMAE and LAMAE+. `CLF` denotes the classifier module (Sect. 3.1.1). `LA-M` denotes the label-assisted memory (Sect. 3.1.2). `CN` denotes a normalizer to refine the reconstruction (Sect. 3.2).

## 3.1 Label-Assisted Memory AutoEncoder (LAMAE)

As shown in Fig. 2, the network architecture of Label-Assisted Memory AutoEncoder (LAMAE) consists of four components: (a) an encoder (`Encoder`) to compress the intrinsic data features, (b) a classifier (`CLF`) to regularize the memory for a better targeted reconstruction, (c) a label-assisted memory (`LA-M`) reserving class-conditional memory chunks and their associated weights, and (d) a decoder (`Decoder`) to recreate the input based on the information stored in (c). Components (a) and (d) have been explained in Sect. 2.1, so this section focuses on the newly proposed components (b) and (c).

### 3.1.1 Classifier Module

Our preliminary experiment shows that when ID data consist of multiple classes, the performance of existing AE-based detectors can deteriorate significantly. Figure 3 provides an illustrative example, where digit “7” is OoD and the rest nine digits are ID. We can see that MemAE can reconstruct both ID and OOD examples very well (Fig. 3), such that it would be difficult to identify the OoD examples based on reconstruction error. Figure 4(a) shows the histograms of the reconstruction errors of ID and OoD data, further confirming the difficulty of achieving good OoD performance by using MemAE under this circumstance.

A potential reason is that the latent space learned from the multi-class ID dataset allows for a combination of features from various ID classes to reconstruct unseen OoD examples. This combination may not have much effect on the reconstruction of ID examples, but can be detrimental for OoD detection since the reconstruction error is no longer distinguishable. To tackle this issue,

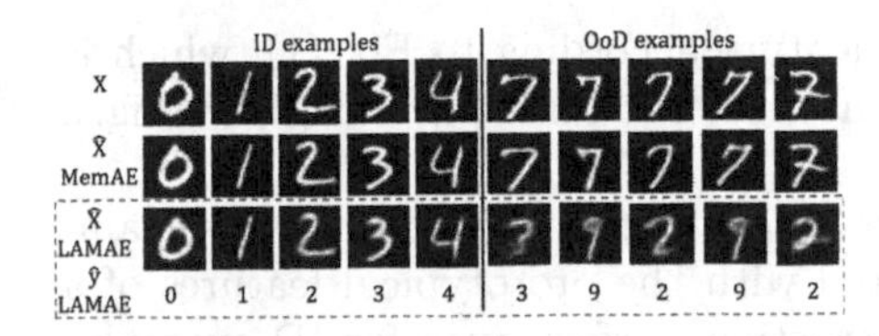


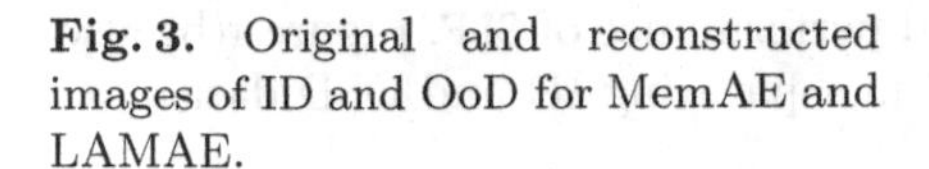


**Fig. 3.** Original and reconstructed images of ID and OoD for MemAE and LAMAE.

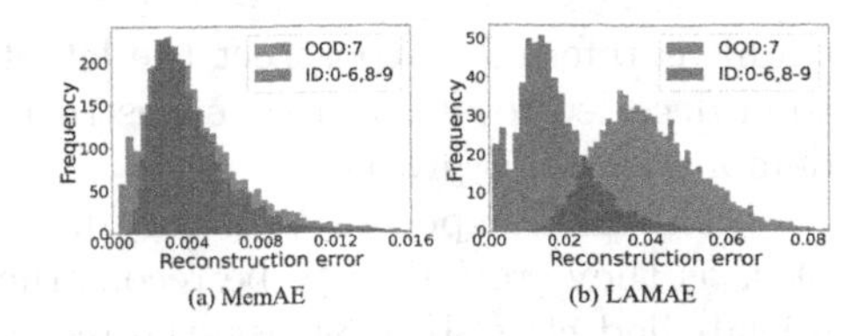


**Fig. 4.** Density of reconstruction error of ID and OoD for MemAE and LAMAE.

we propose to regulate the reconstruction of test images by exploiting their class labels, which is implemented by placing a classifier (CLF) and a label-assisted memory (LA-M) in the AE framework as shown in Fig. 2. The two modules are explained in this section and Sect. 3.1.2, respectively.

A classifier $f_c(\cdot)$ is incorporated into the MemAE architecture by connecting the latent space $\mathcal{Z}$ and memory $\mathbf{M}$ as shown in Fig. 2. $f_c(\cdot)$ can be a single-layered or multi-layered network, depending on the complexity of the application. The predicted label $\hat{y} = f_c(\mathbf{z}; \theta_c)$ of a training image $\mathbf{x}$, where $\theta_c$ denotes the classifier parameter, can then be used to guide the learning of the label-assisted memory. Given a test image, the latent representation induced by the encoder is forwarded to the classifier for the predicted label. We can see from the dotted box in Fig. 3 that the classifier (trained purely on ID data) always assigns the OoD example with one of the existing ID labels.

### 3.1.2 Label-Assisted Memory Module

The Label-Assisted Memory module (LA-M) aims to record the most representative prototypical patterns for each individual class. Therefore, the whole memory $\mathbf{M}$ of size $N$ is divide into $S$ mutually exclusive class-conditional memory chunks $\{\mathbf{M}^s|s = 1, \cdots, S\}$ where $S$ is the number of ID classes, i.e., $\mathbf{M} = \cup_{s=1}^{S}\mathbf{M}^s$. We use $N^s$ and $\mathbf{w}^s$ to denote the size and associated weight vector of $\mathbf{M}^s$, respectively. This study assigns the same size for $\mathbf{M}^s$, i.e., $N^1 = \cdots = N^S$.

Similar to MemAE, cosine similarity is used to calculate the weights (see Sect. 2.1). However, thanks to the information from $f_c(\cdot)$, the latent feature $\mathbf{z}$ is only compared with each memory item $\{\mathbf{m}_i^{\hat{y}} \in \mathbf{M}^{\hat{y}}|i = 1, \cdots, N^{\hat{y}}\}$. The associated weight $\mathbf{w}^{\hat{y}}$ is calculated based on the similarity of the memory items and $\mathbf{z}$. Only $\mathbf{M}^{\hat{y}}$ and $\mathbf{w}^{\hat{y}}$ are used to formulate $\hat{\mathbf{z}}$ as

$$\hat{\mathbf{z}} = \mathbb{1}_{s=\hat{y}} \sum_{i=1}^{N^s} w_i^s \mathbf{m}_i^s. \tag{1}$$

Note that weight vector $\mathbf{w}$ is rather sparse since $\mathbf{w}^s = 0 \;\forall\; s \neq \hat{y}$, so we do not need to apply the hard shrinkage technique as in MemAE.

In the testing phase, $\mathbf{M}$ is fixed. Given a test image, one employs the classifier to predict its label, so that only the predicted-class-conditional memory chunk is referred to construct the latent feature according to Eq. (1), which is then decoded to obtain the reconstruction error. Modules (c) and (d) in Fig. 3 demonstrate this process.

LAMAE is expected to induce higher reconstruction error for OoD examples, as they are forced to be reconstructed with the prototypical features of a mislabelled class. In contrast, the reconstruction performance for ID examples can be maintained, given the typically good performance of CLF. Figure 4(b) also demonstrates the effectiveness of LAMAE in separating ID and OoD data.

#### 3.1.3 Training Objective

The loss function is formulated as the sum of reconstruction error and classification error on the training data as

$$\mathcal{L} = \frac{1}{T}\sum_{t=1}^{T}[R(\mathbf{x}^t, \hat{\mathbf{x}}^t) + \beta L(y^t, \hat{y}^t|\mathbf{x}^t)], \tag{2}$$

where $T$ is the training set size, $\beta$ is a tuning parameter, $R(\mathbf{x}^t, \hat{\mathbf{x}}^t) = ||\mathbf{x}^t - \hat{\mathbf{x}}^t||_2^2$ is the mean-squared reconstruction error, and $L(y^t, \hat{y}^t|\mathbf{x}^t) = -\sum_{s=1}^{S} \mathbb{1}_{\hat{y}^t=s} log(p(y^t = s))$ is cross entropy classification error. In this work, softmax activation is adopted and $\beta$ is set to 1 without tuning.

In the training process, ID examples, along with their true labels, are used to minimize the overall loss during which the predictive performance of the classifier module is also guaranteed.

### 3.2 LAMAE with Complexity Normalizer (LAMAE+)

Further exploration into our experiments shows that although LAMAE generally achieves better performance than other AE-based methods, it may still fail on some specific types of semantic OoDs. For instance, the detection performance is 26% lower when digit "1" is taken as OoD compared with the case when "0" or "2" is taken. In fact, almost all existing AE-based detectors suffer in such a scenario. This section aims to investigate the reason why the detection of some types of OoD is of greater difficulty. After that, we propose a new way to categorize OoD examples. Finally, we design an image complexity-based metric (module (e) in Fig. 2) to upgrade LAMAE, inducing LAMAE+.

#### 3.2.1 Image Reconstruction and Complexity

According to the taxonomy in [11], our experimental setting based on handwritten digits belongs to the *semantic* OoD scenario. Nevertheless, our experimental results show that the detection performances can still vary by a large extent when different digit is treated as OoD, suggesting that it is not adequate to explain the performance variation purely from the perspective of semantics.

Based on this, we hypothesize that the inherent complexity of the image is positively correlated to its reconstruction difficulty, which impacts detection performance. To test this hypothesis, we train and test two AEs on two datasets with very different complexities (handwritten digits with lower complexity and natural images with higher complexity). We adopt Shannon entropy [31,32], which has a well-established information-theoretic basis, to measure the image complexity as

$$H(S) = -\sum_{S_i=0}^{n-1} p(S_i)log(p(S_i)), \tag{3}$$

where $n$ is the number of grey levels, $S_i$ are the grey level pixel values contained in image $S$ and $p(S_i)$ is the probability of pixel having level $S_i$. A Pearson correlation of 0.7952 between image entropy and reconstruction error is derived from a total of 20,000 test images (10,000 from each dataset), suggesting a strong positive correlation. Our hypothesis has been verified.

### 3.2.2 A Characterization of OoD

Experimentally, we found that image complexity played an important role in OoD detection. Hence, we propose to further characterize OoD according to the complexity of OoD examples as compared to the ID ones. When OoD examples have a lower image complexity, such as images with constant pixels, we categorize them as "*plain*". For OoD examples with a higher image complexity, such as images with random pixels, we categorize them as "*fancy*". Altogether, OoDs can be categorized into six classes: *S+P*, *S+F*, *NS+P*, *NS+F*, *NS+S+P*, and *NS+S+F*, where *P*, *F*, *S*, *NS* stand for *plain*, *fancy*, *semantic* and *non-semantic*, respectively.

By nature, the reconstruction for *plain* images should be easier than that of the *fancy* images, since fewer features are required for their description, leading to lower errors. This property would probably mislead OoD detectors towards classifying *plain* images as ID even when they are actually OoD. Therefore, *semantic-and-plain (S+P)* OoD is the hardest to detect among all types of OoD and image complexity should be catered for when making OoD detection based on the criterion of reconstruction error.

### 3.2.3 Complexity-Normalized Test Statistic

To tackle the challenge of detecting *S+P* OoDs, we propose a new metric called Complexity Normalizer (`CN`) to adjust reconstruction error for detection. Indeed, the mechanism of CN can be used in combination with any AEs. When CN is equipped with LAMAE, we form LAMAE+.

To perform LAMAE+ for each test image, we calculate an entropy-based normalizer $CN = log(H(S) + 1)$ and re-scale the reconstruction error derived from LAMAE as:

$$\widehat{Err} = \frac{||\mathbf{x}^t - \hat{\mathbf{x}}^t||_2^2}{CN + \gamma}, \tag{4}$$

where $\gamma > 0$ is a tiny value to avoid the numerical problem of zero division (fixed as 1e–9). $\widehat{Err}$ is the correction of the reconstruction error, taking image complexity under consideration for OoD detection.

# 4 Experimental Studies

This section carries out two sets of experiments. Experiment 1 validates the proposed LAMAE+ by comparing with SOTA OoD detectors. Experiment 2 examines the effectiveness of each component in LAMAE+. Comparisons between LAMAE and LAMAE+ can also be found in Experiment 2.

## 4.1 Experimental Setup

Our experiments are based on the following benchmark datasets with each image standardized to [0,1] channel-wise.

1. **MNIST** [16] contains gray-scale images of handwritten digits 0–9.
2. **Fashion MNIST** (FMNIST) [34] contains gray-scale images of Zalando's article images from 10 classes including sneakers, trousers, pullover, etc.
3. **CIFAR10** [15] contains natural color images from 10 classes including airplane, ship, dog, cat, etc.
4. **CelebA** [20] contains face images of 10,177 celebrities.
5. **notMNIST** [5] contains gray-scale images of English letters from A to J.
6. **Constant** contains images of plain color. All pixels of an image has the same value uniform-randomly drawn from the set $\{0, \cdots, 255\}$.
7. **Noise** contains images of uniform noise. Pixel values are independently drawn from the uniform distribution on the set $\{0, \cdots, 255\}$.

To show the generality and applicability of the proposed detectors, we conduct experiments on three different settings. Setting 1 is built with MNIST dataset. In each experiment, one class is used as the OoD class, and the rest are seen as ID. The procedure is repeated for all classes. Within this setting, *S+P* OoDs and *S+F* OoDs exist. In Setting 2 and 3, FMNIST and CIFAR10 are used as the ID dataset respectively. Various OoD datasets including CelebA, notMNIST, FMNIST, Constant and Noise are adopted. Within this setting, both *NS+S+P* and *NS+S+F* OoDs exist.

Due to page limit, we report the area under the receiver operating characteristic curve (AUROC) which plots the true positive rate (TPR) of ID against the false positive rate (FPR) of OoD data by a varying threshold. The average detection performance and the standard deviation of 10 repetitive experiments are reported. Performance measured by area under the precision-recall curve (AUPRC) shows similar trends.

## 4.2 Experiment 1: Comparative Studies with SOTA Detectors

In this section we validate the proposed methods for the detection of various types of OoD. We compare LAMAE+ with unsupervised detectors including traditional AE, VAE [2], SSVAE [4], MemAE [9] and the latest non-reconstruction-based detector GODIN [11].

**Table 1.** AUROC detection performance for MNIST in Experiment 1. (Each time the model is trained on 9 of the 10 classes and the left-out class is considered to be the OoD class. **Bold** indicates the best scores.)

| OoD Class | AE | VAE(e) | VAE(p) | SSVAE(p) | MemAE | GODIN | LAMAE+ |
|---|---|---|---|---|---|---|---|
| 0 | 81.4 ± 0.7 | 87.0 ± 1.6 | 94.9 ± 0.1 | 96.9 ± 0.1 | 83.0 ± 1.4 | 86.3 ± 8.0 | **98.5 ± 0.2** |
| 1 | 12.7 ± 0.1 | 27.2 ± 1.3 | 47.0 ± 3.0 | 9.5 ± 0.6 | 14.2 ± 1.0 | 86.4 ± 5.0 | **90.1 ± 2.5** |
| 2 | 92.9 ± 0.3 | 96.0 ± 0.6 | 96.1 ± 0.1 | 97.2 ± 0.0 | 94.8 ± 0.4 | 88.7 ± 4.4 | **99.0 ± 0.1** |
| 3 | 82.0 ± 0.4 | 94.2 ± 0.5 | 84.8 ± 0.3 | 90.2 ± 0.2 | 83.1 ± 1.2 | 70.6 ± 6.8 | **97.7 ± 0.3** |
| 4 | 76.6 ± 0.8 | 91.4 ± 0.6 | 70.8 ± 0.4 | 75.1 ± 0.3 | 75.7 ± 0.6 | 79.4 ± 6.0 | **95.5 ± 0.6** |
| 5 | 82.6 ± 0.3 | 92.2 ± 0.7 | 86.0 ± 0.3 | 89.4 ± 0.1 | 84.1 ± 0.8 | 64.8 ± 9.9 | **97.5 ± 0.3** |
| 6 | 83.6 ± 0.6 | 86.1 ± 1.1 | 92.3 ± 0.01 | 96.0 ± 0.2 | 85.9 ± 0.9 | 83.1 ± 7.6 | **97.0 ± 0.5** |
| 7 | 56.9 ± 1.0 | 67.2 ± 1.7 | 66.9 ± 0.2 | 75.5 ± 0.5 | 57.7 ± 0.8 | 79.8 ± 8.7 | **94.7 ± 1.1** |
| 8 | 90.5 ± 0.3 | 95.3 ± 0.5 | 89.1 ± 0.4 | 92.2 ± 0.2 | 90.1 ± 0.6 | 85.8 ± 3.8 | **97.5 ± 0.3** |
| 9 | 59.2 ± 0.7 | 67.3 ± 0.8 | 62.0 ± 2.0 | 67.7 ± 0.5 | 56.5 ± 0.9 | 79.1 ± 9.1 | **89.7 ± 2.0** |

On MNIST and FMNIST, we implement the encoder using three convolution layers as in MemAE [9]. For GODIN [11] where MNIST and FMNIST are not used for training, we experimented with the same structure as the setting for encoder-and-classifier component adopted for LAMAE+. On CIFAR10, with higher data complexity, deeper encoder and decoder are constructed for MemAE and LAMAE+. A skip connection from $\mathbf{z}$ to $\hat{\mathbf{z}}$ with dimension 16 is added to further assist reconstruction. Except for the last layer, each layer is followed by a batch normalization (BN) [13] and a Rectified Linear (ReLU) activation [22]. Batch size is set to 128 and we use an Adam optimization procedure.

For AE-based detectors, the maximum number of training epochs is set to 200, 200 and 500 and the class-conditional memory size $N^s$ in LAMAE and LAMAE+ is set to 10, 10 and 50 for MNIST, FMNIST and CIFAR10 respectively. Later we demonstrate experimentally that performance is insensitive to the selection of memory size. An extra fully connected layer with softmax output is taken as the classifier component. A 10% validation set is extracted from the ID training dataset. Early stopping is adopted to choose the model that achieves the lowest loss on the validation dataset. Note that this validation dataset is still ID so the models are trained without access to any information about OoD.

### 4.2.1 Performance on Semantic OoD Only

Table 1 reports the results of MNIST. Results of VAE (p) and SSVAE (p) based on reconstruction probability are taken from the original papers [2,4]. We also report the results based on reconstruction error (VAE(e)).

We can see that LAMAE+ achieves the best AUROC in all 10 cases. The improvement is especially substantial for digits "1", "4" , "7" and "9". Detailed explanations of how exactly each component in LAMAE contributed to this outcome is presented later in Sect. 4.3. It is also worth noting that the stan-

**Table 2.** AUROC detection performance for FMNIST and CIFAR10 in Experiment 1. (**Bold** indicates the best scores.)

| ID | OoD | AE | VAE (e) | SSVAE (e) | MemAE | GODIN | LAMAE+ |
|---|---|---|---|---|---|---|---|
| FMNIST | MNIST | 96.2 ± 0.2 | 99.0 ± 0.1 | 98.9 ± 0.1 | 97.1 ± 0.1 | 79.0 ± 4.3 | **99.9 ± 0.0** |
| | notMNIST | 99.6 ± 0.0 | 99.8 ± 0.0 | **99.9 ± 0.0** | 99.8 ± 0.0 | 64.0 ± 5.8 | **99.9 ± 0.0** |
| | Constant | 68.1 ± 2.2 | 63.2 ± 1.7 | 82.0 ± 7.0 | 72.3 ± 1.1 | 84.4 ± 9.6 | **100.0 ± 0.0** |
| | Noise | **100.0 ± 0.0** | **100.0 ± 0.0** | **100.0 ± 0.0** | **100.0 ± 0.0** | 86.5 ± 6.3 | 99.9 ± 0.0 |
| CIFAR10 | FMNIST | 71.7 ± 0.7 | 69.4 ± 0.9 | 84.8 ± 1.9 | **98.5 ± 0.0** | 94.2 ± 1.7 | 95.0 ± 0.6 |
| | CelebA | 55.0 ± 0.5 | 58.2 ± 0.3 | 60.0 ± 1.2 | 70.0 ± 0.0 | **75.7 ± 2.2** | 59.5± 1.0 |
| | Constant | 0.0 ± 0.0 | 0.0 ± 0.0 | 0.0 ± 0.0 | 51.2 ± 0.0 | 92.7 ± 2.0 | **100.0 ± 0.0** |
| | Noise | 100.0 ± 0.0 | **100.0 ± 0.0** | **100.0 ± 0.0** | **79.6 ± 0.0** | 91.0 ± 8.8 | **100.0 ± 0.0** |

**Table 3.** AUROC detection performance for MNIST in Experiment 2. The second column lists the average image complexity of each OoD class on 1000 images measured by Eq. (3). **Bold** indicates the best scores among each subgroup.

| | Detector | 1 | 2 | 3 | 4 | 5 | 6 |
|---|---|---|---|---|---|---|---|
| OOD class | Complexity | AE | AE+ | MemAE | MemAE+ | LAMAE | LAMAE+ |
| 0 | 1.91 ± 0.27 | **81.4 ± 0.7** | 78.3 ± 0.8 | **83.0 ± 1.4** | 79.9 ± 1.6 | **98.8 ± 0.1** | 98.5 ± 0.2 |
| 1 | 0.94 ± 0.19 | 12.7 ± 0.1 | **27.0 ± 0.3** | 14.2 ± 1.0 | **32.6 ± 1.8** | 72.8 ± 4.2 | **90.1 ± 2.5** |
| 2 | 1.76 ± 0.29 | **92.9 ± 0.3** | **92.9 ± 0.3** | **94.8 ± 0.4** | **94.8 ± 0.4** | 98.9 ± 0.1 | **99.0 ± 0.1** |
| 3 | 1.74 ± 0.30 | **82.0 ± 0.4** | **82.0 ± 0.4** | **83.1 ± 1.2** | **83.1 ± 1.3** | 97.5 ± 0.4 | **97.7 ± 0.3** |
| 4 | 1.55 ± 0.25 | 76.6 ± 0.8 | **79.2 ± 0.7** | 75.7 ± 0.6 | **78.7 ± 0.7** | 93.9 ± 0.8 | **95.5 ± 0.6** |
| 5 | 1.67 ± 0.29 | 82.6 ± 0.3 | **83.1 ± 0.3** | 84.1 ± 0.8 | **84.6 ± 0.9** | 97.2 ± 0.3 | **97.5 ± 0.3** |
| 6 | 1.68 ± 0.29 | 83.6 ± 0.6 | **84.3 ± 0.6** | 85.9 ± 0.9 | **86.6 ± 0.9** | 96.5 ± 0.5 | **97.0 ± 0.5** |
| 7 | 1.42 ± 0.24 | 56.9 ± 1.0 | **62.1 ± 1.0** | 57.5 ± 0.8 | **63.4 ± 0.8** | 91.7 ± 1.5 | **94.7 ± 1.1** |
| 8 | 1.86 ± 0.30 | **90.5 ± 0.3** | 89.1 ± 0.4 | **90.1 ± 0.6** | 88.5 ± 0.7 | **98.0 ± 0.3** | 97.5 ± 0.3 |
| 9 | 1.58 ± 0.26 | 59.2 ± 0.7 | **60.9 ± 0.7** | 56.5 ± 0.9 | **58.2 ± 1.0** | 87.4 ± 2.1 | **89.8 ± 2.0** |

dard deviation of GODIN is much larger than that of AE reconstruction-based approaches, signifying that this type of approaches are more stable than softmax-based approaches.

#### 4.2.2 Performance on Both Semantic and Non-semantic OoD

Table 2 reports the results of FMNIST and CIFAR10 where various OoD datasets are selected. VAE and SSVAE based on reconstruction likelihood have not been tested within these settings and the exact formulation of reconstruction likelihood is not provided. Hence, we report only VAE (e) and SSVAE (e) based on reconstruction error. We can see that LAMAE+ ranked the first in 5 out of the 8 cases. In particular, it is capable of detecting the OoD examples belonging to the Constant dataset better than the other methods, demonstrating its effectiveness in identifying the *plain* OoDs.

The performance is not as good when CIFAR10 is used as the training dataset. This may be due to the fact that the network structure is not deep enough to account for the complicated details within the CIFAR10 dataset. In addition, a single classifier layer may also be inadequate for this dataset. For instance, the backbone classifier used by GODIN is Resnet-34 [11]. Increasing the complexity of network may lead to improvements in detection performance at a cost of an increasing computational burden.

### 4.3 Experiment 2: Analysis of LAMAE+

In this section, we analyse the functionality of each component in LAMAE+. We experimentally demonstrate that the combination of label-assisted memory and CN-adjusted test statistic helps the detector to achieve better results on the most difficult OoD type, i.e., *S+P* OoDs, with the MNIST dataset.

#### 4.3.1 Effect of the Classifier and the Label-Assisted Memory

To illustrate the effectiveness of the classifier and the label-assisted memory, we compare the results for AE, MemAE and LAMAE on the MNIST dataset in Table 3 (Detectors 1, 3 and 5).

We can see that our results for AE and MemAE confirmed the benefit of establishing a memory component as discussed in MemAE [9], for which MemAE achieved higher AUROC than AE in 7 out of the 10 cases. Furthermore, in all 10 cases, LAMAE achieved a significant improvement in AUROC when compared with both MemAE and AE, demonstrating the dominant advantage of using a classifier and a label-assisted memory for detecting *semantic* OoDs when the ID dataset contains multiple classes. Moreover, for OoD digits "1", "4", "7" and "9" whose image complexity measured with Shannon entropy (Eq. 3) ranked the lowest four, there is still a gap when compared with the rest cases. We will address this issue with *plain* OoDs with the CN component in the following section.

#### 4.3.2 Effect of CN-Adjustment

This section demonstrates that the CN-adjustment can further improve the detection performance, especially for the most difficult OoD type *S+P*. As discussed earlier, CN can be used with any AE reconstruction-based OoD detectors and an improvement in detection performance can be anticipated. We verify this conjecture experimentally by taking AE, MemAE and LAMAE as the base detectors and modify only the reconstruction error-based test statistic. We rename the CN-adjusted detectors by suffixing "+". Results are presented in Table 3.

It can be noted that in 8 of the 10 cases, using a CN-adjusted test statistic indeed leads to a significant improvement in detection performance for the digits "1" , "4" , "7" and "9" , which are the hardest ones to detect among all digits [4] and can be characterized as *S+P* by us. This is true for various types of AEs. For digits "0" and "8" , CN caused a slight decrease in AUROC. This is due to the fact that these two digits are already the most complex 2 with the highest entropy values. Regarding this, we suggest that better complexity measurements may be created in the future so that the detection performance on slightly more complicated images can be maintained.

Examining the overall average performance, we conclude that the improvement in detection performance is attributed to the combination of the classifier component, the label-assisted memory and the complexity normalizer.

#### 4.3.3 Sensitivity to Memory Size

This section provides a sensitivity analysis of the detection performance for LAMAE. We present the performance under different memory size settings for the MNIST experiment. Figure 5 suggests that LAMAE is robust to different memory sizes and for simple datasets such as MNIST, even a small memory size can achieve satisfactory performance.

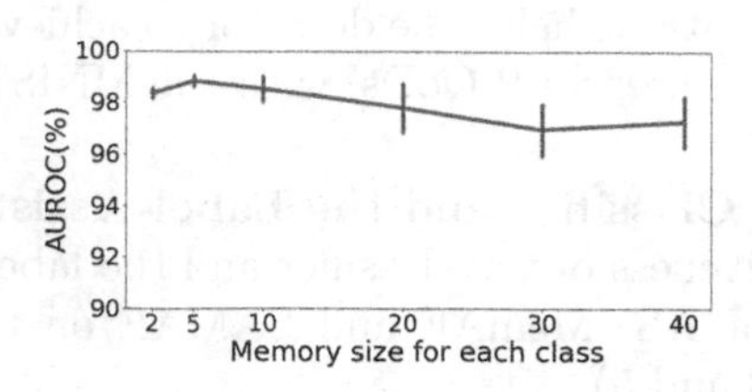


**Fig. 5.** Sensitivity of detection performance to class-conditional memory size on MNIST when digit "0" is held as OoD. Similar trends can be observed for others.

## 5 Conclusion

We proposed LAMAE, a novel AE-based OoD detector with a label-assisted memory. Specifically, we injected a classifier and a class-conditional memory into the AE network architecture to avoid combination of features from different ID classes and thus, constrain the reconstruction of OoD examples while retaining the generalization on ID examples. The detection performance of semantic OoD examples improved significantly. We also proposed a new way to characterize OoD based on image complexity and a new metric, CN, to eliminates the bias associated with the reconstruction error induced by inherent image complexity. Thereby, the refined detector LAMAE+ is capable of detecting the most difficult type of OoD that previous work cannot handle well. It is also worth pointing out that both detectors are purely unsupervised.

In the current work, we only used the basic Shannon entropy to measure image complexity. More suitable measures may also exist [8]. Besides, the performance of the classifier component is of crucial importance to the results. Potential improvements can be made to further improve the detection performance on more complex datasets. Various sizes for each class-conditional memory can also be considered.

**Acknowledgement.** This work was supported by National Natural Science Foundation of China (Grant No. 62002148), the Guangdong Provincial Key Laboratory (Grant No. 2020B121201001), the Program for Guangdong Introducing Innovative and Entrepreneurial Teams (Grant No. 2017ZT07X386), Shenzhen Science and Technology Program (Grant No. KQTD2016112514355531), Shenzhen Fundamental Research Program (Grant No. JCYJ20190809121403553), Research Institute of Trustworthy Autonomous Systems, and Huawei.

## References

1. Abdelzad, V., Czarnecki, K., Salay, R., Denounden, T., Vernekar, S., Phan, B.: Detecting out-of-distribution inputs in deep neural networks using an early-layer output. arXiv preprint arXiv:1910.10307 (2019)

2. An, J., Cho, S.: Variational autoencoder based anomaly detection using reconstruction probability. Spec. Lect. IE **2**(1), 1–18 (2015)
3. Andrews, J.T., Morton, E.J., Griffin, L.D.: Detecting anomalous data using autoencoders. Int. J. Mach. Learn. Comput. **6**(1), 21 (2016)
4. Berkhahn, F., Keys, R., Ouertani, W., Shetty, N., Geißler, D.: Augmenting variational autoencoders with sparse labels: A unified framework for unsupervised, semi-(un) supervised, and supervised learning. arXiv preprint arXiv:1908.03015 (2019)
5. Bulatov, Y.: notMNIST dataset. http://yaroslavvb.blogspot.com/2011/09/notmnist-dataset.html (2020)
6. Chen, J., Sathe, S., Aggarwal, C., Turaga, D.: Outlier detection with autoencoder ensembles. In: SIAM International Conference on Data Mining, pp. 90–98. (2017)
7. Denouden, T., Salay, R., Czarnecki, K., Abdelzad, V., Phan, B., Vernekar, S.: Improving reconstruction autoencoder out-of-distribution detection with mahalanobis distance. arXiv preprint arXiv:1812.02765 (2018)
8. Gao, P., Li, Z., Zhang, H.: Thermodynamics-based evaluation of various improved Shannon entropies for configurational information of gray-level images. Entropy **20**(1), 19 (2018)
9. Gong, D., et al.: Memorizing normality to detect anomaly: memory-augmented deep autoencoder for unsupervised anomaly detection. In: IEEE/CVF International Conference on Computer Vision, pp. 1705–1714 (2019)
10. Hendrycks, D., Mazeika, M., Dietterich, T.: Deep anomaly detection with outlier exposure. In: International Conference on Learning Representations (2018)
11. Hsu, Y.C., Shen, Y., Jin, H., Kira, Z.: Generalized ODIN: detecting out-of-distribution image without learning from out-of-distribution data. In: IEEE/CVF Conference on Computer Vision and Pattern Recognition, pp. 10951–10960 (2020)
12. Huang, Y., Dai, S., Nguyen, T., Baraniuk, R.G., Anandkumar, A.: Out-of-distribution detection using neural rendering generative models. arXiv preprint arXiv:1907.04572 (2019)
13. Ioffe, S., Szegedy, C.: Batch normalization: accelerating deep network training by reducing internal covariate shift. In: International Conference on Machine Learning, pp. 448–456 (2015)
14. Kingma, D.P., Welling, M.: Auto-encoding variational Bayes. In: International Conference on Learning Representations (2014)
15. Krizhevsky, A., Hinton, G., et al.: Learning multiple layers of features from tiny images. Technical Report TR-2009, University of Toronto, Toronto (2009)
16. LeCun, Y.: The MNIST database of handwritten digits. http://yann.lecun.com/exdb/mnist/ (1998)
17. Lee, K., Lee, H., Lee, K., Shin, J.: Training confidence-calibrated classifiers for detecting out-of-distribution samples. In: International Conference on Learning Representations (2018)
18. Lee, K., Lee, K., Lee, H., Shin, J.: A simple unified framework for detecting out-of-distribution samples and adversarial attacks. In: Advances in Neural Information Processing Systems, pp. 7167–7177 (2018)
19. Liang, S., Li, Y., Srikant, R.: Enhancing the reliability of out-of-distribution image detection in neural networks. In: International Conference on Learning Representations (2018)
20. Liu, Z., Luo, P., Wang, X., Tang, X.: Deep learning face attributes in the wild. In: IEEE International Conference on Computer Vision, pp. 3730–3738 (2015)
21. Masana, M., Ruiz, I., Serrat, J., van de Weijer, J., Lopez, A.M.: Metric learning for novelty and anomaly detection. In: British Machine Vision Conference 64 (2018)

22. Nair, V., Hinton, G.E.: Rectified linear units improve restricted Boltzmann machines. In: International Conference on Machine Learning, pp. 807–814 (2010)
23. Nalisnick, E., Matsukawa, A., Teh, Y.W., Gorur, D., Lakshminarayanan, B.: Do deep generative models know what they don't know? In: International Conference on Machine Learning (2019)
24. Perera, P., Patel, V.M.: Learning deep features for one-class classification. IEEE Trans. Image Process. **28**(11), 5450–5463 (2019)
25. Ren, J., et al.: Likelihood ratios for out-of-distribution detection. In: Advances in Neural Information Processing Systems, pp. 14680–14691 (2019)
26. Ruff, L., et al.: Deep one-class classification. In: International Conference on Machine Learning, pp. 4393–4402 (2018)
27. Rumelhart, D.E., Hinton, G.E., Williams, R.J.: Learning internal representations by error propagation. Technical report La Jolla Inst for Cognitive Science (1985)
28. Sarafijanovic-Djukic, N., Davis, J.: Fast distance-based anomaly detection in images using an inception-like autoencoder. In: International Conference on Discovery Science, pp. 493–508 (2019)
29. Shafaei, A., Schmidt, M., Little, J.: A less biased evaluation of OOD sample detectors. In: British Machine Vision Conference (2019)
30. Shalev, G., Adi, Y., Keshet, J.: Out-of-distribution detection using multiple semantic label representations. In: Advances in Neural Information Processing Systems, pp. 7375–7385 (2018)
31. Shannon, C.E.: A mathematical theory of communication. Bell Syst. Tech. J. **27**(3), 379–423 (1948)
32. Tsai, D.Y., Lee, Y., Matsuyama, E.: Information entropy measure for evaluation of image quality. J. Digit. Imaging **21**(3), 338–347 (2008)
33. Tuluptceva, N., Bakker, B., Fedulova, I., Schulz, H., Dylov, D.V.: Anomaly detection with deep perceptual autoencoders. arXiv preprint arXiv:2006.13265 (2020)
34. Xiao, H., Rasul, K., Vollgraf, R.: Fashion-MNIST: a novel image dataset for benchmarking machine learning algorithms. arXiv preprint arXiv:1708.07747 (2017)
35. Xu, D., Ricci, E., Yan, Y., Song, J., Sebe, N.: Learning deep representations of appearance and motion for anomalous event detection. In: British Machine Vision Conference 8 (2015)
36. Yu, Q., Aizawa, K.: Unsupervised out-of-distribution detection by maximum classifier discrepancy. In: IEEE/CVF International Conference on Computer Vision, pp. 9518–9526. (2019)
37. Yuan, Y., Wang, D., Wang, Q.: Anomaly detection in traffic scenes via spatial-aware motion reconstruction. IEEE Trans. Intell. Transp. Syst. **18**(5), 1198–1209 (2016)
38. Zhao, Y., Deng, B., Shen, C., Liu, Y., Lu, H., Hua, X.S.: Spatio-temporal autoencoder for video anomaly detection. In: ACM International Conference on Multimedia, pp. 1933–1941 (2017)
39. Zhou, C., Paffenroth, R.C.: Anomaly detection with robust deep autoencoders. In: ACM SIGKDD International Conference on Knowledge Discovery and Data Mining, pp. 665–674 (2017)
40. Zong, B., et al.: Deep autoencoding Gaussian mixture model for unsupervised anomaly detection. In: International Conference on Learning Representations (2018)

# Quantized Gromov-Wasserstein

Samir Chowdhury[1], David Miller[2], and Tom Needham[3](✉)

[1] Stanford University, Stanford, CA 94305, USA
samirc@stanford.edu
[2] University of Utah, Salt Lake City, UT 84112, USA
dmiller@cs.utah.edu
[3] Florida State University, Tallahassee, FL 32306, USA
tneedham@fsu.edu

**Abstract.** The Gromov-Wasserstein (GW) framework adapts ideas from optimal transport to allow for the comparison of probability distributions defined on different metric spaces. Scalable computation of GW distances and associated matchings on graphs and point clouds have recently been made possible by state-of-the-art algorithms such as S-GWL and MREC. Each of these algorithmic breakthroughs relies on decomposing the underlying spaces into parts and performing matchings on these parts, adding recursion as needed. While very successful in practice, theoretical guarantees on such methods are limited. Inspired by recent advances in the theory of quantization for metric measure spaces, we define Quantized Gromov Wasserstein (qGW): a metric that treats parts as fundamental objects and fits into a hierarchy of theoretical upper bounds for the GW problem. This formulation motivates a new algorithm for approximating optimal GW matchings which yields algorithmic speedups and reductions in memory complexity. Consequently, we are able to go beyond outperforming state-of-the-art and apply GW matching at scales that are an order of magnitude larger than in the existing literature, including datasets containing over 1M points.

**Keywords:** Gromov-Wasserstein distance · Optimal transport · Metric space registration

## 1 Introduction

It is frequently convenient to represent geometric data, such as the set of points in a point cloud or the set of nodes in a network, as a finite metric space. Such a representation naturally enjoys invariances to symmetries such as permutations or rigid Euclidean motions, which typically serve as nuisances in data analysis tasks. The well known Gromov-Hausdorff (GH) distance provides a

**Electronic supplementary material** The online version of this chapter (https://doi.org/10.1007/978-3-030-86523-8_49) contains supplementary material, which is available to authorized users.


N. Oliver et al. (Eds.): ECML PKDD 2021, LNAI 12977, pp. 811–827, 2021.
https://doi.org/10.1007/978-3-030-86523-8_49

mathematical framework for comparing metric spaces, but it is hard to handle computationally, as it inherently requires a *point correspondence* between the spaces [19,20]. Gromov-Wasserstein (GW) distance [15,17] is a relaxation of Gromov-Hausdorff distance that compares *metric measure (mm) spaces*—metric spaces endowed with probability measures—by optimizing a nonconvex loss over a convex domain. We precisely define GW distance below, but the idea is that one compares mm-spaces by finding a *probabilistic correspondence* (also referred to as a *matching*) between their points. Matchings can be approximated via standard optimization tools in this relaxed setting. Such matchings have numerous applications, as they give alignments between spaces which are not directly comparable; e.g., protein-protein interaction networks [36], text corpora from different languages [1] or single-cell multi-omics data [10].

While exact computation of GW distance is NP-Hard, recent algorithmic advances toward estimating it have made GW distance a viable tool for machine learning tasks on metric space-valued data at an increasingly large scale [3,25,36]. A common theme in recent approximation algorithms is to break the GW problem into smaller subproblems. A rough template for using this approach to compare two large mm-spaces $X$ and $Y$ is as follows: partition $X$ and $Y$ into smaller blocks of manageable size, find a matching between representatives of the blocks, then recurse this process to find matchings between the paired blocks. This recursive approach to estimating GW distance increases the feasible scale of $X$ and $Y$ by an order of magnitude ($\sim$1K points to $\sim$10K points). In this paper, we treat this partitioned matching paradigm more formally: we develop a theoretical formulation inspired by the duality between quantization and clustering developed recently in [21], which in turn suggests a new approximation algorithm that scales to metric spaces with $\sim$100K or even $\sim$1M points[1]. Our main contributions are:

1. we define a new metric on the space of partitioned mm-spaces called *quantized Gromov-Wasserstein (qGW) distance*;
2. we give a novel algorithm for estimating qGW distance and show that (for certain parameter choices) its complexity is *nearly linear* in the sizes of the spaces being compared (Proposition 3);
3. we give theoretical error bounds comparing qGW and GW (Theorems 5 and 6)—these error bounds give heuristics for the types of spaces on which qGW should be expected to perform well;
4. we demonstrate empirically that qGW gives state-of-the-art performance for matching large scale shape datasets (Sect. 4).

## 2 Quantized Gromov-Wasserstein

**Gromov-Wasserstein Distance.** A *metric measure space*, or *mm-space* for short, is a triple $(X, d_X, \mu_X)$ consisting of a compact metric space $(X, d_X)$

[1] https://github.com/trneedham/QuantizedGromovWasserstein.

endowed with a Borel probability measure $\mu_X$. We abuse notation and abbreviate the triple simply as $X$. In practice, we deal with finite mm-spaces $X = \{x_1, \ldots, x_n\}$, where the metric can be represented as the distance matrix $(d_X(x_i, x_j))_{i,j}$ and the probability measure can be represented as a vector $(\mu_X(x_1), \ldots, \mu_X(x_n))$. Let $\mathscr{C}(\mu_X, \mu_Y)$ denote the set of *couplings* of $\mu_X$ and $\mu_Y$; i.e., Borel probability measures $\mu$ on the product space $X \times Y$ whose marginals are $\mu_X$ and $\mu_Y$.

For finite mm-spaces $X$ and $Y$, we define the *Gromov-Wasserstein loss* of a coupling $\mu \in \mathcal{C}(\mu_X, \mu_Y)$ as

$$\mathrm{GW}(\mu) = \sum_{i,j,k,\ell} (d_X(x_i, x_k) - d_Y(y_j, y_\ell))^2 \mu(x_i, y_j) \mu(x_k, y_\ell). \tag{1}$$

The *Gromov-Wasserstein (GW) distance* between $X$ and $Y$ is

$$d_{\mathrm{GW}}(X, Y) := \inf_{\mu \in \mathcal{C}(\mu_X, \mu_Y)} \mathrm{GW}(\mu)^{1/2}. \tag{2}$$

For finite spaces, $\mathcal{C}(\mu_X, \mu_Y) \subset \mathbb{R}^{|X| \times |Y|}$ is a convex polytope whose elements are represented as matrices. In this setting, a minimizer $\mu$ of (1) can be considered as a *soft alignment* of the spaces $X$ and $Y$: the row of $\mu$ (considered as a matrix) indexed by $x_i$ gives a probabilistic assignment to each of the points of $Y$. We frequently refer to a minimizer of (1) as a *matching* of the mm-spaces.

We remark that Eq. 2 can be extended to treat not necessarily finite spaces. It is one of two different notions of GW distance [15,17,31] in the literature, and the main theoretical properties of both these GW distances were worked out by Mémoli [15–17] and Sturm [30,31]. The formulation in Eq. (2) is the one used more often in computational settings [26], and we focus on this case throughout the current work. After initial theoretical development, large scale applications of GW to machine learning and graphics problems were later explored in [25,29], and a vast literature on GW distance has since developed, focusing on both theoretical [7–9,18,28,35] and applications-driven [1,6,10] aspects.

For finite mm-spaces $X$ and $Y$ of size $\approx n$, naive evaluation of the objective function (1) incurs cost $O(n^4)$. A less naive implementation brings this down to $O(n^3 \log(n))$ [25], but this is still prohibitively expensive for even medium scale tasks with on the order of several thousand points. We introduce a metric which operates on partitioned spaces, and present an efficient algorithm for approximating this metric further below. These developments formalize ideas which have appeared in more ad hoc forms in the recent literature [3,33,36]; we make these connections precise at the end of this section.

**The Quantized Gromov-Wasserstein Metric.** Recent approaches to approximating GW distance have used a divide-and-conquer strategy, where the spaces are partitioned into blocks and the blocks are matched recursively. In this section, we develop a formal framework for matching partitioned spaces. We treat a partition of a space as input data for a metric, obtained in a preprocessing step; practical methods for finding good partitions are described below.

Let $X$ be a finite mm-space. A *pointed partition* of $X$ is a structure consisting of a partition of $X$ into disjoint, nonempty sets $U^1, \dots, U^m$ together with a representative point $x^p \in U^p$ for each $p = 1, \dots, m$. We denote this structure as $\mathcal{P}_X = \{(x^1, U^1), \dots, (x^m, U^m)\}$. When a pointed partition has $m$ points, we will refer to it as an *m-pointed partition.* We will typically work in the regime $m \ll |X|$. A set $U^p$ in $\mathcal{P}_X$ will be referred to as a *partition block* and the distinguished point $x^p \in U^p$ will be referred to as the *representative* of the partition block. Note that we use superscripts $x^p$ to denote partition block representatives, to distinguish from generic indexed points $x_i$ in the metric space. An *m-pointed metric measure space* is a quadruple $(X, d_X, \mu_X, \mathcal{P}_X)$; this will be abbreviated as $(X, \mathcal{P}_X)$.

To an $m$-pointed mm-space $(X, \mathcal{P}_X)$ we associate several related mm-spaces. Let $X^m := \{x^1, \dots, x^m\}$ denote the set of partition block representatives. There is a well-defined projection map $X \to X^m$ induced by the partition, i.e. the map $x \mapsto x^p$ when $x \in U^p$. We endow $X^m$ with the measure $\mu_{\mathcal{P}_X}$ given by the pushforward of $\mu_X$ by this projection map. Then $X^m$ has an mm-space structure given by restricting $d_X$. We refer to $X^m = (X^m, d_X|_{X^m}, \mu_{\mathcal{P}_X})$ as a *quantized representation* of $X$. For each $(x^p, U^p) \in \mathcal{P}_X$, we also obtain a new mm-space $(U^p, d_X|_{U^p}, \mu_{U^p})$, where $\mu_{U^p} := (\mu_X(U^p))^{-1} \mu_X|_{U^p}$. This can be extended to a probability measure $\bar{\mu}_{U^p}$ on all of $X$ via the formula

$$\bar{\mu}_{U^p}(A) := \mu_{U^p}(A \cap U^p) \qquad \forall\, A \subset X. \tag{3}$$

Let $(Y, d_Y, \mu_Y, \mathcal{P}_Y)$ be another finite $m$-pointed mm-space with quantized representation $Y^m$. A *quantization coupling* is a measure $\mu$ on $X \times Y$ of the form

$$\mu(x, y) = \sum_{p,q} \mu_m(x^p, y^q) \bar{\mu}_{x^p, y^q}(x, y), \tag{4}$$

where $\mu_m \in \mathcal{C}(\mu_{\mathcal{P}_X}, \mu_{\mathcal{P}_Y})$, each $\mu_{x^p, y^q} \in \mathcal{C}(\mu_{U^p}, \mu_{V^q})$ and $\bar{\mu}_{x^p, y^q} \in \mathcal{C}(\bar{\mu}_{U^p}, \bar{\mu}_{V^q})$ is an extension of $\mu_{x^p, y^q}$ using the same trick as (3). We moreover assume that $\mu_{x^p, y^q}(x^p, y^q) > 0$ for each $p, q$. Let $\mathcal{C}_{\mathcal{P}_X, \mathcal{P}_Y}(\mu_X, \mu_Y)$ denote the set of quantization couplings of $\mu_X$ and $\mu_Y$ with respect to $\mathcal{P}_X$ and $\mathcal{P}_Y$. We have the following proposition, which says that quantization couplings are couplings in the usual sense. The proof is provided in the supplementary materials.

**Proposition 1.** *Any quantization coupling is a coupling; that is, the set of quantization couplings $\mathcal{C}_{\mathcal{P}_X, \mathcal{P}_Y}(\mu_X, \mu_Y)$ is a subset of the set of couplings $\mathcal{C}(\mu_X, \mu_Y)$.*

We define the *quantized Gromov-Wasserstein (qGW) distance* between finite $m$-pointed mm-spaces as

$$d_{\mathrm{qGW}}((X, \mathcal{P}_X), (Y, \mathcal{P}_Y)) := \inf_{\mu \in \mathcal{C}_{\mathcal{P}_X, \mathcal{P}_Y}(\mu_X, \mu_Y)} \mathrm{GW}(\mu)^{1/2}. \tag{5}$$

Intuitively, qGW compresses partition blocks into their representative points, matches the representatives across mm-spaces, and then unpacks to give an alignment of the full mm-spaces.

An *isomorphism* of mm-spaces $X$ and $Y$ is a measure-preserving isometry $X \to Y$. Let $\mathcal{M}$ denote the collection of all isomorphism classes of finite mm-spaces. This notion can be specialized to pointed mm-spaces: an *isomorphism* of $m$-pointed mm-spaces $(X, \mathcal{P}_X)$ and $(Y, \mathcal{P}_Y)$ is an isomorphism $X \to Y$ which takes $X^m$ to $Y^m$. We use $\mathcal{M}^m$ to denote the collection of all finite $m$-pointed mm-spaces, considered up to isomorphism.

**Theorem 2.** *The quantized GW distance $d_{\mathrm{qGW}}$ is a metric on $\mathcal{M}^m$.*

The interesting part of the proof is establishing the triangle inequality, which is an application of the gluing lemma from optimal transport theory [34, Lemma 7.6]. The idea is to produce a coupling of $(X, \mathcal{P}_X)$ and $(Z, \mathcal{P}_Z)$ by applying the gluing lemma to quantization couplings of $(X, \mathcal{P}_X)$ and $(Y, \mathcal{P}_Y)$ and $(Y, \mathcal{P}_Y)$ and $(Z, \mathcal{P}_Z)$. A lengthy computation shows that the result is automatically a quantization coupling, and the triangle inequality follows. A full proof of the theorem is provided in the supplementary materials.

Exact computation of quantized GW distance $d_{\mathrm{qGW}}$ is intractable, but the structure of the metrics suggests natural heuristics. We now give an algorithm for approximating $d_{\mathrm{qGW}}$. Theoretical estimates for the quality of this approximation with respect to the standard GW distance between underlying mm-spaces are presented in Sect. 3.

**The Quantized Gromov-Wasserstein Algorithm.** Let $(X, d_X, \mu_X, \mathcal{P}_X)$ and $(Y, d_Y, \mu_Y, \mathcal{P}_Y)$ be $m$-pointed mm-spaces and let $X^m$ and $Y^m$ denote their quantized representations. In this subsection, we present an efficient method for approximating quantized GW distance $d_{\mathrm{qGW}}$ between these spaces, which in turn gives an estimate of GW distance between the underlying mm-spaces. The algorithm proceeds in three steps:

1. **Global Alignment:** The first step in the algorithm is to compute $\mu_m$, an optimal coupling of the quantized representations of $X$ and $Y$—$m \ll |X|, |Y|$ is chosen so that this optimal coupling can be feasibly approximated via existing methods, such as those implemented in the Python Optimal Transport library [12]. This serves to give a *global alignment* of the spaces.
2. **Local Alignment:** The second step is to produce a collection of *local alignments*. For each $x^p \in X^m$ and $y^q \in Y^m$, we obtain a coupling $\mu_{x^p,y^q}$ of the partition block mm-spaces $(U^p, d_X|_{U^p}, \mu_{U^p})$ and $(V^q, d_Y|_{V^q}, \mu_{V^q})$ by solving the optimal transport problem

$$\min_{\mu_{x^p,y^q} \in \mathcal{C}(\mu_{U^p}, \mu_{V^q})} \sum_{x \in U^p, y \in V^q} (d_X(x, x^p) - d_Y(y, y^q))^2 \mu_{x^p,y^q}(x, y). \qquad (6)$$

The solution $\mu_{x^p,y^q}$ is referred to as a *local linear matching* of $U_p$ and $V_q$. The simplified matching problem (6) can be solved efficiently, as it is equivalent to finding an optimal transport plan between distributions on the real line—this is made precise in Proposition 3 below.

3. **Create Coupling:** The final step is to create a quantization coupling from the global and local alignments,

$$\mu = \sum_{p,q} \mu_m(x^p, y^q)\overline{\mu}_{x^p,y^q}. \tag{7}$$

This is treated as an approximate solution to the $d_{\mathrm{qGW}}$ optimization problem.

We remark that the local linear matching obtained by solving (6) is generally *not* a solution of the GW optimization problem (2), applied to the partition blocks $U^p$ and $V^q$. An alternative approach to approximating $d_{\mathrm{qGW}}$ would be to replace the local alignment step above with a local alignment which actually solves the GW subproblems—indeed, this procedure is similar to what is done in the Scalable Gromov-Wasserstein [37] and MREC [3] frameworks. We show below that our simplified algorithm drastically reduces the computational complexity over computing several GW subproblems.

A subroutine in this procedure is a heuristic for generating good partitions. In the graph setting, we applied the Fluid community detection algorithm [24]—available in the Python package `networkx` [14]—to choose partition blocks, and we chose block representatives with maximal PageRank [4]. In the point cloud settings we simply chose uniform iid samples without replacement and computed a Voronoi partition; other approaches such as $k$-means are possible.

**Computational Complexity.** We now give bounds on the computational complexity of finding a quantization coupling via local linear matchings, as was described in (7). The key observation is that the optimization problem (6) can be solved extremely efficiently. The next proposition follows from a result on pushforward measures [7, Lemma 27] and from the well known fact that one-dimensional optimal transport can be solved efficiently [26, Section 2.6]; see the supplementary material for details.

**Proposition 3.** *The optimization problem* (6) *is equivalent to a one dimensional optimal transport problem and can therefore be solved in* $O(k\log(k))$ *time, with* $k$ *the max number of points in* $U^p$ *or* $V^q$.

We can therefore estimate the computational complexity of the qGW approximation algorithm as follows. Suppose that $|X|$ and $|Y|$ are of order $N$ and that they are each partitioned into $m$ blocks. Also suppose that the blocks are of roughly equal size, so that there are approximately $N/m$ points per block. The worst-case complexity of the quantized GW algorithm is the maximum of an iterative $O(m^3\log(m))$ term coming from approximation of the global GW alignment (via some variant of gradient descent) [25] and a $O(m^2 \cdot N/m\log(N/m))$ term coming from performing $m^2$ local linear matchings. However, it has been observed empirically [8,36] that optimal GW couplings tend to have supports whose sizes scale *linearly* in the number points in the spaces being matched (rather than the worst-case quadratic scaling)—in fact, this order of scaling can be theoretically guaranteed when comparing symmetric positive definite kernel matrices [9]. Since local linear matchings only need to be computed for $x^p, y^q$ such that $\mu_m(x^p, y^q) \neq 0$, the expected computational complexity for the algorithm is therefore the maximum of an iterative $O(m^3 \log(m))$ term and an $O(N \log(N/m))$ term; taking $m$ on the order of $N^{1/3}$, for example, gives an algorithm with computational cost $O(N \log(N))$ (iterative).

Memory complexity of GW computations becomes a serious issue at scale. For example, storing and manipulating distance matrices on 50K points (cf. the graph matching experiment in Sect. 4) with 64-bit floats requires 20 GB memory. We resolve this by observing that qGW never requires the full $O(N^2)$ distance matrices: we only require storing a dense $O(m^2)$ matrix of distances between representatives, and a sparse $O(Nm)$ matrix of distances between each block representative to the points *in the same block*. In addition to enabling qGW computations on datasets with ~1M points (cf. the large scale matching experiment in Sect. 4), this observation becomes especially useful when preprocessing distance matrices on graphs: instead of incurring $O(N|E|\log(N))$ cost of computing a full matrix of graph geodesic distances via Dijkstra's algorithm, we simply incur $O(m|E|\log(N))$ cost. Here $E$ denotes the set of edges.

Finally, the quantization approach allows for fast computation of individual queries, i.e. individual rows of the coupling matrix. Given a point $x \in X$ with block representative $x^p$, we can compute $\mu(x, \cdot)$ (i.e. the target of $x$ in $Y$) by only accessing the $m^2$ matrices of distances between representatives in $X, Y$, the distances from $x^p$ to all other points in its block, and likewise for all $y^q$ such that $\mu_m(x^p, y^q) > 0$ (typically $\mu_m$ is sparse).

**Quantized Fused Gromov-Wasserstein.** It is common that data, represented as a finite metric space $(X, d_X)$, comes endowed with attributions; i.e., with a function $f : X \to Z$ valued in another metric space $(Z, d_Z)$. This is the case, for example, when $X$ represents (the nodes of a) network and $f : X \to Z$ represents node attributes, which can be data-driven or crafted from local network features. This extra complexity is handled elegantly in the GW framework by the Fused Gromov-Wasserstein (FGW) distance of Vayer et al. [32]. We give a brief formulation of FGW distance in the setting of finite metric spaces.

Let $(Z, d_Z)$ be a metric space. A *finite $Z$-structured mm-space* is a quadruple $(X, d_X, \mu_X, f_X)$, where $(X, d_X, \mu_X)$ is a finite mm-space and $f_X : X \to Z$ is an arbitrary function. We denote this structure as $(X, f_X)$ when the existence of a metric and measure on $X$ is clear from context. Let $(X, f_X)$ and $(Y, f_Y)$ be $Z$-structured mm-spaces and let $\alpha > 0$. The *Fused Gromov-Wasserstein loss*, with parameter $\alpha$, of a coupling $\mu \in \mathcal{C}(\mu_X, \mu_Y)$ is

$$\mathrm{FGW}_\alpha(\mu) := (1-\alpha)\mathrm{GW}(\mu) + \alpha \mathrm{W}(\mu),$$

where GW is Gromov-Wasserstein loss (1) and

$$\mathrm{W}(\mu) := \sum_{i,j} d_Z(f_X(x_i), f_Y(y_j))^2 \mu(x_i, y_j)$$

is classical Wasserstein loss. One then defines *FGW distance* as

$$d_{\mathrm{FGW},\alpha}((X,f_X),(Y,d_Y)) := \min_{\mu\in\mathcal{C}(\mu_X,\mu_Y)} \mathrm{FGW}_\alpha(\mu)^{1/2}.$$

We now describe a quantized algorithm for approximating FGW distance. Let $(X, d_X, \mu_X, \mathcal{P}_X, f_X)$ and $(Y, d_Y, \mu_Y, \mathcal{P}_Y, f_Y)$ be $m$-partitioned mm-spaces endowed with $Z$-structures. The approximation algorithm proceeds in steps similar to the qGW algorithm, and involves an additional parameter $\beta$.

1. **Global Alignment:** The first step in the approximation algorithm is to determine a global registration. This is done by computing a coupling $\mu_m \in \mathcal{C}(\mu_X, \mu_Y)$ as a minimizer of $\mathrm{FGW}_\alpha$ for the $Z$-structured mm-spaces $(X^m, d_X|_{X^m}, \mu_{\mathcal{P}_X}, f_X|_{X^m})$ and $(Y^m, d_Y|_{Y^m}, \mu_{\mathcal{P}_Y}, f_Y|_{Y^m})$.
2. **Local Alignment:** Next we find a $\mu_{x^p,y^q} \in \mathcal{C}(\mu_{U^p}, \mu_{V^q})$ for each $x^p, y^q$. We first solve the local linear matching problem (6) to get a coupling $\mu^{(0)}_{x^p,y^q}$. Next, we solve another local linear matching problem with respect to $Z$-valued features to obtain a second coupling $\mu^{(1)}_{x^p,y^q}$. Finally, we define $\mu_{x^p,y^q}$ by a simple weighted average with respect to our parameter $\beta$, and set

$$\mu_{x^p,y^q} = (1-\beta)\mu^{(0)}_{x^p,y^q} + \beta\mu^{(1)}_{x^p,y^q}.$$

3. **Create Coupling:** We create a coupling, as in the qGW algorithm.

The parameters in this matching algorithm can therefore be described intuitively: $\alpha$ controls the preference to globally match based on metric structure or feature structure, while $\beta$ controls this preference locally.

**Related Work.** We now describe in more detail other approaches to scaling the GW framework and the relationship between our algorithm and others in the literature. The first serious attempt to scaling GW distance computation is the *scalable Gromov-Wasserstein (sGW)* framework of Xu et al. [36]. This framework is designed specifically to generate GW matchings between graphs, represented as adjacency matrices, by leveraging the observations of [7,25] that the GW framework gives a sensible way to compare arbitrary square matrices (not just distance matrices). Xu et al. introduced a recursive partitioning scheme, allowing a "divide-and-conquer" approach to the matching problem. A similar approach to approximating GW matchings is the basis of the MREC algorithm of Blumberg et al. [3], which finds matchings between Euclidean point clouds or more general metric spaces via a scheme which recursively partitions the data and defines smaller subproblems by matching partition block representatives. The MREC framework is quite general, also allowing matchings based on classical Wasserstein distance. In the Euclidean setting with $L^2$ Wasserstein cost, Mérigot introduced a multiscale approach that solves an optimal transport problem via iterative refinement [22]. Our algorithm qGW fits into the general mold of sGW and MREC, but we replace the recursive definition of submatching problems with a simpler local linear matching problem. More broadly, these methods based on partitioning are related, at least in the metric space setting, to a duality between quantization and clustering as studied in [21]. We use this connection explicitly when obtaining error bounds for qGW.

The ideas presented in this paper are related to other common themes appearing in the GW and broader optimal transport literature. Recently introduced by Vayer et al. as a fast approximation of GW distance, the *sliced Gromov-Wasserstein distance* [33] computes a dissimilarity between Euclidean point clouds by taking an expectation of Gromov-Wasserstein distance between random 1-dimensional projections. By its nature, this method is limited to Euclidean point clouds, but extra optimization steps can be incorporated to compare point clouds lying in different dimensions and to make the dissimilarity rotation-invariant. Our qGW algorithm also speeds up the GW computation by using 1-dimensional projections; here, we are invoking a 1-dimensional problem by "slicing" radially from prealigned anchor points. This approach to slicing means that our algorithm works on general metric spaces and that it is naturally invariant to isometries such as rigid motions. The idea of using distances to anchor points to compare mm-spaces is used in computable *lower* bounds on GW distance derived in [17] (whereas our method always produces an *upper* bound). We remark that efficient algorithms for computing variants of this lower bound have recently been introduced by Sato et al. [27]. Although not directly related to this work, similar ideas involving finding optimal anchor points for simplified graph representations via methods of optimal transport have appeared in [13].

## 3 Theoretical Error Bounds

**Quantized Eccentricity.** Let $(X, \mathcal{P}_X)$ be an $m$-pointed mm-space. As above, we let $X^m = \{x^1, \dots, x^m\}$ denote the set of partition block representatives. We treat $X^m$ and each of the blocks $U^p$ as a mm-space, as described in Sect. 2. The goal of this section is to obtain estimates on the approximation quality of the qGW algorithm with respect to the true GW distance.

Let $x \in X$. The *eccentricity* [17] of $x$ is

$$s_X(x) := \left( \sum_{x'} d_X(x, x')^2 \mu_X(x') \right)^{1/2}.$$

We define the *quantized eccentricity* of $\mathcal{P}_X$ to be the quantity

$$q(\mathcal{P}_X) := \left( \sum_{p} \mu_X(U^p) s_{U^p}(x^p)^2 \right)^{1/2}$$

measuring the expected eccentricity of a partition block. The *m-quantized eccentricity* of $X$, $q_m(X)$, is the minimum of $q(\mathcal{P}_X)$ over all $m$-pointed partitions.

**Lemma 4.** *For a finite mm-space $X$ and $m$-pointed partition $\mathcal{P}_X$ with partition block representatives $X^m$, $d_{GW}(X, X^m) \leq 2q(\mathcal{P}_X)$. It follows that*

$$\min_{\mathcal{P}_X} d_{GW}(X, X^m) \leq 2q_m(X),$$

*where the minimum is over $m$-pointed partitions $\mathcal{P}_X$.*

Intuitively, the lemma says that a mm-space with small $m$-quantized eccentricity is well-approximated by an $m$-point subset. The proof is given in the supplementary materials. The strategy of the proof is inspired by the proof of [21, Theorem 1.12], which compares other abstract measures (related to quantization and clustering) for coarsely representing a mm-space. Combining the lemma with the reverse triangle inequality immediately yields the following result.

**Theorem 5.** *For finite mm-spaces $X$ and $Y$,*

$$\min_{\mathcal{P}_X, \mathcal{P}_Y} |d_{GW}(X, Y) - d_{GW}(X^m, Y^m)| \leq 2\left(q_m(X) + q_m(Y)\right). \tag{8}$$

*where the minimum is taken over $m$-pointed partitions $\mathcal{P}_X$ and $\mathcal{P}_Y$.*

We note that neither side of the estimate (8) is explicitly computable. The theorem should be interpreted as giving intuition about which types of mm-spaces are amenable to accurate comparison by the qGW algorithm. Since the qGW algorithm described above begins by finding a global alignment of mm-spaces using partition block representatives, we would like to understand how well this global alignment reflects the true distance between mm-spaces. The result says that one can only reasonably hope for an accurate representation when the mm-spaces in question have low quantized eccentricity. As a heuristic, this is the case for dense point clouds in low dimensions or for graphs with rigid geometric structure. Low quantized eccentricity is less likely in high-dimensional point clouds due to the concentration of measure phenomenon or in graphs such as social networks, due to typical "small world" structure.

**Error Bound for the qGW Algorithm.** Let $(X, \mathcal{P}_X)$ and $(Y, \mathcal{P}_Y)$ be $m$-pointed mm-spaces and let

$$\delta((X, \mathcal{P}_X), (Y, \mathcal{P}_Y)) := \mathrm{GW}(\mu)^{1/2},$$

where $\mu$ is the coupling obtained by the qGW algorithm with locally linear matchings described in Sect. 2.

**Theorem 6.** *Let $(X, \mathcal{P}_X)$ and $(Y, \mathcal{P}_Y)$ be $m$-pointed mm-spaces such that the metric diameter of each partition block in both $\mathcal{P}_X$ and $\mathcal{P}_Y$ is bounded above by $\epsilon > 0$. Then*

$$|d_{GW}(X, Y) - \delta((X, \mathcal{P}_X), (Y, \mathcal{P}_Y))| \leq 2\left(q(\mathcal{P}_X) + q(\mathcal{P}_Y)\right) + 8\epsilon.$$

The proof uses the quantized eccentricity results of the previous subsection together with some careful estimates on the $\delta$ dissimilarity; details are provided in the supplementary materials. The value of this result is that it once again gives a heuristic for when quantized GW will be most useful: when the spaces being compared admit partitions with small quantized energy and with partition blocks of small diameter.

## 4 Experiments

**Table 1.** Distortion scores (lower is better) and runtimes for variants of Gromov-Wasserstein matchings. The average number of points in each shape class is provided under the shape class name. Results are listed for several parameter choices of each method—see the text for details. Text in **bold** is the best score across scalable GW methods and underlined text is the best score among the methods in the top 50% of methods in terms of compute time. Blank entries failed to complete in 10 h.

| Method | Param | Humans 1926 | Planes 2144 | Spiders 2664 | Cars 5220 | Dogs 8937 | Trees 10433 | Vases 15828 |
|---|---|---|---|---|---|---|---|---|
| GW | | .07 (8.47) | .08 (19.64) | .005 (28.29) | .16 (99.83) | .003 (512.80) | .015 (835.46) | – |
| erGW | 0.2 | **.03 (15.42)** | .09 (17.03) | .090 (10.54) | **.16 (85.89)** | .153 (920.77) | .200 (2490.14) | – |
| | 5 | .63 (2.93) | .87 (6.23) | .182 (9.76) | .67 (41.03) | .687 (181.94) | .714 (275.73) | .61 (1312.87) |
| MREC | (.1, .01) | .32 (1.45) | .52 (2.37) | .060 (1.88) | .87 (3.12) | .459 (8.89) | .472 (9.05) | .43 (23.49) |
| | (5, .01) | .76 (.29) | .76 (.43) | .131 (0.52) | .98 (1.75) | .653 (5.31) | .549 (6.48) | .58 (15.66) |
| | (.1, .1) | .25 (2.31) | .39 (3.37) | .064 (2.61) | .38 (8.00) | .378 (15.59) | .316 (23.20) | .48 (43.05) |
| | (5, .1) | .67 (.54) | .83 (.82) | .166 (1.02) | .63 (2.94) | .674 (8.36) | .580 (10.87) | .62 (29.07) |
| | (.1, .2) | .32 (1.54) | .10 (2.42) | .065 (2.99) | .20 (11.24) | .415 (30.86) | .391 (40.58) | .50 (130.86) |
| | (5, .2) | .53 (1.31) | .89 (2.10) | .160 (2.80) | .72 (8.88) | .686 (26.97) | .720 (37.48) | .60 (100.51) |
| | (.1, .5) | .18 (9.89) | .10 (15.16) | .063 (18.17) | .20 (71.26) | .411 (240.53) | .375 (298.27) | .42 (1337.00) |
| | (5, .5) | .65 (7.55) | .87 (11.94) | .180 (18.29) | .70 (60.56) | .694 (198.40) | .723 (282.02) | .62 (767.81) |
| mbGW | (50, 5$K$) | .22 (19.17) | .44 (17.95) | .043 (19.90) | .74 (22.13) | .494 (25.88) | .325 (26.60) | .51 (31.51) |
| | (50, 0.1) | .30 (.71) | .61 (.72) | .048 (1.02) | .78 (2.23) | .506 (4.51) | .334 (5.35) | .52 (10.18) |
| qGW | .01 | .36 (.06) | .53 (.09) | .044 (.13) | .18 (.34) | .330 (.93) | .161 (1.22) | .26 (5.96) |
| | .1 | .28 (.32) | .08 (.70) | .016 (.78) | .28 (1.48) | .002 (4.20) | .026 (5.96) | **.18 (25.23)** |
| | .2 | .14 (.70) | **.03 (1.15)** | .020 (1.56) | .22 (4.10) | **.001 (11.37)** | .002 (15.38) | .21 (74.43) |
| | .5 | .06 (2.71) | .11 (4.82) | **.010 (6.94)** | .19 (26.02) | .001 (89.82) | **.001 (122.26)** | .21 (642.09) |

**Point Cloud Matching.** In this experiment we investigate the power of several scalable variants of GW distance to uncover "ground truth" matchings between 3D point clouds. We use point clouds from the CAPOD dataset [23], which contains several classes of 3D meshes (we use the vertices as point clouds) of various sizes. For each class, we choose 10 shape samples, each treated as a mm-space by endowing it with Euclidean distance and uniform measure. For each shape sample, we create a copy whose vertices are permuted and perturbed (randomly within 1% of the diameter of the shape). The task is to use a GW matching to get correct matches between points in the original shape $X$ and its noisy, permuted copy $\widetilde{X}$. Given a matching $\mu$ (a probabilistic correspondence between points in $X$ and $\widetilde{X}$), we compute the *distortion* for each point $x_i \in X$ as the distance from its ground truth copy $\widetilde{x}_i$ and its *matched point* $y_j := \mathrm{argmax}_{y_j} \mu(x_i, y_j)$. The *distortion score* for the matching is the mean squared distortion.

For each class and each sample, matchings are produced by several algorithms; in Table 1 we report class average distortion scores and compute times. We test the qGW algorithm, where we choose an $m$-pointed partition of each shape sample $X$ by randomly sampling $\lfloor p \cdot |X| \rfloor$ points, $p \in \{.01, .1, .2, .5\}$, as partition block representatives and then taking a Voronoi partition with respect to these representatives. We also test against several baselines: standard GW is used as an overall baseline, where we see that compute times quickly become infeasible for any large scale task. For scalable baselines, we compare against entropy regularized GW (erGW) [25], MREC [3] and minibatch GW (mbGW) [11]. Entropy regularized GW has a regularization weight parameter $\epsilon$ and we test $\epsilon \in \{0.2, 5\}$ to check results in low and high regularization regimes. The MREC algorithm is very flexible and can incorporate several clustering and matching methods into its overall architecture. Our comparison does not reflect the full capabilities of MREC, and we only used parameters giving a direct comparison to other GW-based methods; that is, we use the GW module for matching and the random Voronoi partitioning module for clustering. With these choices, MREC has two additional parameters $(\epsilon, p)$, where $\epsilon$ is a regularization weight parameter and $p$ is the percentage of points used when creating partitions for recursion. We used $\epsilon \in \{0.1, 5\}$—we ran a larger grid search, and the reported results are qualitatively similar to those we obtained for other parameters—and $p \in \{.01, .1, .2, .5\}$. Minibatch GW has parameters $(n, k)$, where $n$ the number of samples per batch and $k$ is the number of batches—either an integer or a fraction of the size of each dataset. We use $n = 50$, and $k = 5K$ or 10% of the size of the datasets (following the method of [11, Figure 16]). We remark that we are not aware of an official implementation for obtaining mbGW matchings, and the results here were created by our own implementation. Representative matchings for several methods are shown in Fig. 1.

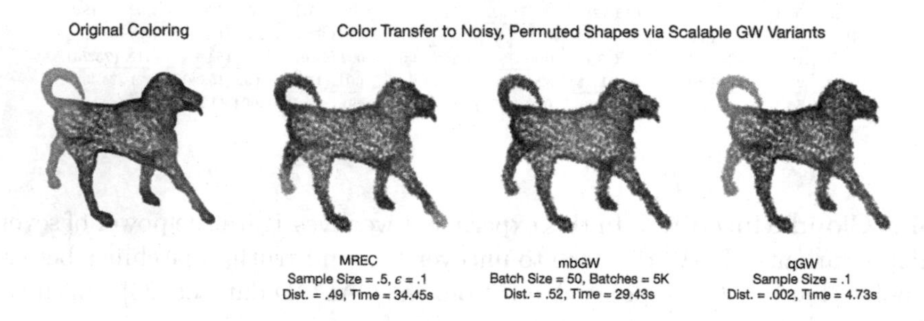


**Fig. 1.** We match the dog point cloud on the left (~9K points) to a copy whose points have been perturbed and whose order has been permuted. Matchings are computed via scalable variants of GW distance: MREC [3], minibatch GW [11] and our qGW algorithm. The distortion (see text) and compute time of each matching are provided. We visualize the matching by transferring a coloring of the points of the original shape to the new shapes via the respective matchings. Each algorithm provides a probabilistic correspondence between points; the color of a point in the target shape is a weighted average of the colors of the points in the source shape, with the weights given by the probability that a source shape matches to the query target shape.

**Table 2.** Distortion percentage (lower is better) and runtimes (s) on graph matching. Blank entries correspond to experiments that did not complete in 1 h or ran out of memory for storing distance matrices.

| Method | Param | Centaur 1<br>15768 | Centaur 2<br>15768 | Centaur 3<br>15768 | Centaur 4<br>15768 | Centaur 5<br>15768 | Cat<br>27894 | David<br>52565 |
|---|---|---|---|---|---|---|---|---|
| erGW | $10^3$ | 92.2 (1059) | 92.0 (1080) | 91.7 (1074) | 91.7 (1060) | 92.0 (1054) | – | – |
| mbGW | $(400, 2K)$ | 48.3 (788) | 48.1 (784) | 46.3 (778) | 46.4 (783) | 48.0 (779) | 42.9 (712) | – |
| MREC | $(750, 10^{-3})$ | 78.6 (273) | 13.3 (267) | 14.5 (288) | 13.8 (255) | 22.7 (272) | 70.1 (707) | – |
| qFGW | $(0.5, 0.75)$ | **6.76** (4.53) | **6.62** (4.54) | **6.65** (4.56) | **6.55** (4.52) | **6.71** (4.68) | **8.28** (7.83) | **82.5** (8.62) |

**Graph Matching.** Graph matching is a fundamental GW application for which [36] produced state-of-the-art scalability results, including matching source and target graphs with 2K and 9K nodes, respectively. Here we match graphs coming from meshes in the TOSCA dataset [5]. We choose multiple models (number of meshes in parentheses) from the "Centaur" (6), "Cat" (2), and "David" (2) mesh families, having approximately 16K, 30K, and 50K vertices, respectively. Meshes in each category correspond to different poses of the same object, and the underlying vertices are numbered in a compatible way to provide for ground truth labels. For the six Centaur graphs, we wished to compute GW matchings between $(G_1, G_2), \ldots, (G_5, G_6)$. We retained the structure of the Point Cloud Matching experiment with three variations. First, for evaluating the quality of a matching, we computed the ratio of the summed distortions of a matching $\mu$ (graph distance from each matched point to its ground truth copy) to the distortion of a *random* matching (averaged over five random matchings), and converted this into a percentage. Thus a lower distortion score is better, and these are the scores reported in Table 2. Second, we used the observation of [32] that adding node features via Weisfeiler-Lehman (WL) leads to superior performance, and devised a WL scheme to apply qFGW. Third, to demonstrate how to use these methods in a cross-validation pipeline, we took the two fastest methods—qFGW and MREC—and optimized parameters for them using leave-one-out cross-validation on the Centaur comparisons. For MREC this cross-validation produced $\epsilon = 10^{-3}$ and #clusters=750, and for qFGW we fixed $m = 1000$ and obtained $\alpha = 0.5, \beta = 0.75$. Results on the five testing folds are presented in Table 2; these optimized parameters are also used for the Cat and David comparisons.

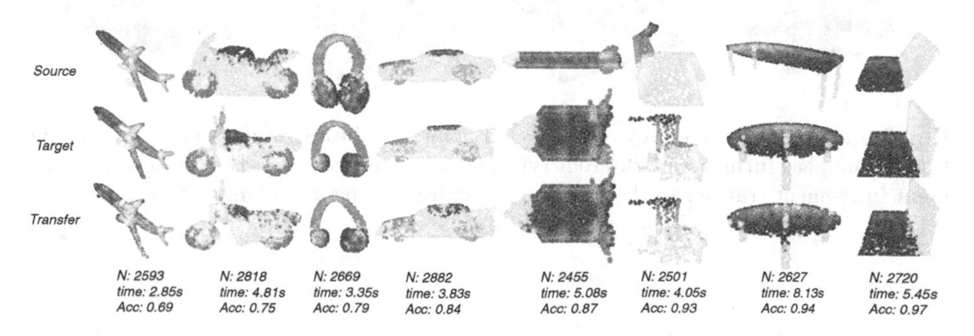


**Fig. 2.** Semantic segmentation transfer on ShapeNet. Colors denote part annotations. (Color figure online)

**Application to Segmentation Transfer.** Given point cloud datasets with semantic labels, segmentation transfer is the problem of constructing correspondences that preserve segment category labels. We demonstrate the applicability of qFGW to segmentation transfer using the ShapeNet CAD model dataset. This dataset contains 16 categories, each category having objects with approximately 3K points split into 2–6 parts. Point features are chosen to be surface normals. We choose 12 models each from the eight categories {Airplane, Car, Earphone, Guitar, Laptop, Motorbike, Rocket, Table}. We optimize parameters over a simple grid of $\alpha, \beta$ parameters as in the Graph Matching experiment, and illustrate results with optimal parameters in Fig. 2. For evaluation, we obtain matchings $\mu$ via an argmax as in the preceding experiments, and then count the fraction of matches between source and target part labels normalized by the number of points.

**Application to Large Scale Segment Transfer.** As further demonstration of segment transfer on extremely large metric spaces, we use the Stanford 3D Indoor Scene Dataset (S3DIS) [2]. This point cloud dataset comprises six areas containing 271 rooms, totaling to 215 million points where each point is labeled with one of 13 semantic categories and comes with an RGB color vector. We choose two Lobby rooms in Area 4, one containing 1,155,072 points and the other containing 909,312 points. We carry out matching using qFGW, using point colors as features. For evaluation we obtain a matching $\mu$, count the number of points that get matched to a point of the same part category, and normalize by the number of points in the source room. Because the rooms may contain parts belonging to different semantic categories, making direct evaluation difficult, we compare against a random matching (higher is better): random matching obtains 10.0%, matching with $m = 1000$ obtains 26.2%, and matching with $m = 5000$ obtains 41.0%. Notably, the overall computation is completed in just 10 min (for $m = 1000$) on a standard Macbook Pro with 8 GB RAM (Fig. 3).

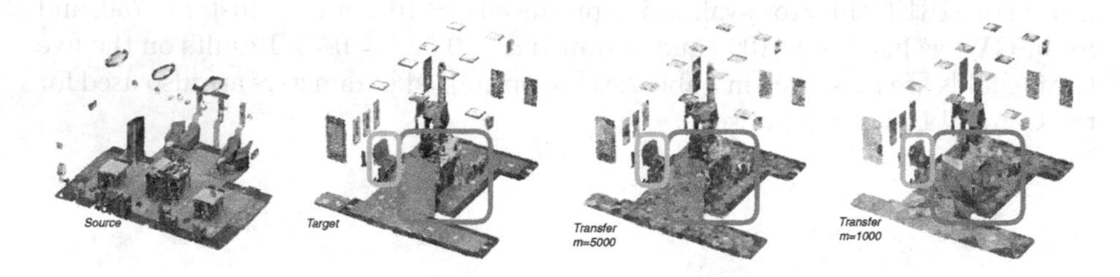


**Fig. 3.** Partial render of Lobby rooms (~1M points) in the S3DIS dataset. Note that the target room has furniture of different types than the source room. Boxes show improvement in segment transfer to the chair and desk from using more landmarks.

## 5 Discussion

By combining the crucial insight of partitioning for scalable GW [3,36], the notion of slicing a mm-space by distance to anchor points [17,33] and ideas regarding the duality between quantizing and clustering a metric space [21], we presented qGW: a theoretical framework for scalable GW computation with error bounds as well as new algorithmic improvements in time and space complexity. These error bounds complement those stated in [3] via the language of *doubling dimension* in metric spaces, and we related these notions using the concept of quantized eccentricity and associated bounds. Future theoretical work could validate this dependence on dimension by examining other intrinsic properties that could potentially tighten these bounds. This in turn motivates us to study how to extract local structure to aid with GW alignment even in high dimensional spaces.

**Acknowledgements.** We would like to thank Mathieu Carrière for help with the MREC code, Vikas Garg for sharing code from [13], Facundo Mémoli for providing useful feedback, and the anonymous reviewers for helpful comments.

## References

1. Alvarez-Melis, D., Jaakkola, T.: Gromov-Wasserstein alignment of word embedding spaces. In: Proceedings of the 2018 Conference on Empirical Methods in Natural Language Processing, pp. 1881–1890 (2018)
2. Armeni, I., et al.: 3D semantic parsing of large-scale indoor spaces. In: Proceedings of the IEEE Conference on Computer Vision and Pattern Recognition, pp. 1534–1543 (2016)
3. Blumberg, A.J., Carriere, M., Mandell, M.A., Rabadan, R., Villar, S.: MREC: a fast and versatile framework for aligning and matching point clouds with applications to single cell molecular data. arXiv preprint arXiv:2001.01666 (2020)
4. Brin, S., Page, L.: The anatomy of a large-scale hypertextual web search engine. Comput. Netw. ISDN Syst. **30**(1–7), 107–117 (1998)
5. Bronstein, A.M., Bronstein, M.M., Kimmel, R.: Numerical Geometry of Non-Rigid Shapes. MCS, Springer, New York (2009). https://doi.org/10.1007/978-0-387-73301-2
6. Bunne, C., Alvarez-Melis, D., Krause, A., Jegelka, S.: Learning generative models across incomparable spaces. In: International Conference on Machine Learning, pp. 851–861 (2019)
7. Chowdhury, S., Mémoli, F.: The Gromov-Wasserstein distance between networks and stable network invariants. Inf. Inference J. IMA **8**(4), 757–787 (2019)
8. Chowdhury, S., Needham, T.: Gromov-Wasserstein averaging in a Riemannian framework. In: Proceedings of the IEEE/CVF Conference on Computer Vision and Pattern Recognition Workshops, pp. 842–843 (2020)
9. Chowdhury, S., Needham, T.: Generalized spectral clustering via Gromov-Wasserstein learning. In: International Conference on Artificial Intelligence and Statistics, pp. 712–720. PMLR (2021)
10. Demetci, P., Santorella, R., Sandstede, B., Noble, W.S., Singh, R.: Gromov-Wasserstein optimal transport to align single-cell multi-omics data. BioRxiv (2020)

11. Fatras, K., Zine, Y., Majewski, S., Flamary, R., Gribonval, R., Courty, N.: Minibatch optimal transport distances; analysis and applications. arXiv preprint arXiv:2101.01792 (2021)
12. Flamary, R., Courty, N.: POT Python Optimal Transport library (2017). https://pythonot.github.io/
13. Garg, V., Jaakkola, T.: Solving graph compression via optimal transport. In: Advances in Neural Information Processing Systems, vol. 32 (2019)
14. Hagberg, A., Swart, P., Chult, D.S.: Exploring network structure, dynamics, and function using NetworkX. Technical report, Los Alamos National Lab. (LANL), Los Alamos, NM (United States) (2008)
15. Mémoli, F.: On the use of Gromov-Hausdorff distances for shape comparison. The Eurographics Association (2007)
16. Mémoli, F.: Gromov-Hausdorff distances in Euclidean spaces. In: IEEE Computer Society Conference on Computer Vision and Pattern Recognition Workshops (2008)
17. Mémoli, F.: Gromov-Wasserstein distances and the metric approach to object matching. Found. Comput. Math. **11**(4), 417–487 (2011)
18. Mémoli, F., Needham, T.: Gromov-Monge quasi-metrics and distance distributions. arXiv preprint arXiv:1810.09646 (2018)
19. Mémoli, F., Sapiro, G.: Comparing point clouds. In: SGP 2004: Proceedings of the 2004 Eurographics/ACM SIGGRAPH symposium on Geometry Processing, pp. 32–40. ACM, New York (2004). https://doi.org/http://doi.acm.org/10.1145/1057432.1057436
20. Mémoli, F., Sapiro, G.: A theoretical and computational framework for isometry invariant recognition of point cloud data. Found. Comput. Math. **5**(3), 313–347 (2005)
21. Mémoli, F., Sidiropoulos, A., Singhal, K.: Sketching and clustering metric measure spaces. arXiv preprint arXiv:1801.00551 (2018)
22. Mérigot, Q.: A multiscale approach to optimal transport. In: Computer Graphics Forum, vol. 30, pp. 1583–1592. Wiley Online Library (2011)
23. Papadakis, P.: The canonically posed 3D objects dataset. In: Eurographics Workshop on 3D Object Retrieval, pp. 33–36 (2014)
24. Parés, F., et al.: Fluid communities: a competitive, scalable and diverse community detection algorithm. In: Cherifi, C., Cherifi, H., Karsai, M., Musolesi, M. (eds.) COMPLEX NETWORKS 2017 2017. SCI, vol. 689, pp. 229–240. Springer, Cham (2018). https://doi.org/10.1007/978-3-319-72150-7_19
25. Peyré, G., Cuturi, M., Solomon, J.: Gromov-Wasserstein averaging of kernel and distance matrices. In: International Conference on Machine Learning (2016)
26. Peyré, G., Cuturi, M., et al.: Computational optimal transport: with applications to data science. Found. Trends® Mach. Learn. **11**(5–6), 355–607 (2019)
27. Sato, R., Cuturi, M., Yamada, M., Kashima, H.: Fast and robust comparison of probability measures in heterogeneous spaces. arXiv:2002.01615 (2020)
28. Séjourné, T., Vialard, F.X., Peyré, G.: The unbalanced Gromov Wasserstein distance: conic formulation and relaxation. arXiv preprint arXiv:2009.04266 (2020)
29. Solomon, J., Peyré, G., Kim, V.G., Sra, S.: Entropic metric alignment for correspondence problems. ACM Trans. Graph. (TOG) **35**(4), 1–13 (2016)
30. Sturm, K.T.: The space of spaces: curvature bounds and gradient flows on the space of metric measure spaces. arXiv preprint arXiv:1208.0434 (2012)
31. Sturm, K.T., et al.: On the geometry of metric measure spaces. Acta Math. **196**(1), 65–131 (2006)

32. Vayer, T., Courty, N., Tavenard, R., Flamary, R.: Optimal transport for structured data with application on graphs. In: International Conference on Machine Learning, pp. 6275–6284 (2019)
33. Vayer, T., Flamary, R., Tavenard, R., Chapel, L., Courty, N.: Sliced Gromov-Wasserstein. In: NeurIPS 2019-Thirty-third Conference on Neural Information Processing Systems, vol. 32 (2019)
34. Villani, C.: Topics in optimal transportation. American Mathematical Soc. (2003)
35. Weitkamp, C.A., Proksch, K., Tameling, C., Munk, A.: Gromov-Wasserstein distance based object matching: asymptotic inference. arXiv:2006.12287 (2020)
36. Xu, H., Luo, D., Carin, L.: Scalable Gromov-Wasserstein learning for graph partitioning and matching. In: Advances in Neural Information Processing Systems (2019)
37. Xu, H., Luo, D., Zha, H., Carin, L.: Gromov-Wasserstein learning for graph matching and node embedding. In: International Conference on Machine Learning (2019)

# Author Index

Printed in the United States
by Baker & Taylor Publisher Services